April 8–11, 2013
Philadelphia, PA, USA

Association for
Computing Machinery

Advancing Computing as a Science & Profession

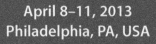

HSCC'13

Proceedings of the 16th International Conference on
Hybrid Systems: Computation and Control

Sponsored by:

ACM SIGBED

Supported by:

NEC

Part of:

CPSWeek 2013

Association for
Computing Machinery

Advancing Computing as a Science & Profession

The Association for Computing Machinery
2 Penn Plaza, Suite 701
New York, New York 10121-0701

ISBN: 978-1-4503-1567-8

Additional copies may be ordered prepaid from:

ACM Order Department
PO Box 30777
New York, NY 10087-0777, USA

Phone: 1-800-342-6626 (USA and Canada)
+1-212-626-0500 (Global)
Fax: +1-212-944-1318
E-mail: acmhelp@acm.org
Hours of Operation: 8:30 am – 4:30 pm ET

Printed in the USA

HSCC'13 Chairs' Welcome

It is our great pleasure to welcome you to the *2013 ACM International Conference on Hybrid Systems: Computation and Control – HSCC'13*. This year's conference continues its tradition of being the premier forum for presentation of basic and applied research that combines computer science and control theory for analysis and controls of dynamical systems that exhibit continuous, discrete, or combined (hybrid) dynamics. The mission of the conference is to bring together researchers from academia and industry to discuss computational challenges arising in both man-made, cyber-physical systems (ranging from mixed signal circuits and small robots to global infrastructure networks) and natural systems (ranging from biochemical networks to physiological models).

The call for papers attracted 86 submissions from Asia, Europe, and the United States. The program committee accepted 30 regular papers and 10 tool and case study papers that cover a variety of topics, including analysis, control, scheduling, formal verification, synthesis, abstraction, and learning for hybrid systems. The applications include robotics, networks, biology, and medicine. In addition, the program includes a poster session for tool and case study papers and work-in-progress discussions and an invited presentation by Aaron Ames from Texas A&M University.

We would like to thank the CPSWeek organizing committee, in particular Oleg Sokolsky and Sriram Sankaranarayanan for their help with financial arrangements and publicity. Special thanks go to Adrienne Griscti from ACM and those at Sheridan Communications for processing the papers in a timely manner.

Putting together *HSCC'13* was a team effort. We first thank the authors for providing the content of the program. We are grateful to the program committee and the additional reviewers, who worked very hard in reviewing papers and providing feedback for authors. We express our gratitude to the HSCC steering committee consisting of Rajeev Alur, Werner Damm, Bruce Krogh, Oded Maler, and Claire Tomlin for the trust and guidance provided throughout the organization of HSCC'13. The HSCC'12 chairs Thao Dang and Ian Mitchell were very supportive by sharing their perspective and experience. We are also indebted to the additional members of the HSCC Best Paper Award committee consisting of Thao Dang, Antoine Girard, Mircea Lazar, Ian Mitchell, André Platzer, and Sriram Sankaranarayanan for selecting a *Best Paper* and a *Best Student Paper*. Finally, we thank our sponsor, ACM, our technical co-sponsor IEEE CSS, and our corporate supporter NEC.

We hope that you will find this program interesting and thought-provoking and that the conference will provide you with a valuable opportunity to share ideas with other researchers and practitioners from institutions around the world.

Calin Belta
HSCC'13 co-Chair
Boston University, USA

Franjo Ivančić
HSCC'13 co-Chair
NEC Laboratories America, USA

Table of Contents

Session 8: Scheduling

Session 9: Verification

Session 10: Stochastic Hybrid Systems

Session 11: Control II

HSCC'13 Conference Organization

Program Chairs: Calin Belta *(Boston University, USA)*
Franjo Ivančić *(NEC Laboratories America, USA)*

Publicity Chair: Sriram Sankaranarayanan *(University of Colorado at Boulder, USA)*

Steering Committee Chair: Werner Damm *(Carl von Ossietzky Universität Oldenburg, Germany)*

Steering Committee: Rajeev Alur *(University of Pennsylvania, USA)*
Bruce Krogh *(Carnegie Mellon University, USA)*
Oded Maler *(Verimag, France)*
Claire Tomlin *(University of California at Berkeley, USA)*

Program Committee: Alessandro Abate *(Delft University of Technology, The Netherlands)*
Erika Ábrahám *(RWTH Aachen University, Germany)*
Aaron D. Ames *(Texas A&M University, USA)*
Manfred Broy *(Technische Universität München, Germany)*
Patrick Cousot *(École normale supérieure, France)*
Thao Dang *(Verimag, France)*
Domitilla Del Vecchio *(MIT, USA)*
Stefano Di Cairano *(Mitsubishi Electric Research Laboratories, USA)*
Alexandre Donzé *(University of California at Berkeley, USA)*
Magnus Egerstedt *(Georgia Institute of Technology, USA)*
Georgios Fainekos *(Arizona State University, USA)*
Eric Feron *(Georgia Institute of Technology, USA)*
Goran Frehse *(Verimag, France)*
Antoine Girard *(Université Joseph Fourier, France)*
Radu Grosu *(Technische Universität Wien, Austria)*
Klaus Havelund *(NASA Jet Propulsion Laboratory, USA)*
Holger Hermanns *(Universität des Saarlands, Germany)*
A. Agung Julius *(Rensselaer Polytechnic Institute, USA)*
Xenofon Koutsoukos *(Vanderbilt University, USA)*
Kim G. Larsen *(Aalborg Universitet, Denmark)*
Mircea Lazar *(Technische Universiteit Eindhoven, The Netherlands)*
Jie Liu *(Microsoft Research, USA)*
Rupak Majumdar *(Max Planck Institute for Software Systems, Germany)*
Ian Mitchell *(University of British Columbia, Canada)*
Sayan Mitra *(University of Illinois at Urbana-Champaign, USA)*
Pieter J. Mosterman *(The MathWorks, USA)*
Jens Oehlerking *(Robert Bosch GmbH, Germany)*
André Platzer *(Carnegie Mellon University, USA)*
Andreas Podelski *(University of Freiburg, Germany)*
Pavithra Prabhakar *(IMDEA Software Institute, Spain)*
S. Ramesh *(General Motors R&D, India)*

Program Committee (continued):

Stefan Ratschan *(Czech Academy of Sciences, Czech Republic)*
Ricardo G. Sanfelice *(University of Arizona, USA)*
Sriram Sankaranarayanan *(University of Colorado at Boulder, USA)*
Paulo Tabuada *(University of California Los Angeles, USA)*
Ashish Tiwari *(SRI International, USA)*
Verena Wolf *(Universität des Saarlands, Germany)*
Paolo Zuliani *(Newcastle University, UK)*

Additional reviewers:

Yongga A
Houssam Abbas
Dieky Adzkiya
Matthias Althoff
Eugene Asarin
Nikolaos Athanasopoulos
Laurent Bako
Ezio Bartocci
Veronika Bauer
Spring Berman
Amit Bhatia
Sergiy Bogomolov
Alessandro Borri
Olivier Bouissou
Bernard Brogliato
Mireille Broucke
Samuel Burden
Alarico Campetelli
Souymodip Chakraborty
Krishnendu Chatterjee
Jian-Jia Chen
Xin Chen
Alessandro Colombo
David Copp
Nikolaus Correll
Florian Corzilius
Eric Cousineau
Sergey Dashkovskiy
Claudio De Persis
Yi Deng
Marco Diciolla
Xu Chu Ding
Deepak D'Souza
Parasara Sridhar Duggirala
Tomas Dzetkulic
Christian Ebenbauer
Ruediger Ehlers
Martin Elshuber
Chinwendu Enyioha
Uli Fahrenberg
Jérôme Feret
Cameron Finucane
Emilio Frazzoli
Roman Geiselhart
Michael Gerke
Reza Ghaemi
Khalil Ghorbal
Mario Gleirscher
Rafal Goebel
Gregor Goessler
Ebru Aydin Gol
Meng Guo
Mike Hafner
Ernst Moritz Hahn
Zhi Han
Arnd Hartmanns
Vahid Hashemi
David Henriques
Jochen Hoenicke
Zhenqi Huang
Benedikt Huber
Jun-Ichi Imura
Md Ariful Islam
Nils Jansen
Shridhar Jayanthi
Susmit Jha
Shu Jiang
Romain Jobredeaux
Taylor T. Johnson
Maximilian Junker
Tamas Kalmar-Nagy
Aditya Kanade

Additional reviewers (continued):

Sertac Karaman
Shahab Kaynama
Shishir Kolathaya
Thilo Krüger
Jan Kuřátko
Jordan Lack
Colas Le Guernic
Lichun Li
Wenchao Li
Wenlong Ma
João G. Martins
Ion Matei
Manuel Mazo Jr.
Yunsong Meng
Linar Mikeev
Marius Mikucionis
Antoine Miné
Swarup Mohalik
Eike Möhlmann
Michael Monerau
Nader Motee
Abhishek Murthy
Roberto Naldi
Philipp Neubeck
Luigi Palopoli
Ivan Papusha
Steven Peters
Erion Plaku
Romain Postoyan
Matthew Powell
Jan-David Quesel
Lillian Ratliff
David Renshaw
Matthias Rungger
Dorsa Sadigh
Indranil Saha
Prahladavaradan Sampath
Peter Schramme
Di Shi
Yasser Shoukry
Ryan Sinnet
Maria Spichkova
David Spieler
Veaceslav Spinu
Peter Struss
Yuichi Tazaki
Ilya Tkachev
Stephan Trenn
Rajeev Verma
Roberto Vigo
Mahesh Viswanathan
Nicola Vitacolonna
Andreas Vogelsang
Gera Weiss
Andrew Winn
Sadegh Esmaeil Zadeh Soudjani
Majid Zamani
Erik Zawadzki
Haibo Zeng
Wei Zhang
Huihua Zhao

HSCC'13 Sponsor & Supporters

Sponsor: acm SIGBED

Technical Supporter: CSS

Supporter: NEC

Least-violating Control Strategy Synthesis with Safety Rules

Jana Tůmová *
Masaryk University
Brno, Czech Republic
xtumova@fi.muni.cz

Gavin C. Hall
Massachusetts Institute of Technology
Cambridge, MA, USA
gvnhall@mit.edu

Sertac Karaman
Massachusetts Institute of Technology
Cambridge, MA, USA
sertac@mit.edu

Emilio Frazzoli
Massachusetts Institute of Technology
Cambridge, MA, USA
frazzoli@mit.edu

Daniela Rus
Massachusetts Institute of Technology
Cambridge, MA, USA
rus@csail.mit.edu

ABSTRACT

We consider the problem of automatic control strategy synthesis, for discrete models of robotic systems, to fulfill a task that requires reaching a goal state while obeying a given set of safety rules. In this paper, we focus on the case when the said task is not feasible without temporarily violating some of the rules. We propose an algorithm that synthesizes a motion which violates only lowest priority rules for the shortest amount of time. Although the proposed algorithm can be applied in a variety of control problems, throughout the paper, we motivate this problem with an autonomous car navigating in an urban environment while abiding by the rules of the road, such as "always stay in the right lane" and "do not enter the sidewalk." We evaluate the algorithm on a case study with several illustrative scenarios.

Categories and Subject Descriptors

D.2.4 [**Software Engineering**]: Software/Program Verification—*Formal Methods*; I.2.8 [**Artificial Intelligence**]: Problem Solving, Control Methods, and Search—*Control Theory*; I.2.9 [**Artificial Intelligence**]: Robotics

General Terms

Algorithms, Verification

Keywords

least-violating planning; control strategy synthesis; robot path planning; formal specification

*This work was initiated while this author was a visiting student and MIT and Singapore-MIT Alliance for Research and Technology.

1. INTRODUCTION

The motivation for our work comes from the following scenario. We would like to automatically generate a control strategy for a robotic car in a road-like environment given its goal is to reach a target location while obeying a set of safety traffic rules (e.g., "always stay in the right lane", or "do not go on the sidewalk"). In many cases, this goal is not feasible (e.g., due to obstacles present in the environment) and in order to reach the target location, the robot needs to temporarily break some of the rules. We would like to systematically find the strategy that violates the rules as little as possible. In order to do that, we define a fine, quantitative characterization of *how much* a safety rule is obeyed or disobeyed on a particular trace and we aim to find a least-violating strategy with respect to this characterization.

In our approach, we decompose the robot motion planning into high-level planning and low-level control. (In fact, this is a common approach that appears, e.g., in [17, 16, 25] and many other recent publications on robot planning with complex specifications.) This hierarchical approach to control synthesis follows roughly three steps: (1) construct a discrete abstraction of the system's dynamics, (2) synthesize a plan for the resulting discrete abstraction, and (3) compute a controller that implements the discrete plan on the original system. Numerous methods exist for the first step of computing discrete abstractions for hybrid systems; See, e.g., [17, 16, 25] and the references therein. The third step, in particular, computing low-level controllers that can drive a robot from one region to another, regardless of the initial position of the robot within the initial region, has also been largely explored (see, e.g., [3]). In this paper, we focus on the second step of the hierarchical approach: synthesizing a plan for a discrete system. We assume that all dynamical phenomena are captured by a weighted, deterministic, discrete transition system.

The safety rules are expressed in any formal specification language that can be translated into finite automata, such as regular expressions or Linear Temporal Logic (LTL) over finite traces. Each safety rule is assigned a priority. We define the *level of unsafety* of a vehicle trace based on the number of steps made that caused violation of each safety rule weighted by the respective rule priority. We present an algorithm for finding a finite trace of the transition system that (1) leads from the current robot's location into a target location, (2) minimizes the level of unsafety among traces satisfying (1), and (3) minimizes the length of the trace among traces satisfying (1) and (2). Our solution builds on some ideas from automata-based model checking and graph theory. We express the rules specification as a set of finite automata and augment each of them with new transitions and weights. Then, we construct the product automaton between the transition system and the newly obtained finite automata. The shortest path of the product (found e.g., by a modification of the Dijsktra's shortest path algorithm) provably projects onto a trace of the transition system that is the solution to our problem.

Our work in this paper can be seen in context of works aimed at the use of formal methods in optimal robot path planning, e.g., [22, 23, 9, 13], where the authors aim to find an optimal high-level controller for a robot that guarantees satisfaction of a given temporal logic mission.

Other related work includes the recent literature adressing the control strategy synthesis problem with an unsatisfiable set of specifications. In [19, 20], the authors focus on explaining the reasons of unsatisfiability; given an LTL specification and a model that does not satisfy this specification, they answer whether or not the invalidity is limited to the provided model. Related literature includes also [5], where the authors aim to pinpoint the (un)realizable fragments of the specification to reveal causes of the specification violation. Other authors propose to construct control strategies with minimal changes in the input. On one hand, in [10, 15, 14], the authors aim to revise the given specification to (i) make it satisfiable by the given model, and (ii) make it as close as possible to the original specification according to a suitable metric. On the other hand, in [12] the author focuses on finding the least set of constraints (in the model) whose removal results in the satisfaction of the specification and in [4, 2] the authors aim to repair model in the form of a transition system or a Markov chain in order to ensure the satisfaction a given CTL or PCTL formula, respectively. Finally, some others, such as [7, 6], focus on synthesis of the least-violating controllers. They find, according to a proposed metric, the maximal part of the specification that can be satisfied by the model, and they generate the corresponding controller for this part of the specification.

The contribution of this paper can be summarized as follows. We address the problem of least-violating control strategy synthesis with a reachability task and a set of safety rules to be obeyed. We suggest a quantitative metric called level of unsafety to capture "how much" a given trace disobeys the given set of rules and suggest an algorithm to find the optimal trace with respect to the level of unsafety. We demonstrate the proposed algorithm on an extensive case study. Throughout the paper, we motivate the problem by a robotic car navigating in an urban environment while abiding by the rules of the road, although the algorithm can be used to address a wide class of automated synthesis problems.

The paper is organized as follows. In Sec. 2, we introduce necessary notation and prelimiaries. In Sec. 3, we formalize our problem and summarize our approach. Sec. 4 is dedicated to the details of the proposed algorithm and its analysis. In Sec. 5, the results of our case study are presented. Finally, we conclude the paper with remarks and directions for future work in Sec. 6.

2. PRELIMINARIES

In this section, we introduce some necessary notation and preliminaries.

Notation

Given a set $\mathcal{S}$, let $|\mathcal{S}|$, $2^{\mathcal{S}}$, and $\mathcal{S}^*$ denote the cardinality of $\mathcal{S}$, the set of all subsets of $\mathcal{S}$, and set of all finite sequences of elements of $\mathcal{S}$, respectively. A word over $\mathcal{S}$ is a finite sequence of elements from $\mathcal{S}$. Given words $u, v \in \mathcal{S}^*$, we denote by $u \cdot v$ their concatenation.

Transition Systems and Finite Automata

Definition 1 (Weighted transition system)
A weighted labeled deterministic transition system (TS) is a tuple $\mathcal{T} = (S, s_{\text{init}}, R, AP, L, W_{\mathcal{T}})$, where S is a finite set of states; $s_{\text{init}} \in S$ is the initial state; $R \subseteq S \times S$ is a deterministic transition relation; AP is a set of atomic propositions; $L : S \to 2^{AP}$ is a labeling function; $W_{\mathcal{T}} : R \to \mathbb{N}$ is a weight function.

A finite *trace* of $\mathcal{T}$ is a finite sequence of states $\tau = s_0 s_1 \dots s_n$ such that $s_0 = s_{\text{init}}$ and $(s_i, s_{i+1}) \in R$, for all $0 \leq i < n$. A finite trace $\tau = s_0 s_1 \dots s_n$ produces a finite *word* $w(\tau) = L(s_0)L(s_1)\dots L(s_n)$. We say that $\tau = s_0 s_1 \dots s_n$ satisfies "go from A to B" task if and only if $A, B \in AP$, $A \in L(s_0)$, and $B \in L(s_n)$.

We define the weight of a trace $s_0 s_1 \dots s_n$ as the sum of the weights of its transitions. With a slight abuse of notation, this quantity is denoted as follows:

$$W_{\mathcal{T}}(s_0 s_1 \dots s_n) = \sum_{i=0}^{n-1} W_{\mathcal{T}}((s_i, s_{i+1})).$$

Definition 2 (Finite automaton)
A (non-deterministic) finite automaton (FA) is a tuple $\mathcal{A} = (Q, q_{\text{init}}, \Sigma, \delta, F)$, where Q is a finite set of states; $q_{\text{init}} \in Q$ is the initial state; Σ is an input alphabet; $\delta \subseteq Q \times \Sigma \times Q$ is a non-deterministic transition relation; $F \subseteq Q$ is the set of accepting states.

The semantics of finite automata are defined over finite input words over Σ (such as those generated by transition system from Def. 1 if $\Sigma = 2^{AP}$). A *run* of the finite automaton $\mathcal{A}$ *over* an input word $w = w(1)\dots w(n)$ is a sequence $\rho = q_0 q_1 \dots q_n$, such that $q_0 = q_{\text{init}}$, and $(q_{i-1}, w(i), q_i) \in \delta$, for all $1 \leq i \leq n$. If $q_n \in F$, then ρ is accepting and the word w is accepted by the automaton. We say that a trace $\tau = s_0 s_1 \dots s_n$ of $\mathcal{T}$ *satisfies* the property defined by $\mathcal{A}$ (denoted by $\tau \models \mathcal{A}$) if and only if the word produced by τ is accepted by $\mathcal{A}$.

A *weighted (non-deterministic) finite automaton* is a tuple $\mathcal{A} = (Q, q_{\text{init}}, \Sigma, \delta, F, W)$, where $(Q, q_{\text{init}}, \Sigma, \delta, F)$ is a finite automaton (see Def. 2) and $W : \delta \to \mathbb{N}^k$, where $k \in \mathbb{N}$ is a weight function. In other words, each transition is assigned a k-tuple of natural numbers.

A (weighted) finite automaton $\mathcal{A}$ can be viewed as a graph $G_{\mathcal{A}} = (V, E)$, where the set of nodes V is the set of states Q and the set of edges E with their weights are given in the expected way by the transition function δ and the weight function W, respectively. Consider a run $\rho = q_0 q_1 \dots q_n$ over a word $w = w(1)\dots w(n)$ and assume that $(x_{j,1},\dots,x_{j,k}) = W((q_j, w(j), q_{j+1}))$ denotes the weight of the corresponding transition, for all $j \in \{1,\dots,n\}$. The weight of the run ρ over the word w is

$$W(\rho) = (X_1,\dots,X_k),$$

$$\text{where } X_i = \sum_{j=0}^{n} x_{j,i}, \text{ for all } i \in \{1,\dots,k\}.$$

In other words, the weight of a run is the tuple obtained by component-wise sum of the weights associated with the transitions executed along the run.

The *shortest run over* $w(1)\dots w(n)$ is the run ρ minimizing the weight $W(\rho)$ in the lexicographical ordering [1].

Linear Temporal Logic

Finite automata can capture a large class of properties that are exhibited by the traces of a transition system. However, some specification languages with similar expressive power, such as regular expressions or variants of Linear Temporal Logic (LTL) interpreted over finite runs, provide a more user-friendly means to express these properties. For a through exposition to regular expressions and the LTL, the reader is referred to the standard texts [21, 1].

In Section 5, we will demonstrate how the desirable properties of the system, e.g., the rules of the road, can be conveniently captured in Finite LTL (FLTL) [18], a definition for which is given below.

Definition 3 *A FLTL formula ϕ over the set of atomic propositions AP is defined inductively as follows:*

1. *every atomic proposition $\sigma \in 2^{AP}$ is a formula, and*
2. *if ϕ_1 and ϕ_2 are formulas, then $\phi_1 \vee \phi_2$, $\neg\phi_1$, $\mathsf{X}\,\phi_1$, $\overline{\mathsf{X}}\,\phi_1$, $\phi_1 \,\mathsf{U}\, \phi_2$, $\mathsf{G}\,\phi_1$, and $\mathsf{F}\,\phi_1$ are each formulas,*

where $\neg$ (negation) and $\vee$ (disjunction) are standard Boolean connectives, and X, $\overline{\mathsf{X}}$, U, G, and F are temporal operators.

Unlike the well-known LTL (see e.g., [1]), FLTL is interpreted over finite traces, as those generated by the transition system from Def. 1. Informally, $\mathsf{X}\,\pi$ states that there is the next state of a trace and that proposition π is true there (i.e., $\pi \in L(s_1)$). Strong next operator $\overline{\mathsf{X}}\,\pi$ is defined as $\neg\mathsf{X}\neg\pi$, expressing that if there exists a next state of a current state of a trace, then π is true there. In contrast, $\pi_1 \,\mathsf{U}\, \pi_2$ states that there is a future moment when proposition π_2 is true, and proposition π_1 is true at least until π_2 is true. The formula $\mathsf{G}\,\pi$ states that proposition π holds at all states of a finite trace, and $\mathsf{F}\,\pi$ states that π holds at some future time instance.

Similarly as LTL formulas can be algorithmically translated into automata over infinite words, FLTL formulas can be translated into finite automata [11].

3. PROBLEM STATEMENT

In this section, we provide a formal description of the problem, after presenting some intermediate definitions.

[1]The lexicographical ordering $\leq$ is defined as follows: $(x_1,\dots,x_k) \leq (x'_1,\dots,x'_k)$ if $(x_1,\dots,x_k) = (x'_1,\dots,x'_k)$ or if $\exists 1 \leq i \leq k$, such that $\forall j < i : x_j \leq x'_j$ and $x_i < x'_i$.

Let us consider a robot moving in a partitioned environment with its motion capabilities modeled as a weighted transition system $\mathcal{T} = (S, s_{\text{init}}, R, AP, L, W_{\mathcal{T}})$ (see Def. 1). Each region of the environment is modeled as a state of the transition system and the robot's ability to move between the adjacent regions is captured through the transition relation. We assume that there are no transitions leading from and to the states representing regions where obstacles are present. Therefore, each trajectory of the robot in the said subset of the environment is obstacle-free. The weight function assigns each transition a weight, such as distance or travel time. Each region is labeled with a set of atomic propositions (e.g., "region A", "region B", "left lane", "right lane", "safe region", or "sidewalk") that are true in this region. Our robot mission is simply "go from A to B" while abiding by a set of (safety) rules, which we will describe using LTL formulas and assign a priority each.

Consider a trace that starts at the region with label A and reaches that with label B. Suppose for the moment that there is a single safety rule, which we denote by ψ. Although its priority, denoted by $\varpi(\psi)$, is not important for now, its role will become clear shortly when we consider multiple safety rules. Examples of such rules include "never enter region C" and "region D can only be visited after visiting region E." Note that ψ can be given in any form, as long as it can be translated into a finite automaton $\mathcal{A}_\psi$. In particular, it can be a regular expression or an LTL formula that is interpreted over finite words (see Sec. 2). For notational simplicity, we denote by $\tau \models \psi$ the property that the trace τ of $\mathcal{T}$ satisfies property ψ, i.e., that $\tau \models \mathcal{A}_\psi$. Furthermore, in the sequel, we tacitly assume that the empty trace of $\mathcal{T}$ and the empty word ϵ always satisfy any safety rule ψ. That is, by convention, an empty run does not violate any safety rules.

We define the *level of unsafety* $\lambda(\tau, \psi)$ of a finite trace τ of the transition system $\mathcal{T}$ as the minimum number of transitions in τ that must "vanish" so that $\tau \models \psi$ holds, weighted by the priority $\varpi(\psi)$. Intuitively, the level of unsafety can be seen as the number of steps that violate the given safety rule multiplied by $\varpi(\psi)$. For instance, a trace $s_A s_D s_C s_E s_C s_B$, where $s_X \in S$ for all $x \in \{A, B, C, D, E\}$, visits regions A, D, C, E, C, B in this order has level of unsafety 2 for the rule ψ_1 "never enter region C," with $\varpi(\psi_1) = 1$ and it has level of unsafety 3 for the rule ψ_2 "region D can only be visited after visiting region E.", where $\varpi(\psi_2) = 3$. On the other hand, the level of unsafety of a trace visiting regions A, D, B is 0 for the former rule and remains 3 for the latter one. The smaller the level of unsafety of a trace, the closer the trace is to a trace that satisfies the given safety rule. Indeed, the trace satisfies the rule ψ if and only if its level of unsafety is equal to 0.

More formally, let $\mathcal{T} = (S, s_{\text{init}}, R, AP, L)$ be a transition system, and let ψ be a property over AP that can be expressed as a finite automaton.

Definition 4 (Level of unsafety for a safety rule) *Let $\tau = s_0 \dots s_n$ be a finite trace producing a word $w(\tau) = w(0)w(1)\dots w(n)$ and $\{i_1,\dots,i_k\} \subseteq \{0,\dots,n\}$. Define*

$$\text{vanish}(\tau, \{i_1, \dots i_k\}) =$$

$$s_0 \dots s_{i_1 - 1} s_{i_1 + 1} \dots s_{i_k - 1} s_{i_k + 1} \dots s_n,$$

i.e., the finite sequence of states obtained by erasing states indexed with $i_1,\dots,i_k$ from τ.

The level of unsafety $\lambda(\tau, \psi)$ of a finite trace τ with respect to a safety rule ψ is the cardinality-wise minimal $I \subseteq \{0, \ldots n\}$ such that the trace vanish(τ, I) *satisfies the safety rule ψ:*

$$\lambda(\tau, \psi) = \min_{I \subseteq \{0,\ldots,n\}, \text{vanish}(\tau, I) \models \psi} |I| \cdot \varpi(\psi)$$

Level of unsafety of the word $w(\tau)$ is defined analogously.

Now, consider a sequence of nonempty sets of safety rules $\mathbf{\Psi} = (\Psi_1, \ldots, \Psi_n)$. Suppose each rule $\psi \in \Psi_i$, for all $1 \le i \le n$, is given in any form that can be translated into a finite automaton. Define the *priority function* as $\varpi : \mathbf{\Psi} \to \mathbb{N}$, which assigns a *priority* to each rule $\psi \in \Psi_i$, for all $1 \le i \le n$. In the sequel, the ordered set $\mathbf{\Psi}$ together with the priority function ϖ will be called a *set of safety rules with priorities* $(\mathbf{\Psi}, \varpi)$.

Given a transition system $\mathcal{T}$ and a set of rules with priorities, denoted by $(\mathbf{\Psi} = (\Psi_1, \ldots, \Psi_n), \varpi)$, we extend the definition of the level of unsafety for a trace τ of $\mathcal{T}$ to a set of safety rules $\Psi_i \in \mathbf{\Psi}$ and a set of safety rules with priorities $(\mathbf{\Psi}, \varpi)$ as follows.

Definition 5 (Level of unsafety for a set of rules) Level of unsafety $\lambda(\tau, \Psi_i)$ *of a finite trace τ with respect to* a set of safety rules Ψ_i *is defined as*

$$\lambda(\tau, \Psi_i) = \sum_{\psi \in \Psi_i} \lambda(\tau, \psi).$$

Level of unsafety $\lambda(\tau, \mathbf{\Psi})$ *of finite trace τ with respect to* a set of safety rules with priorities $(\mathbf{\Psi}, \varpi)$ *is defined as*

$$\lambda(\tau, \mathbf{\Psi}) = \big(\lambda(\tau, \Psi_1), \ldots, \lambda(\tau, \Psi_n)\big).$$

The measure of unsafety, as given by Definition 5, implies an ordering relation, which we formalize below.

Definition 6 (Unsafety order) *A finite trace τ is said to be* safer *than a finite trace τ' if and only if*

$$\lambda(\tau, \mathbf{\Psi}) \le \lambda(\tau', \mathbf{\Psi})$$

in the standard lexicographical ordering, i.e., iff

- $\lambda(\tau, \mathbf{\Psi}) = \lambda(\tau', \mathbf{\Psi})$, *or*
- $\exists 1 \le i \le n$, *such that* $\lambda(\tau, \Psi_i) < \lambda(\tau', \Psi_i)$ *and* $\lambda(\tau, \Psi_j) \le \lambda(\tau', \Psi_j)$, *for all* $j < i$.

Then, the problem can be formally stated as follows.

Problem 1 Given

- *a weighted labeled deterministic transition system* $\mathcal{T} = (S, s_{\text{init}}, R, AP, L, W_{\mathcal{T}})$;
- *a task "go from A to B," where* $A, B \in AP, A \in L(s_{\text{init}})$; *and*
- *a set of safety rules with priorities* $(\mathbf{\Psi}, \varpi)$ *over* AP,

find *a finite trace τ of $\mathcal{T}$ that*

(i) satisfies the task;

(ii) is safer than all the other traces satisfying the condition (i); and

(iii) minimizes $W_{\mathcal{T}}(\tau)$ among the traces satisfying the conditions (i) and (ii).

Intuitively, the sets $\Psi_1, \ldots \Psi_n$ from $\mathbf{\Psi}$ can be seen as *priority classes*. According to Definition 6, it is always preferred to satisfy the rules in Ψ_i rather than those in Ψ_j, for all $i < j$. However, one can trade off when violating rules that are in the same priority class. For an illustrative example, see Sec. 5.

4. ALGORITHM AND ANALYSIS

In this section, we first describe the proposed approach, and subsequently we prove its correctness and analyze its computational complexity. Before closing this section, we also consider a more general problem description, and point out how the proposed algorithm can be modified to address the resulting problem.

4.1 The Proposed Approach

In this section, we outline the proposed algorithm, which leverages ideas from automata theory, graph algorithms and reachability analysis of transition systems. Along the way, we provide the necessary intermediate results to prove the correctness and analyze the computational complexity of our algorithm.

The proposed algorithm can be outlined as follows. First, we translate each specification $\psi_{i,j}$ in $\Psi_i \in \mathbf{\Psi}$ into a finite automaton $\mathcal{A}_{i,j}$. We propose a method to augment this automaton with new transitions and weights, such that the resulting weighted automaton also accepts all words w that do *not* satisfy the rule $\psi_{i,j}$; however, the weights are picked such that the weight of this accepting run over w determines the level of unsafety of w with respect to $\psi_{i,j}$ (see Definition 7). Second, we combine all the weighted automata into a single automaton $\overline{\mathcal{A}}_{\mathbf{\Psi}}$, using its weights to capture the level of unsafety of all words (and thus traces of $\mathcal{T}$) with respect to the given set of safety rules with priorities $(\mathbf{\Psi}, \varpi)$ (see Definition 8). Third, we build a weighted product of the transition system $\mathcal{T}$ and the automaton $\overline{\mathcal{A}}_{\mathbf{\Psi}}$ (see Definition 9), and we search for the shortest path from the initial state to a final state. We show that this path projects onto a trace of $\mathcal{T}$ that satisfies the conditions *(i)–(iii)* put forth in the statement of Problem 1. In the rest of this section, we describe each of these steps in detail in this order.

First, let us describe the construction of $\overline{\mathcal{A}}_{\psi_{i,j}}$. Let $\mathcal{T} = (S, s_{\text{init}}, R, AP, L, W_{\mathcal{T}})$ be a weighted transition system, and let $(\mathbf{\Psi} = (\Psi_1, \ldots, \Psi_n), \varpi)$ be a set of safety rules with priorities, where $\Psi_i = \{\psi_{i,1}, \ldots, \psi_{i,m_i}\}$. Furthermore, let $\mathcal{A}_{\psi_{i,j}} = (Q_{i,j}, q_{\text{init},i,j}, 2^{AP}, \delta_{i,j}, F_{i,j})$ denote a finite automaton capturing the safety rule specification $\psi_j \in \Psi_i$ over AP. We augment $\mathcal{A}_{\psi_{i,j}}$ as follows.

Definition 7 (Automaton $\overline{\mathcal{A}}_{\psi_{i,j}}$) *Given an automaton $\mathcal{A}_{\psi_{i,j}}$, a weighted finite automaton*

$$\overline{\mathcal{A}}_{\psi_{i,j}} = (\overline{Q}_{i,j}, \overline{q}_{\text{init},i,j}, 2^{AP}, \overline{\delta}_{i,j}, \overline{F}_{i,j}, \overline{W}_{i,j})$$

is built as follows:

- $\overline{Q}_{i,j} = Q_{i,j}$;
- $\overline{q}_{\text{init},i,j} = q_{\text{init},i,j}$;
- $\overline{\delta}_{i,j} = \delta_{i,j} \cup \{(q, \sigma, q) \mid q \in Q$ *and* $\sigma \in 2^{AP}\}$;
- $\overline{F}_{i,j} = F_{i,j}$; *and*
- $$\overline{W}((q, \sigma, q')) = \begin{cases} 0 & \text{if } (q, \sigma, q') \in \delta_{i,j} \\ \varpi(\psi_{i,j}) & \text{if } (q, \sigma, q') \in \overline{\delta}_{i,j} \setminus \delta_{i,j}. \end{cases}$$

Lemma 1 *Let w be a word over 2^{AP}. Then w is accepted by $\overline{\mathcal{A}}_{\psi_{i,j}}$ and the weight of the shortest accepting run of $\overline{\mathcal{A}}_{\psi_{i,j}}$ over w is equal to the level of unsafety $\lambda(w, \psi_{i,j})$.*

PROOF. Note that, thanks to our assumption that an empty word and an empty trace satisfy $\psi_{i,j}$, each finite word w has a finite level of unsafety. The proof is via induction with respect to the level of unsafety $\lambda(w, \psi_{i,j})$. We prove that for each w there exists an accepting run with its length equal to $\lambda(w, \psi_{i,j})$ and we prove, that no accepting run over w is shorter than $\lambda(w, \psi_{i,j})$.

First, let $\lambda(w, \psi_{i,j}) = 0$. Then w is accepted by $\mathcal{A}_{\psi_{i,j}}$ and therefore, from the construction of $\overline{\mathcal{A}}_{\psi_{i,j}}$ there exists an accepting run of $\overline{\mathcal{A}}_{\psi_{i,j}}$ over w with its weight equal to 0. At the same time, clearly no accepting run over w is shorter than 0.

Second, let $k \in \mathbb{N}$ and let us assume that the lemma holds for each w, such that $\lambda(w, \psi_{i,j}) \leq k \cdot \varpi(\psi_{i,j})$. Consider a word w', such that $\lambda(w', \psi_{i,j}) = (k+1) \cdot \varpi(\psi_{i,j})$. Necessarily, there exist $u, v \in (2^{AP})^*$, and $\sigma \in 2^{AP}$, such that $\lambda(u \cdot v, \psi_{i,j}) = k \cdot \varpi(\psi_{i,j})$ and $w' = u \cdot \sigma \cdot v$. Thus, from the induction assumption, there exists an accepting run $q_0 \ldots q_{|u \cdot v|}$ over $u \cdot v$ with its weight equal to $k \cdot \varpi(\psi_{i,j})$. From the construction of $\overline{\mathcal{A}}_{\psi_{i,j}}$ there is a transition $(q_{|u|}, \sigma, q_{|u|}) \in \overline{\delta}_{i,j}$ with $W\big((q_{|u|}, \sigma, q_{|u|})\big) = \varpi(\psi_{i,j})$. Thus, there exists an accepting run over $w' = u \cdot \sigma \cdot v$ with its weight equal to $k \cdot \varpi(\psi_{i,j}) + \varpi(\psi_{i,j}) = (k+1) \cdot \varpi(\psi_{i,j})$.

On the other hand, assume that there exists an accepting run $q_0 q_1 \ldots q_{|w'|-1} q_{|w'|}$ over w' with its weight equal to $l \cdot \varpi(\psi_{i,j})$, where $l < k+1$. Then, the run $q_0 q_1 \ldots q_{|w'|-1} q_{|w'|}$ has the property that $(q_{x_y}, w'(x_y), q_{x_y+1}) \in \overline{\delta}_{i,j} \setminus \delta_{i,j}$ for exactly l different indexes x_y, where $y \in \{1, \ldots, l\}$. Hence, the word vanish$(w', \{x_1, \ldots, x_l\}) = w'(0) \ldots w'(x_1-1) w'(x_1+1) \ldots w'(x_l - 1) w'(x_l + 1) \ldots w'(|w'|)$ is accepted by automaton $\mathcal{A}_{\psi_{i,j}}$ and thus the level of unsafety $\lambda(w', \psi_{i,j}) \leq l$ which contradicts the assumption. The proof is thus complete. □

Second, we combine all automata $\overline{\mathcal{A}}_{\psi_{i,j}}$, where $\psi_{i,j} \in \Psi_i \in \mathbf{\Psi}$ into a single weighted automaton $\overline{\mathcal{A}}_{\mathbf{\Psi}}$ capturing, through its weight function, the level of unsafety with respect to the whole set of safety rules with priorities $(\mathbf{\Psi}, \varpi)$.

Definition 8 (Automaton $\overline{\mathcal{A}}_{\mathbf{\Psi}}$) *The weighted finite automaton*

$$\overline{\mathcal{A}}_{\mathbf{\Psi}} = (\overline{Q}, \overline{q}_{\text{init}}, 2^{AP}, \overline{\delta}, \overline{F}, \overline{W})$$

is defined as follows:

- $\overline{Q} = Q_{1,1} \ldots \times \ldots Q_{1,m_1} \cdots \times \ldots Q_{n,1} \ldots \times \ldots Q_{n,m_n}$;
- $\overline{q}_{\text{init}} = (q_{\text{init},1,1}, \ldots, q_{\text{init},n,m_n})$;
- $(p, \sigma, p') \in \overline{\delta}$ *and* $\overline{W}((p, \sigma, p')) = (x_1, \ldots, x_n)$, *if* $p = (q_{1,1}, \ldots, q_{n,m_n}), p' = (q'_{1,1}, \ldots, q'_{n,m_n})$, $(q_{i,j}, \sigma, q'_{i,j}) \in \overline{\delta}_{i,j}$, *for all* $i \in \{1, \ldots, n\}, j \in \{i, \ldots m_i\}$, *and* $x_i = \sum_{j=1}^{m_i} \overline{W}_{i,j}(q_{i,j}, \sigma, q'_{i,j})$.
- $\overline{F} = \{(q_{1,1}, \ldots, q_{n,m_n}) \mid q_{i,j} \in \overline{F}_{i,j}$, *for all* $i \in \{1, \ldots, n\}, j \in \{i, \ldots m_i\}\}$

Lemma 2 *Let w be a word over 2^{AP}. Then w is accepted by $\overline{\mathcal{A}}_{\mathbf{\Psi}}$ and the weight of the shortest accepting run of $\overline{\mathcal{A}}_{\mathbf{\Psi}}$ over w is equal to the level of unsafety $\lambda(w, \mathbf{\Psi})$.*

PROOF. Consider a word $w \in (2^{AP})^*$. From Lemma 1, we have that w is accepted by $\overline{\mathcal{A}}_{\psi_{i,j}}$ and the weight of the shortest accepting run $q_{0,i,j} \ldots q_{|w|,i,j}$ of $\overline{\mathcal{A}}_{\psi_{i,j}}$ over w is $\lambda(w, \psi_{i,j})$, for all $\psi_{i,j} \in \mathbf{\Psi}$. Thus there exists an accepting run $(q_{0,1,1}, \ldots q_{0,n,m_n}) \ldots (q_{|w|,1,1}, \ldots q_{|w|,n,m_n})$ over w and its weight is $(\sum_{k=1}^{|w|} x_{1,k}, \ldots, \sum_{k=1}^{|w|} x_{n,k})$, where $x_{i,k} = \sum_{j=1}^{m_i} \overline{W}_{i,j}(q_{k-1,i,j}, w(k), q_{k,i,j})$, for all $k \in \{1, \ldots |w|\}$. On the other hand, each accepting run $\rho = (q_{0,1,1}, \ldots q_{0,n,m_n}) \ldots (q_{|w|,1,1}, \ldots q_{|w|,n,m_n})$ of $\overline{\mathcal{A}}_{\mathbf{\Psi}}$ over w maps to an accepting run of $q_{0,i,j} \ldots q_{|w|,i,j}$ of $\overline{\mathcal{A}}_{\psi_{i,j}}$ over w, such that

$$W(\rho) \leq (\textstyle\sum_{k=1}^{|w|} x_{k,1}, \ldots, \sum_{k=1}^{|w|} x_{k,n}), \text{ where}$$

$x_{k,i} = \sum_{j=1}^{m_i} \overline{W}_{i,j}(q_{k-1,i,j}, w(k), q_{k,i,j})$, for all $k \in \{1, \ldots |w|\}$ and all $\psi_{i,j} \in \mathbf{\Psi}$. □

Third, following the ideas of automata-based approach to model checking [24], we build a product automaton that captures the traces of $\mathcal{T}$ satisfying the specification given by $\overline{\mathcal{A}}_{\mathbf{\Psi}}$ while preserving the purpose of the weight function.

Definition 9 (Product automaton $\mathcal{P}$)
We build the weighted product automaton

$$\mathcal{P} = \mathcal{T} \otimes \overline{\mathcal{A}}_{\mathbf{\Psi}} = (Q_{\mathcal{P}}, q_{\text{init},\mathcal{P}}, \delta_{\mathcal{P}}, F_{\mathcal{P}}, W_{\mathcal{P}})$$

as follows:

- $Q_{\mathcal{P}} = (S \cup \{init \mid init \notin S\}) \times \overline{Q}$
- $q_{\text{init},\mathcal{P}} = (init, \overline{q}_{\text{init}})$
- $\big((s,q),(s',q')\big) \in \delta_{\mathcal{P}}$ *and* $W_{\mathcal{P}}\big((s,q),(s',q')\big) = (\overline{W}(q, L(s'), q'), W_{\mathcal{T}}(s,s'))$ *if* $(s,s') \in R$, *and* $(q, L(s'), q') \in \overline{\delta}$
- $\big((init,q),(s_{\text{init}},q')\big) \in \delta_{\mathcal{P}}$ *and* $W_{\mathcal{P}}\big((init,q),(s_{\text{init}},q')\big) = (\overline{W}(q, L(s'), q'), 0)$ *if* $(q, L(s'), q') \in \overline{\delta}$
- $F_{\mathcal{P}} = S \times \overline{F}$

The following lemma summarizes the role of the product automaton $\mathcal{P}$.

Lemma 3 *The shortest run (in the lexicographical ordering with respect to $W_{\mathcal{P}}$) $p_0 \ldots p_n$ in $\mathcal{P}$ from the state $p_0 = q_{\text{init},\mathcal{P}}$ to a state $p_n \in F_{\mathcal{P}}$, such that $B \in L(p_n)$ projects onto a trace $\tau = \alpha(p_0 \ldots p_n)$ of $\mathcal{T}$ that is the solution to Problem 1.*

PROOF. The proof follows directly from Lemma 2 the following two facts. (1) Given a trace $\tau = s_0, \ldots, s_n$ of $\mathcal{T}$ that produces a word $w(\tau)$, such that $s_0 = s_{\text{init}}$ and $B \in L(s_n)$, there exists an accepting run of P over $w(\tau)$. Furthermore, the weight of the shortest accepting run of P over $w(\tau)$ is equal to $(\lambda(w(\tau), \mathbf{\Psi}), W_{\mathcal{T}}(\tau))$. (2) Given an accepting run $p_0 \ldots p_n = (q_0, s_0) \ldots (q_n, s_n)$ of $\mathcal{P}$, such that $p_0 = q_{\text{init},P}$, $p_n \in F_{\mathcal{P}}$, and $B \in L(p_n)$ with the weight equal to $(\lambda(w(\tau), \mathbf{\Psi}), W_{\mathcal{T}}(\tau))$, the trace $\tau = s_0 \ldots s_n$ is a trace of $\mathcal{T}$ that produces a word $w(\tau)$ and $q_0 \ldots q_n$ is a run over $w(\tau)$ with its weight equal to $\lambda(w(\tau), \mathbf{\Psi}))$. □

Computation of the shortest path in $\mathcal{P}$ can be done with the use of a slight modification of the well-known Dijsktra's shortest path algorithm [8], where all the comparisons in the algorithm are based on the lexicographical ordering of $(n+1)-$tuples.

For convenience, the proposed algorithm that solves Problem 1 is summarized in Alg. 1.

Algorithm 1 Solution to Problem 1

Input: A TS $\mathcal{T} = (S, s_{\text{init}}, R, AP, L, W_{\mathcal{T}})$; a region label $B \in AP$; a set of safety rules with priorities $(\boldsymbol{\Psi} = (\Psi_1, \ldots, \Psi_n), \varpi)$ expressed as an FA $\mathcal{A}_{\psi_{i,j}} = (Q_{i,j}, q_{\text{init},i,j}, \Sigma, \delta_{i,j}, F_{i,j})$ for each rule $\psi_j \in \Psi_i$.

Output: a trace of $\mathcal{T}$ that is solution to Problem 1

1: **for all** $i \in \{1, \ldots, n\}, \psi_j \in \Psi_i$ **do**
2: build modified automaton $\overline{\mathcal{A}}_{\psi_{i,j}}$ (Def. 7)
3: **end for**
4: build modified automaton $\overline{\mathcal{A}}_{\Psi}$ (Def. 8)
5: build product automaton $\mathcal{P} = \mathcal{T} \otimes \overline{\mathcal{A}}_{\Psi}$ (Def. 9)
6: $path :=$ shortest_path$(\mathcal{P}, B)$
7: $trace :=$ project_onto_first_component$(path)$
8: **return** $trace$

4.2 Correctness and Complexity

The correctness of the algorithm is stated in the following theorem. The computational complexity of the algorithm analyzed subsequently.

Theorem 1 *Alg. 1 returns a solution to Problem 1.*

PROOF. Follows directly from the proof of Lemma 3 and the correctness of the Dijkstra's shortest path algorithm. □

Let $|\mathcal{T}|$ denote the size of the input transition system, and $|\mathcal{A}_{\Psi_{i,j}}|$ the size of the input automaton $\mathcal{A}_{\Psi_{i,j}}$. The size of the automaton $\overline{\mathcal{A}}_{\psi_{i,j}}$ is $|\mathcal{A}_{\Psi_{i,j}}| \cdot |\Sigma|$ in the worst case and the size of $\overline{\mathcal{A}}_{\Psi}$ is then up to $\prod_{\Psi_{i,j} \in \Psi} |\mathcal{A}_{\Psi_{i,j}}| \cdot |\Sigma|$. The size of the product automaton $\mathcal{P}$ is $|\mathcal{P}| = |\mathcal{T}| \cdot \prod_{\Psi_{i,j} \in \Psi} |\mathcal{A}_{\Psi_{i,j}}| \cdot |\Sigma|$ and the modified Dijkstra's algorithm runs in $\mathcal{O}(|\mathcal{P}| \log |\mathcal{P}|)$.

4.3 An Alternative Problem Statement

In case a weighted transition system is used as a model, it is questionable, whether the definition of level of unsafety fits our needs as it does not take into consideration the length of the transitions leading into states that need to be vanished from a trace in order to make the trace satisfying. For instance, if a safety rule says "do not *stay* in the left lane," one expects that the length of the transitions leading into states violating these rules, i.e., distances or travel times spent in the left lane, should impact the level of unsafety of the trace.

Let us introduce an alternative definition for the level of unsafety.

Definition 10 (Weighted level of unsafety) Weighted level of unsafety $\lambda_W(\tau, \psi)$ *of finite trace* τ *with respect to a safety rule* ψ *is*

$$\lambda_W(\tau, \psi) = \min_{I \subseteq \{0,\ldots,n\} | \text{vanish}(\tau, I) \models \psi} \sum_{i \in I} W(s_{i-1}, s_i) \cdot \varpi(\psi).$$

The definition of weighted level of unsafety with respect to a set of safety rules with priorities remains analogous to the one in Def. 5. Problem 1 can be now formulated using the weighted level of unsafety (Def. 10) instead of "regular" level of unsafety (Def. 4). The proposed solution can also be easily adapted to the resulting problem. In fact, the only modification lies in the definition of the weight function of the product automaton: For each

$$W_{\mathcal{P}}((s,q),(s',q')) = \Big(\big(\overline{W}(q_1, L(s'), q_1') \cdot W_{\mathcal{T}}(s,s'), \ldots \ldots, \overline{W}(q_n, L(s'), q_n') \cdot W_{\mathcal{T}}(s,s') \big), W_{\mathcal{T}}(s,s') \Big),$$

for all $((s,q),(s',q')) \in \delta_{\mathcal{P}}$, where $q = (q_1, \ldots, q_n) \in \overline{Q}$.

However, this approach also does not fit all the scenarios. For instance, the safety rule formulated as "never *enter* the left lane." intuitively should not be influenced by the weights of the transitions. Ideally, we would like to combine the above two approaches and associate each safety rule with an indicator determining whether the "regular" level of unsafety (Def. 4) or the weighted level of unsafety (Def. 10) should be used. Therefore, we interconnect both definitions and introduce combined level of unsafety.

Definition 11 (Combined level of unsafety) Combined level of unsafety $\lambda_C(\tau, \Psi_i)$ *of finite trace* τ *with respect to the set of safety rules* Ψ_i *is*

$$\lambda_C(\tau, \Psi_i) = \sum_{j \in J_R} \lambda(\tau, \psi_j) + \sum_{j \in J_W} \lambda_W(\tau, \psi_j),$$

where $\psi_{i,j}$ *such that* $j \in J_R \subseteq \{1, \ldots m_i\}$ *are the safety rules associated with the level of unsafety (Def. 5) and* $\psi_{i,j}$ *such that* $j \in J_W \subseteq \{1, \ldots m_i\}$ *are the safety rules associated with the weighted level of unsafety (Def. 10). Note, that* $J_R \cap J_W = \emptyset$ *and* $J_R \cup J_W = \{1, \ldots, |\Psi_i|\}$.

Combined level of unsafety $\lambda(\tau, \boldsymbol{\Psi})$ *of finite trace* τ *with respect to the set of rules with priorities* $(\boldsymbol{\Psi}, \varpi)$ *is a tuple*

$$\lambda_C(\tau, \boldsymbol{\Psi}) = \big(\lambda_C(\tau, \Psi_1), \ldots, \lambda_C(\tau, \Psi_n)\big).$$

Then, the problem differs from Problem 1 with the change in objective (ii), where we aim to find a trace minimizing the *combined* level of unsafety. To address the new problem, we make only a single modification in the proposed algorithm, which is to change the definition of the weight function of the product

$$W_{\mathcal{P}}((s,q),(s',q')) = ((x_1, \ldots, x_n), W_{\mathcal{T}}((s,s'))),$$

with the property that

$$x_i = \sum_{j \in J_R} \overline{W}_{i,j}(q_{i,j}, \sigma, q_{i,j}') + \sum_{j \in J_W} \overline{W}_{i,j}(q_{i,j}, \sigma, q_{i,j}') \cdot W_{\mathcal{T}}((s,s')),$$

such that $\mathcal{A}_{i,j}$, where $j \in J_R \subseteq \{1, \ldots m_i\}$, are the automata for rules associated with the level of unsafety, and $\mathcal{A}_{i,j}$, where $j \in J_W \subseteq \{1, \ldots m_i\}$, are the automata for rules associated with the weighted level of unsafety, for all $i \in \{1, \ldots, n\}$. Since such a product automaton is built, the rest of the our algorithm remains the same.

5. CASE STUDY

Consider a robotic vehicle in the partitioned environment depicted in Fig. 1. The robot's motion capabilities are modeled as a transition system. Each state of the transition of the system is given by (i) the cell the robot occupies and (ii) the robot's direction within the cell. For simplicity, only four directions N, E, S, W, i.e., North, East, South and

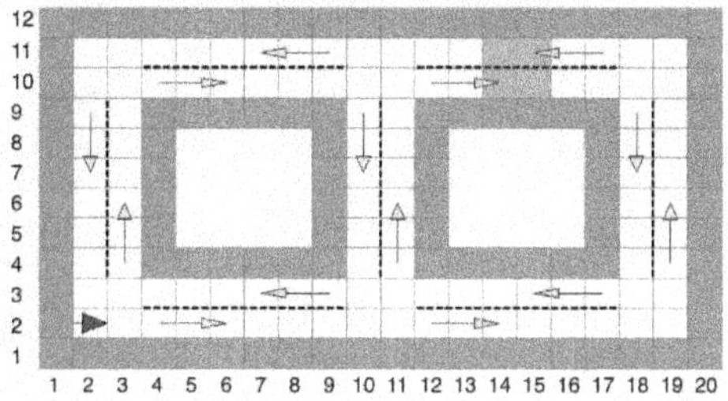

Figure 1: A road-like environment partitioned into cells. Each road segment has two lanes and the purple-shaded area illustrates a sidewalk. The robot's position and direction are depicted as the blue arrow. The white arrows in the lane illustrate the direction the vehicle should be heading when present there. Note that the vehicle is allowed to head in two different directions in each intersection cell. The green-shaded area is the goal region.

West, respectively, are considered. The simulation environment depicted in Fig. 1 has 240 cells and 960 corresponding states. Define $Dir = \{N, E, S, W\}$, $Col = \{1, \ldots, 20\}$, and $Row = \{1, \ldots 12\}$, where Col and Row represent the column number as measured from left to right and the row number as measured from bottom to top in Fig. 1, respectively. Then, the set of states of the transition system can be described conveniently as

$$Q = \{q^d_{c,r} \mid c \in Col, r \in Row, d \in Dir\}.$$

For example, in this notation, the current state of the robot, as depicted in Fig. 1, is $q^E_{2,2}$. Starting from any state, the robot can move into the adjacent cell in front of it by keeping its current direction or into the cell on its left-hand side or its right-hand side by changing it's direction accordingly (if such cells exist in the environment). For instance, when the current state is $q^E_{2,2}$, these transitions would lead into $q^E_{3,2}$, $q^N_{2,3}$ and $q^S_{2,1}$, respectively. The weight of each transition is set to the distance between the adjacent cells, which equals to 1 in this case.

Throughout the case studies, the set of atomic propositions is defined as

$$AP = \{road\_X, car\_X, sidewalk, goal\},$$

where $X \in \{N, E, S, W\}$. Each state $q^X_{c,r}$ is labeled with proposition car_X. The states of the TS that correspond to the road cells of the environment where the vehicle should be positioned in direction X are labeled with $road\_X$. For instance, $road\_E \in L(q^Y_{2,r})$, for all $r \in \{2, \ldots, 19\}, Y \in Dir$ and similarly $road\_S \in L(q^Y_{c,2})$, for all $c \in \{2, \ldots, 11\}, Y \in Dir$. The states that correspond to sidewalk cells (in purple) of the TS are labeled with the proposition $sidewalk$. The states corresponding to the goal region (shown in green in Fig. 1) are labeled with the proposition $goal$. For instance, the state $q^W_{15,11}$ is labeled with $L(q^W_{15,11}) = \{road\_W, car\_W, goal\}$, the state $q^N_{15,11}$ with $L(q^N_{15,11}) = \{road\_W, car\_N, goal\}$, the state $q^N_{20,1}$ with $L(q^N_{20,1}) = \{sidewalk, car\_N\}$, and the state $q^E_{19,2}$ with $L(q^E_{19,2}) = \{road\_N, road\_E, car\_E\}$.

As in Problem 1, the mission is to drive from the current state into a goal state. At the same time, however, we require the vehicle to follow the rules of the road: (i) always head in one of the allowed directions (*direction rule*) and (ii) do not drive on a sidewalk (*sidewalk rule*). This might be not possible if there are some (static) obstacles present on the road. In such a case, we focus on finding a path that is, in some sense as close as possible to obeying the rules, where closeness is defined in terms of level of unsafety discussed in Sec. 3. Here, driving on the sidewalk is considered less safe than driving in a lane while heading in an incorrect direction. That is, the sidewalk rule belongs to a higher priority class when compared to the direction rule.

Furthermore, we would like to prevent the vehicle from performing dangerous maneuvers. For instance, consider two obstacles present in the right lane close to each other. The vehicle could pass the first obstacle, move back to the right lane and then immediately switch back to the wrong lane again in order to pass the second obstacle. However, rather than that, we expect the vehicle to stay in the wrong lane while passing both of the obstacles. Similar behavior is expected when passing a sequence of obstacles using a sidewalk. We specifically set the *dangerous maneuver* as coming back from heading in a wrong direction (or from a sidewalk) into the right lane, staying in the right lane for up to 3 more steps and entering the wrong lane (or a sidewalk, respectively) again. Such a maneuver is forbidden with a higher priority as the direction rule (or the sidewalk rule, respectively). However, spending 10 steps heading in an incorrect direction (or going on the sidewalk) is considered to be less safe than performing the dangerous maneuver. Therefore, we equip the dangerous maneuver rule with priority 3 and the direction rule (or the sidewalk rule, respectively) with priority 1. The level of unsafety of a trace with a single dangerous maneuver is then 3,6, or 9, depending whether the vehicle spends during the maneuver 1,2, or 3 steps in the right lane, respectively. Note, that we could easily distinguish even between unsafety of different dangerous maneuvers by defining a dangerous maneuver of staying in the right lane for up to 2, or even only 1 step and setting their priority accordingly.

The rules that we describe above can be expressed formally using the LTL as follows.

- Never be on the sidewalk.

$$\psi_{1,1} = \mathsf{G}(\neg sidewalk)$$

- Do not perform the dangerous sidewalk maneuver. In other words, if coming back from a sidewalk to a road, stay in a road for at least 4 steps.

$$\begin{aligned}\psi_{1,2} = \mathsf{G}(&sidewalk \wedge \mathsf{X}\neg sidewalk \Rightarrow \\ &\mathsf{X}\,(\neg sidewalk \wedge \\ &\mathsf{X}\,(\neg sidewalk \wedge \\ &\mathsf{X}\,(\neg sidewalk \wedge \\ &\mathsf{X}\,(\neg sidewalk)))))\end{aligned}$$

- Never head in a wrong direction. Heading in an incorrect direction is formalized as a Boolean combination of atomic propositions $car\_X \wedge \neg road\_X$, for some $X \in Dir$, i.e., as the vehicle heading in a different direction than expected in the current cell. For instance, the vehicle heads in a wrong direction in states $q^W_{11,2}$, $q^N_{11,2}$, $q^E_{10,4}$, or $q^S_{11,4}$.

$$\psi_{2,1} = \bigwedge_{X \in Dir} \mathsf{G}\,(car\_X \Rightarrow road\_X)$$

- Do not do the dangerous wrong lane maneuver. In other words, if coming back from heading in a wrong direction to the right one, stay in the right lane for at least 4 steps.

$$\psi_{2,2} = \bigwedge_{X,Y_1,Y_2,Y_3,Y_4 \in Dir} \mathsf{G}\Big(\big(\neg(car\_X \wedge road\_X) \wedge \mathsf{X}\,(car\_Y_1 \wedge road\_Y_1)\big) \Rightarrow \mathsf{X}\,(car\_Y_1 \wedge road\_Y_1 \wedge \mathsf{X}\,(car\_Y_2 \wedge road\_Y_2 \wedge \mathsf{X}\,(car\_Y_3 \wedge road\_Y_3 \wedge \mathsf{X}\,(car\_Y_4 \wedge road\_Y_4))))\Big)$$

Altogether, the set of rules with priorities is

$$(\Psi = (\Psi_1, \Psi_2,), \varpi) = ((\{\psi_{1,1}, \psi_{1,2}\}, \{\psi_{2,1}, \psi_{2,2}\}), \varpi),$$

where $\varpi(\psi_{1,1}) = 1$, $\varpi(\psi_{1,2}) = 3$, $\varpi(\psi_{2,1}) = 1$,$\varpi(\psi_{2,2}) = 3$.

The finite automata corresponding to the rules $\psi_{1,1}$ and $\psi_{1,2}$ are depicted in Fig. 2 and 3, respectively. The finite automata for the rules $\psi_{2,1}$ and $\psi_{2,2}$ are constructed analogously.

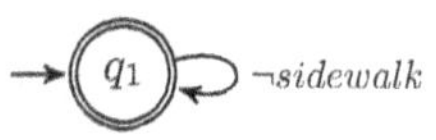

Figure 2: Finite automaton $\mathcal{A}_{1,1}$ for the safety rule $\psi_{1,1}$. The transition labeled with *¬sidewalk* in fact represents $2^{|AP|-1}$ transitions labeled with all the subsets of atomic propositions that do not contain proposition *sidewalk*.

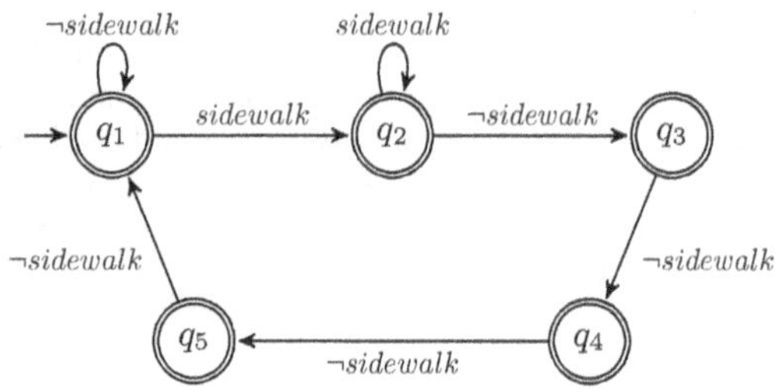

Figure 3: Finite automaton $\mathcal{A}_{1,2}$ for the safety rule $\psi_{1,2}$. Each transition labeled with *sidewalk* in fact represents $2^{|AP|-1}$ transitions labeled with all the subsets of atomic propositions that contain proposition *sidewalk* and each transition labeled with *¬sidewalk* represent $2^{|AP|-1}$ transitions labeled with all the subsets of atomic propositions that do not contain proposition *sidewalk*.

Simulation Results

Below six different scenarios, each of which feature a different set of obstacles, are considered. In each simulation example, the vehicle is started from the initial state, $q^E_{2,2}$. This condition allows for the overall distance and maneuvering penalties to be compared uniformly across all examples.

A weighted product automaton was built for each of the cases, whose transitions are weighted by 3-tuples. The shortest run of the product automaton (in the lexicographical ordering) projects onto an optimal trace of the transition system. The first two elements of the weight tuple represent the level of unsafety of the trace with respect to $\mathbf{\Psi}$ and can be viewed also as penalties collected for violation of the both sidewalk and both direction rules, respectively. The third element is the trace weight, i.e., the total distance travelled.

The results for the considered cases are captured in Fig. 4-9. The red cells depict the obstacles and the blue arrows illustrate the computed robotic traces in the given environments.

A. No obstacles.

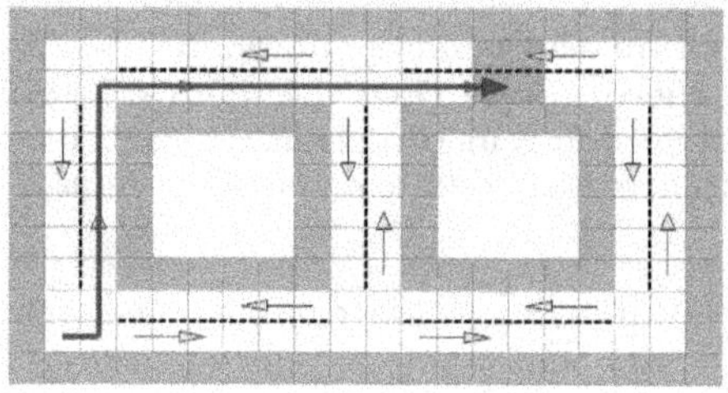

Figure 4: Case A

In this case, there were no obstacles present in this environment. The algorithm simply picked the trace of shortest distance, which can be either the path shown in Fig. 4, or a trace that travels in the North lane of the middle S-N road. The weight assigned to either of these paths was W_A=(0,0,20), i.e., the computed trace satisfies all formulas in $\mathbf{\Psi}$ and its weight, i.e., the total distance traveled is 20.

B. Obstacles present in the right lane in both leftmost and middle S-N roads.

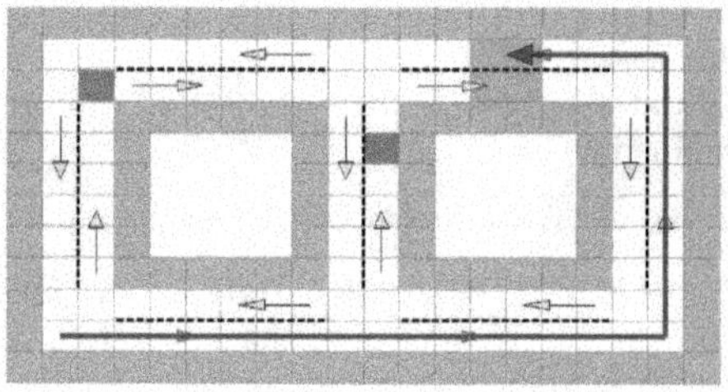

Figure 5: Case B

The obstacles present in this environment removed the options of traveling in the North lane of the middle S-N road and in the North lane of the leftmost S-N road, the shortest paths available for the environment. Considering these restrictions, the algorithm instead chose to travel in the North lane of the rightmost S-N road. This decision was made in lieu of attempting to switch lanes to get around one of the obstacles. While this was not the trace of shortest distance, it was the path of smallest weight, with W_B=(0,0,30). This path was unique in that there were no others paths whose weight was equivalent.

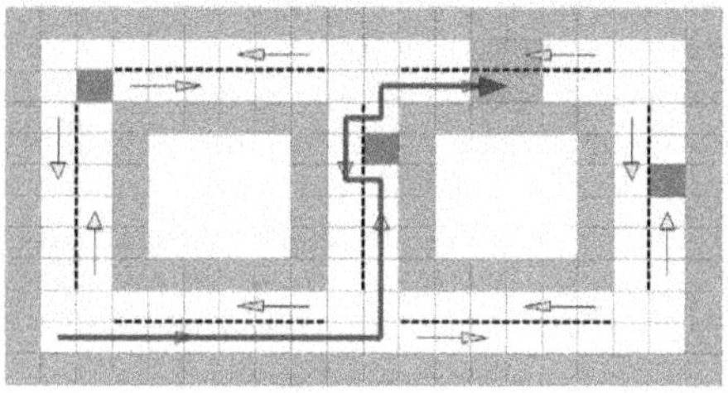

Figure 6: Case C

C. Obstacles present in each S-N road.

The three obstacles in the environment effectively blocked all paths which did not incur any sidewalk or direction rules violation. In this case, the algorithm chose a lane switching maneuver with the resulting weight equal to W_C=(0,4,22). The level of unsafety of the computed trace is 0 with respect to the sidewalk rules Ψ_1 and 4 with respect to the wrong direction rules Ψ_2. The four penalties accrued during this path were a result of two lane switches which resulted in an incorrect (West and East) orientation in the South and North lanes, respectively, and the two North transitions while in the South lane. By inspection, this type of maneuvering on the rightmost S-N road, would have incurred an identical lane penalty, but would have had a larger weight due to the overall distance metric.

D. Obstacles totally blocking the leftmost and the middle S-N roads and two obstacles close to each other in the rightmost S-N road.

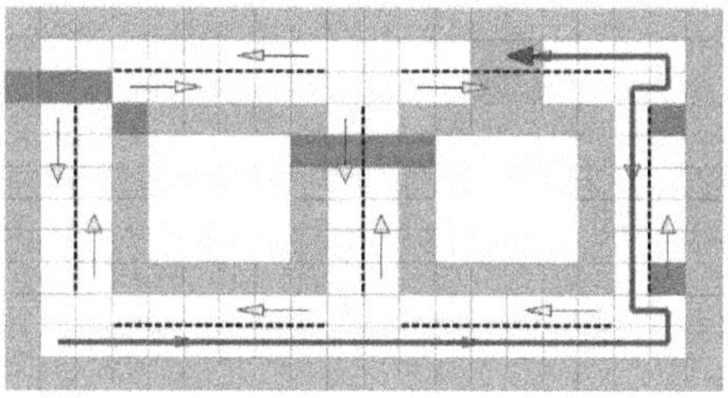

Figure 7: Case D

In this case, the algorithm found a trace that involved lane switching and no sidewalks. As expected from the definition of the dangerous maneuver rule $\Psi_{2,2}$ and its priority, the vehicle does not switch lanes in the presence of obstacles in the same lane that are spaced four units of distance or less apart. The path as shown in Fig. 7 has weight $W_D = (0, 7, 32)$. The seven lane penalties were result of seven transitions with an incorrect orientation in the South lane. In this case, the lane switching does not cause increase in the level of unsafety as the lane switching maneuvers happen in the intersection areas, where heading East and West is allowed in North and South lanes, respectively.

E. Sidewalk needed to pass the obstacles.

With the addition of extra obstacles in the middle S-N column, the vehicle was unable to finish any path without entering the sidewalks. The rightmost S-N road was chosen, because it only required the vehicle to go on the sidewalk to avoid the obstacles, as opposed to both switching lanes and going on the sidewalk required by the middle S-N road. The weight of the path along rightmost S-N road was W_E=(3,1,32), because the vehicle was on the sidewalk for three units of distance and it was penalized once it returned to the North lane from the sidewalk with an incorrect orientation.

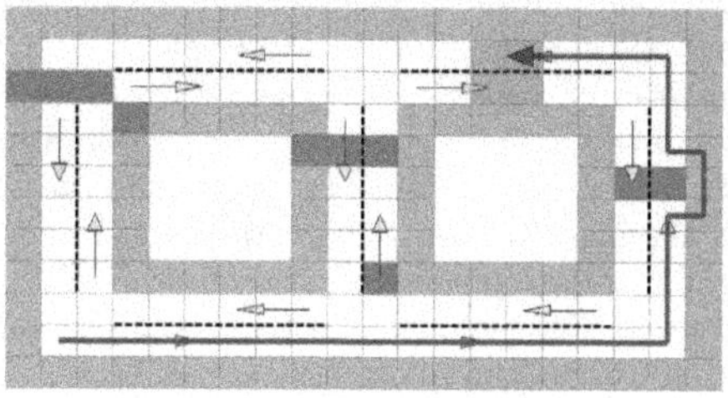

Figure 8: Case E

F. Sidewalk needed to pass the obstacles.

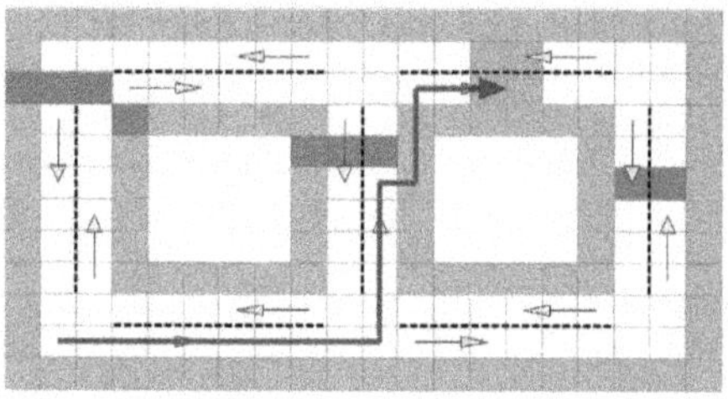

Figure 9: Case F

Because all roads were blocked by obstacles, in this case it was required once again that the vehicle travel on the sidewalk. Instead of returning to the North lane on the middle S-N road, the vehicle chose to re-enter the highway on the uppermost E-W road. The weight of the chosen path, W_F=(3,1,20) was a result of traveling on the sidewalk for three units of distance and entering the uppermost E-W road with an incorrect initial orientation. If the vehicle were to return to the North lane of the middle S-N road instead of directly enter the East lane of the uppermost E-W road, the resulting weight would be $(3, 1, 22)$.

The proposed algorithm was implemented in MATLAB for experimental evaluation on this case study. In each of the scenarios, the product automaton had 24000 states and was built in approx. 8 minutes for each case. The time of the computation of the Dijkstra's shortest path algorithm varied from approx. 1 minute to 20 minutes depending on the total weight of the computed run of the product automaton. The reported times were measured on a Macbook Pro laptop with 2.3 GHz processor and 4GB memory.

6. CONCLUSIONS

In this paper, we have considered the problem of control strategy synthesis where the mission specification is to fulfill the task of reach a goal state while abiding by a given set of rules. We have focused on the case when the task is not feasible without temporarily violating a subset of the rules. We have proposed an algorithm that provably violates on the lowest-priority rules for the shortest amount of time. We have shown the correctness of the proposed algorithm, and we have analyzed its complexity. Furthermore, we have evaluated the algorithm in a case study involving a robotic car navigating in an urban environment.

Future work will include addressing more complex task specifications, other than "reaching a goal state." Examples may include liveness specifications, such as those that describe surveillance missions. Furthermore, we would like to extend our techniques to non-deterministic systems that are often obtained as discrete abstractions of hybrid systems.

Acknowledgments

This work is supported in part by the National Research Foundation of Singapore, through the Future Urban Mobility SMART IRG, by the US National Science Foundation, grant CNS-1016213, ONR-MURI Award N00014-09-1-1051, and grant LH11065 at Masaryk University, Czech Republic.

7. REFERENCES

[1] C. Baier and J.-P. Katoen. *Principles of Model Checking.* MIT Press, 2008.

[2] E. Bartocci, R. Grosu, P. Katsaros, C. R. Ramakrishnan, and S. A. Smolka. Model repair for probabilistic systems. In *International Conference on Tools and Algorithms for the Construction and Analysis of Systems (TACAS)*, pages 326–340. Springer-Verlag, 2011.

[3] C. Belta and L. C. G. J. M. Habets. Control of a class of nonlinear systems on rectangles. *IEEE Transactions on Automatic Control*, 51(11):1749–1759, 2006.

[4] F. Buccafurri, T. Eiter, G. Gottlob, and N. Leone. Enhancing model checking in verification by AI techniques. *Artificial Intelligence*, 112(1-2):57 – 104, 1999.

[5] A. Cimatti, M. Roveri, V. Schuppan, and A. Tchaltsev. Diagnostic information for realizability. In *Proceedings of the International Conference on Verification, Model Checking, and Abstract Interpretation (VMCAI)*, pages 52–67, Berlin, Heidelberg, 2008. Springer-Verlag.

[6] C. Courcoubetis and M. Yannakakis. Markov decision processes and regular events. *IEEE Transactions on Automatic Control*, 43:1399 – 1418, 1998.

[7] W. Damm and B. Finkbeiner. Does it pay to extend the perimeter of a world model? In *Proceedings of the International Symposium on Formal Mehods (FM)*, pages 12–26, Berlin, Heidelberg, 2011. Springer-Verlag.

[8] E. W. Dijkstra. A note on two problems in connexion with graphs. *Numerische Mathematik*, 1:269–271, 1959.

[9] X. C. Ding, S. Smith, C. Belta, and D. Rus. Mdp optimal control under temporal logic constraints. In *Proceedings of the IEEE Conference on Decision and Control and European Control Conference (CDC-ECC)*, pages 532 –538, 2011.

[10] G. E. Fainekos. Revising temporal logic specifications for motion planning. In *Proceedings of the IEEE International Conference on Robotics and Automation (ICRA)*, 2011.

[11] E. L. Gunter and D. Peled. Temporal debugging for concurrent systems. In *International Conference on Tools and Algorithms for the Construction and Analysis of Systems (TACAS)*, pages 431–444. Springer-Verlag, 2002.

[12] K. Hauser. The minimum constraint removal problem with three robotics applications. In *Proceedings of the International Workshop on the Algorithmic Foundations of Robotics (WAFR)*, 2012.

[13] Karaman and Frazzoli. Sampling-based optimal motion planning with deterministic μ-calculus specifications. In *Proceedings of the American Control Conference (ACC)*, 2012.

[14] K. Kim and G. Fainekos. Approximate solutions for the minimal revision problem of specification automata. In *Proceedings of the IEEE/RSJ International Conference on Intelligent Robots and Systems (IROS)*, 2012.

[15] K. Kim, G. Fainekos, and S. Sankaranarayanan. On the revision problem of specification automata. In *Proceedings of the IEEE International Conference on Robotics and Automation (ICRA)*, 2012.

[16] M. Kloetzer and C. Belta. A Fully Automated Framework for Control of Linear Systems from Temporal Logic Specifications. *IEEE Transactions on Automatic Control*, 53(1):287–297, 2008.

[17] H. Kress-Gazit, G. E. Fainekos, and G. J. Pappas. Temporal Logic-based Reactive Mission and Motion Planning. *IEEE Transactions on Robotics*, 25(6):1370–1381, 2009.

[18] Z. Manna and A. Pnueli. *Temporal Verification of Reactive Systems: Safety.* Springer, 1995.

[19] V. Raman and H. Kress-Gazit. Analyzing unsynthesizable specifications for high-level robot behavior using ltlmop. In *Proceedings of International Conference on Computer Aided Verification (CAV)*, pages 663–668, 2011.

[20] V. Raman and H. Kress-Gazit. Automated feedback for unachievable high-level robot behaviors. In *Proceedings of the IEEE International Conference on Robotics and Automation (ICRA)*, pages 5156–5162, 2012.

[21] M. Sipser. *Introduction to the Theory of Computation.* Course Technology, 3rd edition, 2012.

[22] S. L. Smith, J. Tumova, C. Belta, and D. Rus. Optimal Path Planning for Surveillance with Temporal Logic Constraints. *International Journal of Robotics Research*, 30(14):1695–1708, 2011.

[23] A. Ulusoy, S. L. Smith, X. C. Ding, and C. Belta. Robust multi-robot optimal path planning with temporal logic constraints. In *Proceedings of the IEEE International Conference on Robotics and Automation (ICRA)*, pages 4693–4698, 2012.

[24] M. Y. Vardi and P. Wolper. An automata-theoretic approach to automatic program verification. In *Logic in Computer Science*, pages 322–331, 1986.

[25] T. Wongpiromsarn, U. Topcu, and R. M. Murray. Receding Horizon Control for Temporal Logic Specifications. In *Hybrid systems: Computation and Control (HSCC)*, pages 101–110, 2010.

Limited-Information Control of Hybrid Systems via Reachable Set Propagation

Daniel Liberzon*
Coordinated Science Laboratory
University of Illinois at Urbana-Champaign
Urbana, IL 61821, USA
liberzon@uiuc.edu

ABSTRACT

This paper deals with control of hybrid systems based on limited information about their state. Specifically, measurements being passed from the system to the controller are sampled and quantized, resulting in finite data-rate communication. The main ingredient of our solution to this control problem is a novel method for propagating over-approximations of reachable sets for hybrid systems through sampling intervals, during which the discrete mode is unknown. In addition, slow-switching conditions of the (average) dwell-time type and multiple Lyapunov functions play a central role in the analysis.

Categories and Subject Descriptors

H.1.1 [**Models and Principles**]: Systems and Information Theory—*general systems theory, information theory*

Keywords

Hybrid systems, quantized control, reachable sets

1. INTRODUCTION

The topic of this paper is control of hybrid systems based on limited information about their state. More precisely, by "limited information" here we mean that measurements being passed from the system to the controller are sampled and quantized using a finite alphabet, resulting in finite data-rate communication. Our aim is to bring together two research areas—hybrid systems and control with limited information—which have both enjoyed a lot of activity in the last two decades and made great impact on applications, but synergy between which has been lacking.

*This work was supported by the NSF grants CNS-1217811 and ECCS-1231196 and by a National Research Foundation Grant funded by the Korean Government (NRF-2011-220-D00043).

HSCC'13, April 8–11, 2013, Philadelphia, Pennsylvania, USA.

Hybrid systems are ubiquitous in realistic system models, because of their ability to capture the presence of two types of dynamical behavior within the system: continuous flow and discrete transitions. Amidst the large body of research on hybrid and switched systems, particularly relevant here is the work on stability analysis and stabilization of such systems, covered in the books [36, 19], the survey [32], and the many references therein. Among the specific technical tools typically employed to study these problems, common and multiple Lyapunov functions and slow-switching assumptions are prominently featured. There is also recent work that combines these Lyapunov-based approaches with tools from program analysis like ranking functions and abstraction [28, 5].

Feedback control problems with limited information have been an active research area for some time now, as surveyed in [25] (several specifically relevant works will be cited below). Information flow in a feedback control loop is an important consideration in many application-related scenarios. One notable example is *networked control systems*, where the controller is not collocated with the plant and both control and measurement signals are transmitted through a network. Even though in modern applications a lot of communication bandwidth is usually available, there are also multiple resources competing for this bandwidth (due to many control loops sharing a network cable or wireless medium). In the networked control systems literature, Lyapunov analysis and data-rate/transmission-interval bounds are commonly employed tools [27, 26]. In many other applications, one has constraints on the sensors dictated by cost concerns or physical limitations, or constraints on information transmission dictated by security considerations. Besides multiple practical motivations, the questions of how much information is really needed to solve a given control problem, or what interesting control tasks can be performed with a given amount of information, are quite fundamental from the theoretical point of view.

Control problems with limited information do not seem to have received much attention so far in the context of hybrid systems. (One exception we are aware of is some existing work on quantized Markov jump linear systems [24, 39, 22], but these systems are quite different from the models we consider, and the information structure considered in these references implies that the discrete mode is always known to the controller, which would remove most of the difficulties present in our problem formulation. On the other hand, control of hybrid systems with unknown discrete state was also considered in [37] but there the continuous state was not quantized.) In view of the commonality of the technical tools employed for the analysis of hybrid systems and for control design with limited information, evident from the previous discussion, we contend that a marriage of these two research areas is quite natural. In particular, multiple Lyapunov functions and slow-switching assumptions of the (average) dwell-time type will play a crucial role in our work.

In order to understand how much information is needed—and how this information should be used—to control a given system, we must understand how the uncertainty about the system's state evolves over time along its dynamics. In more precise terms, we need to be able to characterize propagation of reachable sets or their suitable over-approximations during the sampling interval (for a known control input). The reason is that at each sampling time, the quantizer singles out a bounded set which contains the continuous state and the controller determines the control signal to be applied to the system over the next sampling interval; no further information about the state is available during this interval, and at the next sampling time a bounded set containing the state must be computed to generate the next quantized measurement. The system will be stabilized if the factor by which the state estimation error is reduced at the sampling times is larger than the factor by which it grows between the sampling times. Thus propagation of reachable set bounds is a crucial ingredient in the available results on quantized sampled-data control of non-hybrid systems (such as our earlier work [18] which provides the basis for some of the ideas presented here), and the bulk of the effort required to handle the hybrid system scenario will be concentrated in implementing this step and analyzing its consequences.

If the discrete state were precisely known to the controller at each time, then the problem of reachable set propagation would be just a sequence of corresponding problems for the individual continuous modes, and as such would pose very little extra difficulty. (This would essentially correspond to the situation considered, in a discrete-time stochastic setting, in [24].) On the other hand, if the discrete state were completely unknown, then the set of possible trajectories of the hybrid system would be too large to hope for a reasonable (not overly conservative) solution. To strike a balance between these two situations, we assume here that we have a partial knowledge of the discrete state of the hybrid system; specifically, we assume that the discrete state is known at each sampling time, whereas its behavior between switching times is unknown except for an upper bound on the number of discrete transitions. Under these assumptions we implement a novel method for computing over-approximations of reachable sets at the end of each sampling interval. If in addition the allowed data rate is large enough, then we are able to design a provably correct communication and control strategy.

Our plan of attack on the problem described above is as follows. As a first step, we consider *switched systems* which can be viewed as high-level abstractions of hybrid systems. Instead of having a precise model of the discrete dynamics governing the transitions of the discrete state, in a switched system the continuous dynamics are given explicitly (by ordinary differential equations) while discrete dynamics are captured more abstractly by a class of possible switching signals which specify the discrete state as a function of time. In other words, we basically first consider systems with time-dependent switching (switched systems) instead of systems with state-dependent switching (hybrid systems). This system class allows us to focus our attention on formulating an appropriate "slow-switching" assumption; the relevant definitions are given in Section 2. We then develop a novel method for propagating over-approximations of reachable sets for such switched systems; this method is described in Sections 3 and 4. With these two ingredients—a slow-switching condition and an algorithm for reachable set propagation—we are able to define and validate a control strategy based on sampled and quantized measurements of the continuous state; the analysis is sketched in Section 5 followed by a short simulation example in Section 6. After this, we build a bridge to hybrid systems by returning to a detailed description of the discrete dynamics in Section 7.

2. PROBLEM FORMULATION

2.1 Switched system

The system to be controlled is the switched linear control system

$$\dot{x} = A_\sigma x + B_\sigma u, \qquad x(0) = x_0 \tag{1}$$

where $x \in \mathbb{R}^n$ is the state, $u \in \mathbb{R}^m$ is the control input, $\{(A_p, B_p) : p \in \mathcal{P}\}$ is a collection of matrix pairs defining the individual control systems ("modes") of the switched system, $\mathcal{P}$ is a finite index set, and $\sigma : [0, \infty) \to \mathcal{P}$ is a right-continuous, piecewise constant function called the *switching signal* which specifies the active mode at each time. The solution $x(\cdot)$ is absolutely continuous and satisfies the differential equation away from the discontinuities of σ (in particular, we assume for now that there are no state jumps, but state jumps can also be handled as explained in Section 7.3). The switching signal σ is fixed but not known to the controller a priori. The discontinuities of σ are called "switching times" or simply "switches" and we let $N_\sigma(t, s)$ stand for their number on a semi-open interval $(s, t]$:

$$N_\sigma(t, s) := \text{number of switches on } (s, t].$$

Our first basic assumption is that the switching is not too fast, in the following sense.

Assumption 1 (Slow Switching)

1. There exists a number $\tau_d > 0$ (called a *dwell time*) such that any two switches are separated by at least τ_d, i.e., $N_\sigma(t, s) \le 1$ when $t - s \le \tau_d$;
2. There exist numbers $\tau_a > \tau_d$ (called an *average dwell time*) and $N_0 \ge 1$ such that

$$N_\sigma(t, s) \le N_0 + \frac{t - s}{\tau_a} \qquad \forall\, t > s \ge 0.$$

The concept of average dwell time was introduced in [13] and has since then become standard; it includes dwell time as a special case (for $N_0 = 1$). Note that if the constraint $\tau_a > \tau_d$ were violated, the average dwell-time condition (item 2) would be implied by the dwell-time condition (item 1). Switching signals satisfying Assumption 1 were considered in [38], where they were called "hybrid dwell-time" signals.

Our second basic assumption is stabilizability of all individual modes.

Assumption 2 (Stabilizability) For each $p \in \mathcal{P}$ the pair (A_p, B_p) is stabilizable, i.e., there exists a state feedback gain matrix K_p such that $A_p + B_p K_p$ is Hurwitz (all eigenvalues have negative real parts).

In the sequel, we assume that a family of such stabilizing gain matrices K_p, $p \in \mathcal{P}$ has been selected and fixed. We understand that (at least some of) the open-loop matrices A_p, $p \in \mathcal{P}$ are not Hurwitz. Note, however, that even if all the individual modes are stabilized by state feedback (or stable without feedback), stability of the switched system is not guaranteed in general.

2.2 Information structure

The task of the controller is to generate a control input $u(\cdot)$ based on limited information about the state $x(\cdot)$ and about the switching signal $\sigma(\cdot)$. The information to be communicated to the controller is subject to the following two constraints.

Sampling: State measurements are taken at times $t_k := k\tau_s$, $k = 0, 1, 2, \dots$, where $\tau_s > 0$ is a fixed *sampling period.*

Quantization: Each state measurement $x(t_k)$ is encoded by an integer from 0 to N^n, where N is an odd positive integer, and sent to the controller. In addition, the value of $\sigma(t_k) \in \mathcal{P}$ is also sent to the controller.

As a consequence, data is transmitted to the controller at the rate of $(\log_2(N^n + 1) + \log_2 |\mathcal{P}|)/\tau_s$ bits per time unit, where $|\mathcal{P}|$ is the number of elements in $\mathcal{P}$. We assume the data transmission to be noise-free and delay-free. We take the sampling period τ_s to be no larger than the dwell time from Assumption 1 (item 1):

$$\tau_s \leq \tau_d. \tag{2}$$

This guarantees that at most one switch occurs within each sampling interval.[1] Since the average dwell time τ_a in Assumption 1 (item 2) is larger than τ_d, we know that switches actually occur less often than once every sampling period. The reason for taking the integer N to be odd is to ensure that the control strategy described later preserves the equilibrium at the origin.

Throughout the paper, we work with the ∞*-norm* $\|x\|_\infty = \max_{1\leq i\leq n} |x_i|$ *on* $\mathbb{R}^n$ *and the corresponding induced matrix norm* $\|A\|_\infty = \max_{1\leq i\leq n} \sum_{j=1}^n |A_{ij}|$ *on* $\mathbb{R}^{n\times n}$*, both of which we denote simply by* $\|\cdot\|$. To formulate our final basic assumption, we define

$$\Lambda_p := \|e^{A_p \tau_s}\|, \qquad p \in \mathcal{P}. \tag{3}$$

Assumption 3 (Data Rate) $\Lambda_p < N$ for all $p \in \mathcal{P}$.

We can view the above inequality as a data-rate bound because it requires N to be sufficiently large relative to τ_s, thereby imposing (indirectly) a lower bound on the available data rate. A very similar data-rate bound but for the case of a single mode appears in [18], where it is shown to be sufficient for stabilizing a non-switched linear system. That bound is slightly conservative compared to known bounds that characterize the minimal data rate necessary for stabilization (see, e.g., [34], [15]). However, the control scheme of [18] can be refined by tailoring it better to the structure of the system matrix A, and then the data rate that it requires will approach the minimal data rate (see also the discussion in [31, Section V]). Therefore, it is fair to say that Assumption 3 does not introduce a significant conservatism beyond requiring that the data rate be sufficient to stabilize each individual mode of the switched system (1).

[1] This assumption is made for simplicity. It could be relaxed to allow multiple switches, up to a fixed number, per sampling interval. This would make our formulas more complicated but would not cause conceptual difficulties.

2.3 Main objective

The control objective is to asymptotically stabilize the system defined in Section 2.1 while respecting the information constraints described in Section 2.2. More concretely, we want to provide a constructive proof of the following result.

Theorem 1 (Main Result) Consider the switched linear system (1) and let Assumptions 1–3 and the inequality (2) hold. If the average dwell time τ_a is large enough, then there exists an encoding and control strategy that yields the following two properties:

Exponential convergence: There exist a number $\lambda > 0$ and a function $g : [0, \infty) \to (0, \infty)$ such that for every initial condition x_0 and every time $t \geq 0$ we have

$$\|x(t)\| \leq e^{-\lambda t} g(\|x_0\|). \tag{4}$$

Lyapunov stability: For every $\varepsilon > 0$ there exists a $\delta > 0$ such that

$$\|x_0\| < \delta \quad \Rightarrow \quad \|x(t)\| < \varepsilon \quad \forall t \geq 0. \tag{5}$$

A precise lower bound on the average dwell time τ_a is derived in the course of the proof (see the formula (24) in Section 5). As for the function g in the exponential convergence property, $g(r)$ does not go to 0 as $r \to 0$ and, in general, g grows faster than any linear function at infinity (see the formula (25) in Section 5 and the discussion at the end of Section 4.3). For this reason, Lyapunov stability needs to be established separately, and the two properties (exponential convergence and Lyapunov stability) combined still do not give the standard global exponential stability, but rather just global asymptotic stability with an exponential convergence rate.

The control strategy that we will develop to prove Theorem 1 is a dynamic one: it involves an additional state denoted by $\hat{x}$. Theorem 1 only discusses the behavior of the state x, which is the main quantity of interest, but it can be deduced from the proof that the controller state $\hat{x}$ satisfies analogous bounds. We will also see that $\hat{x}$ is potentially discontinuous at the sampling times t_k (which are not synchronized with the switching times of the original system); in other words, our controller is a hybrid one.

3. BASIC ENCODING AND CONTROL SCHEME

In this section we outline our encoding and control strategy, assuming for now that the state x satisfies known bounds at the sampling times. The problem of generating such state bounds is solved in the next section.

First, suppose that at some sampling time t_{k_0} we have

$$\|x(t_{k_0})\| \leq E_{k_0}$$

where $E_{k_0} > 0$ is a number known to the controller. (In Section 4.3 we will show how such a bound can be generated for an arbitrary initial state x_0, by using a "zooming-out" procedure.) At the first such sampling time our controller is initialized. The encoder works by partitioning the hypercube $\{x \in \mathbb{R}^n : \|x\| \leq E_{k_0}\}$ into N^n equal hypercubic boxes, N per each dimension, and numbering them from 1 to N^n in some specific way. It then records the number of the box that contains[2] $x(t_{k_0})$ and sends it to the controller, along with the value of $\sigma(t_{k_0})$. We assume that the controller knows the box numbering system used by the encoder, so it can decode the box number. It lets $c_{k_0} \in \mathbb{R}^n$ be the center of the box containing $x(t_{k_0})$. We then have

$$\|x(t_{k_0}) - c_{k_0}\| \leq \frac{E_{k_0}}{N}.$$

For $t \in [t_{k_0}, t_{k_0+1})$, the control is set to

$$u(t) = K_{\sigma(t_{k_0})}\hat{x}(t)$$

where $\hat{x}$ is defined to be the solution of

$$\dot{\hat{x}} = (A_{\sigma(t_{k_0})} + B_{\sigma(t_{k_0})}K_{\sigma(t_{k_0})})\hat{x} = A_{\sigma(t_{k_0})}\hat{x} + B_{\sigma(t_{k_0})}u$$

with the boundary condition

$$\hat{x}(t_{k_0}) = c_{k_0}.$$

At a general sampling time t_k, $k \geq k_0 + 1$, suppose that a point $x_k^* \in \mathbb{R}^n$ and a number $E_k > 0$ are known such that

$$\|x(t_k) - x_k^*\| \leq E_k. \tag{6}$$

Of course the encoder has precise knowledge of x; the quantities x_k^* and E_k have to be obtainable on the decoder/controller side, based on the knowledge of the system matrices (but not the switching signal) and previously received measurements. We explain later how such x_k^* and E_k can be generated. The encoder also computes x_k^* and E_k in the same way, to ensure that the encoder and the decoder are synchronized. The encoding is then done as follows. Partition the hypercube $\{x \in \mathbb{R}^n : \|x - x_k^*\| \leq E_k\}$ into N^n equal hypercubic boxes, N per each dimension. Send the number of the box to the controller, along with the value of $\sigma(t_k)$. On the decoder/controller side, let c_k be the center of the box containing $x(t_k)$. This gives

$$\|x(t_k) - c_k\| \leq \frac{E_k}{N} \tag{7}$$

and also

$$\|c_k - x_k^*\| \leq \frac{N-1}{N}E_k. \tag{8}$$

Note that the formula (8) is also valid for $k = k_0$ if we set $x_{k_0}^* := 0$, a convention that we follow in the sequel. For $t \in [t_k, t_{k+1})$ define the control, along the same lines as before, by

$$u(t) = K_{\sigma(t_k)}\hat{x}(t)$$

where $\hat{x}$ is the solution of

$$\dot{\hat{x}} = (A_{\sigma(t_k)} + B_{\sigma(t_k)}K_{\sigma(t_k)})\hat{x} = A_{\sigma(t_k)}\hat{x} + B_{\sigma(t_k)}u \tag{9}$$

with the boundary condition

$$\hat{x}(t_k) = c_k. \tag{10}$$

[2]In case $x(t_{k_0})$ lies on the boundary of several boxes, either one of these boxes can be chosen.

The above procedure is to be repeated for each subsequent value of k. Note that $\hat{x}$ is, in general, discontinuous (only right-continuous) at the sampling times, and we will use the notation $\hat{x}(t_k^-) := \lim_{t \nearrow t_k} \hat{x}(t)$. In the earlier work [18], x_k^* was obtained directly from $\hat{x}$ via $x_k^* := \hat{x}(t_k^-)$. On sampling intervals containing a switch this construction no longer works, and the task of defining x_k^* as well as E_k becomes more challenging.

4. STATE BOUNDS: REACHABLE SET OVER-APPROXIMATIONS

Proceeding inductively, we start with known x_k^* and E_k satisfying (6), where $k \geq k_0$, and show how to find x_{k+1}^* and E_{k+1} such that

$$\|x(t_{k+1}) - x_{k+1}^*\| \leq E_{k+1}. \tag{11}$$

Generation of E_{k_0} is addressed at the end of the section.

4.1 Sampling interval with no switch

We first consider the simpler case when $\sigma(t_k) = \sigma(t_{k+1}) = p \in \mathcal{P}$. By (2) we know that no switch has occurred on $(t_k, t_{k+1}]$, since two switches would have been impossible. So, we know that on the whole interval $[t_k, t_{k+1}]$ mode p is active. We can then proceed as in [18]. It is clear from (1) and (9) that the error $e := x - \hat{x}$ satisfies $\dot{e} = A_p e$ on $[t_k, t_{k+1})$, and we know from (10) and (7) that $\|e(t_k)\| \leq E_k/N$, hence

$$\|e(t_{k+1}^-)\| \leq \Lambda_p \frac{E_k}{N} =: E_{k+1} \tag{12}$$

where Λ_p was defined in (3). It remains to let

$$x_{k+1}^* := \hat{x}(t_{k+1}^-) = e^{(A_p + B_p K_p)\tau_s} c_k \tag{13}$$

and recall that x is continuous to see that (11) indeed holds.

4.2 Sampling interval with a switch

Suppose now that $\sigma(t_k) = p$ and $\sigma(t_{k+1}) = q \neq p$. Then again by (2) the controller knows that exactly one switch, from mode p to mode q, has occurred somewhere on the interval $(t_k, t_{k+1}]$, but it does not know exactly where. This case is more challenging.

Let the (unknown) time of the switch from p to q be $t_k + \bar{t}$, where $\bar{t} \in (0, \tau_s]$.

4.2.1 Analysis before the switch

On $[t_k, t_k + \bar{t})$ mode p is active, and we can derive as before that

$$\|x(t_k + \bar{t}) - \hat{x}(t_k + \bar{t})\| \leq \|e^{A_p \bar{t}}\| \frac{E_k}{N}.$$

But $\hat{x}(t_k + \bar{t})$ is unknown, so we need to describe a set that contains it. Choose an arbitrary $t' \in [0, \tau_s]$ (which may vary with k). By (9) and (10) we have[3]

$$\hat{x}(t_k + t') = e^{(A_p + B_p K_p)t'} c_k \tag{14}$$

and

$$\hat{x}(t_k + \bar{t}) = e^{(A_p + B_p K_p)(\bar{t} - t')} \hat{x}(t_k + t')$$

[3]In case either $t_k + t'$ or $t_k + \bar{t}$ equals t_{k+1}, the value of $\hat{x}$ at that time should be replaced by the left limit $\hat{x}(t_k + t'^-)$ or $\hat{x}(t_k + \bar{t}^-)$, respectively.

hence

$$\|\hat{x}(t_k+\bar{t})-\hat{x}(t_k+t')\|$$
$$\le \|e^{(A_p+B_pK_p)(\bar{t}-t')}-I\|\|\hat{x}(t_k+t')\|$$
$$\le \|e^{(A_p+B_pK_p)(\bar{t}-t')}-I\|\|e^{(A_p+B_pK_p)t'}\|\|c_k\|.$$

We also have from (8) that

$$\|c_k\| \le \|x_k^*\| + \frac{N-1}{N}E_k. \tag{15}$$

By the triangle inequality, we obtain

$$\|x(t_k+\bar{t})-\hat{x}(t_k+t')\|$$
$$\le \|e^{(A_p+B_pK_p)(\bar{t}-t')}-I\|\|e^{(A_p+B_pK_p)t'}\|$$
$$\times\Big(\|x_k^*\|+\frac{N-1}{N}E_k\Big)+\|e^{A_p\bar{t}}\|\frac{E_k}{N} =: D_{k+1}(\bar{t}).$$

4.2.2 *Analysis after the switch*

On the interval $[t_k+\bar{t}, t_{k+1})$, the closed-loop dynamics are

$$\begin{pmatrix}\dot{x}\\ \dot{\hat{x}}\end{pmatrix}=\begin{pmatrix}A_q & B_qK_p\\ 0 & A_p+B_pK_p\end{pmatrix}\begin{pmatrix}x\\ \hat{x}\end{pmatrix}. \tag{16}$$

Letting

$$z:=\begin{pmatrix}x\\ \hat{x}\end{pmatrix},\qquad \bar{A}_{pq}:=\begin{pmatrix}A_q & B_qK_p\\ 0 & A_p+B_pK_p\end{pmatrix} \tag{17}$$

we can write (16) in the more compact form

$$\dot{z}=\bar{A}_{pq}z. \tag{18}$$

The previous analysis shows that

$$\Big\|z(t_k+\bar{t})-\begin{pmatrix}\hat{x}(t_k+t')\\ \hat{x}(t_k+t')\end{pmatrix}\Big\|\le D_{k+1}(\bar{t})$$

(noting the property $\|(a^T,b^T)^T\|\le\max\{\|a\|,\|b\|\}$ of the ∞-norm). Consider the auxiliary system copy (on $\mathbb{R}^{2n}$)

$$\dot{\bar{z}}=\bar{A}_{pq}\bar{z},\qquad \bar{z}(0)=\begin{pmatrix}\hat{x}(t_k+t')\\ \hat{x}(t_k+t')\end{pmatrix}.$$

We have

$$\|z(t_{k+1}^-)-\bar{z}(\tau_s-\bar{t})\|\le\|e^{\bar{A}_{pq}(\tau_s-\bar{t})}\|D_{k+1}(\bar{t}).$$

We now need to generate a bound for the unknown $\bar{z}(\tau_s-\bar{t})$. Similarly to what we did before, pick a $t''\in[0,\tau_s]$. Then $\bar{z}(t'')=e^{\bar{A}_{pq}t''}\bar{z}(0)$ and $\bar{z}(\tau_s-\bar{t})=e^{\bar{A}_{pq}(\tau_s-\bar{t}-t'')}\bar{z}(t'')$, hence

$$\|\bar{z}(\tau_s-\bar{t})-\bar{z}(t'')\|\le\|e^{\bar{A}_{pq}(\tau_s-\bar{t}-t'')}-I\|\|\bar{z}(t'')\|$$
$$\le\|e^{\bar{A}_{pq}(\tau_s-\bar{t}-t'')}-I\|\|e^{\bar{A}_{pq}t''}\|\|\bar{z}(0)\|$$
$$=\|e^{\bar{A}_{pq}(\tau_s-\bar{t}-t'')}-I\|\|e^{\bar{A}_{pq}t''}\|\|\hat{x}(t_k+t')\|$$
$$\le\|e^{\bar{A}_{pq}(\tau_s-\bar{t}-t'')}-I\|\|e^{\bar{A}_{pq}t''}\|\|e^{(A_p+B_pK_p)t'}\|$$
$$\times\Big(\|x_k^*\|+\frac{N-1}{N}E_k\Big)$$

where we used (14) and (15) in the last step. By the triangle inequality,

$$\|z(t_{k+1}^-)-\bar{z}(t'')\|\le\|e^{\bar{A}_{pq}(\tau_s-\bar{t}-t'')}-I\|$$
$$\times\|e^{\bar{A}_{pq}t''}\|\|e^{(A_p+B_pK_p)t'}\|\Big(\|x_k^*\|+\frac{N-1}{N}E_k\Big)$$
$$+\|e^{\bar{A}_{pq}(\tau_s-\bar{t})}\|D_{k+1}(\bar{t}) =: E_{k+1}(\bar{t}).$$

To eliminate the dependence on the unknown $\bar{t}$, we take the maximum over $\bar{t}$ (with t' and t'' fixed as above):

$$E_{k+1}:=\max_{0\le\bar{t}\le\tau_s}E_{k+1}(\bar{t})=\max_{0\le\bar{t}\le\tau_s}\Big\{\|e^{\bar{A}_{pq}(\tau_s-\bar{t}-t'')}-I\|$$
$$\times\|e^{\bar{A}_{pq}t''}\|\|e^{(A_p+B_pK_p)t'}\|\Big(\|x_k^*\|+\frac{N-1}{N}E_k\Big)$$
$$+\|e^{\bar{A}_{pq}(\tau_s-\bar{t})}\|\Big(\|e^{(A_p+B_pK_p)(\bar{t}-t')}-I\|\|e^{(A_p+B_pK_p)t'}\|$$
$$\times\Big(\|x_k^*\|+\frac{N-1}{N}E_k\Big)+\|e^{A_p\bar{t}}\|\frac{E_k}{N}\Big)\Big\}.$$

We can use the inequalities

$$\|M-I\|\le\|M\|+1,\qquad \|e^{As}\|\le e^{\|A\|s} \tag{19}$$

to obtain a more conservative upper bound which is more useful for computations:

$$E_{k+1}\le\big(e^{\|\bar{A}_{pq}\|\max\{t'',\tau_s-t''\}}+1\big)\|e^{\bar{A}_{pq}t''}\|\|e^{(A_p+B_pK_p)t'}\|$$
$$\times\Big(\|x_k^*\|+\frac{N-1}{N}E_k\Big)+e^{\|\bar{A}_{pq}\|\tau_s}$$
$$\times\Big(\big(e^{\|A_p+B_pK_p\|\max\{t',\tau_s-t'\}}+1\big)\|e^{(A_p+B_pK_p)t'}\|$$
$$\times\Big(\|x_k^*\|+\frac{N-1}{N}E_k\Big)+e^{\|A_p\|\tau_s}\frac{E_k}{N}\Big).$$

Note that setting $t'=t''=0$ simplifies the formulas but does not necessarily minimize E_{k+1}. Finally, x_{k+1}^* is defined by projecting $\bar{z}(t'')$ onto the x-component:

$$x_{k+1}^* := \begin{pmatrix}I_{n\times n} & 0_{n\times n}\end{pmatrix}\bar{z}(t'')$$
$$=\begin{pmatrix}I_{n\times n} & 0_{n\times n}\end{pmatrix}e^{\bar{A}_{pq}t''}\begin{pmatrix}\hat{x}(t_k+t')\\ \hat{x}(t_k+t')\end{pmatrix} \tag{20}$$
$$=\begin{pmatrix}I_{n\times n} & 0_{n\times n}\end{pmatrix}e^{\bar{A}_{pq}t''}\begin{pmatrix}I_{n\times n}\\ I_{n\times n}\end{pmatrix}e^{(A_p+B_pK_p)t'}c_k.$$

4.3 Initial state bound E_{k_0}

Initially, set the control to $u\equiv 0$. At time 0, choose an arbitrary $E_0>0$ and partition the hypercube $\{x\in\mathbb{R}^n : \|x\|\le E_0\}$ into N^n equal hypercubic boxes, N per each dimension. If x_0 belongs to one of these boxes, then send the number of the box to the controller. Otherwise send 0 (the "overflow" symbol). Choose an increasing sequence $E_1, E_2, \dots$ that grows fast enough to dominate the rate of growth of the open-loop dynamics. For example, we can pick a small $\varepsilon>0$ and let

$$E_k := e^{(2+\varepsilon)\max_{p\in\mathcal{P}}\|A_p\|t_k}E_0,\qquad k=1,2,\dots \tag{21}$$

There are other options but for concreteness we assume that the specific "zooming-out" sequence (21) is implemented. Repeat the above encoding procedure at each step. (As long as the quantization symbol is 0, there is no need to send the value of σ to the controller.) Then we claim that there will be a time t_{k_0} such that, for the corresponding value E_{k_0}, the symbol received by the controller will not be 0. At this time, the encoding strategy described in Section 3 can be initialized.

We skip the proof that k_0 is well defined but mention that there exist functions $\eta:[0,\infty)\to\mathbb{Z}_{\ge 0}$ and $\gamma:[0,\infty)\to(0,\infty)$ such that

$$k_0\le\eta(\|x_0\|),\qquad E_{k_0}\le\gamma(\|x_0\|) \tag{22}$$

and

$$\|x(t)\| \le \gamma(\|x_0\|) \qquad \forall\, t \in [0, t_{k_0}]. \tag{23}$$

Both functions depend on the initial choice of E_0. We can pick them so that $\eta(r) = 0$ and $\gamma(r) = E_0$ for all $r \le E_0$. For large values of its argument, $\gamma(\cdot)$ is in general super-linear. In fact, we can calculate that $\gamma(r)$ is of the order of r^2/E_0, and $\eta(r)$ is of the order of $(\max_{p\in\mathcal{P}} \|A_p\|\tau_s)^{-1}\log(r/E_0)$, for large values of r.

5. STABILITY VERIFICATION

We only sketch very briefly the main steps of the stability proof. First, consider a sampling interval with no switch: $\sigma \equiv p$ on $[t_k, t_{k+1}]$. Rewrite (13) as

$$x^*_{k+1} = e^{(A_p+B_pK_p)\tau_s} c_k = e^{(A_p+B_pK_p)\tau_s}(x^*_k + (c_k - x^*_k)).$$

This is an exponentially stable discrete-time system with input $\Delta_k := c_k - x^*_k$. We know from (8) that $\|\Delta_k\| \le E_k(N-1)/N$, whereas (12) and Assumption 3 give us $E_{k+1} = E_k\Lambda_p/N < E_k$. We see that, as long as there are no switches, E_k decays exponentially, hence the overall "cascade" system describing the joint evolution of x^*_k and E_k is exponentially stable. This fact can be formally proved by constructing a Lyapunov function in the form of a sum of a positive definite quadratic form in x^*_k and a positive multiple of E_k^2:

$$V_p(x^*_k, E_k) := (x^*_k)^T P_p x^*_k + \rho_p E_k^2$$

which can be shown to satisfy, on a sampling interval with no switch that we are considering,

$$V_p(x^*_{k+1}, E_{k+1}) \le \nu V_p(x^*_k, E_k)$$

for some number $\nu < 1$ that can be precisely computed.

Next, if a sampling interval $[t_k, t_{k+1}]$ contains a switch from $\sigma = p$ to $\sigma = q$, then the above mode-dependent Lyapunov function satisfies

$$V_q(x^*_{k+1}, E_{k+1}) \le \mu V_p(x^*_k, E_k)$$

for some $\mu > 1$ which again can be computed. It can now be shown that $V_{\sigma(t_k)}(x^*_k, E_k)$ converges to 0 exponentially fast if the average dwell time τ_a introduced in Assumption 1 satisfies the lower bound

$$\tau_a \ge m\tau_s, \qquad m > 1 + \frac{\log \mu}{\log \frac{1}{\nu}}. \tag{24}$$

From this, the same exponential convergence property holds for $x(t_k)$ and, with a bit of extra effort needed to analyze the intersample behavior, can be established for $x(t)$. Specifically, recalling (22) and (23), we can establish the desired exponential convergence property (4) with

$$\lambda := \frac{1}{2\tau_s}\log\frac{1}{\theta}, \qquad \theta := \mu^{1/m}\nu^{(m-1)/m} < 1$$

and

$$g(r) := \bar{c}\Big(\frac{1}{\sqrt{\theta}}\Big)^{1+\eta(r)}\gamma(r). \tag{25}$$

Finally, the proof of Lyapunov stability proceeds along the lines of [18, 20].

6. SIMULATION EXAMPLE

We simulated the above control strategy with the following data: $\mathcal{P} = \{1,2\}$, $A_1 = \begin{pmatrix}1 & 0\\ 0 & -1\end{pmatrix}$, $B_1 = \begin{pmatrix}1\\0\end{pmatrix}$, $K_1 = \begin{pmatrix}-2 & 0\end{pmatrix}$, $A_2 = \begin{pmatrix}0 & 1\\ -1 & 0\end{pmatrix}$, $B_2 = \begin{pmatrix}0\\1\end{pmatrix}$, $K_2 = \begin{pmatrix}0 & -1\end{pmatrix}$, $x_0 = (2,2)^T$, $E_0 = 0.5$, $\tau_s = 0.5$, $N = 5$ (Assumption 3 is satisfied), $\tau_d = 1.05$, $\tau_a = 7.55$, and $N_0 = 5$. Figure 1 plots a typical behavior of the first component x_1 of the continuous state (in solid red) and the corresponding component $\hat{x}_1$ of the state estimate (in dashed green) versus time; switches are marked by blue circles. Observe the initial "zooming-out" phase and the nonsmooth behavior of x when $\hat{x}$ experiences a jump (causing a jump in the control u). The above value of the average dwell time τ_a was picked empirically to be just large enough to provide consistent convergence in simulations. For this example, the theoretical lower bound on the average dwell time from the formula (24) is about 85.5 which is, not surprisingly, quite conservative.

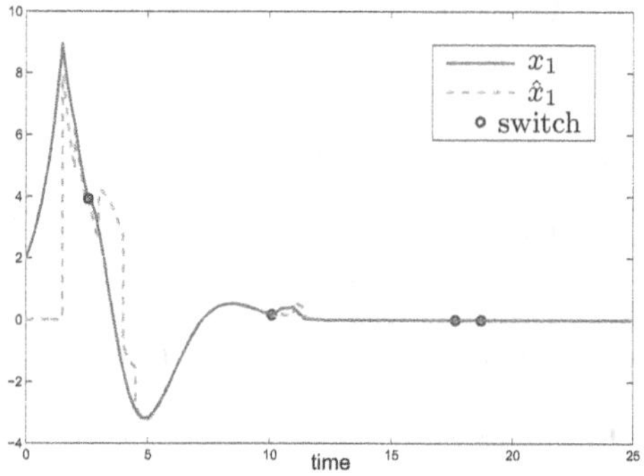

Figure 1: Simulation example

7. HYBRID SYSTEMS

In a hybrid system, the abstract notion of the switching signal σ that we used to define the switched system (1) is replaced (or, we may say, realized) by a discrete dynamics model which generates the sequence of modes, based typically on the evolution of the continuous state. Many specific modeling formalisms for hybrid systems exist in the literature, but a common paradigm which we also have in mind here is that each mode corresponds to a region in the continuous state space (sometimes called the *invariant* for that mode) where the corresponding continuous dynamics are active, and transitions (or switchings) between different modes take place when the continuous state x crosses boundaries (called *switching surfaces*, or *guards*) between these regions. At the times of these discrete transitions, the value of x in general can also jump to a new value according to some *reset map*.

Thus, compared to the switched system model (1), the two main new aspects that must be incorporated are *switching surfaces* and *state jumps*. We will address both these aspects in what follows. However, since we saw that propagating (over-approximations of) reachable sets is a key ingredient of our control strategy, we first discuss some relevant prior work on reachable set computation for hybrid systems in order to put our present developments in a proper context.

7.1 Comparison with existing reachable set algorithms

Without aiming for completeness, we give here an overview of some representative results. We classify them roughly according to the type of dynamics in the considered hybrid system model and the shapes of the sets used for reachable set approximation.

Early work by Puri, Borkar, and Varaiya on differential inclusions [30] approximates a general nonlinear differential inclusion by a piecewise constant one, and computes over-approximations of reachable sets which are unions of polyhedra. Henzinger et. al. [12] and Preußig et. al. [29] approximate hybrid systems by *rectangular automata* (hybrid systems whose regions and flow in each region are defined by constant lower and upper bounds on state and velocity components, respectively) and base reachable set computation on the tool `HyTech`; Frehse later developed a refined tool, `PHAVer` [6], for a similar purpose. Also related to this is reachability analysis using "face lifting" [4]. Asarin et. al. [1, 2] work with linear dynamics and rectangular polyhedra and develop the tool called `d/dt`. They reduce the conservatism due to the so-called "wrapping effect" by combining propagation of exact reachable sets at sampling instants with convex over-approximation during intersample intervals. Similar ideas appeared in the earlier work of Greenstreet and Mitchell [10] who also handle nonlinear models and non-convex polyhedra by using two-dimensional projections. Mitchell et. al. [23, 35] work with general nonlinear dynamics and compute reachable sets as sublevel sets of value functions for differential games, which are solutions of Hamilton-Jacobi PDEs. Kurzhanski and Varaiya [17] work with affine open-loop dynamics and use ellipsoids for reachable set approximation (based on ellipsoidal methods for continuous systems developed in their prior work). They handle discrete transitions by taking the union of reachable sets over possible switching times and covering it with one bounding ellipsoid. Chutinan and Krogh [3] compute optimal polyhedral approximations of continuous flow pipes for general nonlinear dynamics, using the tool `CheckMate`. Stursberg and Krogh [33] work with nonlinear dynamics and "oriented rectangular hulls" relying on principal component analysis. Girard et. al. [8, 9] use a procedure similar to that of Asarin et. al. mentioned earlier, but work with *zonotopes* (affine transforms of hypercubes) which allow more efficient computation for linear dynamics. More recently, this approach was refined with the help of support-function representations [11] and the accompanying tool `SpaceEx` [7] was developed. Another example of very recent work in this area is the result of [16] on computation of ε-reach sets.

There are several similarities between our method and the previous ones just mentioned. Like `d/dt` and related techniques, we also reduce the conservatism due to the "wrapping effect" by making a distinction between sampling and intersample approximations (although we do not present the analysis of our method at sufficient level of detail to fully demonstrate this point here, it should be clear that the bounds derived in Section 4 are valid at sampling times only). Also, similarly to Kurzhanski and Varaiya, we handle discrete transitions by taking the union of reachable sets over possible switching times and covering it with one bounding set, except we work with hypercubes rather than ellipsoids. On the other hand, in spite of the multitude of available methods, these methods were designed for reachability verification and are not directly tailored to control problems of the kind considered here. There are at least two important reasons why we prefer to build on the method from Section 4 rather than just adopt one of the above methods for dealing with hybrid systems:

i) The methods just mentioned are computational (online) in nature; by this we mean that approximations of reachable sets are generated in real time as the system evolves. By contrast, the method from Section 4 is analytical (off-line). Indeed, the size of the reachable set bound E_k at each time step, as well as the center point x_k^*, are obtained iteratively from the formulas given in Section 4. In other words, knowing the system data (the matrices A_p and B_p as well as the control gains K_p), we can pre-compute these bounds; there is no need to synchronize their computation with the evolution of the system. Consequently, the corresponding lower bound on the data rate required for stabilization can be obtained *a priori*, which makes more sense in the context of applications where communication strategies are designed separately from on-line control tasks. (On the negative side, this makes the bounds on reachable sets that our method provides more conservative.)

ii) Our method is tailored specifically to linear dynamics and to sets in the shapes of hypercubes. Our choice of hypercubes as bounding sets is very natural from the point of view of quantizer design with rectilinear quantization boxes, such as those arising from simple sensors. (However, in other application contexts it may be possible to work with different set shapes. For example, zonotopes—which are affine transforms of hypercubes—would correspond to pre-processing the continuous state by an affine transformation before passing it to a digital encoder; this generalization appears to be quite promising for more efficient computations.)

7.2 Switching surfaces

With regards to hybrid systems where mode switching occurs on switching surfaces, the first observation is that our Theorem 1 already covers such systems, because our reachable set over-approximation is computed by taking the union over all possible switching times $\bar{t}$ (see Section 4). Indeed, a switched system admits more solutions than a hybrid system (for which it serves as a high-level abstraction), and so our stabilization result conservatively captures the hybrid system solutions. The main issue is to verify that a given hybrid system fulfills the slow-switching assumption (Assumption 1), i.e., that all solutions satisfy the dwell-time and average dwell-time properties specified there. This can be difficult, but is possible in some cases. Notable examples are hybrid systems whose switching surfaces are concentric circles with respect to some norm, or lines through the origin in the plane. (Average dwell time is not directly helpful but in these cases we can compute dwell time, assuming linear dynamics in each mode.) Some more interesting examples where time-dependent properties (of dwell-time type) are established a posteriori for control systems with state-dependent switching can be found in [14], [21]. Thus, translating a hybrid system to a switched system and applying our previous result off-the-shelf via verifying the slow-switching condition can actually be a reasonable route to follow. In fact, since our strategy guarantees containment of the reachable set at each sampling time within a bounding hypercube, we can just run it and verify empirically whether or not the switching is slow enough for convergence. This is what we actually did in the simulation example given in Section 6. This in some sense moves us closer to the on-line computational methods cited above.

A better approach, however, is to improve our reachable set bounds by explicitly incorporating the information available in a hybrid system about where in the continuous state space the switching can occur. Recall that our information structure makes the current mode available to the controller at each sampling time t_k. So, for example, if we know as in Section 4.1 that no switch has occurred on an interval $(t_k, t_{k+1}]$ and $\sigma(t) = p$ there, then the hypercube $\{x \in \mathbb{R}^n : \|x - x^*_{k+1}\| \leq E_{k+1}\}$, which contains the reachable set at time $t = t_{k+1}$, can be reduced by intersecting it with the invariant for mode p. In other words, if a guard passes through this hypercube then we keep only the portion lying on that side of the guard on which mode p is active; the point x^*_{k+1} can also be redefined at this step. The resulting reduction in the size of the bounding set can be quite significant, especially if the set $\{x : \|x - x^*_k\| \leq E_k\}$ at time $t = t_k$ was close to some of the switching surfaces. (Note, however, that if the reachable set over-approximation at time t_{k+1} must be a hypercube, then some or all of this size reduction might become undone when passing to a bounding hypercube.) Or, consider the situation of Section 4.2 where a sampling interval $(t_k, t_{k+1}]$ contains a switch from $\sigma = p$ to $\sigma = q$ at an unknown time $\bar{t}$. The bounding set before the switch, $\{x : \|x - \hat{x}(t_k + t')\| \leq D_{k+1}(\bar{t})\}$ (see Section 4.2(a)) can be reduced in the same way as above by intersecting it with the invariant for mode p. (Since $\bar{t}$ is unknown, we should either treat it as a parameter for this computation or take the maximum over $\bar{t}$ first; (19) is helpful for doing the latter.) Then, when this possibly reduced intermediate bounding set is used to calculate the bounding set after the switch, which we previously defined as $\{x : \|x - x^*_{k+1}\| \leq E_{k+1}\}$ (see Section 4.2(b)), we may reduce it again, this time intersecting it with the invariant for mode q. Overall, this can lead to a significant reduction in the size of the reachable set over-approximation compared to the method of Section 4 which does not assume any relation between the continuous state x and the switching signal (but again, working with hypercubes would not allow us to take full advantage of this size reduction). Additionally, the knowledge of switching surfaces can be used to obtain some information about the unknown switching time $\bar{t}$: for example, if at time t_k we are far from any switching surface, then using the system dynamics we can calculate a lower bound on the time that must pass before a switch can occur.

7.3 State jumps

The reachable set propagation method of Section 4 assumes that there are no state jumps at the switching times, i.e., the reset map is the identity. However, it is not very difficult to augment it to nontrivial reset maps. Specifically, if we have a reset map $R_{pq} : \mathbb{R}^n \to \mathbb{R}^n$ which defines the new state $R_{pq}(x)$ to which x jumps at the time of mode transition from p to q, all we need is a knowledge of some affine Lipschitz bound of the form $\|R_{pq}(x_1) - R_{pq}(x_2)\| \leq a\|x_1 - x_2\| + b$. Then, we can apply the transformations $c \mapsto R_{pq}(c)$ and $D \mapsto aD + b$ to the reachable set over-approximations of the form $\{x : \|x - c\| \leq D\}$ obtained at each time that corresponds to a switch (these times are $t_k + \bar{t}$ on sampling intervals containing a switch, see Section 4.2). We can continue working with hypercubes because after incorporating resets in this way we still obtain hypercubes. We see that accounting for state jumps does not lead to substantial complications in our reachable set algorithm. (The same claim is true for most of the other existing reachable set algorithms from the literature: many of them assume the identity reset map but can be generalized with not much difficulty.) The stability analysis can proceed similarly, with the constants a, b affecting the evolution of the Lyapunov function and leading to a modified average dwell time bound.

8. CONCLUSIONS

We presented a result on sampled-data quantized state feedback stabilization of switched linear systems, which relies on a slow-switching condition and a novel method for propagating over-approximations of reachable sets. We explained how this result can be applied in the setting of hybrid systems, where it can actually be improved. Future work will focus on refining the reachable set bounds (possibly by combining our method with other known reachable set algorithms for hybrid systems) and on addressing more general systems with external disturbances, output measurements, and nonlinear dynamics.

9. ACKNOWLEDGMENTS

The author thanks Aneel Tanwani for helpful comments on an earlier draft and Sayan Mitra for useful discussions of the literature.

10. REFERENCES

[1] E. Asarin, O. Bournez, T. Dang, and O. Maler. Approximate reachability analysis of piecewise-linear dynamical systems. In *Proc. 3rd Int. Workshop on Hybrid Systems: Computation and Control*, pages 20–31, 2000.

[2] E. Asarin, T. Dang, and O. Maler. The d/dt tool for verification of hybrid systems. In *Computer Aided Verification: Proc. 14th Intl. Conf. (CAV 2002)*, pages 746–770, 2002.

[3] A. Chutinan and B. H. Krogh. Computational techniques for hybrid system verification. *IEEE Trans. Automat. Control*, 48:64–75, 2003.

[4] T. Dang and O. Maler. Reachability analysis via face lifting. In *Proc. 1st Int. Workshop on Hybrid Systems: Computation and Control*, pages 96–109, 1998.

[5] P. S. Duggirala and S. Mitra. Abstraction-refinement for stability. In *Proc. Int. Conf. Cyber-Physical Systems (ICCPS 2011)*, 2011.

[6] G. Frehse. PHAVer: algorithmic verification of hybrid systems past HyTech. In *Proc. 8th Int. Workshop on Hybrid Systems: Computation and Control*, pages 258–273, 2005.

[7] G. Frehse, C. Le Guernic, A. Donzé, S. Cotton, R. Ray, O. Lebeltel, R. Ripado, A. Girard, T. Dang, and O. Maler. SpaceEx: scalable verification of hybrid systems. In *Computer Aided Verification: Proc. 23rd Intl. Conf. (CAV 2011)*, page 379Ű395, 2011.

[8] A. Girard. Reachability of uncertain linear systems using zonotopes. In *Proc. 8th Int. Workshop on Hybrid Systems: Computation and Control*, pages 291–305, 2005.
[9] A. Girard and C. Le Guernic. Zonotope/hyperplane intersection for hybrid systems reachability analysis. In *Proc. 11th Int. Workshop on Hybrid Systems: Computation and Control*, pages 215–228, 2008.
[10] M. R. Greenstreet and I. Mitchell. Integrating projections. In *Proc. 1st Int. Workshop on Hybrid Systems: Computation and Control*, pages 159–174, 1998.
[11] C. Le Guernic and A. Girard. Reachability analysis of hybrid systems using support functions. In *Computer Aided Verification: Proc. 21st Intl. Conf. (CAV 2009)*, page 540Ű554, 2009.
[12] T. A. Henzinger, P.-H. Ho, and H. Wong-Toi. Algorithmic analysis of nonlinear hybrid systems. *IEEE Trans. Automat. Control*, 43:540–554, 1998.
[13] J. P. Hespanha and A. S. Morse. Stability of switched systems with average dwell-time. In *Proc. 38th IEEE Conf. on Decision and Control*, pages 2655–2660, 1999.
[14] J. P. Hespanha and A. S. Morse. Stabilization of nonholonomic integrators via logic-based switching. *Automatica*, 35:385–393, 1999.
[15] J. P. Hespanha, A. Ortega, and L. Vasudevan. Towards the control of linear systems with minimum bit-rate. In *Proc. 15th Int. Symp. on Mathematical Theory of Networks and Systems (MTNS)*, 2002.
[16] K.-D. Kim, S. Mitra, and P. R. Kumar. Computing bounded ϵ-reach set with finite precision computations for a class of linear hybrid automata. In *Proc. 14th ACM Int. Conf. Hybrid Systems: Computation and Control*, 2011.
[17] A. B. Kurzhanski and P. Varaiya. Ellipsoidal techniques for hybrid dynamics: the reachability problem. In W. P. Dayawansa, A. Lindquist, and Y. Zhou, editors, *New Directions and Applications in Control Theory*, volume 321 of *Lecture Notes in Control and Information Sciences*, pages 193–205. Springer, Berlin, 2005.
[18] D. Liberzon. On stabilization of linear systems with limited information. *IEEE Trans. Automat. Control*, 48:304–307, 2003.
[19] D. Liberzon. *Switching in Systems and Control*. Birkhäuser, Boston, 2003.
[20] D. Liberzon and J. P. Hespanha. Stabilization of nonlinear systems with limited information feedback. *IEEE Trans. Automat. Control*, 50:910–915, 2005.
[21] D. Liberzon and S. Trenn. The bang-bang funnel controller. In *Proc. 49th IEEE Conf. on Decision and Control*, pages 690–695, 2010.
[22] Q. Ling and H. Lin. Necessary and sufficient bit rate conditions to stabilize quantized Markov jump linear systems. In *Proc. 2010 American Control Conf.*, pages 236–240, 2010.
[23] I. Mitchell and C. Tomlin. Level set methods for computation in hybrid systems. In *Proc. 3rd Int. Workshop on Hybrid Systems: Computation and Control*, pages 310–323, 2000.
[24] G. N. Nair, S. Dey, and R. J. Evans. Infimum data rates for stabilising Markov jump linear systems. In *Proc. 42nd IEEE Conf. on Decision and Control*, pages 1176–1181, 2003.
[25] G. N. Nair, F. Fagnani, S. Zampieri, and R. J. Evans. Feedback control under data rate constraints: An overview. *Proc. IEEE*, 95:108–137, 2007.
[26] D. Nešić and D. Liberzon. A unified framework for design and analysis of networked and quantized control systems. *IEEE Trans. Automat. Control*, 54:732–747, 2009.
[27] D. Nešić and A. R. Teel. Input-output stability properties of networked control systems. *IEEE Trans. Automat. Control*, 49:1650–1667, 2004.
[28] A. Podelski and S. Wagner. Model checking of hybrid systems: from reachability towards stability. In *Proc. 9th Int. Workshop on Hybrid Systems: Computation and Control*, pages 507–521, 2006.
[29] J. Preußig, O. Stursberg, and S. Kowalewski. Reachability analysis of a class of switched continuous systems by integrating rectangular approximation and rectangular analysis. In *Proc. 2nd Int. Workshop on Hybrid Systems: Computation and Control*, pages 209–222, 1999.
[30] A. Puri, V. Borkar, and P. Varaiya. ϵ-approximations of differential inclusions. In *Hybrid Systems III: Verification and Control*, pages 362–376, 1996.
[31] Y. Sharon and D. Liberzon. Input-to-state stabilizing controller for systems with coarse quantization. *IEEE Trans. Automat. Control*, 57:830–844, 2012.
[32] R. Shorten, F. Wirth, O. Mason, K. Wulff, and C. King. Stability criteria for switched and hybrid systems. *SIAM Review*, 49:545–592, 2007.
[33] O. Stursberg and B. H. Krogh. Efficient representation and computation of reachable sets for hybrid systems. In *Proc. 6th Int. Workshop on Hybrid Systems: Computation and Control*, pages 482–497, 2003.
[34] S. Tatikonda and S. K. Mitter. Control under communication constraints. *IEEE Trans. Automat. Control*, 49:1056–1068, 2004.
[35] C. Tomlin, I. Mitchell, A. Bayen, and M. Oishi. Computational techniques for the verification of hybrid systems. *Proc. IEEE*, 91:986–1001, 2003.
[36] A. van der Schaft and H. Schumacher. *An Introduction to Hybrid Dynamical Systems*. Springer, London, 2000.
[37] R. Verma and D. Del Vecchio. Safety control of hidden mode hybrid systems. *IEEE Trans. Automat. Control*, 57:62–77, 2012.
[38] L. Vu and D. Liberzon. Supervisory control of uncertain linear time-varying plants. *IEEE Trans. Automat. Control*, 56:27–42, 2011.
[39] C. Zhang, K. Chen, and G. E. Dullerud. Stabilization of Markovian jump linear systems with limited information—a convex approach. In *Proc. 2009 American Control Conf.*, pages 4013–4019, 2009.

Learning Nonlinear Hybrid Systems: From Sparse Optimization to Support Vector Regression

Van Luong Le
Université de Lorraine, CRAN
CNRS
van-luong.le@univ-lorraine.fr

Fabien Lauer
Université de Lorraine, LORIA
CNRS
Inria
fabien.lauer@loria.fr

Laurent Bako
Univ Lille Nord
Mines Douai – URIA
laurent.bako@mines-douai.fr

Gérard Bloch
Université de Lorraine, CRAN
CNRS
gerard.bloch@univ-lorraine.fr

ABSTRACT

This paper deals with the identification of hybrid systems switching between nonlinear subsystems of unknown structure and focuses on the connections with a family of machine learning algorithms known as support vector machines. In particular, we consider a recent approach to nonlinear hybrid system identification based on a convex relaxation of a sparse optimization problem. In this approach, the submodels are iteratively estimated one by one by maximizing the sparsity of the corresponding error vector. We extend this approach in several ways. First, we relax the sparsity condition by introducing robust sparsity, which can be optimized through the minimization of a modified ℓ_1-norm or, equivalently, of the ε-insensitive loss function. Then, we show that, depending on the choice of regularizer, the method is equivalent to different forms of support vector regression. More precisely, the submodels can be estimated by iteratively solving a classical support vector regression problem, in which the sparsity of support vectors relates to the sparsity of the error vector in the considered hybrid system identification framework. This allows us to extend theoretical results as well as efficient optimization algorithms from the field of machine learning to the hybrid system framework.

Categories and Subject Descriptors

G.1 [**Mathematics of Computing**]: Numerical Analysis—*Approximation, Optimization*; I.2.6 [**Artificial Intelligence**]: Learning

Keywords

switched nonlinear systems; system identification; switched regression; support vector machine; robustness to noise; sparsity; convex optimization

1. INTRODUCTION

This paper deals with the identification of hybrid dynamical systems and in particular with the connections between this problem and the fields of sparse optimization and machine learning. More precisely, we consider switched nonlinear systems that can be written in input–output Nonlinear ARX (NARX) form as

$$y_i = f_{q_i}(\boldsymbol{x}_i) + v_i, \qquad (1)$$

where v_i is a noise term, $q_i \in \{1, \dots, s\}$ and, at time step i, the output y_i is given by the q_ith function of the collection of submodels $\{f_j\}_{j=1}^s$ and the vector of regressors

$$\boldsymbol{x}_i = [y_{i-1}, \dots, y_{i-n_a}, u_{i-n_k}, \dots, u_{i-n_k-n_b+1}]^T$$

built from past inputs u_{i-k} and outputs y_{i-k}. In this setting, we call q_i the mode of $\boldsymbol{x}_i$. The goal of the identification is to estimate the submodels $\{f_j\}_{j=1}^s$ from a training set of input–output data, $\{(\boldsymbol{x}_i, y_i)\}_{i=1}^N$, without knowledge of the corresponding sequence of modes $\{q_i\}_{i=1}^N$. The paper further focuses on the case where the submodels are nonlinear functions of arbitrary and unknown structure.

Related work. Many approaches have been proposed over the last decade for the case where the submodels f_j are linear (or affine) [24]. These include the algebraic approach [36] and various convex [22, 1, 23, 18] and nonconvex [9, 26, 4, 13, 17, 15] optimization-based approaches, to name a few. But far fewer works have considered the nonlinear case, since the problem was formally stated more recently in [16] with a preliminary solution suffering from strong limitations, particularly regarding the number of data that could be processed by the identification algorithm. In [19], the approach of [17] for linear hybrid systems is extended to estimate the s nonlinear submodels at once via the solution of a continuous, but nonconvex, optimization problem. The quality of the estimates thus obtained rely on the capabilities of global optimization solvers, and thus cannot be guaranteed for models with many parameters. On the other hand, the approach of [2], which extends the method of [1], relies on a sequence of convex optimizations to estimate the submodels one by one, and thus does not suffer from local minima issues. More precisely, the method relies on the formulation of the identification as a sparse optimization problem and its convex relaxation. However, the analysis of the method provided in [2] is only valid for noiseless data. This is all the more unsatisfactory in the nonlinear setting, since the uncertainty on the model structure can be interpreted as a form of noise.

Contribution. The present work considers the sparse optimization framework of [1, 2], which we extend in several ways. First, we introduce the notion of robust sparsity to relax the conditions on the noise under which the method can yield optimal estimates. Then, a convex relaxation is proposed to allow for the optimization of the robust sparsity through the minimization of a modified ℓ_1-norm. Finally, we show that the resulting convex optimization programs can be equivalently formulated as the minimization of the ε-insensitive loss function proposed in the machine learning community for Support Vector Regression (SVR) [32], a particular instance of the Support Vector Machine (SVM) [35]. Depending on the choice of regularizer, a formal equivalence between sparse optimization based hybrid system identification and SVR can be obtained, in which the sparsity of support vectors relates to the sparsity of the error vector in the considered identification framework. Algorithmically, the submodels can be estimated by iteratively solving a classical SVR learning problem, which allows the method to benefit from the numerous advances on SVMs for machine learning, for instance regarding the tuning of the hyperparameters. In addition, efficient optimization algorithms dedicated to SVMs can be applied as off-the-shelf solvers to nonlinear hybrid system identification with very large data sets, which typically cannot be handled by general purpose convex optimization solvers.

Notations. All vectors and matrices are written with bold symbols. In particular, $\mathbf{1}$ denotes a vector of appropriate dimension filled with ones and $\boldsymbol{I}$ is the identity matrix. Inequality symbols applied to vectors are to be understood entry-wise. The notation $\langle\cdot,\cdot\rangle_E$ stands for the inner product in E, while $|I|$ denotes the cardinality of I whenever I is a set.

Paper organization. The rest of the paper starts in Sect. 2 with some background in nonlinear model estimation by focusing more particularly on recent machine learning approaches. Then, Section 3 recalls the sparse optimization approach to hybrid system identification and its extension to the case of nonlinear submodels (Sect. 3.2), including a discussion on the choice of the regularizer (Sect. 3.3). Robust sparsity is introduced in Sect. 4, where the connection with SVMs is explicitly stated (Sect. 4.1–4.3) and its benefits discussed (Sect. 4.4 and 4.6). The paper ends with numerical examples in Sect. 5 and conclusions in Sect. 6.

2. LEARNING NONLINEAR MODELS

In this section we provide the necessary background on nonlinear model estimation. In particular, we consider the case of arbitrary and unknown nonlinearities and focus on the estimation of a single (non-hybrid) nonlinear model. In such a setting, both the structure and the parameters of the model must be estimated from the data. While this clearly constitutes a difficulty, recent approaches developed in machine learning allow both subproblems to be solved simultaneously through convex optimization.

One issue which must be considered with care in nonlinear regression compared to the linear case is overfitting. While linear models are constrained within a restricted function class of low capacity, nonlinear function classes can include very complex functions. The typical functions classes provide sufficient flexibility for the model to yield a perfect fit of the data. Thus, if we were to minimize the error on a data set, the model would learn the noise as well as the target function, i.e., overfit the training data. Hence, most approaches to nonlinear modeling include a regularization scheme to control the complexity (or flexibility) of the model.

Formally, the considered approach to nonlinear modeling can be stated as follows. Assume we are given a training set $\mathcal{D}$ of N pairs $(\boldsymbol{x}_i, y_i) \in (\mathcal{X} \subset \mathbb{R}^d) \times (\mathcal{Y} \subset \mathbb{R})$, $i = 1, \dots, N$, with the general goal of learning a function $f \in \mathcal{H}$ such that this function minimizes, over some function class $\mathcal{H}$, a regularized functional representing a trade-off between the fit to the data and some regularity conditions of f:

$$\min_{f\in\mathcal{H}} \sum_{i=1}^{N} \ell(f(\boldsymbol{x}_i), y_i) + \lambda\mathcal{R}(f), \tag{2}$$

where the data term is defined through a loss function ℓ of $\mathbb{R}^2$ to $\mathbb{R}^+$, $\mathcal{R}(f)$ is a general regularization term and $\lambda \geq 0$ tunes the trade-off between the two terms.

Though searching for f within a specific function class $\mathcal{H}$ can be related in some cases to a particular choice of structure for the nonlinear model f, this can also be more general. In particular, by assuming that f is an expansion over some functional basis, a single function $f \in \mathcal{H}$ can have multiple representations (and parametrizations) depending on the choice of the basis. In addition, we will see below that $\mathcal{H}$ can be an infinite dimensional function space with the universal approximation capacity while still allowing for learning from a finite set of data. As a practical consequence, arbitrary nonlinearities can be learned without introducing a bias due to an arbitrary choice of unsuitable or insufficiently flexible structure for f.[1]

2.1 Learning in RKHS

We start with some formal definitions. Let K be a real-valued positive type (or positive definite) function [5] on $\mathcal{X}^2$ and $(\mathcal{H}, \langle\cdot,\cdot\rangle_{\mathcal{H}})$ the corresponding reproducing kernel Hilbert space (RKHS), i.e., K is the reproducing kernel of $\mathcal{H}$ with the reproducing property: $\forall \boldsymbol{x} \in \mathcal{X}$, $\forall f \in \mathcal{H}$, $\langle f, K(\boldsymbol{x}, \cdot)\rangle_{\mathcal{H}} = f(\boldsymbol{x})$, and in particular

$$\langle K(\boldsymbol{x},\cdot), K(\boldsymbol{x}',\cdot)\rangle_{\mathcal{H}} = K(\boldsymbol{x},\boldsymbol{x}').$$

In this case, the class of functions $\mathcal{H}$ can be written as

$$\mathcal{H} = \left\{ f : f = \sum_{i=1}^{\infty} \alpha_i K(\boldsymbol{x}_i, \cdot),\ \alpha_i \in \mathbb{R}, \boldsymbol{x}_i \in \mathcal{X}, \|f\|_{\mathcal{H}} < +\infty \right\}, \tag{3}$$

where the norm $\|\cdot\|_{\mathcal{H}}$ is the norm in $\mathcal{H}$ induced by the inner

[1] Note that, due to the bias-variance dilemma, this does not imply the optimal recovery of the target function.

product defined as

$$\|f\|_{\mathcal{H}}^2 = \langle f, f\rangle_{\mathcal{H}} = \sum_{i=1}^{\infty}\sum_{j=1}^{\infty} \alpha_i\alpha_j K(\boldsymbol{x}_i, \boldsymbol{x}_j). \quad (4)$$

Typical examples of kernel functions include the Gaussian Radial Basis Function (Gaussian RBF) kernel, $K(\boldsymbol{x}, \boldsymbol{x}') = \exp(-\|\boldsymbol{x} - \boldsymbol{x}'\|_2^2/2\sigma^2)$, the polynomial kernel, $K(\boldsymbol{x}, \boldsymbol{x}') = (\boldsymbol{x}^T\boldsymbol{x}'+1)^\gamma$, and the linear kernel, $K(\boldsymbol{x}, \boldsymbol{x}') = \boldsymbol{x}^T\boldsymbol{x}'$. For the Gaussian RBF kernel, the space $\mathcal{H}$ consists of all infinitely differentiable functions of $\mathcal{X}$ and thus enjoys the so-called universal approximation capacity, i.e., an arbitrary function can be arbitrarily well approximated by a function in $\mathcal{H}$.

When learning in an RKHS, a natural choice for $\mathcal{R}(f)$ is based on the RKHS norm:

$$\mathcal{R}(f) = \frac{1}{2}\|f\|_{\mathcal{H}}^2. \quad (5)$$

Such a regularizer is a measure of the function smoothness[2] and is particularly suitable for cases without prior information on the shape of the target function. In addition, with (5), the representer theorem [27] provides an explicit structure for the solution to (2). This theorem is recalled below, where $\mathcal{D}_x$ denotes the set of all points $\boldsymbol{x}_i$ in the training set $\mathcal{D}$ (a sketch of the proof is given in Appendix A).

THEOREM 2.1 (REPRESENTER THEOREM, [27]). *The solution f^* to* (2), *with $\mathcal{H}$ defined as in* (3), $\mathcal{R}(f) = g(\|f\|_{\mathcal{H}})$ *and a monotonically increasing function $g : \mathbb{R}^+ \rightarrow \mathbb{R}^+$, is a kernel expansion over the training set, i.e., f^* is in the span of $\{K(\boldsymbol{x}_i, \cdot) \ : \ \boldsymbol{x}_i \in \mathcal{D}_x\}$.*

This result shows that minimizing any regularized functional of the form (2) over an RKHS leads to finite linear combinations of kernel functions computed at the training points:

$$f(\boldsymbol{x}) = \sum_{i=1}^{N} \alpha_i K(\boldsymbol{x}_i, \boldsymbol{x}). \quad (6)$$

Note that a semiparametric version of Theorem 2.1 is also provided in [27] to allow for a bias term in the model. This is obtained by considering a model $\tilde{f} = f + b$, with $f \in \mathcal{H}$ and $b \in \mathbb{R}$, regularized only in f. In most of the paper, we omit this straightforward substitution and focus on models in the form of f in (6).

Finally, given the structure of the model (6), solving (2) with a convex loss function ℓ amounts to a finite-dimensional and convex optimization problem.

2.2 ℓ_1-norm regularization

Another typical regularization scheme for models based on kernel functions is to penalize the ℓ_1-norm of the parameters, i.e., to penalize $\|\boldsymbol{\alpha}\|_1$, with $\boldsymbol{\alpha} = [\alpha_1 \ \dots \ \alpha_N]^T$, in (6). However, this scheme cannot apply to (2) with $\mathcal{H}$ defined as in (3), since $\boldsymbol{\alpha}$ depends on the particular choice of basis functions through $\{\boldsymbol{x}_i\}_{i=1}^{\infty}$ and need not be uniquely defined. Therefore, this scheme is usually applied with the structure of f fixed a priori. With f chosen as in (6), this leads to

$$\min_{\boldsymbol{\alpha}\in\mathbb{R}^N} \sum_{i=1}^{N} \ell(f(\boldsymbol{x}_i), y_i) + \lambda\|\boldsymbol{\alpha}\|_1 \quad (7)$$

$$\text{s.t.} \quad f(\boldsymbol{x}) = \sum_{i=1}^{N} \alpha_i K(\boldsymbol{x}_i, \boldsymbol{x}).$$

While this learning strategy also provides control over the complexity of the model, as detailed in Appendix B, it is often chosen in order to favor sparse solutions with few nonzero α_i. Indeed, the sparsity of $\boldsymbol{\alpha}$ directly defines the number of operations required to compute the output of the model (6) for a given $\boldsymbol{x}$.

3. SPARSE OPTIMIZATION FOR HYBRID SYSTEM IDENTIFICATION

We now recall the sparse optimization framework of [1] in the case of switched linear systems, while its extension to the case of switched nonlinear systems will be detailed in Sect. 3.2.

Consider a switched linear system, i.e., of the form (1) with linear submodels, $f_j(\boldsymbol{x}_i) = \boldsymbol{x}_i^T\boldsymbol{\theta}_j$. As proposed in [1], switched linear systems can be identified via sparse optimization. In this approach, a single parameter vector $\boldsymbol{\theta}$ is first estimated by maximizing the sparsity of the error vector, $\boldsymbol{e} = \boldsymbol{y} - \boldsymbol{X}\boldsymbol{\theta}$, where $\boldsymbol{X} = [\boldsymbol{x}_1 \ \dots \ \boldsymbol{x}_N]^T$ and $\boldsymbol{y} = [y_1 \ \dots \ y_N]^T$. In the noiseless case, each entry e_i in $\boldsymbol{e}$ can be zero by choosing $\boldsymbol{\theta}$ as the vector of parameters $\boldsymbol{\theta}_{q_i}$ that generated the corresponding data point. Therefore, by searching for the pair $(\boldsymbol{\theta}, \boldsymbol{e})$ leading to the sparsest vector $\boldsymbol{e}$, we recover the parameters of the submodel that generated the largest percentage of data. Formally, the sparsity is measured through the ℓ_0-pseudo norm,

$$\|\boldsymbol{e}\|_0 = |\{i : e_i \neq 0\}|,$$

where a vector $\boldsymbol{e}$ with small norm $\|\boldsymbol{e}\|_0$ is said to be sparse, and one solves

$$\min_{\boldsymbol{\theta},\boldsymbol{e}} \|\boldsymbol{e}\|_0 \quad (8)$$

$$\text{s.t.} \quad \boldsymbol{e} = \boldsymbol{y} - \boldsymbol{X}\boldsymbol{\theta}$$

to obtain the first parameter vector. Then, the data points with corresponding $e_i = 0$ are removed from the data set, and (8) is solved again to obtain the second parameter vector. Applying this procedure iteratively until all data are correctly approximated and removed from the training set yields all the submodels.

3.1 Convex relaxation

Since (8) is a nonconvex optimization problem, we instead consider a convex relaxation based on the best convex approximation to the ℓ_0-pseudo norm, i.e., the ℓ_1-norm. In order to improve the sparsity of the solution, an iteratively reweighted scheme is employed, as proposed by [7]. Thus, each parameter vector is recovered by iteratively solving

$$\min_{\boldsymbol{\theta}} \|\boldsymbol{W}^k(\boldsymbol{y} - \boldsymbol{X}\boldsymbol{\theta})\|_1, \quad (9)$$

where $\boldsymbol{W}^0 = \boldsymbol{I}$ and $\boldsymbol{W}^k$ is a diagonal matrix of entries $(W^k)_{ii} = 1/(|y_i - \boldsymbol{x}_i^T\boldsymbol{\theta}^{k-1}| + \delta)$, with δ a small positive number, and $\boldsymbol{\theta}^{k-1}$ the solution at iteration $k-1$.

In [1], the following sparse recovery conditions are stated.

[2]For smooth kernel functions K, all $f \in \mathcal{H}$ are smooth functions with infinite order of differentiability. However, here, a large measure of smoothness refers to functions with derivatives of small magnitude rather than a high order of differentiability. The RKHS norm in (5) provides an upper bound on these magnitudes (see, e.g., Corollary 4.36 in [33]).

THEOREM 3.1 (THEOREM 11 IN [1]). *If there is a vector $\boldsymbol{\theta}$ such that*

$$\|\boldsymbol{y} - \boldsymbol{X}\boldsymbol{\theta}\|_0 < \frac{1}{2}\left(1 + \frac{1}{m(\boldsymbol{X})}\right),$$

with

$$m(\boldsymbol{X}) = \max_{1 \le t,k \le N, t \ne k} \frac{|M_{tk}|}{\sqrt{(1 - M_{tt})(1 - M_{kk})}},$$

where $\boldsymbol{M} = \boldsymbol{X}(\boldsymbol{X}^T\boldsymbol{X})^{-1}\boldsymbol{X}^T$ and M_{tk} is the (t,k)th entry of $\boldsymbol{M}$, then $\boldsymbol{\theta}$ is the unique solution to both (8) and (9) with $\boldsymbol{W}^k = \boldsymbol{I}$.

Note that this result directly applies only to the first iteration of the reweighted scheme. However, it provides sufficient ground for the method, while the convergence analysis of the reweighted scheme remains a difficult open issue.

3.2 Extension to nonlinear submodels

When the submodels f_j are nonlinear, the procedure above can be extended to estimate nonlinear submodels. The basic idea is to replace $\boldsymbol{x}_i^T\boldsymbol{\theta}$ by an expansion over a set of basis functions, e.g., a kernel expansion as in (6). As discussed in Sect. 2, depending on the regularizer $\mathcal{R}(f)$, this either corresponds to an arbitrary choice of nonlinear structure for the model or to the explicit form of the solution. We first describe the complete procedure for a general regularizer and nonlinear model before detailing the typical choices.

For a given function class $\mathcal{H}$, the nonlinear submodel f of a single mode is estimated by solving

$$\min_{f \in \mathcal{H}} \sum_{i=1}^{N} w_i |y_i - f(\boldsymbol{x}_i)| + \lambda\mathcal{R}(f), \tag{10}$$

with $w_i = (W^k)_{ii} > 0$. By defining the error vector $\boldsymbol{e} \in \mathbb{R}^N$ with components $e_i = y_i - f(\boldsymbol{x}_i)$, we see that the first term in (10) is merely $\|\boldsymbol{W}^k\boldsymbol{e}\|_1$.

3.3 Choice of the regularizer

In machine learning, it is a well-known fact that one cannot learn without a minimal set of assumptions on the target function. In the most general case, where no prior knowledge is available, the less informative assumption concerns the smoothness of the target function. Indeed, without assuming that function values should be close for two points that are close in the regression space $\mathcal{X}$, one cannot learn from a finite set of points and generalize to others. In practice, the smoothness assumption is typically implemented by regularization as in (2). We now discuss two particular choices for the regularizer $\mathcal{R}(f)$.

3.3.1 Sparsity inducing regularization

In [2], a regularization term based on the ℓ_0-norm of the parameter vector $\boldsymbol{\alpha}$ is introduced, before being relaxed to the convex ℓ_1-norm. While the ℓ_1-norm is a typical choice for regularization, which is also known for its sparsity inducing feature, ℓ_0-norm regularization is more ambiguous regarding the resulting smoothness of f. Therefore, in this case, the aim of minimizing the ℓ_1-norm is not to recover the smallest ℓ_0-norm solution through a convex relaxation, and we will not delve into theoretical guarantees of convergence of the ℓ_1-solution to the ℓ_0-solution.

Let $\boldsymbol{K}$ be the so-called Gram matrix of the kernel K with respect to the sample $\mathcal{D}_x$, i.e., with all components given by $(\boldsymbol{K})_{ij} = K(\boldsymbol{x}_i, \boldsymbol{x}_j)$. Then, with ℓ_1-norm regularization, the submodels are estimated by solving

$$\min_{\boldsymbol{\alpha} \in \mathbb{R}^N} \|\boldsymbol{W}^k(\boldsymbol{y} - \boldsymbol{K}\boldsymbol{\alpha})\|_1 + \lambda\|\boldsymbol{\alpha}\|_1, \tag{11}$$

where the classical reweighting scheme applies.

3.3.2 Capacity control regularization

The typical approach used in machine learning to estimate nonlinear functions is to control the capacity of the model by penalizing the nonsmoothness of f. This can be measured through a norm of f.

Using the natural RKHS squared norm defined in (4), $\mathcal{R}(f) = \frac{1}{2}\|f\|_{\mathcal{H}}^2$, the nonlinear submodels are estimated by solving the convex optimization problem

$$\min_{f \in \mathcal{H}} \sum_{i=1}^{N} w_i |y_i - f(\boldsymbol{x}_i)| + \frac{\lambda}{2}\|f\|_{\mathcal{H}}^2, \tag{12}$$

whose solution is in the form of (6) by application of Theorem 2.1 with the convex loss function[3] $\ell(f(\boldsymbol{x}_i), y_i) = |y_i - f(\boldsymbol{x}_i)|$ in (2).

4. ROBUST SPARSITY

A preliminary condition to the sparse recovery conditions derived in Sect. 3.1 (see Theorem 3.1) is that the data must be noiseless. Indeed, with noisy data, no (or very few) entries of the error vector can be zero[4], hence breaking the sparsity of the optimal solution.

In order to circumvent the issue of the lack of zeros in the error vector, we introduce robust sparsity as defined through the pseudo-norm

$$\|\boldsymbol{e}\|_{0,\varepsilon} = |\{i \;:\; |e_i| > \varepsilon\}|.$$

Under a bounded noise assumption of the type $\|\boldsymbol{v}\|_\infty = \max_{i \in \{1,\dots,N\}} |v_i| \le \varepsilon$, the error vector, $\boldsymbol{e} = [y_1 - f(\boldsymbol{x}_1), \dots, y_N - f(\boldsymbol{x}_N)]^T$, can be *robustly sparse*, i.e., with a small value of $\|\boldsymbol{e}\|_{0,\varepsilon}$, if f is a sufficiently good approximation of one of the target submodels f_j.

Instead of the nonconvex pseudo-norm above, we consider the following convex relaxation based on a modified ℓ_1-norm:

$$\|\boldsymbol{e}\|_{1,\varepsilon} = \sum_i (|e_i| - \varepsilon)_+ = \sum_i \max\{0, |e_i| - \varepsilon\},$$

which is defined as a sum of pointwise maximum of convex functions of e_i and hence is convex with respect to all components e_i. In the following, we will refer to the pseudo norm above as the $\ell_{1,\varepsilon}$-norm.

With these definitions, the procedure to estimate the submodels under noisy conditions is similar to the one presented in Sect. 3 for the noiseless case, with the $\ell_{1,\varepsilon}$-norm substituted for the ℓ_1-norm. Similarly, after the estimation of a submodel f, the data points correctly approximated by f are removed, where "correctly approximated" is now implemented by the test $|y_i - f(\boldsymbol{x}_i)| \le \varepsilon$.

[3]For the sake of brevity, we omitted the weights w_i in (2), but Theorem 2.1 in its original version found in [27] equivalently applies to a weighted sum of losses.

[4]With nonlinear models of sufficient capacity, the error vector can actually be zero. But, as already discussed, this is not a desirable case, since this would clearly indicate overfitting. Here, we focus on sufficiently regularized (and desirable) solutions, for which the error vector cannot be sparse.

4.1 ℓ_1-norm regularization

By assuming submodels in the form of (6), estimating one of the nonlinear submodels by maximizing the robust sparsity of the error vector can be set as the convex optimization problem:

$$\min_{\boldsymbol{\alpha}\in\mathbb{R}^N} \|\boldsymbol{y} - \boldsymbol{K\alpha}\|_{1,\varepsilon} + \lambda\|\boldsymbol{\alpha}\|_1, \tag{13}$$

where the convexity of the first term is due to the convexity of the $\ell_{1,\varepsilon}$-norm and the linearity of f wrt. the parameters $\boldsymbol{\alpha}$. Note that robust sparsity is only considered for the error vector, and that the standard ℓ_1-norm is used for regularization.

Connection with Support Vector Machines.

Problem (13) can be written as the linear program

$$\begin{aligned} \min_{(\boldsymbol{\alpha},\boldsymbol{a},\boldsymbol{\xi})\in\mathbb{R}^{3N}} \quad & \boldsymbol{1}^T\boldsymbol{a} + C\boldsymbol{1}^T\boldsymbol{\xi} \\ \text{s.t.} \quad & -\boldsymbol{\xi} - \varepsilon\boldsymbol{1} \leq \boldsymbol{y} - \boldsymbol{K\alpha} \leq \varepsilon\boldsymbol{1} + \boldsymbol{\xi} \\ & -\boldsymbol{a} \leq \boldsymbol{\alpha} \leq \boldsymbol{a}, \end{aligned} \tag{14}$$

with $C = 1/\lambda$. Here, the objective function has been divided by λ in order to emphasize the equivalence with the training algorithm of the so-called Linear Programming Support Vector Regression (LP-SVR) proposed in [21] for nonlinear function approximation.

4.2 RKHS norm regularization

Introducing robust sparsity in (12) yields

$$\min_{f\in\mathcal{H}} \frac{1}{2}\|f\|_{\mathcal{H}}^2 + C\sum_{i=1}^{N} \ell_\varepsilon(y_i, f(\boldsymbol{x}_i)), \tag{15}$$

with $C = 1/\lambda$ and the ε-insensitive loss function defined as in [35] by

$$\ell_\varepsilon(y_i, f(\boldsymbol{x}_i)) = \max\{0,\ |y_i - f(\boldsymbol{x}_i)| - \varepsilon\}.$$

With these definitions, the connection with Support Vector Regression (SVR) [32] becomes apparent, as detailed below.

Connection with Support Vector Machines and explicit solution.

It is known (see, e.g., [32]) that a kernel function K implicitly defines a (nonlinear) feature map, $\Phi : \mathcal{X} \to \mathcal{F}$, mapping the regressors $\boldsymbol{x}$ into a feature space $\mathcal{F}$, where the model f becomes linear, i.e.,

$$f(\boldsymbol{x}) = \langle \boldsymbol{w}, \Phi(\boldsymbol{x})\rangle_{\mathcal{F}}, \tag{16}$$

with parameters $\boldsymbol{w} \in \mathcal{F}$. In order to emphasize the relationship with SVMs below, we further consider the affine model $\tilde{f} = f + b$, with $b \in \mathbb{R}$. With these notations, and a simple substitution of $\tilde{f}$ for f in the computation of the loss, problem (15) can be written as

$$\begin{aligned} \min_{\boldsymbol{w},b,\xi_i,\xi_i'} \quad & \frac{1}{2}\|\boldsymbol{w}\|^2 + C\sum_{i=1}^{N}(\xi_i + \xi_i') \\ \text{s.t.} \quad & y_i - \langle \boldsymbol{w}, \Phi(\boldsymbol{x}_i)\rangle_{\mathcal{F}} - b \leq \varepsilon + \xi_i \\ & y_i - \langle \boldsymbol{w}, \Phi(\boldsymbol{x}_i)\rangle_{\mathcal{F}} - b \geq -\varepsilon - \xi_i' \\ & \xi_i \geq 0,\ \xi_i' \geq 0, \end{aligned} \tag{17}$$

which is the primal form of the training algorithm of a support vector machine for nonlinear regression (SVR) [32]. Note that, Φ and $\mathcal{F}$ are only implicit and need not be known nor finite-dimensional, and so does $\boldsymbol{w}$. What is known however is that, by construction, $\langle\Phi(\boldsymbol{x}), \Phi(\boldsymbol{x}')\rangle_{\mathcal{F}} = K(\boldsymbol{x}, \boldsymbol{x}')$. Thus, by Lagrangian duality, this problem can be reformulated as the finite-dimensional quadratic program

$$\begin{aligned} \max_{\boldsymbol{\beta}\in\mathbb{R}^N, \boldsymbol{\beta}'\in\mathbb{R}^N} \quad & -\frac{1}{2}(\boldsymbol{\beta} - \boldsymbol{\beta}')^T\boldsymbol{K}(\boldsymbol{\beta} - \boldsymbol{\beta}') - \varepsilon\boldsymbol{1}^T(\boldsymbol{\beta} + \boldsymbol{\beta}') \\ & + \boldsymbol{y}^T(\boldsymbol{\beta} - \boldsymbol{\beta}') \\ \text{s.t.} \quad & \boldsymbol{1}^T(\boldsymbol{\beta} - \boldsymbol{\beta}') = 0 \\ & 0 \leq \boldsymbol{\beta} \leq C,\ 0 \leq \boldsymbol{\beta}' \leq C, \end{aligned} \tag{18}$$

which involves Φ only through the (computable) matrix $\boldsymbol{K}$. Then, the solution of the primal is given by $\boldsymbol{w} = \sum_{i=1}^N \alpha_i\Phi(\boldsymbol{x}_i)$, where $\alpha_i = \beta_i - \beta_i'$. The reader is referred to [32] for more details on the derivation of (18) and on the computation of b.

From these results, the connection with RKHS theory (Sect. 2.1) can readily be seen by choosing Φ as the most natural feature map for a kernel function, i.e., the one that maps $\mathcal{X}$ to the corresponding RKHS: $\Phi(\boldsymbol{x}) = K(\boldsymbol{x}, \cdot)$ and $\mathcal{F} = \mathcal{H}$. Indeed, this yields $f = \boldsymbol{w}$ and $f(\boldsymbol{x}) = \langle f, K(\boldsymbol{x}, \cdot)\rangle_{\mathcal{H}} = \langle \boldsymbol{w}, \Phi(\boldsymbol{x})\rangle_{\mathcal{F}} = \sum_{i=1}^N \alpha_i\langle\Phi(\boldsymbol{x}_i), \Phi(\boldsymbol{x})\rangle_{\mathcal{F}} = \sum_{i=1}^N \alpha_i K(\boldsymbol{x}_i, \boldsymbol{x})$.

4.3 Sparsity, support vectors and outliers

We now detail the connections between nonlinear hybrid system identification and support vector regression at the sparsity level.

In the robust sparsity optimization framework, the error vector, $\boldsymbol{e}$, is (robustly) sparse and the data points with large errors, i.e., $|y_i - f(\boldsymbol{x}_i)| > \varepsilon$, are considered as outliers. In the SVR framework, these points are known as *support vectors* (SVs) and the model f is said to be sparse in the sense that the vector of parameters $\boldsymbol{\alpha}$ is sparse. Formally, a SV is a regression vector $\boldsymbol{x}_i$ from the training set which is retained in the model f after training, i.e., the set of SVs is the set of points $\boldsymbol{x}_i$ for which $\alpha_i \neq 0$. Sparsity of the model is an advantageous feature of SVR which leads to faster computations of outputs $f(\boldsymbol{x})$ by reducing the number of terms in the sum (6). The connection between the two forms of sparsity, measured by $\|\boldsymbol{\alpha}\|_0$ for the model and by $\|\boldsymbol{e}\|_{0,\varepsilon}$ for the error, is given by the following classical result (see, e.g., [32]):

$$\|\boldsymbol{\alpha}\|_0 = \|\boldsymbol{e}\|_{0,\varepsilon} + \|\boldsymbol{e} - \varepsilon\boldsymbol{1}\|_0 + \|\boldsymbol{e} + \varepsilon\boldsymbol{1}\|_0.$$

In words, the set of SVs coincides with the set of points that are not inside the ε-tube of insensitivity, i.e., points with $|y_i - f(\boldsymbol{x}_i)| \geq \varepsilon$. This set differs from the set of outliers only by the points that lie exactly on the boundary of the ε-tube.

4.4 Tuning of the threshold ε

Regarding the tuning of the threshold ε, the following different cases must be considered.

4.4.1 *Bounded-error approach*

In [4], a bounded-error approach is proposed for linear hybrid system identification, in which the number of submodels is estimated in order to satisfy, for a predefined threshold δ, a bound of the form $|y_i - f(\boldsymbol{x}_i)| < \delta$, for all data points. Such a bound is optimal in the bounded noise case, with $\delta = \|\boldsymbol{v}\|_\infty$, where $\boldsymbol{v}$ is the concatenation of all noise terms. But it is also more general in the sense that it does not require a noise model. Indeed, the aim is to obtain a set of submodels which approximate the data with a given tolerance. Thus, the parameter δ allows one to tune the trade-off between the model complexity (measured as the number of submodels) and the fit to the data. The original sparse optimization approach of [1] is of a similar flavor, but uses a data-dependent threshold $\delta(\boldsymbol{x}_i, y_i, f)$. In addition, its analysis focuses on the noiseless case, in which the true number of modes can be recovered.

Following these works, a similar strategy applies to the proposed method, where ε plays a similar role as δ.

4.4.2 With assumptions on the noise model

Optimal values for ε have been investigated in the context of SVR under various noise models by different authors, where "optimal" is to be understood with respect to the precise setting of each author's analysis. A common result of these works shows a linear dependency between the optimal value of ε and the noise standard deviation, σ_v. While this is rather intuitive for a uniform noise model, in which case the optimum is $\varepsilon = \sigma_v$, this is more intricate for Gaussian noise. In particular, [30] obtained a first estimate at $\varepsilon = 0.621\sigma_v$, in a maximum likelihood estimation framework. More precisely, they considered the asymptotically optimal ε as the maximizer of the statistical efficiency of the estimator of a single location parameter, which is an oversimplification of the regression setting. A better estimate of the optimal value of ε for Gaussian noise (also in better accordance with experimental observations) was obtained in [14] by considering the complete regression problem in the maximum a posteriori setting. In this case, they estimated the optimal value of ε with a type 2 maximum likelihood method from Bayesian statistics. Another intuitive result developed in [14] concerns Laplacian noise, for which $\varepsilon = 0$ is the optimal choice. Indeed, with $\varepsilon = 0$, the $\ell_{1,\varepsilon}$-norm coincides with the ℓ_1-norm, and the ε-insensitive loss with the absolute loss which is known to yield a maximum likelihood estimator for Laplacian noise in linear regression. These results are summarized in Table 1.

Table 1: Optimal values of ε with a noise model of standard deviation equal to σ_v.

Noise model	Uniform	Laplacian	Gaussian
Optimal ε	σ_v	0	$1.0043\sigma_v$

An additional difficulty with Gaussian or Laplacian noise is that the criterion, $|y_i - f(\boldsymbol{x}_i)| \leq \varepsilon$, used to remove data points after the estimation of a submodel becomes suboptimal: data points with larger noise terms are not removed. In this case, the complete procedure must be stopped when a sufficiently small, but not too small, number of data points remain in the data set. The rationale here is that points with a low noise magnitude are used to estimate the submodels, while the others are considered as outliers. We further assume that these outliers represent a small fraction of the data set. Since at each iteration of the sparse optimization approach, a submodel is estimated from a set of points representing a large fraction of the remaining data, it is expected that outliers are left unused until the end of the procedure.

4.4.3 Automatic tuning

In the SVM literature, problem (17) is sometimes referred to as ε-SVR to distinguish it from the alternative ν-SVR [28] which allows for the automatic tuning of ε. In the derivation of ν-SVR, the trick is to add a term in the objective function of (17) in order to minimize ε while learning the model. This leads to

$$\min_{f\in\mathcal{H},\varepsilon\in\mathbb{R}^+} \frac{1}{2}\|f\|_{\mathcal{H}}^2 + \frac{C}{N}\sum_{i=1}^N \ell_\varepsilon(y_i, f(\boldsymbol{x}_i)) + C\nu\varepsilon, \qquad (19)$$

where $\nu \geq 0$ is a new hyperparameter tuning the trade-off between the minimization of ε and the minimization of the errors larger than ε. As for the ε-SVR, the solution to this new fomulation is obtained in the form of (6) by solving the dual. However, in this case, the hyperparameter ν enjoys a number of properties which can ease its tuning when compared to ε in (17). In particular, it is shown in [28] that $\nu > 1$ yields $\varepsilon = 0$ and that, if $\varepsilon > 0$, $\nu \in [0, 1]$ can be interpreted as the fraction of data points outside of the ε-tube of insensitivity, i.e., $\nu \approx \|\boldsymbol{e}\|_{0,\varepsilon}$.

A similar approach can be followed in the case of ℓ_1-norm regularization. This leads to a formulation of the LP-SVR allowing for the automatic tuning of ε via linear programming as proposed in [31] or [21].

4.5 Iteratively reweighted scheme

As in the classical case for sparse optimization, a reweighted scheme can be used to improve the recovery of robustly sparse solutions with low sparsity. This leads for instance to

$$\min_{f\in\mathcal{H}} \frac{1}{2}\|f\|_{\mathcal{H}}^2 + C\sum_{i=1}^N w_i\ \ell_\varepsilon(y_i, f(\boldsymbol{x}_i)),$$

with w_i defined as in (10). Such a formulation corresponds to a Weighted-SVR, which has been proposed (with fixed weights) by [34] and others in order to deal with a varying confidence in the data points or to introduce various forms of prior knowledge.

4.6 Algorithmic and implementation issues

The theoretical equivalence between nonlinear hybrid system identification via sparse optimization and support vector regression also yields direct algorithmic benefits. In particular, this means that the problem can be solved efficiently even for large data sets (e.g., with more than ten thousand points).

First, note that all the considered convex formulations, e.g., (11) or (13), are theoretically simple optimization problems due to their convexity. However, despite the possibility to write them as linear programs, e.g., as in (14), solving large-scale instances of such problems requires much more care in practice. In particular, a major issue concerns the memory requirements: the data of the problem, including the (typically dense) N-by-N Gram matrix $\boldsymbol{K}$, simply cannot be stored in the memory of most computers. This basic limitation prevents any subsequent call to a general purpose optimization solver in many cases.

On the other hand, dedicated optimization algorithms have been proposed to train SVMs and benefit from numerous advances in this active field of research, see, e.g., [6]. SVM algorithms typically use decomposition techniques such as sequential minimal optimization (SMO) [25, 29] to avoid the storage of the matrix $\boldsymbol{K}$ in memory. With a proper working set selection strategy, the solution can even be found without having to compute all the elements of the matrix

$\boldsymbol{K}$, thus reducing both the memory and computing load. Good SVM solvers implementing these ideas are for instance SVM^{light} [12] or LibSVM [8]. The latter also implements the Weighted-SVR and can be used in the iteratively reweighted version of the procedure for hybrid system identification, as discussed in Sect. 4.5. For ℓ_1-norm regularization, efficient algorithms are developed in [21]. Finally, these solvers also apply to the original sparse optimization approach of [2] (without robust sparsity) simply by setting $\varepsilon = 0$.

Thus, by showing the equivalence between the robust sparsity optimization approach and support vector regression, we also make the problem tractable for off-the-shelf (and usually freely available) solvers.

5. NUMERICAL SIMULATIONS

We now turn to illustrative examples of application with static functions (Sect. 5.1) and with switching dynamical systems (Sect. 5.2).

5.1 Static example

The first illustrative example considers the approximation of two overlapping nonlinear functions (a sinusoid and a quadratic) from a set of 3000 data points with Gaussian noise of standard deviation $\sigma_v = 0.5$. The submodels are estimated by solving (18) with LibSVM for a Gaussian RBF kernel ($\sigma = 0.5$), $C = 100$ and ε set as in Table 1 for the Gaussian noise model ($\varepsilon = 0.50215$). The first row of Figure 1 shows the first submodel obtained when either one of the two functions dominates the other in terms of the fraction of data points. In both cases, the method correctly estimates the submodel corresponding to the dominating mode. Then, after removing the points close to this submodel, a second submodel is estimated. By thresholding the absolute error, $|y_i - f(\boldsymbol{x}_i)|$, at either 3ε (2nd row of Fig. 1) or ε (last row) in the test for removing points, a sufficient fraction of data are eliminated to allow for the recovery of the second submodel. However, with a threshold of ε, a significant fraction (a bit less than 1/3) of the data remains at the end of the procedure. Then, either the number of submodels is assumed fixed to 2 and the algorithm returns the 2 submodels, or the bounded-error approach is applied and the algorithm continues to estimate additional submodels until all the data are removed.

In this example, we observed that the reweighted scheme of Sect. 4.5 slightly improves the submodels, while the first iteration already yields a satisfactory discrimination between the two modes due to the large fraction of points associated to the dominating one (about 66%). Figure 2 shows the influence of reweighting when this fraction is closer to 50%. For 50.25%, the first iteration is not very accurate, but 10 iterations of the reweighted scheme provide a good approximation of the target submodel. For exactly 50% of data of each mode, the estimated model switches between the two target submodels and cannot discriminate between the modes.

5.2 Switched nonlinear system examples

We now consider the switched nonlinear system example from [19], where the aim is to identify a dynamical system

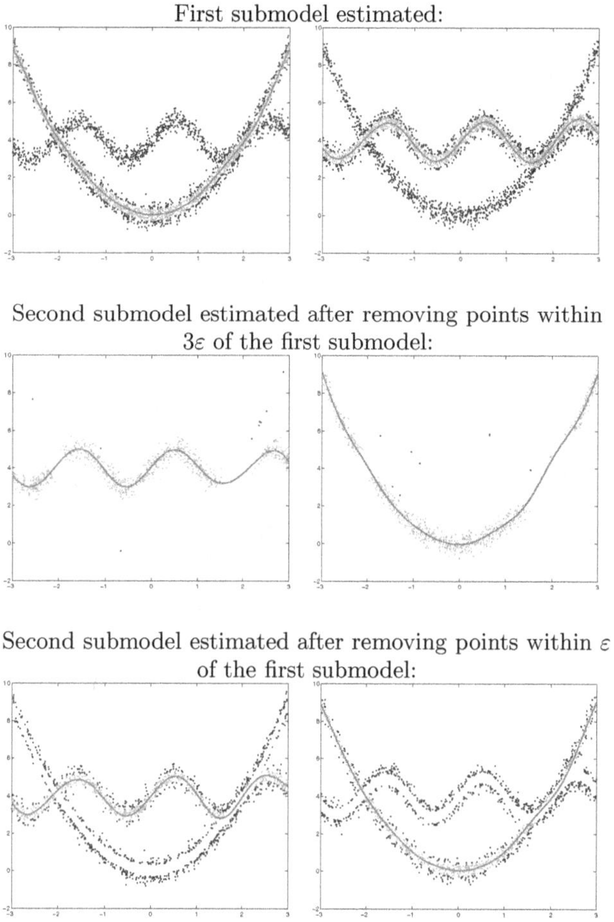
First submodel estimated:
Second submodel estimated after removing points within 3ε of the first submodel:
Second submodel estimated after removing points within ε of the first submodel:

Figure 1: Illustration of the procedure depending on which one of the quadratic (left column) or the sinusoidal (right column) mode dominates the data set.

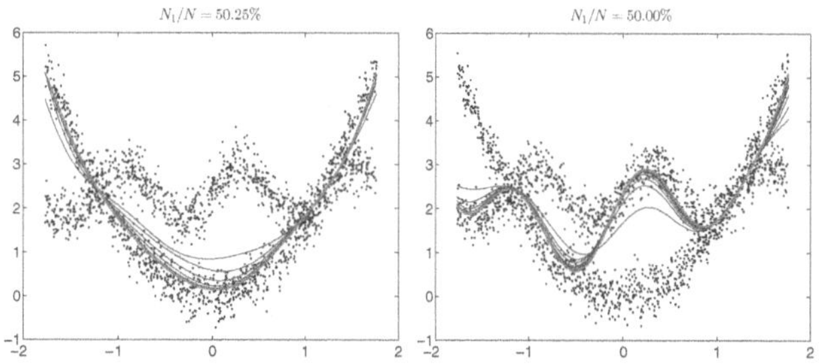
$N_1/N = 50.25\%$
$N_1/N = 50.00\%$

Figure 2: Iterations of the reweighting process for N_1/N = 50.25% (left) and N_1/N = 50% (right), where N_1 is the number of points generated by the quadratic.

arbitrarily switching between two modes as

$$y_i = \begin{cases} 0.9y_{i-1} + 0.2y_{i-2} + v_i, & \text{if } q_i = 1, \\ (0.8 - 0.5\exp(-y_{i-1}^2))y_{i-1} - \\ \quad (0.3 + 0.9\exp(-y_{i-1}^2))y_{i-2} + & \text{if } q_i = 2. \\ \quad 0.4\sin(2\pi y_{i-1}) + 0.4\sin(2\pi y_{i-2}) + v_i, \end{cases} \tag{20}$$

A training set of $N = 3000$ points is generated by (20) with a random sequence of q_i ($P(q_i = 1) = 2/3$ and $P(q_i = 2) = 1/3$), initial conditions $y_0 = y_{-1} = 0.1$, and an additive zero-mean Gaussian noise v_i of standard deviation $\sigma_v = 0.1$.

In this example, the linearity of the first mode is assumed to be known. Thus, a first submodel, f_1, is estimated with a linear kernel and $\varepsilon = 1.0043\sigma_v$, yielding the parameter estimates[5] reported in Table 2 (first column). Then, the points with $i \in I_R = \{i : |y_i - f_1(\boldsymbol{x}_i)| \leq 3\varepsilon\}$ are removed and a nonlinear submodel with a Gaussian RBF kernel ($\sigma = 0.5$) is estimated.

Similar experiments with the reversed order (nonlinear submodel estimated first) are also conducted on a data set with $P(q_i = 1) = 1/3$ (results in Table 2, second column).

The quality of the estimation is evaluated for each mode j in terms of the FIT criterion computed on a test set of 2000 data (generated without noise from the initial conditions $y_0 = 0.4$, $y_{-1} = -0.3$) as

$$\text{FIT}_j = \left(1 - \frac{\sqrt{\sum_{i\in I_j}(y_i - f_{q_i}(\boldsymbol{x}_i))^2}}{\sqrt{\sum_{i\in I_j}(y_i - m_j)^2}}\right),$$

where $I_j = \{i : q_i = j\}$ and m_j is the mean of y_i over all $i \in I_j$. Two additional performance indexes are used to evaluate the ability of the method to discriminate between the two modes during training: the fraction of data that must be removed and have been,

$$\text{D1} = \frac{|I_1 \cap I_R|}{|I_1|},$$

and the fraction of data that must be removed among those that have been,

$$\text{D2} = \frac{|I_1 \cap I_R|}{|I_R|}.$$

Note that these numbers are computed on the training data. The results, shown in Table 2, emphasize the accuracy of the estimated submodels and the fact that the proposed method correctly discriminates between the two modes, independently of the dominating mode.

Table 3 shows similar results for a switched NARX system with two nonlinear modes given by

$$y_i = \begin{cases} 0.4y_{i-1}^2 + 0.2y_{i-2}, & \text{if } q_i = 1, \\ (0.8 - 0.5\exp(-y_{i-1}^2))y_{i-1} - \\ \quad (0.3 + 0.9\exp(-y_{i-1}^2))y_{i-2} + & \text{if } q_i = 2. \\ \quad 0.4\sin(2\pi y_{i-1}) + 0.4\sin(2\pi y_{i-2}), \end{cases} \tag{21}$$

For this example, training trajectories of $N = 16000$ points are generated with 6000 points for mode 1 and 10000 points for mode 2 ($P(q_i = 1) = 0.375$). On these large-scale data sets, the average computing time was about one minute for each SVR training by LibSVM, i.e., for each iteration of the reweighted scheme.

[5]With a linear kernel, the parameters of a linear submodel (6) are recovered by $\boldsymbol{\theta} = \sum_{i=1}^{N} \alpha_i \boldsymbol{x}_i$.

Table 2: Estimation of the system (20) switching between a linear mode (with parameters θ_1, θ_2) and a nonlinear mode. Numbers are averages and standard deviations over 100 trials with different noise, v_i, and mode, q_i, sequences.

$P(q_i = 1)$	2/3	1/3
θ_1 (= 0.9)	0.9008 ± 0.0092	0.9000 ± 0.0070
θ_2 (= 0.2)	0.1824 ± 0.0111	0.2019 ± 0.0068
FIT_1 (%)	97.9148 ± 0.8136	99.1472 ± 0.3261
FIT_2 (%)	83.7052 ± 5.3668	84.8237 ± 6.2603
D1 (%)	99.6840 ± 0.1383	98.7180 ± 0.4904
D2 (%)	88.4713 ± 0.6781	85.7150 ± 0.7850

Table 3: Estimation of the system (21). Numbers are averages and standard deviations over 100 trials with different noise, v_i, and mode, q_i, sequences.

FIT_1 (%)	73.2515 ± 4.2561
FIT_2 (%)	88.6945 ± 1.0960
D1 (%)	99.1160 ± 0.1618
D2 (%)	78.6506 ± 0.2756

6. CONCLUSIONS

This paper discussed the identification of hybrid systems involving arbitrary and unknown nonlinearities in the submodels, particularly focusing on the sparse optimization approach. Conditions of application of this approach were relaxed with the introduction of robust sparsity as a means to deal with noise in the data. We then emphasized the connections between this approach and the support vector machines developed in the field of machine learning. In particular, we have shown that nonlinear hybrid systems can be identified efficiently from large data sets by a sequence of SVM trainings. In addition, this formal equivalence allowed for the derivation of a modified algorithm for the automatic determination of the main hyperparameter (the threshold ε) in the robust sparsity approach. This modified algorithm introduces a new parameter, ν, which can be interpreted as the fraction of data considered as outliers for the model. The precise relationship between this parameter and the fraction of data generated by each mode, which is also involved in the sparse recovery conditions, is the subject of ongoing investigations. In particular, the characterization of the influence of the reweighting scheme on the choice of ν remains an open issue.

An alternative direction of future research concerns the computation of the full solution paths with respect to the regularization constant (λ) and the hyperparameters ε or ν. Here, the aim is to obtain the models for all possible values of the hyperparameters at a low computational cost. In this respect, we should once again take advantage of the equivalence with support vector regression and the large collection of results on this topic [11, 10, 37, 20].

The paper focused on systems with arbitrary switching mechanisms. While this provides an algorithm for a very general class of hybrid systems, this also implies that the active mode can only be predicted a posteriori (after the observation of the actual output), which limits the applicability of the model for predictive purposes. In this respect, the proposed approach should be adapted to deal with piecewise smooth (PWS) systems, where the mode depends on a partition of the regression space. Meanwhile, the classical approach to this issue is to use a classification algorithm in a postprocessing step to recover the partition from the training points grouped into modes, as explained, e.g., in [24]. Note that SVMs were originally developed for classification and provide state-of-the-art solutions to such problems.

Finally, the concept of robust sparsity has also been introduced in the context of switched *linear* systems and more particularly for multi-input multi-output (MIMO) systems in state-space form [3].

7. REFERENCES

[1] L. Bako. Identification of switched linear systems via sparse optimization. *Automatica*, 47(4):668–677, 2011.

[2] L. Bako, K. Boukharouba, and S. Lecoeuche. An ℓ_0-ℓ_1 norm based optimization procedure for the identification of switched nonlinear systems. In *Proc. of the 49th IEEE Int. Conf. on Decision and Control (CDC)*, pages 4467–4472, 2010.

[3] L. Bako, V. L. Le, F. Lauer, and G. Bloch. Identification of MIMO switched state-space models. In *Proc. of the American Control Conference (ACC), Washington, DC, USA*, 2013.

[4] A. Bemporad, A. Garulli, S. Paoletti, and A. Vicino. A bounded-error approach to piecewise affine system identification. *IEEE Transactions on Automatic Control*, 50(10):1567–1580, 2005.

[5] A. Berlinet and C. Thomas-Agnan. *Reproducing Kernel Hilbert Spaces in Probability and Statistics.* Kluwer Academic Publishers, Boston, 2004.

[6] L. Bottou, O. Chapelle, D. DeCoste, and J. Weston, editors. *Large Scale Kernel Machines.* MIT Press, Cambridge, MA, 2007.

[7] E. Candès, M. Wakin, and S. Boyd. Enhancing sparsity by reweighted ℓ_1 minimization. *Journal of Fourier Analysis and Applications*, 14(5):877–905, 2008.

[8] C. Chang and C. Lin. LibSVM: a library for support vector machines, 2001. Available at http://www.csie.ntu.edu.tw/~cjlin/libsvm/.

[9] G. Ferrari-Trecate, M. Muselli, D. Liberati, and M. Morari. A clustering technique for the identification of piecewise affine systems. *Automatica*, 39(2):205–217, 2003.

[10] L. Gunter and J. Zhu. Computing the solution path for the regularized support vector regression. In *Advances in Neural Information Processing Systems 18*, pages 483–490, 2006.

[11] T. Hastie, S. Rosset, R. Tibshirani, and J. Zhu. The entire regularization path for the support vector machine. *Journal of Machine Learning Research*, 5:1391–1415, 2004.

[12] T. Joachims. SVM^{light}, 1998. Available at http://svmlight.joachims.org/.

[13] A. L. Juloski, S. Weiland, and W. Heemels. A Bayesian approach to identification of hybrid systems. *IEEE Transactions on Automatic Control*, 50(10):1520–1533, 2005.

[14] J. Kwok and I. Tsang. Linear dependency between ε and the input noise in ε-support vector regression. *IEEE Transactions on Neural Networks*, 14(3):544–553, 2003.

[15] F. Lauer. Estimating the probability of success of a simple algorithm for switched linear regression. *Nonlinear Analysis: Hybrid Systems*, 8:31–47, 2013. Supplementary material available at `http://www.loria.fr/~lauer/klinreg/`.

[16] F. Lauer and G. Bloch. Switched and piecewise nonlinear hybrid system identification. In *Proc. of the 11th Int. Conf. on Hybrid Systems: Computation and Control (HSCC)*, volume 4981 of *LNCS*, pages 330–343, 2008.

[17] F. Lauer, G. Bloch, and R. Vidal. A continuous optimization framework for hybrid system identification. *Automatica*, 47(3):608–613, 2011.

[18] F. Lauer, V. L. Le, and G. Bloch. Learning smooth models of nonsmooth functions via convex optimization. In *Proc. of the IEEE Int. Workshop on Machine Learning for Signal Processing (MLSP), Santander, Spain*, 2012.

[19] V. L. Le, G. Bloch, and F. Lauer. Reduced-size kernel models for nonlinear hybrid system identification. *IEEE Transactions on Neural Networks*, 22(12):2398–2405, 2011.

[20] G. Loosli, G. Gasso, and S. Canu. Regularization paths for ν-SVM and ν-SVR. In *Advances in Neural Networks - ISNN*, volume 4493 of *LNCS*, pages 486–496, 2007.

[21] O. Mangasarian and D. Musicant. Large scale kernel regression via linear programming. *Machine Learning*, 46(1):255–269, 2002.

[22] H. Ohlsson and L. Ljung. Identification of piecewise affine systems using sum-of-norms regularization. In *Proc. of the 18th IFAC World Congress*, pages 6640–6645, 2011.

[23] N. Ozay, M. Sznaier, C. Lagoa, and O. Camps. A sparsification approach to set membership identification of switched affine systems. *IEEE Transactions on Automatic Control*, 57(3):634–648, 2012.

[24] S. Paoletti, A. L. Juloski, G. Ferrari-Trecate, and R. Vidal. Identification of hybrid systems: a tutorial. *European Journal of Control*, 13(2-3):242–262, 2007.

[25] J. Platt. Fast training of support vector machines using sequential minimal optimization. In B. Schölkopf, C. Burges, and A. Smola, editors, *Advances in Kernel Methods: Support Vector Learning*, pages 185–208. MIT Press, 1999.

[26] J. Roll, A. Bemporad, and L. Ljung. Identification of piecewise affine systems via mixed-integer programming. *Automatica*, 40(1):37–50, 2004.

[27] B. Schölkopf, R. Herbrich, and A. Smola. A generalized representer theorem. In *Proc. of COLT/EuroCOLT*, volume 2111 of *LNAI*, pages 416–426, 2001.

[28] B. Schölkopf, A. Smola, R. Williamson, and P. Bartlett. New support vector algorithms. *Neural computation*, 12(5):1207–1245, 2000.

[29] S. Shevade, S. Keerthi, C. Bhattacharyya, and K. Murthy. Improvements to the SMO algorithm for

SVM regression. *IEEE Transactions on Neural Networks*, 11(5):1188–1193, 2000.

[30] A. Smola, N. Murata, B. Schölkopf, and K. Müller. Asymptotically optimal choice of ε-loss for support vector machines. In *Proc. of the 8th Int. Conf. on Artificial Neural Networks (ICANN)*, pages 105–110, 1998.

[31] A. Smola, B. Schölkopf, and G. Rätsch. Linear programs for automatic accuracy control in regression. In *Proc. of the 9th Int. Conf. on Artificial Neural Networks (ICANN)*, pages 575–580, 1999.

[32] A. J. Smola and B. Schölkopf. A tutorial on support vector regression. *Statistics and Computing*, 14(3):199–222, 2004.

[33] I. Steinwart and A. Christmann. *Support Vector Machines*. Springer, 2008.

[34] F. Tay and L. Cao. Modified support vector machines in financial time series forecasting. *Neurocomputing*, 48(1):847–861, 2002.

[35] V. Vapnik. *The Nature of Statistical Learning Theory*. Springer-Verlag, 1995.

[36] R. Vidal, S. Soatto, Y. Ma, and S. Sastry. An algebraic geometric approach to the identification of a class of linear hybrid systems. In *Proc. of the 42nd IEEE Conf. on Decision and Control (CDC)*, pages 167–172, 2003.

[37] G. Wang, D. Yeung, and F. Lochovsky. Two-dimensional solution path for support vector regression. In *Proc. of the 23rd Int. Conf. on Machine Learning (ICML)*, pages 993–1000, 2006.

APPENDIX

A. SKETCH OF PROOF OF THEOREM 1

Any function $f \in \mathcal{H}$ can be decomposed into a part in the span of $\{K(\boldsymbol{x}_i, \cdot) \ : \ \boldsymbol{x}_i \in \mathcal{D}_x\}$ and a part which is orthogonal to it, i.e., $f = u + v$, with, for all $\boldsymbol{x}_i \in \mathcal{D}_x$, $\langle K(\boldsymbol{x}_i, \cdot), v\rangle_{\mathcal{H}} = 0$. Thus, and by the reproducing property of K, for all $\boldsymbol{x}_i \in \mathcal{D}_x$, $f(\boldsymbol{x}_i) = \langle K(\boldsymbol{x}_i, \cdot), u + v\rangle_{\mathcal{H}} = \langle K(\boldsymbol{x}_i, \cdot), u\rangle_{\mathcal{H}} = u(\boldsymbol{x}_i)$. As a consequence, the first term in (2), which computes the error over the training set, does not depend on v. On the other hand, we have $\|f\|_{\mathcal{H}} = \|u + v\|_{\mathcal{H}} = \sqrt{\|u\|^2_{\mathcal{H}} + \|v\|^2_{\mathcal{H}} + 2\langle u, v\rangle_{\mathcal{H}}}$. Since $u \perp v$, $\langle u, v\rangle_{\mathcal{H}} = 0$. Thus, for any f with $v \neq 0$, there is a function $u \in Span\{K(\boldsymbol{x}_i, \cdot) \ : \ \boldsymbol{x}_i \in \mathcal{D}_x\}$ with smaller norm and which, in the conditions of Theorem 2.1, leads to a lower value of the objective function. A more detailed proof can be found in [27].

B. COMPLEXITY CONTROL VIA ℓ_1-NORM REGULARIZATION

Here, we show why the ℓ_1-norm regularization can be used in (7) instead of the RKHS norm to penalize the nonsmoothness of the model (i.e., the magnitude of the derivatives of f) and control its complexity. This can be seen from the fact that any f in the form used in (7) belongs to the RKHS $(\mathcal{H}, \langle\cdot,\cdot\rangle_{\mathcal{H}})$ and that its norm in this RKHS, given by (4), can be bounded by

$$\|f\|_{\mathcal{H}} = \sqrt{\boldsymbol{\alpha}^T \boldsymbol{K} \boldsymbol{\alpha}} \leq \sqrt{\lambda_{\max} \|\boldsymbol{\alpha}\|_2^2} \leq \sqrt{\lambda_{\max}} \|\boldsymbol{\alpha}\|_1,$$

where $\lambda_{\max}$ is the largest eigenvalue of the Gram matrix $\boldsymbol{K}$ (as defined in Sect. 3.3.1). Thus, minimizing $\|\boldsymbol{\alpha}\|_1$ also provides control over $\|f\|_{\mathcal{H}}$ and the complexity of f.

Mining Requirements from Closed-Loop Control Models

Xiaoqing Jin
Univ. of California Riverside
jinx@cs.ucr.edu

Alexandre Donzé
Univ. of California Berkeley
donze@eecs.berkeley.edu

Jyotirmoy V. Deshmukh
Toyota Technical Center
jyotirmoy.deshmukh@tema.toyota.com

Sanjit A. Seshia
Univ. of California Berkeley
sseshia@eecs.berkeley.edu

ABSTRACT

A significant challenge to the formal validation of software-based industrial control systems is that system requirements are often imprecise, non-modular, evolving, or even simply unknown. We propose a framework for mining requirements from the closed-loop model of an industrial-scale control system, such as one specified in the Simulink modeling language. The input to our algorithm is a *requirement template* expressed in Parametric Signal Temporal Logic — a formalism to express temporal formulas in which concrete signal or time values are replaced by parameters. Our algorithm is an instance of counterexample-guided inductive synthesis: an intermediate candidate requirement is synthesized from *simulation traces* of the system, which is refined using counterexamples to the candidate obtained with the help of a *falsification* tool. The algorithm terminates when no counterexample is found. Mining has many usage scenarios: mined requirements can be used to validate future modifications of the model, they can be used to enhance understanding of legacy models, and can also guide the process of bug-finding through simulations. We present two case studies for requirement mining: a simple automobile transmission controller and an industrial airpath control model for an engine.

Categories and Subject Descriptors

D.4.7 [**Organization and Design**]: Real-time Systems and embedded systems; D.2.1 [**Software Engineering**]: Requirements/Specifications; I.2.8 [**Computing Methodologies**]: Problem Solving, Control Methods, and Search

General Terms

Verification, Theory

Keywords

Model-based design; Parametric Temporal Logics; Simulink

HSCC'13, April 8–11, 2013, Philadelphia, Pennsylvania, USA.

1. INTRODUCTION

Industrial-scale controllers used in automobiles and avionics are now commonly developed using a model-based design (MBD) paradigm [23, 28]. The MBD process consists of a sequence of steps. In the first step, the designer captures the *plant model*, i.e., the dynamical characteristics of the physical parts of the system using logical, differential and algebraic equations. Examples of plant models include the rotational dynamics model of the camshaft in an automobile engine, the thermodynamic model of an internal combustion engine, and atmospheric turbulence models. The next step is to design a *controller* that employs specific control laws to regulate the behavior of the physical system. The *closed-loop model* consists of the composition of the plant and the controller.

The designer may then perform extensive *simulations* of such a model. The goal is to analyze the controller design by observing the temporal behavior of the signals of interest by exciting the exogenous, time-varying inputs to the model. An important aspect of this step is *validation*, i.e. checking if the temporal behavior of the system matches a set of *requirements*. Unfortunately, in practice, these requirements are high-level and often vague. Examples of requirements include, "better fuel-efficiency", "signal should eventually settle", and "resistance to turbulence". If the simulation behavior is deemed unsatisfactory, then the designer refines or tunes the controller design and repeats the validation step.

In the formal methods literature, a requirement (also called a specification) is a mathematical expression of the design goals or desirable design properties in a suitable logic. In an industrial design setting, requirements are rarely expressed formally, and it is common to find them written in natural language. Control designers then validate their design manually by comparing experimental time traces to these informal requirements. In some cases, they simply use simulation-data and their domain expertise to determine the quality of the design. Moreover, to date, formal validation tools have been unable to digest the format or scale of industrial-scale models. As a result, widespread adoption of formal tools has been restricted to testing syntactic coverage of the controller code, with the hope that higher coverage implies better chances of finding bugs. It is clear that even simulation-based tools would benefit from the more semantic notions of coverage offered by formal requirements.

In this paper, we propose a scalable technique to systematically mine requirements from the closed-loop model of a control system from observations of the system behavior. In addition to the model being analyzed, our technique takes as input a *template requirement*. The final output is a synthesized requirement matching the template. In our current implementation, we assume that the model is specified in Simulink [21], an industry-wide standard that is able to: (1) express complex dynamics (differential and algebraic equations), (2) capture discrete state-machine behavior by allowing both Boolean and real-valued variables, (3) allow a layered design approach through modularity and hierarchical composition, and (4) perform high-fidelity simulations. We remark that our technique is not restricted to Simulink models; in principle, it is applicable in any setting where the closed-loop system can be simulated, e.g., hardware-in-the-loop simulations, and tests on the physical system.

Formalisms such as Metric Temporal Logic (MTL) [2,17], and later Parametric Signal Temporal Logic (PSTL) [6] have emerged as logics well-suited to capture both the real-valued and time-varying behaviors of hybrid control systems. As PSTL is equipped with *parameters*, properties in PSTL naturally express template requirements. As an example, consider the following natural language specification: "eventually between time 0 and some unspecified time τ_1, the signal $\mathbf{x}$ is less than some value π_1, and from that point for some τ_2 seconds, it remains less than some value π_2". In PSTL the above property would be expressed as:

$$\Diamond_{[0,\tau_1]}(\mathbf{x} < \pi_1 \ \wedge \ \Box_{[0,\tau_2]}(\mathbf{x} < \pi_2)).$$

Here, we interpret the unspecified values $\tau_1, \tau_2, \pi_1, \pi_2$ as parameters. The subset of PSTL with no parameters is referred to as STL. Robust satisfaction of MTL formulas [13] and quantitative semantics for PSTL [11] allow reasoning about how "close" a system behavior is to satisfying a given specification. Intuitively, a lower *satisfaction value* corresponds to a stronger property, making it easier for a behavior to violate the property.

The proposed mining algorithm is an iterative procedure; in each iteration, it performs the following steps:

1. In the first step, the algorithm synthesizes a *candidate requirement* from a given PSTL template and a set of simulation traces of the model. The candidate requirement is the strongest STL property satisfied by the given set of traces. It is obtained by instantiating the PSTL template with the parameter values that minimize the satisfaction value of the PSTL property over the given traces.

2. It then tries to falsify the candidate requirement using a global optimization-based search, such as using stochastic-search within the tool S-TALIRO [5].

3. If the falsification tool finds a counterexample, we add this trace to the existing set of simulation traces, and go to Step 1 of the next iteration. If no counterexample is found, the algorithm terminates.

At the heart of Step 1 is an efficient search over the space defined by the parameters in the PSTL property in order to generate a candidate requirement. For this purpose, we use the BREACH tool [9]. If the number of parameters is n, a naïve search strategy in the parameter space would have an exponential cost in n.

However, we observe that the satisfaction values of certain PSTL properties are monotonic in their parameter values. For example, consider the property $\phi = \Diamond_{[0,\tau]}(x > \pi)$. Suppose that the minimum value of a given trace $x(t)$ is 3, then, starting from a value less than 3, as π increases, the property ϕ becomes a *stronger* assertion for the trace $x(t)$, i.e., its satisfaction value decreases. Finally, when π exceeds 3, the satisfaction value becomes negative, i.e., ϕ no longer holds for $x(t)$. Thus, we can say that the satisfaction value of ϕ monotonically decreases in the parameter π. Similarly, the satisfaction value of ϕ monotonically increases in τ. When monotonicity holds, we can get exponential savings when searching over the parameter-space by using methods like binary search. Though syntactic rules for *polarity* of a PSTL property identified in previous work [6] ensure satisfaction monotonicity, these rules are not complete. Hence, we provide a general way of reasoning about monotonicity of arbitrary PSTL properties using Satisfiability-Modulo-Theories (SMT) solving [7].

In this paper, we explore two applications for requirement mining. The first application is the obvious one: to generate requirements that serve as high-level specifications for the closed-loop model. In an industrial setting, formalized requirements that can be used for design validation are often unavailable. For example, consider the case of legacy controller code. Such code usually goes through several years of refinement, is developed in a non-formal setting, and is not very easy to understand for any engineers other than its original developers. In this context, mined requirements can enhance understanding of the code and help future code maintenance. The second application explores the use of mining as an enhanced bug-finding procedure. Suppose we wish to check if the model behavior ever has a signal that oscillates with an amplitude greater than a threshold. Considering the huge space of input signals, simply running tests on the closed-loop model is unlikely to detect such behavior. We instead attempt to mine the requirement, "the signal settles to a steady value π in time τ" (roughly corresponding to the negation of the original property). In each step, our algorithm pushes the trajectory-space exploration of the falsification tool in a region not already subsumed by existing traces. Hence, the search for a counterexample is guided by the intermediate candidate requirements. Note that state-of-the-art falsifiers such as S-TALIRO would require a *concrete* STL property encoding the oscillation behavior, which would require tedious manual effort given many possible expressions of such behavior arising from unknowns such as the oscillation amplitude, frequency, and the time at which oscillations start.

To summarize, our contributions are as follows:

1. We propose a novel counterexample-guided iterative procedure for mining temporal requirements satisfied by signals of interest of an industrial-scale closed-loop control model. Specifically, we target the mining of properties expressible in PSTL.

2. We extend BREACH to support Simulink models and the falsification of STL formulas. In addition we enhance the BREACH tool framework with efficient strategies for synthesizing parameters of monotonic PSTL properties. To extend the range of formulas for which we can prove monotonicity, and hence apply these strategies, we formulate the query for monotonicity in a fragment of first order logic with quantifiers, real arithmetic and uninterpreted functions, and use an SMT solver to answer the query.

3. We demonstrate the practical applicability of our technique in two case studies: (a) a simple automatic transmission controller, and (b) an industrial closed-loop model of the airpath-control in an automobile engine model. We also demonstrate the use of the mining technique as a bug-finding tool, showing how it found a bug in the industrial model that was confirmed by a designer.

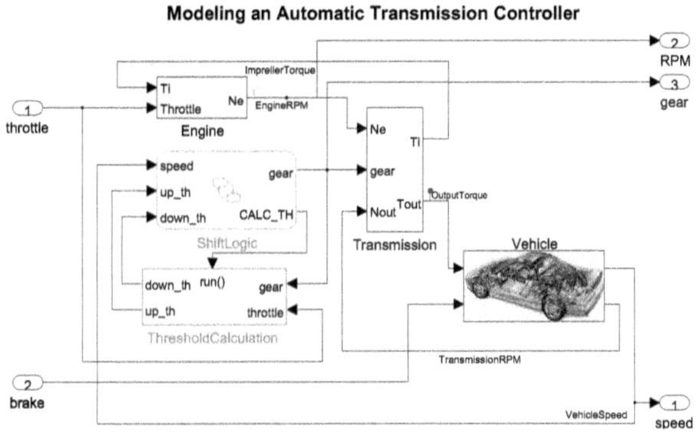

Figure 1: The closed-loop Simulink model of an automatic transmission controller. The input to the model is the throttle position and the brake torque.

The rest of the paper is as follows: In Sec. 2, we present a transmission controller as a running example. Sec. 3 presents the background, the problem formulation and an overview of our technique. We discuss our approach for finding counterexamples to candidate requirements in Sec. 4, and synthesizing parameter values for templates in Sec. 5 and Sec. 6. We present two case studies and experimental results for each in Sec. 7, and conclude with related work in Sec. 8.

2. A RUNNING EXAMPLE

As an illustrative example throughout the paper, we consider a closed-loop model designed for a four-speed automatic transmission controller of a vehicle (shown in Fig. 1). Although this model is not a real industrial model, it has all necessary mechanical components: models for the engine, the transmission, and the vehicle. The transmission block computes the transmission ratio (Ti) using the current gear status, and computes the output torque from the engine speed (Ne), the gear status and the transmission RPM. The other two blocks represent the gear shift logic and the related threshold speed calculation. The model has two inputs: (1) the percentage of the `throttle` position, and (2) the `brake` torque.

We are interested in the following signals: the vehicle `speed`, transmission `gear` position, and engine speed measured in `RPM` (rotations per minute). Suppose we want to use this controller to ensure the requirement that the engine speed never exceeds 4500 rpm, and that the vehicle never drives faster than 120 mph. After simulating the closed-loop system we can show that these requirements are not met, as illustrated in Fig. 2.

However, this negative result does not provide further insight into the model. If a requirement does not hold, we would like to know what *does* hold for the controller, and how narrowly the controller misses the requirement. Such a characterization would shed more light on the working of the system, especially in the context of legacy systems and for reverse engineering the behavior of a very complex system. In the context of this example, it would help to know the maximum speed and RPM that the model can reach, or the minimum dwell time that the transmission enforces to avoid frequent gear shifts. In the next section, we present a technique to automatically obtain such requirements from the model.

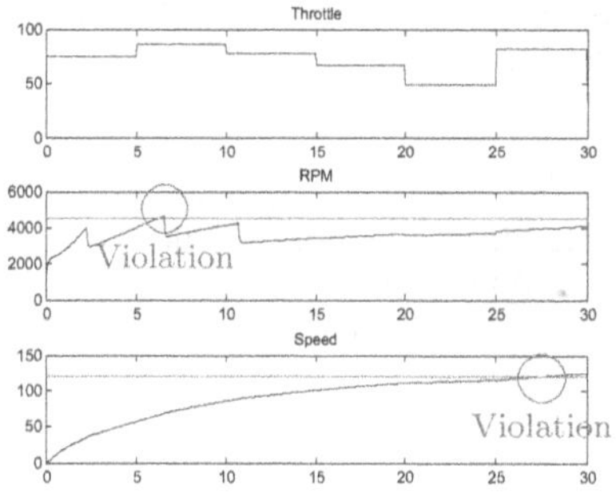

Figure 2: Falsifying trace for the automatic transmission controller and the requirement that RPM never goes beyond 4500 or speed beyond 120 mph.

3. PRELIMINARIES AND OVERVIEW

3.1 Signals and Systems

The systems considered in this paper are hybrid dynamical systems, that is systems mixing discrete dynamics (such as the shifting logic of gears) and continuous dynamics (such as the rotational dynamics of the car engine). Additionally, the systems are closed-loop, meaning that they are obtained by composing a controller and a plant in a loop.[1]

We define a *signal* as a function mapping the time domain $\mathbb{T} = \mathbb{R}^{\geq 0}$ to the reals $\mathbb{R}$. *Boolean signals*, used to represent discrete dynamics, are signals whose values are restricted to *false* (denoted $\bot$) and *true* (denoted $\top$). Vectors in $\mathbb{R}^n$ with $n > 1$ are denoted in bold fonts and their components are indexed from 1 to n, e.g., $\mathbf{p} = (p_1, \cdots, p_n)$. Likewise, a multi-dimensional signal $\mathbf{x}$ is a function from $\mathbb{T}$ to $\mathbb{R}^n$ such that $\forall t \in \mathbb{T}$, $\mathbf{x}(t) = (x_1(t), \cdots, x_n(t))$. A *system* $\mathcal{S}$ (such as a Simulink model) is an input-output state machine: it takes as input a signal $\mathbf{u}(t)$ and computes an output signal $\mathbf{x}(t) = \mathcal{S}(\mathbf{u}(t))$. It is common to drop time t, and say $\mathbf{x} = \mathcal{S}(\mathbf{u})$. A *trace* is a collection of output signals resulting from the simulation of a system, i.e., it can be viewed as a multi-dimensional signal. In the following, we use interchangeably the words trace and signal.

3.2 Signal Temporal Logic

Temporal logics were introduced in the late 1970s [24] to reason formally about the temporal behaviors of *reactive* systems – originally input-output systems with Boolean, discrete-time signals. Temporal logics to reason about real-time signals, such as Timed Propositional Temporal Logic [2], and Metric Temporal Logic (MTL) [17] were introduced later to deal with dense-time signals. More recently, Signal Temporal Logic [20] was proposed in the context of analog and mixed-signal circuits as a specification language for constraints on real-valued signals. These constraints, or *predicates* can be reduced to the form $\mu = f(\mathbf{x}) \sim \pi$, where f is a scalar-valued function over the signal $\mathbf{x}$, $\sim \in \{<, \leq, \geq, >, =, \neq\}$, and π is a real number.

Temporal formulas are formed using temporal operators, "always" (denoted as $\Box$), "eventually" (denoted as $\Diamond$) and "until" (denoted as $\mathbf{U}$). Each temporal operator is indexed by intervals of the form (a, b), $(a, b]$, $[a, b)$, $[a, b]$, (a, ∞) or $[a, \infty)$ where each of a, b is a non-negative real-valued con-

[1]Note that such systems can have exogenous inputs, e.g. a human controlling brakes provides inputs to the vehicle engine and controller system. The term "closed-loop" differs from "closed systems," which are systems with no inputs.

stant. If I is an interval, then an STL formula is written using the following grammar:

$$\varphi := \top \mid \mu \mid \neg\varphi \mid \varphi_1 \wedge \varphi_2 \mid \varphi_1 \, \mathbf{U}_I \, \varphi_2$$

The always and eventually operators are defined as special cases of the until operator as follows: $\Box_I \varphi \triangleq \neg \Diamond_I \neg \varphi$, $\Diamond_I \varphi \triangleq \top \, \mathbf{U}_I \, \varphi$. When the interval I is omitted, we use the default interval of $[0, +\infty)$. The semantics of STL formulas are defined informally as follows. The signal $\mathbf{x}$ satisfies $f(\mathbf{x}) > 10$ at time t (where $t \geq 0$) if $f(\mathbf{x}(t)) > 10$. It satisfies $\varphi = \Box_{[0,2)} \, (x > -1)$ if for all time $0 \leq t < 2$, $x(t) > -1$. The signal x_1 satisfies $\varphi = \Diamond_{[1,2)} \, x_1 > 0.4$ iff there exists time t such that $1 \leq t < 2$ and $x_1(t) > 0.4$. The two-dimensional signal $\mathbf{x} = (x_1, x_2)$ satisfies the formula $\varphi = (x_1 > 10) \, \mathbf{U}_{[2.3,4.5]} \, (x_2 < 1)$ iff there is some time u where $2.3 \leq u \leq 4.5$ and $x_2(u) < 1$, and for all time v in $[2.3, u)$, $x_1(u)$ is greater than 10. Formally, the semantics are given as follows:

$$\begin{array}{lll}
(\mathbf{x}, t) \models \mu & \text{iff} & \mathbf{x} \text{ satisfies } \mu \text{ at time } t \\
(\mathbf{x}, t) \models \neg\varphi & \text{iff} & (\mathbf{x}, t) \not\models \varphi \\
(\mathbf{x}, t) \models \varphi_1 \wedge \varphi_2 & \text{iff} & (\mathbf{x}, t) \models \varphi_1 \text{ and } (\mathbf{x}, t) \models \varphi_2 \\
(\mathbf{x}, t) \models \varphi_1 \, \mathbf{U}_{[a,b]} \, \varphi_2 & \text{iff} & \exists t' \in [t+a, t+b] \text{ s.t.} \\
 & & \quad (\mathbf{x}, t') \models \varphi_2 \text{ and} \\
 & & \quad \forall t'' \in [t+a, t'], (\mathbf{x}, t'') \models \varphi_1
\end{array}$$

Extension of the above semantics to other kinds of intervals (open, open-closed, and closed-open) is straightforward. We write $\mathbf{x} \models \varphi$ as a shorthand of $(\mathbf{x}, 0) \models \varphi$.

Parametric Signal Temporal Logic (PSTL) [6] is an extension of STL introduced to define *template formulas* containing unknown parameters. Syntactically speaking, a PSTL formula is an STL formula where numeric constants, either in the constraints given by the predicates μ or in the time intervals of the temporal operators, can be replaced by symbolic parameters divided into two types:

- A *Scale* parameter π is a parameter appearing in predicates of the form $\mu = f(\mathbf{x}) \sim \pi$,
- A *Time* parameter τ is a parameter appearing in an interval of a temporal operator.

An STL formula is obtained by pairing a PSTL formula with a valuation function that assigns a value to each symbolic parameter. For example, consider the PSTL formula $\varphi(\pi, \tau) = \Box_{[0,\tau]} x > \pi$, with symbolic parameters π (scale) and τ (time). The STL formula $\Box_{[0,10]} x > 1.2$ is an instance of φ obtained with the valuation $v = \{\tau \mapsto 10, \ \pi \mapsto 1.2\}$.

EXAMPLE 3.1. *For the example from Sec. 2, suppose we want to specify that the* `speed` *never exceeds* 120 *and* `RPM` *never exceeds* 4500. *The predicate specifying that the speed is above 120 is:* `speed`$>$120 *and the one for RPM is* `RPM`$>$4500. *The STL formula expressing these to be always false is:*

$$\psi = \Box(\texttt{speed} \leq 120) \wedge \Box(\texttt{RPM} \leq 4500). \qquad (3.1)$$

To turn this into a PSTL formula, we rewrite by introducing parameters π_{speed} and π_{RPM}:

$$\varphi(\pi_{speed}, \pi_{rpm}) = \Box(\texttt{speed} \leq \pi_{speed}) \wedge \Box(\texttt{RPM} \leq \pi_{rpm}). \quad (3.2)$$

The STL formula ψ expressed in (3.1) *is then obtained by using the valuation $v = (\pi_{\texttt{speed}} \mapsto 120, \pi_{rpm} \mapsto 4500)$. formulation.*

PROBLEM 3.1. *Given (a) a system $\mathcal{S}$ with a set $\mathcal{U}$ of inputs, and, (b) a PSTL formula with n symbolic parameters $\varphi(p_1, \ldots, p_n)$ where p_i could either be scale parameter π or time parameter τ, the objective is to find a* tight *valuation function v such that*

$$\forall \mathbf{u} \in \mathcal{U} : \mathcal{S}(\mathbf{u}) \models \varphi(v(p_1), \ldots, v(p_n)).$$

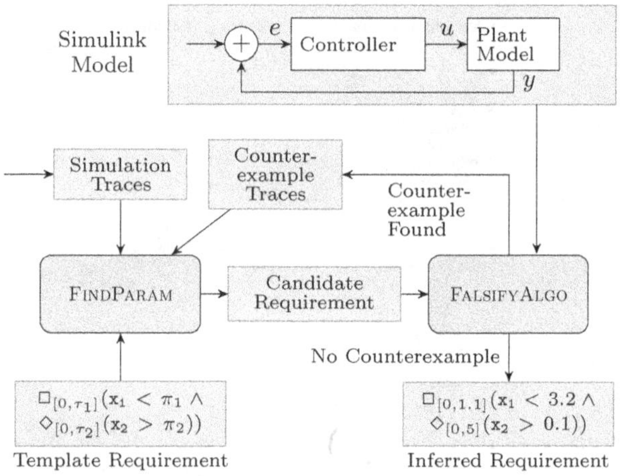

Figure 3: Flowchart for Requirement Mining

Our focus on "*tight* valuations" is to avoid mining trivial requirements or requirements that are overly conservative, e.g. "the car cannot go faster than the speed of light." We make this notion more precise in Section 5.

3.3 Requirement Mining Algorithm: Overview

Our algorithm for mining STL requirements from the closed-loop model in Simulink is an instance of a counterexample-guided inductive synthesis procedure [29], shown in Fig. 3. It consists of two key components:

1. A falsification engine, which, given a formula φ generates an input u such that $\mathbf{x}(t) = \mathcal{S}(u)(t) \not\models \varphi$ if there exists such a u, and returns $\perp$ otherwise. We denote this functionality by FALSIFYALGO.
2. A synthesis function denoted FINDPARAM that given a set of traces $\mathbf{x}_1, \ldots, \mathbf{x}_k$, finds parameters $\mathbf{p}$ such that $\forall i$, $\mathbf{x}_i \models \varphi(\mathbf{p})$. We denote this function by FINDPARAM.

4. FALSIFICATION PROBLEM

Recall that we need to implement a function

$$\mathbf{x} = \text{FALSIFYALGO}(\mathcal{S}, \varphi)$$

such that $\mathbf{x}$ is a valid output signal of a system $\mathcal{S}$ and $\mathbf{x} \not\models \varphi$. Unfortunately, this is an undecidable problem for general hybrid systems; letting φ be a simple safety property establishes a reduction from the reachability problem for general hybrid systems, which is undecidable except for subclasses such as initialized rectangular hybrid automata [16]. For the latter subclasses the mining technique can be *complete*, i.e., absence of a counterexample means that we have found the strongest requirement. However, in general the falsification tool may not be able to find a counterexample though one exists. We argue that a requirement mined in this fashion is still useful as it is something that FALSIFYALGO is unable to disprove even after extensive simulations, and is thus likely to be close to the actual requirement. An alternative is to use a sound verification tool that employs abstraction [15, 30]; however, in our experience, these tools have not scaled to the complex control systems that we consider here. In this paper, we follow the approach taken by the developers of the tool S-TALIRO [5] and propose a falsification algorithm based on the minimization of the quantitative satisfaction of a temporal logic formula.

4.1 Quantitative Semantics of STL

The *quantitative semantics* of STL are defined using a real-valued function ρ of a trace $\mathbf{x}$, a formula φ, and time t satisfying the following property:

$$\rho(\varphi, \mathbf{x}, t) \geq 0 \text{ iff } (\mathbf{x}, t) \models \varphi. \tag{4.1}$$

Quantitative semantics capture the notion of *robustness of satisfaction* of φ by a signal $\mathbf{x}$, i.e., whenever the absolute value of $\rho(\varphi, \mathbf{x}, t)$ is large, a change in $\mathbf{x}$ is less likely to affect the Boolean satisfaction (or violation) of φ by $\mathbf{x}$. In [11], different quantitative semantics for STL have been proposed. We recall the most commonly used semantics defined inductively from the quantitative semantics for predicates and inductive rules for each STL operator.

Without loss of generality, an STL predicate μ can be identified to an inequality of the form $f(\mathbf{x}) \geq 0$ (the use of strict or non strict inequalities is a matter of choice and other inequalities can be trivially transformed into this form). From this form, a straightforward quantitative semantics for predicate μ is defined as

$$\rho(\mu, \mathbf{x}, t) = f(\mathbf{x}(t)). \tag{4.2}$$

Then ρ is defined inductively for every STL formula using the following rules:

$$\rho(\neg\varphi, \mathbf{x}, t) = -\rho(\varphi, \mathbf{x}, t) \tag{4.3}$$

$$\rho(\varphi_1 \wedge \varphi_2, \mathbf{x}, t) = \min(\rho(\varphi_1, \mathbf{x}, t), \rho(\varphi_2, \mathbf{x}, t)) \tag{4.4}$$

$$\rho(\varphi_1 \mathbf{U}_I \varphi_2, \mathbf{x}, t) = \sup_{t' \in t+I} \left(\min(\rho(\varphi_2, \mathbf{x}, t'), \inf_{t'' \in [t,t')} \rho(\varphi_1, \mathbf{x}, t'')\right) \tag{4.5}$$

Then it can be shown [11] that ρ satisfies (4.1) and thus defines a quantitative semantics for STL. Additionally, by combining (4.5), and $\Box_I \varphi \triangleq \neg\Diamond_I \neg\varphi$, we get

$$\rho(\Box_I \varphi, \mathbf{x}, t) = \inf_{t' \in t+I} \rho(\varphi, \mathbf{x}, t') \tag{4.6}$$

For $\Diamond$, we get a similar expression using sup instead of inf.

EXAMPLE 4.1. *Consider again the STL property:*

$$\varphi = \Box(\texttt{speed} \leq 120) \wedge \Box(\texttt{RPM} \leq 4500).$$

It has two predicates, say $\mu_1 : \texttt{speed} \leq 120$ *and* $\mu_2 : \texttt{RPM} \leq 4500$. *To put them into the standard form* $\mu_i : f_i(\mathbf{x}) \geq 0$, *we define* $\mathbf{x} = (\texttt{speed}, \texttt{RPM})$, $f_1(\mathbf{x}) = 120 - \texttt{speed}$ *and* $f_2(\mathbf{x}) = 4500 - \texttt{RPM}$. *From* (4.2), *we get*

$$\rho(\texttt{speed} \leq 120, \mathbf{x}, t) = 120 - \texttt{speed}(t).$$

Applying rule (4.6) *for the semantics of* $\Box$, *we get:*

$$\rho(\Box(\texttt{speed} \leq 120), \mathbf{x}, t) = \inf_{t \in \mathbb{T}}(120 - \texttt{speed}(t)).$$

Similarly for μ_2,

$$\rho(\Box(\texttt{RPM} \leq 4500), \mathbf{x}, t) = \inf_{t \in \mathbb{T}}(4500 - \texttt{RPM}(t)).$$

Finally, by applying rule (4.4):

$$\rho(\varphi, \mathbf{x}, t) = \min(\inf_{t \in \mathbb{T}}(120 - \texttt{speed}(t)), \inf_{t \in \mathbb{T}}(4500 - \texttt{RPM}(t)).$$

Informally, the satisfaction function ρ *looks for the maximum speed and RPMs over time and returns the minimum of the differences with the thresholds* 120 *and* 4500.

4.2 STL vs. MTL Robust Satisfaction

In this section, we clarify the connection between the quantitative semantics of STL defined above and the notion of robust satisfaction of MTL as defined in [13] and used in S-TALIRO. The main difference between STL and MTL lies in the definition of predicates, and so does the difference between quantitative semantics. The robust semantics of MTL is based on the definition of a *metric* on the state space of signals and the fact that each predicate is identified with a set where it holds true. Formally, let d be a metric on $\mathbb{R}^n$ with the usual extension to the *signed distance* from a point $\mathbf{x} \in \mathcal{X}$ to a set $\mathcal{X}' \subseteq \mathcal{X}$:

$$d(\mathbf{x}, \mathcal{X}') = \begin{cases} -\inf_{\mathbf{x}' \in \mathcal{X}} d(\mathbf{x}, \mathbf{x}') & \text{if } \mathbf{x} \notin \mathcal{X}' \\ \inf_{\mathbf{x}' \in \mathcal{X} \setminus \mathcal{X}'} d(\mathbf{x}, \mathbf{x}') & \text{otherwise} \end{cases}$$

For each MTL predicate μ, define its truth set $\mathcal{O}(\mu)$ as:

$$\mathbf{x}, t \models \mu \text{ iff } \mathbf{x}(t) \in \mathcal{O}(\mu) \tag{4.7}$$

and let

$$\rho_d(\mu, \mathbf{x}, t) = d(\mathbf{x}(t), \mathcal{O}(\mu)). \tag{4.8}$$

Finally, the robust satisfaction ρ_d is defined using the same inductive rules (4.3-4.5) as for an STL formula φ. From there, it is clear that the quantitative semantics of STL subsumes the robust semantics of MTL. Indeed, as each predicate in STL is associated with an arbitrary function f, this function can implement the distance d. Then, if ρ_d and ρ coincide on a set of predicates, by induction they coincide on all formulas defined on those predicates.

4.3 Solving the Falsification Problem

The objective of the falsification problem is: given an STL formula φ, find a signal $\mathbf{u}$ such that $\mathcal{S}(\mathbf{u}) \not\models \varphi$. Following the above definitions, this is equivalent to finding a trace $\mathbf{x}$ such that $\rho(\varphi, \mathbf{x}, 0) < 0$. Hence FALSIFYALGO can be implemented by solving

$$\text{Solve } \rho^* = \min_{\mathbf{u} \in \mathcal{U}} \rho(\varphi, \mathcal{S}(\mathbf{u}), 0) \tag{4.9}$$

Then if $\rho^* < 0$, we return $\mathbf{u}^* = \arg\min_{\mathbf{u} \in \mathcal{U}} \rho(\varphi, \mathcal{S}(\mathbf{u}), 0)$, otherwise, $\mathcal{S} \models \varphi$. The undecidability of the falsification problem is reflected here in the fact that the minimization problem (4.9) is a general non-linear optimization problem for which no solver can guarantee convergence, uniqueness or even existence of a solution. On the other hand, many heuristics can be used to find an approximate solution. In a series of recent papers, the authors of S-TALIRO proposed and implemented different strategies, such as Monte-Carlo [22], and the cross-entropy method [25]. In our implementation, we first instrumented S-TALIRO as a falsification tool (made possible by the connection between STL and MTL described in the previous section) and then extended BREACH with a new falsification engine which attacks (4.9) as follows:

1. Define the space of permissible input signals with the help of m *input parameters* $\mathbf{k} = (k_1, \ldots, k_m)$ that take values from a set $\mathcal{P}_u$, and a generator function g such that $\mathbf{u}(t) = g(v(\mathbf{k}))(t)$ is a permissible input signal for $\mathcal{S}$ for any valuation $v(\mathbf{k}) \in \mathcal{P}_u$.

Data: A trace $\mathbf{x}$, a PSTL Formula φ, and parameter set $\mathcal{P}$, $\delta > 0$
Result: A valuation v s.t. $\mathbf{x} \models_\delta \varphi(v)$
Find $v_\top$ s.t. $\mathbf{x} \models \varphi(v_\top)$ or **return** φ *unsat.*;
Find $v_\bot$ s.t. $\mathbf{x} \not\models \varphi(v_\bot)$ or **return** v *maybe not tight*;
Let $v = v_\top$;
for $i = 1$ *to* n **do**
| Find v_i and set $v(p_i) = v_i$ s.t. $\mathbf{x} \models_\delta^i \varphi(v)$
end

Algorithm 1: FINDPARAM algorithm.

2. Sample signal-parameters in a uniform, random fashion to obtain N_{init} distinct valuations $v_i(\mathbf{k}) \in \mathcal{P}_u$.
3. For $i \leq N_{\text{init}}$, solve $\min_{v(\mathbf{k}) \in \mathcal{P}_u} \rho(\varphi, \mathcal{S}(g(v(k))), 0)$ using Nelder-Mead non-linear optimization algorithm and $v_i(\mathbf{k})$ as an initial guess.
4. Return the minimum ρ thus found.

For example, if permissible input signals are step functions, then the input parameters would characterize the amplitude of the step, and the time at which the step input is applied. Note that g does not necessarily generate all possible inputs to the system. However, it is useful in a very generic way to restrict the search space of possible input signals. One motivation for implementing a falsification module in Breach has been to get more flexibility in the definition of input parameters than available in the version of S-TALIRO that we used. In the experimental section, we discuss some results using both S-TALIRO-based falsification and the above algorithm. We found in particular that the choice of input parameters, of N_{init} and the tuning of Nelder-Mead algorithm (which provides a trade-off between global randomized exploration and local optimization) were crucial for the performance of the falsifier.

5. PARAMETER SYNTHESIS

We now discuss the function FINDPARAM. Recall that given a trace[2] $\mathbf{x}$, we need to find a valuation v for the parameters $p_1, \ldots, p_n$, of φ such that $\mathbf{x}$ satisfies $\varphi(v(p_1), \ldots, v(p_n))$ (abbreviated as $\varphi(v)$ in the following). This problem is a dual of the falsification problem (4.9) formulated as:

$$\max_v \rho(\varphi(v), \mathbf{x}, 0). \tag{5.1}$$

However, there is an important difference that the cost function can be expressed as a closed-form expression of the decision variable v whereas for (4.9) as a function of $\mathbf{u}$. By taking advantage of this knowledge, (5.1) can be solved more efficiently, in particular as we will see, if formulas satisfy the important property of *monotonicity*:

DEFINITION 5.1. *A PSTL formula $\varphi(p_1, \cdots, p_n)$ is monotonically increasing with respect to p_i if for every signal $\mathbf{x}$,*

$$\forall v, v' : \begin{pmatrix} \mathbf{x} \models \varphi(\ldots, v(p_i), \ldots) \\ \wedge\ v'(p_i) \geq v(p_i) \end{pmatrix} \Rightarrow \mathbf{x} \models \varphi(\ldots, v'(p_i), \ldots)$$

It is monotonically decreasing if this holds when replacing $v'(p_i) \geq v(p_i)$ with $v'(p_i) \leq v(p_i)$.

[2]We restrict our attention to one trace though in the mining process, FINDPARAM has to work on a set of traces. The generalization to multiple traces is straightforward.

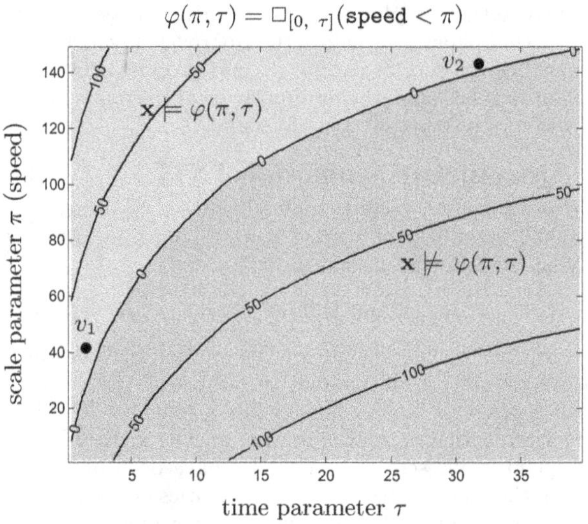

Figure 4: Validity domain of a simple formula for a trace x obtained from the automatic transmission model. The FindParam algorithm will return valuation v_1 (resp. v_2) depending if time (resp. scale) parameter is optimized first. The contour lines are isolines for the satisfaction function ρ.

In the second part of this section, we characterize this notion more precisely. We impose an additional constraint to the parameter synthesis problem: we require that the STL formula mined be "tightly" satisfied by the system up to a given precision $\delta > 0$. Formally,

DEFINITION 5.2. *The signal $\mathbf{x}$ δ-satisfies $\varphi(v)$ for p_i denoted by $\mathbf{x} \models_\delta^i \varphi(v)$ iff $\mathbf{x} \models \varphi(v)$ and there exists a valuation v' such that $\forall j \neq i$, $v'(p_j) = v(p_j)$, $|v(p_i) - v'(p_i)| \leq \delta$ and $\mathbf{x} \not\models \varphi(v')$. The signal $\mathbf{x}$ δ-satisfies $\varphi(v)$, denoted by $\mathbf{x} \models_\delta \varphi(v)$ if $\forall i$, $\mathbf{x} \models_\delta^i \varphi(v)$.*

The rationale is that for a specification to be useful it should not be too conservative. The implication is that it is not enough to find a satisfying valuation, we also need to optimize it for each parameter to get δ-satisfaction. If there is more than one parameter, then the solution is not unique. In fact, all valuations that are at a distance δ from the boundary of the *validity domain* of φ and $\mathbf{x}$ (the set of valuations v for which $\mathbf{x} \models \varphi(v)$) are valid solutions. In [6], the authors note that if the formula is monotonic, then this boundary has the properties of a Pareto surface for which there are efficient computational methods, basically equivalent to multi-dimensional binary search. Here we propose an algorithm (Algorithm 1) for monotonic formulas that takes advantage of this property, and implement it in the BREACH tool. Algorithm 1 starts by trying to find a valuation $v_\top$ that satisfies the property and a valuation $v_\bot$ that violates it in a parameter range $\mathcal{P}$ provided by the user. By property of monotonicity, it is sufficient to check the corners of $\mathcal{P}$ for the existence of $v_\top$ and $v_\bot$. Then, each parameter i is adjusted using a binary search initialized with $v_\top(p_i)$ and $v_\bot(p_i)$. The user can choose which parameters to optimize first by specifying a priority ordering for the input parameters. Note that different orderings can give drastically different results.

EXAMPLE 5.1. *Consider $\varphi(\pi, \tau) = \Box_{[0,\ \tau]}(\texttt{speed} < \pi)$ and the scenario that the vehicle constantly accelerates with the value of* `throttle` *set at* 100. *The validity domain of φ*

Formula	Monot.	Time
$\Box_{(0,\infty)}(x < \pi)$	+	<0.09 s
$\Box_{[s,s+1]}(x \geq 3 \Rightarrow \Diamond_{(0,\infty)} x < 3)$	–	0.1 s
$\Box_{(0,100)}((x < \pi) \Rightarrow \Diamond_{(0,5)}(x > \pi))$	–	<0.09 s
$\mathbf{gear}_i \mathbf{U}_{(s,s+5)} \mathbf{gear}_{i+1}$	*	0.13 s

Table 1: Proving monotonicity with an SMT solver.

is plotted on Fig. 4. The algorithm will return different values depending on the tightness parameter δ and if we order the parameters as (π,τ) or (τ,π). Here, the order represents the preference in optimizing a parameter over the other when mining for a tight specification.

6. SATISFACTION MONOTONICITY

We first show that checking if an arbitrary PSTL formula is monotonic in a given parameter is undecidable.

THEOREM 6.1. *The problem of checking if a PSTL formula $\varphi(\mathbf{p})$ is monotonic in a given parameter p_i is undecidable.*

PROOF. First, we observe that STL is a superset of MTL. We know from [1] that the satisfiability problem for MTL is undecidable. Thus, it follows that the satisfiability problem for STL is also undecidable. This, in turn, implies undecidability of the satisfiability problem of PSTL with at most one parameter (denoted as PSTL-1-SAT). We now show that PSTL-1-SAT can be reduced to a special case of the problem of checking monotonicity of a PSTL formula.

Let $\varphi(\mathbf{p})$ be an arbitrary PSTL formula where the set of parameters $\mathbf{p}$ is the singleton set with one time parameter τ (thus, $\tau \geq 0$). Construct the formula $\psi(\mathbf{p}) \doteq (\tau = 0) \vee \varphi(\mathbf{p})$. Consider the monotonicity query for $\psi(\mathbf{p})$ in parameter τ:

$$\forall v, v', \mathbf{x} : \; [\mathbf{x} \models \psi(v(\tau)) \;\wedge\; v(\tau) \leq v'(\tau)] \;\Rightarrow\; \mathbf{x} \models \psi(v'(\tau)).$$

Consider the specialization of this formula for the case $v(\tau) = 0$. Note that, in this case, $\psi(0) = \top$, and that $v'(\tau) \geq 0$ for all v'. Thus, the query simplifies to $\forall v', \mathbf{x} : \; \mathbf{x} \models \psi(v'(\tau))$, i.e., checking the validity of the PSTL formula $\psi(\tau)$.

Thus, to check monotonicity of PSTL formula φ in one parameter τ one needs to check that the negation of $\psi(\tau)$ is unsatisfiable. Thus the above specialization of the problem of checking the monotonicity of PSTL formulas is also undecidable, implying undecidability of the general case. □

Monotonicity is closely related to the notion of polarity introduced in [6], in which syntactic deductive rules are given to decide whether a formula is monotonic based on the monotonicity of its subformulae. Thus, one way to tackle undecidability is to first query if the given PSTL formula belongs to the syntactic class described in [6]. Unfortunately, the syntactic rules described therein are not complete; there are monotonic PSTL formulas that do not belong to this syntactic class, for instance, formulas with intervals in which both end-points are parameterized, such as the following:

$$\Box_{[\tau,\tau+1]}((x \geq 3) \Rightarrow \Diamond_{(0,\infty)}(x < 3)) \qquad (6.1)$$

Next, we show how we can use SMT solving to query monotonicity of a formula. If the SMT solver succeeds, it tells us that the formula is monotonic and allows us to use a more efficient search in the parameter space. For instance, we were able to show that the PSTL formula represented in (6.1) is monotonically decreasing in the parameter τ.

Encoding PSTL as constraints. Given a PSTL formula φ, we define the SMT encoding of φ in a fragment of first-order logic with real arithmetic and uninterpreted functions. Let $\mathcal{E}(\varphi)$ denote the encoding of φ, which we define inductively as follows:

• Consider a constraint $\mu \doteq g(\mathbf{x}) > \tau$, where $\mathbf{x} = (x_1, \ldots, x_n)$. We model each signal x_i as an uninterpreted function χ_i from $\mathbb{R}$ to $\mathbb{R}$. We create a new free variable t of the type `Real` and replace each instance of the signal x_i in $g(\mathbf{x})$ by $\chi_i(t)$. We assume that the function g itself has a standard SMT encoding. For example, consider the formula $g(\mathbf{x}) > \tau$, where $\mathbf{x} = \{x_1, x_2\}$, and $g(\mathbf{x}) = 2 * x_1 + 3 * x_2$. Then $\mathcal{E}(\mu)$ is: $2 * \chi_1(t) + 3 * \chi_2(t) \;\; > \tau$.

• For Boolean operations, the SMT encoding is inductively applied to the subformulas, i.e., if $\varphi = \neg\varphi_1$, then $\mathcal{E}(\varphi) = \neg\mathcal{E}(\varphi_1)$. If $\varphi = \varphi_1 \wedge \varphi_2$, then first we ensure that if $\mathcal{E}(\varphi_1)$ and $\mathcal{E}(\varphi_2)$ both have a free time-domain variable, then we make it the same variable, and then, $\mathcal{E}(\varphi) = \mathcal{E}(\varphi_1) \wedge \mathcal{E}(\varphi_2)$. Note that as a consequence, there is at most one free time-domain variable in any subformula.

• Consider $\varphi = \mathrm{H}_{(a,b)}(\varphi_1)$, where a, b are constants or parameters, and H is a unary temporal operator (i.e., $\Diamond, \Box$). There are two possibilities:

(1) The SMT encoding $\mathcal{E}(\varphi_1)$ has one free variable t. In this case, we bound the variable t over the interval (a, b) using a quantifier that depends on the type of the temporal operator H. With $\Diamond$ we use $\exists$ as the quantifier, and with $\Box$ we use $\forall$. E.g., let $\varphi = \Diamond_{(2.3,\tau)}(x > \pi)$, then $\mathcal{E}(\varphi)$ is:

$$\exists t : \;\; (2.3 < t < \tau) \wedge (\chi(t) > \pi).$$

(2) The SMT encoding $\mathcal{E}(\varphi_1)$ has no free variable. This can only happen if φ_1 is $\top$ or $\bot$, or if all variables in φ_1 are bound. In the former case, the encoding is done exactly as in Case 1. In the latter case, the encoding proceeds as before, but all bound variables in the scope are *additionally offset* by the top-level free variable. Suppose, $\varphi = \Box_{(0,\infty)}\Diamond_{(1,2)}(x > 10)$. Then, the encoding of the inner $\Diamond$-subformula has no free variable. Note how the bound variable of this formula is offset by the top-level free variable in the underlined portion in $\mathcal{E}(\varphi)$ below:

$$\forall t : [\exists u : [\underline{(t+1 < u < t+2)} \wedge (\chi(u) > 10)]].$$

• Consider $\varphi = \varphi_1 \mathbf{U}_{(a,b)} \varphi_2$, where a, b are constants or parameters. For simplicity, consider the case where φ_1 and φ_2 have no temporal operators, i.e., $\mathcal{E}(\varphi_1)$ and $\mathcal{E}(\varphi_2)$ both have exactly one free variable each. Let t_1 be the free variable in $\mathcal{E}(\varphi_1)$ and t_2 the free variable in $\mathcal{E}(\varphi_2)$. Then $\mathcal{E}(\varphi)$ is given by the formula:

$$\exists t_2 : [(t_2 \in (a,b)) \wedge \mathcal{E}(\varphi_2) \wedge \forall t_1 : [(t_1 \in (a, t_2)) \Rightarrow \mathcal{E}(\varphi_1))].$$

If φ_1, φ_2 contain no free variables, then t_1, t_2 are respectively used to offset all bound variables in their scope as before.

Using an SMT solver to check monotonicity. To check monotonicity, we check the two assertions below:

$$\mathcal{E}(\varphi(\tau)) \wedge (\tau > \tau') \wedge \neg\mathcal{E}(\varphi(\tau'))$$
$$\mathcal{E}(\varphi(\tau)) \wedge (\tau < \tau') \wedge \neg\mathcal{E}(\varphi(\tau'))$$

	S-Taliro-based falsification					Breach-based falsification				
Template	Parameter values	Fals.	Synth.	#Sim.	Sat./x	Parameter values	Fals.	Synth.	#Sim.	Sat./x
$\varphi_{\texttt{sp\_rpm}}(\pi_1, \pi_2)$	(155 mph, 4858 rpm)	55 s	12 s	255	0.004 s	(155 mph, 4858 rpm)	197.2 s	23.1 s	496	0.043 s
$\varphi_{\texttt{rpm100}}(\pi, \tau)$	(3278.3 rpm, 49.91 s)	6422 s	26.5 s	9519	0.327 s	(3273 rpm, 49.92 s)	267.7 s	10.51 s	709	0.026 s
$\varphi_{\texttt{rpm100}}(\tau, \pi)$	(4997 rpm, 12.20 s)	8554 s	53.8 s	18284	0.149 s	(4997 rpm, 12.20 s)	147.8 s	5.188 s	411	0.021 s
$\varphi_{\texttt{stay}}(\pi)$	1.79 s	18886 s	0.868 s	130	147.2 s	0.102 s	430.9 s	2.157 s	1015	0.032 s

Table 2: Results on mining for the automatic transmission control model. We compare runs of the mining algorithm using either S-Taliro or Breach as falsifiers. In each case and for each template formula, we give the parameters valuations found, the time spent in falsification and in parameter synthesis, the number of simulations and the averaged time spent computing the quantitative satisfaction of the formula by one trace.

If either of these queries is unsatisfiable, then it means that satisfaction of φ is indeed monotonic in τ. If both queries are satisfiable, then it means that there is an interpretation for the (uninterpreted) function representing the signal $\mathbf{x}$ and valuations for τ, τ' which demonstrate the non-monotonicity of φ. We conclude by presenting a small sample of formulas for which we could prove or disprove monotonicity using the Z3 SMT solver [8] in Table 1. The symbols $+$, $-$, and $*$ represent monotonically increasing, decreasing and non-monotonic formulas respectively.

7. CASE STUDIES

In what follows, we present our evaluation of both S-TALIRO and an extension of BREACH for falsification. We also show the performance of the parameter synthesis algorithm implemented with the robust satisfaction engine of BREACH. We use the transmission controller model to benchmark the different options within our approach.

7.1 Automatic Transmission Model

For the model described in Sec. 2, we tested different template requirements:

1. Requirement $\varphi_{\texttt{sp\_rpm}}(\pi_1, \pi_2)$ specifying that always the `speed` is below π_1 and `RPM` is below π_2 :

$$\Box\left((\texttt{speed} < \pi_1) \wedge (\texttt{RPM} < \pi_2) \right).$$

2. Requirement $\varphi_{\texttt{rpm100}}(\tau, \pi)$ specifying that the vehicle cannot reach the speed of 100 mph in τ seconds with `RPM` always below π:

$$\neg(\Diamond_{[0,\tau]}(\texttt{speed} > 100) \wedge \Box(\texttt{RPM} < \pi)).$$

3. Requirement $\varphi_{\texttt{stay}}(\tau)$ specifying that whenever the system shifts to gear 2, it dwells in gear 2 for at least τ seconds:

$$\Box\left(\left(\begin{array}{c}\texttt{gear} \neq 2 \;\wedge \\ \Diamond_{[0,\varepsilon]}\texttt{gear} = 2\end{array}\right) \Rightarrow \Box_{[\varepsilon,\tau]}\texttt{gear} = 2\right).$$

Here, the left-hand-side of the implication captures the *event* of the transition from gear 2 to another gear. The operator $\Diamond_{[0,\varepsilon]}$ here is an MTL substitute for a *next-time* operator. With dense time semantics, ε should be an *infinitesimal* quantity, but in practice, we use a value close to the simulation time-step.

The above requirements have strong correlation with the quality of the controller. The first is a safety requirement characterizing the operating region for the engine parameters `speed` and `RPM`. The second is a measure of the performance of the closed loop system. By mining values for τ, we can determine how fast the vehicle can reach a certain speed, while by mining π we find the lowest `RPM` needed to reach this speed. The third requirement encodes undesirable transient shifting of gears. Rapid shifting causes abrupt output torque changes leading to a jerky ride.

Results on the mined specifications are given in Table 2. We used the Z3 SMT solver [8] to show that all of the requirements were monotonic. As expected, the FINDPARAM algorithm takes only a fraction of the total time in the entire mining process. For the second template, we tried two possible orderings for the parameters. By prioritizing the time parameter τ, we obtained the δ-tight requirement that the vehicle cannot reach 100 mph in less than 12.2s (we set δ to 0.1). As the requirement mined is δ-tight, it means that we found a trace for which the vehicle reaches 100 mph in 12.3 s. Similarly, by prioritizing the scale parameter π, we found that the vehicle could reach 100 mph in 50s keeping the `RPM` below 3278 ($\delta = 5$ in that case). For the third requirement, we found that the transmission controller could trigger a transient shift as short as 0.112s. This corresponds to the up-shifting sequence 1-2-3. Using a variant of the requirement (not shown here), we verified that a (definitely undesirable) short transient sequence of the form 1-2-1 or 3-2-3 was not possible.

The comparison between S-TALIRO and BREACH falsifiers shows better overall performance with the extended BREACH-based falsifier, in the sense that it found stronger requirements using less number of simulations and computational time. However, we cannot conclude that the new falsifier will always outperform S-TALIRO, due to the stochastic nature of the problem and a lack of thorough comparison with the different flavors of optimization used within S-TALIRO. Based on results shown in Table 2 and our experience, we make some observations:

- The space of input signals needs to be parameterized with a sensible number of signal-parameters. If too many parameters are used, the search space is too big and falsification becomes difficult. For instance, the short transient shifting of $\varphi_{\texttt{stay}}$ was found by introducing a signal-parameter controlling the time of initial acceleration, and by preventing acceleration and braking at the same time. We remark that extending BREACH to enforce such constraints over the input signal space is a key reason for its better performance, and a fair comparison would be possible only after repeating these steps for S-TALIRO.
- Requirements involving discrete modes are challenging because they induce "flat" quantitative satisfaction functions that are challenging to optimizers and thus have limited value in guiding the falsifier. This is related to the problem of finding a good metric between discrete states in hybrid systems. This was particularly an issue when mining

the $\varphi_{\texttt{stay}}$ requirement. We were able to tune our falsifier by turning off its local optimization phase, and using uniform random sampling, which led us to obtaining a tighter requirement than with S-TALIRO.

• We found that while both falsifiers are expected to exhibit run-times linear in the size of the traces and the formula [11,14], in some cases, BREACH runs faster. In particular, S-TALIRO is more sensitive to parameter priorities. For the same template $\varphi_{\texttt{rpm100}}$, depending on which parameter τ or π is prioritized, S-TALIRO performs differently. This can be explained by the fact that τ affects the horizon of the temporal operator $\Diamond$. We conjecture that the difference in run-times and mined parameter values for the $\varphi_{\texttt{stay}}$ template is due to our inability to express signal-parameterization in S-TALIRO, but these comparisons require more dedicated studies on various benchmarks before drawing firm conclusions.

7.2 Diesel Engine Model

Next, we consider an industrial-scale, closed-loop Simulink model of an experimental airpath controller for a diesel engine. It has more than 4000 Simulink blocks such as data store memories, integrators, 2D-lookup tables, functional blocks with arbitrary Matlab functions, S-Function blocks, and blocks that induce switching behavior such as level-crossing detectors, multiports, and saturation blocks. The models takes two signals as input: the fuel injection rate and the engine speed. The output signal is the intake manifold pressure denoted by x. For proprietary reasons, we suppress the mined values of the parameters and the time-domain constants from our requirements. We replace the time-domain constants by symbols such as c_1 and c_2.

Discussions with control designers revealed that characterizing overshoot behavior of the intake manifold pressure is important. The inputs to the closed-loop model are a step function to the fuel injection rate at time c_1, and a constant value for the engine speed. The first requirement is:

$$\varphi_{\texttt{overshoot}}(\pi) = \Box_{(c_1,\infty)}(\mathtt{x} < \pi).$$

This template characterizes the requirement that the signal x never exceeds π during the time interval (c_1, ∞), i.e., it finds the maximum peak value (i.e., π) of the step response. Our mining algorithm obtained 7 intermediate candidate requirements that were falsified by S-TALIRO, till we found a requirement that it could not falsify in its 8th iteration. The total number of simulations was 7000 over a period of 13 hours.

Next, we chose to mine the settling behavior of the signal. The *settling time* is the time after which the amplitude of signal is always within a small *error* from its calculated ideal reference value. We wish to mine both the error and how fast the signal settles. Such a template requirement is given by the following PSTL formula:

$$\varphi_{\texttt{settling\_time}}(\tau, \pi) = \Box_{[\tau,\infty)}(|\mathtt{x}| < \pi).$$

It specifies that the absolute value of $\mathbf{x}$ is always less than π starting from the time τ to the end of the simulation. The smaller the settling time and the error, the more stable is the system. We found out from the control designer that a smaller settling time needs to be prioritized over the error (as long as the error lies within the 10% of the signal amplitude), so we prioritize minimizing τ over minimizing π.

After 4 iterations, the procedure stops as the inferred value for τ is very close to the end of the simulation trace,

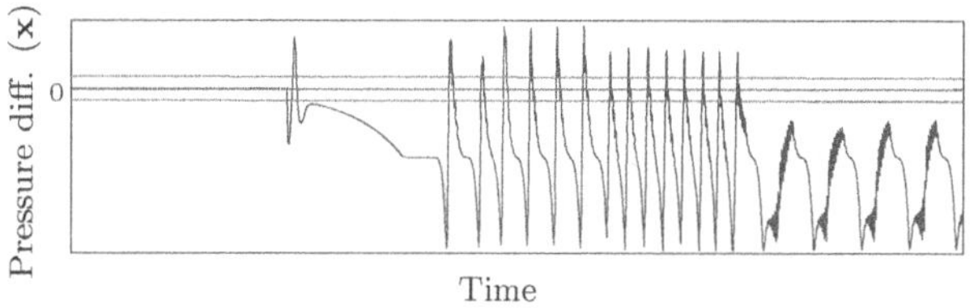

Figure 5: The simulation trace x (in blue) denoting the difference between the intake manifold pressure and its reference value[4] found when mining $\varphi_{\texttt{settling\_time}}(\tau, \pi)$ displays unstable behavior. The maximum error threshold is depicted in red.

but the error is still larger than the tolerance. The implication here is that the algorithm pushed the falsifier to finding a behavior in the model that exhibits *hunting behavior*, or oscillations of magnitude exceeding the tolerance. This output signal is shown in Fig. 5. This behavior was unexpected; discussions with the designers revealed that it was a real bug. Investigating further, we traced the root-cause to an incorrect value in a lookup table; such lookup tables are commonly used to speed up the computation time by storing pre-computed values approximating the control law.

This experiment demonstrates the use of requirement mining as an advanced, guided debugging strategy. Instead of verifying correctness with a concrete formal requirement, the process of trying to infer what requirement a model must satisfy can reveal erroneous behaviors that could be otherwise missed. In the course of our experiments, we encountered other suspicious (for instance Zeno-like) behaviors, which we suspect to be either an error in the model, or an improper tuning of the numerical solver leading to discontinuities in the dynamics.

8. RELATED WORK

Mining requirements from programs and circuits is well-studied in the field of computer science [3, 4, 12, 18, 19, 26, 27, 32]. In computer science, the word "requirement" is often synonymous with "specification". Specification mining techniques vary based on the kind of specifications mined; examples include automata, temporal rules, and sequence diagrams. They also vary based on the input to the miner; techniques based on static analysis or model checking operate on the source code, while dynamic techniques mine from execution traces. Mining temporal rules [3,27] involves learning an automaton that captures the temporal behavior and typically focusses on API usage in libraries. The individual components within such libraries are often *terminating* programs, and specification automata encode legal interaction-patterns between components. In contrast to the software world, where most programs have discrete-time semantics, the behavioral requirements that we mine are for systems with both continuous and discrete-time semantics. It may be worthwhile to see if automata-based mining could be adapted to the hybrid systems domain. The work closest to the proposed approach appears in [33], in which the authors introduce Parametric MTL (PMTL), which adds a *single* time or scale parameter to MTL formulas. This parameter is then estimated using stochastic optimization

[4]Note that the values along the axes have been suppressed for proprietary reasons. We remark that the actual values are irrelevant and the intention is to show an oscillating behavior arising from a real bug in the design.

within S-TALIRO. We remark that we provide a way to reason about monotonicity of PSTL formulas with arbitrary number of parameters, and also allow mining non-monotonic PSTL formulas (albeit less efficiently).

To the best of our knowledge, this work is among the first to address the specification mining problem for cyber-physical systems. From a broader perspective, the literature reports several attempts to apply formal methods to industrial-scale block-based design tools such as Simulink. In [10], the authors verify simple safety properties using sensitivity analysis. Other approaches try to transform Simulink diagrams into models amenable to formal verification [31, 34]. When successful, such approaches provide very strong guarantees. However the type of blocks that can be handled is usually limited and we are not aware of scalable analysis tools for models representing hybrid systems.

Acknowledgment

We thank the anonymous referees for their comments, and James Kapinski, Koichi Ueda, and Ken Butts for providing Simulink models, help with experiments, and insightful discussions. The first author was supported in part by National Science Foundation grant CCF-1018057. The second and fourth authors were funded in part by Toyota Motor Corporation via the Center for Hybrid and Embedded Software Systems (CHESS) at UC Berkeley, and by the MultiScale Systems Center (MuSyC), one of six research centers funded under the Focus Center Research Program, a Semiconductor Research Corporation program.

9. REFERENCES

[1] R. Alur, T. Feder, and T. A. Henzinger. The benefits of relaxing punctuality. *J. ACM*, 43(1):116—-146, Jan. 1996.

[2] R. Alur and T. A. Henzinger. A really temporal logic. In *Symposium on Foundations of Computer Science*, pages 164–169, 1989.

[3] R. Alur, P. Černý, P. Madhusudan, and W. Nam. Synthesis of interface specifications for Java classes. *ACM SIGPLAN Notices*, 40:98–109, 2005.

[4] G. Ammons, R. Bodík, and J. R. Larus. Mining specifications. *ACM SIGPLAN Notices*, 37:4–16, 2002.

[5] Y. Annpureddy, C. Liu, G. E. Fainekos, and S. Sankaranarayanan. S-TaLiRo: A Tool for Temporal Logic Falsification for Hybrid Systems. In *Tools and Algorithms for the Construction and Analysis of Systems*, pages 254–257, 2011.

[6] E. Asarin, A. Donzé, O. Maler, and D. Nickovic. Parametric identification of temporal properties. In *Runtime Verification*, pages 147–160, 2011.

[7] C. Barrett, R. Sebastiani, S. A. Seshia, and C. Tinelli. Satisfiability Modulo Theories. In A. Biere, H. van Maaren, and T. Walsh, editors, *Handbook of Satisfiability*, volume 4, chapter 8. IOS Press, 2009.

[8] L. De Moura and N. Bjørner. Z3: An efficient SMT,solver. page 337–340, 2008.

[9] A. Donzé. Breach: A Toolbox for Verification and Parameter Synthesis of Hybrid Systems. In *Computer-Aided Verification*, pages 167–170, 2010.

[10] A. Donzé, B. Krogh, and A. Rajhans. Parameter synthesis for hybrid systems with an application to Simulink models. In *Hybrid Systems: Computation and Control*, pages 165–179, 2009.

[11] A. Donzé and O. Maler. Robust satisfaction of temporal logic over real-valued signals. In *Formal Modeling and Analysis of Timed Systems*, pages 92–106, 2010.

[12] M. D. Ernst, J. H. Perkins, P. J. Guo, S. McCamant, C. Pacheco, M. S. Tschantz, and C. Xiao. The Daikon,system for dynamic detection of likely invariants. *Science of Computer Programming*, 69(1-3):35–45, 2007.

[13] G. E. Fainekos and G. J. Pappas. Robustness of temporal logic specifications for continuous-time signals. *Theoretical Computer Science*, 410(42):4262–4291, 2009.

[14] G. E. Fainekos, S. Sankaranarayanan, K. Ueda, and H. Yazarel. Verification of Automotive Control Applications using S-TaLiRo. In *American Control Conference*, 2012.

[15] G. Frehse, C. Le Guernic, A. Donzé, R. Ray, O. Lebeltel, R. Ripado, A. Girard, T. Dang, and O. Maler. Spaceex: Scalable verification of hybrid control systems. In *Computer-Aided Verification*, 2011.

[16] T. Henzinger, P. Kopke, A. Puri, and P. Varaiya. What's decidable about hybrid automata? In *Symposium on Theory of Computing*, pages 373–382, 1995.

[17] R. Koymans. Specifying real-time properties with metric temporal logic. *Real-Time Syst.*, 2(4):255–299, 1990.

[18] C. Lee, F. Chen, and G. Rosu. Mining parametric specifications. In *International Conference on Software Engineering*, page 591–600, 2011.

[19] W. Li, A. Forin, and S. A. Seshia. Scalable specification mining for verification and diagnosis. In *Design Automation Conference*, page 755–760, 2010.

[20] O. Maler and D. Nickovic. Monitoring temporal properties of continuous signals. In *Formal Modeling and Analysis of Timed Systems*, pages 152–166, 2004.

[21] The MathWorks Inc., Natick, Massachusetts. *Simulink version 8.0 (R2012b)*, 2012.

[22] T. Nghiem, S. Sankaranarayanan, G. E. Fainekos, F. Ivancic, A. Gupta, and G. J. Pappas. Monte-carlo techniques for falsification of temporal properties of non-linear hybrid systems. In *Hybrid Systems: Computation and Control*, pages 211–220, 2010.

[23] G. Nicolescu and P. J. Mosterman. *Model-Based Design for Embedded Systems*. CRC,Press, 2009.

[24] A. Pnueli. The temporal logic of programs. In *Symposium on Foundations of Computer Science*, pages 46–57, 1977.

[25] S. Sankaranarayanan and G. E. Fainekos. Falsification of temporal properties of hybrid systems using the cross-entropy method. In *Hybrid Systems: Computation and Control*, pages 125–134, 2012.

[26] S. Sankaranarayanan, F. Ivancic, and A. Gupta. Mining library specifications using inductive logic programming. In *International Conference on Software Engineering*, page 131–140, 2008.

[27] S. Shoham, E. Yahav, S. J. Fink, and M. Pistoia. Static specification mining using automata-based abstractions. *IEEE Trans. on Software Eng.*, 34(5):651–666, 2008.

[28] S. Skogestad and I. Postlethwaite. *Multivariable feedback control: Analysis and Design*. Wiley, 2007.

[29] A. Solar-Lezama, L. Tancau, R. Bodík, S. A. Seshia, and V. A. Saraswat. Combinatorial sketching for finite programs. In *Architectural Support for Programming Languages and Operating Systems*, pages 404–415, 2006.

[30] A. Tiwari. HybridSal relational abstracter. In *Computer-Aided Verification*, pages 725–731, 2012.

[31] S. Tripakis, C. Sofronis, P. Caspi, and A. Curic. Translating discrete-time Simulink to Lustre. *ACM Trans. Embedded Comput. Syst.*, 4(4):779–818, 2005.

[32] W. Weimer and G. Necula. Mining temporal specifications for error detection. In *Tools and Algorithms for the Construction and Analysis of Systems*, page 461–476, 2005.

[33] H. Yang, B. Hoxha, and G. Fainekos. Querying parametric temporal logic properties on embedded systems. In *Int. Conference on Testing Software and Systems*, pages 136–151, 2012.

[34] C. Zhou and R. Kumar. Semantic translation of Simulink diagrams to input/output extended finite automata. *Discrete Event Dynamic Systems*, 22:223–247, 2012.

On the Decidability of Stability of Hybrid Systems

Pavithra Prabhakar
IMDEA Software Institute, Madrid, Spain
Madrid, Spain
pavithra.prabhakar@imdea.org

Mahesh Viswanathan
University of Illinois at Urbana-Champaign
Urbana, IL, USA
vmahesh@illinois.edu

ABSTRACT

A rectangular switched hybrid system with polyhedral invariants and guards, is a hybrid automaton in which every continuous variable is constrained to have rectangular flows in each control mode, all invariants and guards are described by convex polyhedral sets, and the continuous variables are *not* reset during mode changes. We investigate the problem of checking if a given rectangular switched hybrid system is stable around the equilibrium point 0. We consider both Lyapunov stability and asymptotic stability. We show that checking (both Lyapunov and asymptotic) stability of planar rectangular switched hybrid systems is decidable, where by planar we mean hybrid systems with at most 2 continuous variables. We show that the stability problem is undecidable for systems in 5 dimensions, i.e., with 5 continuous variables.

Categories and Subject Descriptors

D.2 [**Software**]: Software Engineering

General Terms

Verification

Keywords

Stability, Decidability, Verification

1. INTRODUCTION

Stability is a key property required of many dynamical and hybrid systems. Intuitively, it says that when a system is started somewhere close to its desired operating behavior, it will stay close to its desired operating behavior at all times (Lyapunov stability) and will converge to its desired operating behavior in the limit (asymptotic stability). Of particular importance is the stability of equilibrium points, which requires that a system starting close to an equilibrium point stays close to the equilibrium point at all times (Lyapunov stability) and converges to the equilibrium point in the limit (asymptotic stability).

Given the importance of system designs to be stable, a number of proof techniques have been discovered to establish the stability of a system under a controller. These rely on extensions of Lyapunov's second or direct method of establishing stability, wherein one identifies an "energy" function that decreases over the trajectories of the system. In the context of hybrid systems, this could be either a common function for all control modes, or separate functions for each control mode that nonetheless decrease over finite sequences of control mode changes. For a comprehensive overview of these techniques, see surveys [5, 7, 6] and textbooks [15, 11]. Such Lyapunov functions are either constructed through the ingenuity of the designer or by algorithmically searching for Lyapunov functions of a special form using convex programming. Automated techniques have also been developed to check a weaker, "qualitative" version of stability, wherein one only requires that all trajectories starting from an initial state eventually enter a "good" region [12, 13, 8, 9]. All the above techniques typically only check for sufficient conditions for a system to be stable. In other words, if the above approaches (automatic and manual) succeed then one can conclude that the system is stable. On the other hand, if they fail then one *cannot* conclude the instability of the system.

Unfortunately, despite the undeniable importance of stability, very little is known about the *computational difficulty* of deciding stability of a hybrid system. This is quite unlike other important properties like safety and liveness. In a series of papers [3, 4, 2], Blondel, Tsitsiklis, and their co-authors, present some results concerning discrete-time, switched hybrid systems with linear dynamics in each control mode, i.e., in each control mode the state at time instant $k+1$ (x_{k+1}) is given by $x_{k+1} = Ax_k$, where A depends on the current mode, and mode switches don't change the continuous state x. They show that stability is undecidable for saturated linear dynamical systems [2], is NP-hard for various restricted forms of discrete time linear switched hybrid systems with mode switches governed by polyhedral guards [4] (the upper bound in these cases is not known, and the problems may in fact be undecidable), and undecidable for certain simple systems with a control input [4].

In this paper, we present some results concerning decidability of stability of equilibrium points in the continuous-time, hybrid setting; we hope this will spur much needed fur-

HSCC'13, April 8–11, 2013, Philadelphia, Pennsylvania, USA.

ther research into the computational complexity of stability. We consider *rectangular switched hybrid systems with polyhedral guards and invariants.* These are hybrid automata that are switched (i.e., the continuous state does not change when the control mode changes), the continuous dynamics in each control mode is governed by rectangular constraints which require that the time derivative of any variable's continuous trajectory lie in an interval at all times, and the invariants for each control mode and guards on all control switches are given by convex polyhedra. Thus, these systems are generalizations of piecewise constant derivative (PCD) systems [1], and related to rectangular hybrid automata in the sense that guards and invariants can be more general than rectangular sets, but restricted in that control mode switches don't reset the continuous state.

Our main result is that the stability (both Lyapunov and asymptotic) of planar (i.e., systems with only two continuous variables) rectangular switched hybrid systems with polyhedral invariants and guards can be effectively decided. Our proof relies on the following sequence of observations. We assume that 0 is the desired equilibrium point whose stability we wish to establish. We first observe that a hybrid system $\mathcal{H}$ is stable at 0 iff for some $\gamma > 0$, $\mathcal{H}_\gamma$ is stable, where $\mathcal{H}_\gamma$ is a system that is identical to $\mathcal{H}$ in the open ball of radius γ around 0. For the class systems we consider, $\mathcal{H}_\gamma$ has a special structure wherein all the guards and invariants can be seen to be unions of "wedge"-like sets. Based on this observation, we show that stability questions about $\mathcal{H}_\gamma$ can be reduced to questions about an "equivalent" finite, weighted graph $G_\mathcal{H}$. The crux of the decidability proof is showing that this graph $G_\mathcal{H}$ can be effectively constructed from $\mathcal{H}$. This involves showing that some restricted forms of reachability predicates can be answered for the class of planar switched systems we consider. These observations about the computability of limited reachability predicates for our systems may be of independent interest, since the decidability of the general reachability problem for our class of hybrid systems is not known. We conclude our results by showing that stability is indeed likely to be a difficult problem: we show that determining the Lyapunov stability (at 0) of a rectangular switched hybrid system (with polyhedral invariants and guards) with at least 5 continuous variables, is undecidable.

Our undecidability result for 5 dimensions may seem unsurprising and disappointing in the light of the fact that reachability is undecidable for 3-dimensional PCD systems (which is a special case of our systems). This disappointment stems from a popular belief that stability is "harder" than reachability. We provide some evidence to support this belief. Our undecidability proof reduces reachability in certain special cases to Lyapunov stability. On the other hand, our paper also provides some evidence to suggest that establishing the hardness of stability relative to reachability (in general) may not be as obvious as the common belief seems to suggest. First, the reachability problem is not known to be decidable for the class of planar systems we consider, even though we show stability to be decidable; for special subclasses of the hybrid systems considered here like PCD [1] and monotonic linear systems [14], reachability is known to be decidable. Next, while our stability algorithm does use certain types of reachability queries, it does so for special "pie shaped" hybrid systems. The reason is because we show that for stability one can always focus on the behavior of the system close to 0, and in all systems sufficiently close to 0, the invariants and guards look like unions of "wedge-like" sets. Thus, the reachability problem must be hard on such instances for the hardness to be lifted to stability.

2. PRELIMINARIES

2.1 Notations

Let $\mathbb{R}$, $\mathbb{R}_{\geq 0}$, $\mathbb{Q}$ and $\mathbb{N}$ denote the set of reals, non-negative reals, rationals and natural numbers, respectively. Given a function F, we use $Dom(F)$ to denote the domain of F. Given function $F : A \to B$ and a set $A' \subseteq A$, $F(A')$ will denote the set $\{F(a) \mid a \in A'\}$.

Sequences.

Let *SeqDom* denote the set of all subsets of $\mathbb{N}$ which are prefix closed, where a set $S \subseteq \mathbb{N}$ is prefix closed if for every $m, n \in \mathbb{N}$ such that $n \in S$ and $m < n$, $m \in S$. For a finite $S \in SeqDom$, we use $|S|$ to denote the largest element of S, that is, for $S = \{0, 1, \cdots, n\}$, $|S| = n$. A *sequence* over a set A is a mapping from an element of *SeqDom* to A. Given a sequence $\pi : S \to A$, we also denote the sequence by enumerating its elements in the order, that is, $\pi(0), \pi(1), \cdots$.

Convex polyhedral sets.

Let X denote the d-dimensional Euclidean space $\mathbb{R}^d$ for some d. Given $\mathbf{x}, \mathbf{y} \in X$, we use the standard notation $\mathbf{x} \cdot \mathbf{y}$ to denote the dot product of vectors $\mathbf{x}$ and $\mathbf{y}$ and $|\mathbf{x}|$ to denote the Euclidean norm. For a vector $\mathbf{x} = (x_1, \cdots, x_d) \in \mathbb{R}^d$ and $1 \leq i \leq d$, $(\mathbf{x})_i$ will denote the projection of $\mathbf{x}$ to the i-th component, that is, x_i. Given $\epsilon \in \mathbb{R}_{\geq 0}$, we use $B_\epsilon(\mathbf{x})$ to denote an open ball around $\mathbf{x}$ of radius ϵ, that is, $B_\epsilon(\mathbf{x}) = \{\mathbf{y} \mid |\mathbf{x} - \mathbf{y}| < \epsilon\}$. A set $S \subseteq X$ is *open* if for every $\mathbf{x} \in S$, there exists a $\delta > 0$ such that $B_\delta(\mathbf{x}) \subseteq S$.

A set $S \subseteq X$ is *convex* if for every $\mathbf{x}, \mathbf{y} \in S$ and $\alpha \in [0, 1]$, $\alpha\mathbf{x} + (1-\alpha)\mathbf{y} \in S$. Let *Conv*($X$) denote the set of all convex subsets of X. Given a set S, *ConvHull*(S) is the smallest convex set containing S. A *half-space* in X is the set which can be expressed as the set of all points $\mathbf{x} \in X$ satisfying a linear constraint, $\mathbf{a} \cdot \mathbf{x} + \mathbf{b} \sim 0$, for some $\mathbf{a}, \mathbf{b} \in X$ and $\sim \in \{<, \leq\}$. A *convex polyhedral set* is an intersection of finitely many half-spaces. We will use *ConvPolyhed*(X) to denote the set of all convex polyhedral subsets of X. Given a convex polyhedral set S and a point $s \in S$, we call the cone of S at s, the set of all vectors $v \neq \mathbf{0}$ such that there exists a $t > 0$, for which $s + vt \in S$. We denote this set by *Cone*(S, s).

A *partition* $\mathcal{P}$ of $\mathbb{R}^d$ into convex polyhedral sets is a finite set of convex polyhedral sets $\{P_1, \cdots, P_k\}$ such that $\cup_{i=1}^k P_i = \mathbb{R}^d$ and for each $i \neq j$, $P_i \cap P_j = \emptyset$.

Intervals and Rectangular sets.

An interval is a convex subset of $\mathbb{R}$ and is represented using the standard notation ($[a, b]$, (a, b), and so on). An interval I is said to be compact if $I = [a, b]$ for some $a, b \in \mathbb{R}$. We use *TimeDom* to represent the finite and infinite closed intervals starting from 0, that is, *TimeDom* consists of the intervals $I = [0, T]$ for some $T \in \mathbb{R}_{\geq 0}$ and the interval $[0, \infty)$. Given $I \in TimeDom$, we define the size of the interval I, denoted *Size*(I), to be an element in $\mathbb{R}_{\geq 0} \cup \{+\infty\}$ such that for a compact interval $I = [0, T]$, *Size*(I) $= T$, otherwise *Size*(I) $= \infty$.

A *rectangular set* over X is a set $R \subseteq X$ which can be expressed as the Cartesian product of compact intervals, that is, R is given by a set $I_1 \times \cdots \times I_d$, where I_i for $1 \leq i \leq d$ are compact intervals. Let $Rect(X)$ denote the set of all rectangular subsets of X. Given $R \in Rect(\mathbb{R}^d)$, we define the extreme vectors in R, denoted $Extremes(R)$, as $Extremes(R) = \{\mathbf{v} \in R \mid \forall i, (\mathbf{v})_i = min((R)_i) \text{ or } max((R)_i)\}$.

Planar Elements.

Two dimensional objects will be referred to using the prefix "planar". A planar convex polyhedral set can be decomposed into an interior region and boundary elements which are the vertices and the straight line segments. Formally, given a planar convex polyhedral set S, *region* of S is the set of all points in the interior of S, that is, all point $s \in S$ such that $B_\delta(s) \subseteq S$ for some $\delta > 0$. *Boundary* of S is the set of all points $s \in S$ such that s is not in the interior of S. A *line* of S is a maximal convex subset l contained in the boundary of S, such that for every $s \in I$, there exist $s_1, s_2 \in I$ and $0 < \lambda < 1$ with $s = \lambda s_1 + (1-\lambda)s_2$. (The last condition is to eliminate the end-points of the line, if any). A *vertex* of S is a singleton set $\{s\}$ such that s belongs to the boundary of S but is not contained in any line of S. The closure of S, denoted $Closure(S)$, is the smallest closed set containing S. Elements of S, denoted $Elements(S)$, is the set of all vertices, lines and regions of S.

Let $\mathcal{P} = \{P_1, \cdots, P_k\}$ be a convex polyhedral partition of $\mathbb{R}^2$. Elements of $\mathcal{P}$, denoted by $Elements(\mathcal{P})$, is the set of elements of the convex polyhedral sets in $\mathcal{P}$, that is, $Elements(\mathcal{P}) = \cup_{i=1}^{k} Elements(P_i)$.

Graphs.

A *graph* G is a triple (V, E, V_0), where V is a finite set of nodes, $E \subseteq V \times V$ is a finite set of edges, and $V_0 \subseteq V$ is a finite set of initial nodes. A *path* π of a graph $G = (V, E, V_0)$ is a finite sequence of nodes $v_0, \cdots, v_k$ such that $(v_i, v_{i+1}) \in E$ for $0 \leq i < k$. The length of a path π, denoted $|\pi|$, is the number of edges occurring in it. A path π is *simple* if all the nodes occurring in the path are distinct. A *cycle* in a graph $G = (V, E, V_0)$ is a path $\pi = v_0, \cdots, v_k$ in which the first and the last nodes are the same, that is, $v_0 = v_k$. A cycle is *simple* if all the nodes except the last one are distinct. A node v is *reachable* from a node u if there exists a path whose first element is u and the last element is v, that is, there exists a π such that $\pi(0) = u$ and $\pi(|Dom(\pi)|) = v$.

We associate weights with the edges of a graph using weighting functions. A *weight function* w of a graph $G = (V, E, V_0)$ is a function $w : E \to \mathbb{R} \cup \{-\infty, +\infty\}$. We extend the weight function to a path as follows. Weight of a path π of G, denoted $w(\pi)$, is $\sum_{0 \leq i < |Dom(\pi)|} w((\pi(i), \pi(i+1)))$.

A *strongly connected component SCC* in a graph $G = (V, E)$ is a set of nodes $V' \subseteq V$ such that for every $v_1, v_2 \in V'$, there exists a path from v_1 to v_2 in G such that all elements in the path are in V'. Let $SCC(G)$ denote the set of maximal strongly connected components of G. Note that $SCC(G)$ represents a partition of the nodes in V, and there is a partial order on the elements of $SCC(G)$, where an element C_1 is less than C_2 if there exists a path from some node in C_1 to some node in C_2. In particular, we can construct a graph $G/_{SCC}$ whose nodes are the elements of $SCC(G)$ and whose edges are pairs (C_1, C_2) such that there exists $u \in C_1$ and $v \in C_2$ with (u, v) being an edge of G. We call this the quotient graph of G with respect to strongly connected components.

2.2 Switched Hybrid Systems

In this section, we define a formal model of systems which exhibit discrete-continuous behaviors. We focus on a class of systems, which exhibit different continuous behaviors in different modes of the system, and in which the switching between the modes is dictated by a controller which can read the continuous state of the system. In particular, the controller cannot force resets in the system state. We use the hybrid automaton model of [10], with some restrictions on the class of constraints used to define the elements of the model.

A *switched hybrid system (SHS)* of dimension d is given by $\mathcal{H} = (Q, Q_0, \Delta, X, Fl, Inv, Gd)$, where

- Q is a finite set of locations or modes,
- $Q_0 \subseteq Q$ is a set of initial locations,
- $\Delta \subseteq Q \times Q$ is a set of mode changes or transitions,
- $X = \mathbb{R}^d$ is the set of continuous states,
- $Fl : Q \to Conv(X)$ specifies vector fields for locations,
- $Inv : Q \to Conv(X)$ specifies invariants for locations, and
- $Gd : \Delta \to Conv(X)$ specifies guards for transitions.

In this paper, we focus on only switched hybrid systems, hence, we take the liberty to drop the prefix "switched" when referring to this class of systems.

Notation. The components of a hybrid system $\mathcal{H}_r$ are denoted using appropriate subscripts, for example, the set of locations of $\mathcal{H}_r$ is denoted by Q_r.

The hybrid system $\mathcal{H}$ starts in a state in X with the control in a mode in Q_0. The state of the system then evolves such that the derivative of the evolution belongs to the flow Fl of the current mode, and simultaneously satisfies the invariant of the mode. The system can change mode at any time from the current mode q_1 to a new mode q_2 provided there is a transition (q_1, q_2) in Δ and the current state satisfies the guard associated with the transition.

The semantics of a switched hybrid system $\mathcal{H}$ is given by the set of executions of the system. An *execution* of $\mathcal{H}$ is a triple $\eta = (\sigma, \tau, \gamma)$ such that there exists a $D \in TimeDom$ and an $S \in SeqDom$, which satisfy the following:

- $\sigma : D \to X$ is a continuous function,
- $\tau : S \to D$ is a non-decreasing function, such that
  - $\tau(0) = 0$,
  - if $S = \mathbb{N}$, then $\lim_{i \to \infty} \tau(i) = \infty$,
  - σ is differentiable in the interval $(\tau(i), \tau(i+1))$ for every i such that $\tau(i) < \tau(i+1)$, and in the interval $(\tau(|S|), Size(D))$ if S is finite, and
- $\gamma : S \to Q$, such that
  - for every i such that $\tau(i) = \tau(i+1)$, $e = (\gamma(i), \gamma(i+1)) \in \Delta$ and $\sigma(\tau(i)) \in Gd(e)$,
  - for every i such that $\tau(i) < \tau(i+1)$, $\gamma(i) = \gamma(i+1)$, $\frac{d\sigma}{dt}(t) \in Fl(\gamma(i))$ and $\sigma(t) \in Inv(\gamma(i))$, for every $t \in (\tau(i), \tau(i+1))$, and

- if S is finite and $\tau(|S|) < Size(D)$, then again $\frac{d\sigma}{dt}(t) \in Fl(\gamma(|S|))$ and $\sigma(t) \in Inv(\gamma(|S|))$, for every $t \in (\tau(|S|), Size(D))$.

Further, if $\gamma(0) \in Q_0$, then we call $\eta = (\sigma, \tau, \gamma)$ an *initial execution* of $\mathcal{H}$.

Notation. Given an execution η_r, we refer to its components with appropriate subscripts, that is, $\eta_r = (\sigma_r, \tau_r, \gamma_r)$.

An execution η of $\mathcal{H}$ is called *complete* if $Size(Dom(\sigma)) = \infty$, and *incomplete* or *finite* otherwise. We denote the set of executions of $\mathcal{H}$ by $Exec(\mathcal{H})$, the set of initial executions of $\mathcal{H}$ by $IExec(\mathcal{H})$ and the set of complete executions of $\mathcal{H}$ by $CExec(\mathcal{H})$.

η is said to be an execution from $\mathbf{x}$ to $\mathbf{y}$ if σ is an incomplete execution with $\sigma(0) = \mathbf{x}$ and $\sigma(Size(Dom(\sigma))) = \mathbf{y}$. Further, it is an execution from $(q_1, \mathbf{x})$ to $(q_2, \mathbf{y})$, if in addition $\gamma(0) = q_1$ and $\gamma(|Dom(\gamma)|) = q_2$. Also, η *reaches* $\mathbf{y}$ ($(q, \mathbf{y})$) if η is an execution from some $\mathbf{x}$ to $\mathbf{y}$ (and $\gamma(|Dom(\gamma)|) = q$), and η *touches* $\mathbf{y}$ ($(q, \mathbf{y})$) if there exists a time $t \in Dom(\sigma)$ (and $i \in Dom(\tau)$) such that $\sigma(t) = \mathbf{y}$ (and $\tau(i) = t$ and $\gamma(i) = q$).

Splitting an execution.

Next we define splitting an execution into two executions such that executing the second from the end-point of the first results in the original execution. Two executions η_1, η_2 *split* an execution η if the following hold:

- There exists $c \in \mathbb{R}_{\geq 0}$ such that $c = Size(Dom(\sigma_1))$, and $Dom(\sigma_1) \cup \{c + t \mid t \in Dom(\sigma_2)\} = Dom(\sigma)$.
- $|Dom(\tau)| = |Dom(\tau_1)| + |Dom(\tau_2)| - 1$, such that, for every $i \in Dom(\tau_1)$, $\tau_1(i) = \tau(i)$, for every $j > |Dom(\tau_1)|$ such that $j \in Dom(\tau)$, $\tau_2(j - |Dom(\tau_1)| + 1) = \tau(j) - c$, and $\tau_2(0) = c$.
- $\gamma_1(i) = \gamma(i)$ for all $i \in Dom(\gamma_1)$, for every $j > |Dom(\gamma_1)|$ such that $j \in Dom(\gamma)$, $\gamma_2(j - |Dom(\gamma_1)| + 1) = \gamma(j)$, and $\gamma(0) = \gamma_1(|Dom(\gamma_1)|)$.

A finite sequence $\eta_1, \cdots, \eta_n$ splits η, if there exists a sequence of sequences starting from the sequence η and ending with the sequence $\eta_1, \cdots, \eta_n$ such that every successive sequence is obtained from the previous one by splitting some execution. An infinite sequence $\eta_1, \eta_2, \cdots$ splits η, if for every $i \geq 1$, there exists an η'_i such that the finite sequence $\eta_1, \eta_2, \cdots, \eta_i, \eta'_i$ splits η.

Note that in general to split an η uniquely we need to specify the time of splitting σ and a element in the $Dom(\tau)$ which maps to the time of splitting, if one exists. $\eta[t_1, t_2]$ will used to denote η_2 in some splitting η_1, η_2, η_3 of η such that $t_1 = Size(Dom(\sigma_1))$ and $t_2 - t_1 = Size(Dom(\sigma_2))$. A prefix of η is an execution $\eta[0, t]$ for some $t \in Dom(\sigma)$.

Remark. Note that if η is an execution of $\mathcal{H}$, then $\eta[t_1, t_2]$ exists and is also an execution of $\mathcal{H}$.

Scaling and Weight.

Given a finite execution η, we define the *scaling* and *weight* of an execution η, denoted $Scaling(\eta)$ and $Weight(\eta)$, to be $|\sigma(Size(Dom(\sigma)))|/|\sigma(0)|$ and $\ln(Scaling(\eta))$, respectively.

PROPOSITION 1. *Given a splitting $\eta_1, \cdots, \eta_n$ of a finite execution η, weight of η is the sum of the weights of η_i for $1 \leq i \leq n$.*

Restrictions of Hybrid Systems.

Let $\mathcal{H}$ be a hybrid system. Given a set of locations $S \subseteq Q$, the restriction of the $\mathcal{H}$ to S, is the system $(Q \cap S, Q_0 \cap S, \Delta \cap S \times S, X, Fl[S], Inv[S], Gd[S])$, where $f[A]$ is the function whose domain is $A \subseteq Dom(f)$ and agrees with f on all points in the domain. We will denote this system as $\mathcal{H}[Q \to S]$. Similarly, given a set $S \subseteq X$, we define the restriction of $\mathcal{H}$ to the set S, denoted $\mathcal{H}[X \to S]$ as the system $(Q, Q_0, \Delta, X, Fl', Inv', Gd')$, where $Inv'(q) = Inv(q) \cap S$ and $Gd'(e) = Gd(e) \cap S$ for every location $q \in Q$ and transition $e \in \Delta$. We use $\mathcal{H}[F_1 \to F_2]$ to denote the hybrid system which is same as $\mathcal{H}$ except that the component F_1 ranging over Fl, Inv and Gd is replaced by F_2.

PROPOSITION 2. *Given a hybrid system $\mathcal{H}$ and $S \subseteq X$, an execution $\eta = (\sigma, \tau, \gamma)$ in $Exec(\mathcal{H})$ belongs to $Exec(\mathcal{H}[X \to S])$ if and only if $\sigma(t) \in S$ for every $t \in Dom(\sigma)$.*

3. STABILITY: LYAPUNOV AND ASYMPTOTIC

In this section, we define two classical notions of stability for hybrid systems, and state some general results about stability of hybrid systems.

We consider stability of the system with respect to an equilibrium point, which in our setting will be the origin.

Definition. $\mathbf{0}$ is an *equilibrium point* of a hybrid system $\mathcal{H}$ if any initial execution of $\mathcal{H}$ starting at $\mathbf{0}$ remains at $\mathbf{0}$.

Intuitively, Lyapunov stability captures the notion that an execution starting close to the equilibrium point remains close to it, and asymptotic stability, in addition, enforces converges to the equilibrium point.

Definition. A hybrid system $\mathcal{H}$ is said to be *Lyapunov stable*, if for every $\epsilon > 0$, there exists a $\delta > 0$ such that for every initial execution $\eta \in IExec(\mathcal{H})$ with $\sigma(0) \in B_\delta(\mathbf{0})$, $\sigma(t) \in B_\epsilon(\mathbf{0})$ for every $t \in Dom(\sigma)$.

If $\mathcal{H}$ is Lyapunov stable, we use $Lyap(\mathcal{H}, \epsilon, \delta)$ to denote the fact that for every execution $\eta \in IExec(\mathcal{H})$ with $\sigma(0) \in B_\delta(\mathbf{0})$, $\sigma(t) \in B_\epsilon(\mathbf{0})$ for every $t \in Dom(\sigma)$.

In fact, we do not need to consider all possible values for ϵ in the definition of Lyapunov stability but only values in a small neighborhood around 0.

Definition. A hybrid system $\mathcal{H}$ is said to be *asymptotically stable*, if it is Lyapunov stable and there exists a $\delta > 0$ such that every complete initial execution $\eta \in CExec(\mathcal{H}) \cap IExec(\mathcal{H})$ with $\sigma(0) \in B_\delta(\mathbf{0})$ converges to $\mathbf{0}$, that is, for every $\epsilon > 0$, there exists a $T \in Dom(\sigma)$, such that $\sigma(t) \in B_\epsilon(\mathbf{0})$ for every $t \geq T$.

If $\mathcal{H}$ is asymptotically stable, we use $Asymp(\mathcal{H}, \delta)$ to denote the fact that every complete execution $\eta \in CExec(\mathcal{H})$ with $\sigma(0) \in B_\delta(\mathbf{0})$ converges to $\mathbf{0}$.

The next lemma says that to ensure Lyapunov stability it suffices to consider the restriction of $\mathcal{H}$ to a small enough neighborhood of $\mathbf{0}$. Let us use the shorthand $\mathcal{H}_\epsilon$ to represent $\mathcal{H}[X \to B_\epsilon(\mathbf{0})]$. The next lemma says that the stability of the system can be determined by examining its behavior in a small neighborhood around $\mathbf{0}$.

4. PLANAR RECTANGULAR SYSTEMS AND NORMAL FORM

In this section, we present some preliminary results which we will use later.

We consider a subclass of hybrid systems with rectangular flows and polyhedral invariants and guards, which we call rectangular hybrid systems. A hybrid system $\mathcal{H}$ with a partition $\mathcal{P}$ of X into convex polyhedral sets is a *rectangular hybrid system* if $Fl: Q \rightarrow Rect(X)$, $Inv: Q \rightarrow ConvPolyhed(X)$ and $Gd: Q \rightarrow ConvPolyhed(X)$, and for each $q \in Q$ and $e \in \Delta$, $Inv(q)$ and $Gd(e)$ are a finite union of the elements of $\mathcal{P}$. We will denote a rectangular system as a pair $(\mathcal{H}, \mathcal{P})$, however, when $\mathcal{P}$ is clear from the context we drop it from the notation. We use $Elements(\mathcal{H})$ to denote $Elements(\mathcal{P})$.

In this paper, we further focus on planar or 2-dimensional hybrid systems. We present certain structural properties of planar hybrid systems, and present a normal form for this class with respect to stability analysis.

A set S is said to be *ϵ-pie-shaped* if either

1. S is $\{\mathbf{0}\}$ or $\emptyset$,
2. S is a set $\{\alpha\mathbf{x} \mid 0 < \alpha < \epsilon\}$ for some $\mathbf{x}$ such that $|\mathbf{x}| = 1$, or
3. S is expressed as a set of points $(x, y) \in \mathbb{R}^2$ satisfying the constraints $a_1x + b_1y < 0$, $a_2x + b_2y < 0$ and $x^2 + y^2 < \epsilon$ for some $a_1, b_1, a_2, b_2 \in \mathbb{R}$.

We say that a planar hybrid system is *ϵ-pie-shaped* if for every location q and every transition e, $Inv(q)$ and $Gd(e)$ can be expressed as the union of a finite number of pie-shaped sets. The next lemma states that a rectangular hybrid system is a pie-shaped hybrid system when restricted to a small enough neighborhood around the origin.

LEMMA 1. *For a rectangular hybrid system $\mathcal{H}$, there exists $\epsilon > 0$, such that $\mathcal{H}_\epsilon$ is ϵ-pie-shaped.*

Given an ϵ-pie-shaped hybrid system, we construct a rectangular hybrid system by replacing each pie-shaped set by an unbounded conical set. Given an ϵ-pie-shaped set S, which is not $\{\mathbf{0}\}$, we define its infinite extension, denoted $InfExt(S)$, as follows. If $S = \{\alpha\mathbf{x} \mid 0 < \alpha < \epsilon\}$ for some $\mathbf{x}$ such that $|\mathbf{x}| = 1$, then $InfExt(S) = \{\alpha\mathbf{x} \mid \alpha \in \mathbb{R}_{\geq 0}/\{0\}\}$. If S is expressed as a set of points $(x, y) \in \mathbb{R}^2$ satisfying the constraints $a_1x + b_1y < 0$, $a_2x + b_2y < 0$ and $x^2 + y^2 < \epsilon$ for some $a_1, b_1, a_2, b_2 \in \mathbb{R}$, then $InfExt(S)$ is the set of points satisfying the constraints $a_1x + b_1y < 0$ and $a_2x + b_2y < 0$. Given an ϵ-pie-shaped system $\mathcal{H}$, we define $InfExt(\mathcal{H})$ to be the hybrid system obtained by replacing each invariant and guard given as the union of a finite set R of ϵ-pie-shaped sets, by the set which is the union of the sets $InfExt(S)$ for $S \in R$. Note that the result does not depend on the particular representation R of the invariant or guard.

Remark. Given an ϵ-pie-shaped hybrid system $\mathcal{H}$, the restriction to the ϵ-ball of its extension gives back $\mathcal{H}$, that is, $InfExt(\mathcal{H})_\epsilon = \mathcal{H}$.

The next proposition says that for a pie-shaped hybrid system, its stability is equivalent to that of its infinite extension.

PROPOSITION 3. *Let $\mathcal{H}$ be an ϵ-pie-shaped hybrid system. Then:*
$\mathcal{H}$ is Lyapunov (asymptotically) stable iff $InfExt(\mathcal{H})$ is Lyapunov (asymptotically) stable.

A planar rectangular hybrid system $\mathcal{H}$ is said to be in *normal form* if there exists an $\epsilon > 0$ such that $\mathcal{H}_\epsilon$ is ϵ-pie-shaped and $InfExt(\mathcal{H}_\epsilon)$ is $\mathcal{H}$.

PROPOSITION 4. *Given a planar rectangular hybrid system $\mathcal{H}$, there exists a planar rectangular hybrid system $\hat{\mathcal{H}}$ which is in normal form such that:*
$\mathcal{H}$ is Lyapunov (asymptotically) stable if and only if $\hat{\mathcal{H}}$ is Lyapunov (asymptotically) stable.

5. DECIDABILITY OF PLANAR RECTANGULAR HYBRID SYSTEMS

The algorithm for decidability consists of constructing a weighted graph and analyzing the graph for the absence of certain positive weight cycles. We need a primitive for constructing the graph, which is the maximum and minimum scaling of the state with respect to the start and end states of certain execution fragments. We formalize this primitive, present the construction of the graph and characterize the stability in terms of properties of this graph.

Elements of $\mathcal{H}$.

Let us fix a planar rectangular system $\mathcal{H}$ in normal form, as obtained by first taking its restriction to a small enough ϵ and then performing an infinite extension of its elements. The elements of $\mathcal{H}$, $Elements(\mathcal{H})$, are as follows:

1. The vertex $\{\mathbf{0}\}$. (We will abuse notation and represent $\{\mathbf{0}\}$ as just $\mathbf{0}$).
2. Lines $p_0, \cdots, p_{n-1}$, where p_i is an infinite ray starting from $\mathbf{0}$ but not including $\mathbf{0}$. (We will assume that the lines are numbered in the order in which they appear in the plane).
3. $R_{i,j}$, $0 \leq i, j \leq n-1$, $|i - j| = 1$, represents the open region between the lines p_i and p_j. (We assume that the addition of the indices i, j is modulo n).

Neighbors and elements between them.

We will define the concept of neighboring elements and elements between two neighboring elements as follows. The neighbors of a line p_i is $Neigh(p_i) = \{p_i, p_{i-1}, p_{i+1}, R_{i,i-1}, R_{i,i+1}, \mathbf{0}\}$. The elements between a line p_i and a neighboring element S of p_i, denoted $Between(p_i, S)$, is defined as follows:

1. If S is p_i, then $Between(p_i, S)$ is $\{p_i, R_{i,i+1}, R_{i-1,i}\}$.
2. If S is p_j and $|i - j| = 1$, then $Between(p_i, S)$ is $\{R_{i,j}\}$.
3. If S is $R_{i,j}$ for $|i - j| = 1$, then $Between(p_i, S)$ is $\{R_{i,j}\}$.
4. If S is $\mathbf{0}$, then $Between(p_i, S)$ is $\{p_i, R_{i,i+1}, R_{i-1,i}\}$.

Element and Location execution.

We will break up an execution into smaller fragments each of which either remains inside an element of $\mathcal{H}$ or inside a location of $\mathcal{H}$.

Definition. Given a line p_i, a neighbor S of p_i and an element $R \in Between(p_i, S)$, we say that an execution η of $\mathcal{H}$ is an *element execution* with respect to p_i, S and R if η is an execution from some point in p_i to some point in S such that $\sigma(t) \in R$ for every $0 < t < Size(Dom(\sigma))$.

We call η an element execution if it is an element execution with respect to some p_i, S and R.

Definition. Given a line p_i and a location q, we say that η is a *location execution* with respect to q and p_i if η is an execution from some point in p_i to some point in $R = p_i \cup R_{i,i-1} \cup R_{i,i+1}$ such that $\sigma(t) \in R$ for every $t \in Dom(\sigma)$, and $\gamma(i) = q$ for all $i \in Dom(\gamma)$.

We call η a location execution if it is a location execution with respect to some location q and some line p_i. Note that a location execution need not be an element execution because it could move in both the sets $R_{i,i-1}$ and $R_{i,i+1}$, where as an element execution requires it to stay in one of these sets.

Max and Min Scalings.

Next, we define four primitives which denote the maximum and minimum scaling between the starting and ending points of an element execution and a location execution.

Given a neighbor S of a line p_i and locations q_1, q_2 of $\mathcal{H}$, we define $MaxSclE(q_1, p_i, q_2, S)$ and $MinSclE(q_1, p_i, q_2, S)$ as follows. Let Σ_E denote the set of all executions η of $\mathcal{H}$ such that η is an element execution with respect to p_i, S, and R for some $R \in Between(p_i, S)$ and $\gamma(0) = q_1$ and $\gamma(|Dom(\gamma)|) = q_2$. Then:

- $MaxSclE(q_1, p_i, q_2, S) = MinSclE(q_1, p_i, q_2, S) = \perp$ if Σ_E is empty.
- $MaxSclE(q_1, p_i, q_2, S) = \sup_{\eta \in \Sigma_E} Scaling(\eta)$, and $MinSclE(q_1, p_i, q_2, S) = \inf_{\eta \in \Sigma_E} Scaling(\eta)$, otherwise.

Given a line p_i and a location q, let Σ_L denote the set of all location executions with respect to q and p_i.

- $MaxSclL(q, p_i) = MinSclL(q, p_i) = \perp$ if Σ_L is empty.
- Otherwise, $MaxSclL(q, p_i) = \sup_{\eta \in \Sigma_L} Scaling(\eta)$, and $MinSclL(q, p_i) = \inf_{\eta \in \Sigma_L} Scaling(\eta)$.

Construction of the graph.

Next, we present the construction of the graph. We construct the graph $G_{\mathcal{H}} = (V, E, V_0)$ with two weight functions W_{max} and W_{min} as follows:

1. $V = Q \times Elements(\mathcal{H})$.
2. $V_0 = \{(q, S) \in V \mid q \in Q_0, S \subseteq Inv(q)\}$.
3. $E = \{((q, p_i), (q', S)) \in V \times V \mid S \in Neigh(p_i), MaxSclE(q, p_i, q', S) \neq \perp\} \cup \{((q, p_i), (q, p_i)) \in V \times V \mid p_i \in P\}$.
4. $W_{max}((q, p_i), (q', S)) = \ln(MaxSclE(q, p_i, q', S))$ if $(q, p_i) \neq (q', S)$, $max(\ln(MaxSclE(q, p_i, q', S)), \ln(MaxSclL(q, p_i)))$ otherwise.
5. $W_{min}((q, p_i), (q', S)) = \ln(MinSclE(q, p_i, q', S))$ if $(q, p_i) \neq (q', S)$, $max(\ln(MinSclE(q, p_i, q', S)), \ln(MinSclL(q, p_i)))$ otherwise.

We will assume that the invariant associated with an initial location is a line. If the invariant is a region, then we can consider the lines of its closure (which are reachable) as the invariant for the purpose of the analysis, along with a test to check if the execution can blow up while remaining in the region.

Definition. Given an edge $e = ((q, p), (q', p'))$, we say that execution η *realizes* it, denoted $\eta \in Realize(e)$, if η is an element or location execution and satisfies $q = \gamma(0)$, $q' = \gamma(|Dom(\gamma)|)$, $\sigma(0) \in p$ and $\sigma(Size(Dom(\sigma))) \in p'$.

Definition. An execution η is said to *realize* an edge e of $G_{\mathcal{H}}$ with precision $\epsilon \geq 0$, denoted $\eta \in Realize_{\epsilon}(e)$, if it realizes e and $Weight(\eta) \in [W_{max}(e) - \epsilon, W_{max}(e)]$.

PROPOSITION 5. *Let $e = ((q_1, r_1), (q_2, r_2))$ be an edge of $G_{\mathcal{H}}$. Then:*

1. *Every execution η realizing an e satisfies $Weight(\eta) \leq W_{max}(e)$.*
2. *For every $\epsilon > 0$ and every $v_1 \in r_1$, there exists an η from (q_1, v_1) realizing e with precision ϵ.*

Characterization of Lyapunov Stability.

We need the notion of a q-cycle, negative cycle and an exploding node.

Definition. A cycle π of $G_{\mathcal{H}}$ is said to be a *q-cycle*, if for every $i \in Dom(\pi)$, $\pi(i) = (q, S)$ for some S.

Definition. A cycle π of $G_{\mathcal{H}}$ is a *negative* cycle if $W_{min}(\pi) < 0$.

Definition. A node $(q, \mathbf{0})$ of $G_{\mathcal{H}}$ is said to be *exploding* if there exists a location q' such that $Trans(q, q', \mathbf{0})$ and $Fl(q') \cap Cone(Inv(q'), \mathbf{0}) \neq \emptyset$, where $Trans(q_s, q_f, r_c)$ holds iff q_f is reachable from q_s by a series of transitions whose guards contain r_c.

LEMMA 2. *Any finite execution η of $\mathcal{H}$ which starts on some line p_i and does not reach $\mathbf{0}$ can be split into finitely many fragments $\eta_1, \cdots, \eta_n$ such that each η_i is either an element execution or a location execution.*

LEMMA 3. *If an execution η of $\mathcal{H}$ starting at some point in p_i at location q reaches $\mathbf{0}$ and then leaves $\mathbf{0}$, then in the graph $G_{\mathcal{H}}$, either there exists a path from the node (q, p_i) to an exploding node or there exists a simple negative q' cycle for some q' such that $(q', \{\mathbf{0}\})$ is an exploding node.*

THEOREM 1. *$\mathcal{H}$ is Lyapunov stable if and only if the following conditions hold:*

1. *There does not exist a simple cycle π in $G_{\mathcal{H}}$ such that $W_{max}(\pi) > 0$ and $\pi(0)$ is reachable from an initial node in V_0.*
2. *No edge (u, v) of $G_{\mathcal{H}}$ with $W_{max}((u, v)) = +\infty$ is reachable from an initial node.*
3. *For every exploding node $v = (q, \mathbf{0})$, v is not reachable from an initial node and there does not exist a simple negative q-cycle π such that $\pi(0)$ is reachable from an initial node.*

Characterization of Asymptotic Stability.

We need the notions of a locally stable region and a time bounded edge.

Definition. A region R in $Elements(\mathcal{H})$ is said to be *locally stable* with respect to a location q if every complete execution starting in (q, R) and remaining within R converges to $\mathbf{0}$.

PROPOSITION 6. *A region R in $Elements(\mathcal{H})$ is locally stable with respect to a location q iff $\mathbf{0} \notin ConvHull(S)$ and $S \setminus Cone(Closure(R), \mathbf{0}) = \emptyset$, where S is given by $\cup_{q'} Fl(q')$ such that q' is reachable from q by a series of transitions whose guards contain R.*

Further, if R is not locally stable with respect to q, then there exists a complete execution from every point in R starting at q which does not converge.

Definition. An edge $e = ((q_1, p_i), (q_2, p_i))$ with $W_{max}(e) = 0$ is said to be *bounded in time*, if for every $v_i \in p_i$, there exists a time bound $T \in \mathbb{R}_{\geq 0}$ such that all executions η from (q_1, v_i) realizing e with precision 1 satisfy $Size(Dom(\sigma)) \leq T$.

PROPOSITION 7. *Suppose $e = ((q_1, p_i), (q_2, p_i))$ is bounded in time and T is the bound on the time for a point $v_i \in p_i$, then for every $\epsilon > 0$, the time domain of any execution η realizing e from any point $\alpha v_i \in p_i$, $\alpha > 0$ with precision ϵ is bounded by $\epsilon\alpha T$.*

LEMMA 4. *Any complete execution η of $\mathcal{H}$ which starts on some line p_i and does not reach $\mathbf{0}$ can be split into either infinitely many fragments $\eta_1, \eta_2, \cdots$ such that each η_i is either an element execution or a location execution, or into finitely many fragments $\eta_1, \cdots, \eta_n, \eta'$, where η_i are element or location executions and η' is an execution which always remains in a particular region R.*

THEOREM 2. *$\mathcal{H}$ is asymptotically stable if and only if $\mathcal{H}$ is Lyapunov stable and the following conditions hold:*

1. *Every simple cycle π in $G_{\mathcal{H}}$ where $\pi(0)$ is reachable from an initial node in V_0 by a path with finite weight satisfies one of the following:*
   - *(a) $W_{max}(\pi) < 0$, or*
   - *(b) $W_{max}(\pi) = 0$, for every edge e of π, $W_{max}(e) = 0$, all the nodes of π correspond to a single p_i, and every edge of π is bounded in time.*
2. *Every node (q, R), where R is a region, which is reachable in $G_{\mathcal{H}}$ from an initial node, is locally stable.*

Algorithm for checking Theorem 1 and Theorem 2.

We will assume that all the constants appearing in the specification of the hybrid automaton are rational. The algorithm for checking Lyapunov and asymptotic stability involves constructing the graph $G_{\mathcal{H}}$ and checking the conditions in Theorem 1 and Theorem 2. Constructing the graph requires computation of the primitives *MaxSclE*, *MinSclE*, *MaxSclL* and *MinSclL*, which we will explain in the next section.

In terms of Lyapunov stability, we need to show that the primitives $Trans(q, q', r)$ can be computed and whether a node is exploding can be decided. One can compute the primitive $Trans(q, q', r)$ by first identifying the edges in the underlying graph of $\mathcal{H}$ which contain r and checking reachability using those edges. Containment, emptiness and intersection of sets expressed as first order logic formulas can be effectively computed, and so is the primitive $Cone(S, s)$ when S is a convex set. Hence, one can compute both primitives. Once the graph is constructed, the conditions involving existence of reachable simple cycles with positive and negative weights, reaching an edge with $+\infty$ weight or reaching an exploding node can be computed in time polynomial in the number of nodes of the graph. Hence, all the conditions in Theorem 1 can be effectively checked.

For checking asymptotic stability, the conditions on the graph can again be effectively checked. Determining local stability of a region involves checking for intersection of two sets, computing $Cone(S, s)$ for a convex polyhedral set S, and computing the convex hull of a finite number of convex sets, each of which can be effectively carried out using the decision procedures for first-order logic over $(\mathbb{R}, 0, +, <)$. It remains to show for deciding asymptotic stability that the property of an edge to be bounded in time can be decided, which we will show in the next section.

6. COMPUTING THE GRAPH ELEMENTS

In this section, we outline the procedure for computing Max and Min Scaling. Essentially, we perform a reachability computation and extract the min and max scaling from the set.

6.1 Reachability in a Convex Polyhedral Set

We compute the reachable points in a convex polyhedral set by executions which remain within the set.

Problem 1. *Let us fix a convex polyhedral set R which is open. Let $\mathcal{H}$ be a hybrid system whose invariants and guards are R. Let l be a line and r an element, both of which are in $Closure(R)$. Let q_1 and q_2 be locations in $\mathcal{H}$. Compute the set $Reach_{\mathcal{H}}(R, l, r, q_1, q_2)$, which is the set of all (v_1, v_2, t) such that $v_1 \in l$ and $v_2 \in r$ and there exists an incomplete execution η of $\mathcal{H}$ from (q_1, v_1) to (q_2, v_2) with $\sigma(Dom(\sigma)) = t$ and which remains within R, that is, $\sigma(t') \in R$ for $0 < t' < t$.*

The structure of the computation is similar to that in [14], but there are a few differences due to the more general dynamics we consider. Below are some series of reductions to a simpler problem that we need to solve.

Procedure 1. *Procedure for reducing to a hybrid system whose underlying graph is a strongly connected component.*

- *Compute $G/_{SCC}$, where $G = (Q, \Delta)$ is the underlying graph of $\mathcal{H}$.*
- *Note that $Reach_{\mathcal{H}}(R, l, r, q_1, q_2)$ is equivalent to computing the union of $Reach_{\mathcal{H}_\pi}(R, l, r, q_1, q_2)$ for every path $\pi = C_1 \cdots C_n \in G$, where $\mathcal{H}_\pi$ is the restriction of $\mathcal{H}$ to the locations in components of π, that is, $\cup_{1 \leq i \leq n} C_i$. Since the set of paths of G/SCC is finite, we can compute $Reach_{\mathcal{H}_\pi}(R, l, r, q_1, q_2)$ for each π.*
- *Let us fix a $\pi = D_1, \cdots, D_k$. We can further reduce the problem to reachability with respect to one strongly connected component as follows. We can assume w.l.o.g that $q_1 \in D_1$ and $q_2 \in D_k$.*
  - *First we compute the set of points (q, v') reached in R from (q_1, v) in $\mathcal{H}[Q \to D_1]$. Let us call this set S_1.*
  - *Next, for $i = 2, \cdots, k-1$, we iteratively compute S_i from S_{i-1} as follows. S_i is the set of points (q, v') reached in R starting from some point in S_{i-1} in the hybrid system $\mathcal{H}[Q \to D_i]$.*
  - *Finally, we compute the set of points (q, v') reached in r starting from some point in S_{k-1} in the hybrid system $\mathcal{H}[Q \to D_k]$.*

We will show that each of the sets S_i can be expressed as a formula of the first order logic over $(\mathbb{R}, 0, +, <)$, and hence is a finite union of convex polyhedral sets. From the above sketch of the algorithm, it follows that the primitive that we need to compute is the following.

Problem 2. *Let R be an open convex polyhedral set, and r_i, $i = 1, 2$ be a convex polyhedral sets each of which is a subset of some element of Closure(R). Let $\mathcal{H}$ be a hybrid system all of whose invariants and guards are R and its underlying graph is strongly connected. Let q_1 and q_2 be two locations in $\mathcal{H}$. We want to find the set of all triples (v_1, v_2, t) such that (q_2, v_2) is reached in r_2 from points (q_1, v_1) in r_1 by an executions of $\mathcal{H}$ whose domain is $[0, t]$.*

In the rest of the section, we show how to compute this. We focus on a single starting point $v \in r_1$. We prove a series of properties which reduce the problem to a simple computation. First, we consider a hybrid system $\mathcal{H}_1$ which is the same as $\mathcal{H}$ except that the invariants and guards are replaced by the whole Euclidean space, that is, $\mathbb{R}^2$. Note that solving Problem 2 for $\mathcal{H}_1$ gives us a set which is a superset of the set we would obtain for $\mathcal{H}$. We will show that they are equivalent modulo some exceptions.

The next property states that in $\mathcal{H}_1$ we only need to consider executions which are piecewise linear and whose derivative belong to a finite set of vectors. Let $\mathcal{H}_2$ be the hybrid system which is same as $\mathcal{H}_1$ except that for each location q, its flow $Fl_2(q) = Extremes(Fl_1(q))$.

PROPOSITION 8. *If v_2 is reachable in $\mathcal{H}_1[Q \to \{q\}]$ from v_1 by an execution η such that σ is differentiable everywhere in the interval $(0, T)$, then there exists an $a \in Fl_1(q)$ such that $v_2 = v_1 + aT$, where $T = Size(Dom(\sigma))$).*

LEMMA 5. *v_2 is reachable from v_1 in the hybrid system $\mathcal{H}_1[Q \to \{q\}]$ by an execution in time T iff v_2 is reachable from v_1 in the hybrid system $\mathcal{H}_2[Q \to \{q\}]$ by an execution in time T.*

Remark. The above lemma states that reachability in a strongly connected component with $\mathbb{R}^2$ as invariants and guards can be interpreted as a single location with $\mathbb{R}^2$ as invariant and the union of the extremes of the flows of the locations in the strongly connected component as its flow. Note that the locations do not play a role since starting in any location one can take a series of transitions to any other location without any time elapse, due to the strongly connected nature of the graph.

Next we want to lift the results in the setting where invariants and guards are $\mathbb{R}^2$ to one where they are given by a convex polyhedral open set R. For that we use a result from [14] which shows that if some point $v_2 \in Closure(R)$ is reachable from $v_1 \in Closure(R)$ by following at most two flows, then v_2 can be reached from v_1 by switching finitely many times between the two flows while remaining within R as long as neither of them is a vertex. And if one of them is a vertex, then there is an additional test involving the flows of the locations which determines if it is reachable. We summarize the result from [14] below.

LEMMA 6. *Let R be a convex polyhedral open set. Suppose $v_2 \in Closure(R)$ is reachable from $v_1 \in Closure(R)$ using flows a_1 and a_2, that is, $v_2 = a_1t_1 + a_2t_2$ for some $t_1, t_2 \geq 0$. Let us associate a set S_i with the vertices v_i, $i = 1, 2$ as follows. The set associated with v_i is empty if v_i is not a vertex. Otherwise, let S_i be a singleton set with a non-zero vector v such that there exists $t > 0$ with $v_i + tv \in R$ if $i = 1$ and $v_i - tv \in R$ otherwise. Then:*

- *There exists a piecewise continuous function $f : [0, T] \to \mathbb{R}^2$ such that $f(0) = v_1$, $f(T) = v_2$, $T = t_1 + t_2$ and the derivative in each piece belongs to $\{a_1, a_2\} \cup S_1 \cup S_2$ such that $f(t) \in R$ for every $t \in (0, T)$.*

In fact, we can generalize Lemma 6 to any number of flows. This tells us that reachability in $\mathcal{H}_2$ is almost equivalent to reachability in $\mathcal{H}$, except for the condition that we need to be able to enter R from our initial state and exit to the final state. We formalize this fact below.

THEOREM 3. *Let R be an open convex polyhedral set. Let r_1 and r_2 be elements of Closure(R). Let $\mathcal{H}$ be a hybrid system whose underlying graph is strongly connected and all of whose guards and invariants are R. Let $t > 0$. The following are equivalent:*

1. *A point v_2 in r_2 at location q_2 is reachable from a point v_1 in r_1 at location q_1 by an execution η of $\mathcal{H}$ with $Size(Dom(\sigma)) = t$ which remains within R in the interval $(0, t)$.*

2. *Let $\cup_{q \in Q} Extremes(Fl(q)) = \{a_1, \cdots, a_n\}$. There exist times $t_0, t_1, \cdots, t_n, t_{n+1} \geq 0$ with $t_0, t_n > 0$ and $\sum_{i=0}^{n+1} t_i = t$, and flow $a_0 \in Fl(q_1')$ and $a_{n+1} \in Fl(q_2')$ for some location q_1' and q_2' such that $Trans(q_1, q_1', r_1)$ and $Trans(q_2', q_2, r_2)$ holds, $v_2 = v_1 + a_0t_0 + \cdots + a_nt_n + a_{n+1}t_n$ and $v_1 + a_0t', v_2 - a_{n+1}t'' \in R$ for every $t' \in (0, t_0)$ and $t'' \in (0, t_{n+1})$, where $Trans(q_s, q_f, r_c)$ holds iff q_f is reachable from q_s by a series of transitions whose guards contain r_c.*

The above theorem summarizes the algorithm for solving Problem 2. One can write a first order logic formula over $(\mathbb{R}, 0, +, <)$ with free variables for v_1, v_2 and t. Let us call this formula $Reach_{\mathcal{H},R,r_1,r_2,q_1,q_2}(v_1, v_2, t)$. This formula can be used in Procedure 1 to iteratively compute a formula for each $\pi \in G_{SCC}$, and then for $\mathcal{H}$ itself.

6.2 Computing the primitives

Computing reachability by element executions.

Given p_i, a neighbor S of p_i, an element $R \in Between(p_i, S)$ and locations q_1 and q_2 of the hybrid system $\mathcal{H}$, we want to compute the set of points reached in S at location q_2 from the points in p_i at location q_1 by executions which remain within R except at the end-points. When R is an open set, it is equivalent to computing $Reach_{\mathcal{H}}(R, l, r, q_1, q_2)$. When R is a line or a point, we essentially follow Procedure 1 except that the analogue of Problem 2 can be solved by much simpler tests. For example, if R is a line, then essentially, $p_i = R$ and $S = p_i$ or $\mathbf{0}$. The set of points computed is the union of a subset of elements from the set $\{U, L, \mathbf{0}\}$, where U is the set of all points "above" v_i in p_i and L is the set of all points "below" v_i in p_i. And the inclusion of each of the set U, L and $\mathbf{0}$ can be tested easily. For example, U is included iff there exists a vector along p_i in the positive direction in one of the locations reachable from the given location.

Computing reachability by location executions.

Given p_i and q, we want to compute the set of all points reached by finite executions starting from some point in p_i at location q which remain within $R = p_i \cup R_{i,i-1} \cup R_{i,i+1}$ and in location q all the time. Note that R is an open set and p_i is a convex subset of R. So, the problem reduces to Problem 2, with $\mathcal{H}[Q \to \{q\}]$.

Determining edges which are bounded in time.

It is straightforward to write a first order logic formula over $(\mathbb{R}, 0, <, +)$ which determines if an edge $((q_1, p_i), (q_2, p_i))$ with $W_{max}(e) = 0$ is bounded in time, since we can not only compute which point is reachable from which but also the time taken by the execution, as given by the predicate $Reach_{\mathcal{H},R,r_1,r_2,q_1,q_2}(v_1, v_2, t)$.

Max and Min Scaling.

First, we prove a property which says that the set of scalings associated with any point on a line p_i are the same.

PROPOSITION 9. *Let p_i be a line, S a neighbor of p_i and $R \in Between(p_i, S)$. Let q be a location. Then for every $v_1, v_2 \in p_i$,*

1. *if there exists an element execution η_1 from v_1 with respect to p_i, S and R, then there exists an element execution η_2 with respect to p_i, S and R with $Scaling(\eta_1) = Scaling(\eta_2)$, and*
2. *if there exists a location execution η_1 from v_1 with respect to q and p_i, then there exists a location execution η_2 with respect to q and p_i such that $Scaling(\eta_1) = Scaling(\eta_2)$.*

The above lemma states that *MaxSclE*, *MinSclE*, *MaxSclL* and *MinSclL* can be computed by focusing on a point v_i on the line p_i. To compute *MaxSclE* and *MinSclE*, we compute the reachability by element execution with a fixed point v_i as the initial state and compute the maximum and minimum scaling from the set (by constructing a formula with one free variable for max or min using the predicate *Reach* and performing quantifier elimination). Similarly, to compute *MaxSclL* and *MinSclL*, we compute reachable sets with respect to location executions and find the maximum and minimum scaling.

7. UNDECIDABILITY OF STABILITY OF PCD IN 5 DIMENSIONS

In this section, we show that Lyapunov and asymptotic stability are undecidable for rectangular hybrid systems. In particular, we show that for a subclass of rectangular hybrid systems called piecewise constant derivative systems, Lyapunov and asymptotic stability are undecidable in 5 dimensions.

Piecewise constant derivatives is a subclass of rectangular hybrid systems with the additional constraint that the invariants in distinct locations are disjoint and the guard associated with a transition is the common boundary, if any, of the closure of the invariants associated with the two locations in the transition. Further, and the flow associated with every location is a singleton set. We will use a simpler notation to represent such systems as defined below. A *piecewise constant derivative* (*PCD*) $\mathcal{H}$ of dimension d is a triple $(\mathcal{P}, \varphi, P_0)$, where $\mathcal{P}$ is a partition of $\mathbb{R}^d$ into convex polyhedral sets, $\varphi : \mathcal{P} \to \mathbb{R}^d$ associates a vector with each region of the partition, and P_0 is an initial region.

In [1], Asarin, Maler and Pnueli show that a pushdown automaton with 2 stacks (2-*PDA*) can be simulated by a 3-dimensional *PCD*, such that the control state reachability problem of 2-*PDA* is equivalent to "point-to-point" reachability of 3-dimensional *PCD*, where point-to-point reachability is the problem of deciding, given points v_0 and v_f in the state space of the *PCD*, if there exists an execution from v_0 to v_f. Since the control state reachability problem for 2-*PDA* is undecidable, the point-to-point reachability for 3-dimensional *PCD* is undecidable as well. Below we highlight some of the properties of the *PCD*s which are output by the reduction.

1. The initial partition P_0 is given by the initial vector $\{v_0\}$.
2. Since *PCD* are deterministic, there exists a unique maximal execution starting from the initial state v_0. If the 2-PDA is halting, then the unique maximal initial execution of the *PCD* is incomplete and reaches v_f in finite time, otherwise it has a complete execution which does not reach v_f.
3. The reachable state space of the *PCD* is bounded, that is, the set of points reached by initial executions is contained within a bounded region.

Below we summarize the undecidability result for *PCD* in [1].

PROPOSITION 10 ([1]). *The problem of point-to-point reachability from a point v_0 to a point v_f for the class of 3-dimensional PCD, whose reachable state space is bounded and in which the unique maximal trajectory from v_0 is either incomplete and reaches v_f, or is complete and does not reach v_f, is undecidable.*

We reduce the above problem to the problem of deciding Lyapunov and asymptotic stability. We take a *PCD* $\mathcal{H}_3$ of the above form and construct a 5-dimensional *PCD* such that point-to-point reachability of $\mathcal{H}_3$ is equivalent to Lyapunov (asymptotic) stability of $\mathcal{H}_5$. This exhibits the undecidability of Lyapunov and asymptotic stability for *PCD* in 5 dimensions.

We explain the construction from $\mathcal{H}_3$ to $\mathcal{H}_5$. First, we construct a 4-dimensional *PCD* $\mathcal{H}_4$. Intuitively, $\mathcal{H}_4$ is obtained by stacking uncountable many copies of a scaled version of $\mathcal{H}_3$ in a fourth dimension. That is, for each value α of the fourth dimension, we place at α a copy of $\mathcal{H}_3$ scaled by a factor of α. Concretely, we replace each element P of the partition $\mathcal{P}_3$ associated with $\mathcal{H}_3$, by a 4 dimensional polyhedral set as follows. We can assume without loss of generality that P is expressed as the conjunction of finitely many linear constraints of the form $ax+by+cz \sim d$ for some $a, b, c, d \in \mathbb{R}$ and $\sim \in \{<, \leq\}$. We obtain the 4 dimensional *PCD* P' by replacing each constraint $ax + by + cz \sim d$ in the variables x, y, z by the constraint $ax + by + cz \sim hd$ in the variables x, y, z and h. The vector associated with the element P' in $\mathcal{H}_4$ is the extension of the vector associated with P by the value 0 in the fourth dimension.

Remark. Observe that the only points reachable from each other in $\mathcal{H}_4$ are those whose fourth components (corresponding to the fourth dimension) have the same value. Note that if $y \in \mathbb{R}^3$ is reached from $x \in \mathbb{R}^3$ in time t in $\mathcal{H}$, then for every h, $(hy, h) \in \mathbb{R}^4$ is reached in time th from $(hx, h) \in \mathbb{R}^4$ in $\mathcal{H}_4$.

The last dimension we add corresponds to time. We add a variable which is a clock, that is, evolves at rate 1. More precisely, every partition in $\mathcal{H}_5$ is obtained by interpreting the constraints of the corresponding partition in $\mathcal{H}_4$ as a constraint over 5 variables, where the coefficient associated with the 5-th variable is always 0. The vector associated with a partition is extended in the 5-th dimension by the value 1. The initial partition of $\mathcal{H}_5$ is the set of points $\{(v_0, h, 0) \mid h \geq 0\}$.

Next, we present a proof of the correctness of the reduction.

THEOREM 4. *The following are equivalent:*

1. *v_f is reachable from v_0 in $\mathcal{H}_3$.*
2. *$\mathcal{H}_5$ is Lyapunov stable.*
3. *$\mathcal{H}_5$ is asymptotically stable.*

PROOF. (1) to (2): We show that if v_f is reachable from v_0, then $\mathcal{H}_5$ is Lyapunov stable. Let us fix an $\epsilon > 0$. Since the reachable state space of $\mathcal{H}_3$ is bounded, assume that it is contained in $B_d(\mathbf{0})$ for some $d > 0$. Then we know that all the points starting with value h for the 4-th variable in $\mathcal{H}_4$ are contained within a dh-ball around $(\mathbf{0}, h)$, or equivalently in a $\sqrt{d^2h^2 + h^2}$-ball around the origin. Since every execution of $\mathcal{H}_3$ is a prefix of the maximal initial execution η which is incomplete, the maximum time elapsed to reach v_f from v_0 is bounded by the domain of σ, say, T. Observe that any initial execution of $\mathcal{H}_4$ starting at (v_0, h) and ending at (v_f, h) takes time at most Th, since the flows evolve at the same rate as in $\mathcal{H}_3$ but the regions are scaled by a factor of h. Therefore, the execution starting from $(v_0, h, 0)$ in $\mathcal{H}_5$ at any time is in a $\sqrt{d^2h^2 + h^2 + T^2h^2}$-ball around $\mathbf{0}$. Choosing $\delta < \epsilon/\sqrt{d^2 + T^2 + 1}$, ensures that every point starting at $(v_0, \delta', 0)$ for some $\delta' \leq \delta$ remains within the ϵ-ball. But all the points starting within a δ-ball are a subset of the points in $\{(v_0, \delta', 0) \mid \delta' \leq \delta\}$. Hence, every point starting within the δ-ball around the origin remains within an ϵ-ball around the origin. Therefore, $\mathcal{H}_5$ is Lyapunov stable.

(1) to (3): (1) implies that $\mathcal{H}_5$ is Lyapunov stable. But then the system does not have any complete executions, therefore the system is trivially asymptotically stable.

(2) to (1): Next, we show that if $\mathcal{H}_5$ is Lyapunov stable, then v_f is reachable from v_0 in $\mathcal{H}_3$. Suppose not. Then the execution η of $\mathcal{H}_3$ is complete. So the corresponding execution starting at any initial point in $\mathcal{H}_5$ evolves unboundedly in the 5-th dimension, which is time. This implies that $\mathcal{H}_5$ is not Lyapunov stable, a contradiction to our assumption.

(3) to (2): If follows from the definition of asymptotic stability. This completes the proof. □

8. CONCLUSIONS

We proved two main results. First we showed that checking (Lyapunov and asymptotic) stability of planar rectangular switched hybrid systems is decidable. Second, we showed that the same problem is undecidable for systems in 5 dimensions. There are a number of questions left open by our investigations. First, there are a couple of questions left unresolved that are intimately tied to our paper: can we reduce the dimensionality gap between our decidability and undecidability results for rectangular switched systems; and, is reachability decidable for planar rectangular switched systems. Next, more generally, one would like to study the stability problem for other classes of hybrid systems, and in particular, identify other decidable subclasses. Finally, one would like to understand the formal relationship between reachability and stability from a computational standpoint: can one of these problems be reduced to the other in general or are these problems computationally incomparable.

9. ACKNOWLEDGEMENT

This work was partially supported by a postdoctoral fellowship from the Center for Mathematics of Information, California Institute of Technology, USA and by the US National Science Foundation grants CCF 1016989 and CNS 1016791.

10. REFERENCES

[1] E. Asarin, O. Maler, and A. Pnueli. Reachability analysis of dynamical systems have piecewise-constant derivatives. *Theoretical Computer Science*, 138:35–66, 1995.

[2] V. Blondel, O. Bournez, P. Koiran, and J. Tsitsiklis. The stability of saturated linear dynamical systems is undecidable. *Journal of Computer and System Science*, 62:442–462, 2001.

[3] V. Blondel and J. Tsitsiklis. Overview of complexity and decidability results for three classes of elementary nonlinear systems. In *Proeedings of Learning, Control and Hybrid Systems*, pages 46–58, 1998.

[4] V. Blondel and J. Tsitsiklis. Complexity of stability and controllability of elementary hybrid systems. *Automatica*, 35:479–489, 1999.

[5] M. Branicky. Multiple Lyapunov functions and other analysis tools for switched and hybrid systems. *IEEE Transactions on Automatic Control*, 43:475–482, 1998.

[6] M. Branicky. Stability of hybrid systems. In H. Unbehauen, editor, *Encylopedia of Life Support Systems*, volume Theme 6.43:Control Sytems, Robotics and Automation, chapter Article 6.43.28.3. UNESCO Publishing, 2004.

[7] G. Davrazos and N. Koussoulas. A review of stability results for switched and hybrid systems. In *Proceedings of the Mediterranean Conference on Control*, 2001.

[8] P. Duggirala and S. Mitra. Abstraction Refinement for stability. In *International Conference on Cyber-Physical Systems*, pages 22–31, 2011.

[9] P. Duggirala and S. Mitra. Lyapunov abstractions for inevitability of hybrid systems. In *ACM Conference on Hybrid Systems: Computation and Control*, pages 115–124, 2012.

[10] T. A. Henzinger. The Theory of Hybrid Automata. In *Logic In Computer Science*, pages 278–292, 1996.

[11] D. Liberzon. *Switching in Systems and Control.* Birkhauser, 2003.

[12] A. Podelski and S. Wagner. Model checking of hybrid systems: From reachability towards stability. In *ACM Conference on Hybrid Systems: Computation and Control*, pages 507–521, 2006.

[13] A. Podelski and S. Wagner. A sound and complete proof rule for region stability of hybrid systems. In *ACM Conference on Hybrid Systems: Computation and Control*, pages 750–753, 2007.

[14] P. Prabhakar, V. Vladimerou, M. Viswanathan, and G. Dullerud. A decidable class of planar linear hybrid systems. In *ACM Conference on Hybrid Systems: Computation and Control*, pages 401–414, 2008.

[15] A. van der Schaft and H. Schumacher. *An introduction to hybrid dynamical systems.* Springer, 2000.

Lyapunov Analysis of Rigid Body Systems with Impacts and Friction via Sums-of-Squares

Michael Posa
Massachusetts Institute of Technology
77 Massachusetts Avenue
Cambridge, MA, USA
mposa@mit.edu

Mark Tobenkin
Massachusetts Institute of Technology
77 Massachusetts Avenue
Cambridge, MA, USA
mmt@mit.edu

Russ Tedrake
Massachusetts Institute of Technology
77 Massachusetts Avenue
Cambridge, MA, USA
russt@mit.edu

ABSTRACT

Many critical tasks in robotics, such as locomotion or manipulation, involve collisions between a rigid body and the environment or between multiple bodies. Sums-of-squares (SOS) based methods for numerical computation of Lyapunov certificates are a powerful tool for analyzing the stability of continuous nonlinear systems, which can play a powerful role in motion planning and control design. Here, we present a method for applying sums-of-squares verification to rigid bodies with Coulomb friction undergoing discontinuous, inelastic impact events. The proposed algorithm explicitly generates Lyapunov certificates for stability, positive invariance, and reachability over admissible (non-penetrating) states and contact forces. We leverage the complementarity formulation of contact, which naturally generates the semialgebraic constraints that define this admissible region. The approach is demonstrated on multiple robotics examples, including simple models of a walking robot and a perching aircraft.

Categories and Subject Descriptors

I.2.8 [**Computing Methodologies**]: Problem Solving, Control Methods, and Search—*control theory*; I.1.2 [**Computing Methodologies**]: Algorithms—*algebraic algorithms*; I.2.9 [**Computing Methodologies**]: Robotics—*kinematics and dynamics, manipulators*

Keywords

Lyapunov analysis and stability verification, rigid body dynamics with impacts and friction, sums-of-squares

1. INTRODUCTION

Many tasks in robotics require making and breaking contact with objects in the robot's environment. For highly dynamic tasks, such as walking [7, 35], and perching [8], this drives the need for control design techniques capable of handling impulsive dynamics and realistic friction models. Recent work has demonstrated that, for smooth nonlinear systems, techniques for stability verification can play a pivotal role in incremental motion planning and control design strategies, [30], and direct optimization over feedback laws [18]. Motivated by these developments, this paper presents a numerical approach for analyzing questions of stability, invariance, and reachability for rigid body systems subject to inelastic collisions and friction.

Our central observation is that the complementarity framework for modeling such systems (e.g. [4, 27, 12]) is compatible with recent advances in polynomial optimization, in particular sums-of-squares (SOS) optimization [23]. For a polynomial to be non-negative, it is sufficient that it be expressible as the sum of squares of polynomials. Optimizing over such polynomials can be formulated as a semidefinite program (SDP), a form of convex optimization. In the controls community, SOS has been widely applied, particularly in automating Lyapunov analysis of polynomial dynamical systems [10]. A major advantage of the complementarity framework over traditional approaches using hybrid automata is that non-smooth and impulsive dynamics can be expressed without suffering from the combinatorial explosion of "modes" resulting from distinct combinations of contacts. This approach also naturally encompasses instances of Zeno phenomena, which pose a challenge in some frameworks. Instead, the dynamics are described by conditions expressed jointly in the coordinates, velocities and feasible contact forces, the number of which grows linearly in the number of contact points.

This work demonstrates how Lyapunov analysis can be performed by testing sufficient semialgebraic conditions in these variables. We present algorithms based on testing these semialgebraic conditions using SOS optimization. Procedures for maximizing the size of regions of positive invariance or guaranteed regions of "safety" by solving a convex program (or a sequence of such programs) are provided. We apply the algorithms detailed here on multiple passive systems of interest to the robotics community.

1.1 Related Work

In this paper, we adopt the complementarity formulation for modeling rigid bodies with frictional impacts. This framework and its historical development are reviewed in [27] and a more comprehensive bibliography and discussion are provided in [4]. Notions of equilibria, stability, and extensions of Lyapunov analysis to such systems are presented in [4] and [12] (see also the related article [13]). Section 2 will summarize the aspects of these works used in this paper.

An alternative formalism for modeling and control of non-smooth mechanical systems is that of hybrid automata [3]. Example applications of this framework to the control and analysis of hybrid mechanical systems can be found in [17], [26], and [21]. A number of numerical techniques have been presented for addressing verification, stability, and control design of general hybrid automata. These include methods based on approximate solutions of Hamilton-Jacobi equations, [31], and the construction of discrete abstractions [1]. SOS optimization has also been used to find polynomial barrier certificates, [24], as well as Lyapunov functions about equilibria, [22], and of transverse dynamics about hybrid limit cycles [19].

The fundamental difference between the approaches in [24], [22], and [19] and in this work stems from the choice of modeling framework. For a general rigid body with m possible contact points, 2^m different hybrid modes are possible, each with a distinct associated differential inclusion. By contrast, this work simultaneously reasons over the set of system trajectories and feasible contact force trajectories.

2. BACKGROUND

Here, we introduce a notion of solutions to discontinuous rigid body systems and describe the friction and impact laws used in this paper.

2.1 Measure Differential Inclusions

We consider systems whose state is given by a set of generalized coordinates $q \in \mathbb{R}^n$ and generalized velocities $v \in \mathbb{R}^n$ and we let $x = [q^T\ v^T]^T$. For mechanical systems, $q(t)$ will evolve continuously whereas $v(t)$ may have discontinuities due to impacts which present an obstacle to applying classical Lyapunov analysis.

Recently, a number of authors have provided extensions of Lyapunov analysis to discontinuous dynamics using the framework of *measure differential inclusions* (MDIs), see [12] Ch. 4 or [4] for more details. This framework addresses both the discontinuities and non-smoothness of the system evolution that arise from impacts and standard friction force laws. We provide a high-level overview of MDIs here, focused on autonomous Lagrangian mechanical systems.

A solution of a measure differential inclusion will be taken to be a pair of functions, $q(t)$ and $v(t)$, such that $q(t)$ is absolutely continuous and $v(t)$ is of locally bounded variation, allowing for countably many discontinuities. The left and right limits of $v(t)$, denoted $v^+(t)$ and $v^-(t)$, are guaranteed to exist and we require that solutions satisfy:

$$q(t) - q(t_0) = \int_{t_0}^{t} v(\tau)d\tau, \tag{1}$$

$$v^+(t) - v^-(t_0) = \int_{t_0}^{t} \dot{v}(\tau)d\tau + \int_{t_0}^{t} v^+(\tau) - v^-(\tau)d\eta(\tau). \tag{2}$$

Here $\dot{v}(t)$ is an integrable function and η is a sum of Dirac measures centered at times $\{t_k\}_{k=1}^{\infty}$, which model the continuous evolution of and jumps in the velocity respectively. By assumption, $v(t)$ has no singular part (see [12], Ch. 3). To describe the dynamics we must give rules for specifying legal values of $\dot{v}(t)$, the locations $\{t_k\}_{k=1}^{\infty}$, and the values of jumps $v^+(t_k) - v^-(t_k)$. Specific rules are given in the next section, but we briefly note the following: we require $\dot{v}(t) \in \mathfrak{F}(q(t), v(t))$ for almost all t, where $\mathfrak{F}(q, v)$ is a function which provides a *set* of possible values. This use of differential inclusions (i.e. set-valued laws) instead of equations is a standard approach for addressing Coulomb friction. Similarly, the value of jumps will be drawn from a set which generally depends on $q(t)$ and $v^-(t)$. The locations of impacts will be defined implicitly by the locations where $v^+(t)$ and $v^-(t)$ disagree. Finally, we take $v(t)$ to be undefined at points of discontinuity.

Our problems center around systems where solutions must lie in an *admissible set*, $\mathcal{A}$, defined by a finite family of functions $\phi_i : \mathbb{R}^n \to \mathbb{R}$:

$$\mathcal{A} = \{(q, v) \in \mathbb{R}^{2n} \mid \phi_i(q) \geq 0\, \forall\, i \in \{1, \dots, m\}\}. \tag{3}$$

Here the functions $\phi_i(\cdot)$ represent non-penetration constraints for the rigid body. We will focus on MDIs which are *consistent* (see [12], Ch. 4).

DEFINITION 1. *A measure differential inclusion is consistent if every solution defined for t_0 is defined for almost all $t \geq t_0$, all such solutions remain within $\mathcal{A}$, and for each $x_0 \in \mathcal{A}$ there exists at least one solution passing through x_0.*

An *equilibrium point* for such a system is defined as any point $x_0 \in \mathcal{A}$ such that such that $x(t) = x_0$ is a solution. In general we do not expect to have unique solutions from the systems covered by this work, particularly for models with dry friction ([4, 27]).

For systems governed by MDIs, natural extensions exist to the notions of stability and positive invariance ([12] Ch. 6). An equilibrium point $x_0 \in \mathcal{A}$ of a consistent MDI is *stable in the sense of Lyapunov* if, for each $\epsilon > 0$, there exists a $\delta > 0$ such that every solution $x(t)$ satisfying $|x_0 - x(t_0)| < \delta$ satisfies $|x_0 - x(t)| < \epsilon$ for almost all $t \geq t_0$. A set $B \subset \mathcal{A}$ is positively invariant if each solution $x(t)$ satisfying $x^-(t_0) \in B$ satisfies $x(t) \in B$ for almost all $t \geq t_0$.

In order to apply Lyapunov analysis to MDIs we make note of the following fact (see [12] Proposition 6.3): if $V : D \to \mathbb{R}$ is a continuously differentiable function on a compact set $D \subset \mathbb{R}^{2n}$, and $x(t)$ is of locally bounded variation, then $V(x(t))$ is of locally bounded variation and

$$\begin{aligned} V(x^+(t)) - V(x^-(t_0)) = &\int_{t_0}^{t} \frac{\partial V}{\partial x}\dot{x}(\tau)d\tau \\ &+ \int_{t_0}^{t} V(x^+(\tau)) - V(x^-(\tau))d\eta(\tau), \end{aligned} \tag{4}$$

where $\dot{x}(t) = [v(t)^T\ \dot{v}(t)^T]^T$, as in (1) and (2). For the remainder of this paper, when we write $dV(x) \leq 0$ for certain $x \in A$ we mean that, for any solution satisfying $x^-(t) = x$, we have $\frac{\partial V}{\partial x}\dot{x}(t) \leq 0$ and $V(x^+(t)) - V(x^-(t)) \leq 0$.

2.2 Rigid Body Dynamics

Many robotic systems are appropriately modeled as a set of rigid links connected through some combination of joints. The continuous dynamics of these rigid body systems subject to frictional forces are well described by the manipulator equations

$$H(q)\dot{v} + C(q, v) = J(q)^T\lambda_N + J_f(q)^T\lambda_T, \tag{5}$$

where the dependence of q, v, λ_N, and λ_T on time has been suppressed for clarity. Here, $H(q)$ is the inertia matrix and $C(q, v)$ is the combined Coriolis and gravitational terms. In this paper, we primarily consider planar models, though the extension to three dimensions is straightforward. For m potential contacts, the constraint forces are split into those normal to the contacts, $\lambda_N \in \mathbb{R}^m$, $\lambda_N \geq 0$, and the frictional forces tangential to the contact surface, $\lambda_T \in \mathbb{R}^m$. We will also write λ to be the stacked vector $\begin{bmatrix}\lambda_N^T & \lambda_T^T\end{bmatrix}^T$. The Jacobian matrices $J(q), J_f(q) \in \mathbb{R}^{m\times n}$ project the normal and frictional contact forces into joint coordinates. We will also refer to $J_i(q)$ and $J_{f,i}(q)$ as the ith row of the Jacobians associated with the particular contact forces $\lambda_{N,i}$ and $\lambda_{T,i}$.

We will focus on the dynamics of a planar rigid body contacting a fixed environment (such as the ground or a wall) at a finite number of contact points. Here, the velocity of the ith contact point has components $J_i v(t)$ and $J_{f,i}v(t)$ normal to and tangential to the contact surface. We use a simple Coulomb friction model to represent our contact forces:

$$J_{f,i}v(t) = 0 \Rightarrow |\lambda_{T,i}(t)| \leq \mu\lambda_{N,i}(t),$$
$$J_{f,i}v(t) \neq 0 \Rightarrow \lambda_{T,i}(t) = -\text{sgn}(J_{f,i}v(t))\mu\lambda_{N,i}(t).$$

When the tangential velocity vanishes, $\lambda_{T,i}$ can take on any value within the friction cone. If the contact point is sliding, then $\lambda_{T,i}$ directly opposes the direction of slip.

2.3 Inelastic Collisions with Friction

Rigid body impacts are often modeled as instantaneous events where an impulse causes a discontinuity in the velocity. Impacts occur when there is contact, $\phi_i(q(t)) = 0$, and when the velocity normal to the contact would cause penetration, $J_i v^-(t) < 0$. As with the continuous case, we let $\Lambda_N, \Lambda_T \in \mathbb{R}^m$ be the normal and tangential impulses. Derived from the manipulator equations, the pre- and post-impact velocities for a collision at the ith contact point are related by $v^+(t) = v^-(t) + H^{-1}(J_i^T\Lambda_{N,i} + J_{f,i}^T\Lambda_{T,i})$. In the special case of frictionless inelastic collisions, we observe that the inelastic condition is $J_i v^+(t) = 0$ and so we can explicitly solve for the normal impulse and post-impact state:

$$\Lambda_{N,i} = -(J_i H^{-1} J_i^T)^{-1} J_i v^-(t), \tag{6}$$
$$v^+(t) = \left(I - H^{-1}J_i^T(J_i H^{-1}J_i^T)^{-1}J_i\right)v^-(t). \tag{7}$$

However, when considering Coulomb friction, we have no explicit formula. We provide a model for inelastic impacts into a single contact surface with friction. We additionally assume that simultaneous collisions, a well studied problem in both simulation and analysis[5], can be modeled as (potentially non-unique) sequences of single surface impacts.

To model frictional collisions, we adopt an impact law first proposed by Routh[25] that is described in detail in [34, 2]. Originally a graphical approach, this method constructs a path in impulse space that will fit naturally into our Lyapunov analysis. To briefly summarize Routh's technique for computing the net impulses and the post-impact state:

1. Monotonically increase the normal impulse $\Lambda_{N,i}$.
2. Increment the tangential impulse $\Lambda_{T,i}$ according to the friction law:
$$J_{f,i}\bar{v} = 0 \Rightarrow |\Lambda'_{T,i}| \leq \mu\Lambda'_{N,i},$$
$$J_{f,i}\bar{v} \neq 0 \Rightarrow \Lambda'_{T,i} = -\text{sgn}(J_{f,i}\bar{v})\mu\Lambda'_{N,i},$$
where $\bar{v} = v^-(t) + H^{-1}(J_i^T\Lambda_{N,i} + J_{f,i}^T\Lambda_{T,i})$.
3. Terminate when the normal contact velocity vanishes, $J_i\bar{v} = 0$, and take $v^+(t) = \bar{v}$.

This method amounts to following a piecewise linear path in the impulse space where the slopes of the impulses are $\Lambda'_{N,i}$ and $\Lambda'_{T,i}$. Along each linear section, these slopes must satisfy the Coulomb friction constraints. Solutions may transition from sliding to sticking and vice versa and the direction of slip may even reverse as a result of each impact.

3. CONDITIONS FOR STABILITY

The highly structured nature of rigid body dynamics and the complementarity formulation of contact allow us to construct semialgebraic conditions for stability in the sense of Lyapunov and positive invariance.

3.1 Lyapunov Conditions for MDIs

We begin by describing sufficient conditions for stability in the sense of Lyapunov and positive invariance stated in terms of Lyapunov functions. Say that a function $\alpha : [0,\infty) \to [0,\infty)$ belongs to class $\mathcal{K}$ if it is strictly increasing and $\alpha(0) = 0$. The following theorem is adapted from [12], Theorem 6.23, and stated without proof.

THEOREM 1. *Let x_0 be an equilibrium point for a consistent MDI and let $V : \mathbb{R}^{2n} \to \mathbb{R}$ be a continuously differentiable function. If there exists a neighborhood U of x_0 and a class $\mathcal{K}$ function α such that $x \in U \cap \mathcal{A}$ implies $\mathrm{d}V \leq 0$ and $V(x) \geq \alpha(\|x - x_0\|)$ then x_0 is stable in the sense of Lyapunov.*

For a candidate Lyapunov function $V(q,v)$, define the c-sublevel set

$$\Omega_c = \{(q,v) \in \mathbb{R}^{2n} \mid V(q,v) < c\}.$$

For a system whose solutions are continuous functions of time, $\mathrm{d}V \leq 0$ on $\Omega_c \cap \mathcal{A}$ would be sufficient to show that each connected component of $\Omega_c \cap \mathcal{A}$ is positively invariant. However, where $v(t)$ is discontinuous, the pre- and post-impact states may be in disjoint connected components. The following lemma provides stronger conditions which guarantee positive invariance of such a connected component. The proof can be found in the Appendix.

LEMMA 2. *Let $V : \mathbb{R}^{2n} \to \mathbb{R}$ be a continuously differentiable function, and $\mathcal{C}$ be a connected component of $\Omega_c \cap \mathcal{A}$ with $\mathrm{d}V \leq 0$ on $\mathcal{C}$. Then $\mathcal{C}$ is positively invariant if, for every solution $(q(t), v(t))$ with $(q(t), v^-(t)) \in \mathcal{C}$, there exists a path $\bar{v}(s)$ from $v^-(t)$ to $v^+(t)$ such that $V(q(t), \bar{v}(s))$ is a non-increasing function of s.*

Lemma 2 holds for general Lyapunov functions and systems where the unilateral constraints are defined by the generalized positions q. While we are generally interested in systems with friction, we briefly consider the special structure implied by rigid body dynamics and frictionless, inelastic collisions. The following lemma, whose proof is in the Appendix, shows that for such systems, and for V a convex function in v for each fixed q, the above sufficient condition for positive invariance is also necessary. In particular, no additional conservatism is added by requiring $\bar{v}$ to be the chord connecting $v^-(t)$ to $v^+(t)$.

LEMMA 3. *For a rigid body system undergoing frictionless, inelastic collisions, let $V : \mathbb{R}^{2n} \to \mathbb{R}$ be a continuously differentiable function, and $\mathcal{C}$ be a connected component of $\Omega_c \cap \mathcal{A}$ such that $\mathrm{d}V \leq 0$ on $\mathcal{C}$. If V is convex in v for each fixed q the following conditions are equivalent:*

(i) $\mathcal{C}$ is positively invariant.

(ii) For each solution $(q(t), v(t))$, when $(q(t), v^-(t)) \in \mathcal{C}$, $V(q(t), \bar{v}(s))$ is non-increasing along the path $\bar{v}(s) = sv^+(t) + (1-s)v^-(t)$ for $s \in [0,1]$

The proof for Lemma 3 fails for contacts with friction as a result of the piecewise linear resolution of collisions. Solutions which transition from slip to stick or where the direction of slip reverses may have intermediate points during the Routh solution which leave Ω_c. The frictionless assumption does, however, cover a class of interesting problems including collisions due to impacting hard joint limits.

3.2 Conditions For Complementarity Systems

We now focus on Lagrangian mechanical systems with impacts and friction described by complementarity conditions. This section contains sufficient conditions for demonstrating $dV \le 0$ and the additional constraint on jump discontinuities in the statement of Lemma 2. We partition the admissible set, $\mathcal{A}$, into $\mathcal{F}$ and $\mathcal{A} \setminus \mathcal{F}$, where $\mathcal{F} = \bigcap_{i=1}^{m} \mathcal{F}_i$ and

$$\mathcal{F}_i = \{(q,v) | \phi_i(q) > 0\} \cup \{(q,v) | \phi_i(q) = 0,\ J_i(q)v > 0\}.$$

Intuitively, $\mathcal{F}$ is the region where there is no contact or all contacts are being broken which forces the contact forces to vanish. On $\mathcal{F}$, we know that $\lambda = \Lambda = 0$ and so that $dV \le 0$ is equivalent to

$$\nabla V(q,v)^T \begin{bmatrix} v \\ -H(q)^{-1}C(q,v) \end{bmatrix} \le 0. \tag{8}$$

On $\mathcal{A} \setminus \mathcal{F}$, there may be frictional forces and collisions. The condition on continuous state evolution is simply

$$\nabla V(q,v)^T \begin{bmatrix} v \\ \dot{v} \end{bmatrix} \le 0, \tag{9}$$

where (5) gives an expression for $\dot{v}$. We now provide conditions on V and a path for each jump discontinuity such that the requirements of Lemma 2 are satisfied. Furthermore, this path ensures that the jump conditions for $dV \le 0$ are met. We explicitly construct this path between pre- and post-impact states. Recall that the Routh method of Section 2.3 constructs a piecewise linear path through the space of contact impulses. We take $\bar{v}(s)$ to be the velocities defined in step 2 of the Routh method, where s is the path parameter varying the impulses. As $\bar{v}$ depends linearly on the forces, we can find the derivative of $\bar{v}$ with respect to s, defined along each path segment:

$$\frac{d\bar{v}(s)}{ds} = H^{-1}(J_i(q)^T \Lambda'_{N,i} + J_{f,i}(q)^T \Lambda'_{T,i}),$$

where $\Lambda'_{N,i}$ and $\Lambda'_{T,i}$ satisfy the Coulomb friction conditions. To show V is non-increasing along the path, we require

$$\left.\frac{\partial V(q,v)}{\partial v}\right|_{(q,\bar{v}(s))} H^{-1}(J_i^T \Lambda'_{N,i} + J_{f,i}^T \Lambda'_{T,i}) \le 0. \tag{10}$$

Since the Routh method for resolving impacts is memoryless, any point $(q, \bar{v}(s))$ is also a possible pre-impact state. So the set of all possible $(q, \bar{v}(s))$ is precisely equivalent to $\mathcal{A} \setminus \mathcal{F}$ and it is equivalent to enforce (10) for all $(q,v) \in \mathcal{A} \setminus \mathcal{F}$ instead of along potential paths. This constraint must hold for all i, so we construct a single condition that encompasses all contact points:

$$\frac{\partial V(q,v)}{\partial v} H(q)^{-1}(J(q)^T \Lambda'_N + J_f(q)^T \Lambda'_T) \le 0. \tag{11}$$

Both (9) and (11) are defined in terms of permissible contact forces λ and slopes of the impulse path Λ' when resolving a collisions. Complementarity conditions can be used to describe the set of feasible contact normal forces [4, 27, 12]:

$$\phi_i(q), \lambda_{N,i} \ge 0, \tag{12}$$
$$\phi_i(q)\lambda_{N,i} = 0, \tag{13}$$
$$(J_i(q)v)\lambda_{N,i} \le 0. \tag{14}$$

These constraints prohibit contact at a distance and ensure that the contact normal is a compressive and dissipative force. Note that (12-14) apply not only to the continuous force λ, but they also describe the set of feasible impulse slopes Λ'. Observing that the friction constraints on both are also identical, we write the additional set of constraints:

$$(J_{f,i}(q)v)\lambda_{T,i} \le 0, \tag{15}$$
$$\mu^2 \lambda_{N,i}^2 - \lambda_{T,i}^2 \ge 0, \tag{16}$$
$$(\mu^2 \lambda_{N,i}^2 - \lambda_{T,i}^2)(J_{f,i}(q)v) = 0. \tag{17}$$

Here, we diverge from the standard linear complementarity description of Coulomb friction to avoid introducing additional slack variables. However, for any (q,v), this full set of conditions is exactly equivalent to our formulations of both frictional, inelastic collisions and Coulomb friction.

We now have three separate positivity conditions for stability. On $\mathcal{F}$ we have (8) and on $\mathcal{A} \setminus \mathcal{F}$ we have (9) and (11), with the contact forces and impulses subject to (12-17). However, since the conditions on λ and Λ' are identical, observe that (9) is equal to the sum of (8) and (11), so it is a redundant condition. Note that since all of these conditions are continuous and all points in $\mathcal{A} \setminus \mathcal{F}$ are in the closure of $\mathcal{F}$, (8) holding on $\mathcal{F}$ implies that it will hold on all of $\mathcal{A}$. This precise overlap between the constraints on the forces and impulses allows us to restrict our attention to the conditions on (q, v, λ) for the remainder of the paper.

3.3 Semialgebraic Representation

Rigid body dynamics and the manipulator equations offer a great deal of structure that we can exploit to make the problems of control and verification more amenable to algebraic methods. For many rigid body systems, especially those of interest in robotics, trigonometric substitutions can reduce the task of kinematics to an algebraic problem[33]. Concretely, for rotational joints, substituting c_i and s_i for $\cos(q_i)$ and $\sin(q_i)$ respectively, with the constraint that $s_i^2 + c_i^2 = 1$, will often result in polynomial kinematics in c_i and s_i. For simple contact surfaces, the various terms of the manipulator equations and constraints (H, C, B, J, J_f, and ϕ) are also polynomial in c_i, s_i, v and the remaining translational coordinates of q. Several methods can be used to accommodate the appearance of $H(q)^{-1}$ in the conditions of the previous section. First we note that by explicitly introducing an additional variable $\dot{v} \in \mathbb{R}^n$, the condition (5) is algebraic in $\dot{v}, v, \lambda$, the translational components of q and any introduced trigonometric variables. Alternatively, as $H(q)$ is positive definite and polynomial, its inverse is a rational function and we can find equivalent conditions by clearing the denominator. These facts imply that semialgebraic conditions can be posed that are equivalent to those in Section 3.2.

4. APPROACH

For our systems of interest, the Lyapunov conditions in Section 3 amount to non-negativity constraints on polynomials over basic semialgebraic sets. This formulation is amenable to SOS based techniques, which provide certificates that a polynomial can be written as a sum of squares of polynomials, a clearly sufficient condition for non-negativity. Searching over polynomials which satisfy these sufficient conditions can be cast as an SDP, allowing for the application of modern convex optimization tools. For the examples in this paper we use the YALMIP[15, 16] and SPOT[20] toolboxes to generate programs for the semidefinite solver SeDuMi[29]. For a portion of our approach, we exploit bilinear alternation (related to the techniques of DK-iteration [14] and also referred to as coordinate-wise descent). We briefly review these concepts in the Appendix.

4.1 Global Verification

For some dynamic systems, we can verify the Lyapunov conditions over the entire admissible set. Define $\mathcal{D}$ to be the set of all (q, v, λ) that satisfy the complementarity conditions (12-17). Note that this also implies $(q, v) \in \mathcal{A}$. If $(0, 0)$ is an equilibrium of the system, we can then pose the global feasibility SOS program:

$$\begin{aligned} \text{find} \quad & V(q,v) && (18)\\ \text{subj. to} \quad & V(0,0) = 0, \\ & V(q,v) \geq \alpha(||q|| + ||v||) && \text{for } (q,v) \in \mathcal{A}, \\ & \nabla V^T \begin{bmatrix} v \\ \dot{v} \end{bmatrix} \leq 0 && \text{for } (q,v) \in \mathcal{A}, \\ & \frac{\partial V}{\partial v} H^{-1}(J^T \lambda_N + J_f^T \lambda_T) \leq 0 && \text{for } (q,v,\lambda) \in \mathcal{D}, \end{aligned}$$

with $\alpha(\cdot)$ is in class $\mathcal{K}$ (see Section 3.1). SOS allows us to search over a parameterized family of polynomial Lyapunov functions via SDP. By finding such a function, we verify that every sublevel set of V is positively invariant and that the origin is stable in the sense of Lyapunov. This certificate of a nested set of invariant regions is weaker than asymptotic stability but stronger than invariance of a single set.

4.2 Regional Verification

For many problems of interest, we would like to maximize the verifiable region about an equilibrium. Specifically, find a Lyapunov function that maximizes the volume of a connected component $\mathcal{C} \subseteq \Omega_1 \cap \mathcal{A}$, which is positively invariant and, for all $\rho \leq 1$, $\mathcal{C} \cap \Omega_\rho$ is also positively invariant. This leads to the following optimization problem:

$$\begin{aligned} \max_{V} \quad & \text{Volume}(\mathcal{C}) && (19)\\ \text{subj. to} \quad & V(0,0) = 0, \\ & V(q,v) \geq \alpha(||q|| + ||v||) && \text{for } (q,v) \in \mathcal{C}, \\ & \nabla V^T \begin{bmatrix} v \\ \dot{v} \end{bmatrix} \leq 0 && \text{for } (q,v) \in \mathcal{C}, \\ & \frac{\partial V}{\partial v} H^{-1}(J^T \lambda_N + J_f^T \lambda_T) \leq 0 && \text{for } (q,v,\lambda) \in \mathcal{D} \\ & && \text{and } (q,v) \in \mathcal{C}. \end{aligned}$$

As currently posed, this problem is not amenable to convex optimization techniques. It is difficult to directly measure the volume of $\mathcal{C}$ and, as $\mathcal{C}$ is only one connected component of $\Omega_1 \cap \mathcal{A}$, it is not naturally described as a semialgebraic set. We approximate these regions by finding polynomials $g_i(q, v)$ and $g_o(q, v)$ such that their one sublevel sets ($\mathcal{G}_i$ and $\mathcal{G}_0$ resp.) are inner and outer approximations of $\mathcal{C}$, i.e.

$$(\mathcal{G}_i \cap \mathcal{A}) \subseteq \mathcal{C} \subseteq (\mathcal{G}_o \cap \mathcal{A}). \quad (20)$$

By containing $\mathcal{C}$ within the semialgebraic set $\mathcal{G}_o$ and verifying the Lyapunov conditions on $\mathcal{G}_o$, we provide sufficient conditions on $\mathcal{C}$. The inner approximation $\mathcal{G}_i$ is used to estimate the volume of the verified region. In practice, we parameterize g_i and g_o as quadratic forms. For $g_i(q, v) = \begin{bmatrix} q^T & v^T \end{bmatrix} G_i \begin{bmatrix} q^T & v^T \end{bmatrix}^T$, we will use $-\text{Trace}(G_i)$ as a proxy for the volume of $\mathcal{C}$. Given this, we pose the problem:

$$\begin{aligned} \min_{V, G_i, G_o} \quad & \text{Trace}(G_i) && (21)\\ \text{subj. to} \quad & V(0,0) = 0, \quad G_i, G_o \succeq 0, \\ & V(q,v) \geq \alpha(||q|| + ||v||) && \text{for } (q,v) \in \mathcal{A} \cap \mathcal{G}_o, \\ & \nabla V^T \begin{bmatrix} v \\ \dot{v} \end{bmatrix} \leq 0 && \text{for } (q,v) \in \mathcal{A} \cap \mathcal{G}_o, \\ & \frac{\partial V}{\partial v} H^{-1}(J^T \lambda_N + J_f^T \lambda_T) \leq 0 && \text{for } (q,v,\lambda) \in \mathcal{D} \\ & && \text{and } (q,v), \in \mathcal{G}_o, \\ & V(q,v) \geq 1 && \text{for } (q,v) \in \mathcal{A} \\ & && \text{and } g_o(q,v) = 1, \\ & V(q,v) \leq 1 && \text{for } (q,v) \in \mathcal{A} \cap \mathcal{G}_i. \end{aligned}$$

This problem verifies the Lyapunov conditions on the outer approximation $\mathcal{G}_o$ and ensures the containment in (20). It is now posed in the familiar form of an optimization over polynomials that are positive on a basic semialgebraic set. As described in the Appendix, we use a bilinear alternation technique to solve this problem. One of the potentially difficult aspects of this alternation is that we must typically supply an initial feasible Lyapunov candidate. Previous sums-of-squares based methods have used local linearizations of the dynamics to find initial candidates [32, 30], but this approach fails when the dynamics are discontinuous. However, since the passive rigid body dynamics and inelastic collisions are energetically conservative, taking V to be the total energy provides a viable starting point for most mechanical systems. Solutions to (21) are guaranteed to be feasible Lyapunov functions to the original problem (19), although they will generally be suboptimal. This method, however, provides a tractable technique to synthesize useful regional certificates through contact discontinuities.

An alternate approach to bilinear alternations is to fix $\mathcal{G}_o$ and to fix the form of $\mathcal{G}_i$ to within a scalar factor and pose (21) as a feasibility problem. The optimal scaling of G_i can then be found by binary search. Though it only searches over a subset of the solutions to the first formulation, this SDP may be better conditioned numerically for some applications.

4.3 Reachability

The algorithm above for verifying stability and invariance can be easily adapted to address questions of dynamic reachability. For instance, we might wish to determine the largest set of initial conditions such that the infinite horizon reachable set does not intersect some unsafe semialgebraic set $\mathcal{U}$. We pose this problem in a manner similar (21), although here we do not require that V be positive definite:

$$\min_{V,G_i,G_o} \quad \text{Trace}(G_i) \tag{22}$$

$$\begin{aligned}
\text{subj. to} \quad & G_i, G_o \succeq 0, \\
& \nabla V^T \begin{bmatrix} v \\ \dot{v} \end{bmatrix} \leq 0 && \text{for } (q,v) \in \mathcal{A} \cap \mathcal{G}_o, \\
& \frac{\partial V}{\partial v} H^{-1}(J^T \lambda_N + J_f^T \lambda_T) \leq 0 && \text{for } (q,v,\lambda) \in \mathcal{D} \\
& && \text{and } (q,v) \in \mathcal{G}_o, \\
& V(q,v) \geq 1 && \text{for } (q,v) \in \mathcal{A} \\
& && \text{and } g_o(q,v) = 1, \\
& V(q,v) \leq 1 && \text{for } (q,v) \in \mathcal{A} \cap \mathcal{G}_i, \\
& V(q,v) \geq 1 && \text{for } (q,v) \in \mathcal{A} \cap \mathcal{U}.
\end{aligned}$$

The optimization program in (22) verifies that $\mathcal{C}$ is positively invariant and that $\mathcal{C} \cap \mathcal{U} = \emptyset$, so no trajectory that originates in $\mathcal{C}$ can leave the safe region.

5. EXAMPLES

5.1 Bean Bag Toss

We first examine the simple problem of a bean bag, modeled as a planar point mass, colliding with the ground. This example serves to demonstrate the method on a system simple enough where the calculations can be easily verified by hand. With $q = \begin{bmatrix} x & z \end{bmatrix}^T$ and $v = \begin{bmatrix} \dot{x} & \dot{z} \end{bmatrix}^T$, we define $\phi(q) = z$ and the dynamics are given by

$$m\ddot{x} = \lambda_T, \quad m\ddot{z} = -mg + \lambda_N.$$

For this simple system, the dynamics are invariant under x so we consider the equilibrium set where the mass rests on the ground, $\{(x,z) \in \mathbb{R}^2 | z = 0\}$. Choosing our Lyapunov candidate function be equal to the total energy of the system, we will show stability in the sense of Lyapunov and invariance of a series of nested sets. That is, each sublevel set of energy is positively invariant. Substituting $V(q,v) = .5m\dot{x}^2 + .5m\dot{z}^2 + mgz$ and the dynamics into (18), we have the two conditions:

$$-\nabla V^T \begin{bmatrix} v \\ \dot{v} \end{bmatrix} = -mg\dot{z} + mg\dot{z} \geq 0 \text{ for } (q,v,\lambda) \in \mathcal{D}, \tag{23}$$

$$-\frac{\partial V}{\partial v} H(q)^{-1}(J(q)^T \lambda_N + J_f(q)^T \lambda_T) =$$
$$-\dot{x}\lambda_T - \dot{z}\lambda_N \geq 0 \text{ for } (q,v,\lambda) \in \mathcal{D}. \tag{24}$$

The first condition is trivially true. Observing that $Jv = \dot{z}$ and $J_f v = \dot{x}$, we use S-procedure type multipliers to verify the second condition. Generating sums-of-squares multipliers $\sigma_i(q,v,\lambda)$ for the relevant unilateral constraints (14) and (15), replace (24) with

$$-\dot{x}\lambda_T - \dot{z}\lambda_N + \sigma_1 \dot{x}\lambda_T + \sigma_2 \dot{z}\lambda_N \text{ is SOS.} \tag{25}$$

Choosing $\sigma_1 = \sigma_2 = 1$, the equation above vanishes and is trivially non-negative. Thus, we have used our methods to demonstrate the rather obvious fact that every sublevel set of energy will be positively invariant.

Note that there also exists a quartic Lyapunov function that satisfies (18) but where we can additionally verify that $\dot{V} < -\alpha(z^2 + \dot{x}^2 + \dot{z}^2)$ for some class $\mathcal{K}$ function α. Combined with the condition that $dV \leq 0$, this is sufficient to verify asymptotic stability of the equilibrium set, though we do not prove this here. In general, it is difficult to find such Lyapunov functions for discontinuous mechanical systems.

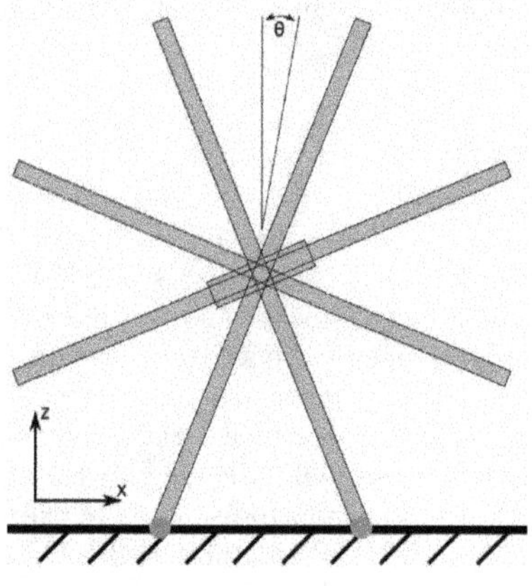

Figure 1: The rimless wheel shown in an equilibrium state, with two feet on the ground. Verified trajectories pass through four possible contact states (no contact, double-support, and both single-support phases).

Table 1: Rimless Wheel Parameters

mass	1 kg
moment of inertia	0.25 kgm^2
center to foot distance	1 m

5.2 Rimless Wheel

The rimless wheel model is a single rigid body composed of a number of equally-spaced spokes about a simple mass. This simple model has been used extensively as a proxy for a passive-dynamic walking robot[6]. Though previous works have primarily been interested in analyzing the limit-cycle behavior of the rimless wheel, here we focus on the stability of a single, static configuration of the system. We allow for frictional contacts between two of the spokes and the ground, highlighted in Fig. 1, and we consider the equilibrium set where both of these spokes rest on a flat ground. We differentiate between resting on these two particular spokes and any other equilibrium state. Trajectories of the rimless wheel that come to rest in the equilibrium set may undergo an infinite number of collisions rocking back and forth between the two feet, in an example of Zeno phenomena.

The planar floating base model of the rimless wheel has three degrees of freedom, $q = \begin{bmatrix} x & z & \theta \end{bmatrix}^T$ and $v = \begin{bmatrix} \dot{x} & \dot{z} & \dot{\theta} \end{bmatrix}^T$. With the trigonometric substitutions $s = \sin(\theta)$ and $c = \cos(\theta)$, the dynamics of the rimless wheel and the contact related elements $\phi_i(q)$, $J_i(q)$, and $J_{f,i}(q)$ are all polynomial functions of the redundant state variables $(x, z, s, c, \dot{x}, \dot{z}, \dot{\theta})$ and the contact forces $(\lambda_{N,i}, \lambda_{T,i})$. As with the point mass example, the dynamics are invariant under x and so the equilibrium set is defined as $\{(x,z,\theta) \in \mathbb{R}^3 | z = 0, \theta = 0\}$. The parameters of the rimless wheel model are given in Table 1.

Taking the total mechanical energy of the system as an initial candidate Lyapunov function, we search for a solution to (21) to find a nested set of invariant regions and verify

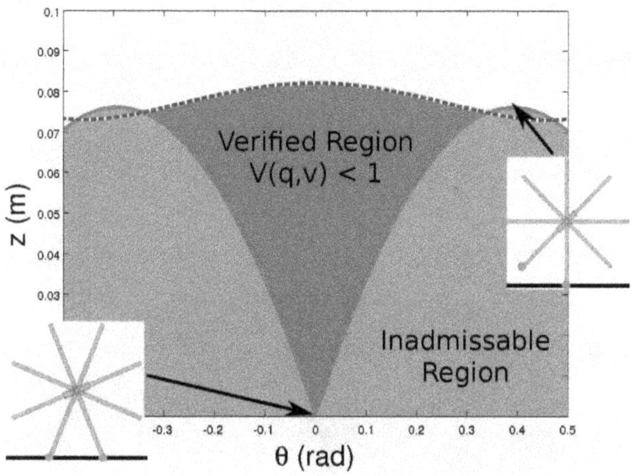

(a)

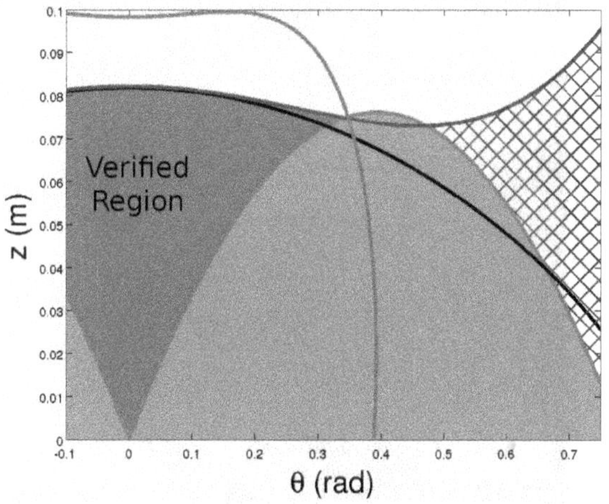

(b)

Figure 2: A slice of the rimless wheel state space is shown where all velocities are zero. (a) The red region below the solid line indicates the inadmissible set, where at least one of the contact points is penetrating the ground. For reference, two particular states are indicated: the stable equilibrium in double-support at the bottom and the unstable equilibrium in single-support on the upper right. The blue region below the dashed line is the connected component $\mathcal{C} \subseteq \Omega_1 \cap \mathcal{A}$ that contains the equilibrium (the verified region). (b) Two additional curves are shown. Under the black curve is $\mathcal{G}_i$, tight to $\Omega_1 \cap \mathcal{A}$, which we use to approximate the volume of the verified region. Under the magenta curve is $\mathcal{G}_o$, which must contain $\mathcal{C}$. $\mathcal{G}_o$ is parameterized as an ellipsoid in the redundant state variables, including s and $(1-c)$, which is why it is not convex when plotted against θ. The hatched region, while a subset of $\Omega_1 \cap \mathcal{A}$, is not connected to $\mathcal{C}$ and is unverified by our algorithm (and also outside the true invariant region).

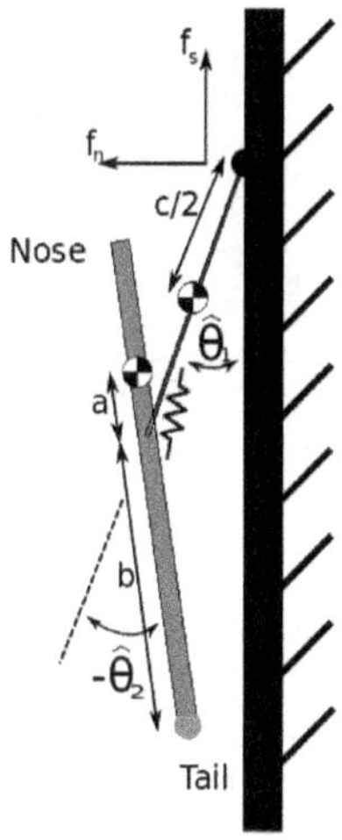

Figure 3: A simple model of a perching aircraft using two rigid links. The foot of the aircraft is pinned to the wall surface and there is a contact point at the tail that can collide with and slide along the wall. A torsional spring damper connects the main body of the aircraft to the foot.

stability in the sense of Lyapunov. When we parameterize V as a quartic polynomial, we verify a significant region of state space about the origin. A slice of this region is shown in Fig. 2(a) where the verified region is the connected component of $\Omega_1 \cap \mathcal{A}$ containing the origin. Fig. 2(b) illustrates the use of $\mathcal{G}_i$ and $\mathcal{G}_o$ to provide inner and outer bounds on $\mathcal{C}$.

It is interesting to note that if we parameterize V as a quadratic polynomial, the alternations quickly converge to verify a region that is nearly identical to the maximal sub-level set of energy that does not contain any additional equilibrium points. Additionally, we recover a scaled version of energy as our Lyapunov candidate. The quartic parameterization, however, verifies a larger region with a Lyapunov function significantly different from energy.

Note that the true region of attraction of this model is unbounded. For instance, for any x, z, take $q = \begin{bmatrix} x & z & 0 \end{bmatrix}^T$ and all velocities to be zero. A trajectory starting from any height will fall and then come to rest in the equilibrium set. By our parameterizations of $\mathcal{G}_i$ and $\mathcal{G}_o$, our regional approach is limited to ellipsoidal volumes and so will not recover the entire region of attraction. We do find a significant volume about the equilibrium set that would be relevant to planning or control applications.

5.3 Perching Glider

We lastly examine the problem of verifying a safe set of initial conditions for a glider perching against a wall, by adapting a model first presented in [8]. We consider the instant after the glider feet, which have adhesive microspines, have first impacted the wall and so we treat this contact as a pin joint. The glider is then modeled as a two-link body, with a spring damper connecting the bodies as shown in Fig. 3. We allow the tail of the glider to impact the surface of the wall and slide along it. The specific problem of verification was earlier addressed in [9], although the authors there used a model with a single joint and fixed the tail of the glider to slide along the wall, disallowing collisions. In

Table 2: Glider Properties

Body Mass	0.4 kg	Leg Length, c	0.15 m
Body Inertia	0.0164 kgm^2	Spring Const.	4.1e-3 Nm/°
Foot Mass	0.05 kg	Damper Const.	1.2e-4 Nms/°
Foot Inertia	0.001 kgm^2	Friction Coef. μ	0.3
Body CG, a	0.03 m	Nose Clearance	$\hat{\theta}_1 \geq 0$
Foot Dist., b	0.15 m	Force Limit	$f_s + f_n \geq 0$

this paper, we verify a significantly larger region than in our previous work, though a direct comparison is impossible since our model is higher dimension and uses Coulomb friction instead of viscous damping at the tail contact.

There are two relevant failure modes for the perching behavior of the glider, described in more detail in [9]. In one, the nose of the glider impacts the wall, which would be a potentially damaging event. In the other, the force limit of the feet microspines is exceeded and the glider falls from the wall. The force at the feet is a rational function of the state variables, and so the force limit can be easily expressed as a semialgebraic constraint. The full parameters of the glider are given in Table 2.

This is a two degree of freedom model, with $q = \begin{bmatrix}\hat{\theta}_1 & \hat{\theta}_2\end{bmatrix}^T$ and we again use a trigonometric substitution for both angles. We also change coordinates so that $(0,0)$ is an equilibrium point with the tail resting against the wall, substituting $\theta_1 = \hat{\theta}_1 - .2604$ and $\theta_2 = \hat{\theta}_2 + .5207$. With the rimless wheel, $H(q)$ was a constant matrix and so $H^{-1}(q)$ was also constant. Here, $H(q)^{-1}$ is rational and so, for our dynamical constraints, we clear the denominator to ensure that our conditions are algebraic.

We search for a solution to (22), to maximize the set of initial conditions of trajectories that do not violate either constraint. Here, we let $\mathcal{G}_o$ be all of $\mathbb{R}^{2n}$ and define

$$\mathcal{G}_i(\rho) = \{(q, v) | 2(1 - c_1) + 2(1 - c_2) + 0.1||v||^2 < \rho\}.$$

We seek to maximize ρ through a binary search, observing that $2(1 - c_i)$ well approximates θ_i^2 near $\theta_i = 0$. As with the rimless wheel, if we restrict our search to parameterizations of V of equal degree to energy, we recover the maximal sublevel set of energy that does not intersect either constraint boundary. If we expand our search to include all quartic polynomials, we find a Lyapunov function which verifies a visibly larger region. Two slices of this region are shown in Fig. 4(a) and 4(b). Our binary search terminates finding $\rho = 0.25$ and we can verify that an upper bound on the true optimal value is $\rho = 0.327$, since there $\mathcal{G}_i(\rho)$ intersects the constraint boundary. Of course, the true optimal value might be lower still since there is no claim that any $\mathcal{G}_i(\rho)$ is invariant. Here, we find a significant invariant region that usefully approximates the safe set of initial conditions.

6. CONCLUSIONS

Reliable planning and control through contact, as in locomotion and manipulation tasks, are critical challenges in robotics. The natural structure of rigid body dynamics and the complementarity formulation of frictional impacts provide a framework for posing questions of stability and invariance as sums-of-squares optimization problems. This paper presents a class of algorithms for numerical computation of Lyapunov certificates for such systems. By invok-

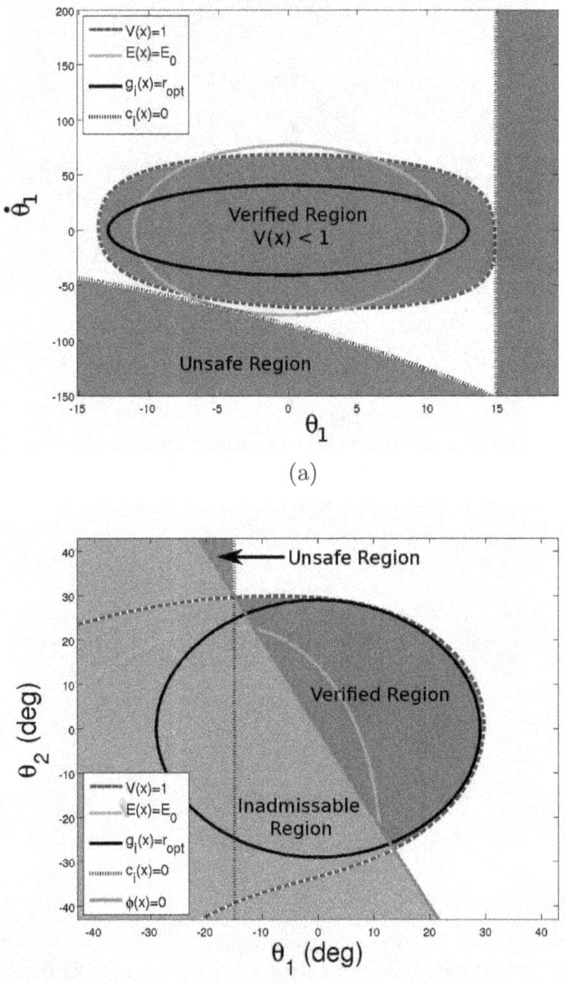

(a)

(b)

Figure 4: (a) A slice of the glider state space where the tail is restricted to the surface of the wall. The shaded blue region within the dashed line indicates the verified region $\mathcal{C} \cap \mathcal{A}$ for a quartic Lyapunov function and the solid black ellipse outlines $\mathcal{G}_i$. The green ellipse indicates the maximal sublevel set of energy that does not intersect the unsafe region, shown in gray. (b) A second slice of state space where the joint velocities are zero is also shown. In this slice, the general quartic Lyapunov function vastly outperforms energy and $\mathcal{G}_i$ is tight to $\mathcal{C}$.

ing the complementarity model of contact, we avoid directly reasoning about both the complexity of Zeno phenomena and the combinatorial number of hybrid modes associated with the set of potential contact states. Initial experiments have found physically significant certificates of stability and invariance for multiple problems of interest to the robotics community.

In this work, we have been primarily interested in stability in the sense of Lyapunov and positive invariance, but we hope to extend these methods to asymptotic stability. Here, we briefly discuss some of the challenges posed by this extension. Two common approaches to verifying asymptotic stability are to find a Lyapunov candidate where $\dot{V}$ is strictly negative or to apply LaSalle's Invariance Principle. With the former method, energy no longer provides an initial feasible candidate to begin bilinear alternation. In [12], Theorem 6.31 gives a generalization of LaSalle to discontinuous systems with the additional condition that the limit sets of trajectories also be positively invariant. The algorithms presented here might be extended to meet this condition and search for certificates of asymptotic stability.

Future work in this area will also center on extending these algorithms to more complex tasks. In particular, we are interested in the analysis of trajectories and limit cycles of robotics systems, where previous work has demonstrated the effectiveness of SOS based methods [19]. Numerical Lyapunov analysis has also been experimentally demonstrated to be an effective tool for controller synthesis, both at fixed points and along trajectories [18]. Furthermore, a natural extension of this work would be to include elastic impacts, where many models exist which are amenable to complementarity formulations such as in [28]. Recent work on the computation of regions of attraction for polynomial systems has included methods for approximating the volume of semi-algebraic sets [11]. A similar approach here might eliminate the requirement for bilinear alternations and allow the problem to be posted as a single convex optimization program.

7. ACKNOWLEDGMENTS

This work was supported by the National Science Foundation [Contract IIS-1161679] and Defense Advanced Research Projects Agency program [BAA-10-65-M3-FP-024].

APPENDIX

A. PROOFS

The following proof is for Lemma 2.

PROOF. Fix a solution $x(t) = (q(t), v(t))$ with $x^-(t_0) \in \mathcal{C}$. Let $\bar{\tau}$ be the supremum over all τ such that $x^-(t) \in \mathcal{C}$ for all $t \in [t_0, \tau]$. Assume for contradiction $\bar{\tau}$ is finite. We see $x^-(\bar{\tau}) \in \mathcal{C}$ as it is the limit of a sequence in a connected component. The function $s \mapsto (q(\bar{\tau}), \bar{v}(s))$ is a path from $x^-(\bar{\tau})$ to $x^+(\bar{\tau})$. This path lies in Ω_c as $\mathrm{d}V \leq 0$ on $\mathcal{C}$ implies $V(x^-(\bar{\tau})) \leq V(x^-(t_0)) < c$ and, by assumption, $V(q(\bar{\tau}), \bar{v}(s))$ is a non-increasing function of s. The path lies in $\mathcal{A}$ as only the generalized velocities vary (recall our definition of $\mathcal{A}$ in (3)). Thus $x^+(\bar{\tau})$ belongs to $\mathcal{C}$. As $V(x^+(\bar{\tau})) < c$ and V is continuous, there exists an $r > 0$ such that:

$$U_r = \{x \in \mathbb{R}^{2n} \mid \|x - x^+(\bar{\tau})\|_\infty < r\}$$

is contained in Ω_c (where $\|\cdot\|_\infty$ is the maximum norm). As $x^+(\tau)$ is a right limit there exists an $\epsilon > 0$ such that $x(t) \in U_r$ for $t \in (\bar{\tau}, \bar{\tau} + \epsilon)$.

We show that $x(t) \in \mathcal{C}$ for almost all $t \in (\bar{\tau}, \bar{\tau} + \epsilon)$, contradicting the definition of $\bar{\tau}$. Fix t such that $v(t)$ is defined and examine the following functions:

$$\sigma_1 \mapsto (q(\sigma_1), v^+(\bar{\tau})),$$
$$\sigma_2 \mapsto (q(t), \sigma_2 v(t) + (1 - \sigma_2) v^+(\bar{\tau})),$$

the first defined for $\sigma_1 \in [\bar{\tau}, t]$ and the second for $\sigma_2 \in [0, 1]$. The second function is clearly continuous, and the first is continuous as $q(t)$ is continuous. We see the range of both maps lie in $\mathcal{A}$ as $\{q(t)\} \times \mathbb{R}^n \subset \mathcal{A}$ for all $t \geq t_0$. We see the ranges of the functions also lie in U_r: the first lies in U_r as $(q_1, v_1), (q_2, v_2) \in U_r$ implies $(q_2, v_1) \in U_r$ and the second as U_r is convex. Together these functions provide a path from $x^+(\bar{\tau})$ to $x(t)$ that lies in $\Omega_c \cap \mathcal{A}$, thus $x(t) \in \mathcal{C}$. □

The following proof is for Lemma 3.

PROOF. That (ii) implies (i) is the content of Lemma 2. Now assume (i) holds and fix a solution $(q(t), v(t))$ with $(q(t), v^-(t)) \in \mathcal{C}$. For convenience, let q,v^+, and v^- denote $q(t), v^+(t), v^-(t)$. Take the path $\bar{v}(s) = sv^+ + (1-s)v^-$. Since V is convex in v and $\mathrm{d}V \leq 0$, we know

$$V(q, v^-) \geq (1-s)V(q, v^-) + sV(q, v^+) \geq V(q, \bar{v}(s)),$$

so that the chord $(q, \bar{v}(s))$ lies in Ω_c, and clearly lies in $\mathcal{A}$.

We show that $\frac{dV(q, \bar{v}(s))}{ds} \leq 0$. Observe that $\{(q, \bar{v}(s) | s \in [0, 1)\}$ are all possible pre-impact states since $\phi(q) = 0$ and the impact conditions $Jv^- < 0, Jv^+ = 0$ imply that $J\bar{v}(s) < 0$. Let Λ_N be a feasible impulse such that $v^+ = v^- + H^{-1}J^T\Lambda_N$. Substituting into the definition of $\bar{v}(s)$,

$$\bar{v}(s) = v^- + sH^{-1}J^T\Lambda_N.$$

Since the constraints on Λ_N are linear, we know that the impulse $(1-s)\Lambda_N > 0$ will also be feasible. Applying this impulse to $(q, \bar{v}(s))$ we get the post-impact velocity

$$\bar{v}^+(s) = \bar{v}(s) + (1-s)H^{-1}J^T\Lambda_N = v^+.$$

And so v^+ is a possible post-impact velocity from an impact at any point along the chord. Since $\mathrm{d}V \leq 0$, we then know $V(q, \bar{v}(s)) \geq V(q, v^+)$. This implies that the minimum of V along the chord is achieved at (q, v^+) and, since V is convex, the derivative along the chord must be non-positive. □

B. BILINEAR ALTERNATION

Sum-of-squares optimization enables optimization over linearly parameterized polynomials that are guaranteed to be non-negative. This can be extended to guarantee non-negativity on basic semialgebraic sets in the following fashion[23]. To demonstrate that $g(x) \geq 0$ implies $f(x) \geq 0$ we introduce multiplier polynomials $\sigma_i(x)$ and require:

$$\sigma_1(x)f(x) - \sigma_2(x)g(x) \geq 0, \quad \sigma_1(x) - 1 \geq 0, \quad \sigma_2(x) \geq 0.$$

In general, we may wish to simultaneously search over the free coefficients of the multiplier polynomials $\sigma_i(x)$ and those of $f(x)$ and $g(x)$. However, such coefficients enter the condition in a bilinear fashion. Fixing the free coefficients of f and g, we can search over the free coefficients of σ_1 and σ_2, and vice versa. These allows for a coordinate-descent strategy where each step consists of a single convex optimization problem and is guaranteed to improve the objective value. Similar approaches have been applied to determine region-of-attraction estimates for smooth dynamical systems [32].

C. REFERENCES

[1] R. Alur, T. Henzinger, G. Lafferriere, and G. Pappas. Discrete abstractions of hybrid systems. *Proceedings of the IEEE*, 88(7):971 –984, july 2000.

[2] V. Bhatt and J. Koechling. Three-dimensional frictional rigid-body impact. *Journal of applied mechanics*, 62:893, 1995.

[3] M. Branicky, V. Borkar, and S. Mitter. A unified framework for hybrid control: model and optimal control theory. *Automatic Control, IEEE Transactions on*, 43(1):31 –45, jan 1998.

[4] B. Brogliato. *Nonsmooth mechanics: models, dynamics, and control.* Springer Verlag, 1999.

[5] A. Chatterjee and A. Ruina. A new algebraic rigid-body collision law based on impulse space considerations. *Journal of Applied Mechanics*, 65(4):939–951, 1998.

[6] M. J. Coleman, A. Chatterjee, and A. Ruina. Motions of a rimless spoked wheel: a simple 3d system with impacts. *Dynamics and Stability of Systems*, 12(3):139–160, 1997.

[7] S. H. Collins, A. Ruina, R. Tedrake, and M. Wisse. Efficient bipedal robots based on passive-dynamic walkers. *Science*, 307:1082–1085, February 18 2005.

[8] A. L. Desbiens, A. Asbeck, and M. Cutkosky. Hybrid aerial and scansorial robotics. *ICRA*, May 2010.

[9] E. L. Glassman, A. L. Desbiens, M. Tobenkin, M. Cutkosky, and R. Tedrake. Region of attraction estimation for a perching aircraft: A Lyapunov method exploiting barrier certificates. In *Proceedings of the 2012 IEEE International Conference on Robotics and Automation (ICRA)*, 2012.

[10] D. Henrion and A. Garulli, editors. *Positive Polynomials in Control.* Lecture Notes in Control and Information Sciences. Springer-Verlag, 2005.

[11] D. Henrion, J. Lasserre, and C. Savorgnan. Approximate volume and integration for basic semialgebraic sets. *SIAM Review*, 51(4):722–743, 2009.

[12] R. Leine and N. van de Wouw. *Stability and convergence of mechanical systems with unilateral constraints.* Springer Verlag, 2008.

[13] R. Leine and N. van de Wouw. Stability properties of equilibrium sets of non-linear mechanical systems with dry friction and impact. *Nonlinear Dynamics*, 51(4):551–583, 2008.

[14] R. Lind, G. Balas, and A. Packard. Evaluating D-K iteration for control design. In *American Control Conference, 1994*, volume 3, pages 2792 – 2797 vol.3, june-1 july 1994.

[15] J. Löfberg. YALMIP : A toolbox for modeling and optimization in MATLAB. In *Proceedings of the CACSD Conference*, Taipei, Taiwan, 2004.

[16] J. Lofberg. Pre- and post-processing sum-of-squares programs in practice. *IEEE Transactions On Automatic Control*, 54(5):1007–, May 2009.

[17] J. Lygeros, K. Johansson, S. Simic, J. Zhang, and S. Sastry. Dynamical properties of hybrid automata. *Automatic Control, IEEE Transactions on*, 48(1):2 – 17, jan 2003.

[18] A. Majumdar, A. A. Ahmadi, and R. Tedrake. Control design along trajectories with sums of squares programming. *Under review*, 2013.

[19] I. R. Manchester. Transverse dynamics and regions of stability for nonlinear hybrid limit cycles. *Proceedings of the 18th IFAC World Congress, extended version available online: arXiv:1010.2241 [math.OC]*, Aug-Sep 2011.

[20] A. Megretski. Systems polynomial optimization tools (SPOT), available online: `http://web.mit.edu/ameg/www/`. 2010.

[21] Y. Or and A. Ames. Stability and completion of zeno equilibria in lagrangian hybrid systems. *Automatic Control, IEEE Transactions on*, 56(6):1322 –1336, june 2011.

[22] A. Papachristodoulou and S. Prajna. Robust stability analysis of nonlinear hybrid systems. *Automatic Control, IEEE Transactions on*, 54(5):1035 –1041, may 2009.

[23] P. A. Parrilo. Semidefinite programming relaxations for semialgebraic problems. *Mathematical Programming*, 96(2):293–320, 2003.

[24] S. Prajna and A. Rantzer. Convex programs for temporal verification of nonlinear dynamical systems. *SIAM Journal of Control and Optimization*, 46(3):999–1021, 2007.

[25] E. Routh. *Dynamics of a system of rigid bodies.* MacMillan and co. London, 1891.

[26] A. S. Shiriaev and L. B. Freidovich. Transverse linearization for impulsive mechanical systems with one passive link. *IEEE Transactions On Automatic Control*, 54(12):2882–2888, December 2009.

[27] D. Stewart. Rigid-body dynamics with friction and impact. *SIAM Review*, 42(1):3–39, 2000.

[28] W. Stronge. Rigid body collisions with friction. *Proceedings of the Royal Society of London. Series A: Mathematical and Physical Sciences*, 431(1881):169–181, 1990.

[29] J. F. Sturm. Using SeDuMi 1.02, a Matlab toolbox for optimization over symmetric cones. *Optimization Methods and Software*, 11(1-4):625 – 653, 1999.

[30] R. Tedrake, I. R. Manchester, M. M. Tobenkin, and J. W. Roberts. LQR-Trees: Feedback motion planning via sums of squares verification. *International Journal of Robotics Research*, 29:1038–1052, July 2010.

[31] C. Tomlin, I. Mitchell, A. Bayen, and M. Oishi. Computational techniques for the verification of hybrid systems. *Proceedings of the IEEE*, 91(7):986 – 1001, july 2003.

[32] U. Topcu, A. Packard, and P. Seiler. Local stability analysis using simulations and sum-of-squares programming. *Automatica*, 44(10):2669 – 2675, 2008.

[33] C. W. Wampler and A. J. Sommese. Numerical algebraic geometry and algebraic kinematics. *Acta Numerica*, 20:469–567, 2011.

[34] Y. Wang and M. T. Mason. Two-dimensional rigid-body collisions with friction. *ASME Journal of Applied Mechanics*, 59:635–642, 1992.

[35] E. R. Westervelt, J. W. Grizzle, C. Chevallereau, J. H. Choi, and B. Morris. *Feedback Control of Dynamic Bipedal Robot Locomotion.* CRC Press, Boca Raton, FL, 2007.

Hybrid Control Lyapunov Functions for the Stabilization of Hybrid Systems

Sterano Di Cairano
Mechatronics
Mitsubishi Electric Research Laboratories
Cambridge, MA 02139
dicairano@ieee.org

Maurice Heemels
Dept. Mechanical Engineering
Technical University of Eindhoven
P.O. Box 513
5600 MB Eindhoven, The Netherlands
m.heemels@tue.nl

Mircea Lazar
Dept. Electrical Engineering
Technical University of Eindhoven
P.O. Box 513
5600 MB Eindhoven, The Netherlands
m.lazar@tue.nl

Alberto Bemporad
IMT Institute for Advanced Studies
Piazza San Ponziano 6 55100
Lucca, Italy
alberto.bemporad@imtlucca.it

ABSTRACT

The design of stabilizing controllers for hybrid systems is particularly challenging due to the heterogeneity present within the system itself. In this paper we propose a constructive procedure to design stabilizing dynamic controllers for a fairly general class of hybrid systems. The proposed technique is based on the concept of a hybrid control Lyapunov function (hybrid CLF) that was previously introduced by the authors. In this paper we generalize the concept of hybrid control Lyapunov function, and we show that the existence of a hybrid CLF guarantees the existence of a standard control Lyapunov function (CLF) for the hybrid system. We provide a constructive procedure to design a hybrid CLF and the corresponding dynamic control law, which is stabilizing because of the established connection to a standard CLF that becomes a Lyapunov function for the closed-loop system. The obtained control law can be conveniently implemented by constrained predictive control in the form of a receding horizon control strategy. A numerical example highlighting the features of the proposed approach is presented.

HSCC'13, April 8–11, 2013, Philadelphia, Pennsylvania, USA.

Categories and Subject Descriptors

I.2.8 [**Artificial Intelligence**]: Problem Solving, Control Methods, and Search—*Control theory*; G.1.6 [**Numerical Analysis**]: Optimization—*Constrained Optimization*

General Terms

Theory, Algorithms

Keywords

Hybrid system, control Lyapunov function, dynamic controller, receding horizon control

1. INTRODUCTION

Hybrid systems [1,47] are powerful models for describing physical processes interacting with computer systems and embedded controllers, since they allow to formulate continuous and discrete dynamics in a unified framework. However, such modeling power is also associated with an inherent complexity in analyzing and manipulating the hybrid system, due to its heterogenous nature. While the concept of hybrid system was already introduced by Witsenhausen in the sixties [48], only in recent years tools to address stability analysis and control of certain classes of hybrid systems [8,21,23,25,34,38] have been developed. For the problem of stabilizing a hybrid system equilibrium, several techniques have been proposed for particular classes of hybrid systems, see, e.g., [11]. However, most of the available techniques can be applied only to switched systems [23,34], and piecewise affine systems [25,38,43], where the discrete dynamics are trivial, since the discrete state is uniquely determined by the current input, continuous state, and possibly external disturbances. In order to allow control applications such as robot tasks [5,17], industrial batch processes [18,33], and program executions in embedded and software-enabled control systems [4,42] stabilizing controller design techniques for hybrid systems with more general discrete dynamics are needed. Unfortunately, the existing methodologies for designing stabilizing controllers in this case are still limited.

Some previous works on stability and stabilization of hybrid systems with discrete dynamics are the following. A stability analysis based on the *hybrid distance* was developed in [35] for hybrid automata [22]. In [7] an approach based on model predictive control guaranteeing attractivity of the equilibrium is presented. More recently, in [19] a novel perspective on the stability of hybrid systems is presented, based on the concepts of *hybrid time* and *graphic convergence*, together with conditions that guarantee stability of the closed-loop system. Existence results for stabilizing controllers are given in [19], but the problem on how to synthesize stabilizing controllers for hybrid systems still remains widely open.

In [14], the authors have proposed a design approach for stabilizing controllers for a hybrid system based on receding horizon control and on the concept of *hybrid control Lyapunov function*, that resulted in a specific notion of closed-loop stability. The control design was implementable for classes of systems that have been used in practical applications [6,9,13,15,40,41]. In this paper we extend the ideas of [14] towards a general control design framework based on a formal notion of hybrid control Lyapunov function. We derive a fundamental result that shows that a hybrid CLF, which is significantly simpler to construct than a standard CLF, guarantees the existence of a standard control Lyapunov function [2,26,44] for the hybrid system. As a result, a controller synthesized based on the hybrid control Lyapunov function achieves (asymptotic) Lyapunov stability of the hybrid system in closed loop. A major advantage of the hybrid CLF approach is that we can derive a systematic procedure to construct the hybrid CLF and the related control law, thereby obtaining a systematic design procedure for stabilizing controllers for hybrid systems, without the need of constructing a standard CLF, which may be extremely difficult. By combining the hybrid CLF conditions with ideas from receding horizon control, a feasible implementation of the controller for a large class of systems [12,20,46] is also obtained.

The paper is structured as follows. In Section 2, we briefly review the basic notions of stability, control Lyapunov function and some notions of graph theory, and we formulate the stabilizing control design problem. In Section 3 we introduce the concept of hybrid control Lyapunov function, and we show that its existence guarantees the existence of a standard control Lyapunov function for the closed-loop system. In Section 4, starting from our previous results in [14], we propose a construction for the hybrid CLF and the consequent design of the stabilizing control law, and we implement the stabilizing controller by using a receding horizon constrained control strategy. In Section 5 we present a numerical example that highlights the features of the proposed approach, and in Section 6 we summarize the conclusions.

2. PRELIMINARIES AND PROBLEM DEFINITION

$\mathbb{R}$, $\mathbb{R}_+$, $\mathbb{R}_{0+}$, $\mathbb{Z}$, $\mathbb{Z}_+$, $\mathbb{Z}_{0+}$ denote the set of real, positive real, and non-negative real, integer, positive integer, and non-negative integer numbers, respectively. For a countable set $\mathcal{S}$, $|\mathcal{S}|$ denotes its cardinality. We use the notation $\mathbb{Z}_{(c_1,c_2]}$, where $c_1, c_2 \in \mathbb{Z}$, (and similarly with $\mathbb{R}$) to denote the set $\{k \in \mathbb{Z} : c_1 < k \leq c_2\}$. The Hölder p-norm of a vector $x \in \mathbb{R}^n$ is defined as $\|x\|_p \triangleq (|[x]_1|^p + \ldots + |[x]_n|^p)^{\frac{1}{p}}$, if $p \in \mathbb{Z}_{[1,\infty)}$ and $\|x\|_\infty \triangleq \max_{i=1,\ldots,n} |[x]_i|$, where $[x]_i$, $i = 1,\ldots,n$, is the i-th component of x and $|\cdot|$ is the absolute value. By $\|\cdot\|$ we denote an arbitrary p-norm.

We denote a function from set $\mathcal{A}$ to set $\mathcal{B}$ by $\phi : \mathcal{A} \to \mathcal{B}$, and a set-valued function by $\phi : \mathcal{A} \rightrightarrows \mathcal{B}$. Given a system $x(k+1) = \phi(x(k), u(k))$, an initial state $x(0)$ and an input sequence $\mathbf{u}_N = (u_0, \ldots, u_{N-1})$, $N \in \mathbb{Z}_+$, $\mathbf{x}_N = (x_0, \ldots, x_N)$ is the sequence of states obtained from $x(0)$ following the application of the input sequence $\mathbf{u}_N$. For the simplicity of notation, we denote $\phi^j(x(0), \mathbf{u}_N) \triangleq x(j)$ for $j \in \mathbb{Z}_{[0,N]}$, where we use the convention $\phi^0(x(0), \mathbf{u}_N) \triangleq x(0)$. For two vectors $u \in \mathbb{R}^{n_u}$ and $v \in \mathbb{R}^{n_v}$, we sometimes write $(u, v) = [u' \ v']' \in \mathbb{R}^{n_u+n_v}$. In addition, with a little abuse of notation, we sometimes separate the discrete valued and the real (continuous) valued arguments of a function $f(x, u)$, i.e., given $x = [x_c' \ x_d']'$, $u = [u_c' \ u_d']'$, where x_c, u_c are the continuous components, and x_d, u_d are the discrete components of x and u, respectively, we write sometimes $f(x_c, x_d, u_c, u_d) \triangleq f(x, u)$.

2.1 Stability Notions

A function $\varphi : \mathbb{R}_{0+} \to \mathbb{R}_{0+}$ *belongs to class* $\mathcal{K}$ if it is continuous, strictly increasing and $\varphi(0) = 0$. It *belongs to class* $\mathcal{K}_\infty$ if $\varphi \in \mathcal{K}$ and $\varphi(s) \to \infty$ when $s \to \infty$. A function $\beta : \mathbb{R}_{0+} \times \mathbb{R}_{0+} \to \mathbb{R}_{0+}$ *belongs to class* $\mathcal{KL}$ if for each $k \in \mathbb{R}_{0+}$, $\beta(\cdot, k) \in \mathcal{K}$, for each $s \in \mathbb{R}_{0+}$, $\beta(s, \cdot)$ is decreasing, and $\lim_{k\to\infty} \beta(s, k) = 0$.

Consider the hybrid system

$$x(k+1) = \begin{bmatrix} x_c(k+1) \\ x_d(k+1) \end{bmatrix} \in \begin{bmatrix} \Phi_c(x(k)) \\ \Phi_d(x(k)) \end{bmatrix} = \Phi(x(k)), \quad (1)$$

where $x(k) = [x_c(k)' \ x_d(k)']' \in \mathbb{X} \subseteq \mathbb{X}_c \times \mathcal{E}$ is the hybrid state at time $k \in \mathbb{Z}_{0+}$, with $x_c(k) \in \mathbb{R}^n$ the continuous part and $x_d(k) \in \mathcal{E}$ the discrete part, and $\mathcal{E} \triangleq \{\epsilon_1, \ldots, \epsilon_{n_d}\}$ is a finite set of symbols. The mapping $\Phi_c : \mathbb{X} \rightrightarrows \mathbb{X}_c$ is an arbitrary, possibly discontinuous, nonlinear set-valued function, that defines the continuous state dynamics of the hybrid system (1), and $\Phi_d : \mathbb{X} \rightrightarrows \mathcal{E}$ is an arbitrary set-valued function that defines the discrete state dynamics of the hybrid system (1)

Let $x^e = [x_c^{e\prime} \ x_d^{e\prime}]' \in \mathbb{R}^n \times \mathcal{E}$. If $\Phi(x^e) = \{x^e\}$, $x^e \in \mathbb{X}$ is an equilibrium for (1). In order to define asymptotic stability for discrete-time hybrid systems that exhibit both discrete and continuous dynamics, we introduce a distance function d_h in a hybrid state space. We first introduce a discrete distance function d_d for purely discrete state spaces.

Definition 1 [35] *Given a finite set* $\mathcal{E}$ *the* discrete distance *is the function* $d_d : \mathcal{E} \times \mathcal{E} \to \mathbb{R}_{0+}$ *defined by*

$$d_d(x_d, y_d) \triangleq \begin{cases} 0 & \text{if} \quad x_d = y_d \\ 1 & \text{if} \quad x_d \neq y_d \ . \end{cases} \quad (2)$$

for $x_d, y_d \in \mathcal{E}$.

Definition 2 [35] *Given a hybrid state space* $\mathbb{X}$ *the* hybrid distance *is the function* $d_h : \mathbb{X} \times \mathbb{X} \to \mathbb{R}_{0+}$ *defined by*

$$d_h(x, \chi) = \|x_c - \chi_c\| + d_d(x_d, \chi_d), \quad (3)$$

for $x = \left[\begin{smallmatrix} x_c \\ x_d \end{smallmatrix}\right] \in \mathbb{X}$ and $\chi = \left[\begin{smallmatrix} \chi_c \\ \chi_d \end{smallmatrix}\right] \in \mathbb{X}$.

Definition 3 *Consider hybrid system* (1) *and let* $x^e \in \mathbb{X}$ *satisfy* $\Phi(x^e) = \{x^e\}$. *The equilibrium* x^e *is called* asymptotically stable (AS) *in* $\mathbb{X}$ *for* (1), *if there exists a* $\mathcal{KL}$-*function* β *such that for any* $x(0) \in \mathbb{X}$ *all the trajectories generated by* (1) *satisfy*

$$d_h(x(k), x^e) \le \beta(d_h(x(0), x^e), k), \ \forall k \in \mathbb{Z}_+. \quad (4)$$

Definition 3 is consistent with [35], and it coincides with the stability definition of purely continuous or purely discrete systems in case (1) is a purely continuous or purely discrete system, respectively.

2.2 Lyapunov Functions and Control Lyapunov Functions

Definition 4 *A set* $\mathcal{P} \subseteq \mathbb{X}_c \times \mathcal{E}$ *is called* positively invariant (PI) *for system* (1) *if for all* $x \in \mathcal{P}$, $\Phi(x) \subseteq \mathcal{P}$.

Theorem 1 *Let* $\mathbb{X}$ *be a PI set for* (1) *with* $x^e \in \mathbb{X}$. *Let* $\alpha_1, \alpha_2 \in \mathcal{K}_\infty$, $\rho \in \mathbb{R}_{[0,1)}$, *and let* $\mathcal{V} : \mathbb{R}^n \to \mathbb{R}_{0+}$ *be a function such that*

$$\alpha_1(d_h(x, x^e)) \le \mathcal{V}(x) \le \alpha_2(d_h(x, x^e)) \quad (5a)$$

$$\mathcal{V}(x^+) \le \rho\mathcal{V}(x) \quad (5b)$$

for all $x \in \mathbb{X}$, *and all* $x^+ \in \Phi(x)$. *Then,* x^e *is AS for* (1) *in* $\mathbb{X}$.

The proof of Theorem 1 is similar in nature to the proof given in [29,32] by replacing the continuous difference *equation* with the hybrid difference *inclusion* (1), and hence it is omitted here. The proof can also be obtained by following [27], which discusses robust stability of discrete-time difference inclusions. A function $\mathcal{V}$ that satisfies the hypothesis of Theorem 1 is called a *Lyapunov function* for hybrid system (1).

Consider now the discrete-time hybrid system with control inputs described by the difference equation

$$\begin{aligned} x(k+1) &= \begin{bmatrix} x_c(k+1) \\ x_d(k+1) \end{bmatrix} \quad (6) \\ &= \begin{bmatrix} \phi_c(x(k), u(k)) \\ \phi_d(x(k), u(k)) \end{bmatrix} = \phi(x(k), u(k)), \end{aligned}$$

where $x(k) \in \mathbb{X} \subseteq \mathbb{X}_c \times \mathcal{E}$, $u(k) \in \mathbb{U} \subseteq \mathbb{U}_c \times \mathcal{E}_u$ are the state and input at $k \in \mathbb{Z}_{0+}$, and $\mathcal{E}_u \triangleq \{\epsilon_{u_1}, \ldots, \epsilon_{u_{m_d}}\}$ is a finite set of input symbols. In (6), $\phi : \mathbb{X} \times \mathbb{U} \to \mathbb{X}$ is an arbitrary, possibly discontinuous, nonlinear function. Assume that for a desired equilibrium $x^e = [x_c^{e\prime} \ x_d^{e\prime}]' \in \mathbb{X}$ there exists $u^e = [u_c^{e\prime} \ u_d^{e\prime}]' \in \mathbb{U}$ such that $\phi(x^e, u^e) = x^e$.

Definition 5 *A function* $\mathcal{V}_h : \mathbb{X} \to \mathbb{R}_{0+}$ *that satisfies* (5a) *for some* $\alpha_1, \alpha_2 \in \mathcal{K}_\infty$ *and for which there exists* $\rho \in \mathbb{R}_{[0,1)}$ *such that for all* $x \in \mathbb{X}$, *there exists* $u \in \mathbb{U}$ *such that* $\phi(x, u) \in \mathbb{X}$ *and*

$$\mathcal{V}_h(\phi(x, u)) \le \rho\mathcal{V}_h(x), \quad (7)$$

is called a control Lyapunov function *(CLF) for* x^e *in* $\mathbb{X}$ *for* (6).

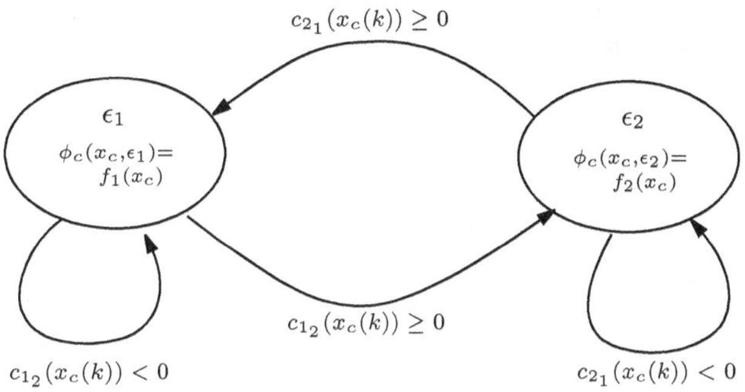

Figure 1: Graphical representation of a simple hybrid system with a graph associated to the discrete dynamics.

Given the CLF $\mathcal{V}_h$, define the control law

$$u(k) \in R(x(k)), \ k \in \mathbb{Z}_{0+}, \quad (8)$$

where $R : \mathbb{X} \rightrightarrows \mathbb{U}$ satisfies for all $x \in \mathbb{X}$,

$$\emptyset \ne R(x) \subseteq \Gamma(x) := \{u \in \mathbb{U} : \phi(x, u) \in \mathbb{X}, \text{ and (7) hold}\}. \quad (9)$$

By using (8) in (6), we obtain the closed-loop system

$$x(k+1) \in \phi(x(k), R(x(k))) \triangleq \{\phi(x(k), u) : u \in R(x(k))\}. \quad (10)$$

Theorem 2 *Consider* (6) *and* $x^e = [x_c^{e\prime} \ x_d^{e\prime}]' \in \mathbb{X}$, *where there exists* $u^e \in \mathbb{U}$ *such that* $\phi_c(x^e, u^e) = x^e$. *Suppose that there exists a CLF for* x^e *in* $\mathbb{X}$ *for* (6). *Then,* x^e *is asymptotically stable in* $\mathbb{X}$ *for* (10).

Theorem 2 is a consequence of Theorem 1 as $\mathbb{X}$ is PI for (10) by the definition of R. Theorem 2 shows that once a CLF is found, the construction of controller (8) satisfying (9) for all $x \in \mathbb{X}$ is immediate. If (6) consists only of continuous dynamics, we obtain a continuous CLF as defined in [26].

2.3 Graph Notions

A directed graph $G = (V, E)$ is described by the set of *nodes* $V = \{v_1, \ldots, v_s\}$ and the set of *edges* $E \subseteq (V \times V)$, where $e_{ij} \in E$, $e_{ij} = (v_i, v_j)$ is the edge from node $v_i \in V$ to node $v_j \in V$.

In this paper, the discrete dynamics of (6) are considered from the perspective of an automaton (see the example in Figure 1) with states $\mathcal{E} = \{\epsilon_1, \ldots, \epsilon_{n_d}\}$ and transitions $T \subseteq \mathcal{E} \times \mathcal{E}$, where $(\epsilon_i, \epsilon_j) \in T$ if and only if there exists $u \in \mathbb{U}$ and $x = [x_c' \ x_d']'$, such that $x_d = \epsilon_i$, $\phi(x, u) \in \mathbb{X}$ and $\phi_d(x, u) = \epsilon_i$. A graph is associated to the automaton, where each node $v \in V$ is bijectively associated to a state $\epsilon \in \mathcal{E}$ and each edge $e \in E$ is bijectively associated to a transition $t \in T$. The discrete distance (2) can be used as a distance measure between the nodes of the graph. However, for the graph associated to ϕ_d, when using the discrete distance (2) all the states appear to be equally far from the target state x_d^e, except the target state itself. Thus, this distance is less useful for measuring how close (in terms of discrete transitions) to the target state a certain state is. Therefore, we introduce a different notion of distance, the *graph distance*.

Definition 6 *Given a graph $G = (V, E)$, a* graph path *from $v_r \in V$ to $v_t \in V$, is a sequence of vertices $\tau = (\nu^{(0)}, \dots, \nu^{(\ell)})$, $\ell \in \mathbb{Z}_{0+}$, where $\nu^{(j)} \in V$ for $j \in \mathbb{Z}_{[0,\ell]}$, $(\nu^{(j)}, \nu^{(j+1)}) \in E$ for $j \in \mathbb{Z}_{[0,\ell-1]}$, and $\nu^{(0)} = v_r$, $\nu^{(\ell)} = v_t$. The length of the path is $\mathcal{L}(\tau) \triangleq \ell$, i.e., the number of edges traversed along τ from v_r to v_s.*

For $v_r, v_t \in V$, let $\mathcal{T}_{r,t}$ denote the set of all graph paths from v_r to v_t.

Definition 7 *Given a directed graph $G = (V, E)$, the* graph distance *between $v_r, v_t \in V$ is the length of the shortest graph path between them, i.e., for $v_r \neq v_t$, if $\mathcal{T}_{r,t} \neq \emptyset$, $d(v_r, v_t) \triangleq \min_{\tau \in \mathcal{T}_{r,t}} \mathcal{L}(\tau)$, and if $\mathcal{T}_{r,t} = \emptyset$, $d(v_r, v_t) \triangleq \infty$. For $v_r = v_t$, $d(v_r, v_t) \triangleq 0$.*

The graph distance, which represents the minimum number of edges to travel between two nodes is a proper distance function on undirected graphs, but it lacks the symmetry property on directed graphs, since in general $d(v_r, v_t) \neq d(v_t, v_r)$. However, this does not impact our use of the graph distance. For a given graph $G(V, E)$, for all $v_r, v_t \in V$ the graph distance $d(v_r, v_t)$ can be computed using, for instance, Dijkstra's algorithm [16].

2.4 Problem Definition

Consider hybrid system (6), where, for $k \in \mathbb{Z}_{0+}$, $x(k) \in \mathbb{X} \subseteq \mathbb{X}_c \times \mathcal{E}$, $\mathbb{X}_c \subseteq \mathbb{R}^n$ and $\mathcal{E} = \{\varepsilon_1, \dots, \varepsilon_{n_d}\}$, and $u(k) \in \mathbb{U} \subseteq \mathbb{U}_c \times \mathcal{E}_u$ with $\mathbb{U}_c \subseteq \mathbb{R}^m$ and $\mathcal{E}_u = \{\epsilon_{u_1}, \dots, \epsilon_{u_{m_d}}\}$. The sets $\mathbb{X}$ and $\mathbb{U}$ define the ranges of states and inputs, respectively, which possibly model system constraints. While state and input constraints defined by $\mathbb{X}$ and $\mathbb{U}$ are independent from each other, this condition is introduced here only to simplify the notation, and it can be easily relaxed. Given $\epsilon \in \mathcal{E}$, $\mathcal{X}_h(\epsilon) \triangleq \{x \in \mathbb{X} : x_d = \epsilon\}$ is the set of hybrid states where the discrete state is ϵ, and obviously $\mathbb{X} = \bigcup_{\epsilon_i \in \mathcal{E}} \mathcal{X}_h(\epsilon)$. Furthermore, $\mathcal{X}_h(\epsilon) = \mathcal{X}_c(\epsilon) \times \{\epsilon\}$, where $\mathcal{X}_c(\epsilon) \triangleq \{x_c \in \mathbb{R}^n : [\begin{smallmatrix} x_c \\ \epsilon \end{smallmatrix}] \in \mathbb{X}\}$ is the set of continuous states compatible with discrete state ϵ, sometimes referred to as the *domain of* ϵ.

We consider the stabilization of the desired (closed-loop) equilibrium $x^e = [x_c^{e\prime}\ x_d^{e\prime}]' \in \mathbb{X}$, for which there exists $u^e \in \mathbb{U}$ such that $\phi(x^e, u^e) = x^e$. The general problem that this paper addresses is to provide a *constructive* procedure to design a controller such that x^e is asymptotically stable for (6) in closed loop with the designed controller. In Section 2.2 we have described how such a control law can be obtained from a CLF. However, the direct derivation of a CLF for (6) is far from trivial.

In order to stabilize the desired equilibrium x^e of hybrid system (6), we consider the following class of *dynamic* controllers

$$z(k+1) = \psi(x(k), z(k), u(k), v(k)), \tag{11a}$$

$$\begin{bmatrix} u(k) \\ v(k) \end{bmatrix} \in R(x(k), z(k)), \tag{11b}$$

where $z \in \mathcal{Z} \subseteq \mathbb{R}^{n_z}$ is the controller state with dynamics defined by (11a), $v \in \mathbb{V}$ is an additional (endogenous) control input, and (11b) defines the set-valued command as a function of x and z. The problem addressed in this paper is formulated as follows.

Problem: Stabilizing Feedback Control Design. Given a desired equilibrium $x^e \in \mathbb{X}$ for (6) with $u^e \in \mathbb{U}$ satisfying $\phi(x^e, u^e) = x^e$, synthesize (11) such that there exist a non-trivial set $\Xi \subseteq \mathbb{X} \times \mathcal{Z}$, and $z^e \in \mathcal{Z}$, such that (x^e, z^e) is an asymptotically stable equilibrium in Ξ for the closed-loop system (6), (11).

At a conceptual level, the approach that we take in this paper is first to appropriately select the controller dynamics (11a) such that the interconnection of (6) and (11a) allows for a CLF $\mathcal{V}_h : \Xi \to \mathbb{R}_{0+}$, and then to choose the feedback R such that (9) is satisfied for (6), (11). The choice of z and the construction of the dynamics (11a) are major contributions of the paper, next to crafting the CLF in a systematic manner. In fact, the CLF is built in a compositional manner based on a so-called "hybrid CLF". Before presenting the hybrid CLF approach in the next section, note that the proposed approach is different from the standard CLF-based stabilization, which typically results in a *static* state feedback law $u(k) = \gamma(x(k))$, while (11) is a dynamic controller.

3. HYBRID CONTROL LYAPUNOV FUNCTIONS

Given a desired equilibrium $x^e \in \mathbb{X}$, we construct (11) using a so-called hybrid control Lyapuonv function (hybrid CLF). In [14] we introduced a specific type of hybrid CLF for the first time. Here, we generalize the concept of hybrid CLF and we show that it induces a CLF $\mathcal{V}_h$ consistent with Definition 5 for the interconnection of (6) and (11a), and that can be used for constructing R in (11b).

3.1 Definition of Hybrid CLF

A hybrid control Lyapunov function is defined as follows.

Definition 8 *A hybrid CLF for (6), (11a) for $(x^e, z^e) \in \Xi \subset \mathbb{X} \times \mathcal{Z}$ is a triple $(\mathcal{V}_c, \mathcal{V}_d, \mathcal{V}_z)$, where $\mathcal{V}_c : \mathbb{X}_c \to \mathbb{R}_{0+}$, $\mathcal{V}_d : \mathcal{E} \to \mathbb{R}_{0+}$ and $\mathcal{V}_z : \mathcal{Z} \to \mathbb{R}_{0+}$ satisfy the bounds*

$$\alpha_1^c(\|x_c - x_c^e\|) \leq \mathcal{V}_c(x_c) \leq \alpha_2^c(\|x_c - x_c^e\|), \ \forall x_c \in \mathbb{X}_c, \tag{12a}$$

$$\alpha_1^d(d_d(x_d, x_d^e)) \leq \mathcal{V}_d(x_d) \leq \alpha_2^d(d_d(x_d, x_d^e)), \ \forall x_d \in \mathcal{E}, \tag{12b}$$

$$\alpha_1^z(\|z - z^e\|) \leq \mathcal{V}_z(z) \leq \alpha_2^z(\|z - z^e\|), \ \forall z \in \mathcal{Z}, \tag{12c}$$

for some $\alpha_1^c, \alpha_2^c, \alpha_1^d, \alpha_2^d, \alpha_1^z, \alpha_2^z \in \mathcal{K}_\infty$. Moreover, for each $(x, z) \in \Xi$ there must exist $(u, v) \in \mathbb{U} \times \mathbb{V}$ such that

$$(\phi(x, u), \psi(x, z, u, v)) \in \Xi \tag{13}$$

and

$$\begin{cases} \mathcal{V}_c(\phi_c(x, u)) \leq \rho_c \mathcal{V}_c(x_c) + M_c \\ \mathcal{V}_z(\psi(x, z, u, v)) \leq \mathcal{V}_z(z) - 1 \\ \mathcal{V}_d(\phi_d(x, u)) \leq \mathcal{V}_d(x_d) \end{cases} \quad \text{if} \quad x_d \neq x_d^e \tag{14a}$$

$$\begin{cases} \mathcal{V}_c(\phi_c(x, u)) \leq \rho_c \mathcal{V}_c(x_c) \\ \mathcal{V}_z(\psi(x, z, u, v)) \leq \rho_z \mathcal{V}_z(z) \\ \mathcal{V}_d(\phi_d(x, u)) \leq \mathcal{V}_d(x_d) \end{cases} \quad \text{if} \quad x_d = x_d^e, \tag{14b}$$

for some constants $\rho_c, \rho_z \in [0, 1)$, $M_c \in \mathbb{R}_{0+}$.

Roughly speaking, (14) imposes that $\mathcal{V}_c$ is a local CLF for the continuous dynamics of (6) once the discrete state is equal to the desired discrete state (as in (14b)), $\mathcal{V}_z$ is a CLF for the controller dynamics (11a), and $\mathcal{V}_d$ for the discrete dynamics of (6), although no strict decrease is required. The three components of a hybrid CLF can be combined to obtain a standard CLF $\mathcal{V}_h$ for (6) and (11a) in the sense of Definition 5, hence justifying the name "hybrid CLF". In this way, by obtaining a procedure for constructing a hybrid CLF, which will be shown later to be easier than obtaining a standard CLF due to its compositional nature, a constructive procedure for the design of stabilizing controllers is defined.

3.2 Existence of a Hybrid CLF Guarantees Existence of a Standard CLF

Before proving that a hybrid CLF induces a standard CLF, we need the following technical lemma[1].

Lemma 1 *Let a hybrid CLF* $(\mathcal{V}_c, \mathcal{V}_d, \mathcal{V}_z)$ *for* $(x^e, z^e) \in \Xi \subset \mathbb{X} \times \mathcal{Z}$ *be given for system* (6), (11a), *and assume* $\mathcal{Z}$ *is a bounded set. Consider the function* $\mathcal{V}_D : \mathcal{E} \times \mathcal{Z} \to \mathbb{R}_{0+}$,

$$\mathcal{V}_D(x_d, z) = \mathcal{V}_d(x_d) + \mathcal{V}_z(z), \ (x_d, z) \in \mathcal{E} \times \mathcal{Z}.$$

Then, there exist $0 < \lambda_1 < 1$ *and* $0 < \lambda_2 < 1$ *such that for all* $(x, z) \in \Xi$ *with* $x_d \neq x_d^e$ *there exists* $(u, v) \in \mathbb{U} \times \mathbb{V}$ *such that*

$$\mathcal{V}_D(\phi_d(x, u), \psi(x, z, u, v)) \leq \lambda_1 \mathcal{V}_D(x_d, z) - \lambda_2.$$

Theorem 3 *Let a hybrid CLF* $(\mathcal{V}_c, \mathcal{V}_d, \mathcal{V}_z)$ *for* $(x^e, z^e) \in \Xi \subseteq \mathcal{X} \times \mathcal{Z}$ *be given, and assume* $\mathcal{Z}$ *is bounded. Then, for a sufficiently large* $\alpha \in \mathbb{R}_+$, *the function* $\mathcal{V}_h : \Xi \to \mathbb{R}_{0+}$, *given by*

$$\mathcal{V}_h(x, z) = \alpha \mathcal{V}_D(x_d, z) + \mathcal{V}(x_c), \tag{15}$$

where $(x, z) \in \Xi$ *and* $\mathcal{V}_D$ *is as in Lemma 1, is a CLF for* (6), (11) *for* (x^e, z^e) *in* Ξ.

From Theorem 3, the next corollary follows immediately.

Corollary 1 *Let a hybrid CLF* $(\mathcal{V}_c, \mathcal{V}_d, \mathcal{V}_z)$ *for* $(x^e, z^e) \in \Xi \subseteq \mathbb{X} \times \mathcal{Z}$ *be given, and assume* $\mathcal{Z}$ *is bounded. Consider a CLF* $\mathcal{V}_h$ *for* (6) *and* (11) *for* $(x^e, z^e) \in \Xi$ *obtained as in Theorem 3 for a sufficiently large* $\alpha > 0$. *Then, there exists* $0 \leq \rho_h < 1$ *such that if* $(u, v) \in \mathbb{U} \times \mathbb{V}$ *satisfies* (13)-(14) *for* $(x, z) \in \Xi$, *then* $(u, v) \in \mathbb{U} \times \mathbb{V}$ *satisfies*

$$\mathcal{V}_h(\phi(x, u), \psi(x, z, u, v)) \leq \rho_h \mathcal{V}_h(x, z)), \tag{16a}$$

$$(\phi(x, u), \psi(x, z, u, v)) \in \Xi. \tag{16b}$$

Corollary 1 is instrumental for designing $R : \Xi \to \mathbb{U} \times \mathbb{V}$ in (11b), since it guarantees that for $(x, z) \in \Xi$, if $(u, v) \in \mathbb{U} \times \mathbb{V}$ is chosen such that the hybrid CLF conditions (13)-(14) are satisfied, the standard CLF conditions (16) are satisfied for the standard CLF $\mathcal{V}_h$.

3.3 Stabilizing Dynamic Controller

Due to Theorem 2 and Theorem 3, if $R : \Xi \to \mathbb{U} \times \mathbb{V}$ is chosen according to (9) for $\mathcal{V}_h$, $(x^e, z^e) \in \Xi$ is asymptotically stable for (6), (11). For $(x, z) \in \Xi$, Corollary 1 shows that if R is chosen as

$$R(x, z) := \{(u, v) \in \mathbb{U} \times \mathbb{V} \mid (13) - (14)\}, \tag{17}$$

[1]In this paper the technical proofs are omitted due to space limitations. The statements of lemmas, theorems and corollaries provide an effective guidance towards the rationale of the control design properties.

then (9) is satisfied, and hence $(x^e, z^e) \in \Xi$ is asymptotically stable for (6), (11).

Thus, we obtain the following corollary.

Corollary 2 *Let a hybrid CLF* $(\mathcal{V}_c, \mathcal{V}_d, \mathcal{V}_z)$ *for* $(x^e, z^e) \in \Xi \subseteq \mathbb{X} \times \mathcal{Z}$ *be given for the system* (6),(11a), *and assume* $\mathcal{Z}$ *is bounded. If* $R : \Xi \to \mathbb{U} \times \mathbb{V}$ *is chosen as in* (17), *then* (x^e, z^e) *is asymptotically stable in* Ξ *for the closed-loop system* (6), (11).

4. DESIGN PROCEDURE FOR STABILIZING CONTROLLERS

In this section we provide a systematic method to design (11) that stabilizes (x^e, z^e) in Ξ, based on appropriately selecting (11a) and choosing the hybrid CLF for the hybrid dynamics (6), (11a). The procedure results in a hybrid CLF similar to the one originally proposed in [14], with some important modifications. Then $R : \Xi \to \mathbb{U} \times \mathbb{V}$ as in (11b) will follow automatically via Corollary 2. The ingredients we have to select for specifying (11) according to this procedure are z, z^e, $\mathcal{Z}$, ψ, Ξ, $\mathcal{V}_c$,$\mathcal{V}_d$, and $\mathcal{V}_z$ in (11).

4.1 Controller Dynamics

To specify the controller dynamics (11a), and in particular z, $\mathcal{Z}$, v, and $\mathbb{V}$, consider a desired equilibrium $x^e \in \mathbb{X}$ with equilibrium input $u^e \in \mathbb{U}$, i.e., $\phi(x^e, u^e) = x^e$, and an input sequence $\mathbf{u}_N(k) = (u_0(k), \ldots, u_{N-1}(k)) \in \mathbb{U}^N$ at time $k \in \mathbb{Z}_{0+}$. In (11), define $v(k)$ such that $(u(k), v(k)) = \mathbf{u}_N(k)$, so that $u(k) = u_0(k) \in \mathbb{U}$, and $v(k) = (u_1(k), \ldots, u_{N-1}(k)) \in \mathbb{U}^{N-1}$. Hence, $\mathbb{V} \triangleq \mathbb{U}^{N-1}$. Then, define the controller dynamics as

$$\psi(x, z, u, v) \triangleq \sum_{j=1}^{N} d(\phi_d^j(x, \mathbf{u}_N), x_d^e). \tag{18}$$

Hence, at time $k \in \mathbb{Z}_{0+}$, the controller state update is

$$z(k+1) = \psi(x(k), u(k), v(k)) = \psi(x(k), \mathbf{u}_N(k)). \tag{19}$$

Equation (18) defines the next controller state $z(k+1)$ as the cumulated graph distance from step 1 to step N along the trajectory generated from $x(k)$ following the application of $\mathbf{u}_N(k)$. Note that $\mathcal{Z} = \mathbb{R}_{[0,c_z]}$, $c_z = N \max_{x_d \in \mathcal{E}} d(x_d, x_d^e)$.

For the subsequent discussion it is important to notice that, by (19), the first element of the summation in (18) for $z(k)$ is $d(x_d(k), x_d^e)$, when $k \in \mathbb{Z}_+$. Hence, if $z(k) = 0$ for $k \in \mathbb{Z}_+$, then necessarily $x_d(k) = x_d^e$. In fact, we take $z^e = 0$, which satisfies for $(u^e, v^e) = \mathbf{u}_N = (u^e, \ldots, u^e)$ that $\psi(x^e, z^e, u^e, v^e) = 0 = z^e$. Hence, (x^e, z^e) serves as the desired equilibrium for (6), (11a), as we already have that $\phi(x^e, u^e) = x^e$.

4.2 Construction of Hybrid CLF

The component $\mathcal{V}_d$ of the hybrid CLF (14) related to the discrete state of the hybrid system is defined by using the discrete distance (2) as

$$\mathcal{V}_d(x_d) = d_d(x_d, x_d^e). \tag{20}$$

Thus, the constraints on $\mathcal{V}_d$ in (14) require now that

$$d_d(\phi_d(x, u), x_d^e) \leq d_d(x_d, x_d^e). \tag{21}$$

This condition is guaranteed by the following.

Assumption 1 *For any $x \in \mathcal{X}_h(x_d^e)$ there exists $u \in \mathbb{U}$ such that $\phi(x,u) \in \mathcal{X}_h(x_d^e)$.*

Assumption 1 requires that for any hybrid state where the discrete state is at the target, there exists an input that keeps the discrete state at the target state.

The component $\mathcal{V}_z$ of the hybrid CLF is defined as

$$\mathcal{V}_z(z) = \|z\| = z. \tag{22}$$

where the second equality holds due to z being always non-negative. Given the controller dynamics (18), the conditions (14) on the hybrid CLF impose the constraints

$$\psi(x,z,\mathbf{u}_N) \leq z - 1 \quad \text{if} \quad x_d \neq x_d^e \tag{23a}$$
$$\psi(x,z,\mathbf{u}_N) \leq \quad \rho_z z \quad \text{if} \quad x_d = x_d^e, \tag{23b}$$

for all $(x,z) \in \Xi$ and some constant $0 < \rho_z < 1$. Constraint (23) is called the *cumulative graph distance contraction* (CGDC) requirement, and it can be seen as a relaxation of

$$d(\phi_d(x,u), x_d^e) \leq \rho_d d(x_d, x_d^e),\ 0 \leq \rho_d < 1, \tag{24}$$

that would require the discrete state to come closer to the equilibrium at every time step. Constraint (24) would be difficult to enforce in most practical systems the discrete state cannot change at *every* step. In order to guarantee feasibility of (23) we introduce the following assumption.

Assumption 2 *Let $x^e \in \mathbb{X}$. For any discrete state $x_d \in \mathcal{E}\setminus\{x^e\}$ there exists $n_g \in \mathbb{Z}_{0+}$ such that for any $x \in \mathcal{X}_h(x_d)$, there exists $\bar{x}_d \in \mathcal{E}$, where $d(\bar{x}_d, x_d^e) < d(x_d, x_d^e)$, and an input sequence $\mathbf{u}_\ell \in \mathbb{U}^\ell$, such that: (i) $\ell \leq n_g$; (ii) $\phi^q(x,\mathbf{u}_\ell) \in \mathbb{X}$, $\phi_d^q(x,\mathbf{u}_\ell) = x_d$, $q \in \mathbb{Z}_{[1,\ell-1]}$; (iii) $\phi_d^\ell(x,\mathbf{u}_\ell) = \bar{x}_d$.*

Assumption 2 requires the existence of a horizon ℓ such that the discrete state gets closer to x_d^e, when in discrete state x_d. In general, if Assumption 2 is satisfied, the horizon n_g required is shorter than the one required by the approach in [7], because reachability of a discrete state closer to the equilibrium, rather than reachability of the equilibrium itself is required.

Definition 9 *Given $x_d \in \mathcal{E}$, the minimum graph distance progress horizon $n(x_d) \in \mathbb{Z}_{0+}$ for $x_d \in \mathcal{E}$ is the minimum value n_g for which Assumption 2 holds for x_d, where we use $n(x_d^e) \triangleq 0$.*

In Definition 9, $n(x_d)$ is the minimum horizon needed for the discrete state to get closer to x_d^e, with respect to the graph distance. The value $n(x_d)$ can be computed by offline backward reachability analysis (see, e.g., [3,10,39,45]). For the proposed approach backward reachability analysis is computationally viable since by Assumption 2 the discrete state remains constant, hence we only have to verify reachability for a constrained continuous system.

The final component in the hybrid CLF $(\mathcal{V}_c, \mathcal{V}_d, \mathcal{V}_z)$ is

$$\mathcal{V}_c : \mathbb{X}_c \to \mathbb{R}_{0+}, \tag{25}$$

which by (14), should satisfy that for x with $(x,z) \in \Xi$ there exists $u \in \mathbb{U}$ such that

$$\mathcal{V}_c(\phi_c(x,u)) \leq \rho_c \mathcal{V}_c(x_c) + M_c \quad \text{if} \quad x_d \neq x_d^e \tag{26a}$$
$$\mathcal{V}_c(\phi_c(x,u)) \leq \rho_c \mathcal{V}_c(x_c) \quad \text{if} \quad x_d = x_d^e, \tag{26b}$$

where $\rho_c \in \mathbb{R}_{[0,1)}$ and $M_c \in \mathbb{R}_+$ are appropriately selected constants. In fact, (26a) implies that $\mathcal{V}_c$ is a standard CLF of the continuous dynamics locally around the equilibrium x^e, and only for the dynamics associated to $x_d = x_d^e$. Finding CLFs for continuous dynamics is a well-studied problem [2,26,44], and is significantly simpler than the design of a (global) CLF for the hybrid system. Techniques for calculating local CLFs based on infinity norms are discussed, for instance, in [30,31], while techniques for computing local CLFs based on quadratic forms are discussed, for instance, in [24,28,31]. We adopt the following assumption regarding $\mathcal{V}_c$.

Assumption 3 *There exists $\mathcal{V}_c$ as in (25) and $0 \leq \rho_c < 1$, such that the bounds (12a) are satisfied for some $\alpha_1^c, \alpha_2^c \in \mathcal{K}_\infty$, and for all $x \in \mathcal{X}_h(x_d^e)$ there exists $u \in \mathbb{U}$ such that*

$$\phi(x,u) \in \mathcal{X}_h(x_d^e), \tag{27a}$$
$$\mathcal{V}(\phi_c(x,u)) \leq \rho_c \mathcal{V}(x_c). \tag{27b}$$

In addition, we assume that $\sup_{x\in\mathbb{X}} \mathcal{V}_c(x_c) < \infty$.

Two observations are in order. First of all, note that Assumption 3 implies Assumption 1. Second, note that to guarantee (26b) we can set $M_c = \sup_{x\in\mathbb{X}} \mathcal{V}_c(x_c)$.

To prove that Assumptions 2, 3 imply that $(\mathcal{V}_c, \mathcal{V}_d, \mathcal{V}_z)$ is indeed a hybrid CLF for (6), (11a), we first state the following lemma.

Lemma 2 *Suppose Assumptions 1 and 2 hold. Given $x \in \mathbb{X}$ for any $\varsigma \in \mathbb{Z}_+$ there exists $\mathbf{u}_\varsigma \in \mathbb{U}^\varsigma$, such that $\phi^i(x,\mathbf{u}_\varsigma) \in \mathbb{X}$, for $i \in \mathbb{Z}_{[1,\varsigma]}$, and*

$$d(\phi_d^{i+1}(x,\mathbf{u}_\varsigma), x_d^e) \leq d(\phi_d^i(x,\mathbf{u}_\varsigma), x_d^e),\ i \in \mathbb{Z}_{[0,\varsigma-1]}.$$

Theorem 4 *Suppose Assumptions 2, 3 hold and let $N \geq \max_{x_d\in\mathcal{E}} n(x_d)$. Define*

$$\Xi = \{(x,z) \in \mathbb{X} \times \mathbb{R}_{[0,c_z]} : \exists(u,v) \in \mathbb{U}\times\mathbb{V},\ (14) \text{ holds}\}. \tag{28}$$

Then $(\mathcal{V}_c, \mathcal{V}_d, \mathcal{V}_z)$, defined respectively by (25), (20), (22) is a hybrid CLF for (6), (19) for (x^e, z^e) in Ξ.

In order to prove Theorem 4 we need the following technical Lemma.

Lemma 3 *Suppose Assumptions 1 and 2 hold, and let $N \geq \max_{x_d\in\mathcal{E}} n(x_d)$. Let $(\mathcal{V}_c, \mathcal{V}_d, \mathcal{V}_z)$ in (14) be defined respectively by (25), (20), (22). If (14) is feasible for $(x,z) \in \mathbb{X}\times\mathbb{R}_{[0,c_z]}$ for $(u,v) \in \mathbb{U}\times\mathbb{V}$, then there exists $(\tilde{u},\tilde{v}) \in \mathbb{U}\times\mathbb{V}$ such that it is feasible for $(\phi(x,u), \psi(x,u,v)) \in \mathbb{X}\times\mathbb{R}_{[0,c_z]}$.*

We can now prove Theorem 4.

Proof. (Theorem 4) Given $(\mathcal{V}_c, \mathcal{V}_d, \mathcal{V}_z)$ defined respectively by (25), (20), (22), the existence of class $\mathcal{K}$ bounds on $\mathcal{V}_c$ is guaranteed by Assumption 3, while for $\mathcal{V}_d$, $\mathcal{V}_z$, it follows by construction since $\mathcal{V}_z(z) = z = \|z\|$ and $\mathcal{V}_d(x_d) = d_d(x_d, x_d^e)$. We only need to prove that for each $(x,z) \in \Xi$ there exists $(u,v) \in \mathbb{U}\times\mathbb{V}$ such that $(\phi(x,u), \psi(x,z,u,v)) \in \Xi$ and (14) is satisfied. Lemma 3 ensures that if there exists $(u,v) \in \mathbb{U}\times\mathbb{V}$, such that (14) is feasible for $(x,z) \in \mathbb{X}\times\mathbb{R}_{[0,c_z]}$, then there exists $(\tilde{u},\tilde{v}) \in \mathbb{U}\times\mathbb{V}$ such that (14) is feasible for $(\phi(x,u), \psi(x,u,v)) \in \mathbb{X}\times\mathbb{R}_{[0,c_z]}$. Hence, by choosing Ξ as in (28), for any $(x,z) \in \Xi$ there is always $(u,v) \in \mathbb{U}\times\mathbb{V}$ such that (14) holds and $(\phi(x,u), \psi(x,z,u,v)) \in \Xi$. □

The next corollary follows directly.

Corollary 3 *Let Assumptions 1 and 2 hold and* $(\mathcal{V}_c, \mathcal{V}_d, \mathcal{V}_z)$ *be defined respectively by (25), (20), (22). For any* $x \in \mathbb{X}$, *there exists* $0 \leq \bar{z} \leq c_z$, *such that for any* $z \geq \bar{z}$, *(14) is feasible. If (14) is feasible for* $(x(0), z(0)) = (x, z)$, *there exists a finite* $\bar{k} \in \mathbb{Z}_{0+}$ *such that* $z(k) = 0$, *and* $x_d(k) = x_d^e$, *for all* $k \geq \bar{k}$.

Corollary 3 proves that by initializing the controller state appropriately, convergence to the equilibrium is achieved for any initial state, and that the discrete state converges in finite time to the discrete equilibrium state.

The mapping R in (11) can now be designed according to Corollary 2 providing the complete dynamical controller (11) that stabilizes (x^e, z^e) in Ξ.

Next, we discuss a specific implementation of (17) based on receding horizon control.

4.3 Controller Implementation by Receding Horizon Control

The hybrid CLF (14) results in a controller (11) with R as in (17) that generates a sequence of inputs along a future horizon. As previously shown in [14], such controller can be implemented by using a predictive control strategy.

Corollary 3 guarantees that there exists a finite value $\bar{z} \in \mathbb{R}_{0+}$ such that $(x, z) \in \Xi$, for any $x \in \mathbb{X}$. Hence, with an appropriate initialization of the controller state z, Corollary 2 guarantees convergence to the desired equilibrium for any initial state of the hybrid system. The stabilizing properties established in Corollary 2 are guaranteed for any control input that satisfies (14), i.e., for any $(u, v) \in R(x, z)$ in (17). The actual input (u, v) can be chosen by optimizing the set of feasible inputs with respect to a defined performance criterion. In this way a receding horizon predictive control strategy based on the repetitive solution of an optimization problem is obtained. A common definition of the performance criterion in optimization-based predictive control, such as model predictive control [36], is

$$J(x, \mathbf{u}_N) \triangleq F(\phi^N(x, \mathbf{u}_N)) + \sum_{h=0}^{N-1} L(\phi^h(x, \mathbf{u}_N), u_h), \quad (29)$$

where $F(\cdot)$ and $L(\cdot)$ denote suitable terminal and stage costs, respectively, [37]. Cost (29) typically trades off the regulation performance, in terms of distance from the equilibrium, and the actuation effort.

Constraint (23) can be implemented by a single constraint as

$$\begin{aligned} \mathcal{V}_z(\psi(x(k), \mathbf{u}_N(k))) \;\leq\; & (1 - d_d(x_d(k), x_d^e))\rho_z \mathcal{V}_z(z(k)) \\ & - d_d(x_d(k), x_d^e) \end{aligned}$$

Algorithm 1 *(Hybrid CLF-based Stabilizing Receding Horizon Control)*
Initialization. *Set* $k = 0$, *measure* $x(0) \in \mathbb{X}$ *and set* $z(0) \geq N\, d(x_d(0), x_d^e)$.

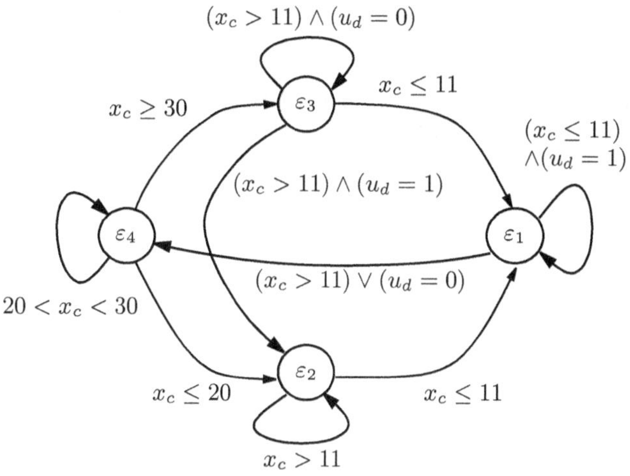

Figure 2: Automaton describing the discrete dynamics of the system in the numerical example.

Step 1. *Solve the optimization problem*

$$\begin{aligned} &\min_{\mathbf{u}_N(k)} && J(x(k), \mathbf{u}_N(k)) && (30a) \\ &\text{s.t.}: \quad x_{h+1} &=& \phi(x_h, u_h(k)), && (30b) \\ &z_1 &=& \psi(x_0, \mathbf{u}_N(k)) && (30c) \\ &\mathcal{V}_c(\phi_c^1(x_0, \mathbf{u}_N(k))) &\leq& \rho \mathcal{V}_c(x_{0c}(k)) + M d_d(x_{0d}, x_d^e) && (30d) \\ &\mathcal{V}_z(\psi(x_0, \mathbf{u}_N(k))) &\leq& (1 - d_d(x_{0d}(k), x_d^e))\rho_z \mathcal{V}_z(z_0) \\ & && -d_d(x_{0d}, x_d^e) && (30e) \\ &\mathcal{V}_d(\phi_d^1(x_0, \mathbf{u}_N(k))) &\leq& \mathcal{V}_d(x_{0d}) && (30f) \\ & && \mathbf{u}_N(k) \in \mathbb{U}^N, && (30g) \\ & && x_h \in \mathbb{X}, \; h \in \mathbb{Z}_{[1,N]} && (30h) \\ & && x_0 = x(k), \; z_0 = z(k). && (30i) \end{aligned}$$

Step 2. *Let* $\bar{\mathbf{u}}_N(k) = (\bar{u}_0(k), \ldots, \bar{u}_{N-1}(k))$ *be a feasible solution of (30), possibly, but not necessarily, the optimal one. Set* $u(k) = \bar{u}_0(k)$, *and* $z(k+1) = \psi(x(k), \bar{\mathbf{u}}_N(k))$.
Step 3 *Measure* $x(k+1)$, *set* $k \leftarrow k+1$, *and go to Step 1.*

Algorithm 1 implements the constraints as required in the definition of the hybrid CLF and minimizes the performance criterion (29). It is worth noticing that the optimization problem (30) is always feasible because of the hybrid CLF existence results. Also, similarly to what demonstrated in [14], (30) can be formulated as a mixed integer linear/quadratic problem for a large and practically useful class of hybrid systems, for which several high performance numerical algorithms are available.

5. NUMERICAL EXAMPLE

In what follows we present the application of the proposed technique to a numerical. We consider a system with continuous state domain $\mathcal{X}_c := [-5, 30]$, discrete state domain $\mathcal{E} := \{\epsilon_1, \epsilon_2, \epsilon_3, \epsilon_4\}$, one continuous input $u_c \in \mathbb{U}_c := [-2.5, 2.5]$ and one discrete input $u_d \in \mathcal{E}_u := \{0, 1\}$. Hence, $\mathbb{U} := [-2.5, 2.5] \times \{0, 1\}$, and $\mathbb{X} \subseteq [-5, 30] \times \{\epsilon_1, \epsilon_2, \epsilon_3, \epsilon_4\}$, where in particular $\mathcal{X}_c(\epsilon_1) = [-5, 11]$. The automaton describing the discrete dynamics ϕ_d of the example system is shown in Figure 2. The continuous dynamics are

$$x(k+1) = A_i x(k) + B_i u(k) \quad \text{if} \quad x_d = \epsilon_i,$$

and $(A_1, B_1) = (1.07, 0.4)$, $(A_2, B_2) = (0.85, 1.25)$, $(A_3, B_3) = (0.7, 1.05)$, $(A_4, B_4) = (1.02, 1)$.

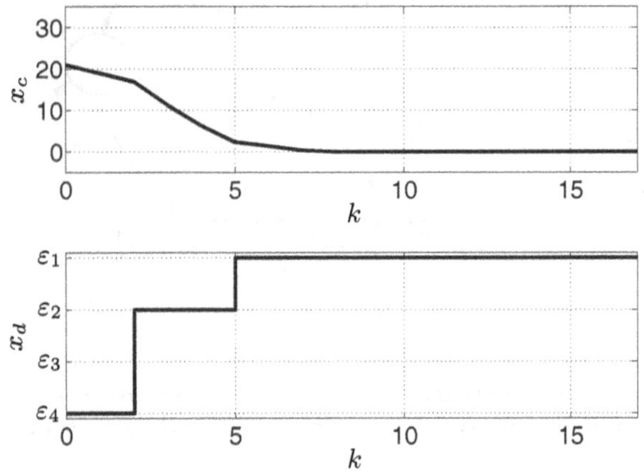

(a) Continuous (top) and discrete (bottom) state evolution.

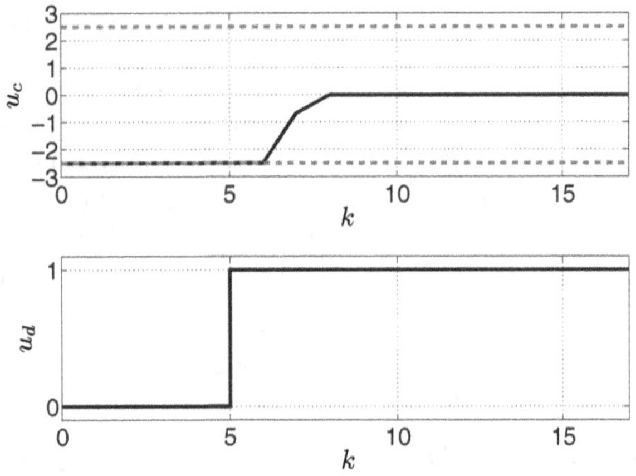

(b) Continuous (top) and discrete (bottom) input evolution.

Figure 3: Simulation results of the numerical example for $(x_c(0), x_d(0)) = (21, \epsilon_4)$.

The desired equilibrium is $(x_c^e,\ x_d^e) = (0,\ \epsilon_1)$ for a steady state input $(u_c^e,\ u_d^e) = (0,\ 1)$. The function $\mathcal{V}(x_c) = \|x_c\|_\infty$ is chosen as the local CLF for the continuous state in $\mathcal{X}_c(\varepsilon_1)$ which can be proved to exists in $\mathcal{X}_c(\varepsilon_1)$ by using the auxiliary controller $(u_c, u_d) = (K_c x_c, 1)$, where $K_c = -0.2250$. From $\mathcal{V}_c$ and the construction of z, $\mathcal{V}_z$ (where we have chosen $N = 4$ which satisfies Assumption 2) and $\mathcal{V}_d$ proposed in Section 4 we obtain the hybrid control Lyapunov function (14).

The stabilizing dynamic controller (11a) is obtained by executing Algorithm 1, where problem (30) is implemented with

$$L(x, u) = \|Q_x(x - x^e)\|_\infty + \|Q_u(u - u^e)\|_\infty,$$
$$Q_x = \begin{bmatrix} 1 & 0 \\ 0 & 1 \end{bmatrix},\ Q_u = \begin{bmatrix} 0.1 & 0 \\ 0 & 0.1 \end{bmatrix},\ \bar{\rho}_c = 0.98,\ N = 4. \quad (31)$$

The model of the hybrid system dynamics (30c), (30b) is formulated as a discrete hybrid automaton using the language in [46], so that problem (30) results in a mixed-integer linear program.

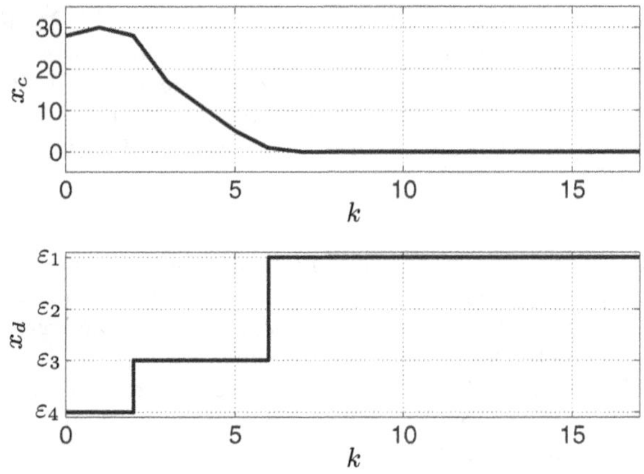

(a) Continuous (top) and discrete (bottom) state evolution.

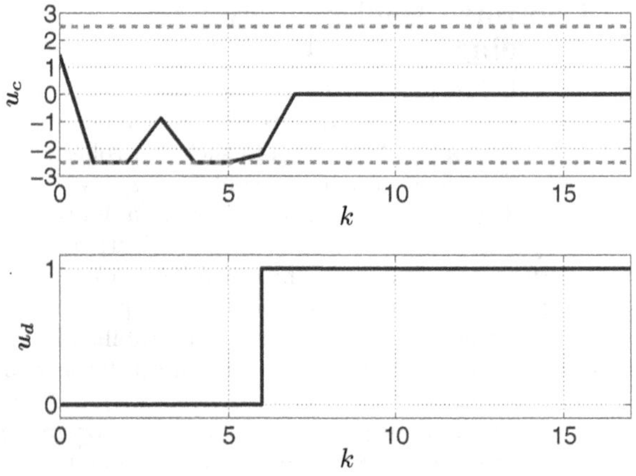

(b) Continuous (top) and discrete (bottom) input evolution.

Figure 4: Simulation results of the numerical example for $(x_c(0), x_d(0)) = (28, \epsilon_4)$.

Figure 3 shows the simulation results for initial condition $x(0) = (21,\ \epsilon_4)$. Note that in this simulation inequality (5b) is enforced at every step, which means that the continuous state component of the hybrid CLF, $\mathcal{V}_c(x_c)$ is decreasing along the entire trajectory. In Figure 3 we show the simulation simulation results for the case when the initial condition is $x(0) = (28,\ \epsilon_4)$ are shown by solid lines. In this case $\mathcal{V}_c(x_c)$ is not monotonically decreasing along the whole trajectory. This is according to (26), where the decrease of $\mathcal{V}_c$ is required only in the set $\mathcal{X}_c(\epsilon_1)$. However, the closed-loop system is asymptotically stable, due to Theorem 3 that guarantees the existence of a control Lyapunov function for the hybrid dynamics, which becomes a Lyapunov function for the closed-loop system, is guaranteed. The continuous state trajectories for many initial conditions are shown in Figure 5.

It is worth to point out that for the same setup, the hybrid controller proposed in [7] that guarantees attractivity, but not Lyapunov stability, is infeasible unless a longer horizon

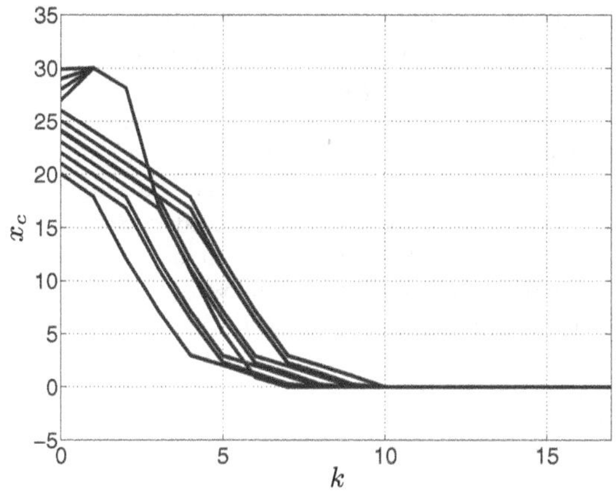

Figure 5: State evolution in the numerical example for different initial conditions.

(at least $N = 9$) is used, due to the required controllability to the equilibrium by the end of the horizon. On the other hand, for the control law proposed here it is enough to enforce a decrease of the cumulated graph distance along the prediction horizon, which is possible for a horizon $N = 4$. Clearly, this indicates also numerical advantages of our novel approach.

6. CONCLUSIONS

In this paper we have provided a general control design framework based on a formal notion of the concept of hybrid control Lyapunov function. We have shown that the existence of a hybrid CLF guarantees the existence of a standard CLF. This result induces a class of dynamic control laws based on the hybrid CLF that stabilize a desired hybrid system equilibrium. Building on our previous research, we have defined a constructive procedure to obtain a hybrid control Lyapunov function and the corresponding control law for a fairly general class of hybrid systems, and we have implemented that by receding horizon constrained control. The proposed approach has been demonstrated on a numerical example.

7. ACKNOWLEDGMENTS

M. Lazar, W.P.M.H. Heemels, and A. Bemporad are partially supported by the European Commission under project FP7-INFSO-ICT-248858 'MOBY-DIC - Model-based synthesis of digital electronic circuits for embedded control'.

8. REFERENCES

[1] P. Antsaklis. A brief introduction to the theory and applications of hybrid systems. *Proc. IEEE, Special Issue on Hybrid Systems: Theory and Applications*, 88(7):879–886, July 2000.

[2] Z. Artstein. Stabilization with relaxed controls. *Nonlinear Analysis*, 7:1163–1173, 1983.

[3] E. Asarin, O. Bournez, T. Dang, and O. Maler. Approximate reachability analysis of piecewise-linear dynamical systems. In B. Krogh and N. Lynch, editors, *Hybrid Systems: Computation and Control*, volume 1790 of *Lecture Notes in Computer Science*, pages 20–31. Springer-Verlag, 2000.

[4] A. Balluchi, L. Benvenuti, M. Di Benedetto, C. Pinello, and A. Sangiovanni-Vincentelli. Automotive engine control and hybrid systems: Challenges and opportunities. *Proceedings of the IEEE*, 88(7):888–912, 2002.

[5] C. Belta, A. Bicchi, M. Egerstedt, E. Frazzoli, E. Klavins, and G. Pappas. Symbolic Planning and Control of Robot Motion. *Robotics & Automation Magazine*, 14:62–66, 2007.

[6] A. Bemporad, S. Di Cairano, E. Henriksson, and K. Johansson. Hybrid model predictive control based on wireless sensor feedback: an experimental study. *Int. J. Robust Nonlinear Control*, 20(2), 2010. special issue "Industrial applications of wireless control".

[7] A. Bemporad and M. Morari. Control of systems integrating logic, dynamics, and constraints. *Automatica*, 35(3):407–427, 1999.

[8] M. S. Branicky, V. S. Borkar, and S. K. Mitter. A unified framework for hybrid control: model and optimal control theory. *IEEE Trans. Aut. Control*, 43(1):31–45, 1998.

[9] M. Braun and J. Shear. A mixed logical dynamic model predictive control approach for handling industrially relevant transportation constraints. In *IEEE Conf. Automation Science and Engineering*, pages 966–971, Toronto, Canada, 2010.

[10] T. Dang. Approximate reachability computation for polynomial systems. In *Hybrid Systems: Computation and Control*, number 3927, pages 93–107. 2006.

[11] R. DeCarlo, M. Branicky, S. Pettersson, and B. Lennartson. Perspectives and results on the stability and stabilizability of hybrid systems. *Proceedings of the IEEE*, 88(7):1069–1082, 2002.

[12] S. Di Cairano and A. Bemporad. Equivalent piecewise affine models of linear hybrid automata. *IEEE Trans. Aut. Control*, 55(2):498–502, 2010.

[13] S. Di Cairano, A. Bemporad, I. Kolmanovsky, and D. Hrovat. Model predictive control of magnetically actuated mass spring dampers for automotive applications. *Int. J. Control*, 80(11):1701–1716, 2007.

[14] S. Di Cairano, M. Lazar, A. Bemporad, and W. Heemels. A control Lyapunov approach to predictive control of hybrid systems. In B. M. M. Egerstedt, editor, *Hybrid Systems: Computation and Control*, Lect. Not. in Computer Science, pages 130–143. Springer-Verlag, 2008.

[15] S. Di Cairano, H. Tseng, D. Bernardini, and A. Bemporad. Vehicle yaw stability control by coordinated active front steering and differential braking in the tire sideslip angles domain. *IEEE Trans. Contr. Systems Technology*, 2012. Preprints on ieeexplore.org.

[16] E. W. Dijkstra. A note on two problems in connexion with graphs. *Numerische Mathematik*, 1(1):269–271, 1959.

[17] M. Egerstedt. Behavior based robotics using hybrid automata. In B. Krogh and N. Lynch, editors, *Hybrid Systems: Computation and Control*, number 1790 in Lecture Notes in Computer Science, pages 103–116. Springer-Verlag, 2000.

[18] S. Engell and O. Stursberg. Hybrid control techniques for the design of industrial controllers. In *Proc. 44th*

IEEE Conf. on Decision and Control, pages 5612–5617, Seville, Spain, 2005.

[19] R. Goebel, R. Sanfelice, and A. Teel. Hybrid dynamical systems. *Control Systems Magazine*, 29(2):28–93, 2009.

[20] W. Heemels, B. de Schutter, and A. Bemporad. Equivalence of hybrid dynamical models. *Automatica*, 37(7):1085–1091, July 2001.

[21] W. Heemels, B. D. Schutter, J. Lunze, and M. Lazar. Stability analysis and controller synthesis for hybrid dynamical systems. *Philosophical Transactions of the Royal Society A*, 368:4937–4960, 2010.

[22] T. Henzinger. The theory of hybrid automata. In *Proceedings of the 11th Annual IEEE Symposium on Logic in Computer Science (LICS '96)*, pages 278–292, New Brunswick, New Jersey, 1996.

[23] J. Hespanha. Uniform stability of switched linear systems: extensions of LaSalle's invariance principle. *IEEE Trans. Aut. Control*, 49(4):470–482, 2004.

[24] M. Johansson. *Piecewise linear control systems*. PhD thesis, Lund Institute of Technology, Sweden, 1999.

[25] M. Johansson and A. Rantzer. Computation of piecewise quadratic Lyapunov functions for hybrid systems. *IEEE Trans. Aut. Control*, 43(4):555–559, 1998.

[26] C. Kellett and A. Teel. Discrete-time asymptotic controllability implies smooth control-Lyapunov function. *Systems & Control Letters*, 52(5):349–359, 2004.

[27] C. M. Kellett and A. R. Teel. On the robustness of $\mathcal{KL}$-stability for difference inclusions: Smooth discrete-time Lyapunov functions. *SIAM Journal on Control and Optimization*, 44(3):777–800, 2005.

[28] M. V. Kothare, V. Balakrishnan, and M. Morari. Robust constrained model predictive control using linear matrix inequalities. *Automatica*, 32(10):1361–1379, 1996.

[29] M. Lazar. *Model predictive control of hybrid systems: Stability and robustness*. PhD thesis, Eindhoven University of Technology, The Netherlands, 2006.

[30] M. Lazar, W. P. M. H. Heemels, B. J. P. Roset, H. Nijmeijer, and P. P. J. van den Bosch. Input-to-state stabilizing sub-optimal NMPC algorithms with an application to DC-DC converters. *International Journal of Robust and Nonlinear Control*, 18:890–904, 2007.

[31] M. Lazar, W. P. M. H. Heemels, S. Weiland, and A. Bemporad. Stabilizing model predictive control of hybrid systems. *IEEE Trans. Aut. Control*, 51(11):1813–1818, 2006.

[32] M. Lazar, D. Muñoz de la Peña, W. P. M. H. Heemels, and T. Alamo. On input-to-state stability of min-max nonlinear model predictive control. *Systems & Control Letters*, 57:39–48, 2008.

[33] B. Lennartson, M. Tittus, B. Egardt, and S. Pettersson. Hybrid systems in process control. *Control Systems Magazine*, 16(5):45–56, 1996.

[34] D. Liberzon and A. Morse. Basic problems in stability and design of switched systems. *Control Systems Magazine*, 19(5):59–70, 2002.

[35] J. Lygeros, K. Johansson, S. Simic, J. Zhang, and S. Sastry. Dynamical properties of hybrid automata. *IEEE Trans. Aut. Control*, 48:2–17, Jan. 2003.

[36] D. Mayne, J. Rawlings, C. Rao, and P. Scokaert. Constrained model predictive control: Stability and optimality. *Automatica*, 36(6):789–814, June 2000.

[37] D. Q. Mayne, J. B. Rawlings, C. V. Rao, and P. O. M. Scokaert. Constrained model predictive control: Stability and optimality. *Automatica*, 36(6):789–814, 2000.

[38] D. Mignone, G. Ferrari-Trecate, and M. Morari. Stability and stabilization of piecewise affine and hybrid systems: An LMI approach. In *Proc. 39th IEEE Conf. on Decision and Control*, volume 1, pages 504–509, 2002.

[39] I. Mitchell, A. Bayen, and C. Tomlin. A time-dependent hamilton-jacobi formulation of reachable sets for continuous dynamic games. *IEEE Trans. Aut. Control*, 50(7):947–957, 2005.

[40] B. Potocnik, A. Bemporad, F. Torrisi, G. Music, and B. Zupancic. Hybrid modelling and optimal control of a multiproduct batch plant. *Control Engineering Practice*, 12(9):1127–1137, 2004.

[41] G. Ripaccioli, A. Bemporad, F. Assadian, C. Dextreit, S. Di Cairano, and I. Kolmanovsky. Hybrid Modeling, Identification, and Predictive Control: An Application to Hybrid Electric Vehicle Energy Management. In *Hybrid Systems: Computation and Control*, volume 5469, pages 321–335. 2009.

[42] M. Roozbehani, E. Feron, and A. Megrestki. Modeling, optimization and computation for software verification. In L. T. M. Morari, editor, *Hybrid Systems: Computation and Control*, number 3414 in Lecture Notes in Computer Science, pages 606–622. Springer-Verlag, 2005.

[43] E. D. Sontag. Nonlinear regulation: the piecewise linear approach. *IEEE Transactions on Automatic Control*, 26(2):346–357, 1981.

[44] E. D. Sontag. A Lyapunov-like characterization of asymptotic controllability. *SIAM Journal of Control and Optimization*, 21:462–471, 1983.

[45] C. Tomlin, I. Mitchell, A. Bayen, and M. Oishi. Computational techniques for the verification of hybrid systems. *Proceedings of the IEEE*, 91(7):986–1001, 2003.

[46] F. Torrisi and A. Bemporad. HYSDEL — A tool for generating computational hybrid models. *IEEE Trans. Contr. Systems Technology*, 12(2):235–249, Mar. 2004.

[47] A. J. van der Schaft and J. M. Schumacher. *An introduction to hybrid dynamical systems*, volume 251 of *Lecture Notes in Control and Information Sciences*. Springer, 2000.

[48] H. Witsenhausen. A class of hybrid-state continuous-time dynamic systems. *IEEE Trans. Aut. Control*, 11(2):161–167, 1966.

CoSyMA: A Tool for Controller Synthesis using Multi-scale Abstractions*

Sebti Mouelhi
POP ART project
INRIA Grenoble – Rhône-Alpes
38334 St Ismier, France

Antoine Girard
Laboratoire Jean Kuntzmann
Université de Grenoble
B.P. 53, 38041 Grenoble, France

Gregor Gössler
POP ART project
INRIA Grenoble – Rhône-Alpes
38334 St Ismier, France

ABSTRACT

We introduce CoSyMA, a tool for automatic controller synthesis for incrementally stable switched systems based on multi-scale discrete abstractions. The tool accepts a description of a switched system represented by a set of differential equations and the sampling parameters used to define an approximation of the state-space on which discrete abstractions are computed. The tool generates a controller — if it exists — for the system that enforces a given safety or time-bounded reachability specification. We illustrate by examples the synthesized controllers and the significant performance gains during their computation.

Categories and Subject Descriptors

I.2.8 [**Artificial Intelligence**]: Problem Solving, Control Methods and Search—*Control theory*

Keywords

Switched systems, Multi-scale abstractions, Controller synthesis, Symbolic Algorithms.

1. INTRODUCTION

Controller synthesis for hybrid systems using discrete abstractions has become an established approach (see [12] and the references therein). In [6], it has been demonstrated that the (sampled) continuous behavior of incrementally stable [1] switched systems are *approximately bisimilar* to some discrete abstractions. The states of these abstractions are elements of lattices that approximate the continuous state-space. Time and space sampling parameters are chosen to achieve a desired precision; the smaller the time sampling parameter, the finer the lattice used for approximating the state-space, and consequently, the larger the number of states in the abstraction. The approach detailed in [4, 3] based on *multi-scale* discrete abstractions can be used to cope with this problem. These abstractions are defined over a set of embedded lattices. The finer lattices are only explored when the specification cannot be met at the coarsest level.

In this paper, we present CoSyMA (*COntroller SYnthesis using Multi-scale Abstractions*), a tool implementing symbolic approaches based on multi-scale abstractions to synthesize controllers for incrementally stable switched systems. CoSyMA accepts as input a switched system defined by differential equations indexed by a set of modes, time and space sampling parameters used to set an approximation of the continuous state-space, and a safety or a time-bounded reachability specification. If it exists, it computes a controller satisfying the specification. The tool is implemented using OCaml [8] and it is available (with documentation) for download at `multiscale-dcs.gforge.inria.fr`.

Approximately bisimilar abstractions are also used in the tool PESSOA [9]. PESSOA handles arbitrary switched or continuous systems (not only incrementally stable ones) and applies thus to a more general class of systems, however it does not feature the multi-scale discrete abstractions, which constitute the core of CoSyMA for which we implemented dedicated algorithms allowing us to reduce the computational effort demanded by controller synthesis.

2. THEORETICAL BACKGROUND

2.1 Incrementally stable switched systems

DEFINITION 2.1. *A* switched system *is a quadruple* $\Sigma = \langle \mathbb{R}^n, P, \mathcal{P}, F\rangle$ *where* $\mathbb{R}^n$ *is the state-space;* $P = \{1, \ldots, m\}$ *is a finite set of modes;* $\mathcal{P}$ *is the set of piecewise constant functions from* $\mathbb{R}^+$ *to* P*, continuous from the right and with a finite number of discontinuities on every bounded interval of* $\mathbb{R}^+$*;* $F = \{f_1, \ldots, f_m\}$ *is a collection of smooth vector fields indexed by* P*. For all modes* $p \in P$*,* $f_p : \mathbb{R}^n \to \mathbb{R}^n$ *is a locally Lipschitz continuous map.*

A *switching signal* of Σ is a function $\mathbf{p} \in \mathcal{P}$. A piecewise $\mathcal{C}^1$ function $\mathbf{x} : \mathbb{R}^+ \to \mathbb{R}^n$ is said to be a *trajectory* of Σ if it is continuous and there exists a switching signal $\mathbf{p} \in \mathcal{P}$ such that, at each $t \in \mathbb{R}^+$ where the function $\mathbf{p}$ is continuous, $\mathbf{x}$ is continuously differentiable and satisfies $\dot{\mathbf{x}}(t) = f_{\mathbf{p}(t)}(\mathbf{x}(t))$. We denote by $\mathbf{x}(t, x, p)$ the point reached at time t, starting from the state x and applying the constant switching

*This work was supported by the French project VEDECY number ANR 2009 SEGI 015 01 and the project UJF-MSTIC SYMBAD.

signal $\mathbf{p}(s) = p$, for all $s \in [0, t]$. A switched system is δ-GUAS (i.e. globally uniformly asymptotically incrementally stable [1, 6]) if all the trajectories associated with the same switching signal converge asymptotically to the same reference trajectory independently of their initial states.

2.2 Approximate bisimulation

In this section we provide a brief introduction of the notion of approximate bisimulation that relates a switched system to the specific discrete abstraction we construct. Let us consider a class of transition systems of the form $T = \langle Q, L, r, O, H, I\rangle$ consisting of a set of states Q; a set of labels L; a transition relation $r \subseteq Q \times L \times Q$; an output set O; an output function $H : Q \to O$; a set of initial states $I \subseteq Q$. T is said to be *metric* if the output set O is equipped with a metric d, *discrete* if Q and L are finite or countable sets. For $q \in Q$ and $l \in L$ let $succs_l(q) = \{q' \in Q \mid (q, l, q') \in r\}$. An action $l \in L$ belongs to the set of *enabled actions* at the state q, denoted $Enab(q)$, if $succs_l(q) \neq \emptyset$. The transition system is said to be *deterministic* if for all $q \in Q$ and $l \in Enab(q)$, $succs_l(q)$ has only one element denoted by $succ_l(q)$. A *trajectory* of the transition system is a finite or infinite sequence of transitions $\sigma = q_0 l_0 q_1 l_1 q_2 l_2 \ldots$, it is *initialized* if $q_0 \in I$. A state $q \in Q$ is *reachable* if there exists an initialized trajectory reaching q. We denote by $\Phi(T)$ the set of all the trajectories of T.

Transition systems can describe the dynamics of switched systems. Given a switched system $\Sigma = \langle \mathbb{R}^n, P, \mathcal{P}, F\rangle$, let $T(\Sigma) = \langle Q, L, r, O, H, I\rangle$ be the transition system where the set of states is $Q = \mathbb{R}^n$; the set of labels is $L = P \times \mathbb{R}^+$; the transition relation is given by $(x, (p, \tau), x') \in r$ iff $\mathbf{x}(\tau, x, p) = x'$, i.e. the switched system Σ goes from state x to state x' by applying the constant mode p for a duration τ; the set of outputs is $O = \mathbb{R}^n$; the observation map H is the identity map over $\mathbb{R}^n$; the set of initial states is $I = \mathbb{R}^n$. $T(\Sigma)$ is deterministic and metric when the set of outputs $O = \mathbb{R}^n$ is equipped with the metric $d(x, x') = \|x - x'\|$. The relation between the discrete abstractions and $T(\Sigma)$ can be defined by an approximate bisimulation [5].

DEFINITION 2.2. *Let $T_i = \langle Q_i, L_i, r_i, O_i, H_i, I_i\rangle$, for $i \in \{1, 2\}$, be metric transition systems where $L_1 = L_2$ and $O_1 = O_2$, equipped with the metric d, and a precision $\varepsilon \geq 0$. A relation $R \subseteq Q_1 \times Q_2$ is said to be an* ε-approximate bisimulation relation *between T_1 and T_2 if for all $(q_1, q_2) \in R$:*

- $d(H_1(q_1), H_2(q_2)) \leq \varepsilon$;
- $\forall (q_1, l, q_1') \in r_1, \exists (q_2, l, q_2') \in r_2$, *such that* $(q_1', q_2') \in R$;
- $\forall (q_2, l, q_2') \in r_2, \exists (q_1, l, q_1') \in r_1$, *such that* $(q_1', q_2') \in R$.

T_1 and T_2 are said to be approximately bisimilar with precision ε, denoted $T_1 \sim_\varepsilon T_2$, if for all $q_1 \in I_1$, there exists $q_2 \in I_2$, such that $(q_1, q_2) \in R$, and for all $q_2 \in I_2$, there exists $q_1 \in I_1$, such that $(q_1, q_2) \in R$.

2.3 Multi-scale abstractions

In applications where the switching has to be fast, uniform approximately bisimilar abstractions, as defined in [6], approximate the state-space using fine lattices which results in a huge number of abstract states. In practice, fast switching is generally necessary only on a restricted part of the state space. For instance, for safety specifications, fast switching is needed only when the system gets close to unsafe regions. In order to enable fast switching while using abstractions with a reasonable number of states, we consider discrete abstractions enabling transitions of different durations. For transitions of long duration, it is sufficient to consider abstract states on the coarse lattice. The finer ones are reached by shorter transitions only when the specification cannot be met at the coarsest level.

Let us consider a switched system Σ whose switching is determined by a time-triggered controller with time-periods in the finite set $\Theta_\tau^N = \{2^{-s}\tau \mid s = 0, \ldots, N\}$ that consists of dyadic fractions of a time sampling parameter $\tau > 0$ up to some scale parameter $N \in \mathbb{N}$. The dynamics of a switched system Σ is then described by the transition system $T_\tau^N(\Sigma) = \langle Q_1, P \times \Theta_\tau^N, r_1, O, H_1, I_1\rangle$ where $Q_1 = O = I_1 = \mathbb{R}^n$, H_1 is the identity map over $\mathbb{R}^n$, and $(x, (p, 2^{-s}\tau), x') \in r_1$ iff $\mathbf{x}(2^{-s}\tau, x, p) = x'$.

The discrete abstraction of $T_\tau^N(\Sigma)$ is defined on an approximation of $Q_1 = \mathbb{R}^n$ by a set of embedded lattices $[\mathbb{R}^n]_\eta^s$ defined by

$$[\mathbb{R}^n]_\eta^s = \left\{ q \in \mathbb{R}^n \;\middle|\; q[i] = k_i \frac{2^{-s+1}\eta}{\sqrt{n}},\ k_i \in \mathbb{Z},\ i = 1, ..., n \right\}$$

where $s = 0, \ldots, N$, $q[i]$ is the i-th coordinate of q and $\eta > 0$ is a state space discretization parameter. By simple geometrical considerations, we can check that for all $x \in \mathbb{R}^n$ and $s = 0, \ldots, N$, there exists $q \in [\mathbb{R}^n]_\eta^s$ such that $\|x - q\| \leq 2^{-s}\eta$. Then, we can define the abstraction of $T_\tau^N(\Sigma)$ as the transition system $T_{\tau,\eta}^N(\Sigma) = \langle Q_2, P \times \Theta_\tau^N, r_2, O, H_2, I_2\rangle$, where the set of states is $Q_2 = [\mathbb{R}^n]_\eta^N$; the set of actions remains $L = P \times \Theta_\tau^N$; r_2 is defined such that $(q, (p, 2^{-s}\tau), q') \in r_2$ iff $q' = \arg\min_{m \in [\mathbb{R}^n]_\eta^s}(\|\mathbf{x}(2^{-s}\tau, q, p) - m\|)$. The approximation principle is illustrated in Figure 1. The observation map H_2 is the natural inclusion map from $[\mathbb{R}^n]_\eta^N$ to $\mathbb{R}^n$; the set of initial states is $I_2 = [\mathbb{R}^n]_\eta^0$.

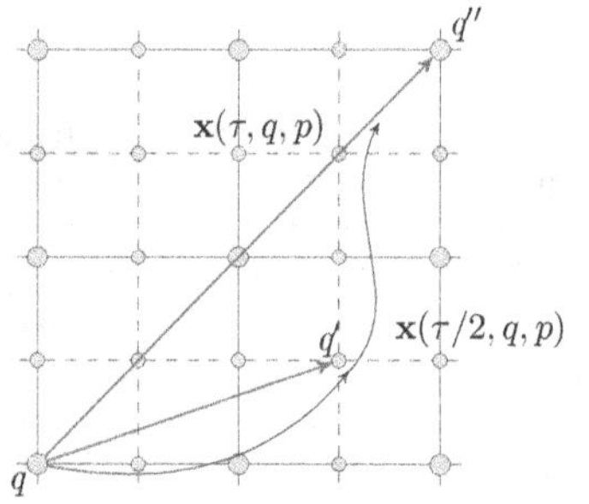

Figure 1: Computation of the discrete abstraction: $q' = succ_2(q, (p, \frac{\tau}{2})) = \arg\min_{m \in [\mathbb{R}^n]_\eta^1}(\|\mathbf{x}(\frac{\tau}{2}, q, p) - m\|)$ **and** $q'' = succ_2(q, (p, \tau)) = \arg\min_{m \in [\mathbb{R}^n]_\eta^0}(\|\mathbf{x}(\tau, q, p) - m\|)$

The resulting abstraction $T_{\tau,\eta}^N(\Sigma)$ is discrete and deterministic, its set of states and its set of actions are respectively countable and finite. For $N = 0$, we recover the "uniform" abstractions introduced in [6]. In [4], it was proved that for a switched system Σ admitting a common δ-GUAS Lyapunov function, $T_\tau^N(\Sigma) \sim_\varepsilon T_{\tau,\eta}^N(\Sigma)$ where the precision ε can be made arbitrarily small by reducing the state sampling parameter η.

2.4 Controller synthesis using multi-scale abstractions

Before presenting the tool details and our experimental results, we explain briefly how we use multi-scale abstractions for synthesizing safety and time-bounded reachability controllers. Let $T = \langle Q, L, r, O, H, I\rangle$ be a deterministic transition system, a *controller* for T is a map $\mathcal{S} : Q \to 2^L$ such that for all $q \in Q$, $\mathcal{S}(q) \subseteq Enab(q)$. The system T controlled by $\mathcal{S}$ is $T/\mathcal{S} = \langle Q, L, r_{\mathcal{S}}, O, H, I\rangle$ where the transition relation is given by $(q, l, q') \in r_{\mathcal{S}}$ iff $(l \in \mathcal{S}(q) \wedge (q, l, q') \in r)$. The *support* of $\mathcal{S}$ is defined by $\mathsf{supp}(\mathcal{S}) = \{q \in Q \mid \mathcal{S}(q) \neq \emptyset\}$.

Safety controller synthesis

Given a safety specification $Q_S \subseteq Q$ (obtained from a subset $O_S \subseteq O$ of safe outputs), a state q of T is *controllable with respect to a safety specification* Q_S if $q \in Q_S$ and there exists an infinite trajectory $\sigma \in \Phi(T)$ starting from q and remaining in Q_S. We denote the set of controllable states of T with respect to the safety specification Q_S by $SCont(T, Q_S)$. A *safety controller* $\mathcal{S}$ for T and Q_S is defined such that $\mathsf{supp}(\mathcal{S}) \subseteq SCont(T, Q_S)$ and for all $q \in \mathsf{supp}(\mathcal{S})$: (1) $q \in Q_S$ (safety) and (2) $\forall l \in \mathcal{S}(q)$, $succ_l(q) \in \mathsf{supp}(\mathcal{S})$ (deadend freedom). The set $SCont(T, Q_S)$ is computable for discrete abstractions. However, the larger the number of states, the more expensive the computation. For that reason, we want to capitalize on multi-scale abstractions to propose an efficient algorithm for safety controller synthesis.

The lazy safety synthesis problem consists in controlling a system so as to keep any trajectory starting from some initial state in I within the safe subset of states Q_S, while applying at each state transitions of the longest possible duration for which safety can be guaranteed. For that purpose we define a priority relation on the set of labels $L = P \times \Theta_\tau^N$ giving priority to transitions of longer duration: for $l, l' \in L$ with $l = (p, \tau)$, $l' = (p', \tau')$, $l \preceq l'$ iff $\tau \leq \tau'$, $l \prec l'$ iff $\tau < \tau'$ and $l \cong l'$ iff $\tau = \tau'$. Given a subset of labels $L' \subseteq L$, we define $\max_{\preceq}(L') = \{l' \in L' \mid \forall l \in L', l \preceq l'\}$.

DEFINITION 2.3. *A maximal lazy safety (MLS) controller $\mathcal{S} : Q \to 2^L$ for T and Q_S is a safety controller such that $I \cap SCont(T, Q_S) \subseteq \mathsf{supp}(\mathcal{S})$ and for all states $q \in \mathsf{supp}(\mathcal{S})$, we have: (1) if $l \in \mathcal{S}(q)$, then for any $l \prec l'$, $succ_{l'}(q) \notin SCont(T, Q_S)$ (laziness), and (2) if $l \in \mathcal{S}(q)$, then for any $l \cong l'$, $l' \in \mathcal{S}(q)$ iff $succ_{l'}(q) \in SCont(T, Q_S)$ (maximality).*

The MLS controller exists and is unique as proved in [3].

Time-bounded reachability controller synthesis

Given a transition system $T = \langle Q, L, r, O, H, I\rangle$, for all transitions $(q, l, q') \in r$, let $\delta(l)$ be the time needed by T to reach q' from q by action l. For all finite trajectories $\sigma = q_0 l_0 q_1 l_1 \dots l_{n-1} q_n$ in $\Phi(T)$, we define its *duration* by $\Delta(\sigma) = \delta(l_0) + \delta(l_1) + \dots + \delta(l_{n-1})$. For instance, for the transition systems $T_\tau^N(\Sigma)$ and $T_{\tau,\eta}^N(\Sigma)$, we have $L = P \times \Theta_\tau^N$ and for all $l = (p, 2^{-s}\tau) \in L$, we have $\delta(l) = 2^{-s}\tau$.

To formally define a time-bounded reachability controller, we define $C(T) = \langle Q_c, L, r_c, O, H_c, I_c\rangle$ the *transition system with clock* of T where $Q_c = Q \times \mathbb{R}^+$ is the set of states Q extended by a clock; for all $((q, c), l, (q', c')) \in r_c$, $(q, l, q') \in r$ and $c' = c + \delta(l)$; $H_c((q, c)) = H(q)$ for all $(q, c) \in Q_c$; $I_c = I \times \{0\}$. The set of reachable states of $C(T_{\tau,\eta}^N(\Sigma))$ is countable and defined by $Q \times 2^{-N}\tau\mathbb{N}$. Given a maximal time bound $B \in \mathbb{R}^+$, the state (q, c) of $C(T)$ is controllable with respect to a time-bounded reachability specification (Q_S, Q_T, B), where $Q_T \subseteq Q_S$ if (1) $q \in Q_S$ and there exists a finite trajectory σ of T starting from q, eventually reaching Q_T, and remaining in Q_S until reaching Q_T, such that $c + \Delta(\sigma) \leq B$. The set of these states is denoted by $RCont(C(T), Q_S, Q_T, B)$.

Next we define time-bounded reachability controllers using the notion of safety controllers. We start by defining the notion of *stuttering* ($\circ$) actions. An outgoing transition from a state q labeled by a stuttering action loops on the same state ($succ_\circ(q) = q$ and $\delta(\circ) = 0$).

Let us now define $T_{\circlearrowleft Q'} = \langle Q, L \cup \{\circ\}, r^{\circlearrowleft}, O, H, I\rangle$ for $Q' \subseteq Q$ such that

$$(q, l, q') \in r^{\circlearrowleft} \text{ iff } \begin{cases} q = q' & \text{if } l = \circ \text{ and } q \in Q'; \\ (q, l, q') \in r & \text{if } l \neq \circ \text{ and } q \in Q \setminus Q'. \end{cases}$$

$T_{\circlearrowleft Q'}$ is the transition system derived from T where the only actions enabled from a state in Q' are stuttering.

Since we are not concerned with the evolution of the system after reaching the target Q_T, we will use the transition system $C(T_{\circlearrowleft Q_T}) = \langle Q_c, L, r_c^{\circlearrowleft}, O, H_c, I_c\rangle$ rather than $C(T)$. We easily show that $RCont(C(T_{\circlearrowleft Q_T}), Q_S, Q_T, B) = RCont(C(T), Q_S, Q_T, B) = SCont(C(T_{\circlearrowleft Q_T}), Q_S \times [0, B])$.

A *time-bounded reachability controller* for the transition system $C(T_{\circlearrowleft Q_T})$ and (Q_S, Q_T, B) is a safety controller for $C(T_{\circlearrowleft Q_T})$ and $Q_S \times [0, B]$. We define the maximal lazy time-bounded reachability controller based on Definition 2.3 as follows.

DEFINITION 2.4. *The* maximal lazy time-bounded reachability (MLBR) controller *$\mathcal{R}^m : Q \times \mathbb{R}^+ \to 2^L$ for $C(T_{\circlearrowleft Q_T})$ and (Q_S, Q_T, B) is the MLS controller for $C(T_{\circlearrowleft Q_T})$ and $Q_S \times [0, B]$.*

It is clear, based on the previous definition, that $\mathcal{R}^m$ is unique and that Q_T is reachable within the specified time bound starting from controllable initial states. We can use the algorithm synthesizing MLS controllers proposed in [3] to synthesize MLBR controllers. However, their computation is expensive because dealing with problems of n dimensions amounts to handle the equivalents of $n+1$ dimensions by adding a clock. To avoid this constraint, we can settle for a sub-controller $\mathcal{V}$ of $\mathcal{R}^m$ such that all controllable initial states of $\mathcal{R}^m$ are also controllable by $\mathcal{V}$.

Let $\mathcal{R}^m$ be the MLBR controller for $C(T_{\circlearrowleft Q_T})$ and (Q_S, Q_T, B). A *sub-controller* $\mathcal{V}$ of $\mathcal{R}^m$ is a time-bounded reachability controller for $C(T_{\circlearrowleft Q_T})$ and (Q_S, Q_T, B) such that for all $(q, c) \in \mathsf{supp}(\mathcal{V})$, $\mathcal{V}((q, c)) \subseteq \mathcal{R}^m((q, c))$. The sub-controller $\mathcal{V}$ is *complete* if $\{i \in I \mid (i, 0) \in \mathsf{supp}(\mathcal{R}^m)\} = \{i \in I \mid \exists 0 \leq c \leq B \mid (i, c) \in \mathsf{supp}(\mathcal{V})\}$. We can now define static reachability controllers based on the previous definition.

DEFINITION 2.5. *Consider a complete sub-controller $\mathcal{V}$ of $\mathcal{R}^m$. The* static reachability controller *$\mathcal{R}_{\mathcal{V}}^{ls} : Q \to 2^L$ for $T_{\circlearrowleft Q_T}$ and (Q_S, Q_T, B) obtained from $\mathcal{V}$ is the controller such that for all $(q, c) \in \mathsf{supp}(\mathcal{V})$, $q \in \mathsf{supp}(\mathcal{R}_{\mathcal{V}}^{ls})$ and for all $q \in \mathsf{supp}(\mathcal{R}_{\mathcal{V}}^{ls})$, $\mathcal{R}_{\mathcal{V}}^{ls}(q) = \mathcal{V}((q, c_{max}(q)))$ where $c_{max}(q) = max\{c' \mid (q, c') \in \mathsf{supp}(\mathcal{V})\}$.*

It can be shown that using $\mathcal{R}_{\mathcal{V}}^{ls}$, Q_T is reachable within the specified time bound starting from all controllable initial states (see [10]). For time-bounded reachability specifications CoSyMA synthesizes static reachability controllers. In the next section, we present some details about the tool.

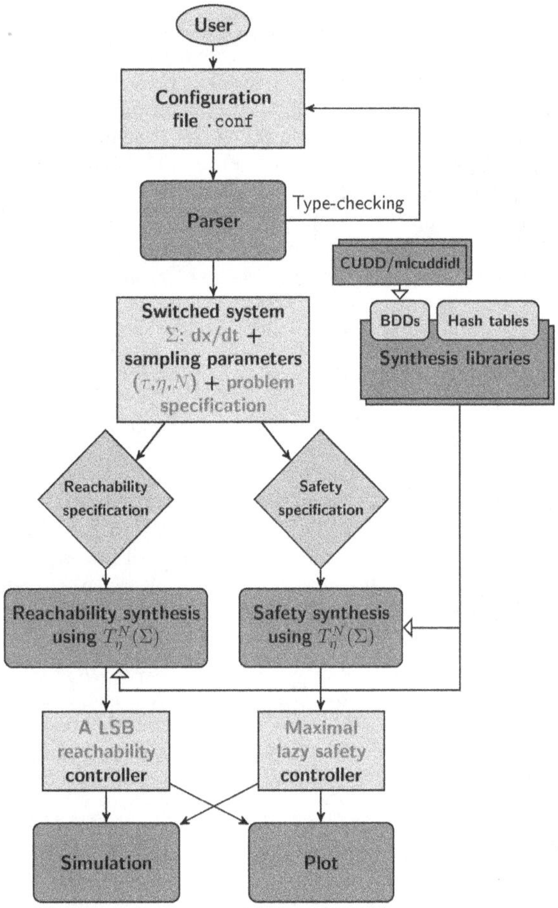

Figure 2: Execution flow of CoSyMA

3. TOOL DETAILS

In this section, we present the internal architecture of CoSyMA, its description language, and our implementation choices. CoSyMA accepts a configuration (`.conf`) file describing the system and the synthesis parameters. It contains in the order: the description of a switched system Σ in terms of differential equations $\dot{\mathbf{x}}(t) = f_{\mathbf{p}(t)}(\mathbf{x}(t))$; time τ and space η sampling parameters, and the scale N used to compute the finer lattice $[\mathbb{R}]_\eta^N$ that approximates the continuous state-space; a safety specification Q_S or a time-bounded reachability specification (Q_S, Q_T, B); the plot parameters. The grammar is detailed in the reference manual of the tool.

The execution flow of the tool, shown in Figure 2, represents the different steps by which the tool synthesizes the controllers of the described system according to the given safety or a time-bounded reachability specification. After the parsing of the configuration file, the tool represents the vector fields f_p for each switching mode p by an OCaml function of type `float -> float array -> float array`. It computes the successor of a state q under a label $l = (p, 2^{-s}\tau)$ by solving $\dot{\mathbf{x}}(t) = f_p(\mathbf{x}(t))$ for $t \in [0, \delta(l)]$ and $\mathbf{x}(0) = q$ using the common fourth-order Runge-Kutta method. The tool synthesizes the controllers based on $T_{\tau,\eta}^N$ approximately bisimilar to T_τ^N (cf. Section 2.3). For safety specifications Q_S, the tool synthesizes the MLS controller based on the algorithm presented in [3] which computes the abstraction $T_{\tau,\eta}^N$ on the fly.

For time-bounded reachability specifications (Q_S, Q_T, B), it uses a new algorithm computing the static reachability controller $\mathcal{R}_\mathcal{V}^{ls}$ where $\mathcal{V}$ is a complete sub-controller of the MLBR controller $\mathcal{R}^m$ for $C(T_{\circlearrowright Q_T})$ and (Q_S, Q_T, B) (with $T = T_{\tau,\eta}^N(\Sigma)$). The algorithm is a depth-first traversal of paths $\sigma \in \Phi(T_{\tau,\eta}^N(\Sigma))$ starting from initial states I until reaching Q_T to keep track of the clocks of the states reached by σ. From an operational point of view, the complete sub-controller $\mathcal{V} \subseteq \mathcal{R}^m$, according to which the static reachability controller $\mathcal{R}_\mathcal{V}^{ls}$ is defined, is computed according to the order of exploration of initial states I. By keeping the problem at its original dimension, the synthesis complexity is significantly reduced. The algorithm details are given in the reference manual [10].

The user has the choice to represent the system abstractions either by *enumerated types* or *boolean functions*. The tool uses *hash tables* as an enumerated type to operate on the system abstractions. Hash tables turn out to be more efficient than search trees or other table lookup structures, especially for large numbers of entries. Alternatively, the tool can use BDDs (*Binary Decision Diagrams*) [2] to represent the system abstractions. BDDs are able to represent sets and relations compactly in memory as boolean functions. We implement BDDs using the modules **`Cudd.Man`** and **`Cudd.Bdd`** of the OCaml IDL interface MLCUDDIDL [7] of the CUDD (*CU Decision Diagram*) package [11]. However, using BDDs makes the controller synthesis algorithms presented above more costly than hash tables since the symbolic abstraction is constructed by enumerating the states and computing their successors using the Runge-Kutta method.

The Plot backend of CoSyMA uses its own TikZ/Pgf scripts [13] used to generate plots for controllers and their simulation.

4. EXPERIMENTAL RESULTS

DC-DC Converter

As a first case study, we apply our approach to a boost DC-DC converter. It is a switched system with two modes, the two dimensional dynamics associated with both modes are affine of the form $\dot{\mathbf{x}}(t) = a_p\mathbf{x}(t) + b$ for $p \in \{1, 2\}$ (see [6] for numerical values). It can be shown that it is incrementally stable and thus approximately bisimilar discrete abstractions can be computed. We consider the problem of keeping the state of the system in a desired region of operation given by the safe set $O_S = [1.15, 1.55] \times [5.45, 5.85]$.

We use approximately bisimilar abstractions to synthesize MLS controllers for the DC-DC converter. We compare the cost of controller synthesis for the uniform abstraction T_{τ_1,η_1}^0 for parameters $\tau_1 = 0.5s$ and $\eta_1 = 10^{-3}\sqrt{2}/4$ (containing transitions of duration 0.5s) and the multi-scale abstractions T_{τ_2,η_2}^2 for parameters $\tau_2 = 4\tau_1$ and $\eta_2 = 4\eta_1$ (containing transitions of durations in $\Theta_\tau^2 = \{2s, 1s, 0.5s\}$). These two abstractions have the same precision. Table 1 details the experimental results obtained for the synthesis of the controllers for T_{τ_1,η_1}^0 and T_{τ_2,η_2}^2. We can see that there is a noteworthy reduction of the time used to compute the controller using multi-scale abstractions instead of using uniform ones (up to a 86% improvement between T_{τ_1,η_1}^0 and T_{τ_2,η_2}^2). This is due to the fact that the size of uniform abstractions grows

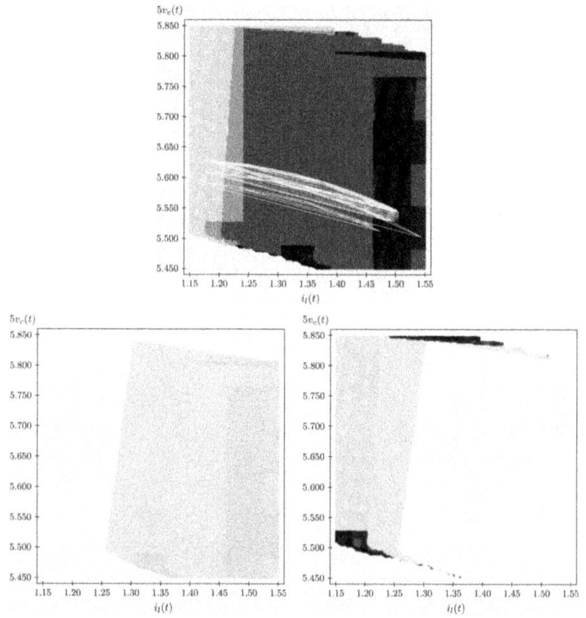

Figure 3: The MLS controller for $T^2_{\tau_2,\eta_2}$ and Q_S. Top: mode 1 is activated (light gray); mode 2 is activated (black); modes 1 and 2 (gray); Bottom (Left): actions of 2s are enabled (light gray); Bottom (right): actions of 1s are enabled (light gray), actions of 0.5s are enabled (black).

exponentially with higher resolutions, whereas using multi-scale abstractions are refined only when we get close to unsafe regions (reduction of more than 91% between $T^0_{\tau_1,\eta_1}$ and $T^2_{\tau_2,\eta_2}$). Interestingly, this reduction in computation time and size does not affect the performance of the multi-scale controllers, which yield a ratio of controllable initial states[1] (CR) over the safety specification comparable to that of their its uniform counterparts. It is worth emphasizing that using CoSyMA there is a remarkable reduction of the computation times compared to those reported in [3] obtained by a prototype implementation of the algorithm. Figure 3 depicts the maximal lazy safety controller for $T^2_{\tau_2,\eta_2}$ and Q_S and the trace of its simulation starting from the state (1.15,5.6).

	Abstractions $T^N_{\tau,\eta}$	
	$N=0, \tau=0.5s$, $\eta=10^{-3}\sqrt{2}/4$, $\varepsilon=0.1$	$N=2, \tau=2s$, $\eta=10^{-3}\sqrt{2}$, $\varepsilon=0.1$
Time	8.32s	1.10s
Size	599 294	53 479
$\delta(l)$	0.5s (100%)	2s (63.61%) 1s (31.67%) 0.5s (4.72%)
CR	93.52%	93.51%

Table 1: Experimental results for the MLS controller synthesis for the boost DC-DC converter

Now, we consider the time-bounded reachability specification (Q_S, Q_T, B) where $Q_S = [0.65, 1.65] \times [4.95, 5.95]$,

[1]The ratio of controllable initial states for a controller $\mathcal{S} : Q \to 2^L$ and a system $T = (Q, L, \to, O, H, I)$ is computed as $|\{q \in I | \mathcal{S}(q) \neq \emptyset\}|/|I|$.

$Q_T = [1.1, 1.6] \times [5.4, 5.9]$ and $B = 20s$. We synthesize static reachability controllers respectively for $T^0_{\tau_1,\eta_1}$ where $\tau_1 = 0.25$ and $\eta_1 = 10^{-3}\sqrt{2}/4$, and $T^2_{\tau_2,\eta_2}$ where $\tau_2 = 4\tau_1$ and $\eta_2 = 4\eta_1$, and the specification (Q_S, Q_T, B).

As shown in Section 2.4, the MLBR controller for the abstraction $C(T_{\circlearrowleft Q_T})$ and (Q_S, Q_T, B) where T equal to $T^0_{\tau_1,\eta_1}$ or $T^2_{\tau_2,\eta_2}$ is the maximal lazy safety controller for $C(T_{\circlearrowleft Q_T})$ and the safety specification $Q_S \times [0, B]$. Its computation is costly because the problem is grown from 2 to 3 dimensions by considering the supplementary clock parameter. Synthesizing a static reachability controller rather than an MLB reachability controller significantly reduces complexity. In Table 2, we observe a considerable reduction of the size of the controlled abstraction of $T^2_{\tau_2,\eta_2}$ of more than 91.46% compared to that of $T^0_{\tau_1,\eta_1}$ with comparable controllability ratios of initial states. Also, the computation time of the controller of $T^2_{\tau_2,\eta_2}$ is slightly shorter than that of $T^0_{\tau_1,\eta_1}$.

	Abstractions $T^N_{\tau,\eta}$	
	$N=0, \tau=0.5s$, $\eta=10^{-3}\sqrt{2}/4$, $\varepsilon=0.1$	$N=2, \tau=2s$, $\eta=10^{-3}\sqrt{2}$, $\varepsilon=0.1$
Time	658 s	223 s
Size	3 149 538	262 593
$\delta(l)$	0.5s (100%)	2s (72.26%) 1s (15.77%) 0.5s (11.97%)
CR	89.97%	90.30%

Table 2: Experimental results for the static reachability controller for the boost DC-DC converter

Building Temperature Regulation

The second case study deals with temperature regulation in a circular building. Each room is equipped with a heater and at a given instant at most one heater is switched on. The temperature t_i of the room i is defined by the differential equation $\dot{t}_i = \alpha(t_{i+1} + t_{i-1} - 2t_i) + \beta(t_e - t_i) + \gamma(t_h - t_i)u_i(t)$ where t_{i-1} is the temperature of the room $i-1$; t_{i+1} is the temperature of the room $i+1$; t_e is the temperature of the external environment of the building; t_h is the temperature of the heater; α is the temperature transfer ratio between the rooms $i \pm 1$ and the room i; β is the temperature transfer ratio between the external environment and the room i; γ is the temperature transfer ratio between the heater and the room i; $u_i(t)$ equals to 1 if the room i is heated, or 0 otherwise. Given a number $n \geq 2$ of rooms, we distinguish $n+1$ switching modes. For $1 \leq i \leq n$, the mode p_i represents the mode of activating the heater of room i. The mode p_{n+1} represents that no heater is activated. The values of α, β, γ, t_e, and t_h are respectively 1/20, 1/200, 1/100, 10, and 50. We will increase the system dimension to test the limits of the tool in terms of memory usage and computation time. Given the safety specification $Q_S = [20.0, 22.0]^n$ for $n \in \{3, 4, 5\}$, we synthesize safety controllers for buildings of three, four, and five rooms. The values of τ and η are given in Table 3. By looking to the results, we can see the combinatorial explosion of the size of abstractions by increasing the system dimension from 3 to 5. Also, it makes sense that the ratio of controllability of initial states decreases by increasing the number of rooms. On our machine equipped with a Core i5-2430M and 4GB of RAM, synthesis fails for the 6-dimensional instance due to running out of memory.

	Abstractions $T^N_{\tau,\eta}$		
	$n=3$, $N=2$, $\eta = 50 \times 10^{-3}$ $\tau = 20s$ $\varepsilon = 0.2$	$n=4$, $N=2$, $\eta = 50 \times 10^{-3}$ $\tau = 20s$ $\varepsilon = 0.2$	$n=5$, $N=1$ $\eta = 0.1$ $\tau = 10s$ $\varepsilon = 0.4$
Time	2.40s	595 s	571 s
Size	55 564	3 927 564	6 135 218
$\delta(l)$	20s (20.06%) 10s (79.94%) 5s (0%)	20s (8.77%) 10s (86.99%) 5s (4.24%)	10s (86.44%) 5s (13.56%)
CR	99.99%	99.89%	99.79%

Table 3: Comparison of experimental results for the safety synthesis for the temperature regulator system of three, four, and five dimensions

Figure 4 shows the MLS controller for the transition system $T^2_{20,0.05}$ of three dimensions and the safety specification $Q_S = [20.0, 22.0]^3$. The plots are slices of the state space in the dimensions (t_1, t_2) for a fixed $t_3 \approx 20°$ (left) and $t_3(t) \approx 22°$ (right), respectively. The plots on the top depict scales and those in the middle and the bottom depicts modes. We can remark the predominance of the mode p_4 (no heater is activated) by increasing the temperature of third room.

5. CONCLUSION

In this paper we have introduced CoSyMA, a tool that automatically synthesizes controllers for incrementally stable switched systems based on multi-scale discrete abstractions. We have illustrated by examples the synthesized controllers for safety and time-bounded reachability problems. The benchmarks provide evidence that the use of multi-scale abstractions leads to a substantial reduction of synthesis time and size of the obtained controller while maintaining coverage of the state space.

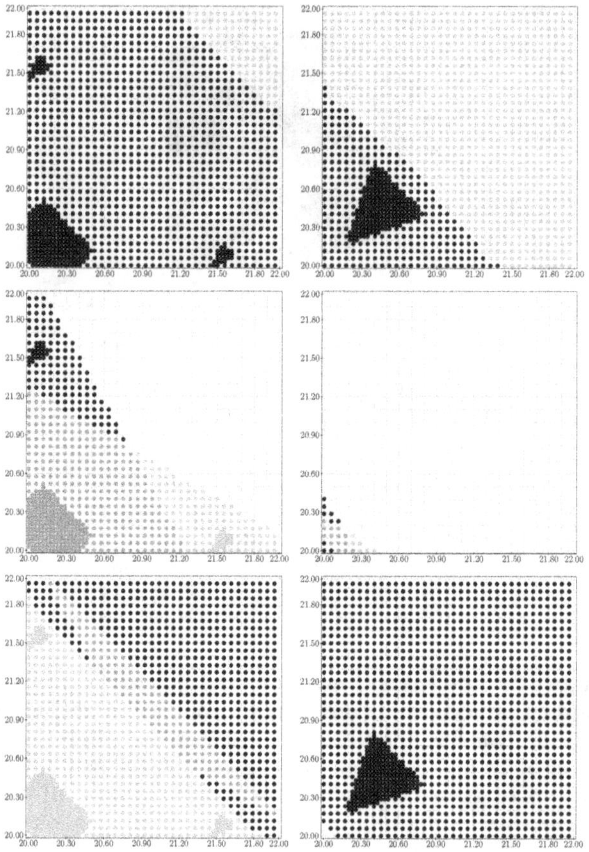

$t_3(t) \approx 20$ $t_3(t) \approx 22$

Figure 4: MLS controller for $T^2_{20,0.05}$ of three dimensions and Q_S; Horizontal axis: $t_1(t)$; Vertical axis: $t_2(t)$; Top: actions of 20s are enabled (black); actions of 10s are enabled (gray); Middle: mode p_1 (black); mode p_2 (light gray); p_1 and p_2 (gray); Bottom: mode p_4 (black); mode p_3 (light gray); p_3 and p_4 (gray).

6. REFERENCES

[1] D. Angeli. A Lyapunov approach to incremental stability properties. *IEEE Transactions on Automatic Control*, 47(3):410 –421, 2002.

[2] R. E. Bryant. Graph-based algorithms for Boolean function manipulation. *IEEE Transactions on Computers*, 35(8):677–691, 1986.

[3] J. Cámara, A. Girard, and G. Gössler. Safety controller synthesis for switched systems Using multi-scale symbolic models. In *IEEE Conference on Decision and Control and European Control Conference*, pages 520–525. IEEE, 2011.

[4] J. Cámara, A. Girard, and G. Gössler. Synthesis of switching controllers using approximately bisimilar multiscale abstractions. In *Hybrid Systems: Computation and Control*, pages 191–200. ACM, 2011.

[5] A. Girard and G. J. Pappas. Approximation metrics for discrete and continuous systems. *IEEE Transactions on Automatic Control*, 52(5):782 –798, 2007.

[6] A. Girard, G. Pola, and P. Tabuada. Approximately bisimilar symbolic models for incrementally stable switched systems. *IEEE Transactions on Automatic Control*, 55(1):116 –126, 2010.

[7] B. Jeannet. MLCUDDIDL: OCaml interface for CUDD library, version 2.2.0, 2011. `http://pop-art.inrialpes.fr/~bjeannet/mlxxxidl-forge/mlcuddidl/html/Cudd.html`.

[8] X. Leroy, D. Doligez, A. Frisch, J. Garrigue, D. Rémy, and J. Vouillon. The Objective Caml system release 3.12. Documentation and user manual, 2010.

[9] M. Mazo Jr., A. Davitian, and P. Tabuada. Pessoa: A tool for embedded controller synthesis. In *Computer Aided Verification*, volume 6174 of *LNCS*, pages 566–569. Springer, 2010.

[10] S. Mouelhi, G. Gössler, and A. Girard. Synthesizing controllers for switched systems using COSYMA. Technical report, INRIA, 2012.

[11] F. Somenzi. CUDD: CU decision diagram package. `http://vlsi.colorado.edu/~fabio/CUDD`.

[12] P. Tabuada. *Verification and control of hybrid systems: a symbolic approach.* Springer, 2009.

[13] T. Tantau. The TikZ and PGF packages, manual for version 2.10, 2010.

An Aircraft Electric Power Testbed for Validating Automatically Synthesized Reactive Control Protocols *

Robert Rogersten†, Huan Xu§, Necmiye Ozay§, Ufuk Topcu‡, and Richard M. Murray§

†KTH Royal Inst. of Tech.
rrog@kth.se

§California Inst. of Tech.
{mumu, necmiye, murray}
@cds.caltech.edu

‡University of Pennsylvania
utopcu@seas.upenn.edu

ABSTRACT

Modern aircraft increasingly rely on electric power for subsystems that have traditionally run on mechanical power. The complexity and safety-criticality of aircraft electric power systems have therefore increased, rendering the design of these systems more challenging. This work is motivated by the potential that correct-by-construction reactive controller synthesis tools may have in increasing the effectiveness of the electric power system design cycle. In particular, we have built an experimental hardware platform that captures some key elements of aircraft electric power systems within a simplified setting. We intend to use this platform for validating the applicability of theoretical advances in correct-by-construction control synthesis and for studying implementation-related challenges. We demonstrate a simple design workflow from formal specifications to autogenerated code that can run on software models and be used in hardware implementation. We show some preliminary results with different control architectures on the developed hardware testbed.

Categories and Subject Descriptors

D.2.4 [**Software Engineering**]: Software/Program Verification—*Formal methods*

Keywords

reactive synthesis; testbed; aircraft electric power system

1. INTRODUCTION AND MOTIVATION

Aircraft electric power systems have become increasingly important over the years because they support various subsystems and essential services on aircraft. These electrical services and subsystems are commonly referred to as system loads. System loads are of two categories, namely, primary loads (some of these are safety- or mission-critical) and secondary (noncritical) loads. The system needs to ensure that the primary loads are supplied with power at all times; that is, if a fault affects a part of the system that powers a primary load, the system must be able to reconfigure and provide power to the load through another path. In order to reconfigure a system, it is necessary to reroute power, which is accomplished with high power electromagnetic devices called contactors. The contactors are arranged such that they are magnetically held in a preferred state by an applied signal. The state is either open or closed. To reconfigure the contactors to react to faults and modes of operation, the system uses control logic that can sense system conditions and environmental conditions under which the system operates. The electric power system, therefore, includes voltage and current sensors connected to the control logic. In current practice, the control logic is often designed by hand, resulting in lengthy design and verification cycles. As an alternative approach, [12] and [13] explored the application of correct-by-construction reactive controller synthesis techniques.

In this paper, we report on our recently developed simulation models and a hardware testbed for validating reactive controllers synthesized using `TuLiP` [12], a temporal logic planning toolbox, in order to investigate the validity of the assumptions made in controller synthesis. `TuLiP` is a collection of Python-based code used for automatic synthesis of correct-by-construction embedded control software. Automatic synthesis of reactive centralized and distributed controllers of aircraft electric power systems is described in detail in [13]. The particular distributed synthesis method adopted in this study is introduced in [4] and [5].

University-scale testbeds for research on correct-by-construction controller synthesis are fairly limited. An advanced diagnostics and prognostics testbed is described in [9]. Some applications of this testbed to the electric power systems of spacecraft and aircraft are detailed in [3]. However, the experiments focused on diagnostic queries of the system, while our work is focused on the implementation of correct-by-construction control protocols for fault-tolerant operations. A robotics testbed implementing correct-by-construction controllers is described in [2].

`TuLiP` can be used to synthesize logic so that the satisfaction of certain safety requirements is guaranteed. The synthesized logic enables the contactors to react to changes in system conditions such as the status of generators and rectifier units. This is commonly referred to as a reactive system. The safety requirements used in our simulation models and hardware testbed stipulate that the alternating current generators should never be paralleled and that the duration for which the bus is not powered should never

*An extended version of this paper is available at [10].

HSCC'13, April 8–11, 2013, Philadelphia, Pennsylvania, USA.

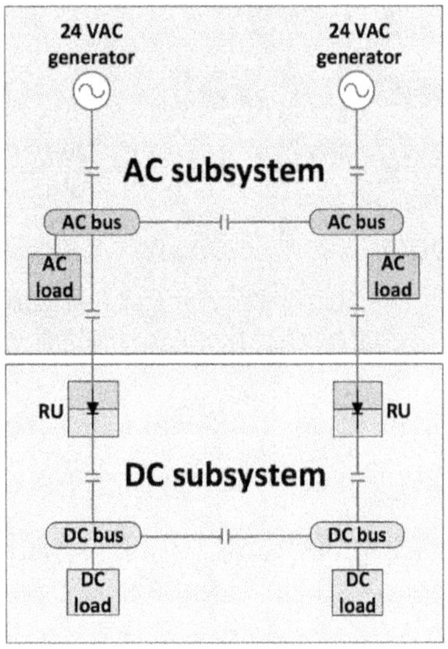

Figure 1: Single-line diagram of the power system testbed. Contactors are represented by double bars. The AC and DC sides of the system are separated by rectifier units (RU).

exceed a certain limit. They also include the environment-related assumption that at least a subset of the generators and rectifier units must be working at all times. The simulation models were built with the physical modeling software SIMPOWERSYSTEMS, an extension of SIMULINK [11]. In order to validate the controller on the experimental hardware platform, we synthesized and tested it using TuLiP and SIMPOWERSYSTEMS, respectively. Thereafter, we investigated the validity of the assumptions used for controller synthesis on the experimental hardware platform.

An aircraft electric power system uses different voltage levels, which can broadly be divided into four categories, namely, high-voltage AC, high-voltage DC, low-voltage AC, and low-voltage DC. The topology in Figure 1 is of specific interest because it is representative of some of the key features of aircraft electric power systems in simplified settings. Therefore, the hardware testbed was built based on the above mentioned topology.

2. THEORETICAL BACKGROUND

We now discuss the formal specification language utilized for the synthesis of control protocols and how these protocols are implemented in the software models and on the testbed.

2.1 Linear Temporal Logic

In reactive systems, correctness depends, not only on inputs and outputs of a computation, but on execution of the system. Temporal logic is a branch of logic that incorporates temporal aspects to reason about propositions in time. In this paper, we consider a version of temporal logic called linear temporal logic (LTL) [1].

LTL includes Boolean connectors like negation ($\neg$), disjunction ($\vee$), conjunction ($\wedge$), material implication ($\rightarrow$), and two basic temporal modalities *next* ($\bigcirc$) and *until* ($\mathcal{U}$). By combining these operators, it is possible to specify a wide range of requirements. Formulas involving other operators can be derived from these basic ones, including *eventually* ($\Diamond$) and *always* ($\Box$).

An *atomic proposition* is a statement on system variables v that has a unique truth value (*True* or *False*) for a given value v. For a set π of atomic propositions, any atomic proposition $p \in \pi$ is an LTL formula. Given a propositional formula describing properties of interest, widely used temporal specifications can be defined in terms of their corresponding LTL formulas as follows. A *safety formula* asserts that a property will remain true throughout the entire execution (i.e., nothing bad will happen). A *response formula* states that at some point in the execution following a state where a property is true, there exists a point where a second property is true. A response formula is used to describe how systems need to react to changes in environment or operating conditions. A response property, for example, can be used to describe how the system should react to a generator failure: if a generator fails, then at some point a corresponding contactor should open [12], [13].

2.2 Reactive Synthesis

A system consists of a set V of variables. The domain of V, denoted by $dom(V)$, is the set of valuations of V. Let E and P be sets of environment and controlled variables, respectively. Let $s = (e, p) \in dom(E) \times dom(P)$ be a state of the system. Consider a LTL specification φ of assume-guarantee form $\varphi = \varphi_e \rightarrow \varphi_s$, where, roughly speaking, φ_e characterizes the assumptions on the environment and φ_s characterizes the system requirements. LTL formulas are interpreted over infinite sequences of states, where $s_0 s_1 s_2 \ldots$ is an infinite sequence of valuations of environment and controlled variables. The synthesis problem is then concerned with constructing a strategy in the form of a partial function $f : (s_0 s_1 \ldots s_{t-1}, e_t) \mapsto p_t$, which chooses the move of the controlled variables based on the state sequence so far and the behavior of the environment so that the system satisfies φ_s as long as the environment satisfies φ_e.

For general LTL, the synthesis problem has a doubly exponential complexity [7]. A subset of LTL, namely generalized reactivity (GR(1)), can be solved in polynomial time (polynomial in the number of valuations of the variables in E and P) [6]. GR(1) specifications restrict φ_e and φ_s to take the following form, for $\alpha \in \{e, s\}$,

$$\varphi_\alpha := \varphi_{\text{init}}^\alpha \wedge \bigwedge_{i \in I_1^\alpha} \Box \varphi_{1,i}^\alpha \wedge \bigwedge_{i \in I_2^\alpha} \Box \Diamond \varphi_{2,i}^\alpha,$$

where $\varphi_{\text{init}}^\alpha$ is a propositional formula characterizing the initial conditions; $\varphi_{1,i}^\alpha$ are transition relations characterizing safe, allowable moves and propositional formulas characterizing invariants; and $\varphi_{2,i}^\alpha$ are propositional formulas characterizing states that should be attained infinitely often. For the specifications considered in this paper, the safety fragment of GR(1) suffices.

Given a GR(1) specification, the digital design synthesis tool implemented in JTLV (a framework for developing temporal verification algorithm) [8] generates a finite-state automaton that represents a switching strategy for the system. TuLiP provides an interface to JTLV.

2.3 Testbed Specifications

Consider the single-line diagram in Figure 1 in which environment variables are health statuses of generators and rectifier units, and controlled variables are the state of contactors. Consider also two different controller implementations: a *centralized logic* that runs the system with a single automaton and a *distributed logic* that has two different automata, one for the AC subsystem and one for the DC subsystem, running sequentially.

For the centralized logic, the environment assumptions are: (i) at least one generator must always be healthy, and (ii) at least one rectifier unit must always be healthy. In LTL, this can be written as

$$\Box(((gen_1 = healthy) \vee (gen_2 = healthy)) \wedge ((ru_1 = healthy) \vee (ru_2 = healthy))), \quad (1)$$

where gen_1, gen_2, ru_1, and ru_2 are health statuses of the two generators and the two rectifier units, respectively. To ensure non-paralleling of AC sources, we disallow any configuration of contactors in which a path may be created between the two generators. The contactors c_1 and c_2 are below the generators in Figure 1, and c_3 is between the AC buses. Therefore, contactors c_1, c_2, and c_3 can never be closed at the same time. This is written as

$$\Box\neg((c_1 = closed) \wedge (c_2 = closed) \wedge (c_3 = closed)).$$

The last specification ensures that all buses can be unpowered for no more than a time T. The limit that unpowered time can be set to depends on timing characteristics of the testbed, witch is explained in Section 4.1. To synthesize centralized logic, we used the assumption that this time is zero; thus, the specifications that all buses b_i fulfill $\Box(b_i = powered)$, for $i \in \{1,2,3,4\}$ can be set.

To synthesize distributed logic, we separate the system into two subsystems, seen in Figure 1. The AC subsystem contains all AC components (generators, AC contactors, AC buses, and loads). The DC subsystem contains all rectifier units, DC contactors, buses, and loads. All specifications from the centralized case decompose and carry over to the distributed case. However, in order to ensure that the overall specification is realizable, we impose additional restrictions on the components located at the interface between subsystems. The rectifier units contain capacitors that can be chosen so that they create a delay T_{RU}, in which the DC buses stays powered even after that an AC bus gets unpowered.

If $T_{RU} > T$ the additional interface refinement comes in the form of a guarantee specification that all DC buses b_i, for $i \in \{1,2\}$ will always be powered $\Box(b_i = powered)$, provided that both rectifier units stay healthy, i.e.,

$$\Box((ru_1 = healthy) \wedge (ru_2 = healthy)).$$

This guarantee is written as an environment for the DC subsystem. With this refinement, both subsystems can be synthesized independently, and the overall system specifications are satisfied when they are implemented together. We assume that the time a generator remains healthy is not arbitrarily short so that the AC bus powered time (i.e., the time between two intervals when AC bus is unpowered) is large enough to keep the capacitors on rectifier units charged.

2.4 Implementing Formal Specifications

`TuLiP` generates finite-state automata in the form of a text file that enumerates the possible states of the system and how the transitions could be carried out according to the current state. It also generates a text file that specifies environment variables (e.g., generators and rectifier units) and system variables (e.g., contactors). In order to implement the control logic in SimPowerSystems, we automatically translate these files into a Matlab-compatible script. A preliminary solution uses a Python script for this translation. A

```
State 0 <gen1:1, gen2:1, c1:1, c2:1, c3:0>
With successors: 1, 2, 3, 0
State 1 <gen1:0, gen2:0, c1:0, c2:0, c3:0>
With no successors
State 2 <gen1:0, gen2:1, c1:1, c2:0, c3:1>
With successors: 1, 2, 3, 0
State 3 <gen1:1, gen2:0, c1:0, c2:1, c3:1>
With successors: 1, 2, 3, 0
```

Figure 2: Sample of a `TuLiP` output in two-generator and three-contactor case. The generator status variables are gen1 and gen2, and the contactor status variables are c1, c2, and c3. Each state has successors, which define where the controller can transit depending on current state. In addition, no-successor states exist.

```
function [c1, c2, c3] = mscript(gen1, gen2)
global state;
switch (state)
case 0:
  if gen1 == 1 and gen2 == 1 then
    state = 0; c1 = 1; c2 = 1; c3 = 0;
  else if gen1 == 0 and gen2 == 0 then
    state = 1; c1 = 0; c2 = 0; c3 = 0;
    ...
  end if
case 1:
  ...
end switch
```

Figure 3: Sample code generated using `TuLiP` controller shown in Figure 2.

Python script generating the Matlab code is released with `TuLiP` version 0.3c under the tools directory[1].

Figure 2 shows an example four-state `TuLiP` generated controller for the two-generator and three-contactor case. A few lines of the auto-generated code that corresponds to this controller is shown in Figure 3. The auto-generated code can be inserted in SimPowerSystems as a Matlab function block. It can also be connected to the board with the code shown in Figure 4.

3. DESIGN AND IMPLEMENTATION

The single-line diagram in Figure 1 is a simplified notation for representing a three-phase power system. However, as described in Section 3.1, power supply to the hardware testbed is not three-phase. In order to represent the installations of the sensors, circuit protection devices, and fault injection switches, we present a detailed schematic of the testbed in Figure 6. Descriptions of the components shown in Figure 6 are given in Figure 7.

The hardware testbed has two different voltage levels: 24 VAC and 2.5 VDC. The DC section is connected to the AC section by rectifier units. Aircraft contactors are designed to switch three-phase electric power with relatively high currents. Relays are generally used for switching lower currents. These operate in a similar fashion to contactors but are lighter, simpler, and less expensive. Therefore, it was more convenient to handle the switching in the hardware

[1] `http://tulip-control.sf.net`

```
global state;
while 1 do
  gen1 = readgen1();
  gen2 = readgen2();
  [c1, c2, c3] = mscript(gen1, gen2);
  writeboard(c1, c2, c3);
end while
```

Figure 4: Code that implements the control software running on hardware model.

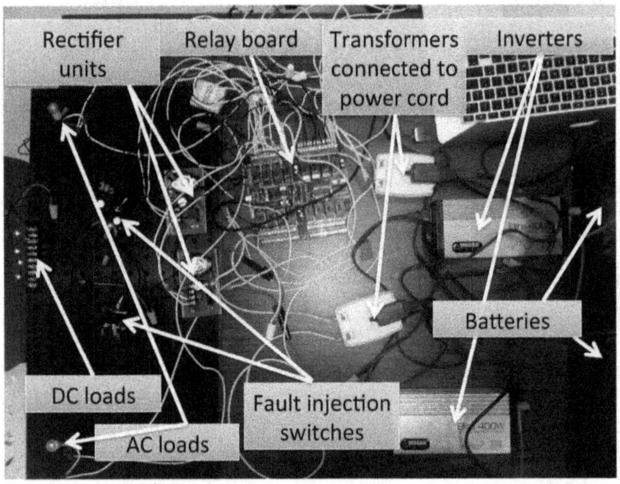

Figure 5: Hardware setup corresponding to the single-line diagram shown in Figure 1.

model with relays. It was possible to connect the control logic to the relays with the use of a relay board[2], which is a set of computer-controlled relays that can communicate with programming languages supporting serial communications, e.g., MATLAB. Analog-to-digital (A/D) connections on the relay board are used to monitor the system conditions. A photo of the setup[3] is shown in Figure 5. The transformers in Figure 5 are connected to power cords; these can be unplugged to simulate a generator failure. The rectifier units are connected to a switch, which can be used to generate a fault on the DC subsystem. Next, we describe how we monitor and sense the status of generators and rectifier units.

3.1 Generation and Circuit Protection

Each generation unit consists of a 12 V battery connected to an inverter that generates 120 VAC; that is then transformed down to 24 VAC to ensure safety. If the controller violates one of the safety requirements and connects these two sources in parallel, it would result in a short-circuit and cause the fuses installed next to the generators, shown in Figure 6(a), to blow. This observation makes it possible to monitor the correctness of the controllers at run time.

3.2 Sensing

The relay board needs to react consistently to faults injected into the system; this requirement implies that sensor

[2]A company called RelayPROS sells such relay boards. For more information, visit `www.relaypros.com`.

[3]A photo of the relay board can be found online at `assets.controlanything.com/photos/usb_relay/ZADSR165DPDTPROXR_USB-900.jpg`.

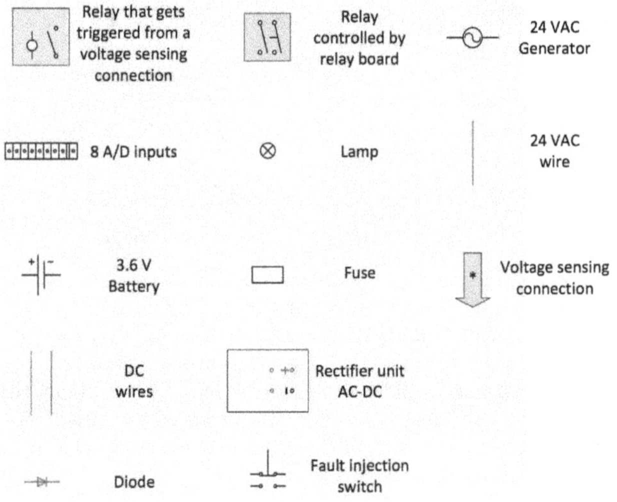

Figure 7: Description of the components used in Figure 6.

	T_c [ms]	T_c' [ms]
Mean	303.7	187.5
Max	333.3	234.1
Min	282.5	166.6

Table 1: Control cycle time, both when relay configuration changes, i.e., T_c and without any change, i.e., T_c'. The values with and without change were calculated from 20 and 250 measurements, respectively.

placement, functionality, accuracy, and time delay play crucial roles in design. Two types of faults can be injected in the system, namely, rectifier unit failures and generator failures. Voltage sensing for generator failures is handled using additional relays. These relays close a 3.6 V circuit to a battery when triggered by the voltage from the transformers. If a fault occurs and a generator does not work properly, the 3.6 V circuit opens and the system reacts accordingly. The voltage sensors of the rectifier units are directly connected to the A/D ports of the relay board because the voltage can be tuned to the appropriate value using an adjustable output on the rectifier units. Figure 6(b) illustrates the sensing configuration on the testbed.

4. EXPERIMENTS

We next describe the characteristics of the hardware testbed and show some preliminary test runs with different control architectures.

4.1 Testbed Characteristics

The first step before the implementation and testing of different controllers is characterizing the timing properties of the hardware testbed. Every relay has a time delay between the time a command is sent by the computer and the time an action (i.e., relay opening or closing) is taken, this is referred to as the *relay delay time*, T_d. Furthermore, the system has delays resulting from *control cycle times*, T_c and T_c', defined as

$$\begin{aligned} T_c &= T_r + T_I + T_w \\ T_c' &= T_r + T_I, \end{aligned} \quad (2)$$

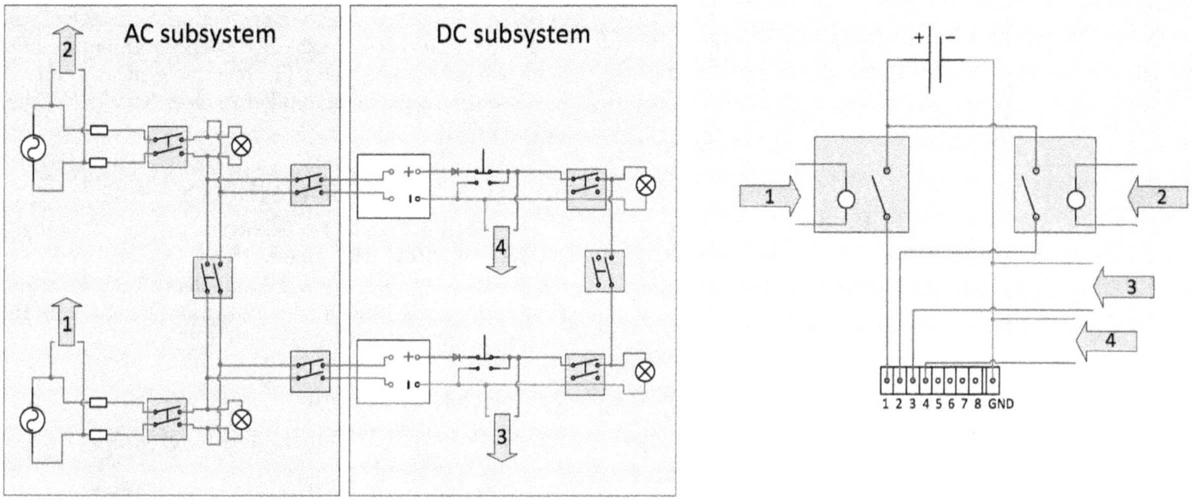

(a) Circuit schematic (b) Sensing configuration

Figure 6: Circuit schematic of the hardware testbed, which corresponds to the single-line diagram shown in Figure 1. The numbered arrows in (a) denote voltage sensing connections to the corresponding numbered arrows in (b).

where T_r is the time it takes to read the health statuses from all of the four environment variables, T_I is the time it takes to run the logic (the time can be interpreted as the time taken to run the code shown in Figure 3), and T_w is the time it takes to write information to the board (see Figure 4). Writing information to the board is not needed in every iteration (for instance, if the system state remains the same), therefore the control cycle time also include T_c'.

The control cycle times T_c and T_c' are listed in Table 1. The relay delay time can be found from the board specifications and shall be less than 20 ms.

An important safety requirement in an aircraft is that a bus should never lose power for more than a certain duration, e.g., typically 50 ms. In the hardware testbed, the time for which the bus is unpowered depends on the control cycle times and the relay delay time, and because the control cycle times exceed 50 ms, we cannot use the typically specified time for which an aircraft can be unpowered. Therefore, it was necessary to adopt a suitable limit. As illustrated with two environment variables in Figure 4 the relay board read the health status from each environment variable in a specified order. It is therefore necessary to include a part of T_c' from the previous control cycle in this limit. The time T_I in Equation (2) is negligible compared to T_r and T_w, the time taken to read the health status from one environment variable can therefore be approximated as $T_c'/4$. A reasonable value of an acceptable unpowered time for the hardware testbed can be

$$T \approx \max(T_d) + \max(T_c) + \frac{4-n}{4}\max(T_c'), \qquad (3)$$

where $n \in \{1, 2, 3, 4\}$ is the number which denotes the order of when the environment variable that is faulty is read in the code.

4.2 Controller Tests

Two controllers were tested, one with distributed logic and one with centralized logic. The controller with centralized logic had a 16-state automaton synthesized as explained in Section 2.3. The controller with distributed logic had two four-state automata that run on each subsystem. Both of these automata were synthesized in a similar fashion to the 16-state controller.

If the environment-related assumption is violated, the controller may end up in a state with no outgoing transitions, referred to as the *no-successor* state. The environment-related assumptions for the testbed are expressed in Equation (1) of Section 2.3. A violation of Equation (1) results in the controller entering a no-successor state, which happens when both generators or both rectifier units are faulty. If a centralized controller senses that both rectifier units are faulty, the whole system stops working because a no-successor state has been reached. This is not the case when distributed logic is used, because the AC system continues working even if the DC environment assumption is violated and the DC part reaches a no-successor state. The distributed logic implementation has two different automata that represent the logic, one for each subsystem, with coupling between them. However, the distributed logic is centralized in that it consists of single control software running on a single computer and communicating with the hardware through a single channel.

Figure 8 shows the voltage measurement for the centralized 16-state controller. The measurement was taken on the AC bus when the generator, which health status is read at second place ($n = 2$ in Equation (3)) of the four environment variables in the code, was switched off and then on again. The generator was switched off at $t = 2.83$ s, at which point the bus becomes unpowered. The second vertical line from the left indicates when the controller reacts and power up the bus using the other generator, which happens at $t = 3.1$ s. The generator was switched on again at $t = 3.73$ s; this was accompanied by a discernible change in the sine curve. Once a generator is switched on again after a fault, the time for which the bus is without power is not noticeable because the controller sends simultaneous commands to two relays.

The measured bus-unpowered times are listed in Table 2, which show a maximum value of $T_{max} = 414.9$ ms. An acceptable unpowered time when $n = 2$ and $\max(T_d) = 20$ ms

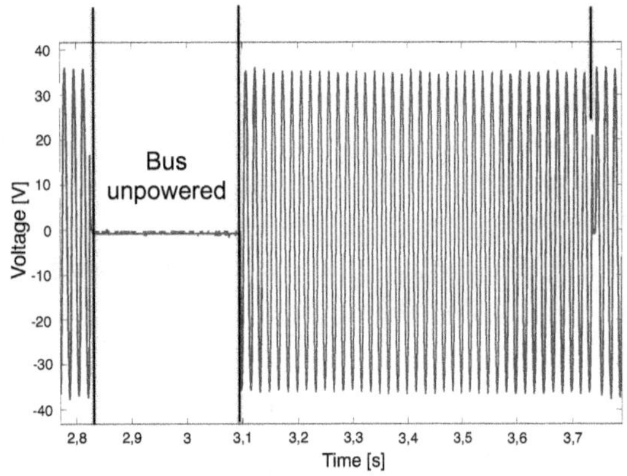

Figure 8: Bus voltage measurement when a generator is switched off and then turned back on. The first vertical line indicates the fault, the second vertical line is when the controller reacts, and the third line is when the generator is turned back on.

	Bus-unpowered time [ms]
Mean	333.9
Max	414.9
Min	232.7

Table 2: Time for which bus is unpowered after a fault is injected. These values are calculated using measurements from 10 fault injections.

can be calculated with Equation (3). It follows that $T \approx \max(T_d) + \max(T_c) + \frac{1}{2}\max(T_c') = 470.35\,\text{ms}$ and hence, $T_{max} < T$. We used a digital storage oscilloscope (Rigol DS1052E 50 MHz) for measurements. The measurement data are imported into MATLAB for analysis and to estimate the unpowered times.

5. LIMITATIONS AND EXTENSIONS

As discussed earlier, when we implemented distributed logic with the hardware model, it was still centralized in that only one relay board was connected to one computer. However, it is possible to use two relay boards connected to two different computers, with each of them controlled by different automata. The part that contributes the most to the control cycle times (T_c and T_c'), is the time it takes to read data from the board (T_r); if the controller is operated with two relay boards, T_r would be split in half, which would cause T_c and T_c' to decrease significantly. The distributed control architecture would also be more like that of an aircraft.

We injected faults in the hardware testbed by unplugging the power cords and changing the switches; however, a more accurate approach to generate faults would be the use of an additional relay board. Using an additional fault injection board, we can systematically study synchronous, correlated, and cascaded failures and their influence on controller performance; with the current method of fault injection, it could be difficult to switch off a generator and a rectifier unit within the same control cycle.

On an aircraft, the controller is an embedded system designated for a specific task. To increase its reliability and performance, the hardware model could be adapted to run the relay boards through microcontrollers. Embedded code for these microcontrollers can be generated using MATLAB.

At last, we want to emphasize the fact that it is entirely possible to synthesize the controller with another synthesis tool and test it on the testbed. It was convenient as an initial demonstration to choose LTL, reactive synthesis, and `TuLiP` because they have been applied to electric power systems in the past [13].

6. ACKNOWLEDGMENTS

The authors wish to acknowledge the funding from MuSyC, iCyPhy, the Boeing Corporation, and AFOSR (award # FA9550-12-1-0302), and thank Rich Poisson from Hamilton-Sundstrand for helpful discussions about the development of the hardware testbed.

7. REFERENCES

[1] C. Baier and J. Katoen. *Principles of Model Checking.* MIT press, 1999.

[2] M. Lahijanian, M. Kloetzer, S. Itani, C. Belta, and S. B. Andersson. Automatic deployment of autonomous cars in a robotic urban-like environment. In *IEEE Intl. Conf. on Robotics and Automation,* pages 2055–2060, Kobe, Japan, 2009.

[3] O. J. Mengshoel, A. Darwiche, K. Cascio, M. Chavira, S. Poll, and S. Uckun. Diagnosing faults in electrical power systems of spacecraft and aircraft. In *Innovative Applications of Artificial Intelligence Conference,* pages 1699–1705, Chicago, IL, 2008.

[4] N. Ozay, U. Topcu, and R. M. Murray. Distributed power allocation for vehicle management systems. In *IEEE Conference on Decision and Control,* 2011.

[5] N. Ozay, U. Topcu, T. Wongpiromsarn, and R. M. Murray. Distributed synthesis of control protocols for smart camera networks. In *ACM/IEEE Intl. Conf. on Cyber-Physical Systems,* Chicago, IL, 2011.

[6] N. Piterman, A. Pneuli, and Y. Sa'ar. Synthesis of reactive(1) designs. *Verification, Model Checking and Abstract Interpretation,* 3855, 2006.

[7] A. Pnueli and R. Rosner. Distributed reactive systems are hard to synthesize. In *IEEE Symposium on Foundations of Computer Science,* 1990.

[8] A. Pnueli, Y. Sa'ar, and L. Zuck. *JTLV* a framework for developing verification algorithms. In *Intl. Conf. on Computer Aided Verification,* 2010.

[9] S. Poll, A. Patterson-hine, J. Camisa, D. Garcia, D. Hall, C. Lee, et al. Advanced diagnostics and prognostics testbed. In *International Workshop on Principles of Diagnosis,* pages 178–185, 2007.

[10] R. Rogersten, H. Xu, N. Ozay, U. Topcu, and R. M. Murray. *An Aircraft Electric Power Testbed for Validating Automatically Synthesized Reactive Control Protocols.* Caltech, Tech. Rep. Id. 36376, 2013.

[11] SIMULINK. *version 7.7 (R2011a).* The MathWorks Inc., Natick, Massachusetts, 2011.

[12] T. Wongpiromsarn, U. Topcu, N. Ozay, H. Xu, and R. Murray. `TuLiP`: a software toolbox for receding horizon temporal logic planning. In *Intl. Conf. on Hybrid Systems: Computation and Control,* 2011.

[13] H. Xu, U. Topcu, and R. Murray. A case study on reactive protocols for aircraft electric power distribution. In *IEEE Conference on Decision and Control,* 2012.

Towards Sensitivity Analysis of Hybrid Systems Using Simulink

Zhi Han
MathWorks
zhi.han@mathworks.com

Pieter J. Mosterman
MathWorks
pieter.mosterman@mathworks.com

ABSTRACT

In the design of engineered systems two types of models are used: (i) *analysis models* and (ii) *system models*. The system models are primary deliverables between design stages whereas analysis models are employed within a design stage. Sensitivity analysis studies the behavior of the system under small parameter variations which proves to be useful in design. To enable sensitivity analysis in verification of hybrid dynamic systems that model industry-size problems, support for simulation-based methods is desired. The computational semantics for simulation of corresponding analysis models must then be consistent with the computational semantics of the system models. A method is presented that enables direct sensitivity analysis on system models via an implementation in the Simulink® software. The approach relies on the existing ordinary differential equation solver of Simulink and the block-by-block analytic Jacobian computation to provide the analytic Jacobian for solving the sensitivity equations. Results of a prototype implementation show that sensitivity analysis can be applied to moderate size Simulink models of continuous and hybrid systems.

Categories and Subject Descriptors

I.6.3 [**Computing Methodologies**]: Simulation and ModelingApplications

Keywords

Sensitivity analysis; simulation; verification

1. INTRODUCTION

Design of complex technical systems such as automobiles, airplanes, satellites, etc., relies on the principles of (i) separation of concerns and (ii) divide and rule [15, 29]. Where the latter is the motivation to partition and modularize an overall system, the former presages the utility of various abstractions such as linearized models, architecture models, performance models, requirement models, trade-off models, etc. Generally, these models can be classified in (i) *system models*, which are included as deliverables in specifications between design stages and (ii) *analysis models*, which are utilized within design stages to deliver a specification based on incoming requirements.

With the advent of personal computing, the documents storing the various models increasingly became electronic by using office applications such as spreadsheet software, word processing software, graphics software, etc. The availability in an electronic modality then unlocked the potential to automate the generation of a representation amenable to computational simulation. While initially it eliminated the task performed by 'simulationists' to transform a model into computer code, such automation proceeded to enable sophisticated and domain specific formalism with semantics defined by their corresponding computational representation. By employing the same computational semantics for system models as they progress throughout the various design stages, in due time, the formal meaning of models became defined by the computational semantics [22].

As a consequence, for models that include continuous-time dynamics, such as *hybrid dynamic systems*, the approximations made by the resulting computational semantics (e.g., numerical integration solvers, algebraic equation solvers, root-finding solvers, etc.) have effectively come to define the meaning of the models. Though to an extent because of uninformed use of numerical methods, engineers working on the development of complex technical systems have come to accept and even rely on the computational approximations introduced by their modeling tool of choice. To these engineers, consistency of computational results between design stages supersedes correctness as a measure of computational closeness to the underlying theoretically analytic solutions that is typically not practically available. For tool developers, when developing features for computational models it is imperative to preserve the precise computational semantics even in the face of their approximative fidelity [23].

For verification of hybrid systems, the necessity to preserve computational semantics is critical in simulation-based approaches. Such approaches are developed to tackle scalability issues in reachability analysis for formal verification (*e.g.* [9, 11, 28]) as based on a combination of methods from discrete systems, such as model checking and algorithmic abstractions, and computational methods for continuous systems, such as computational geometry and numerical simulations. Here, for industry problems, the state space that must be analyzed often becomes prohibitively large and solu-

HSCC'13, April 8–11, 2013, Philadelphia, Pennsylvania, USA.

tions resort to simulation-based reachability analysis, which uses sensitivity analysis of the simulated trajectories as an intermediate result (*e.g.* [1, 4, 6, 7, 13]). Thus solving industry problems renders it important to support sensitivity analyses and so computing sensitivities as part of the *analysis models* must preserve the computational semantics and not affect the simulation results of the *system models*.

Sensitivity analysis (SA) is the study of how variation in the output of a model can be apportioned to different sources of variation, and of how the given model depends upon the information input into it. SA has been used in engineering prior to the use in reach set computation, for instance, to find the most influential parameters in a model for optimization and parameter tuning [8]. In the hybrid systems domain, SA methods have been used to analyze hybrid dynamic systems and applied in computing critical parameter values of electric power networks [14, 24]. Although success cases have been reported in the literature, there is still a lack of tool support to enable SA in the software tools for the design of engineered system. For example, as a popular industry tool, Simulink® [20] provides a modeling environment for control system design engineers to build complex, hybrid systems using hierarchical block diagrams. Although SA functionality has been implemented in ordinary differential equation (ODE) solvers (*e.g.*, [26]), this functionality cannot be integrated in Simulink by simply replacing the Simulink ODE solver because the computational semantics of Simulink models include various additional numerical algorithms to support not only continuous-time behavior (*e.g.*, an algebraic equation solver) but, perhaps more importantly, also discrete and hybrid behavior (*e.g.*, root-finding).

Ideally one would like to have the capability to perform SA in analysis models without affecting the numerical computations performed in simulation of the corresponding system models. If implemented in a separate tool, this would require capturing the complete computational semantics of industry-strength simulation software such as Simulink, which presents a fundamental language engineering challenge that is yet to be solved [21]. Instead, the approach taken in this paper *embeds* SA directly into the existing simulation engine and relies on the existing ODE solver combined with the block-by-block analytic Jacobian computation of Simulink to solve the sensitivity equations. By implementing the majority of the SA capability in a block based on the standard Simulink block interface (S-function) API, a model can be easily configured for SA by including the block in the model diagram without diagrammatic complications. The prototype SA for Simulink hybrid system models is based on the research results of [8] and [14] and applied to a small set of test models. Results show that without affecting the computational semantics, SA can be performed on moderate size Simulink models with hundreds of blocks and over 60 continuous states, illustrating the scalability of the approach.

2. SENSITIVITY EQUATIONS

Most of the discussion in this section can be found in [8] and [14] with a slightly different formulation.

Consider an ODE system given in the following form $\dot{x} = f(x, p, t)$, $x(0) = x_0$ where $x \in \mathbb{R}^n$ is the state vector, $p \in \mathbb{R}^m$ is a vector of constant parameters, and $f : \mathbb{R}^n \to \mathbb{R}^n$ is a function that is continuously differentiable with respect to p, x, and t. Also assume that the system has a unique solution for any given initial value x_0 and parameter value p. Denote the unique solution by $\xi(t)$ and call it the *trajectory* of the system. The trajectory sensitivity of the system is

$$s_p^x \equiv \left.\frac{\partial \xi}{\partial p}\right|_{p,x_0}(t). \tag{1}$$

where $s_p^x \in \mathbb{R}^{n\times m}$ is the trajectory sensitivity matrix. The trajectory sensitivity can be used as a linear approximation for a *perturbed* trajectory $\xi|_{p+\Delta p,x_0}(t) \approx \xi|_{p,x_0}(t) + s_p^x|_{p,x_0}(t)\Delta p$, where $\Delta p \in \mathbb{R}^m$ and $\|\Delta p\| \ll \|p\|$.

The dynamics of s satisfies the *sensitivity equations* [8]:

$$\dot{s}_p^x = \frac{\partial f}{\partial x} s_p^x + \frac{\partial f}{\partial p}, \text{ and } s_p^x(0) = \frac{\partial x_0}{\partial p} \tag{2}$$

In formal verification of hybrid systems, a useful intermediate step is to compute the sensitivity matrix with respect to the initial state, where the sensitivity equation is: ([6])

$$\dot{s}_{x_0}^x = \frac{\partial f}{\partial x} s_{x_0}^x, \; s_{x_0}^x(0) = I_n. \tag{3}$$

Solving sensitivity equations for continuous systems involves solving the ODEs in Eq. (2) or Eq. (3). The sensitivity equations are time-varying linear ODE systems, which can be solved by general-purpose nonlinear ODE solvers or exponential ODE solvers [25]. Note that since $\frac{\partial f}{\partial x}$ must be evaluated, it is necessary to solve the system equations in Eq. (1) simultaneously, which results in a problem that consists of $n \times m + n$ continuous states.

For hybrid dynamic systems, special attention must be paid to the discrete events that are triggered by the continuous dynamics. Since the time of a discrete event t_e is dependent on the parameter values, the sensitivity of t_e must be calculated with respect to p [14]. To find t_e, Simulink employs a root-finding mechanism called zero-crossing detection [30]. Now define a continuously differentiable function $g : \mathbb{R}^n \to \mathbb{R}$ and call it a *zero-crossing* function. Let t_e denote the time of the *zero-crossing event*, that is, the instance of time where the function g changes its sign. Under the same assumptions provided in [14], the sensitivity of the zero-crossing event time can be derived as follows. Since the zero-crossing event is described by $y = g(\xi(t_e)) = 0$ it follows that

$$\frac{\partial y}{\partial t_e}\Delta t + \frac{\partial y}{\partial p}\Delta p = 0$$

With some manipulation, the sensitivity of the time of zero-crossing event can be derived as

$$s_p^{t_e} \equiv \frac{\partial t_e}{\partial p} = -\frac{1}{\frac{\partial g}{\partial x}\cdot f(x,p,t)} \frac{\partial g}{\partial x} s_p^x(t_e) \tag{4}$$

Assuming that there is no state reset after the zero-crossing event, and let f^+ denote the right-hand side of the system equations after the zero-crossing event, the trajectory sensitivity matrix is reset at the zero-crossing event according to the following *jump condition* [14]:

$$s_p^x(t_e^+) = s_p^x(t_e^-) + \left(f^-(x,p,t_e) - f^+(x,p,t_e)\right) s_p^{t_e}. \tag{5}$$

3. IMPLEMENTATION

Numerical solvers for ODEs are necessary to solve Eq. (2) and Eq. (3) along with means to evaluate the partial derivative terms (Jacobians) in those equations. The evaluation of Jacobians can be performed either analytically or by numerical approximations such as finite differences. Analytic computation of the Jacobian is usually preferred as it can be more accurate and efficient using algorithmic differentiation (AD) tools [12]. In Simulink, block-by-block analytic Jacobian evaluations have been used in tools such as Simulink® Control Design™ [19] to evaluate the model Jacobian.

3.1 Simulation loop in Simulink®

The simulation of a continuous dynamic system model in Simulink consists of a model compilation phase, a link phase, and a simulation loop phase [20]. In the compilation phase, Simulink models are compiled into an operational form. In the linking phase, the resources necessary for simulation are allocated. The simulation loop phase depicted in Fig. 1 is the step where the dynamic equations of the model are solved by computing state and output signal values at each consecutive time step. At each time step, Simulink first invokes the **Output** method to compute the output signals of each block, then Simulink invokes the *ODE solver* to compute the next time step and the state values at the next time step. For zero-crossing functions Simulink then checks if the signs of the zero-crossing functions have changed for the new state values. If so, Simulink invokes the *zero-crossing detector* to further reduce the time step to locate the time instances of the zero-crossing events. Once an appropriate time step has been found and the corresponding state vector values are computed, Simulink advances to the next time step. This loop is repeated till the simulation is completed.

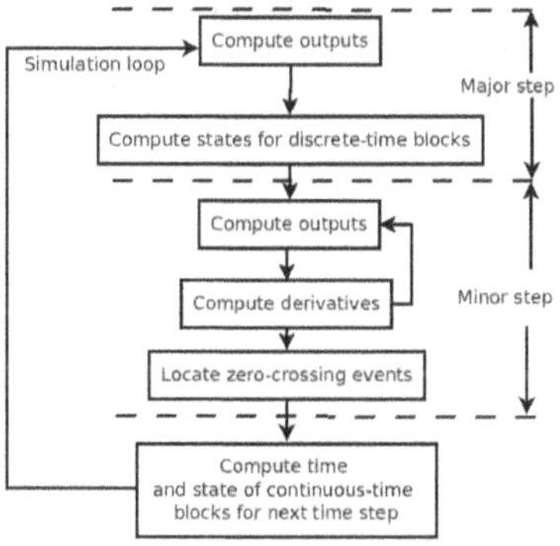

Figure 1: Simulation loop of Simulink®

The ODE solver and zero-crossing detector of Simulink invoke the **Output** and **Derivative** methods of the model to compute signal values and the state derivative vector. In variable step solvers, the **Output** method and **Derivative** method may be invoked multiple times in one step. And some results from these method calls may be discarded based on estimations of the approximation error. To distinguish the **Output** calls from the simulation loop and the calls originating from the solver and the zero-crossing detector, the former is referred to as the *major* step and the latter is called the *minor* step [20].

After the **Output** method completes, Simulink may optionally compute the analytic model Jacobian before invoking the ODE solvers for the sensitivity analysis. The computation of the model Jacobian consists of three steps. First, the Jacobian of each block is computed by invoking their **Jacobian** method and a set of matrices is computed to represent the linear relationship corresponding to block connections. Second, the analytic Jacobians of all blocks are concatenated into a set of Jacobian matrices along with the matrices for block connections combined into the *open-loop model Jacobian* data structure. Third, the linearization algorithm performs *derivative accumulation*: it uses the open-loop model Jacobian to compute the closed-loop model Jacobian $\frac{\partial f}{\partial x}$ [12]. The derivative accumulation of Simulink is implemented as linear fractional transformations (LFT) [31].

3.2 The sensitivity blocks

The SA for continuous-time Simulink blocks is implemented in a number of S-function blocks. To perform SA with respect to initial conditions, the **Sensitivity** block must be included in the root level of a Simulink model. If a block contains parameters for which the sensitivity is computed, changes must also be made to the block. The solution of the sensitivity vector is computed in the callback function of the block. At the major **Output** steps, the **Sensitivity** block notifies Simulink that the block-by-block analytic Jacobian of the model must be computed. After completing the **Output** of all blocks, Simulink computes the open-loop model Jacobian and invokes the block callback function with the open-loop model Jacobian as an argument.

Figure 2 shows the control flow of the simulation loop with the computation of the sensitivity. Once control has reached the block callback function, it uses the open-loop model Jacobian to compute the closed-loop model Jacobian. Then the state value of the model is used to compute the sensitivity matrix. Note that since the block callback function is invoked at the end of the current step (*i.e.*, the solver has not yet computed the next time step), the sensitivity is computed for the current time step. To handle sensitivity of the zero-crossing event time, another block that is implemented as an S-function and only invoked when a zero-crossing event is detected computes the jump conditions.

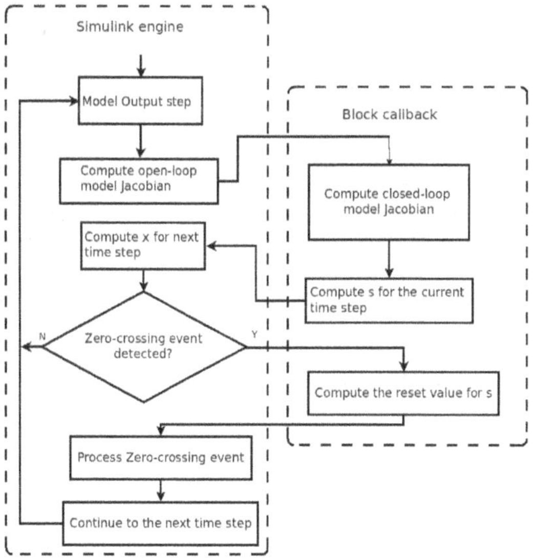

Figure 2: Block callback for SA

To compute the sensitivity with respect to a block parameter the *linear analysis point* specification of Simulink [19] is utilized. When a signal y is marked as a linear analysis point, the Simulink engine treats the output port of the signal as a special port and marks the location of the signal in the open-loop model Jacobian.

For example, suppose the block implements $y = \phi(p, i)$ where y is the output signal of the block, p is a constant parameter, and i is the input signal of the block. Now suppose a state-less block provides the input to a block with a derivative function to form the system equation $\dot{x} = f(x, \phi(p, i), t)$. With u the input to the connected blocks the following relation holds $\frac{\partial f}{\partial p} = \frac{\partial f}{\partial y}\frac{\partial \phi(i,p)}{\partial p}$. To obtain $\frac{\partial f}{\partial p}$, the signal y is marked as a linear analysis point, which enables Simulink to compute the partial derivative $\frac{\partial f}{\partial y}$ from the open-loop model Jacobian using LFT. Also $\frac{\partial \phi}{\partial u}$ can be computed from the function ϕ using u and p. For the **Gain** blocks used in this section, and for which $\phi(u) = Ku$, the partial derivative is $\frac{\phi}{K} = u$. For blocks with complicated algorithms, the source code of the block must be modified to implement the partial derivative.

Figure 3 illustrates this by the Van der Pol example augmented with the blocks to perform SA. The changes to the model are highlighted in green. It has a masked **Gain** block that specifies the gain value as the parameter sensitivity input. It also has a **Sensitivity** block that implements the block callback function for SA. The masked Gain block is a subsystem that has two blocks as its children: one block computes the output and the other block computes $\frac{\partial \phi}{\partial p}$ and passes the value to the **Sensitivity** block.

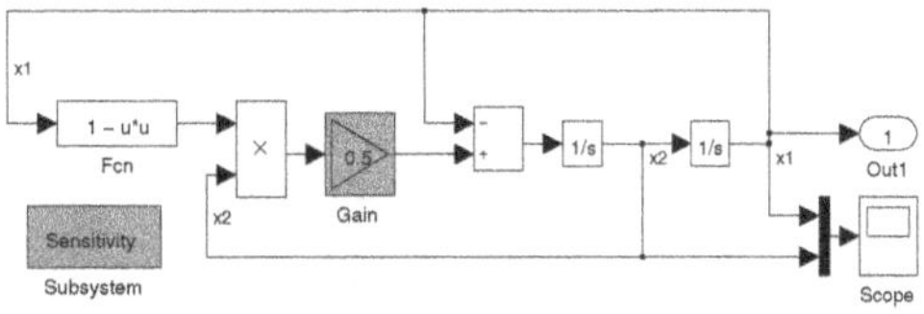

Figure 3: A Simulink® model with SA

Sensitivity for systems with zero-crossing events is implemented in a few Simulink blocks. The **Reset Sensitivity** block and **Sensitivity** block share data by reading and writing to a shared workspace. Figure 4 shows the block diagram used for the computation of sensitivity for a model with zero-crossing events. Figure 4b shows the content of the masked subsystem **Relay Block With Zero-crossing Sensitivity** in Fig. 4a. The subsystem consists of a **Relay** block and a triggered subsystem block to compute the jump condition at the zero crossing event. The computation of the zero-crossing event requires computing $\frac{\partial g}{\partial x}$ in Eq. (4), which can be computed from the open-loop model Jacobian once the signal corresponding to $g(x)$ is marked for analytic Jacobian as linear analysis points. The upward arrow icon on the output port of the **Inport** block in Fig. 4b shows that indeed this signal is marked as linear analysis point. The **Sensitivity** block performs the computation to solve the sensitivity equations.

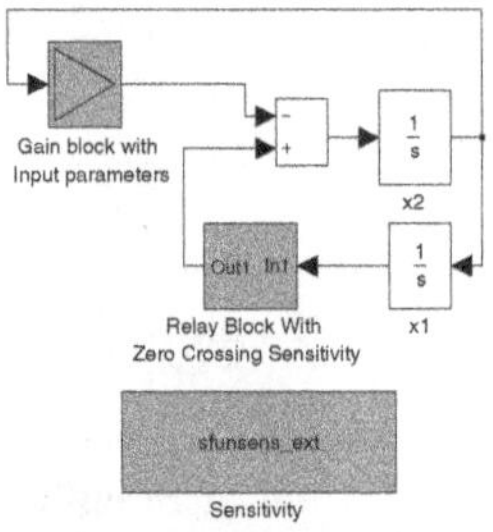

(a) System model

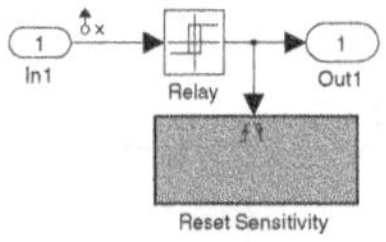

(b) Relay subsystem

Figure 4: SA for models with zero-crossing events

Figure 5a shows the SA results for the model with zero-crossing events in Fig. 4 where the sensitivity of the state trajectory is computed with respect to the parameter value of the **Gain** block. Note that the state sensitivity has discontinuities for both states because of the jump conditions at the zero-crossing events.

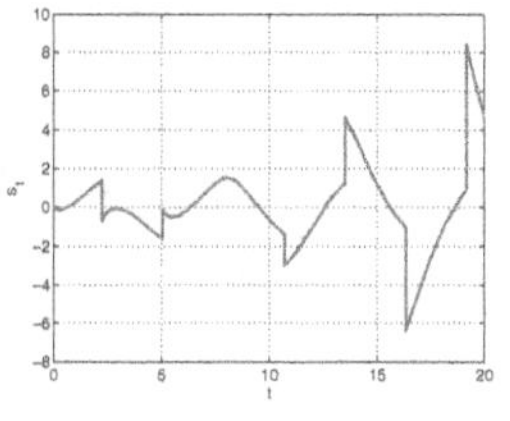

(a) $s_K^{x_1}$

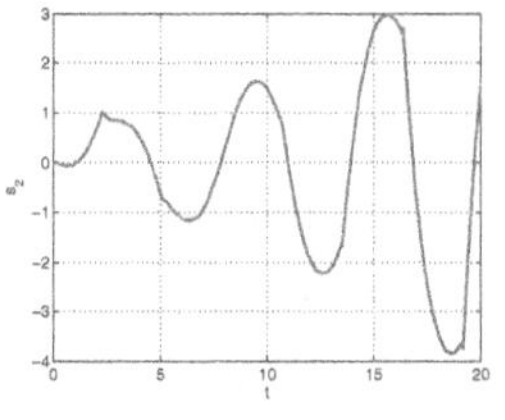

(b) $s_K^{x_2}$

Figure 5: SA for systems with zero-crossing events

4. COMPUTATION RESULTS

The prototype implementation is tested on a number of Simulink models of continuous-time dynamic systems. The first is a Simulink model of a voltage controlled oscillator (VCO) taken from [6, 10]. The initial values of the simulation are chosen to be $V_{d_1} = 1.5$ Volts, $V_{d_2} = 0.5$ Volts and $\dot{V}_{d_1} = \dot{V}_{d_2} = 0$ Volts/s. Using a variable-step solver **ODE45**, the simulation of a Simulink model of the VCO system takes 253 steps to simulate from 0 to 0.2 seconds. The time steps and state trajectories with SA are identical to the results without SA which confirms that the computational semantics are preserved. Figure 6 shows the simulation result of the VCO model together with the approximation of the perturbed responses. The simulation results are shown in solid lines and the approximation of the perturbed trajectories for $\pm\Delta V_{d_1}$ using SA are shown as the two dashed curves.

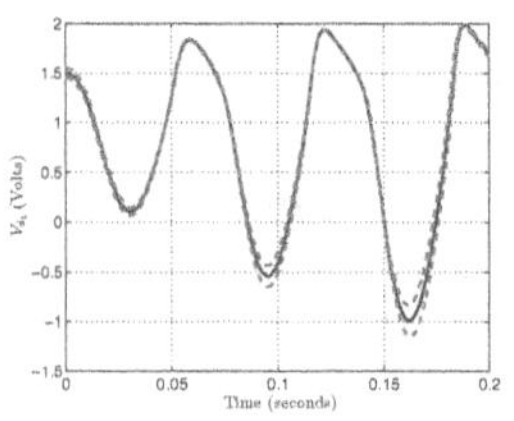

(a) Response of V_{d_1}

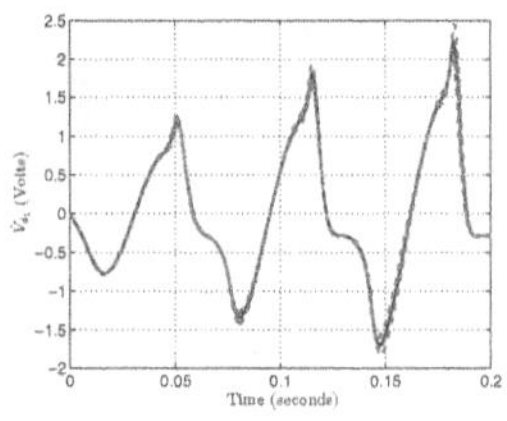

(b) Response of $\dot{V}_{d_1}$

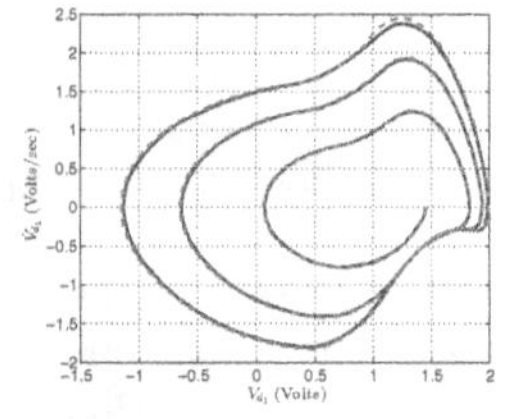

(c) State trajectory

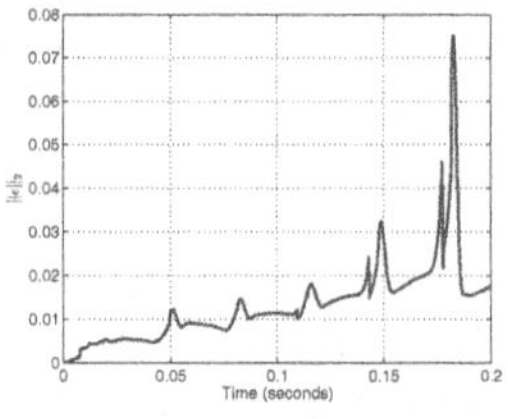

(d) Approximation error

Figure 6: Simulation results of VCO model

To evaluate the approximation errors, the linear approximation using SA is compared with a simulation of the perturbed state trajectory. Figure 6c shows the approximation of a perturbed simulation for $V_{d_1} = 1.45$ Volts overlaid with the actual simulation results of the perturbed initial state. The approximation using SA is shown as dashed lines and the actual simulation result is shown in solid lines.

The approximation error for state trajectory using SA is

$$\epsilon(t) = \xi|_{p+\Delta p, x_0}(t) - \xi|_{p,x_0}(t) - s_p^x|_{p,x_0}(t)\Delta p.$$

Figure 6d shows the two-norm of the approximation error for the perturbed state trajectory. The approximation error is small compared to the state values. The approximation error accumulates as the simulation progresses, leading to noticeable approximation errors in the state trajectory.

The prototype implementation is further used to compute trajectory sensitivity for a number of Simulink models. All SA are performed for sensitivity to initial conditions. The experiments are all performed on a computer with Intel® Xeon® CPU X5650 at 2.67GHz and 24 GiB memory running Debian Linux 6. Table 1 shows the computational results for a number of Simulink example models and Table 2 shows the computational results for a number of mechanical system models using SimMechanics™ [18]. The models are listed in the order of number of blocks in the block diagram. The results show that SA can be applied to Simulink models of continuous dynamic systems with moderate complexity. For example, for mechanical system Model 6, where the system consists of 268 blocks and 61 continuous states, the total computation time for SA is slightly more than twice that of the computation time for a simulation, even though the combined sensitivity equation and system equation comprises an ODE with 3782 continuous state variables.

Test Model	# of blocks	# of states	Simulation time (seconds)	Time for SA (seconds)
`sldemo_ dblcart1`	20	8	0.50	19.73
`sldemo_ f14`	75	10	0.54	18.01
`sldemo_ engine`	85	4	1.67	20.53
`sldemo_ hydrod`	128	6	1.75	4.52
`sldemo_ hydcyl4`	136	4	2.94	6.30

Table 1: Results for Simulink® example models

Test Model	# of blocks	# of states	Simulation time (seconds)	Time for SA (seconds)
Scotch yoke	47	8	1.15	6.56
Double pendulum	49	4	2.54	5.42
Flexible four bar mechanism	56	8	1.38	6.55
Cross slider	66	8	10.48	118.75
Crank slider	79	8	1.93	2.24
Offset slider	117	16	14.75	216.75
Stewart platform	268	61	5.49	11.8
Walking gait mechanism	414	32	12.83	86.05

Table 2: Results for SimMechanics™ models

There are a number of cases where the computational overhead of SA is significant. For example, the total time of SA for the double-cart example model is around 40 times the simulation time. Profiling the computation reveals that most of the computation time is spent in derivative accumulation and numerical integration of the sensitivity equations.

5. DISCUSSION

Because of their popularity in embedded systems design, Simulink models have attracted attention from researchers in the hybrid dynamic system domain. There have been numerous efforts to apply research results in hybrid dynamic systems to Simulink products, particularly in the area of formal verification (*e.g.*, [2, 4, 16, 27]). Most methods limit the scope to a fairly restricted subset of Simulink blocks. Some methods rely heavily on translation mechanisms that translate Simulink models into equivalent models that are amenable to analysis with a developed method, so the analysis can be performed externally on the translated model.

An existing software package that provides SA directly for Simulink is `DiffEdge` [3]. The software provides SA capability by augmenting the system model, that is, by creating new block diagrams on top of the existing block diagram model. The newly created block diagram models include the sensitivity equations, which can be simulated to provide SA results. The approach has a number of shortcomings, though. First, the newly created block diagram model has significantly increased complexity as it consists of at least twice the number of blocks and twice the number of signal connections compared to the original block diagram model. Second, the method creates a new model from an original model for sensitivity analysis and it is unclear how close the trajectories of the new model and those of the original model are. Third, the method relies on analyzing the structure of a given model and creating blocks, subsystems, and adding connections to create the new model. During model compilation the Simulink software may transform the model by removing blocks, inserting blocks, and rerouting signals [20] and it is unclear whether the method is compatible with the transformations performed by Simulink.

The approach taken in this paper is direct: it does not require intrusive changes to the original block diagram for simulation. It uses the existing computation procedures in Simulink. Although the implementation is still a prototype, it demonstrates that the functionality required by SA is available in the Simulink software and the method holds promise to handle industry models. The computational results show that SA for hybrid systems can be a feasible and useful enhancement to the Simulink software.

The software tool `Breach` performs SA and parameter synthesis [5]. It uses Real-Time Workshop® [17] to generate C code for Simulink models and uses CVODES for solving ODEs and computing trajectory sensitivity [26]. The approach in this paper is different because it applies to the Simulink execution engine and solver directly, which does not require the additional step of code generation.

The prototype implementation also reveals some limitations of the approach. Preliminary results showed that the computational overhead involved in the SA can be large. One of the computational steps that require a significant amount of time is the computation of the closed-loop Jacobians. Because for SA the model Jacobian must be evaluated many times during a simulation, it is crucial to improve the efficiency to compute Jacobians and some work in progress in this direction has been made.

Another limitation of the method is that it does not consider discrete-time systems. A straightforward manner to support them is to consider discrete-time dynamic systems described as difference equations with fixed sampling rate and extend the sensitivity equations to difference systems.

$$y_k = g(x_k, u_k) \quad (6)$$

$$x_{k+1} = f(x_k, u_k) \quad (7)$$

This method is adopted by Simulink Control Design and used in its linearization functionality [19]. When multiple sampling rates are encountered, a system with one sampling rate is converted to an equivalent system with another sampling rate. Since the sampling rate has changed, the converted system is not semantically identical to the original. The method is not used in the implementation of SA because, first, it does not support Simulink capabilities such as sample time offsets and, second, it does not consider how rate transitions are handled in Simulink.

Another difficulty with difference equations is the compatibility with implementations of discrete-time blocks. A Simulink block must have Eq. (6) and Eq. (7) implemented separately for the method to apply. When modeling discrete-time dynamic systems, the block author is allowed to implement an `Output` method with the combined effects of Eq. (6) and Eq. (7). To compute the Jacobian of Eq. (6) and Eq. (7) the block author must first analyze the implementation of the dynamics and extract the equations for Eq. (6) and Eq. (7) and re-implement these methods for the block.

Finally, in addition to discrete-time dynamic systems, the method developed in this paper does not support state reset functions because of their implementation in Simulink. In Simulink, state reset is modeled by configuring a state port, initial condition port, and reset port for blocks with continuous states. The implementation makes it difficult to isolate and extract the parts that contribute to the reset function.

6. CONCLUSION

This paper presents a prototype implementation of sensitivity analysis in Simulink that preserves the computational semantics between analysis models and systems models. The implementation employs existing Simulink algorithms for solving ODEs and sensitivity equations. Computational results using the implementation showed that it can be applied to Simulink models with hundreds of blocks and less than 100 continuous states with reasonable computational overhead. Limitations of the current implementation have been discussed along with reference to recent progress on improving the efficiency.

Acknowledgments

We thank Dr. Fu Zhang at MathWorks® for the numerous discussions and contributions to the ideas that are published in this paper. We thank professor Dr. Goran Frehse for the original implementation of the VCO model.

References

[1] M. Althoff, O. Stursberg, and M. Buss. Reachability analysis of nonlinear systems with uncertain parameters using conservative linearization. In *IEEE Conference on Decision and Control (CDC)*, pages 4042–4048, 2008.

[2] R. Alur, A. Kanade, K. C. Shashidhar, and S. Ramesh. Generating and analyzing symbolic traces of Simuink/Stateflow models. In *International Conference on Computer-Aided Verification*, pages 430–445, 2009.

[3] AppEdge Consulting and Engineering. DiffEdge: Differentiation, sensitivity analysis and identification of hybrid models under simulink. http://pagesperso-orange.fr/appedge/diffedge_intro.htm.

[4] T. Dang, A. Donzé, O. Maler, and N. Shalev. Sensitive state-space exploration. In *Proceedings of the IEEE Conference on Decision and Control*, pages 4049 –4054, 2008.

[5] A. Donzé, B. Krogh, and A. Rajhams. Parameter synthesis for hybrid systems with an application to simulink models. In *HSCC*, pages 165 – 179, 2009.

[6] A. Donzé and O. Maler. Systematic simulation using sensitivity analysis. In *Hybrid Systems: Computation and Control (HSCC 2007)*, pages 174–189, 2007.

[7] A. Donzé and O. Maler. Robust satisfaction of temporal logic over real-valued signals. In *Formal modeling and analysis of timed systems*, pages 92 – 106, 2010.

[8] P. M. Frank. *Introduction to System Sensitivity Theory*. Academic Press Inc., 1978.

[9] G. Frehse. PHAVer: Algorithmic verification of hybrid systems past HyTech. In *Hybrid Systems: Computation and Control (HSCC'05)*, pages 258–273, 2005.

[10] G. Frehse, B. H. Krogh, and R. A. Rutenbar. Verifying analog oscillator circuits using forward/backward abstraction refinement. In *Design, Automation and Test in Europe (DATE)*, pages 257 – 262, 2006.

[11] A. Girard. Reachability of uncertain linear systems using zonotopes. In *Hybrid Systems: Computation and Control (HSCC'05)*, pages 291–305, March 2005.

[12] A. Griewank and A. Walther. *Evaluating Derivatives: Principle and Techniques of Algorithmic Differentiation*. SIAM, 2008.

[13] Z. Han and B. H. Krogh. Reachability analysis of nonlinear systems using trajectory piecewise linearized models. In *American Control Conference*, pages 1505–1510, 2006.

[14] I. A. Hiskens and M. A. Pai. Trajectory sensitivity analysis of hybrid systems. *IEEE Transaction on Circuits and Systems*, 47(2):204 – 220, 2000.

[15] M. Jackson. Some complexities in computer-based systems and their implications for system development. In *Proceedings of the 1990 IEEE International Conference on Computer Systems and Software Engineering*, pages 344–351, Tel-Aviv, Israel, May 1990.

[16] K. Manamcheri, S. Mitra, S. Bak, and M. Caccamo. A step towards verification and synthesis from Simuink/Stateflow models. In *Hybrid systems computation and control (HSCC)*, pages 317–318, 2011.

[17] MathWorks. *Real-Time Workshop*. MathWorks, Inc., Natick, MA, Mar. 2009.

[18] MathWorks. *SimMechanics*. MathWorks, Inc., Natick, MA, Sept. 2012.

[19] MathWorks. *Simulink Control Design*. MathWorks, Inc., Natick, MA, Sept. 2012.

[20] MathWorks. *Simulink User's Guide*. The MathWorks, Inc., Natick, MA, Sept. 2012.

[21] P. J. Mosterman and H. Vangheluwe. Computer automated multi-paradigm modeling: An introduction. *SIMULATION: Transactions of The Society for Modeling and Simulation International*, 80(9):433–450, Sept. 2004.

[22] P. J. Mosterman and J. Zander. Advancing model-based design by modeling approximations of computational semantics. In *Proceedings of the 4th International Workshop on Equation-Based Object-Oriented Modeling Languages and Tools*, pages 3–7, Zürich, Switzerland, Sept. 2011.

[23] P. J. Mosterman, J. Zander, G. Hamon, and B. Denckla. A computational model of time for stiff hybrid systems applied to control synthesis. *Control Engineering Practice*, 20(1):2–13, 2012.

[24] T. B. Nguyen, M. A. Pai, and I. A. Hiskens. Computation of critical values of parameters in power systems using trajectory sensitivities. In *Power System Computation Conference (PSCC)*, pages 24 – 28, 2002.

[25] D. A. Pope. An exponential method of numerical integration of ordinary differential equations. *Communications of the ACM*, 6(8):491–493, 1963.

[26] R. Serban and A. C. Hindmarsh. CVODES: An ODE solver with sensitivity analysis capabilities. Technical Report UCRL-JP-200039, LLNL, 2003.

[27] B. I. Silva, K. Richeson, B. H. Krogh, and A. Chutinan. Modeling and verifying hybrid dynamic systems using CheckMate. In *ADPM 2000 Conference Proceedings: The 4th International Conference on Automation of Mixed Processes - Hybrid Dynamic Systems*, 2000.

[28] P. Varaiya. Reach set computation using optimal control. Available online at http://paleale.eecs.berkeley.edu/ ~varaiya/ papers_ps.dir/reachset.ps, 1998.

[29] K. Wijbrans. *Twente Hierarchical Embedded Systems Implementation by Simulation: a structured method for controller realization*. PhD dissertation, University of Twente, Enschede, The Netherlands, 1993. ISBN 90-9005933-4.

[30] F. Zhang, M. Yeddanapudi, and P. J. Mosterman. Zero-crossing location and detection algorithm for hybrid system simulation. In *IFAC World Congress*, pages 7967 – 7972, 2008.

[31] K. Zhou and J. C. Doyle. *Essentials of Robust Control*. Prentice Hall, 1998.

A Toolbox for Simulation of Hybrid Systems in Matlab/Simulink

[Hybrid Equations (HyEQ) Toolbox] *

Ricardo G. Sanfelice
University of Arizona
USA
sricardo@u.arizona.edu

David A. Copp
University of Arizona
USA
dacopp@email.arizona.edu

Pablo Ñañez
Universidad de Los Andes
Colombia
pa.nanez49@uniandes.edu.co

ABSTRACT

This paper describes the Hybrid Equations (HyEQ) Toolbox implemented in Matlab/Simulink for the simulation of hybrid dynamical systems. This toolbox is capable of computing approximations of trajectories to hybrid systems given in terms of differential and difference equations with constraints, called *hybrid equations*. The toolbox is suitable for the simulation of hybrid systems with different type of trajectories, including those that are Zeno and that have multiple jumps at the same instant. It is also capable of simulating hybrid systems without inputs, with inputs, as well as interconnections of hybrid systems. The structure, components, and usage of the simulation scripts within the toolbox are described. Examples are included to illustrate the main capabilities of the toolbox.

Categories and Subject Descriptors

I.6.7 [**Simulation and Modeling**]: Simulation Support Systems - Environments; G.4 [**Mathematical Software**]: Reliability and robustness; I.2.8 [**Problem Solving, Control Methods, and Search**]: Control theory.

Keywords

Simulation; Hybrid Systems; Matlab/Simulink

1. INTRODUCTION

Simulation is a key tool in the validation of results in engineering and science. Complete theories of numerical simulation have been documented in several textbooks (see, e.g., [17, 3]) and have permitted the development of integration schemes for accurate computation of solutions to differential equations. Integration schemes for such systems are widely available in simulation packages and include one-step methods, such as forward/backward Euler and Runge-Kutta, multi-step methods, such as Adams method and backward differentiation, and their variable step versions. When, in addition to continuous behavior, the model incorporates variables that exhibit jumps, advanced integration methods are needed. Numerous software packages have been recently developed for such class of systems, including Modelica [6], Ptolemy [13], Charon [2], HYSDEL [18], and HyVisual [12], to just list a few. Significant progress has been also made in the development of theory for hybrid systems, with results including the definition of semantics for simulation [10, 15, 12], event detection [14, 7, 5], solvers and error control [7, 4, 1], and structural properties of simulators [16].

Exploiting the robustness properties of hybrid systems modeled as in [9, 8] and the structural properties of simulators in [16], we present a toolbox for the simulation of hybrid systems in Matlab/Simulink. A hybrid system $\mathcal{H}$ is given by the *hybrid equations*

$$\mathcal{H}: \quad x \in \mathbb{R}^n,\ u \in \mathbb{R}^m \quad \begin{cases} \dot{x} &= f(x,u) \quad (x,u) \in C \\ x^+ &= g(x,u) \quad (x,u) \in D \end{cases} \tag{1}$$

with the following objects defining its data: the set $C \subset \mathbb{R}^n \times \mathbb{R}^m$ called the *flow set*; the function $f : \mathbb{R}^n \times \mathbb{R}^m \to \mathbb{R}$ called the *flow map*; the set $D \subset \mathbb{R}^n \times \mathbb{R}^m$ called the *jump set*; and the function $g : \mathbb{R}^n \times \mathbb{R}^m \to \mathbb{R}$ called the *jump map* [1]. These objects define the *data* of $\mathcal{H}$, which is explicitly denoted as $\mathcal{H} = (f, C, g, D)$. Examples in Section 3. illustrate the type of systems that can be modeled within the hybrid equations framework; see [9, 8] for more examples.

The *Hybrid Equations (HyEQ) Toolbox* is capable of simulating hybrid systems given as in (1). The toolbox consists of a set of Matlab/Simulink scripts to numerically compute and plot the system's trajectories. The HyEQ Toolbox has the following main components and features:

1) A Matlab script for simulation of hybrid equations without inputs (*Lite HyEQ Simulator*);

2) a Simulink library for simulation of hybrid equations with inputs (*HyEQ Simulator*);

*The current version of the toolbox can be found at Matlab Central and at the author's website http://www.u.arizona.edu/~sricardo/. This research has been partially supported by the National Science Foundation under CAREER Grant no. ECS-1150306 and by the Air Force Office of Scientific Research under Grant no. FA9550-12-1-0366.

[1]In other works, the terms *reset map* and *impact map* are used for g while the term *switching surface* is sometimes used for D.

HSCC'13, April 8–11, 2013, Philadelphia, Pennsylvania, USA.

3) Computation of trajectories that are Zeno and that have multiple jumps at the same instants;

4) Basic event detection and capability to implement advanced crossing detection algorithms;

5) Simulation of interconnections of hybrid systems.

In Section 2.2, the main components of the toolbox are described. In Section 3, examples illustrate the main features of the toolbox listed above as well as its broad applicability. The examples feature a system with Zeno behavior, a system with zero-crossing event detection, and interconnections of hybrid systems with multiple jumps at the same instant.

2. HYEQ: A TOOLBOX FOR SIMULATION OF HYBRID EQUATIONS

The HyEQ Toolbox provides a set of Matlab/Simulink scripts to numerically compute and plot the trajectories to hybrid systems given in terms of hybrid equations as in (1). Given an input u, a trajectory (or solution) to $\mathcal{H}$ is conveniently defined as a function[2]

$$x : \operatorname{dom} x \to \mathbb{R}^n \tag{2}$$

parameterized by ordinary time $t \in \mathbb{R}_{\geq 0}$ and discrete time $j \in \{0, 1, \ldots\} =: \mathbb{N}$. The set $\operatorname{dom} x \subset \mathbb{R}_{\geq 0} \times \mathbb{N}$ defines the *hybrid time domain* of x and has the following structure: there exists a sequence of times $0 = t_0 \leq t_1 \leq \ \ldots \ \leq t_j \leq \ldots$ such that $\operatorname{dom} x$ is the union of (possibly infinitely) many intervals of flow time $[t_j, t_{j+1}]$ indexed by j when $t_{j+1} > t_j$ (i.e., of intervals of the form $[t_j, t_{j+1}] \times \{j\}$ with $[t_j, t_{j+1}]$ of nonzero length) and of jump instants $[t_j, t_{j+1}] \times \{j\}$ when $t_{j+1} = t_j$, with the last interval of flow possibly of the form $[t_j, t_{j+1}) \times \{j\}$. During flows, the trajectory x and the input u satisfy the continuous dynamics

$$\dot{x} = f(x, u) \quad (x, u) \in C \tag{3}$$

over the intervals of flow while, at jumps, they satisfy the discrete dynamics

$$x^+ = g(x, u) \quad (x, u) \in D. \tag{4}$$

Over a finite amount of flow and a finite number of jumps, the HyEQ Toolbox computes an approximation of the trajectory x in (2) by evaluating the flow condition $(x, u) \in C$ and the jump condition $(x, u) \in D$, and, according to the result of this evaluation, by appropriately discretizing the differential equation defining the flows in (3) or computing the new value of the state after jumps using (4). In this way, the HyEQ Toolbox returns a discrete version of x and its hybrid time domain $\operatorname{dom} x$. More precisely, given an input u, the computed version of the trajectory x is denoted

$$x_s : \operatorname{dom} x_s \to \mathbb{R}^n,$$

which we call *a simulated trajectory* of $\mathcal{H}$, and satisfies

$$x_s^+ = f_s(x_s, u_s) \quad (x_s, u_s) \in C \tag{5}$$

over the intervals of flow and, at jumps, satisfies the discrete dynamics

$$x_s^+ = g(x_s, u_s) \quad (x_s, u_s) \in D. \tag{6}$$

[2]See [9] for a formal definition of this function as a hybrid arc.

The input u_s is the discretization of u. The function f_s is the resulting discretized flow map obtained when employing an integration scheme for the differential equation $\dot{x} = f(x, u)$. For instance, when the integration scheme is given by forward Euler, $f_s(x_s, u_s) = x_s + sf(x_s, u_s)$ with $s > 0$ denoting the step size for integration. Formal definitions of simulated trajectories (or solutions) and dynamical properties of the discretization (5)-(6) of $\mathcal{H}$ can be found in [16].

Within this framework for simulation, we describe the Lite HyEQ Simulator in Section 2.1, the HyEQ Simulator in Section 2.2, and common initialization and plotting functions in Section 2.3.

2.1 Lite HyEQ Simulator

A way to compute the trajectories of hybrid equations is to use ODE function calls with events that reset the state according to the discrete dynamics of the system. The Lite HyEQ Simulator computes x_s in this manner and uses four Matlab functions to implement the data of the hybrid system (1) without external input u. The Matlab functions employed by this simulator are defined as follows:

i) The flow map is defined in the Matlab function `f.m`. The input to this function is a vector with components defining the state of the system x_s. Its output is the value of the flow map f evaluated at x_s.
ii) The flow set is defined in the Matlab function `C.m`. The input to this function is a vector with components defining the state of the system x_s. Its output is equal to 1 if the state belongs to the set C or equal to 0 otherwise.
ii) The jump map is defined in the Matlab function `g.m`. Its input is a vector with components defining the state of the system x_s. Its output is the value of the jump map g evaluated at x_s.
iv) The jump set is defined in the Matlab function `D.m`. Its input is a vector with components defining the state of the system x_s. Its output is equal to 1 if the state belongs to D or equal to 0 otherwise.

The Matlab script `HyEQsolver.m` implements the simulation of a hybrid equation defined by functions `f.m`, `C.m`, `g.m`, and `D.m` implementing the data. `HyEQsolver.m` uses these functions to integrate the differential equation during continuous evolution and to execute the jump map when jumps shall occur. To do this, the algorithms in `HyEQsolver.m` check at each integration step if x_s is in the set C, D, or neither. Depending on which set x_s is in, the simulation is accordingly reset following the dynamics given in f or g, or the simulation is stopped, respectively. `HyEQsolver.m` is configured to call ODE45, but a different ODE solver can be configured similarly.

The syntax to perform a simulation using `HyEQsolver.m` is as follows:

```
[t j xs] = HyEQsolver(@f,@g,@C,@D,x0,TSPAN,
                       JSPAN,rule,options);
```

The function call has arguments given by the data implemented as Matlab functions as well as simulation parameters. The $n \times 1$ vector `x0` defines the initial condition. The 2×1 parameter `TSPAN = [TSTART TFINAL]` defines the initial and final values of the flow variable t, i.e., the *continuous horizon*, and the 2×1 parameter `JSPAN = [JSTART JFINAL]` defines the initial and final values of the jump index

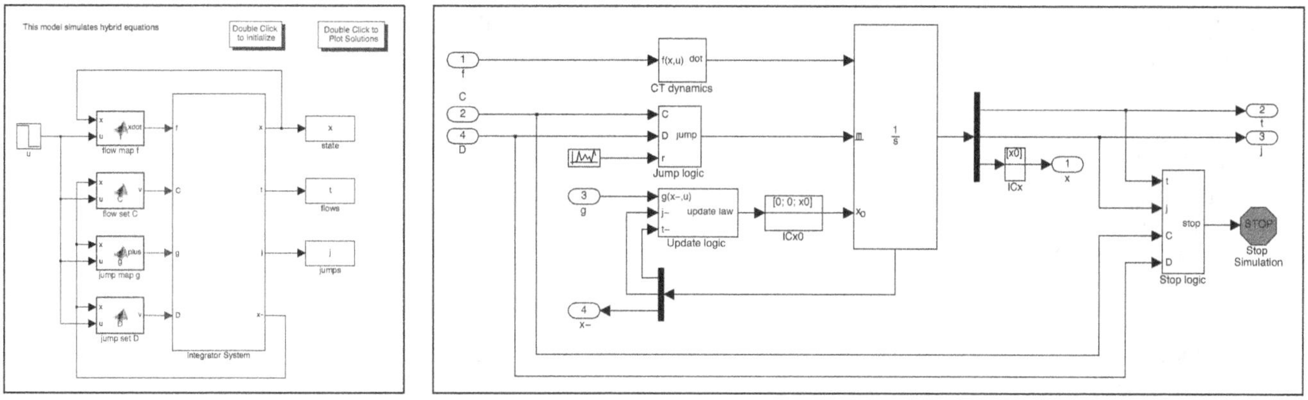

Figure 1: Matlab/Simulink implementation of a hybrid system $\mathcal{H} = (C, f, D, g)$ with inputs (left). Internals of integrator system (right).

j, i.e., the *discrete horizon*. The simulation stops when either `TFINAL` or `JFINAL` is reached (or exceeded). The scalar parameter `rule` defines whether the simulator gives priority to jumps (rule $= 1$), priority to flows (rule $= 2$), or no priority (rule $= 3$) when both $x \in C$ and $x \in D$ hold. When no priority is selected, then the simulator selects flowing or jumping randomly. The parameter `options` configures the relative tolerance, maximum integration step allowed, and other knobs of the ODE solver. The Matlab function `odeset` can be used to define this parameter; e.g., `options = odeset('RelTol',1e-6,'MaxStep',.1)` sets the relative tolerance to 10^{-6} and the maximum integration step to 0.1.

The Matlab script `HyEQsolver.m` returns the computed state x_s along with the (discretized) hybrid time domain $\operatorname{dom} x_s$. The Matlab script `run.m` is provided to initialize these parameters and run the Lite HyEQ solver. This script can also be used to plot the computed trajectories after the simulation is complete.

2.2 HyEQ Simulator

Trajectories to hybrid equations with inputs are computed by implementing the ideas in the Lite HyEQ simulator of Section 2.1 within Simulink. This permits the use of Simulink libraries for the generation of signals to be used as inputs. It also permits the computation of trajectories to hybrid equations with interconnections defined using Simulink's user interface. Figure 1 depicts a diagram of the HyEQ Simulator (left) and its internals (right). The HyEQ Simulator computes x_s using an integrator block with external resets and state port enabled, and uses four Embedded Matlab function blocks to implement the data of the hybrid system (1) with external input u. More precisely:

i) The flow map f is implemented in the Embedded Matlab function block "flow map f." Its input is a vector with components defining the state of the system x and the input u. Its output is the value of the flow map f which is connected to the input of an integrator.
ii) The flow set C is implemented in the Embedded Matlab function block "flow set C." Its input is a vector with components x^- and input u [3]. Its output is equal to "true" if the state belongs to the set C or equal to "false" otherwise.
iii) The jump map g is implemented in the Embedded Matlab function block "jump map g." Its input is a vector with components x^- and input u of the *Integrator system*. Its output is the value of the jump map g.
iv) The jump set is implemented in an *Embedded Matlab function block* executing the function `D.m`. Its input is a vector with components x^- and input u. Its output is equal to "true" if the state belongs to the set D or equal to "false" otherwise.

[3]The minus notation denotes the previous value of the state (before integration). In Simulink, the value x^- is obtained from the state port of the integrator.

The integrator block integrates $\dot{x} = f(x, u)$ to get the state trajectory from a given initial condition `x0`. This block also computes the (discretized) hybrid time domain associated with the state trajectory. The internal components of this block are shown in Figure 1 (right) and are described next.

2.2.1 CT Dynamics

This block defines the continuous-time (CT) dynamics by assembling the time derivative of t, j, and x. The parameters t and j are considered as states of the integrator and, in this way, are updated throughout the simulation so as to keep track of the flow time and number of jumps. The "CT dynamics" block implements the differential equation

$$\dot{t} = 1, \qquad \dot{j} = 0, \qquad \dot{x} = f(x, u),$$

where u is provided by an external signal generator (see left of Figure 1).

2.2.2 Jump Logic

This block uses the outputs of the Embedded Matlab function blocks "flow set C" and "jump set D," which indicate whether the state and input are in those sets or not, and a random signal r with uniform distribution in $[0, 1]$. Figure 2 shows the Simulink blocks used to implement the logic for jumps. As defined in Section 2.1, the variable *rule* defines whether the simulator gives priority to jumps, priority to flows, or no priority. The output of the "Jump logic" block is equal to one when either

- the output of the "jump set D" block is equal to one and `rule`$= 1$;

- the output of the "flow set C" block is equal to zero, the output of the "jump set D" block is equal to one, and **rule**= 2;
- the output of the "flow set C" block is equal to zero, the output of the "jump set D" block is equal to one, and **rule**= 3; or
- the output of the "flow set C" block is equal to one, the output of the "jump set D" block is equal to one, **rule**= 3, and the random signal r is larger or equal than 0.5.

Under any of these events, the output of the "Jump logic" block, which is connected to the integrator's external reset input, triggers a reset of the integrator, that is, a jump of the hybrid equation being simulated. The reset or jump is activated since the configuration of the reset input is set to "level hold," which executes resets when this external input is equal to one (if the next input remains set to one, multiple resets would be triggered). Otherwise, the output is equal to zero.

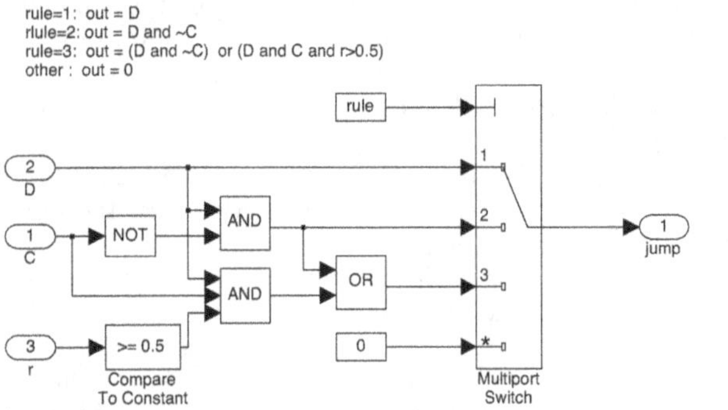

Figure 2: Implementation of the logic for jumps within "Jump logic" block.

2.2.3 *Update Logic*

The "Update logic" block makes use of the *state port* of the integrator to reset the state. This port reports the value of the state of the integrator at the exact instant that the reset condition becomes true. Notice that x^- differs from x since at a jump, x^- indicates the value of the state that triggers the jump, but it is never assigned as the output of the integrator. In other words, "$(x, u) \in D$" is checked using x^- and, if true, x is reset to $g(x^-, u)$. Notice, however, that u is the same because at a jump, u indicates the next evaluated value of the input, and it is assigned as the output of the integrator. The flow time t is kept constant at jumps and j is incremented by one. More precisely, the "Update logic" block executes the following reset law

$$t^+ = t^-, \qquad j^+ = j^- + 1, \qquad x^+ = g(x^-, u),$$

where (x^-, u) is the value that triggers the jump.

2.2.4 *Stop Logic*

This block stops the simulation under any of the following events:

- The flow time is larger than or equal to the maximum flow time specified by `TFINAL`.
- The jump time is larger than or equal to the maximum number of jumps specified by `JFINAL`.
- The state of the hybrid system x is neither in C nor in D.

Under any of these events, the output of the logic operator connected to the "Stop logic" block becomes one, stopping the simulation. Note that the inputs C and D are routed from the output of the blocks computing whether the state is in C or D and use the value of x^-.

2.3 Initialization and postprocessing functions

The following functions are used to generate plots:

- **`plotflows(t,j,x)`**: plots (in blue) the projection of the trajectory x onto the flow time axis t. The value of the trajectory for intervals $[t_j, t_{j+1}]$ with empty interior is marked with $*$ (in blue). Dashed lines (in red) connect the value of the trajectory before and after the jump.
- **`plotjumps(t,j,x)`**: plots (in red) the projection of the trajectory x onto the jump time j. The initial and final value of the trajectory on each interval $[t_j, t_{j+1}]$ is denoted by $*$ (in red) and the continuous evolution of the trajectory on each interval is depicted with a dashed line (in blue).
- **`plotHybridArc(t,j,x)`**: plots (in black) the trajectory x on hybrid time domains. The intervals $[t_j, t_{j+1}]$ indexed by the corresponding j are depicted in the $t - j$ plane (in red).

An initialization file is used to define variables needed for a simulation. Also, a post-processing file is used to plot the simulated solutions after a simulation is complete. These two `.m` files are called by double-clicking the *Double Click to...* blocks at the top of the HyEQ Simulator (left in Figure 1).

3. EXAMPLES

Next, we illustrate the HyEQ Toolbox in several hybrid systems modeled as hybrid equations. The files associated with the examples below are available at the software entry at http://www.u.arizona.edu/~sricardo/.

3.1 A system with Zeno behavior

Consider the model of a bouncing ball given by a hybrid equation with data

$$f(x) := \begin{bmatrix} x_2 \\ -\gamma \end{bmatrix}, \; C := \{x \in \mathbb{R}^2 \mid x_1 \geq 0\}$$

$$g(x) := \begin{bmatrix} 0 \\ -\lambda x_2 \end{bmatrix}, \; D := \{x \in \mathbb{R}^2 \mid x_1 \leq 0 \, , \, x_2 \leq 0\}$$

where $\gamma > 0$ is the gravity constant and $\lambda \in [0, 1)$ is the restitution coefficient. This system models a ball bouncing on a floor at zero height. The following procedure is used to simulate this system in the Lite HyEQ Simulator: 1) Inside the Matlab script **`run.m`**, initial conditions, simulation horizons, a rule for jumps, ode solver options, and a step size coefficient are defined. The function **`HyEQsolver`** is called in order to run the simulation, and a script for plotting simulated solutions is included; 2) Then, the Matlab functions **`f.m, C.m, g.m, D.m`** are edited according to the data given above; 3) Finally, the simulation is run by calling **`run.m`** in the Matlab command window. A simulated

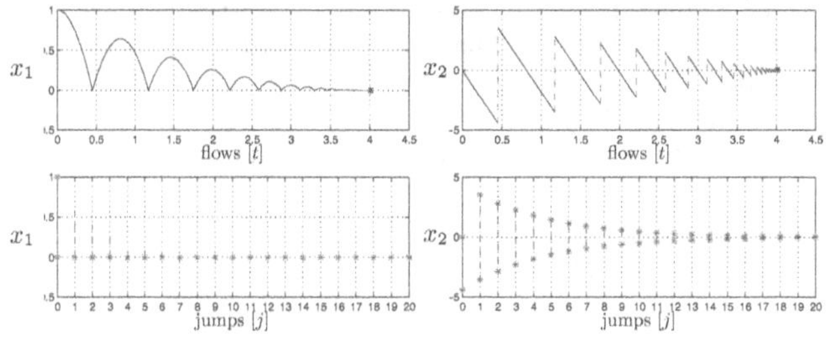

(a) Height (b) Velocity

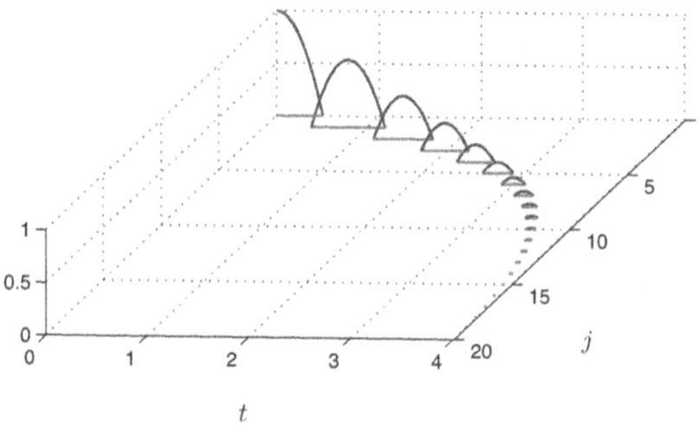

(c) Simulated solution vs. (t, j)

Figure 3: Simulated solution to system in Section 3.1.

solution to the bouncing ball system from x0= $[1, 0]^\top$ and with TSPAN= [0 10], JSPAN= [0 20], rule= 1, $\gamma = 9.81$, and $\lambda = 0.8$ is depicted in Figure 3(a) (height) and Figure 3(b) (velocity). Both the projection onto t using plotflows and onto j using plotjumps are shown. Figure 3(c) depicts the corresponding hybrid arc for the position state using plotHybridArc.

3.2 A system with special event detection

Consider the hybrid equation with data

$$f(x) := \begin{bmatrix} -x_2 \\ 0 \end{bmatrix}, \quad C := \{x \in \mathbb{R} \times \{-1, 1\} \mid x_1 > 0\} \cup \{x \in \mathbb{R} \times \{-1, 1\} \mid x_1 < 0\},$$

$$g(x) := \begin{bmatrix} -x_2 \\ -x_2 \end{bmatrix}, \quad D := \{x \in \mathbb{R} \times \{-1, 1\} \mid x_1 = 0\}.$$

The trajectories to this system are such that x_1 flows according to $\dot{x}_1 = 1$ when $x_2 = -1$ and according to $\dot{x}_1 = -1$ when $x_2 = 1$. When x_1 hits zero, x_1 is reset to x_2, which is either 1 or -1. Unfortunately, the jump condition is fragile and numerical errors would prevent from it being satisfied.

So that the simulated solution jumps when x_1 crosses zero, a zero-crossing detection algorithm could be used. Figure 4 shows a construction of the jump set that exploits the capabilities of Simulink's block *Hit Crossing*, which will not miss the crossing of x_1 by zero and, hence, trigger a jump. A trajectory for this system from x0= [2, 1] is depicted in Figure 4(b) (projected onto t and j) with jumps at the correct instants.

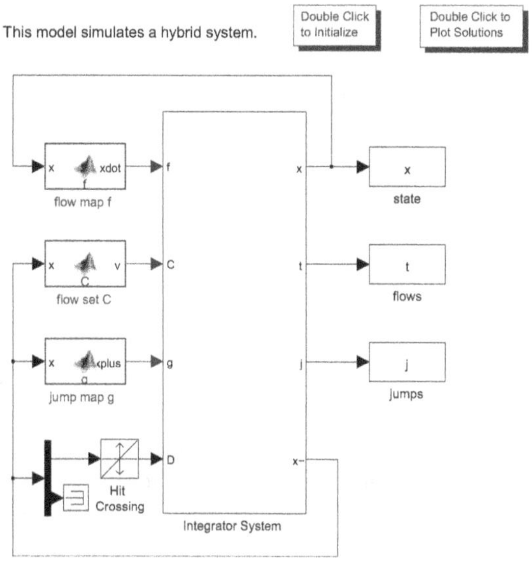

(a) HyEQ Simulator with *Hit Crossing* block.

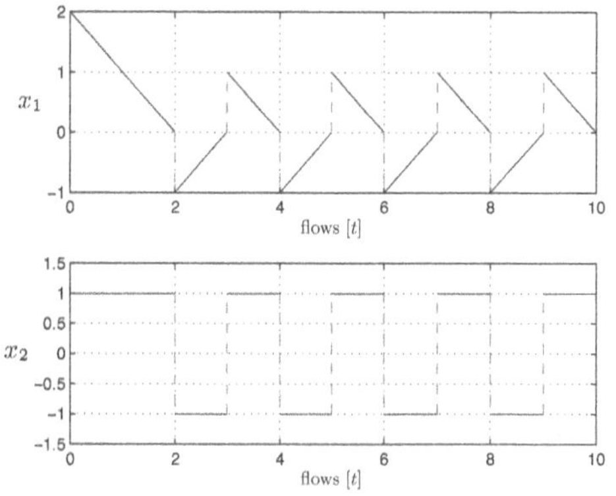

(b) Simulated solution with *Hit Crossing* block.

Figure 4: HyEQ Simulator using zero-crossing detection to detect jumps and a trajectory.

3.3 An interconnected system w/multiple jumps

Consider the synchronization of two impulse-coupled oscillators. The timers within each oscillator are modeled as periodic oscillators with timer state x_i, $i = 1, 2$, evolving on the unitary interval $[0, 1]$. When the i-th state reaches $x_i = 1$, the said state is reset to zero but the state of the other timer is reset via $x_j^+ = \max\{1, (1+\varepsilon)x_j\}$, where $\varepsilon > 0$ is a constant coefficient. It can be shown that the timers converge to each other asymptotically, for almost every initial condition [9]. This model has been used in the literature to model impulse-coupled oscillators in nature (fireflies, crickets, Parkinson's disease, etc.). Each timer can be modeled by the hybrid equations given by $f_i(x_i, u_i) := 1$,

$$\begin{aligned} C_i &:= \left\{(x_i, u_i) \in \mathbb{R}^2 \mid 0 \le x_i \le 1\right\} \cap \\ &\qquad \left\{(x_i, u_i) \in \mathbb{R}^2 \mid 0 \le u_i \le 1\right\}, \\ g_i(x_i, u_i) &:= \begin{cases} (1+\varepsilon)x_i & \text{if } (1+\varepsilon)x_i < 1 \\ 0 & \text{if } (1+\varepsilon)x_i \ge 1, \end{cases} \\ D_i &:= \left\{(x_i, u_i) \in \mathbb{R}^2 \mid x_i = 1\right\} \cup \\ &\qquad \left\{(x_i, u_i) \in \mathbb{R}^2 \mid u_i = 1\right\}. \end{aligned}$$

The interaction between the timers can be modeled as the interconnection between a copy of this system for $i = 1$ and another copy for $i = 2$, with the interconnection relationship $u_1 = x_2$ and $u_2 = x_1$. In this way, the reset of one timer affects the dynamics of the other timer. A trajectory to this interconnection is obtained by coding the system above in the HyEQ Simulator, converting it into a subsystem, and then duplicating the system and interconnecting them within a common Simulink file; see Figure 5(b)[4]. A simulated solution for $\varepsilon = 0.3$ is depicted in Figure 5(a). It confirms the expected behavior of the interconnected hybrid equations: the timers initially evolve out of phase and progressively synchronize their resets. Once the timers are synchronized, two jumps at the same instant occur.

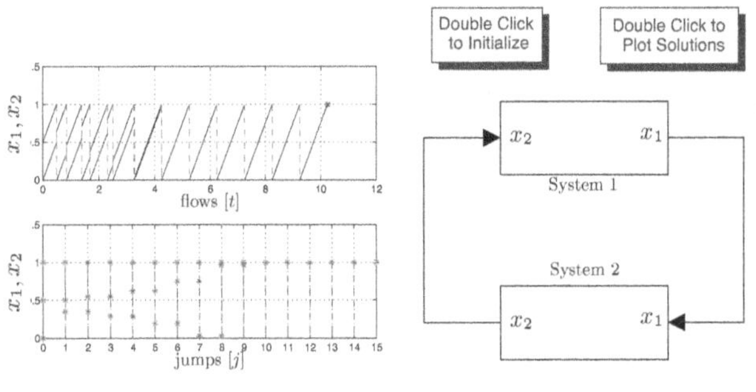

(a) Simulated solution (b) Interconnection

Figure 5: Simulated solution to system and the interconnected system in Section 3.3.

4. CONCLUSIONS

The Hybrid Equations (HyEQ) Toolbox for simulation of hybrid systems modeled as hybrid equations was described and illustrated in examples. A webinar presenting this toolbox and illustrating it in examples has been prepared and is currently under review by Mathworks before it goes live at their website. The toolbox has been tested for the past five years, including graduate courses taught at the University of Arizona and at the University of California Santa Barbara, and at short courses and workshops at the International EECI Graduate School on Control, CDC, and ACC.

5. REFERENCES

[1] A. Abate, A. D. Ames, and S. S. Sastry. Error bounds based stochastic approximations and simulations of hybrid dynamical systems. In *Proc. 25th American Control Conference*, pages 4742–4747, 2006.

[4]Furthermore, a complex system with up to twenty five impulse-coupled oscillators is included in the examples that come with the toolbox.

[2] R. Alur, T. Dang, J. Esposito, Y. Hur, F. Ivancic, V. Kumar, I. Lee, P. Mishra, G. J. Pappas, and O. Sokolsky. Hierarchical modeling and analysis of embedded systems. *Proc. of the IEEE*, 91:11–28, 2003.

[3] U. M. Ascher and L. R. Petzold. *Computer Methods for Ordinary Differential Equations and Differential-Algebraic Equations.* SIAM, 1998.

[4] M. K. Camlibel, W.P.M.H. Heemels, and J.M.H. Schumacher. Consistency of a time-stepping method for a class of piecewise-linear networks. *IEEE Trans. Circ. Syst. I*, 49(3):349–357, 2002.

[5] D. A. Copp and R. G. Sanfelice. On the effect and robustness of zero-crossing detection algorithms in simulation of hybrid systems jumping on surfaces. In *Proceedings of the American Control Conference*, pages 2449–2454, 2012.

[6] H. Elmqvist, S.E. Mattsson, and M. Otter. Modelica: The new object-oriented modeling language. In *Proc. of 12th European Simulation Multiconference*, 1998.

[7] J.M. Esposito, V. Kumar, and G.J. Pappas. Accurate event detection for simulating hybrid systems. In *Hybrid Systems: Computation and Control: 4th International Workshop*, pages 204–217, 2001.

[8] R. Goebel, R. G. Sanfelice, and A. R. Teel. *Hybrid Dynamical Systems: Modeling, Stability, and Robustness.* Princeton University Press, 2012.

[9] R. Goebel, R.G. Sanfelice, and A.R. Teel. Hybrid dynamical systems. *IEEE Control Systems Magazine*, pages 28–93, 2009.

[10] K.H. Johansson, J. Lygeros, S. Sastry, and M. Egerstedt. Simulation of Zeno hybrid automata. In *Proc. 38th IEEE Conference on Decision and Control*, volume 4, pages 3538–3543, 1999.

[11] E. Kofman. Discrete event simulation of hybrid systems. *SIAM Journal on Scientific Computing*, 25(5):1771–1797, 2004.

[12] E. A. Lee and H. Zheng. Operational semantics for hybrid systems. In *Hybrid Systems: Computation and Control: 8th International Workshop*, pages 25–53, 2005.

[13] J. Liu and E. A. Lee. A component-based approach to modeling and simulating mixed-signal and hybrid systems. *ACM Transactions on Modeling and Computer Simulation*, 12(4):343–368, 2002.

[14] J. Liu, X. Liu, T. J. Koo, B. Sinopoli, S.Sastry, and E. A. Lee. A hierarchical hybrid system model and its simulation. In *Proc. 38th IEEE Conf. Dec. Contr.*, volume 4, pages 3508 – 3513, 1999.

[15] P.J. Mosterman and G. Biswas. A hybrid modeling and simulation methodology for dynamic physical systems. *Simulation*, 78:5–17, 2002.

[16] R. G. Sanfelice and A. R. Teel. Dynamical properties of hybrid systems simulators. *Automatica*, 46(2):239–248, 2010.

[17] A. M. Stuart and A.R. Humphries. *Dynamical Systems and Numerical Analysis.* Cambridge University Press, 1996.

[18] F.D. Torrisi and A. Bemporad. HYSDEL - A tool for generating computational hybrid models for analysis and synthesis problems. *IEEE Trans. Control Systems Technology*, 12:235–249, 2004.

Stabhyli – A Tool for Automatic Stability Verification of Non-Linear Hybrid Systems*

Eike Möhlmann and Oliver Theel
Carl von Ossietzky University of Oldenburg
Department of Computer Science
D-26111 Oldenburg, Germany
{eike.moehlmann, theel}@informatik.uni-oldenburg.de

ABSTRACT

We present Stabhyli, a tool that automatically proves stability of non-linear hybrid systems. Hybrid systems are systems that exhibit discrete as well as continuous behavior. The stability property basically ensures that a system exposed to a faulty environment (e.g. suffering from disturbances) will be able to regain a "good" operation mode as long as errors occur not too frequently. Stabilizing Hybrid systems are omnipresent, for instance in control applications where a discrete controller is controlling a time-continuous process such as a car's movement or a particular chemical reaction. We have implemented a tool to automatically derive a certificate of stability for non-linear hybrid systems. Certificates are obtained by Lyapunov theory combined with decomposition and composition techniques.

Categories and Subject Descriptors

D.2.4 [**Software Engineering**]: Software/Program Verification

General Terms

Design, Verification

Keywords

Hybrid Systems; Automatic Verification; Stability; Lyapunov Theory; LMIs; Sums-of-Squares; Computer-Aided Design

1. INTRODUCTION

We present a tool for the automatic verification of stability of non-linear hybrid systems. Stability, in general and for hybrid systems in particular, is a very desirable property, since stable systems are inherently fault-tolerant: after the occurrence of faults leading to, for example, a changed environment, the system will automatically "drive back" to the set of desired (i.e., stable) states. Stable systems are therefore particularly suited for contexts where autonomy is important or in the context of dependable assistance systems.

*This work has been partly supported by the German Research Foundation (DFG) as part of the Transregional Collaborative Research Center "Automatic Verification and Analysis of Complex Systems" (SFB/TR 14 AVACS).

HSCC'13, April 8–11, 2013, Philadelphia, Pennsylvania, USA.

Often, modeling real world systems involves the interaction of embedded systems (e.g., a controller) and its surrounding environment (e.g., a plant). Examples for such systems are automatic cruise controllers, engine control units, or unmanned powerhouses. In all these examples, an optimal operating range should be maintained. Although it is sometimes possible to discretize physical relations (using sampling) or to fluidize discrete steps (having a real-valued count of objects) it is more natural and less error-prone to use hybrid systems for modeling and verification. This is due to the fact that hybrid systems allow both, the representation of discrete and continuous behavior.

As real world systems tend to become very large and complicated, automated computer-aided verification and assistance in the design is very important. In this paper, we present a tool that addresses both, verification and design.

Oehlerking et al. implemented a very powerful state space partitioning scheme which is restricted to linear behavior [8]. In case their tool is not able to prove stability of a hybrid system, it cannot give a hint about the reason, as it would not terminate and continue to partition the state space. Stabhyli, on the contrary, overcomes these drawbacks by using the (de-)compositional proof schemes presented in [9, 3]. We have fully automatized these schemes and combined them with pre- and postprocessing steps that simplify the design and counteract numerical problems. Furthermore, in case stability cannot be proven, our tool returns a hint to the user. This information can then be used for system redesign. In addition, Stabhyli handles non-linear systems whose behavior is expressible in forms of polynomials.

To the best of our knowledge, no other tool with those capabilities does currently exist. Podelski and Wagner have presented a tool that automatically proves region stability by discretization of the system using snapshot sequences and then computing a decreasing relation on these snapshots [11]. But to compute the snapshots, their tool relies on sets of states which usually are overapproximations of the actual reachable states. In cases where the overapproximations are to coarse, no decreasing relation between the snapshots can be found. A tool that combines Lyapunov functions with the search for a well-foundedness relation for symmetric linear hybrid systems can be found in [4]. Another tool is RSolver which is restricted to purely continuous systems [13]. All these tools including Stabhyli are incomplete i.e., not all stable hybrid systems can be proven stable. This is due to the fact, that stability is in general undecidable. Therefore, sufficient conditions are checked instead.

This paper is organized as follows. In Section 2, we define

the hybrid system model, the stability property, and give an adaptation of the Lyapunov Theorem which is the bases of the proof schemes. Section 3 describes the proof schemes which Stabhyli supports as well as the preprocessing, constraint solving, and postprocessing steps that we have integrated. In Section 4, we prove stability for two examples using Stabhyli. Both examples cannot be proven stable using the tools mentioned above, since either the state space partitioning scheme is not suitable or due to non-linear constraints and the hybrid nature of the examples. Finally, in Section 5, we give a short summary.

2. PRELIMINARIES

A hybrid automaton models both continuous evolution and discrete actions over time. It is often represented as a graph made of vertices (called *modes*) and edges (called *jumps* or *transitions*). The discrete actions (i.e. *guards* and *resets*) are annotated at the transitions between the modes and the continuous evolution in form of *invariants, differential equations*, or *differential inclusions* within the modes.

A hybrid automaton is defined as follows

Definition 1. A **Hybrid Automaton** is a quintuple

$$\mathcal{H} = (\mathcal{V}, \mathcal{M}, \mathcal{T}, Flow, Inv) \text{ where}$$

- $\mathcal{V}$ is a finite set of *variables* and $\mathcal{S} = \mathbb{R}^{|\mathcal{V}|}$ is the corresponding *continuous state space*,
- $\mathcal{M}$ is a finite set of *modes*,
- $\mathcal{T}$ is a finite set of *transitions* (m_1, G, R, m_2) where
  - $m_1, m_2 \in \mathcal{M}$ are the *source and target mode* of the transition, respectively,
  - $G \subseteq \mathcal{S}$ is a *guard* which restricts the valuations of the variables for which this transition can be taken,
  - $R : \mathcal{S} \to \mathcal{S}$ is the *reset function* which might reset some valuations of the variables,
- $Flow : \mathcal{M} \to [\mathcal{S} \to \mathcal{P}(\mathcal{S})]$ is the *flow function* which assigns a *flow* to every mode. A flow $f \subseteq \mathcal{S} \to \mathcal{P}(\mathcal{S})$ in turn assigns a closed subset of $\mathcal{S}$ to each $x \in \mathcal{S}$, which can be seen as the right hand side of a differential inclusion $\dot{x} \in f(x)$,
- $Inv : \mathcal{M} \to \mathcal{S}$ is the *invariant function* which assigns a closed subset of the continuous state space to each mode $m \in \mathcal{M}$, and therefore restricts valuations of the variables for which this mode can be active.

A *trajectory* of $\mathcal{H}$ is an infinite solution in form of a function $x(t)$ over time. Each solution has an associated (possibly infinite) sequence of modes visited by the trajectory.

Intuitively, stability is a property basically expressing that all trajectories of the system eventually reach an equilibrium point of the sub-state space and stay in that point forever given the absence of errors. For technical reasons the equilibrium point is usually assumed to be the origin of the continuous state space, i.e. 0. This is not a restriction, since a system can always be shifted such that the equilibrium is 0 via a coordinate transformation. In the sequel, we focus on *asymptotic stability* which does not require the equilibrium point to be reached in finite time but only requires every trajectory to "continuously approach" it (in contrast to *exponential stability* where additionally the existence of an exponential rate of convergence is required).

In the following, we refer to $x_{|\mathcal{V}'} \in \mathbb{R}^{|\mathcal{V}'|}$ as the sub-vector of a vector $x \in \mathbb{R}^{\mathcal{V}}$ containing only values of variables in $\mathcal{V}' \subseteq \mathcal{V}$.

Definition 2. **Global Asymptotic Stability with Respect to a Subset of Variables [7].** Let $\mathcal{H} = (\mathcal{V}, \mathcal{M}, \mathcal{T}, Flow, Inv)$ be a hybrid automaton, and let $\mathcal{V}' \subseteq \mathcal{V}$ be the set of variables that are required to converge to the equilibrium point 0. A continuous-time dynamic system $\mathcal{H}$ is called *globally stable (GS) with respect to* $\mathcal{V}'$ if for all functions $x_{|\mathcal{V}'}(t)$,

$$\forall \epsilon > 0 : \exists \delta > 0 : \forall t \geq 0 : ||x(0)|| < \delta \Rightarrow ||x_{|\mathcal{V}'}(t)|| < \epsilon.$$

$\mathcal{H}$ is called *globally attractive (GA) with respect to* $\mathcal{V}'$ if for all functions $x_{|\mathcal{V}'}(t)$,

$$\lim_{t\to\infty} x_{|\mathcal{V}'}(t) = 0, \text{ i.e.}, \forall \epsilon > 0 : \exists t_0 \geq 0 : \forall t > t_0 : ||x_{|\mathcal{V}'}(t)|| < \epsilon,$$

where 0 is the origin of $\mathbb{R}^{|\mathcal{V}'|}$. If a system is both globally stable with respect to $\mathcal{V}'$ and globally attractive with respect to $\mathcal{V}'$, then it is called *globally asymptotically stable (GAS) with respect to* $\mathcal{V}'$.

Intuitively, GS is a boundedness condition, i.e. each trajectory starting δ-close to the origin will remain ϵ-close to the origin. GA ensures progress, i.e. for each ϵ-distance to the origin, there exists a point in time t_0 such that a trajectory always remains within this distance. By induction it follows, that every trajectory is eventually always approaching the origin. This property can be proven using Lyapunov Theory [6]. Lyapunov Theory was originally restricted to continuous systems but has been lifted to hybrid systems.

THEOREM 1. **Discontinuous Lyapunov Functions for a subset of variables [7].** Let $\mathcal{H} = (\mathcal{V}, \mathcal{M}, \mathcal{T}, Flow, Inv)$ be a hybrid automaton and let $\mathcal{V}' \subseteq \mathcal{V}$ be the set of variables that are required to converge. If for each $m \in \mathcal{M}$, there exists a set of variables $\mathcal{V}_m$ with $\mathcal{V}' \subseteq \mathcal{V}_m \subseteq \mathcal{V}$ and a continuously differentiable function $V_m : \mathcal{S} \to \mathbb{R}$ such that

1. for each $m \in \mathcal{M}$, there exist two class K^∞ functions α and β such that
$$\forall x \in Inv(m) : \alpha(||x_{|\mathcal{V}_m}||) \preccurlyeq V_m(x) \preccurlyeq \beta(||x_{|\mathcal{V}_m}||),$$
2. for each $m \in \mathcal{M}$, there exists a class K^∞ function γ such that
$$\forall x \in Inv(m) : \dot{V}_m(x) \preccurlyeq -\gamma(||x_{|\mathcal{V}_m}||)$$
for each $\dot{V}_m(x) \in \left\{ \left\langle \frac{dV_m(x)}{dx} \middle| f(x) \right\rangle \middle| f(x) \in Flow(m) \right\}$,
3. for each $(m_1, G, R, m_2) \in \mathcal{T}$,
$$\forall x \in G : V_{m_2}(R(x)) \preccurlyeq V_{m_1}(x),$$

then $\mathcal{H}$ is globally asymptotically stable with respect to $\mathcal{V}'$ and V_m is called a *Local Lyapunov Function (LLF)* of m.

In Theorem 1, $\left\langle \frac{dV(x)}{dx} \middle| f(x) \right\rangle$ denotes the inner product between the gradient of a Lyapunov function V and a flow function $f(x)$.

3. THE NEW TOOL

Our tool generates Lyapunov functions for a given hybrid system and thereby proves stability of this system. Figure 1 gives an overview of the steps performed. They will be described in this section. To prove stability of a hybrid system given as a hybrid automaton, the tool reads a Hybrid Automaton Language (HAL) file and parses the description into internal graph data structures. It then shifts the given hybrid system such that the equilibrium point is at the origin

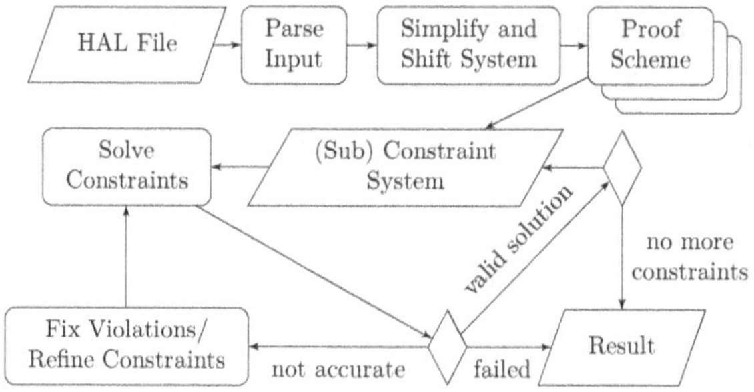

Figure 1: Overview of Stabhyli's internal steps.

of the state space, i.e., 0. The preprocessing steps are described in more detail in Section 3.1. To actually obtain the Lyapunov functions, and thereby deriving a proof, one out of four currently available proof schemes can be used. We call a *proof scheme* a method that takes a hybrid system as an input and returns either that the system is stable or that the method was not successful. In the following, the proof schemes are explained in more detail. All proof schemes have in common that they generate constraint systems that will be handed over to an external solver. The outcome of the solver is then double-checked whether there are violated constraints, since the solver might only be able to satisfy the constraints up to a certain threshold. If a violated constraint is found, then we refine the constraint system and rerun the solver until either no further refinement is possible or a valid solution of the constraint system has been found. The steps for solving constraints are described in Section 3.2.

The currently available proof schemes are (1) a proof scheme using a common Lyapunov function, (2) a proof scheme using a piecewise Lyapunov function, (3) a decompositional proof scheme, and (4) a compositional proof scheme.

(1) Common Lyapunov Function. This proof scheme tries to find a single *common Lyapunov function* V such that $\forall m \in \mathcal{M} : V_m = V$. The proof scheme directly corresponds to Lyapunov's Second Method [6] lifted to hybrid systems by requiring a single Lyapunov function for all modes. An advantage is that constraints of Type 3 in Theorem 1 can be omitted iff the reset function R is the identity function. As a drawback this approach is very restrictive since all modes have to share the same LLF.

(2) Piecewise Lyapunov Function. This proof scheme tries to find a *piecewise Lyapunov function* such as in Theorem 1. The scheme is not that restrictive as it allows the use of a LLF for each mode. These LLFs together form a piecewise Lyapunov function valid for all feasible trajectories. Here, all constraints of Type 3 are needed to ensure non-increase of the piecewise Lyapunov function when switching between modes.

(3) Piecewise Lyapunov Function via Decomposition. The third proof scheme tries to find a piecewise Lyapunov function using the decomposition proof scheme proposed by Oehlerking and Theel [9]. This proof scheme seeks to render the sequence of modes, that a trajectory may visit, "more explicit," thereby allowing multiple LLFs for the same mode depending on the path. This is done by disjoining paths within the graph's structure of the hybrid automaton. Thus, a single mode might have different LLFs depending on preceding or succeeding modes. For example, a trajectory might enter a mode m_b either coming from m_a or $m_{a'}$

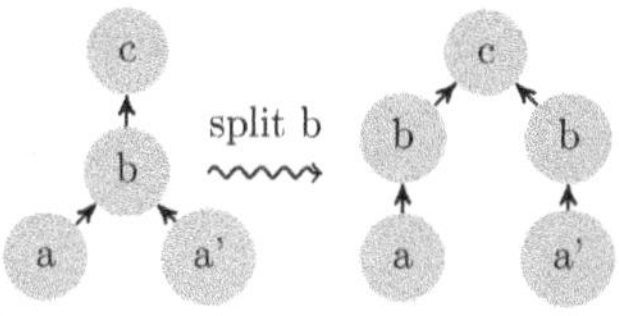

Figure 2: Splitting mode b.

with $a \neq a'$ and visiting m_c next. Generating the constraint system exactly as described in Theorem 1 leads to four constraints of Type 3 involving mode m_b. This might be unnecessarily restrictive since m_b might operate differently depending on the preceding mode. In this case, the decompositional proof scheme will disjoin the paths by creating a copy of m_b for each path, and thereby allowing every instance to have its own – possibly different – LLF. This step is called *splitting* and it is visualized in Figure 2. As a consequence, this approach is more powerful.

The decomposition procedure is as follows: search for cycles that are connected to the rest of the graph by at exactly one vertex (called "border mode"). Generate a sub-proof for the cycle and thereby obtain Lyapunov function candidates for the "border mode." Replace the cycle by a new mode which has the set of Lyapunov function candidates attached. This set serves as a "summary of the sub-proof" in following proofs and using conic combinations of the candidates only, ensures the existence of Lyapunov functions for all modes in the cycle. If no such cycle exists, then select a single vertex and split it into one vertex for each pair of incoming and outgoing edges. After splitting, restart with a search for a cycle. Note that termination of this procedure depends on the selection of vertices for splitting. Furthermore, the order of the splitting steps affects the total number of steps required to complete the proof and even the chance of success. For controlling the order, we have implemented three heuristics:

Selection by product. Sort the vertices by their product of incoming and outgoing edges. Select the first vertex with a product greater than 1. Thus, this heuristic always chooses a vertex that will be split into the least possible number of vertices.

Prioritization of zippers. A vertex is a *zipper* if and only if it either has exactly one incoming edge and multiple outgoing edges or vice versa. Perform a "selection by product" on zippers. If no zipper exists, then perform a "selection by product" on all vertices. This heuristic aims to separate cycles as fast as possible.

Selection by pairwise degree. Sort the vertices according to the vector of incoming and outgoing edges and select the first vertex having a vector greater than $\begin{bmatrix}1 & 1\end{bmatrix}^T$. The vector order is a partial order defined as

$$\begin{bmatrix}a_1\\a_2\end{bmatrix} \leq \begin{bmatrix}b_1\\b_2\end{bmatrix} \overset{def}{=} a_1 \leq b_1 \wedge a_2 \leq b_2.$$

This heuristic is a tradeoff between the others.

Another benefit of the decompositional proof scheme is that, if the hybrid automaton cannot be proven stable, then the particular part of the automaton, that let the proof fail, can easily be identified. This is due to the fact that the proof scheme successively generates sub-proofs to derive a complete proof. On one hand, the use of sub-proofs creates smaller constraint systems, and thus, will likely attract less numerical problems. On the other hand, a failed sub-proof will always identify parts of the automaton that might require a redesign. By using this knowledge, a guided analysis, refinement, or redesign of this part can be performed by the user. This can in some cases be seen as a debugging process for hybrid systems.

(4) Piecewise Lyapunov Function via Composition. The incremental proof scheme enables the design of a hybrid automaton while incrementally deriving a proof. This is done by adding components to the automaton in a bottom-up fashion, while simultaneously assuring that a proof exists. A possible design flow could then be to start with some small basic sub-automata (called *submodules*) and combine them to derive an automaton that fulfills a desired specification. To guarantee the existence of LLFs that together form a piecewise Lyapunov function, Stabhyli generates submodule interfaces. These interfaces describe how a submodule might be entered (the control is transfered to the submodule) and exited (the control is transfered from the submodule to another one). Furthermore, the interface is annotated with a set of valid Lyapunov function candidates. These candidates can then be combined conically, but excluding zero, in the constraints of Type 3 between submodules. By allowing only conical combinations, the existence of LLFs for the submodule's internal modes is guaranteed. The techniques used by this proof scheme are broadly described in [3].

3.1 Preprocessing

The preprocessing consists of multiple steps. The first step is parsing the HAL file. The second step is to shift the hybrid automaton such that the assumed equilibrium point becomes the origin of the state space. The third step is to transform the terms defining the flows, guards, invariants, and resets into canonical forms.

HAL. The Hybrid Automaton Language (HAL) is a language similar to HLang [5] but specialized to describe hybrid automata. It consists of five declaration parts.

1. A **variable part**. A variable can additionally be marked as *convergent* or *constant*. It also allows to create variables that evaluate to constant expressions.
2. An **interface part**. This part is only available when using the proof scheme (4). It serves as a template for the generated interface and contains a list of variables that are propagated to other submodules, a list of entries, and a list of exits.
3. A **submodule part**. Again, this part is only available when using the proof scheme (4) and contains a generated interface description of the submodules.
4. A **mode part**. A mode has a unique name and optionally an associated flow and an invariant.
5. A **transition part**. A transition has a source, an optional guard, a reset function which defaults to the identity function, and a target.

Variable Shifting. HAL allows the user to define variables that are required to converge to a fixed value other than 0. This means that Stabhyli must prove convergence wrt. this equilibrium point. Although this simplifies the modeling process, it requires a tool to shift the system. A successful proof then guarantees that the difference between the value of the variable and the equilibrium point converges to zero which in turn guarantees the desired property. Such shifting can be done by replacing each occurrence of a variable x with a new variable $x' := x - x_e$ for an equilibrium point x_e.

Forms of Flows, Guards, Invariants, and Resets. HAL allows to define flows and resets in terms of arbitrary mathematical expressions of the following form

$$\texttt{expr} ::= \texttt{number} \mid \texttt{variable} \mid (\texttt{expr}) \mid \\ -\texttt{expr} \mid \texttt{expr} * \texttt{expr}$$

with operators $* \in \{-, +, \cdot, /, \wedge\}$ having their usual meaning. Stabhyli symbolically evaluates such expressions and checks whether these expressions are polynomials since it currently handles only polynomials. For doing so, it tries to rewrite every such expression as a sum of monomials, i.e. $\sum_{j=0}^{n} c_j \prod_{v \in \mathcal{V}} v^{i_{vj}}$, and checks whether all exponents i_{vj} evaluate to non-negative integers. During this step, any constant occurring in an expression will also be replaced by its value. We refer to systems only using mathematical expression rewritable as a sum of monomials as *polynomial systems*. For invariants and guards, HAL allows to define arbitrary logical formulae of the form

$$\texttt{formula} ::= \texttt{expr} \sim \texttt{expr} \mid (\texttt{formula}) \mid \\ \neg\texttt{formula} \mid \texttt{formula} \triangle \texttt{fromula}$$

with comparators $\sim \in \{=, <, >, \leq, \geq, \neq\}$ and logical operators $\triangle \in \{\wedge, \vee, \Rightarrow, \Leftrightarrow\}$ also having their usual meaning. Such logical formulae will be evaluated by Stabhyli and rewritten in disjunctive normal form.

Variables that are Required for Convergence. HAL allows to globally mark variables as *required to converge*, that is, $v \in \mathcal{V}'$ according to Definition 2. In order to determine the set of variables that require to converge locally to a mode $m \in \mathcal{M}$, we define the relation "required to converge:" a variable $v \in \mathcal{V}$ is *required to converge* if and only if the variable is marked as required to converge or it occurs on the right-hand side of a differential inclusion of a variable that is required to converge. Thus, the set of *variables* $\mathcal{V}_m$ *that require to converge for a mode* m is:

$$\mathcal{V}_m \stackrel{def}{=} \mathcal{V}' \cup \left\{x \middle| \exists y \in \mathcal{V}_m : x \in RHS\Big(Flow(m)_{|y}\Big)\right\},$$

where $Flow(m)_{|y}$ is the projection of $Flow(m)$ onto variable y. This projection is the differential inclusion for $\dot{y}$. The set $RHS(f(y))$ is the set of all variables occurring in $f(y)$ with a non-zero coefficient, defined as:

$$RHS(f(y)) \stackrel{def}{=} \mathcal{V} \setminus \{v \in \mathcal{V} | f(y) = f(y_{|\mathcal{V} \setminus \{v\}})\}$$

Note that the recursive definition of $\mathcal{V}_m$ is transitive and, thus, we can generate the set by the transitive closure, which can be compute in $\mathcal{O}(|\mathcal{V}|^3)$ using the Floyd-Warshall algorithm [14].

For the set of variables $\mathcal{V} = \{x, y\}$ with $\mathcal{V}' = \{y\}$ and a flow

$$\begin{bmatrix}\dot{x}\\ \dot{y}\end{bmatrix} \in \text{convex}\left(\begin{bmatrix}-x\\ -x - 2.1y\end{bmatrix}, \begin{bmatrix}-1.5x\\ -1.1x - 2.1y\end{bmatrix}\right)$$

for a mode m, the set of variables, that are "required to converge," is $\mathcal{V}_m = \{x, y\}$ since

$$\mathcal{V}' \cup RHS(\text{convex}(-x - 2.1y, -1.1x - 2.1y)) = \mathcal{V}' \cup \{x, y\}.$$

Remark: This is only a heuristic; for a monomial $x \cdot y$ that occurs on the right-hand side, it would be sufficient that either x or y converges to 0 but our heuristic demands both.

3.2 Solving Constraints

A constraint system defined by Theorem 1 can directly be solved using *Linear Matrix Inequalities (LMI)* for linear flows, guards, invariants and resets [10]. For polynomial systems, it is required to transform the constraint system using the so-called *sums-of-squares (SOS)* method [12] into an LMI. For an inequality $\sum_{j=1}^{n} c_j \prod_{v \in \mathcal{V}} v^{i_{v_j}} = f(x) \succcurlyeq 0$, the SOS method requires to find a vector of monomials z and a matrix F such that $z^T F z = f(x)$. Now, $F \succcurlyeq 0$ implies $f(x) \succcurlyeq 0$. Note that the converse is not always true and deciding positivity of polynomials belongs to the class of NP-hard problems.
The S-Procedure [2] can be used to restrict certain constraints to some regions. The *S-Procedure* exploits the fact that finding a solution for

$$F \succcurlyeq \sum_{1 \le k \le l} a_k \cdot F_k \text{ with } a_k \ge 0 \text{ implies } \bigwedge_{1 \le k \le l} F_k \succcurlyeq 0 \Rightarrow F \succcurlyeq 0.$$

By automatically generating the constraint system for Theorem 1 and successively applying the SOS method and the S-Procedure, we obtain an LMI problem, which – in turn – can be solved by a *semi-definite program (SDP)*. An SDP has the form

$$\min. \quad c^T \cdot x \qquad \text{s.t.} \quad F_0 + \sum_{i=1}^{n} x_i \cdot F_i \succcurlyeq 0.$$

Such optimization problems have been well-studied and tools that efficiently solve such problems have been implemented. See CSDP [1] as an example, which is the solver that Stabhyli uses. These tools usually use some kind of interior point method and, thus, are very fast. Unfortunately, numerical solvers suffer from numerical instability.
A typical situation where numerical problems arise is that a constraint requires free parameters to have specific values (such as zero). This happens if for a mode $m \in \mathcal{M}$, the invariant contains the origin i.e., $0 \in Inv(m)$ (here, it is required that $V(x) = 0$ for $x = 0$).

Obtaining Lyapunov Function Templates. To find a solution for a constraint system as defined by Theorem 1, one needs to provide templates for Lyapunov functions. Such a template is a polynomial involving free parameters, and a solver's task is to find a good valuation of these parameters. Having an additive constant in the Lyapunov function is only valid for modes whose invariant does not contain the origin, because otherwise the value of the Lyapunov function must be zero at the origin. Thus, Stabhyli will not add an additive free parameter iff the invariant contains the origin. Furthermore, Stabhyli allows the Lyapunov function template to involve only combinations of variables v that have an associated flow function i.e., $Flow(m)_{|v} \neq 0$. The combinations are restricted by a user-specified degree.

Refining Constraints. As mentioned above, Stabhyli uses a numerical solver as a backend. To further counteract numerical problems, we have implemented a refinement loop. This loop exploits the fact that due to technical reasons, numerical solvers – if not able to solve a constraint system – report valuations of free parameters that violate the least number of constraints. Our refinement loop therefore double-checks the result of the solver and reruns the solver using a refined problem. To obtain this refinement, Stabhyli replaces each occurrence of a parameter by

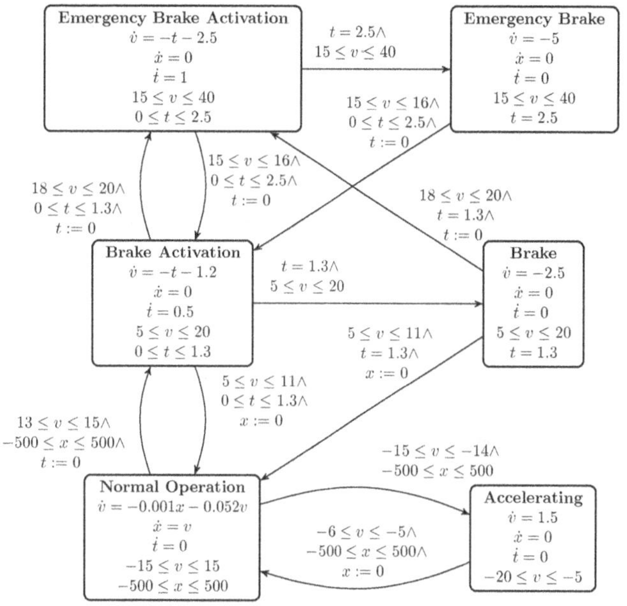

Figure 3: Cruise Controller Automaton[7]

- zero if the parameter is supposed to be non-negative and the valuation is negative, or by
- a small positive number if the parameter is supposed to be positive and its valuation is non-positive.

In order not to narrow down the solution space too fast, Stabhyli eliminates those parameters first which do not occur in the objective, and which have the largest distance to zero.
Stabhyli runs the check even if the solver reports success since it might happen that the solver returns solutions that satisfy the constraints only up to some threshold. Accordingly, we use the fact that a matrix is positive semi-definite iff it has only non-negative eigenvalues. Through this refinement, Stabhyli incrementally tries to build a valid solution even if the solver was not able to find one at first try.

4. EXPERIMENTS

In this section, we will give some examples that were proven stable using Stabhyli. The first example is a simplified automatic cruise controller that might be used to adjust the velocity of a vehicle. The second example demonstrates that higher order Lyapunov functions allow stability proofs even if no quadratic Lyapunov function exists [12].

Example 1. Figure 3 shows an automatic cruise controller. This controller regulates the velocity (in $^m/_s$) of a vehicle. Its objective is to approach a desired velocity. In this example, the variable v is the difference between this desired velocity and the current velocity. The automaton contains six modes; four of them model the brake system, one mode models the acceleration, and one mode implements a fine-tuning of the current velocity using a PI controller in non-critical situations. The brake system of the vehicle has two levels and, for convenience, slowly increases the deceleration towards a maximum. In our experiments Stabhyli was able to obtain a certificate of stability for this controller via decomposition with all three heuristics in twelve steps (sum of splitting and reduction steps). While the sequences of steps performed by the "prioritization of zippers" and the "selection by product" heuristic are identical, the heuristic "selection by pairwise degree" has chosen a different sequence of steps but obtained the same result. Note that neither applying proof scheme (1) nor (2) was successful. RSolver and [8] cannot be applied to this example since the system class of the example is not supported.

Intentionally "sabotaging" the controller in a way that it cannot stabilize causes Stabhyli to recognize and report the problem. E.g., changing the differential equation in mode "Emergency Brake" to $\dot{v} = 5$ leads to an instable cycle containing the modes "Emergency Brake", "Brake Activation", and "Emergency Brake Activation." In this case, Stabhyli terminates the search for a proof and blames the responsible cycle. This information allows guided redesign of this particular part of the automaton. The other proof schemes simply fail without any information about the reason.

Example 2. Consider the example from [12],

$$Flow(m_1)(x) = \begin{bmatrix} -5x_1 - 4x_2 \\ -x_1 - 2x_2 \end{bmatrix} \qquad Flow(m_2)(x) = \begin{bmatrix} -2x_1 - 4x_2 \\ 20x_1 - 2x_2 \end{bmatrix},$$

with $x = \begin{bmatrix} x_1 & x_2 \end{bmatrix}^T$ which has two modes m_1, m_2 and allows arbitrary switching.

For this system, Stabhyli cannot find a quadratic common Lyapunov function, which is not a deficiency since no such function exists [12]. Note, that one cannot conclude that the system is not stable. In fact, it is indeed stable and can be proven stable by allowing Stabhyli to search for a sextic common Lyapunov function. The normalized result, Stabhyli returns, is

$$\begin{aligned} V(x) =& 19.576x_1^6 + 11.627x_1^5x_2 + 15.267x_1^4x_2^2 \\ &+ 3.0857x_1^3x_2^3 + 8.9471x_1^2x_2^4 - 1.3629x_1x_2^5 \\ &+ 1.0539x_2^6 \end{aligned}$$

which is nearly the same Lyapunov function as the one presented in [12]. By default, Stabhyli uses quadratic Lyapunov function templates. For this system, the default setting is not sufficient, but the tool user can allow Stabhyli to retry using higher degrees and thereby allowing more flexible templates. In our experiments Stabhyli returned the above result within few seconds, while RSolver was not able to return any solution within nine hours. Relaxing the problem to region stability let RSolver report "Unknown."

5. SUMMARY

A tool, called Stabhyli, for automatically deriving proofs of stability of hybrid systems, has been presented. Stabhyli uses state-of-the-art techniques like sums-of-squares to cast polynomial constraint systems to LMIs. Furthermore, Stabhyli features four different proof schemes of which one allows the use of three different heuristics. These proof schemes are combined with powerful preprocessing and refinement steps. The decompositional proof scheme allows a designer to identify proof-critical parts of the hybrid system. The incremental proof scheme, in contrast, allows a bottom-up design of hybrid systems while Stabhyli ensures that a proof exists.

Future work will be the automatic detection of invariants that exhibit special shapes; such as ellipsoids. Such a detection will facilitate the process of finding a solution for a constraint system, thus, making it more likely that a system can be proven stable. Another improvement will be methods – possibly symbolic – that identify implicit equality constraint. Such situations often requires different free parameters to have identical valuations and numerical solvers usually do not recognize this. Furthermore, we will focus on finding additional ways to decompose and to transform hybrid systems based on their graph structure. This will increase the chance that a hybrid system can be proven stable automatically.

6. REFERENCES

[1] B. Borchers. Csdp, a C library for semidefinite programming. *Optim. Met. Softw.*, 10:613–623, 1999.

[2] S. Boyd and L. Vandenberghe. *Convex Optimization.* Cambridge University Press, Mar. 2004.

[3] W. Damm, H. Dierks, J. Oehlerking, and A. Pnueli. Towards Component Based Design of Hybrid Systems: Safety and Stability. In *Essays in Memory of Amir Pnueli*, volume 6200 of *LNCS*, pages 96–143. Springer, 2010.

[4] P. S. Duggirala and S. Mitra. Lyapunov abstractions for inevitability of hybrid systems. In *HSCC*, pages 115–124. ACM, 2012.

[5] M. Fränzle, H. Hungar, C. Schmitt, and B. Wirtz. Hlang: Compositional representation of hybrid systems via predicates. Reports of SFB/TR 14 AVACS 20, SFB/TR 14 AVACS, July 2007. ISSN: 1860-9821, http://www.avacs.org.

[6] M. Lyapunov. Problème général de la stabilité du movement. In *Ann. Fac. Sci. Toulouse, 9*, pages 203–474. Université Paul Sabatier, 1907. (Translation of a paper published in Comm. Soc. Math. Kharkow, 1893, reprinted Ann. Math. Studies No. 17, Princeton Univ. Press, 1949).

[7] J. Oehlerking. *Decomposition of Stability Proofs for Hybrid Systems.* PhD thesis, Carl von Ossietzky University of Oldenburg, Department of Computer Science, Oldenburg, Germany, 2011.

[8] J. Oehlerking, H. Burchardt, and O. E. Theel. Fully Automated Stability Verification for Piecewise Affine Systems. In *HSCC*, volume 4416 of *LNCS*, pages 741–745. Springer, 2007.

[9] J. Oehlerking and O. E. Theel. Decompositional Construction of Lyapunov Functions for Hybrid Systems. In *HSCC*, volume 5469 of *LNCS*, pages 276–290. Springer, 2009.

[10] S. Pettersson. *Analysis and Design of Hybrid Systems.* PhD thesis, CTH, Gothenburg, Sweden, 1999.

[11] A. Podelski and S. Wagner. Region Stability Proofs for Hybrid Systems. In *FORMATS*, volume 4763 of *LNCS*, pages 320–335. Springer, 2007.

[12] S. Prajna and A. Papachristodoulou. Analysis of Switched and Hybrid Systems - Beyond Piecewise Quadratic Methods, 2003.

[13] S. Ratschan and Z. She. Providing a Basin of Attraction to a Target Region of Polynomial Systems by Computation of Lyapunov-Like Functions. *SIAM J. Control and Optimization*, 48(7):4377–4394, 2010.

[14] S. Warshall. A Theorem on Boolean Matrices. *Journal of the ACM*, 9:11–12, 1962.

Zélus: A Synchronous Language with ODEs

Tool Paper

Timothy Bourke
NICTA, Sydney
INRIA Paris-Rocquencourt
DI, École normale supérieure
Timothy.Bourke@ens.fr

Marc Pouzet
Univ. Pierre et Marie Curie
DI, École normale supérieure
INRIA Paris-Rocquencourt
Marc.Pouzet@ens.fr

ABSTRACT

ZÉLUS is a new programming language for modeling systems that mix discrete logical time and continuous time behaviors. From a user's perspective, its main originality is to extend an existing LUSTRE-like synchronous language with Ordinary Differential Equations (ODEs). The extension is conservative: any synchronous program expressed as data-flow equations and hierarchical automata can be composed arbitrarily with ODEs in *the same source code*.

A dedicated type system and causality analysis ensure that all discrete changes are aligned with zero-crossing events so that no side effects or discontinuities occur during integration. Programs are statically scheduled and translated into sequential code that, by construction, runs in bounded time and space. Compilation is effected by source-to-source translation into a small synchronous subset which is processed by a standard synchronous compiler architecture. The resultant code is paired with an off-the-shelf numeric solver.

We show that it is possible to build a modeler for explicit hybrid systems *à la Simulink/Stateflow* on top of an existing synchronous language, using it both as a semantic basis and as a target for code generation.

Categories and Subject Descriptors

C.3 [**Special-Purpose and Application-Based Systems**]: Real-time and embedded systems; D.3.2 [**Programming Languages**]: Language Classifications—*Data-flow languages*; I.6.8 [**Simulation and Modeling**]: Types of Simulation—*Continuous, Discrete Event*; D.3.4 [**Programming Languages**]: Processors—*Code generation, Compilers*

General Terms

Algorithms, Languages

Keywords

Hybrid systems; Hybrid automata; Synchronous languages; Block diagrams; Type systems;

HSCC'13, April 8–11, 2013, Philadelphia, Pennsylvania, USA.

1. INTRODUCTION

Hybrid systems modelers are used not only in the high-level design and simulation of complex embedded systems, but also as development platforms in which the same source is used for formal verification, testing, simulation, and the generation of target executables. The quintessential example is the SIMULINK/STATEFLOW suite,[1] but there are also LABVIEW,[2] MODELICA,[3] and several others [6].

These tools are distinguished from the formal model of hybrid automata [14] by their focus on modular programming and simulation of both physical models and their controllers. In this context, reproducible, efficient simulations and the ability to generate statically scheduled code are essential features. As a major consequence, programming constructs are typically deterministic whereas hybrid automata are essentially non-deterministic and oriented toward specification and formal verification through the over-approximation of piecewise continuous behaviors.

The underlying mathematical model of hybrid modelers is the synchronous parallel composition of stream equations, differential equations, and hierarchical automata. But even with a carefully chosen numeric solver, actual simulations only ever approximate the ideal behaviors of such models. A formal semantics exists for discrete subsets [13, 20], but mixes of discrete and continuous-time signals often have unpredictable and mathematically weird behaviors [2, 3]. For instance, a continuous STATEFLOW state may be triggered repeatedly if a transition guard remains true, and, although transitions are instantaneous, the amount of simulation time that elapses between triggerings is non-zero and depends on when the solver decides to stop, which in turn is influenced by simulation parameters, global numerical error, and the occurrence of other zero-crossings. The behavior of a model may even change if it is placed in parallel with an independent block, for example one that tests the zero-crossings of a sinusoid signal. In this case, changing the sinusoid frequency may radically change the output of the other model. This time leak is not due to numerical artifacts but to a more fundamental reason: discrete time is not logical but global, it exposes the internal steps of the simulation engine.

Synchronous languages [5] differ from this approach by providing a logical notion of time, independent of a physical implementation. For example, the LUSTRE equations

$$\mathsf{x} = 0 \rightarrow \mathsf{pre}\ \mathsf{y} \quad \text{and} \quad \mathsf{y} = \mathsf{if}\ \mathsf{c}\ \mathsf{then}\ \mathsf{x} + 1\ \mathsf{else}\ \mathsf{z}$$

[1]http://www.mathworks.com/products/simulink/
[2]http://www.ni.com/labview/
[3]http://www.modelica.org/

define the two sequences $(x_n)_{n \in I\!N}$ and $(y_n)_{n \in I\!N}$ which are computed sequentially with $x_0 = 0$, $x_n = y_{n-1}$, and for all $n \in I\!N$, $y_n = \text{if } c_n \text{ then } x_n + 1 \text{ else } z_n$.[4] Time is logical, that is, nothing can be inferred about the actual time that passes between instants i and $i+1$. Synchronous programs see the environment as a source of input and output sequences and ignore intervening gaps. They are thus only suitable for the design of discrete controllers. In contrast, a model expressed with Ordinary Differential Equations (ODEs) or Differential Algebraic Equations (DAEs) continues to evolve during such gaps and a variable-step numeric solver is necessary to approximate continuous trajectories efficiently and faithfully.

So, what is the best way to combine the precision of synchronous languages for programming discrete components with the extra expressiveness afforded by ODEs approximated by numeric algorithms? Any combination must be *conservative*: the behavior of a synchronous program should not change if ODEs are placed in parallel, it should have the same logical-time semantics and compile to the same code; in particular to avoid inconsistencies between simulation and execution. Furthermore, discrete computations and side effects should not occur during numerical approximation.

In order to avoid the aforementioned time leak and to have a clear separation between discrete and continuous-time signals, we proposed the following convention, quoting [3]:

> "A signal is *discrete* if it is activated on a discrete clock. A clock is termed discrete if it has been declared so or if it is the result of a zero-crossing or a sub-sampling of a discrete clock. Otherwise, it is termed *continuous*."

This means that any synchronous program can be paired with ODEs as long as it is activated on a discrete clock. The definition is sufficiently general to model, for example, a discrete controller activated on a timer (periodic or not) or a deterministic hybrid automata with dynamic conditions.

Previously, we proposed the basis of a LUSTRE-like language extended with ODEs and following the above discipline. We defined the semantics of a minimal language using non-standard analysis [4], and proposed a type system that ensures the absence of time leaks, and a compilation method [3]. We later added hierarchical automata [2]. These techniques have been implemented in a new language, called ZÉLUS, which allows arbitrary combinations of data-flow equations, hierarchical automata, and ODEs. A type system and causality analysis statically ensure that discrete computations are aligned with zero-crossings. Compilation is by source-to-source translation into synchronous code which is then compiled to sequential code and paired with an off-the-shelf numeric solver. This paper describes key aspects of the language and implementation. We use synchronous programming as both a semantical foundation, where we are strongly influenced by the work of Lee et al. [17] and Mosterman et al. [11, 19], and as a target for code generation.

2. AN OVERVIEW OF ZÉLUS

ZÉLUS[5] is a first-order synchronous dataflow programming language extended with resettable ODEs and hierarchical automata. Rather than define the abstract syntax (available in [2, §3.1]), we show its main features through examples.

A ZÉLUS program is a sequence of definitions. The following declares a discrete function that counts the occurrences of a Boolean v and detects when there have been n:

```
let node after (n, v) = (c = n) where
  rec c = 0 → pre(min(tick, n))
  and tick = if v then c + 1 else c
```

where pre(·) is the non-initialized unit delay, . → . is the initialization operator of LUSTRE, min computes the minimum of its arguments and if/then/else is the mux operator that executes both branches and takes the value of one. The semantics in terms of infinite sequences is

$$\forall i \in I\!N^*, c_i = min(tick_{i-1}, n_{i-1}) \text{ and } c_0 = 0$$
$$\forall i \in I\!N, tick_i = \text{if } v_i \text{ then } c_i + 1 \text{ else } c_i,$$
$$\forall i \in I\!N, (after(n, v))_i = (c_i = n_i)$$

and the compiler infers the signature:

```
val after : int × bool ⇒ bool
```

This node can be used in a two state automaton,

```
let node blink (n, m, v) = x where
  automaton
  | On → do x = true until (after(n, v)) then Off
  | Off → do x = false until (after(m, v)) then On
```

which returns a value for x that alternates between true for n occurrences of v and false for m occurrences of v. The keyword until stands for a *weak* preemption, that is, x equals true when after(n, v) is true and becomes false the following instant. The semantics and compilation of automata, defined in [8], is that of SCADE 6[6] and LUCID SYNCHRONE.[7] The blinking behavior can be reset on a boolean condition r by nesting it inside a one state automaton that tests r, which we write using the reset/every syntactic sugar:

```
let node blink_reset (r, n, m, v) = x where
  reset
    automaton
    | On  → do x = true until (after(n, v)) then Off
    | Off → do x = false until (after(m, v)) then On
  every r
```

The type signatures inferred by the compiler are:

```
val blink : int × int × bool ⇒ bool
val blink_reset : bool × int × int × bool ⇒ bool
```

Up to syntactic details, these ZÉLUS programs could have been written *as is* in SCADE 6 or LUCID SYNCHRONE.

But ZÉLUS goes beyond discrete dataflow programming and allows the definition of continuous-time variables. For instance, consider a boom turning on a fixed pivot. The boom's angle can be expressed as a differential equation with initial value i and derivative v using the der keyword:

```
der angle = v init i
```

Its (ideal) value at model time t is:

$$\text{angle}(t) = i(0) + \int_0^t v(x)\, \mathrm{d}x,$$

It is compiled into a continuous state variable whose value is approximated by a numeric solver as described in §3.2.

[4]The unit delay initialized to 0, 0 → pre(.), is $\frac{1}{z}$ in SIMULINK.
[5]Available at `http://www.di.ens.fr/~pouzet/zelus`.
[6]`http://www.esterel-technologies.com`
[7]`http://www.di.ens.fr/~pouzet/lucid-synchrone`

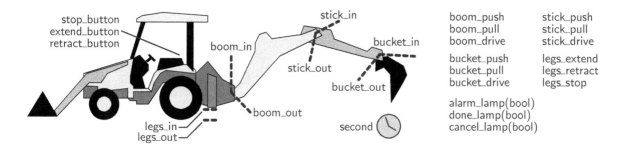

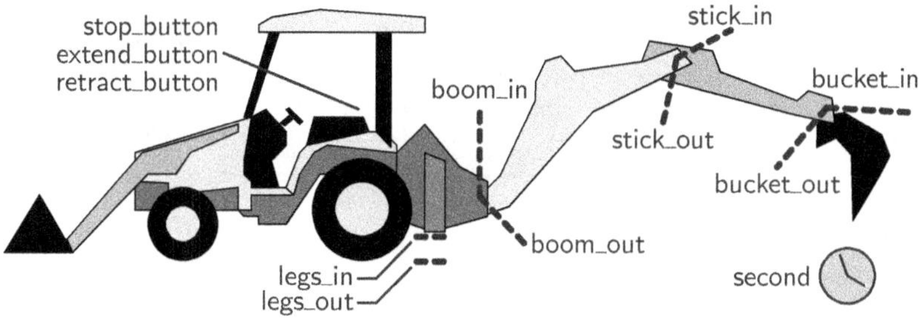

Figure 1: Idealized Backhoe Loader model

Now, if we wanted to achieve a reference velocity v_r using Proportional-Integral (PI) control, we need only add three equations in parallel:[8]

```
... and error = v_r −. v
    and der v = 0.7 ∗. error +. 0.3 ∗. z init 0.0
    and der z = error init 0.0
```

It could be that we also want to reset the controller state z when the angle reaches or exceeds a limit max. Two new constructs are needed: a way of detecting such interesting events and a way of directly setting the value of a continuous state. The standard way to detect events in a numeric solver is via zero-crossings where a solver monitors expressions for changes in sign and then, if they are detected, searches for a more precise instant of crossing. We introduce an up(.) operator to monitor an expression for a (rising) zero-crossing; with this, the definition of z becomes:

```
der z = error init 0.0 reset up(angle −. max) → 0.0
```

which says to reset the value of z to 0.0 the instant when angle−max becomes zero or positive. A der definition defines two values: a state (z) initially and in response to discrete events, and its derivative ($\dot{z}$) during continuous phases.

More complicated behaviors are better described as automata where defining equations and events being monitored change depending on mode. For instance, if the direction of the boom's motion changes in response to the input signals push and pull, and if the boom becomes stuck when it reaches a limit of motion, we may define v_r as follows:

```
automaton
| Pushing → do v_r = maxf
              until up(angle −. max) then Stuck
              else pull(_) then Pulling
| Pulling → do v_r = −. maxf
             until up(min −. angle) then Stuck
             else push(_) then Pushing
| Stuck → do v_r = 0.0 done
```

The automaton construct is effectively an equation and each mode contains a set of definitions which, naturally, may themselves include derivatives and automata. Here, the value of v_r is defined as either maxf, −maxf, or 0.

Transitions are ordered by priority and their guards have type α signal; events of type α signal are emitted by discrete components or introduced by up(.) : float ⇝ unit signal. As a consequence of this typing rule, boolean expressions cannot serve directly as guards in continuous automata. While it would be possible to compile an expression up(e : bool) into up(if e then 1.0 else −1.0), as effectively occurs in Stateflow, the search for zero-crossings then degenerates into binary search, and, more unsatisfying still, boolean complement, like signal absence, is not closed on discrete signals. Ultimately, we think we can better analyze and execute hybrid programs by restricting the form of triggering expressions.

[8] −., ∗., +. are floating-point arithmetic operators.

```
let hybrid segment ((min, max, i), maxf, (push, pull, go))
    = ((segin, segout), angle) where

 rec der angle = v init i
 and error = v_r −. v
 and der v = (0.7 /. maxf) ∗. error +. 0.3 ∗. z init 0.0
               reset hit(v0) → v0
 and der z = error init 0.0 reset hit(_) → 0.0
 and v_r = if go then rate else 0.0
 and (segin, segout) = (angle <= min, angle >= max)

 and automaton
   | Stuck →
     do rate = 0.0
     until push() on (not segout) then Pushing
     else pull() on (not segin) then Pulling

   | Pushing → local atlimit in
     do rate = maxf and atlimit = up(angle −. max)
     until atlimit() on (last v > 0.3 ∗. maxf) then
          do emit hit = −0.8 ∗. last v in Pushing
     else atlimit() then Stuck
     else pull() then Pulling

   | Pulling → local atlimit in
     do rate = −. maxf and atlimit = up(min −. angle)
     until atlimit() on (last v < −0.3 ∗. maxf) then
          do emit hit = −0.8 ∗. last v in Pulling
     else atlimit() then Stuck
     else push() then Pushing
```

Figure 2: Model of segment (boom, stick, or bucket)

One of ZÉLUS's strengths is the way that larger models can be constructed using abstraction and instantiation. For example, the various fragments discussed so far are readily combined into a more interesting model: that of the idealized backhoe loader that we use to teach discrete reactive programming. A labeled screen capture from its graphical simulator is shown in Figure 1. Using input signals from buttons and *_in/*_out sensors, and the outputs listed at right, students must write a discrete controller to operate the three backhoe segments. The simulator must, of course, approximate the movement of the segments.

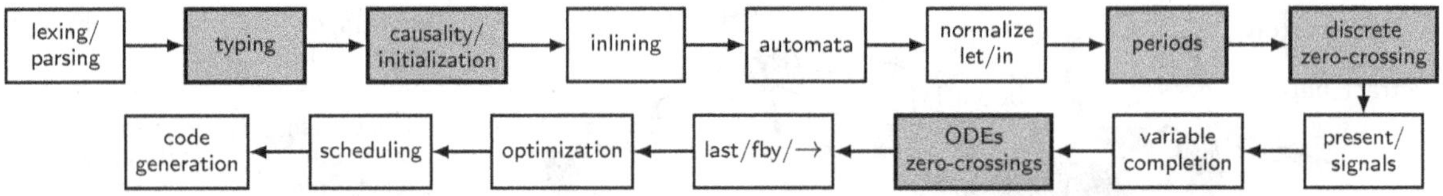

Figure 3: Compiler architecture

The node declaration for a single segment is shown in Figure 2. It declares a hybrid node called segment taking three inputs: a triple of movement parameters (min, max, i), the maximum force maxf, the control signals (push, pull, go); and giving as output the sensor values (segin, segout) and the segment position, angle.

The body of segment combines the elements already discussed with some minor modifications. For one, the reference velocity v_r is 0 when go is false and rate otherwise. The value of rate depends on the direction of motion, which in reality is determined by a hydraulic valve but which we model as a hybrid automaton. The automaton differs in its initial state, but also because of the self-loop transitions that model bouncing at the limits of motion:

```
until (atlimit() on (last v > 0.3 *. maxf)) then do ...
```

The operator on : α signal $\times$ bool $\rightsquigarrow$ α signal filters signals through boolean expressions. When atlimit occurs and the expression is satisfied, the transition resets the value of v and emits the signal hit. We are obliged to write last v in the guard and do/in equations to respect causality: the value of v cannot be tested before it is defined! Semantically last v is the left-limit of v. The hit signal resets the controller integrator, z, so that the sudden spikes on v are ignored.

A complete simulation is constructed as the parallel composition of instantiations of the segment node (three times: boom, stick, and bucket), a similar node that moves the legs, a function for updating the visualization, and a node implementing a discrete controller. Whereas a hybrid node like segment may be executed repeatedly to approximate continuous states, activations of the visualization function and controller node must be triggered more conservatively: the former because it has side-effects (it draws in a window); the latter because it may update internal discrete states. For instance, we call the update function with

```
present period (0.1) →
  showupdate (leg_pos, boom_ang, stick_ang, bucket_ang,
              alarm_lamp, done_lamp, cancel_lamp)
```

3. COMPILER ARCHITECTURE

Our starting point in developing a compiler for ZÉLUS was to recycle existing synchronous techniques so that a language like SCADE could be extended without disturbing its existing semantics and compilation. After a year of trial and error, we arrived at the architecture depicted in Figure 3. ZÉLUS is implemented in OCaml;[9] the size of each stage is given in Table 1. The compiler is a pipeline of stages that only aborts early if one of the front-end passes fails:

1. Parsing turns the program into an abstract syntax tree.
2. Typing annotates expressions with types (refer §3.1).

[9] `http://caml.inria.fr`

Compiler	*LOC*
Main driver (incl. main data structures)	1769
Abstract syntax and pretty printers	767
Lexer & parser	1002
Typing	1696
Initialization and causality analysis	839
Transformation of hybrid features	974
Transformation of automata	337
Transformation of synchronous features	940
Inlining and other optimizations	597
Code generation	852
Runtime	
Simulation algorithm	317
Solver interface (generic)	340
Solver interface (Sundials, compiler specific)	200
Zero-crossing detection (Illinois)	151

Table 1: ZÉLUS in numbers

3. The causality analysis checks for causality loops (refer §3.1). Then, the initialization analysis checks for reads from uninitialized delays [9]; ODEs are readily treated.

After the front-end stages, a series of source-to-source transformations are applied, each yielding a valid program.

4. 'Small' functions are inlined as an optimization.
5. Each automaton is replaced with a pair of switch-like statements [8]; ODEs remain unchanged.
6. Local declarations are un-nested to simplify later steps. For example, let x = (let $y = e_1$ in e_2) in e_3 is transformed into let $x = e_2$ and $y = e_1$ in e_3.
7. A primitive exists for periodic clocks, like period 0.2(3.4) which has a phase of 0.2 and a period of 3.4 and is mathematically equivalent to z and the sawtooth s:

   z = up(s)   der s = 1.0 init −0.2 reset z → −3.4

   Nonetheless, this direct translation with its costly continuous state and zero-crossing is avoided in favor of an output that returns the next horizon to the solver.
8. Each hybrid function is augmented with a boolean flag go to signal when a weak transition has occurred and thus that a subsequent discrete reaction is required.
9. The present and emit constructs are replaced, respectively, by a switch statement and the pairing of a value with an enable bit [7].
10. ODEs and up(.) operators are removed (refer §3.2).

(At this point, the code is purely synchronous.)

11. All last, fby, and → operators are replaced by pre delays.

12. Simple optimizations occur: dead-code removal, elimination of copy variables and common sub-expressions.
13. Equations are statically scheduled according to dataflow dependencies.
14. The code is modularly translated into sequential code.

The architecture is mainly that of the LUCID SYNCHRONE compiler[10] and only the highlighted boxes are really novel. We detail them in the following sections.

3.1 Typing and Causality

As the backhoe example demonstrates, ZÉLUS allows liberal combinations of combinational, discrete, and continuous elements. Nevertheless, discontinuities and side-effects must only occur on discrete clocks; programs that violate this rule are rejected. The principles and formal rules underlying the type system of ZÉLUS are presented elsewhere [2, 3]; here we focus on the pragmatic motivations and implementation.

Every function definition, equation or expression is associated to a kind $k \in \{\mathsf{A}, \mathsf{D}, \mathsf{C}\}$: A if combinatorial, D if discrete-time, and C if continuous-time. Kinds are related by the minimal subtyping relation such that $\mathsf{A} \le \mathsf{D}$ and $\mathsf{A} \le \mathsf{C}$. They are ascribed to entire 'blocks' rather than to individual 'wires'—each node or set of equations has a single kind which is inherited by individual inputs and outputs. The system extends naturally to automata: all the states of a 'continuous' automaton are also of kind C and may thus contain ODEs. The signature of a function f with input type t_1 and output type t_2 is thus of the form:

$$f : \forall \beta_1, ..., \beta_n . t_1 \xrightarrow{k} t_2 \quad \text{(where the } \beta_i \text{ are type variables)}$$

(We write $\xrightarrow{\mathsf{A}}$ as $\rightarrow$, $\xrightarrow{\mathsf{D}}$ as $\Rightarrow$, and $\xrightarrow{\mathsf{C}}$ as $\leadsto$). This block-based scheme greatly simplifies the formal rules, implementation, and type-related messages (interface files and errors) and so far we have not found it hinders writing programs.

A combinational function is defined by writing:

```
let square(x) = x *. x
```

Its inferred type is float → float. A declaration with the keyword node gives a function that executes in discrete time and which may thus contain unit delays, and the keyword hybrid gives a function that executes in continuous time.

As an example of the typing analysis, consider a program that tries to place an ODE in parallel with a discrete counter:

```
let hybrid wrong(x) = o where
  rec der o = 1.0 init -1.0
  and cpt = 0.0 fby cpt +. o
```

The compiler rejects it with the message:

```
File "ex.ls", line 3, characters 12-25:
> and cpt = 0 fby cpt +. o
>           ^^^^^^^^^^^^
Type error: this is a discrete expression and is
expected to be continuous.
```

As wrong is defined with keyword hybrid, it must not contain a discrete-time computation which is not triggered on a discrete clock. It could be made valid by writing, for example,

```
let hybrid good(x) = o where
  rec der o = 1.0 init -1.0 reset z() -> -1.0
  and cpt = present z() -> (0 fby cpt + o) init 0
  and z = up(last o)
```

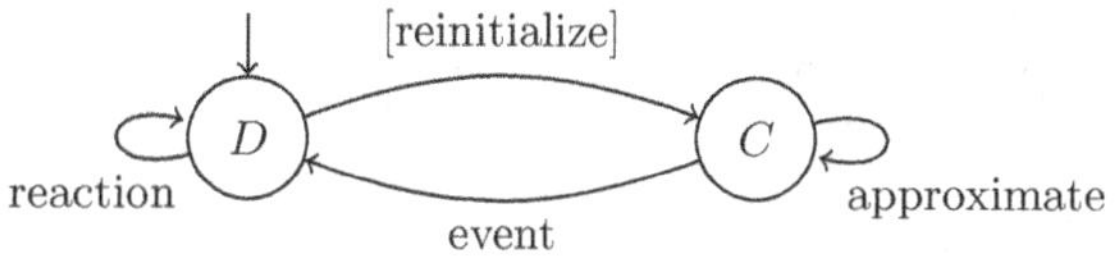

Figure 4: The basic structure of a hybrid simulation

After typing, a causality analysis is performed to ensure the absence of instantaneous loops. It follows two simple principles: every loop on a discrete signal must be broken by a unit delay and every loop on a continuous-time signal must be broken by an integrator. This is essentially what existing tools like SIMULINK do.

3.2 Compilation of ODEs and Runtime

The typing and causality analyzes help fulfill three fundamental requirements: *(a)* to simulate continuous processes with state-of-the-art numeric solvers, *(b)* to import existing synchronous code without modification, and, *(c)* to use existing tools to compile everything.

A hybrid simulation alternates between two phases as depicted in Figure 4. In state D, some code *next* is executed,

$$y', go' = next(y, \vec{x}),$$

to compute the next values of the discrete state variables y and go from the current value of y and a continuous state $\vec{x}$. This computation occurs when any of the zero-crossings $z_1, ..., z_n$ or go are true. Side effects and changes of state variables (continuous or discrete) may occur at such instants. A simulation iterates in this state, in which physical time does not progress, until go becomes false. Then, integration in a numeric solver begins. A solver takes functions f_y and g_y parameterized by y and a horizon h with:

$$\dot{\vec{x}} = f_{y,h}(t, \vec{x}) \quad \vec{z} = g_{y,h}(t, \vec{x})$$

and approximates $\vec{x}(t_i + h)$ from the current time t_i while monitoring the elements of $\vec{z}$ for changes of sign. The solver stops when it reaches time $t_i + h$ or when one or more zero-crossings has been detected. State D is then entered and the code is executed and may respond to detected events.

It is up to the compiler to construct *next*, f, g, the discrete state y, and the continuous state $\vec{x}$. This is done by the source-to-source transformations which turn hybrid functions into discrete nodes with additional inputs and outputs as shown in this extract from the compilation of der z = ... and up(angle -. max) in segment:

```
let node segment((iz1, iz2), (lz, lv, langle), d,
          (min, max, i), maxf, (push, pull, go)) =
  ((upz1, upz2), (nz, nv, nangle),
   go1 or go2 or go3, ((segin, segout), angle))
  ...
  and lastz = 0.0 -> lz
  and dz = error
  and match hit with
      | (_, true) -> do go1 = true and z = 0.0 done
      | _ ->         do go1 = false and z = lastz done
  and nz = if d then z else dz
  ...
  and atlimit = iz1 and upz1 = angle -. max
  ...
```

[10]Through it has been rewritten entirely.

Each up(e) is replaced with a boolean input iz$i$ to signal detection of the corresponding zero-crossing and a floating-point output upz$i$ to transmit the value of the expression e. The equation der z = error init 0.0 reset hit(_) → 0.0 is translated into dz = error, a match statement on the concrete representation of the signal hit (when hit is present z is set directly to 0.0), and equation nz = if d then z else dz, where d is a boolean flag that is true in D and false in C, and nz is an output. The lastz variable replaces all occurrences of last z, whether implicit or explicit, and lz is an input that contains the value of the state variable z estimated by the solver (an element of $\vec{x}$). This synchronous function is then compiled to sequential code.

Once the source program has been compiled into an executable, it is possible to choose the numeric solver and zero-crossing detection algorithm, and to set their parameters from the command-line. We have implemented a modular framework based on OCaml functors and first-class modules to integrate solvers. In addition to an interface to the SUNDIALS CVODE solver [15], we have implemented several numeric solvers and the 'Illinois' false position method for zero-crossing detection using standard techniques [10] (Butcher tables, Hermite interpolation, and error estimation).

We have experimented with various examples from the literature, including the sticky masses [16, Example 6.8] and air traffic control [21], and from SIMULINK, including the bang-bang temperature controller and clutch model.

4. COMPARISON WITH OTHER TOOLS

The ZÉLUS language is most distinguished from SIMULINK by the type system that regulates compositions of discrete and continuous elements and the compilation by source-to-source transformation. While the basic semantics of a SIMULINK model follow simple principles, the behavior of important corner-cases can really only be understood by careful operational reasoning, and such intricacies must be faithfully reproduced by compilers and analysis tools. Users intending to compile their models into controllers are advised to avoid certain features, or to favor others; for instance, to use function call triggers to explicitly determine the execution order of blocks. In contrast, ZÉLUS has a consistent and simple semantics, the same source-to-source translations generate code for simulation and for embedded targets. Only the last step, that turns basic dataflow assignments into imperative code, requires customization for specific targets.

ZÉLUS shares the same basic model of executions, that alternate between continuous phases and sequences of 'run-to-completion' discrete actions, as specification frameworks like Charon [1], SpaceEx [12], and Hybrid I/O Automata [18], and like them, concerns itself with modular composition. The biggest difference is the use of synchronous, rather than interleaved, parallelism, which enforces a strong discipline on communication through shared variables, which we consider as clocked streams, and on causality, since the language ensures a single value per variable per instant. Furthermore, we insist on externalizing all non-determinism from models.

Acknowledgments

We warmly thank Cyprien Lecourt for his work on solver modules and the development of various examples.

5. REFERENCES

[1] R. Alur, R. Grosu, Y. Hur, V. Kumar, and I. Lee. Modular specification of hybrid systems in CHARON. In *HSCC'00*, pages 6–19, 2000.

[2] A. Benveniste, T. Bourke, B. Caillaud, and M. Pouzet. A Hybrid Synchronous Language with Hierarchical Automata: Static Typing and Translation to Synchronous Code. In *EMSOFT'11*, Taiwan, Oct. 2011.

[3] A. Benveniste, T. Bourke, B. Caillaud, and M. Pouzet. Divide and recycle: types and compilation for a hybrid synchronous language. In *LCTES'11*, USA, Apr. 2011.

[4] A. Benveniste, T. Bourke, B. Caillaud, and M. Pouzet. Non-Standard Semantics of Hybrid Systems Modelers. *J. Computer and System Sciences*, 78:877–910, May 2012.

[5] A. Benveniste, P. Caspi, S. Edwards, N. Halbwachs, P. Le Guernic, and R. de Simone. The synchronous languages 12 years later. *Proc. IEEE*, 91(1), Jan. 2003.

[6] L. Carloni, R. Passerone, A. Pinto, and A. Sangiovanni-Vincentelli. Languages and tools for hybrid systems design. *Foundations & Trends in Electronic Design Automation*, vol. 1, 2006.

[7] J.-L. Colaço, G. Hamon, and M. Pouzet. Mixing Signals and Modes in Synchronous Data-flow Systems. In *EMSOFT'06*, South Korea, Oct. 2006.

[8] J.-L. Colaço, B. Pagano, and M. Pouzet. A Conservative Extension of Synchronous Data-flow with State Machines. In *EMSOFT'05*, USA, Sept. 2005.

[9] J.-L. Colaço and M. Pouzet. Type-based initialization analysis of a synchronous data-flow language. *J. Software Tools for Technology Transfer*, 6(3):245–255, Aug. 2004.

[10] G. Dahlquist and Å. Björck. *Numerical Methods in Scientific Computing: Volume 1*. SIAM, 2008.

[11] B. Denckla and P. Mosterman. Stream- and state-based semantics of hierarchy in block diagrams. In *17th IFAC World Congress*, pages 7955–7960, South Korea, 2008.

[12] G. Frehse, C. Le Guernic, A. Donzé, S. Cotton, R. Ray, O. Lebeltel, R. Ripado, A. Girard, T. Dang, and O. Maler. SpaceEx: Scalable verification of hybrid systems. In *23rd Conf. CAV*, pages 379–395, USA, July 2011.

[13] G. Hamon. A denotational semantics for Stateflow. In *EMSOFT'05*, pages 164–172, 2005.

[14] T. Henzinger. The theory of hybrid automata. *NATO ASI Series F: Comp. & Systems Sciences*, 170:265–292, 2000.

[15] A. Hindmarsh, P. Brown, K. Grant, S. Lee, R. Serban, D. Shumaker, and C. Woodward. SUNDIALS: Suite of nonlinear and differential/algebraic equation solvers. *ACM Trans. Math. Soft.*, 31(3):363–396, Sept. 2005.

[16] E. A. Lee and P. Varaiya. *Structure and Interpretation of Signals and Systems*. `http://LeeVaraiya.org`, second edition, 2011.

[17] E. A. Lee and H. Zheng. Leveraging synchronous language principles for heterogeneous modeling and design of embedded systems. In *EMSOFT'07*, Austria, Sept. 2007.

[18] N. A. Lynch, R. Segala, and F. W. Vaandrager. Hybrid I/O automata. *Info. & Comp.*, 185(1):105–157, Aug. 2003.

[19] P. Mosterman, J. Zander, G. Hamon, and B. Denckla. Towards computational hybrid system semantics for time-based block diagrams. In *3rd IFAC Conf. Analysis & Design of Hybrid Sys.*, pages 376–385, Spain, Sept. 2009.

[20] N. Scaife, C. Sofronis, P. Caspi, S. Tripakis, and F. Maraninchi. Defining and translating a "safe" subset of Simulink/Stateflow into Lustre. In *EMSOFT'04*, pages 259–268, 2004.

[21] C. Tomlin, G. J. Pappas, and S. Sastry. Conflict resolution for air traffic management: A study in multiagent hybrid systems. *IEEE Trans. Automatic Control*, 43(4):509–521, Apr. 1998.

A Stochastic Hybrid System Model of Collective Transport in the Desert Ant *Aphaenogaster cockerelli*

Ganesh P. Kumar*
School of Computing, Informatics and Decision Systems Engineering
Arizona State University
Tempe, AZ, USA
Ganesh.P.Kumar@asu.edu

Aurélie Buffin
School of Life Sciences
Arizona State University
Tempe, AZ, USA
baurelie@asu.edu

Theodore P. Pavlic
School of Life Sciences
Arizona State University
Tempe, AZ, USA
tpavlic@asu.edu

Stephen C. Pratt
School of Life Sciences
Arizona State University
Tempe, AZ, USA
Stephen.Pratt@asu.edu

Spring M. Berman
School for Engineering of Matter, Transport and Energy
Arizona State University
Tempe, AZ, USA
Spring.Berman@asu.edu

ABSTRACT

Collective food transport in ant colonies is a striking, albeit poorly understood, example of coordinated group behavior in nature that can serve as a template for robust, decentralized multi-robot cooperative manipulation strategies. We investigate this behavior in *Aphaenogaster cockerelli* ants in order to derive a model of the ants' roles and behavioral transitions and the resulting dynamics of a transported load. In experimental trials, *A. cockerelli* are induced to transport a rigid artificial load to their nest. From video recordings of the trials, we obtain time series data on the load position and the population counts of ants in three roles. From our observations, we develop a stochastic hybrid system model that describes the time evolution of these variables and that can be used to derive the dynamics of their statistical moments. In our model, ants switch stochastically between roles at constant, unknown probability rates, and ants in one role pull on the load with a force that acts as a proportional controller on the load velocity with unknown gain and set point. We compute these unknown parameters by using standard numerical optimization techniques to fit the time evolution of the means of the load position and population counts to the averaged experimental time series. The close fit of our model to the averaged data and to data for individual trials demonstrates the accuracy of our proposed model in predicting the ant behavior.

*Corresponding author.

HSCC'13, April 8–11, 2013, Philadelphia, Pennsylvania, USA.

Categories and Subject Descriptors

G.3 [**Probability and Statistics**]: Markov processes, Stochastic processes, Time series analysis; I.6.3 [**Simulation and Modeling**]: Applications; I.6.4 [**Simulation and Modeling**]: Model Validation and Analysis; I.2.9 [**Artificial Intelligence**]: Robotics—*autonomous vehicles, kinematics and dynamics*; I.2.11 [**Artificial Intelligence**]: Distributed Artificial Intelligence—*coherence and coordination, intelligent agents, multiagent systems*; J.2 [**Physical Sciences and Engineering**]: Mathematics and statistics; J.3 [**Life and Medical Sciences**]: Biology and Genetics

Keywords

stochastic hybrid system, collective transport, social insect behavior modeling, distributed robot systems, bio-inspired robotics, biomimicry

1. INTRODUCTION

Recent advances in technologies for swarm robotic systems, consisting of hundreds to thousands of autonomous, relatively expendable robots with limited capabilities, are facilitating the development of robotic teams to collectively manipulate and transport a variety of objects in their environment. These multi-robot transport teams can be used to amplify productivity in construction, manufacturing, and automated warehouse applications, as well as to aid in disaster scenarios and search-and-rescue missions. The problem of controlling swarm robotic transport teams in such applications presents certain challenges. The control approach should be scalable to arbitrary team sizes and should accommodate limitations on the robot platform's sensing, communication, and computation abilities. In addition, the control strategy should be robust to robot failures and not rely on detailed *a priori* information about the payload or environment so that it is generalizable to a wide range of scenarios.

As a step toward synthesizing a control approach for an adaptable, resource-constrained robotic transport team, we develop a model of collective transport by a group of agents with minimal capabilities. Toward this end, we look to an analogous system in nature: the food retrieval teams formed in colonies of the desert ant *Aphaenogaster cockerelli*. Although most ant species are relatively unskilled at group transport, *A. cockerelli* has evolved the impressive coordination skills necessary for this task. This behavior is an example of a fully decentralized cooperative manipulation strategy that is scalable in the number of transporters and successful for a wide range of payloads in environments with uneven terrain and obstacles.

Berman et al. [1] and Czaczkes and Ratnieks [2] present overviews of the incidence, advantages, and organization of group retrieval teams in ants. Group transport requires individual ants to coordinate their movement to carry a bulky food item to the nest. How this is achieved remains very poorly understood [2, 3, 11]. Many have argued that coordination proceeds entirely by indirect interactions through the item itself, known as stigmergy [7], but more direct interactions and signaling among carriers may also play a role. A new approach to this problem is to describe the range of behavioral states occupied by transporters, the rates at which ants change states, and the contextual influences on these changes. Models of this type can link individual behavior to the dynamics of group transport, ultimately allowing the identification of behavioral rules crucial to successful coordination. Furthermore, these models can be adapted to describe other species and to explain why some species are better at collective transport than others. Thus, modeling group transport in ants not only provides a template for the engineering of multi-robot transport systems, but such models can also assist biologists in the analysis of natural transport behavior.

In previous work of S.M.B. and S.C.P. [1], qualitative observations of *A. cockerelli* transporting elastic vision-based force sensors were used to develop a model of ants collectively dragging a load with compliant attachment points. The model consisted of a behavioral component, comprised of a hybrid system with probabilistic transitions between two modes, and a dynamic component, described by a quasi-static planar manipulation model that incorporated friction on the load surface. In the current work, we quantitatively derive a model of collective retrieval that combines the stochastic ant behavioral transitions and the continuous load dynamics in a single framework, a *polynomial stochastic hybrid system (pSHS)* [4, 6], that is amenable to analysis and control. Using this framework, we can derive the time evolution of the model variables' statistical moments, which can be fit to experimental data. The pSHS framework has been recently applied to problems of stochastic task allocation in multi-robot teams [8] and the control of self-assembly of stochastically interacting robots [10].

We conduct experiments on group transport in *A. cockerelli* using a rigid load, similar to one that would be encountered in nature, which the ants can lift as well as pull. This type of load allows us to develop a simpler dynamical model than our original one, and we expand the previous behavioral model of Berman et al. [1] with an additional mode to explicitly capture the directionality of the ants' efforts. We propose that the ants switch stochastically between roles at constant probability rates and that their pulling force on the load acts as a proportional controller on the load velocity. The unknown parameters of our model, consisting of a set of behavioral transition rates, the load velocity set point, and the gain on the load velocity regulator, are estimated from the experimental data using a weighted-least-squares procedure that fits theoretical to sampled means. The resulting model generates trajectories that not only closely match in the mean, but also accurately predict the load dynamics as a function of measured counts of ants in different roles in several individual trials.

2. EXPERIMENTAL TRIALS

We filmed colonies of *A. cockerelli* collectively retrieving a standardized artificial load. A total of 17 colonies were located in South Mountain Park in Phoenix, AZ. Experiments were carried out during the activity period of the colony in the early morning (0600–0830 hours) and in the late afternoon (1700–1900 hours) in May 2012.

A new colony was located each day. A Plexiglas® sheet with dimensions 61 cm × 46 cm × 0.5 cm was positioned such that one edge, Edge A, was 50 cm south of the main nest entrance, and the opposite edge, Edge B, was 111 cm south of the nest. The sheet was covered with white paper and leveled to avoid inclination in any direction. An artificial load was constructed by gluing a dime with mass $m_L = 2.30$ g and radius 0.90 cm to an ethylene vinyl acetate (EVA) foam disk (0.2 mm thickness, 1.0 cm radius), which was rubbed with fig paste to attract ants. We filmed the transport with a Canon G12 camera positioned above the sheet. The camera's field of view was 1280 pixels × 720 pixels, centered on the sheet. Foragers were recruited to a whole fig placed at Edge B. Once 10 workers were feeding on the fig, we replaced the fruit with the artificial load. Ants were able to manipulate the load by gripping the excess 0.1 cm of foam around the perimeter of the dime. Ants carrying the load were filmed until they reached Edge A.

From the video recording of each experimental trial, we selected a segment of duration 145 s during which the ants were smoothly transporting the load; i.e., the load moved a nonzero distance during each consecutive 5-second interval of the segment. From this segment, we extracted the positions of the ants around the load and the position of the center of the load using ImageJ [12] and the Mtrack plug-in [9]. This information was obtained from single frames at 5-second intervals.

We observed that during these segments, the ants moved the load along an approximately straight path (left column of Fig. 1), allowing us to model the load movement as one-dimensional. In addition, we found that the transport teams move the load at approximately the same speed across trials (right column of Fig. 1). Figure 2(a) shows an overhead snapshot of the ants and load during one trial. We observed that ants switched at random times between three behavioral states. The *Detached* state describes ants that are not attached to the load; two ants in Fig. 2(a) are in this state at the instant of the snapshot. To classify ants that were attached to the load, we divided the load in half by a line perpendicular to the direction of the load motion. Ants that gripped the half of the load in the direction of travel were labeled as being in state *Front*, and ants that gripped the other half were assigned state *Back*. In Fig. 2(a), three ants are in state *Front* and three are in state *Back*. From side-view videos of transport (Fig. 2(b)), we observed that ants on both sides lift the load off the ground in addition to exerting forces parallel to the surface that drive the load across the substrate.

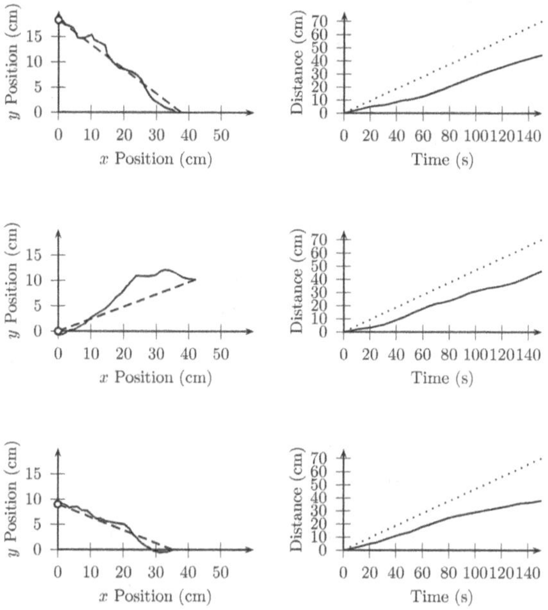

Figure 1: The x–y trajectory of the artificial load center (left column) and the distance traveled by the load over time (right column) for three experimental trials. The dashed lines in the left column show the ideal straight-line path. The dotted reference lines in the right column all have the same slope.

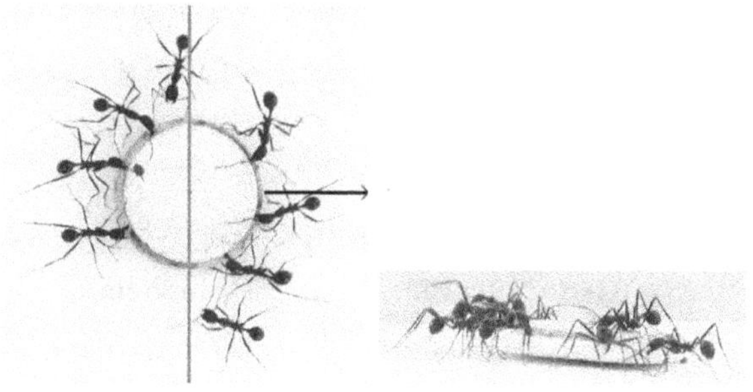

(a) Top view (b) Side view

Figure 2: *Aphaenogaster cockerelli* ants transporting an artificial load: (a) top view, with the arrow indicating the direction of the load motion and the red line dividing the load into *front* (right) and *back* (left) halves; (b) side view. The views are from different trials.

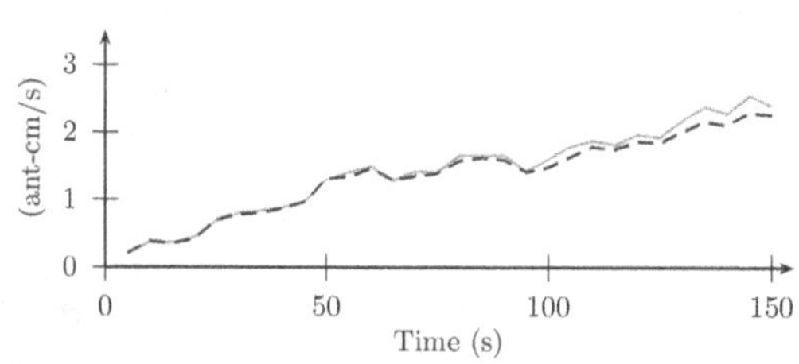

Figure 3: The solid line shows the sample mean of the product $N_F v_L$ from the data; the dashed line shows the product of the sample mean of N_F with the sample mean of v_L from the data.

The average pulling force of a single ant on an elastic load was previously measured by Berman et al. [1] to be 10.5 ± 5.0 mN, with 99.1% of the 10906 samples less than 30.0 mN. To further characterize the dynamics of the artificial load, we estimated the coefficient of kinetic friction μ of the load on the surface by measuring the angle θ_s at which the load started to slide down an inclined plane covered with the same paper used in the transport experiments. We measured $\theta_s = 30°$, yielding the value $\mu = \tan(\theta_s) = 0.58$. Finally, the force F_l with which an individual ant lifts the load was estimated as $F_l = 2.653$ mN by averaging the peak lifting forces that several ants applied individually to a rigid plastic disk (0.5 mm thickness, 5 mm radius) glued to the pin of a 10 gram capacity load cell (Transducer Techniques GSO series).

3. COLLECTIVE TRANSPORT MODEL

In this section, we model collective transport as a stochastic hybrid system (SHS) [6]. This SHS is a cascade connection of a chemical reaction network representing stochastic ant behavioral transitions followed by the deterministic dynamics of a load transported along a surface with friction. In Section 4, we derive moment dynamics of the model that are used in Section 5 to fit model parameters to statistics from the experimental ant data.

Discrete Behavioral Modes.

We represent the stochastic switching of ants between behavioral states in the form of a set of chemical reactions. The species X_i denotes an ant in behavioral state $i \in \{F, B, D\}$, where i signifies the states *Front*, *Back*, and *Detached*, respectively. Each ant is assumed to switch from state i to state $j \neq i$ at a constant probability per unit time r_{ij}, which we call the *transition rate*. The six reactions representing these transitions take the form

$$X_i \xrightarrow{r_{ij}} X_j, \quad i,j \in \{F,B,D\}, \quad i \neq j. \tag{1}$$

We define $N_i(t)$ as the number of ants in state i at time t. The instantaneous probability rate of the reaction in Eq. (1) occurring within the group of ants is called the *transition intensity*, λ_{ij} [5]. Because all reactions are unimolecular, this quantity is given by $\lambda_{ij} = r_{ij} N_i$. Hence, although each transition rate is constant, the transition intensities vary with the number of ants in each state. Consequently, when the number of ants in a given state drops to zero, so does the probability per unit time of further reactions out of that state. We note that the total ant population, $N_F(t) + N_B(t) + N_D(t)$, is conserved at all times.

Load Dynamics.

As stated in Section 2, we model the load movement as one-dimensional. We specify that the load is initially located at the origin and then travels only in the positive direction along the x axis toward the nest. The load position and velocity at time t are denoted by $x_L(t)$ and $v_L(t)$, respectively. We assume that each ant in state *Front* walks backward in the positive x direction toward the nest and pulls on the load with force F_p. Because the ant teams in our experiments moved the load at an approximately constant velocity (see Fig. 1), we assume a proportional velocity regulation policy for F_p of the form

$$F_p = K(v_L^d - v_L(t)), \tag{2}$$

where the positive gain K and velocity set point v_L^d are parameters fit from the experimental data. Ants in state *Back* are assumed to only lift the load, and so the force applied by the ants to the load in the x direction at time t is $N_F(t)F_p$. We assume that ants in both states *Front* and *Back* lift the load with the upward force $F_l = 2.653\,\text{mN}$ from the measurements described in Section 2. The load is subject to a kinetic frictional force μF_n, where the normal force F_n on the load is determined by assuming that the load is in static equilibrium in the vertical direction. That is,

$$F_n = m_L g - (N_F(t) + N_B(t))F_l,$$

where g is the acceleration due to gravity. The net force F on the load in the x direction is then given by

$$F = N_F(t)F_p - \mu F_n.$$

Stochastic Hybrid System.

Our models of the ant behavioral dynamics and the load dynamics together constitute a polynomial stochastic hybrid system (pSHS) [4, 5]. The pSHS is characterized by the state vector $\mathbf{x} = [N_F\ N_B\ N_D\ x_L\ v_L]^\top$ with the continuous flow

$$\dot{\mathbf{x}} = \begin{bmatrix}0 & 0 & 0 & v_L & F/m_L\end{bmatrix}^\top \tag{3}$$

and a set of discrete *reset maps* $\phi_{ij}(\mathbf{x})$, each representing the stochastic transition $(N_i, N_j) \mapsto (N_i - 1, N_j + 1)$ corresponding to the reaction in Eq. (1) that occurs with intensity λ_{ij}. The fact that the λ_{ij}, $\phi_{ij}(\mathbf{x})$, and components of $\dot{\mathbf{x}}$ are all finite polynomial functions of the continuous state variables establishes our SHS model as a pSHS [5]. The stochastic variation in $N_F(t)$, $N_B(t)$, and $N_D(t)$ over time causes $x_L(t)$ and $v_L(t)$ to vary randomly over time as well.

4. MOMENT DYNAMICS OF THE MODEL

The *extended generator* L of an SHS can be used to predict the time evolution of the statistical moments of the SHS continuous state [4, 5]. For any function $\psi(\mathbf{x}) : \mathbb{R}^n \to \mathbb{R}$ that is continuously differentiable, the dynamics of the expected value of ψ are given by $\mathrm{d}\,\mathrm{E}(\psi)/\mathrm{d}t = \mathrm{E}(L\psi)$. In our case, L is defined as

$$L\psi(\mathbf{x}) \triangleq \frac{\partial \psi}{\partial x_L}\dot{x}_L + \frac{\partial \psi}{\partial v_L}\dot{v}_L + \sum_{\substack{i,j \in \{F,B,D\}\\ i \neq j}} \left(\psi(\phi_{ij}(\mathbf{x})) - \psi(\mathbf{x})\right) r_{ij} N_i.$$

Hence, by setting $\psi = N_i$, we can derive the dynamics of the mean number of ants in state i, $\mathrm{E}(N_i)$, as

$$\frac{\mathrm{d}}{\mathrm{d}t}\mathrm{E}(N_i) = \sum_{\substack{j \in \{F,B,D\}\\ j \neq i}} \left(r_{ji}\,\mathrm{E}(N_j) - r_{ij}\,\mathrm{E}(N_i)\right). \tag{4a}$$

Likewise, using $\psi = x_L$, the mean load position is such that

$$\frac{\mathrm{d}}{\mathrm{d}t}\mathrm{E}(x_L) = \mathrm{E}(v_L), \tag{4b}$$

and, using $\psi = v_L$ and the approximation that the instantaneous N_F and v_L are uncorrelated (see Fig. 3), the mean load velocity is such that

$$\frac{\mathrm{d}}{\mathrm{d}t}\mathrm{E}(v_L) = c_g + c_F\,\mathrm{E}(N_F) + c_B\,\mathrm{E}(N_B) + c_{Fv}\,\mathrm{E}(N_F)\,\mathrm{E}(v_L) \tag{4c}$$

where $c_g \triangleq -\mu g$, $c_F \triangleq (Kv_L^d + \mu F_l)/m_L$, $c_B \triangleq \mu F_l/m_L$, and $c_{Fv} \triangleq K/m_L$.

5. PARAMETER ESTIMATION RESULTS

In this section, we discuss the fitting procedure used to estimate the pSHS model parameters: the transition rates r_{ij} of the reactions in Eq. (1) and the gain K and velocity set point v_L^d in Eq. (2). We compute parameters that best fit the sample means of the experimental data to the first-order moment dynamics derived in Section 4.

As described in Section 2, the experimental data consist of the load position and velocity and the counts of ants in each behavioral state sampled at 5 s intervals during 17 trials. The number of *Detached* ants at each sampling time was generated by subtracting the number of attached ants during that sampling period from the maximum number of attached ants across all periods. Thus, the total number of ants is constant within each trial. The load velocity was estimated numerically from the position data. For the purpose of fitting model parameters to the data, we computed the across-trial means of load position and velocity as well as the across-trial mean numbers of *Front*, *Back*, and *Detached* ants at each sampling time. Weighted-least-squares (WLS) optimization was then used to fit the theoretical moment dynamics Eq. (4) at these times to the corresponding empirical mean values. In particular, the numerical optimizer minimized the sum of weighted squared errors between each theoretical and empirical mean, where each weight was the sample variance of the corresponding data set. The numerical operation was performed in MATLAB using the `fmincon` tool for active-set optimization; the theoretical mean-field trajectories were integrated using `ode15s` with initial conditions chosen to match the initial means from the data.

The best-WLS-fit parameters were the transition rates

$$r_{DB} = 0.0197\,\text{s}^{-1}, \qquad r_{BD} = 0.0205\,\text{s}^{-1},$$
$$r_{DF} = 0, \qquad r_{FD} = 0,$$
$$r_{BF} = 0.0301\,\text{s}^{-1}, \qquad r_{FB} = 0.0184\,\text{s}^{-1},$$

and the gain and velocity set point parameters

$$K = 0.0035\,\text{N/(cm/s)}, \qquad v_L^d = 0.3185\,\text{cm/s}.$$

Based on the rates that best fit the model to the data, each ant in the transport team appears much more likely to attach to the back of the moving load than the front (i.e., $r_{DB} > r_{DF} = 0$). This hypothesis can be tested in future experiments and in a more detailed analysis of the video data. If it is supported, we can speculate as to reasons for the trend. For example, during movement, the front of the load may have near-maximal occupancy, hindering new attachment to the front even by detached ants downstream of the load motion. Alternatively, it may be easier to attach to the back of the moving object because that end is vertically stationary and moving horizontally in the same direction as a forward-walking ant approaching it. When an *A. cockerelli* transports a small object individually, she lifts the seed off the ground and carries it forward toward her nest. Hence, it is possible that backward carrying in teams results from the inability to carry the load forward in the normal posture. Moreover, the model infers a relatively high *Back*-to-*Front* transition rate r_{BF}, which is supported by the observation that backward walking is a contingency when loads are difficult to move. Even if these speculations are not biologically accurate, they may assist in the design of grasping mechanisms and approach postures for robots performing collective transport.

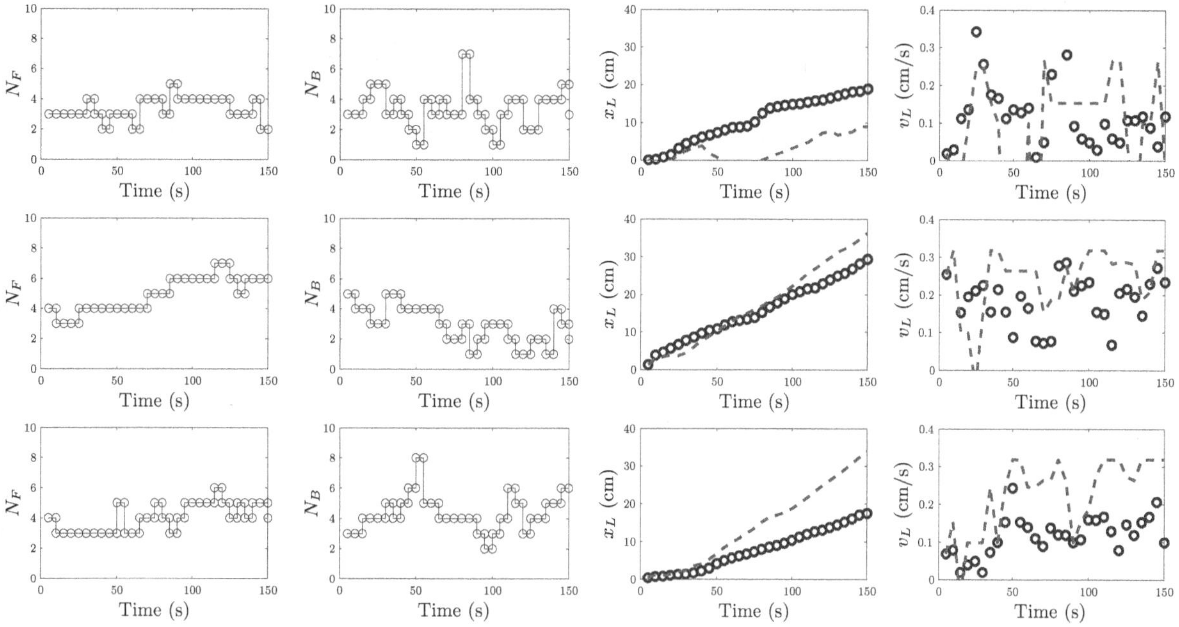

Figure 4: Observed numbers of ants in states *Front* (first column) and *Back* (second column) over time and observed (circles) and predicted (dashes) load position (third column) and velocity (fourth column) over time for three selected experimental trials (one per row).

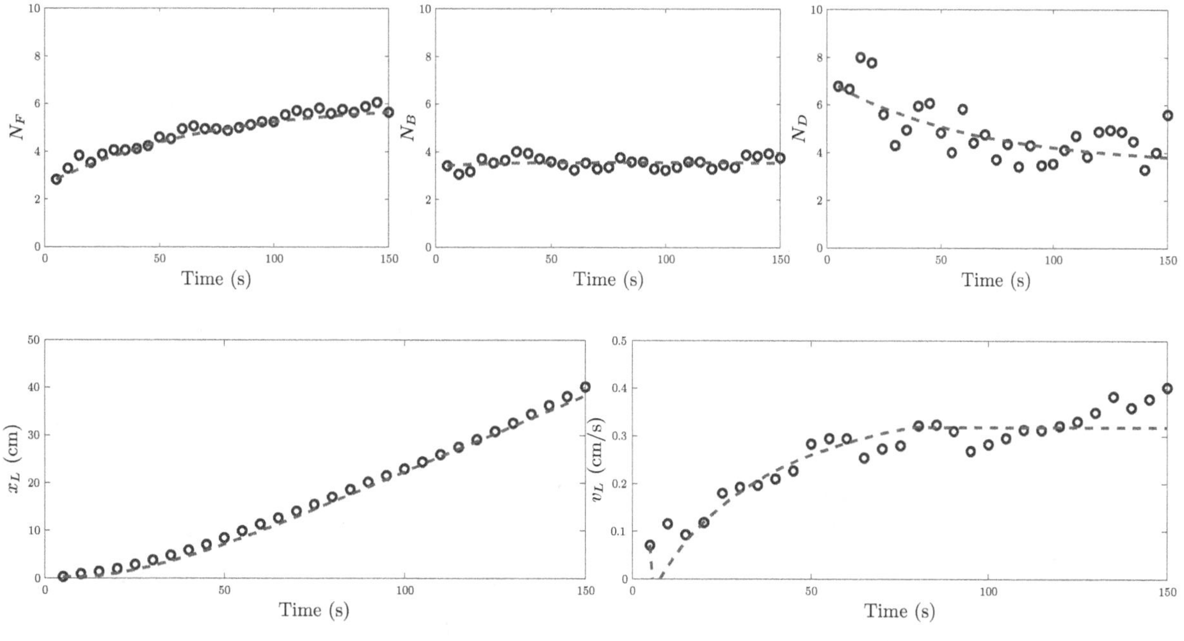

Figure 5: Observed (circles) and predicted (dashes) time evolution of the mean numbers of ants in each state (top row) and the mean load position and velocity (bottom row).

Figures 4 and 5 compare the experimental data with the model predictions. Figure 4 shows this comparison for three selected experimental trials: the first two columns show the measured numbers of *Front* and *Back* ants, and the third and fourth columns show the measured load position and velocity and their predictions based on the measured ant counts and the deterministic feedback policy from Eq. (2) instantiated with the WLS-fit K and v_L^d parameters in the load dynamics Eq. (4b), Eq. (4c). In Fig. 5, the first row displays the mean numbers of ants in each state measured across trials against the predicted mean-field trajectories from Eq. (4a) with the WLS-fit transition rates r_{ij}, and the second row shows the analogous plots for load position and velocity. The parameter v_L^d may be viewed as a desired traveling speed of the ant transport teams that can only be achieved in the absence of friction. That is, some amount of steady-state velocity error is required to balance the frictional force. Additional experiments can further validate the model for a higher load mass and coefficient of friction. The model should then predict a shift in the steady-state load velocity.

6. CONCLUSION

In this work, we conducted an experimental investigation of collective transport of a rigid artificial load by *Aphaenogaster cockerelli* ants. A stochastic hybrid system (SHS) model was developed to describe the dynamics of the load and the behavioral transitions of the ants during transport. The model was fit to the data to minimize the difference in mean behavior, and the resulting best-fit parameters are presented as reduced-order metrics of collective transport.

In future work, we plan to further validate our model by fitting both first-order and second-order moments to statistics from experimental data. We will also investigate how the best-fit parameters vary from optimized parameters that minimize criteria including path variance, load transport time, and transport team size.

We also plan to expand the model to incorporate other features of collective transport by ants. Additional behavioral states as well as heterogeneity and stochasticity within states will be included. The modeled behaviors will include teams of individuals pulling and lifting with different time-varying forces. An important future direction is to adjust the transition rates so that they depend on factors such as the load position, load velocity, ant force applied to the load, and the number of ants attached. Especially in the uncoordinated phase before smooth transport, it is likely that the probability of an ant attaching to the front or back of the load is strongly determined by the load's nascent motion as well as the number of ants gathered around it. State-dependent transition rates can capture this initial behavior and still allow for the smooth motion that is the focus of this paper. As the uncoordinated phase has less directionality than the smooth phase discussed here, it will require augmenting the model for two-dimensional load motion.

In general, we hope to catalyze bio-inspired research on multi-robot transport teams. SHS frameworks are utilized in robotics, and so they have potential to be substrates for trans-disciplinary knowledge transfer.

Acknowledgments

We acknowledge the support of ONR Grant N00014-08-1-0696. We thank Jessica D. Ebie, Ti Eriksson, and Kevin L. Haight for their help in ant collection and care, and we are grateful for the help of Denise Wong and Vijay Kumar in the measurement of forces exerted by the ants.

7. REFERENCES

[1] S. Berman, Q. Lindsey, M. S. Sakar, V. Kumar, and S. C. Pratt. Experimental study and modeling of group retrieval in ants as an approach to collective transport in swarm robotic systems. *Proc. IEEE*, 99(9):1470–1481, September 2011. doi: 10.1109/JPROC.2011.2111450.

[2] T. J. Czaczkes and F. L. W. Ratnieks. Cooperative transport in ants (Hymenoptera: Formicidae) and elsewhere. *Myrmecol. News*, 18:1–11, 2013.

[3] N. R. Franks. Teams in social insects: group retrieval of prey by army ants (*Eciton burchelli*, Hymenoptera: Formicidae). *Behav. Ecol. Sociobiol.*, 18(6):425–429, 1986. doi: 10.1007/BF00300517.

[4] J. P. Hespanha. Polynomial stochastic hybrid systems (extended version). Technical report, University of California, October 2004. URL `http://www.ece.ucsb.edu/~hespanha/published/tr_pshs_hscc05.pdf`.

[5] J. P. Hespanha and A. Singh. Stochastic models for chemically reacting systems using polynomial stochastic hybrid systems. *Int. J. Robust Nonlinear Control.*, 15(15):669–689, October 2005. doi: 10.1002/rnc.1017.

[6] J. Hu, J. Lygeros, and S. Sastry. Towards a theory of stochastic hybrid systems. In N. A. Lynch and B. H. Krogh, editors, *Hybrid Systems: Computation and Control Third International Workshop Proceedings*, volume 1790 of *Lecture Notes in Computer Science*, pages 160–173, Pittsburgh, PA, March 23–25, 2000. doi: 10.1007/3-540-46430-1.

[7] C. R. Kube and E. Bonabeau. Cooperative transport by ants and robots. *Robot. Auton. Syst.*, 30(1–2):85–101, January 2000. doi: 10.1016/S0921-8890(99)00066-4.

[8] T. W. Mather and M. A. Hsieh. Distributed robot ensemble control for deployment to multiple sites. In *Proceedings of Robotics: Science and Systems*, Los Angeles, CA, USA, June 2011.

[9] E. Meijering, O. Dzyubachyk, and I. Smal. *Methods for Cell and Particle Tracking*, volume 504 of *Methods in Enzymology*, chapter 9, pages 183–200. Elsevier, February 2012.

[10] N. Napp, S. Burden, and E. Klavins. Setpoint regulation for stochastically interacting robots. In *Proceedings of Robotics: Science and Systems*, Seattle, WA, USA, June 2009.

[11] S. K. Robson and J. F. A. Traniello. Resource assessment, recruitment behavior, and organization of cooperative prey retrieval in the ant *Formica schaufussi* (Hymenoptera: Formicidae). *J. Insect Behav.*, 11(1): 1–22, 1998. doi: 10.1023/A:1020859531179.

[12] C. A. Schneider, W. S. Rasband, and K. W. Eliceiri. NIH Image to ImageJ: 25 years of image analysis. *Nat. Methods*, 9(7):671–675, July 2012. doi: 10.1038/nmeth.2089.

Formal Verification of Distributed Aircraft Controllers*

Sarah M. Loos
Computer Science Dept.
Carnegie Mellon University
Pittsburgh, PA, USA
sloos@cs.cmu.edu

David Renshaw
Computer Science Dept.
Carnegie Mellon University
Pittsburgh, PA, USA
dwrensha@cs.cmu.edu

André Platzer
Computer Science Dept.
Carnegie Mellon University
Pittsburgh, PA, USA
aplatzer@cs.cmu.edu

ABSTRACT

As airspace becomes ever more crowded, air traffic management must reduce both space and time between aircraft to increase throughput, making on-board collision avoidance systems ever more important. These safety-critical systems must be extremely reliable, and as such, many resources are invested into ensuring that the protocols they implement are accurate. Still, it is challenging to guarantee that such a controller works properly under every circumstance. In tough scenarios where a large number of aircraft must execute a collision avoidance maneuver, a human pilot under stress is not necessarily able to understand the complexity of the distributed system and may not take the right course, especially if actions must be taken quickly. We consider a class of distributed collision avoidance controllers designed to work even in environments with arbitrarily many aircraft or UAVs. We prove that the controllers never allow the aircraft to get too close to one another, even when new planes approach an in-progress avoidance maneuver that the new plane may not be aware of. Because these safety guarantees always hold, the aircraft are protected against unexpected emergent behavior which simulation and testing may miss. This is an important step in formally verified, flyable, and distributed air traffic control.

Categories and Subject Descriptors

F.3.1 [**Logics and Meanings of Programs**]: Specifying and Verifying and Reasoning about Programs

Keywords

formal verification; distributed aircraft controllers

1. INTRODUCTION

Verification of air traffic control is particularly challenging because it lies in the intersection of many fields which already give tough verification problems when examined independently. It is a distributed system, with a large number of aircraft interacting over an unbounded time horizon. Each aircraft has nonlinear continuous dynamics combined with complex discrete controllers. And finally, every protocol must be flyable (i.e. not cause the aircraft to enter a stall, bank too sharply, or require it to turn on sharp corners).

In this paper, we investigate the safety of collision-avoidance controllers for aircraft systems. We want to prove safety not just for a single aircraft or a pair of aircraft, but for all aircraft operating simultaneously in the sky. Because this system is composed of multiple independent computational agents that interact with the physical world, it is called a *distributed hybrid system*. It is this combination of continuous flight dynamics, discrete flight control decisions, and distributed communication that causes verification of aircraft control protocols to be extremely challenging.

Aircraft control systems are safety-critical, so they must be designed with a high assurance of correctness. When the costs of failure are high, system designers must be able to guarantee ahead of time that their systems work as intended. Many methods, such as testing and simulation, are used in combination to improve reliability. While testing and simulation may reveal software bugs and increase safety assurance, they are not able to prove safety guarantees over the continuous and infinite state-spaces characteristic of hybrid systems like flight control, where the aircraft move continuously through space and time. The complexity of curved flight dynamics has been difficult for many analysis techniques [1–8], which often resort to unflyable approximations of flight trajectories that require aircraft to turn on corners. However, the formal verification techniques described in this paper are able to provide guarantees for flyable maneuvers over the entirety of this continuous state-space and therefore over all evolutions of all aircraft movement.

These strong guarantees are especially important in a distributed system with a large number of interacting participants. As in [1, 4, 7, 9, 10], many previous approaches to aircraft control have looked into a relatively small number of agents. But with thousands of aircraft flying through commercial airspace daily, this system is already far too complex for humans to predict every scenario by looking at interactions of only a few aircraft. And this challenge increases when we examine controllers for Unmanned Aerial Vehicles (UAVs), which are becoming increasingly autonomous and fly even closer together, with less direct supervision by humans. As a result, we must provide a good argument for why a controller will always take the right action, even in extremely crowded airspace.

In this paper we specify and verify two control policies for planar aircraft avoidance maneuvers using automated theorem prover KeYmaeraD to produce a proof of safety for each of them. We design these policies such that all aircraft adhere to a simple and easy-to-implement separation principle: associated with each aircraft is

*This material is based upon work supported by the National Science Foundation under NSF CAREER Award CNS-1054246, NSF EXPEDITION CNS-0926181, grant no. CNS-0931985, and grant no. CNS-1035800. Sarah M. Loos was supported by a DOE Computational Science Graduate Fellowship.

HSCC'13, April 8–11, 2013, Philadelphia, Pennsylvania, USA.

a disc, within which the aircraft must remain. In this way, the problem reduces to proving that i) sufficient separation is maintained between pairs of discs, and ii) individual aircraft always remain inside their associated disc. We model 2D flight dynamics since they are the relevant dynamics for planar maneuvers, but investigating 3D maneuvers and dynamics may make interesting future work.

The complexities which arise from the curved flight trajectories of an arbitrary number of aircraft interacting in a distributed manner, along with the tight coupling of discrete control and continuous dynamics presently make KeYmaeraD the only verification tool capable of proving safety for this system. Our contributions are:

- We provide the *first formally* verified distributed system of aircraft with *curved flight* dynamics.
- Our controller requires only *flyable* aircraft trajectories with no corners or instantaneous changes of ground speed.
- We prove our controller is safe for an *arbitrarily large number of aircraft*. This guarantee is necessary for high-traffic applications such as crowded commercial airspace, unmanned aerial vehicle maneuvers, and robotic swarms.
- Other aircraft may enter an avoidance maneuver already *in progress* and safety for all aircraft is guaranteed still.
- We use *Arithmetic coding* to reduce proof complexity and branching.
- We prove that even when the interactions of many aircraft cause unexpected *emergent behaviors*, all resulting control choices are still safe.
- We present *hierarchical and compositional* techniques to reduce a very complex system into smaller, provable pieces.

2. RELATED WORK

Many methods for ensuring correctness have been researched, each having different strengths in dealing with the various challenges posed by air traffic control. Pallottino et al. [11] proposed a distributed collision avoidance policy that is closely related to the systems we examine here. They provide a thorough empirical description of the system's behavior, emphasizing simulation and physical experiment. They formulate a liveness property and give probabilistic evidence for it using Monte Carlo methods. They also provide an informal proof of safety that is similar in high-level ideas to our proofs, but does not consider a model for flight dynamics. However, since we provide *formal* proofs of safety based directly on the control protocols and working with a continuous model of flight dynamics, we provide a higher degree of assurance and a clearer avenue to safely extend the systems.

Verification methods for systems with an arbitrary number of agents behaving under distributed control fall primarily into one of two categories: theorem proving and parameterized verification. Johnson and Mitra [8] use parameterized verification to guarantee that a distributed air traffic landing protocol (SATS) is collision free. Using backward reachability, they prove safety in the SATS landing maneuver given a bound on the number of aircraft that can be engaged in the landing maneuver. The protocol divides the airspace into regions and models the aircraft flight trajectory within each region by a clock. We consider the complementary problem of free flight instead of airport landing traffic, and we show that in free space, *arbitrarily* many aircraft can join our maneuver, and we model aircraft movement using *flyable*, curved flight dynamics.

Other provably safe systems with a specific (usually small) number of agents are presented in [1, 4, 7, 9]. The work by Umeno and Lynch [4, 7] is complementary to ours; however while they consider real-time properties of airport protocols using Timed I/O Automata, we prove local properties of the actual hybrid system flight dynamics. Duperret et al. [9] verify a roundabout maneuver with three vehicles. Each vehicle is constrained to a specific, pre-defined path, so physical dynamics are simplified to one dimension. Tomlin et al. [1] analyze competitive aircraft maneuvers game theoretically using numerical approximations of partial differential equations. As a solution, they propose roundabout maneuvers and give bounded-time verification results for up to four aircraft using straight-line approximations of flight dynamics.

Flyability is identified as a major challenge in Košecká et al. [12], where planning based on superposition of potential fields is used to resolve air traffic conflicts. This planning does not guarantee flyability but, rather, defaults to classical vertical altitude changes whenever a nonflyable path is detected. The resulting maneuver has not yet been verified. The planning approach has been pursued by Bicchi and Pallottino [13] with numerical simulations.

Numerical simulation algorithms approximating discrete-time Markov Chain approximations of aircraft behavior have been proposed by Hu et al. [2]. They approximate bounded-time probabilistic reachable sets for one initial state. We consider hybrid systems combining discrete control choices and continuous dynamics instead of uncontrolled, probabilistic continuous dynamics. Hwang et al. [6] have presented a straight-line aircraft conflict avoidance maneuver involving optimization over complicated trigonometric computations, and validate it using random numerical simulation and informal arguments. The work of Dowek et al. [3] and Galdino et al. [5] shares many goals with ours. They consider unflyable, straight-line maneuvers and formalize geometrical proofs in PVS.

Our approach has a different focus from complementary work:

- Our maneuver directly involves curved flight unlike [1–8]. This makes our maneuver more realistic since it is *flyable*, but much more difficult to analyze.
- Unlike [2,6,12], we do not give results for a finite (sometimes small) number of initial flight positions (as in simulation). Instead, we verify uncountably many initial states and give unbounded-time horizon verification results.
- Unlike [1, 2, 6, 12–14], we use symbolic computation so that numerical and floating point errors can not violate soundness.
- Unlike [2–8, 13, 15], we analyze hybrid system dynamics directly, not approximations like clocks.
- Unlike [1,2,6,11–13,15] we produce *formal*, deductive proofs.
- In [3–7, 11], it is not proved that the hybrid dynamics and flight equations follow the geometrical thoughts. In contrast, our approach directly works for the hybrid flight dynamics.
- Unlike [1–3,5,6,9,10,12,13], we verify the case of *arbitrarily many* aircraft, which is crucial for dense airspace.
- Unlike [13, 14], we do not guarantee optimality of the resulting maneuver.

3. PRELIMINARIES

Quantified Hybrid Programs.

QHPs [16, 17] are defined by the following grammar (α, β are QHPs, θ terms, i a variable of sort C, f a function symbol, s a term with sort compatible to f, and H is a formula of first-order logic):

$$\alpha, \beta \ ::= \ \forall i\!:\!C\ \mathcal{A} \mid \forall i\!:\!C\ \{\mathcal{D}\ \&\ H\} \mid ?H \mid \alpha \cup \beta \mid \alpha;\beta \mid \alpha^*$$

where $\mathcal{A}$ is a list of assignments of the form $f(s) := \theta$ and nondeterministic assignments of the form $f(s) := *_C$, and $\mathcal{D}$ is a list of differential equations of the form $f(s)' = \theta$. When an assignment list does not depend on the quantified variable i, we may elide the quantification for clarity.

The effect of *assignment* $f(s) := \theta$ is a discrete jump assigning θ to $f(s)$. The effect of *nondeterministic assignment* $f(s) := *_C$ is a discrete jump assigning *any value* in C to $f(s)$. The effect of *quantified assignment* $\forall i\!:\!C\ \mathcal{A}$ is the simultaneous effect of all assignments in $\mathcal{A}$ for all objects i of sort C. For example, the QHP $\forall i\!:\!C\ \omega(i) := \omega(i) + 1$ expresses that all aircraft i of sort C simultaneously increase their angular velocity. The effect of *quantified differential equation* $\forall i : C\{\mathcal{D}\&H\}$ is a continuous evolution where, for all objects i of sort C, all differential equations in $\mathcal{D}$ hold and formula H holds throughout the evolution (i.e. the state remains in the region described by *evolution domain constraint* H). The dynamics of QHPs changes the interpretation of terms over time: for an $\mathbb{R}$-valued function symbol f, $f(\vec{s})'$ denotes the derivative of the interpretation of the term $f(\vec{s})$ over time during continuous evolution, not the derivative of $f(\vec{s})$ by its argument $\vec{s}$. For this paper, f does not occur in $\vec{s}$. In most cases, $\vec{s}$ is just i. For instance, the following QHP expresses that all aircraft i of sort C fly by $\forall i\!:\!C\ x(i)' = d(i), d(i)' = \omega(i)d(i)$ such that their position $x(i)$ changes continuously according to their respective direction $d(i)$ and their direction changes according to their angular velocity $\omega(i)$.

The effect of *test* $?H$ is a *skip* (i.e., no change) if formula H is true in the current state and *abort* (blocking the system run by a failed assertion), otherwise. *Nondeterministic choice* $\alpha \cup \beta$ is for alternatives in the behavior of the distributed hybrid system. In the *sequential composition* $\alpha;\beta$, QHP β starts after α finishes (β never starts if α continues indefinitely). *Nondeterministic repetition* α^* repeats α an arbitrary number of times, possibly zero times.

Quantified Differential Dynamic Logic.

The formulas of Qd$\mathcal{L}$ [16, 17] are defined as in first-order dynamic logic plus many-sorted first-order logic by the following grammar (ϕ, ψ are formulas, θ_1, θ_2 are terms of the same sort, i is a variable of sort C, and α is a QHP):

$$\begin{aligned}\phi, \psi \ ::= &\ \theta_1 = \theta_2 \mid \theta_1 \geq \theta_2 \mid \neg\phi \mid \phi \wedge \psi \mid \phi \vee \psi \\ &\mid \forall i\!:\!C\ \phi \mid \exists i\!:\!C\ \phi \mid [\alpha]\phi \mid \langle\alpha\rangle\phi\end{aligned}$$

We use standard abbreviations to define $\leq, >, <, \rightarrow$. The real numbers $\mathbb{R}$ form a distinguished sort, upon which are defined the rigid functions $+$ and $\times$. Sorts $C \neq \mathbb{R}$ have no ordering and hence $\theta_1 = \theta_2$ is the only relation allowed on them. For sort $\mathbb{R}$, we abbreviate $\forall x\!:\!\mathbb{R}\ \phi$ by $\forall x\,\phi$. In the following, all formulas and terms have to be well-typed. Qd$\mathcal{L}$ formula $[\alpha]\phi$ expresses that *all states* reachable by QHP α satisfy formula ϕ. Likewise, $\langle\alpha\rangle\phi$ expresses that *there is at least one state* reachable by α for which ϕ holds.

Proof Calculus and Prover.

The Qd$\mathcal{L}$ proof calculus [16, 17] consists of *proof rules* that operate on *sequents*, which are syntactic objects of the form $\Gamma \Rightarrow \Delta$ where Γ and Δ are finite sets of Qd$\mathcal{L}$ formulas. Loos et al. [18] use Qd$\mathcal{L}$ to verify adaptive cruise control for arbitrarily many cars on a highway, with simple continuous dynamics on a straight lane.

KeYmearaD is a theorem prover which mechanizes the use of the Qd$\mathcal{L}$ proof calculus. It has previously been used to verify a simple car control system [19]. KeYmaeraD further implements quantified differential invariants [20]. KeYmaeraD constructs proofs by following a user-created *tactic script*. When KeYmaeraD runs a tactic script, it applies proof rules to the sequent based on tactics specified in the script which will ultimately reduce the sequent to several problems in first-order real arithmetic. These simpler problems are then sent to a backend decision procedure in Mathematica.

4. CASE STUDY SYSTEMS

In this section, we present two classes of aircraft controllers, each of which maintains a guaranteed minimum distance p between all aircraft. We then prove that both of these controller classes are safe, (i.e. that the minimum distance p between aircraft is never violated). Each aircraft has a disc-shaped zone large enough to fly a circle within and which no other aircraft will be allowed to enter.

In the first controller class (Section 4.1), each aircraft maintains a larger buffer disc with the aircraft at its center. This disc allows pilots some freedom during an avoidance maneuver, including the choice of circling direction. We imagine this controller will be useful when passenger comfort is a factor, as in commercial airlines. The second class of controllers (Section 4.2) uses smaller buffer discs centered to the left or right of the aircraft. The smaller discs allow the aircraft to fly closer together, but there may be little choice in how a maneuver is executed. This is well suited to UAVs which may fly very close together and are concerned only with flyability, not passenger comfort. Additionally, since many UAVs may be monitored and managed remotely by a small group of people, it may be more desirable to have a specific collision avoidance maneuver with little freedom and high predictability. Because the first controller class requires a larger disc than the second, we call it *Big Disc*, and appropriately we name the second class *Small Discs*.

We use two levels of abstraction to analyze the controllers. At the higher abstraction level we model the buffer discs, which can freeze instantaneously when they get within p distance of each other. At the lower level, we model the movement of aircraft within their discs, ensuring they always stay within the buffer zone while following flyable trajectories. In the proof, these two levels of abstraction are joined so that safety is assured for the system as a whole.

We model airspace as $\mathbb{R}^2$ and aircraft as points moving in this space. Each aircraft i steers by adjusting its angular velocity $\omega(i)$. When $\omega(i)$ is zero, the plane flies in a straight line. As angular velocity increases, the plane flies in a tighter and tighter circle, so we put an upper bound on the angular velocity $\Omega(i)$ based on the smallest circle that aircraft i can fly while maintaining constant linear speed $v(i)$. We keep linear speed $v(i)$ constant for each aircraft. We can determine the radius of each aircraft's smallest flyable circle by the equation $minr(i) = v(i)/\Omega(i)$. This model is known as the *Dubins vehicle* [21] and has been used previously for aircraft verification [1]. We allow an arbitrarily large number of aircraft to be present in airspace, so long as there is enough space to pack their discs. To our knowledge, no other method has been able to verify a protocol or controller safe for an arbitrary number of aircraft using a continuous model of their flight dynamics. This is not surprising, since safety must be guaranteed even for unpredictable emergent behaviors and in crowded, worst-case scenarios. Models written as QHPs inherently have a compositional and hierarchical structure which makes them easier to decompose into smaller, provable pieces by using sound proof rules. We also use nondeterminism in the model of the controller, which means that our proof is robust to variations in implementation on individual aircraft.

4.1 Big Disc

During normal free flight (i.e. whenever the aircraft is not engaged in a collision avoidance maneuver), the buffer zone for an aircraft i is a disc of radius $2minr(i)$ centered around the aircraft, which is at planar position $x(i) = (x_1(i), x_2(i))$. So long as the aircraft does not enter a collision avoidance maneuver, its buffer disc remains centered on the aircraft. However, should aircraft i come too close to another plane, it will enter collision avoidance mode and begin circling at radius $minr(i)$ to either the left or the right. The disc allows just enough room for this maneuver; however, the disc is big in the sense that it allows a considerable amount of freedom once the aircraft has gone halfway around this initial circle. The beginning of one possible trajectory of a collision avoidance

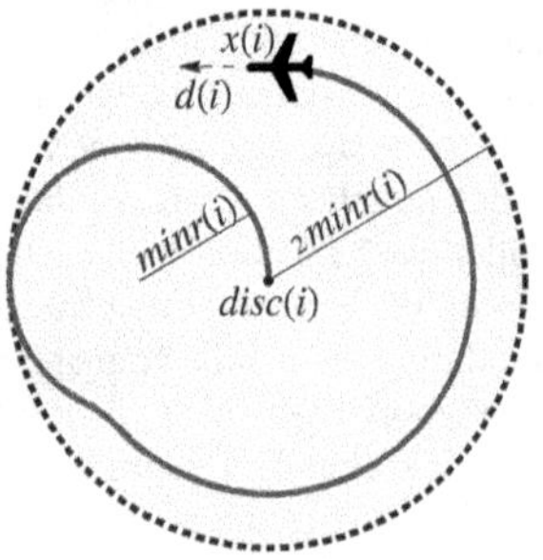

Figure 1: A possible collision avoidance trajectory of BigDisc.

maneuver is illustrated in Figure 1. The current direction of flight of aircraft i is given by the indexed variable $d(i)$ as a 2D unit vector. The variable $disc(i)$ stores the position of the center of i's buffer disc. The aircraft need not always turn at the maximum angular velocity $\Omega(i)$; we require only that the aircraft remain within the disc by circling in its original direction. (Note: while it is possible for the aircraft to change its circling direction while staying within the disc by flying a figure eight or an 'S' shape, we disallow this behavior since it would increase the complexity of the controller.)

4.1.1 *Formal Model*

The Big Disc policy is presented formally in QHP 1 and we describe it in this section. The variable $ca(i)$, indicates whether aircraft i is in a collision avoidance maneuver. If $ca(i) = 0$ then i is in free flight; if $ca(i) = 1$ then i is in an avoidance maneuver and is circling within its disc. Each aircraft has the ability to enter collision avoidance independently and asynchronously. This simplifies collision avoidance maneuvers with more than two aircraft and improves reliability, since no perfect synchronization is required, which would be difficult to implement in a distributed system.

The variable $side(i)$ indicates i's circling direction when it enters an avoidance maneuver. If $side(i) = 1$, i will circle counterclockwise; if $side(i) = -1$, i circles clockwise. We use $\|y\|$ to denote the Euclidean norm and we use $y^{\perp}$ to denote the vector obtained by rotating y ninety degrees counter-clockwise, $y^{\perp} := (-y_2, y_1)$.

The quantified hybrid program `BigDisc` is a loop, which is represented in line 1 by *, the nondeterministic repetition operator. Each iteration is either a control action as represented by `Control` or an evolution of physics as represented by `Plant`. This loop may repeat arbitrarily many times. In the `Control` branch, the program nondeterministically selects an aircraft k ($k := *_{\mathbb{A}}$ assigns an arbitrary aircraft into k) and then allows k to perform some action. The allowed actions depend on whether k is in a collision avoidance maneuver. If it is (case `CA`), then k may either adjust its angular velocity in the `Steer` branch, or exit the maneuver with the `Exit` branch. The new angular velocity in the `Steer` branch is arbitrary ($\omega(k) := *_{\mathbb{R}}$, where $*_{\mathbb{R}}$ is an arbitrary real number) but bounded by $-\Omega(k)$ and $\Omega(k)$ due to the subsequent test. The aircraft may only `Exit` the collision avoidance circling maneuver when $x(k) = disc(k)$, i.e., the aircraft must return to the center of the disc before exiting the maneuver. If k is not in a collision avoidance maneuver (case `NotCA`), then it may once again `Steer`, or it may switch its circling direction with the `Flip` branch, or it may enter collision avoidance with the `Enter` branch. In the `Enter` branch, the aircraft sets its angular velocity so that it will circle with radius $minr(k)$, thereby entering a collision avoidance. It also sets the $ca(k)$ flag to indicate internally that it has entered this maneuver.

The other branch in `BigDisc`'s main loop is `Plant`. The position of the aircraft, $x(i)$, changes according to its direction, $d(i)$, which in turn changes according to its angular velocity, $\omega(i)$. This

Quantified Hybrid Program 1 Big Disc

$$\texttt{BigDisc} \equiv (\texttt{Control} \cup \texttt{Plant})^* \tag{1}$$

$$\texttt{Control} \equiv k := *_{\mathbb{A}};\ (\texttt{CA} \cup \texttt{NotCA}) \tag{2}$$

$$\texttt{CA} \equiv ?(ca(k) = 1);\ (\texttt{Steer} \cup \texttt{Exit}) \tag{3}$$

$$\texttt{NotCA} \equiv ?(ca(k) = 0);\ (\texttt{Steer} \cup \texttt{Flip} \cup \texttt{Enter}) \tag{4}$$

$$\texttt{Steer} \equiv \omega(k) := *_{\mathbb{R}};\ ?(-\Omega(k) \le \omega(k) \le \Omega(k)) \tag{5}$$

$$\texttt{Exit} \equiv ?(disc(k) = x(k));\ ca(k) := 0 \tag{6}$$

$$\texttt{Enter} \equiv \omega(k) := side(k) \cdot \Omega(k);\ ca(k) := 1 \tag{7}$$

$$\texttt{Flip} \equiv side(k) := -side(k) \tag{8}$$

$$\texttt{Plant} \equiv \forall i : \mathbb{A}\ \big(x(i)' = v(i) \cdot d(i),\ d(i)' = \omega(i) \cdot d(i)^{\perp}, \tag{9}$$

$$disc(i)' = (1 - ca(i)) \cdot v(i) \cdot d(i)\ \&\ \texttt{EvDom}\big) \tag{10}$$

$$\texttt{EvDom} \equiv \forall j : \mathbb{A} \tag{11}$$

$$((j \ne i \wedge (ca(i) = 0 \vee ca(j) = 0)) \rightarrow \mathsf{Sep}(i, j) \tag{12}$$

$$\wedge\ \|disc(i) - (x(i) + minr(i) \cdot side(i) \cdot d(i)^{\perp})\|$$

$$\le minr(i)) \tag{13}$$

$$\mathsf{Sep}(i, j) \equiv \|disc(i) - disc(j)\| \ge 2minr(i) + 2minr(j) + p \tag{14}$$

makes $\omega(i)$ our primary control variable. These physical dynamics are modeled by the differential equation in line 9. The center of the disc, $disc(i)$, is stationary during a collision avoidance maneuver, but otherwise it is equal to the aircraft position. This case distinction is achieved by using arithmetic coding, whereby we multiply by $(1 - ca(i))$, in line 10. This reduces a branching of the system which would be incurred if we were to use traditional if-else coding style, and thereby reduces the complexity of the safety proof. When the aircraft is in free flight, $ca(i) = 0$, which causes $disc(i)'$ to equal $x(i)'$. But, when the aircraft is in a collision avoidance maneuver, $ca(i) = 1$, causing $disc(i)' = 0$, so the disc is stationary. The evolution domain, `EvDom`, has two purposes. First, it monitors the disc positions of other aircraft (line 12). Recall that in order for the system to be considered safe, no aircraft can pass closer than distance p to another. So, if the aircraft's disc comes within p of another disc (as quantified in line 14), it forces both aircraft to enter collision avoidance. Second, while the aircraft has a great amount of freedom in how it maneuvers during collision avoidance, it must always be able to flyably remain within its buffer disc. We use the inequality in line 13 to quantify this condition. It states that if the aircraft turns in a tight circle with radius $minr(i)$, the origin of that tight circle is no more than $minr(i)$ away from the point $disc(i)$.

Our model allows for a huge amount of nondeterminism, both in the discrete dynamics of the controller (e.g. which aircraft are controlled ($k := *_{\mathbb{A}}$, line 2), how the aircraft steer ($\omega(k) := *_{\mathbb{R}}$, line 5), and whether to enter a collision avoidance maneuver), and in the continuous dynamics of the plant (e.g. how long to wait between control choices). This nondeterminism is a beneficial property of our collision avoidance protocol, since it allows for each aircraft to implement slightly different control algorithms without violating the proof of safety for the entire system. As a result, we have verified a class of controllers, rather than one specific implementation.

4.1.2 *Theorem Statement*

In order to guarantee safety, we must prove that for all pairs of distinct aircraft i, j, the distance between i and j is greater than or equal to p. We can express this condition formally as

$$\mathsf{Safe} \equiv \forall\, i, j : \mathbb{A}\ \ (i \ne j \rightarrow \|x(i) - x(j)\| \ge p).$$

We must show that Safe holds during all executions of BigDisc. The Qd$\mathcal{L}$ formula expressing this property is [BigDisc]Safe. We must also ensure that the aircraft begin in a controllable state. This means each aircraft must have a buffer disc (InitA), which is empty (InitB), and within which it may flyably maneuver (InitC). For aircraft i that are not in an avoidance maneuver, we ensure that i's disc is at the same point as i. Since $ca(i) = 1$ for aircraft in a maneuver and $ca(i) = 0$ otherwise, we may write this property as

$$\texttt{InitA} \equiv \forall i : \mathbb{A}\ (1 - ca(i)) \cdot disc(i) = (1 - ca(i)) \cdot x(i).$$

By again using this arithmetic coding style rather than an if-else statement, we eliminate a significant branching factor in the resulting proof. We then show that the discs are empty by ensuring sufficient separation between the discs as defined in QHP 1 line 14.

$$\texttt{InitB} \equiv \forall i, j : \mathbb{A}\ \ (i \neq j \rightarrow \mathsf{Sep}(i, j))$$

Finally, we ensure that aircraft in collision avoidance maneuvers are able to flyably remain within their discs. We do this by proving that the following formula is an invariant of our system. It states that if i tightly circles in its current circling direction, the center of this tight circle will be within distance $minr(i)$ of $disc(i)$.

$$\texttt{InitC} \equiv \forall i : \mathbb{A}\ \ \|disc(i) - (x(i) + minr(i) \cdot side(i) \cdot d(i)^{\perp})\| \leq minr(i)$$

Note that these initial conditions all hold trivially if the aircraft begin far enough apart that none is in a collision avoidance maneuver.

THEOREM 1 (SAFETY OF BigDisc). *If the aircraft are initially in a controllable state, then no two aircraft will come closer than distance p while all aircraft follow* Control*; therefore safety of the* BigDisc *controller is expressed by the provable* Qd$\mathcal{L}$ *formula:*

$$(\texttt{InitA} \wedge \texttt{InitB} \wedge \texttt{InitC}) \rightarrow [\texttt{BigDisc}]\mathsf{Safe}$$

We proved Theorem 1 for all parameter values by showing that InitA, B and C are maintained as invariants. This proof was generated using KeYmaeraD from a 330 line user-generated tactic script. The tactic file and the KeYmaeraD theorem prover are available online [22]. A discussion of the critical techniques needed to complete this proof is presented in [23, Appendix A.1].

4.2 Small Discs

One drawback of the Big Disc policy is that it may trigger collision avoidance maneuvers that are not strictly necessary; the buffer zones are larger than the required circling space of the aircraft. Our second policy, *Small Discs*, aims to decrease the size of the disc. The only way to do this is to abandon the assumption that the disc must be centered on the aircraft during free flight. Instead, the buffer zone, a disc of radius $minr(i)$, is centered at a point with distance $minr(i)$ away from $x(i)$, in a direction perpendicular to i's motion either to the left or the right. Thus the aircraft is always on the edge of its disc, and during collision avoidance, the aircraft follows the circumference of its disc. As with BigDisc, an aircraft may flip its circling direction during free flight. This now makes the disc jump to the other side of the aircraft, and before an aircraft can flip its circling direction, it must check that it may do so safely.

Fig. 2 illustrates a situation where flipping the disc to the opposite side prevents an unnecessary collision avoidance maneuver. Each aircraft has an *active disc* (solid discs in Fig. 2) that it will use for collision avoidance if needed. We also illustrate the *inactive disc* (dotted discs) as an alternative choice for the disc if the aircraft decides to flip its circling direction. In Fig. 2, the active discs of aircraft i and aircraft j are on a collision course. If nothing is done, at the latest when the edges of the discs are separated by distance p, both aircraft will enter collision avoidance by circling to the right

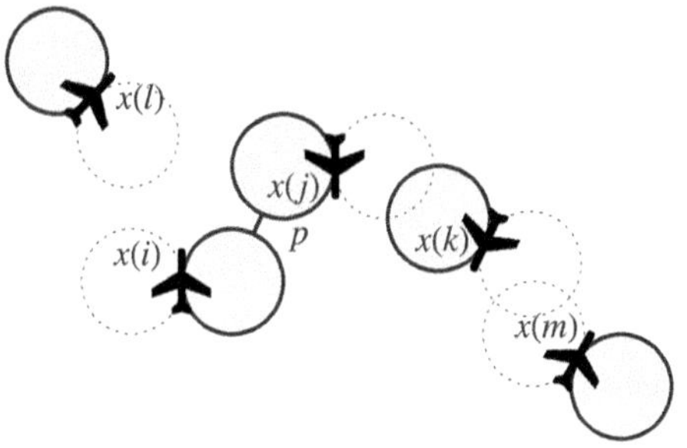

Figure 2: One possible scenario in the Small Discs policy

around the circumference of their respective discs. Notice that in this case, the aircraft may pass as close as the minimum separation distance p when aircraft i is at the top of its disc while aircraft j is at the bottom. However, since aircraft i has free space to its left, it may flip its circling disc to the opposite side and no collision avoidance maneuvers are necessary. Aircraft j is unable to make such a flip, since aircraft k's disc occupies the necessary space. Only collisions of active discs are a problem. The fact that the inactive discs of k and m overlap is immaterial, because if collision avoidance is necessary, every aircraft will follow the circumference of its *active* disc. This also illustrates that aircraft must synchronize disc flipping. If, to enable j to flip its disc, k flips its disc, but, at the same time, and unaware of this, m flips its disc, then k and m would have incompatible collision avoidance discs. Since purely discrete standard solutions exist for ensuring consistency in such discrete mode changes, our model simply uses sequentialized flipping decisions.

4.2.1 Formal Model

The Small Discs policy is presented formally as SmallDiscs in QHP 2. The overall structure is similar to that of BigDisc. One notable difference is that SmallDiscs no longer uses the state variable $disc(i)$. During free flight, the center of an aircraft's disc moves with dynamics that are not easy to express in terms of other variables; it is certainly not as easy as setting $disc(i)' = x(i)'$, as we did in BigDisc. The point $disc(i)$ moves faster or slower than $x(i)$, depending on whether the aircraft is veering away from its active disc, or towards it. As a result, $disc(i)$ has very involved continuous dynamics. Fortunately, however, the position of aircraft i's disc can be simply expressed in terms of other state variables. By using differential-algebraic equations as in [24], we equate

$$disc(i) = x(i) + minr(i) \cdot side(i) \cdot d(i)^{\perp}. \qquad (15)$$

In order to simplify the mathematics of the system, we directly reduce the system to ordinary differential equations, which also makes the connection to BigDisc more apparent. Thus, instead of using (15) as part of a differential-algebraic equation [24], we consider (15) as a definition and statically replace all occurrences of $disc(i)$ by the right-hand side of (15). The subsequent model should be read with this in mind. The other major change in QHP 2 is the separation condition Sep and the newly introduced separation condition FlipSep for flipping the disc.

4.2.2 Theorem Statement

Here again we want to prove that under safe initial conditions, the SmallDiscs controller is always Safe, where Safe is exactly as we defined it for BigDisc. We need to modify the initial conditions for SmallDiscs. There are two properties that we want to hold: first, the discs are separated (InitD), and second, when an aircraft is engaged in collision avoidance it is flying along the circum-

Quantified Hybrid Program 2 Small Discs

$$\texttt{SmallDiscs} \equiv (\texttt{Control} \cup \texttt{Plant})^* \quad (16)$$

$$\texttt{Control} \equiv k := *_{\mathbb{A}};\ (\texttt{CA} \cup \texttt{NotCA}) \quad (17)$$

$$\texttt{CA} \equiv ?(ca(k) = 1);\ (\texttt{Exit} \cup \texttt{Skip}) \quad (18)$$

$$\texttt{NotCA} \equiv ?(ca(k) = 0);\ (\texttt{Steer} \cup \texttt{Flip} \cup \texttt{Enter}) \quad (19)$$

$$\texttt{Skip} \equiv ?true \quad (20)$$

$$\texttt{Steer} \equiv \omega(k) := *_{\mathbb{R}};\ ?(-\Omega(k) \leq \omega(k) \leq \Omega(k)) \quad (21)$$

$$\texttt{Exit} \equiv ca(k) := 0 \quad (22)$$

$$\texttt{Enter} \equiv (\omega(k) := side(k) \cdot \Omega(k));\ ca(k) := 1 \quad (23)$$

$$\texttt{Flip} \equiv ?(\forall j : \mathbb{A}\ (j \neq k \rightarrow \mathsf{FlipSep}(j, k))); \quad (24)$$

$$side(k) := -side(k) \quad (25)$$

$$\mathsf{FlipSep}(i, j) \equiv \|(x(i) + minr(i) \cdot side(i) \cdot d(i)^{\perp}) \quad (26)$$

$$- (x(j) - minr(j) \cdot side(j) \cdot d(j)^{\perp})\| \quad (27)$$

$$\geq minr(i) + minr(j) + p \quad (28)$$

$$\texttt{Plant} \equiv \forall i : \mathbb{A}\ \big(x(i)' = v(i) \cdot d(i),\ d(i)' = \omega(i)d(i)^{\perp} \quad (29)$$

$$\&\ \forall j : \mathbb{A}\ ((j \neq i \wedge (ca(i) = 0 \vee ca(j) = 0)) \quad (30)$$

$$\rightarrow \mathsf{Sep}(i, j))\big) \quad (31)$$

$$\mathsf{Sep}(i, j) \equiv \|(x(i) + minr(i) \cdot side(i) \cdot d(i)^{\perp}) \quad (32)$$

$$- (x(j) + minr(j) \cdot side(j) \cdot d(j)^{\perp})\| \quad (33)$$

$$\geq minr(i) + minr(j) + p \quad (34)$$

ference of its active disc (`InitE`). `InitD` is similar to `InitB` for `BigDisc`, but it uses our new definition of Sep for `SmallDiscs`:

$$\texttt{InitD} \equiv \forall i, j : \mathbb{A}\ \ (i \neq j \rightarrow \mathsf{Sep}(i, j)).$$

If i is in a collision avoidance maneuver, then i is turning at maximal angular velocity. This implication is expressed with arithmetic coding by multiplying both sides with the indicator $ca(i)$:

$$\texttt{InitE} \equiv \forall i : \mathbb{A}\ \ (\omega(i) \cdot ca(i) = \Omega(i) \cdot side(i) \cdot ca(i)).$$

THEOREM 2 (SAFETY OF `SmallDiscs`). *If the aircraft are initially in a controllable state (i.e. where* `InitD` *and* `InitE` *hold), then no aircraft will come closer than distance p to any other aircraft so long as each aircraft follows* `Control`*; therefore safety of the* `SmallDiscs` *controller is expressed by the provable* Qd$\mathcal{L}$ *formula:*

$$(\texttt{InitD} \wedge \texttt{InitE}) \rightarrow [\texttt{SmallDiscs}]\mathsf{Safe}$$

We proved Theorem 2 in KeYmaeraD by showing `InitD` and `InitE` are invariant. The accompanying tactic script is 309 lines in length and is available online [22].

5. REFERENCES

[1] Tomlin, C., Pappas, G.J., Sastry, S.: Conflict resolution for air traffic management. IEEE T. Automat. Contr. **43** (1998) 509–521

[2] Hu, J., Prandini, M., Sastry, S.: Probabilistic safety analysis in three-dimensional aircraft flight. In: CDC. (2003)

[3] Dowek, G., Muñoz, C., Carreño, V.A.: Provably safe coordinated strategy for distributed conflict resolution. In: AIAA-2005-6047. (2005)

[4] Umeno, S., Lynch, N.A.: Proving safety properties of an aircraft landing protocol using I/O automata and the PVS theorem prover. In Misra, J., Nipkow, T., Sekerinski, E., eds.: FM. Volume 4085 of LNCS., Springer (2006) 64–80

[5] Galdino, A.L., Muñoz, C., Ayala-Rincón, M.: Formal verification of an optimal air traffic conflict resolution and recovery algorithm. In Leivant, D., de Queiroz, R., eds.: WoLLIC. Volume 4576 of LNCS., Springer (2007) 177–188

[6] Hwang, I., Kim, J., Tomlin, C.: Protocol-based conflict resolution for air traffic control. Air Traffic Control Quarterly **15** (2007) 1–34

[7] Umeno, S., Lynch, N.A.: Safety verification of an aircraft landing protocol: A refinement approach. In Bemporad, A., Bicchi, A., Buttazzo, G., eds.: HSCC. Volume 4416 of LNCS., Springer (2007) 557–572

[8] Johnson, T., Mitra, S.: Parameterized verification of distributed cyber-physical systems:an aircraft landing protocol case study. In: ACM/IEEE ICCPS. (2012)

[9] Duperret, J.M., Hafner, M.R., Del Vecchio, D.: Formal design of a provably safe roundabout system. (In: IEEE/RSJ IROS) 2006–2011

[10] Platzer, A., Clarke, E.M.: Formal verification of curved flight collision avoidance maneuvers: A case study. In: FM. Volume 5850 of LNCS., Springer (2009) 547–562

[11] Pallottino, L., Scordio, V., Frazzoli, E., Bicchi, A.: Decentralized cooperative policy for conflict resolution in multi-vehicle systems. IEEE Trans. on Robotics **23** (2007)

[12] Košecká, J., Tomlin, C., Pappas, G., Sastry, S.: 2-1/2D conflict resolution maneuvers for ATMS. In: CDC. Volume 3., Tampa, FL, USA (1998) 2650–2655

[13] Bicchi, A., Pallottino, L.: On optimal cooperative conflict resolution for air traffic management systems. IEEE Trans. ITS **1** (2000) 221–231

[14] Hu, J., Prandini, M., Sastry, S.: Optimal coordinated motions of multiple agents moving on a plane. SIAM Journal on Control and Optimization **42** (2003) 637–668

[15] Massink, M., Francesco, N.D.: Modelling free flight with collision avoidance. In Andler, S.F., Offutt, J., eds.: ICECCS, Los Alamitos, IEEE (2001) 270–280

[16] Platzer, A.: Quantified differential dynamic logic for distributed hybrid systems. In Dawar, A., Veith, H., eds.: CSL. Volume 6247 of LNCS., Springer (2010) 469–483

[17] Platzer, A.: A complete axiomatization of quantified differential dynamic logic for distributed hybrid systems. Logical Methods in Computer Science **8** (2012) 1–44

[18] Loos, S.M., Platzer, A., Nistor, L.: Adaptive cruise control: Hybrid, distributed, and now formally verified. In Butler, M., Schulte, W., eds.: FM. LNCS, Springer (2011) 42–56

[19] Renshaw, D.W., Loos, S.M., Platzer, A.: Distributed theorem proving for distributed hybrid systems. In Qin, S., Qiu, Z., eds.: ICFEM. Volume 6991 of LNCS., Springer (2011) 356–371

[20] Platzer, A.: Quantified differential invariants. In Frazzoli, E., Grosu, R., eds.: HSCC, ACM (2011) 63–72

[21] Dubins, L.E.: On curves of minimal length with a constraint on average curvature, and with prescribed initial and terminal positions and tangents. Am J Math **79** (1957) pp. 497–516

[22] Electronic proofs: `www.ls.cs.cmu.edu/discworld`.

[23] David Renshaw, Sarah M. Loos, A.P.: Formal verification of distributed aircraft controllers. Technical Report CMU-CS-12-132, Carnegie Mellon (2012)

[24] Platzer, A.: Differential-algebraic dynamic logic for differential-algebraic programs. J. Log. Comput. **20** (2010) 309–352

A Simulink Hybrid Heart Model for Quantitative Verification of Cardiac Pacemakers

Taolue Chen
Department of Computer Science
University of Oxford, UK

Marco Diciolla
Department of Computer Science
University of Oxford, UK

Marta Kwiatkowska
Department of Computer Science
University of Oxford, UK

Alexandru Mereacre
Department of Computer Science
University of Oxford, UK

ABSTRACT

We develop a novel hybrid heart model in Simulink that is suitable for quantitative verification of implantable cardiac pacemakers. The heart model is formulated at the level of cardiac cells, can be adapted to patient data, and incorporates stochasticity. It is inspired by the timed and hybrid automata network models of Jiang *et al* and Ye *et al*, where probabilistic behaviour is not considered. In contrast to our earlier work, we work directly with action potential signals that the pacemaker sensor inputs from a specific cell, rather than ECG signals. We validate the model by demonstrating that its composition with a pacemaker model can be used to check safety properties by means of approximate probabilistic verification.

Categories and Subject Descriptors

I.6.4 [**Computing Methodologies**]: SIMULATION AND MODELING—*Model Validation and Analysis*

General Terms

Verification

Keywords

Hybrid systems, Quantitative verification, Pacemaker

1. INTRODUCTION

Today's implantable medical devices are increasingly often controlled by embedded software and rigorous software design methodologies are needed to ensure their safe operation and to avoid costly device recalls. The natural models for medical devices, such as cardiac pacemakers [6], GPCA infusion pumps [8] and continuous glucose monitors [16], are stochastic hybrid systems: they involve discrete mode switching and nonlinear continuous flows, e.g., electrical signal or glucose level, while at the same time allowing for stochasticity that arises from randomness of the timing of events. Therefore, developing effective methodologies to provide safety assurance in this setting by means of *quantitative verification* is an important challenge.

Regarding cardiac pacemakers, a number of model-based frameworks have been proposed, to mention the Virtual Heart Model (VHM) of Jiang *et al* [4, 5]. Though mainly intended for simulation and testing, its timed automata pacemaker model [6] has been verified using UPPAAL [9] against a *random heart model*. The random heart model can capture the timing delays between events, but is unable to model the stochasticity in the timing that is characteristic in a heart rhythm and varies from patient to patient. Following a suggestion in [6] that physiologically-relevant heart models are needed to establish the correctness of more complex properties for pacemakers, we earlier proposed a realistic heart model that addresses this issue [1]. The model was adapted from a sophisticated model that generates multichannel electrocardiogram (ECG) based on nonlinear ordinary differential equations (ODEs) due to Clifford *et al* [2]. To transfer to our setting, where we need to consider that the pacemaker is implanted in the heart tissue, we convert external ECG signals into action potential signals read by implanted sensors. A unique feature of the model of [1] is that the heart can probabilistically switch between normal and abnormal beat types, in a manner that can be learnt from patient data. We performed quantitative, probabilistic verification by analysing the composition of the pacemaker model of [6] and the heart model against typical correctness properties such as (i) whether the pacemaker corrects faulty heart beats, maintaining normal heart rhythm of 60-100 beats per minute (BPM), and (ii) that the pacemaker does not induce erroneous heart behaviours (that is, it does not overpace the heart unless necessary). These were implemented based on the probabilistic model checker PRISM [7] and MATLAB.

One of the shortcomings of the heart model in [1] is that it does not capture the electrical conduction system of the heart, and specifically the delays in the action potential signal as it is propagated from cell to cell. In this paper, we propose an accurate, fine-grained, heart model which is a network of cells, and which can therefore model the conduction delays. Moreover, we work directly with the action potentials that the pacemaker can read from a specific cell. Our model is inspired by the VHM system developed in [5], except that we model cells as *hybrid automata*, in the style of [17], and add *stochasticity* not considered in [5, 17].

HSCC'13, April 8–11, 2013, Philadelphia, Pennsylvania, USA.

The model of [5] can reproduce the timing of the action potential signals, but they do not address the voltage change when the signal is propagated through the cells. Our model is more precise, in that we work directly with cell action potential, explicitly representing the voltage of the cardiac cell signal as a hybrid automaton of [17]. As a result, the timing of the *effective refractory period* (ERP), i.e., the time of non-responsiveness for the cell to further stimulus, and the *relative refractory period* (RRP), i.e., the period of time when an altered secondary excitation stimulus is possible, can be naturally expressed.

Contribution. The contribution of the paper is as follows. We propose a physiologically-relevant heart model built as a network of communicating input-output hybrid automata which features stochasticity. The model enables the modelling of both diseased and normal rhythms, and can be adapted to exhibit random delays in the timing of events that are patient-specific. We implement the heart model in Simulink and validate it against the pacemaker models of [5], demonstrating basic safety properties of the pacemaker by means of probabilistic approximate model checking, with encouraging results.

Related work. [6] formulate a timed automata model for a cardiac pacemaker and verify it using UPPAAL against a simple random heart model. Tuan *et al* [12] develop a real-time formal model for a pacemaker and verify it with the PAT model checker. Networks of timed automata are employed to devise the Virtual Heart Model [4, 5] and hybrid automata are used in the model of [17], both analysed through simulation. Macedo *et al* [13] develop and analyse a concurrent and distributed real-time model for pacemakers through a pragmatic incremental approach using VDM and a scenario-based approach. Gomes *et al* [3] present a formal specification of the pacemaker using the Z notation and employ theorem proving, whereas Mert *et al* [15] use Event-B and the ProB tool, to validate their models in different situations. None of the above approaches considers stochastic behaviours and properties. Risk analysis of glucose infusion pumps is performed with physiological models using statistical model checking in [16], but there is no stochasticity in the models.

Organisation. The rest of the paper is organised as follows. Sect. 2 presents the necessary background on the function of human heart, its modelling and a pacemaker model. Sect. 3 introduces the electrical conduction system of the heart. There, it is discussed how single cells are implemented, how the SA node differs from other cells and how the conduction system works. Sect. 4 gives an overview of the pacemaker model and some of its characteristic features. Sect. 5 describes the composition of the heart and the pacemaker and how probabilistic approximate verification on such a model is performed. Sect. 6 presents experimental results for basic safety properties for pacemakers. Finally, Sect. 7 includes conclusion and possible future directions.

2. PRELIMINARIES

In this section we describe the working of the heart, including its electrical system. The main function of the human heart is to maintain blood circulation of the body. This rhythmic, pump-like function is driven by muscle contractions, in particular, the contraction of the atria and ventricles which are triggered by electrical signals.

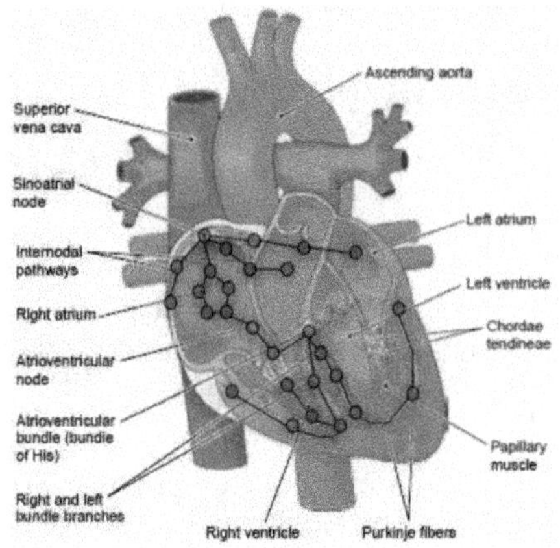

Figure 1: Electrical conduction system of the heart.

The *sinoatrial* (SA) node (a special tissue in the heart, see Fig. 1) spontaneously produces an electrical signal, which is the primary pacemaker of the heart. On each heart beat, it generates the control electrical signal which is conducted through prescribed *internodal pathways* into the atrium causing its contraction. The signal then passes through the slow conducting *atrioventricular (AV) node*, allowing the blood to empty out the atria and fill the ventricles. The fast conducting *Purkinje* system spreads the electricity through the ventricles, causing all tissues in both ventricles to contract simultaneously and to force blood out of the heart. This electrical system is called the *natural pacemaker* (in contrast to the artificial pacemaker) of the heart. At the cellular level, the electrical signal is a change in the potential across the cell membrane, which is caused by the flow of ions between the inside and outside of the cell.

Abnormalities in the electrical signal generation and propagation can cause different types of arrhythmias, such as *Tachycardia* (fast heart beat) and *Bradycardia* (slow heart beat), which require medical intervention in the form of medication, surgery or implantable pacemakers.

Action Potential. At the cellular level, the heart tissue is activated by an external voltage applied to the cell. After the activation, a transmembrane voltage change over time can be sensed due to ion channel activities, which is referred to as an *action potential* (AP). The AP is fired as an all-or-nothing response to a supra-threshold electrical signal, and each AP follows roughly the same sequence of events and has the same magnitude regardless of the applied stimulus. This is also the signal that an implantable pacemaker will receive or generate (see Sect. 2 of [1] for more detail).

3. ELECTRICAL CONDUCTION SYSTEM

In this section, we propose a model for the electrical conduction system of the heart which is tailored for the verification of pacemakers.

3.1 The Cardiac Cell

In this paper, we adapt the so called Luo-Rudy Guinea Pig Ventricular Cell model (LRd), which is presented in [17] as a hybrid automaton and is depicted in Fig. 2. In a nutshell, there are four (discrete) locations and each of them is associated with an AP phase: *resting and final repolarisation* (q_0), *stimulated* (q_1), *upstroke* (q_2), and *plateau and early repolarisation* (q_3). The variables introduced in the model are: the membrane voltage v, which controls switches among locations; a memory variable v_n which is used to modify next ERP phases upon new rounds of excitation; and the excitation current i_{st}.

The memory variable v_n captures the proper response of AP to pacing frequency, which is an essential feature of cardiac excitation. Following [17] we define $\theta = \frac{v_n}{V_R}$ and incorporate the function $f(\theta) = 1 + 13\sqrt[6]{\theta}$ into location q_3. The function $g(\vec{v})$ denotes the voltage contribution from the neighbouring cells. Assuming the total number of cells connected to the current cell k is $N - 1$, we define $g_k(\vec{v})$, the function for the k'th cell to be $g_k(\vec{v}) = \sum_{i=1, i \neq k}^{N} v_i(t - \delta_{ki}) \cdot a_{ki} - v_k \cdot d_k$, where a_{ki} is the gain applied to the potential v_i from cell i, δ_{ki} is the time it takes for the potential to reach cell k, and d_k is the distance coefficient. Moreover, the mode invariants of each location are given as linear inequalities which constraint the membrane voltage. They depend on three model-specific constants: threshold voltage V_T, overshoot voltage V_O, and repolarisation voltage V_R.

The cell starts at location q_0 where two different scenarios are possible. If the cell is externally stimulated with the event $\mathsf{V_s}$?, it enters the stimulated mode updating its voltage according to the *stimulus current* (i_{st}). When the stimulus is terminated, via event $\overline{\mathsf{V}}_\mathsf{s}$?, with a sub-threshold voltage, the cell returns to resting without firing an AP. If the stimulus is supra-threshold, i.e., $v \geq V_T$ holds, the excited cell will generate an AP by progressing to mode *upstroke.* Similarly, without any external stimulus, the cell can move from q_0 to q_2 if the voltage (due to the contribution of the neighbouring cells) is supra-threshold $v \geq V_T$. The recovery course of the cell follows transitions to mode *plateau and ER* and then to *resting and FR.* The jump conditions on the control switches monitor the transmembrane potential v, rather than imposing a rigid timing scheme. This approach allows for AP adaptation (response to various pacing frequencies).

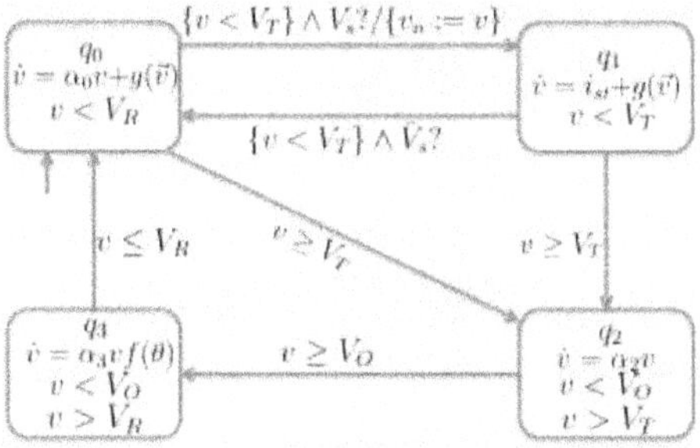

Figure 2: Hybrid automaton for the cardiac cell model.

3.2 The Conduction System

Modelling every single cell of the electrical conduction system is computationally intensive. Thus, we abstract the conduction system as a network of cells in order to achieve a good trade-off between the complexity of the model and the running time of the experiments.

In our experiments each cell is connected to neighbouring cells forming a graph as shown in Fig. 1 (black circles connected with lines). There are 33 cells in our graph. The electrical conduction system of the heart consists of conduction pathways with different *conduction delays.* Cells are connected by pathways. The delays of the pathways depend on the physiology of the tissue considered. Moreover, it is possible to use the pathway delays to model various known tissue diseases.

More specifically, our model consists of the SA node, the AV node and 31 cells that share similar properties. The SA node generates sequences of AP which are propagated through the electrical conduction system of the heart (see Sect. 3.3). The AV node is a special cell with the role of slowing down the signal coming from the atria to the ventricles and 31 cells. The 31 cells are connected together forming the graph structure presented in Fig. 1.

Cells communicate with their neighbours through input and output actions. Output actions are propagated to neighbouring cells. For each output action there is a corresponding input action. The set of input actions is {AP?, $\overline{\mathsf{AP}}$?, VP?, $\overline{\mathsf{VP}}$?, $\mathsf{V_s}$? and $\overline{\mathsf{V}}_\mathsf{s}$?}, and the set of output actions is defined accordingly. AP? and $\overline{\mathsf{AP}}$? are the start and end actions that the pacemaker generates when it paces the atrium (and similarly for VP? and $\overline{\mathsf{VP}}$? in the ventricle). Cells that are not stimulated by the pacemaker are stimulated by the voltage of their neighbours.

3.3 The SA Node

In this section we present a model for the SA node, which is known to control the normal rhythm of the heart. The SA node is the impulse-generating tissue located in the right atrium of the heart.

The heart rate is composed of two main periodic components: *respiratory sinus arrhythmia* (RSA) and *Mayer wave* (MW). The RSA oscillation is located between $0.15 - 0.4$ Hz (HF band), while the MW oscillation is at approximately 0.1 Hz between $0.04 - 0.15$ Hz (LF band). The heart rate is measured by monitoring the electrocardiogram (ECG) of the heart. An ECG is a signal recorded from the surface of the human chest. Typically, an ECG signal describes a cardiac cycle, which has three main waves, P, QRS and T. The P wave denotes the atrial depolarisation. The QRS wave reflects the rapid depolarisation of the right and left ventricles. The T wave denotes the repolarisation of the ventricle. The RR-interval is the time between successive R-peaks of the QRS wave, and is the inverse of this time interval that determines the instantaneous heart rate. Analysis of variations in the instantaneous heart rate time series using the beat-to-beat RR-intervals is known as HRV analysis, which has been shown to provide an assessment of cardiovascular diseases.

The RR time series can be generated by first constructing the power spectrum $S(f)$ as a sum of two Gaussian distributions for the LF band and HF band

$$S(f) = \frac{\sigma_1^2}{\sqrt{2\pi c_1^2}} \exp\left(\frac{(f - f_1)^2}{2\pi c_1^2}\right) + \frac{\sigma_2^2}{\sqrt{2\pi c_2^2}} \exp\left(\frac{(f - f_2)^2}{2\pi c_2^2}\right),$$

with means f_1, f_2, standard deviations c_1, c_2 and power σ_1^2, σ_2^2. Then the spectrum is mapped into the time domain by inverse Fourier transform. More details on the construction of the RR time series are given in [14].

In the sequel, we use the RR time series to create the firing times for the events $\mathsf{V_s}$? and $\overline{\mathsf{V}}_\mathsf{s}$? corresponding to the SA node cell (see Fig. 2). We define RR time series as a

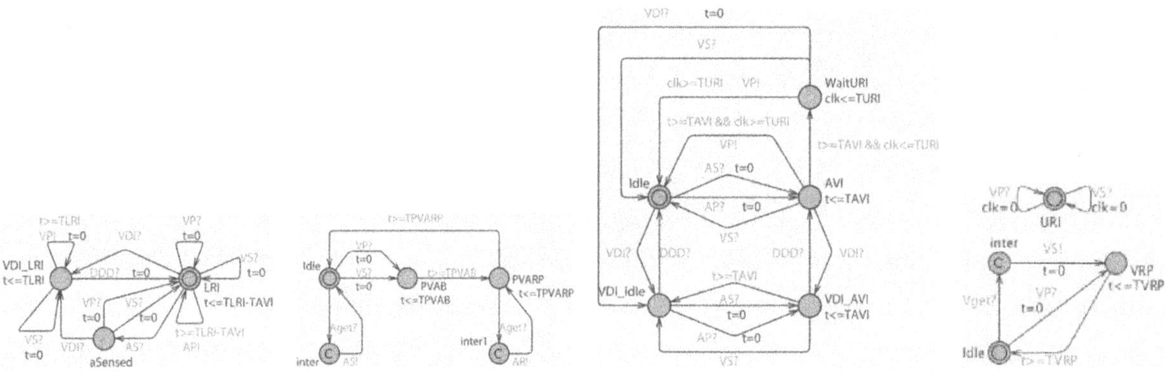

(a) LRI component (b) PVARP component (c) AVI component (d) URI and VRP components

Figure 3: LRI, PVARP, AVI, URI and VRP components [6].

sequence $\{r_i\}_{i\in\mathbb{N}}$. Intuitively, each r_i denotes the period of two consecutive SA node stimulations, i.e., it marks the beginning and end of the stimulus. If the time interval r_i is small the SA node is not stimulated, i.e., the $\overline{\mathsf{V}}_s$? happens before $v \geq V_T$. Otherwise, the SA node is stimulated and the stimulus is propagated through the heart.

4. PACEMAKER MODEL

The basic pacemaker model in [6] consists of five timed automata (TA) components: the lower rate interval (LRI) component, the atrio-ventricular interval (AVI) component, the upper rate interval (URI) component, the post ventricular atrial refractory period (PVARP) component and the ventricular refractory period (VRP) component. The LRI component (see Fig. 3(a)) has the function to keep the heart rate above a given minimum value. The AVI component (see Fig. 3(c)) has the purpose to maintain the synchronisation between the atrial and the ventricular events. An event is when the pacemaker senses or generates an action. The AVI component also defines the longest interval between an atrial event and a ventricular event. The PVARP component (see Fig. 3(b)) notifies all other components that an atrial event has occurred. Notice that there is no AR signal as we are not using the advanced algorithms given in [6]. The URI component (see Fig. 3(d) top) sets a lower bound on the times between consecutive ventricular events. The VRP component (see Fig. 3(d) bottom) filters noise and early events that may cause undesired behaviour.

There are four actions in the pacemaker model that are used to communicate with the heart model. The input actions Aget and Vget will notify the pacemaker when there is an action potential from the atrium or from the ventricle, respectively. The output actions AP and VP are responsible for pacing the atrium and the ventricle, respectively. Notice that in a real pacemaker device the input will be a signal. The pacemaker will have a voltage threshold that will be used to decide whether the signal yields an Aget or a Vget action. It is important to remark that all transitions from the pacemaker model that are not labelled with an input or output action are assumed to be labelled with the internal action τ. The locations that have transitions labelled with τ, as well as the locations labelled with **C**, do not allow the time to elapse.

5. QUANTITATIVE VERIFICATION

In this section, we show how quantitative, probabilistic verification can be performed for a given heart model and the pacemaker model which exhibit real-time, hybrid and stochastic features.

5.1 Generation of Abnormal Heart Behaviour

The heart exhibits abnormal behaviours due to many different reasons and in this paper we consider the malfunction of the SA node. Modelling this aspect plays a crucial role in the verification of pacemakers, as the pacemaker's function is to correct the heart behaviour for such scenarios. We now describe our approach. For the disease of SA nodes, we consider a model based on Markov chains. In order to verify the pacemaker we generate different behaviours, including Bradycardia and Tachycardia, as well as a normal rhythm, and allow switching between them. To generate a behaviour where, for instance, the SA node goes from Bradycardia to Tachycardia, we construct a Markov chain with three states (modes) as shown in Fig. 4, labelled as Normal (**N**), Tachycardia (**T**) and Bradycardia (**B**). An important observation is that the probability to switch between states can be learnt from patient data [2].

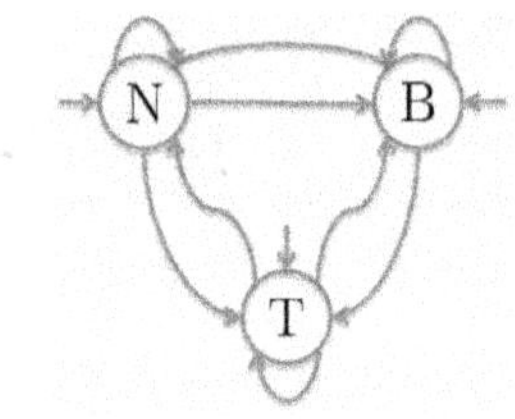

Figure 4: Markov chain.

5.2 Property Specification

The composition of the heart and the pacemaker models gives rise to a timed sequence which records the voltage of certain cells of the heart, for which we need to specify a property that checks whether the sequence corresponds to a normal heart behaviour. Intuitively, we define as a "normal path" any path for which *there are between 60 and 100 heart beats (ventricular events) in any interval of window time of 60 seconds.*

5.3 Approximate Model Checking

Technically speaking, the composition of the heart model introduced in Sect. 3 and the pacemaker model yields a stochastic hybrid system. Automated verification of such systems presents significant research challenges. We exploit techniques from *approximate model checking* (AMC) to deliver verification results with high confidence. AMC is an approach to verify quantitative properties of stochastic systems by simulating the system for finitely many runs, and then analysing the drawn samples (simulation trajectories) to obtain statistical evidence for the satisfaction or violation of the specification. Note that it is closely related to the *statistical model checking* technique [11], which relies on hypothesis testing.

Let T be the time bound, h be the simulation step, and $k = \frac{T}{h}$ be the path length. To apply AMC, one selects two parameters: $0 < \varepsilon < 1$ and $0 < \delta < 1$. Intuitively, ε is the error bound while $1-\delta$ is the confidence level. We randomly draw $N = \log(\frac{2}{\delta})/2\varepsilon^2$ paths of length k. For each path, we check whether it is "normal". Suppose there are N' normal paths. By the Chernoff bound, one can conclude that the probability $Prob\left[\left|\frac{N'}{N} - p\right| \leq \varepsilon\right] \geq 1-\delta$, where p is what we want to estimate, i.e., the probability of all "normal paths" of length k. Intuitively this means that, with a very high probability (i.e., $1 - \delta$), the ratio $\frac{N'}{N}$ that we compute is ε-close to the real probability p. We refer the readers to [10] for further details.

6. EXPERIMENTAL RESULTS

We implement both the heart model and the pacemaker model in Simulink. We run the experiments on a 2.83GHz 4 Core(TM)2 Quad CPU with 3.7GB of memory. All model files can be accessed via `http://www.veriware.org`.

Fig. 5(a) shows the Simulink implementation of the SA node. The cell is implemented by means of three Simulink blocks: *Event generator*, *Hybrid set* and *Subsystem*. The *Event generator* block is responsible to generate the input events to the cell. The *Hybrid set* implements the cell hybrid automaton model described in Sec. 3.1. The *Subsystem* block performs the integration procedure to compute the voltage level of the cell. Fig. 5(b) shows a network of six cells. Each cell block is composed from the three sub-blocks shown in Fig. 5(a) and connected to other cells through delay and gain components.

Basic Safety Analysis. In the first set of experiments we induce Bradycardia from the SA node and verify that the pacemaker corrects the faulty behaviour by restoring a normal heart beat. In Fig. 6(a) we depict two signals. The first one (in blue) denotes the action potential generated by the SA node which is running in Bradycardia mode. More precisely, we have three beats in six seconds, which is approximately 30 beats per minute. The number of heart beats is thus too slow and needs the intervention of the pacemaker. The second signal (in red) denotes the action potential from one of the cells of the His bundle situated in the ventricle. This is the signal which is captured and paced by the pacemaker. Note that the pacemaker increases the number of beats per minute by first delivering a beat to the ventricle after approximately one second.

Probabilistic Analysis. The second experiment depicts the relation between the probability to generate Bradycardia and the number of pacemaker beats to the ventricle. We range probability from 0.05 to 0.95. The results are presented in Fig. 6(b). We run 40 experiments, each representing 8 minutes of the heart beat. As expected, by increasing the probability, the pacemaker delivers more beats to the ventricle.

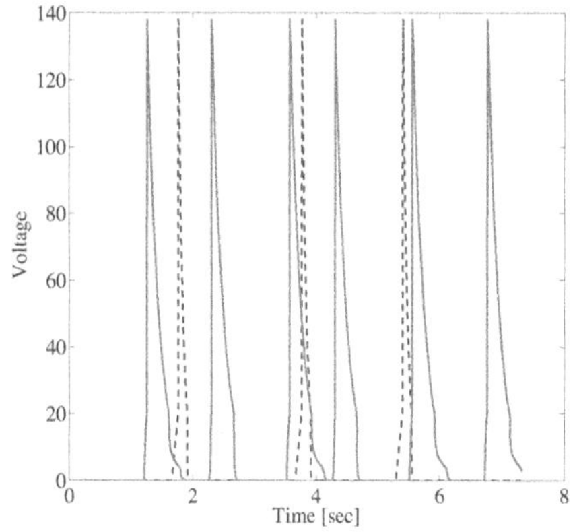

(a) Pacemaker beats

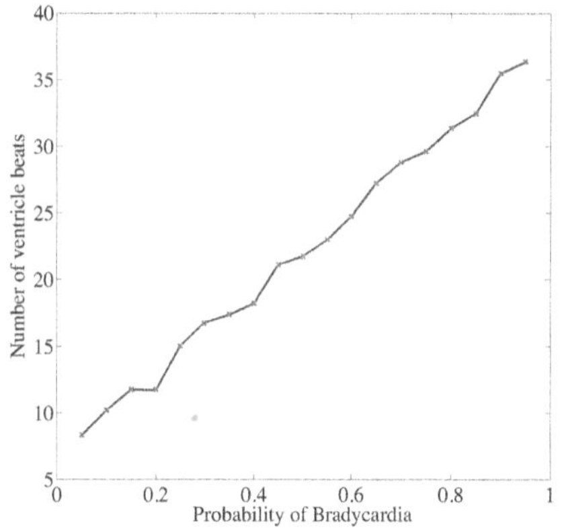

(b) Bradycardia experiment

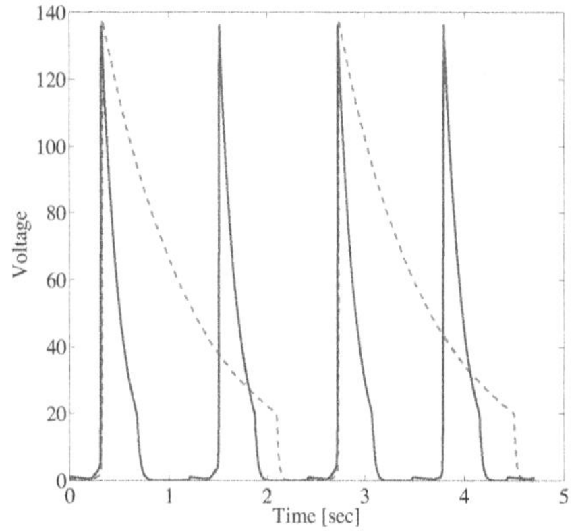

(c) AV node block

Figure 6: Experiments.

AV node block. In Fig. 6(c) we depict the case when the ERP value of the AV node is long enough, so that it filters out the signal from the SA node. Since a cell cannot be stimulated during its ERP phase, increasing the ERP value of the AV node results in filtering some of the signal that comes from the atrium. In this case, the SA node signal (in blue) is being blocked by a high ERP value of the AV node

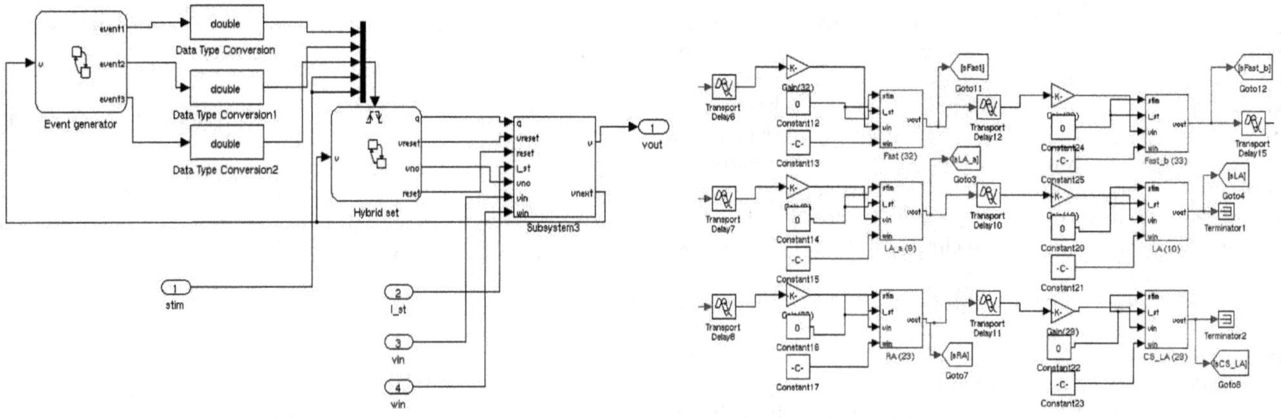

(a) Cell block

(b) Cell connection

Figure 5: Simulink models

(signal in red). The factor is 2 : 1 (two beats in the atrium result in one beat in the ventricle). A long ERP value for the AV node induces Bradycardia in the ventricle.

7. CONCLUSION

In this paper we have proposed a model for the electrical conduction system of the heart and performed quantitative verification of pacemaker models composed with the conduction system model. We worked directly with action potential signals that the pacemaker sensor inputs from specific cardiac cells. We have implemented our heart model in Simulink and evaluated it via approximate model checking using the timed automata pacemaker models of Jiang *et al*, with appropriate extensions.

There are several interesting directions for future work. For instance, we plan to explore the parameter synthesis problem of the pacemakers. Moreover, considering a failure model for the pacemaker seems to be a promising direction.

Acknowledgements

This work is supported by the ERC Advanced Grant VERI-WARE and Oxford Martin School Institute for the Future of Computing.

8. REFERENCES

[1] T. Chen, M. Diciolla, M. Kwiatkowska, and A. Mereacre. Quantitative Verification of Implantable Cardiac Pacemakers. *RTSS*, pp. 263-272. IEEE, 2012.

[2] G. Clifford, S. Nemati, and R. Sameni. An Artificial Vector Model for Generating Abnormal Electrocardiographic Rhythms. *Physiological Measurements*, 31(5):595–609, May 2010.

[3] A. Gomes and M. Oliveira. Formal Specification of a Cardiac Pacing System. *FM '09*, pp. 692–707, 2009.

[4] Z. Jiang, M. Pajic, A. Connolly, S. Dixit, and R. Mangharam. Real-Time Heart model for implantable cardiac device validation and verification. *ECRTS*, pp. 239–248. IEEE, 2010.

[5] Z. Jiang, M. Pajic, and R. Mangharam. Cyber-Physical Modeling of Implantable Cardiac Medical Devices. *Proceedings of the IEEE*, 100(1):122–137, 2012.

[6] Z. Jiang, M. Pajic, S. Moarref, R. Alur, and R. Mangharam. Modeling and Verification of a Dual Chamber Implantable Pacemaker. *TACAS*, pp. 188–203, 2012.

[7] M. Kwiatkowska, G. Norman and D. Parker. PRISM 4.0: Verification of Probabilistic Real-time Systems *CAV'11*, pp. 585–591, 2011.

[8] B. Kim, A. Ayoub, O. Sokolsky, I. Lee, P. L. Jones, Y. Zhang, and R. P. Jetley. Safety-assured development of the gpca infusion pump software. *EMSOFT'11*, pp. 155–164. ACM, 2011.

[9] K. G. Larsen, P. Pettersson, and W. Yi. UPPAAL in a Nutshell, 1997.

[10] R. Lassaigne and S. Peyronnet. Probabilistic Verification and Approximation. *Ann. Pure Appl. Logic*, 152(1-3):122–131, 2008.

[11] A. Legay, B. Delahaye, and S. Bensalem. Statistical Model Checking: An Overview. *RV'10*, LNCS 6418, pp. 122–135. Springer, 2010.

[12] A. T. Luu, M. C. Zheng, and Q. T. Tho. Modeling and Verification of Safety Critical Systems: A Case Study on Pacemaker. *SSIRI*, pp. 23–32. IEEE, 2010.

[13] H. Macedo, P. Larsen, and J. Fitzgerald. Incremental development of a distributed real-time model of a cardiac pacing system using vdm. *FM'8*, LNCS 5014, pp. 181–197. 2008.

[14] P. McSharry, G. Clifford, L. Tarassenko, and L. Smith. A Dynamical Model for Generating Synthetic Electrocardiogram Signals. *Biomedical Engineering, IEEE Transactions on*, 50(3):289 –294, 2003.

[15] D. Méry and N. K. Singh. Pacemaker's Functional Behaviors in Event-B. Rapport de recherche, 2009.

[16] S. Sankaranarayanan and G. E. Fainekos. Simulating Insulin Infusion Pump Risks by In-Silico Modeling of the Insulin-Glucose Regulatory System. *CMSB*, pp. 322–341, 2012.

[17] P. Ye, E. Entcheva, R. Grosu, and S. A. Smolka. Efficient Modeling of Excitable Cells Using Hybrid Automata. *CMSB*, 2005.

A Partially Observable Hybrid System Model for Bipedal Locomotion for Adapting to Terrain Variations

Koushil Sreenath
University of Pennsylvania
3300 Walnut St.
Philadelphia, PA 19104
koushils@seas.upenn.edu

Connie R Hill Jr.
Lockheed Martin Advanced Technology Laboratories
3 Executive Campus
Cherry Hill, NJ 08002
connie.r.hill.jr@lmco.com

Vijay Kumar
University of Pennsylvania
3300 Walnut St.
Philadelphia, PA 19104
kumar@seas.upenn.edu

ABSTRACT

We propose a methodology of applying PoMDPs at a sufficiently high abstraction of a high-dimensional continuous-time partially observable hybrid system. In particular, we develop a two-layer hybrid controller, where the higher-level PoMDP-based hybrid controller learns the boundaries between various modes and appropriately switches between them. The modes partition the state-space and represent a closed-loop hybrid system with a lower-level hybrid controller. We apply this methodology onto the problem of bipedal walking on varying terrain, where the gradient change in the terrain is only partially observable (due to poor and noisy sensors.) We develop three lower-level hybrid controllers that result in robust walking on level ground, up and down ramps. The higher-level PoMDP-based hybrid controller then learns the boundary between these controllers and is used to perform appropriate controller switching. With only a coarse, discrete estimate of walking speed, the controller enables traversing terrain both with long sustained constant slopes, and with rapid changes in slope. Simulation results are presented on a 26-dimensional planar bipedal robot model that incorporates contact forces and friction.

Categories and Subject Descriptors

H.4 [**Information Systems Applications**]: Miscellaneous

General Terms

Theory

1. INTRODUCTION

Partially Observable Markov Decision Processes (PoMDPs) are good for solving problems with partial observability of

HSCC'13, April 8–11, 2013, Philadelphia, Pennsylvania, USA.

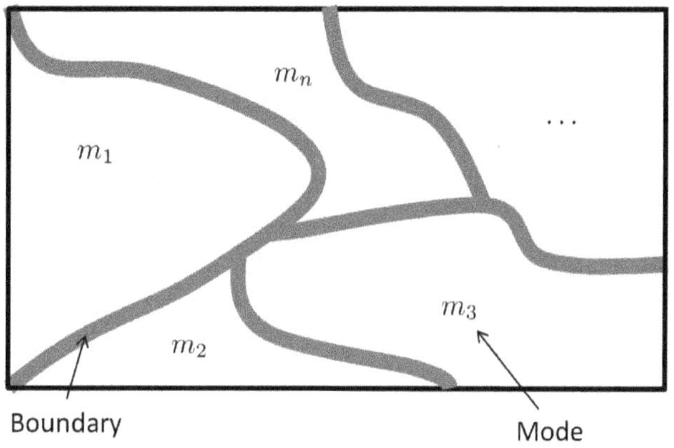

Figure 1: An abstract illustration of a two-layer hybrid controller with different modes separated by boundaries. Each mode is a closed-loop hybrid system with a lower-level hybrid controller, comprised of an inner-loop continuous-time controller and a discrete-time event-based outer-loop. Additionally, a higher-level hybrid controller learns the boundaries between the modes, by solving the postulated PoMDP problem, and then achieves switching between them.

systems with uncertainty in the states [11, 4, 1], however they are plagued by the problem of state explosion. Here we apply PoMDPs at a sufficiently high abstraction of a high-dimensional continuous-time partially observable hybrid system. In particular, we develop a two-layer hybrid controller, where the higher-level PoMDP-based hybrid controller learns the boundaries between various modes and appropriately switches between them. The modes partition the state-space and represent a closed-loop hybrid system with a lower-level hybrid controller. Similar ideas exist for simpler systems [5, 6].

Essentially, our method proposes a two-layer hybrid controller hierarchy, with multiple discrete modes that partition the state space; see Figure 1. Each mode is a closed-loop hybrid system with a *lower-level* hybrid controller, comprised of an inner-loop continuous-time controller and a discrete-time event-based outer-loop. Additionally, an *higher-level* hybrid controller learns the boundaries between the modes, by solving the postulated PoMDP problem, and then achieves switching between them. We have a nested hybrid system and the proposed methodology can be applied at any level.

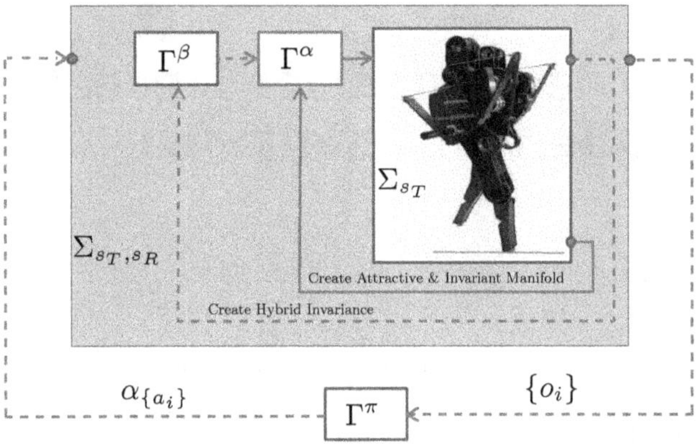

Figure 2: Feedback diagram illustrating the PoMDP-based hybrid controller structure. Continuous lines represent signals in continuous time; dashed lines represent signals in discrete time. The controllers Γ^α and Γ^β create an attractive and invariant hybrid zero dynamics [15]. Σ_{s_T}, and Σ_{s_T,s_R} are open-loop, and closed-loop partially observable hybrid systems respectively. The controller Γ^π is an hybrid controller that implements a PoMDP policy to switch the walking gait (specified by the α parameters) based on the observations that are being received from the hybrid system (marked by the blue box).

We chose to apply it at the highest level because, as we will see, this is where our state is not fully observable.

We particularize the method for studying the problem of designing feedback controllers for structured terrain where the gradient changes discretely, but is not accurately perceivable either by a tactile sensor through forward kinematics or a vision system. For instance, indoor corridors and sidewalks with gradual changes in slope are examples of such terrain. Walking on varying slopes has been considered in [14, 3], however these assume that the slope can be perfectly known or can be inferred at each step. Bipedal walking on rough terrain has been primarily addressed by developing controllers that are robust to bounded variation in step heights of the terrain [7, 2]. These controllers are typically hybrid, with a continuous-time inner-loop controller and a discrete-time event-based outer-loop controller designed to provide robustness to changes in the terrain.

However, a single controller can not be easily designed to achieve walking at multiple ground slopes, but rather multiple locally stable controllers that are robust about different specific ground slopes are sought, along with some higher-level sequential decision making to switch between the controllers. Partially observable Markov decision process (PoMDP), [1], provide a natural model for sequential decision making under uncertainty, and is particularly applicable to situations where a robot cannot reliably identify the state of the underlying environment. Although switching controllers have been demonstrated in the literature for rough terrain walking [3, 9], these approaches do not consider any uncertainty of when to switch.

We apply the two-layer PoMDP hybrid controller framework to a special case of three simple discrete modes for walking on level ground, walking on a up ramp and a down ramp of specified gradient. For each of these modes, there is a continuous-time controller for tracking, and a discrete-time event-based controller to provide additional robustness to adapt to small changes in terrain gradient. The higher-level PoMDP-based hybrid controller learns the boundaries between the discrete modes based on the stochastic distributions specified as part of the PoMDP problem, and provides a way to switch between the low-level hybrid controllers. Note that we are not considering the quasi-static case [12], but we are using a PoMDP formulation to solve a problem in which the dynamics and real-time control are critical for operation of the system.

Employing only three specific walking gaits, our proposed method is able to successfully walk on level ground and sustained up and down slopes of 10° while also working in situations where the slope changes every step. Moreover, this method can be easily extended to have additional modes (rather than the 3 illustrated here) corresponding to walking on different ground slopes.

Note that most work on rough terrain walking assumes the availability of precise ground height, and/or walking step speed. However, for physical legged robots, computing small changes in ground height at impact using forward kinematics is error-prone due to sensor and calibrations errors that add up as the kinematic chain is traversed. Moreover, relying on visual or laser sensors for estimating small changes in ground slope is also not feasible. Furthermore, accurate forward velocity, usually obtained by fusing odometry and visual information, is not available for legged robots due to the error-prone odometry. The proposed method sidesteps these issues by only requiring a coarse estimate of the walking speed (speed roughly equal to, greater than, or less than nominal speed) for estimating the terrain gradient.

The rest of the paper is organized as follows. Section 2 briefly presents the hybrid model for walking and the nominal controller used for walking. Section 3 develops and formulates a PoMDP problem as a Kronecker product of separate PoMDPs for modeling the terrain and the robot. Section 4 presents simulation results by evaluating the postulated PoMDP-based hybrid controller on a high-dimensional planar bipedal model that captures unilateral ground constraints and stick-slip friction, to demonstrate walking on stochastically varying terrain. Finally, Section 5 presents concluding remarks.

2. DYNAMICAL MODEL AND NOMINAL CONTROL DESIGN FOR WALKING

We have seen several methods for handling rough terrain in the previous section, with researchers considering simple systems for demonstrating their method. We will illustrate the PoMDP-based control design on a dynamical model of a real-life complex experimental system called MABEL, a planar bipedal robot at The University of Michigan, which has an underactuated, compliant transmission [15].

A dynamical model for walking can be developed by modeling the single support phase, when one foot is assumed to be pinned to the ground, and the subsequent double support phase that occurs when the swing foot makes contact with the ground. The single support phase is modeled as the continuous-time dynamics of a pinned, planar, kinematic chain with revolute joints and rigid links, while the double support phase is modeled as an instantaneous impact. The hybrid model for walking is then as follows,

$$\Sigma_{s_T}: \begin{cases} \dot{x} = f(x) + g(x)u, & x^- \notin \mathcal{G}_{s_T} \\ x^+ = \Delta(x^-), & x^- \in \mathcal{G}_{s_T}. \end{cases} \tag{1}$$

where $x = [q \;\; \dot{q}] \in \mathcal{X}$ is the state of the system, f, g are vector fields that captures the continuous-time dynamics, Δ represents the impact map that maps the pre-impact state x^- to the post-impact state x^+, and $\mathcal{G}_{s_T} = \{x \in \mathcal{X} \mid H_{s_T}(x) = 0\}$ is the guard surface representing contact of the swing foot with the ground, with $s_T \in \mathcal{S}_T := \{L, U, D\}$ representing if the terrain is level, a up ramp, or a down ramp. In the absence of sensors for precisely measuring small variations in the terrain either through forward kinematics, or through vision, the terrain state is only partially observable, making this a *partially observable hybrid model.*

To coordinate and synchronize the links of a robot to achieve the objectives of walking, such as keeping the torso upright, advancing the swing foot to take a step, maintaining a foot clearance for a specific slope of the terrain, etc., we use *virtual* constraints [15]. Virtual constraints are holonomic constraints on the robot's configuration variables that are asymptotically imposed through feedback control. One virtual constraint is typically chosen per actuator in the form of an output, such that when a feedback controller drives the output to zero, the constraint is enforced. Virtual constraints can be written as,

$$y = h(q, \alpha) = H_0 q - h_d(\theta(q), \alpha), \tag{2}$$

where, H_0 is a selection matrix for choosing the variables to be controlled, $h_d(\theta, \alpha)$ are the desired trajectories, expressed as Bézier polynomials, parametrized by θ, a monotonous function of the joint variables, and α the Bézier polynomial coefficients. The α-parameters are chosen through a constrained nonlinear optimization to obtain a periodic gait for walking. For the purpose of this paper, we design a set of periodic walking gaits, represented by the parameters α_{s_R}, with $s_R \in \mathcal{S}_R := \{WL, WU, WD\}$, designed for walking on level ground, and up, down ramps of a 10° gradient respectively.

Next we present the lower-level hybrid controller, comprised of a continuous-time inner-loop controller and a discrete-time outer-loop controller, that locally, exponentially stabilizes the periodic walking gaits. Assuming the output (2) has vector relative degree two, the continuous-time controller Γ^α is given by the input-output linearizing controller,

$$\Gamma^\alpha: \quad u = -L_g L_f h(q, \alpha)^{-1} \left(L_f^2 h + K_p y + K_d \dot{y} \right). \tag{3}$$

This controller locally drives the output to zero exponentially and creates the invariant manifold $\mathcal{Z} = \{x \in \mathcal{X} \mid y = 0, L_f y = 0\}$.

The discrete-time event-based controller Γ^β serves to perform step-to-step parameter updates by adding an additional term to the virtual constraint defined in (2) to get a new output,

$$\Gamma^\beta: \quad y_b = h(q, \alpha, \beta) = H_0 q - h_d(\theta(q), \alpha) - h_b(\theta(q), \beta). \tag{4}$$

The event-based control is used to select β so as to perform corrections to the virtual constraint to obtain hybrid invariance, i.e., $x^- \in \mathcal{Z} \implies x^+ \in \mathcal{Z}$ [15, Sec. IV-B], at each step, thereby smoothly handling transients.

Finally, we present a coarse speed sensor that indicates if the walking step speed is greater than, equal to, or lesser than the nominal speed, by providing an observation $o_R \in \mathcal{O}_R := \{Sp, S0, Sm\}$, according to the following rule,

$$o_R = \begin{cases} Sp, & v > (1+\delta)\, v_{s_R} \\ S0, & (1-\delta)\, v_{s_R} < v < (1+\delta)\, v_{s_R} \\ Sm, & v < (1-\delta) v_{s_R}, \end{cases} \tag{5}$$

where v_{s_R} corresponds to the steady-state speed of the walking gait α_{s_R}, for $s_R \in \mathcal{S}_R$, and δ a small positive number used as a threshold for discretization by the sensor.

$A_T = \{\}$	$s' = L$	U	D
$s = L$	0.8	0.1	0.1
U	0.35	0.6	0.05
D	0.35	0.05	0.6

(a)

$A_T = \{\}$	$O_T = \{\}$
$s = L$	1
U	1
D	1

(b)

Table 1: (a) State transition, $T_T(s, \{\}, s')$, and (b) observation functions, $Z_T(\{\}, s, \{\})$, for the terrain PoMDP, $\mathcal{P}_T$.

3. POMDP FORMULATION FOR WALKING ON SLOPES

The hybrid system Σ_{s_T} in (1) under the action of the lower-level hybrid controller (3), (4) is another (closed-loop) hybrid system Σ_{s_T, s_R}, represented by the blue box in Figure 2. In this section, we will develop a higher-level hybrid controller for this system. Note that we are able to analyze the lower-level controller developed in the previous section, and the higher-level hybrid controller developed here separately. This is primarily due to the loose coupling between the two that occurs through the incoming discrete speed observations (5), which in turn indirectly depend on the continuous states and interactions with the ground.

Here, we will assume that we neither have access to precise foot height at contact, nor accurate estimates of step speed (due to poor and noisy sensors.) Rather, we will assume we have a coarse estimate of step speed that only indicates if the speed is roughly equal to, greater than, or lesser than nominal speed as in (5). It must be noted that just looking at the speed for a single step is not sufficient to immediately determine the state of terrain so as to switch to an appropriate walking controller. Instead, we will need to pose the problem as a PoMDP, that will enable us to learn the boundaries between when each of the three walking controllers, developed in Section 2, for walking on level ground, up and down ramps, are active and when to switch to the appropriate one, based on a sequence of observations of the coarse speed estimate.

A partially observable Markov decision process (PoMDP) is a Markov decision process that does not make an assumptions that the states are directly observable [11, 4, 1]. A PoMDP can be characterized by a 6-tuple $\mathcal{P} = (\mathcal{S}, \mathcal{A}, \mathcal{O}, T, Z, R)$, where $\mathcal{S} = \{s_1, s_2, ..., s_n\}$ is a set of states that the system can be in, $\mathcal{A} = \{a_1, a_2, ..., a_m\}$ is a set of actions that can be applied to affect the system, $\mathcal{O} = \{o_1, o_2, ..., o_l\}$ is a set of observations that describe the perception of the system which partially reflects the current state of the system, $T: \mathcal{S} \times \mathcal{A} \to \Pi(\mathcal{S})$ is the state transition function that maps each state-action pair into a probability distribution over the state space, such that $T(s, a, s') = Pr(S_{t+1} =$

$a = WL$	$s' = WL$	WU	WD	$a = WU$	$s' = WL$	WU	WD	$a = WD$	$s' = WL$	WU	WD
$s = WL$	1	0	0	$s = WL$	0	1	0	$s = WL$	0	0	1
WU	1	0	0	WU	0	1	0	WU	0	0	1
WD	1	0	0	WD	0	1	0	WD	0	0	1

(a)

$a = WL$	$o = Sp$	$S0$	Sm	$a = WL$	$o = Sp$	$S0$	Sm	$a = WL$	$o = Sp$	$S0$	Sm
$s = WL$	0.1	0.8	0.1	$s = WL$	0.04	0.06	0.9	$s = WL$	0.9	0.06	0.04
WU	0.7	0.2	0.1	WU	0.1	0.7	0.2	WU	0.95	0.03	0.02
WD	0.1	0.2	0.7	WD	0.02	0.03	0.95	WD	0.2	0.7	0.1

(b)

Table 2: (a) State transition, $T_R(s, a, s')$, and (b) observation functions, $Z_R(a, s, o)$, for the robot PoMDP, $\mathcal{P}_R$.

$s' \mid S_t = s, A_t = a)$, $Z : \mathcal{A} \times \mathcal{S} \rightarrow \Pi(\mathcal{O})$ is an observation function that maps the current state and the previous action to a distribution over the observations, such that $Z(a, s, o) = Pr(O_t = o \mid S_t = s, A_{t-1} = a)$, and finally $R : \mathcal{S} \times \mathcal{A} \rightarrow \mathbb{R}$ is the immediate reward function.

At any given point in time the system is in state s_t, which is partially observable through observation o_t, with probability $b_t = Pr(s_t \mid o_t, a_t, o_{t-1}, a_{t-1}, \cdots, o_0, a_0)$. This *belief* distribution represents the entire history of the interaction of the system. The goal of a PoMDP is to learn an optimal policy describing action selection, that maximizes the expected discounted cumulative reward, i.e.,

$$V_t^*(s) = \max_{a \in \mathcal{A}} \left[R(s, a) + \gamma \sum_{s' \in \mathcal{S}} T(s, a, s') V_{t-1}(s') \right], \quad (6)$$

where γ is the discount, with V being the value function. The *policy* is defined as a mapping from the belief state to action state.

We model two PoMDPs, one for the terrain, $\mathcal{P}_T = (\mathcal{S}_T, \mathcal{A}_T, \mathcal{O}_T, T_T, Z_T, R_T)$, and the other for the robot, $\mathcal{P}_R = (\mathcal{S}_R, \mathcal{A}_R, \mathcal{O}_R, T_R, Z_R, R_R)$, and then form a composite PoMDP by taking the Kronecker product of the two, $\mathcal{P} = \mathcal{P}_T \otimes \mathcal{P}_R$. We model the terrain PoMDP as follows, $\mathcal{S}_T = \{L, U, D\}$ as defined earlier represents the state of the terrain to be either level, a up ramp, or a down ramp, and with no actions or observations, i.e., $\mathcal{A}_T = \{\}, \mathcal{O}_T = \{\}$. Further, we choose the state transition and observation functions as given in Table 1. These probabilities were chosen arbitrarily and do not reflect true values of terrain distributions occurring in the real world. However, it would be pretty straightforward to collect real-world data and compute these probabilities.

Next, we model the robot PoMDP as follows, $\mathcal{S}_R = \{WL, WU, WD\}$ as defined earlier represents the state of the robot controller to be either a level walking controller, a up ramp walking controller, or a down ramp walking controller. In a similar way, we choose the actions $\mathcal{A}_R = \{WL, WU, WD\}$ to represent the transition to these controllers (Note that we have abused notation and have used the same symbols for states and actions. The state represents the current controller being used on the robot, while the action represents what controller should be used next.) The observations, $\mathcal{O}_R = \{Sp, S0, Sm\}$ as defined earlier, indicate if the speed is greater than (Sp = speed plus), equal to, or lesser than (Sm = speed minus) nominal walking speed. These observations provide an indirect estimate of the change in the terrain and do not require precise accurate sensors for measuring the speed of walking. Further, we choose the state transition and observation functions as given in Table 2. To keep the reward structure simple, we choose unit reward for both PoMDPs, i.e., $R_T \equiv 1, R_R \equiv 1$.

Remark 1. *Decomposing our PoMDP into the terrain and robot PoMDPs makes presentation of the state transition and observation functions in Tables 1, 2 compact. The state transition and observation functions for the composite PoMDP are then easily obtained from these. Moreover, in the future, we can easily model a visual sensor as part of the observations in $\mathcal{P}_T$, to provide a direct estimate of the terrain. This enables, easily fusing observations from a visual sensor providing direct estimates of the terrain in $\mathcal{P}_T$, and a discrete speed sensor providing an indirect estimate of the terrain in $\mathcal{P}_R$.*

The composite PoMDP thus formed is then solved using a solver such as ZMDP, Perseus, Cassandras, or Pegasus [13, 1, 8]. Figure 4 illustrates the resulting policy graph. The obtained policy is essentially a finite state controller that describes what actions to take based on the observations received. The received observations are used to update the belief, and when the belief of a particular state is high enough, an appropriate action is issued. Thus, the outer-loop control Γ^π is specified as,

$$\Gamma^\pi : \quad \alpha = \alpha_{\pi(o_t)}, \quad (7)$$

where o_t is the observation received at discrete time instant t, and $a_t = \pi(o_t)$ is the action specified by the policy, and the α-parameters corresponding to the action are picked.

4. RESULTS

We now evaluate the proposed controller, comprising of the continuous-time controller with the discrete-time outer-loop controller as described in Section 2, and the PoMDP hybrid controller developed in Section 3, on the complex model of MABEL developed in [10]. This model is 26-dimensional and captures unilateral constraints through a compliant ground and stick-slip friction model. This captures the robot-ground interaction more realistically and is important for testing controllers especially on terrain that is varying, where foot slippage is bound to happen and is an important source of failure. This is in contrast to papers that do not model stick-slip friction while evaluating proposed rough terrain controllers.

As an initial test of validity of the control design, we consider a deterministically generated terrain that consists of segments of a level ground, a up ramp at 10°, a level ground and a down ramp at 10°; see Figure 3a. This sample terrain will test if the PoMDP policy correctly switches to an appropriate walking controller.

In this simulation, the robot is able to take over 65 steps to traverse the entire terrain, with the PoMDP correctly switching the controller at appropriate instants in time. The PoMDP policy is able to transition from $WL \to WU$, $WU \to WL$, $WL \to WD$, $WD \to WL$. Step speeds, observations received, which are computed using the rule in (5), and the actions generated by the PoMDP policy-based controller Γ^{π} are also illustrated. Specifically, the coarse discrete speed measurement indicating if the speed is approximately equal to, greater than or lesser than the nominal speed, is used to build a belief in the state of the partially observable hybrid model and take an appropriate action based on the belief, enabling switching to an appropriate controller, thereby adapting to and traversing the terrain.

Next, we test the controller on a suite of stochastically generated terrains based on the terrain state transition in Table 1. It must be noted that the PoMDP only captures the transitions of the terrain. Neither the length of terrain segments nor the specific slope to choose for the up/down ramps are specified by the PoMDP. Instead of picking just one particular value, we randomly choose a segment length from a uniform distribution in $\{2m, 4m, 6m\}$ and a slope for up ramp and another slope for down ramp from $\{5°, 7.5°, 10°\}$. With the segment length and the slopes for the up / down ramps fixed, we generate a terrain based on the terrain transition function. This forms one sample path to test the controller on, and we repeat the above procedure to obtain additional sample paths. The proposed PoMDP controller is able to successfully complete walking on over 85% of the generated terrain, some of which involve over 100 steps of walking down at a 10° incline. The most common reason for failure is an inability of the controller to accommodate changes in terrain slope of over 10°, for instance when the up ramp changes to a down ramp involving a change in slope of 20°. A small percentage of failures can be attributed to the foot slipping while going up a ramp, which by itself does not cause the robot to fall, but instead causes the swing foot to advance rapidly to catch the robot from falling, causing subsequent steps to be taken quickly, which increases the speed of walking. This results in the PoMDP estimating a wrong terrain gradient thereby not transitioning to, the correct controller.

Finally, we test the controller on a stochastically generated terrain, where the slope of the terrain changes every 1m with the terrain state transition as in Table 1. The slopes of ramp for each segment are selected from an uniform distribution of slopes in $\{2°, 4°, 6°, 8°, 10°\}$. Simulation results shown in Figures 3b, illustrate that the robot is able to traverse terrain not only when the slope is constant for a long duration, but also when the slope changes rapidly.

5. CONCLUSION

A two-layer hybrid controller methodology for applying PoMDPs at a sufficiently high abstraction of a high-dimensional continuous-time partially observable hybrid system has been presented. This has been particularized for application onto planar bipedal walking on stochastically varying slopes, where neither an assumption of requiring an estimate of step height, nor requiring an accurate estimate of the step speed at each step is made. With only a coarse estimate of step speed, the presented PoMDP-based hybrid controller is able to appropriately switch between three hybrid controllers designed for robust walking on level ground, up and down ramps of 10°. Simulations carried out on a model of MABEL illustrate the controller being able not only to traverse terrain with long and sustained constant slopes, but also ones where the slope changes more rapidly. The gradient of the ramps addressed in this paper can easily be increased. The method itself can easily be extended to multiple ground slopes and the controllers can be stitched together by appropriate switching. The only assumption is that a periodic walking gait at that slope is available. As future work, we are exploring extending this framework to address rough and difficult terrain in addition to the presented gentle ramps.

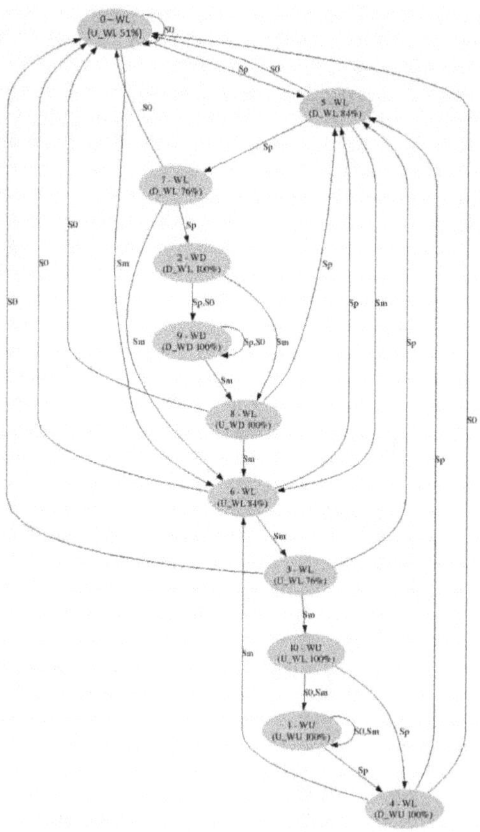

Figure 4: The policy graph obtained by solving the postulated PoMDP. The obtained policy has 11 nodes, with each node specifying the action to be taken, the state of the composite PoMDP, $\mathcal{P}$, that has the highest belief, along with the belief expressed as a percentage. An appropriate node transition occurs when the specified observation is received from the system. From the policy it can be seen that to switch from a level walking gait to a walking up (or down) controller requires three consecutive observations of S_p, speed up (or S_m, speed down), as seen in the node transitions $0 \to 5 \to 7 \to 2$ (or $0 \to 6 \to 3 \to 10$). This sequence is optimally required for the belief to build-up, and is specific to the chosen transition, observation, and reward functions.

6. REFERENCES

[1] A. R. Cassandra, "Exact and Approximate Algorithms

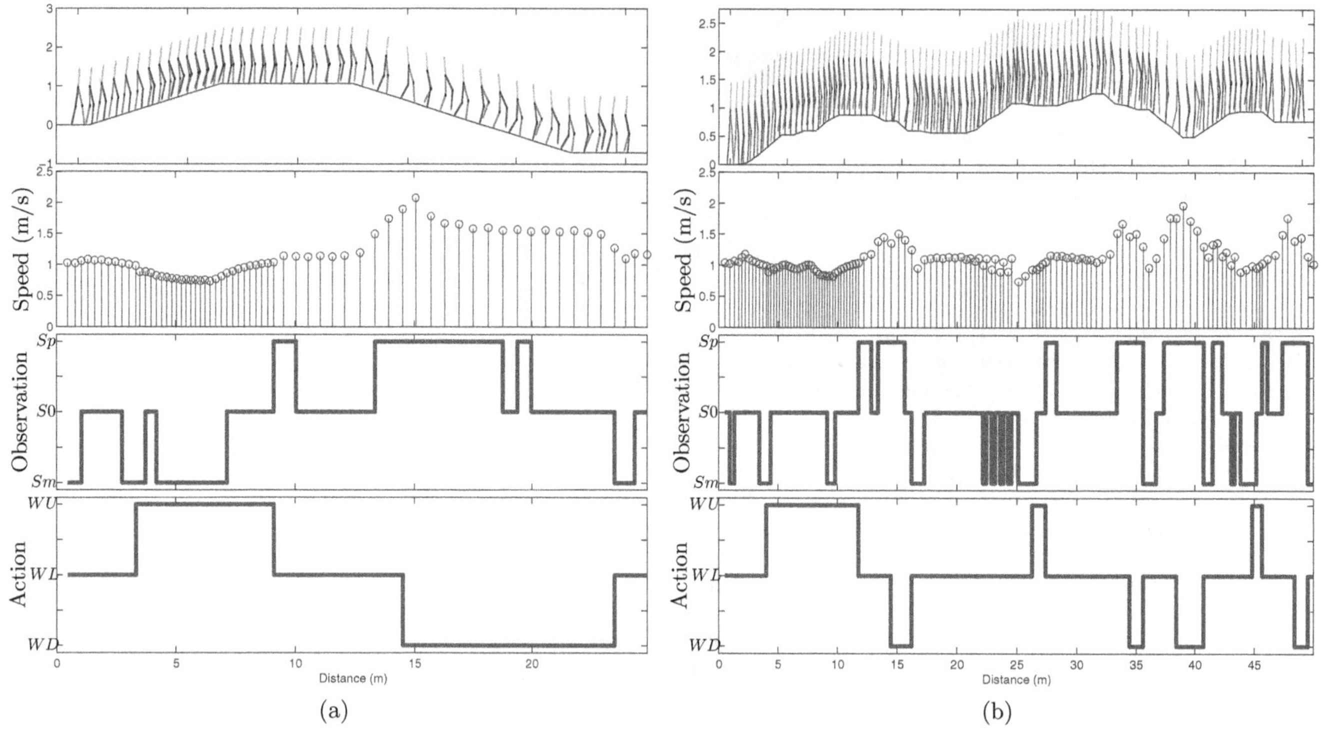

Figure 3: Stick figure plot, step speeds, observations received, and the actions taken by the PoMDP-based hybrid controller Γ^{π} for (a) 65 step walk over a sample deterministic terrain with the up/down ramp slopes set at 10°, (b) a 125 step walk over a sample stochastically generated terrain. The slopes for the up and down ramp gradients are uniformly selected from $\{2°, 4°, 6°, 8°, 10°\}$.

for Partially Observable Markov Decision Processes," Ph.D., Brown University, 1998.

[2] H. Dai and R. Tedrake, "Optimizing Robust Limit Cycles for Legged Locomotion on Unknown Terrain," in *International Conference on Robotics and Automation*, 2012.

[3] T. Erez, W. D. Smart, and S. Louis, "Bipedal Walking on Rough Terrain Using Manifold Control," in *Intelligent Robots and Systems*, 2007, pp. 1539 - 1544.

[4] A. Foka and P. Trahanias, "Real-time hierarchical POMDPs for autonomous robot navigation," *Robotics and Autonomous Systems*, no. January, 2007.

[5] G. Grudic, V. Kumar, and L. Ungar, "Using policy gradient reinforcement learning on autonomous robot controllers," *Robotics: Science and Systems*, pp. 406–411, 2003.

[6] M. Lahijanian, S. B. Andersson, and C. Belta, "A probabilistic approach for control of a stochastic system from LTL specifications," in *IEEE Conference on Decision and Control*, December 2009, pp. 2236–2241.

[7] I. R. Manchester, U. Mettin, F. Iida, and R. Tedrake, "Stable Dynamic Walking over Rough Terrain Theory and Experiment," in *Proceedings of the International Symposium on Robotics Research*, 2009, pp. 1–16.

[8] A. Y. Ng, "PEGASUS : A policy search method for large MDPs and POMDPs," in *Proceedings of the Sixteenth Conference on Uncertainty in Artificial Intelligence*, 2000.

[9] H.-W. Park, A. Ramezani, and J. W. Grizzle, "A Finite-state Machine for Accommodating Unexpected Large Ground Height Variations in Bipedal Robot Walking," *IEEE Transactions on Robotics*, to appear, 2013.

[10] H.-W. Park, K. Sreenath, J. W. Hurst, and J. W. Grizzle, "Identification of a Bipedal Robot with a Compliant Drivetrain," *IEEE Control Systems Magazine (CSM)*, vol. 31, no. 2, pp. 63–88, April 2011.

[11] J. Pineau and S. Thrun, "Hierarchical POMDP decomposition for a conversational robot," *ICML Workshop on Hierarchy and Memory*, 2001.

[12] G. Singh and A. Ng, "Quadruped robot obstacle negotiation via reinforcement learning," in *IEEE International Conference on Robotics and Automation*, 2006, pp. 3003–3010.

[13] T. Smith and R. Simmons, "Heuristic search value iteration for POMDPs," *Proceedings of the 20th conference on Uncertainty*, 2004.

[14] M. Spong and G. Bhatia, "Further results on control of the compass gait biped," *Proceedings 2003 IEEE/RSJ International Conference on Intelligent Robots and Systems*, vol. 2, pp. 1933–1938, 2003.

[15] K. Sreenath, H.-W. Park, I. Poulakakis, and J. W. Grizzle, "A Compliant Hybrid Zero Dynamics Controller for Stable, Efficient and Fast Bipedal Walking on MABEL," *The International Journal of Robotics Research*, vol. 30, no. 9, pp. 1170–1193, Sep. 2010.

State Estimation for Polyhedral Hybrid Systems and Applications to the Godunov Scheme

Jérôme Thai
Electrical Engineering and Computer Science
University of California, Berkeley
jerome.thai@berkeley.edu

Alexandre M. Bayen
Systems Engineering
University of California, Berkeley
bayen@berkeley.edu

ABSTRACT

In this article, the problem of estimating the state of a discretized hyperbolic scalar partial differential equation is studied. The discretization of the *Lighthill-Whitham-Richards* equation with a triangular flux function using the Godunov scheme is shown to lead to a hybrid linear system or *Switched Linear Systems* (SLS) with a number of modes exponential in the size of the discretized model. Some geometric properties of the partition of the space into polyhedra (in which a mode is active) are exploited to find heuristics to reduce the number of modes to a representative set. This motivates a new approach inspired from a well established technique for hybrid system estimation, namely the *interactive multiple model* (IMM). qThe performance of this new variant of the IMM is compared to the *extended Kalman filter* and the *ensemble Kalman filter* using the *Mobile Millennium* data set.

Categories and Subject Descriptors

G.1.8 [Numerical Analysis]: Partial Differential Equations

Keywords

Godunov Scheme, Hybrid Systems, Interactive Multiple Model

1. INTRODUCTION

Partial Differential Equations (PDEs) have been extensively used in the scientific literature because they provide a concise mathematical model to capture essential properties of a wide variety of phenomena such as fluid flow, heat, and electrodynamics. They are often used in traffic as density based traffic models. The *Lighthill-Whitham-Richards* (LWR) PDE [20, 23] and its discretization using the Godunov scheme [14, 17, 25] is a well established model for traffic dynamics, also known as the *Cell Transmission Model* (CTM) [6, 7] in the transportation literature. State of the art traffic estimation techniques for this model include the application of the *extended Kalman filter* (EKF) to the LWR PDE by Schreiter et al. [24], and to non-scalar traffic model by Papageorgiou [22]. The application of the EKF to the LWR PDE model is problematic due to the non-differentiability of its discretization, a problem which has been partially adressed in [3] and [27]. The *ensemble Kalman filter* (EnKF) has also been applied to a velocity-based model in [29], in order to circumvent the difficulties of non-differentiability of numerical solutions to these PDEs such as the one presented in this article.

For a triangular flux function, the Godunov scheme applied to the LWR model can be proven to lead to a *piecewise affine* (PWA) hybrid system. The resulting switching-mode dynamical system combines continuous dynamics in the form of linear discretized dynamical systems, and discrete dynamics modeled by a finite automaton for the *transitions* between the modes. The problem of estimation of hybrid systems has been widely studied in past work [15, 28, 16, 1]. In particular, such techniques have been successfully used for aircraft tracking in [13] in which Bar-Shalom's *interacting multiple model* (IMM) algorithm was used [2]. Similar hybrid estimation algorithms and their applications are described in [21, 26, 12]. In the context of traffic estimation, computational difficulties arise when the IMM algorithm is applied to the highway model due to the exponential number of modes. A priori, each cell of the discretized model can be in seven different modes, which lead to 7^n modes, where n is the dimension of the state thus creating serious computational challenges in the estimation problem. One possible way to address this is with the *mixture Kalman filter* algorithm [5] which handles this complexity by randomly sampling in the space of modes.

In the present work, we approach this difficulty differently, leading to the following contributions:

- The article uses an explicit formulation of the Godunov scheme to express the evolution of the discretized LWR PDE as a PWA hybrid system

- It develops methods for reducing the number of modes to a tractable number using geometric results derived from the PWA representation of the system, and using *k-means clustering*.

- It demonstrates the performance of the method on field experiment data, and compares it to a traditional EnKF approach on the discretized LWR PDE.

Thus, the contribution of the article can be viewed as the construction of a method in which the estimation of the state of a discretized LWR PDE can be done in two steps; (1) using Kalman filtering on each of the modes written explicitly in a linear manner (which does not constitute a linearization of the dynamics like the EKF), (2) using a new framework for the mode estimation step, so it becomes tractable, which enables the use of various techniques, in particular the IMM.

The rest of the article is organized as follow: Section 2 presents the mathematical model used in the rest of the article. Section 3 unravels the PWA expression of the Godunov scheme. Section 4 presents feasible heuristics inspired from IMM using the PWA character of the Godunov scheme, and Section 4.5 presents some numerical results.

2. A HYBRID MODEL FOR THE GODUNOV SCHEME

This section briefly summarizes results derived earlier on hybrid formulations of the Godunov scheme [27]. We use the first order Godunov scheme for the discretization of the *Lighthill-Whitham-Richards* equation with a triangular flux function, and proves that the resulting nonlinear dynamical system can be decomposed in a *piecewise affine* manner. Using this explicit representation, the resulting hybrid system is shown to have an exponential number of modes.

2.1 Discretization of the LWR equation

We consider the *Lighthill-Whitham-Richards* hydrodynamic model used earlier in [27] as the distributed parameter system model of interest:

$$\frac{\partial \rho(x,t)}{\partial t} + \frac{\partial Q(\rho(x,t))}{\partial x} = 0 \tag{2.1}$$

where $Q(\rho(x,t))$ is the flux function historically introduced by Greenshields [10]. For the rest of the work, we will instantiate the results on a triangular flux function introduced by Daganzo [7]:

$$Q_T(\rho) = \begin{cases} v_f \rho & \text{if } \rho \leq \rho_c \\ -\omega_f(\rho - \rho_{\text{jam}}) & \text{if } \rho > \rho_c \end{cases} \tag{2.2}$$

where $\omega_f = v_f \rho_c / (\rho_{\text{jam}} - \rho_c)$ is the backwards propagation wave speed.

A seminal numerical method to solve the above equations is given by the Godunov scheme, which is based on exact solutions to Riemann problems [8, 9]. This leads to the construction of a nonlinear discrete time dynamical system. The Godunov discretization scheme is applied on the LWR PDE, where the discrete time step Δt is indexed by t, and the discrete space step Δx is indexed by i:

$$\rho_i^{t+1} = \rho_i^t - \frac{\Delta t}{\Delta x}\left(G(\rho_i^t, \rho_{i+1}^t) - G(\rho_{i-1}^t, \rho_i^t)\right) \tag{2.3}$$

In order to ensure numerical stability, the time and space steps are coupled by the CFL condition [17]: $c_{\max}\frac{\Delta t}{\Delta x} \leq 1$ where $c_{\max}$ denotes the maximal characteristic speed.

For a triangular flux function, the Godunov flux can be expressed as the minimum of the *sending flow* $S(\rho)$ from the upstream cell and the *receiving flow* $R(\rho)$ from the downstream cell (2.4,2.5,2.6) through a boundary connecting two cells of a homogeneous road (i.e. the upstream and downstream cells have the same characteristics). For the triangular flux function:

$$G(\rho_1, \rho_2) = \min(S(\rho_1), R(\rho_2)) \tag{2.4}$$

$$S(\rho) = \begin{cases} Q(\rho) = v_f \rho & \text{if } \rho \leq \rho_c \\ q_c & \text{if } \rho > \rho_c \end{cases} \tag{2.5}$$

$$R(\rho) = \begin{cases} q_c & \text{if } \rho \leq \rho_c \\ Q(\rho) = -\omega_f(\rho - \rho_{\text{jam}}) & \text{if } \rho > \rho_c \end{cases} \tag{2.6}$$

where ρ_1 is the density of the cell upstream and ρ_2 is the density of the cell downstream.

The explicit values taken by $G(\rho_1, \rho_2)$ for a partition of the space in different regions of the space (ρ_1, ρ_2) **W**, **L**, and **D** are shown in Figure 2.1 and defined by equations 2.8. In the triangular case:

$$G_T(\rho_1, \rho_2) = \begin{cases} R(\rho_2) = -\omega_f(\rho_2 - \rho_{\text{jam}}) & \text{if } (\rho_1, \rho_2) \in \mathbf{W} \\ q_c & \text{if } (\rho_1, \rho_2) \in \mathbf{L} \\ S(\rho_1) = v_f \rho_1 & \text{if } (\rho_1, \rho_2) \in \mathbf{D} \end{cases} \tag{2.7}$$

$$\begin{aligned} \mathbf{W} &= \{(\rho_1, \rho_2) \mid \rho_2 > h(\rho_1)\ ,\ \rho_2 > \rho_c\} \\ \mathbf{L} &= \{(\rho_1, \rho_2) \mid \rho_1 > \rho_c\ ,\ \rho_2 \leq \rho_c\} \\ \mathbf{D} &= \{(\rho_1, \rho_2) \mid \rho_2 \leq h(\rho_1)\ ,\ \rho_1 \leq \rho_c\} \end{aligned} \tag{2.8}$$

The boundary between the **W** and **D** regions follows the $(\rho_1, \rho_2) = (\rho_1, h(\rho_1))$ trajectory for $\rho_1 \leq \rho_c$, with:[1]

$$h(\rho_1) = \bar{R}^{-1}(\bar{S}(\rho_1)) = -\frac{v_f}{\omega_f}\rho_1 + \rho_{\text{jam}} \tag{2.9}$$

where $\bar{S}$ and $\bar{R}$ respectively denote the restrictions of the sending and receiving flows S and R to the sub-regions $[0, \rho_c)$ and $(\rho_c, \rho_{\text{jam}}]$ respectively, which also correspond to the left and right parts of the flux function (w.r.t. ρ_c), as shown in Figure 2.1.

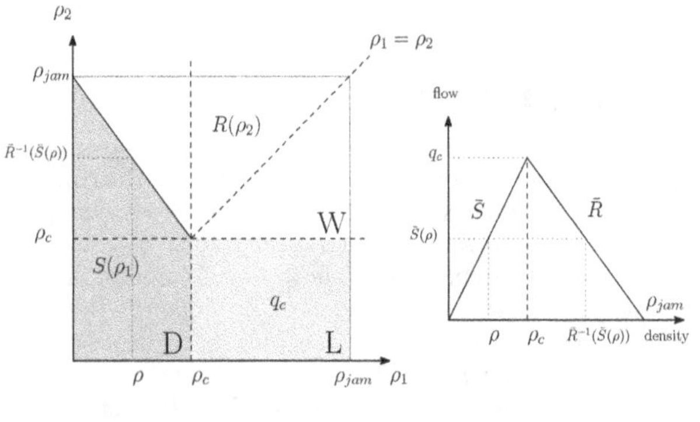

Figure 2.1: Values of $G(\rho_1, \rho_2)$ in the space (ρ_1, ρ_2).

2.2 Piecewise affine model

In the Godunov scheme (2.3), the update of the density ρ_i^{t+1} at cell i depends on the triplet $(\rho_{i-1}^t, \rho_i^t, \rho_{i+1}^t)$. With $\frac{\Delta t}{\Delta x} = \alpha$, the Godunov scheme reads:

[1]Here, we suppose that $\bar{R}$ is a strictly monotonic function on $(\rho_c, \rho_j]$, hence invertible, and $\bar{R}^{-1}$ denotes its inverse, which is the case for the triangular flux function.

$$\rho_i^{t+1} = \rho_i^t - \alpha\left(G(\rho_i^t, \rho_{i+1}^t) - G(\rho_{i-1}^t, \rho_i^t)\right) \tag{2.10}$$

The expression (2.10) is non-linear but can be formulated as a *piecewise affine* hybrid system. Depending on whether (ρ_{i-1}^t, ρ_i^t) and (ρ_i^t, ρ_{i+1}^t) are in $\mathbf{W}$, $\mathbf{L}$, or $\mathbf{D}$, there are nine possible combinations at cell i, which can be reduced to seven "modes" since the pairs (ρ_{i-1}^t, ρ_i^t) and (ρ_i^t, ρ_{i+1}^t) have ρ_i^t in common. Let us denote by $f(\rho_{i-1}^t, \rho_i^t, \rho_{i+1}^t)$ the vector function for the possible values of ρ_i^{t+1}. Table 2.1 lists these seven possibilities, which can be derived from Figure 2.1.

Mode	(ρ_{i-1}^t, ρ_i^t)	(ρ_i^t, ρ_{i+1}^t)	$f(\rho_{i-1}^t, \rho_i^t, \rho_{i+1}^t)$
1	W	W	$(1-\alpha\omega_f)\rho_i^t + \alpha\omega_f\rho_{i+1}^t$
2	W	L	$(1-\alpha\omega_f)\rho_i^t + \alpha\omega_f\rho_c$
3	L	W	$\rho_i^t + \alpha\omega_f\rho_{i+1}^t - \alpha\omega_f\rho_c$
4	L	D	$(1-\alpha v_f)\rho_i^t + \alpha v_f\rho_c$
5	D	W	$\alpha v_f\rho_{i-1}^t + \rho_i^t + \alpha\omega_f\rho_{i+1}^t - \alpha\omega_f\rho_{\text{jam}}$
6	D	L	$\alpha v_f\rho_{i-1}^t + \rho_i^t - \alpha v_f\rho_c$
7	D	D	$\alpha v_f\rho_{i-1}^t + (1-\alpha v_f)\rho_i^t$

Table 2.1: 7×1-dimensional column vector $f(\rho_{i-1}^t, \rho_i^t, \rho_{i+1}^t)$ of the different values of ρ_i^{t+1} depending on the mode.

For example, for the first mode, (ρ_{i-1}^t, ρ_i^t) and (ρ_i^t, ρ_{i+1}^t) are both in $\mathbf{W}$ (see Figure 2.1), thus $G(\rho_{i-1}^t, \rho_i^t) = R(\rho_i^t)$ and $G(\rho_i^t, \rho_{i+1}^t) = R(\rho_{i+1}^t)$, and then $\rho_i^{t+1} = \rho_i^t - \alpha(R(\rho_{i+1}^t) - R(\rho_i^t))$. By extending this result to an entire link with discrete state space indexed by $i = 1, \cdots, n$, where n is the number of space steps, we have an exhaustive description of the space of "modes" along the link.

We define J, the Jacobian matrix of f with respect to $(\rho_{i-1}^t, \rho_i^t, \rho_{i+1}^t)$ in each of the seven modes above (which are all linear):

$$J = \left(\frac{\partial f_j}{\partial \rho_k}\right)_{j=1,\cdots,7,k=i-1,i,i+1} \tag{2.11}$$

Where f_j is the j-th entry of the vector function f defined in Table 2.1. It is useful to make the Jacobian matrix J explicit with respect to $(\rho_{i-1}^t, \rho_i^t, \rho_{i+1}^t)$, and the constant term w:

$$J = \begin{pmatrix} 0 & 1-\alpha\omega_f & \alpha\omega_f \\ 0 & 1-\alpha\omega_f & 0 \\ 0 & 1 & \alpha\omega_f \\ 0 & 1-\alpha v_f & 0 \\ \alpha v_f & 1 & \alpha\omega_f \\ \alpha v_f & 1 & 0 \\ \alpha v_f & 1-\alpha v_f & 0 \end{pmatrix}, \quad w = \begin{pmatrix} 0 \\ \alpha\omega_f\rho_c \\ -\alpha\omega_f\rho_c \\ \alpha v_f\rho_c \\ -\alpha\omega_f\rho_{\text{jam}} \\ -\alpha v_f\rho_c \\ 0 \end{pmatrix} \tag{2.12}$$

Since f is a *linear function* of $(\rho_{i-1}^t, \rho_i^t, \rho_{i+1}^t)$ as shown in Table 2.1, we note that J is constant. More notably, the seven possible values of ρ_i^{t+1} in Table 2.1 can be rewritten in vector form as follows:

$$f(\rho_{i-1}^t, \rho_i^t, \rho_{i+1}^t) = J\begin{pmatrix} \rho_{i-1}^t \\ \rho_i^t \\ \rho_{i+1}^t \end{pmatrix} + w \tag{2.13}$$

In the next section, we will show that the decomposition in "modes" as shown in Table 2.1 leads to a piecewise affine formulation of the Godunov scheme in the case of the triangular flux function.

Let us consider a link with discrete time step indexed by $t \geq 0$ and discrete space step indexed by $i = 1, \cdots, n$, and let us denote $\boldsymbol{\rho}^t = (\rho_0^t, \rho_1^t, \cdots, \rho_n^t, \rho_{n+1}^t)$ an $n+2$ dimensional vector which describes the state of the link at time t in the space $\mathcal{S} = [0, \rho_{\text{jam}}]^{n+2}$, where ρ_i^t is the density at time t and cell i. Note that the ghost cells 0 and $n+1$ are included in the state of the link.[2]

Definition of the space of modes: Let us denote by $\mathcal{M}_n$ the space of modes of the system ($\mathcal{M}_n \subset \{1, \cdots, 7\}^n$, see Table 2.1). For $\boldsymbol{m} \in \mathcal{M}_n$, $\boldsymbol{m}$ is a vector of dimension n for which the i-th entry $m_i \in \{1, \cdots, 7\}$ is the mode at cell i. Equivalently, each element of $\mathcal{M}_n$ can be described as a sequence of regions in which the pair (ρ_i, ρ_{i+1}) is, for $i = 0, \cdots, n$. Hence, we define the equivalent space of modes $\tilde{\mathcal{M}}_n \subset \{w, l, d\}^{n+1}$, and for $\boldsymbol{s} \in \tilde{\mathcal{M}}_n$, $\boldsymbol{s}$ is a vector of dimension $n+1$ for which the i-th entry s_i is equal to l if $(\rho_i, \rho_{i+1}) \in \mathbf{L}$, for $i = 0, \cdots, n$, and similar definitions for w and d. As will be seen later, this second definition gives a description of the *partition of the space $\mathcal{S}$ into different polyhedra $\mathbf{P}_{\boldsymbol{m}}$ in which the mode is $\boldsymbol{m}$*. See Figure 2.2 for an illustration.

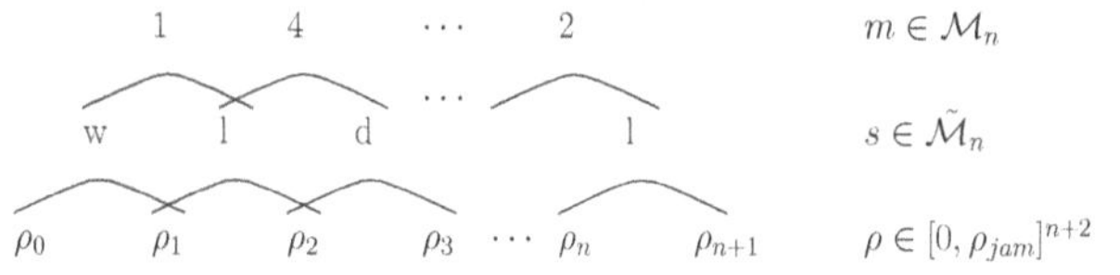

Figure 2.2: Illustration of the vectors $\boldsymbol{\rho} \in [0, \rho_{\text{jam}}]^{n+2}$, $\boldsymbol{s} \in \tilde{\mathcal{M}}_n \subset \{w, l, d\}^{n+1}$, and $\boldsymbol{m} \in \mathcal{M}_n \subset \{1, \cdots, 7\}^n$ for n cells.

The n-dimensional vector $\boldsymbol{m} \in \mathcal{M}_n$ describes the mode of the link at any time, as defined in the previous section. At each time step, the state of the link is updated through the following nonlinear dynamical system:

$$\boldsymbol{\rho}^{t+1} = F_{\boldsymbol{m}}[\boldsymbol{\rho}^t] \quad \text{if} \quad \boldsymbol{\rho}^t \in \mathbf{P}_{\boldsymbol{m}} \tag{2.14}$$

with $F_{\boldsymbol{m}}[\cdot]$ an $n+2$ dimensional function vector, and $\boldsymbol{m}$ the mode at time t. With u^t and d^t the boundary conditions upstream and downstream at time step t, the i-th entry $\rho_i^{t+1} = F_{\boldsymbol{m},i}[\boldsymbol{\rho}^t]$ is:

$$\rho_i^{t+1} = \begin{cases} f_{m_i}(\rho_{i-1}^t, \rho_i^t, \rho_{i+1}^t) & \text{for} \quad i = 1, \cdots, n \\ u^t & \text{for} \quad i = 0 \\ d^t & \text{for} \quad i = n+1 \end{cases} \tag{2.15}$$

where m_i denotes the i-th entry of $\boldsymbol{m} \in \mathcal{M}_n$, i.e. the mode of cell i at time step t, and $f_{m_i}(\rho_{i-1}^t, \rho_i^t, \rho_{i+1}^t)$ is the m_i-th entry of the vector-valued function f evaluated at $(\rho_{i-1}^t, \rho_i^t, \rho_{i+1}^t)$. We note that $\rho_0^{t+1} = u^t$ and $\rho_{n+1}^{t+1} = d^t$, which means that the ghost cells are the boundary conditions of the Godunov

[2] The values of ρ_0^t and ρ_{n+1}^t are given by the prescibed boundary conditions to be imposed on the in left and right side of the domain respectively. Note that these boundary values do not always affect the physical domain because of the nonlinear operator (2.7), which causes the boundary conditions to be implemented in the weak sense. For more details, see [29] and [25].

scheme. For a triangular flux function, with L_{m_i} the m_i-th line of J_T and w_{m_i} the m_i-th entry of w, the update operator of the dynamical system is:

$$\rho_i^{t+1} = \begin{cases} L_{m_i} \cdot \begin{pmatrix} \rho_{i-1}^t \\ \rho_i^t \\ \rho_{i+1}^t \end{pmatrix} + w_{m_i} & \text{for } i = 1, \cdots, n \\ u^t & \text{for } i = 0 \\ d^t & \text{for } i = n+1 \end{cases} \tag{2.16}$$

When $\boldsymbol{\rho}^t \in \mathbf{P}_{\boldsymbol{m}}$, the $(n+2) \times (n+2)$-dimensional state-transition matrix $A_{\boldsymbol{m}}$ is obtained by concatenating the 3×1 row vectors L_{m_i} along the diagonal. It is tridiagonal with diagonal elements $\{0, J_{m_1,2}, \cdots, J_{m_n,2}, 0\}$, lower diagonal elements $\{J_{m_1,1}, J_{m_2,1}, \cdots, J_{m_n,1}, 0\}$, and upper diagonal elements $\{0, J_{m_1,3}, J_{m_2,3}, \cdots, J_{m_n,3}\}$ where J is defined in equation (2.12). Equivalently:

$$A_{\boldsymbol{m}} = \begin{pmatrix} 0 & \cdots & 0 \\ L_{m_1} & & \\ & \ddots & \\ & & L_{m_n} \\ 0 & \cdots & 0 \end{pmatrix} \tag{2.17}$$

Let us denote $b_{\boldsymbol{m}}$ and c_t the two vectors of dimension $(n+2)$ with entries $\{0, w_{m_1}, \cdots, w_{m_n}, 0\}$ and $\{u^t, 0, \cdots, 0, d^t\}$ repectively, and $\mathbf{P}_{\boldsymbol{m}}$ the subset of space $\mathcal{S}$ where the mode is $\boldsymbol{m}$. The update operator of the dynamical system is *piece-wise affine*:

$$\boldsymbol{\rho}^{t+1} = A_{\boldsymbol{m}} \boldsymbol{\rho}^t + b_{\boldsymbol{m}} + c^t \quad \text{if} \quad \boldsymbol{\rho}^t \in \mathbf{P}_{\boldsymbol{m}} \tag{2.18}$$

We now provide a description of the partition of the space into the polyhedra $\mathbf{P}_{\boldsymbol{m}}$ in which the mode is $\boldsymbol{m}$. Note that in this formula, $A_{\boldsymbol{m}} \boldsymbol{\rho}^t$ represents the local (affine) discretization of the PDE, and c_t the boundary condition.

2.3 Number of modes

A priori, for n cells, the number of possible modes at any given time is equal to 7^n (cf. Table 2.1). Since two consecutive indices are constrained by the evolution of equation (2.10) as derived before, the number of modes for the entire link is less than $3 \cdot (2.5)^n$.

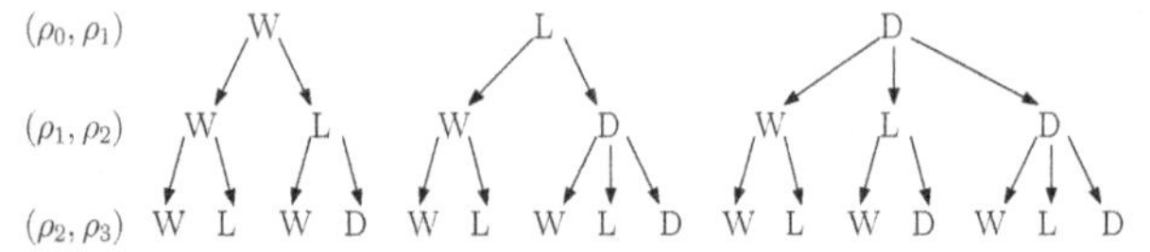

Figure 2.3: The sixteen possible modes for the first three pairs (ρ_0, ρ_1), (ρ_1, ρ_2), and (ρ_2, ρ_3).

Let n_k be the number of modes for a discretized model with k cells. Then we can recursively compute n_k with respect to k. Let us denote by w_k, l_k, and d_k the number of modes for which (ρ_k, ρ_{k+1}) is in $\mathbf{W}$, $\mathbf{L}$, and $\mathbf{D}$ respectively $(n_k = w_k + l_k + d_k)$. Then these equations can be derived:

$$\begin{array}{lllllll} w_0 & = & l_0 & = & d_0 & = & 1 \\ w_{k+1} & = & w_k & + & l_k & + & d_k \\ l_{k+1} & = & w_k & + & d_k & & \\ d_{k+1} & = & l_k & + & d_k & & \end{array} \quad \text{for } k \geq 0 \tag{2.19}$$

Using matrix notations and equation (2.19):

$$\begin{bmatrix} w_k \\ l_k \\ d_k \end{bmatrix} = A^k \times \begin{bmatrix} w_0 \\ l_0 \\ d_0 \end{bmatrix} \text{ where } A = \begin{bmatrix} 1 & 1 & 1 \\ 1 & 0 & 1 \\ 0 & 1 & 1 \end{bmatrix} \tag{2.20}$$

It is possible to compute A^k explicitly by diagonalizing the matrix A, to obtain an explicit expression for w_k, l_k, and d_k in the form of $a.\beta^k + b.\gamma^k + c.\delta^k$. However, this analytical expression is unwieldy, so we will just derive lower and upper bounds to n_k. It can be proved that $d_k \leq n_k/2$ for $k \geq 0$, then we can prove recursively that $3 \cdot 2^k \leq n_k \leq 3 \cdot (2.5)^k$.

number of cells	1	2	5	10	20
number of modes	7	16	182	10426	34206521
bound without analysis	7	49	16807	282475249	$8 \cdot 10^{16}$

Table 2.2: Number of modes for a homogeneous road.

3. HYBRID ESTIMATION ALGORITHMS

In this section, we first introduce a geometric framework for the description of the *polyhedral structure* of the hybrid system (2.18). We then develop an *interactive multiple model Kalman filter* (IMM KF) algorithm to the resulting *Polyhedral Piecewise Affine* (PPWA) hybrid system.

3.1 PPWA hybrid system

We now provide a description of the partition of the space into the polyhedra $\mathbf{P}_{\boldsymbol{m}}$ in which the mode is $\boldsymbol{m}$. Note that in this formula, $A_{\boldsymbol{m}} \boldsymbol{\rho}^t$ represents the local (affine) discretization of the PDE, and c_t the boundary condition.

For a discretization into n cells, we chose to describe the ensemble of modes $\tilde{\mathcal{M}}_n$ in sequences $\boldsymbol{s} \in \{w, l, d\}^{n+1}$ and define $\mathbf{P}_{\boldsymbol{s}}$ the corresponding polyhedron for each sequence. Let us define 3^{n+1} polyhedra $\mathbf{W}_i$, $\mathbf{L}_i$, and $\mathbf{D}_i$ for $i = 0, \cdots, n$ in the space $\mathcal{S}$ obtained by instantiating $h(\rho_1)$ with (2.9):

$$\begin{array}{ll} \mathbf{W}_i & = \{(\rho_i, \rho_{i+1}) \mid \rho_{i+1} + \frac{v_f}{w_f}\rho_i > \rho_{\text{jam}} \ , \ \rho_{i+1} > \rho_c\} \\ \mathbf{L}_i & = \{(\rho_i, \rho_{i+1}) \mid \rho_i > \rho_c \ , \ \rho_{i+1} \leq \rho_c\} \\ \mathbf{D}_i & = \{(\rho_i, \rho_{i+1}) \mid \rho_{i+1} + \frac{v_f}{w_f}\rho_i \leq \rho_{\text{jam}} \ , \ \rho_i \leq \rho_c\} \end{array} \tag{3.1}$$

The polyhedron $\mathbf{P}_{\boldsymbol{s}}$, in which the mode is $\boldsymbol{s} \in \tilde{\mathcal{M}}_n$, can be described as an intersection of $n+1$ polyhedra $\mathbf{Q}_i$:

$$\mathbf{P}_{\boldsymbol{s}} = \bigcap_{i=0}^{n} \mathbf{Q}_i \quad \text{with} \quad \mathbf{Q}_i = \begin{cases} \mathbf{W}_i & \text{if } s_i = w \\ \mathbf{L}_i & \text{if } s_i = l \\ \mathbf{D}_i & \text{if } s_i = d \end{cases} \tag{3.2}$$

Moreover, for two different modes $\boldsymbol{s}$ and $\boldsymbol{s}'$, and corresponding polyhedra $\mathbf{P}_{\boldsymbol{s}} = \bigcap_{i=0}^n \mathbf{Q}_i$ and $\mathbf{P}_{\boldsymbol{s}'} = \bigcap_{i=0}^n \mathbf{Q}'_i$, we can find an index i for which $\mathbf{Q}_i$ and $\mathbf{Q}'_i$ are disjoint. For instance, suppose without loss of generality that $\mathbf{Q}_i = \mathbf{W}_i$ and $\mathbf{Q}'_i = \mathbf{D}_i$, and we know that $\mathbf{W}_i$ and $\mathbf{D}_i$ are disjoint.

Then in this case, the hyperplan $\{\boldsymbol{\rho} \mid \rho_{i+1} + \frac{v_f}{w_f}\rho_i = \rho_{\text{jam}}\}$ is a seperating hyperplan between $\mathbf{P}_{\boldsymbol{s}}$ and $\mathbf{P}_{\boldsymbol{s}'}$. Hence, $\mathbf{P}_{\boldsymbol{s}}$ and $\mathbf{P}_{\boldsymbol{s}'}$ are disjoint and the family $\{\mathbf{P}_{\boldsymbol{s}}\}_{\boldsymbol{s}\in\mathcal{M}_n}$ is a partition of $\tilde{\mathcal{M}}_n$.

Each polyhedron $\mathbf{P}_{\boldsymbol{s}}$ can be defined as an intersection of a finite number of half-spaces. Such definition is called a *halfspace representation* or *H-representation* [11]. Let us introduce the following indicator functions:

$$\begin{aligned} \alpha_i(\boldsymbol{\rho}) &= 1_{\{\rho_{i+1}+\frac{v_f}{w_f}\rho_i>\rho_{\text{jam}}\}} && \text{for } i = 0,1,\cdots,n \\ \beta_i(\boldsymbol{\rho}) &= 1_{\{\rho_i>\rho_c\}} && \text{for } i = 0,1,\cdots,n+1 \end{aligned} \tag{3.3}$$

We define the corresponding half-spaces $\mathbf{H}_{\alpha_i}$ and $\mathbf{H}_{\beta_i}$. The complementary half-spaces $\mathcal{S}\backslash\mathbf{H}$ are denoted by $\mathbf{H}^c_{\alpha_i}$ and $\mathbf{H}^c_{\beta_i}$ and the corresponding indicator functions are $1 - \alpha_i(\boldsymbol{\rho})$ and $1 - \beta_i(\boldsymbol{\rho})$. Since we have:

$$\begin{aligned} \mathbf{W}_i &= \mathbf{H}_{\alpha_i} \cap \mathbf{H}_{\beta_{i+1}} \\ \mathbf{L}_i &= \mathbf{H}_{\beta_i} \cap \mathbf{H}^c_{\beta_{i+1}} \\ \mathbf{D}_i &= \mathbf{H}^c_{\alpha_i} \cap \mathbf{H}^c_{\beta_i} \end{aligned} \tag{3.4}$$

for $i = 0,1,\cdots,n$ the polyhedra defined in (3.1), their respective indicator functions are:

$$\begin{aligned} w_i(\boldsymbol{\rho}) &= \alpha_i(\boldsymbol{\rho})\beta_{i+1}(\boldsymbol{\rho}) \\ l_i(\boldsymbol{\rho}) &= \beta_i(\boldsymbol{\rho})(1-\beta_{i+1}(\boldsymbol{\rho})) \qquad \text{for } i = 0,1,\cdots,n \\ d_i(\boldsymbol{\rho}) &= (1-\alpha_i(\boldsymbol{\rho}))(1-\beta_i(\boldsymbol{\rho})) \end{aligned} \tag{3.5}$$

Hence, evaluating the indicator functions $\alpha_i(\boldsymbol{\rho})$ for $i = 0,\cdots,n$, and $\beta_i(\boldsymbol{\rho})$ for $i = 0,\cdots,n+1$ gives the mode $\boldsymbol{m}$ of state $\boldsymbol{\rho}$. Equations (3.2, 3.4) give an H-representation of $\mathbf{P}_{\boldsymbol{s}}$. With $\mathcal{H}_{\boldsymbol{s}}$ the set of half-spaces in this representation:

$$\mathbf{P}_{\boldsymbol{s}} = \bigcap_{\mathbf{H}\in\mathcal{H}_{\boldsymbol{s}}} \mathbf{H} \tag{3.6}$$

3.2 Kalman filter algorithm

In order to use the *Kalman filter* to estimate the state of the link given a sequence of noisy observations, we model the process by adding a white noise to the underlying dynamical system model. The "true" state at time $t+1$, namely $\boldsymbol{\rho}^{t+1}$, is then given by the update equation:

$$\boldsymbol{\rho}^{t+1} = A_{\boldsymbol{m}}\boldsymbol{\rho}^t + b_{\boldsymbol{m}} + c^t + \boldsymbol{\eta}^t \quad \text{if} \quad \boldsymbol{\rho}^t \in \mathbf{P}_{\boldsymbol{m}} \tag{3.7}$$

where $\boldsymbol{\eta}^t \sim N(0, Q^t)$ is the Gaussian zero-mean, white state noise with covariance Q^t. To apply the *control update* of the Kalman filter, it is then necessary to know the mode $\boldsymbol{m}$ of the state $\boldsymbol{\rho}^t$ (i.e. $\boldsymbol{m}$ such that $\boldsymbol{\rho}^t \in \mathbf{P}_{\boldsymbol{m}}$) by evaluating the indicator functions (3.5).

Additionally, the *observation model* for the link is given by:

$$\boldsymbol{z}^t = H^t\boldsymbol{\rho}^t + \boldsymbol{\chi}^t \tag{3.8}$$

where H^t is the $p_t \times n$-dimensional linear observation matrix which encodes the p^t observations (each one of them being at a discrete cell on the discretization domain) for which the density is observed during discrete time step t, and n is the number of cells along the link. The last term in equation (3.8) is the white, zero mean observation noise $\boldsymbol{\chi}^t \sim N(0, R^t)$ with covariance matrix R^t.

Let $\hat{\boldsymbol{\rho}}^{t:t}$ and $P^{t:t}$ be the *a posteriori* state estimate and error covariance matrix at time t. The *predicted* state estimate $\hat{\boldsymbol{\rho}}_j^{t+1:t}$ and covariance estimate $P_j^{t+1:t}$ of the *prediction step* in *mode* $\boldsymbol{m}_j$ are:[3]

$$\begin{aligned} \hat{\boldsymbol{\rho}}_j^{t+1:t} &= A_{\boldsymbol{m}_j}\hat{\boldsymbol{\rho}}^{t:t} + b_{\boldsymbol{m}_j} + c^t \\ P_j^{t+1:t} &= A_{\boldsymbol{m}_j}P^{t:t}(A_{\boldsymbol{m}_j})^T + Q^t \end{aligned} \tag{3.9}$$

The *measurement residual* $\boldsymbol{r}_j^{t+1}$, *residual covariance* S_j^{t+1}, *Kalman gain* K_j^{t+1}, *updated state estimate* $\hat{\boldsymbol{\rho}}_j^{t+1:t+1}$, and *updated estimate covariance* $P_j^{t+1:t+1}$ of the *update step* in *mode* j are:

$$\begin{aligned} \boldsymbol{r}_j^{t+1} &= \boldsymbol{z}^{t+1} - H^{t+1}\hat{\boldsymbol{\rho}}_j^{t+1:t} \\ S_j^{t+1} &= H^{t+1}P_j^{t+1:t}(H^{t+1})^T + R^{t+1} \\ K_j^{t+1} &= P_j^{t+1:t}(H^{t+1})^T(S_j^{t+1})^{-1} \\ \hat{\boldsymbol{\rho}}_j^{t+1:t+1} &= \hat{\boldsymbol{\rho}}_j^{t+1:t} + K_j^{t+1}\boldsymbol{r}_j^{t+1} \\ P_j^{t+1:t+1} &= (I - K_j^{t+1}H^{t+1})P_j^{t+1:t} \end{aligned} \tag{3.10}$$

In [18], a measure of the likelihood of the Kalman filter in mode j is given by the *mode likelihood function* Λ_j^{t+1}, where $\mathcal{N}(x; a, b)$ is the probability density function of the normal distribution with mean a and variance b:

$$\Lambda_j^{t+1} = \mathcal{N}(\boldsymbol{r}_j^{t+1};\ 0,\ S_j^{t+1}) \tag{3.11}$$

3.3 Interactive multiple model KF

Let us denote by $\boldsymbol{m}_j^t$ the event that the system is in the mode j at time t. We then assume that the model is a discrete-time stochastic linear hybrid system in which the mode evolution is governed by the finite state Markov chain

$$\mu^{t+1} = \Pi\mu^t \tag{3.12}$$

where $\Pi = \{\pi_{ij}\} = P(\boldsymbol{m}_j^{t+1}|\boldsymbol{m}_i^t)$ is the mode transition matrix and $\mu^t = \{\mu_j^t\} = P(\boldsymbol{m}_j^t)$ is the mode probability at time t.

Effective estimation techniques for stochastic hybrid systems are based in multiple models since it is natural to apply a statistical filter for each of the modes. The *Interactive Multiple Model* (IMM) algorithm [2, 4, 19] is a cost-effective (in terms of performance versus complexity) estimation scheme in which there is a *mixing/interacting* step at the beginning of the estimation process, which computes new initial conditions for the Kalman filters matched to the individual modes at each time step as illustrated in Figure 3.1.

Let $\mathcal{M}^t$ be the set of modes for which the Kalman filter is applied at time step t. For the IMM algorithm, $\mathcal{M}^t = \mathcal{M}^n$ for all $t \geq 0$ since we apply the filter to every mode. The components of the *mixing* step are the *mixing probability* $\mu_{ij}^{t|t+1}$ of being in mode i at time t given that the mode at time $t+1$ is j, the *mixed condition* $\hat{\boldsymbol{\rho}}_{0j}^{t:t}$ and $P_{0j}^{t:t}$ for the state estimate and covariance of mode j at time t, and the "spread-of-the-means" X_j in the expression of $P_{0j}^{t:t}$. They are computed for $j \in \mathcal{M}^{t+1}$ w.r.t. $\hat{\boldsymbol{\rho}}_i^{t:t}$ and $P_i^{t:t}$, the state estimate and its covariance of Kalman filter i at time t:

[3]Unlike in Section 2 where ρ_i and m_i denote the i-th entry of the state vector $\boldsymbol{\rho}$ and the mode $\boldsymbol{m}$ respectively, $\hat{\boldsymbol{\rho}}_j$ is the state estimate in mode $\boldsymbol{m}_j$.

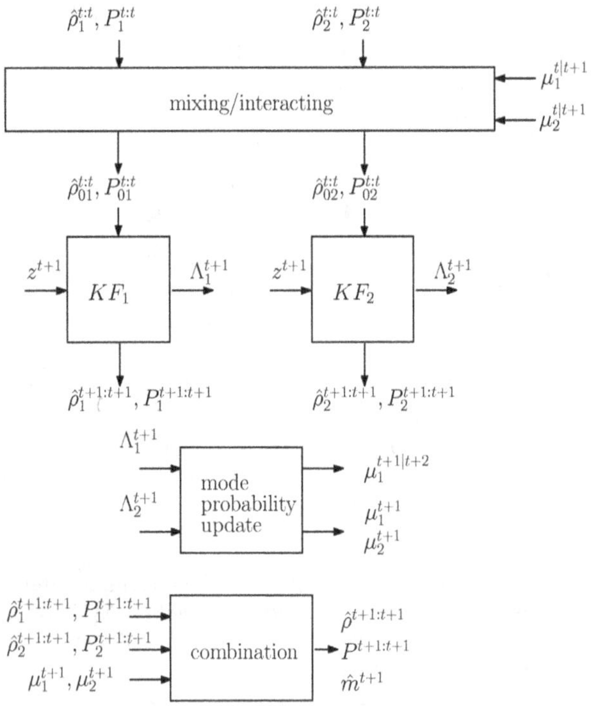

Figure 3.1: Illustration of the structure of IMM algorithm for a two-mode system from [18].

$$
\begin{array}{ll}
\mu_{ij}^{t|t+1} & = \frac{1}{Z_j}\pi_{ij}\mu_i^t \text{ for } i \in \mathcal{M}^t \\
Z_j & = \sum_{i\in\mathcal{M}^t} \pi_{ij}\mu_i^t \\
\hat{\rho}_{0j}^{t:t} & = \sum_{i\in\mathcal{M}^t} \hat{\rho}_i^{t:t}\mu_{ij}^{t|t+1} \\
P_{0j}^{t:t} & = \sum_{i\in\mathcal{M}^t} P_i^{t:t}\mu_{ij}^{t|t+1} + X_j \\
X_j & := \sum_{i\in\mathcal{M}^t} (\hat{\rho}_i^{t:t} - \hat{\rho}_{0j}^{t:t})(\hat{\rho}_i^{t:t} - \hat{\rho}_{0j}^{t:t})^T \mu_{ij}^{t|t+1}
\end{array}
\tag{3.13}
$$

We apply the Kalman filter in each mode $j \in \mathcal{M}^{t+1}$ (KF_j) as described with equations (3.9,3.10) and the resulting mode likelihood functions Λ_j^{t+1} are obtained from $\hat{\rho}_j^{t+1:t+1}$ and $P_j^{t+1:t+1}$ with equation (3.11). The mode probability $\mu^t = \{\mu_j^t\}$ is then updated through:

$$
\mu_j^{t+1} = \frac{1}{Z}\Lambda_j^{t+1} \sum_{i\in\mathcal{M}^t} \pi_{ij}\mu_i^t \text{ for } j \in \mathcal{M}^{t+1} \tag{3.14}
$$

where Z is a normalization constant. The output of the IMM algorithm are the state estimate $\hat{\rho}^{t+1:t+1}$ which is a weighted sum of the estimates from the Kalman filters in each mode and its covariance $P^{t+1:t+1}$, and the mode estimate $\hat{m}^{t+1}$ is the mode which has the highest mode probability. They are given by the *combination* step:

$$
\begin{array}{l}
\hat{\rho}^{t+1:t+1} = \sum_{j\in\mathcal{M}^{t+1}} \hat{\rho}_j^{t+1:t+1}\mu_j^{t+1} \\
P^{t+1:t+1} = \sum_{j\in\mathcal{M}^{t+1}} P_j^{t+1:t+1}\mu_j^{t+1} + X \\
X := \sum_{j\in\mathcal{M}^{t+1}} (\hat{\rho}_j^{t+1:t+1} - \hat{\rho}^{t+1:t+1})(\hat{\rho}_j^{t+1:t+1} - \hat{\rho}^{t+1:t+1})^T \mu_j^{t+1} \\
\hat{m}^{t+1} := \text{argmax}_{j\in\mathcal{M}^{t+1}} \mu_j^{t+1}
\end{array}
\tag{3.15}
$$

In [18, 13], the IMM algorithm is used as a hybrid estimator for Air Traffic Control (ATC) tracking. The models used include one for the uniform motion and one (or more) for the maneuver. However, the discretized PDE model described in section 2 has an exponential number of modes, as shown in Section 2.3, which induces an exponential time complexity of the IMM algorithm. Thus, the straight application of the IMM algorithm [18] as presented earlier is not tractable. The next section provides a reduced version of the algorithm, which is tractable.

4. REDUCED IMM

A solution to the tractability problem presented in the previous section consists in selecting a "representative" sample of modes following an algorithm based on the polyhedral structure of the model. Specifically, we only consider the mode in which the state estimate is and its *adjacent modes*.

4.1 Geometric properties

We extend the geometric setting presented in section 3 with definitions and the concept of *adjacent polyhedra*. We do not make any distinction between an open and closed half-space which are both denoted by $\mathbf{H}$, and the associated hyperplane is $\partial\mathbf{H}$.

Faces of a polyhedron: A supportive hyperplane of a closed convex set $\mathbf{C}$ is a hyperplane $\partial\mathbf{H}$ such as $\mathbf{C}\cap\partial\mathbf{H} \neq \emptyset$ and $\mathbf{C} \subseteq \mathbf{H}$, where $\mathbf{H}$ is one of the two half-spaces (associated to the hyperplane). Given a polyhedron $\mathbf{P}$, the inersection with any supportive hyperplane is a face of $\mathbf{P}$. Moreover, a vertex is a zero-dimension face, an edge a one-dimension face, and a facet is a face of dimension $d-1$ if $\mathbf{P}$ is of dimension d. For a full-dimensional polyhedron, a facet is of dimension $n+1$ (recall that the space $\mathcal{S} = [0, \rho_j]^{n+2}$ is of dimension $n+2$).

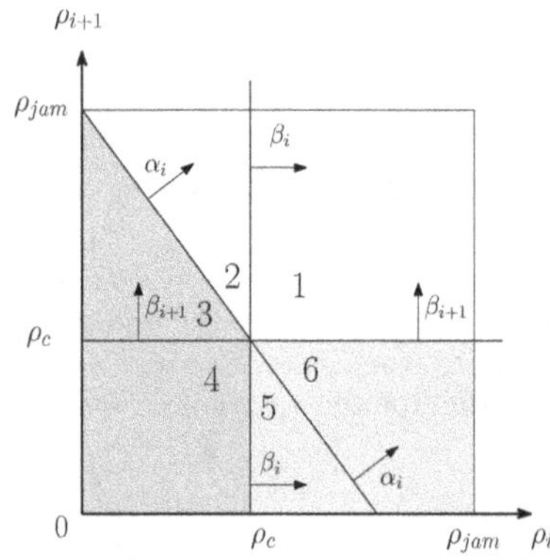

Figure 4.1: Projection of the half-spaces $\mathbf{H}_{\alpha_i}$, $\mathbf{H}_{\beta_i}$, $\mathbf{H}_{\beta_{i+1}}$, $\mathbf{H}^c_{\alpha_i}$, $\mathbf{H}^c_{\beta_i}$, $\mathbf{H}^c_{\beta_{i+1}}$ on the plan (ρ_i, ρ_{i+1}).

Minimal H-representation: There exist infinitely many H-descriptions of a convex polytope. However, for a full-dimensional convex polytope, the minimal H-description is in fact unique and is given by the set of the facet-defining halfspaces [11]. The following procedure gives the minimal H-representation of a polyhedron of the partition of $\mathcal{S}$ in our highway model.

Proposition 1: Let $\mathcal{H}_{\boldsymbol{s}}^{\min}$ be the set of half-spaces in the minimal H-representation of $\mathbf{P}_{\boldsymbol{s}}$. We have:

$$
\mathcal{H}_{\boldsymbol{s}}^{\min} \subset \left(\cup_{i=0}^{n}\{\mathbf{H}_{\alpha_i}, \mathbf{H}^c_{\alpha_i}\}\right) \cup \left(\cup_{i=0}^{n+1}\{\mathbf{H}_{\beta_i}, \mathbf{H}^c_{\beta_i}\}\right) \tag{4.1}
$$

where the previous formula means family of mathematical objects as opposed to union of half-spaces. With $\mathcal{W} := \{k|s_k = w, 0 \le k \le n-1\}$, $\mathcal{L} := \{k|s_k = l, 0 \le k \le n-1\}$, $\mathcal{D} := \{k|s_k = d, 0 \le k \le n-1\}$, and $|\cdot|$ the cardinality:

$$|\mathcal{H}_{\boldsymbol{s}}^{\min}| \le 2+|\mathcal{W}|+|\mathcal{L}|+2\times|\mathcal{D}| = n+2+|\mathcal{D}| \le 2n+2 \quad (4.2)$$

Proof of proposition 1: The following proof gives an algorithm to find the minimal H-representation of $\mathbf{P}_{\boldsymbol{s}}$. We have seen in section 3 that equations (3.2, 3.4) give an H-representation of $\mathbf{P}_{\boldsymbol{s}}$ as an intersection of $2 \times (n+1)$ half-spaces, equivalently a set of $2\times(n+1)$ inequalities. However, some of them do not define a facet of the polytope, equivalently the associated linear inequality is redundant. We want to remove such half-spaces from the H-representation $\mathcal{H}_{\boldsymbol{s}}$ and get the set $\mathcal{H}_{\boldsymbol{s}}^{\min}$ of half-spaces in the minimal H-representation of $\mathbf{P}_{\boldsymbol{s}}$:

$$\mathbf{P}_{\boldsymbol{s}} = \bigcap_{\mathbf{H}\in\mathcal{H}_{\boldsymbol{s}}^{\min}} \mathbf{H} \quad (4.3)$$

For this purpose, we list all the scenarios in which a half-space is redundant. They exactly happen when the intersection of two half-spaces is included in another half-space in the H-representation. There are $6 \times (n+1)$ of them as illustrated in Figure 4.1:

$$\begin{array}{llll} 1 & \mathbf{H}_{\beta_i} \cap \mathbf{H}_{\beta_{i+1}} & \subset & \mathbf{H}_{\alpha_i} \\ 2 & \mathbf{H}_{\alpha_i} \cap \mathbf{H}^c_{\beta_i} & \subset & \mathbf{H}_{\beta_{i+1}} \\ 3 & \mathbf{H}^c_{\alpha_i} \cap \mathbf{H}_{\beta_{i+1}} & \subset & \mathbf{H}^c_{\beta_i} \\ 4 & \mathbf{H}^c_{\beta_i} \cap \mathbf{H}^c_{\beta_{i+1}} & \subset & \mathbf{H}^c_{\alpha_i} \\ 5 & \mathbf{H}^c_{\alpha_i} \cap \mathbf{H}_{\beta_i} & \subset & \mathbf{H}^c_{\beta_{i+1}} \\ 6 & \mathbf{H}_{\alpha_i} \cap \mathbf{H}^c_{\beta_{i+1}} & \subset & \mathbf{H}_{\beta_i} \end{array} \quad \text{for } i = 0, \cdots, n \quad (4.4)$$

Each triangular domain indexed from 1 to 6 in Figure 4.1 represents the intersection of the two half-spaces in each line of (4.4) respectively.projected on the plan (ρ_i, ρ_{i+1}). Then the six inclusions in (4.4) projected on (ρ_i, ρ_{i+1}) become clear and they hold in the entire space since the constraints only span on the variables ρ_i and ρ_{i+1}.

Starting from $k = 0$, we construct the sequence of sets $\{\mathcal{H}_k\}_{k=0,\dots,n}$ such that $\mathcal{H}_k$ is the set of half-spaces in the minimal H-representation of the polyhedron $\mathbf{P}_{\boldsymbol{s},k} = \bigcap_{i=0}^{k} \mathbf{Q}_i$ for $k = 0, \dots, n$. We have $\mathcal{H}_0 = \{\mathbf{H}_0, \mathbf{H}'_0\}$ where $\mathbf{H}_0$ and $\mathbf{H}'_0$ are the two half-spaces in the minimal H-reprentation of $\mathbf{Q}_0$ given by equations (3.4). Suppose we know $\mathcal{H}_k$ and we construct $\mathcal{H}_{k+1}$ by adding only the non-redundant constraints in $\mathbf{Q}_{k+1}$. There are seven cases depending on $s(k)$ and $s(k+1)$:

Case	$s(k)$	$s(k+1)$	$\mathcal{H}_{k+1}$	$\|\mathcal{H}_{k+1}\|$
1	w	w	$\mathcal{H}_k \cup \{\mathbf{H}_{\beta_{k+2}}\}$	$= \|\mathcal{H}_k\| + 1$
2	w	l	$\mathcal{H}_k \cup \{\mathbf{H}^c_{\beta_{k+2}}\}$	$= \|\mathcal{H}_k\| + 1$
3	l	w	$\mathcal{H}_k \cup \{\mathbf{H}_{\alpha_{k+1}}\}$	$= \|\mathcal{H}_k\| + 1$
4	l	d	$\mathcal{H}_k \cup \{\mathbf{H}^c_{\alpha_{k+1}}\}$	$= \|\mathcal{H}_k\| + 1$
5	d	w	$\mathcal{H}_k \cup \{\mathbf{H}_{\alpha_{k+1}}, \mathbf{H}_{\beta_{k+2}}\}$	$= \|\mathcal{H}_k\| + 2$
6	d	l	$\mathcal{H}_k \backslash \{\mathbf{H}^c_{\beta_k}\} \cup \{\mathbf{H}_{\beta_{k+1}}, \mathbf{H}^c_{\beta_{k+2}}\}$	$\le \|\mathcal{H}_k\| + 2$
7	d	d	$\mathcal{H}_k \backslash \{\mathbf{H}^c_{\alpha_k}\} \cup \{\mathbf{H}^c_{\alpha_{k+1}}, \mathbf{H}^c_{\beta_{k+1}}\}$	$\le \|\mathcal{H}_k\| + 2$

Table 4.1: Construction of $\mathcal{H}_{k+1}$ given $\mathcal{H}_k$, and cardinality of $\mathcal{H}_{k+1}$.

For case 1 (repectively 3), the non-redundant constraint when adding $\mathbf{W}_{k+1}$ is $\mathbf{H}_{\beta_{k+2}}$ (respectively $\mathbf{H}_{\alpha_{k+1}}$) from scenario 1 (respectively 2) in (4.4). In case 2 (respectively 4), the constraint $\mathbf{H}_{\beta_{k+1}}$ (respectively $\mathbf{H}^c_{\beta_{k+1}}$) in $\mathbf{L}_{k+1}$ (respectively $\mathbf{D}_{k+1}$) is already implied by $s(k) = w$ (respectively $s(k) = l$). For case 6 (repectively 7), the constraint $\mathbf{H}^c_{\beta_k}$ (respectively $\mathbf{H}^c_{\alpha_k}$) in $\mathbf{D}_k$ becomes redundant from scenario 3 (respectively 4) in (4.4).

We have sequentially constructed $\mathcal{H}_k$ for $k = 0, \cdots, n$ and $\mathcal{H}_n = \mathcal{H}_{\boldsymbol{s}}^{\min}$ is the sef of half-spaces in the minimal H-representation of $\mathbf{P}_{\boldsymbol{s}}$. Since the minimal H-representation is unique, this completes the proof of (4.1) and (4.2). And Table 4.1 also shows that $|\mathcal{H}_{k+1}| \le |\mathcal{H}_{k+1}|+1$ if $s(k) \in \{w, l\}$ and $|\mathcal{H}_{k+1}| \le |\mathcal{H}_{k+1}| + 2$ if $s(k) = d$, which gives (4.2) by induction. □

The polyhedra are assumed to be closed in the following definitions.

Adjacent polyhedra: Two polyhedra $\mathbf{P}$ and $\mathbf{P}'$ in a polyhedral partition of the space $\mathcal{S}$ are said to be *k-adjacent* if they have a face of dimension k in common. Formally, this is when there exists a supportive hyperplane $\partial\mathbf{H}$ for both $\mathbf{P}$ and $\mathbf{P}'$ and the intersection $\mathbf{P} \cap \mathbf{P}' \cap \partial\mathbf{H}$ is of dimension k.

$\mathbf{Q}_k$	Transformation 1	Transformation 2
$\mathbf{W}_k$	$\mathbf{H}^c_{\alpha_k} \cap \mathbf{H}_{\beta_{k+1}} \subset \mathbf{D}_k$	$\mathbf{H}_{\alpha_k} \cap \mathbf{H}^c_{\beta_{k+1}} \subset \mathbf{L}_k$
$\mathbf{L}_k$	$\mathbf{H}^c_{\beta_k} \cap \mathbf{H}^c_{\beta_{k+1}} \subset \mathbf{D}_k$	$\mathbf{H}_{\beta_k} \cap \mathbf{H}_{\beta_{k+1}} \subset \mathbf{W}_k$
$\mathbf{D}_k$	$\mathbf{H}_{\alpha_k} \cap \mathbf{H}^c_{\beta_k} \subset \mathbf{W}_k$	$\mathbf{H}^c_{\alpha_k} \cap \mathbf{H}_{\beta_k} \subset \mathbf{L}_k$

Table 4.2: Transformation of the polyhedra $\mathbf{Q}_k = \mathbf{H} \cap \mathbf{H}'$ into $\mathbf{H}^c \cap \mathbf{H}'$ for transformation 1, and into $\mathbf{H} \cap (\mathbf{H}')^c$ for transformation 2.

Proposition 2: Recall that the space $\mathcal{S} = [0, \rho_j]^{n+2}$ is of dimension $n+2$. Given a polyhedron $\mathbf{P}_{\boldsymbol{s}_0}$ of the partition of $\mathcal{S}$, let us define the function $\mathcal{F}_{\boldsymbol{s}_0}$ from $\mathcal{H}_{\boldsymbol{s}_0}^{\min}$ the set of half-spaces in its minimal H-representation to $\mathcal{A}_{\boldsymbol{s}_0}$ the set of polyhedra of the partition $(n+1)$-adjacent to $\mathbf{P}_{\boldsymbol{s}_0}$:

$$\mathcal{F}_{\boldsymbol{s}_0} : \begin{array}{ccc} \mathcal{H}_{\boldsymbol{s}_0}^{\min} & \rightarrow & \mathcal{A}_{\boldsymbol{s}_0} \\ \mathbf{H} & \mapsto & \mathbf{P}_{\boldsymbol{s}} \end{array} \quad (4.5)$$

where $\boldsymbol{s}$ in $\mathcal{F}_{\boldsymbol{s}_0}[\mathbf{H}] = \mathbf{P}_{\boldsymbol{s}}$ is the unique mode $\boldsymbol{s}'$ such that

$$\left(\cap_{\mathbf{H}'\in\mathcal{H}_{\boldsymbol{s}_0}\backslash\mathbf{H}} \mathbf{H}'\right) \cap \mathbf{H}^c \subset \mathbf{P}_{\boldsymbol{s}'} \quad (4.6)$$

Then $\mathcal{F}_{\boldsymbol{s}_0}$ is a *bijection*.

Proof of proposition 2: Let us show that $\mathcal{F}_{\boldsymbol{s}_0}$ is *well-defined*. Without loss of generality, let us assume that $s_{i_0-1} = w$, $s_{i_0} = l$, and $\mathbf{H} = \mathbf{H}_{\beta_{i_0}}$, since all the other cases can be treated in a similar manner. Hence, $\mathbf{W}_{i_0-1} = \mathbf{H}_{\alpha_{i_0}-1} \cap \mathbf{H}_{\beta_{i_0}}$ and $\mathbf{L}_{i_0} = \mathbf{H}_{\beta_{i_0}} \cap \mathbf{H}^c_{\beta_{i_0}+1}$ are the only two polyhedra of the H-representation of $\mathbf{P}_{\boldsymbol{s}}$ which have $\mathbf{H}_{\beta_{i_0}}$ as a facet-defining half-space. Therefore:

$$\begin{aligned} &\left(\cap_{\mathbf{H}'\in\mathcal{H}_{\boldsymbol{s}_0}\backslash\mathbf{H}} \mathbf{H}'\right) \cap \mathbf{H}^c \\ &= (\cap_{i \neq i_0-1, i_0} \mathbf{Q}_i) \cap (\mathbf{H}_{\alpha_{i_0-1}} \cap \mathbf{H}^c_{\beta_{i_0}}) \cap (\mathbf{H}^c_{\beta_{i_0}} \cap \mathbf{H}^c_{\beta_{i_0+1}}) \\ &\subset (\cap_{i \neq i_0-1, i_0} \mathbf{Q}_i) \cap \mathbf{L}_{i_0-1} \cap \mathbf{D}_{i_0} \end{aligned} \quad (4.7)$$

The inclusion is obtained from Table 4.2. Hence the subset $\left(\cap_{\mathbf{H}' \in \mathcal{H}_{s_0} \setminus \mathbf{H}} \mathbf{H}'\right) \cap \mathbf{H}^c$ is included in a polyhedron of the partition $\mathbf{P}_{\boldsymbol{s}}$. Since the subset is non empty, $\mathbf{P}_{\boldsymbol{s}}$ is the only polyhedron containing it, and $\mathcal{F}_{\boldsymbol{s}_0}(\mathbf{H}) = (\cap_{i \neq i_0-1, i_0} \mathbf{Q}_i) \cap \mathbf{L}_{i_0-1} \cap \mathbf{D}_{i_0}$ is uniquely defined. This is the only polyhedron of the partition $(n+1)$-adjacent to $\mathbf{P}_{\boldsymbol{s}}$ with common supportive hyperplane $\partial\mathbf{H} = \partial\mathbf{H}_{\beta_{i_0}}$.

Let us show that $\mathcal{F}_{\boldsymbol{s}_0}$ is *bijective*. The function $\mathcal{F}_{\boldsymbol{s}_0}$ is *surjective* since it maps each of the facet-defining half-space $\mathbf{H}$ of $\mathbf{P}_{\boldsymbol{s}}$ to the $(n+1)$-adjacent polyhedron of the partition sharing the supportive hyperplane $\partial\mathbf{H}$. And $\mathcal{F}_{\boldsymbol{s}_0}$ is *injective* since for a given i and a given polyhedron $\mathbf{Q}_i$ of the H-representation, the two possible transformations of $\mathbf{Q}_i$ yield a different inclusion as shown in Table 4.2, and therefore a different adjacent polyhedron. □

4.2 Reduction to adjacent modes

We have presented in the previous section an algorithm to construct the minimal H-representation of $\hat{\boldsymbol{s}}$, which enables us to find the adjacent modes of $\hat{\boldsymbol{s}}$. We note that it follows from Equation (4.2) and *Proposition 2* that there are less than $2n+2$ modes adjacent to a given mode $\boldsymbol{m}$, where n is the number of cells of the discretized model. Moreover, two adjacent modes only differ by at most two entries. Hence, when the discretized model is in quasi-steady state (with only small variations between consecutive time steps), every pair of most likely active modes at time t must have adjacent elements. This suggests different heuristics for reducing the number of modes at each time step t of the IMM algorithm to a set $\mathcal{M}^t$ of cardinality linear in the dimension. In the update equations, one can exclusively consider the mode $\hat{\boldsymbol{m}}$ (or $\hat{\boldsymbol{s}}$) of the state estimate $\hat{\boldsymbol{\rho}}^{t:t}$ and its adjacent modes. Hence, the number of modes considered is less than $2n+2$, and we will call this variation of the IMM algorithm the *reduced* IMM (RIMM1) algorithm.

We can further reduce the number of modes by taking into account the covariance P of the estimate and the distance between the state estimate and the facets of the polyhedron $\mathcal{P}_{\hat{\boldsymbol{s}}}$. Let $\bigcap_{\mathbf{H} \in \mathcal{H}_{\boldsymbol{s}}^{\min}} \mathbf{H}$ be the minimal H-representation of $\mathbf{P}_{\boldsymbol{s}}$, $\partial\mathbf{H}$ the associated hyperplane, and $\mathbf{F}_H := \partial\mathbf{H} \cap \mathbf{P}_{\boldsymbol{s}}$ the facet associated to $\mathbf{H}$. The half-spaces in the set $\mathcal{H}_{\boldsymbol{s}}$ can be written as follows:

$$\mathbf{H} = \{\boldsymbol{\rho} \mid a_H.\boldsymbol{\rho} - b_H \leq 0\} \text{ for } \mathbf{H} \in \mathcal{H}_{\boldsymbol{s}} \qquad (4.8)$$

Since $\mathbf{F}_H \subset \partial\mathbf{H}$, the distance between $\hat{\boldsymbol{\rho}}$ and each of the facet $\mathbf{F}_H$ is such that:

$$d(\hat{\boldsymbol{\rho}}, \mathbf{F}_H) \geq d(\hat{\boldsymbol{\rho}}, \partial\mathbf{H}) = \frac{|b_H - a_H.\hat{\boldsymbol{\rho}}|}{||a_H||} \text{ for } \mathbf{H} \in \mathcal{H}_{\boldsymbol{s}} \qquad (4.9)$$

With P the state covariance, let us define the ratio:

$$d_H(\hat{\boldsymbol{\rho}}) = \frac{d(\hat{\boldsymbol{\rho}}, \partial\mathbf{H})}{(a_H)^T P a_H} \text{ for } \mathbf{H} \in \mathcal{H}_{\boldsymbol{s}} \qquad (4.10)$$

Thus, we only look at the adjacent modes for which $d_H(\hat{\boldsymbol{\rho}})$ is less than a given threshold *thres* (RIMM2). Intuitively, when there is a high variance $(a_H)^T P a_H$ along the direction a_H orthogonal to $\partial\mathbf{H}$, there is a higher probability that the state at the next time step is in the half-space $\mathbf{H}^c$, which is the complementary of $\mathbf{H}$, and therefore in the adjacent mode $\boldsymbol{s}_H$ with common supportive hyperplane $\partial\mathbf{H}$. We note that this is an approximation since the projection of $\boldsymbol{\rho}$ on $\partial\mathbf{H}$ is not always on a facet of the adjacent mode $\boldsymbol{s}_j$.

This variation on the IMM algorithm is closely related to the EKF since the Kalman filter applied in the mode $\hat{\boldsymbol{m}}$ of the state estimate $\hat{\boldsymbol{\rho}}^{t:t}$ is equivalent to linearizing around the state estimate. There is a refinement since we include the modes adjacent to $\hat{\boldsymbol{m}}$. Instead of relying on one possible active mode, we represent a set of possible modes active at time t and apply the KF to each one of them. However, the adjacent modes differ from $\hat{\boldsymbol{m}}$ by only one or two entries, and so they only represent a very restricted set of close possibilities centered around the mode estimate. As will be seen in the next section, the reduced IMM based on this reduction of modes is very similar to the EKF. An alternative approach consists in taking the most likely mode $\tilde{\boldsymbol{m}}^t = \text{argmax}_{j \in \mathcal{M}^t} \mu_j^t$, and its adjacent modes. However, this variation of the IMM algorithm fails to give a good representation of the dynamics of the discretized system. Since the mode estimate $\hat{\boldsymbol{m}}^t$ is chosen among the adjacent modes of the mode estimate $\hat{\boldsymbol{m}}^{t-1}$ at the previous time step, it differs from $\hat{\boldsymbol{m}}^{t-1}$ by at most one or two entries. However, the physical system have empirically larger variations, resulting in relatively different active modes between consecutive time steps.

4.3 Clustering of the space of modes

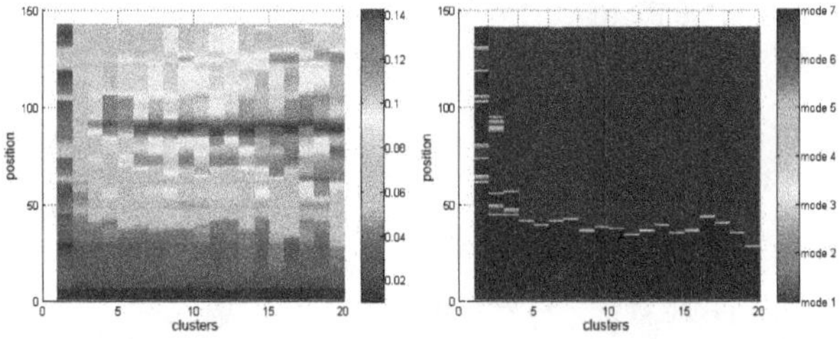

Figure 4.2: 20 clusters of the density space using *k-means* (left) and their respective modes (right).

An intuitive method to solve the latter problem consists in partitioning the space of modes $\mathcal{M}_n$ itself into K clusters with the widely-used *k-means*[4]algorithm, and transfer the adjacencies between modes studied above to the adjacencies between clusters. However, the clusters are no longer seperated by hyperplanes and additional geometrical analysis is needed to find the clusters that are close to each other. Carried further, the IMM algorithm can be applied to each of the clusters if their number K is small, and gives reasonable results (RIMM3).

Figure 4.2 shows 20 clusters of the density space. They are given by *k-means* applied to samples output by the EKF estimator using density easurements from loop detectors on March 1st, 2012 between 7am and 8am. Hence we used "historic" data to reduce the space of modes for estimation on March 5th as described in 4.5. The centroids of the partitioning of the space of modes is then given by the mode of the 20 clusters of the density space.

4.4 Analysis

It is possible to compute the *predicted state estimate* $\hat{\boldsymbol{\rho}}^{t+1:t}$ and the *predicted covariance estimate* $P^{t+1:t}$ given by (3.9)

[4]The *k-means* seems more adapted since it finds clusters with comparable spatial extent contrary to the *expectation-maximization* which partitions in clusters of different shapes.

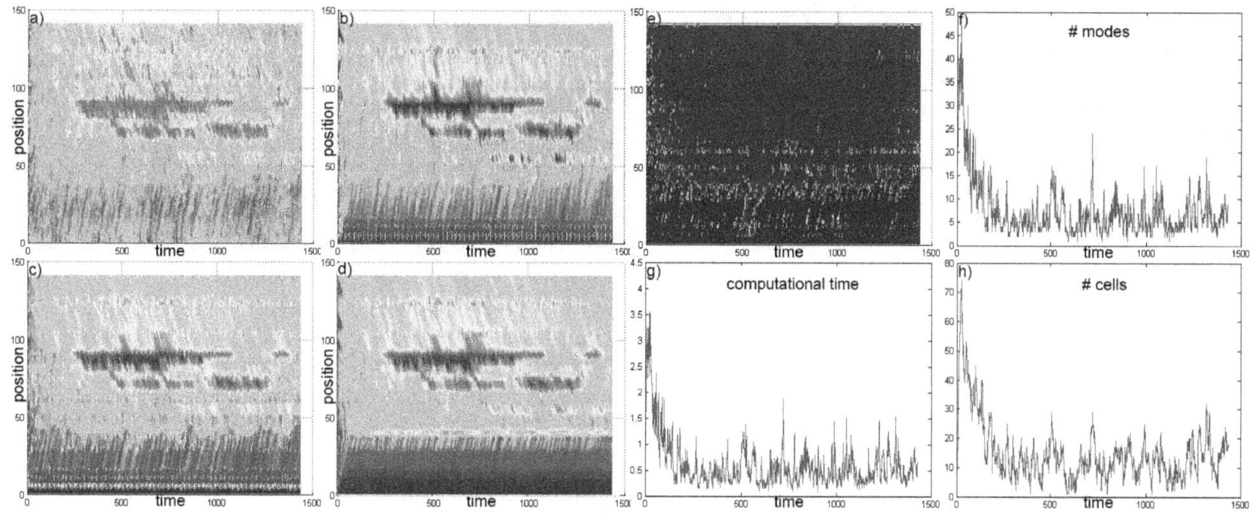

Figure 4.3: Contour plot of the density given by (a) the EnKF with 100 ensembles, (b) the EKF, (c) the RIMM2 with $thres = 1$, (d) the RIMM3 with 20 clusters using the k-means algorithm. Analysis of each time step of the RIMM2 with $thres = 1$: (e) plot of the mode estimate, (f) number of modes selected via the method of ratios, (g) computational time, (h) number of cells with density close to ρ_c.

in linear time and quadratic time respectively as shown in [27]. Hence, both time and space complexities of the *prediction step* of the Kalman filter are $O(n^2)$.

A second analysis shows that the time complexity of the *update step* of the Kalman filter is $O(mn^2 + m^3 + nm^2)$ with m the number of observations. As the density measurements along the highway are sparse, the complexity becomse $O(n^2)$.[5]Hence, the total complexity of the Kalman filter is $O(n^2)$.

Since the number of modes is reduced to $O(n)$, the complexity of the reduced IMM is $O(n^3)$ for both reduction heuristics presented in section 4.2. This is a significant gain from the exponential complexity of the original IMM algorithm, which considers all the modes. And if we further reduce the number of modes by restricting to the ones which have a ratio less than a threshold *thres* small enough (see equation (4.10)), then the number of.modes can be bounded. This gives a complexity $O(n^2)$.

4.5 Numerical results

The previous method is implemented on an 18-mile section of I-880 Northbound in the Bay Area, California combined with two variants of the reduced IMM estimation algorithms (RIMM2, RIMM3) presented above. We use density measurements along the I-880 from 29 loop detectors (PeMS) every 30s on March 5th, 2012 between 7am and 8am to compute density values and integrate them in the model. Each cell has a length of 198m and the time step is 5s.

The results are compared with the output of the EKF which was implemented in [27], and the output of the EnKF which is commonly used in the traffic monitoring community with this class of discretized models [29].

Figures 4.3 (a, b, c, d) present the output of the four estimators which consist in the density in the time-space domain. The regions with high density, average density, and low density are represented in red, yellow, and blue respectively. The estimators give similar higher resolution scalar fields of the density (1440 time steps by 141 cells) by assimilating sparse density measurements (240 time steps by 29 PeMS stations. The shock wave propagation is more noticeable in the output of both RIMM estimators in the congested regions. Hence, these estimators are more tuned to the discretized physical system which is PWA.

Figure 4.3 (e) shows the mode estimate $\hat{\mathbf{s}}$ computed in the *combination step* of the IMM algorithm and presented by a contour plot in the time-space domain. The regions in which $\hat{s}_i^t = w$ are in congestion and they are colored in red and the regions in which $\hat{s}_i^t = d$ are in free flow and they are represented in blue. The points of the time-space domain where $\hat{s}_i^t = l$ are colored in white and Figure 4.3 (e) shows that they are at the boundaries between the congested and free flow regions. They correspond to a regime of transition from free flow to congestion and vice versa. The estimate of the mode provided by the IMM algorithm at each point of the discretized time-space domain is thus accessible, as shown here.

Finally, Figures 4.3 (f, g, h) show that the number of modes in $\mathcal{M}^t$, the computational time, and the number of cells for which the density is close to ρ_c are highly correlated. These results show that the computational time is proportional to the number of modes. This was predictable since the application of the KF to each of the modes is the most expensive step of the IMM algorithm. This underlines the importance of reducing the number of modes to a bounded number.

5. CONCLUSION AND FUTURE WORK

A new approach in estimation of discretized hyperbolic PDEs is developed. It uses a Godunov scheme to discretize the LWR PDE with a triangular flux function. The resulting non-linear dynamical system can be decomposed in PWA components evolving with linear constraints. They partition the state space into an exponential number of polyhedra.

[5]The measurements are sparse for our experimental data location (cf. section 4.5), but this may not be true in the general case.

The initially intractable IMM becomes tractable by reducing the modes to a set of adjacent modes centered around the mode of the state estimate.

The implementation of the reduced IMM algorithm shows that it is tuned to the discretized model due to its PWA structure. The RIMM also provides an estimate of the congestion or free flow state in the time-space domain. More importantly, we have constructed a framework for the estimation of the discretized LWR PDE which enables: (1) the use of Kalman filtering on each of the linear modes, (2) the use of statistical analysis in the space of modes to make the IMM tractable.

6. ACKNOWLEDGEMENT

The authors would like to thank Professor Claire Tomlin from University of California, Berkeley for insightful discussions on hybrid systems and the members of the staff of the California Institute for Innovative Transportation for its contributions to develop, build, and deploy the *Mobile Millennium* system on which this article relies.

7. REFERENCES

[1] A. Balluchi, L. Benvenutti, M. D. di Benedetto, C. Pinello, and A. L. Sangiovanni-Vincentelli. Automative engine control and hybrid systems: challenges and opportunities. *Proc. IEEE*, 88 (7):888–912, 2000.

[2] Y. Bar-Shalom and X. R. Li. Estimation and Tracking: Principles, Techniques, and Software. *Norwood, MA: Artech House*, 1993.

[3] S. Blandin, A. Couque, A. Bayen, and D. Work. On sequential data assimilation for scalar macroscopic traffic flow models. *Physica D*, 2012.

[4] H. A. P. Blom and Y. Bar-Shalom. The interacting multiple model algorithm for systems with Markovian switching coefficients. *IEEE Trans. Autom. Control*, AC-33:780–783, August 1988.

[5] R. Chen and J. S. Liu. Mixture Kalman filters. *Royal Statistical Society*, 62:493–508, 2000.

[6] C. F. Daganzo. The cell transmission model: a dynamic representation of highway traffic consistent with the hydrodynamic theory. *Transportation Research Part B 28, no. 4*, 28:269–287, 1994.

[7] C. F. Daganzo. The cell transmission model, part II: Network traffic. *Transportation Research Part B 29, no. 2*, 29:79–93, 1995.

[8] E. Godlewski and P-A. Raviart. *Numerical approximation of hyperbolic systems of conservation laws.* Applied Mathematical Sciences, 1996.

[9] S.K. Godunov. A finite difference method for the numerical computation of discontinuous solutions of the equations of fluid dynamics. *Math. Sbornik*, 47:271–306, 1959.

[10] B. D. Greenshields. A study of traffic capacity. *Proceedings of the 14th annual meeting of the Highway Research Board*, 14:448–477, 1934.

[11] Branko Grünbaum. *Convex Polytopes.* Springer, 2003.

[12] R. M. Hawkes and J. B. Moore. Performance bounds for adaptive estimation. *Proc. IEEE*, 64:1143–1150, 1976.

[13] I. Hwang, H. Balakrishnan, and C. Tomlin. State estimation for hybrid systems: applications to aircraft tracking. *IEE Proc. Control Theory and Applications*, 153(5):556–566, Sep. 2006.

[14] J. P. Lebacque. The Godunov scheme and what it means for first order traffic flow models. *13th International Symposium on Transportation and Traffic Theory*, pages 647–77, 1996.

[15] M. D. Lemmon, K. X. He, and I. Markovsky. Supervisory hybrid systems. *IEEE Control Systems Magazine*, 19 (4):42–55, 1999.

[16] B. Lennartson, M. Tittus, B. Egardt, and S. Petterson. Hybrid systems in process control. *IEEE Control Systems Magazine*, 16 (5):45–56, 1996.

[17] R. J. LeVeque. *Numerical Methods for Conservation Laws.* Birkhäuser Basel, 1992.

[18] X. R. Li and Y. Bar-Shalom. Design of an Interacting Multiple Model Algorithm for Air Traffic Control Tracking. *IEEE Transactions on Control Systems Technology*, 1(3), September 1993.

[19] X. R. Li and Y. Bar-Shalom. Performance prediction of the interacting multiple model algorithm. *IEEE Trans. Aerosp. Electron. Syst*, 29:755–771, July 1993.

[20] M. J. Lighthill and G. B. Whitham. On Kinematic Waves II. A Theory of Traffic Flow on Long Crowded Roads. *Proceedings of the Royal Society of London. Series A, Mathematical and Physical Sciences*, 229:317–345, 1955.

[21] P. S. Maybeck. Stochastic models, estimation, and control. *Academic Press*, 2, 1982.

[22] M. Papageorgiou, J.-M. Blosseville, and H. Hadj-Salem. Modelling and real-time control of traffic flow on the southern part of Boulevard Peripherique in Paris: Part I: Modelling. *Transportation Research*, 24 (5):345–359, 1990.

[23] P. I. Richards. Shock Waves on the Highway. *Operations Research*, 4:42–51, 1956.

[24] T. Schreiter, C. van Hinsbergen, F. Zuurbier, H. van Lint, and S. Hoogendoorn. Data-model synchronization in extended Kalman filters for accurate online traffic state estimation. *2010 Proceedings of the Traffic Flow Theory Conference, Annecy, France*, 2010.

[25] I. S. Strub and A. M. Bayen. Weak formulation of boundary conditions for scalar conservation laws: an application to highway traffic modeling. *Int. J. Robust Nonlinear Control*, 16:733–748, 2006.

[26] D. D. Sworder and J. E. Boyd. Estimation problems in hybrid systems. *Cambridge University Press*, 1999.

[27] J. Thai, B. Prodhomme, and A. M. Bayen. State Estimation for the discretized LWR PDE using explicit polyhedral representations of the Godunov scheme. *In review, American Control Conference*, 2013.

[28] C. Tomlin, G. Pappas, and S. Sastry. Conflict resolution for air traffic management: A study in multiagent hybrid systems. *IEEE Trans. Autom. Control*, 43 (4), 1998.

[29] D. B. Work, S. Blandin, O. Tossavainen, B. Piccoli, and A. M. Bayen. A Traffic Model for Velocity Data Assimilation. *Applied Mathematics Research eXpress*, 2010.

Observer Design for a Class of Piecewise Affine Hybrid Systems

D.A.J. van Zwieten
Eindhoven University of Technology
P.O. Box 513
5600 MB Eindhoven, The Netherlands
D.A.J.v.Zwieten@tue.nl

E. Lefeber
Eindhoven University of Technology
P.O. Box 513
5600 MB Eindhoven, The Netherlands
A.A.J.Lefeber@tue.nl

W.P.M.H. Heemels
Eindhoven University of Technology
P.O. Box 513
5600 MB Eindhoven, The Netherlands
M.Heemels@tue.nl

ABSTRACT

A methodology for the design of observers is proposed for a special class of hybrid dynamical systems, which are motivated by traffic and manufacturing applications. The class of hybrid systems is characterized as switched system models with constant drift and constant output, rendering all subsystems unobservable by themselves. However, an observer can still be derived due to the fixed switching pattern, even though the switching times may be unknown. A main step in the observer design methodology is the usage of a discrete-time linear observer based on the discretized hybrid dynamics at the event times that are visible. Using this step, a continuous-time observer is built that incorporates additional modes compared to the original hybrid system. This continuous-time observer is shown to asymptotically reconstruct the state of the original system under suitable assumptions. Manufacturing and traffic applications are used to illustrate the proposed observer design methodology.

Categories and Subject Descriptors

D.4.8 [**Performance**]: Modeling and prediction

Keywords

Observer design, Piecewise affine hybrid systems, Manufacturing systems, Traffic applications

1. INTRODUCTION

Nowadays, logic decision making and control actions are combined with continuous (physical) processes in many technological (cyber-physical) systems. Such systems are labeled *hybrid* as they have interacting continuous and discrete dynamics. Hybrid models are not only important in such man-made systems, but also in describing the behavior of many mechanical, biological, electrical and economical systems. Therefore, in the past decades, the structural properties of hybrid systems have been investigated by many researchers. This led to various techniques for controller synthesis, see e.g. [14, 18] for recent overviews. However, many of these controller synthesis methods are based on the assumption that the full state variable of the hybrid system is available, which is hardly ever the case in practice. This renders the *design of observers* for hybrid systems, providing good estimates of both continuous and discrete states, of crucial importance. Despite this high practical relevance, there are only a few results on hybrid observer design, see, e.g., [2, 5, 7, 12, 17, 21, 24] and the references therein. Theoretical results regarding the fundamental question of the existence of an observer are related to notions such as final state observability, reconstructability and final state determinability on which also some work in the area of hybrid systems appeared, see, for instance, [4, 6, 8, 9, 11, 23].

In this paper we are interested in designing observers for a special class of piecewise affine hybrid systems (*PWAHS*) [11], motivated by switching servers in manufacturing systems serving multiple products consecutively, and traffic applications such as signalized intersections. The considered hybrid system is autonomous with the mode dynamics consisting of constant drift and the output within a mode being constant. Only at some times the output reveals information about the currently active mode. In particular, this implies that all subsystems are unobservable, eliminating many currently available solutions for synthesizing hybrid observers proposed in the literature. Furthermore, the mode transitions are state-dependent, which is a property that will turn out to be useful, and are therefore a priori unknown. In addition, during a mode transition some specific state variables might exhibit jumps. The order in which the modes are traversed is fixed and periodic.

For this class of PWAHS, of which the full details are specified later, we propose a methodology for designing continuous-time observers. This methodology consists of a few main steps. First, the system is sampled (with varying sampling periods) at so-called visible event times, i.e., times at which the output changes during a mode transition, resulting in a linear time-varying periodic system. Based on the resulting sampled system a periodic discrete-time observer is derived with the guarantee that the observer's state converges asymptotically to the (original) system's state. Next, this observer is used as a stepping stone for designing an observer

HSCC'13, April 8–11, 2013, Philadelphia, Pennsylvania, USA.

in continuous time. This requires the inclusion of additional modes in the observer structure and additional reset laws at visible event times to ensure the asymptotic recovery of the system's state. A formal proof of the asymptotic recovery of the system's state is provided. Via an example of a traffic application we demonstrate the effectiveness of the proposed observer.

The remainder of this paper is organized as follows. Section 2 introduces the considered class of PWAHS and presents a two-buffer switching server as an introductory example. Section 3 presents the sampled system at the event times. In Section 4 the method for the observer design is presented. First, a discrete-time observer is presented for the sampled system. Next, the continuous-time observer is presented. In Section 5, a signalized traffic intersection with three flows is presented for which an observer is derived. Conclusions are provided in Section 6.

Nomenclature

In this paper we use $\{e_1, e_2, \ldots, e_M\}$ as the standard orthonormal basis in $\mathbb{R}^M$ in which e_i is the vector which contains a 1 at the i-th entry, and zeros elsewhere. By $\mathbb{R}_+$ we denote the set of non-negative reals, i.e., $\mathbb{R}_+ := [0, \infty)$. Furthermore, the product of matrices is considered as a left multiplication, i.e., $\prod_{i=1}^{3} A_i = A_3 A_2 A_1$.

2. CLASS OF PIECEWISE AFFINE HYBRID SYSTEMS

In this section, we present the dynamics of the class of piecewise affine hybrid systems (PWAHS) studied in this paper. Before doing so, we first present an illustrative example of a manufacturing system to motivate the structure of the class. In fact, this manufacturing system is used as a running example throughout the paper.

2.1 Illustrative example

Consider a single server that serves two different job types denoted by $n = 1, 2$, see Figure 1. Each job type n has a separate buffer in which x_n jobs are stored. Jobs arrive at buffer 1 with a constant arrival rate denoted by $\lambda_1 > 0$. After service in buffer 1 the jobs move to buffer 2. The server can only serve one job type at a time and operates based on a clearing policy, i.e., it completely empties the buffer of one job type before it switches to the next job type. The processing speed of job type n is denoted by $\mu_n > 0$. Switching to job type n requires a setup time with duration $\gamma_n \geq 0$ and at least one setup time is non-zero, i.e., $\gamma_1 + \gamma_2 > 0$. The only (measurement) information we get from the server is when the server is processing job type 2.

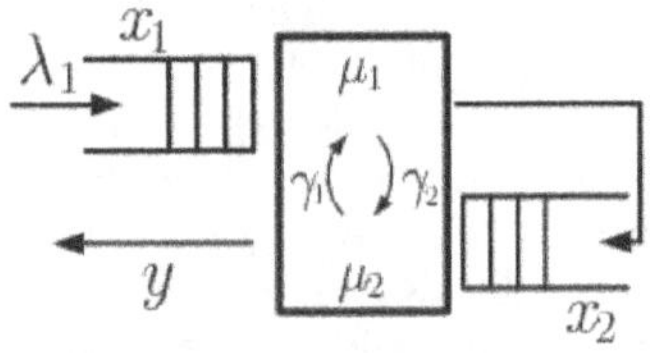

Figure 1: Two-product switching server.

To model this server system, we use a continuous state that consists next to the buffer contents x_n, $n = 1, 2$, also of the remaining setup time at the server, denoted by x_0. Therefore, $x = \begin{bmatrix} x_0 & x_1 & x_2 \end{bmatrix}^\top \in \mathbb{R}_+^{N+1}$ with $N = 2$ being the number of job types and $\mathbb{R}_+ = [0, \infty)$. For each job type n the system has two modes, one for setting up to serve the job type and the other for serving the job type. Hence, this results in $Q = 4$ modes (discrete states) in this case and the modes are denoted by $q \in \mathcal{Q} := \{1, 2, \ldots, Q\}$. The modes 1, 2, 3, and 4 represent setting up the server to serve job type 1, serving job type 1, setting up the server to serve job type 2, and serving job type 2, respectively. Note that the order in which the modes are traversed is fixed. The system evolves from mode 1 via modes 2, 3, and mode 4 back to mode 1 after which the cycle is repeated. In each mode $q \in \mathcal{Q}$ the continuous state x has a constant drift vector, denoted by f_q, i.e., $\dot{x} = f_q$. For the example system, these drift vectors are given by

$$f_1 = f_3 = \begin{bmatrix} -1 \\ \lambda_1 \\ 0 \end{bmatrix}, \quad f_2 = \begin{bmatrix} 0 \\ \lambda_1 - \mu_1 \\ \mu_1 \end{bmatrix}, \quad f_4 = \begin{bmatrix} 0 \\ \lambda_1 \\ -\mu_2 \end{bmatrix}.$$

Furthermore, a transition occurs in modes 1 and 3 to the next modes 2 and 4, respectively, when $x_0 = 0$ indicating that the setup time has elapsed. A transition from mode 2 to mode 3 occurs when $x_1 = 0$ (buffer of job type 1 is empty) and from mode 4 to mode 1 when $x_2 = 0$ (buffer of job type 2 is empty).

Due to the cyclic behavior in the way the nodes are traversed, it holds at the k-th event time t_k, $k \in \mathbb{N}_{\geq 1}$, that the system switches from mode $q = ((k-1) \bmod Q) + 1$ to the next mode $\sigma(q)$ where $\sigma : \mathcal{Q} \to \mathcal{Q}$ is given by

$$\sigma(q) := 1 + (q \bmod Q). \tag{1}$$

In addition, at the event time t_k the setup time x_0 instantaneously increases with constant $\alpha_q(t_k^-) \in \mathbb{R}^+$, $q \in \mathcal{Q}$, given by

$$\alpha_1 = \alpha_3 = 0, \quad \alpha_2 = \gamma_2, \quad \alpha_4 = \gamma_1,$$

i.e., we have a reset of the continuous state variable given by

$$x(t_k^+) = \begin{bmatrix} x_0(t_k^+) & x_1(t_k^+) & x_2(t_k^+) \end{bmatrix}^\top = \begin{bmatrix} \alpha_{q(t_k^-)} & x_1(t_k^-) & x_2(t_k^-) \end{bmatrix}^\top = x(t_k^-) + \alpha_q e_1 \tag{2}$$

(as $x_0(t_k^-) = 0$, see also Lemma 2.7 below), where t_k^- denotes the time just before the k-th event. Note that e_1 is the unit vector with a 1 at the first entry and zeros elsewhere.

This reset law shows that discontinuities only appear in x_0, while x_n, $n = 1 \ldots, N$, evolve continuously in time. Combining the above, leads to the following overall dynamics that can be compactly written as

$$\left.\begin{aligned} \dot{x} &= f_q \\ \dot{q} &= 0 \\ y &= h_q \end{aligned}\right\} \quad \text{if } e_{k_q}^\top x \geq 0, \tag{3a}$$

$$\left.\begin{aligned} x^+ &= x + \alpha_q e_1 \\ q^+ &= \sigma(q) \end{aligned}\right\} \quad \text{if } e_{k_q}^\top x = 0 \tag{3b}$$

with $k_1 = k_3 = 1$, $k_2 = 2$ and $k_4 = 3$ selecting the flow and jump sets (recall that e_{k_q} is the k_q-th unit vector in

$\{e_1, e_2, \ldots, e_{N+1}\}$). Note that we expressed the example in terms of the modeling framework of jump-flow systems advocated in [13]. In fact, solutions/executions of the system under study can be interpreted in the sense of [13]. Initial conditions for this system are given by $x(0) \in \mathbb{R}_+^{N+1}$ with $x_0(0) = \alpha_Q$, and $q(0) = 1$. Besides we use the convention that $t_0 = 0$.

Measurement information regarding the knowledge of when the server is processing job type 2 is included via the output y, with

$$h_1 = h_2 = h_3 = 0, \qquad h_4 = 4.$$

Hence, as long as the system is in mode 4, this is directly seen in the output y. When the system is in one of the other modes $q \in \mathcal{Q} \setminus \{4\}$, the output y is equal to 0 and no information is available from the server.

2.2 General dynamics

In this section we provide the general description of the class of PWAHS under study, which includes the single server system discussed in the previous section as a particular case. Essentially, the general dynamics of the class of systems is given by (3) with continuous state

$$x = \begin{bmatrix} x_0 & x_1 & \ldots & x_N \end{bmatrix}^\top \in \mathbb{R}_+^{N+1}$$

with $N \in \mathbb{N}_{\geq 1}$, discrete state $q \in \mathcal{Q} = \{1, 2, \ldots, Q\}$ with $Q \in \mathbb{N}_{\geq 1}$ and output $y \in \mathcal{Q}_0 := \mathcal{Q} \cup \{0\}$. The *data* of the system are given by the drift vectors $f_q \in \mathbb{R}^{N+1}$, the outputs $h_q \in \mathcal{Q}_0$, and $k_q \in \{1, 2, \ldots, N+1\}$ for each mode $q \in \mathcal{Q}$ together with the reset parameters α_q, $q \in \mathcal{Q}$. In addition, we have for outputs h_q, $q \in \mathcal{Q}$ that $h_q \in \{0, q\}$ for all $q \in \mathcal{Q}$. As in the single server system example, the general dynamics exhibits a *cycle*, i.e., a sequence of Q consecutive modes being repeated over time. Some other special characteristics in the data being inherited from server-like systems in manufacturing and traffic applications are summarized below.

ASSUMPTION 2.1. *For all $q \in \mathcal{Q}$ it holds that $e_{k_q}^\top f_q < 0$ and $e_i^\top f_q \geq 0$ when $i \in \{1, \ldots, N+1\} \setminus \{k_q\}$.*

This assumption guarantees that only one continuous state component decreases (being the one that also triggers the mode transition).

ASSUMPTION 2.2. *For all $q \in \mathcal{Q}$*

$$e_1^\top f_q = -1, \qquad \text{if } k_q = 1, \tag{4a}$$

$$e_1^\top f_q = 0, \qquad \text{if } k_q \neq 1. \tag{4b}$$

This assumption expresses that x_0 is indeed a timer-related variable with only 0 and -1 as slopes. In case x_0 acts as a timer that triggers the next event (i.e., $k_q = 1$) then $e_1^\top f_q = -1$, otherwise it is 0.

ASSUMPTION 2.3. *For all $q \in \mathcal{Q}$ it holds that*

$$\alpha_q > 0 \Leftrightarrow k_{\sigma(q)} = 1. \tag{5}$$

This assumption states that if a mode transition in mode q governs a jump in the state x_0, this state x_0 decreases in mode $\sigma(q)$ and triggers the next mode transition (and vice versa).

ASSUMPTION 2.4.

$$\sum_{q=1}^{Q} \alpha_q > 0. \tag{6}$$

If translated in terms of the server system example, this assumption states that during a cycle of modes at least one setup of non-zero duration is present.

ASSUMPTION 2.5. *For all $q \in \mathcal{Q}$ it holds that*

$$k_q = 1 \Rightarrow k_{\sigma(q)} \neq 1. \tag{7}$$

This assumption expresses that a setup mode is followed by an operational mode ($k_q \neq 1$).

ASSUMPTION 2.6. *There is at least one $q \in \mathcal{Q}$ such that $h_q = q$.*

This assumption states that we get at least some information from the system.

Throughout the paper we assume that all the mentioned assumptions are true (without further reference).

2.3 Basic results

In this section we derive some basic results for the class of PWAHS under study. To do so, let us denote, as before, by t_k the k-th time occurrence of a transition and $t_0 = 0$. Then we can prove the following lemma.

LEMMA 2.7. *For any trajectory of the PWAHS (3) it holds that $x_0(t_k^-) = 0$ for all $k \in \mathbb{N}_{\geq 1}$.*

PROOF. We prove this statement using induction. Suppose the statement holds for some $k \in \mathbb{N}_{\geq 1}$, i.e., $x_0(t_k^-) = 0$. At event time t_k, when going from mode q to mode $\sigma(q)$, two situations can occur, being $\alpha_q > 0$ or $\alpha_q = 0$. In the first case, it holds that $x_0(t_k^+) = x_0(t_k^-) + \alpha_q$. Then due to (3), (4a) and (5) it follows that $\dot{x}_0 = -1$ for $t \in [t_k, t_{k+1})$ until $x_0 = 0$. Hence $x_0(t_{k+1}^-) = 0$. In the second case, i.e., $\alpha_q = 0$, no jump occurs and $x_0(t_{k+1}^-) = x_0(t_k^-)$ and $\dot{x} = 0$ for $t \in [t_k, t_{k+1})$ due to (4b) and (5). Therefore, $x_0(t_{k+1}^-) = x_0(t_k^-) = 0$. To complete the proof we need $x_0(t_1^-) = 0$. If $\alpha_Q > 0$, $x_0(t_1^-) = 0$ follows by the same reasoning as in the first case above. If $\alpha_Q = 0$, we immediately have $x_0(t_0) = 0$ and can use the reasoning in the second case to conclude $x_0(t_1^-) = 0$, thereby completing the proof. □

LEMMA 2.8. *Consider $q \in \mathcal{Q}$ such that $\alpha_q > 0$. The dwell time in mode $\sigma(q)$ is equal to α_q.*

PROOF. From Assumption 2.2 and Assumption 2.3 we know that if $\alpha_q > 0$ we have $e_1^\top f_{\sigma(q)} = -1$ and $k_{\sigma(q)} = 1$. Due to (3b) and Lemma 2.7 it holds that $x_0(t_k^+) = \alpha_q$ for the event at time t_k where the system switches from mode q to mode $\sigma(q)$. In addition, in mode $\sigma(q)$ state x_0 decreases according to $\dot{x}_0 = -1$ until $x_0(t_{k+1}^-) = 0$. Hence, the dwell time in mode $\sigma(q)$ is therefore given by

$$\frac{\alpha_q}{-e_1^\top f_{\sigma(q)}} = \alpha_q.$$

□

For the class of PWAHS the following statements can be made regarding Zeno behavior and fixed points:

LEMMA 2.9. *Zeno behavior is not present in system (3).*

PROOF. Assumption 2.3 and Assumption 2.4 imply that there exists at least one mode in the cycle with $\alpha_q > 0$, $q \in \mathcal{Q}$. Lemma 2.8 shows that the dwell time in mode $\sigma(q)$ is α_q. Therefore, the dwell time of a cycle is bounded away from zero given the cyclic behavior, and no Zeno behavior occurs. □

LEMMA 2.10. *System (3) does not contain any fixed points.*

PROOF. The condition $e_{k_q}^\top f_q < 0$ in Assumption 2.1 guarantees that $e_{k_q}^\top x$ decreases at constant rate, until it reaches the jump criterion $e_{k_q}^\top x = 0$. Therefore, every mode is left in finite time. Since Zeno behavior is excluded, the system has no fixed points. □

LEMMA 2.11. *The system (3) is a positive system in the sense that if $x(0) \in \mathbb{R}_+^{N+1}$ with $x_0(0) = \alpha_Q$, and $q(0) = 1$, then $x(t) \in \mathbb{R}_+^{N+1}$ for all $t \in \mathbb{R}_+$.*

PROOF. The proof follows similar reasoning as the proof of Lemma 2.7 using Assumptions 2.1 and 2.2. □

2.4 Visible and invisible modes and event times

In the remainder of this paper we use the following notations in which we make a distinction between visible and invisible modes, and visible and invisible events.

A mode $q \in \mathcal{Q}$ is called *visible* if $h(q) = q$ and otherwise it is called *invisible.* The set of visible modes is denoted by $\mathcal{Q}_v$, i.e,

$$\mathcal{Q}_v = \{q \in \mathcal{Q} \mid h_q = q\}. \tag{8}$$

Transitions to and from visible modes are called *visible events* and transitions from invisible modes to invisible modes are called *invisible events.* To describe the visible events we define the set

$$\mathcal{V} = \mathcal{Q}_v \cup \{q \in \mathcal{Q} \mid \sigma(q) \in \mathcal{Q}_v\},$$

which is enumerated as $\{v_1, v_2, \ldots, v_V\} \subseteq \mathcal{Q}$ with $1 \le v_1 < v_2 < \ldots < v_V \le Q$. Using the above notation, if at time t_k the k-th event occurs jumping from mode $q(t_k^-)$ to mode $\sigma(q(t_k^-))$, this event is visible if and only if $q(t_k^-) \in \mathcal{V}$. Hence, loosely speaking, we know when the system enters or leaves visible modes and we know when the corresponding events occur.

In the remainder we will use j as the visible event counter and denote visible event times by $t_j^v = t_{k(j)}$, $j \in \mathbb{N}_{\ge 1}$. where $k(j)$ translates the visible event j into the corresponding ordinary event k, i.e.,

$$k(j) = v_{1+(j-1 \bmod V} + \lfloor (j-1)/V \rfloor \cdot Q, \tag{9}$$

where $\lfloor r \rfloor$ denotes the largest integer smaller than $r \in \mathbb{R}$. Due to the cyclic behavior in the way the (visible) nodes are traversed, it holds that visible event $v_{1+(j-1 \bmod V)}$ has successive visible event $\sigma_v(v_{1+(j-1 \bmod V)})$ where $\sigma_v : \mathcal{V} \to \mathcal{V}$, given by

$$\sigma_v(v_l) := v_{1+(l \bmod V)}, \qquad l = 1, 2, ..., V. \tag{10}$$

3. SAMPLING THE HYBRID SYSTEM

One of the main ideas in the observer design methodology is to derive the desired continuous-time observer for system (3) from a discrete-time observer. To that end, we sample the system at the visible event times t_j^v, $j \in \mathbb{N}_{\ge 1}$. To easily derive the sampled data, we split its computation into three parts. First the system is sampled at all events t_k, $k \in \mathbb{N}_{\ge 1}$. Second, the state dimension is reduced by removing the timer variable x_0. In the third step the sampled system description is limited to only the visible events.

3.1 Sampling at all event times

Let $x(t_k^-)$ denote the state at time t_k just before the jump from mode $q(t_k^-)$ to mode $\sigma(q(t_k^-))$. Sampling at all event times t_k, $k \in \mathbb{N}_{\ge 1}$, results in the system

$$x(t_{k+1}^-) = \tilde{A}_{q(t_k^-)} x(t_k^-) + \tilde{a}_{q(t_k^-)}, \tag{11a}$$

$$t_{k+1} = t_k + \tilde{C}_{q(t_k^-)} x(t_k^-) + \tilde{c}_{q(t_k^-)} \tag{11b}$$

in which

$$\tilde{A}_{q(t_k^-)} = I + \frac{f_{\sigma(q(t_k^-))} e_{k_{\sigma(q(k))}}^\top}{-e_{k_{\sigma(q(t_k^-))}}^\top f_{\sigma(q(t_k^-))}}, \qquad \tilde{a}_{q(t_k^-)} = \alpha_{q(t_k^-)} A_{q(t_k^-)} e_1, \tag{12a}$$

$$\tilde{C}_{q(t_k^-)} = \frac{e_{k_{\sigma(q(t_k^-))}}^\top}{-e_{k_{\sigma(q(t_k^-))}}^\top f_{\sigma(q(t_k^-))}}, \qquad \tilde{c}_{q(t_k^-)} = \alpha_{q(t_k^-)} C_{q(t_k^-)} e_1 \tag{12b}$$

as can be derived from system (3). Note that the system (11) is periodic with period Q in the sense that $\tilde{A}_{k+Q} = \tilde{A}_k$ for $k \in \mathbb{N}_{\ge 1}$, and similar expressions hold for $\tilde{a}_k$, $\tilde{C}_k$ and $\tilde{c}_k$.

REMARK 3.1. *3 Notice that from Assumptions 2.1 and 2.2 we obtain that $\tilde{A}_k$, $\tilde{a}_k$, $\tilde{C}_k$, and $\tilde{c}_k$ only contain non-negative elements, resulting in a positive system, which is to be expected given Lemma 2.11.*

The sampled system (11) can be used to write the system as a timed automaton [3] with a single clock, due to the linear dynamics and cyclic behavior. Then, the continuous state can be derived via a linear combination of the clock and the sampled state. However, observability and observer design for timed automata considers the discrete state reconstruction, see e.g., [20], which is straightforward for the class of systems under consideration. Furthermore, we are also interested in reconstructing the continuous state.

Example

For the illustrative system in Section 2.1, we obtain a 4-periodic system for which the matrices in (12) are given as

$$\tilde{A}_1 = \begin{bmatrix} 1 & 0 & 0 \\ 0 & 0 & 0 \\ 0 & \frac{\mu_1}{\mu_1 - \lambda_1} & 1 \end{bmatrix}, \qquad \tilde{a}_1 = \begin{bmatrix} 0 \\ 0 \\ 0 \end{bmatrix},$$

$$\tilde{C}_1 = \begin{bmatrix} 0 & \frac{1}{\mu_1 - \lambda_1} & 0 \end{bmatrix}, \qquad \tilde{c}_1 = 0,$$

$$\tilde{A}_2 = \begin{bmatrix} 0 & 0 & 0 \\ \lambda_1 & 1 & 0 \\ 0 & 0 & 1 \end{bmatrix}, \qquad \tilde{a}_2 = \begin{bmatrix} 0 \\ \gamma_2 \lambda_1 \\ 0 \end{bmatrix},$$

$$\tilde{C}_2 = \begin{bmatrix} 1 & 0 & 0 \end{bmatrix}, \qquad \tilde{c}_2 = \gamma_2,$$

$$\tilde{A}_3 = \begin{bmatrix}1 & 0 & 0\\ 0 & 1 & \frac{\lambda_1}{\mu_2}\\ 0 & 0 & 0\end{bmatrix}, \qquad \tilde{a}_3 = \begin{bmatrix}0\\0\\0\end{bmatrix},$$

$$\tilde{C}_3 = \begin{bmatrix}0 & 0 & \frac{1}{\mu_2}\end{bmatrix}, \qquad \tilde{c}_3 = 0,$$

$$\tilde{A}_4 = \begin{bmatrix}0 & 0 & 0\\ \lambda_1 & 1 & 0\\ 0 & 0 & 1\end{bmatrix}, \qquad \tilde{a}_4 = \begin{bmatrix}0\\ \gamma_1\lambda_1\\ 0\end{bmatrix},$$

$$\tilde{C}_4 = \begin{bmatrix}1 & 0 & 0\end{bmatrix}, \qquad \tilde{c}_4 = \gamma_1.$$

3.2 State reduction

From Lemma 2.7 we have $x_0(t_k^-) = 0$ for all $k \in \mathbb{N}_{\geq 1}$. Therefore, we can consider the reduced state $\bar{x} = \begin{bmatrix}x_1 & \dots & x_N\end{bmatrix}^\top = \begin{bmatrix}0 & I\end{bmatrix} x \in \mathbb{R}_+^N$, for which the dynamics at the event times becomes

$$\bar{x}(t_{k+1}^-) = \bar{A}_{q(t_k^-)}\bar{x}(t_k^-) + \bar{a}_{q(t_k^-)}, \tag{13a}$$

$$t_{k+1} = t_k + \bar{C}_{q(t_k^-)}\bar{x}(t_k^-) + \bar{c}_{q(t_k^-)}. \tag{13b}$$

where

$$\bar{A}_{q(t_k^-)} = \begin{bmatrix}0 & I\end{bmatrix}\tilde{A}_{q(t_k^-)}\begin{bmatrix}0 & I\end{bmatrix}^\top, \quad \bar{a}_{q(t_k^-)} = \begin{bmatrix}0 & I\end{bmatrix}\tilde{a}_{q(t_k^-)},$$

$$\bar{C}_{q(t_k^-)} = \tilde{C}_{q(t_k^-)}\begin{bmatrix}0 & I\end{bmatrix}^\top, \qquad \bar{c}_{q(t_k^-)} = \tilde{c}_{q(t_k^-)}.$$

Note that due to this reduction no discontinuities in $\bar{x}(t)$ occur, as they only appear in x_0 (2). Hence, $\bar{x}(t_k^-) = \bar{x}(t_k^+) = \bar{x}(t_k)$.

Example
For the illustrative system in Section 2.1, these reduced matrices are as follows:

$$\bar{A}_1 = \begin{bmatrix}0 & 0\\ \frac{\mu_1}{\mu_1-\lambda_1} & 1\end{bmatrix}, \ \bar{a}_1 = \begin{bmatrix}0\\0\end{bmatrix}, \quad \bar{C}_1 = \begin{bmatrix}\frac{1}{\mu_1-\lambda_1} & 0\end{bmatrix}, \ \bar{c}_1 = 0,$$

$$\bar{A}_2 = \begin{bmatrix}1 & 0\\ 0 & 1\end{bmatrix}, \quad \bar{a}_2 = \begin{bmatrix}\gamma_2\lambda_1\\ 0\end{bmatrix}, \ \bar{C}_2 = \begin{bmatrix}0 & 0\end{bmatrix}, \quad \bar{c}_2 = \gamma_2,$$

$$\bar{A}_3 = \begin{bmatrix}1 & \frac{\lambda_1}{\mu_2}\\ 0 & 0\end{bmatrix}, \quad \bar{a}_3 = \begin{bmatrix}0\\0\end{bmatrix}, \quad \bar{C}_3 = \begin{bmatrix}0 & \frac{1}{\mu_2}\end{bmatrix}, \quad \bar{c}_3 = 0,$$

$$\bar{A}_4 = \begin{bmatrix}1 & 0\\ 0 & 1\end{bmatrix}, \quad \bar{a}_4 = \begin{bmatrix}\gamma_1\lambda_1\\ 0\end{bmatrix}, \ \bar{C}_4 = \begin{bmatrix}0 & 0\end{bmatrix}, \quad \bar{c}_4 = \gamma_1.$$

3.3 Sampling at visible event times

Let $\bar{x}(t_j^v)$ denote the reduced state vector at t_j^v, the jth visible event. From (13) it is clear that sampling at the visible events results in the system

$$\bar{x}(t_{j+1}^v) = A_{q(t_j^{v-})}\bar{x}(t_j^v) + a_{q(t_j^{v-})}, \tag{14a}$$

$$t_{j+1}^v = t_j^v + C_{q(t_j^{v-})}\bar{x}(t_j^v) + c_{q(t_j^{v-})}. \tag{14b}$$

Since we have V different visible events, the system (14) is a periodic linear system with period V. The system matrices in (14) for all $j \in \mathbb{N}_{\geq 1}$ are presented below. For $j \neq lV$, $l \in \mathbb{N}_{\geq 1}$, we have

$$A_j = \prod_{q=v_j}^{\sigma_v(v_j)-1} \bar{A}_q, \tag{15a}$$

$$a_j = \sum_{q=v_j}^{\sigma_v(v_j)-1}\left(\prod_{r=q+1}^{\sigma_v(v_j)-1} \bar{A}_r\right)\bar{a}_q, \tag{15b}$$

$$C_j = \sum_{q=v_j}^{\sigma_v(v_j)-1} \bar{C}_q \prod_{r=v_j}^{q-1} \bar{A}_r, \tag{15c}$$

$$c_j = \sum_{q=v_j}^{\sigma_v(v_j)-1} \bar{c}_q + \sum_{q=v_j+1}^{\sigma_v(v_j)-1} \bar{C}_q \sum_{r=v_j}^{q-1}\left(\prod_{s=r+1}^{q-1} \bar{A}_s\right)\bar{a}_r \tag{15d}$$

For $j = lV$, $l \in \mathbb{N}_{\geq 1}$, we have

$$A_j = \prod_{q=1}^{\sigma_v(v_j)-1} \bar{A}_q \prod_{q=v_j}^{Q} \bar{A}_q, \tag{15e}$$

$$a_j = \sum_{q=v_j}^{Q+\sigma_v(v_j)-1}\left(\prod_{r=q+1}^{Q+\sigma_v(v_j)-1} \bar{A}_r\right)\bar{a}_q, \tag{15f}$$

$$C_j = \sum_{q=v_j}^{Q+\sigma_v(v_j)-1} \bar{C}_q \prod_{r=v_j}^{q-1} \bar{A}_r, \tag{15g}$$

$$c_j = \sum_{q=v_j}^{Q+\sigma_v(v_j)-1} \bar{c}_q + \sum_{q=v_j+1}^{Q+\sigma_v(v_j)-1} \bar{C}_q \sum_{r=v_j}^{q-1}\left(\prod_{s=r+1}^{q-1} \bar{A}_s\right)\bar{a}_r \tag{15h}$$

with $A_{j+V} = A_j$, $j, \mathbb{N}_{\geq 1}$, and similar expressions for a_j, C_j, c_j and v_j, $j \in \mathbb{N}_{\geq 1}$.

Example
For the illustrative system in Section 2.1, we obtain a 2-periodic system (14) for which the matrices above are given by

$$A_1 = \begin{bmatrix}1 & \frac{\lambda_1}{\mu_2}\\ 0 & 0\end{bmatrix}, \qquad a_1 = \begin{bmatrix}0\\0\end{bmatrix},$$

$$C_1 = \begin{bmatrix}0 & \frac{1}{\mu_2}\end{bmatrix}, \qquad c_1 = 0,$$

$$A_2 = \begin{bmatrix}0 & 0\\ \frac{\mu_1}{\mu_1-\lambda_1} & 1\end{bmatrix}, \qquad a_2 = \begin{bmatrix}\gamma_2\lambda_1\\ \frac{\lambda_1\mu_1\gamma_1}{\mu_1-\lambda_1}\end{bmatrix},$$

$$C_2 = \begin{bmatrix}\frac{1}{\mu_1-\lambda_1} & 0\end{bmatrix}, \qquad c_2 = \frac{\gamma_1\mu_1}{\mu_1-\lambda_1} + \gamma_2.$$

4. OBSERVER DESIGN

Notice that we only receive information about the system's state when visible events occur. Therefore, from the occurrence of visible events we need to reconstruct the system's state.

We therefore build an observer in two steps. First, starting from the dynamics (14) we build a linear time-varying (periodic) discrete-time observer to reconstruct the system's state at visible event times. Next, we use the dynamics (3) to make an open-loop prediction of the system's state, which is corrected (if necessary) at the next visible event time.

4.1 Discrete-time observer at visible event times

Our first goal is to build an observer which reconstructs the state at visible event times t_j^v, as described by the dynamics (14). To that end, we can use a Luenberger observer

$$\hat{\bar{x}}(t_{j+1}^v) = A_{q(t_j^{v-})}\hat{\bar{x}}(t_j^v) + a_{q(t_j^{v-})} + L_{q(t_j^{v-})}(t_{j+1}^v - \hat{t}_{j+1}^v), \tag{16a}$$

$$\hat{t}_{j+1}^v = t_j^v + C_{q(t_j^{v-})}\hat{\bar{x}}(t_j^v) + c_{q(t_j^{v-})}, \tag{16b}$$

where $L_1, L_2, ..., L_V$ are the observer gains. Evolution of the observation error $\bar{e}(t_j^v) = \bar{x}(t_j^v) - \hat{\bar{x}}(t_j^v)$ is given by

$$\bar{e}(t_{j+1}^v) = \left[A_{q(t_j^{v-})} - L_{q(t_j^{v-})}C_{q(t_j^{v-})}\right]\bar{e}(t_j^v).$$

ASSUMPTION 4.1. *There exist $L_1, L_2, ..., L_V$ such that the observer* (16) *leads to*

$$\lim_{j\to\infty} \|\bar{e}(t_j^v)\| = 0.$$

Finding observer gains for this periodic system satisfying (4.1) is a known observer design problem, cf. [16, 19, 22]. For the example presented in the next section, we exploit the periodicity of this system and use a simple sequential algorithm presented in [15] for determining the time-varying observer gains to guarantee deadbeat convergence to zero at visible event times.

4.2 Continuous-time observer

Starting from the *reduced* state estimates at visible event times, as generated by the observer (16), we will now provide a state estimate $\hat{x} \in \mathbb{R}^{N+1}$ of the full state x of (3) (i.e., including x_0). In fact, we will guarantee that the estimated states $\hat{x}(t_j^{v+})$ just after visible events satisfy

$$[0\ I]\hat{x}(t_j^{v+}) = \hat{\bar{x}}(t_j^v), \tag{17}$$

$j \in \mathbb{N}_{\geq 1}$, indicating that just after the visible events the estimated states (without timer x_0) and the estimates $\hat{\bar{x}}(t_j^v)$ of the discrete-time observer (16) coincide. In addition, we use the dynamics (3) to make an open-loop prediction of the system's state after visible events. This open-loop prediction is updated using the observer (16) as soon as a new visible event happens.

However, notice that due to estimation errors, the predicted occurrence of the next visible event, $\hat{t}_{j+1}^v$ based on (16b), can be either sooner or later than the actual occurrence of the next visible event t_{j+1}^v. In the latter case ($t_{j+1}^v \leq \hat{t}_{j+1}^v$) we can simply update the observer state according to (16), but in the former case we cannot (yet) use (16), as the duration to the next visible event $t_{j+1}^v - \hat{t}_{j+1}^v$ is not known yet. Hence in this case ($t_{j+1}^v > \hat{t}_{j+1}^v$), we have to determine the continuous-time observer dynamics for the period from predicted to actual occurrence of the next visible event. To do so, we introduce additional modes (called waiting modes) for the continuous-time observer. We therefore extend the set $\mathcal{Q}$ of modes, by defining the set of observer modes as

$$\hat{\mathcal{Q}} := \mathcal{Q} \cup \left\{ v_l + \frac{1}{2} \middle| l = 1, 2, \ldots, V \right\},$$

where $v_l + \frac{1}{2}$, $l = 1, 2, \ldots, V$ are labels to denote the waiting modes. Furthermore, we define the mode transition map $\hat{\sigma} : \hat{\mathcal{Q}} \to \hat{\mathcal{Q}}$, as

$$\hat{\sigma}(\hat{q}) := \begin{cases} \hat{q} + \frac{1}{2} & \text{if } \hat{q} \in \mathcal{V} \\ \sigma(\hat{q}) & \text{if } \hat{q} \in \mathcal{Q} \setminus \mathcal{V} \\ \sigma\left(\hat{q} - \frac{1}{2}\right) & \text{if } \hat{q} \in \hat{\mathcal{Q}} \setminus \mathcal{Q} \end{cases}$$

The waiting modes will only be used after an expected visible event which has not yet happened, i.e., when $\hat{t}_{j+1}^v \leq t < t_{j+1}^v$. For the additional waiting modes of the observer, we have to determine a drift vector for the state estimate. Notice from the observer (16) that at the visible event times we update the state estimate according to (16a). That is, we add $L_{q(t_j^{v-})}$ times the amount of time that the actual visible event is later than predicted. From this we can derive a continuous evolution of the state estimate keeping $\hat{x}_0$ constant (at zero) and using a drift vector of $L_{q(t_j^{v-})}$ for the remaining state, i.e., $\dot{\hat{x}} = [0\ L_{q(t_j^{v-})}^\top]^\top$, to avoid the need to reset the states at t_{j+1}^v, $j \in \mathbb{N}_{\geq 1}$ (although other choices guaranteeing (17) are fine as well). Furthermore, the output in the waiting mode is identical to the output of the preceding mode, i.e., $h_{\hat{q}} = h_{\hat{q}-\frac{1}{2}}$ for $\hat{q} \in \hat{\mathcal{Q}} \setminus \mathcal{Q}$.

Using the above reasoning, we can deduce a continuous-time observer in the form of a jump-flow system [13]. The discrete-time observer (16) will be embedded in the observer by including the state variables $\tilde{x} \in \mathbb{R}^N$ and $\tilde{t} \in \mathbb{R}$ in the continuous-time observer, which will satisfy $\tilde{x}(t) = \hat{\bar{x}}(t_j^v)$, $t \in [t_j^v, t_{j+1}^v)$ (solutions to (16)), and $\tilde{t}(t) = t_j^v$, $t \in [t_j^v, t_{j+1}^v)$. In between visible events $\tilde{x}$ and $\tilde{t}$ will be constant. This leads to the following flow expressions for the continuous-time observer. When the measurement information equals the estimated output, i.e., $y = \hat{y}$ with $y = h_q$ and $\hat{y} = h_{\hat{q}}$ see (3), the flow expressions are given by

$$\left.\begin{aligned} \dot{\hat{x}} &= f_{\hat{q}} \\ \dot{\hat{q}} &= 0 \\ \dot{\tilde{x}} &= 0 \\ \dot{\tilde{t}} &= 0 \\ \hat{y} &= h_{\hat{q}} \end{aligned}\right\} \quad \text{if } e_{k_{\hat{q}}}^\top \hat{x} \geq 0 \wedge \hat{q} \in \mathcal{Q} \wedge y = \hat{y}, \tag{18a}$$

$$\left.\begin{aligned} \dot{\hat{x}} &= \begin{bmatrix} 0 \\ L_{\sigma_v^{-1}(\hat{q}-\frac{1}{2})} \end{bmatrix} \\ \dot{\hat{q}} &= 0 \\ \dot{\tilde{x}} &= 0 \\ \dot{\tilde{t}} &= 0 \\ \hat{y} &= h_{\hat{q}} \end{aligned}\right\} \quad \text{if } \hat{q} \in \hat{\mathcal{Q}} \setminus \mathcal{Q} \wedge y = \hat{y}. \tag{18b}$$

Note that (18a) describes the normal flow predictions based on model (3), while (18b) corresponds to the predictions in the waiting modes. The jump expressions for the normal flow predictions are given by

$$\left.\begin{aligned} \hat{x}^+ &= \hat{x} + \alpha_{\hat{q}} e_1 \\ \tilde{x}^+ &= \tilde{x} \\ \hat{q}^+ &= \hat{\sigma}(\hat{q}) \\ \tilde{t}^+ &= \tilde{t} \end{aligned}\right\} \quad \text{if } e_{k_{\hat{q}}}^\top \hat{x} = 0 \wedge \hat{q} \in \mathcal{Q} \wedge y = \hat{y}, \tag{18c}$$

following the normal jump expressions based on model (3). Furthermore, (18c) also describes that if a predicted visible event occurs before the actual visible event ($\hat{t}_{j+1}^v < t_{j+1}^v$), the discrete state jumps to a waiting mode.

If the measurement information differs from the estimated output, i.e., $y \neq \hat{y}$, a jump is required, since the system is in a different mode than the observer. Apart from an initial mode mismatch, the measurement difference $y \neq \hat{y}$ occurs via a change in y. In this case, a distinction is made between jumping from a mode without measurement information to a mode with measurement information, i.e., $y \neq \hat{y}$ and $y \neq 0$, and jumping from a mode with measurement information to a mode without measurement information, i.e., $y \neq \hat{y}$ and $y = 0$. In the former case, the measurement information provides information about the mode after the jump ($\hat{q}^+ = y$). In the latter case, the mode after the jump is derived by using the cyclic mode transitions ($\hat{q}^+ = \sigma(\hat{y})$). The jump expressions are given by

$$\left.\begin{aligned}
\xi &= \tilde{t} + C_{p(y)}\tilde{x} + c_{p(y)} \\
\hat{x}^+ &= \begin{bmatrix} 0 \\ A_{p(y)} \end{bmatrix}\tilde{x} + \begin{bmatrix} \alpha_{\sigma^{-1}(y)} \\ a_{p(y)} \end{bmatrix} + \begin{bmatrix} 0 \\ L_{p(y)} \end{bmatrix}(t-\xi) \\
\hat{q}^+ &= y \\
\tilde{x}^+ &= \hat{\bar{x}}^+ \\
\tilde{t}^+ &= t
\end{aligned}\right\}\ldots$$

$$\ldots\text{if } y \neq \hat{y} \wedge y \neq 0, \tag{18d}$$

$$\left.\begin{aligned}
\xi &= \tilde{t} + C_{\sigma^{-1}(\hat{y})}\tilde{x} + c_{\sigma^{-1}(\hat{y})} \\
\hat{x}^+ &= \begin{bmatrix} 0 \\ A_{\sigma^{-1}(\hat{y})} \end{bmatrix}\tilde{x} + \begin{bmatrix} \alpha_{\sigma(\hat{y})} \\ a_{\sigma^{-1}(\hat{y})} \end{bmatrix} + \begin{bmatrix} 0 \\ L_{\sigma^{-1}(\hat{y})} \end{bmatrix}(t-\xi) \\
\hat{q}^+ &= \sigma(\hat{y}) \\
\tilde{x}^+ &= \hat{\bar{x}}^+ \\
\tilde{t}^+ &= t
\end{aligned}\right\}\ldots$$

$$\ldots\text{if } y \neq \hat{y} \wedge y = 0, \tag{18e}$$

where we denoted $p(y) = \sigma_v^{-1}(\sigma^{-1}(y))$ and introduced ξ for ease of exposition. Recall that $\hat{\bar{x}}(t) = [0\ I]\hat{x}(t)$. Note that (18d) and (18e) describe the jump expressions at occurrence of a visible event. Based on some knowledge of initial conditions of the original system (3), we initialize the observer as $\hat{x}(0) \in \mathbb{R}_+^{N+1}$ with $\hat{x}_0(0) = \alpha_Q$, $\hat{q}(0) = 1$, $\tilde{x}(0) = \hat{\bar{x}}(0)$ and $\tilde{t}(0) = 0$.

PROPOSITION 4.2. *Suppose Assumption 4.1 holds. Then the continuous-time observer* (18) *asymptotically reconstructs the states $\bar{x}$ of the system* (3), *i.e.,*

$$\lim_{t\to\infty} \|\bar{x}(t) - \hat{\bar{x}}(t)\| = 0.$$

PROOF. Since the continuous-time observer was designed to satisfy $\|[0\ I]x(t_j^{v+}) - [0\ I]\hat{x}(t_j^{v+})\| = \|\bar{x}(t_j^v) - \hat{\bar{x}}(t_j^v)\|$, $j \in \mathbb{N}_{\geq 1}$, it suffices to show that there exists a constant P such that $\|\bar{x}(t) - \hat{\bar{x}}(t)\| \leq P\|\bar{x}(t_j^v) - \hat{\bar{x}}(t_j^v)\|$ for $t \in [t_j^v, t_{j+1}^v)$, $j \in \mathbb{N}_{\geq 1}$. The latter will follow from the observation that the observation error only changes when the observer is in a different mode than the actual system, as we will show below.

Notice that (11) describes the evolution of the mode changes of the system, i.e., the state updates at the switching times t_k. Let $\hat{t}_k$ denote the switching times as predicted by the continuous-time observer. Then we have from (11b):

$$t_{k+1} - \hat{t}_{k+1} = t_k - \hat{t}_k + \tilde{C}_{q(t_k^-)}[x(t_k^-) - \hat{x}(t_k^-)]. \tag{19}$$

Since at time t_j^v we have $\hat{t}_{k(j)} = t_{k(j)}$, it follows from (19) that the duration of the mode difference is a linear function of the initial observer error, i.e., and so is the sum of the durations of mode differences. Finally, since during mode differences the rate of increase is bounded, and the number of invisible modes between two consecutive visible modes is finite, the observation error $\|\bar{x}(t) - \hat{\bar{x}}(t)\|$ on the interval $t \in [t_j, t_{j+1})$ can be upperbounded by $\|\bar{x}(t) - \hat{\bar{x}}(t)\| \leq P\|\bar{x}(t_j^v) - \hat{\bar{x}}(t_j^v)\|$.

Notice that from (19) we also have that $\lim_{k\to\infty} |t_k - \hat{t}_k| = 0$.
□

REMARK 4.3. *Using a deadbeat observer the complete state can be asymptotically reconstructed, i.e.,* $\lim_{t\to\infty} \|x(t) - \hat{x}(t)\| = 0$. *For non-deadbeat observers peaking occurs in* $\|x_0(t) - \hat{x}_0(t)\|$ *due to mismatch in event times combined with system jumps* (2). *However, in most manufacturing and traffic applications one is interested in the buffer sizes or queue lengths, i.e.,* $\hat{\bar{x}}(t)$.

REMARK 4.4. *As we observed in Lemma 2.11 and Remark 3.1, we are essentially dealing with a positive system. However, the observer* (16) *does not necessarily guarantee positivity of the state estimates* $\hat{x}(t)$, $t \in \mathbb{R}$. *Though the observer* (18) *is well defined and asymptotically recovers the state of the original system, from a physical point of view it would be better to have non-negative state estimates. As we show in the next section by means of an example, it is possible to derive observers that respect the positivity property by generating non-negative state estimates. Note that observer gains satisfying* $(A_v - L_vC_v) \geq 0$ *and* $L_vC_v \geq 0$, $v \in \mathcal{V}$, *suffice since*

$$\begin{aligned}
\hat{\bar{x}}(t_{j+1}^{v-}) =& (A_{q(t_j^{v-})} - L_{q(t_j^{v-})}C_{q(t_j^{v-})})\bar{x}(t_j^{v-}) + \ldots \\
&\ldots L_{q(t_j^{v-})}C_{q(t_j^{v-})}x(t_j^{v-}) + a_{q(t_j^{v-})}.
\end{aligned}$$

In fact, designing directly positive observers is one of the questions for future research.

One direction to pursue in this context is considering the dynamics (14) *only once every J time-instances and lift the system (see, for instance, [10]) leading to a linear time-invariant positive system. The positive observation problem for linear discrete time-invariant positive systems has been dealt with in [1], where a necessary and sufficient condition for the existence and the design of a positive linear observer of Luenberger form has been given by means of the feasibility of a linear program (LP). This could form an interesting starting point to obtain positive discrete-time observers of the form* (16a) *(which is doable under certain assumptions). The step towards a continuous-time observers could follow then* mutatis mutandis *the line of reasoning as indicated above.*

An alternative solution leading to positive estimates is to use a projection of $\hat{\bar{x}}(t)$ *on the positive cone* $\Omega = \mathbb{R}_+^N$ *as the estimated state instead of* $\hat{\bar{x}}(t)$, *i.e., use* $P_\Omega\hat{\bar{x}}(t)$ *with* $P_\Omega : \mathbb{R}^N \to \mathbb{R}_+^N$, *given for* $z \in \mathbb{R}^N$

$$(P_\Omega z)_i = \max(0, z_i).$$

This projected estimate also asymptotically recovers the true state, i.e.,

$$\lim_{t\to\infty} \|P_\Omega\hat{\bar{x}}(t) - \bar{x}(t)\| = 0.$$

5. TRAFFIC INTERSECTION

To demonstrate the observer design, we introduce a signalized T-junction consisting of three flows of cars, see Figure 2. Each flow can go into two directions. For this example we assume that each flow has a single signal, i.e., if a car receives a green light it can move in two different directions. Therefore, it is not possible to give multiple flows a green light simultaneously. The flows are served in order 1, 2, 3, and then back to flow 1 after which the cycle is repeated. The intersection uses a clearing policy, i.e., it completely empties the queue of a flow before switching to serve the next flow. Switching to serve cars from flow i requires a clearing/setup time γ_i to make sure all vehicles from the previously served flow have cleared the intersection. Vehicles arrive at flow i with arrival rate λ_i and are served with process rates μ_i, $i = 1, 2, 3$. A sensor measures the crossing of vehicles in lane 1.

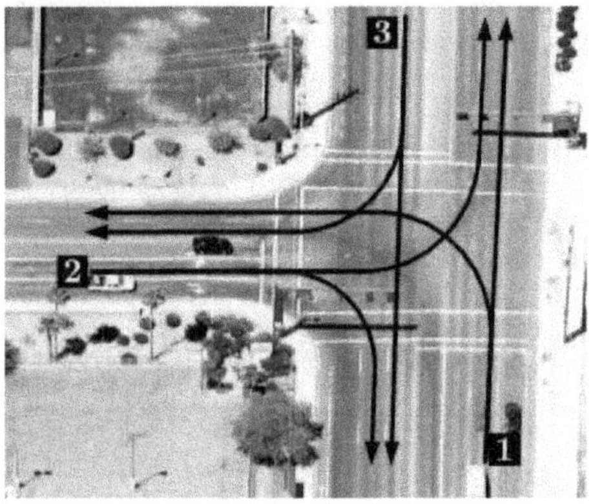

Figure 2: Signalized T-junction containing three vehicle flows 1-3.

The dynamics can be written in the form (3) with

$$f_1 = f_3 = f_5 = \begin{bmatrix} -1 \\ \lambda_1 \\ \lambda_2 \\ \lambda_3 \end{bmatrix},$$

$$f_2 = \begin{bmatrix} 0 \\ \lambda_1 - \mu_1 \\ \lambda_2 \\ \lambda_3 \end{bmatrix}, \quad f_4 = \begin{bmatrix} 0 \\ \lambda_1 \\ \lambda_2 - \mu_2 \\ \lambda_3 \end{bmatrix}, \quad f_6 = \begin{bmatrix} 0 \\ \lambda_1 \\ \lambda_2 \\ \lambda_3 - \mu_3 \end{bmatrix}$$

$$\alpha_1 = \alpha_3 = \alpha_5 = 0, \quad \alpha_2 = \gamma_2, \quad \alpha_4 = \gamma_3, \quad \alpha_6 = \gamma_1,$$
$$k_1 = k_3 = k_5 = 1, \quad k_2 = 2, \quad k_4 = 3, \quad k_6 = 4,$$
$$h_1 = h_3 = h_4 = h_5 = h_6 = 0, \quad h_2 = 2.$$

The modes 1,3 and 5 denote setting up to serve flow 1,2 and 3, respectively. Modes 2,4 and 6 denote serving flow 1,2 and 3, respectively. Writing this system in the form (14) gives

$$A_1 = \begin{bmatrix} 0 & 0 & 0 \\ \frac{\lambda_2}{\mu_1 - \lambda_1} & 1 & 0 \\ \frac{\lambda_3}{\mu_1 - \lambda_1} & 0 & 1 \end{bmatrix}, \quad C_1 = \begin{bmatrix} \frac{1}{\mu_1 - \lambda_1} & 0 & 0 \end{bmatrix},$$

$$A_2 = \begin{bmatrix} 1 & \frac{\lambda_1 \mu_3}{(\mu_2 - \lambda_2)(\mu_3 - \lambda_3)} & \frac{\lambda_1}{\mu_3 - \lambda_3} \\ 0 & \frac{\lambda_2 \lambda_3}{(\mu_2 - \lambda_2)(\mu_3 - \lambda_3)} & \frac{\lambda_2}{\mu_3 - \lambda_3} \\ 0 & 0 & 0 \end{bmatrix},$$

$$C_2 = \begin{bmatrix} 0 & \frac{\mu_3}{(\mu_2 - \lambda_2)(\mu_3 - \lambda_3)} & \frac{1}{\mu_3 - \lambda_3} \end{bmatrix}.$$

we omitted a_i and c_i as they cancel out in the observer error dynamics. Note that though the pairs (A_1, C_1) and (A_2, C_2) are unobservable, we can build a periodic deadbeat observer by using the observer gains

$$L_1 = \begin{bmatrix} 0 & \lambda_2 & \lambda_3 \end{bmatrix}^\top, \tag{20a}$$

$$L_2 = \begin{bmatrix} \lambda_1 & \frac{\lambda_2 \lambda_3}{\mu_3} & 0 \end{bmatrix}^\top, \tag{20b}$$

Using these gains, the matrices $A_v - L_v C_v$ and $L_v C_v$ ($v = 1, 2$) are positive matrices, yielding a positive observer. In this example the observer estimation starts at $t = 50$. For parameters

$$\lambda = \begin{bmatrix} 1 & 2 & 3 \end{bmatrix}^\top, \qquad \mu = \begin{bmatrix} 8 & 10 & 12 \end{bmatrix}^\top,$$
$$\gamma = \begin{bmatrix} 5 & 10 & 15 \end{bmatrix}^\top,$$

and initial estimated state $\hat{x}(50) = \begin{bmatrix} 70 & 20 & 30 \end{bmatrix}^\top$ and mode $\hat{q}(50) = 4$, i.e., serving vehicles from flow 2, a simulation result is presented in Figure 3. The queue lengths at the intersection are presented by dashed lines and the estimated queue lengths are presented by solid lines. Since the measurement information equals the estimated output, i.e., $y(50) = \hat{y}(50)$, the observer dynamics are given by (18a)–(18e) until measurement and estimated output differ. This occurs at $t = 70.7$, when the system starts serving vehicles from queue 1. At this time instant $y \neq \hat{y}$ and the system jumps according to (18d). According to the initial estimated state $\hat{x}(50)$, this event was predicted at $\tilde{t} = 81.7$ with $\tilde{x}(\tilde{t}) = \begin{bmatrix} 101.7 & 58.3 & 15 \end{bmatrix}^\top$. Therefore, $\hat{x}(70.7^+) = \begin{bmatrix} 90.7 & 52.8 & 15 \end{bmatrix}^\top$, causing jumps in $\hat{x}_2$ and $\hat{x}_3$. Also, the estimated mode changes to serving products from flow 1.

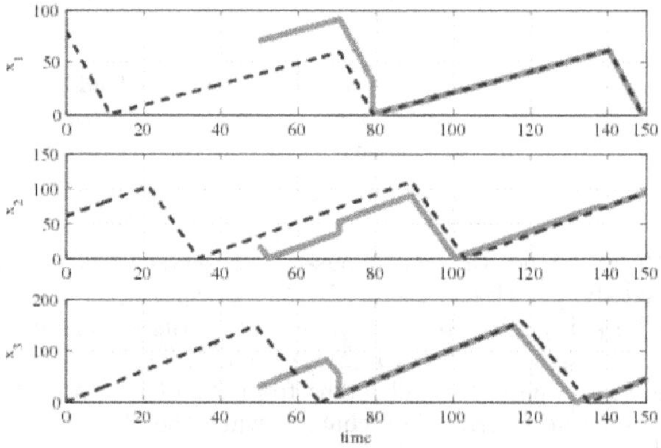

Figure 3: Actual (dashed lines) and estimated (solid lines) queue lengths of the 3-flow T-junction.

The second visible event occurs at $t = 79.2$ when $x_1 = 0$, also earlier than predicted since $\hat{x}_1$ is estimated too large. At this moment $y \neq \hat{y}$ and $y = 0$, therefore jump (18e) results in $\hat{x}_1(79.2^+) = 0$. Note that at this moment both $\hat{x}_1(t)$ and $\hat{x}_3(t)$ are correctly estimated. The observer predicts the next visible event at $\tilde{t} = 137$. However, since $\hat{x}_2(79.2^+) < x_2(79.2^+)$ the actual event occurs later. Therefore the observer switches to the additional waiting mode (18b) via jump (18c). At $t = 140.3$, detection of the next visible event, the estimated buffer levels have converged exactly to the actual buffer levels, due to the deadbeat observer.

6. CONCLUSIONS

This paper presented a methodology to design observers for a special class of piecewise affine hybrid systems (PWAHS), being highly relevant in the context of manufacturing and traffic applications. Although all subsystems are unobservable and not all events are visible, a continuous-time observer was constructed which guarantees that the estimate converges to the state of the plant under suitable conditions. One of the main ideas in the construction of the continuous-time observer was sampling of the system at the visible events leading to a discrete-time periodic linear system, for which an observer can be designed using standard techniques from control theory. If this step is successfully performed, a continuous-time observer that asymptotically recovers the true state of the original hybrid systems can be synthesized. Indeed, the discrete-time observer can then be used as a blueprint for the continuous-time observer, where besides the plant dynamics additional 'waiting' modes are assigned to the observer. Occurrence of visible events before the time that the event was predicted to occur results in a discrete state switch and an update of the continuous states. The observer switches to a 'waiting' mode if an event occurs later than predicted. These principles are formally shown to result in a successful observer design.

The results in this paper lead to various other questions that will be considered in future work. First of all, it would be of interest to take the positivity of the state variable into account leading to observers that always create positive state estimates as well (positive observers). Some first hints were already provided in this direction. In addition, it is of interest to formulate necessary and sufficient conditions in terms of the data of the original PWAHS system when the proposed design is indeed successful and how this relates to fundamental observability and detectability properties. Finally, it is of interest to investigate to what extent the observer design principles put forward in this paper can be applied to more general classes of hybrid systems. As such, this paper provides ideas that might be fruitfully exploited into various future research directions.

7. ACKNOWLEDGMENTS

This work is supported by the Netherlands Organization for Scientific Research (NWO-VIDI grant 639.072.072).

8. REFERENCES

[1] M. Ait Rami and F. Tadeo. Positive observation problem for linear discrete positive systems. In *Proceedings of the 45th IEEE Conference on Decision and Control*, pages 4729–4733, San Diego, CA, USA, 2006.

[2] A. Alessandri and P. Coletta. Design of Luenberger observers for a class of hybrid linear systems. In *Proc. of Hybrid Systems: Computation and Control*, volume 2034 of *Lecture Notes in Computer Science*, pages 7–18, Rome, 2001. Springer.

[3] R. Alur and D. L. Dill. A theory of timed automata. *Theoretical Computer Science*, 126:183–235, 1994.

[4] M. Babaali and M. Egerstedt. Observability for switched linear systems. In *Hybrid Systems: Computation and Control*, volume 2623 of *Lecture Notes in Computer Science. Hybrid Systems: Computation and Control.* Springer, 2004.

[5] A. Balluchi, L. Benvenuti, M. D. Di Benedetto, and A. L. Sangiovanni-Vincentelli. Design of observers for hybrid systems. In C. Tomlin and J. Greenstreet, editors, *Hybrid Systems: Computation and Control*, volume 2289 of *Lecture Notes in Computer Science*, pages 76–89. Springer-Verlag, Stanford, CA, 2002.

[6] A. Balluchi, M. Di Benedetto, L. Benvenuti, and A. Sangiovanni-Vincentelli. Observability for hybrid systems. In *Proceedings of the 42nd IEEE Conference on Decision and Control*, volume 2, pages 1159–1164, 2003.

[7] G. Bara, J. Daafouz, F. Kratz, and J. Ragot. Parameter dependent state observer design for affine LPV systems. *International Journal of Control*, 74(16):1601–1611, 2001.

[8] A. Bemporad, G. Ferrari-Trecate, and M. Morari. Observability and controllability of piecewise affine and hybrid systems. *IEEE Transactions on Automatic Control*, 45(10):1864–1876, 2000.

[9] M. Camlibel, J. Pang, and J. Shen. Conewise linear systems: non-zenoness and observability. *SIAM J. Control Optimizaton*, 45:1769–1800, 2006.

[10] T. Chen and B. Francis. *Optimal sampled-data control systems.* Springer-Verlag, London, 1995.

[11] P. Collins and J. van Schuppen. *Observability of Piecewise-Affine Hybrid Systems*, volume 2993 of *Lecture Notes in Computer Science. Hybrid Systems: Computation and Control*, pages 265–279. Springer Berlin - Heidelberg, 2004.

[12] G. Ferrari-Trecate, D. Mignone, and M. Morari. Moving horizon estimation for hybrid systems. *IEEE Transactions on Automatic Control*, 47:1663–1676, 2002.

[13] R. Goebel, R. Sanfelice, and A. Teel. Hybrid dynamical systems. *IEEE Control Systems Magazine*, 29(2):28–93, 2009.

[14] W. Heemels, B. De Schutter, J. Lunze, and M. Lazar. Stability analysis and controller synthesis for hybrid dynamical systems. *Philosophical Transactions of the Royal Society A: Mathematical, Physical and Engineering Sciences*, 368(1930):4937–4960, 2010.

[15] G. H. Hostetter. Ongoing deadbeat observers for linear time-varying systems. In *American Control Conference*, pages 1099–1101, 1982.

[16] J. Hu. Discrete-time linear periodically time-varying systems: Analysis, realization and model reduction. Master's thesis, Rice University, July 2003.

[17] A. Juloski, W. Heemels, and S. Weiland. Observer design for a class of piecewise linear systems. *Intern. J. Robust and Nonlinear Control*, 17(15):1387–1404, 2007.

[18] J. Lunze and F. Lamnabhi-Lagarrigue. *The HYCON Handbook of Hybrid Systems Control: Theory, Tools, Applications.* Cambridge University Press, Cambridge, 2009.

[19] J. O'Reilly. *Observers for linear systems.* Academic Press: Mathematics in Science & Engineering, London, 1983.

[20] C. Ozveren and A. Willsky. Observability of discrete

event dynamic systems. *Automatic Control, IEEE Transactions on*, 35(7):797–806, 1990.

[21] S. Petterson. Switched state jump observers for switched systems. In *Proc. IFAC World Congress*, Prague, Czech Republic, 2005.

[22] W. Rugh. *Linear System Theory (2nd ed.)*. Prentice-Hall, Inc., Upper Saddle River, NJ, USA, 1996.

[23] A. Tanwani, H. Shim, and D. Liberzon. Observability implies observer design for switched linear systems. *Proceedings of the 2011 Hybrid Systems: Computation and Control*, pages 3–12, 2011.

[24] N. van de Wouw and A. Pavlov. Tracking and synchronisation for a class of PWA systems. *Automatica*, 44(11):2909–2915, 2008.

Automated Analysis of Real-Time Scheduling using Graph Games

Krishnendu Chatterjee
IST Austria
(Institute of Science and Technology Austria)
Klosterneuburg, Austria
krish.chat@ist.ac.at

Alexander Kößler and Ulrich Schmid
Embedded Computing Systems Group
Vienna University of Technology
Vienna, Austria
{koe,s}@ecs.tuwien.ac.at

ABSTRACT

In this paper, we introduce the powerful framework of graph games for the analysis of real-time scheduling with firm deadlines. We introduce a novel instance of a partial-observation game that is suitable for this purpose, and prove decidability of all the involved decision problems. We derive a graph game that allows the automated computation of the competitive ratio (along with an optimal witness algorithm for the competitive ratio) and establish an NP-completeness proof for the graph game problem. For a given on-line algorithm, we present polynomial time solution for computing (i) the worst-case utility; (ii) the worst-case utility ratio w.r.t. a clairvoyant off-line algorithm; and (iii) the competitive ratio. A major strength of the proposed approach lies in its flexibility w.r.t. incorporating additional constraints on the adversary and/or the algorithm, including limited maximum or average load, finiteness of periods of overload, etc., which are easily added by means of additional instances of standard objective functions for graph games.

Categories and Subject Descriptors

I.2.8 [**Problem Solving, Control Methods, and Search**]: Scheduling

Keywords

Game Theory; Competitive Analysis; Real-Time Scheduling

1. INTRODUCTION

Competitive analysis [6] allows to compare the performance of an on-line algorithm $\mathcal{A}$, processing a sequence of inputs without knowing the future, with what can be achieved by an optimal off-line algorithm C that does know the future (a *clairvoyant* algorithm). Its primary focus is determining the *competitive ratio*, which gives the worst-case performance ratio of $\mathcal{A}$ vs. C taken over all input scenarios.

This paper is devoted to the automated competitive analysis of deterministic real-time scheduling algorithms. Their purpose is to schedule a sequence of dynamically arriving instances of *tasks*, which have specific execution time requirements and firm deadlines, on a single processor: A task that is completed by its deadline contributes some positive utility value to the system; a task that does not meet its deadline does not harm, but does not add any utility. The goal of the scheduling algorithm is to maximize the cumulated utility. Tasks that may sometimes miss deadlines arise in several application domains, such as machine scheduling, multimedia and video streaming, QoS management in data networks and other systems that may suffer from overload [14].

In their seminal paper [3], Baruah et. al. proved that no on-line algorithm can achieve a competitive ratio better than 1/4 over a clairvoyant algorithm. The proof is based on constructing a specific task sequence, which takes into account the on-line algorithm's actions and thereby forces it to deliver a sub-optimal cumulated utility. By also giving an on-line algorithm TD1 with competitive ratio 1/4, they proved that 1/4 is indeed the best competitive ratio one can achieve for this real-time scheduling problem.

Obviously, the above situation can be viewed as an instance of a game between the on-line algorithm (Player 1) and an adversary (Player 2) that determines the task sequence. The question addressed in this paper is whether and how the powerful "framework" of *graph games* [15, 20] can be utilized for competitive analysis of real-time scheduling algorithms.

Related work: Following [3], real-time systems research has studied the competitive factor of several extended real-time scheduling algorithms. Among the additional issues taken into consideration are energy consumption [2, 9] (including dynamic voltage scaling), imprecise computation tasks (having both a mandatory and an optional part and associated utilities) [5], lower bounds on slack time [4], and fairness [17]. Note that dealing with these extensions involved considerable ingenuity and efforts w.r.t. identifying and analyzing appropriate worst case scenarios, which do not necessarily carry over even to minor variants of the problem. Maximizing cumulated utility while satisfying multiple resource constraints is also the purpose of the Q-RAM (QoS-based Resource Allocation Model) [19] approach. However, rather than with individual task instances, utility is associated with applications (classes of tasks), and the problem is to (statically) assign schedulable resource(s) to applications such that the cumulated utility is maximized.

Algorithmic game theory [16] has also been applied to classic scheduling problems since decades, primarily in economics and operations research, see e.g. [13] for just one

HSCC'13, April 8–11, 2013, Philadelphia, Pennsylvania, USA.

example of some more recent work. Game theoretic methods have also been utilized successfully for the competitive analysis of on-line algorithms [6], which can typically be expressed as two-person zero-sum games. However, most of this research deals with games in strategic, rather than behavioral, form and focuses on equilibria and related approximation factors. In addition, to the best of our knowledge, algorithmic game theory has never been applied to the competitive analysis of real-time scheduling with firm deadlines before. It has been applied for real-time scheduling of hard real-time tasks [1], though, as an instance of a controller synthesis problem dedicated to meeting all deadlines. This approach does not generalize to firm deadlines, however.

Contributions: This paper shows that, in principle, graph games provide a very powerful and flexible framework for the competitive analysis of real-time scheduling of tasks with firm deadlines. The major strength of this framework lies in its *flexibility*: Additional constraints on the adversary and/or the algorithm, including limited maximum or average load, finiteness of periods of overload, etc., are easily incorporated by adding instances of certain standard objectives for graph games. We provide a new instance of a partial-information game that gives the competitive ratio, and prove that all involved decision problems are decidable. The detailed contributions are as follows:

(a) In Sect. 2, we consider the game models (a partial-observation game with memoryless strategies for Player 1 with mean-payoff and ratio objectives) that are suitable for competitive analysis of real-time scheduling algorithms. The mean-payoff resp. ratio objective allows to compute the cumulated utility resp. the competitive ratio of the best on-line algorithm under the worst-case task sequence. In Sect. 3, we establish that the relevant decision problems are NP-complete in the size of the underlying game graph.

(b) In Sect. 4, we present a reduction of the real-time scheduling algorithms of interest, namely, (i) the particular algorithm TD1 from [3]; and (ii) a general non-deterministic one that covers all possible algorithms; to a finite-state *labeled transition system* (LTS). Those LTS naturally lead to the corresponding game graphs, and suitable reward functions are used to express utility values. This establishes a reduction of the real-time scheduling problem at hand to the graph games framework.

(c) In Sect. 5.1, we show that worst-case average cumulated utitities can be derived from a perfect-information game variant of the above model. Deciding whether there is an on-line algorithm with worst-case average cumulated utility above some threshold is in NP $\cap$ coNP in general, and polynomial in the size of the game graph for reasonable choices of task utility values. Automatically determining the optimal values of the worst-case average cumulated utility, as well as constructing an optimal real-time scheduling algorithm achieving it, can be done efficiently as well. Given some on-line algorithm, we present polynomial time algorithms in the size of the LTS for computing (i) the worst-case utility of the algorithm; and (ii) the worst-case utility ratio, which relates the worst-case cumulated utility of the given on-line algorithm to the worst-case cumulated utility of a clairvoyant off-line algorithm.

(d) In Sect. 5.2, we show that the general partial-information game model with ratio objectives is suitable for competitive analysis. It follows that deciding whether there is an on-line algorithm with competitive ratio above some threshold is in NP w.r.t. the size of the underlying game graph; for a given on-line algorithm, the computation can be done in polynomial time.

(e) In Sect. 6, we show that additional constraints on the adversary, like limited maximum load, finite duration of overload, as well as limited averages of additional constraints, can easily be accommodated: Since any safety objective (prohibiting reachability of certain bad states), any Büchi objective (ensuring repeated reachability of certain target states) and multiple mean-payoff objectives (securing desired limit-average behaviors) can be added without unduly increasing the complexity of the decision problems involved, adding such extensions comes at no extra cost.

2. PARTIAL-OBSERVATION GAMES

In this section, we will first present the basic definitions of partial-observation games with mean-payoff and ratio objectives, and then present the complexity results.

2.1 Basic definitions

We start with the basic definitions of partial-observation games and their subclasses, and the notions of strategies and objectives. We focus on partial-observation games, where one player has partial observation and the other player has perfect observation.

Partial-observation games. A *partial-observation game* (or simply a *game*) is a tuple $\mathsf{G} = \langle S^G, \Sigma_1^G, \Sigma_2^G, \delta^G, \mathcal{O}_S, \mathcal{O}_\Sigma \rangle$ with the following components:

1. *(State space).* The set S^G is a finite set of states.
2. *(Actions).* Σ_i^G $(i = 1, 2)$ is a finite set of actions for Player i.
3. *(Transition function).* The transition function $\delta^G : S^G \times \Sigma_1^G \times \Sigma_2^G \to S^G$ given the current state $s \in S^G$, an action $\sigma_1 \in \Sigma_1^G$ for Player 1, and an action $\sigma_2 \in \Sigma_2^G$ for Player 2, gives the next (or successor) state $s' = \delta^G(s, \sigma_1, \sigma_2)$. A shorter form to depict the previous transition is to write $(s\sigma_2\sigma_1 s')$.
4. *(Observations).* The set $\mathcal{O}_S \subseteq 2^{S^G}$ is a finite set of observations for Player 1 that partitions the state space S^G. The partition uniquely defines a function $\mathsf{obs}_S : S^G \to \mathcal{O}_S$ that maps each state to its observation such that $s \in \mathsf{obs}_S(s)$ for all $s \in S^G$. In other words, the observation partitions the state space according to equivalence classes. Similarly, $\mathcal{O}_\Sigma$ is a finite set of observations for Player 1 that partitions the action set Σ_2^G, and analogously defines the function obs_Σ. Intuitively, Player 1 will have partial observation, and can only obtain the current observation of the state (not the precise state but only the equivalence class the state belongs to) and current observation of the action of Player 2 (but not the precise action of Player 2) to make her choice of action.

Perfect-information games. *Games of complete-observation* (or perfect-information games) are a special case of partial-observation games where $\mathcal{O}_S = \{\{s\} \mid s \in S^G\}$ and $\mathcal{O}_\Sigma = \{\{\sigma_2\} \mid \sigma_2 \in \Sigma_2^G\}$, i.e., every individual state and action is fully visible to Player 1, and thus she has perfect information. For perfect-information games, for the sake of simplicity, we will omit the corresponding observation sets from the description of the game.

Plays. In a game, in each turn, first Player 2 chooses an action, then Player 1 chooses an action, and given the current state and the joint actions, we obtain the next state following the transition function δ^G. A *play* in G is an infinite sequence of states and actions $\rho = s^0\sigma_2^0\sigma_1^0 s^1\sigma_2^1\sigma_1^1 s^2\sigma_2^2\sigma_1^2 s^3 \dots$ such that, for all $j \geq 0$, we have $\delta^G(s^j, \sigma_1^j, \sigma_2^j) = s^{j+1}$. The *prefix up to* s^n of the play ρ is denoted by $\rho(n)$, its *length* is $|\rho(n)| = n+1$ and its *last element* is $\mathsf{Last}(\rho(n)) = s^n$. The set of plays in G is denoted by $\mathsf{Plays}(\mathsf{G})$, and the set of corresponding finite prefixes is denoted $\mathsf{Prefs}(\mathsf{G})$.

Strategies. A *strategy* for a player is a recipe that specifies how to extend finite prefixes of plays. We will consider *memoryless* strategies for Player 1 (where its next action depends only on the current state, but not on the entire history) and general history-dependent strategies for Player 2. A strategy for Player 1 is a function $\pi : \mathcal{O}_S \times \mathcal{O}_\Sigma \to \Sigma_1^G$ that given the current observation of the state and the current observation on the action of Player 2, selects the next action. A strategy for Player 2 is a function $\sigma : \mathsf{Prefs}(\mathsf{G}) \to \Sigma_2^G$ that given the current prefix of the play chooses an action. Observe that the strategies for Player 1 are both observation-based and memoryless; i.e., depend only on the current observations (rather than the whole history), whereas the strategies for Player 2 depend on the history. A memoryless strategy for Player 2 only depends on the last state of a prefix. We denote by Π_{G}^M, Σ_{G}, Σ_{G}^M the set of all observation-based memoryless Player 1 strategies, the set of all Player 2 strategies, and the set of all memoryless Player 2 strategies, respectively. In sequel when we write strategy for Player 1 we consider only observation-based memoryless strategies. Given a strategy π and a strategy σ for Player 1 and Player 2, and an initial state s^0, we obtain a unique play $\rho(s^0, \pi, \sigma) = s^0\sigma_2^0\sigma_1^0 s^1\sigma_2^1\sigma_1^1 s^2\sigma_2^2\sigma_1^2 s^3 \dots$ such that, for all $n \geq 0$, we have $\sigma(\rho(n)) = \sigma_2^n$ and $\pi(\mathsf{obs}_S(s^n), \mathsf{obs}_\Sigma(\sigma_2^n)) = \sigma_1^n$.

Objectives. In this work we will consider *mean-payoff* (or long-run average or limit-average) objectives, as well as *ratio* objectives. For mean-payoff objectives we will consider games with a reward function $r : S^G \times \Sigma_1^G \times \Sigma_2^G \times S^G \to \mathbb{Z}$ that maps every transition to an integer valued reward. In case of ratio objectives, we will consider games with two reward functions $r_1 : S^G \times \Sigma_1^G \times \Sigma_2^G \times S^G \to \mathbb{N}$ and $r_2 : S^G \times \Sigma_1^G \times \Sigma_2^G \times S^G \to \mathbb{N}$ that map every transition to a non-negative valued reward. Given a reward function r and a play $\rho = s^0\sigma_2^0\sigma_1^0 s^1\sigma_2^1\sigma_1^1 s^2\sigma_2^2\sigma_1^2 s^3 \dots$, for $n \geq 0$, let $\mathsf{Sum}(r, \rho(n)) = \sum_{i=0}^{n-1} r(s^i, \sigma_1^i, \sigma_2^i, s^{i+1})$ denote the sum of the rewards for the prefix $\rho(n)$. We define the mean-payoff and ratio objectives as follows.

1. The mean-payoff-sup (resp. mean-payoff-inf) objective $\mathsf{MPsup}(r) : \mathsf{Plays}(\mathsf{G}) \to \mathbb{R}$ (resp. $\mathsf{MPinf}(r) : \mathsf{Plays}(\mathsf{G}) \to \mathbb{R}$) for a reward function r, assigns a real-valued reward to every play ρ as follows:
$$\mathsf{MPsup}(r)(\rho) = \limsup_{n\to\infty} \frac{1}{n} \cdot \mathsf{Sum}(r, \rho(n));$$
$$\mathsf{MPinf}(r)(\rho) = \liminf_{n\to\infty} \frac{1}{n} \cdot \mathsf{Sum}(r, \rho(n)).$$
2. The ratio objective $\mathsf{Ratio}(r_1, r_2) : \mathsf{Plays}(\mathsf{G}) \to \mathbb{R}$ for two reward functions r_1 and r_2 also assigns a real-valued reward to every play as follows: for a play ρ we have
$$\mathsf{Ratio}(r_1, r_2)(\rho) = \liminf_{n\to\infty} \frac{1 + \mathsf{Sum}(r_1, \rho(n))}{1 + \mathsf{Sum}(r_2, \rho(n))}.$$

Note that adding 1 in both numerator and denominator of the ratio avoids division by 0 issues, yet is irrelevant for sums that go to infinity for $n \to \infty$.

Decision problems. The relevant decision problems we are interested in are: given a game G, a starting state s^0, reward functions r, r_1, r_2, and a rational threshold $\nu \in \mathbb{Q}$, whether it holds that $\sup_{\pi\in\Pi_{\mathsf{G}}^M} \inf_{\sigma\in\Sigma_{\mathsf{G}}} \mathsf{MPinf}(r)(\rho(s^0, \pi, \sigma)) \geq \nu$; and $\sup_{\pi\in\Pi_{\mathsf{G}}^M} \inf_{\sigma\in\Sigma_{\mathsf{G}}} \mathsf{Ratio}(r_1, r_2)(\rho(s^0, \pi, \sigma)) \geq \nu$.

2.2 Perfect information games

We now present the fundamental results about *perfect-information* games with mean-payoff objectives that follow from the classical results of [10, 22, 7, 11].

THEOREM 1. *(Determinacy and complexity of perfect-information mean-payoff games [10, 22, 7, 11]). The following assertions hold for perfect-information games with reward function r:*

1. (Determinacy). *We have*
$$\sup_{\pi\in\Pi_{\mathsf{G}}^M} \inf_{\sigma\in\Sigma_{\mathsf{G}}} \mathsf{MPinf}(r)(\rho(s^0, \pi, \sigma))$$
$$= \inf_{\sigma\in\Sigma_{\mathsf{G}}} \sup_{\pi\in\Pi_{\mathsf{G}}^M} \mathsf{MPsup}(r)(\rho(s^0, \pi, \sigma))$$
$$= \inf_{\sigma\in\Sigma_{\mathsf{G}}^M} \sup_{\pi\in\Pi_{\mathsf{G}}^M} \mathsf{MPsup}(r)(\rho(s^0, \pi, \sigma)).$$

2. Whether $\sup_{\pi\in\Pi_{\mathsf{G}}^M} \inf_{\sigma\in\Sigma_{\mathsf{G}}} \mathsf{MPinf}(r)(\rho(s^0, \pi, \sigma)) \geq \nu$ can be decided in NP ∩ CONP, *for a rational threshold ν.*

3. The time complexity for computing the optimal value $v^* = \sup_{\pi\in\Pi_{\mathsf{G}}^M} \inf_{\sigma\in\Sigma_{\mathsf{G}}} \mathsf{MPinf}(r)(\rho(s^0, \pi, \sigma))$ and an optimal memoryless strategy $\pi^* \in \Pi_{\mathsf{G}}^M$, satisfying $v^* = \inf_{\sigma\in\Sigma_{\mathsf{G}}} \mathsf{MPinf}(r)(\rho(s^0, \pi^*, \sigma))$, is in $O(n \cdot m \cdot W)$, where n is the number of states, m is the number of transitions, and W is the maximal value of all the rewards (i.e., the algorithm is pseudo-polynomial time, and if the maximal value W of rewards is polynomial in the size of the game, then the algorithm is polynomial).

4. For any given memoryless strategy $\pi \in \Pi_{\mathsf{G}}^M$, computing $\inf_{\sigma\in\Sigma_{\mathsf{G}}} \mathsf{MPinf}(r)(\rho(s^0, \pi, \sigma))$ can be done in polynomial time $O(n \cdot m \cdot \log W)$.

3. COMPLEXITY RESULTS

In this section, we establish the complexity of the decision problems we consider for partial-observation mean-payoff and ratio objectives. In particular, we will show that for partial-observation games with memoryless strategies for Player 1 all the decision problems are NP-complete.

Transformation. We start with a simple transformation that will allow us to simplify the technical results we prove. In the definition of games, we defined them in such a way that in every state every action was available for the players for simplicity. For technical development, we will consider games where, at certain states, some actions are not allowed for a player. The transformation of such games to games where all actions are allowed is as follows: we add two absorbing dummy states (with only a self-loop), one for Player 1 and the other for Player 2, and assign rewards in a way such that the objectives are violated for the player. For example, for mean-payoff objectives with threshold $\nu > 0$, we assign reward 0 for the only out-going (self-loop) transition of the Player 1 dummy state, and a reward strictly greater than ν for the self-loop of the Player 2 dummy state; and in case of ratio-objectives we assign the reward pairs similarly. Given a state s, if Player 1 plays an action that is not allowed at s, we go to the dummy Player 1 state; and if Player 2 plays an action that is not allowed we go to the Player 2 dummy state. Thus we have a simple linear time transformation. Hence, for technical convenience, we can assume in the sequel that different states have different sets of available actions for the players. We first start with the hardness result.

LEMMA 1. *The decision problems for partial observation games with mean-payoff objectives and ratio objectives, i.e, whether* $\sup_{\pi \in \Pi_{\mathsf{G}}^{M}} \inf_{\sigma \in \Sigma_{\mathsf{G}}} \mathsf{MPinf}(r)(\rho(s^0, \pi, \sigma)) \geq \nu$ *(respectively* $\sup_{\pi \in \Pi_{\mathsf{G}}^{M}} \inf_{\sigma \in \Sigma_{\mathsf{G}}} \mathsf{Ratio}(r_1, r_2)(\rho(s^0, \pi, \sigma)) \geq \nu$*), are* NP*-hard in the strong sense.*

PROOF. We present a reduction from the 3-SAT problem, which is NP-hard in the strong sense [18]. Let Φ be a 3-SAT formula over n variables $x_1, x_2, \ldots, x_n$ in conjunctive normal form, with m clauses $\mathfrak{c}_1, \mathfrak{c}_2, \ldots, \mathfrak{c}_m$ consisting of a disjunction of 3 literals (a variable x_k or its negation $\overline{x}_k$) each. We will construct a game graph G_Φ as follows:

1. *(State space).* $S^G = \{s_{\mathsf{init}}\} \cup \{s_{i,j} \mid 1 \leq i \leq m, 1 \leq j \leq 3\} \cup \{\mathsf{dead}\}$; i.e., there is an initial state s_{init}, a dead state dead, and there is a state $s_{i,j}$ for every clause $\mathfrak{c}_i$ and a literal j of i.
2. *(Actions).* The set of actions applicable for Player 1 is $\{\mathsf{true}, \mathsf{false}, \bot\}$ and for Player 2 is $\{1, 2, \ldots, m\} \cup \{\bot\}$.
3. *(Transition).* In the initial state s_{init}, Player 1 has only one action $\bot$ available, and Player 2 has actions $\{1, 2, \ldots, m\}$ available, and given action $1 \leq i \leq m$, the next state is $s_{i,1}$. In all other states Player 2 has only one action $\bot$ available. In states $s_{i,j}$ Player 1 has two actions available, namely, true and false. The transitions are as follows:
   - If the action of Player 1 is true in $s_{i,j}$, then (i) if the j-th literal in $\mathfrak{c}_i$ is x_k, then we have a transition back to the initial state; and (ii) if the j-th literal in $\mathfrak{c}_i$ is $\overline{x}_k$ (negation of x_k), then we have a transition to $s_{i,j+1}$ if $j \in \{1, 2\}$, and if $j = 3$, we have a transition to dead.
   - If the action of Player 1 is false in $s_{i,j}$, then (i) if the j-th literal in $\mathfrak{c}_i$ is $\overline{x}_k$ (negation of x_k), then we have a transition back to the initial state; and (ii) if the j-th literal in $\mathfrak{c}_i$ is x_k, then we have a transition to $s_{i,j+1}$ if $j \in \{1, 2\}$, and if $j = 3$, we have a transition to dead.

   In state dead both players have only one available action $\bot$, and dead is a state with only a self-loop (transition only to itself).
4. *(Observations).* First, Player 1 does not observe the actions of Player 2 (i.e., Player 1 does not know which action is played by Player 2). The observation mapping for the state space for Player 1 is as follows: the set of observations is $\{0, 1, \ldots, n\}$ and we have $\mathsf{obs}_S(s_{\mathsf{init}}) = \mathsf{obs}_S(\mathsf{dead}) = 0$ and $\mathsf{obs}_S(s_{i,j}) = k$ if the j-th variable of $\mathfrak{c}_i$ is either x_k or its negation $\overline{x}_k$, i.e., the observation for Player 1 corresponds to the variables.

A pictorial description is shown in Fig 1. The intuition for the above construction is as follows: Player 2 chooses a clause from the initial state s_{init}, and an observation-based memoryless strategy for Player 1 corresponds to a non-conflicting assignment to the variables. Note that Player-1 strategies are observation-based memoryless; hence, for every observation (i.e., a variable), it chooses a unique action (i.e., an assignment) and thus non-conflicting assignments are ensured. We consider G_Φ with reward functions r, r_1, r_2 as follows: r_2 assigns reward 1 to all transitions; r and r_1 assigns reward 1 to all transitions other than the self-loop at state dead, which is assigned reward 0. We ask the decision questions with $\nu = 1$. Observe that the answer to the decision problems for both mean-payoff and ratio objectives is "Yes" iff the state dead can be avoided by Player 1 (because if dead is reached, then the game stays in dead forever, violating both the mean-payoff as well as the ratio objective). We now present the two directions of the proof.

Satisfiable implies dead *is not reached.* We now show that if Φ is satisfiable, then Player 1 has an observation-based memoryless strategy π^* to ensure that dead is never reached. Consider a satisfying assignment A for Φ, then the strategy π^* for Player 1 is as follows: given an observation k, if A assigns true to variable x_k, then the strategy π^* chooses action true for observation k, otherwise it chooses action false. Since the assignment A satisfies all clauses, it follows that for every $1 \leq i \leq m$, there exists $s_{i,j}$ such that the strategy π^* for Player 1 ensures that the transition to s_{init} is chosen. Hence the state dead is never reached, and both the mean-payoff and ratio objectives are satisfied.

If dead *is not reached, then* Φ *is satisfiable.* Consider an observation-based memoryless strategy π^* for Player 1 that ensures that dead is never reached. From the strategy π^* we obtain an assignment A as follows: if for observation k, the strategy π^* chooses true, then the assignment A chooses true for variable x_k, otherwise it chooses false. Since π^* ensures that dead is not reached, it means for every $1 \leq i \leq m$, that there exists $s_{i,j}$ such that the transition to s_{init} is choosen (which ensures that $\mathfrak{c}_i$ is satisfied by A). Thus since π^* ensures dead is not reached, the assignment A is a satisfying assignment for Φ.

Thus, it follows that the answers to the decision problems are Yes iff Φ is satisfiable, and this establishes the NP-hardness result. ∎

REMARK 1. *Note that our reduction used only weight values* 0 *and* 1*. This implies that* NP*-hardness holds also for the case where weight values are bounded by a constant.*

The NP *upper bounds.* We now present the NP upper bounds for the problem. In both cases, the polynomial witness for the decision problem is a memoryless strategy (i.e., if the answer to the decision problem is Yes, then there is a witness memoryless strategy π for Player 1, and the NP algorithm just guesses the witness strategy π). Once the memoryless strategy is guessed and fixed, we need to show that we have a polynomial time verification procedure. The polynomial time verification procedures are as follows:

1. *(Mean-payoff objectives).* Once the memoryless strategy for Player 1 is fixed, the game problem reduces to a problem when there is only Player 2 (this is interpreted as a path problem in directed graphs). The polynomial time

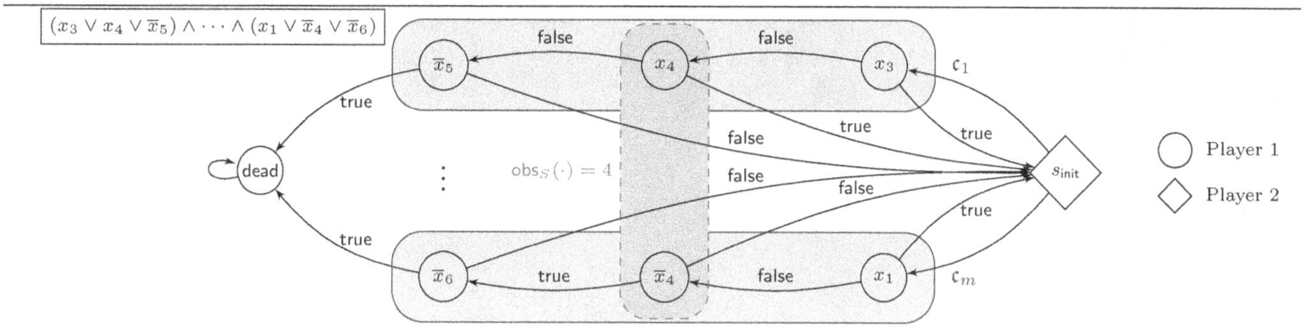

Figure 1: Illustration of construction of the game from a formula.

algorithm to solve directed graphs with mean-payoff objectives is obtained from the well-known minimum mean-cycle algorithm of [11]. Thus we have a polynomial time verification procedure.

2. *(Ratio objectives).* Again once the memoryless strategy for Player 1 is fixed, the game problem reduces to a problem on directed graphs. It follows from the results of [12] that in directed graphs with ratio objectives memoryless optimal strategies exist (i.e., once the memoryless strategy for Player 1 is fixed, if there is a counter strategy for Player 2, then there is a memoryless one)[1]. Given this fact, we present a simple reduction of the decision problem of ratio objectives to mean-payoff objectives as follows: given reward function r_1 and r_2 and the rational threshold $\nu = p/q$, where $p, q \in \mathbb{N}$; we consider the reward function $r = p \cdot r_1 - q \cdot r_2$, i.e., for every transition $(s\sigma_2\sigma_1 s')$ we have $r(s, \sigma_1, \sigma_2, s') = p \cdot r_1(s, \sigma_1, \sigma_2, s') - q \cdot r_2(s, \sigma_1, \sigma_2, s')$. We consider the reward function r with mean-payoff objective and threshold 0. The mean-payoff objective with reward r and threshold 0 is satisfied iff the ratio objective with rewards r_1 and r_2 is satisfied with threshold ν. Thus from the minimum mean-cycle algorithm of [11], we also obtain a polynomial time verification algorithm for ratio objectives.

We summarize the result in the following theorem.

THEOREM 2. *(The complexity of the decision problems). 1. The decision problems for partial-observation games with mean-payoff objectives and ratio objectives, i.e., whether* $\sup_{\pi \in \Pi_G^M} \inf_{\sigma \in \Sigma_G} \mathsf{MPinf}(r)(\rho(s^0, \pi, \sigma)) \geq \nu$ *(resp.* $\sup_{\pi \in \Pi_G^M} \inf_{\sigma \in \Sigma_G} \mathsf{Ratio}(r_1, r_2)(\rho(s^0, \pi, \sigma)) \geq \nu$*), are* NP*-complete.*
2. For any given memoryless strategy $\pi \in \Pi_G^M$*, whether* $\inf_{\sigma \in \Sigma_G} \mathsf{Ratio}(r_1, r_2)(\rho(s^0, \pi, \sigma)) \geq \nu = \frac{p}{q}$*, for a rational number* ν*, can be decided in polynomial time* $O(n \cdot m \cdot \log(W \cdot p \cdot q))$.

4. REAL-TIME TASK SCHEDULING

Consider a finite set of tasks $\mathcal{T} = \{\tau_1, \ldots, \tau_n\}$, $n < \infty$, which are all executed on a single processor. Every task τ_i releases countably many task instances (called *jobs*) $J_{i,j}$, $j \geq 1$, over time. All jobs, of all tasks, are independent of each other and can be preempted and resumed during execution without any overhead. We assume a discrete notion of real-time $t = k\varepsilon$, $k \geq 1$, where $\varepsilon > 0$ is both the unit time and the smallest unit of preemption (called a *slot*). Since both task releases and scheduling activities occur at slot boundaries only, all timing values are specified as integers.

[1]It is straightforward to verify that ratio objectives satisfy the criteria of [12] for memoryless optimal strategies.

4.1 Basic definitions

Every task τ_i, $1 \leq i \leq n$, is characterized by its non-zero *worst-case execution time* $C_i \in \mathbb{N}$ (slots), its non-zero *relative deadline* $D_i \in \mathbb{N}$ (slots) and its non-zero *utility value* $V_i \in \mathbb{N}$.[2] We will thus sometimes write $\tau_i = (C_i, D_i, V_i)$ to completely specify task τ_i. Every job $J_{i,j}$ of task τ_i needs the processor for C_i (not necessarily consecutive) slots exclusively to execute to completion. All tasks have firm deadlines: Only a job $J_{i,j}$ that completes within D_i slots, as measured from its release time, provides utility to the system. A job that misses its deadline does not do harm but provides zero utility. The real-time scheduling problem considered in this paper is to maximize the *cumulated utility* $V \in \mathbb{N}$ (abbreviated utility), which is the sum of V_i times the number of jobs $J_{i,j}$ that can be completed by their deadlines.

Since there may be slots where no job can be scheduled since no task has been released, we add a special idle task $\tau_0 = (1, 1, V_0)$ with some non-zero $V_0 \in \mathbb{N}$. We assume that a job $J_{0,\ell}$ of τ_0 is released internally at the beginning of every slot $\ell \geq 1$, which may only be executed, however, if no other unfinished job with absolute deadline $> \ell$ has been released earlier. This way, the idle task contributes to the cumulated utility whenever there is no "real" work to be done. This way, the adversary cannot minimize the cumulated utility by not releasing any task.[3]

More formally, let $\Sigma = 2^{\mathcal{T}}$ and consider an infinite *task sequence* $\sigma = (\sigma)^{\ell \geq 1} \in \Sigma^\infty$. For every $\ell \geq 1$, σ^ℓ specifies the subset of tasks a (single) new job of which is released at the beginning of slot ℓ. For $\tau_i \in \sigma^\ell \cup \{\tau_0\}$, job $J_{i,j}$ with $j = |\{k | \tau_i \in \sigma^k \cup \{\tau_0\} \text{ and } 1 \leq k \leq \ell\}|$ is released: $r_{i,j} = \ell$ denotes the *release time* of $J_{i,j}$, which is the earliest time when $J_{i,j}$ can be scheduled and executed, and $d_{i,j} = r_{i,j} + D_i$ denotes its absolute *deadline*.

Given a task sequence σ, the *schedule* $\pi = (\pi)^{\ell \geq 1} \in (\mathcal{T} \cup \{\tau_0\})^\infty$ computed by a real-time scheduling algorithm in the

[2]Rational utility values $V_1, \ldots, V_n$ can be mapped to integers by proper scaling.

[3]Alternatively, we could omit the idle task and force the adversary to create infinitely many jobs by adding an appropriate Büchi objective, see Sect. 6.

presence of σ is a function that assigns at most one job for execution to every slot $\ell \geq 1$: π^ℓ is either $\emptyset$ (meaning no job is executed) or else some job $J_{i,j}$, which must satisfy the following constraints:

(i) $r_{i,j} \leq \ell$ (execute only released jobs),

(ii) $\ell < d_{i,j}$ (execute only jobs that could still complete by their deadline),

(iii) $|\{k|\pi^k = J_{i,j} \text{ and } 1 \leq k \leq \ell\}| < C_i$ (execute only jobs that have not been completed),

(iv) $J_{i,j} = J_{0,\ell}$ only if there is no unfinished job $J_{i',j'}$ with $i' > 0$ and $d_{i',j'} > \ell$ (execute idle jobs only if there is no work to be done).[4]

Given some π, let $\pi(k)$ denote the prefix of length $k \geq 1$ of π and $\gamma_i(\pi, k)$ the number of jobs of task τ_i that are completed by their deadlines in $\pi(k)$. The cumulated utility $V(\pi, k)$, also called utility for brevity, achieved in $\pi(k)$ is defined as $V(\pi, k) = \sum_{i=0}^{n} \gamma_i(\pi, k)V_i$. Since typically $\lim_{k\to\infty} V(\pi, k) = \infty$, however, the utility is not a suitable performance measure for infinite task sequences. We will hence typically focus on the limiting average of the utility (called average utility) $\overline{V}(\pi) = \liminf_{k\to\infty} \frac{1}{k}V(\pi, k)$ instead. Note that $\overline{V}(\pi)$ is meaningful for characterizing the worst-case cumulated utility of finite task sequences as well, since one can always form an infinite task sequence from a finite one with the same average utility by infinite repetition.

4.2 Labeled Transition Systems

We focus on deterministic *on-line* scheduling algorithms $\mathcal{A}$, which, at time ℓ, do not know any of the σ^k for $k > \ell$. By contrast, an *off-line* algorithm knows the entire task sequence σ beforehand. Note that, due to our firm deadline assumption, we can restrict our attention to finite-history algorithms here: Let $D_{\max} = \max_{1\leq i\leq n} D_i$ (analogous definitions hold for $C_{\max}$ and $V_{\max}$). Obviously, any job $J_{i,j}$ released by $\ell - D_{\max}$ can be discarded by an algorithm at time ℓ, as scheduling this job would violate (ii). Consequently, algorithms do not need to maintain the full history.

It is easy to express the above setting for a particular on-line algorithm $\mathcal{A}$ as a deterministic labeled transition system $(S, \Sigma_2, \nabla \subseteq S \times \Sigma_2 \times S)$ with reward function r: We choose $\Sigma_2 = \Sigma$ as the set of possible new task releases per slot. Every finite-history algorithm $\mathcal{A}$ can be encoded as a deterministic finite-state automaton with state space S and transition function ∇, which takes the current state and the new jobs released at the beginning of the current slot to compute the new state; the latter also encodes the job $J_{i,j}$ to be executed in the current slot. The reward function $r(\delta)$ for $\delta \in \nabla$ is V_i if $J_{i,j}$ will be completed by the end of the current slot, or 0 otherwise. Given a run $\pi = (\delta)^{\ell\geq 1}$ of the LTS, the average utility is $\overline{V}(\pi) = \liminf_{k\to\infty} \frac{1}{k}\sum_{\ell=1}^{k} r(\delta^\ell)$.

[4]Note carefully that this definition does not allow the idle task to create utility as long as there is a regular task in the system with a deadline that has not yet expired - even if this task will eventually not make its deadline because of excessive remaining execution time. We cannot allow the idle task to already contribute utility when it becomes clear that the remaining tasks will not make their deadlines, though: An optimal on-line algorithm would then just discard any arriving job immediately, and still create 100% of the achievable cumulated utility by means of the idle task.

Example: LTS for TD1. We illustrate this construction by means of the simple version of the algorithm TD1 introduced in [3, Fig. 1], which is optimal for zero-laxity tasks (i.e., $D_i = C_i$) with uniform value density, i.e., utility values $V_i = C_i$. It maintains three variables

- $v \in \{0, V_1, \ldots, V_n\}$, the utility value of the currently executed job,
- $\Delta \in \{0, \ldots, 4V_{\max}\}$, the time interval between the start of the current busy period and the latest deadline (the range follows from the proof of [3, Lem. 5], which reveals $v \geq \Delta/4$),
- $k \in \{0, \ldots, V_{\max} - 1\}$, the number of slots remaining for executing the current job to completion,

all initialized to 0 when the system is/becomes (pre-)idle (see below). Whenever a new job of task $(C_i = V_i, D_i = V_i, V_i)$ is released, where V_i is maximal among all simultaneously released jobs (all others are discarded right away), TD1 updates Δ to $\Delta := \max\{\Delta, \Delta - k + V_i\}$ and then checks whether $v < \Delta/4$. If so, it discards the current job [if any] and schedules the new job by setting $v := V_i$ and $k = V_i - 1$; otherwise, the current job is not changed and k is just decremented once per slot (if $k > 0$).

If k has reached 0, the current job $J_{i,j}$ has been completed. However, TD1 is not allowed to just switch to the idle state (where the idle task τ_0 adds 1 to the cumulated utility in every slot) in our setting: It may be the case that some jobs have been discarded during $J_{i,j}$'s execution, which have a deadline past $J_{i,j}$'s finishing time. In this case, property (iv) above does not allow the idle task to contribute to the cumulated utility until the deadline of the last discarded job has expired. Hence, our TD1 switches to a *pre-idle* state first, which is the same as the idle state except that the idle task does not add 1 to the cumulated utility. The original TD1 must hence slightly be extended to accommodate this: We add another variable d, $0 \leq d \leq V_{\max} - 1$, initially 0, which is set to $\max\{d, V_i - 1\}$ upon the arrival of a new job of task τ_i (i.e., when TD1 updates Δ). Like k, d is decremented by 1 in every slot (if $d > 0$); our modified TD1 switches from pre-idle to idle when d reaches 0.

If we denote the states in our state set S as $s(v, \Delta, k, d)$, the corresponding LTS includes e.g. the following states:

- $s(0, 0, 0, 0)$, the idle state,
- $s(0, 0, 0, d)$ with $d > 0$, the pre-idle state with d remaining slots until making the transition to the idle state,
- $s(v, \Delta, 1, d)$ with $v > 0$ and $\Delta > 0$, a state where the current job only needs one additional slot for being completed.

Recalling the ranges of the involved variables, the size of the state space S is $|S| = (1 + 4nV_{\max}^2)V_{\max}$. For $C_i = D_i = V_i = i$ and hence $V_{\max} = n$, which is the task set that allows comparison of our results with [3], the above version of TD1 hence needs at most $4n^4 + n = O(n^4)$ states.[5]

Every transition $\delta \in S \times \Sigma \times S$ of our LTS corresponds to the execution of a single slot, and depends on the jobs arriving at the beginning of this slot. In general, transitions

[5]There is certainly room for improvement here, by using a more efficient encoding of the state.

must be labeled with the *sets* of newly arriving jobs σ^ℓ. In case of TD1, however, (cf. the update-rule for Δ) it suffices to label transitions just with the task τ_i with maximum utility V_i in σ^ℓ. We list a few example transitions of TD1 (we write $s \xrightarrow{\tau_i} s'$ for $(s, \tau_i, s') \in \nabla$ for clarity):

- For $d \geq 0$: $s(0,0,0,d) \xrightarrow{\tau_i} s(V_i, V_i, V_i - 1, \max\{d, V_i - 1\})$: τ_i initiates new busy period.
- For $k > 1$, $d > 1$, $V_i \geq (\Delta - k + V_j)/4$: $s(V_i, \Delta, k, d) \xrightarrow{\tau_j} s(V_i, \Delta - k + V_j, k-1, d-1)$: TD1 retains current task but does not complete it.
- For $d > 0$, $V_i \geq (\Delta - k + V_j)/4$: $s(V_i, \Delta, 1, d) \xrightarrow{\tau_j} s(0,0,0,d-1)$: TD1 retains current task and does complete it (adds V_i to cumulated utility).
- $s(0,0,0,1) \rightarrow s(0,0,0,0)$: Move from pre-idle to idle (do not add 1 to cumulated utility).
- $s(0,0,0,0) \rightarrow s(0,0,0,0)$: Execute idle task (add 1 to cumulated utility).
- For $k > 1$, $d > 1$, $V_i < (\Delta - k + V_j)/4$: $s(V_i, \Delta, k, d) \xrightarrow{\tau_j} s(V_j, \Delta - k + V_j, V_j - 1, V_j - 1)$: TD1 discards current task and starts new one.

Since at most $2n+1$ transitions lead away from any state (no release or one of n tasks, w/o replacing the current task), the total number of transitions is hence at most $O(n^5)$.

The non-deterministic LTS. Devising an algorithm-specific LTS is already sufficient for computing its average utility and competitive ratio. For a general competitive analysis, a *non-deterministic* LTS that can simulate all possible on-line algorithms in a memoryless way is required. In case of our real-time scheduling problem, such an LTS for a given task set $\mathcal{T} = \{\tau_1, \ldots, \tau_n\}$ can be easily constructed, since property (ii) allows it to discard all jobs that have been released $\geq D_{\max}$ slots ago. Its state space hence only has to encode, for the at most $D_{\max}$ jobs of each $\tau_1, \ldots, \tau_n$ arriving during the $D_{\max}$ most recent slots, every job's current state.

In our non-deterministic LTS, the state of any job $J_{i,j}$ consists of

- $k \in \{0 \ldots, C_i\}$, the number of slots of $J_{i,j}$ that still remain to be executed,
- $d \in \{0, \ldots, D_i - 1\}$, the number of slots of $J_{i,j}$ remaining until $d_{i,j}$ expires (the job's laxity).

Note that $(0, d)$ denotes a state where $J_{i,j}$ has been completed, with d slots remaining until its deadline expires. Similarly, $(k, 0)$ with $k > 0$ depicts a job that did not make its deadline.

The entire state space S of our LTS consists of an array $J[i, \ell]$ of D_{max} columns and n rows. Row i, $1 \leq i \leq n$, represents task τ_i, and column ℓ, $1 \leq \ell \leq D_{\max}$, represents a slot. The columns are organized as a circular buffer for the jobs arriving in the most recent $D_{\max}$ slots: S also contains an index variable $\ell \in \{1, \ldots, D_{\max}\}$, initially 1, which is incremented (with $D_{\max}$ followed by 1) at the end of every slot. If $\sigma \subseteq 2^{\mathcal{T}}$ denotes the tasks a new job of which is released at the beginning of the current slot, every state transition atomically performs the following actions:

1. $\forall \tau_i \in \sigma : J[i, \ell] := (C_i, D_i)$ (record newly arriving jobs, if any).
2. Non-deterministically choose some index $[i', \ell']$ where $J[i', \ell'] = (k', d')$ with both $k' > 0$ and $d' > 0$, and set $J[i', \ell'] := (k' - 1, d')$ (schedule a slot of $J[i', \ell']$ with non-zero remaining processing time and laxity for execution).
3. For all $i, 1 \leq i \leq n, 1 \leq m \leq D_{\max}$ where $J[i, m] = (., d)$ with $d > 0$, set $J[i, m] := (., d - 1)$ (decrease laxity of all jobs with non-zero laxity).

The transition relation of our LTS contains *all* transitions adhering to the above, for all σ. A transition δ where $J[i', \ell']$ is set to $(0, d)$ represents the completion of a job of $\tau_{i'}$; its reward is $r(\delta) = V_{i'}$. All other transitions (except idle transitions, see below) have reward 0.

As in the LTS for TD1, we must distinguish between the real idle states of the system (one for every ℓ), where all entries of J are $(0, 0)$, and pre-idle states, where at least one entry of J is of the form (k, d) with $d > 0$. Transitions δ from an idle state to an idle state have reward $r(\delta) = 1$, representing the utility value of our idle task τ_0. Transitions from a pre-idle state to a pre-idle or idle state have reward 0.

Figure 2 shows two (non-deterministic) transitions $\delta' = (s, \sigma, s')$ and $\delta'' = (s, \sigma, s'')$ from state s into successor states s' and s'', respectively, including the three internal actions performed during these transitions. We assume a task set consisting of three tasks here:

$$\begin{aligned} \tau_1 &: \; C_1 = 2, D_1 = 2, V_1 = 2, \\ \tau_2 &: \; C_2 = 1, D_2 = 1, V_2 = 1, \\ \tau_3 &: \; C_3 = 2, D_3 = 3, V_3 = 2. \end{aligned}$$

State s is such that the first job of τ_1 and τ_3 have already been released in the slot before (where $J_{1,1}$ has been chosen for execution), such that $\ell = 2$. In the slot corresponding to δ' resp. δ'', we assume that two additional jobs arrive: the second instance of task τ_1 and the first instance of task τ_2, i.e., $\sigma = \{\tau_1, \tau_2\}$. In the second action, the non-deterministic choice happens. There are actually four possibilities here, as there are four jobs ready for being scheduled in the system. The label on each arrow gives the entry $[i', \ell']$ of the job being chosen. In the third action, the remaining laxity is decreased for all jobs in the system, which concludes the state transition. Transition δ'', leading to s'', chooses entry $[2, 2]$ in the second action and thus gains reward $r(\delta'') = V_2 = 1$, as the job is completed. On the other hand, transition δ', leading to s', chooses job $[1, 2]$ and thus gains reward $r(\delta') = 0$.

Obviously, this non-deterministic finite-state LTS can simulate any possible real-time scheduling algorithm with a memoryless strategy, i.e., one that does not need to refer to the history of Σ_2 actions/transitions, as all required history information is encoded in the state. Taking into account that every of the $nD_{\max}$ job entries in J can take on at most $(C_{max}+1)D_{\max}$ different states, and the fact that ℓ can take on $D_{\max}$ different values, the overall size of the state space is $|S| = D_{\max}((C_{max} + 1)D_{\max})^{nD_{\max}}$. We hence arrive at the following result.

THEOREM 3. *(LTS reduction of real-time scheduling problems). Given a real-time scheduling problem with firm deadlines, there is a reduction that constructs a non-deterministic LTS with $|S| = D_{\max}((C_{max} + 1)D_{\max})^{nD_{\max}}$ states, such that, for every real-time scheduling algorithm*

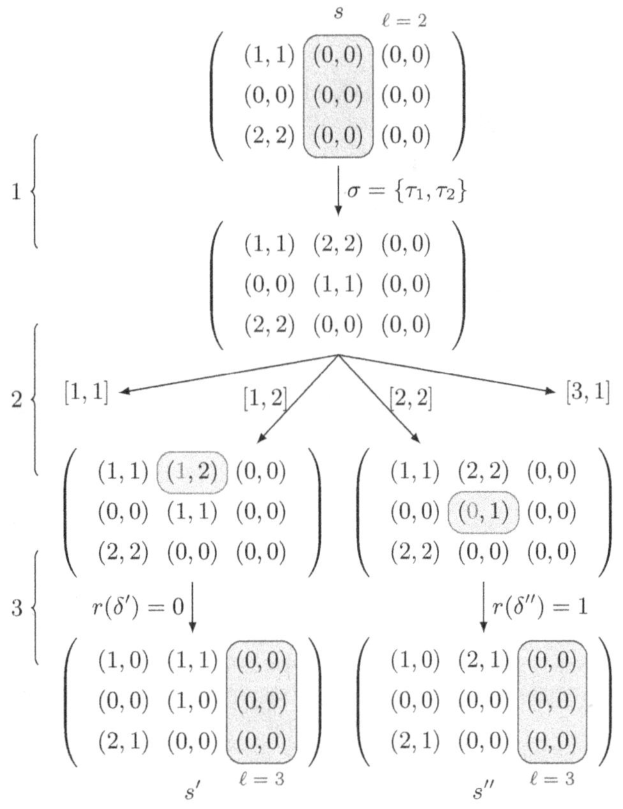

Figure 2: Example of non-deterministic LTS transitions $\delta' = (s, \sigma, s')$ and $\delta'' = (s, \sigma, s'')$, from state s into successor states s' and s'', including the internal actions performed during these transitions, for a task set consisting of three tasks. δ'' chooses job $[2, 2]$ and thus gains reward $r(\delta'') = V_2 = 1$, δ' chooses job $[1, 2]$ and thus gains reward $r(\delta') = 0$.

for this problem adhering to properties (i)–(iv), there is a memoryless strategy in the LTS to simulate this algorithm.

5. GAMES FOR SCHEDULING ANALYSIS

5.1 Worst-case average utility

Before turning our attention to the competitive analysis of our real-time scheduling algorithms, i.e., the competitive ratio [3], we first establish general results on worst-case average utilities. They follow easily from the powerful framework of perfect-information graph games (Theorem 1).

We have already shown that our scheduling problem can be reduced to a non-deterministic finite-state LTS $(S, \Sigma_2, \nabla \subseteq S \times \Sigma_2 \times S)$ along with a reward function r, such that the mean-payoff function $\mathsf{MPinf}(r)$ gives the average utility. We simply interpret the transition system as a game, where the choices of transitions δ^G correspond to the actions for Player 1. Thus we have a perfect-information game, and every memoryless strategy correspond to a scheduling algorithm and vice-versa (i.e., every scheduling algorithm is a memoryless strategy of the game). The worst-case utility of a given on-line algorithm, corresponding to a memoryless strategy $\pi \in \Pi_{\mathsf{G}}^M$, is

$$\inf_{\sigma \in \Sigma_{\mathsf{G}}} \mathsf{MPinf}(r)(\rho(s^0, \pi, \sigma)). \tag{1}$$

Note that it can be computed from the LTS of the particular on-line algorithm. The worst-case utility of the optimal on-line algorithm is given by

$$\sup_{\pi \in \Pi_{\mathsf{G}}^M} \inf_{\sigma \in \Sigma_{\mathsf{G}}} \mathsf{MPinf}(r)(\rho(s^0, \pi, \sigma)); \tag{2}$$

its computation requires the non-deterministic LTS of the real-time scheduling problem. Using the results of Theorem 1, we obtain the following result.

THEOREM 4. *The following assertions hold for the class of real-time scheduling problems defined in Sect. 4:*

1. *The worst-case average utility for a given on-line algorithm can be determined in polynomial time $O(|S| \cdot m \cdot \log V_{\max})$, where $|S|$ resp. m is the number of states resp. transitions of the deterministic LTS of the algorithm and $V_{\max}$ is the maximum utility value of any task.*
2. *Whether there exists an on-line algorithm with worst-case average utility at least ν can be decided in* NP ∩ coNP *in general; and if $V_{\max}$ is bounded by the size of the non-deterministic LTS, then the decision problem can be solved in polynomial time.*
3. *An on-line algorithm with optimal worst-case average utility can be constructed in time $O(|S| \cdot m \cdot V_{\max})$, where $|S|$ resp. m is the number of states resp. transitions of the non-deterministic LTS.*

Theorem 1 also allows us to derive an interesting result on the worst-case utility *ratio.*

Worst-case utility ratio. The worst-case utility ratio is the worst-case limiting average utility of the best on-line algorithm over the worst-case limiting average utility achievable by the best off-line algorithm.

Worst-case utility means the smallest cumulated utility that can be guaranteed by the best (and hence by any) algorithm in the presence of the worst-case task sequence. Note that defining the worst-case *utility* ratio in terms of a ratio of limiting *average utilities* is possible, as the leading factor of $1/k$ in the latter cancels out in the ratio. In contrast to the competitive ratio, the worst-case utility ratio relates the average utilities of the on-line and off-line algorithm for possibly *different* task sequences. It is not difficult to prove that the worst-case utility ratio is always an upper bound on the competitive factor of an optimal on-line algorithm.

Recalling (2), the worst-case utility of an off-line algorithm is given by

$$\inf_{\sigma \in \Sigma_{\mathsf{G}}} \sup_{\pi \in \Pi_{\mathsf{G}}^M} \mathsf{MPsup}(r)(\rho(s^0, \pi, \sigma)). \tag{3}$$

Observe that the on-line algorithm makes the choice of its strategy without knowing in advance the strategy of the adversary, whereas the off-line algorithm makes the choice after the input sequence has been chosen by the opponent. The first item of Theorem 1 also provides the following result for the ratio of (2) and (3).

THEOREM 5. *The worst-case utility ratio for the class of real-time scheduling problems defined in Sect. 4 is always 1.*

Finally, according to Theorem 4, it is also possible to (efficiently) compute the worst-case utility ratio for a given on-line algorithm, by computing the ratio of (1) and (3).

5.2 Competitive analysis

We now show how the competitive ratio analysis can be reduced to the partial-observation game problem with ratio objectives. It follows from Theorem 2 that the decision problem for the competitive ratio can be solved in NP.

Labeled transition system to partial-observation game. Given our non-deterministic LTS $(S, \Sigma_2, \nabla \subseteq S \times \Sigma_2 \times S)$ with reward function r, we first construct the equivalent perfect-information game $(S, \Sigma_1, \Sigma_2, \delta)$. Σ_1 represent the choice of the transition function as actions and $\delta : S \times \Sigma_1 \times \Sigma_2 \to S$ gives the transition. We now construct a partial-observation game G as follows: $\mathsf{G} = (S^G = S \times S, \Sigma_1^G = \Sigma_1, \Sigma_2^G = \Sigma_2 \times \Sigma_1, \delta^G, \mathcal{O}_S, \mathcal{O}_\Sigma)$. Intuitively, we construct the product game with two components, where Player 1 only observes the first component and makes the choice of the transitions there; and Player 2 is in charge of choosing the input and also the transitions of the second component. However, due to partial observation, Player 1 does not observe the choice of transitions in the second component. We describe the transition and the observation mapping to capture this: (i) the transition function $\delta^G : S^G \times \Sigma_1^G \times \Sigma_2^G \to S^G$ is as follows: $\delta^G((s_1, s_2), \sigma_1, (\sigma_2, \overline{\sigma}_1)) = (\delta(s_1, \sigma_1, \sigma_2), \delta(s_2, \sigma_2, \overline{\sigma}_1))$; (ii) the observation for states for Player 1 maps every state to the first component, i.e., $\mathsf{obs}_S((s_1, s_2)) = s_1$ and the observation for actions for Player 1 maps every state to the first component as well (i.e., the input from Player 2), i.e., $\mathsf{obs}_\Sigma((\sigma_2, \overline{\sigma}_1)) = \sigma_2$. The two reward functions are as follows: the reward function r_1 gives reward according to r and the transitions of the first component and the reward function r_2 assigns reward according to r and the transitions of the second component. Note that the construction ensures that we compare the utility of an on-line algorithm and an off-line algorithm (chosen by Player 2 using the second component) that operate on the *same* input sequence. It follows that there is an on-line algorithm with competitive-ratio at least ν iff $\sup_{\pi \in \Pi_{\mathsf{G}}^M} \inf_{\sigma \in \Sigma_{\mathsf{G}}} \mathsf{Ratio}(r_1, r_2)(\rho(s^0, \pi, \sigma)) \geq \nu$. Note that in the ratio objective we add the value 1 in the denominator and numerator to ensure that a division by zero cannot occur. As our choice of rewards in conjunction with the idle task ensures that the sum of the rewards in the denominator goes to infinity, adding the constant value 1 does not affect the competitive ratio analysis. By Theorem 2, the decision problem is in NP in the size of the LTS; by Theorem 2, given a particular on-line algorithm the competitive ratio computation is polynomial in the size of the LTS.

THEOREM 6. *For the class of scheduling problems defined in Section 4, the competitive ratio analysis decision problem is in* NP *in the size of the LTS constructed from the scheduling problem. Given an on-line algorithm, whether the competitive ratio is at least a rational number ν can be achieved in time polynomial in the size of the LTS.*

6. EXTENSIONS

In the previous section, we presented results of applying partial-observation games and perfect-observation games with mean-payoff and ratio objectives to our real-time scheduling problem. In this section, we show that it is easy to add various constraints on the adversary to the original setting, without unduly increasing the complexity of the involved decision problems. This particularly attractive feature of our approach follows from the very flexible framework of games, which allows to combine mean-payoff and ratio objectives with other classes of Boolean and quantitative objectives.

1. *(Safety objectives).* In general, safety objectives ensure that certain states of the game are never reached in any play. This is accomplished by penalizing the transitions to/from a forbidden state by a prohibitively large weight, such that the adversary can never visit it without losing the game. All our complexity results continue to hold, as the underlying (polynomial time) graph algorithms also work with additional safety conditions.

In our real-time scheduling context, safety objectives can e.g. be used to constrain the maximum load created by the adversary by some threshold u. All that is needed to accomplish this is to keep track of the actual load $u(t)$ within the state of the LTS, and to go to a winning state for Player 1 whenever $u(t) > u$. Enforcing sporadic and even periodic releases of certain tasks, rather than purely aperiodic ones, can be achieved in a similar way. A more elaborate application of safety objectives can be found in energy-constrained real-time scheduling, where the goal is to maximize the cumulated utility achievable with some fixed energy budget E. In [9], the adversary must be constrained to only generate task sequences that can be feasibly scheduled (without any deadline violation) for $E = \infty$. Since EDF is known to be an optimal scheduling algorithm for the latter problem, it suffices to also incorporate an instance of EDF into our LTS that leads to a winning state for Player 1 in case of a deadline violation.

2. *(Büchi objectives).* We can extend our results with additional Büchi objectives, which ensures that certain states are visited infinitely often in every play. Again, our results hold, as the underlying graph algorithms just need to first consider strongly connected components with Büchi states and then to apply the mean-payoff graph algorithm to ensure that the mean-payoff cycle contains at least one Büchi state. In a strongly connected component C with a Büchi state, the mean-payoff objective and the Büchi objective is satisfied as follows: the path first visits the optimal mean-payoff cycle C' for a long time, then visits a Büchi state s (in the same component); then again visits the cycle C' for long, then s, then the cycle C' and so on. Thus both the mean-payoff and Büchi objectives are satisfied.

In our real-time scheduling context, Büchi objectives can e.g. be used to constrain the adversary to generate only finite periods of overload. This can be achieved by adding the requirement that idle states of the LTS are visited infinitely often. Another application of Büchi objectives is for modeling semi-online algorithms, which assume that the scheduling algorithm knows something about the future behavior. A typical example is that the adversary will eventually generate an instance of a particular task (e.g., with some given execution time C_i [9]).

3. *(Multi-dimensional objectives).* We can also extend our standard setting with multi-dimensional objectives, where instead of one mean-payoff objective there are multiple ones. Again, all our results also extend to such objectives, since graphs with multiple mean-payoff objectives can also be solved in polynomial time [8, 21].

This extension could be used in real-time scheduling context for e.g. constraining the average load—rather than the maximum load—generated by the adversary during some time interval. Multi-dimensional objectives also allow to incorporate additional cost functions. E.g., minimizing energy consumption, in addition to maximizing cumulated utility, is an important concern in modern real-time systems [2]. Due to lack of space, we cannot elaborate on these extensions.

7. CONCLUSIONS

We showed how to employ the powerful framework of algorithmic graph games for competitive analysis of real-time scheduling algorithms for tasks with firm deadlines. We developed a novel partial-observation game suitable for this purpose, and showed that all involved problems of interest are decidable. Comparison with [3] reveals the following advantages/disadvantages of our approach: On the positive side, as an algorithmic approach, our solution does not involve human ingenuity. Moreover, apart from the fact that it deals with arbitrary task utility values V_i right from the start, i.e., not just uniform densities $V_i = C_i$, our method can easily accommodate various additional constraints on the adversary. Part of our future work will be devoted to exploring these extensions in detail.

On the negative side, whereas computing the worst-case cumulated utility etc. of a *given* on-line algorithm can usually be done in polynomial time, the utility of our approach for automated construction of an optimal on-line algorithm and its competitive analysis is currently severely impaired by the exponential time complexity involved in the LTS reduction (Theorem 3) and the competitive analysis decision problem (Theorem 2). While the present work sets up the basic foundational framework to use graph games for competitive analysis, and presents all the required theoretical results, a central part of our future work will be devoted to coping with practical issues, both identifying relevant problem sub-classes with better computational complexity; and trying to design heuristics and other practical techniques such as abstraction of games that allow to efficiently deal with (most) instances of the general problem.

Another limitation of our approach, though not relevant in practice, follows from the required finiteness of the state space of the LTS (without which all problems of interest are undecidable). In particular, we cannot allow the number of tasks n to become infinite. Consequently, we could not prove $1/4$ to be the competitive factor for uniform utilities $V_i = C_i$, as this effectively requires infinitely many tasks [3]. What we can prove, however, is a competitive factor of $1/4 + \epsilon(n)$, for some $\epsilon(n) > 0$, by applying our approach to the task set $\mathcal{T} = \{\tau_i = (i, i, i) | \; 1 \leq i \leq n\}$.

Overall, our findings reveal that automated competitive analysis using algorithmic graph games is theoretically feasible and opens up an interesting direction of further real-time systems research.

8. ACKNOWLEDGMENTS

This work has been supported by the Austrian Science Foundation (FWF) under the NFN RiSE (S11405 and S11407), FWF Grant P23499-N23, ERC Start grant (279307: Graph Games), and Microsoft faculty fellows award.

9. REFERENCES

[1] K. Altisen, G. Gößler, and J. Sifakis. Scheduler modeling based on the controller synthesis paradigm. *Real-Time Systems*, 23(1-2):55–84, 2002.

[2] H. Aydin, R. Melhem, D. Mossé, and P. Mejía-Alvarez. Power-aware scheduling for periodic real-time tasks. *IEEE Trans. Comput.*, 53(5):584–600, May 2004.

[3] S. Baruah, G. Koren, D. Mao, B. Mishra, A. Raghunathan, L. Rosier, D. Shasha, and F. Wang. On the competitiveness of on-line real-time task scheduling. *Real-Time Syst.*, 4:125–144, May 1992.

[4] S. Baruah and J. R. Haritsa. Scheduling for overload in real-time systems. *IEEE Trans. Comput.*, 46:1034–1039, September 1997.

[5] S. Baruah and M. E. Hickey. Competitive on-line scheduling of imprecise computations. *IEEE Trans. Comp.*, 47:1027–1032, 1998.

[6] A. Borodin and R. El-Yaniv. *Online Computation and Competitive Analysis*. Cambridge Univ. Press, 1998.

[7] L. Brim, J. Chaloupka, L. Doyen, R. Gentilini, and J-F. Raskin. Faster algorithms for mean-payoff games. *FMSD*, 38:97–118, 2011.

[8] K. Chatterjee, L. Doyen, T. A. Henzinger, and J-F. Raskin. Generalized mean-payoff and energy games. In *FSTTCS*, pages 505–516, 2010.

[9] V. Devadas, F. Li, and H. Aydin. Competitive analysis of online real-time scheduling algorithms under hard energy constraint. *Real-Time Syst.*, 46(1):88–120, September 2010.

[10] A. Ehrenfeucht and J. Mycielski. Positional strategies for mean payoff games. *Int. J. of Game Theory*, 8:109–113, 1979.

[11] R. M. Karp. A characterization of the minimum cycle mean in a digraph. *Discr. Math.*, 23:309–311, 1978.

[12] E. Kopczynski. Half-positional determinacy of infinite games. In *ICALP (2)*, pages 336–347, 2006.

[13] E. Koutsoupias. Scheduling without payments. In *SAGT* pages 143–153, 2011.

[14] G. Koren and D. Shasha. D^{over}: An optimal on-line scheduling algorithm for overloaded uniprocessor real-time systems. *SIAM J. on Comp.*, 24:318–339, 1995.

[15] D. A. Martin. Borel determinacy. *Annals of Math.*, 102:363–371, 1975.

[16] N. Nisan, T. Roughgarden, E. Tardos, and V. Vazirani. *Algorithmic Game Theory*. Cambridge Univ. Press, New York, NY, USA, 2007.

[17] M. A. Palis. Competitive algorithms for fine-grain real-time scheduling. *RTSS*, pages 129–138, 2004.

[18] C. H. Papadimitriou. *Computational Complexity*. Addison-Wesley, 1993.

[19] R. Rajkumar, C. Lee, J. Lehoczky, and D. Siewiorek. A resource allocation model for qos management. In *RTSS*, pages 298 -307, 1997.

[20] L. S. Shapley. Stochastic games. *PNAS USA*, 39:1095–1100, 1953.

[21] Y. Velner. The complexity of mean-payoff automaton expression. In *ICALP*, 2012.

[22] U. Zwick and M. Paterson. The complexity of mean payoff games on graphs. *TCS*, 158:343–359, 1996.

Reachability Analysis of Nonlinear Systems using Conservative Polynomialization and Non-Convex Sets

Matthias Althoff
Ilmenau University of Technology
98693 Ilmenau, Germany
matthias.althoff@tu-ilmenau.de

ABSTRACT

A new technique for computing the reachable set of hybrid systems with nonlinear continuous dynamics is presented. Previous work showed that abstracting the nonlinear continuous dynamics to linear differential inclusions results in a scalable approach for reachability analysis. However, when the abstraction becomes inaccurate, linearization techniques require splitting of reachable sets, resulting in an exponential growth of required linearizations. In this work, the nonlinearity of the dynamics is more accurately abstracted to polynomial difference inclusions. As a consequence, it is no longer guaranteed that reachable sets of consecutive time steps are mapped to convex sets as typically used in previous works. Thus, a non-convex set representation is developed in order to better capture the nonlinear dynamics, requiring no or much less splitting. The new approach has polynomial complexity with respect to the number of continuous state variables when splitting can be avoided and is thus promising when a linearization technique requires splitting for the same problem. The benefits are presented by numerical examples.

Categories and Subject Descriptors

G.1.0 [**Numerical Analysis**]: General; I.6.4 [**Simulation and Modeling**]: Model Validation and Analysis

Keywords

Reachability Analysis, Hybrid Systems, Nonlinear Dynamics, Difference Inclusion, Polynomial Zonotopes

1. INTRODUCTION

Formal verification of hybrid systems has enormous practical relevance since in almost all engineering fields, complex systems have a mixed discrete/continuous dynamics due to the interplay of physical behavior and digital control. Those systems are difficult to analyze, especially since disturbances and other uncertainties can lead to completely different behaviors. For this reason, formal verification techniques have been developed in order to mathematically guarantee that a system model satisfies a formalized specification.

Over the past years, a variety of formal methods for hybrid system verification have been developed: Reachability analysis [6], theorem proving [33], barrier certificates [34], simulation-based verification [23], abstraction to discrete systems [14], constraint propagation [35], and many more. Related topics are the falsification of systems [27,38], i.e. finding solutions that violate a specification, and probabilistic model checking [10,15], where a probability for satisfying a specification is computed. This work is about reachability analysis so that the remainder of the literature review is on this topic. Reachability analysis is concerned with the problem of computing the set of discrete and continuous states that a system can reach, making it possible to verify if a state can avoid a set of unsafe states.

Early works on reachability analysis considered timed automata, i.e., automata with time as the only continuous variable [5,8]. This concept has been extended to linear hybrid automata, where in each discrete mode, the derivative of the continuous state vector is bounded by a hyperrectangle [11] or a polytope [19]. Linear automata can also be used to overapproximate the solutions of more complex systems, such as linear continuous systems ($\dot{x} = Ax(t) + u(t)$, $A \in \mathbb{R}^{n\times n}$, $x, u \in \mathbb{R}^n$) [24]. However, this approach causes a wrapping effect, i.e., the overapproximation increases since reachable sets are computed based on sets of previous times and the error accumulates for each iteration. A wrapping-free approach for linear systems has first been published in [22]. When using special set representations such as zonotopes or support functions, linear systems with more than 1000 continuous state variables can be verified [21,22].

For hybrid systems with nonlinear continuous dynamics, no wrapping-free algorithm exist, except when the dynamics can be rewritten as a linear system in a new coordinate system [37]. Approaches for nonlinear systems can be categorized into approaches that (1) reformulate the reachability analysis as an optimization problem, (2) use the Picard iteration in combination with Taylor models, or (3) construct mappings to propagate the set of reachable states. Reformulation as an optimization problem is performed in [13] and [31], where the first approach uses optimization to obtain the outwards translation of halfspaces confining polytopical reachable sets and the latter rewrites the entire reachable set computation in the form of Hamilton-Jacobi equations.

HSCC'13, April 8–11, 2013, Philadelphia, Pennsylvania, USA.

Typically, approaches involving optimization techniques do not scale well with the number of continuous state variables.

The Picard iteration in combination with rigorous Taylor models is first proposed by Berz and Makino and adopted by other researchers [25, 30, 32]. The main idea is to use the Picard iteration to obtain a polynomial of given degree that approximates the solution over time with respect to varying initial states and other parameters. In order to guarantee that the exact solution is captured, an uncertain multidimensional interval is added to the polynomial solution, resulting in a so-called *Taylor model.* A Taylor model is acceptable if it is contractive, i.e., running the Picard iteration with the suggested Taylor model has to result in a Taylor model enclosed by the previous one. The approach is extended to hybrid systems in [12].

Construction of mappings for the propagation of reachable sets is well developed for linear systems with a large variety of convex set representations (polytopes [13], zonotopes [22], ellipsoids [29], support functions [21], oriented hyperrectangles [39], and others). Those approaches do not require to formulate solutions over time by polynomials as done for Taylor model approaches. When the dynamics is nonlinear, most approaches linearize the dynamics. Earlier approaches define regions in which the original dynamics is linearized to which a compensating linearization error is added, resulting in linear differential inclusions [7]. When the nonlinear dynamics is a multi-affine system and the regions are hyperrectangles, it is sufficient to only consider the flow field at the vertices to determine which cells of the partition are reachable [28]. The disadvantage of fixed partitions is that the number of required regions grows exponentially with respect to the number of continuous state variables and most approaches require intersection at the borders of the regions, which is computationally expensive. More recent approaches overcome this problem by defining overlapping linearization regions that move along with the reachable set [4, 18]. The main disadvantage of those approaches is that for large linearization errors, the reachable set has to be split and one has to continue with several reachable sets in parallel, causing exponential complexity in the number of variables contributing to the linearization error.

In this work, the problem of avoiding or at least reducing splitting of map-based reachable set approaches is addressed. Instead of applying linearization, the system dynamics is abstracted to a nonlinear system with polynomial right-hand side plus additive uncertainty, resulting in *polynomial differential inclusions.* Unlike for linear systems, there exists no closed-form solution for polynomial differential inclusions, but a new tight overapproximation is presented in the form of a *polynomial difference inclusion* $x(t_{k+1}) \in f(x(t_k), u(t_k)) \oplus \mathcal{W}$, where $\mathcal{W}$ is an additive uncertainty with proper dimension and $\oplus$ denotes the Minkowski addition. Thus, reachable sets which are represented by a convex set, are now possibly mapped to non-convex sets. If one uses a convex set representation as done by almost all previous approaches, the benefit of capturing the nonlinearity by a polynomial differential inclusion is lost. For this reason, a new non-convex set representation called *polynomial zonotope* is proposed, which extends the definition of zonotopes, and is as expressive as Taylor models. We demonstrate the approach by numerical examples, compare the results to the Taylor model tool *flow** [12], and show the benefits compared to map-based linearization procedures. Since the overall complexity of the new approach is polynomial in the number of continuous state variables, it is preferable over linearization techniques for many problems which require splitting.

In order to focus on the continuous dynamics of hybrid systems, the approach is first developed for purely continuous systems. In Sec. 3, the solution of a nonlinear differential equation is overapproximated by a polynomial difference inclusion. The abstraction is used in Sec. 4 to develop the reachability algorithm. Polynomial zonotopes are introduced as the set representation in Sec. 5. The paper finishes with numerical examples in Sec. 6, where one example also briefly presents the integration of the approach into hybrid system reachability analysis.

2. BASIC PRINCIPLE

In this paper, nonlinear systems of the form

$$\dot{x}(t) = f(x(t), u(t)), \quad x(t) \in \mathbb{R}^n, \quad u(t) \in \mathbb{R}^m, \qquad (1)$$

are considered, where x is the state vector and u is the input vector. The differential equation is required to be Lipschitz continuous and the input trajectory $u(\cdot)$ is required to be piecewise continuous so that a solution is guaranteed to exist. Let $\chi(t; x_0, u(\cdot))$ denote the solution to (1) for an initial state $x(0) = x_0$ and the input trajectory $u(\cdot)$. For a set of initial states $\mathcal{R}(0) \subset \mathbb{R}^n$ and a set of possible input values $\mathcal{U} \subset \mathbb{R}^m$, the set of reachable states is

$$\mathcal{R}^e([0,r]) := \Big\{ \chi(t; x_0, u(\cdot)) \Big| x_0 \in \mathcal{R}(0), t \in [0,r], \\ \forall \tau \in [0,t]\, u(\tau) \in \mathcal{U} \Big\}.$$

The superscript e on $\mathcal{R}^e([0,r])$ denotes the exact reachable set, which cannot be computed for general nonlinear systems. Therefore, an overapproximation $\mathcal{R}([0,r]) \supseteq \mathcal{R}^e([0,r])$ is computed as accurately as possible, while at the same time ensuring that the computations are efficient and scale well with the system dimension n. As in many other works (see e.g. [13, 16, 20, 39]), the reachable set is computed for consecutive time intervals $\tau_s := [t_s, t_{s+1}]$, where $t_s = s \cdot r$, $r \in \mathbb{R}^+$ is the time increment, and $s \in \mathbb{N}$ is the time step. The reachable set for a user-defined time horizon t_f is $\mathcal{R}([0, t_f]) = \bigcup_{s=0}^{t_f/r-1} \mathcal{R}(\tau_s)$, where t_f is a multiple of r.

As later shown, the wrapping effect in this work is almost entirely determined by the accuracy of auxiliary sets $\mathcal{R}(t_s)$ at points in time t_s since those sets are computed based on the accuracy of the previous ones. In contrast to this, the sets of time intervals $\mathcal{R}(\tau_s)$ fill the "time gaps" based on the sets $\mathcal{R}(t_s)$ as shown in Fig. 1, and are not used again in the computation.

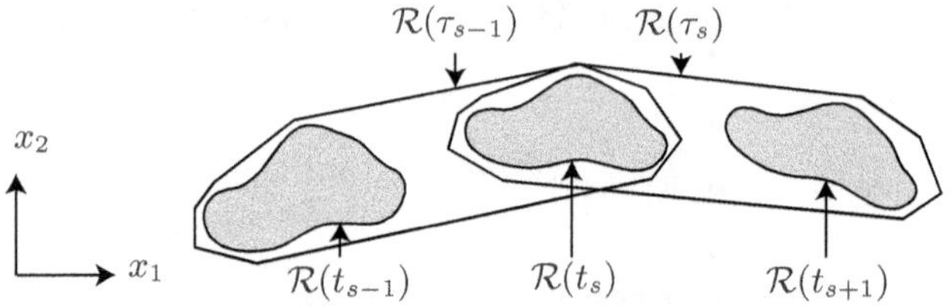

Figure 1: Stepwise computation of the reachable set.

3. ABSTRACTION OF THE NONLINEAR DYNAMICS

Nonlinear systems are hard to analyze since almost all of them do not have a closed-form solution. In this work, their solution is tightly overapproximated by first abstracting the nonlinear differential equations to a polynomial differential inclusion using a Taylor expansion. In a second step, a novel approach is used to obtain a polynomial difference equation.

3.1 Abstraction to Polynomial Differential Inclusions

For a concise notation, the combined state/input vector $\tilde{z} = \begin{bmatrix} x^T & u^T \end{bmatrix}^T \in \mathbb{R}^o$ $(o = n + m)$ and the Nabla symbol $\nabla = \sum_{i=1}^{o} e^{(i)} \frac{\partial}{\partial \tilde{z}_i}$, using orthogonal unit vectors $e^{(i)}$, are introduced. Note that the superscript i is in parentheses to avoid confusion with powers, which is a notation used for other variables in this work, too. The nonlinear system in (1) is abstracted by a Taylor expansion of order κ at point z^* with Lagrange remainder $\mathcal{L}$ (see [9]):

$$\dot{x}_i = f_i(\hat{z}(t)) \in \sum_{\nu=0}^{\kappa} \frac{\left((\hat{z}(t) - z^*)^T \nabla\right)^{\nu} f_i(\tilde{z})}{\nu!}\Bigg|_{\tilde{z}=z^*} \oplus \mathcal{L}_i(t) \tag{2}$$

$$\mathcal{L}_i(t) = \left\{ \frac{\left((\hat{z}(t) - z^*)^T \nabla\right)^{\kappa+1} f_i(\tilde{z})}{(\kappa+1)!}\Bigg| \tilde{z} = z^* + \alpha(\hat{z}(t) - z^*),\, \alpha \in [0,1] \right\},$$

where $\mathcal{A} \oplus \mathcal{B} := \{a + b | a \in \mathcal{A}, b \in \mathcal{B}\}$ is a Minkowski addition and for later derivations, set-based multiplication $\mathcal{A} \otimes \mathcal{B} := \{a\, b | a \in \mathcal{A}, b \in \mathcal{B}\}$ is introduced, too. Note that set-based multiplication has precedence over set-based multiplication, expressions are evaluated from left to right, and the symbol $\otimes$ is sometimes omitted as for classical multiplications. For subsequent derivations, the alternative notation of (2)

$$\begin{aligned} \dot{x}_i \in & w_i + \frac{1}{1!} \sum_{j=1}^{o} C_{ij} z_j(t) + \frac{1}{2!} \sum_{j=1}^{o} \sum_{k=1}^{o} D_{ijk} z_j(t) z_k(t) \\ & + \frac{1}{3!} \sum_{j=1}^{o} \sum_{k=1}^{o} \sum_{l=1}^{o} E_{ijkl} z_j(t) z_k(t) z_l(t) + \ldots \oplus \mathcal{L}_i(t) \end{aligned} \tag{3}$$

is used, where $z(t) = \hat{z}(t) - z^*$ and

$$w_i = f_i(z^*),\, C_{ij} = \frac{\partial f_i(\tilde{z})}{\partial \tilde{z}_j}\Big|_{\tilde{z}=z^*},\, D_{ijk} = \frac{\partial^2 f_i(\tilde{z})}{\partial \tilde{z}_j \partial \tilde{z}_k}\Big|_{\tilde{z}=z^*}, \ldots$$

Note that z^* is changed at times t_s so that we have $w(t_s)$, $C(t_s)$, $D(t_s)$, ..., where the dependency on time is omitted in the notation since for the remainder of this section it is always assumed that $t \in \tau_s$. For reachability analysis we require a difference inclusion that encloses the solution of the differential inclusion in (3).

3.2 Abstraction to Difference Inclusions

As a first step, all higher order terms in (3) are interpreted as an input v to a linear system, where the matrices $A \in \mathbb{R}^{n \times n}$ and $B \in \mathbb{R}^{n \times m}$ are obtained from $C = \begin{bmatrix} A & B \end{bmatrix}$ and z is partially substituted by $\begin{bmatrix} x^T & u^T \end{bmatrix}^T$:

$$\begin{aligned} \dot{x} &\in Ax(t) + v(z(t), u(t)) \oplus \mathcal{L}(t) \\ v_i(z, u) &= w_i + \sum_{j=1}^{m} B_{ij} u_j + \frac{1}{2!} \sum_{j=1}^{o} \sum_{k=1}^{o} D_{ijk} z_j z_k + \ldots \end{aligned} \tag{4}$$

In order to obtain a tight overapproximation, the auxiliary variables $u^{\Delta}(t) = u(t) - u^c$, $z^{\Delta}(t) = z(t) - z(t_s)$ are introduced to split the input $v(z(t), u(t))$ for $t \in \tau_s$ into a constant part $v(z(t_s), u^c)$ fixed at the specific point in time t_s and a time-varying part $v^{\Delta}(z^{\Delta}(t), z(t_s), u^{\Delta}(t))$:

$$\begin{aligned} v_i(z(t), u(t)) &= w_i + \sum_{j=1}^{m} B_{ij}(u_j^c + u_j^{\Delta}(t)) \\ &+ \frac{1}{2!} \sum_{j=1}^{o} \sum_{k=1}^{o} D_{ijk} \underbrace{(z_j(t_s) + z_j^{\Delta}(t))(z_k(t_s) + z_k^{\Delta}(t))}_{\substack{= z_j(t_s) z_k(t_s) + z_j(t_s) z_k^{\Delta}(t) \\ + z_j^{\Delta}(t) z_k(t_s) + z_j^{\Delta}(t) z_k^{\Delta}(t)}} + \ldots \\ &= v(z(t_s), u^c) + v^{\Delta}(z^{\Delta}(t), z(t_s), u^{\Delta}(t)) \end{aligned}$$

where

$$\begin{aligned} v(z(t_s), u^c) &= w_i + \sum_{j=1}^{m} B_{ij} u_j^c + \\ & \frac{1}{2!} \sum_{j=1}^{o} \sum_{k=1}^{o} D_{ijk} z_j(t_s) z_k(t_s) + \ldots \\ v^{\Delta}(z^{\Delta}(t), z(t_s), u^{\Delta}(t)) &= \sum_{j=1}^{m} B_{ij} u_j^{\Delta}(t) + \frac{1}{2!} \sum_{j=1}^{o} \sum_{k=1}^{o} D_{ijk} \\ & \left(z_j(t_s) z_k^{\Delta}(t) + z_j^{\Delta}(t) z_k(t_s) + z_j^{\Delta}(t) z_k^{\Delta}(t)\right) + \ldots \end{aligned} \tag{5}$$

After defining $\mathcal{U}^{\Delta} := \mathcal{U} \oplus (-u^c)$ and assuming that the reachable set $\mathcal{R}(\tau_s)$ and

$$\begin{aligned} \mathcal{R}^{\Delta}(\tau_s) := \Big\{ & \chi(t; x(t_s), u(\cdot)) - x(t_s) \Big| t \in \tau_s, \\ & x(t_s) \in \mathcal{R}(t_s), \forall t \in \tau_s\, u(t) \in \mathcal{U} \Big\} \end{aligned} \tag{6}$$

are already known ($\chi()$ was defined as the solution of (1)), the set of possible values of $v^{\Delta}(z^{\Delta}(t), z(t_s), u^{\Delta}(t))$ is bounded by

$$\begin{aligned} \mathcal{V}^{\Delta}(\tau_s) := \Big\{ & v^{\Delta}(z^{\Delta}, z, u^{\Delta}) \Big| \\ & z^{\Delta} \in \mathcal{R}^{\Delta}(\tau_s) \times \mathcal{U}^{\Delta}, z \in \mathcal{R}(\tau_s) \times \mathcal{U},\, u^{\Delta} \in \mathcal{U}^{\Delta} \Big\}. \end{aligned} \tag{7}$$

Using (4) - (7), the linear differential inclusion

$$\dot{x} \in Ax(t) + v(z(t_s), u^c) \oplus \left(\mathcal{V}^{\Delta}(\tau_s) \oplus \mathcal{L}(\tau_s)\right)$$

is obtained for $t \in \tau_s$. Due to the superposition principle of linear systems, the solution is obtained by adding the solution of the homogeneous solution $x^h(t_{s+1})$, the input solution due to constant input $x^{p,c}(r)$, where $r = t_{s+1} - t_s$, and the input solution set due to time-varying inputs $\mathcal{R}^{p,\Delta}(\mathcal{V}^{\Delta}(\tau_s) \oplus \mathcal{L}(\tau_s), r)$ to

$$x(t_{s+1}) \in x^h(t_{s+1}) + x^{p,c}(t) \oplus \mathcal{R}^{p,\Delta}\left(\mathcal{V}^{\Delta}(\tau_s) \oplus \mathcal{L}(\tau_s), r\right). \tag{8}$$

The well-known homogeneous solution is $x^h(t_{s+1}) = e^{Ar}x(t_s)$, the input solution due to constant input is

$$x^{p,c}(r) = \Gamma(r)\, v(z(t_s), u^c), \quad \Gamma(r) := \int_0^r e^{A(r-t)}dt,$$

where $\Gamma(r) = A^{-1}(e^{Ar} - I)$ (I is the identity matrix) and when A is not invertible, the approach in [3] is used. The reachable set due to the set of uncertain time-varying inputs within $\tilde{\mathcal{V}}(\tau_s) := \mathcal{V}^{\Delta}(\tau_s) \oplus \mathcal{L}(\tau_s)$ is computed as in [3] as

$$\mathcal{R}^{p,\Delta}(\tilde{\mathcal{V}}(\tau_s), r) = \bigoplus_{i=0}^{\nu} \frac{t^{i+1}}{(i+1)!}\mathtt{CH}(A^i \otimes \tilde{\mathcal{V}}(\tau_s)) \oplus \mathcal{E}^{p,\Delta}, \quad (9)$$

where CH() is the operator returning the convex hull of a set, and $\mathcal{E}^{p,\Delta}$ is an interval vector which becomes arbitrarily small for $\nu \to \infty$.

The difference to previous approaches (e.g. [4,18]) is that due to the separation in a constant and a time-varying input, nonlinear terms are saved from linearization at times t_s, while within τ_s, an abstracting linear differential inclusion is used. Inserting $v(z(t_s), u^c)$ from (5) into the overall solution (8) results in a nonlinear difference equation that encloses the exact solution:

$$\begin{aligned} x_i(t_{s+1}) \in &\sum_{j=1}^{n}(e^{Ar})_{ij}x_j(t_s) + \sum_{j=1}^{n}\Gamma_{ij}(r)\Big(w_j + \sum_{k=1}^{m} B_{jk}u_k^c \\ &+ \frac{1}{2!}\sum_{k=1}^{o}\sum_{l=1}^{o} D_{jkl}z_k(t_s)z_l(t_s) + \dots\Big) \\ &\oplus \mathcal{R}_i^{p,\Delta}(\mathcal{V}^{\Delta}(\tau_s) \oplus \mathcal{L}(\tau_s), r) \end{aligned} \quad (10)$$

The benefits of the above difference inclusion for reachability analysis do not immediately show. Note that for small time increments r, as typically used in reachability analysis, the set $\mathcal{R}_i^{p,\Delta}$ becomes small, no matter how large the set of $z(t_s)$ becomes during the reachability analysis (proof is omitted due to space limitations). Thus, for large sets of $z(t_s)$, the nonlinearity is well captured by all other terms, while the abstractions in $\mathcal{R}_i^{p,\Delta}$ are not dominant.

4. REACHABLE SET COMPUTATIONS

This section describes Alg. 1 for computing the set of reachable states when using the previously presented abstraction to difference inclusions. In order to focus on the novel aspects, the possibility to split reachable sets is not included. The algorithm consists of 2 parts as indicated in Alg. 1:

① Computing a linearization and the corresponding set of linearization errors denoted by $\Psi(\tau_s)$. The reachable set $\mathcal{R}(\tau_s)$ for the time interval τ_s is obtained as a by-product.

② Computing the reachable set at the next point in time $\mathcal{R}(t_{s+1})$ by a set-based evaluation of (10).

Since the computations of each time interval τ_s are based on the reachable set at points in time t_s, the sets $\mathcal{R}(t_s)$ are foremost contributing to the wrapping effect. The sets $\mathcal{R}(\tau_s)$ are filling the gaps between points in time t_s, which are not used for subsequent computations (see Fig. 1).

For the remainder of the paper we focus on Taylor expansions of second order with third order Lagrange remainder since higher order terms do not require a modification of the approach.

At the beginning of each time interval τ_s, the Taylor terms are re-evaluated according to a new expansion point z^*, which is indicated in line 3 of Alg. 1 by

$$\mathtt{taylor} \to z^*, w, A, B, D, E,$$

where the dependency on the time step is omitted in the notation. The expansion point z^* is chosen heuristically as $z^*(t_s) = [x^*(t_s), u^*]$, where $x^*(t_s) = x^c(t_s) + 0.5 \cdot r \cdot f(x^c(t_s), u^c) \approx x^c(t_s + 0.5 \cdot r)$, $u^* = u^c$ and $x^c(t_s)$, u^c are the volumetric centers of $\mathcal{R}(t_s)$, $\mathcal{U}$. Other linearization points within $\mathcal{R}(\tau_s)$ can be chosen, but better heuristics have not been found so far. Next, the linearization error for the reachable set within time intervals is obtained.

4.1 Overapproximating the Linearization Error

To obtain the set of linearization errors for the time interval solution $\mathcal{R}(\tau_s)$, (4) is abstracted to a linear differential inclusion

$$\begin{aligned} \dot{x} \in Ax(t) \oplus \Psi(\tau_s), \quad \Psi(\tau_s) = \mathcal{V}(\tau_s) \oplus \mathcal{L}(\tau_s), \quad (11) \\ \mathcal{V}(\tau_s) := \{v(z,u) | z \in \mathcal{R}(\tau_s) \times \mathcal{U},\, u \in \mathcal{U}\}. \end{aligned}$$

The computation of the set of linearization errors $\Psi(\tau_s)$ requires the computation of the reachable set $\mathcal{R}(\tau_s)$, which in turn requires $\Psi(\tau_s)$. This mutual dependence is initially resolved by an estimation of the set of linearization errors $\overline{\Psi}(\tau_s)$ with the goal that $\Psi(\tau_s) \subseteq \overline{\Psi}(\tau_s)$. This estimation is used to compute the set of state differences $\mathcal{R}^{\Delta}(\tau_s)$ for (11) using a slight modification of the standard techniques for linear system reachability presented in [1, Chap. 3.2]. The modification involves returning the set of state differences $\mathcal{R}^{\Delta}(\tau_s)$ instead of the set of states $\mathcal{R}(\tau_s)$. We denote this standard operation (see line 6 of Alg. 1) by

$$\mathcal{R}^{\Delta}(\tau_s) = \mathtt{post}^{\Delta}(\mathcal{R}(t_s), \overline{\Psi}(\tau_s), A).$$

Using the definition of $\mathcal{R}^{\Delta}(\tau_s)$ in (6), the overapproximation of the time interval solution is obtained as $\mathcal{R}(\tau_s) = \mathcal{R}(t_s) \oplus \mathcal{R}^{\Delta}(\tau_s)$, where $\mathcal{R}(t_s)$ is later represented by a non-convex set, $\mathcal{R}^{\Delta}(\tau_s)$ by a convex set, and the overapproximation $\mathcal{R}(\tau_s) \subseteq \mathtt{CH}(\mathcal{R}(t_s)) \oplus \mathcal{R}^{\Delta}(\tau_s)$ is used in line 7 of Alg. 1 for the efficient computation of $\mathcal{R}(\tau_s)$, since $\mathcal{R}(\tau_s)$ contributes only marginally to the wrapping effect by enlarging the overapproximation of the linearization error.

For a simple notation of subsequent computations, the operations

$$\mathrm{sq}(D, \mathcal{R}_1) := \Big\{\lambda \Big| \lambda_i = \sum_{j=1}^{o}\sum_{k=1}^{o} D_{ijk}z_jz_k,\, z \in \mathcal{R}_1\Big\},$$

$$\mathrm{mu}(D, \mathcal{R}_1, \mathcal{R}_2) := \Big\{\lambda \Big| \lambda_i = \sum_{j=1}^{o}\sum_{k=1}^{o} D_{ijk}z_j\hat{z}_k,\, z \in \mathcal{R}_1, \hat{z} \in \mathcal{R}_2\Big\}$$

are introduced. Using the definitions of the linearization error in (11) and the function $v(z,u)$ in (4), the set of linearization errors is overapproximated in line 9 of Alg. 1 by

$$\Psi(\tau_s) \subseteq w \oplus B \otimes \mathcal{U} \oplus \frac{1}{2}\mathrm{sq}(D, \mathcal{R}(\tau_s) \times \mathcal{U}) \oplus \mathcal{L}(\tau_s).$$

The computation of $B \otimes \mathcal{U}$ and $\mathrm{sq}(D, \mathcal{R}(\tau_s) \times \mathcal{U})$ is later presented for polynomial zonotopes. The Lagrangian remainder $\mathcal{L}(\tau_s)$ is small compared to the other sets and thus less accurately overapproximated by obtaining the enclosing boxes of all sets and applying interval arithmetic [26]. This is denoted in line 8 of Alg. 1 by $\texttt{lagrangeRemainder}(\mathcal{R}(\tau_s), E(t_s), z^*(t_s))$.

In case $\Psi(\tau_s) \not\subseteq \overline{\Psi}(\tau_s)$, the result $\Psi(\tau_s)$ is uniformly enlarged in each direction by a factor $\lambda > 1$ around its volumetric center, denoted by $\overline{\Psi}(\tau_s) = \texttt{enlarge}\big(\Psi(\tau_s), \lambda\big)$ in line 5 of Alg. 1. Using the enlarged set $\overline{\Psi}(\tau_s)$, the the linearization error computation is started over.

4.2 Reachable Set at the Next Point in Time

The reachable set at the next point in time is obtained by a set-based computation of (10). Thereto, it is first required to compute $\mathcal{V}^{\Delta}(\tau_s)$ in line 12 of Alg. 1 using (5), for which the sets $\mathcal{R}_z^{\Delta}(\tau_s) = \mathcal{R}^{\Delta}(\tau_s) \times \mathcal{U}^{\Delta}$ and $\mathcal{R}_z(t_s) = \mathcal{R}(t_s) \times \mathcal{U}$ are introduced:

$$\begin{aligned}\mathcal{V}^{\Delta}(\tau_s) =& \texttt{varInputs}(\mathcal{R}_z(t_s), \mathcal{R}_z^{\Delta}(\tau_s), \mathcal{U}^{\Delta}, B, D)\\ :=& B \otimes \mathcal{U}^{\Delta} \oplus \frac{1}{2!}\Big(\mathrm{mu}(D, \mathcal{R}_z(t_s), \mathcal{R}_z^{\Delta}(\tau_s))\\ &\oplus \mathrm{mu}(D, \mathcal{R}_z^{\Delta}(\tau_s), \mathcal{R}_z(t_s)) \oplus \mathrm{sq}(D, \mathcal{R}_z^{\Delta}(\tau_s))\Big)\end{aligned}$$

The operator $\texttt{post}()$ computes $\mathcal{R}(t_{s+1})$ in line 13 of Alg. 1 by replacing exact values with sets in (10):

$$\begin{aligned}\mathcal{R}(t_{s+1}) =& \texttt{post}(\mathcal{R}(t_s), \mathcal{R}_z(t_s), w, A, B, D, \mathcal{V}^{\Delta}(\tau_s), \mathcal{L}(\tau_s))\\ :=& \underbrace{e^{Ar}\mathcal{R}(t_s)}_{=:\mathcal{PZ}_1} \oplus \mathcal{R}^{p,\Delta}(\mathcal{V}^{\Delta}(\tau_s) \oplus \mathcal{L}(\tau_s), r)\\ &\underbrace{\oplus \Gamma(r)\Big(w \oplus Bu^c \oplus \frac{1}{2!}\mathrm{sq}(D, \mathcal{R}_z(t_s))\Big)}_{=:\mathcal{PZ}_2}\end{aligned} \tag{12}$$

As previously mentioned, for small time steps r as typically used in reachability analysis, the set $\mathcal{R}^{p,\Delta}$ is small compared to the set $\mathcal{R}$ and $\mathcal{R}_z$. From this follows that the nonlinear terms capturing the original nonlinear dynamics are not marginalized by the abstractions applied to compute $\mathcal{R}^{p,\Delta}$.

The new algorithm includes nonlinear mappings so that in general convex sets are no longer mapped to convex sets as in other works, requiring a new non-convex set representation as presented in the following section.

5. POLYNOMIAL ZONOTOPES

Set representations in most previous works are convex since they are easy to represent and manipulate (see e.g. [4, 6, 7, 13, 21, 22, 29, 39]). However, the convexity property makes the efforts in capturing the nonlinear dynamics obsolete, since convex sets only work well for linear maps. A new non-convex set representation is proposed, which can be efficiently stored and manipulated. The new representation shares many similarities with Taylor models [25] (as shortly discussed later) and is a generalization of zonotopes, which have shown great performance for linear and nonlinear reachability analysis [4, 22].

Definition 1 (Polynomial Zonotope): Given a *starting point* $c \in \mathbb{R}^n$, multi-indexed *generators* $f^{([i],j,k,\ldots,m)} \in \mathbb{R}^n$, and single-indexed *generators* $g^{(i)} \in \mathbb{R}^n$, a polynomial zonotope is defined as

$$\begin{aligned}\mathcal{PZ} = \Big\{c &+ \sum_{j=1}^{p}\beta_j f^{([1],j)} + \sum_{j=1}^{p}\sum_{k=j}^{p}\beta_j\beta_k f^{([2],j,k)}\\ &+ \ldots + \sum_{j=1}^{p}\sum_{k=j}^{p}\ldots\sum_{m=l}^{p}\underbrace{\beta_j\beta_k\ldots\beta_m}_{\eta\text{ factors}} f^{([\eta],j,k,\ldots,m)}\\ &+ \sum_{i=1}^{q}\gamma_i g^{(i)}\Big|\beta_i, \gamma_i \in [-1,1]\Big\}.\end{aligned} \tag{13}$$

The scalars β_i are called *dependent factors*, since changing their value does not only affect the multiplication with one generator, but other generators, too. On the other hand, the scalars γ_i only affect the multiplication with one generator, so that they are called *independent factors*. The number of dependent factors is p, the number of independent factors is q, and the polynomial order η is the maximum power of the scalar factors β_i. The order of a polynomial zonotope is defined as the number of generators ξ divided by the dimension, which is $\rho = \frac{\xi}{n}$. For a concise notation and later derivations, we introduce the matrices

$$\begin{aligned}E^{[i]} &= [\underbrace{f^{([i],1,1,\ldots,1)}}_{=:e^{([i],1)}} \ldots \underbrace{f^{([i],p,p,\ldots,p)}}_{=:e^{([i],p)}}] \text{ (equal indices)},\\ F^{[i]} &= [f^{([i],1,1,\ldots,1,2)}\ f^{([i],1,1,\ldots,1,3)} \ldots f^{([i],1,1,\ldots,1,p)}\\ &\quad f^{([i],1,1,\ldots,2,2)}\ f^{([i],1,1,\ldots,2,3)} \ldots f^{([i],1,1,\ldots,2,p)}\\ &\quad f^{([i],1,1,\ldots,3,3)} \ldots] \text{ (unequal indices)},\\ G &= [g^{(1)} \ldots g^{(q)}],\end{aligned}$$

and $E = \begin{bmatrix}E^{[1]} & \ldots & E^{[\eta]}\end{bmatrix}$, $F = \begin{bmatrix}F^{[2]} & \ldots & F^{[\eta]}\end{bmatrix}$ ($F^{[i]}$ is only defined for $i \geq 2$). Note that the indices in $F^{[i]}$ are ascending due to the nested summations in (13). In short form, a polynomial zonotope is written as $\mathcal{PZ} = (c, E, F, G)$. □

Algorithm 1 $\texttt{reach}(\mathcal{R}(0), t_f, \ldots)$

Require: Initial set $\mathcal{R}(0)$, input set $\mathcal{U}$, time horizon t_f, time step r, factor λ
Ensure: $\mathcal{R}([0, t_f])$

1: $t_0 = 0$, $s = 0$, $\Psi(\tau_0) = \{\mathbf{0}\}$, $\mathcal{R}^{union} = \emptyset$, $\mathcal{U}^{\Delta} = \mathcal{U} \oplus (-u^c)$
2: **while** $t_s < t_f$ **do**
3: $\quad\texttt{taylor} \rightarrow z^*, w, A, B, D, E$ ①
4: $\quad$**repeat**
5: $\quad\quad\overline{\Psi}(\tau_s) = \texttt{enlarge}\big(\Psi(\tau_s), \lambda\big)$
6: $\quad\quad\mathcal{R}^{\Delta}(\tau_s) = \texttt{post}^{\Delta}(\mathcal{R}(t_s), \overline{\Psi}(\tau_s), A)$
7: $\quad\quad\mathcal{R}(\tau_s) = \texttt{CH}\big(\mathcal{R}(t_s)\big) \oplus \mathcal{R}^{\Delta}(\tau_s)$
8: $\quad\quad\mathcal{L}(\tau_s) = \texttt{lagrangeRemainder}(\mathcal{R}(\tau_s), E, z^*)$
9: $\quad\quad\Psi(\tau_s) = w \oplus B \otimes \mathcal{U} \oplus \frac{1}{2}\mathrm{sq}(D, \mathcal{R}(\tau_s) \times \mathcal{U}) \oplus \mathcal{L}(\tau_s)$
10: $\quad$**until** $\Psi(\tau_s) \subseteq \overline{\Psi}(\tau_s)$
11: $\quad\mathcal{R}_z(t_s) = \mathcal{R}(t_s) \times \mathcal{U}$, $\quad\mathcal{R}_z^{\Delta}(\tau_s) = \mathcal{R}^{\Delta}(\tau_s) \times \mathcal{U}^{\Delta}$ ②
12: $\quad\mathcal{V}^{\Delta}(\tau_s) = \texttt{varInputs}(\mathcal{R}_z(t_s), \mathcal{R}_z^{\Delta}(\tau_s), \mathcal{U}^{\Delta}, B, D)$
13: $\quad\mathcal{R}(t_{s+1}) = \texttt{post}(\mathcal{R}(t_s), \mathcal{R}_z(t_s), w, A, B, D,$
14: $\quad\quad\quad\quad\mathcal{V}^{\Delta}(\tau_s), \mathcal{L}(\tau_s))$
15: $\quad\mathcal{R}^{union} = \mathcal{R}^{union} \cup \mathcal{R}(\tau_s)$
16: $\quad t_{s+1} = t_s + r$, $\ s := s + 1$
17: **end while**
18: $\mathcal{R}([0, t_f]) = \mathcal{R}^{union}$

For a given polynomial order i, the total number of generators in $E^{[i]}$ and $F^{[i]}$ is derived using the number $\binom{p+i-1}{i}$ of combinations of the scalar factors β with replacement (i.e. the same factor can be used again). Adding the numbers for all polynomial orders and adding the number of independent generators q, results in $\xi = \sum_{i=1}^{\eta} \binom{p+i-1}{i} + q$ generators, which is in $\mathcal{O}(p^{\eta})$ with respect to p. The non-convex shape of a polynomial zonotope with polynomial order 2 is shown in Fig. 2.

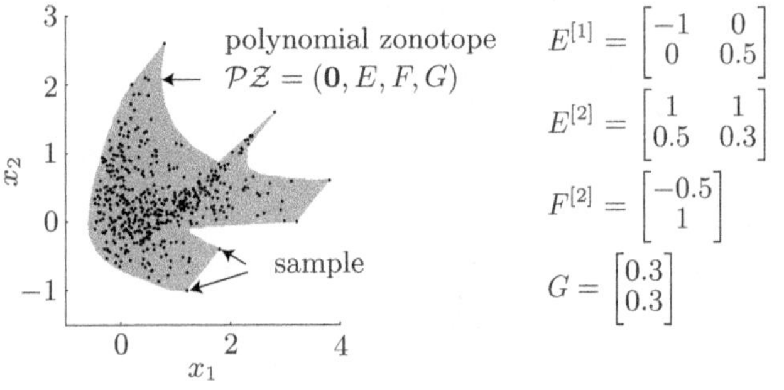

Figure 2: Overapproximative plot of a polynomial zonotope as specified in the figure. Random samples of possible values demonstrate the accuracy of the overapproximative plot.

A zonotope $\mathcal{Z}$ is a special case of a polynomial zonotope that has only generators $g^{(i)}$, which is denoted by $\mathcal{Z} = (c, G)$. Due to the absence of $E^{[i]}, F^{[i]}$ a zonotope is centrally symmetric to c so that for zonotopes, c is referred to as the *center* and not the *starting point*.

Although a Taylor model [25] is not a set, but a multidimensional polynomial plus a multidimensional interval, they can represent exactly the same sets than polynomial zonotopes when the input of the Taylor models is a multidimensional interval. The different organization of polynomial zonotopes, separating dependent from independent variables, makes it easier to overapproximate them by zonotopes or perform the order reduction techniques presented subsequently.

5.1 Operations on Polynomial Zonotopes

It is often required to overapproximate a polynomial zonotope by a zonotope:

Proposition 1 (Overapproximation by a Zonotope): A polynomial zonotope $\mathcal{PZ} = (c, E, F, G)$ can be overapproximated by a zonotope $\mathcal{Z} = \mathtt{zonotope}(\mathcal{PZ}) = (\tilde{c}, \tilde{G})$ so that $\mathcal{PZ} \subseteq \mathcal{Z}$, by choosing

$$\tilde{c} = c + \frac{1}{2} \sum_{i=1}^{\lfloor \eta/2 \rfloor} \sum_{j=1}^{p} e^{([2i],j)},$$

$$\tilde{G} = \left[\tfrac{1}{2}E^{[2]} \quad \tfrac{1}{2}E^{[4]} \dots \quad E^{[1]} \quad E^{[3]} \dots \quad F^{[1]} \quad F^{[2]} \dots \quad G\right],$$

where $\lfloor \eta/2 \rfloor$ returns the lowest integer of $\eta/2$. The computational complexity for a given polynomial zonotope order ρ with respect to n is $\mathcal{O}(n^2)$. □

The proof is omitted due to limited space. A sketch of the proof is as follows: Generators with dependent factors are made independent by moving them into the generator matrix G, which always results in an overapproximation. Dependent factors β_i with even powers are within $[0, 1]$ (e.g. $\beta_1^2 \in [0, 1]$) instead of $[-1, 1]$ so that $E^{[2]}$, $E^{[4]}$, ... can be multiplied by 0.5 and their mean is added to c.

The multiplication of a matrix $M \in \mathbb{R}^{o \times n}$ with a polynomial zonotope $\mathcal{PZ} = (c, E, F, G)$ and the Minkowski addition of a zonotope $\mathcal{Z} = (\tilde{c}, \tilde{G})$ with $\mathcal{PZ}$ follow directly from the definition of polynomial zonotopes:

$$\begin{aligned} M \otimes \mathcal{PZ} &= (Mc, ME, MF, MG), \\ \mathcal{PZ} \oplus \mathcal{Z} &= (c + \tilde{c}, E, F, [G, \tilde{G}]). \end{aligned} \tag{14}$$

For a given polynomial zonotope order ρ, the computational complexity with respect to n is $\mathcal{O}(n^3)$ for the multiplication and $\mathcal{O}(n)$ for the addition. Note that the Minkowski addition of two polynomial zonotopes is never required since $\mathcal{R}^{\Delta}$ and $\mathcal{R}^{p,\Delta}$ are represented by zonotopes. The only addition between two polynomial zonotopes is between $\mathcal{PZ}_1 = (c_1, E_1, F_1, G_1)$ and $\mathcal{PZ}_2 = (c_2, E_2, F_2, G_2)$ in (12). Since both summands have the same dependent factors (proof omitted), one can apply an exact set addition, where the resulting polynomial zonotope is $(c_1 + c_2, E_1 + E_2, F_1 + F_2, [G_1, G_2])$. The generators with independent factors in G_1 and G_2 are added by concatenation as for the Minkowski addition in (14).

The reachability algorithm in Alg. 1 require set-based evaluations of higher order terms, such as the quadratic map $\mathrm{sq}(D, \tilde{\mathcal{R}})$. When $\tilde{\mathcal{R}}$ is a zonotope, the result is exactly enclosed by a polynomial zonotope as shown by the following theorem.

Theorem 1 (Quadratic Map of a Zonotope): Given a zonotope $\mathcal{Z} = (d, g^{(1)}, \dots, g^{(h)})$ and a discrete set of matrices $Q^{(i)} \in \mathbb{R}^{n \times n}$, $i = 1 \dots n$, the set

$$\mathcal{PZ} = \mathrm{sq}(Q, \mathcal{Z}) = \{\varphi | \varphi_i = x^T Q^{(i)} x,\ x \in \mathcal{Z}\}$$

is a polynomial zonotope (c, E, F, G), where the center is computed as $c_i = d^T Q^{(i)} d$, the generators of E and F are computed for $j = 1..h$, $k = j..h$ as

$$e_i^{([1],j)} = d^T Q^{(i)} g^{(j)} + {g^{(j)}}^T Q^{(i)} d, \quad e_i^{([2],j)} = {g^{(j)}}^T Q^{(i)} g^{(j)},$$
$$f_i^{([2],j,k)} = {g^{(j)}}^T Q^{(i)} g^{(k)} + {g^{(k)}}^T Q^{(i)} g^{(j)},$$

and $G = \emptyset$. The complexity of constructing this polynomial zonotope with respect to the dimension n is $\mathcal{O}(n^5)$. □

Proof: Inserting the definition of a zonotope into the set $\mathcal{PZ} = \{\varphi | \varphi_i = x^T Q^{(i)} x,\ x \in \mathcal{Z}\}$ yields

$$\Big\{\varphi \Big| \varphi_i = \Big(d + \sum_{j=1}^{h} \beta_j g^{(j)}\Big)^T Q^{(i)} \Big(d + \sum_{j=1}^{h} \beta_j g^{(j)}\Big),\ \beta_j \in [-1, 1]\Big\},$$

which can be rearranged to

$$\begin{aligned} \Big\{\varphi \Big| \varphi_i = &\underbrace{d^T Q^{(i)} d}_{c_i} + \sum_{j=1}^{h} \beta_j \underbrace{(d^T Q^{(i)} g^{(j)} + {g^{(j)}}^T Q^{(i)} d)}_{e_i^{([1],j)}} \\ &+ \sum_{j=1}^{h-1} \sum_{k=j+1}^{h} \beta_j \beta_k \underbrace{({g^{(j)}}^T Q^{(i)} g^{(k)} + {g^{(k)}}^T Q^{(i)} g^{(j)})}_{f_i^{([2],j,k)}} \\ &+ \sum_{j=1}^{h} \beta_j^2 \underbrace{{g^{(j)}}^T Q^{(i)} g^{(j)}}_{e_i^{([2],j)}},\ \beta_i \in [-1, 1]\Big\}. \end{aligned}$$

Comparing the structure of the above terms with the definition of a polynomial zonotope in Def. 1 shows that the structure is identical and thus a polynomial zonotope.

It remains to derive the complexity. Quadratic operations such as ${g^{(j)}}^T Q^{(i)} g^{(k)}$ have complexity $\mathcal{O}(n^2)$. The number h of generators of $\mathcal{Z}$ can be expressed by its order as ρn, such that the resulting polynomial zonotope has $\binom{(\rho n)+2}{2}-1$ generators, a number which can be bounded by $\mathcal{O}(n^2)$, such that we have $\mathcal{O}(n^4)$ for all generator computations for each dimension and $\mathcal{O}(n^5)$ for all dimensions. □

Corollary 1 (Quadratic Map of a Polynomial Zonotope): Given is a polynomial zonotope $\mathcal{PZ} = (c, E, F, G)$ and the enclosing zonotope $\mathcal{Z} = \texttt{zonotope}(\mathcal{PZ}) = (c, h^{(1)}, \ldots, h^{(\sigma)})$ according to Prop. 1. The quadratic map is overapproximated by

$$\mathrm{sq}(Q, \mathcal{Z}) \subseteq \mathrm{sq}(Q, \mathcal{Z}_{EF}) \oplus \texttt{zonotope}(\mathrm{sq}(Q, \mathcal{Z}_G)),$$

where $\mathcal{Z}_{EF} = (c, h^{(1)}, .., h^{(p)})$ and $\mathcal{Z}_G = (\mathbf{0}, h^{(p+1)}, .., h^{(\sigma)})$, and $\mathbf{0}$ is a vector of zeros of proper dimension. The addition is performed according to (14) and the result has the same dependent factors as $\mathcal{PZ}$. The computational complexity for a given order ρ is identical to Theorem 1 ($\mathcal{O}(n^5)$). □

Proof: The zonotope $\mathcal{Z}$ is split into $\mathcal{Z} = \mathcal{Z}_{EF} \oplus \mathcal{Z}_G$, where the first one has p generators, which equals the number of generators in $E^{[1]}$. Ignoring dependencies always results in an overapproximation, such that

$$\mathrm{sq}(Q, \mathcal{Z}) \subseteq \mathrm{sq}(Q, \mathcal{Z}_{EF}) \oplus \underbrace{\mathrm{sq}(Q, \mathcal{Z}_G)}_{\subseteq \texttt{zonotope}(\mathrm{sq}(Q, \mathcal{Z}_G))}. \quad \square$$

In order to make the above quadratic map accurate, the generators in $\mathcal{Z}_{EF}$ have to be dominant since $\mathrm{sq}(Q, \mathcal{Z}_{EF})$ is computed exactly, while $\mathrm{sq}(Q, \mathcal{Z}_G)$ is overapproximated by a zonotope. This is achieved by the order reduction technique described in the next subsection.

The operation $\mathrm{mu}(D, \tilde{\mathcal{R}})$ is similar to $\mathrm{sq}(D, \tilde{\mathcal{R}})$, which is the reason for omitting a detailed description.

5.2 Order Reduction of Polynomial Zonotopes

Many of the previously presented operations increase the order of polynomial zonotopes due to added generators. As a consequence, an order reduction technique has to be applied to limit the representation size and the computational costs. Most techniques for classical zonotopes remove generators and add new, but fewer ones that capture the spanned set of the removed generators (see e.g. [20]). This results in a reordering of the generators, which is no problem for zonotopes since the ordering of generators is irrelevant. However, generators of polynomial zonotopes can only be reordered within G, where generators are multiplied by independent factors γ_i. For this reason, a new order reduction technique is developed that does not change the ordering of generators in E and F.

The size of E and F is fixed and only G grows after performing the required operations presented in this work. Thus, it is required to remove generators from G and stretch the generators in E and F such that the ones removed from G are compensated in an overapproximative way. As for most applied order reduction techniques of zonotopes, heuristics are used rather than strict optimization techniques due to their favorable ratio of computational costs to obtained overapproximation.

Proposition 2 (Overapproximative Generator Removal): Given is $\mathcal{PZ} = (c, E, F, G)$ of which n linearly independent generators with indices $\mathrm{ind}_1, \ldots \mathrm{ind}_n$ are picked from $E^{[1]}$ and stored in $P = [e^{([1],\mathrm{ind}_1)} \ldots e^{([1],\mathrm{ind}_n)}]$ ($\det(P) \neq 0$). The overapproximating polynomial zonotope $\widehat{\mathcal{PZ}} = (c, \hat{E}, F, \hat{G})$ from which the generator $g^{(i)}$ is removed, is computed as

$$\hat{G} = \left[g^{(1)} \ldots g^{(i-1)}, g^{(i+1)}, \ldots g^{(q)}\right],$$

$$\hat{E} = \left[\hat{E}^{[1]}\, E^{[2]} \ldots E^{[\eta]}\right],$$

$$\hat{e}^{[1],j} = \begin{cases} (1 + (P^{-1} g^{(i)})_j) e^{([1],j)} & \text{for } j \in \{\mathrm{ind}_1, \ldots, \mathrm{ind}_n\}, \\ e^{([1],j)} & \text{otherwise.} \end{cases}$$

The computational complexity is $\mathcal{O}(n^3)$ due to the matrix inversion when using the Gauss-Jordan elimination. □

Proof: The generator $g^{(i)}$ can be composed from the generators in P:

$$g^{(i)} = e^{([1],\mathrm{ind}_1)}\phi_1 + \ldots + e^{([1],\mathrm{ind}_n)}\phi_n = P\phi \rightarrow \phi = P^{-1} g^{(i)}.$$

Note that P^{-1} can always be computed since $\det(P) \neq 0$. Thus, $g^{(i)}$ can be replaced by n new generators $e^{([1],\mathrm{ind}_1)}\phi_1$, …, $e^{([1],\mathrm{ind}_n)}\phi_n$, which causes an overapproximation since each generator has an independent factor γ_{q+j}:

$$\left\{\gamma_i g^{(i)} \middle| \gamma_i \in [-1, 1]\right\} \subseteq \left\{\sum_{j=1}^{n} \gamma_{q+j} e^{([1],\mathrm{ind}_j)}\phi_j \middle| \gamma_{q+j} \in [-1, 1]\right\}$$

The n new generators are aligned with the corresponding generators in $E^{[1]}$ and can be removed by stretching each $e^{([1],\mathrm{ind}_j)}$ by the factor

$$\frac{\|e^{([1],\mathrm{ind}_j)}\|_2 + \|e^{([1],\mathrm{ind}_j)}\phi_j\|_2}{\|e^{([1],\mathrm{ind}_j)}\|_2} = 1 + \phi_j = 1 + (P^{-1} g^{(i)})_j. \quad \square$$

In this work, the longest generators $g^{(i)}$ (maximum 2-norm) are removed from G so that the generators in E are more dominant than the ones in G, which is required since only the generators in E are used for the exact computation of quadratic maps. The heuristic for choosing the set of picked generators P in Prop. 2 is as follows: The first generator is the one in $E^{[1]}$ that is best aligned with the removed generator $g^{(i)}$, and the other $n-1$ generators are the ones which have the least alignment with $g^{(i)}$ and among each other. The alignment is measured by the normalized scalar product $\frac{|{g^{(i)}}^T e^{([1],\mathrm{ind}_i)}|}{\|g^{(i)}\|_2 \|e^{([1],\mathrm{ind}_i)}\|_2}$, where a value of 1 occurs when the vectors are aligned and 0 for perpendicular vectors. The generators $g^{(i)}$ are removed until the order is less than a user-defined order. Without giving the proof, the complexity of the order reduction heuristic is $\mathcal{O}(n^3)$.

For a given order ρ of the polynomial zonotopes, no required operation has a complexity exceeding $\mathcal{O}(n^5)$ when a second order Taylor expansion with third order Lagrangian remainder is used. The scalability of the approach is demonstrated by the subsequent numerical examples.

6. NUMERICAL EXAMPLES

The approach is demonstrated for three examples. In all examples, the nonlinear dynamics is abstracted by a difference inclusion of second order with third order Lagrange remainder. The first example is a Van-der-Pol oscillator, which is a standard example for nonlinear systems that have a limit cycle:

$$\dot{x}_1 = x_2$$
$$\dot{x}_2 = (1 - x_1^2)x_2 - x_1$$

The reachable sets are computed with a time step of $r = 0.005$, $p = 20$ dependent factors (resulting in 230 generators with dependent factors), and 100 generators with independent factors. The initial set is a rectangle, where $x_1 \in [1.25, 1.55]$ and $x_2 \in [2.28, 2.32]$. When using polynomial zonotopes of polynomial order $\eta = 2$, a complete cycle can be computed without splitting, while with conventional zonotopes, the reachable set grows over all limits when splitting is not enforced, see Fig. 3. For a fair comparison, the maximum number of used generators are equal for the polynomial and the classical zonotope. We additionally plotted the results of the tool flow* [12] based on Taylor models, which has similar accuracy than the approach using polynomial zonotopes. The computational time in MATLAB is 23.2 seconds for polynomial zonotopes and similar for zonotopes since the total number of generators is equal. The computational time in flow* is 46.1 seconds, but an improved and faster version will soon be released. The computations have been performed on an Intel XEON X5690 processor with 3.47 Ghz. An advantage of this approach compared to Taylor models is that unlike Taylor models, arbitrarily time-varying uncertain inputs can be considered.

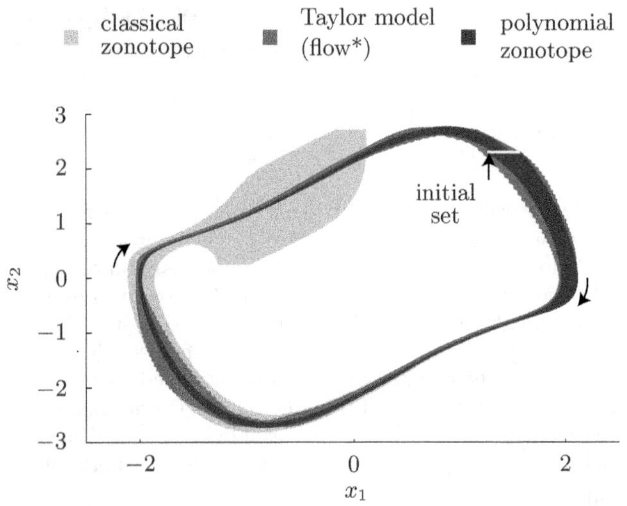

Figure 3: Reachable sets of the Van-der-Pol oscillator.

The second example is a biological model taken out of [17] with strong nonlinearity measure for studying the mitochondrial theory of aging. Due to the strong nonlinear nature, the auxiliary variables

$$S = \frac{x_9}{x_9 + ATP_c} \frac{k_1}{1 + (\frac{x_9}{ATP_c})^3} \frac{1}{x_1 + \frac{2x_2}{GDF+1} + \frac{x_3+x_4+x_5+x_6}{GDF}}$$
$$B = \frac{PAO_x}{x_8(x_1 + x_2 + x_3) + RDF\, x_8(x_4 + x_5 + x_6)}$$

are introduced. The differential equations of the model are

$$\begin{aligned}
\dot{x}_1 =& Sx_1 + \frac{2Sx_2}{GDF+1} - (\alpha + (k_M + k_D)x_8)x_1,\\
\dot{x}_2 =& \frac{-2Sx_2}{GDF+1} + \frac{2Sx_3}{GDF} + k_M x_8 x_1 - (\beta + (k_M + k_D)x_8)x_1,\\
\dot{x}_3 =& \frac{-2Sx_3}{GDF} + k_M x_8 x_2 - (\gamma + k_D x_8)x_3,\\
\dot{x}_4 =& \frac{S(x_4 + x_5)}{GDF} + k_D x_8 x_1 - (\alpha + k_M RDF x_8)x_4,\\
\dot{x}_5 =& \frac{-Sx_5}{GDF} + \frac{2Sx_6}{GDF} + k_D x_8 x_2\\
&+ k_M RDF x_8 x_1 - (\beta + k_M RDF x_8)x_5,\\
\dot{x}_6 =& \frac{-2Sx_6}{GDF} + k_D x_8 x_3 + k_M RDF x_8 x_5 - \gamma x_6,\\
\dot{x}_7 =& \frac{x_9}{x_9 + ATP_c} \frac{k_2}{1+B} - \delta x_7,\\
\dot{x}_8 =& k_R - \frac{k_3(x_7 x_8)}{x_1 + x_2 + x_3 + x_4 + x_5 + x_6},\\
\dot{x}_9 =& k_{ATP} x_1 + 0.5 k_{ATP} x_2 - \frac{x_9}{x_9 + ATP_c}\\
&\left(\frac{k_{EM} k_1}{1 + (\frac{x_9}{ATP_c})^3} + k_{EC} + \frac{k_{EP} k_2}{1+B}\right),
\end{aligned}$$

where x_1-x_9 are the continuous state variables and the remaining variables are parameters whose values are as listed in [17]. The initial set has the same center as the one in [17], but the size of the initial set is 20 times larger for each coordinate, so that the volume is 20^9 times larger than in [17]. The initial set is bounded by a hyperrectangle, where $x_1 - x_3 \in [481, 521]$, $x_4 - x_6 \in [81, 121]$, $x_7 \in [181, 221]$, $x_8 \in [481, 521]$, and $x_9 \in [0, 40]$. The time increment is chosen as $r = 1.5 \cdot 10^{-5}$ and the time horizon $t_f = 0.01$ is as in [17]. For the reachable set computations, polynomial zonotopes with $p = 9$ dependent factors (resulting in 54 generators with dependent factors) and 90 generators with independent factors are used.

The results are compared for 3 approaches: Polynomialization with polynomial zonotopes, polynomialization with classical zonotopes, and linearization with classical zonotopes, which is the approach in [4]. For this example, it is not possible to compare the results with flow* since the tool does not yet support non-polynomial differential equations. When using polynomialization with polynomial zonotopes, the reachable set for the entire time horizon is computed without splitting. For polynomialization in combination with conventional zonotopes, the reachable set has to be split for the first time at $t = 0.0050$ and 6 parallel computations are required in the end. The linearization with classical zonotopes already requires the first split at time $t = 9 \cdot 10^{-5}$ and 98 parallel reachable set computations are performed in the end. Due to the possibility of splitting reachable sets, all approaches provide similar accuracy. However, the approach using polynomial zonotopes results in the tightest overapproximation.

Selected projections of the reachable set using polynomial zonotopes are shown in Fig. 4. For a fair comparison, the total number of used generators for the polynomial and the classical zonotope are chosen equal. The computation time is 1180 s using polynomial zonotopes, while almost all the time (1121 s) is spent on evaluating the third order Lagrange remainder using the interval arithmetic toolbox INTLAB

[36] for MATLAB. Since this toolbox is only efficient when matrix operations are used, the performance can be drastically improved when the interval computations are performed by precompiled code using e.g. C++. When using polynomialization in combination with classical zonotopes, the computation takes 4121 s, which is more than 3 times longer compared to polynomial zonotopes. For the linearization approach, the total computation time is even 18316 s, which is more than 15 times longer compared to polynomialization in combination with polynomial zonotopes. Note that for the specific example, the variable x_9 dominates the linearization error and thus splitting is mainly performed in one direction only. Otherwise, splits in many more directions are required, resulting in an exponential complexity with respect to the variables that have to be split.

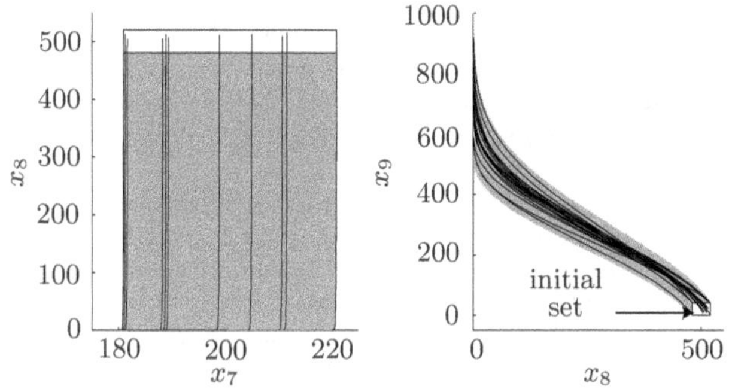

Figure 4: Reachable sets (gray area) of the biological aging model starting from a set of initial states (white area). Black lines show possible trajectories.

Finally, it is demonstrated by a third example that the proposed approach can be directly applied to hybrid systems. Thereto, the aging model is made hybrid by changing the parameter k_{ATP} from 1200 to 120 when the value $x_9 = 100$ is reached. As a consequence, the state x_9 converges to a different steady state, while the other states are only marginally affected, see Fig. 5. The extension to hybrid systems is performed as in [1], which computes the intersection with the guard set $x_9 = 100$ geometrically. Integrating the nonlinear reachability into a more efficient approach [2], which avoids computationally expensive guard intersections, is part of future work. The computation time is 1296 s, where again almost all the time (1155 s) is spent on interval arithmetic. The computation has been performed on the same machine as used for the previous experiments.

7. CONCLUSIONS

This work presents a new approach for reachability analysis of hybrid systems with nonlinear continuous dynamics. The new method successfully improves the major problem of linearization-based approaches: When the linearization error becomes too large, the linearization region has to be split, resulting in many regions that have to be considered simultaneously. Splitting results in an exponential complexity in the number of continuous state variables contributing to the linearization error.

By improving the accuracy of the approach, due to abstracting to polynomial difference inclusions instead of linear difference inclusions, splitting is avoided in the presented examples, saving computation time and improving accuracy.

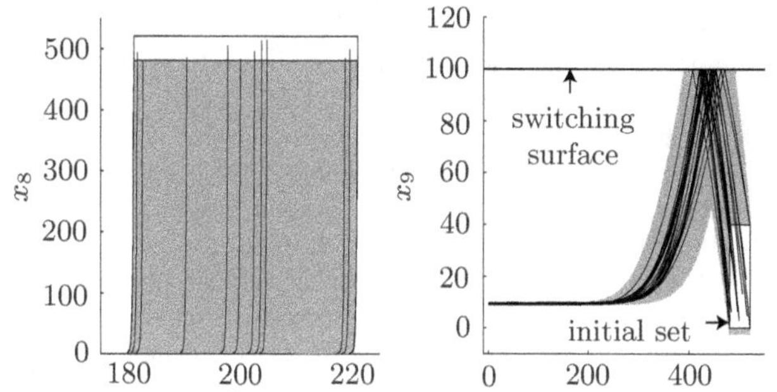

Figure 5: Reachable sets (gray area) of the hybrid biological aging model starting from a set of initial states (white area). Black lines show possible trajectories.

The new approach requires the use of a non-convex set representations, which is newly developed in this work. Since many aspects are implemented for the first time, there is a lot of potential for future improvements. For instance, one could compute with Taylor expansions of order greater than two and also use polynomial zonotopes with polynomial order $\eta > 2$, making it possible to compute with even larger sets of initial states in the future.

Acknowledgment

The author likes to thank Xin Chen for support with the tool flow* and discussions on Taylor models.

8. REFERENCES

[1] M. Althoff. *Reachability Analysis and its Application to the Safety Assessment of Autonomous Cars.* Dissertation, Technische Universität München, 2010. http://nbn-resolving.de/urn/resolver.pl?urn:nbn:de:bvb:91-diss-20100715-963752-1-4.

[2] M. Althoff and B. H. Krogh. Avoiding geometric intersection operations in reachability analysis of hybrid systems. In *Hybrid Systems: Computation and Control*, pages 45–54, 2012.

[3] M. Althoff, C. Le Guernic, and B. H. Krogh. Reachable set computation for uncertain time-varying linear systems. In *Hybrid Systems: Computation and Control*, pages 93–102, 2011.

[4] M. Althoff, O. Stursberg, and M. Buss. Reachability analysis of nonlinear systems with uncertain parameters using conservative linearization. In *Proc. of the 47th IEEE Conference on Decision and Control*, pages 4042–4048, 2008.

[5] R. Alur and D. L. Dill. A theory of timed automata. *Theoretical Computer Science*, 126:183–235, 1994.

[6] E. Asarin, T. Dang, G. Frehse, A. Girard, C. Le Guernic, and O. Maler. Recent progress in continuous and hybrid reachability analysis. In *Proc. of the 2006 IEEE Conference on Computer Aided Control Systems Design*, pages 1582–1587, 2006.

[7] E. Asarin, T. Dang, and A. Girard. Hybridization methods for the analysis of nonlinear systems. *Acta Informatica*, 43:451–476, 2007.

[8] G. Behrmann, A. David, K. G. Larsen, P. Pettersson, and W. Yi. Developing UPPAAL over 15 years. *Softw., Pract. Exper.*, 41(2):133–142, 2011.

[9] M. Berz and G. Hoffstätter. Computation and application of Taylor polynomials with interval remainder bounds. *Reliable Computing*, 4:83–97, 1998.

[10] P. Bulychev, A. David, K. G. Larsen, A. Legay, G. Li, D. B. Poulsen, and A. Stainer. Monitor-based statistical model checking for weighted metric temporal logic. In *Proc. of 18th International Conference on Logic for Programming Artificial Intelligence and Reasoning*, pages 168–182, 2012.

[11] X. Chen, E. Ábráham, and G. Frehse. Efficient bounded reachability computation for rectangular automata. In *5th Workshop on Reachability Problems*, LNCS 6945, pages 139–152. Springer, 2011.

[12] X. Chen, S. Sankaranarayanan, and E. Ábrahám. Taylor model flowpipe construction for non-linear hybrid systems. In *Proc. of the 33rd IEEE Real-Time Systems Symposium*, 2012.

[13] A. Chutinan and B. H. Krogh. Computational techniques for hybrid system verification. *IEEE Transactions on Automatic Control*, 48(1):64–75, 2003.

[14] E. Clarke, A. Fehnker, Z. Han, B. H. Krogh, J. Ouaknine, O. Stursberg, and M. Theobald. Abstraction and counterexample-guided refinement in model checking of hybrid systems. *International Journal of Foundations of Computer Science*, 14(4):583–604, 2003.

[15] E. M. Clarke and P. Zuliani. Statistical model checking for cyber-physical systems. In *Proc. of the 9th International Symposium on Automated Technology for Verification and Analysis*, LNCS 6996, pages 1–12, 2011.

[16] T. Dang. *Vérification et synthèse des systèmes hybrides*. PhD thesis, Institut National Polytechnique de Grenoble, 2000.

[17] T. Dang, C. Le Guernic, and O. Maler. Computing reachable states for nonlinear biological models. In *Computational Methods in Systems Biology*, LNCS 5688, pages 126–141. Springer, 2009.

[18] T. Dang, O. Maler, and R. Testylier. Accurate hybridization of nonlinear systems. In *Hybrid Systems: Computation and Control*, pages 11–19, 2010.

[19] G. Frehse. PHAVer: Algorithmic verification of hybrid systems past HyTech. *International Journal on Software Tools for Technology Transfer*, 10:263–279, 2008.

[20] A. Girard. Reachability of uncertain linear systems using zonotopes. In *Hybrid Systems: Computation and Control*, LNCS 3414, pages 291–305. Springer, 2005.

[21] A. Girard and C. Le Guernic. Efficient reachability analysis for linear systems using support functions. In *Proc. of the 17th IFAC World Congress*, pages 8966–8971, 2008.

[22] A. Girard, C. Le Guernic, and O. Maler. Efficient computation of reachable sets of linear time-invariant systems with inputs. In *Hybrid Systems: Computation and Control*, LNCS 3927, pages 257–271. Springer, 2006.

[23] A. Girard and G. J. Pappas. Verification using simulation. In *Hybrid Systems: Computation and Control*, LNCS 3927, pages 272–286. Springer, 2006.

[24] T. A. Henzinger, B. Horowitz, R. Majumdar, and H. Wong-Toi. Beyond HyTech: Hybrid systems analysis using interval numerical methods. In *Hybrid Systems: Computation and Control*, LNCS 1790, pages 130–144. Springer, 2000.

[25] J. Hoefkens, M. Berz, and K. Makino. *Scientific Computing, Validated Numerics, Interval Methods*, chapter Verified High-Order Integration of DAEs and Higher-Order ODEs, pages 281–292. Springer, 2001.

[26] L. Jaulin, M. Kieffer, and O. Didrit. *Applied Interval Analysis*. Springer, 2006.

[27] A. A. Julius, G. E. Fainekos, M. Anand, I. Lee, and G. J. Pappas. Robust test generation and coverage for hybrid systems. In *Hybrid Systems: Computation and Control*, LNCS 4416, pages 329–342. Springer, 2007.

[28] M. Kloetzer and C. Belta. Reachability analysis of multi-affine systems. In *Hybrid Systems: Computation and Control*, LNCS 3927, pages 348–362. Springer, 2006.

[29] A. B. Kurzhanski and P. Varaiya. Ellipsoidal techniques for reachability analysis. In *Hybrid Systems: Computation and Control*, LNCS 1790, pages 202–214. Springer, 2000.

[30] K. Makino and M. Berz. Rigorous integration of flows and ODEs using Taylor models. In *Proc. of Symbolic-Numeric Computation*, pages 79–84, 2009.

[31] I. M. Mitchell, A. M. Bayen, and C. J. Tomlin. A time-dependent Hamilton–Jacobi formulation of reachable sets for continuous dynamic games. *IEEE Transactions on Automatic Control*, 50:947–957, 2005.

[32] M. Neher, K. R. Jackson, and N. S. Nedialkov. On Taylor model based integration of ODEs. *SIAM Journal on Numerical Analysis*, 45(1):236–262, 2007.

[33] A. Platzer. *Logical Analysis of Hybrid Systems: Proving Theorems for Complex Dynamics*. Springer, 2010. ISBN 978-3-642-14508-7.

[34] S. Prajna. Barrier certificates for nonlinear model validation. *Automatica*, 42(1):117–126, 2006.

[35] S. Ratschan and Z. She. Constraints for continuous reachability in the verification of hybrid systems. In *Proc. of Artificial Intelligence and Symbolic Computation*, LNCS 4120, pages 196–210. Springer, 2006.

[36] S. M. Rump. *Developments in Reliable Computing*, chapter INTLAB - INTerval LABoratory, pages 77–104. Kluwer Academic Publishers, 1999.

[37] S. Sankaranarayanan. Automatic abstraction of non-linear systems using change of bases transformations. In *Hybrid Systems: Computation and Control*, pages 143–152, 2011.

[38] S. Sankaranarayanan and G. Fainekos. Falsification of temporal properties of hybrid systems using the cross-entropy method. In *Hybrid Systems: Computation and Control*, pages 125–134, 2012.

[39] O. Stursberg and B. H. Krogh. Efficient representation and computation of reachable sets for hybrid systems. In *Hybrid Systems: Computation and Control*, LNCS 2623, pages 482–497. Springer, 2003.

One-Shot Computation of Reachable Sets for Differential Games

Insoon Yang
Department of Electrical Engineering and Computer Sciences
University of California
Berkeley, CA 94720
iyang@eecs.berkeley.edu

Sabine Becker-Weimann
Life Sciences Division
Lawrence Berkeley National Laboratory
Berkeley, CA 94720
sbecker-weimann@lbl.gov

Mina J. Bissell
Life Sciences Division
Lawrence Berkeley National Laboratory
Berkeley, CA 94720
mjbissell@lbl.gov

Claire J. Tomlin
Department of Electrical Engineering and Computer Sciences
University of California
Berkeley, CA 94720
tomlin@eecs.berkeley.edu

ABSTRACT

We present a numerical method for computing backward reachable sets in differential games. A backward reachable set for time t is captured by the t sublevel set of the lower value function of the game, which coincides with the viscosity solution of a stationary Hamilton-Jacobi-Isaacs (HJI) equation. We solve the stationary HJI equation in a computationally efficient way that does not involve any numerical integration over time, which would otherwise be required for time-dependent HJI equations. Backward reachable sets for all time points can simultaneously be extracted from the solution. The performance of the method is demonstrated by investigating the growth of multicellular structures of non-malignant and malignant breast cells as a proof of principle.

Categories and Subject Descriptors

G.1.0 [**Numerical Analysis**]: General

Keywords

Reachability Analysis, Differential Games, Hamilton-Jacobi-Isaacs Equations, Cancer Cells Systems Biology

1. INTRODUCTION

Reachable set computation is a core component of reachability analysis, which is an important research problem in the formal verification of, and controller synthesis for discrete, continuous and hybrid systems. A number of reachability algorithms have been developed and extensively studied for several classes of dynamical systems. One broad class of the methods uses polyhedral over-approximations of reachable sets, such as polytopes [4, 11, 13, 5], oriented rectangular hulls [37], zonotopes [19], or ellipsoidal over-approximations [25]. Recent progress in this approach has demonstrated their scalability, accuracy and capability for handling parameter uncertainties [21, 20, 2, 3]. A second approach utilizes interval numerical method-based algorithms to track multi-dimensional axis-aligned intervals of a state to over-approximate reachable sets for systems with (or without) uncertain parameters [22, 33]. A third class of reachability algorithms is based on partitions of the (uncountable) continuous state space [39, 24]. This approach then utilizes an abstraction function that maps the continuous state space into the countable elements of the partition to simplify reachability analysis.

Most of the aforementioned algorithms in the three classes are designed to compute forward reachable sets for systems with (or without) control inputs. However, we often want to compute *backward reachable sets* for systems with adversarial disturbances as well as the control inputs e.g. safety guaranteed-design of controllers. To be more precise, we consider the situation in which we want to control a safety-critical system under the existence of a disturbance. Let Γ be the target set representing the set of unsafe states that we want to avoid. Then the t-backward reachable set consists of initial states from which the system can reach the unsafe set Γ in time t. Therefore, the t-backward reachable set information is useful for predictive threat assessment because it provides us with the set of states that will cause the system to be potentially unsafe at time t.

A natural approach for computing backward reachable sets with two competing inputs in continuous systems is to formulate a differential game. One general method for this approach uses techniques based on set-valued analysis and viability theory [12]. Another method is based on a Hamilton-Jacobi-Isaacs (HJI) partial differential equation (PDE). By applying the dynamic programming principle, we

HSCC'13, April 8–11, 2013, Philadelphia, Pennsylvania, USA.

can derive an HJI equation whose viscosity solution is the (lower or upper) value function of the differential game of interest. The time-dependent formulation of the HJI equation [29] is a method for computing backward reachable sets in the presence of competing inputs and is widely applicable to both continuous and hybrid systems [40, 26, 41]. Well-studied numerical techniques [27, 29, 28] based on level set methods [31, 35, 32] have been developed to solve for the backward reachable set. In this time-dependent method, the backward reachable set for time t is the zero sublevel set of the viscosity solution of the HJI equation at time t. In other words, we need to integrate the HJI equation backward in time to obtain the solution, and the solution only contains the information of the reachable set for a single time t.

In this paper, we focus a zero-sum differential game problem with the value function whose t sublevel set contains the information of the backward reachable set for time t. That is, we save the information of the reachable sets for all time t in one value function, which represents the time to reach a target of interest when two inputs are competing each other. This value function is indeed the viscosity solution of a stationary HJI equation. Therefore, once we numerically solve the stationary HJI equation to obtain the solution, we can extract the backward reachable set for any time t by capturing the t sublevel set. Unlike the time-dependent method, the method is highly practical when we want to obtain the reachable set for large time t because the computational time is independent of t, unlike the time-dependent method. Furthermore, the method uses only $O(N)$ memory to save the reachable sets for all time, where N is the number of nodes at which we evaluate the solution. This method is also flexible enough to handle nonlinear time-invariant dynamical systems. We apply our method to formation of structure of non-malignant and malignant mammary epithelial cells in 3D cultures: we assume a dynamic game of cellular growth potential and cell-cell adhesion potential.

2. REACHABILITY PROBLEM

Consider the following dynamical system in $\mathbb{R}^n$ with two competing inputs

$$\dot{y}(t) = f(y(t), u(t), d(t)) \quad (1a)$$

$$y(0) = \boldsymbol{x}, \quad (1b)$$

where $u(\cdot)$ and $d(\cdot)$ are control and disturbance injected into the system, respectively, and $U \subseteq \mathbb{R}^{n_u}$ and $D \subseteq \mathbb{R}^{n_d}$ are feasible sets of control and disturbance values, respectively, and assumed to be compact. The two competing inputs $u(\cdot)$ and $d(\cdot)$ are considered as controls for Player I and Player II, respectively. We also introduce sets of admissible controls $\mathcal{U} = \{u : [0, +\infty) \to U \mid u(\cdot) \text{ measurable}\}$ and $\mathcal{D} = \{d : [0, +\infty) \to D \mid d(\cdot) \text{ measurable}\}$. We assume that $f : \mathbb{R}^n \times U \times D \to \mathbb{R}^n$ is continuous and that there exists $L > 0$ such that $\|f(x_1, a, b) - f(x_2, a, b)\| \leq L\|x_1 - x_2\|$ for all $x_1, x_2 \in \mathbb{R}^n$, $a \in U$, $b \in D$. Under these assumptions, the dynamical system (1) has a unique solution.

To clarify the objectives of Player I and Player II in a reachability problem that we are interested in, we introduce the following two subproblems:

(A) (*Backward reachable sets*) Find the set $\mathcal{B}(t)$ of initial states that can be driven to the target by Player II no matter how Player I chooses her control input, under the assumption that Player II knows Player I's current and past decisions.

(B) (*Time-to-reach*) Find the time to reach the target from a starting point $\boldsymbol{x}$ when Player I wants to maximize the time, while Player II uses a strategy to minimize the time with knowledge of Player I's current and past decisions.

Note that the *Time-to-reach* problem (B) is a *two-player zero-sum differential game*, in which Player II uses a *nonanticipating strategy*. Once we obtain the time-to-reach function $\phi : \mathbb{R}^n \to \mathbb{R}$ in problem (B), we can compute the backward reachable set $\mathcal{B}(t)$ as $\{\boldsymbol{x} \mid \phi(\boldsymbol{x}) \leq t\}$ as we will see later. Extracting the reachable sets from the time-to-reach function, $\phi(\cdot)$, provides advantages in terms of computational memory and complexity because we save all of the information of reachable sets in one time-to-reach function and the complexity of the algorithm to compute $\phi(\cdot)$ is invariant in time. We first address problem (B), which is valuable in itself, and then solve problem (A). To this end, we begin by introducing the mathematical notion of the time to reach a closed target $\Gamma \in \mathbb{R}^n$ with a compact boundary, given $u(\cdot)$ and $d(\cdot)$:

$$T_{\boldsymbol{x}}[u, d] = \min \{t \mid y(t) \in \Gamma\},$$

which is the *payoff* functional of the game (B). Following Varaiya [43] and Elliott-Kalton [16], we define the set of nonanticipatory strategies for Player II by

$$\Theta := \{\theta : \mathcal{U} \to \mathcal{D} \mid u(\tau) = \hat{u}(\tau)\ \forall \tau \leq t \text{ implies } \theta[u](\tau) = \theta[\hat{u}](\tau)\ \forall \tau \leq t\}.$$

Here, $\theta[u] \in \mathcal{D}$ denotes the control of Player II with strategy θ for the control u of Player I. Then, problem (B) is equivalent to the differential game problem

$$\phi(\boldsymbol{x}) := \min_{\theta \in \Theta} \max_{u \in \mathcal{U}} T_{\boldsymbol{x}}[u, \theta[u]]. \quad (2)$$

We call $\phi(\cdot)$ the *lower value function* of the differential game problem (B). In this paper, we assume that the minimum and the maximum in (2) exist. Let us introduce the (lower) capturability set $\mathcal{R}^* = \{\boldsymbol{x} \in \mathbb{R}^n \mid \phi(\boldsymbol{x}) < +\infty\}$. Under the assumption that the minimum and the maximum in (2) exist, the capturability set $\mathcal{R}^*$ coincides with the state space $\mathbb{R}^n$. Applying the dynamic programming principle, we can obtain the following stationary HJI equation with viscosity solution ϕ:

$$H(\boldsymbol{x}, \nabla\phi(\boldsymbol{x})) = 0, \quad \text{in } \mathcal{R}^* \setminus \Gamma \quad (3a)$$

$$\phi(\boldsymbol{x}) = 0, \quad \text{on } \Gamma \quad (3b)$$

where H denotes the Hamiltonian defined by

$$H(\boldsymbol{x}, \boldsymbol{p}) = \min_{a \in U} \max_{b \in D} \{-\boldsymbol{p}^\top f(\boldsymbol{x}, a, b) - 1\}. \quad (4)$$

Detailed derivations and discussions are presented in [10, 8, 6] and [17] describes the time-dependent version. The viscosity solution of (3) is a weak solution that is consistent and unique in the domain where an associated free boundary problem yields a viscosity solution bounded below [14, 36]. Having the lower value function ϕ as the viscosity solution of an HJI equation offers practical advantages. Once we have a numerical solution that approximates the viscosity solution, then it is automatically guaranteed that the solution approximates the lower value function due to the uniqueness and consistency of the viscosity solution. Before presenting a numerical method for approximating the viscosity solution of (3), we need to determine how to extract the t-backward reachable set from the lower value function ϕ, which represents the time-to-reach, in the differential game setting.

2.1 Backward reachable sets from time-to-reach

The backward reachable set for a certain time t is the collection of all initial states that can be driven to the target set Γ in time t with some disturbance $d(\cdot) = \theta[u(\cdot)]$ no matter how we choose the control input $u(\cdot)$. A mathematical definition of the backward reachable set for time s is as follows:

Definition 1. An s-backward reachable set $\mathcal{B}(s)$ of (1a) is defined by

$$\mathcal{B}(s) = \{\boldsymbol{x} \mid \exists \theta \in \Theta \text{ such that } \forall u \in \mathcal{U}, \exists \tau \in [0, s] \text{ satisfying } y(0) = \boldsymbol{x}, y(\tau) \in \Gamma\},$$

where $y(\cdot)$ evolves with (1a).

Note that this definition is consistent with the description of the reachable set $\mathcal{B}(t)$ in problem (A). The following proposition suggests a connection between the backward reachable set and lower value function ϕ of the differential game (B).

PROPOSITION 1. *Given the lower value function ϕ in (2), we can obtain the t-reachable set as*

$$\mathcal{B}(t) = \{\boldsymbol{x} \mid \phi(\boldsymbol{x}) \leq t\}. \tag{5}$$

PROOF. We first claim that $\mathcal{B}(t) \subseteq \{\boldsymbol{x} \mid \phi(\boldsymbol{x}) \leq t\}$. To see this, we fix a point $\boldsymbol{x} \in \mathcal{B}(t)$. By definition, there exists $\theta^* \in \Theta$ such that for each $u \in \mathcal{U}$, there exists $\tau \in [0, t]$ with $y(\tau) \in \Gamma$ and $y(0) = \boldsymbol{x}$. Let τ^* be the supremum of these τ's. Then we have

$$\max_{u \in \mathcal{U}} T_{\boldsymbol{x}}[u, \theta^*[u]] \leq \tau^* \leq t.$$

Due to the definition of the lower value function in (2), we have $\phi(\boldsymbol{x}) \leq \max_{u \in \mathcal{U}} T_{\boldsymbol{x}}[u, \theta^*[u]] \leq t$, which implies that $\mathcal{B}(t) \subseteq \{\boldsymbol{x} \mid \phi(\boldsymbol{x}) \leq t\}$, as desired.
Now, we prove that $\{\boldsymbol{x} \mid \phi(\boldsymbol{x}) \leq t\} \subseteq \mathcal{B}(t)$. Fix a point $\boldsymbol{x}$ such that $\phi(\boldsymbol{x}) \leq t$. Then,

$$\min_{\theta \in \Theta} \max_{u \in \mathcal{U}} T_{\boldsymbol{x}}[u, \theta[u]] \leq t.$$

This inequality means that there exists a nonanticipatory strategy $\hat{\theta} \in \Theta$ such that, for any $\hat{u} \in \mathcal{U}$, $T_{\boldsymbol{x}}[\hat{u}, \hat{\theta}[\hat{u}]] \leq t$. Hence, for any control $\hat{u} \in \mathcal{U}$, the system starting at $\boldsymbol{x}$ can always be steered to target Γ in time t with the control $\hat{d} = \hat{\theta}[\hat{u}]$ of Player II. Therefore, $\boldsymbol{x} \in \mathcal{B}(t)$. This completes the proof. □

Proposition 1 is the key motivation for designing a *one-shot method* for computing t-reachable sets for all t at once. This consists of the following two steps:

(i) Compute $\phi(\cdot)$ by solving the HJI equation (3).

(ii) Extract $\mathcal{B}(t)$ for any t by capturing the t sublevel set of $\phi(\cdot)$.

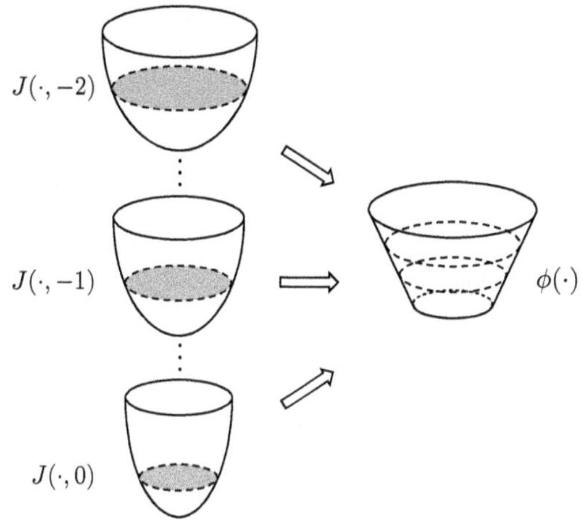

Figure 1: Compression of zero sublevel sets: the zero sublevel set (gray) of $J(\cdot, -t)$ corresponds to the set $\{\boldsymbol{x} \mid \phi(\boldsymbol{x}) \leq t\}$. In this way, we compress the information of reachable sets contained in multiple value functions $J(\cdot, -t)$'s into a single *time-to-reach* function $\phi(\cdot)$.

Note that a single evaluation of $\phi(\cdot)$ on the state space is enough to obtain all t-reachable sets. In practice, a subset of the state space is chosen as the computational domain Ω in which we evaluate ϕ. In this case, the t sublevel set of ϕ coincides with the intersection of the t-backward reachable set and the computational domain, i.e., $\mathcal{B}(t) \cap \Omega$. The computational domain Ω should be chosen large enough in the event that one wants to retrieve a whole region of the t-backward reachable set. Even if Ω is chosen to not include the t-backward reachable set, because we do not know the size and shape of the reachable set *a priori*, the set $\{\boldsymbol{x} \mid \phi(\boldsymbol{x}) \leq t\}$ is equal to $\mathcal{B}(t) \cap \Omega$, which is still valuable information.

2.2 Comparison with a time-dependent HJI method

It is important to note that our one-shot method is *time-independent* in the sense that the HJI equation (3) does not depend explicitly on time. In other words, (3) is a *boundary value* problem, whereas an HJI equation in a time-dependent formulation for computing reachable set is an *initial (or final) value* problem of the form

$$\begin{aligned} &\frac{\partial J(\boldsymbol{x}, t)}{\partial t} + \min[0, H(\boldsymbol{x}, \nabla J(\boldsymbol{x}, t))] = 0, \quad \text{in } \Omega \times [-T, 0) \\ &J(\boldsymbol{x}, 0) = l(\boldsymbol{x}), \quad \text{on } \Omega \times \{t = 0\}, \end{aligned} \tag{6}$$

where the zero sublevel set of the initial value l coincides with the target Γ. Here, the Hamiltonian is given by $H(\boldsymbol{x}, \boldsymbol{p}) = \max_{a \in U} \min_{b \in D} \boldsymbol{p}^\top f(\boldsymbol{x}, a, b)$. In this formulation [40, 29], we compute the t-backward reachable set as the zero sublevel set of $J(\cdot, -t)$. The time-dependent method is somewhat more flexible than our stationary formulation because it is capable of dealing with time-varying dynamics.

The time-dependent formulation is limited in scalability because the memory and computational cost grow exponentially as the system dimension increases. Our stationary formulation also intrinsically exhibits the scalability problem because we need to evaluate the solution of the stationary HJI equation at all nodes that appropriately discretize the state space. However, the one-shot method alleviates the cost on computational complexity and memory costs in the following sense. Suppose that we discretize the domain Ω with N nodes and time interval $[0, T]$ with N_T number of points. To solve the stationary HJI (3), we use the Lax-Friedrichs sweeping method [23] with complexity independent of time. However, the complexity of any grid-based numerical method for the time-dependent HJI (6) linearly increases as N_T increases. Hence, we can reduce the computational complexity by using the one-shot method when N_T is large. This complexity result is linked to the improvement of computational time when we use the one-shot method to compute reachable sets for a long time horizon. Once we solve the stationary HJI equation, then the t-reachable set (intersected with the computational domain) can be automatically extracted from the solution no matter how large t is. However, the time-dependent method requires us to integrate the non-stationary HJI equation (6) over time with a sufficiently small time step to satisfy the Courant-Friedrichs-Lewy (CFL) condition for the stability of the method. Hence, it is quite computationally intensive to determine a reachable set for a long time horizon using the time-dependent method. The comparison between the computational times of the two methods is presented in an example in Section 4.

Another important advantage of the one-shot method over the time-dependent method is the improvement in computational memory when we are in a situation that requires holding all t-reachable sets in a given time interval. As depicted in Figure 1, we need N_T value functions to hold the reachable sets at N_T time points, because only the zero sublevel set of each value function contains the reachability information. However, in the one-shot method, every t sublevel set of one value function $\phi(\cdot)$ provides the t reachable set. That is, the one-shot method uses only $O(N)$ memory since it saves the information of reachable sets for all time in one value function $\phi(\cdot)$, while the time-dependent method requires $O(N \times N_T)$ memory when we want to save the information of t-reachable sets for $t \in \{t_1, t_2, \cdots, t_{N_T}\}$. One can reformulate the stationary HJI into a time-dependent HJI (with solution $\tilde{J}$) if the Hamiltonian satisfies the *non-characteristic criterion* [30]. When this reformulation is valid, one may want to use a thresholding formula $\phi(\boldsymbol{x}) = \{t \mid \tilde{J}(\boldsymbol{x}, t) = 0\}$ at each time step to construct the time-to-reach function $\phi(\cdot)$ from the time-dependent HJI in order to achieve $O(N)$ memory. However, the resolution of ϕ obtained by this approach can be poor unless high-order interpolations in both time and space are employed. High-order interpolations are not desirable in terms of the memory usage because we must save solution copies at multiple time points. Another downside of this reformulation is that it precludes gaining the advantages of the one-shot method in computational time and complexity.

The information on the time-to-reach (2) that we obtain in the stationary formulation is another benefit. In the time-dependent HJI formulation, the solution rather works as an indicator for distinguishing the reachable set from another region in the computational domain. However, the information on the time-to-reach can be used for synthesizing optimal controllers under adversarial disturbances. Ideally, if the value function in the time-dependent method is a signed-distance function, then its value is related to the distance to the reachable set. To obtain this information, we need to frequently reinitialize the value function as a signed distance function. However, the reinitialization procedure imposes additional computational burden.

Various numerical techniques and supporting theory allow us to compute solutions of time-dependent Hamilton-Jacobi equations with high accuracy and convergence rate [31, 35, 32, 29]. A MATLAB toolbox for solving (6) is also publicly available [1]. The numerical algorithms for the time-dependent methods enjoy the continuity of the value function [29]. In the stationary HJI, however, the viscosity solution is likely to have the set of discontinuities, which has zero measure. To approximate the discontinuous viscosity solution, one can use a method that directly discretizes the system dynamics and formulate the discrete dynamic programming equation [7, 9]. The numerical solution obtained by this method converges to the discontinuous viscosity solution. In this article, we use the *Lax-Friedrichs sweeping method* [23], which directly discretizes (3) to compute the solution. This method, which will be discussed in the next section, is very simple, easy to implement and handles any number of dimensions. Although the convergence of the Lax-Friedrichs sweeping method for a solution with discontinuities has not been shown, the method seems to work well in practice. When ϕ has severe discontinuities, we recommend that readers use the convergent fully discrete method that can handle the discontinuities [7, 9]. We also note that the fully discrete method for approximating the time-to-reach function using the viability theory [12] is a good alternative to the stationary HJI approach because the viability theory-based method does not assume the continuity of the value function.

3. COMPUTATION OF REACHABLE SETS

Let us define a uniform grid on an n-dimensional computational domain Ω with grid spacing h. Let N_i be the number of nodes on the ith axis. If Ω is two-dimensional, for example, we let $\boldsymbol{x}_{i,j}$ denote the position of node (i, j), and let $\phi_{i,j}$ denote $\phi(\boldsymbol{x}_{i,j})$ for $i = 1, \cdots, N_1$ and $j = 1, \cdots, N_2$. A similar notation can be naturally extended for any higher-dimensional Ω.

3.1 Lax-Friedrichs scheme

As we have discussed, the time-to-reach function $\phi(\cdot)$ can be computed as the lower value function (2) of the differential game problem (B), which also corresponds to the viscosity solution of (3). Therefore, our goal is to solve (3) with an appropriate numerical method, which directly approximates (3) without any time-dependent reformulation. We first consider the one-dimensional case for the sake of exposition. One way to discretize (3) is by using the Lax-Friedrichs scheme for (3), i.e.,

$$H\left(\boldsymbol{x}_i, \frac{(\phi_x)_i^- + (\phi_x)_i^+}{2}\right) - \sigma_1 \frac{(\phi_x)_i^- - (\phi_x)_i^+}{2} = 0, \qquad (7)$$

where $(\phi_x)_i^-$ and $(\phi_x)_i^+$ are backward (left-sided) and forward (right-sided) discretizations for ϕ_x at the ith node, respectively. The Lax-Friedrichs numerical Hamiltonian is consistent. Here the bound on a (numerical) artificial viscosity σ_1,

$$\sigma_1 \geq \max_{p \in [p_L, p_U]} \left| \frac{\partial H(\boldsymbol{x}, p)}{\partial p} \right|,$$

is analogous to the monotonicity condition on the Lax-Friedrichs scheme for time-dependent Hamilton-Jacobi equations, where p_L and p_U are the lower and upper bounds of ϕ_x, respectively [15]. If we use the first-order approximation of $(\phi_x)^{\pm}$, then the Lax-Friedrichs Hamiltonian can be written as

$$H\left(\boldsymbol{x}_i, \frac{\phi_{i+1}-\phi_{i-1}}{2h}\right) - \sigma_1 \frac{\phi_{i+1}-2\phi_i+\phi_{i-1}}{2h} = 0, \quad (8)$$

Alternatively, one can use higher-order approximations such as the third-order *essentially non-oscillatory interpolation* (ENO) or the fifth-order *weighted ENO* (WENO). The discretization provides a system of nonlinear equations for the ϕ_i's because the Hamiltonian is, in general, nonlinear. We observe that the ith equation only depends on ϕ_{i-1}, ϕ_i and ϕ_{i+1} in (8). This motivates us to use a *Gauss-Seidel*-type iterative method even for non-convex Hamiltonians as in the fast sweeping method for convex Hamilton-Jacobi equations [42].

3.2 A sweeping algorithm

The Lax-Friedrichs sweeping method [23] approximates the viscosity solution of (3) with the Lax-Friedrichs scheme using the Gauss-Seidel type sweeping idea. In other words, the method uses the following update formula based on (7) for Gauss-Seidel iteration [23]:

$$\phi_i^{\text{new}} = c\left[-H\left(\boldsymbol{x}_i, \frac{(\phi_x)_i^- + (\phi_x)_i^+}{2}\right) - \sigma_1 \frac{(\phi_x)_i^- - (\phi_x)_i^+}{2h}\right] + \phi_i.$$

If we use the first-order approximations of spatial derivatives, we can rewrite the update formula as

$$\phi_i^{\text{new}} = c\left[-H\left(\boldsymbol{x}_i, \frac{\phi_{i+1}-\phi_{i-1}}{2h}\right) + \sigma_1 \frac{\phi_{i+1}+\phi_{i-1}}{2h}\right].$$

For a two-dimensional problem, we can simply extend it to

$$\begin{aligned}\phi_{i,j}^{\text{new}} = c\Bigg[&-H\left(\boldsymbol{x}_{i,j}, \frac{\phi_{i+1,j}-\phi_{i-1,j}}{2h}, \frac{\phi_{i,j+1}-\phi_{i,j-1}}{2h}\right) \\ &+\sigma_1 \frac{\phi_{i+1,j}+\phi_{i-1,j}}{2h} + \sigma_2 \frac{\phi_{i,j+1}+\phi_{i,j-1}}{2h}\Bigg],\end{aligned} \quad (9)$$

where $c = h/(\sigma_1+\sigma_2)$ and

$$\sigma_1 \geq \max_{\substack{p\in[p_L,p_U]\\ q\in[q_L,q_U]}} \left|\frac{\partial H(\boldsymbol{x},p,q)}{\partial p}\right|, \quad \sigma_2 \geq \max_{\substack{p\in[p_L,p_U]\\ q\in[q_L,q_U]}} \left|\frac{\partial H(\boldsymbol{x},p,q)}{\partial q}\right|.$$

The essence of the Gauss-Seidel method is its successive displacement, which accelerates the convergence compared to the *Jacobi* method. For example, if we sweep a two-dimensional grid from SW to NE, then $\phi_{i,j}^{\text{new}}$ is updated with (9) using updated $\phi_{i-1,j}, \phi_{i,j-1}$ and un-updated $\phi_{i+1,j}$, $\phi_{i,j+1}$. To cover all characteristic directions of (3), we alternate the sweeping direction as SW, SE, NW and NE. As presented in Algorithm 1, we initialize ϕ with some prescribed value ϕ_U for all $\boldsymbol{x}\in\Omega$. Here, ϕ_U is supposed to be greater than $\max_{\boldsymbol{x}\in\Omega}\phi(\boldsymbol{x})$. Then, $\phi_{i,j}$ is updated only if $\phi_{i,j}^{\text{new}}$ is less than or equal to its previous value as imposed in lines 8, 13, 18, 23, so as to capture the correct characteristic direction that results in the shortest time to reach the target from $\boldsymbol{x}_{i,j}$. Once the solution at a node reaches the minimal value it can achieve, it remains unchanged at the correct solution value. Algorithm 1 can be extended to any dimension. If the state space is n dimensional, however, we then need to consider 2^n alternating sweeping directions.

Algorithm 1: Pseudocode of the Lax-Friedrichs sweeping-based reachable set computation.

```
1  Initialization:
2  Set φ(x) = { 0    if x ∈ Γ
              { φ_U  otherwise;
3  Solving the HJI PDE (3):
4  while |φ − φ^old| > ε do
5      φ^old ← φ
6      for i = 2 : N1 − 1; j = 2 : N2 − 1 do
7          Compute φ^new_{i,j} using (9).
8          φ_{i,j} ← min(φ^new_{i,j}, φ_{i,j}).
9      end
10     Impose boundary conditions using (10).
11     for i = N1 − 1 : 2; j = 2 : N2 − 1 do
12         Compute φ^new_{i,j} using (9).
13         φ_{i,j} ← min(φ^new_{i,j}, φ_{i,j}).
14     end
15     Impose boundary conditions using (10).
16     for i = 2 : N1 − 1; j = N2 − 1 : 2 do
17         Compute φ^new_{i,j} using (9).
18         φ_{i,j} ← min(φ^new_{i,j}, φ_{i,j}).
19     end
20     Impose boundary conditions using (10).
21     for i = N1 − 1 : 2; j = N2 − 1 : 2 do
22         Compute φ^new_{i,j} using (9).
23         φ_{i,j} ← min(φ^new_{i,j}, φ_{i,j}).
24     end
25     Impose boundary conditions using (10).
26 end
27 Reachable set extraction:
28 B(t) ← {x | φ(x) ≤ t}.
```

3.3 Conditions on computational boundaries

The Lax-Friedrichs scheme requires conditions on computational boundaries, unlike an upwind scheme that automatically assigns a solution value on boundary nodes using the characteristic information. We proceed along the line of linear extrapolation-based conditions [23], which is analogous to the boundary conditions for time-dependent HJI equations with Lax-Friedrichs Hamiltonians [29, 1]. The conditions are as follows:

$$\begin{aligned}\phi_{1,j}^{\text{new}} &= \min\left[\max(2\phi_{2,j}-\phi_{3,j}, \phi_{3,j}), \phi_{1,j}\right]\\ \phi_{N_1,j}^{\text{new}} &= \min\left[\max(2\phi_{N_1-1,j}-\phi_{N_1-2,j}, \phi_{N_1-2,j}), \phi_{N_1,j}\right]\\ \phi_{i,1}^{\text{new}} &= \min\left[\max(2\phi_{i,2}-\phi_{i,3}, \phi_{i,3}), \phi_{i,1}\right]\\ \phi_{i,N_2}^{\text{new}} &= \min\left[\max(2\phi_{i,N_2-1}-\phi_{i,N_2-2}, \phi_{i,N_2-2}), \phi_{i,N_2}\right]\end{aligned} \quad (10)$$

These computational boundary conditions prevent inflow characteristics that can introduce spurious data into the numerical solution. To see this, let us consider the case $\phi_{2,j} < \phi_{3,j}$, in which characteristic passes $(2,j)$ in advance to $(3,j)$, i.e., the characteristics are locally inflows. Then the rules (10) choose $\phi_{1,j} = \phi_{3,j} > \phi_{2,j}$ to enforce characteristics outflowing on the computational boundary. If $\phi_{3,j} \leq \phi_{2,j}$, on the other hand, we already have outflowing characteristics near the computational boundary. Hence it is reasonable to approximate $\phi_{0,j}$ as $2\phi_{2,j} - \phi_{3,j}$, which is a linear extrapolation, as suggested by the rules (10). The

Table 1: Errors, convergence rates and computational times (unit: sec) of the test problem.

h	CPU	$\|\phi - \phi^A\|_1$	r	$\|\phi - \phi^A\|_\infty$	r
0.02	0.12	5.548×10^{-2}	0.83	6.160×10^{-2}	0.82
0.01	0.65	3.096×10^{-2}	0.84	3.434×10^{-2}	0.84
0.005	3.91	1.715×10^{-2}	0.85	1.892×10^{-2}	0.86
0.0025	25.03	9.420×10^{-3}	0.86	1.033×10^{-2}	0.87

outer minimizations in (10) are to ensure the numerical solution can only decrease in the sweeping algorithm.

4. EXAMPLES

4.1 Test problem

To test the convergence rate of our algorithm, we consider a simple pursuit-evasion game problem, where Player I is an evader and Player II is a pursuer. Let V_1 and V_2 be the velocities of Player I and Player II, respectively. Then the dynamics of the relative coordinate y of Player II with respect to Player I can be described by

$$\dot{y} = f(y, u, d) = V_1 u - V_2 d,$$

where u and d are controls of Player I and Player II, respectively. If we set the target Γ at the origin, then the t-backward reachable set $\mathcal{B}(t)$ corresponds to the set of relative initial configurations of Player II with respect to Player I, with which Player II can capture Player I in time t using the nonanticipatory strategy. In other words, $\mathcal{B}(t) = \{\boldsymbol{x} \mid \phi(\boldsymbol{x}) \leq t\}$, where $\phi(\cdot)$ is defined as (2) with $\Gamma = \{0\}$. Let $U = [-\alpha, +\alpha]$ and $D = [-\beta, +\beta]$, then $\phi(\cdot)$ solves the HJI equation (3) with the Hamiltonian (4) given by

$$H(\boldsymbol{p}) = (-V_1\alpha + V_2\beta)\|p\| - 1.$$

Hence, the HJI equation (3) is indeed an Eikonal equation with analytic solution

$$\phi^A(\boldsymbol{x}) = \frac{\|\boldsymbol{x}\|}{-V_1\alpha + V_2\beta}.$$

We choose $(\alpha, \beta, V_1, V_2) = (1, 2, 1, 1)$. The computational domain Ω is chosen as $[-1, 1]^2$. The Lax-Friedrichs sweeping method was implemented in C++. All tests in this and the following examples were performed on a 2.53GHz Intel Core i5 processor with 4 GB RAM. The errors, convergence rates and computational times are presented in Table 1. These convergence rates are consistent with the results reported in [23].

4.2 Structure formation by non-malignant and malignant cells

To illustrate our method, we applied it to the different growth patterns of non-malignant and malignant cells for identifying the characteristics of growth and adhesion potentials. The non-malignant S1 and malignant T4-2 cell lines of the HMT-3522 breast cancer progression series can be distinguished by their growth pattern in a three-dimensional laminin-rich gel: S1 cells form spherical hollow structures ('acini') with a homogeneous layer of basement membrane formed around them (Figure 2 (a)), while T4-2 cells grow in a disorganized pattern, forming large multicellular clusters

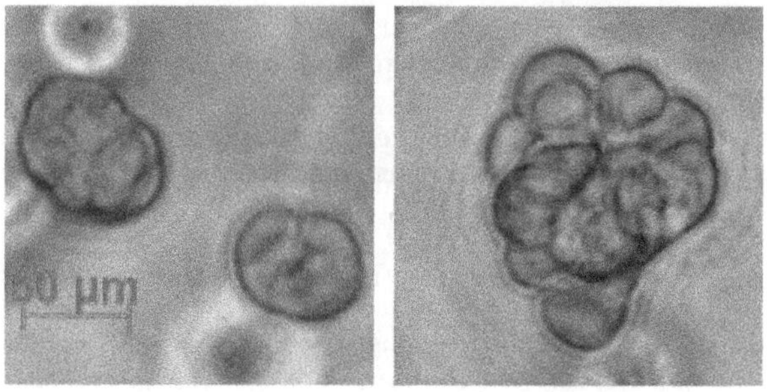

(a) (b)

Figure 2: Multicellular structures formed by (a) non-malignant S1 cells ('acini') and (b) malignant T4-2 cells when grown in a laminin-rich gel.

and patchy basement membranes surrounding the structure (Figure 2 (b)) [45, 34]. In S1 cells, the cell-cell adhesion protein E-Cadherin is localized at the lateral cell membrane together with its binding partner β-catenin; as a complex they mediate cell-cell adhesion. T4-2 cells display decreased cell-cell adhesion with E-Cadherin not being co-localized with β-catenin [45]. Modifications of E-Cadherin expression not only influence cell adhesion but also modulate cellular growth: inhibition of E-Cadherin in S1 cells leads to higher cell numbers per structure [18]; however, overexpression of E-Cadherin in the breast cancer cell line MDA-MB-231 reduces the structure size to 50% [44]. In this example, we propose a hypothesis of a dynamic game between adhesion and growth potentials and use our technique to predict the size of the initial and intermediate structures, where the structures at a final time are given as data. We want to illustrate how comparing the structure size and shape at the final and initial times can help to draw conclusions about the underlying mechanism.

Let $\mathcal{R}(t)$ and $\mathcal{V}(t)$ be the outer boundary and the inner volume of the multicellular structure formed by HMT-3522 cells at time t in a 3D laminin 111-rich gel (lr-ECM), respectively. We assume that $\mathcal{R}(T)$ and $\mathcal{V}(T)$ are given as data, where $T > 0$ is the final time and $\mathcal{V}(t)$ is to be open for all t. We describe the trajectory starting from a point on $\mathcal{R}(s)$ for any $s \leq T$ by the dynamical system $\dot{y} = f(y, u, d)$, where u represents adhesion and d represents the growth potential. More specifically, we set

$$\dot{y}(t) = f(y(t), u(t), d(t)) = \alpha(y(t), u(t))\beta(y(t))d(t),$$

where $f : \mathbb{R}^3 \times \mathbb{R}^3 \times \mathbb{R} \to \mathbb{R}^3$. The first term $\alpha : \mathbb{R}^3 \times \mathbb{R}^3 \to \mathbb{R}^3$, which determines the direction of the growth, represents the effect of adhesion. We assume that cells with strong adhesion try to maximize cell-cell contact. Therefore, to model the effect of adhesion, we define α by

$$\alpha(x, w) = \frac{x/\|x\| + w}{\|x/\|x\| + w\|},$$

with $\|w\| \leq R$. Note that α only determines the direction, and that $\|\alpha(x, w)\| \equiv 1$. As shown in Figure 3 (a), adhesion changes the growth direction from the normal line of the surface of $\mathcal{S}(0, 1)$ at x, where $\mathcal{S}(0, 1)$ is the unit sphere centered at the origin. As the effect of adhesion gets stronger, the growth direction deviates further from the normal line, i.e., R is large in this case.

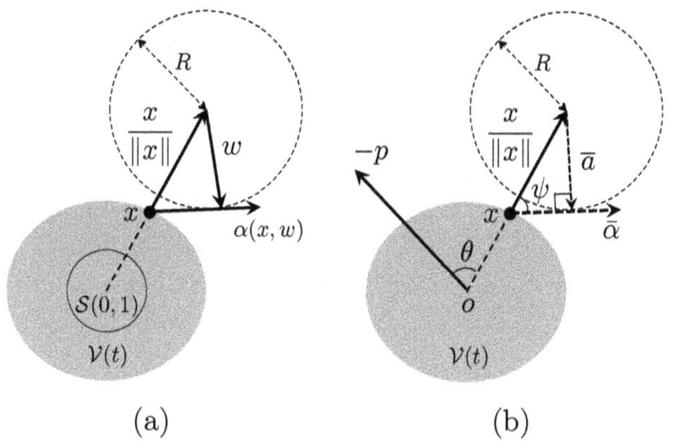

Figure 3: (a) Modeling the effect of adhesion. (b) Angles θ and ψ.

The second term of f models the effect of the growth response (determined by surrounding growth stimulating agents as well as the cellular sensitivity towards those agents). The growth response leads to cell proliferation in the direction determined by α. The given function $\beta : \mathbb{R}^3 \to \mathbb{R}$ represents the inhomogeneity of the cellular microenvironment. To model the inhomogeneity, we synthesize β^* as a noisy map, the value of which is uniformly distributed at all nodes. At $\boldsymbol{x}$, which is not a node, we define $\beta(\boldsymbol{x})$ by the linear interpolation of values at the neighboring nodes. By this definition, β^* is continuous.

We propose a hypothesis that the growth potential minimizes the time-to-reach the boundary of the given final multicellular structure competing with adhesion. We also assume that the growth potential (Player II) has the knowledge of the current and past decision of the adhesion potential (Player I). Then, the t-backward reachable set $\mathcal{B}(t)$ corresponds to the set of initial states that can reach the final structure in time t, no matter how the adhesion potential tries to maximize the time-to-reach. Therefore, the complement of the t-reachable set, $\mathcal{B}(t)^c$, is the set of all initial states that cannot reach the final structure in time t with some controls of adhesion potential. From this observation we have $\mathcal{B}(0) = \mathcal{V}(T)^c$, $\mathcal{R}(T-t) \subseteq \mathcal{B}(t)$ and $\mathcal{B}(t)^c \subseteq \mathcal{V}(T-t)$. We define the bounds of the adhesion and growth potentials as $U = \{u \,|\, \|u\| \leq R\}$ and $D = [\underline{d}, \overline{d}]$ with $\underline{d} \geq 0$, respectively. To compute the backward reachable set, we proceed along the lines of Section 2. We formulate the HJI equation (3) with the Hamiltonian (4),

$$\begin{aligned} H(\boldsymbol{x}, \boldsymbol{p}) &= \min_{a \in U} \max_{b \in D} \{-\boldsymbol{p}^\top \alpha(\boldsymbol{x}, a)\beta(\boldsymbol{x})b - 1\} \\ &= \|\boldsymbol{p}\| \cos(\theta + \psi)\beta(\boldsymbol{x})\overline{d}\mathbf{1}_{\{(\theta+\psi)<\pi/2\}} \\ &+ \|\boldsymbol{p}\| \cos(\theta + \psi)\beta(\boldsymbol{x})\underline{d}\mathbf{1}_{\{\pi/2 \leq (\theta+\psi)<\pi\}} \\ &- \|\boldsymbol{p}\|\beta(\boldsymbol{x})\underline{d}\mathbf{1}_{\{\pi \leq (\theta+\psi)\}} - 1, \end{aligned}$$

where $\mathbf{1}$ is the indicator function. As depicted in Figure 3 (b), $\theta \leq \pi$ is the angle between $-\boldsymbol{p}$ and $\boldsymbol{x}$ and ψ is the angle between $\boldsymbol{x}$ and $\bar{\alpha}$, where α is a unit vector starting from $\boldsymbol{x}$ and tangential to the sphere. We choose $\bar{\alpha}$ so that $-\boldsymbol{p}$, $\boldsymbol{x}$ and $\bar{\alpha}$ lie in the same plane. (There are two candidates for such $\bar{\alpha}$; we choose the one outward from the (convex) conic hulls of $\{o, -\boldsymbol{p}, \boldsymbol{x}\}$, as shown in Figure 3 (b).) Then the angle between $-\boldsymbol{p}$ and $\bar{\alpha}$ is equal to $\theta + \psi = \cos^{-1} \frac{\|\boldsymbol{p}\|^2 + \|\boldsymbol{x}\|^2 - \|-\boldsymbol{p}-\boldsymbol{x}\|^2}{2\|\boldsymbol{p}\|\|\boldsymbol{x}\|} + \sin^{-1} R$. If $\theta + \psi < \pi/2$, then $\cos(\theta + \psi) \geq 0$ and $\cos(\cdot)$ is decreasing. Hence, an optimal a^* can be chosen as $\bar{a}$ so that $\alpha(\boldsymbol{x}, a^*) = \bar{\alpha}$. Furthermore, an optimal b^* must be $\overline{d}$ because $\beta \geq 0$ and $\cos(\theta + \psi) \geq 0$. If $\pi/2 \leq \theta + \psi < \pi$, then $\cos(\theta + \psi) < 0$ and $\cos(\cdot)$ is decreasing. In this case, an optimal a^* is $\bar{a}$ as before so that $\alpha(\boldsymbol{x}, a^*) = \bar{\alpha}$. Due to the fact that $\beta \geq 0$ and $\cos(\theta + \psi) < 0$, an optimal b^* should be $\underline{d}$. If $\theta + \psi \geq \pi$, then an optimal ψ^*, which is determined by an optimal a^*, is $\pi - \theta$ so that $\cos(\theta + \psi^*) = -1$. Then an optimal b^* is $\underline{d}$ because $\beta \geq 0$ and $\cos(\theta + \psi) < 0$.

We compute the backward reachable sets of both non-malignant and malignant cells, where the target is $\Gamma = \mathcal{V}(T)^c$, i.e., the complement of the final structure. For non-malignant S1 cells, we assume strong cell adhesion and therefore a high value of R ($R = 0.5$). For malignant T4-2 cells, we assume weak cell adhesion, which is represented by a low value of R ($R = 0.01$). As malignant cells have a higher proliferation rate than non-malignant cells, we choose the bound on the growth rate per hour for malignant cells to be $(\underline{d}, \overline{d}) = (0.001/2, 0.01/2)$ and $(\underline{d}, \overline{d}) = (0.001/3, 0.01/3)$ for non-malignant cells. Last, the variance of the map of microenvironmental inhomogeneity for the malignant T4-2 cells is chosen to be twice as high as that for non-malignant cells because the formation of the basement membrane surrounding the structures is disturbed in the malignant case.

The complement of the t-backward reachable set $\{\boldsymbol{x} | \phi(\boldsymbol{x}) \leq t\}$ computed by Algorithm 1 is shown in Figure 4. We chose the computational domain Ω as $[-5, 5]^3$ and evaluated the solution at 101^3 nodes with grid spacing $h = 0.1$. Recall that $\mathcal{B}(t)^c \subseteq \mathcal{V}(10 - t) \subseteq \mathcal{B}(0)^c$, we observe that the structure of non-malignant S1 cells with adversarial adhesion has much lower variation than that of malignant T4-2 cells. It was found that multicellular structures formed by S1 cells rotate within the matrigel with a coherent angular motion (CaMo) and display low variation in the structure shapes. However, T4-2 cells show more random lateral movement in the early multicellular stage and thus result in rather high variations in the structures [38].

Although the final structures of non-malignant and malignant cells are widely different in terms of size and shape (Figures 4 (a) and (f)), the smallest possible size of their initial structures are comparable to each other (Figures 4 (e) and (j)), which is significant. This agrees with the experimental data: at an early stage the size of S1 and T4-2 structures is similar *in vitro*. Being able to compare the final, intermediate and initial structures *in vitro* with the final structure and backward reachable sets of the simulation gives us a tool with which to validate hypotheses related to how the structures evolve and which processes are involved. In this particular simulation we find that the competition of adhesion and growth indeed allows for the structure formations observed *in vitro*. In the future, further analysis using our method can clarify the detailed roles of cell-adhesion, growth and the microenvironment for the size and shape of biological multicellular structures.

To compare the one-shot method and the time-dependent method, we also compute the same reachable sets using the C++ implementation of the level set method for time-dependent HJI equations with Lax-Friedrichs Hamiltonians. When solving the time-dependent HJI equation (6), we chose the CFL constant as 0.75 and used the first-order approximations for spatial derivatives and forward Euler method

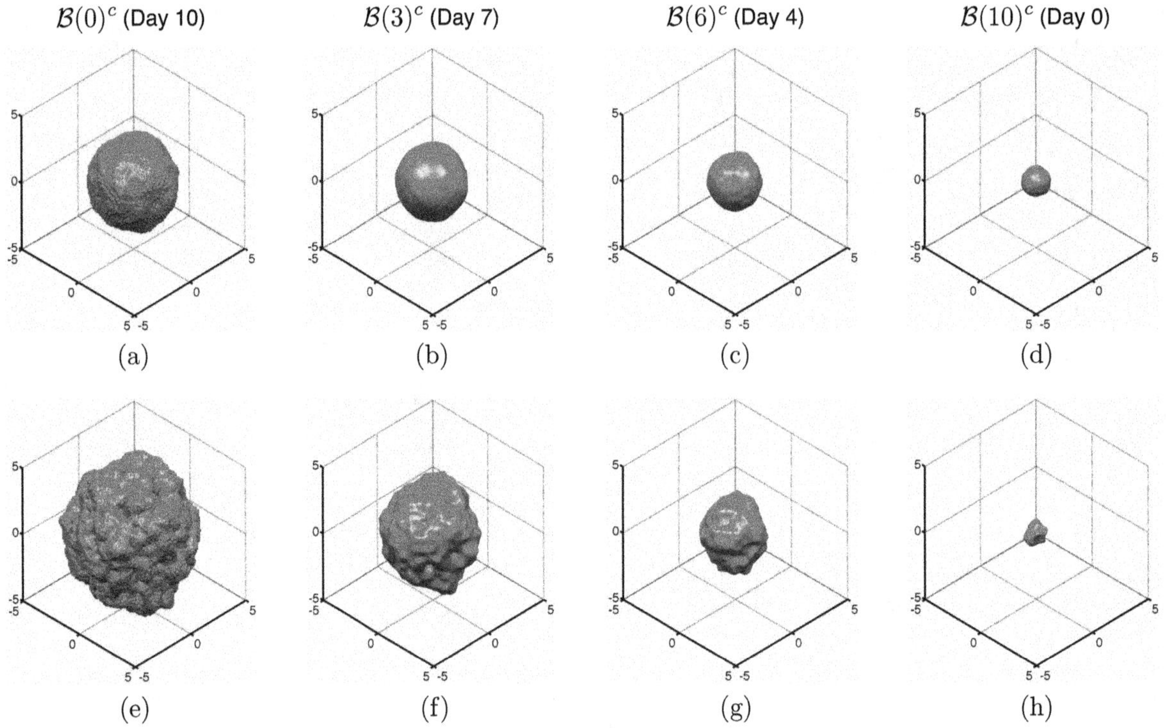

Figure 4: Backward reachable sets obtained by the one-shot method. Structure formation of (a)–(e) non-malignant cells; (f)–(j) malignant cells.

Table 2: Computational times (unit: sec) for the one-shot method and the time-dependent method.

	One-shot			Time-dependent		
h	$\mathcal{B}(5)$	$\mathcal{B}(10)$	$\mathcal{B}(20)$	$\mathcal{B}(5)$	$\mathcal{B}(10)$	$\mathcal{B}(20)$
0.2	2.74	2.74	2.74	1.49	3.29	6.32
0.1	34.22	34.22	34.22	23.78	46.31	93.00
0.05	458.30	458.30	458.30	373.00	729.16	1466.7

for integration over time, which are the fastest among other discretization options. As shown in Figures 4 and 5, the reachable sets generated by the two methods are very similar. Although the Lax-Friedrichs sweeping method is known to introduce somewhat higher numerical diffusion into the solution, we do not observe these effects in this example due to small artificial viscosities σ_i for $i = 1, 2, 3$.

The Lax-Friedrichs sweeping-based one-shot method saves computational time if we compute reachable sets for long time horizons. As presented in Table 2, the one-shot method is faster than the time-dependent method for computing $\mathcal{B}(10)$ and $\mathcal{B}(20)$, although it is slower for $\mathcal{B}(5)$. The computational time gap between the two methods increases as we compute the reachable set for a longer time horizon because the one-shot method takes the same time to compute $\mathcal{B}(t)$ for any t while the computational time for the time-dependent method increases linearly with respect to t. For example, when using $h = 0.05$, the one-shot method is approximately 3.2 times faster than the time-dependent method for computing $\mathcal{B}(20)$. This computational time behavior shows one of the advantages of the one-shot method: because we save the information of the time-to-reach in one level-set function in the one-shot method, it takes the same time to extract the t-reachable set for any t. We also observe that, when we refine the grid twice, the one-shot method takes approximately 12–13 times longer, while the time-dependent method takes approximately 16 times longer. Let h^* and k^* be the nominal grid spacing and time step sizes, respectively. The time-dependent method requires the CFL condition, which restricts the time step size by the grid spacing size. If we decrease the grid size by half, i.e., $h = h^*/2$, for example, then the time step size should be $k = k^*/2$ when using the same CFL constant. Therefore, in theory, the computational complexity of the time-dependent method increases $2^4 = 16$ times when we use the grid with $h = h^*/2$. However, the Lax-Friedrichs sweeping-based one-shot method does not require such a restriction because we do not need to integrate the HJI PDE over time. Recall that the HJI equation (3) is stationary in our formulation. Although we need more sweeping iterations as we refine the grid with $h = h^*/2$, the increase in the number of iterations is much less than a factor of two. Therefore, the computational time of the one-shot method with $h = h^*/2$ is only 12–13 times higher than that with $h = h^*$.

5. CONCLUSION

We have presented a numerical method for computing the (lower) value function of a dynamic game from which we can extract multiple reachable sets at once. The method is very useful when holding all reachable sets in a given time interval or computing reachable sets for a long time horizon. We applied the method to computing backward reachable

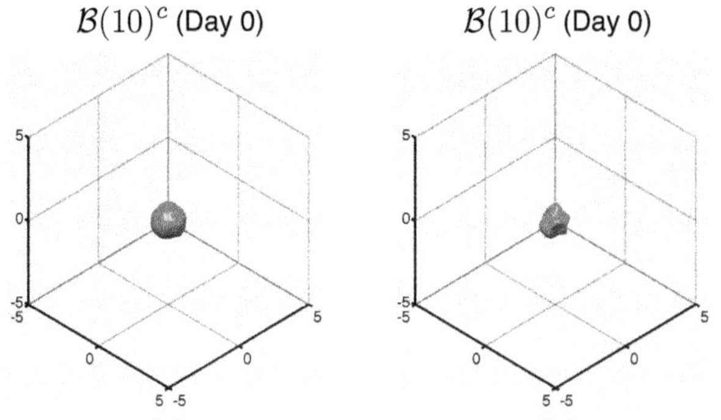

(a) (b)

Figure 5: Backward reachable sets obtained by solving the time-dependent HJI equation (6): (a) non-malignant cells; (b) malignant cells.

sets when growth and adhesion potentials compete with each other in structure formation of non-malignant and malignant cells.

A weakness of the Lax-Fredrichs sweeping method is that it introduces an undesirable numerical diffusion. It would be interesting to minimize this diffusiveness by using the Roe-Fix schemes or the semi-Lagrangian schemes. The most critical disadvantage of the method is its scalability: the complexity of Algorithm 1 increases exponentially as the system dimension increases because we must evaluate the value function at all nodes that fully discretize the state space. To resolve this scalability issue, we are very interested in utilizing approximate dynamic programming with carefully designed basis functions of the value function. We also plan to extend the one-method for computing reachable sets for stochastic differential games.

6. ACKNOWLEDGMENTS

The authors would like to thank Professor Ian Mitchell and Ryo Takei for sharing their insights on the numerical algorithms for Hamilton-Jacobi equations. This work was supported by National Cancer Institute Physical Sciences-Oncology Center (PS-OC) program under grant number 299 49-31150-44-IQPRJ1-IQCLT; by the U.S. Department of Energy, Office of Biological and Environmental Research and Low Dose Scientific Focus Area (contract number DE-AC02-05CH1123); by National Cancer Institute (awards R37CA06 4786, R01CA057621, R01CA140663, U54CA112970, U01CA 143233, and U54CA143836 - PS-OC, UC Berkeley); and by U.S. Department of Defense (W81XWH0810736).

7. REFERENCES

[1] `http://www.cs.ubc.ca/~mitchell/ToolboxLS`.

[2] M. Althoff and B. H. Krogh. Avoiding geometric intersection operations in reachability analysis of hybrid systems. In *Hybrid Systems: Computation and Control*, pages 45–54, Beijing, China, 2012.

[3] M. Althoff, C. Le Guernic, and B. H. Krogh. Reachable set computation for uncertain time-varying linear systems. In *Hybrid Systems: Computation and Control*, pages 93–102, Chicago, IL, 2011.

[4] R. Alur, C. Courcoubetis, H. Halbwachs, T. A. Henzinger, P.-H. Ho, X. Nicollin, A. Olivero, J. Sifakis, and S. Yovine. The algorithmic analysis of hybrid systems. *Theoretical Computer Science*, 3(34):3–34, 1995.

[5] E. Asarin, T. Dang, and A. Girard. Reachability analysis of nonlinear systems using conservative approximation. In *Hybrid Systems: Computation and Control*, LNCS 2623, pages 20–35. Springer-Verlag, 2003.

[6] T. Başar and G. J. Olsder. *Dynamic Noncooperative Game Theory*. SIAM, Philadelphia, 2nd edition, 1999.

[7] M. Bardi, S. Bottacin, and M. Falcone. Convergence of discrete schemes for discontinuous value functions of pursuit-evasion games. In *New Trends in Dynamic Games and Application*, pages 273–304. Birkhäuser, Boston, 1995.

[8] M. Bardi and I. Capuzzo-Dolcetta. *Optimal Control and Viscosity Solutions of Hamilton-Jacobi-Bellman equations*. Birkhäuser, Boston, 1997.

[9] M. Bardi, M. Falcone, and P. Soravia. Numerical methods for pursuit-evasion games via viscosity solutions. In *Stochastic and Differential Games: Theory and Numerical Methods*, volume 4 of *Annals of International Society of Dynamic Games*. Birkhäuser, Boston, 1999.

[10] M. Bardi and P. Soravia. Hamilton-Jacobi equations with singular boundary conditions on a free boundary and applications to differential games. *Transactions of the Americal Mathematical Society*, 325(1):205–229, 1991.

[11] A. Bemporad, F. D. Torrisi, and M. Morari. Optimization-based verification and stability characterization of piecewise affine and hybrid systems. In *Hybrid Systems: Computation and Control*, LNCS 1790, pages 45–58. Springer-Verlag, 2000.

[12] P. Cardaliaguet, M. Quincampoix, and P. Saint-Pierre. Set-valued numerical analysis for optimal control and differential games. In *Stochastic and Differential Games: Theory and Numerical Methods*, volume 4 of *Annals of International Society of Dynamic Games*. Birkhäuser, Boston, 1999.

[13] A. Chutinan and B. H. Krogh. Computational techniques for hybrid system verification. *IEEE Transactions on Automatic Control*, 48(1):64–75, 2003.

[14] M. G. Crandall and P.-L. Lions. Viscosity solutions of Hamilton-Jacobi equations. *Transactions of the Americal Mathematical Society*, 277(1):1–42, 1983.

[15] M. G. Crandall and P. L. Lions. Two approximations of solutions of Hamilton-Jacobi equations. *Mathematics of Computation*, 43(167):1–19, 1984.

[16] R. J. Elliott and N. J. Kalton. *The existence of value in differential games*. Number 126 in Memoirs of The American Mathematical Society. American Mathematical Society, Providence, RI, 1972.

[17] L. C. Evans and P. E. Souganidis. Differential games and representation formulas for solutions of Hamilton-Jacobi-Isaacs equations. *Indiana Univeristy Mathematics Journal*, 33(5), 1984.

[18] M. V. Fournier, J. E. Fata, K. J. Martin, P. Yaswen, and M. J. Bissell. Interaction of E-cadherin and PTEN

regulates morphogenesis and growth arrest in human mammary epithelial cells. *Cancer Research*, 69(10):4545–4552, 2009.

[19] A. Girard. Reachability of uncertain linear systems using zonotopes. In *Hybrid Systems: Computation and Control*, LNCS 3414, pages 291–305. Springer-Verlag, 2005.

[20] A. Girard, C. Le Guernic, and O. Maler. Efficient computation of reachable sets of linear time-invariant systems with inputs. In *Hybrid Systems: Computation and Control*, LNCS 3927, pages 257–271. Springer-Verlag, 2006.

[21] Z. Han and B. H. Krogh. Reachability analysis of large-scale affine systems using low-dimensional polytopes. In *Hybrid Systems: Computation and Control*, LNCS 3927, pages 287–301. Springer-Verlag, 2006.

[22] T. A. Henzinger, B. Horowitz, R. Majumdar, and H. Wong-Toi. Beyond HyTech: Hybrid systems analysis using interval numerical methods. In *Hybrid Systems: Computation and Control*, LNCS 1790, pages 130–144. Springer-Verlag, 2000.

[23] C. Y. Kao, S. Osher, and J. Qian. Lax-Friedrichs sweeping scheme for static Hamilton-Jacobi equations. *Journal of Computational Physics*, 196:367–391, 2004.

[24] M. Kloetzer and C. Belta. Reachability analysis of multi-affine systems. In *Hybrid Systems: Computation and Control*, LNCS 3927, pages 348–362. Springer-Verlag, 2006.

[25] A. B. Kurzhanski and P. Varaiya. Reachability analysis for uncertain systems–the ellipsoidal techinique. *Dynamics of Continuous, Discrete and Impulsive Systems Series B: Applications and Algorithms*, 9(3):347–367, 2002.

[26] J. Lygeros, C. Tomlin, and S. S. Sastry. Controllers for reachability specifications for hybrid systems. *Automatica*, 35:349–370, 1999.

[27] I. Mitchell and C. J. Tomlin. Level set methods for computation in hybrid systems. In *Hybrid Systems: Computation and Control*, LNCS 1790, pages 310–323. Springer-Verlag, 2000.

[28] I. M. Mitchell. The flexible, extensible and efficient toolbox of level set methods. *Journal of Scientific Computing*, 35:300–329, 2008.

[29] I. M. Mitchell, A. M. Bayen, and C. J. Tomlin. A time-dependent Hamilton-Jacobi formulation of reachable sets for continuous dynamics games. *IEEE Transactions on Automatic Control*, 50(7):947–957, 2005.

[30] S. Osher. A level set formulation for the solution of the Dirichlet problem for Hamilton-Jacobi equations. *SIAM Journal on Mathematical Analysis*, 24(5):1145–1152, 1993.

[31] S. Osher and J. A. Sethian. Fronts propagating with curvature-dependent speed: algorithms based on Hamilton-Jacobi formulations. *Journal of Computational Physics*, 79:12–49, 1988.

[32] S. J. Osher and R. P. Fedkiw. *Level Set Methods and Dynamic Implicit Surfaces*. Springer, New York, 1st edition, 2002.

[33] N. Ramdani, N. Meslem, and Y. Candau. Reachability of uncertain nonlinear systems using a nonlinear hybridization. In *Hybrid Systems: Computation and Control*, LNCS 4981, pages 415–428. Springer, New York, 2008.

[34] A. Rizki, V. M. Weaver, S.-Y. Lee, G. I. Rozenberg, K. Chin, C. A. Myers, J. L. Bascom, J. D. Mott, J. R. Semeiks, L. R. Grate, I. Saira Mian, A. D. Borowsky, R. A. Jensen, M. O. Idowu, F. Chen, D. J. Chen, O. W. Petersen, J. W. Gray, and M. J. Bissell. A human breast cell model of preinvasive to invasive transition. *Cancer Research*, 68(5):1378–1387, 2008.

[35] J. A. Sethian. *Level Set Methods and Fast Marching Methods: Evolving Interfaces in Computational Geometry, Fluid Mechanics, Computer Vision, and Materials Science*. Cambridge University Press, New York, 2nd edition, 1999.

[36] P. Soravia. Pursuit-evasion problems and viscosity solutions of Isaacs equations. *SIAM Journal on Control and Optimization*, 31(3):604–623, 1992.

[37] O. Stursberg and B. H. Krogh. Efficient representation and computation of reachable sets for hybrid systems. In *Hybrid Systems: Computation and Control*, LNCS 2623, pages 482–497. Springer-Verlag, 2003.

[38] K. Tanner, H. Mori, R. Mroue, A. Bruni-Cardoso, and M. J. Bissell. Coherent angular motion in the establishment of multicellular architecture of glandular tissues. *Proceedings of the National Academy of Sciences*, 109(6):1973–1978, 2012.

[39] A. Tiwari and G. Khanna. Series of abstractions for hybrid automata. In *Hybrid Systems: Computation and Control*, LNCS 2289, pages 465–478. Springer-Verlag, 2002.

[40] C. Tomlin, G. J. Pappas, and S. Sastry. Conflict resolution for air traffic management: A study in multiagent hybrid systems. *IEEE Transactions on Automatic Control*, 43(4):509–521, 1998.

[41] C. J. Tomlin, I. Mitchell, A. M. Bayen, and M. Oishi. Computational techniques for the verification of hybrid systems. *Proceedings of the IEEE*, 91(7):986–1001, 2003.

[42] Y.-H. R. Tsai, L.-T. Cheng, S. Osher, and H.-K. Zhao. Fast sweeping algorithms for a class of Hamilton-Jacobi equations. *SIAM Journal on Numerical Analysis*, 41(2):673–694, 2003.

[43] P. P. Varaiya. On the existence of solutions to a differential game. *SIAM Journal on Control*, 5(1):153–162, 1967.

[44] F. Wang, R. K. Hansen, D. Radisky, T. Yoneda, H. Barcellos-Hoff, O. W. Petersen, E. A. Turley, and M. J. Bissell. Phenotypic reversion or death of cancer cells by altering signaling pathways in three-dimensional contexts. *Journal of the National Cancer Institute*, 94(19):1494–1503, 2002.

[45] V. M. Weaver, O. W. Petersen, F. Wang, C. A. Larabell, P. Briand, C. Damsky, and M. J. Bissell. Reversion of the malignant phenotype of human breast cells in three-dimensional culture and in vivo by integrin blocking antibodies. *The Journal of Cell Biology*, 137(1):231–245, 1997.

Tracking Differentiable Trajectories across Polyhedra Boundaries*

Massimo Benerecetti
Elect. Eng. and Information Technologies
Università di Napoli "Federico II", Italy
massimo.benerecetti@unina.it

Marco Faella
Elect. Eng. and Information Technologies
Università di Napoli "Federico II", Italy
m.faella@unina.it

ABSTRACT

We analyze the properties of differentiable trajectories subject to a constant differential inclusion which constrains the first derivative to belong to a given convex polyhedron. We present the first exact algorithm that computes the set of points from which there is a trajectory that reaches a given polyhedron while avoiding another (possibly non-convex) polyhedron. We discuss the connection with (Linear) Hybrid Automata and in particular the relationship with the classical algorithm for reachability analysis for Linear Hybrid Automata.

Categories and Subject Descriptors

D.2.4 [**Software Engineering**]: Software/Program Verification

Keywords

Reachability, Differentiability, Exact algorithm, Linear Hybrid Automata

1. INTRODUCTION

Hybrid Automata are a mathematical abstraction of systems that feature both discrete and continuous dynamics. Linear Hybrid Automata (LHAs) [2] were introduced as a computationally tractable model of hybrid systems that still allows for non-trivial dynamics. In particular, LHAs can approximate complex dynamics up to an arbitrary precision [11].

In an LHA, discrete dynamics is represented by a finite set of control modes called *locations*, while the continuous dynamics is ensured by a finite set of real-valued variables. In each discrete location, the continuous dynamics is constrained by a differential inclusion of the type $\dot{x} \in F$, where $\dot{x}$ is the vector of the time-derivatives of all the variables in the system, and $F \subseteq \mathbb{R}^n$ is a convex polyhedron. The main decision problem that was considered for LHAs is *reachability*, i.e., given two system configurations, say an initial state and an error state, establish whether there is a system behavior that leads from the first to the second. A more complex task consists in verifying whether a given LHA can be modified (i.e., *controlled*) in such a way that a given error configuration (or region) is *not* reached by any behavior. This problem can be called *safety control* and is analogous to a game with a safety objective. Both problems require an algorithm for the following sub-problem, which applies to a single discrete location: given a region G (for *goal*) and a region A (for *avoid*) of system configurations, find the set of points from which there is a trajectory that reaches G while avoiding A at all times. We denote this set by $RWA(G, A)$ for *reach while avoiding*. In reachability problems, the goal region G can be thought of as error states, and the avoidance region A is the complement of the *invariant* of the automaton, which is the set of configurations that make physical sense for the system. Hence, $RWA(G, A)$ is the set of states that reach an error state while remaining in the invariant. In a safety control problem, the goal region G is taken to be a set of uncontrollable states (such as, states outside the safe region) and A is a set of controllable states (included in the invariant). Then, $RWA(G, A)$ identifies the region in which the environment can reach an error state while avoiding the good, controllable states.

The RWA operator is recognized as a central tool in the analysis of various kinds of hybrid systems: it corresponds to the *Reach* operator in Tomlin et al. [13] and *Unavoid_Pre* in Balluchi et al. [4]; it was also used in the synthesis of controllers for reachability objectives [8].

Computing reach-while-avoiding. Computing of $RWA(G, A)$ is simple when A is co-convex, corresponding to the case of reachability analysis for LHAs with convex invariants. In that case, RWA can be expressed in the first-order theory of reals and computed using a constant number of basic polyhedra operations.

When A is not co-convex, one may adapt the procedure that is presented in one of the early papers on LHAs, in the context of reachability analysis in presence of non-convex invariants [2]. The idea of the algorithm is simple: consider a partition of the non-convex invariant I into a finite set of convex polyhedra $P_1, \ldots, P_n$; then, split the location with invariant I into n different locations, each with convex invariant P_i; finally, connect these new locations with

*This work was supported by the European Union 7th Framework Program project 600958-SHERPA.

HSCC'13, April 8–11, 2013, Philadelphia, Pennsylvania, USA.

virtual transitions corresponding to the boundaries between two adjacent convex polyhedra P_i, P_j.

The problem with this approach is that the discrete transitions that model the crossing from one convex polyhedron to an adjacent one do not preserve the differentiability of the trajectories. In other words, there are trajectories in the modified system that do not correspond to any trajectory in the original system.

Consider for example the situation depicted in Figure 1. Assume that the invariant for the current location is $P \cup Q \cup G$ and the goal is to reach G. Dashed lines identify topologically open sides of polyhedra. The flow constraint F is also depicted in the figure: it allows trajectories to move in a range of directions going from straight right to straight up, and it forbids stopping (i.e., it does not include the origin).

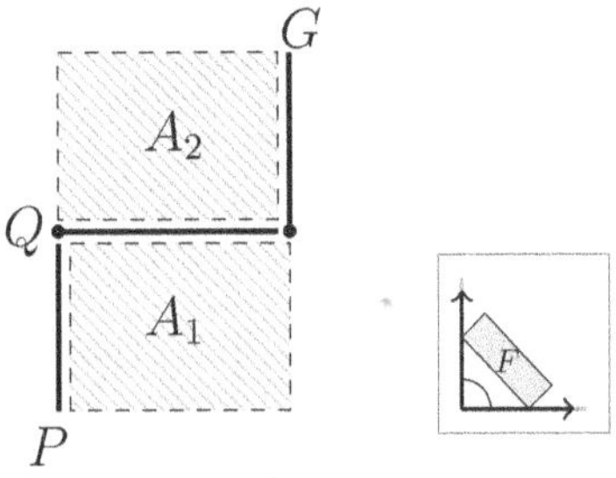

Figure 1: Can the points in P reach the target region G while remaining in $P \cup Q \cup G$? The flow constraint F is shown on the r.h.s.

The above procedure splits the invariant into three convex polyhedra, and then performs a backward reachability analysis which starts from the goal G and progressively enlarges the set of "good" states W by including the states that can reach W while remaining in one of the convex parts of the invariant.

In our example, the points in the line segment Q can reach the target by moving straight to the right, while remaining in one convex part of the invariant. Hence, in the first iteration the set Q is added to the set of good states W. The points in the line segment P can reach the extreme point of Q by moving straight up. Hence, in the second iteration the above procedure puts P in W. On the other hand, no differentiable trajectory can start in a point of P and reach G while remaining in $P \cup Q \cup G$. In other words, the classical algorithm for reachability analysis of LHAs is based on trajectories that can take a finite number of sharp turns, i.e., a trajectory that is differentiable almost everywhere.

By rotating the polyhedra in Figure 1 (including the flow constraint F) by 45°, it becomes apparent that the issue is also present with *rectangular* flow constraints. However, Rectangular Hybrid Automata and Games [12] do not exhibit the above issue, due to the presence of multiple restrictions, such as the fact that guards and invariants are convex and topologically closed.

In this paper, we present an exact algorithm for computing $RWA(G, A)$ for general polyhedra G and A, assuming that trajectories are differentiable everywhere. Our solution is related to the one proposed by Henzinger et al. [11], which however does not apply to general non-convex polyhedra like the one presented in Figure 3.

The rest of the paper is organized as follows. Section 2 is devoted to preliminary definitions, including the problem statement. In Section 3.2 we recall the previous attempts at computing *RWA* and remark that they only work for trajectories that are almost everywhere differentiable. We also lay the basis for our algorithm, by introducing an operator (called *Cross*) that contains the points that can move in a differentiable fashion from a given convex polyhedron P to another one P' and spend a positive amount of time in the latter. In Section 4 we present a series of steps that lead from *Cross* (which deals with arbitrary differentiable trajectories) to a computationally simpler operator (called *ClosCross*) that only reasons about straight-line trajectories. Then, in Section 5 we show how the latter operator can be computed using basic operations on polyhedra. Section 6 argues that a finite number of repeated applications of *Cross* (plus a preliminary step) captures exactly *RWA*. Finally, we provide some conclusions in Section 7.

2. DEFINITIONS

Let $\mathbb{R}$ (respectively, $\mathbb{R}_{\geq 0}$) denote the set of real numbers (resp., non-negative real numbers). Throughout the paper we consider a fixed ambient space $\mathbb{R}^n$. A *convex polyhedron* is a subset of $\mathbb{R}^n$ that is the intersection of a finite number of open and closed half-spaces. A *polyhedron* is a subset of $\mathbb{R}^n$ that is the union of a finite number of convex polyhedra. For a general (i.e., not necessarily convex) polyhedron $G \subseteq \mathbb{R}^n$, we denote by $cl(G)$ its topological closure, by $\overline{G}$ its complement, and by $[\![G]\!] \subseteq 2^{\mathbb{R}^n}$ its representation as a finite set of convex polyhedra. We assume w.l.o.g. that $[\![G]\!]$ contains mutually disjoint convex polyhedra, called *patches* of G.

Let $\mathcal{C} \subseteq [\mathbb{R}_{\geq 0} \to \mathbb{R}^n]$ be a class of functions from the time domain to our ambient space. Given a convex polyhedron F, called the *flow constraint*, an $(F, \mathcal{C})$*-trajectory* is a function $f \in \mathcal{C}$ such that $\dot{f}(t) \in F$ for all $t \geq 0$ such that f is differentiable in t. Given a point $x \in \mathbb{R}^n$, let $Adm^{\mathcal{C}}_F(x)$ (for *admissible*) denote the set of $(F, \mathcal{C})$-trajectories f starting from x (i.e., such that $f(0) = x$). We henceforth consider two classes of functions: the class $\mathcal{C}_{\text{d}}$ of functions that are continuous and differentiable everywhere, and the class $\mathcal{C}_{\text{ae}}$ of functions that are continuous everywhere and differentiable almost everywhere (i.e., always except for a finite set of time points).

Given two disjoint polyhedra G (for *goal*) and A (for *avoid*), we denote by $RWA^{\mathcal{C}}_F(G, A)$ (for *reach while avoiding*) the set of points from which there is an $(F, \mathcal{C})$-trajectory that reaches G while avoiding A. Formally, we have:

$$RWA^{\mathcal{C}}_F(G, A) = \Big\{ x \in \mathbb{R}^n \;\Big|\; \exists f \in Adm^{\mathcal{C}}_F(x),\, \delta_f \geq 0 : f(\delta_f) \in G \text{ and } \forall 0 \leq \delta < \delta_f : f(\delta) \notin A \Big\}.$$

For every $x \in RWA^{\mathcal{C}}_F(G, A)$, any pair (f, δ_f) satisfying the condition in the definition of $RWA^{\mathcal{C}}_F$ will be called a *witness* for x. Notice that

$$RWA^{\mathcal{C}}_F(G_1 \cup G_2, A) = RWA^{\mathcal{C}}_F(G_1, A) \cup RWA^{\mathcal{C}}_F(G_2, A),$$

whereas $RWA^{\mathcal{C}}_F$ does not distribute over unions in the second argument (see [7]). Therefore, in the following we can assume, w.l.o.g., that the goal G is a convex polyhedron.

3. TRACKING DIFFERENTIABLE TRAJECTORIES

Starting from this section, we consider a fixed flow constraint F, and we omit the F parameter from all notations.

3.1 The Previous Algorithm

Let us briefly recall the previous approach to the problem, which can be used to compute *RWA*, assuming that trajectories are differentiable everywhere except for a finite set of time points.

Given two convex polyhedra P, P', let $\mathit{Reach}(P, P')$ be the set of points in P that can reach P' via a trajectory that remains within $P \cup P'$ at all times (until P' is reached). Formally,

$$\mathit{Reach}(P, P') = \{x \in P \mid \exists f \in \mathit{Adm}^{\mathcal{C}_{\mathrm{ae}}}(x), \delta \geq 0 : \\ f(\delta) \in P' \text{ and } \forall \delta' \in (0, \delta) : f(\delta') \in P \cup P'\}.$$

It can be shown that $\mathit{Reach}(P, P')$ is a convex polyhedron which can be computed from P, P', and F using basic polyhedra operations [5, 7]. Then, Theorem 1 shows how *RWA* can be computed by iterative application of *Reach*.

THEOREM 1. *[5, 7] For all polyhedra G, A, and W, let* $\tau_{\mathrm{ae}}(G, A, W) =$

$$G \cup \bigcup_{P \in [\![A]\!]} \bigcup_{P' \in [\![W]\!]} \mathit{Reach}(P, P').$$

We have $\mathit{RWA}^{\mathcal{C}_{\mathrm{ae}}}(G, A) = \mu W \,.\, \tau_{\mathrm{ae}}(G, A, W)$*, where* μW *denotes the least fixed point operator. Moreover, the fixed point is reached within a finite number of iterations.*

By Knaster-Tarski fixed point theorem, the fixed point equation in Theorem 1 suggests the semi-algorithm consisting in repeated applications of $\tau_{\mathrm{ae}}(G, A, W)$, starting from $W = \emptyset$, i.e.: $W_0 = \emptyset$ and $W_{i+1} = \tau_{\mathrm{ae}}(G, A, W_i)$ for all $i \geq 0$. Theorem 1 states that there exists $k > 0$ such that $W_k = W_{k+1} = \mathit{RWA}^{\mathcal{C}_{\mathrm{ae}}}(G, A)$.

The *Reach* and τ_{ae} operators, together with Theorem 1, represent a reformulation of the original algorithm for reachability analysis of LHAs under non-convex invariants, which was expressed in terms of locations of a hybrid automaton. Notice that both we ourselves (in [5], later corrected by [7]) and Alur et al. (in [1, 2]) have claimed that τ_{ae} (or very similar variations thereof) can be used to compute $\mathit{RWA}^{\mathcal{C}_{\mathrm{d}}}$. Those claims are incorrect, as shown in the Introduction and again below. In the rest of the paper we show that significant new developments are required to correctly compute $\mathit{RWA}^{\mathcal{C}_{\mathrm{d}}}$.

3.2 Differentiable Trajectories

Consider again the example in Figure 1, and let $A = A_1 \cup A_2$. It holds $\mathit{Reach}(Q, G) = Q$ and $\mathit{Reach}(P, Q) = P$. On the other hand, when considering differentiable trajectories it holds $\mathit{RWA}^{\mathcal{C}_{\mathrm{d}}}(G, A) = G \cup Q$. The points in P do not belong to $\mathit{RWA}^{\mathcal{C}_{\mathrm{d}}}(G, A)$ because no differentiable trajectory can start from P and turn into Q without hitting either A_1 or A_2.

An obvious first step consists in modifying the definition of *Reach* by replacing $\mathit{Adm}^{\mathcal{C}_{\mathrm{ae}}}$ with $\mathit{Adm}^{\mathcal{C}_{\mathrm{d}}}$. However, this replacement is not sufficient to solve the problem with the example in Figure 1: it would still hold $\mathit{Reach}(P, Q) = P$, because the points in P can reach Q along a straight-line trajectory, which is both in $\mathcal{C}_{\mathrm{ae}}$ and in $\mathcal{C}_{\mathrm{d}}$.

Then, we notice that all trajectories going from P to Q lie within P at all times, except for the final point, which belongs to Q. Hence, we modify the definition of $\mathit{Reach}(P, P')$ by requiring not only that there exists a (differentiable) trajectory from P to P' contained in $P \cup P'$, but also that this trajectory *spends a positive amount of time in* P':

$$\mathit{Reach}'(P, P') = \{x \in P \mid \exists f \in \mathit{Adm}^{\mathcal{C}_{\mathrm{d}}}(x), 0 \leq \delta_1 < \delta_2 : \\ \forall \delta \in (0, \delta_1] : f(\delta) \in P \cup P' \text{ and} \\ \forall \delta \in (\delta_1, \delta_2] : f(\delta) \in P'\}.$$

Unfortunately, *Reach'* still suffers from a shortcoming. Consider again the example in Figure 1, but this time let the avoidance region be $A = A_1$. Notice that the status of the immediate neighborhood of P is identical to the previous case: Q is still a "good" neighbor (w.r.t. reaching G while avoiding A) and A_1 is still a "bad" neighbor. However, we have $P \subseteq \mathit{RWA}^{\mathcal{C}_{\mathrm{d}}}(G, A)$, because a differentiable trajectory can start from P, pass instantaneously through the left vertex of Q, then curve into A_2 and finally reach G. The fact that P is a set of good points is essentially due to A_2, which is not an immediate neighbor of P (in our terminology, it is a *weak* neighbor, since $\mathit{cl}(P) \cap \mathit{cl}(A_2) \neq \emptyset$).

Therefore, we realize that *Reach'* cannot solve this example because it is constrained to consider pairs of adjacent convex polyhedra. In particular, it holds $\mathit{Reach}'(P, A_2) = \emptyset$ because P and A_2 are not adjacent, and $\mathit{Reach}'(P, Q) = \emptyset$ because differentiable trajectories cannot start from P and spend a positive amount of time in Q while remaining in $P \cup Q$.

Guided by this example, we devise a third version of *Reach*, called *Cross*, which carries three arguments: $\mathit{Cross}(P, \hat{P}, P')$ contains the points of P that can reach and spend a positive amount of time in P', via a trajectory that remains within $P \cup P'$ at all times, except for an intermediate time instant in which the trajectory may be in $\hat{P}$. Formally,

$$\mathit{Cross}(P, \hat{P}, P') = \{x \in P \mid \exists f \in \mathit{Adm}^{\mathcal{C}_{\mathrm{d}}}(x), 0 \leq \delta_1 < \delta_2 : \\ \forall \delta \in (0, \delta_1) : f(\delta) \in P \text{ and} \\ f(\delta_1) \in P \cup \hat{P} \cup P' \text{ and} \\ \forall \delta \in (\delta_1, \delta_2] : f(\delta) \in P'\}.$$

Notice that the conditions above imply that $f(\delta_1) \in \mathit{cl}(P) \cap \mathit{cl}(P')$. Hence,

$$\mathit{Cross}(P, \hat{P}, P') = \mathit{Cross}(P, \hat{P} \cap \mathit{cl}(P) \cap \mathit{cl}(P'), P'), \quad (1)$$

and we can assume w.l.o.g. that $\hat{P} \subseteq \mathit{cl}(P) \cap \mathit{cl}(P')$.

We show in Section 6 that *Cross* is the main ingredient in a fixed point characterization of $\mathit{RWA}^{\mathcal{C}_{\mathrm{d}}}$. Sections 4 and 5 prove that $\mathit{Cross}(P, \hat{P}, P')$ is a polyhedron that can be computed using basic operations on convex polyhedra.

In the rest of the paper we always refer to the class of trajectories $\mathcal{C}_{\mathrm{d}}$. Hence, we drop the corresponding superscript and write *Adm* for $\mathit{Adm}^{\mathcal{C}_{\mathrm{d}}}$ and *RWA* for $\mathit{RWA}^{\mathcal{C}_{\mathrm{d}}}$.

4. FROM GENERAL TRAJECTORIES TO STRAIGHT DIRECTIONS

As a first step towards the computation of the operator $\mathit{Cross}(P, \hat{P}, P')$ defined in the previous section, we will show how to reformulate it in terms of straight trajectories only.

The main results of this section and the next one are summarized in Figure 4.

We assume that we can compute the following basic operations on arbitrary convex polyhedra P and P': the Boolean operations $P \cup P'$, $P \cap P'$, and $\overline{P}$; the topological closure $cl(P)$ of P; finally, the *pre-* and *post-flows* of P, defined as follows:

$$P\swarrow = \{x - \delta c \mid x \in P, \delta \geq 0, c \in F\}$$
$$P\swarrow_{>0} = \{x - \delta c \mid x \in P, \delta > 0, c \in F\}$$
$$P\nearrow = \{x + \delta c \mid x \in P, \delta \geq 0, c \in F\}$$
$$P\nearrow_{>0} = \{x + \delta c \mid x \in P, \delta > 0, c \in F\}.$$

Intuitively, the pre- and post-flow operators compute the pre- and post-image, respectively, of a convex polyhedron with respect to the straight directions contained in F. The algorithm for $P\swarrow_{>0}$ and $P\nearrow_{>0}$ can be found in [6].

It is well known that $P\nearrow$ (resp., $P\swarrow$) is not a convex polyhedron when F is non-necessarily closed. The following example shows that the same is true even if both P and F are closed convex polyhedra. This observation contradicts a claim made by Halbachs et al. [10].

THEOREM 2. *Given two closed convex polyhedra P and F, the following hold:*

1. *$P\nearrow$ may not be a convex polyhedron;*
2. *$P\nearrow = P \cup (P\nearrow_{>0})$.*

PROOF. To show that the first property holds it suffices to consider the closed convex polyhedra $P = \{(0,0)\}$ and $F = \{(x,y) \mid x \geq 1\}$. According to the definition, $P\nearrow = P \cup \{(x,y) \mid x > 0\}$, which is a convex set but not a convex polyhedron. As to the second property, it is enough to observe that $P = \{x + \delta c \mid x \in P, \delta = 0, c \in F\}$. ∎

We start by splitting $Cross(P, \hat{P}, P')$ into three simpler operators, according to the properties of the trajectory f and the delay δ_1 occurring in the definition of *Cross*: $Cross^0(P, P')$ takes care of the case when $\delta_1 = 0$; $Cross^+(P, P')$ takes care of the case when $\delta_1 > 0$ and $f(\delta_1) \in P \cup P'$; finally, $Cross^+(P, \hat{P}, P')$ covers the remaining case, i.e., $\delta_1 > 0$ and $f(\delta_1) \in \hat{P}$. Hence, it holds:

$$Cross(P, \hat{P}, P') = Cross^0(P, P') \cup Cross^+(P, P') \cup Cross^+(P, \hat{P}, P').$$

We recall the following result from the literature, which guarantees that every point reachable from a point x along an admissible and differentiable trajectory can also be reached from x along an admissible straight trajectory.

LEMMA 1. [2] *For all points $x \in \mathbb{R}^n$, if there is a trajectory $f \in Adm(x)$ and a time $\delta > 0$ such that $f(\delta) = y$, then there is a slope $c \in F$ such that $y = x + \delta c$.*

The following result shows that $Cross^0$ can immediately be reformulated in terms of straight directions and easily computed with basic polyhedral operations. Given a set $P \subseteq \mathbb{R}^n$, we say that a trajectory f *lingers* in P if there exists $\delta > 0$ such that $f(\delta') \in P$ for all $0 < \delta' < \delta$.

THEOREM 3. *For all disjoint convex polyhedra P, P', it holds:*

$$Cross^0(P, P') = P \cap cl(P') \cap P'\swarrow.$$

PROOF. First, notice that the definition of $Cross(P, \hat{P}, P')$ boils down to the following:

$$Cross^0(P, P') = \{x \in P \mid \exists f \in Adm(x), \delta > 0 : \forall \delta' \in (0, \delta] : f(\delta') \in P'\}.$$

Now, we can prove the two sides of the equivalence.

($\subseteq$) Let $x \in Cross^0(P, P')$ and let $f \in Adm(x)$ be the trajectory whose existence is postulated by the definition of $Cross^0$. Since f lingers in P', in each neighborhood of x there is a point in P'. Hence, $x \in cl(P')$. Moreover, Lemma 1 implies that $x \in P'\swarrow$.

($\supseteq$) Let $x \in P \cap cl(P') \cap P'\swarrow$ and let $y \in P'$ such that $y = x + \bar{\delta}c$, for suitable $\bar{\delta} \geq 0$ and $c \in F$. Since P and P' are disjoint, it holds $\bar{\delta} > 0$. Then, the trajectory $f(\delta) = x + \delta c$ lingers in P' and proves that $x \in Cross^0(P, P')$. ∎

Next, we show how $Cross^+(P, P')$ can be expressed in terms of $Cross^+(P, \hat{P}, P')$ for a suitable $\hat{P}$. Let the *boundary* between P and P' be defined as:

$$bndry(P, P') \triangleq (cl(P) \cap P') \cup (P \cap cl(P')).$$

It has been proved that $bndry(P, P')$ is a convex polyhedron [7].

LEMMA 2. *For all convex polyhedra P, P', it holds:*

$$Cross^+(P, P') = Cross^+(P, bndry(P, P'), P').$$

PROOF. ($\subseteq$) Let $x \in Cross^+(P, P')$, according to the definition there exist a trajectory $f \in Adm(x)$ and two delays δ_1, δ_2 such that f stays in P from 0 to δ_1 (excluded), then $f(\delta_1) \in P \cup P'$, and finally f lies in P' from δ_1 (excluded) to δ_2. So, in each neighborhood of $f(\delta_1)$ there is both a point in P and a point in P'. Hence, $f(\delta_1) \in cl(P) \cap cl(P')$ and therefore $f(\delta_1) \in bndry(P, P')$.

($\supseteq$) Conversely, let $x \in Cross^+(P, bndry(P, P'), P')$. The thesis follows immediately from the fact that $bndry(P, P') \subseteq P \cup P'$. ∎

Lemma 2 implies that the only remaining task that we need to address is a way to compute $Cross^+(P, \hat{P}, P')$. As we shall show, this task turns out to be significantly involved. The first step towards the solution consists in reformulating $Cross^+$ in terms of straight trajectories. If we simply replace the arbitrary trajectory $f \in Adm(x)$ in the definition of $Cross^+$ with a straight trajectory of slope c, we obtain the following operator:

$$StrCross(P, \hat{P}, P') = \{x \in P \mid \exists c \in F, 0 < \delta_1 < \delta_2 : \forall \delta \in (0, \delta_1) : x + \delta c \in P \text{ and } x + \delta_1 c \in \hat{P} \text{ and } \forall \delta \in (\delta_1, \delta_2] : x + \delta c \in P'\}.$$

Intuitively, these are the points of P which can reach a point in P' following a straight direction while remaining in $P \cup P'$ at all times, except in a single point of $\hat{P}$.

Clearly, $StrCross(P, \hat{P}, P') \subseteq Cross^+(P, \hat{P}, P')$. In addition, any point in P which can reach $StrCross(P, \hat{P}, P')$ also belongs to $Cross^+(P, \hat{P}, P')$:

LEMMA 3. *For all convex polyhedra P, $\hat{P}$ and P' the following holds:*

$$P \cap StrCross(P, \hat{P}, P')\swarrow \subseteq Cross^+(P, \hat{P}, P').$$

To prove Lemma 3, we shall need the following result, which shows how to connect, in an admissible and differentiable fashion, two points which can be connected by the concatenation of two straight-line trajectories. Due to space constraints, the proof of the following lemma, as well as those of several other results, can be found in the Appendix.

LEMMA 4 (INTERPOLATION). *Given three points $x_0, x_1, x_2 \in \mathbb{R}^n$, two directions $c_0, c_1 \in F$ and two delays $\delta_0, \delta_1 \geq 0$ such that: $x_1 = x_0 + \delta_0 c_0$ and $x_2 = x_1 + \delta_1 c_1$, there exists a trajectory $f \in Adm(x_0)$ such that $\dot{f}(0) = c_0$, $f(\delta_0+\delta_1) = x_2$, $\dot{f}(\delta_0 + \delta_1) = c_1$, and $f(\delta)$ is a strict convex combination of x_0, x_1, x_2 for all $\delta \in (0, \delta_0 + \delta_1)$.*

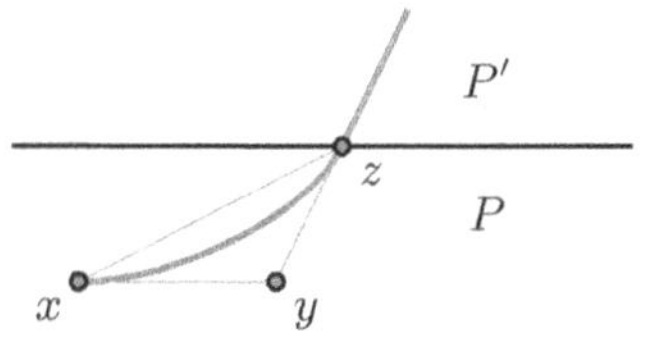

Figure 2: When a point $x \in P$ can reach another point y which is in $StrCross(P, \hat{P}, P')$, x can differentiably cross into P' (see Lemma 3). Here, $\hat{P} = \{z\}$.

PROOF OF LEMMA 3. Let $x \in P \cap StrCross(P, \hat{P}, P')_{\swarrow}$. The proof is illustrated in Figure 2. Let $y \in StrCross(P, \hat{P}, P')$ such that $x \in \{y\}_{\swarrow}$. There exist $c \in F$ and $0 < \delta_1 < \delta_2$ satisfying the definition of $StrCross(P, \hat{P}, P')$ for y. Let $z = y + \delta_1 c$, by construction it holds $z \in \hat{P} \cap cl(P) \cap cl(P')$. By applying Lemma 4 with $x_0 = x$, $x_1 = y$, and $x_2 = z$, we get a differentiable trajectory f in $Adm(x)$ from x to z whose derivative in z is c. Moreover, f is contained in P, with the possible exception of $z \in \hat{P}$. Therefore, the concatenation of f and the straight trajectory $g(\delta) = z + \delta c$ is everywhere differentiable and crosses into P'. As a consequence, $x \in Cross^+(P, \hat{P}, P')$. ∎

While Lemma 3 provides a sound approximation of $Cross^+$ in terms of straight trajectories, it does not, unfortunately, ensure completeness. Indeed, there may be points that belong to $Cross^+(P, \hat{P}, P')$ but not to $StrCross(P, \hat{P}, P')_{\swarrow}$.

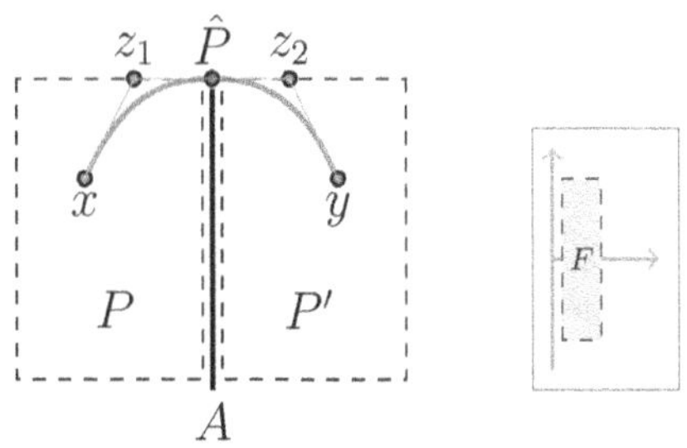

Figure 3: Reaching points in $StrCross$ is not necessary to be in $Cross^+$.

Consider the situation depicted in Figure 3, where P and P' are open rectangles, $\hat{P}$ contains a single point z (the upper right corner of the closure of P), and the line segment A (which does not include $\hat{P}$) is a region to avoid. Given the flow constraint depicted on the right-hand side of the figure, it holds $StrCross(P, \hat{P}, P') = \emptyset$, as no straight trajectory leads from P to P' passing through $\hat{P}$.

Notice, however, that the point z_1, lying in the closure of P, is reachable from x by following a straight trajectory which always remains in the closure of P. Then, a straight trajectory, with derivative c, leads from z_1 to a point z_2, which lies in the closure of P', without ever leaving the closures of the two polyhedra. Finally, z_2 can reach y in P', following a straight trajectory that never leaves the closure of P'. Therefore, Lemma 4, applied with $x_0 = x$, $x_1 = z_1$ and $x_2 = z$, gives a differentiable trajectory from x to z, which never leaves P except for the end point $z \in \hat{P}$ and whose derivative in z is c (the straight direction from z_1 to z_2). Similarly, another application of Lemma 4, this time applied to $x_0 = z$, $x_1 = z_2$ and $x_2 = y$, gives a differentiable trajectory from z to y, which never leaves P' except for the starting point $z \in \hat{P}$ and whose derivative in z is, again, c. The concatenation of these two trajectories in z is depicted in Figure 3 and is a differentiable trajectory from x to y which never leaves $P \cup P'$ except for the single point $z \in \hat{P}$. Hence, we have that $x \in Cross^+(P, \hat{P}, P')$.

This example suggests that the straight trajectory reformulation of Lemma 3, while not complete, can be extended, exploiting Lemma 4, by also allowing certain straight trajectories lying in the closures of P and P'. We, therefore, obtain the following operator:

$$
\begin{aligned}
ClosCross(P, \hat{P}, P') = \{x \in cl(P) \cap P_{\nearrow} \mid \exists c \in F, 0 < \delta_1 < \delta_2 :& \\
\forall \delta \in (0, \delta_1) : x + \delta c \in cl(P) \text{ and}& \\
x + \delta_1 c \in \hat{P} \text{ and}& \\
\forall \delta \in (\delta_1, \delta_2) : x + \delta c \in cl(P') \text{ and}& \\
x + \delta_2 c \in cl(P') \cap P'_{\swarrow} \}.&
\end{aligned}
$$

The following theorem ensures that $ClosCross$ allows us to obtain a sound and complete reformulation of $Cross^+$ in terms of straight trajectories.

THEOREM 4. *For all convex polyhedra P, $\hat{P}$ and P' the following holds:*

$$Cross^+(P, \hat{P}, P') = P \cap ClosCross(P, \hat{P}, P')_{\swarrow}.$$

Before proving Theorem 4, we need an additional result, connecting arbitrary trajectories and straight trajectories. Consider a differentiable trajectory lying within a convex polyhedron P. Along any tangent to the curve one can find a point that is still in the closure of P (if not in P itself) and that is reachable from the initial point of the curve. The following lemma formalizes this property.

LEMMA 5 (TANGENT). *Let P be a convex polyhedron, x a point, $f \in Adm(x)$ a trajectory, and $\hat{\delta} > 0$ a delay such that in all non-empty intervals $(\delta, \hat{\delta})$ there is a time γ such that $f(\gamma) \in P$. Then, there exists $\delta^* > 0$ such that $f(\hat{\delta}) - \delta^* \dot{f}(\hat{\delta}) \in cl(P) \cap \{x\}_{\nearrow_0}$.*

PROOF OF THEOREM 4. *($\supseteq$)* This part of the proof has been illustrated in the above discussion regarding the example in Figure 3. A full proof is provided in the Appendix.

($\subseteq$) Let $x \in Cross^+(P, \hat{P}, P')$ and let $f \in Adm(x)$ and δ_1, δ_2 be the trajectory and the delays whose existence is postulated by the definition of $Cross^+$. By definition, $x \in P$. Let $y = f(\delta_1)$, recall that $y \in \hat{P}$. Moreover, since $f(\delta) \in P$ for all $\delta \in [0, \delta_1)$, we also have that $y \in cl(P)$.

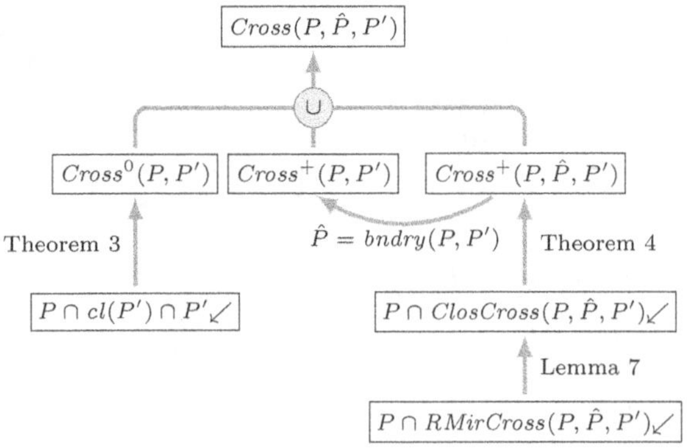

Figure 4: The main steps required for computing *Cross* when $\hat{P} \subseteq cl(P) \cap cl(P')$. Equation 1 in Section 3.2 shows that the latter condition does not restrict generality.

Let $c = \dot{f}(\delta_1) \in F$, by Lemma 5, there exists $\gamma_1 > 0$ such that $u_1 \triangleq y - \gamma_1 c \in cl(P) \cap \{x\}\nearrow$. Since $u_1 \in cl(P)$ and $y \in cl(P)$, then for all $\delta \in (0, \gamma_1)$ we have $u_1 + \delta c \in cl(P)$.

Similarly, by applying Lemma 5 backwards from a point $f(\delta) \in P'$ (it is sufficient to consider any $\delta \in (\delta_1, \delta_2)$), we obtain another value $\gamma_2 > 0$, such that $u_2 \triangleq y + \gamma_2 c \in cl(P') \cap P'\swarrow$. Since $y \in cl(P')$ and $u_2 \in cl(P')$, then for all $\delta \in (\gamma_1, \gamma_1 + \gamma_2)$ we have $y + \delta c \in cl(P')$. As a consequence, we obtain that $x \in P \cap StrCross(P, \hat{P}, P')\swarrow$. ∎

In conclusion, Theorems 3 and 4 allow us to compute $Cross(P, \hat{P}, P')$ provided we can compute $ClosCross(P, \hat{P}, P')$, or at least its pre-flow. Indeed, that is the subject of the following section.

5. COMPUTING THE $Cross^+$ OPERATOR

We start by introducing some preliminary geometric notions.

5.1 Geometric Primitives

An *affine combination* of two points x and y is any point $z = ax + (1-a)y$, for $a \in \mathbb{R}$. An *affine set* is any set of points closed under affine combinations. The empty set, a single point, a line, a hyper-plane, and the whole ambient space are all examples of affine sets. Given a convex polyhedron P, the *affine hull* of P, denoted $ahull(P)$, is the smallest affine set containing P.

Given a polyhedron $P \subseteq \mathbb{R}^n$, the *affine dimension* of P is the natural number $k \leq n$, such that the maximum number of affinely independent points in P is $k + 1$.

The *relative interior* of a convex polyhedron P, denoted $rint(P)$, is the interior of P itself taken in the space corresponding to the affine hull of P itself. More formally, $x \in rint(P)$ if and only if there is a ball $N_\epsilon(x)$ of radius $\epsilon > 0$ centered in x, such that $N_\epsilon(x) \cap ahull(P) \subseteq P$.

Similarly, given a polyhedron P, we call the set N a *relative neighborhood* of a point $x \in P$, if there is a ball $N_\epsilon(x)$ of radius $\epsilon > 0$, such that $N = N_\epsilon(x) \cap ahull(P)$. Intuitively, the relative neighborhood of x is a neighborhood of x relative to the the affine hull of P.

The *point reflection* of a point x w.r.t. a point y, in symbols $mirror(x, y)$, is the point z, beyond y along the line connecting x and y, such that $\|y - x\| = \|z - y\|$. Similarly, one can define the reflection $mirror(x, Q)$ of x w.r.t. an affine set Q. More precisely, $mirror(x, Q)$ is defined as the point reflection of x w.r.t. the *orthogonal projection* of x on Q. Finally, the above definitions can be extended to sets of points P, giving rise to the reflections $mirror(P, y)$ and $mirror(P, Q)$. For a convex polyhedron Q which is not necessarily an affine set, we will abuse the notation and write $mirror(P, Q)$ when we mean $mirror(P, ahull(Q))$.

The affine hull and the relative interior can be easily computed using the standard "double description" of convex polyhedra via constraints and generators [3]. Moreover, it is well known that the reflection of a point w.r.t. a given affine set is a linear transformation, and as such it is exactly computable starting from a representation of the affine set. The details are beyond the scope of the present paper.

5.2 Computation of $Cross^+$

In Section 4 we have reduced the problem of computing the set $Cross^+(P, \hat{P}, P')$ to the problem of computing (the pre-flow of) $ClosCross(P, \hat{P}, P')$, with $\hat{P}$ a convex polyhedron and $\hat{P} \subseteq cl(P) \cap cl(P')$. In the rest of the section, we shall then assume that $\hat{P}$ is any convex polyhedron contained in the closures of P and P'.

The main difficulty we have to face in order to compute $ClosCross(P, \hat{P}, P')$ is to ensure that the points collected in the set can actually cross from P to P' following a "single" admissible straight direction in F. Clearly, if a point $y \in cl(P)$ can reach a point $x \in \hat{P}$ along a straight direction $c \in F$, i.e., $x = y + \delta c$ for some $\delta > 0$, then its point reflection w.r.t. x, namely $mirror(y, x)$, is reachable from y along the same admissible direction, indeed $mirror(y, x) = y + 2\delta c$. For instance, in Figure 5 if point y can reach point x along a straight direction, then it can also reach its point reflection y' (w.r.t. x) along the same direction.

The observation above motivates the definition:

$$MirCross(P, x, P') = cl(P) \cap P\nearrow \cap \{x\}\swarrow_{>0} \cap mirror(cl(P') \cap P'\swarrow, x).$$

The set $MirCross(P, x, P')$ collects all the points of the closure of P, reachable from P, that can reach, following a single straight trajectory passing through x, some point of the portion of the closure of P' that can, in turn, reach P'.

In addition, when $x \in cl(P) \cap cl(P')$, any point belonging to $MirCross(P, x, P')$ can reach, following a single straight trajectory, some point in $cl(P') \cap P'\swarrow$, never leaving $cl(P) \cup cl(P')$, as required by the definition of $ClosCross(P, \{x\}, P')$. As a consequence, a point belongs to $MirCross(P, x, P')$ if and only if it belongs to $ClosCross(P, \{x\}, P')$. This approach ensures that, when $\hat{P}$ contains a single point, we can compute the set of points in $ClosCross(P, \hat{P}, P')$ that cross from P to P' along a straight direction passing through $\hat{P}$. By generalizing the operator $MirCross$ to take an arbitrary convex polyhedron Q as second argument, we obtain the following:

$$MirCross(P, Q, P') = cl(P) \cap P\nearrow \cap Q\swarrow_{>0} \cap mirror(cl(P') \cap P'\swarrow, Q).$$

The set $MirCross(P, Q, P')$ is a convex polyhedron[1] containing points of the closure of P which: (*i*) are reachable (along a straight trajectory) from P, (*ii*) can reach a point

[1]It is easy to verify that $cl(P) \cap P\nearrow$ is a convex polyhedron even when $P\nearrow$ is not.

in Q in a positive amount of time along a straight direction, and *(iii)* have a reflection w.r.t. the affine hull of Q which lies in the portion of the closure of P' that can reach P'.

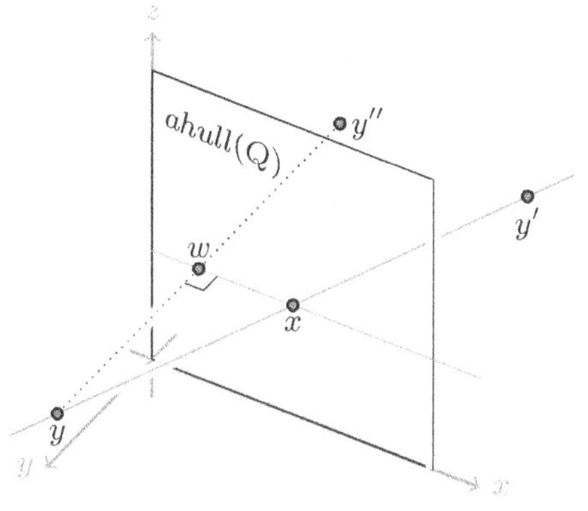

Figure 5: Relationship between the point reflection $y' = mirror(y, x)$ and the reflection $y'' = mirror(y, Q)$ w.r.t. Q. The affine hull of Q is drawn truncated.

It is important to notice that, differently from the case of a single point, the straight trajectory reaching Q from a point $x \in cl(P) \cap P\nearrow$, which is clearly admissible, need not be the same straight line connecting x with its reflection w.r.t. Q. Even worst, this straight line may not even be an admissible trajectory. Figure 5 shows that the point reflection y' of y w.r.t. x may be different from the reflection of y with respect to Q even when $x \in Q$.

The following result, however, ensures that, if Q is an open set w.r.t. its affine hull, namely when $Q = rint(Q)$, and $Q \subseteq cl(P) \cap cl(P')$, then for all $x \in Q$, any point in the set $MirCross(P, x, P')$ can reach some point in the set $MirCross(P, Q, P')$, and vice versa.

LEMMA 6. *For all convex polyhedra P, P', $Q \subseteq cl(P) \cap cl(P')$ such that $Q = rint(Q)$, points $x \in Q$, and directions $c \in F$, let D_Q (resp., D_x) be the set of delays $\delta > 0$ such that $x - \delta c \in MirCross(P, Q, P')$ (resp., $x - \delta c \in MirCross(P, x, P')$). Then, either $D_Q = D_x = \emptyset$ or both D_Q and D_x are non-empty intervals having the infimum 0.*

As a consequence of Lemma 6, whenever $\hat{P}$ is the relative interior of itself and is included in $cl(P) \cap cl(P')$, a point in the closure of P can reach the closure of P' following a single straight trajectory passing through $\hat{P}$ if and only if it can reach a point in $MirCross(P, \hat{P}, P')$. It is now easy to solve the problem for a general convex polyhedron $\hat{P}$ which is not a relative interior. It suffices to recursively decompose $\hat{P}$ into convex polyhedra Q, each of which is the relative interior of itself, and compute $MirCross(P, Q, P')$ for each such component. This leads to the following operator, called $RMirCross$ for *recursive*: let $RMirCross(P, \emptyset, P') = \emptyset$, and

$$RMirCross(P, \hat{P}, P') = MirCross(P, rint(\hat{P}), P') \cup \bigcup_{Q \in [\![\hat{P} \setminus rint(\hat{P})]\!]} RMirCross(P, Q, P').$$

The operator can easily be computed by a recursive algorithm, whose termination is guaranteed by noticing that for every $Q \in [\![\hat{P} \setminus rint(\hat{P})]\!]$, the affine dimension of Q is strictly lower than the affine dimension of $\hat{P}$, down to the case where Q contains a single point, in which case Q is the relative interior of itself and the recursion terminates.

As an example, Figure 6 depicts sets $RMirCross(P, \hat{P}, P')$, $ClosCross(P, \hat{P}, P')$ and $Cross^+(P, \hat{P}, P')$, assuming that the upper extreme of the segment labeled $\hat{P}$ actually belongs to $\hat{P}$, while the lower one does not. Dashed boundaries identify topologically open sides of polyhedra. It is easy to verify that no admissible differentiable trajectory can cross from a point along the dashed boundaries of $C_1 \cup C_2$ to P' passing through $\hat{P}$, hence they belong neither to $ClosCross(P, \hat{P}, P')$ nor to $Cross^+(P, \hat{P}, P')$. The reason why they do not belong to $RMirCross(P, \hat{P}, P')$ is that the operator considers the relative interior of $\hat{P}$ first, collecting C_1. Then, it recursively considers the upper extreme point of $\hat{P}$, say x. The resulting set $RMirCross(P, x, P')$ is empty, as none of the points of P that can reach x (namely, the points along the upper boundaries of C_1 and C_2 and below) has a point reflection w.r.t. x lying in the closure of P'.

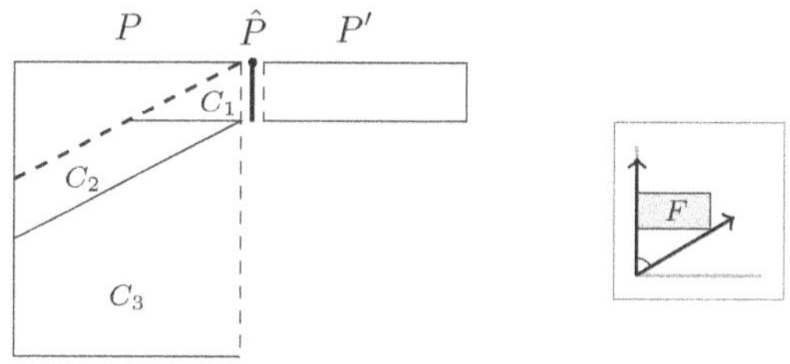

Figure 6: Difference between $ClosCross$ and $RMirCross$. We have $C_1 = RMirCross(P, \hat{P}, P')$, $C_1 \cup C_2 = ClosCross(P, \hat{P}, P')$, and $C_1 \cup C_2 \cup C_3 = Cross^+(P, \hat{P}, P')$.

The following lemma ensures that a point can reach the set $ClosCross(P, \hat{P}, P')$ if and only if it can reach the set $RMirCross(P, \hat{P}, P')$, allowing us to obtain the desired result:

LEMMA 7. *Let P, $\hat{P}$ and P' be convex polyhedra, with $\hat{P} \subseteq cl(P) \cap cl(P')$, then*

$$ClosCross(P, \hat{P}, P')\swarrow = RMirCross(P, \hat{P}, P')\swarrow .$$

PROOF. $(\supseteq)$ Let $x \in RMirCross(P, \hat{P}, P')\swarrow$. Then, there is a point $y \in RMirCross(P, \hat{P}, P')$, which is reachable from x. Hence, there must be a polyhedron $Q \subseteq \hat{P}$, with $Q = rint(Q)$ and $y \in MirCross(P, Q, P')$. By the definition of $MirCross(P, Q, P')$, y must belong to $cl(P)$, must be reachable from P, and must reach a point in $z \in Q$ along a straight direction $c \in F$ in a positive amount of time $\delta > 0$ (i.e., $z = y + \delta c$). Clearly, the set of delays D_Q of Lemma 6 contains at least δ, since $z - \delta c = y \in MirCross(P, Q, P')$. Therefore, Lemma 6 ensures that both D_Q and D_z are non-empty time intervals with infimum 0. Let δ_z be any time instant contained in D_z and $\delta' = \min\{\delta, \delta_z\} > 0$. Then, let $y' \triangleq z - \delta' c \in cl(P)$, it holds $y' \in MirCross(P, z, P')$. As a consequence, $y' \in cl(P) \cap P\nearrow$, it can reach z in a positive amount of time δ' along direction c, and in time $2\delta'$, along the same direction c, it can reach a point in $cl(P') \cap P'\swarrow$. This gives us a point $y' \in ClosCross(P, \hat{P}, P')$. Finally, we obtain the thesis $x \in ClosCross(P, \hat{P}, P')\swarrow$, by observing that, since x reaches y along a direction in F and y, in turn, reaches y' along some direction in F, also x reaches y' along a direction in F.

($\subseteq$) As to the other direction, let $x \in \mathit{ClosCross}(P, \hat{P}, P')\swarrow$. This means that x reaches a point $y \in cl(P) \cap P\nearrow$. In addition, there are two positive delays $0 < \delta_1 < \delta_2$ and a direction $c \in F$, such that $y + \delta c \in cl(P)$ for all $\delta \in (0, \delta_1)$, $y + \delta c \in cl(P')$ for all $\delta \in (\delta_1, \delta_2)$, $z \triangleq y + \delta_1 c \in \hat{P}$ and $y' \triangleq y + \delta_2 c \in cl(P) \cap P'\swarrow$.

Let $\delta' = \min\{\delta_1, \delta_2 - \delta_1\}$. Clearly, the point $z_1 \triangleq z - \delta' c \in cl(P)$ is reachable from y along direction c, hence from x as well. Similarly, the point $z_2 \triangleq z + \delta' c \in cl(P')$ can reach y' along direction c, hence also any point reachable from y' in P'. It is immediate to verify that $z_1 = \mathit{mirror}(z_2, z) \subseteq \mathit{mirror}(cl(P') \cap P'\swarrow, z)$. We, then, obtain that $z_1 \in \mathit{MirCross}(P, z, P')$. Since, however, $z \in \hat{P}$, there is a $Q \subseteq \hat{P}$, with $Q = \mathit{rint}(Q)$ and which contains z. Now, the application of Lemma 6 gives us a point z_1' (possibly different from z_1), such that $z_1' \in \mathit{MirCross}(P, Q, P') \subseteq \mathit{RMirCross}(P, \hat{P}, P')$ which is reachable from z_1 along direction c. Therefore, z_1 is reachable also from y, along the same direction. Once again, since x reaches y, which, in turn, reaches z_1', also x can reach z_1' and we obtain that $x \in \mathit{RMirCross}(P, \hat{P}, P')\swarrow$, as desired. ∎

In conclusion, for all convex polyhedra P, $\hat{P}$ and P', whenever $\hat{P} \subseteq cl(P) \cap cl(P')$ we can compute $\mathit{Cross}^+(P, \hat{P}, P')$ by means of the following:

$$\mathit{Cross}^+(P, \hat{P}, P') = P \cap \mathit{RMirCross}(P, \hat{P}, P')\swarrow .$$

6. FIXPOINT CHARACTERIZATION OF REACH-WHILE-AVOIDING

In this section, we show how to use the *Reach* and *Cross* operators to compute $\mathit{RWA}^{\mathcal{C}_\mathrm{d}}$. Analogously to the τ_{ae} operator, we define the following:

$$\tau_\mathrm{d}(G, A, W) = G \cup \bigcup_{P \in [\![\overline{A}]\!]} \mathit{Reach}(P, G) \cup \bigcup_{P \in [\![\overline{A}]\!]} \; \bigcup_{\hat{P}, P' \in [\![W]\!]} \mathit{Cross}(P, \hat{P}, P').$$

We prove that a finite number of repeated applications of $\tau_\mathrm{d}(G, A, W)$, starting from $W = \emptyset$, capture exactly all points in $\mathit{RWA}^{\mathcal{C}_\mathrm{d}}(G, A)$.

THEOREM 5. *For all convex polyhedra G and polyhedra A, such that $G \cap A = \emptyset$, we have*

$$\mathit{RWA}^{\mathcal{C}_\mathrm{d}}(G, A) = \mu W \,.\, \tau_\mathrm{d}(G, A, W). \qquad (2)$$

Moreover, the fixpoint is reached in a finite number of iterations.

In the following, let $W_0 = \emptyset$ and $W_{i+1} = \tau_\mathrm{d}(G, A, W_i)$, for all $i \geq 0$. In order to prove the above theorem, we proceed as follows. First, in Section 6.1 we prove that there is a normal form for witnesses, which in turn induces a partition of $\mathit{RWA}(G, B)$ into a finite number of classes $R_0, \ldots, R_k$. Then, in Section 6.2, we show that the points belonging to the class R_i are included in the $(i+1)$-th iteration of the fixpoint (2), i.e., to W_{i+1}. Finally, in Section 6.3 we prove that whenever W is a subset of $\mathit{RWA}(G, A)$, $\tau_\mathrm{d}(G, A, W)$ is also a subset of $\mathit{RWA}(G, A)$. Together, these results imply that W_{k+1} coincides with $\mathit{RWA}(G, B)$.

6.1 Canonicity

In this section, we introduce a normal form for witnesses. For a point $x \in \mathit{RWA}(G, A)$ and a witness (f, δ_f) for x, for all δ between 0 and δ_f, $f(\delta)$ lies in one of the convex polyhedra that constitute $\overline{A}$. In general, a witness can enter in and exit from the same polyhedron $P \in [\![\overline{A}]\!]$ an unbounded number of times. However, we prove that there are special witnesses, called *$\overline{A}$-canonical*, that do not spend a positive amount of time more than once in the same polyhedron.

Given a witness $\xi = (f, \delta_f)$ and a convex polyhedron P, let Δ_ξ^P be the set of delays $\delta \leq \delta_f$ such that f lies in P in an open interval around δ. Formally,

$$\Delta_\xi^P = \{0 \leq \delta \leq \delta_f \mid \exists \gamma > 0 \forall \delta' \in (\delta - \gamma, \delta + \gamma) : f(\delta') \in P\}.$$

We say that a witness $\xi = (f, \delta_f)$ for x is *P-canonical* if either $\Delta_\xi^P = \emptyset$ or $f(\delta) \in P$ for all $\delta \in (\inf \Delta_\xi^P, \sup \Delta_\xi^P)$. The definition implies that once a P-canonical witness spends a positive amount of time in P and then exits from it, it can only return to P for instantaneous visits (i.e., in isolated time points).

For a non-convex polyhedron B, we say that the witness ξ is B-canonical if it is P-canonical for all $P \in [\![B]\!]$.

LEMMA 8 (CANONICITY). *For all $x \in \mathit{RWA}(G, A)$ there exists an $\overline{A}$-canonical witness for x.*

If a witness $\xi = (f, \delta_f)$ lies entirely within a polyhedron B (i.e., $f(\delta) \in B$ for all $\delta \in [0, \delta_f]$), it induces a partition of the interval $[0, \delta_f]$ in a possibly infinite[2] sequence of intervals $I_0, I_1, \ldots, I_\alpha, \ldots$ such that f lies in the same patch $P_\alpha \in [\![B]\!]$ during each interval I_α and two subsequent polyhedra are distinct (i.e., we do not partition more often than necessary). We call the sequence of pairs (I_α, P_α) the *B-segmentation* of ξ. Notice that for all pairs of subsequent intervals $I_\alpha, I_{\alpha+1}$, at most one of them is singular (i.e., of length 0).

Consider the B-segmentation of a B-canonical witness. For all patches $P \in [\![B]\!]$, there is at most one non-singular time interval during which the witness lies in P. Since every other interval in the sequence must be non-singular, the segmentation can contain at most $2|[\![B]\!]| + 1$ intervals (i.e., $|[\![B]\!]|$ non-singular intervals interleaved with $|[\![B]\!]| + 1$ singular ones). Now, for a witness ξ define $\mathit{rank}(\xi)$ as the length (i.e., the order type) of its $\overline{A}$-segmentation, and $\mathit{rank}(x)$ the minimum rank of all witnesses for x. In light of Lemma 8 and the above discussion, we obtain the following result.

COROLLARY 1. *For all points $x \in \mathit{RWA}(G, A)$, it holds $\mathit{rank}(x) \leq 2|[\![\overline{A}]\!]| + 1$.*

The above corollary provides the main reason why our fixpoint procedure terminates. However, in order to prove that our fixpoint characterization is complete, we need more than $\overline{A}$-canonicity. We also need witnesses to be canonical w.r.t. the intermediate results of the computation, i.e., the sets W_i. Hence, we introduce the following definition and the corresponding lemma.

We say that a polyhedron B *refines* $\overline{A}$ if for all $P \in [\![B]\!]$ there exists $P' \in [\![\overline{A}]\!]$ such that $P \subseteq P'$. Notice that this implies $B \subseteq \overline{A}$.

LEMMA 9. *Let B be a refinement of $\overline{A}$. Then, for all $x \in \mathit{RWA}(G, A)$ there exists a B-canonical witness ξ such that $\mathit{rank}(\xi) = \mathit{rank}(x)$.*

[2]In fact, the order type of this sequence may be greater than ω, so that the sequence needs to be indexed by ordinals.

B-canonicity is a sort of regularity condition w.r.t. the internal boundaries of B (i.e., the boundaries between different patches of B). The following lemma formalizes this regularity.

LEMMA 10. *For all polyhedra B, if a witness $\xi = (f, \delta_f)$ is contained in B and it is B-canonical then for all $\delta \in [0, \delta_f]$ there exists a non-empty open time interval $(\delta, \delta + \gamma)$ during which f lies in the same polyhedron $P \in [\![B]\!]$.*

6.2 Completeness

We can now prove that all points in $RWA(G, A)$ will be eventually included in the fixed point (2).

THEOREM 6 (COMPLETENESS). *Let $x \in RWA(G, A)$ and $r = rank(x)$. If $r = 0$ then $x \in W_1$ and if $r > 0$ then $x \in W_r$.*

PROOF. First, we assume w.l.o.g. that G is one of the patches of $\overline{A}$. If this is not the case, modify $[\![\overline{A}]\!]$ as follows: replace each $P \in [\![\overline{A}]\!]$ with $[\![P \setminus G]\!]$; then, add G as an extra patch.

Let $x \in RWA(G, A)$. If $rank(x) = 0$ then there exists a witness for x that is entirely contained in $G \in [\![\overline{A}]\!]$. Then, by definition of τ_d we have $x \in \tau_d(G, A, \emptyset) = W_1$.

For the higher values of r we proceed by induction. If $rank(x) = 1$, there exists a witness $\xi = (f, \delta_f)$ for x that is entirely contained in two patches $P, G \in [\![\overline{A}]\!]$. More precisely, $f(\delta_f) \in G$ and $f(\delta) \in P \cup G$ for all $\delta \in [0, \delta_f]$. Hence, it holds $x \in Reach(P, G) \subseteq \tau_d(G, A, \emptyset) = W_1$.

Next, assume that $rank(x) = r + 1 > 1$. We apply Lemma 9 to x and W_r. This is legitimate, as W_r can be represented in a way that refines $\overline{A}$. Indeed, according to the definition of τ_d, W_r is the union of G and several other polyhedra, each of which is included in some $P \in [\![\overline{A}]\!]$.

Then, by Lemma 9 there exists a witness $\xi = (f, \delta_f)$ for x that is W_r-canonical and such that $rank(\xi) = r + 1$. Let $(I_i, P_i)_{i=0,\ldots,r+1}$ be the $\overline{A}$-segmentation of ξ. We distinguish the following cases.

Case1: $I_0 = [0, 0] = \{0\}$, i.e., f immediately leaves the first polyhedron P_0. In this case, for all $\delta \in I_1$ the rank of $f(\delta)$ is at most r because the suffix of f starting from δ is a witness for $f(\delta)$ and its rank is r. By inductive hypothesis, it holds $f(\delta) \in W_r$. Since ξ is W_r-canonical, by Lemma 10 f lingers in some $P' \in [\![W_r]\!]$. Hence, f proves that $x \in Cross^0(P_0, P') \subseteq \tau_d(G, A, W_r) = W_{r+1}$.

Case2: $I_0 \neq \{0\}$, i.e., f lingers in P_0. Let δ_0 be the right extreme of I_0, for all $\delta \in (\delta_0, \delta_f]$ it holds $rank(f(\delta)) \leq r$. By inductive hypothesis, $f(\delta) \in W_r$. By Lemma 10, applied to ξ and δ_0, there exists a non-empty time interval $(\delta_0, \delta_0 + \gamma)$ and a patch $P' \in [\![W_r]\!]$ such that $f(\delta) \in P'$ for all $\delta \in (\delta_0, \delta_0 + \gamma)$. If I_0 is right-closed, i.e., $I_0 = [0, \delta_0]$, then $f(\delta_0) \in P_0$. Then, f proves that $x \in Cross^+(P_0, P_0, P') \subseteq W_{r+1}$.

Otherwise, $rank(f(\delta_0)) \leq r - 1$ and, therefore, by inductive hypothesis $f(\delta_0) \in W_r$. In particular, let $\hat{P} \in [\![W_r]\!]$ be the patch such that $f(\delta_0) \in \hat{P}$. Then, f proves that $x \in Cross^+(P_0, \hat{P}, P') \subseteq W_{r+1}$. ∎

6.3 Correctness

The following auxiliary result states that from each point in $RWA(G, A)$ there is a witness trajectory that follows a straight line for a positive amount of time.

LEMMA 11. *For all $x \in RWA(G, A)$ there exist a witness (f, δ_f), a delay $\delta^* \leq \delta_f$, and a slope $c \in F$ such that $\dot{f}(\delta) = c$ for all $\delta \in [0, \delta^*)$.*

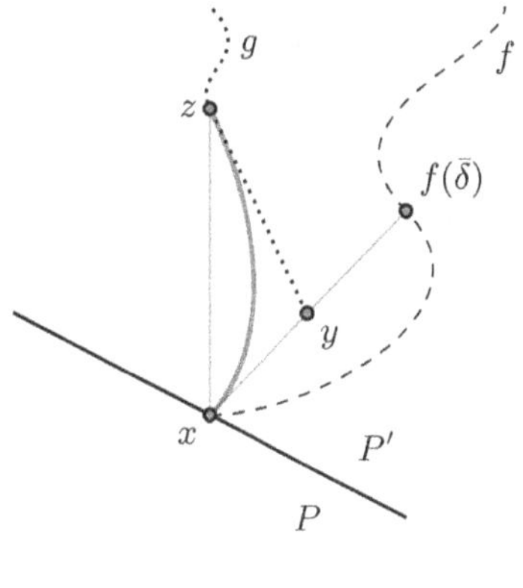

Figure 7: Case 1 of Lemma 7: connecting x to a witness that starts from y. Notice that point z need not be in P'.

With the result given by the lemma above, we are now ready to prove that the fix-point procedure is correct, namely that at any iteration of the procedure the operator τ_d only adds points which belong to $RWA(G, A)$.

THEOREM 7 (INCREMENTAL CORRECTNESS). *For all convex polyhedra G and polyhedra A and W, such that $G \cap A = \emptyset$ and $W \subseteq RWA(G, A)$, it holds $\tau_d(G, A, W) \subseteq RWA(G, A)$.*

PROOF. Let $x \in \tau_d(G, A, W)$. If $x \in G \cup \bigcup_{P \in [\![\overline{A}]\!]} Reach(P, G)$, it is easy to verify that either $x \in G$ or there is a straight-line trajectory that starts from x and reaches G without hitting A. Otherwise, it holds $x \in \bigcup_{P \in [\![\overline{A}]\!]} \bigcup_{\hat{P}, P' \in [\![W]\!]} Cross(P, \hat{P}, P')$. In particular, let $P \in [\![\overline{A}]\!]$ and $\hat{P}, P' \in [\![W]\!]$ be such that $x \in Cross(P, \hat{P}, P')$. We distinguish two cases based on the definition of $Cross$.

Case 1: $x \in Cross^0(P, P')$. This case is illustrated by Figure 7. According to the definition of $Cross^0$, let $f \in Adm(x)$ and $\bar{\delta} > 0$ be such that $f(\delta) \in P'$ for all $\delta \in (0, \bar{\delta}]$. Notice that f shows that $x \in cl(P')$, because in every neighborhood of x there is a point in P'.

By Lemma 1, there is a slope $c \in F$ and a delay $\gamma > 0$ such that $f(\bar{\delta}) = x + \gamma c$. Consider an arbitrary point x' of the type $x + \delta' c$, for $0 < \delta' \leq \gamma$. Being a convex combination of a point in $cl(P')$ and a point in P', different from x, x' belongs to P'. By Lemma 11, x' has a witness that starts with a straight line segment, and this witness must linger in one of the patches in $[\![\overline{A}]\!]$. Hence, there exists a delay $\hat{\delta} < \gamma$ and a patch $P'' \in [\![\overline{A}]\!]$ such that all points $x + \delta' c$, for $0 < \delta' \leq \hat{\delta}$, have a witness that starts with a straight line segment and lingers in P'' (it may be $P'' = P'$). Let $\xi = (g, \delta_g)$ be such a witness for the point $y = x + \hat{\delta} c$. By construction, there exists $\delta^* > 0$ such that $g(\delta') \in P''$ for all $0 < \delta' \leq \delta^*$ and moreover g follows a straight line segment of a given slope $c^* \in F$ at all times between 0 and δ^*. Let $z = g(\delta^*)$. By applying Lemma 4 to points $x_0 = x$, $x_1 = y$, and $x_2 = z$, we obtain a trajectory f' that starts in x and reaches z with final slope c^*. Moreover, each point of f' (except its extremes) is a strict convex combination of two points in $cl(P'')$ (namely, x and y) and a point in P'' (namely, z); hence it belongs to P''. Finally, f' can be connected to g at point z, giving rise to a witness for the fact that $x \in RWA(G, A)$.

Case 2: $x \in Cross^+(P, \hat{P}, P')$. This case is illustrated in

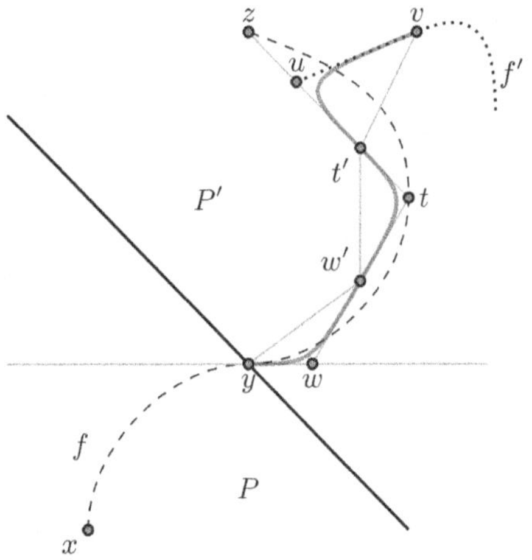

Figure 8: Case 2 of Lemma 7: connecting two witnesses (the dashed line starting from x and the dotted line starting from u) into a single one. Notice that point v need not be in P'.

Figure 8. Let $f \in Adm(x)$ and δ_1, δ_2 be the trajectory and the delays identified by the definition of $Cross_+(P, \hat{P}, P')$. Let $y = f(\delta_1)$, $z = f(\delta_2)$, and t be an arbitrary point between them (i.e., $t = f(\delta')$ for $\delta' \in (\delta_1, \delta_2)$). Let $c = \dot{f}(\delta_1)$, by applying Lemma 5 backwards from t, we obtain that there exists $\delta^* > 0$ such that $w \triangleq f(\delta_1) + \delta^* c \in cl(P') \cap \{t\} \swarrow_{>0}$. Let w' be an intermediate point between w and t.

Next, consider the points lying on the line segment between t and z. Clearly, they belong to P'. Hence, from each of them there is a witness that starts with a straight line segment (Lemma 11) and lingers in a patch in $[\![\overline{A}]\!]$. Let us pick a point u along the segment connecting t and z, in such a way that all points between u (included) and t (excluded) have a witness that lingers in the same patch $P'' \in [\![\overline{A}]\!]$ (it may be $P'' = P'$). Let f' be such a witness for the point u. In detail, f' starts with a straight line segment and lingers in P''. Formally, there exist a slope $c \in F$ and a delay δ' such that $f'(\delta) = u + \delta c$ for all $\delta \in [0, \delta']$. Let $v = f'(\delta')$. Let t' be an intermediate point on the line connecting t and u.

In order to connect f to f', we apply Lemma 4 three times. First, to points $x_0 = y$, $x_1 = w$, and $x_2 = w'$; then, to points $x_0 = w'$, $x_1 = t$, and $x_2 = t'$; finally, to points $x_0 = t'$, $x_1 = u$, and $x_2 = v$. In this way, we obtain three admissible trajectories f_1, f_2, f_3. The trajectory obtained by concatenating f, the three trajectories f_1, f_2, f_3, and f' is a witness for x. ∎

7. CONCLUSIONS

We considered the problem of computing the set of points that can reach a given polyhedron while avoiding another one via a differentiable trajectory that is subject to a polyhedral differential inclusion. We have shown that previous solutions do not guarantee differentiability of the trajectories. We provided a precise formulation of the problem, showing that significant developments are needed to obtain the correct solution. This enabled us to devise the first exact algorithm in the literature to solve the problem.

As future work, we plan to implement the proposed algorithm using a library for polyhedra manipulation. Additionally, it may be of interest to integrate our algorithm in frameworks that approximate complex dynamics with differential inclusions [9].

8. REFERENCES

[1] R. Alur, T. Henzinger, and P.-H. Ho. Automatic symbolic verification of embedded systems. In *RTSS 93: Real-Time Systems Symposium*, pages 2–11, 1993.

[2] R. Alur, T. Henzinger, and P.-H. Ho. Automatic symbolic verification of embedded systems. *IEEE Trans. Softw. Eng.*, 22:181–201, March 1996.

[3] R. Bagnara, P. M. Hill, and E. Zaffanella. Not necessarily closed convex polyhedra and the double description method. *Formal Aspects of Computing*, 17:222–257, 2005.

[4] A. Balluchi, L. Benvenuti, T. Villa, H. Wong-Toi, and A. Sangiovanni-Vincentelli. Controller synthesis for hybrid systems with a lower bound on event separation. *Int. J. of Control*, 76(12):1171–1200, 2003.

[5] M. Benerecetti, M. Faella, and S. Minopoli. Revisiting synthesis of switching controllers for linear hybrid systems. In *Proc. of the 50th IEEE Conf. on Decision and Control*. IEEE, 2011.

[6] M. Benerecetti, M. Faella, and S. Minopoli. Towards efficient exact synthesis for linear hybrid systems. In *GandALF 11: 2nd Int. Symp. on Games, Automata, Logics and Formal Verification*, volume 54 of *Electronic Proceedings in Theoretical Computer Science*, 2011.

[7] M. Benerecetti, M. Faella, and S. Minopoli. Automatic synthesis of switching controllers for linear hybrid systems: Safety control. *Theoretical Computer Science*, 2012. To appear.

[8] M. Benerecetti, M. Faella, and S. Minopoli. Reachability games for linear hybrid systems. In *HSCC 12: Hybrid Systems Computation and Control. 15th Int. Conf.*, pages 65–74. ACM, 2012.

[9] X. Briand and B. Jeannet. Combining control and data abstraction in the verification of hybrid systems. *IEEE Trans. on Computer-Aided Design of Integrated Circuits and Systems*, 29(10):1481 –1494, oct. 2010.

[10] N. Halbwachs, Y.-E. Proy, and P. Roumanoff. Verification of real-time systems using linear relation analysis. *Formal Methods in System Design*, 11:157–185, 1997.

[11] T. Henzinger, P.-H. Ho, and H. Wong-toi. Algorithmic analysis of nonlinear hybrid systems. *IEEE Transactions on Automatic Control*, 43:540–554, 1998.

[12] T. Henzinger, B. Horowitz, and R. Majumdar. Rectangular hybrid games. In *CONCUR 99: Concurrency Theory. 10th Int. Conf.*, volume 1664 of *Lect. Notes in Comp. Sci.*, pages 320–335. Springer, 1999.

[13] C. Tomlin, J. Lygeros, and S. Shankar Sastry. A game theoretic approach to controller design for hybrid systems. *Proc. of the IEEE*, 88(7):949–970, 2000.

Flowpipe Approximation and Clustering in Space-Time

Goran Frehse, Rajat Kateja
Université Grenoble 1 Joseph Fourier - Verimag
Centre Equation, 2 av. de Vignate
38610 Gières, France
goran.frehse@imag.fr

Colas Le Guernic
DGA-Maîtrise de l'Information
La Roche Marguerite route de Laillé BP 7
35998 Rennes, France
colas.le.guernic@gmail.com

ABSTRACT

In this paper, we present an approximation of the set of reachable states, called flowpipe, for a continuous system with affine dynamics. Our approach is based on a representation we call flowpipe sampling, which consists of a set of continuous, interval-valued functions over time. A flowpipe sampling attributes to each time point a polyhedral enclosure of the set of states reachable at that time point, and is capable of representing a nonconvex enclosure of a nonconvex flowpipe. The use of flowpipe samplings allows us to represent and approximate the nonconvex flowpipe efficiently. In particular, we can measure the error incurred by the initial approximation and by further processing such as simplification and convexification. A flowpipe sampling can be efficiently translated into a set of convex polyhedra in a way that minimizes the number of convex sets for a given error bound. When applying flowpipe approximation for the reachability of hybrid systems, a reduction in the number of convex sets spawned by each image computation can lead to drastic performance improvements.

Categories and Subject Descriptors

G.1.7 [**Numerical Analysis**]: Ordinary Differential Equations—*Initial value problems*

Keywords

Hybrid systems, verification, reachability, tools

1. INTRODUCTION

A widely used strategy for computing the reachable states of a continuous or hybrid system is to cover the flowpipes (bundles of trajectories in the state space) with a finite but frequently large number of convex sets, which can be represented, e.g., as polyhedra, zonotopes, ellipsoids and support functions [1, 2, 7, 8, 9, 4]. Each of the convex sets covers the flowpipe on a certain time interval, and the approximation error usually increases rapidly with the size of this interval.

Often, the time step has to be made very small to achieve a desired accuracy. This in turn may lead to a very large number of convex sets that, depending on the processing or further image computation to be performed, can quickly become prohibitive. For reachability analysis of hybrid systems in particular, we have often observed a fatal explosion in the number of sets when more than a few of the convex sets can take a discrete transition, as each will spawn a new flowpipe in the next state, and so on.

The goal of this paper is to address this fundamental problem from two angles: Firstly, we use a representation, which we call flowpipe sampling, that consists of a set of continuous, interval-valued functions over time. A flowpipe sampling attributes to each time point a polyhedral enclosure of the set of states reachable at that time point, thus capable of representing a nonconvex enclosure of a nonconvex flowpipe. This representation helps to decouple, as far as possible, the accuracy from the number of convex sets created in the end. Secondly, we propose a clustering procedure that aims to minimize the number of convex sets produced for a desired accuracy and does so using accuracy bounds established by the flowpipe construction. A-posteriori error measurements help evaluate the distance of the approximation to the actual flowpipe.

The following examples shall illustrate different aspects of the flowpipe approximation problem, as well as showcase the performance of our proposed solution.

EXAMPLE 1.1 (HELICOPTER). *Figure 1(a) show a flowpipe approximation for an affine helicopter model with 28 state variables plus a clock variable. It was obtained using the approach in [4], which constructs for each time-step a convex polyhedron in the 29-dimensional state space. The facet normals of the polyhedra, also called* template directions, *are given by the user. In this case, the axis directions are used so that the polyhedra are boxes. The complex dynamics of the system require using a very small time step throughout the time horizon of 30 s. Note that only the projection on two of the 29 variables is shown (the vertical speed and the clock), while the approximation takes the variation of all variables into account. As a result of the small time step, 1440 convex sets are constructed for a given directional error* estimate *of* $\epsilon = 0.025$. *The construction itself is computationally cheap at 5.9 s CPU time, but the sheer number of sets makes further processing and image computation impractical.*
The approach proposed in this paper combines an enhanced flowpipe approximation with adaptive clustering that guarantees a conservative error bound on the directional distance to

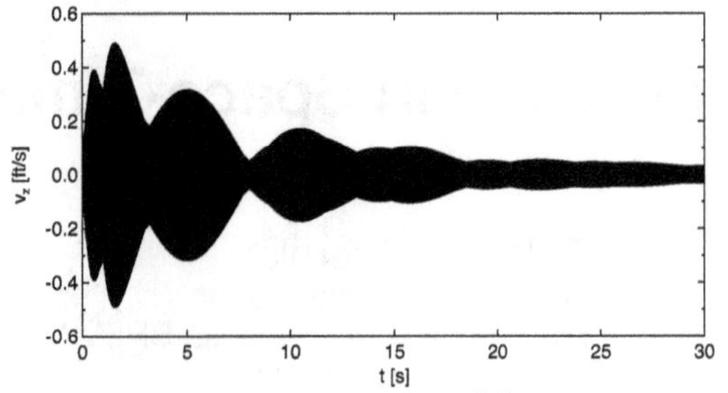

(a) The flowpipe approximation from [4] constructs one convex polyhedra per time-step, in total 1440 polyhedra

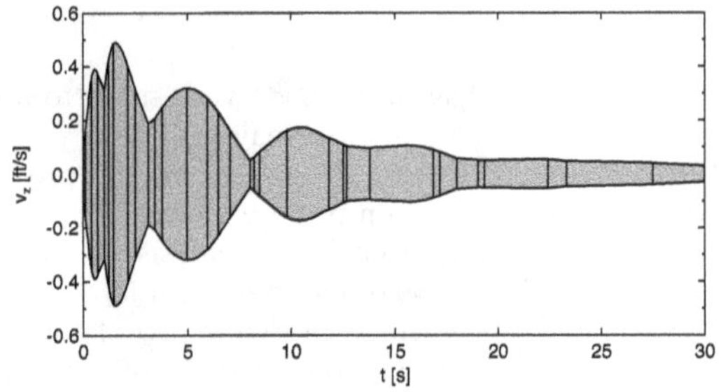

(b) The flowpipe approximation with clustering proposed in this paper constructs 32 convex polyhedra

Figure 1: Flowpipe approximations for the 28-dimensional affine helicopter model from Ex. 1.1, plus a clock. The shown sets are projections of 29-dimensional polyhedra onto two variables, the vertical speed and the clock

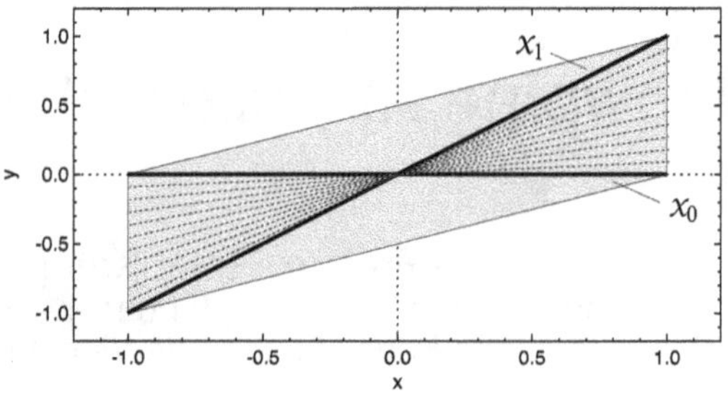

(a) The initial set $\mathcal{X}_0$, the final set $\mathcal{X}_1$, and the smallest convex approximation of the reachable set $\mathcal{X}_{[0,1]}$ (shaded)

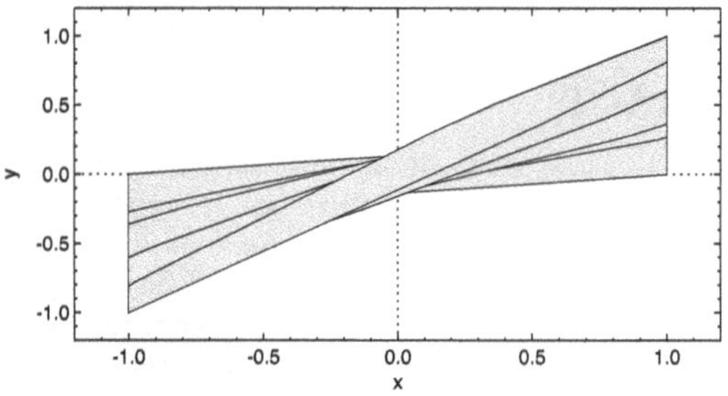

(b) An approximation of $\mathcal{X}_{[0,t]}$ constructed by the proposed technique for a given error bound 0.1

Figure 2: Example 1.2 has a nonconvex reachable set in the shape of an hourglass. It is pointwise convex in time, but every convex set covering a nonsingular interval is forcibly an overapproximation. All trajectories are linear, so the approximation error is impossible to detect based on the dynamics alone

the actual reachable set at each point in time. The time step is adapted separately for each of the template directions and can therefore be considerably larger. In the directions corresponding to the axis of the clock the system evolves linearly, so the time step spans the entire time horizon. The clustering step produces the 32 polyhedra shown in Fig. 1(b) for a given directional error bound *of* $\epsilon = 0.025$. *The construction of the flowpipe sampling takes 9.4 s and the clustering and outer polyhedral approximation 4.8 s.*

As the following example illustrates, the convexification error is by no means restricted to complex dynamics.

EXAMPLE 1.2 (HOURGLASS). *Consider the simple linear ODE system* $\dot{x} = 0$, $\dot{y} = x$, *with initial states* $\mathcal{X}_0 = \{-1 \le x(0) \le 1, y(0) = 0\}$ *as shown in Fig. 2(a). We consider the states reachable up to time* $t = 1$. *The reachable set is pointwise convex in time, but every convex set that covers the reachable states over a nonsingular time interval is forcibly an overapproximation. The time-step adaptation of the LGG-algorithm in [4] decreases the time step until all error terms fall below the desired threshold. In this case, the error terms are zero since the trajectories are linear, i.e.,* $x(t) = x(0), y(t) = t \cdot x(0)$. *The LGG-algorithm therefore covers the whole flowpipe in a single timestep. In the best case (arbitrarily well-chosen template directions), the result is the convex hull of the flowpipe, shown in Fig. 2(a).*

The flowpipe approximation proposed in this paper can produce a result of arbitrary accuracy in terms of a directional error that is measured in the template directions. As more template directions are added, this directional error converges towards the Hausdorff distance between the actual reachable states and the overapproximation. Figure 2(b) shows the result for a given error bound 0.1 *in 64 uniformly distributed template directions.*

The basis of our approach is the representation of a convex set by its support function. Simply put, given the normal vector of a halfspace (direction), the support function tells the position where the halfspace touches the set such that it contains it. If the dynamics are affine and the set of initial states is convex, the states reachable at a given point in time is also convex and can be represented by their support function. The flowpipe can therefore be described by a family of support functions parameterized by time, which has been considered in, e.g., [13, 10]. Our flowpipe sampling builds on this concept, which we refine by accounting for the fact that we can only compute a finite number of values in terms of both direction and time points. There are methods to approximate flowpipes directly with sets that are nonconvex, e.g., with polynomial tubes [12].

The flowpipe construction in this paper builds heavily on previous work in [9, 4]. Notably, this paper provides error measurements that include the convexification error. For more detailed comments, see Sect. 2.4. The clustering approach of Sect. 3 is, to the best of our knowledge, novel. For lack of space, we have omitted several proofs that can be found in [5]. The SpaceEx tool and examples from the paper are available at the SpaceEx website [3].

In the next section, we define flowpipe samplings as a general means to describe and approximate flowpipes and then present our flowpipe approximation algorithm. We also show how a flowpipe sampling can be translated into a set of convex polyhedra. In Sect. 3, we present our clustering approach, which minimizes the number convex polyhedra that a flowpipe sampling defines. Section 4 presents some experimental results.

2. FLOWPIPE APPROXIMATION IN SPACE-TIME

We consider continuous dynamical systems given by differential equations of the form

$$\dot{x}(t) = Ax(t) + u(t), \qquad u(t) \in \mathcal{U}, \tag{1}$$

where $x(t) \in \mathbb{R}^n$ and $A \in \mathbb{R}^n \times \mathbb{R}^n$. The closed, bounded, and convex set $\mathcal{U} \subseteq \mathbb{R}^n$ can represent, e.g., nondeterministic inputs, disturbances or approximation errors. The initial states of the system are given as a convex compact set $\mathcal{X}_0$, i.e., $x(0) \in \mathcal{X}_0$. We refer to the states reachable from $\mathcal{X}_0$ by time elapse as the *flowpipe* of $\mathcal{X}_0$. We compute a sequence of closed and bounded convex sets $\Omega_0, \dots, \Omega_N$ that covers the flowpipe using an extension of the approximation technique in [4]. In our construction, each Ω_i is the result of convex hull and Minkowski sum operations on polytopes. So Ω_i is itself a polytope, but explicitly computing it would be prohibitively expensive in higher dimensions. Its support function can, however, be computed efficiently for any given direction. In the next section we present how we approximate convex sets by computing their support function, and in the following section we extend the approach to flowpipes.

2.1 Approximating Convex Sets with Support Samples

A convex set can be represented by its support function, which attributes to each direction in $\mathbb{R}^n$ the signed distance of the farthest point of the set to the origin. Computing the value of the support function for a given set of directions, one obtains a polyhedron that overapproximates the set. We call this *sampling the support function*. In this paper, we allow this computation to be approximative, i.e., a lower and an upper bound on the support function is computed. In this section, we recall some basics, define support samples and derive error bounds for approximations from support samples.

We use the following standard notation for operations on sets. Let $\mathcal{X}, \mathcal{Y} \subseteq \mathbb{R}^n$ be sets, $\lambda \in \mathbb{R}$, and M be an $m \times n$ matrix of reals. We denote $\lambda\mathcal{X} = \{\lambda x \mid x \in \mathcal{X}\}$, $M\mathcal{X} = \{Mx \mid x \in \mathcal{X}\}$, and $\mathcal{X} \oplus \mathcal{Y} = \{x + y \mid x \in \mathcal{X}, y \in \mathcal{Y}\}$ (Minkowski sum). A *halfspace* $\mathcal{H} \subseteq \mathbb{R}^n$ is the set of points satisfying a linear constraint

$$\mathcal{H} = \{x \mid a^\mathsf{T} x \le b\},$$

where $a = (a_1 \cdots a_n) \in \mathbb{R}^n$ and $b \in \mathbb{R}$. A *polyhedron* $\mathcal{P} \subseteq \mathbb{R}^n$ is the intersection of a finite number of half spaces

$$\mathcal{P} = \Big\{x \;\Big|\; \bigwedge_{i=1}^{m} a_i^\mathsf{T} x \le b_i\Big\},$$

where $a_i \in \mathbb{R}^n$ and $b_i \in \mathbb{R}$. If such a_i and b_i are know, we say the polyhedron is given in *constraint representation.* A *polytope* is a polyhedron that is bounded. The *convex hull* $\mathrm{CH}(\mathcal{X}) \subseteq \mathbb{R}^n$ of a set $\mathcal{X}$ is

$$\mathrm{CH}(\mathcal{X}) = \Big\{\sum_{i=1}^{n+1} \lambda_i v_i \;\Big|\; v_i \in X, \lambda_i \in \mathbb{R}^{\ge 0}, \sum_{i=1}^{n+1} \lambda_i = 1\Big\}.$$

The *support function* of a nonempty, closed and bounded continuous set $\mathcal{X} \subseteq \mathbb{R}^n$ with respect to a direction vector $\ell \in \mathbb{R}^n$ is

$$\rho_\mathcal{X}(\ell) = \max\{\ell^\mathsf{T} x \mid x \in \mathcal{X}\}.$$

The set of *support vectors* (or *maximizers*) of $\mathcal{X}$ in direction ℓ is denoted by

$$\sigma_\mathcal{X}(\ell) = \{x^* \in \mathcal{X} \mid \ell^\mathsf{T} x^* = \rho_\mathcal{X}(\ell)\}.$$

We are interested in support functions because many set operations are computationally cheaper to carry out on support functions than on, say polyhedra [9]. For instance, the operations $M\mathcal{X}$, $\mathcal{X} \oplus \mathcal{Y}$, and $\mathrm{CH}(\mathcal{X} \cup \mathcal{Y})$ can be very expensive for polyhedra in constraint representation, while for support functions they are simple:

$$\begin{aligned} \rho_{M\mathcal{X}}(\ell) &= \rho_\mathcal{X}(M^T \ell), \\ \rho_{\mathcal{X} \oplus \mathcal{Y}}(\ell) &= \rho_\mathcal{X}(\ell) + \rho_\mathcal{Y}(\ell), \\ \rho_{\mathrm{CH}(\mathcal{X} \cup \mathcal{Y})}(\ell) &= \max\{\rho_\mathcal{X}(\ell), \rho_\mathcal{Y}(\ell)\}. \end{aligned}$$

Because support functions are cheap, we would like to use them in our flowpipe approximation. However, in our construction it is not always possible or efficient to compute the exact value of the support function. Instead, we allow for interval bounds on the support function. Furthermore, we consider that those bounds are only computed for a finite number of directions. In the following, we examine how such bounds provide an outer approximation of the actual set and characterize the approximation error. Given a set $\mathcal{X}$, a *support sample* $r = (\ell, [r^-, r^+])$ pairs a direction $\ell \in \mathbb{R}^n$ with a real-valued interval $[r^-, r^+]$ that contains the value of the support function of $\mathcal{X}$, i.e.,

$$\rho_\mathcal{X}(\ell) \in [r^-, r^+]. \tag{2}$$

A *support sampling* is a set of support samples

$$R = \{r_1, \dots, r_K\}, \quad \text{with} \quad r_k = (\ell_k, [r_k^-, r_k^+]).$$

Its *outer approximation* is the polyhedron

$$\lceil R \rceil = \Big\{x \;\Big|\; \bigwedge_k \ell_k^\mathsf{T} x \le r_k^+\Big\}, \tag{3}$$

i.e., given a support sampling R of $\mathcal{X}$, we have that $\mathcal{X} \subseteq \lceil R \rceil$.

From the lower bounds in the support samples we can derive a lower bound on the support function in any direction, which allows us to bound the approximation error of the outer approximation. By definition, a support sample r_k implies that there is at least one point $x \in \mathcal{X}$ such that $\ell_k^\mathsf{T} x \ge r_k^-$, as illustrated in Fig. 3. Let the *facet slab* of r_k be

$$\lfloor R \rfloor_k = \lceil R \rceil \cap \{\ell_k^\mathsf{T} x \ge r_k^-\}, \tag{4}$$

then the support function in direction ℓ cannot be lower than

$$\min\{\ell^\mathsf{T} x \mid x \in \lfloor R \rfloor_k\} = -\rho_{\lfloor R \rfloor_k}(-\ell).$$

Combining the lower bounds from all facet slabs, we obtain the following result:

LEMMA 2.1. *Given a support sampling R of a nonempty compact convex set $\mathcal{X}$, the support function of $\mathcal{X}$ is bounded in any direction ℓ by $\rho_R^-(\ell) \le \rho_\mathcal{X}(\ell) \le \rho_R^+(\ell)$, where*

$$\rho_R^+(\ell) = \rho_{\lceil R \rceil}(\ell), \tag{5}$$

$$\rho_R^-(\ell) = \max_{k=1,\dots,K} -\rho_{\lfloor R \rfloor_k}(-\ell). \tag{6}$$

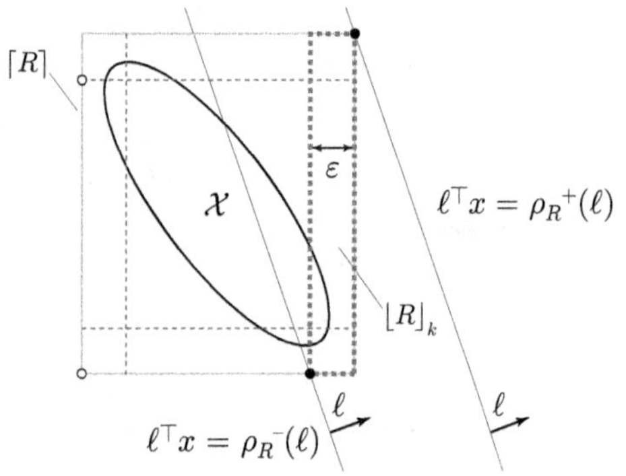

Figure 3: A support sampling R of a set $\mathcal{X}$ (solid black) defined over the axis directions, with lower and upper bound being ε apart. The outer approximation $\lceil R \rceil$ (solid grey) is shown together with the facet slabs $\lfloor R \rfloor_k$ (dashed), each of which contains at least one point of $\mathcal{X}$. For every direction ℓ, $\lceil R \rceil$ provides an upper bound $\rho_R^+(\ell)$ on the support function, and the facet slabs a lower bound $\rho_R^-(\ell)$

For a given direction ℓ, the lower bound (6) can be reformulated as a linear program with $O(Kn)$ variables and $O(K^2)$ constraints by introducing an additional variable z:

$$\rho_R^-(\ell) = \min\Big\{ z \in \mathbb{R} \Big| \bigwedge_{k=1,\ldots,K} z \geq \ell^\top x_k \wedge x_k \in \lfloor R \rfloor_k \Big\}. \quad (7)$$

We consider two ways to measure the error between the actual set and its outer approximation: a directional error and the Hausdorff distance. The *directional error* of a support sampling R is the width of the bounds on the support function,

$$\varepsilon_R(\ell) = \rho_R^+(\ell) - \rho_R^-(\ell). \quad (8)$$

Let $\mathcal{B}_k = \{x \mid \|x\|_k = 1\}$ be the unit ball in the k-norm. The *directed Hausdorff distance* between sets $\mathcal{X}, \mathcal{Y}$ is

$$\vec{d}_H(\mathcal{X}, \mathcal{Y}) = \sup_{x \in \mathcal{X}} \inf_{y \in \mathcal{Y}} \|x - y\|_2 = \inf\{\varepsilon > 0 \mid \mathcal{X} \subseteq \mathcal{Y} \oplus \varepsilon \mathcal{B}_2\},$$

and the *Hausdorff distance* is

$$d_H(\mathcal{X}, \mathcal{Y}) = \max\Big(\vec{d}_H(\mathcal{X}, \mathcal{Y}), \vec{d}_H(\mathcal{Y}, \mathcal{X})\Big).$$

LEMMA 2.2. *Given a support sampling R of $\mathcal{X}$,*

$$d_H(\mathcal{X}, \lceil R \rceil) \leq \max_{\|\ell\|_2 = 1} \varepsilon_R(\ell). \quad (9)$$

Using the LP formulation (7), the above bound on the Hausdorff distance can be rewritten as a quadratic maximization problem with bilinear constraints. This bound is generally quite costly to compute. But it implies that, as more and more directions are sampled, the largest directional error tends towards a bound on the Hausdorff distance (assuming directions are uniformly distributed).

2.2 Approximating Flowpipes with Support Samples over Time

Our space-time construction is a natural extension of the support function representation of sets. For a given convex and bounded set of initial states $\mathcal{X}_0$, we define the flowpipe as the states reachable from this set.

Formally, let $\mathcal{X}_t$ be the states reachable from $\mathcal{X}_0$ after exactly time t,

$$\mathcal{X}_t = \{x(t) \mid x(0) \in \mathcal{X}_0, \forall \tau \in [0,t]\ \exists u(\tau) \in \mathcal{U} : \dot{x}(\tau) = Ax(\tau) + u(\tau)\}. \quad (10)$$

A *flowpipe segment* over a time interval $[t_1, t_2]$ is the set

$$\mathcal{X}_{t_1,t_2} = \bigcup_{t_1 \leq t \leq t_2} \mathcal{X}_t.$$

In this paper, we assume a finite time horizon T and refer to $\mathcal{X}_{0,T}$ as the *flowpipe*. Given that $\mathcal{X}_0$ is convex and that the dynamics are affine, $\mathcal{X}_t$ is convex at any time t. For a fixed value of t, we can approximate $\mathcal{X}_t$ with a support sampling

$$R = \{(\ell_1, r_1), \ldots, (\ell_K, r_K)\},$$

where the ℓ_k given template directions, and the r_k are intervals containing the support function of $\mathcal{X}_t$. Recall that R allows us to construct an outer approximation of $\mathcal{X}_t$ and quantify the approximation error.

We describe the nonconvex flowpipe over the time interval $[0,T]$ in a similar way. Letting t vary in the time interval $[0,T]$, we consider the bounds of the interval $r_k(t) = [r_k^-(t), r_k^+(t)]$ to be continuous functions over time. For every t, $r_k(t)$ contains the support function of $\mathcal{X}_t$ in direction $\ell_k(t)$. A *flowpipe sampling* over K directions is a function F that attributes to each t a support sampling

$$F(t) = \{(\ell_1(t), r_1(t)), \ldots, (\ell_K(t), r_K(t))\}. \quad (11)$$

The pairs $(\ell_k(\cdot), r_k(\cdot))$ are called *flowpipe samples.* In this paper, we consider the directions to be constant over time, and simply write ℓ_k instead of $\ell_k(t)$. By combining the outer approximation of the support sampling $F(t)$ at each time point, we obtain an outer approximation of a flowpipe segment $\mathcal{X}_{t_1,t_2}$. With Lemma 2.2 it is straightforward to derive a bound on the Hausdorff distance between the flowpipe segment and its outer approximation.

LEMMA 2.3. *Let F be a flowpipe sampling (11) and let*

$$\lceil F \rceil_{t_1,t_2} = \bigcup_{t_1 \leq t \leq t_2} \lceil F(t) \rceil, \quad (12)$$

$$\varepsilon_{t_1,t_2} = \max_{t_1 \leq t \leq t_2} \max_{\|\ell\|_2 = 1} \varepsilon_{F(t)}(\ell). \quad (13)$$

Then $\mathcal{X}_{t_1,t_2} \subseteq \lceil F \rceil_{t_1,t_2}$ and the Hausdorff distance between $\lceil F \rceil_{t_1,t_2}$ and $\mathcal{X}_{t_1,t_2}$ is bounded by ε_{t_1,t_2}.

EXAMPLE 2.4. *In Ex. 1.2 (hourglass), the trajectories are $x(t) = x(0), y(t) = t \cdot x(0)$, with initial states $\mathcal{X}_0 = \{-1 \leq x(0) \leq 1, y(0) = 0\}$. The support function over time for a direction vector $\ell = \begin{pmatrix}\alpha & \beta\end{pmatrix}$ is*

$$\rho_{\mathcal{X}_0}(\ell) = \max_{x(0) \in \mathcal{X}_0} \begin{pmatrix}\alpha & \beta\end{pmatrix} \cdot \begin{pmatrix}x(0) & t \cdot x(0)\end{pmatrix} = \max(\alpha + \beta t, -\alpha - \beta t). \quad (14)$$

Let's assume that flowpipe samples have been computed for directions $\ell_1 = \begin{pmatrix}-1 & 4\end{pmatrix}$, $\ell_2 = \begin{pmatrix}-3 & 5\end{pmatrix}$, $\ell_3 = \begin{pmatrix}1 & 0\end{pmatrix}$, as well as their negatives. Assuming the computation is exact, the lower and upper bounds of the flowpipes are identical. The flowpipe samples $r_1(t)$, $r_2(t)$, $r_3(t)$ are shown in Fig. 4(a). The resulting outer approximation $\lceil F \rceil_{0,T}$ is shown in Fig. 4(b) .

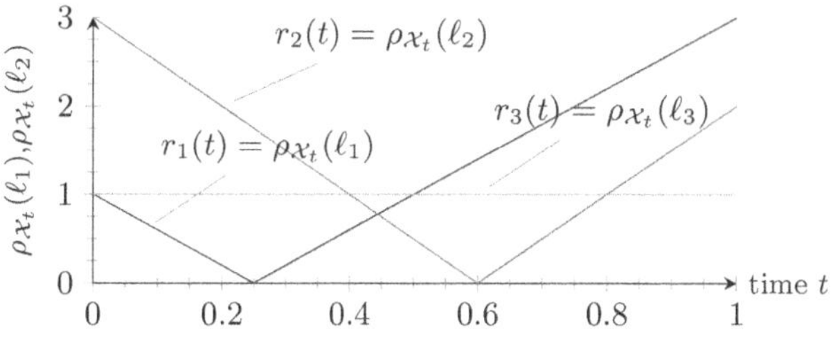

(a) Flowpipe samples for directions ℓ_1, ℓ_2, ℓ_3

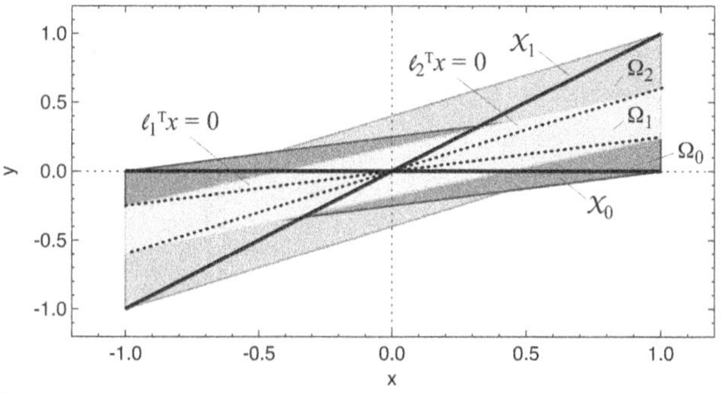

(b) Outer flowpipe approximation $\lceil F \rceil_{0,T}$ in the state space, consisting of convex polyhedra $\Omega_0, \Omega_1, \Omega_2$

Figure 4: Approximating the flowpipe of Ex. 1.2 for a given set of directions ℓ_1, ℓ_2, ℓ_3, and their negatives

2.3 Polyhedral Approximations of Flowpipes

A flowpipe sampling describes a flowpipe in the same way that a support sampling describes a convex set, only that the flowpipe and its outer approximation can be nonconvex. We now show that the outer approximation of a flowpipe sampling with piecewise linear upper bounds is a set of convex polyhedra, one for each segment for which all upper bounds are concave.

Let F be a flowpipe sampling (11), such that for all k, $r_k^+(t), r_k^-(t)$ are piecewise linear. For constructing the polyhedra, we need some technical notation for describing the linear pieces. Let the i-th pieces of $r_k^+(t)$, $r_k^-(t)$ be

$$\begin{aligned} r_k^+(t) &= \alpha_{k,i}^+ t + \beta_{k,i}^+ \text{ for } \tau_{k,i}^+ \le t \le \tau_{k,i+1}^+, \\ r_k^-(t) &= \alpha_{k,i}^- t + \beta_{k,i}^- \text{ for } \tau_{k,i}^- \le t \le \tau_{k,i+1}^-. \end{aligned}$$

In the following, we consider the time interval $[t_1, t_2]$ and will assume for simplicity that $r_k^+(t)$, $r_k^-(t)$ have breakpoints at the boundary of $[t_1, t_2]$, i.e., for some $i' < i''$, $t_1 = \tau_{k,i'}^+$ and $t_2 = \tau_{k,i''}^+$. Let I_k^+, I_k^- be the sets of indices i of the pieces of $r_k^+(t)$, respectively $r_k^-(t)$, that lie completely inside the time interval $[t_1, t_2]$.

If all computed flowpipe samples are concave over $[t_1, t_2]$, their outer approximation is convex:

LEMMA 2.5. *If for all k, $r_k^+(t)$ is concave on the time interval $[t_1, t_2]$, then $\lceil F \rceil_{t_1,t_2}$ is the convex polyhedron*

$$\begin{aligned} \lceil F \rceil_{t_1,t_2} &= M \, [\![F]\!]_{t_1,t_2}, \qquad \textit{where} \qquad (15) \\ [\![F]\!]_{t_1,t_2} &= \Big\{ (x,t) \;\Big|\; t_1 \le t \le t_2 \wedge \bigwedge_{k, i \in I_k^+} \ell_k^{\mathsf{T}} x \le \alpha_{k,i}^+ t + \beta_{k,i}^+ \Big\}, \end{aligned}$$

and the matrix M maps $(x,t) \in \mathbb{R}^{n+1}$ to $x \in \mathbb{R}^n$.

With Lemma 2.5, we can take a flowpipe sampling F and compute the support function of $\lceil F \rceil_{t_1,t_2}$ by solving a single LP with $O(n)$ variables and $O(KZ)$ constraints, where K is the number of template directions and Z is a bound on the number of pieces of the $r_k^+(t)$ in the time interval $[t_j, t_{j+1}]$.

With the above, we can construct a flowpipe approximation consisting of convex polyhedra $\Omega_0, \ldots, \Omega_N$ as follows:

1. Compute a piecewise linear flowpipe sample for each template direction.
2. Identify time intervals $[t_0, t_1], \ldots, [t_N, t_{N+1}]$, with $t_0 = 0$ and $t_{N+1} = T$, such that in each interval all samples have concave upper bounds.
3. Construct for each interval $[t_i, t_{i+1}]$ its convex polyhedron $\Omega_i = \lceil F \rceil_{t_i,t_{i+1}}$ using (15).

EXAMPLE 2.6. *The flowpipe samples of Ex. 2.9, shown in Fig. 4(a), are all concave on the time intervals $[0, 0.25]$, $[0.25, 0.6]$, and $[0.6, 1]$. The outer approximation of the flowpipe consists of three convex polyhedra $\Omega_0 = \lceil F \rceil_{0,0.25}$, $\Omega_1 = \lceil F \rceil_{0.25,0.6}$, and $\Omega_2 = \lceil F \rceil_{0.6,1}$, shown in Fig. 4(b). The facet normals of Ω_0 are ℓ_1, ℓ_3, those of Ω_2 are ℓ_2, ℓ_3, and those of Ω_3 are a linear combination of ℓ_1 and ℓ_2.*

The above approach produces a precise flowpipe approximation, but the number of polyhedra can be very large, especially if the concave intervals of the different flowpipe samples do not coincide. If an upper bound of a flowpipe samples is not concave on an interval, we can replace it by its concave envelope. The concave envelope of a piecewise linear function with N points, sorted along the time axis, can be computed in $O(N)$ with the Graham scan. The approximation error can be measured via the distance to the envelope. In Sect. 3, we will present a clustering technique that establishes the largest concave intervals that can be created by relaxing the upper bounds, under a desired error bound.

Note that a convex outer approximation does not imply that the flowpipe segment is convex. We now derive a bound on the approximation error by using the lower bounds of the flowpipe samples.

LEMMA 2.7. *Let for all k, $r_k^+(t)$ be concave and $r_k^-(t)$ be convex on the time interval $[t_1, t_2]$. Let the k-th* facet slab *of F be*

$$[\![F]\!]_{t_1,t_2}^k = \Big\{ (x_k, t) \in [\![F]\!]_{t_1,t_2} \;\Big|\; \bigwedge_{i \in I_k^-} \ell_k^{\mathsf{T}} x_k \ge \alpha_{k,i}^- t + \beta_{k,i}^- \Big\}, \quad (16)$$

Then the Hausdorff distance between $\lceil F \rceil_{t_1,t_2}$ and $\mathcal{X}_{t_1,t_2}$ is bounded by

$$\varepsilon_{t_1,t_2} = \max_{\|\ell\|_2 = 1} \varepsilon_{t_1,t_2}(\ell), \qquad \textit{where} \qquad (17)$$

$$\begin{aligned} \varepsilon_{t_1,t_2}(\ell) = \max \Big\{ & \ell^{\mathsf{T}} x - z \;\Big|\; (x,t) \in [\![F]\!]_{t_1,t_2} \wedge \\ & \bigwedge_{k=1,\ldots,K} z \ge \ell^{\mathsf{T}} x_k \wedge (x_k, t) \in [\![F]\!]_{t_1,t_2}^k \Big\} \qquad (18) \end{aligned}$$

For a given direction ℓ, (18) can be formulated as a linear program. Consequently, Lemma 2.7, allows us to compute a bound on the directional approximation error $\varepsilon_{t_1,t_2}(\ell)$ by solving a single LP with $O(Kn)$ variables and $O(K^2 Z)$ constraints. If we can solve the program for all ℓ, we obtain a bound on the Hausdorff distance of $\lceil F \rceil_{t_1,t_2}$ to the actual flowpipe segment.

2.4 Computing Flowpipe Samples for Affine Dynamics

We now present a way to construct flowpipe samples for affine dynamics of the form (1), i.e., an interval-valued function that bounds the support function of the reachable states at time t for a given direction. Our construction takes as input the initial set $\mathcal{X}_0$, a time horizon T, a template direction ℓ, and an error bound ϵ. It produces a flowpipe sample $(\ell, [r^-(t), r^+(t)])$, such that for all $0 \leq t \leq T$,

$$r^+(t) - r^-(t) \leq \epsilon.$$

The sample is piecewise quadratic and can easily be approximated by a piecewise linear sample so that the techniques of the previous section can be applied. The construction is based on the approach in [4], from which it differs in two ways: First, we include a lower bound on the support function, which is used to evaluate the approximation error at all stages including clustering. Second, instead of computing forward with a certain time-step, we start with a time step that covers the whole time horizon and recursively refine with smaller steps on subdomains where the difference between upper and lower bound exceeds the error bound.

We exploit the superposition principle to adapt the approximation separately to the autonomous dynamics (created by $\mathcal{X}_0$), and to the non-autonomous dynamics (created by $\mathcal{U}$). $\mathcal{X}_t$ can be decomposed into

$$\mathcal{X}_t = \mathcal{Z}_t \oplus \mathcal{Y}_t, \tag{19}$$

where $\mathcal{Z}_t = e^{At}\mathcal{X}_0$ and $\mathcal{Y}_t$ is the set of states reachable when starting from $x = 0$ instead of $\mathcal{X}_0$:

$$\mathcal{Y}_t = \{x(t) \mid x(0) = 0, \forall \tau \in [0,t]\ \exists u(\tau) \in \mathcal{U} : \dot{x}(\tau) = Ax(\tau) + u(\tau)\}. \tag{20}$$

Note that $\mathcal{Z}_0 = \mathcal{X}_0$ and $\mathcal{Y}_0 = 0$. We now turn to constructing a flowpipe sample $\omega(t) = [\omega^-(t), \omega^+(t)]$ for $\mathcal{Z}_t$. We need the following notation. The *symmetric interval hull* of a set S, denoted $\boxdot(S)$, is $\boxdot(S) = [-\overline{|x_1|}; \overline{|x_1|}] \times \ldots \times [-\overline{|x_d|}; \overline{|x_d|}]$ where for all i, $\overline{|x_i|} = \sup\{|x_i| \mid x \in S\}$. Let $M = (m_{i,j})$ be a matrix, and $v = (v_i)$ a vector. We define as $|M|$ and $|v|$ the absolute values of M and v respectively, i.e., $|M| = (|m_{i,j}|)$ and $|v| = (|v_i|)$. The approximation uses a transformation matrix Φ_2 defined as

$$\Phi_2(A, \delta) = \sum_{i=0}^{\infty} \frac{\delta^{i+2}}{(i+2)!} A^i, \tag{21}$$

which is computed similarly to a matrix exponential [4]. Our starting point is a linear interpolation between $\mathcal{Z}_0$ and $\mathcal{Z}_\delta$. Using a forward, respectively backward, interpolation leads to error terms represented by sets $\mathcal{E}_\Omega^+$,$\mathcal{E}_\Omega^-$. The intersection of both error terms gives $\mathcal{E}_\Omega$. Using a normalized time variable $\lambda = t/\delta$, let

$$\mathcal{E}_\Omega(\delta, \lambda) = \left(\lambda \mathcal{E}_\Omega^+(\delta) \cap (1-\lambda)\mathcal{E}_\Omega^-(\delta)\right)$$

$$\mathcal{E}_\Omega^+(\delta) = \boxdot\left(\Phi_2(|A|, \delta) \boxdot \left(A^2 \mathcal{X}_0\right)\right),$$

$$\mathcal{E}_\Omega^-(\delta) = \boxdot\left(\Phi_2(|A|, \delta) \boxdot \left(A^2 e^{A\delta} \mathcal{X}_0\right)\right).$$

The support function of $\mathcal{E}_\Omega(\delta, \lambda)$ is piecewise linear,

$$\rho(\ell, \mathcal{E}_\Omega(\delta, \lambda)) = \sum_{i=1}^{n} \min(\lambda e_i^+, (1-\lambda)e_i^-)|\ell_i|,$$

where vectors e^+ and e^- are such that $\rho(\ell, \mathcal{E}_\Omega^+) = |\ell|^\mathsf{T} e^+$ and $\rho(\ell, \mathcal{E}_\Omega^-) = |\ell|^\mathsf{T} e^-$.

An upper bound on the support function of $\mathcal{Z}_t$ over a time interval $[0, \delta]$ is easy to derive from the linear interpolation between $\mathcal{Z}_0$, $\mathcal{Z}_\delta$, and the above error terms [4]. For deriving a lower bound, consider a support vector x^- of $\mathcal{Z}_0 = \mathcal{X}_0$ in direction ℓ. Since the support function of $\mathcal{Z}_\delta = e^{A\delta}\mathcal{X}_0$ is the maximum of $\ell^\mathsf{T} x$ over all $x \in \mathcal{Z}_\delta$, it is bounded below by the image of x^- at time δ, i.e., by $\ell^\mathsf{T} e^{A\delta} x^-$. From the linear interpolation between x^- and $e^{A\delta}x^-$ we derive a lower bound by subtracting a suitable error term. A similar argument can be made with the support vector x^+ at the end of the interval, and we take the maximum of both lower bounds. Using the above error terms we obtain a flowpipe sample as follows.

LEMMA 2.8. *We consider the time interval $[t_i, t_{i+1}]$. Let $\delta_i = t_{i+1} - t_i$, $\ell_i = {e^{At_i}}^\mathsf{T}\ell$, $\ell_{i+1} = {e^{A\delta_i}}^\mathsf{T}\ell_i$, let x^- be a support vector of $\rho(\ell_i, \mathcal{X}_0)$, and x^+ be a support vector of $\rho(\ell_{i+1}, \mathcal{X}_0)$. Let $\lambda = (t - t_i)/\delta_i$ and*

$$\omega^+(t) = (1-\lambda)\rho_{\mathcal{X}_0}(\ell_i) + \lambda \rho_{\mathcal{X}_0}(\ell_{i+1}) + \rho_{\mathcal{E}_\Omega(\delta_i, \lambda)}(\ell_i) \tag{22}$$

$$\omega^-(t) = \max\{(1-\lambda)\ell_i^\mathsf{T} x^- + \lambda \ell_{i+1}^\mathsf{T} x^-, (1-\lambda)\ell_i^\mathsf{T} x^+ + \lambda \ell_{i+1}^\mathsf{T} x^+\} - \rho_{\mathcal{E}_\Omega(\delta_i, \lambda)}(\ell_i). \tag{23}$$

Then $\omega^-(t) \leq \rho_{\mathcal{Z}_t}(\ell) \leq \omega^+(t)$ for all $t_i \leq t \leq t_{i+\delta_i}$.

The approximation error of $\omega^-(t), \omega^+(t)$ in the time interval $[t_i, t_i + \delta_i]$ is

$$\varepsilon_\omega(t_i, t_{i+1}) = \max_{t_i \leq t \leq t_{i+1}} \omega^+(t) - \omega^-(t). \tag{24}$$

Note that the approximation error decreases at least linearly with δ_i. To meet the given error bound ϵ_ω, we construct $\omega^-(t), \omega^+(t)$ and the corresponding sequence of time points t_i by establishing a list of suitable intervals. We begin with a single interval $[t_0, t_1] = [0, T]$, which covers the entire time horizon. Each interval $[t_i, t_{i+1}]$ in the list is processed in the following steps:

1. Construct $\omega^-(t), \omega^+(t)$ on the interval $[t_i, t_{i+1}]$ and compute $\varepsilon_\omega(t_i, t_{i+1})$.
2. If $\varepsilon_\omega(t_i, t_{i+1}) > \epsilon_\omega$, split the interval in two. Let $t' = (t_i + t_{i+1})/2$. Replace $[t_i, t_{i+1}]$ with intervals $[t_i, t']$, $[t', t_{i+1}]$, and process each starting with step 1.

EXAMPLE 2.9. *Consider computing a flowpipe sample of Ex. 1.2 (hourglass) for direction ℓ_2 and up to an error bound of $\epsilon = 1$, as illustrated by Fig. 5. There are no inputs, so $r_2(t) = \omega(t)$. We start with the interval $[0, 1]$, which yields as upper bound the linear interpolation between $\omega^+(0) = 3$ and $\omega^+(1) = 2$, shown dashed in Fig. 5. In this example, the lower bound $\omega^-(t)$ happens to coincide with $r_2(t)$. The inital approximation error is $\varepsilon_\omega(0, 1) = 2.4$. Since this exceeds ϵ, the interval is split into two pieces, $[0, 0.5]$ and $[0.5, 1]$. The approximation errors are $\varepsilon_\omega(0, 0.5) = 0$ and $\varepsilon_\omega(0.5, 1) = 0.6$. They satisfy the error bound ϵ, and we obtain the upper bound $\omega^+(t)$ shown in Fig. 5.*

We now establish a flowpipe sample $\psi(t) = [\psi^-(t), \psi^+(t)]$ for $\mathcal{Y}_t$. Computing $\psi(t)$ is more difficult than computing $\omega(t)$ because there is no analytic solution for the integral over the input $u(t)$. Because of the integration, the approximation error accumulates over time. In order to guarantee that for all t the (accumulated) error of $\psi(t)$ is below a given bound

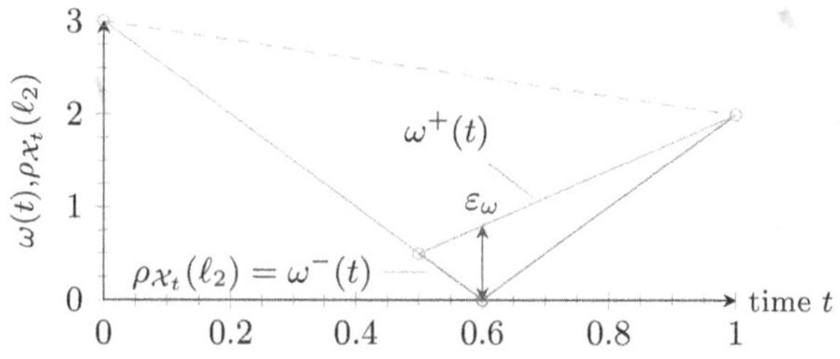

Figure 5: Computing a flowpipe sample of Ex. 1.2 for direction ℓ_2

ϵ_ψ, we impose that the accumulated error at the end of each interval $[t_i, t_{i+1}]$ must lie below the *error rate* $\epsilon_\psi \cdot t_{i+1}/T$. We use a two-step process: We first compute $\psi(t_i)$ at discrete points in time t_i such that the desired error rate is met. Based on these values we then define $\psi(t)$ over continuous time. To bound the approximation error we use the error term

$$\mathcal{E}_\Psi(\mathcal{U}, \delta) = \boxdot\left(\Phi_2(|A|, \delta) \boxdot (A\mathcal{U})\right).$$

For the following two lemmas, we assume the sequence of time points t_i as given. Its construction is presented afterwards, when the required error terms have been defined. We have the following bounds on $\rho_{\mathcal{Y}_t}(\ell)$ at the discrete time points t_i.

LEMMA 2.10. *[4] Let $t_0 = 0, t_1, t_2, \ldots, t_N = T$ be an increasing sequence of time points. Let $\delta_i = t_{i+1} - t_i$, $\ell_i = e^{At_i{}^\mathsf{T}}\ell$, $\ell_{i+1} = e^{A\delta_i{}^\mathsf{T}}\ell_i$, $\psi^+_{t_0} = 0, \psi^-_{t_0} = 0$, and*

$$\begin{aligned} \psi^+_{t_{i+1}} &= \psi^+_{t_i} + \delta_i \rho_\mathcal{U}(\ell_i) + \rho_{\mathcal{E}_\Psi(\mathcal{U},\delta_i)}(\ell_i) && (25)\\ \psi^-_{t_{i+1}} &= \psi^-_{t_i} + \delta_i \rho_\mathcal{U}(\ell_i) - \rho_{-A\Phi_2(A,\delta_i)\mathcal{U}}(\ell_i). && (26)\end{aligned}$$

Then for all t_i, $\psi^-_{t_i} \le \rho_{\mathcal{Y}_{t_i}}(\ell) \le \psi^+_{t_i}$.

Based on the bounds on $\rho_{\mathcal{Y}_t}(\ell)$ at the discrete times t_i, we obtain the following bounds over the intervals $[t_i, t_{i+1}]$.

LEMMA 2.11. *Let $\lambda = (t - t_i)/\delta_i$ and*

$$\begin{aligned} \psi^+(t) =& \psi^+_{t_i} + \lambda \delta_i \rho_\mathcal{U}(\ell_i) + \lambda^2 \rho_{\mathcal{E}_\Psi(\mathcal{U},\delta_i)}(\ell_i), && (27)\\ \psi^-(t) =& \psi^-_{t_i} + \lambda \delta_i \rho_\mathcal{U}(\ell_i) - \lambda \rho_{-A\Phi_2(A,\delta_i)\mathcal{U}}(\ell_i) \\ & - \lambda^2 \rho_{\mathcal{E}_\Psi(\mathcal{U},\delta_i)}(\ell_i). && (28)\end{aligned}$$

Then $\psi^-(t) \le \rho_{\mathcal{Y}_t}(\ell) \le \psi^+(t)$ for all $t_i \le t \le t_{i+\delta_i}$.

We construct the time intervals $[t_i, t_{i+1}]$ by refinement until the error $\psi(t)$ falls below ϵ_ψ. According to the above Lemmas, the error bound on the interval $[t_i, t_{i+1}]$ is defined by the following sequence, starting with $\varepsilon_\psi(t_0) = 0$:

$$\begin{aligned} \varepsilon_\psi(t_i, t_{i+1}) =& \varepsilon_\psi(t_i) + \max_{0 \le \lambda \le 1} 2\lambda^2 \rho_{\mathcal{E}_\Psi(\mathcal{U},\delta_i)}(\ell_i) \\ & + \lambda \rho_{-A\Phi_2(A,\delta_i)\mathcal{U}}(\ell_i), \qquad \text{where} && (29)\\ \varepsilon_\psi(t_{i+1}) =& \varepsilon_\psi(t_i) + \rho_{\mathcal{E}_\Psi(\mathcal{U},\delta_i)}(\ell_i) + \rho_{-A\Phi_2(A,\delta_i)\mathcal{U}}(\ell_i). && (30)\end{aligned}$$

When choosing the time points t_i, we must ensure that largest error in the interval $[t_i, t_{i+1}]$ lies below the error bound, i.e., $\varepsilon_\psi(t_i, t_{i+1}) \le \epsilon_\psi$. To take into account that the error $\varepsilon_\psi(t_i)$ accumulates, we also ensure that the rate of the accumulated error stays below ϵ_ψ/T, i.e., $\varepsilon_\psi(t_{i+1}) \le \epsilon_\psi \cdot t_{i+1}/T$. Beginning with a single interval $[t_0, t_1] = [0, T]$, each interval $[t_i, t_{i+1}]$ is processed in the following steps:

1. Compute $\varepsilon_\psi(t_{i+1})$ and $\varepsilon_\psi(t_i, t_{i+1})$.
2. If $\varepsilon_\psi(t_{i+1}) > \epsilon_\psi \cdot t_{i+1}/T$ or $\varepsilon_\psi(t_i, t_{i+1}) > \epsilon_\psi$, split the interval in two. Let $t' = (t_i + t_{i+1})/2$. Replace $[t_i, t_{i+1}]$ with intervals $[t_i, t']$, $[t', t_{i+1}]$, and process each starting with step 1.

Using the superposition principle (19), we finally combine $\omega(t)$ and $\psi(t)$ to obtain a flowpipe sample for $\mathcal{X}_t$ as

$$r(t) = [r^-(t), r^+(t)] = \omega(t) + \psi(t).$$

The error bound on $r(t)$ is below $\epsilon = \epsilon_\omega + \epsilon_\psi$.

3. CLUSTERING IN SPACE-TIME

Given a flowpipe sampling, our goal is to construct a sequence of convex sets that cover the flowpipe and that are no further than a given distance ϵ from it. For computational efficiency, our distance measure is the directional error in each of the sampled directions, but this implies also a distance in the Hausdorff sense.

We now given an informal description of our clustering algorithm, deferring a formal discussion to the subsections that follow. The algorithm takes as input a flowpipe sampling $F = \{(\ell_1, r_1(t)), \ldots, (\ell_K, r_K(t))\}$ and an error bound ϵ, and produces a flowpipe sampling F' by replacing the upper bounds $r_k^+(t)$ with a piecewise concave envelope with as few pieces as possible for the given error bound. The basic principle is to (over)approximate the upper bounds $r_i^+(t)$ of the flowpipe samples with a set of piecewise concave hulls $y_i(t)$, which are constructed such that they are concave over the same pieces. Recalling from in Sect. 2.3 that an outer approximation in the form of a convex polyhedron can be constructed for each concave piece, this effectively reduces the number of convex sets.

Let $\rho_i(t) = \rho_{\mathcal{X}_t}(\ell_i)$ be the actual value of the support function over time. By definition, $r_i^-(t) \le \rho_i(t) \le r_i^+(t)$. The goal of our clustering is to produce a piecewise concave hull $y_i(t)$ that is no farther than ϵ away from the actual value $\rho_i(t)$, i.e., such that $\rho_i(t) \le y_i(t) \le \rho_i(t) + \epsilon$. Since only $r_i^-(t)$ and $r_i^+(t)$ are known, we must construct the $y_i(t)$ such that

$$r_i^+(t) \le y_i(t) \le r_i^-(t) + \epsilon. \qquad (31)$$

Finding the minimal number of concave pieces for a function between a lower and an upper bound is possible by establishing the *inflection intervals* of $(r_i^+(t), r_i^-(t) + \epsilon)$, which will be presented in Sect. 3.1. The set of inflection intervals has the following property: Any piecewise concave function between $r_i^+(t)$ and $r_i^-(t) + \epsilon$ has at least one inflection point inside every inflection interval. The number of inflection intervals is thus equal to the minimum number of concave pieces of any $y_i(t)$. To have the minimum number of concave pieces, we must find the minimum number of points such that there is at least one point in every inflection interval of every sample. This turns out to be a graph coloring problem that is described in Sect. 3.2. The clustering step itself terminates with the construction of a piecewise concave hull of the flowpipe samples with a minimum number of pieces. Convex polyhedra can be derived from the concave pieces as previously described in Sect. 2.3.

If the clustering results in a number of concave pieces that is still considered too high, one can try to reduce the number further by recomputing the flowpipe samples with higher accuracy. This brings the $r_i^-(t)$ and $r_i^+(t)$ closer together,

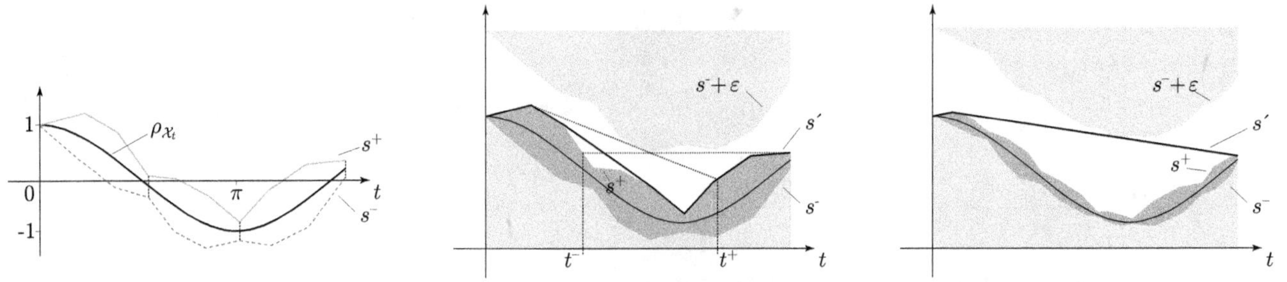

(a) Let $\mathcal{X}_t$ be a unit circle in the x/y-plane, with $\rho_{\mathcal{X}_t}(\ell) = cos(t)$ if ℓ is the x-direction. A flowpipe sample $(\ell, s^-(t), s^+(t))$ encloses this function

(b) Any piecewise linear s' between $s^+(t)$ and $s^-(t)+\varepsilon$ has at least two concave pieces, separated by an inflection point between t^- and t^+

(c) Refining s^-, s^+ leaves more slack between s^+ and $s^- + \varepsilon$, and enables an approximation s' with a single concave piece

Figure 6: Given a flowpipe sample with bounds $s^-(t), s^+(t)$ and a desired approximation error ε, we construct a piecewise concave function $s'(t)$ that lies between $s^+(t)$ and $s^-(t) + \varepsilon$. Fewer concave pieces in $s'(t)$ mean fewer convex sets produced by the clustering

which increases the slack in the bounds (31) used for clustering. As illustrated by Fig. 6, the new bounds may be wide enough to admit a fewer pieces. We can obtain a lower bound on the number of pieces by computing the number of inflection intervals for the bounds

$$r_i^-(t) \leq y_i(t) \leq r_i^+(t) + \epsilon. \qquad (32)$$

3.1 Inflection Intervals of a Flowpipe Sample

Let $l(t)$, $u(t)$ be a pair of piecewise linear functions with domain $[0, T]$ such that $l(t) \leq u(t)$. An *inflection interval* is an interval over t that contains at least one inflection point of any piecewise concave function $y(t)$ lying on or between $l(t)$ and $u(t)$, and no points that are not inflection points of a piecewise concave $y(t)$ with a minimum number of pieces. As a consequence of this definition, the minimum number of pieces of any $y(t)$ is equal to the number of inflection intervals of $(l(t), u(t))$.

Let the breakpoints of $l(t)$ be l_0 to l_N. We propose the following algorithm for finding inflection intervals (for simplicity we omit some special cases), see Fig. 7 for an illustration:

1. Perform a greedy piecewise concave minimal function construction from l_0 to l_N: choose at each step the point on the lower bound farthest towards l_N that is still visible.
2. Denote the breakpoints where the function is convex by $b_0, \dots, b_z$.
3. Perform a greedy piecewise concave minimal function construction in reversed direction, from l_N to l_0.
4. Denote the breakpoints where the function is convex by $a_z, \dots, a_0$.
5. Return the inflection intervals $I_0 = [a_0, b_0], \dots, I_z = [a_z, b_z]$.

Proposition 3.1. *Given intervals $I_0, \dots, I_z$ returned by the above algorithm, there exists a piecewise concave function between $l(t)$ and $u(t)$ with $z+1$ inflection points, one in every interval I_i. There exists no piecewise concave function between $l(t)$ and $u(t)$ with less than $z+1$ inflection points.*

Proof. Let us consider one of the b_i, let us denote it b for simplicity, and l_b the previous vertex in the piecewise concave piecewise linear function. l_b is on $l(t)$ (but not necessarily one of the l_i) otherwise the function would not be minimal. b is on a segment $]l_i, l_{i+1}[$ and there is a point $u' \in u(t)$ on the segment $[l_b, b]$ otherwise the function would not have been greedily constructed. Let us take x in $]b, l_{i+1}[$, any concave function on $[b, x]$ above $l(t)$ and below $u(t)$ must be above l_b and x and below u' which is not possible since u' is below $[l_b, x]$. Thus any piecewise concave function must contain at least one inflection point on $]l_b, x[$, and thus at least one on $]l_b, b]$, and one on each $]l_{b_i}, b_i]$. Since the intervals are disjoint, the greedy algorithm reaches a minimum number of inflexion point. Similarly any piecewise concave function must contain at least one inflection point on each $[a_i, l_{a_i}[$. □

3.2 Combining Inflection Intervals

Having established the inflection intervals for each flowpipe sample, we combine them to find the minimum number of inflection points, as well as their possible positions, for our piecewise cover of all samples. Recall that the pieces of the piecewise cover we seek are common to all samples. We therefore need to pick at least one point from every inflection interval of every sample. To minimize their number, we construct their common sub-intervals, which we call *overlap intervals*.

For each function we have a (possibly empty) set of inflection intervals I_i obtained using the algorithm of Sect. 3.1. For the following it is not relevant that the inflection intervals originate from different functions, so let $I_0, I_1, \dots, I_z$ simply be the set of all inflection intervals. We need to partition the intervals into groups inside which all intervals overlap. The output of the algorithm consists of the groups and for each group their common *overlap intervals* J_j.

Finding maximal groups of overlapping intervals is equivalent to a coloring problem. Each color j corresponds to one of the groups and defines an overlap interval, which consists of the overlap between all members of the group. Two intervals I_1, I_2 need to be colored differently if they do not overlap, i.e., if $I_1^+ < I_2^-$ or $I_2^+ < I_1^-$. This relationship is captured by the *comparability graph*, whose vertices are the intervals I_i. Its edges are given by $I_i \rightarrow I_j \Leftrightarrow I_i^+ < I_j^-$, which is the so-called interval ordering (a strict partial order). Our problem is equivalent to finding a coloring of the

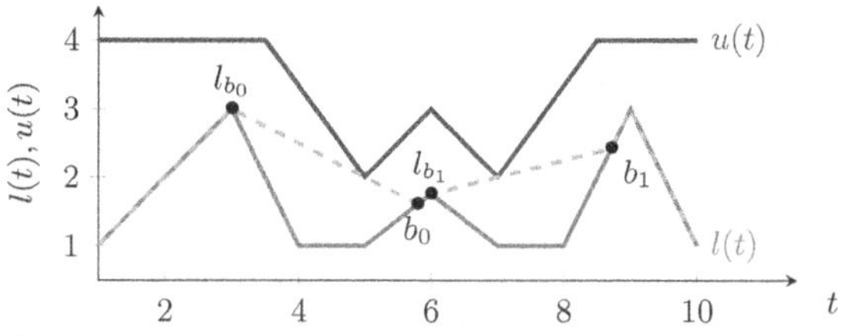

(a) Greedy scan of visible lower bound points from l_0

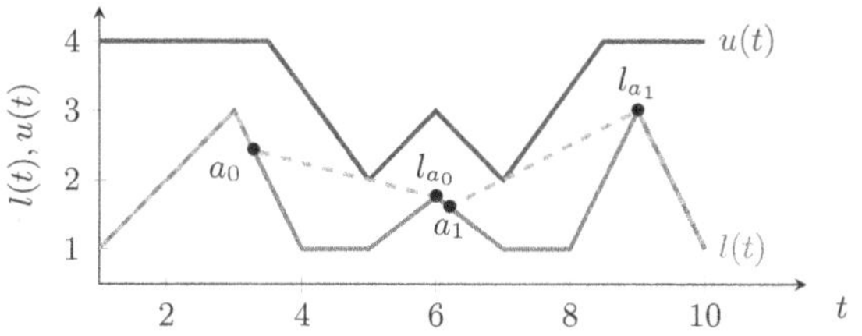

(b) Greedy scan of visible lower bound points from l_N

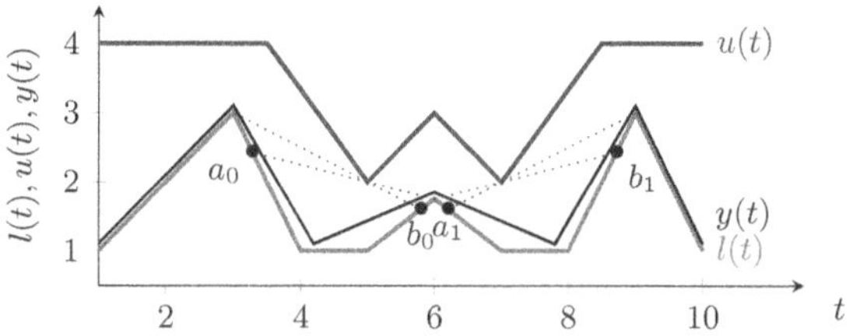

(c) Resulting inflection intervals $I_0 = [a_0, b_0], I_1 = [a_1, b_1]$, and a piecewise concave function $y(t)$ with the minimum number of pieces

Figure 7: Finding the set of inflection intervals, inside each of which any piecewise concave function between $l(t)$ and $u(t)$ has at least one inflection point

comparability graph with the smallest number of colors such that no two adjacent vertices have the same color. Once the intervals have been colored, each color corresponds to a set of intervals that all overlap. We may freely choose an inflection point from inside the common region for that color.

It is known that the interval ordering is a perfect elimination ordering of the comparability graph of a set of intervals. Consequently, a greedy coloring algorithm produces the optimal result if it chooses the vertices in an order that satisfies the interval ordering [11]. Such an order of the vertices can be obtained by a topological sort, i.e., a depth-first search in the graph. The total complexity is determined by the size of the comparability graph and therefore $O(z^2)$, where z is the number of intervals.

Choosing Inflection Points.

Our final set of inflection points consists of one point from each overlap interval J_j. The approximation error of this choice can be measured as the distance of the lower bounds to the resulting piecewise concave functions. The choice in one inflection interval generally influences the approximation error of its neighboring intervals as well, so the optimal choice is a multivariate optimization problem. In our experiments, we have observed that the overlap between the intervals of different directions is usually small, and that choosing inflection points in the middle of each interval yields results that are close to the local optimum.

4. EXPERIMENTAL RESULTS

In this section, we present experiments with the proposed flowpipe approximation and clustering algorithms and compare it with the approximation from [4]. These algorithms are intended to be used for reachability of hybrid systems in the verification tool SpaceEx [4]. There, each flowpipe approximation is followed by the computation of the image of all enabled discrete transitions. This image computation involves intersecting the flowpipe with the invariants of source and target locations, as well as the transition guard, which can be carried out efficiently on polyhedra. The final result of a flowpipe approximation therefore the polyhedral outer approximation as described in Sect. 2.3. Note that other variants of the reachability algorithm avoid polyhedra, e.g., by carrying out the intersection on the support function level through transformation into an optimiziation problem [9, 6].

We compare the following variants:

- LGG (state-space approximation without clustering) variable time-step flowpipe construction in the state-space, then outer polyhedral approximation, both as in [4],
- STA (space-time approximation with all pieces) flowpipe construction as in Sect. 2, then outer polyhedral approximation of all pieces as in Sect. 2.3 (no clustering),
- STC (space-time approximation with clustering) flowpipe construction as in Sect. 2 and clustering as in Sect. 3, then outer polyhedral approximation as in Sect. 2.3.

Note that the STA/STC implementation is still a prototype, and we expect that memory consumption and clustering runtime can be reduced. The parameter settings are not entirely comparable between LGG and STA/STC, since the error bounds in STA/STC are conservative, while in LGG they are mere estimates that do not take the nonconvexity error in account. The error bound in STC measures the total error, including both flowpipe approximation and clustering. We choose that 80% of the error can be taken up by the flowpipe approximation, so that at least 20% of the error bound remain as slack for the clustering step.

To avoid a lengthy description of the models, they are available for download on the SpaceEx website [3]. For illustration, consider a ball in free-fall together with a clock, with 3 variables x, v, t, dynamics $\dot{x} = v$, $\dot{v} = -1$, $\dot{t} = 1$, and initial states $10 \leq x \leq 10.2$, $v = t = 0$. We construct the flowpipe until x falls below 0. The axis directions are used as template directions, so LGG creates bounding boxes of flowpipe segments in the state space. The flowpipe approximation of STA/STC creates a bounding box for each point in time, which projected to the state space yields polyhedra with facet normals other than the template directions, as Fig. 8 illustrates. The error bound $\epsilon = 1$ is fairly large, and the clustering step in STC uses the slack to reduce the number of sets from 8 sets (STA) to 2 sets.

Table 1 shows performance results obtained on a laptop with i7 processor and 8 GB RAM. All examples use the axis directions as template directions. For each algorithm, the table shows the time taken for flowpipe approximation, the time taken for clustering and constructing the polyhedral approximation, and the memory consumption. As an implementation-independent indicator of the computational cost, it shows the total number of times the support function of the initial set has been evaluated. The key column is the total number of convex sets covering the flowpipe, and the

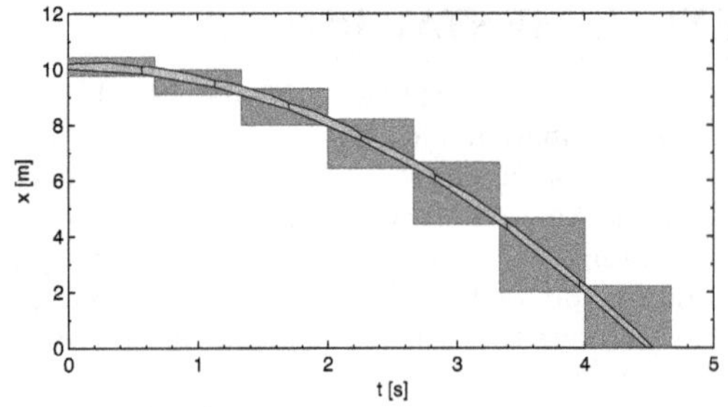

(a) LGG (blue) versus STA (red)

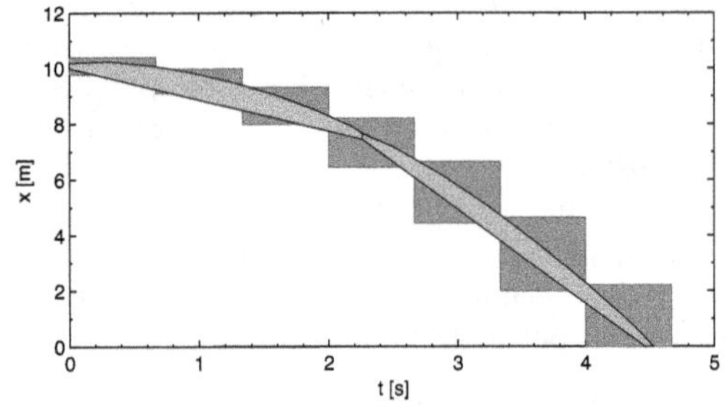

(b) LGG (blue) versus STC (red)

Figure 8: Flowpipe approximation of a ball in free fall (position over time), with axis directions as template directions and a directional error bound $\epsilon = 1$

Table 1: Performance results for different examples

Algo	*Fl.T* [s]	*Cl.T.* [s]	Mem. [MB]	#eval	#sets
Helicopter with controller & clock, 29 variables, $\epsilon = 0.025$					
LGG	5.9	–	14	85144	1440
STA	9.0	3.6	2390	103726	2649
STC	9.4	4.8	203	103726	32
Helicopter with diff. controller & clock, 29 variables, $\epsilon = 0.025$					
LGG	10.4	–	15	150278	2563
STA	14.5	0.0	2620	154026	2568
STC	14.3	19.0	225	154026	10
Ball in free-fall & clock, 3 variables, $\epsilon = 0.001$					
LGG	0.04	–	11	1770	261
STA	0.02	0	14	1453	85
STC	0.02	0	12	1453	56
Ball i. ff. w. input disturbance & clock, 3 variables, $\epsilon = 0.001$					
LGG	6.3	–	94	344676	17208
STA	29.9	0	15	1580834	256
STC	29.9	0	12	1580834	64
Three-tank system, 3 variables, $\epsilon = 0.125$					
LGG	0.03	–	2	1032	105
STA	0.03	0	15	756	137
STC	0.03	0	2	756	9
Overhead crane, 4 variables, $\epsilon = 0.05$					
LGG	0.12	–	11	3944	369
STA	0.17	0	37	4896	633
STC	0.17	0	13	4896	15

Fl.T: flowpipe construction time, *Cl.T.*: clustering and polyhedral approximation time, Mem.: memory consumption, #eval: number of evaluations of the support function of the initial set, #sets: number of convex sets covering the flowpipe

goal of the STC algorithm is to reduce it as much as possible for the given error bound.

The results indicate that the flowpipe approximation in space-time with clustering (STC) can produce a flowpipe cover with a small number of sets, while meeting the desired directional error bounds. The construction uses template directions, with which the reachable set is approximated in space-time, pointwise for every time instant. The projection onto the state space produces polyhedra with facet normals that are linear combinations of the template directions. Compared to our previous work, which approximates the flowpipe directly in the state space, this can improve precision and reduce the number of sets at the same time.

5. ACKNOWLEDGMENTS

The authors thank Oded Maler for valuable discussions and support and the anonymous reviewers for their comments and corrections.

6. REFERENCES

[1] E. Asarin, T. Dang, and O. Maler. The d/dt tool for verification of hybrid systems. In E. Brinksma and K. G. Larsen, editors, *CAV*, volume 2404 of *LNCS*, pages 365–370. Springer, 2002.

[2] A. Chutinan and B. H. Krogh. Computational techniques for hybrid system verification. *IEEE Trans. Automat. Contr.*, 48(1):64–75, 2003.

[3] G. Frehse. SpaceEx state space explorer. Verimag, Grenoble, `http://spaceex.imag.fr`, 2010.

[4] G. Frehse, C. Le Guernic, A. Donzé, S. Cotton, R. Ray, O. Lebeltel, R. Ripado, A. Girard, T. Dang, and O. Maler. SpaceEx: Scalable verification of hybrid systems. In G. Gopalakrishnan and S. Qadeer, editors, *CAV*, volume 6806 of *LNCS*, pages 379–395. Springer, 2011.

[5] G. Frehse, C. Le Guernic, and R. Kateja. Flowpipe approximation and clustering in space-time. Technical Report TR-2013-1, Verimag, February 2013.

[6] G. Frehse and R. Ray. Flowpipe-guard intersection for reachability computations with support functions. In *IFAC ADHS*, pages 94–101, 2012.

[7] A. Girard. Reachability of uncertain linear systems using zonotopes. In M. Morari and L. Thiele, editors, *HSCC*, volume 3414 of *LNCS*, pages 291–305. Springer, 2005.

[8] A. B. Kurzhanski and P. Varaiya. Ellipsoidal techniques for reachability under state constraints. *SIAM J. Control and Optimization*, 45(4):1369–1394, 2006.

[9] C. Le Guernic and A. Girard. Reachability analysis of hybrid systems using support functions. In A. Bouajjani and O. Maler, editors, *CAV*, volume 5643 of *LNCS*, pages 540–554. Springer, 2009.

[10] A. V. Lotov, V. A. Bushenkov, and G. K. Kamenev. *Interactive Decision Maps*, volume 89 of *Applied Optimization*. Kluwer, 2004.

[11] F. Maffray. On the coloration of perfect graphs. In B. A. Reed and C. L. Sales, editors, *Recent Advances in Algorithms and Combinatorics*, CMS Books in Mathematics, pages 65–84. Springer, 2003.

[12] P. Prabhakar and M. Viswanathan. A dynamic algorithm for approximate flow computations. In M. Caccamo, E. Frazzoli, and R. Grosu, editors, *HSCC*, pages 133–142. ACM, 2011.

[13] P. Varaiya. Reach set computation using optimal control. In *Proc. KIT Workshop*, pages 377–383, 1997.

Bounded Model-checking of Discrete Duration Calculus *

Quan Zu and Miaomiao Zhang
School of Software Engineering
Tongji University
Shanghai, China
{7quanzu,miaomiao}@tongji.edu.cn

Jiaqi Zhu and Naijun Zhan*
State Key Lab. of Comp. Sci.
Institute of Software, CAS
Beijing, China
{zhujq,znj}@ios.ac.cn

ABSTRACT

Fränzle and Hansen investigated the model-checking problem of the subset of Duration Calculus without individual variables and quantifications w.r.t. some approximation semantics by reduction to the decision problem of Presburger Arithmetic, thus obtained a model-checking algorithm with 4-fold exponential complexity [6, 7]. As an alternative, inspired by their work, we consider the bounded model-checking problem of the subset in the context of the standard discrete-time semantics in this paper. Based on our previous work [20], we reduce this problem to the reachability problem of timed automata. The complexity of our approach is singly exponential in the size of formulas and quadratic in the number of states of models. We implement our approach using UPPAAL and demonstrate its efficiency by some examples.

Keywords

Model Checking, Duration Calculus, Timed Automata

1. INTRODUCTION

In their seminal work [24], Zhou *et al* introduced the notion of durations of states into Interval Temporal Logic (ITL) [13] for specifying and reasoning about quantitative properties of real-time and hybrid systems and founded Duration Calculus (DC). DC is a very expressive interval-based logic for real-time and hybrid systems at a very abstract level, which is thought as a new trend in formal design of real-time and hybrid systems [8] and has been widely and successfully applied in practice [22]. However, because of its expressiveness, the dark side of DC is the high undecidability of its decision procedure and model-checking issues [23] in general, unless the notion of duration, the use of negation and chop (the only modality in DC), or the models considered are severely constrained, e.g. [23, 21, 4, 5, 9, 12, 15, 14, 17].

*The first two authors are funded by NSFC 61073022 and the Fundamental Research Funds for the Central Universities (No.2100219031), and the last two authors are funded by NSFC-91118007, NSFC-60970031 and National Science and Technology Major Project of China (No. 2012ZX03039-004).
*The corresponding author

HSCC'13, April 8–11, 2013, Philadelphia, Pennsylvania, USA.

Linear duration invariants (LDIs) form an important subset of DC, as many safety properties of real-time systems can be defined with them. For instance, in the gas burner example [16], it is easy to specify the requirement that for any observed interval of length greater than or equal to 60, the duration of the Leak is not greater than one twentieth by an LDI as

$$60 \leq \ell \Rightarrow (19\textstyle\int \mathrm{Leak} - \int \neg\mathrm{Leak}) \leq 0.$$

In [21], the model-checking problem of LDIs is reduced to the linear programming problem and therefore solvable, where models are given by *real-time automata*. While in [3, 11, 17, 19], the authors investigated the model-checking problem of LDIs over *timed automata*.

Whether it is possible to find out a larger subset of DC whose decision and/or model-checking problem are/is decidable is quite interesting. An obvious solution is to investigate the extension of LDIs with Boolean connectives and the chop (we call such extension ELDIs for short). However, unfortunately, according to the results given in [23], the decision problem of ELDIs becomes undecidable both in the discrete time and continuous time settings. Moreover, in [6, 7], Fränzle and Hansen pointed out that the model-checking problem of ELDIs over finite-state Kripke structures turns out to be undecidable also both in the discrete time and continuous time settings. Motivated by this observation, they proposed an approximation semantics for ELDIs, called *doubly situation based semantics*, and showed that its model-checking is decidable in the discrete time setting with cubic complexity in the number of states of the model and linear in the size of the formula. However, further observation indicates that the approximation semantics is too coarse to be useful in practice [7]. So, the authors refined the semantics to another approximation semantics called *counting semantics* and reduced the model-checking problem of ELDIs to Presburger Arithmetic with 4-fold exponential complexity. So, two obstacles hinder the application of their approach:

1. The first one is the approximation semantics. According to their approach, one can only prove/disprove those formulas that can be approximated to be **true/false** over the given model represented by a finite-state Kripke structure, but cannot say anything about other formulas;

2. The second one is the efficiency as explained above.

In this paper, motivated by Fränzle and Hansen's work, as an alternative, we give a more efficient algorithm for model-checking a subset of ELDIs over timed automata in the context of the standard discrete-time semantics. ELDI formulas considered here are of the form $a \leq \ell \leq b \Rightarrow \phi$, where ϕ is an ELDI formula defined in [5, 6] (see the definition given later), a is a natural number, and b is a natural number or ∞. When $a = 0$ and $b = \infty$, this case exactly corresponds to the subset of DC considered in [5, 6]. However, in this paper, we only focus on the case when b is bounded, which means all reference intervals should be with a bounded length. In other words, we just investigate "bounded model-checking" of ELDIs. The solution is based on the technique developed in our previous work [20]. The basic idea is as follows: For a given timed automaton $\mathcal{A}$ and an ELDI formula Φ, we first construct an auxiliary automaton $\mathcal{S}$ to count the observation time of $\mathcal{A}$, and meanwhile, to check whether Φ is satisfied or not at every integral time point whenever the observation time is within the scope; then for the product $\mathcal{A}\|\mathcal{S}$, we use a CTL formula to characterize all the failure states at which the checking procedure returns **false**. So, $\mathcal{A} \models \Phi$ is reduced to verifying that none of these failure states is reachable in $\mathcal{A}\|\mathcal{S}$, i.e., the formula is not satisfied by the product. The hardest part is to design an algorithm called *BMC-DC*, to check at every integral time point whether or not the given formula is satisfied on any reachable execution segment whose length is within the bound. *BMC-DC* is executed as an action when some transitions of the auxiliary automaton $\mathcal{S}$ happen. This allows us to easily implement our approach in the model checker UPPAAL, and we will demonstrate the efficiency of our approach by some examples.

The rest of the paper is organized as follows: Section 2 recalls some basic notions of timed automata and Duration Calculus. Section 3 explains the basic idea of our approach by some running examples, while the detail is given in Section 4. Section 5 reports the implementation and experimental results. Section 6 concludes this paper and discusses the future work.

2. PRELIMINARIES

In this section, we introduce timed automata with discrete time that will be used as models, and ELDIs that will be used as specification language for real-time systems.

For convenience, we fix a finite set of *state variables* $\mathcal{P}$, ranged by $P, Q, \cdots$, and let $\mathcal{L} = 2^{\mathcal{P}}$ in the rest of this paper.

2.1 Timed Automata with Discrete Time

A timed automaton [1] is a finite state machine equipped with a set of clocks. In our case, we use a subset of $\mathcal{P}$ to represent a state (location), and a set X of integer valued variables to represent the clocks. Let $\Delta(X)$ be the set of clock constraints on X, which are conjunctions of formulas of the form $x \leq c$ or $c \leq x$, where $x \in X$ and $c \in \mathbb{N}$. Formally,

Definition 1. A timed automaton $\mathcal{A}$ is a tuple $\mathcal{A} = (L, l_0, \Sigma, X, E, I)$, where $L \subseteq \mathcal{L}$ is a finite set of locations; $l_0 \in L$ is the initial location; Σ is a finite set of actions; X is a finite set of clocks; I is a mapping that assigns each location $l \in L$ with a clock constraint $I(l) \in \Delta(X)$ called the invariant at l; $E \subseteq L \times \Sigma \times \Delta(X) \times 2^X \times L$ is a relation among locations, whose elements are called edges labeled with an action, a guard and a set of clocks to be reset.

A clock interpretation ν for the set of clocks X is a mapping that assigns a natural number to each clock. For $t \in \mathbb{N}$, let $\nu + t$ denote the clock interpretation which maps each clock $x \in X$ to $\nu(x) + t$. For $\lambda \subseteq X$, let $\nu[\lambda = 0]$ denote the clock interpretation which assigns 0 to each $x \in \lambda$ and agrees with $\bar{\nu}$ over the rest of the clocks.

A state of automaton $\mathcal{A}$ is a pair (l, ν), where l is a location of $\mathcal{A}$ and ν is a clock interpretation which satisfies the invariant $I(l)$. State (l_0, ν_0) is the initial state, where $\nu_0(x) = 0$ for any clock $x \in X$.

Let $\mathcal{A}$ be a timed automaton,

1. A *run* or an *execution* r of $\mathcal{A}$ is an infinite sequence of the form

$$r \;:\; (l_0, \nu_0) \xrightarrow[t_1]{a_1} (l_1, \nu_1) \xrightarrow[t_2]{a_2} (l_2, \nu_2) \xrightarrow[t_3]{a_3} \cdots$$

with $l_i \in L$ and ν_i is a clock interpretation, for $i \geq 0$, satisfying the following requirements:

Initiation: (l_0, ν_0) is the initial state.

Consecution: for all $i > 0$, either there is an edge in E of the form $(l_{i-1}, a_i, \delta_i, \lambda_i, l_i)$ such that $(\nu_{i-1} + t_i - t_{i-1})$ satisfies δ_i and ν_i equals $(\nu_{i-1} + t_i - t_{i-1})[\lambda_i = 0]$; or $l_{i-1} = l_i$, $\nu_i = \nu_{i-1} + (t_i - t_{i-1})$, $a_i = t_i - t_{i-1}$ to denote an action to delay $t_i - t_{i-1}$ time units, and for any $0 \leq t < t_i - t_{i-1}$, $\nu_{i-1} + t$ satisfies $I(l_{i-1})$.

2. A *behaviour* ρ corresponding to the above run r, is the infinite sequence of timed locations

$$\rho \;:\; (l_0, t_0)(l_1, t_1) \cdots (l_n, t_n) \cdots$$

satisfying the following conditions: (1) $t_0 = 0$; (2) for any $T \in \mathbb{N}$, there is some $i \geq 0$ such that $t_i \geq T$, which guarantees **divergence** and **nonzeno**.

Note that t_i is the instant that $\mathcal{A}$ enters l_i, for all $i \geq 0$. This means that it stays in l_{i-1} for $t_i - t_{i-1}$ time units and then transits to l_i in the run. Also, in this paper we use a sequence of time-stamped locations to denote a behaviour instead of a sequence of time-stamped switches as in other papers.

The product of several timed automata is defined in a standard way (please refer to [1, 2]). As in UPPAAL [2], each component timed automaton can be associated with a priority in a product. Priorities among the component timed automata are specified on the system level using a partial order $<$ to indicate that the right component timed automaton has a higher priority.

2.2 Extended Linear Duration Invariants

ELDIs with $\mathcal{P}$ investigated in [5, 6] consist of three syntactic categories, which are state expressions S, linear duration formulas (LDFs) $\mathcal{D}$, and ELDI formulas ϕ. The BNFs for them are as follows:

$$\begin{aligned}
S &::= 0 \mid P \mid \neg S \mid S_1 \vee S_2 \\
\mathcal{D} &::= \sum_{i \in \Omega} c_i \textstyle\int S_i \leq c \\
\phi &::= \mathcal{D} \mid \neg\phi \mid \phi_1 \vee \phi_2 \mid \phi_1 ; \phi_2
\end{aligned}$$

where c_is and c are integers, and Ω is a finite set of indices.

As the convention of DC, ℓ is defined as $\int 1$, denoting the length of the reference interval. The Boolean value **true**, denoted by $\top$, is defined by $\ell \geq 0$, falling in ELDIs. Obviously, each ELDI formula can be represented by the form $a \leq \ell \leq b \Rightarrow \phi$, where a is a natural number, b is either a natural number or ∞, and ϕ is defined as above. In this paper, we only focus on the case when b is bounded, and will represent an ELDI of this form by $\Phi, \Psi, \cdots$, possibly with superscript and subscript in the sequel.

Given a timed automaton $\mathcal{A}$, each of its behaviours ρ, defines an interpretation $\mathcal{I}_\rho$ of ELDIs in the following way:

State expressions: $\mathcal{I}_\rho(0)(t) = 0$ for any $t \in \mathbb{N}$;

$$\mathcal{I}_\rho(P)(t) = \begin{cases} 1 & \text{if } t_{i-1} \leq t < t_i \wedge P \in l_{i-1} \text{ for some } i > 0 \\ 0 & \text{otherwise} \end{cases}$$

$\mathcal{I}_\rho(\neg S)(t) = 1 - \mathcal{I}_\rho(S)(t)$;

$\mathcal{I}_\rho(S_1 \vee S_2)(t) = \max\{\mathcal{I}_\rho(S_1)(t), \mathcal{I}_\rho(S_2)(t)\}$.

Durations: given an interval $[t_1, t_2]$, where $t_1, t_2 \in \mathbb{N}$ and $t_1 \leq t_2$, then $\int S$ is interpreted by $\int_{t_1}^{t_2} \mathcal{I}_\rho(S)(t)dt$.

Formulas: given an interval $[t_1, t_2]$ as above, an ELDI formula ϕ is interpreted by

$\mathcal{I}_\rho, [t_1, t_2] \models \sum_{i \in \Omega} c_i \int S_i \leq c$ iff $\sum_{i \in \Omega} c_i \mathcal{I}_\rho(\int S_i, [t_1, t_2]) \leq c$;

$\mathcal{I}_\rho, [t_1, t_2] \models \neg\phi$ iff $\mathcal{I}_\rho, [t_1, t_2] \not\models \phi$;

$\mathcal{I}_\rho, [t_1, t_2] \models \phi_1 \vee \phi_2$ iff $\mathcal{I}_\rho, [t_1, t_2] \models \phi_1$ or $\mathcal{I}_\rho, [t_1, t_2] \models \phi_2$;

$\mathcal{I}_\rho, [t_1, t_2] \models \phi_1; \phi_2$ iff $\mathcal{I}_\rho, [t_1, t] \models \phi_1$ and $\mathcal{I}_\rho, [t, t_2] \models \phi_2$ for some $t \in [t_1, t_2] \cap \mathbb{N}$.

In the above, $(\mathcal{I}_\rho, [t_1, t_2])$ is called an ELDI model of $\mathcal{A}$, and we denote $\mathcal{M}(\mathcal{A})$ the set of all ELDI models of $\mathcal{A}$. We say $\mathcal{A} \models \phi$ iff $M \models \phi$ for any $M \in \mathcal{M}(\mathcal{A})$.

Notice that using the axioms of DC [22], it is easy to transform each ELDI formula to an equivalent one in which all state expressions are of the form $P_1 \wedge \cdots \wedge P_n$, where $P_i \in \mathcal{P}$. Thus, hereafter, we assume all ELDIs are of this form unless otherwise stated.

3. BASIC IDEA AND RUNNING EXAMPLES

Given an ELDI $\Phi = (a \leq \ell \leq b \Rightarrow \phi)$ and a timed automaton $\mathcal{A}$, our approach for checking $\mathcal{A} \models \Phi$ is sketched as follows: Firstly, we construct an auxiliary automaton $\mathcal{S}$ that is parallel with $\mathcal{A}$ and can be triggered at any reachable state of $\mathcal{A}$ to check if ϕ is satisfied on any execution segment of $\mathcal{A}$ starting from the state whose length is in between a and b. Then we define a CTL formula to characterize the set of failure states at which the checking algorithm returns **false**. Thus, $\mathcal{A} \models \Phi$ iff the CTL formula is not satisfied by $\mathcal{A} \| \mathcal{S}$.

The auxiliary automaton $\mathcal{S}$ is given in Figure 1, in which

- there are three states: the initial state, `p0` and `p1`, and the invariants at `p0` and `p1` are both $x \leq 1$;

- there are five transitions: one from the initial state to `p0` with an action represented by the procedure *Init* to analyze the formula to be checked; one from `p0` to `p1` with an action represented by the checking algorithm *BMC-DC* from which the reference interval is counted and the checking algorithm is triggered; one self-transition at `p0` that keeps idle so that the checking algorithm can be triggered on arbitrary reference interval starting from any reachable state; and two self-transitions at `p1`. The first self transition labeled with *BMC-DC* keeps checking the formula at any integral time points of the reference interval whose length is still within the scope, while the second one does nothing whenever the reference interval is beyond the given scope. The details of *Init* and *BMC-DC* will be given in the next section.

- there are two clocks: gc is a local variable to record the length of the observed interval starting from a reachable state; x is a local clock variable with an initial value 1, to be reset to 0 whenever its value is 1, which is used to indicate only integral time points are observed, and the checking algorithm should be triggered and executed only at these points.

Obviously, the set of failure states $\mathcal{F}$ of $\mathcal{A} \| \mathcal{S}$ are the ones in which *BMC-DC* returns **false**, which can be characterized by a CTL formula $\psi \mathrel{\widehat{=}}$ `E<>`$\neg$*BMC-DC*(). When b is finite, it is easy to see that $\mathcal{A} \models \Phi$ iff $A \| \mathcal{S} \not\models \psi$.

In order to guarantee the elapse of clocks is never blocked by other actions in the product $\mathcal{A} \| \mathcal{S}$, it is required that $\mathcal{S}$ has a higher priority in $\mathcal{A} \| \mathcal{S}$, i.e., $\mathcal{S} < \mathcal{A}$.

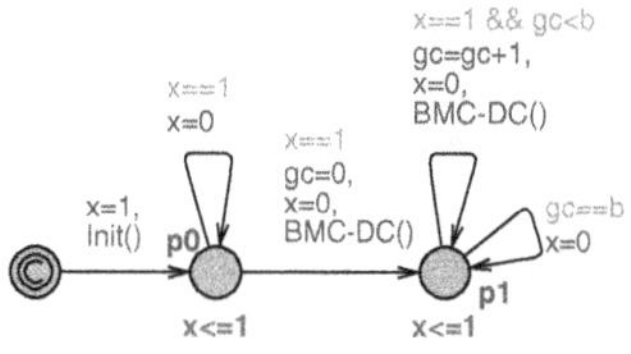

Figure 1: Auxiliary automaton $\mathcal{S}$

3.1 Running Examples

First of all, we use some examples to show how to use *BMC-DC* to check if a given ELDI formula is satisfied on a considered execution segment.

Let's consider an execution segment

$$\rho = (\{P_0\}, 0)(\{P_1\}, 1)(\{P_2\}, 2)(\{P_3\}, 3)(\{P_4\}, 4)(\{P_5\}, 5),$$

and five LDFs $\int P_0 - \int P_1 + \int P_2 + \int P_3 + \int P_4 \leq 0$, $2\int P_1 + \int P_2 - \int P_3 \leq 0$, $-\int P_0 + 2\int P_2 - 2\int P_4 \leq 0$, $\int P_0 \leq 0$ and $\int P_3 \leq 0$, denoted by $\mathcal{D}_1, \cdots, \mathcal{D}_5$ respectively. Moreover, for each $\mathcal{D}_i$, we introduce a variable d_i.

Firstly, consider to check the above segment against a simple formula $5 \leq \ell \leq 5 \Rightarrow \mathcal{D}_1; \mathcal{D}_2$. A natural way is to check its satisfiability by considering $\mathcal{D}_1$ on $[0, 0], [0, 1], [0, 2], [0, 3], [0, 4], [0, 5]$, correspondingly, $\mathcal{D}_2$ on $[0, 5], [1, 5], [2, 5], [3, 5], [4, 5], [5, 5]$, according to the semantics. Thus, the duration expressions of $\mathcal{D}_1$ and $\mathcal{D}_2$ have to be calculated on all the corresponding subintervals. Alternatively, we give a more efficient approach by using the notion of *optimal potential chop point* (OPCP). A time point is called *potential chop point* (PCP) if $\mathcal{D}_1$ is satisfied on the interval from the start point to this point, and a PCP is called *optimal* if the value of the duration expression of $\mathcal{D}_2$ on the interval from this PCP to the current point is minimal among all its values on these intervals from a PCP to the current point. We calculate the values of the duration expressions of $\mathcal{D}_1$ and $\mathcal{D}_2$ step by step from the beginning point to the end point of the execution segment, and record the value of the duration expression of $\mathcal{D}_1$ on the segment from the beginning time point to the current time point by d_1, and that of $\mathcal{D}_2$ on the segment from the OPCP to the current time point by d_2. Meanwhile, reset d_2 to 0 whenever the OPCP is updated. In this example, at the beginning, d_1 is set to 0, and d_2 to 11, which is large enough to guarantee that the duration expressions of $\mathcal{D}_1$ and $\mathcal{D}_2$ on the execution segment are always smaller than this value. At $t = 0$, obviously, $\mathcal{D}_1$ is satisfied as $d_1 \leq 0$, so this point is the current OPCP and the duration expression of $\mathcal{D}_2$ needs to be computed from scratch and hence d_2 is reset to 0. At $t = 1$, d_1 is increased by 1 (staying at $\{P_0\}$ for one time unit), so $t = 1$ cannot be a PCP. At $t = 2$, d_1 is changed to 0 and d_2 is updated to 2. Now, $t = 2$ becomes the new OPCP as $d_2 > 0$, and accordingly, d_2 is reset to 0 again. Repeat the above procedure until $t = 5$. Then we can conclude the formula is satisfied as $d_2 = 0 \leq 0$. In our approach, the duration expressions of $\mathcal{D}_1$ and $\mathcal{D}_2$ are just needed to be calculated one time on the execution segment.

However, in many cases, we cannot distinguish which is optimal among several PCPs. E.g., consider the above model against the formula

$$5 \leq \ell \leq 5 \Rightarrow \mathcal{D}_1; \neg(\neg(\mathcal{D}_2; \mathcal{D}_3); (\mathcal{D}_4 \wedge \mathcal{D}_5)).$$

Obviously, $t = 0$ and $t = 2$ are two PCPs w.r.t. the outmost chop, but we cannot tell which is optimal as we cannot guarantee that the values of the duration expression of $\mathcal{D}_4$ and $\mathcal{D}_5$ achieve *optimal* simultaneously. In this case, the two PCPs have to be checked according to the semantics separately. Thus, we duplicate d_2 at $t = 0$ and $t = 2$, and denote by d_2^0 and d_2^2 respectively. Obviously, d_2^0 is set to 0 at $t = 0$, while d_2^2 is set to 0 at $t = 2$. Moreover, according to the semantics, $\rho, [t_1, t_2] \models \neg(\neg(\mathcal{D}_2; \mathcal{D}_3); (\mathcal{D}_4 \wedge \mathcal{D}_5))$ iff $\rho, [t_1, t] \models \neg(\mathcal{D}_2; \mathcal{D}_3)$ implies $\rho, [t, t_2] \models \neg(\mathcal{D}_4 \wedge \mathcal{D}_5)$, for any $t \in [t_1, t_2]$. Checking the premise part can be done as above thanks to the existence of OPCPs. Accordingly, we also need to duplicate d_3 respectively corresponding to the two PCPs, and denote by d_3^0 and d_3^2 respectively. W.r.t. the PCP $t = 0$ of the outmost chop, it is easy to see that $t = 3$ and $t = 4$ are two PCPs of the second outmost chop as $\neg(\mathcal{D}_2; \mathcal{D}_3)$ is satisfied on both $[0, 3]$ and $[0, 4]$. Regarding the chop point $t = 4$ of the second outmost chop, $\mathcal{D}_4$ and $\mathcal{D}_5$ are both satisfied on $[4, 5]$. This indicates that firstly, $t = 0$ cannot be seen as a chop point of the outmost chop; secondly, the second outmost chop does not have "optimal" property, as we cannot guarantee the duration expressions of $\mathcal{D}_4$ and $\mathcal{D}_5$ achieve "optimal" simultaneously. An analogous analysis indicates the formula is satisfied if $t = 2$ is chosen as the chop point of the outmost chop, because only $t = 3$ is the PCP of the second outmost chop, and $\mathcal{D}_4$ is satisfied on $[3, 5]$, but $\mathcal{D}_5$ not.

4. ALGORITHMS

In this section, we focus on how to implement the procedures *Init* and *BMC-DC* for checking ELDIs.

Clearly, negation of an LDF or a logical combination of LDFs can be easily eliminated and we can obtain another equivalent LDF or logical combination of LDFs in which no negation occurs. Thus, for simplicity, as a convention, we just consider the ELDIs in which any $\neg$ is only applied to a subformula with ; as its outmost operator.

4.1 In a Nutshell

It is a natural way to check the satisfiability of an ELDI Φ against a timed automaton directly by the semantics, possibly with a complexity $O(n^b b^r)$, where n is the number of locations of the considered timed automaton, b is the upper bound of reference intervals, and r is the number of LDIs in Φ. Obviously, it is quite high. However, we found we can dramatically reduce the complexity by using the notion of *optimal potential chop point* (OPCP).

In order to define the notion of OPCP, we introduce some notations first. First of all, let's fix an ELDI $\Phi = a \le \ell \le b \Rightarrow \phi$ for the sequel discussions, $\mathcal{D}_1, \cdots, \mathcal{D}_r$ are all LDFs occurring ϕ, and all other formulas are subformulas of Φ unless otherwise stated. For each $\mathcal{D}_k$, we introduce a variable d_k to denote the value of the duration expression of $\mathcal{D}_k$ on the reference interval, and meanwhile, define an upper bound $d_{\max}$ of d_ks by $b \cdot \max\{ |c_{i_k,k}|, |c_k| \mid i_k \in \Omega_k \wedge k = 1, \ldots, r\} + 1$.

Definition 2. Given an ELDI formula $\phi'; (\mathcal{D}_1 \vee \cdots \vee \mathcal{D}_i)$ and an interval $[t_1, t_2]$, an integral time point t of $[t_1, t_2]$ is called *potential chop point* (PCP) if ϕ' is satisfied on $[t_1, t]$; a PCP t is called *optimal* up to t', if there is $1 \le k \le i$ such that

$$d_k([t, t']) = \min\{d_k([t^*, t']) \mid t^* \le t' \text{ and } t^* \text{ is a PCP}\},$$

where $t_1 \le t \le t' \le t_2$ and $t, t' \in \mathbb{N}$, and $d_k([b, e])$ denotes the value of d_k on $[b, e]$.

For the given $\phi'; (\mathcal{D}_1 \vee \cdots \vee \mathcal{D}_i)$ and $[t_1, t_2]$, we can show that at any time $t' \in [t_1, t_2]$, the formula is satisfied on $[t_1, t']$ iff there exists an OPCP t such that ϕ' is satisfied on $[t_1, t]$ and $\mathcal{D}_1 \vee \cdots \vee \mathcal{D}_i$ is satisfied on $[t, t']$ (see the proof for Theorem 1 in Appendix). Thus, we just need to visit all PCPs from t_1 to t_2, update the current OPCPs t, and keep the values of duration variables in ϕ' on $[t_1, t']$ and values of $d_1, \ldots, d_i$ on $[t, t']$, in order to check the satisfiability of the whole formula over $[t_1, t_2]$. We do not need to consider all possible values of variables on all subintervals of $[t_1, t_2]$. This indeed reduces the complexity of the checking so much. Similar idea is applicable to any ELDI formulas only with chop and disjunction of the form $\phi'; (\mathcal{D}_{1,1} \vee \cdots \vee \mathcal{D}_{1,i_1}); \cdots; (\mathcal{D}_{k,1} \vee \cdots \vee \mathcal{D}_{k,i_k})$ by simultaneously maintaining k interdependent OPCPs.

However, the idea is not applicable to the combinations of chop with conjunction or negation. For example, consider $\mathcal{D}_1; (\mathcal{D}_2 \wedge \mathcal{D}_3)$. It is impossible to guarantee the existence of OPCPs, because if it exists, the duration expressions of $\mathcal{D}_2$ and $\mathcal{D}_3$ both have to be optimal w.r.t. it. So, in order to check the satisfiability of $\mathcal{D}_1; (\mathcal{D}_2 \wedge \mathcal{D}_3)$, all the PCPs need to be maintained and checked for $\mathcal{D}_2 \wedge \mathcal{D}_3$ according to the semantics. Therefore, we duplicate the variable d_2 at each visited point (at most $b + 1$ times), denoted by $d_2^0, d_2^1, \cdots, d_2^b$, which respectively stand for the values of the duration expression of $\mathcal{D}_2$ on the interval from the corresponding time point to the end of the reference interval. Analogously for $\mathcal{D}_3$. In this case, we call the subformula $\mathcal{D}_2 \wedge \mathcal{D}_3$ *duplicated.* Similarly, for the negation of a subformula which is a right operand of chop, we also need to duplicate the corresponding variables for each LDF in the subformula at each visited time point. E.g., in $\mathcal{D}_1; \neg(\mathcal{D}_2; \mathcal{D}_3)$, $\neg(\mathcal{D}_2; \mathcal{D}_3)$ is a *duplicated subformula.* In order to improve the efficiency, we had better exploit the idea of OPCP as many as possible. So, we give a syntactical criterion to tell to which subformulas the OPCP-based approach is not applicable, called *duplicated formulas.*

Definition 3. Given a subformula φ of ϕ, we say φ is *duplicated,* if φ is of the form $\phi_1 \wedge \phi_2$ or $\neg\phi_1$, and $\psi; \varphi$ is a subformula of ϕ for some ψ. Duplicated subformulas can be nested and there is a largest nested depth for each ELDI formula. We use *D-Sub*(χ) to denote the set of duplicated subformulas of ϕ, of which χ is a proper subformula; furthermore denote the minimal one $\min(\textit{D-Sub}(\chi))$ by *MD*(χ).

Example 1. In $\mathcal{D}_1; \neg(\mathcal{D}_2; (\mathcal{D}_3 \wedge \mathcal{D}_4))$, $\neg(\mathcal{D}_2; (\mathcal{D}_3 \wedge \mathcal{D}_4))$ is *duplicated,* and $\mathcal{D}_3 \wedge \mathcal{D}_4$ is *duplicated* too. So the largest nested depth of the formula is 2. $\textit{D-Sub}(\mathcal{D}_3) = \{\mathcal{D}_3 \wedge \mathcal{D}_4, \neg(\mathcal{D}_2; (\mathcal{D}_3 \wedge \mathcal{D}_4))\}$ and $\textit{MD}(\mathcal{D}_3) = \mathcal{D}_3 \wedge \mathcal{D}_4$; $\textit{D-Sub}(\mathcal{D}_3 \wedge \mathcal{D}_4) = \{\neg(\mathcal{D}_2; (\mathcal{D}_3 \wedge \mathcal{D}_4))\}$ and $\textit{MD}(\mathcal{D}_3 \wedge \mathcal{D}_4) = \neg(\mathcal{D}_2; (\mathcal{D}_3 \wedge \mathcal{D}_4))$; while neither $\textit{D-Sub}(\neg(\mathcal{D}_2; (\mathcal{D}_3 \wedge \mathcal{D}_4)))$ nor $\textit{MD}(\neg(\mathcal{D}_2; (\mathcal{D}_3 \wedge \mathcal{D}_4)))$ exists.

According to Definition 3, we design a preprocessing procedure *PreTreat* in Algorithm 1 to analyze the syntactical structure of ϕ, including which subformulas have the "optimal property", which are duplicated, and the relation between them. To this end, ϕ will be represented as a binary syntactical tree, in which each node is organized as a tuple of the form (*Nu*, *Op*, *Left*, *Right*, *V*, *W*, *C*, *S*, *tag*), where *Nu* stands for the number of the subformula in the tree given by a mapping J, *Op* for its outmost operator, *Left* and *Right* for the numbers of the subformulas at the left and the right of *Op* respectively, V for the set of indices of the LDIs, W for the set of indices of the LDIs that are likely to be duplicated, C for the set of indices of the LDIs that have to be duplicated, S for the set of indices of the LDIs of the subformula whose duration expressions need to be initialized to 0 whenever the subformula is considered, and *tag* to indicate if the subformula itself is *duplicated.* Note that if a subformula is an LDF $\mathcal{D}_k$, then its *Op* will be set to its index k, and its *Left* and *Right* will be set to -1; if a subformula is of the form $\neg\psi$, *Left* will be set to the number of ψ, while *Right* will be set to -1. The numbering mapping J follows the convention that all of ϕ's subformulas are numbered by consecutive integers from 0, and different subformulas with different numbers. In addition, ϕ itself is numbered as 0 and the number of a formula is always less than those of its subformulas. We use an array A of the tuple to represent the syntax tree, and each subformula corresponds to an element of A whose index is the same as the subformula's number Nu.

Algorithm 1 *PreTreat*()

Input: ϕ, *flag*
$n := J(\phi)$;
if $\phi = \mathcal{D}_k$ **then**
  $A[n].(Nu, Op, tag, Left, Right) := (n, k, \textbf{false}, -1, -1)$;
  $A[n].(V, W, C, S) := (\{k\}, \emptyset, \emptyset, \{k\})$;
if $\phi = \phi_1 \vee \phi_2$ **then**
  PreTreat(ϕ_1, *flag*); *PreTreat*(ϕ_2, *flag*);
  $A[n].(Nu, Op, tag, Left, Right) := (n, \vee, \textbf{false}, J(\phi_1), J(\phi_2))$;
  $A[n].(V, W, C, S) :=$
    $(A[J(\phi_1)].V \cup A[J(\phi_2)].V, A[J(\phi_1)].W \cup A[J(\phi_2)].W,$
    $A[J(\phi_1)].C \cup A[J(\phi_2)].C, A[J(\phi_1)].S \cup A[J(\phi_2)].S)$;
if $\phi = \phi_1 \wedge \phi_2$ **then**
  PreTreat(ϕ_1, **false**); *PreTreat*(ϕ_2, **false**);
  $A[n].(Nu, Op, tag, Left, Right) := (n, \wedge, flag, J(\phi_1), J(\phi_2))$;
  $A[n].(V, W, C, S) :=$
    $(A[J(\phi_1)].V \cup A[J(\phi_2)].V, A[J(\phi_1)].V \cup A[J(\phi_2)].V,$
    $A[J(\phi_1)].C \cup A[J(\phi_2)].C, A[J(\phi_1)].S \cup A[J(\phi_2)].S)$;
if $\phi = \neg\phi_1$ **then**
  PreTreat(ϕ_1, **false**);
  $A[n].(Nu, Op, tag, Left, Right) := (n, \neg, flag, J(\phi_1), -1)$;
  $A[n].(V, W, C, S) := A[J(\phi_1)].(V, V, C, S)$;
if $\phi = \phi_1; \phi_2$ **then**
  PreTreat(ϕ_1, *flag*); *PreTreat*(ϕ_2, **true**);
  $A[n].(Nu, Op, tag, Left, Right) := (n, ;, \textbf{false}, J(\phi_1), J(\phi_2))$;
  $A[n].(V, W, C, S) := (A[J(\phi_1)].V \cup A[J(\phi_2)].V, A[J(\phi_1)].W,$
    $A[J(\phi_1)].C \cup A[J(\phi_2)].C \cup A[J(\phi_2)].W, A[J(\phi_1)].S)$

Example 2. Still consider the formula above $\mathcal{D}_1; \neg(\mathcal{D}_2; (\mathcal{D}_3 \wedge \mathcal{D}_4))$. Thus, four d_k $(k = 1, \ldots, 4)$ are introduced. The syntactic tree just with the number and operator of each subformula is shown in Figure 2, and the array represented the syntactic tree is shown in the table. Using *PreTreat*, except that *tag* is computed from top to bottom, other information is computed from bottom to top. E.g., 3 and 4 are added to $A[3].C$ as they are contained in $A[5].W$, and $A[5].tag$ is true as node 5 is the right child of node 3 and its operator is $\wedge$.

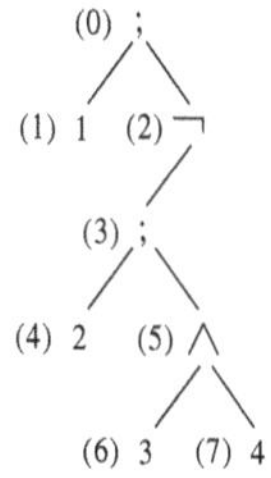

Nu	V	W	C	S	tag
0	{1,2,3,4}	{ }	{2,3,4}	{1}	false
1	{1}	{ }	{ }	{1}	false
2	{2,3,4}	{2,3,4}	{3,4}	{2}	true
3	{2,3,4}	{ }	{3,4}	{2}	false
4	{2}	{ }	{ }	{2}	false
5	{3,4}	{3,4}	{ }	{3,4}	true
6	{3}	{ }	{ }	{3}	false
7	{4}	{ }	{ }	{4}	false

Figure 2: The syntactic tree and the information on the tuples

So, the procedure *Init* in the auxiliary automaton $\mathcal{S}$ of Figure 1 is implemented in Algorithm 2. It first analyzes the syntactical information of ϕ and records it in A by calling *PreTreat*. Then it initializes the corresponding introduced variables.

Then, we implement the action *BMC-DC* in the auxiliary automaton by calling two subroutines after updating the values of the d_is and their duplicates, where *Reset* for resetting the values of the corresponding duration expressions if the considered time point is a new OPCP, and *Sat* for calculating the return value indicating whether a formula is satisfied on the current reference interval. We will explain these two subroutines in detail later.

Algorithm 2 *Init*()

Input: ϕ
PreTreat(ϕ, **false**);
for all $k \in A[0].V - A[0].C$ **do**
  $d_k := d_{\max}$;
for all $k \in A[0].C$ **do**
  for $i := 0$ to b **do**
    $d_k^i := d_{\max}$;
for all $k \in A[0].S$ **do**
  $d_k := 0$;

Algorithm 3 *BMC-DC*()

for all $k \in A[0].V - A[0].C$ **do**
  for all $i \in \Omega_k$ **do**
    if $d_k \neq d_{\max}$ and $S_{i,k}$ is satisfied at the current location[1] and $gc > 0$ **then**
      $d_k := d_k + c_{i,k}$;
for all $k \in A[0].C$ **do**
  for $i := 0$ to b **do**
    for all $i \in \Omega_k$ **do**
      if $d_k^i \neq d_{\max}$ and $S_{i,k}$ is satisfied at the current location and $gc > 0$ **then**
        $d_k^i := d_k^i + c_{i,k}$;
Reset(0, **false**);
if $gc \geq a \wedge \neg Sat(0, -1, -1)$ **then**
  return false;
return true;

4.2 The Subroutine *Reset*

The functions of *Reset* include the following two points:

1. Whenever a PCP is visited, we need to see whether to update the OPCPs, and accordingly reset the values of the duration expressions listed in the set S of the right operand of the chop;

2. For any two immediately nested chops in a nested duplicated subformula, maintain a correspondence relation between the chop points of the outer chop and the ones of the inner chop. That is recorded in a mapping relation T, which is empty at the beginning.

Reset takes as parameters the number n of a formula to be considered and a Boolean variable *flag* to indicate whether the formula is inside a duplicated subformula. For an LDF, *Reset* does nothing but return; for logical connectives, we recursively call the subroutine with its subformulas and the recalculated *flag* as parameters; for the chop, it is needed to reset the values of some duration expressions in the right operand of the chop, which is executed between the recursive invocations of the subroutines to its two operands. That can be categorized into the following two cases:

¬*flag* In this case, the d_ks corresponding to the LDFs of the left operand $A[n].Left$ do not need to be duplicated, therefore there is no need to keep a correspondence with the chop points of any outer chops. The current time point is possibly optimal if $A[n].Left$ is satisfied, and then the d_ks corresponding to the LDFs listed in the set S of the right operand $A[n].Right$ should be reset by case analysis:

- If the index of a d_k is in $A[A[n].Right].S$, but not in $A[A[n].Right].W$, then if its value is greater than 0, it should be reset to 0 as the current point is indeed

[1] This can be easily implemented by checking if each state variable in $S_{i,k}$ occurs in the current location.

optimal; otherwise unchanged, as the previous optimal point is better than the current one.

- If the index of a d_k is in both $A[n].Right$'s S and W, a new duplicated subformula is encountered, so we just simply reset its duplicate at the current point to 0.

flag In this case, the OPCPs of the left operand $A[n].Left$ are related to the PCPs of the immediate outer chop. Thus, the current time point is possibly optimal w.r.t. some outer chop point i if the right operand of the outer chop has been calculated from i (indicated by $d_h^i \neq d_{\max}$) and $A[n].Left$ is satisfied w.r.t. i, so the respective duplicates of the d_ks corresponding to the LDFs listed in the set S of the right operand $A[n].Right$ should be reset by case analysis:

- If the index of a d_k is in $A[A[n].Right].S$, but not in $A[A[n].Right].W$, then if the duplicate d_k^i is greater than 0, it should be reset to 0 as the current point is indeed optimal w.r.t. i; otherwise unchanged.

- If the index of a d_k is in both $A[n].Right$'s S and W, we just simply reset its duplicate at the current point to 0. Meanwhile, the correspondence between the chop point of the outer chop at i and the chop point of the inner chop at gc is recorded in T.

Algorithm 4 $Reset()$

Input: n, *flag*
1: **if** $A[n].Op \in \mathbb{N}$ **then**
2:     **return** ;
3: **if** $A[n].Op == \vee$ or $A[n].Op == \wedge$ **then**
4:     $Reset(A[n].Left, \textit{flag} \vee A[n].tag)$;
      $Reset(A[n].Right, \textit{flag} \vee A[n].tag)$;
5: **if** $A[n].Op == \neg$ **then**
6:     $Reset(A[n].Left, \textit{flag} \vee A[n].tag)$;
7: **if** $A[n].Op ==;$ **then**
8:     $Reset(A[n].Left, \textit{flag})$;
9:     **if** $\neg \textit{flag}$ **then**
10:       **if** $Sat(A[n].Left, -1, -1)$ **then**
11:         **for all** $h \in A[A[n].Right].S - A[A[n].Right].W$ **do**
12:           $d_h := \min(0, d_h)$;
13:         **for all** $h \in A[A[n].Right].S \cap A[A[n].Right].W$ **do**
14:           $d_h^{gc} := 0$;
15:     **else**
16:       choose $k \in A[n].S$;
17:       **for all** $i \in [0, gc]$ **do**
18:         **if** $d_k^i \neq d_{\max} \wedge Sat(A[n].Left, k, i)$ **then**
19:           **for all** $h \in A[A[n].Right].S - A[A[n].Right].W$ **do**
20:             $d_h^i := \min(0, d_h^i)$;
21:           **for all** $h \in A[A[n].Right].S \cap A[A[n].Right].W$ **do**
22:             $d_h^{gc} := 0$;
23:           **for all** $k \in A[n].S$ **do**
24:             $T(d_k^i) := T(d_k^i) \cup \{d_h^{gc}\}$;
25:     $Reset(A[n].Right, \textit{flag})$;

4.3 The Subroutine *Sat*

The subroutine *Sat* is to determine whether a considered formula is satisfied on the reference interval, which is used as the return condition of *BMC-DC* and also as resetting conditions in *Reset*. *Sat* takes three parameters: n is the number corresponding to the formula to be checked; if the minimal duplicated subformula strictly containing n exists, say m, i.e., $m = MD(n)$, then k is the index of some d_k in $A[m].S$ and i is the specific point to duplicate d_k; otherwise, k and i are both assigned with -1.

We compute the satisfiability in a recursive way. For an LDF $\mathcal{D}_k$, the return value is calculated according to the semantics depending

on whether $MD(n)$ exists. If it does not exist, the value of d_k is used; otherwise, the value of d_k's i-th duplicate, i.e., d_k^i, is used. ";" and "$\vee$" can be handled in a standard way as they do not need to be duplicated. Regarding "$\neg$" and "$\wedge$", if the considered formula is not duplicated, then it is handled in a standard way; otherwise, the satisfiability of the formula should be discussed on all possible reference subintervals indicated by the duplicates of a d_h from its S according to the following two cases:

1. the first is when the formula itself is not strictly contained in another duplicated subformula, indicated by $i == -1$, i.e., $MD(n)$ does not exist. Then, we just need to check if the formula is satisfied on some of the considered subintervals (indicated by $d_h^j \neq d_{\max}$);

2. the other is when the formula itself is strictly contained in another duplicated subformula, indicated by $i \neq -1$. Thus, we need to check if the formula is satisfied on some of the subintervals corresponding to the chop point of the outer chop (indicated by $d_h^i \in T(d_k^i)$).

Algorithm 5 Boolean $Sat()$

Input: n, k, i
1: **if** $A[n].Op \in \mathbb{N}$ **then**
2:     **if** $i == -1$ **then**
3:       **return** $d_{A[n].Op} \leq c_{A[n].Op}$;
4:     **else**
5:       **return** $d_{A[n].Op}^i \leq c_{A[n].Op}$;
6: **if** $A[n].Op == \vee$ **then**
7:     **return** $Sat(A[n].Left, k, i) \vee Sat(A[n].Right, k, i)$;
8: **if** $A[n].Op == \wedge$ and $A[n].tag == false$ **then**
9:     **return** $Sat(A[n].Left, k, i) \wedge Sat(A[n].Right, k, i)$;
10: **if** $A[n].Op == \wedge$ and $A[n].tag == true$ **then**
11:     choose a $h \in A[n].S$;
12:     **for** $j := 0$ to gc **do**
13:       **if** $i == -1 \wedge d_h^j \neq d_{\max} \wedge Sat(A[n].Left, h, j) \wedge$
        $Sat(A[n].Right, h, j)$ **then**
14:         **return** **true**;
15:       **if** $i \neq -1 \wedge d_h^j \in T(d_k^i) \wedge Sat(A[n].Left, h, j) \wedge$
        $Sat(A[n].Right, h, j)$ **then**
16:         **return** **true**;
17:     **return** **false**;
18: **if** $A[n].Op == \neg$ and $A[n].tag == false$ **then**
19:     **return** $\neg Sat(A[n].Left, k, i)$;
20: **if** $A[n].Op == \neg$ and $A[n].tag == true$ **then**
21:     choose $h \in A[n].S$;
22:     **for** $j := 0$ to gc **do**
23:       **if** $i == -1 \wedge d_h^j \neq d_{\max} \wedge \neg Sat(A[n].Left, h, j)$ **then**
24:         **return** **true**;
25:       **if** $i \neq -1 \wedge d_h^j \in T(d_k^i) \wedge \neg Sat(A[n].Left, h, j)$ **then**
26:         **return** **true**;
27:     **return** **false**;
28: **if** $A[n].Op ==;$ **then**
29:     **return** $Sat(A[n].Right, k, i)$;

4.4 Correctness and Complexity

The correctness of our approach is guaranteed by the following theorem. The proof is given in the appendix.

THEOREM 1. *Our approach is correct. That is,*

Termination: *Our approach is certain to terminate.*

Soundness: *If none of the failure states is reachable in $\mathcal{A}\|\mathcal{S}$, then the given ELDI formula Φ is satisfied by $\mathcal{A}$.*

Completeness: *If some of the failure states are reachable in $\mathcal{A}\|\mathcal{S}$, then the given ELDI formula Φ is not satisfied by $\mathcal{A}$.*

In order to implement our algorithms on UPPAAL, given a timed automaton $\mathcal{A}$, we reformulate it by replacing each location with a label and using a labeling function f to map each label to the set of state variables corresponding to the location, denoted by $\mathcal{G}$.

As we implement the checking of the reachability of the failure states in the composed automaton $\mathcal{G}\|\mathcal{S}$ using UPPAAL, the complexity of our approach depends on the implementation of UPPAAL, in which on-the-fly checking on the whole system is applied. Here the state space is the product of the locations of the automaton and the values of introduced variables. Regarding the number of variables, first of all, we need to introduce a duration variable d_k for each LDF $\mathcal{D}_k$ $(k = 1, \ldots, r)$; moreover, each d_k could be duplicated at most $b+1$ times; finally, the mapping T is implemented as a $(b+1)r \times (b+1)r$ matrix, in which each entry is a Boolean variable that indicates whether there is a correspondence between the two corresponding introduced variables. So, the total number of introduced variables is at most $(b+1)r \cdot (br + r + 1)$. In addition, in our approach we have to take the cost for executing the action *BMC-DC* on the corresponding transitions into account.

The checking procedure consists of three phases. The first phase is when $\mathcal{S}$ stays in `p0`, in which all the introduced variables as well as gc keep unchanged. So, this phase contains at most $|S_{\mathcal{A}}|$ state changes and $|S_{\mathcal{A}}|^2$ transitions.

The second phase is to do the actual checking in which gc keeps increased from 0 to b. Let the transition from `p0` to `p1` happen at a specific state s. Each execution of *BMC-DC* consists of the following three steps: *updating*, *resetting* and *checking*. Obviously, the cost for updating is $O(br)$. Resetting may change the values of the introduced variables and the entries of the mapping matrix, and recursively call *Reset* and *Sat* many times. Each d_k possibly with a superscript i is reset at most once, so the time cost of reset operations is $O(br)$. Meanwhile, applying *Sat* to a duplicated subformula n could result in recursive invocations of *Sat* of at most $(2(b+1))^h$ times, and each execution of *BMC-DC* checks each node (subformula) of the ELDI ϕ at most once, so the cost of the satisfiability checking in each execution of *BMC-DC* is $O(|\phi|b^h)$, where h is the largest nested depth of duplicated subformulas of ϕ.

Let x be the maximal number of outgoing transitions from the locations of $\mathcal{A}$. In each execution of *BMC-DC*, gc increases by 1, and there are at most $x+1$ possibilities for the next location, which results in at most $2(x+1)^{(b+1)}$ possible states. As shown above, each transition costs time $O(br + |\phi|b^h)$, so the cost for checking intervals starting from s is $O((br + |\phi|b^h)x^b)$. Thus, the cost of the second phase is totally $O(|S_{\mathcal{A}}|(br + |\phi|b^h)x^b)$.

The third phase is when gc is equal to b and the self transition at `p1` is executed. In this phase, the length of the current execution segment has exceeded b, and *BMC-DC* will not be invoked any more, therefore we can conclude that none of the failure states will be reached. Thus, the cost of this phase is zero.

In summary, the number of transitions in the system handled by UPPAAL is $O(|S_{\mathcal{A}}|^2 + |S_{\mathcal{A}}|x^b)$ and the time complexity of our approach is $O(|S_{\mathcal{A}}|^2 + |S_{\mathcal{A}}|(br + |\phi|b^h)x^b)$. We can see that the largest nested depth of duplicated subformulas directly affects the complexity of the algorithm. As it is generally very small, compared with the 4-fold exponential approximation algorithm in [7], we have an essential improvement subject to the constraint of the finite observation time.

5. IMPLEMENTATION AND EXPERIMENTS

Using C++, we develop a tool that can be integrated with UPPAAL to check whether a given system $\mathcal{A}$ satisfies an ELDI Φ (see the dotted box in Figure 3). The input of the tool is Φ and an *XML* file representing $\mathcal{A}$. Then it automatically generates $\mathcal{G}$ in terms of these information. While the generation of $\mathcal{S}$, whose crucial part is the procedure *BMC-DC*, is only dependent on Φ. At last, the composed automaton $\mathcal{G}\|\mathcal{S}$, as well as the CTL formula defined in Section 3 is the input to the model checker UPPAAL.

We now use the following four examples to show the efficiency of our approach, the first three of which are taken from [10] and the fourth is the complex one given in Section 3.1. All the experiments are conducted on a laptop with the Intel Core2 Duo T6400 processor and 2 GB RAM using the operating system Windows 7. Due to the space limitation, we here skip the detailed description of models and model transformation, while only compare the results with those in paper [10].

1. We first check the automaton $\mathcal{A}_N$ obtained by N iterations of the automaton depicted in Figure 4(a) with respect to the property also shown in the figure. The automaton $\mathcal{A}_N$ is constructed by combining N automata $M_1, M_2, \ldots, M_N$, so that there are edges from D_i to A_{i+1}, for $1 \leq i < N$, and edges from D_i to A_j , for $1 \leq j \leq i \leq N$. With regard to different values of N, the checking times (t_1) using our approach are given in the second column of Figure 4(b). Clearly, for each N, the time is less than those listed in the third and the fourth columns (t_2 and t_3), which are needed by two different approaches in [10].

2. The second problem is taken from the experiment 2 of [10][2], where k_1, k_2 and k_3 are coefficients that appear in an ELDI to be checked. We need to solve the problem that for given values of k_1 and k_2, the smallest value of k_3 should be found, so that the ELDI is satisfied. The results of the experiment demonstrate that for each case of different values of k_1 and k_2 given in [10], it takes less time to find the smallest value of k_3 using our approach. For instance, in the first case, it takes $6.3s$, while the two methods proposed in [10] respectively need $38.9s$ and $9.0s$.

3. As discussed in [10], the approximation algorithm fails to check the automaton with respect to the property shown in Figure 4(c) even when the observation interval is finite. While our tool is able to give definite results. For instance, it takes $0.2s$ to verify the satisfiability of the property for $\ell = 4$.

4. The tool can handle complex ELDI formulas as well. For example, for the simple and the complex ones in Section 3.1, it respectively takes $0.16s$ and $0.22s$ to verify the satisfiability of the formulas for the given execution segment.

6. CONCLUSION

In this paper, inspired by Fränzle and Hansen's work [6, 7, 10], we investigated bounded model-checking of ELDIs, which is a very expressive subset of DC. Compared with their work, the advantages of our approach include:

1. instead of using approximation semantics of DC, the standard discrete time semantics of DC is adopted;

2. our approach is much more efficient by case studies. As analyzed in Section 4.4, the complexity of our approach is much

[2]This example actually comes from the full version of [10] thanks to Prof. Michael Hansen for his courtesy of this example.

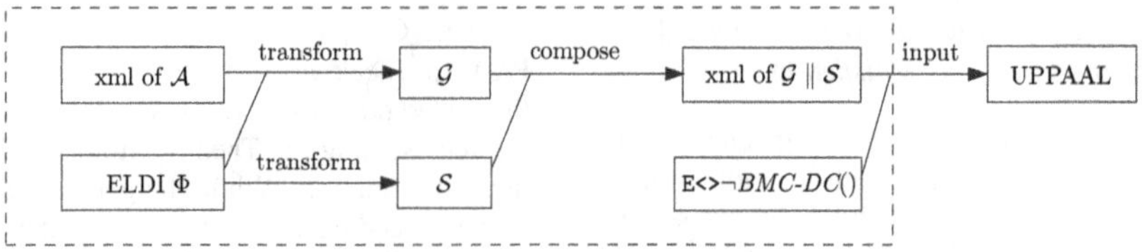

Figure 3: ELDI checking tool

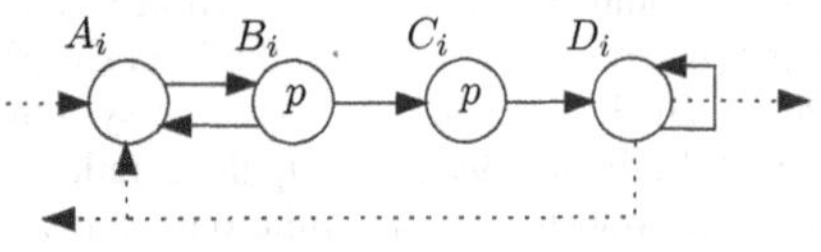

Property: $[](l < 4 \rightarrow \int p < 3)$

(a)

N	$t_1(s)$	$t_2(s)$	$t_3(s)$
3	0.59	5.7	5.7
4	0.64	16.5	15.8
5	0.75	27.9	26.2
6	0.84	55.4	--

(b)

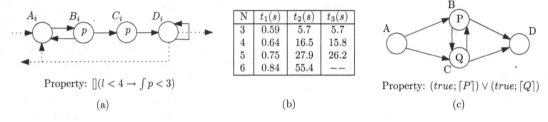

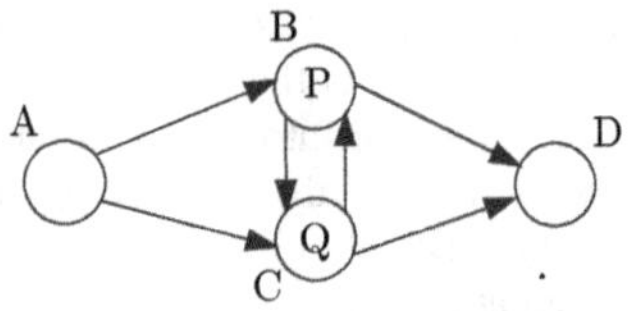

Property: $(true; \lceil P \rceil) \vee (true; \lceil Q \rceil)$

(c)

Figure 4: (a) Models in experiment 1 (b) Execution time of experiment 1 (c) Models in experiment 3

lower than theirs subject to the constraint of the finite observation time.

We implemented the approach using UPPAAL, and showed its efficiency by some examples.

The disadvantage of our approach is that all reference intervals are constrained to be bounded. So, our main future work is to consider how to weaken this constraint.

7. REFERENCES

[1] R. Alur and D. L. Dill. A theory of timed automata. *Theoretical Computer Science*, 126(2):183–235, 1994.

[2] G. Behrmann, A. David, and K. G. Larsen. A tutorial on UPPAAL. In *SFM'04*, pages 200–236, 2004.

[3] V. A. Braberman and D. V. Hung. On checking timed automata for linear duration invariants. In *RTSS'98*, pages 264–273. IEEE Computer Society, 1998.

[4] M. Fränzle. Model-checking dense-time duration calculus. *Formal Aspects of Computing*, 16(2):121–139, 2004.

[5] M. Fränzle and M. R. Hansen. Deciding an interval logic with accumulated durations. In *TACAS'07*, pages 201–215, 2007.

[6] M. Fränzle and M. R. Hansen. Efficient model checking for duration calculus based on branching-time approximations. In *SEFM'08*, pages 63–72, 2008.

[7] M. Fränzle and M. R. Hansen. Efficient model checking for duration calculus. *International Journal of Software and Informatics*, 3(2-3):171–196, 2009.

[8] V. Goranko, A. Montanari, and G. Sciavicco. A road map of interval temporal logics and duration calculi. *Journal of Applied Non-Classical Logics*, 14(1-2):9–54, 2004.

[9] M. R. Hansen. Model-checking discrete duration calculus. *Formal Aspects of Computing*, 6(6):826–845, 1994.

[10] M. R. Hansen and A. W. Brekling. On tool support for duration calculus on the basis of presburger arithmetic. In *TIME'11*, pages 115–122, 2011.

[11] X. Li and D. V. Hung. Checking linear duration invariants by linear programming. In *ASIAN'96*, pages 321–332, 1996.

[12] R. Meyer, J. Faber, J. Hoenicke, and A. Rybalchenko. Model checking duration calculus: a practical approach. *Formal Aspects of Computing*, 20(4-5):481–505, 2008.

[13] B. Moszkowski. A temporal logic for multilevel reasoning about hardware. *Computer*, 18(2):10–19, February 1985.

[14] P. Pandya. Specifying and deciding quantified discrete-time duration calculus formulae using DCVALID. In *RT-TOOLS'01*, 2001.

[15] B. Sharma, P. K. Pandya, and S. Chakraborty. Bounded validity checking of interval duration logic. In *TACAS'05*, pages 301–316, 2005.

[16] E. V. Sorensen, A. P. Ravn, and H. Rischel. Control program for a gas burnew: Part 1: Informal requirements, ProCoS case study 1. Technical report, Department of Computer Science, Technical University of Denmark, 1990.

[17] P. H. Thai and D. V. Hung. Verifying linear duration constraints of timed automata. In *ICTAC'04*, pages 295–309, 2004.

[18] N. Zhan and M. E. Majster-Cederbaum. On hierarchically developing reactive systems. *Inf. Comput.*, 208(9):997–1019, 2010.

[19] M. Zhang, D. V. Hung, and Z. Liu. Verification of linear duration invariants by model checking CTL properties. In *ICTAC'08*, pages 395–409, 2008.

[20] M. Zhang, Z. Liu, and N. Zhan. Model checking linear duration invariants of networks of automata. In *FSEN'09*, pages 244–259, 2009.

[21] C. Zhou. Linear duration invariants. In *FTRTFT'94*, pages 86–109, 1994.

[22] C. Zhou and M. R. Hansen. *Duration Calculus: A Formal Approach to Real-Time Systems*. Springer, 2004.

[23] C. Zhou, M. R. Hansen, and P. Sestoft. Decidability and undecidability results for duration calculus. In *STACS'93*, pages 58–68, 1993.

[24] C. Zhou, C. A. R. Hoare, and A. P. Ravn. A calculus of durations. *Information Processing Letters*, 40(5):269–276, 1991.

Appendix: Proof of Theorem 1

In order to prove Theorem 1, we need the following three lemmas. Lemma 1 certifies the existence of OPCPs for ELDIs of the form $\mathcal{D}_1;\mathcal{D}_2$, which will be a basic case in the proof of the correctness of *BMC-DC*. Based on that, Lemma 2 shows the correctness of the subroutine *Sat*, and Lemma 3 shows the correctness of the procedure *BMC-DC* in general cases.

LEMMA 1. *Let $\rho = (l_0', t_0')(l_1, t_1)\cdots(l_{gc}, t_{gc})$ be the current execution segment in the automaton $\mathcal{A}$ and Φ be $(a \le \ell \le b \Rightarrow \mathcal{D}_1;\mathcal{D}_2)$. Then, the model checking algorithm BMC-DC is correct w.r.t. ρ and Φ, i.e.,*

1. *The algorithm is certain to terminate.*
2. *If the algorithm returns* **true**, *then Φ is satisfied by ρ.*
3. *If the algorithm returns* **false**, *then Φ is not satisfied by ρ.*

PROOF. When checking $\mathcal{D}_1;\mathcal{D}_2$ subject to $a \le \ell \le b$, the effect of *Reset*(0, **false**) is same as

if $d_1 \le c_1$ **then**
  $d_2 := \min(0, d_2)$;

while the return condition by calling *Sat*$(0, -1, -1)$ is $d_2 \le c_2$. So, 1 is obvious.

Regarding to 2, if the algorithm returns **true**, then $gc < a$ or $(gc \ge a \wedge d_2 \le c_2)$ holds. Φ holds trivially for the former case. For the latter case, since $d_2 \le c_2 \ne d_{\max}$, it must have been reset to 0 at some point. Assume the latest reset statement happens at point $t = t_i$ $(0 \le i \le gc)$. As the condition of resetting is $d_1 \le c_1$, $\rho, [t_0, t_i] \models \mathcal{D}_1$ holds. Also, $t = t_i$ is the latest resetting point, so d_2 is exactly the value of the duration expression of $\mathcal{D}_2$ on the interval $[t_i, t_{gc}]$, hence $\rho, [t_i, t_{gc}] \models \mathcal{D}_2$. Therefore, Φ is satisfied by ρ.

As for 3, if the algorithm returns **false**, then both $gc \ge a$ and $d_2 > c_2$ hold. By contraposition, suppose $\rho, [t_0, t_{gc}] \models \mathcal{D}_1;\mathcal{D}_2$, i.e., there exists t_j such that $\rho, [t_0, t_j] \models \mathcal{D}_1$ and $\rho, [t_j, t_{gc}] \models \mathcal{D}_2$. So d_2 must have been reset to 0 at some point. Also, assume the latest reset happens at $t = t_i$ $(0 \le i \le gc)$. Now, we make a case analysis on the relation between t_i and t_j. Firstly, if $t_i = t_j$, then $d_2 \le c_2$ simply holds. Secondly, if $t_i < t_j$, as $t = t_i$ is the latest resetting point and no reset operation happens at point $t = t_j$, the duration expression of $\mathcal{D}_2$ on the subinterval $[t_i, t_j]$ is less than 0. Besides, the duration expression of $\mathcal{D}_2$ on the subinterval $[t_j, t_{gc}]$ is less than or equal to c_2 as $\rho, [t_j, t_{gc}] \models \mathcal{D}_2$ holds. So, d_2 is equal to the summation of the above two parts and thus also less than or equal to c_2. Thirdly, if $t_i > t_j$, the duration expression of $\mathcal{D}_2$ is larger than 0 on $[t_j, t_i]$ as $t = t_i$ is a resetting point. So the duration expression of $\mathcal{D}_2$ is less than or equal to c_2 on $[j, gc]$ implies that the proposition also holds on $[t_i, t_{gc}]$. In summary, in any case the duration expression of $\mathcal{D}_2$ is always less than or equal to c_2 on $[t_i, t_{gc}]$, which contradicts to $d_2 > c_2$ at $t = t_{gc}$. Hence, ϕ is not satisfied by ρ. □

In the following, we give some definitions that will be used in the proof.

Definition 4. We say a subformula ϕ is a *left formula*, if $A[J(\phi)].S \cap A[J(MD(\phi))].S \ne \emptyset$. That is, in $MD(\phi)$, ϕ does not occur as the right operand of any chop operator. Hereafter, conceptually, if *D-Sub*(ϕ) is $\emptyset$, we set $MD(\phi)$ to be the whole formula.

Regarding left formulas, we have

LEMMA 2. *Let $\rho = (l_0', t_0')(l_1, t_1)\cdots(l_{gc}, t_{gc})$ be the current execution segment, ϕ be a left formula with $k \in A[J(MD(\phi))].S$ and $J(\phi) = n$, and $i \ge 0$. Then,*

(1) Sat$(\phi, -1, -1)$ returns **true** *iff $\rho, [t_0, t_{gc}] \models \phi$; and*

(2) Sat(ϕ, k, i) returns **true** *iff $\rho, [t_i, t_{gc}] \models \phi$.*

In order to prove the lemma by induction on formulas, as in [18], we need to define a well-founded order on the formulas of ELDIs, denoted by $<$. To this end, we first define a partial order, denoted by $\prec$ over ELDIs as: $(\phi_1;\phi_2) \prec (\psi_1;\psi_2)$ iff $\phi_1;\phi_2 \Leftrightarrow \psi_1;\psi_2$ and ψ_1 is a proper subformula of ϕ_1. In other words, we assume the left association of ; has a higher precedence. Then, we say $\phi < \psi$ iff either ϕ is a proper subformula of ψ, or $\phi \prec \psi$. It can be proved that $<$ is well-founded, referring to [18] for the detail.

PROOF. By induction on the structure of ϕ w.r.t "$<$".

Base case $\phi = \mathcal{D}_h$. Then for (1)

Sat$(\phi, -1, -1)$ returns **true**
iff $d_h \le c_h$ (line 2 of *Sat*)
iff $\rho, [t_0, t_{gc}] \models \mathcal{D}_h$ ($h \in A[n].S$ and d_h is initialized to 0)

For (2)

Sat(ϕ, h, i) returns **true**
iff $d_h^i \le c_h$ (line 4 of *Sat*)
iff $\rho, [i, gc] \models \mathcal{D}_h$ ($h \in A[n].S$ so is reset to 0 at t_i)

Induction Hypothesis (IH): For any ψ, if $\psi < \phi$, then (1) and (2) hold for ψ.

Inductive steps: We just prove (1) if (2) can be proved similarly for the considered case.

- $\phi = \phi_1 \vee \phi_2$
  Sat$(n, -1, -1)$ returns **true**
  iff *Sat*$(\phi_1, -1, -1)$ returns **true** or
  Sat$(\phi_2, -1, -1)$ returns **true** (line 6)
  iff $\rho, [t_0, t_{gc}] \models \phi_1$ or $\rho, [t_0, t_{gc}] \models \phi_2$
  (IH, as ϕ_1 and ϕ_2 are both left formulas)
- $\phi = \phi_1 \wedge \phi_2$. Because ϕ is a left formula, $A[n]$*.tag* is false. So,
  Sat$(n, -1, -1)$ returns **true**
  iff *Sat*$(\phi_1, -1, -1)$ returns **true** and
  Sat$(\phi_2, -1, -1)$ returns **true** (line 8)
  iff $\rho, [t_0, t_{gc}] \models \phi_1$ and $\rho, [t_0, t_{gc}] \models \phi_2$
  (IH, as ϕ_1 and ϕ_2 are both left formulas)
- $\phi = \neg\phi_1$. Because ϕ is a left formula, $A[n]$*.tag* is false. Thus,
  Sat$(n, -1, -1)$ returns **true**
  iff *Sat*$(\phi_1, -1, -1)$ returns **false** (line 18)
  iff $\rho, [t_0, t_{gc}] \models \neg\phi_1$ (IH, as ϕ_1 is a left formula)
- $\phi = \phi_1;\phi_2$. By induction on the structure of ϕ_2.
  - $\phi = \phi_1;\mathcal{D}_h$
    Sat$(n, -1, -1)$ returns **true**
    iff *Sat*$(\phi_2, -1, -1)$ returns **true** (line 28)
    iff $d_h \le c_h$ (line 2)
    iff $\exists t \bullet (\rho, [t_0, t] \models \phi_1$ and $\rho, [t, t_{gc}] \models \mathcal{D}_h)$
    (d_h has been reset to 0 at t by line 11 of *Reset*, and the property of OPCPs (Lemma 1))
    iff $\exists t \bullet (\rho, [t_0, t] \models \phi_1 \wedge \rho, [t, t_{gc}] \models \mathcal{D}_h)$
    (IH, as ϕ_1 is a left formula)
    iff $\rho, [t_0, t_{gc}] \models \phi_1;\phi_2$
  - $\phi = \phi_1;(\phi_2 \vee \phi_3)$
    Sat$(\phi, -1, -1)$ returns **true**
    iff *Sat*$(\phi_1;\phi_2, -1, -1)$ returns **true** or
    Sat$(\phi_1;\phi_3, -1, -1)$ returns **true**
    (Distributivity of chop over disjunction)
    iff $\rho, [t_0, t_{gc}] \models \phi_1;\phi_2$ or $\rho, [t_0, t_{gc}] \models \phi_1;\phi_3$ (IH)
    iff $\rho, [t_0, t_{gc}] \models \phi_1;(\phi_2 \vee \phi_3)$

- $\phi = \phi_1; (\phi_2 \wedge \phi_3)$. Here $A[A[n].Right].tag$ is true as the conjunction occurs as the right operand of the outmost chop operator. For (1)

$$
\begin{array}{ll}
 & Sat(\phi, -1, -1) \text{ returns } \textbf{true} \\
\text{iff} & Sat(\phi_2 \wedge \phi_3, -1, -1) \text{ returns } \textbf{true} \text{ (line 28)} \\
\text{iff} & \exists j \in [0, gc], h \in A[A[n].Right].S \bullet \\
 & (d_h^j \neq d_{\max} \wedge Sat(\phi_2, h, j) \wedge Sat(\phi_3, h, j)) \text{ (line 13)} \\
\text{iff} & \exists j \in [0, gc], h \in A[A[n].Right].S \bullet \\
 & (d_h^j \neq d_{\max}, \rho, [t_j, t_{gc}] \models \phi_2 \text{ and } \rho, [t_j, t_{gc}] \models \phi_3) \\
 & \text{(IH, as } \phi_2 \text{ and } \phi_3 \text{ are both left formulas)} \\
\text{iff} & \exists j \in [0, gc] \bullet (\rho, [t_0, t_j] \models \phi_1, \\
 & \rho, [t_j, t_{gc}] \models \phi_2 \text{ and } \rho, [t_j, t_{gc}] \models \phi_3) \\
 & (d_h^j \text{ is reset to 0 at } t_j \text{, and line 13 of } Reset) \\
\text{iff} & \exists j \in [0, gc] \bullet (\rho, [t_0, t_j] \models \phi_1, \\
 & \rho, [t_j, t_{gc}] \models \phi_2 \text{ and } \rho, [t_j, t_{gc}] \models \phi_3) \\
 & \text{(IH, as } \phi_1 \text{ is a left formula)} \\
\text{iff} & \rho, [t_0, t_{gc}] \models \phi_1; (\phi_2 \wedge \phi_3)
\end{array}
$$

For (2)

$$
\begin{array}{ll}
 & Sat(\phi, k, i) \text{ returns } \textbf{true} \\
\text{iff} & Sat(\phi_2 \wedge \phi_3, k, i) \text{ returns } \textbf{true} \quad \text{(line 28)} \\
\text{iff} & \exists j \in [i, gc], h \in A[A[n].Right].S \bullet \\
 & (d_h^j \in T(d_k^i) \wedge Sat(\phi_2, h, j) \wedge Sat(\phi_3, h, j)) \quad \text{(line 15)} \\
\text{iff} & \exists j \in [i, gc], h \in A[A[n].Right].S \bullet \\
 & (d_h^j \in T(d_k^i) \text{ and } \rho, [j, gc] \models \phi_2 \text{ and } \rho, [j, gc] \models \phi_3) \\
 & \text{(IH, as } \phi_2 \text{ and } \phi_3 \text{ are both left formulas)} \\
\text{iff} & \exists j \in [i, gc] \bullet (\rho, [t_0, t_j] \models \phi_1 \\
 & \rho, [t_j, t_{gc}] \models \phi_2 \text{ and } \rho, [t_j, t_{gc}] \models \phi_3) \\
 & (d_h^j \text{ is reset to 0 at } t_j \text{, and line 21 of } Reset) \\
\text{iff} & \exists j \in [i, gc] \bullet (\rho, [t_i, t_j] \models \phi_1, \\
 & \rho, [t_j, t_{gc}] \models \phi_2 \text{ and } \rho, [t_j, t_{gc}] \models \phi_3) \\
 & \text{(IH, as } \phi_1 \text{ is a left formula)} \\
\text{iff} & \rho, [t_i, t_{gc}] \models \phi_1; (\phi_2 \wedge \phi_3)
\end{array}
$$

- $\phi = \phi_1; (\neg\phi_2)$. Here $A[A[n].Right].tag$ is also true as the negation occurs as the right operand of the outmost chop operator. For (1)

$$
\begin{array}{ll}
 & Sat(\phi, -1, -1) \text{ returns } \textbf{true} \\
\text{iff} & Sat(\neg\phi_2, -1, -1) \text{ returns } \textbf{true} \text{ (line 28)} \\
\text{iff} & \exists j \in [0, gc], h \in A[A[n].Right].S \bullet \\
 & (d_h^j \neq d_{\max} \wedge \neg Sat(\phi_2, h, j)) \quad \text{(line 23)} \\
\text{iff} & \exists j \in [0, gc], h \in A[A[n].Right].S \bullet \\
 & (d_h^j \neq d_{\max} \text{ and } \rho, [t_j, t_{gc}] \models \neg\phi_2) \\
 & \text{(IH, as } \phi_2 \text{ is a left formula)} \\
\text{iff} & \exists j \in [0, gc] \bullet (\rho, [t_0, t_j] \models \phi_1 \text{ and} \\
 & \rho, [t_j, t_{gc}] \models \neg\phi_2) \\
 & (d_h^j \text{ has to be reset to 0 at } t_j \text{ by line 13 of } Reset) \\
\text{iff} & \exists j \in [0, gc] \bullet (\rho, [t_0, t_j] \models \phi_1 \text{ and } \rho, [t_j, t_{gc}] \models \neg\phi_2) \\
 & \text{(IH, as } \phi_1 \text{ is a left formula)} \\
\text{iff} & \rho, [t_0, t_{gc}] \models \phi_1; \neg\phi_2
\end{array}
$$

For (2)

$$
\begin{array}{ll}
 & Sat(\phi, k, i) \text{ returns } \textbf{true} \\
\text{iff} & Sat(\neg\phi_2, k, i) \text{ returns } \textbf{true} \quad \text{(line 28)} \\
\text{iff} & \exists j \in [i, gc], h \in A[A[n].Right].S \bullet \\
 & (d_h^j \in T(d_k^i) \wedge \neg Sat(\phi_2, h, j)) \quad \text{(line 25)} \\
\text{iff} & \exists j \in [i, gc], h \in A[A[n].Right].S \bullet \\
 & (d_h^j \in T(d_k^i) \text{ and } \rho, [j, gc] \models \neg\phi_2) \\
 & \text{(IH, as } \phi_2 \text{ is a left formula)} \\
\text{iff} & \exists j \in [i, gc] \bullet (\rho, [t_0, t_j] \models \phi_1 \text{ and } \rho, [t_j, t_{gc}] \models \neg\phi_2) \\
 & (d_h^j \text{ has to be reset to 0 at } t_j \text{ by line 21 of } Reset) \\
\text{iff} & \exists j \in [i, gc] \bullet (\rho, [t_i, t_j] \models \phi_1 \text{ and } \rho, [t_j, t_{gc}] \models \neg\phi_2) \\
 & \text{(IH, as } \phi_1 \text{ is a left formula)} \\
\text{iff} & \rho, [t_i, t_{gc}] \models \phi_1; \neg\phi_2
\end{array}
$$

- $\phi = \phi_1; (\phi_2; \phi_3)$.

$$
\begin{array}{ll}
 & Sat(\phi, -1, -1) \text{ returns } \textbf{true} \\
\text{iff} & Sat((\phi_1; \phi_2); \phi_3, -1, -1) \text{ returns } \textbf{true} \\
\text{iff} & \rho, [t_0, t_{gc}] \models (\phi_1; \phi_2); \phi_3 \\
 & \text{(IH as } (\phi_1; \phi_2); \phi_3 \prec \phi_1; (\phi_2; \phi_3)) \\
\text{iff} & \rho, [t_0, t_{gc}] \models \phi_1; (\phi_2; \phi_3) \quad \square
\end{array}
$$

From this lemma, we can conclude the correctness of *BMC-DC* w.r.t. a given execution.

LEMMA 3. *Let $\rho = (l'_0, t'_0)(l_1, t_1) \cdots (l_{gc}, t_{gc})$ be the current execution segment in the automaton $\mathcal{A}$ and Φ be $(a \leq l \leq b \Rightarrow \phi)$. The model checking algorithm BMC-DC is correct, that is*

1. *Termination: The algorithm is certain to terminate.*
2. *Soundness: If the algorithm returns **true**, the ELDI formula Φ is satisfied by ρ.*
3. *Completeness: If the algorithm returns **false**, the ELDI formula Φ is not satisfied by ρ.*

PROOF. From the complexity analysis, we can see that in each execution of *BMC-DC*, the reset operation by *Reset* is executed at most $O(br)$ times, and the basic satisfiability checking by *Sat* is executed at most $O(|\phi| b^h)$ times, so each *BMC-DC* is certain to terminate.

The soundness and the completeness can be directly obtained from Lemma 2 as follows.

$$
\begin{array}{ll}
 & \text{The algorithm returns } \textbf{false} \\
\text{iff} & t_{gc} \geq a \wedge \neg Sat(n, -1, -1) \quad \text{(line 11 of } BMC\text{-}DC) \\
\text{iff} & t_{gc} \geq a \text{ and } \rho, [t_0, t_{gc}] \models \neg\phi \quad \text{(Lemma 2(1))} \\
\text{iff} & \rho, [t_0, t_{gc}] \not\models (a \leq \ell \leq b \Rightarrow \phi) \; (t_{gc} \leq b \text{ always holds}) \\
\text{iff} & \rho, [t_0, t_{gc}] \not\models \Phi \quad \square
\end{array}
$$

Now we can prove Theorem 1, which guarantees the correctness of our approach.

PROOF FOR THEOREM 1.

Termination From the complexity analysis, we can see that the on-the-fly procedure in UPPAAL has at most $O(|S_{\mathcal{A}}|^2 + |S_{\mathcal{A}}| x^b)$ transitions to handle. Moreover, no action is needed for any transition in the first phase, and each transition in the second phase is contained in an execution of *BMC-DC*, which is certain to terminate (Lemma 3(1)). Thus, our approach is also certain to terminate.

Soundness If none of the failure states is reachable, *BMC-DC* always returns **true**. The first phase of the on-the-fly checking can stay on any location of the original automaton, and the second phase checks each execution segment from that location with the length bounded by b, so all the ELDI models of $\mathcal{A}$ have been considered and they all satisfy the formula Φ according to Lemma 3(2). Therefore, Φ is satisfied by the original automaton, which guarantees the soundness of our approach.

Completeness If some failure states are reachable, *BMC-DC* returns **false** at some time, so according to Lemma 3(3), the current execution segment does not satisfy Φ and a counterexample is found. As the current execution segment is an ELDI model of the original automaton, Φ is not satisfied by the original automaton. That guarantees the completeness of our approach. $\square$

Pipelining for Cyclic Control Systems

Arquimedes Canedo
Siemens Corporation
Corporate Technology
Princeton, USA
arquimedes.canedo@siemens.com

Livio Dalloro
Siemens Corporation
Corporate Technology
Princeton, USA
livio.dalloro@siemens.com

Hartmut Ludwig
Siemens Corporation
Corporate Technology
Princeton, USA
hartmut.ludwig@siemens.com

ABSTRACT

This paper presents a novel pipelining technique designed specifically to improve the execution of cyclic control systems and applications in terms of scan cycle time reduction and/or execution of additional workload. Based on the observation that cyclic control systems are tightly coupled with physical processes and that state information (e.g. on/off signals, temperature, pressure) is the most critical data in an application, we present new execution schemes that overlap the execution of multiple cycles in multi-core processors over time. Using an edge detection application benchmark and a software real-time PLC (Programmable Logic Controller) implementation on a quad-core processor, we validate our pipelining method and show the performance scalability of the various schemes. Additionally, we analyze the resilience of our pipelining approach to time delays and propose a speculative execution method to deal with data dependency violations.

Categories and Subject Descriptors

C.3 [**Special-Purpose and Application-Based Systems**]: Real-time and embedded systems; D.1.3 [**Software**]: Concurrent Programming—*Parallel Programming*; D.3.4 [**Programming Languages**]: Processors—*Compilers, Run-time environments*

Keywords

Cyclic control systems; pipelining; scan cycle time; programmable logic controllers; cyber-physical systems; industry automation

1 Introduction

Cyclic control systems are hard real-time systems designed to interact with and control physical processes through sensors and actuators in pharmaceutical, manufacturing, energy, and automotive industries. Inspired by control and information theory, most cyclic control systems execute applications periodically (cyclically). The *scan cycle time* (also known as the sampling rate) refers to the period of time in which the system is expected to (1) read the state of the system under control (plant), (2) compute the corrections to bring the plant to the desired state, and (3) send the corrective commands. Historically, cyclic control systems took advantage of faster uniprocessors to execute more functionality in software and to reduce the scan cycle times for the benefit of control algorithms. Because the entire workload of the application has been executed in a single CPU, sophisticated pre-emptive scheduling algorithms had been developed to guarantee real-time response of these systems [8]. Recent announcements [7, 16, 15, 14] indicate that embedded processor makers are pushing multi-/many-core technology to the industrial automation domain. This has triggered opportunities for researchers and practitioners to explore the benefits of performance, energy efficiency, scalability, consolidation, and redundancy provided by multi-/many-core cyclic control systems. While considering performance, previous work has mainly focused on parallel execution approaches based on application-level parallelization, where different applications are mapped to different cores [2], or in functional parallelization where a compiler identifies data-independent fragments of the program that can be executed at the same time in different cores [4]. Unfortunately, these approaches have fundamental limitations that create unbalanced execution and inhibit a scalable parallelization approach. In this paper, we present a novel highly-scalable parallelization approach for cyclic control systems based on pipelining that significantly reduces the scan cycle time of an application, executes additional workload while maintaining the original configured scan cycle time, or compromises between scan cycle time reduction and additional workload. For the first time, our work presents a general approach for effective application parallelization, considering all the requirements associated with cyclic control systems. The original contributions of this paper are:

1. A new pipelining parallelization method targeting two key performance indicators of cyclic control systems: scan cycle time reduction and execution of additional workload.
2. The characterization of various pipelining execution schemes based on the access of critical state variables in cyclic control systems applications.
3. Analysis and validation of the pipelining for cyclic control systems algorithm and execution schemes based on a software real-time implementation.

The rest of the paper is organized as follows. Section 2 introduces the theory behind cyclic control systems and discusses the requirements imposed on these applications. Section 3 presents our pipelining for cyclic control systems technique and the different execution schemes. Section 4 evaluates the method with a realistic edge detection application and provides insights on its implementation. Section 5 provides the concluding remarks and discusses the future work.

2 Cyclic Control Systems

Modern industrial control systems are hard real-time systems designed to interact with and control physical processes in pharmaceutical, manufacturing, energy, and automotive industries through sensors and actuators. Physical processes typically have continuous states (e.g. temperature, pressure, velocity) that the digital control systems reconstruct to be able to calculate the corrective actions to bring the system to the desired state. From the Nyquist sampling theorem we know that a continuous signal of B Hertz can be reconstructed if the sampling rate exceeds 2B Hertz and therefore

HSCC'13, April 8–11, 2013, Philadelphia, Pennsylvania, USA.

modern control systems must be fast enough to control high-speed physical processes such as chemical and atomic reactions. The sampling time, or frequency between I/O synchronization, of a cyclic control system is known as the *scan cycle time* and it is one of the most important performance metrics because it dictates the cost and ultimately the type of physical process the control system is able to control. The delay in I/O, known as *jitter*, is another critical parameter because it affects the accuracy and therefore the quality of the control. Cyclic control systems are often operating safety-critical systems that are subject to governmental regulations and rigorous validation and verification, and therefore their implementation must guarantee a deterministic execution, jitter-free, deterministic cycle time and I/O behavior of the control software. Cyclic control systems are built on top of a well established computation model and a runtime system.

The computation model of cyclic control systems is based on "automation tasks"[1]. An automation task is the cyclic execution of a control program on regular and equidistant time intervals called *cycles* as shown in Figure 1. This *scan cycle time* is user-configurable and its length is determined by the frequency of the physical system under control. For every cycle, two logical actions are taken: (1) I/O and (2) execution of the program. The I/O event triggers a write/read operations to the *process image* (PI) to update the actuators and read the state of the sensors, respectively. The process image is an addressable memory area that serves as the interface between the control system and the sensors and actuators connected to it and it provides a "snapshot" of the sensors and actuators states. The process image is accessible throughout the execution of a cycle. The actual time required to execute the program within a cycle, including reading inputs and writing outputs, is known as *execution time*. If the execution of the program completes before the next cycle fires (i.e. the execution time is less than the scan cycle time), the program is set into an *idle* state. The automation tasks can be generally programmed and configured in a large variety of proprietary programming languages such as motion control [26, 21], programmable logic controller [13, 19], and robotic languages [23]. The automation task configuration specifies parameters such as priority of the task and the scan cycle time. The hardware configuration describes the type and number of sensors, actuators, I/O modules, and other hardware such as human-machine interfaces (HMI).

The *automation runtime*, or runtime, instantiates the automation tasks defined by the application programmer in the underlying architecture and ensures that the user configuration is honored in terms of priority and scan cycle time of the automation tasks. The runtime is also responsible for monitoring the execution of the program and fire cycle overrun alarms for which the user can write special alarm handling routines. It also monitors and detects bottlenecks such as CPU starvation and provides indication of user configuration errors. The automation runtime is analogous to an operating system in a general purpose computer system, but its implementation must also satisfy the real-time requirements and cyclic execution. In combination, the automation tasks and the automation runtime create a well established model suitable for industrial applications.

The two key performance indicators of any cyclic control system are *scan cycle time* and *jitter*. As a rule of thumb, shorter scan cycle times and absence of jitter are always better. Jitter is defined as the variation in I/O communication that is introduced by hardware and software components. Hardware including the out-of-order execution of the instructions in a CPU, memory access hierarchies (e.g. caches), bus contention, simultaneous multi-threading, and dynamic frequency scaling may generate variable latencies that contribute to jitter [29]. Software including the operating system context switches created by the thread scheduler and interrupt latency can have a significant influence on the jitter of an application [24]. Scan cycle time and jitter have a profound

[1]Motion control task, operational block (OB), IEC 61131-3 tasks.

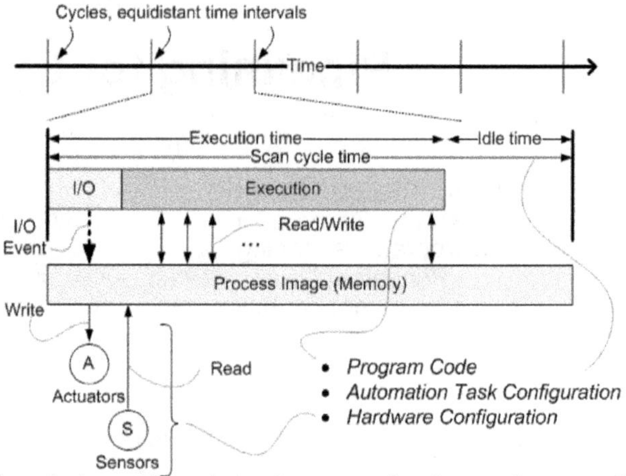

Figure 1: Automation tasks are the essence of cyclic control systems. On every cycle, two logical actions are performed: I/O and Execution. The process image provides to the program a "snapshot" of the state of sensors and actuators.

impact in cyclic control systems. While some applications tolerate jitter effects gracefully, for other applications it is unacceptable and system programmers write subroutines that explicitly handle and compensate for jitter. To understand the effects of scan cycle time and jitter on cyclic control systems it is convenient to distinguish between the types of signals they use to interact with the physical system: analog and discrete signals.

In cyclic control systems processing analog signals (e.g. motion control applications), the velocity with which the actuator (e.g. motor) must bring the system to the desired position is calculated based on the sampled current position, and parameters such as maximum velocity and acceleration. In general, faster scan cycle times allow analog signals, e.g. velocity, to be "more continuous" and therefore to be more accurately approximated. Jitter, however, introduces critical calculation errors of the velocity because the calculation of the velocity relies on precise equidistant sampling points. These systems typically have a scan cycle time of hundreds of microseconds to 1ms and a tolerable jitter between 10-50μs. Cyclic control systems processing digital signals (e.g. programmable logic controllers or PLCs), on the other hand, typically operate on a scan cycle time greater than 1ms and up to 100ms and therefore are more tolerant to jitter. For example, a door security system that opens and closes a hydraulic valve simply cares about a true/false discrete value and jitter of several milliseconds does not affect the correctness of the program. For both, digital and analog control signals, reducing the scan cycle time in the control system is very important because it directly affects its ability to respond faster to inputs and outputs and therefore be more accurate.

The *isochronous* computation principle [25] was introduced to reduce jitter. This model synchronizes all the devices of the control system (sensors actuators, user program, HMI) with the same global clock. This ensures fast and deterministic data exchange between the user program and the I/O to sensors and actuators because a cyclic equidistant clock pulse maintains coherency[2] in the process image, the user program, and the state of the sensors and actuators. Isochronous systems have successfully reduced the jitter to less than 1 microsecond and for this reason we elaborate our ideas based on this principle.

It is important to note that although other alternatives for control systems exist, they are not as dependable as the cyclic control model. One example are the *event-based* control systems that work on the assumption that computation should be performed only when required, in order to avoid unnecessary computation and power consumption when the system under control is in a steady state or not changing. In contrast, the cyclic control

[2]Logical and chronological consistency of data.

model performs computation regardless of the state or changes in the system under control. Although event-based control systems are attractive for saving computation resources, they have been thoroughly investigated [1] and most implementations may suffer from non-deterministic execution [20] and excessive jitter that in certain scenarios, make them not suitable for cyclic control systems.

Historically, the scan cycle time in control systems took advantage of faster clock frequency of microprocessors because the same program is executed in less time and therefore faster scan cycle times were possible. However, due to limitations in the manufacturing process, modern microprocessors provide scalability in terms of multiple processing units (cores) rather than faster clock frequencies. Therefore, some attempts have been made to take advantage of the additional cores in multi-core processors to reduce the scan cycle time in cyclic control systems. An important aspect to observe is that the cyclic control system computation model has the ability to execute multiple automation tasks concurrently and this provides a "natural" source of parallelism that can be exploited by multi-core processors. Figure 2(a) illustrates how the runtime system executes multiple automation tasks (Tasks A, B, C) and system services using pre-emptive time-sharing scheduling algorithms in a single CPU [3]. The addition of multiple cores, as shown in Figure 2(b), allows the execution of automation tasks and system services in dedicated cores, thus reducing the scan cycle time. Notice that the scan cycle time reduction is proportional to the longest automation task (critical path) because the data dependencies is what ultimately limits parallelism [12]. In practice[3], automation tasks in a program are few and not equally balanced in terms of workload. This method is not scalable because modern microprocessors with 8 or more cores are underutilized and the scan cycle time is typically not reduced significantly. The rest of this paper presents a highly-scalable novel parallelization approach that significantly reduces the scan cycle time while effectively utilizing the multiple cores in multi-/many-core processors.

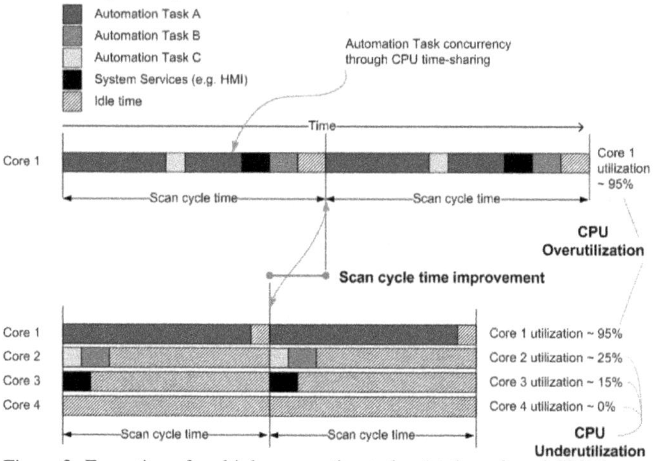

Figure 2: Execution of multiple automation tasks. (a) time-shared CPU for the multiple automation tasks, (b) dedicated execution of automation tasks in multiple cores. Scan cycle time is, in practice, modestly reduced.

3 Pipelining for Cyclic Control Systems

Pipelining refers to the decomposition of computation into *stages* (data processing program fragments) connected in series such that the outputs of a stage are the inputs of the following stage. The key objective of pipelining is to increase the throughput of a system. The fundamental principle behind pipelining is to maximize the utilization of resources by maintaining the stages busy at all times. Immediately after a stage concludes its execution, the result is dispatched to the next stage and computation of the *next* iteration (or data element) begins. If multiple processing units are available, different stages processing contiguous iterations can be executed in parallel. Figure 3 shows the 4-stage (A, B, C, D) pipelined program executed on 4 cores. Computation flows left-to-right and top-down. After stage A reads and processes the inputs, the intermediate results are transferred to stage B, followed by stages C, and D that produces the program outputs. The subscript indicates the iteration number starting from 0 (i.e. A0, B0, C0, D0). Once the *steady state* of the pipeline has been reached in the 4th iteration, all cores are busy processing different stages and iterations of the program (i.e. A3, B2, C1, D0). This type of parallelism is known as *pipeline parallelism* [22][4], and it has been widely exploited in computer architecture [11], signal processing [10, 28], software running on multi-core [6, 27] and other areas [9] because it increases the throughput of a system by allowing for higher sampling rates. Pipeline parallelism does not reduce the execution time of *one program iteration*. In fact, due to the additional data transfer between stages, the execution time may increase. However, the major advantage of pipelining is the increased throughput. The throughput of a pipeline is given by the length of the longest stage, also known as the critical path. If the pipeline stages are equally balanced, the throughput is proportional to the number of stages in the pipeline. In the example given in Figure 3, once the steady state has been reached, the pipelined program produces 4 times more inputs and outputs than the original program in a single core. Notice that while the pipeline scheduling methods in Figure 3(a) and Figure 3(b) are different and may have different performance, both achieve the same results. For example, the scheduling method in Figure 3(a) produces less data transfer but it requires the context switch of the execution of one cycle after every stage of the pipeline completes. The scheduling method in Figure 3(b), on the other hand, does not require context switches and it is easier to read. Therefore, for the rest of the paper we adopt the scheduling method in Figure 3(b) for explaining our concepts but we note that our principles are also applicable to the scheduling method shown in Figure 3(a).

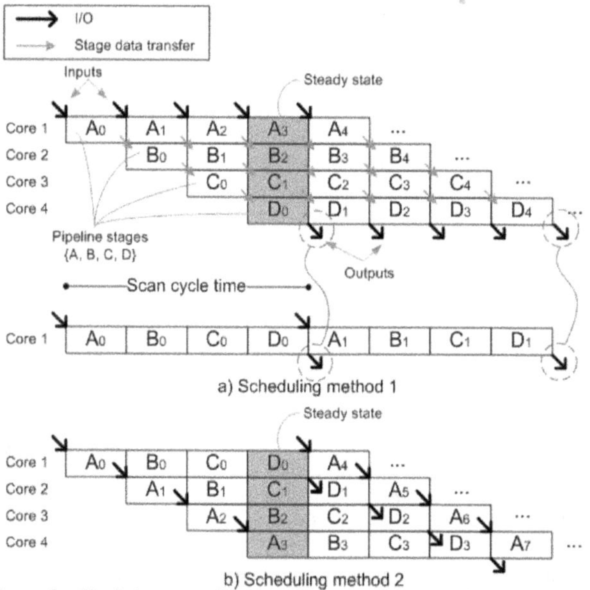

Figure 3: Pipelining a cyclic control system increases the frequency of input/output updates. Different scheduling methods are available, for example, a) dedicates one core for one pipeline stage and b) interleaves the cycles in the available cores.

Delivering small scan cycle times is one of the primary objectives of any cyclic control system (See Section 2). In this Section,

[3]This has not been published but we have identified this pattern through practice.

[4]The granularity of pipeline parallelism we focus on this paper is at the automation task-level, that can be comparable to thread-level pipeline parallelism in general purpose computer systems.

we present a novel run-time execution principle for cyclic control systems based on pipeline parallelism. By using a time-based partitioning approach, we enable the overlapping of contiguous iterations of a cyclic control program in a pipelined manner, without significantly re-architecting the compilers and run-time system. This enables a highly-scalable approach that allows a cyclic control system to:

1. **significantly reduce the scan cycle time of an application,**
2. **execute additional workload while maintaining the original scan cycle time,**
3. **or a hybrid execution of 1) and 2) to compromise between scan cycle time and additional workload.**

The program A shown in Figure 4(a), when pipelined for maximizing the scan cycle time reduction, produces twice as more inputs and outputs as seen in Figure 4(b). This is possible because two contiguous iterations are executed concurrently with an overlap of 50% with respect of each other. Notice that with 2 cores, the pipeline parallelism reduces the scan cycle time down to a factor of 0.5. In general, N cores reduces the scan cycle time at best by the factor 1/N. If the program is pipelined to maximize additional workload, as shown in Figure 4(c), the original scan cycle time remains identical but each iteration is capable of utilizing the otherwise idle time to execute additional instructions. Notice that in the traditional single core execution, increasing workload always results in longer scan cycle time, and therefore fewer updates from/to the I/Os. Furthermore, a compromise between these two objectives (scan cycle time reduction and workload increase) is also possible. Figure 4(d) shows the case where both the scan cycle time can be reduced while increasing the workload per iteration. This approach provides high flexibility for the cyclic control system and for the application programmer because it can be configured according to the requirements of the application. When very small scan cycle times are needed for the reconstruction of analog I/Os, our pipelining method can maximize the scan cycle time reduction (e.g. Figure 4(b)). Similarly, when a scan cycle time is required to stay identical while executing more instructions, our pipelining method can maximize for additional workload. Finally, if both scan cycle time reduction and increased workload are required; our pipelining algorithm can compromise between the two objectives. This configuration is particularly useful for scheduling system services such as HMI in the idle periods.

Existing algorithms [5] for parallelization of cyclic control systems heavily rely on dataflow analysis to discover regions of statements that can be executed in parallel. While this approach works well on applications with high degree of task/data parallelism (e.g. video processing), it is ineffective for the typical applications executed in cyclic control systems. Figure 5 shows the typical distribution of data dependencies in a cyclic control program. Although several tasks are identified (automation task-level parallelism), one single task accounts for more than 90% of the data dependencies of the program[5]. Data dependencies inhibit parallelism and therefore the scalability of existing parallelization approaches because scheduling these programs into multiple cores results in an imbalanced execution where only one core is highly utilized and others are idle for most of the scan cycle time as shown in Figure 2(b). This lack of scalability is also manifested in the limited gains of scan cycle time when using this approach, which is ultimately one of the most important aspects in cyclic control systems.

This paper introduces a novel technique time-based pipelining approach for partitioning cyclic control systems that is easy to introduce in existing control systems, and it is highly scalable across multi-/many-core processors in terms of gains of scan cycle time and/or increased workload. Our pipelining approach addresses the problems in a broad range of industry automation

[5]Breaking a data dependency also corrupts the semantics of a program and this can lead to incorrect execution that may result in severe consequences because cyclic control systems are controlling physical processes.

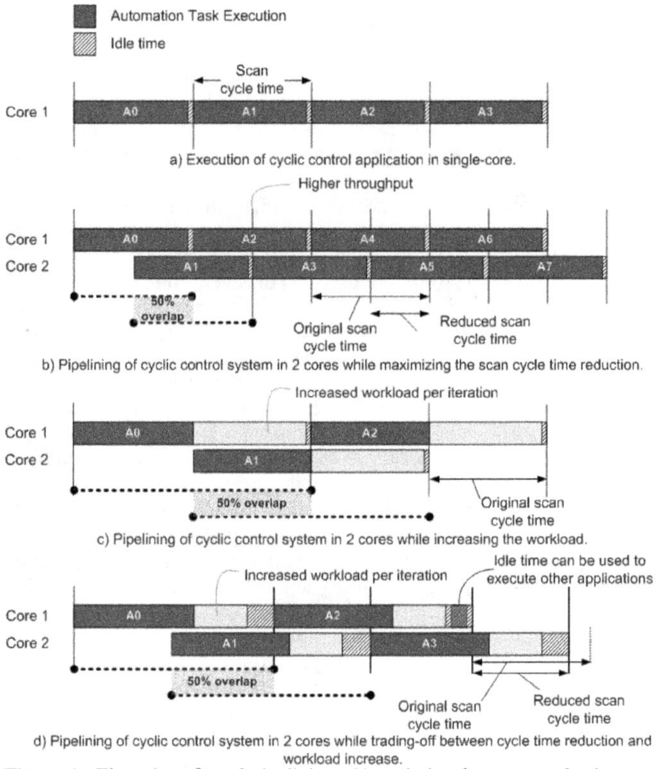

Figure 4: Three benefits of pipelining: b) optimize for scan cycle time reduction; c) optimize for increased workload, d) reduce scan cycle time and increase workload.

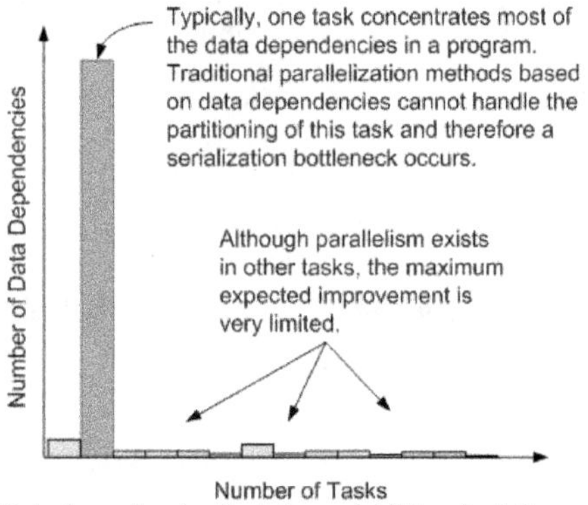

Figure 5: Data dependencies limit the scalability of existing approaches.

applications including analog signal and digital signal processing. Table 1 summarizes the key characteristics of existing application-level parallelization methods and our pipelining approach.

The rest of this section introduces the technical challenges of pipelined execution for cyclic control systems, and we propose an algorithm to maintain correct program semantics and consistency of the process image while overlapping the execution of a cyclic control system to maximize scan cycle time and/or workload.

3.1 Execution Order

The limits of parallelism, including pipeline parallelism, are the data dependencies in a program. Explicit data dependencies are easy to compute by dataflow analysis techniques at compile-time because causality is expressed within the defined local and global variables, and the process image. Implicit data dependencies, on the other hand, are impossible to track at compile-time because of the indirect memory references that are resolved only at run-time. Although run-time data dependence analysis techniques exist, the overhead is high and it directly increases the jitter of the system because there is unforeseeable computation necessary to calculate

Table 1: Qualitative comparison between application-level and pipelining parallelization of cyclic control systems.

	Application-level Parallelization	Pipelining
Scan cycle time reduction	**Limited** by the longest automation task, regardless of the number of available cores.	**Scalable** up to the number of available cores in the system.
CPU Utilization	**Unbalanced**, depends on the load distribution among automation tasks. Some cores are busy while others are idling for long periods of time.	**High**, processing elements execute multiple cycles in multiple cores and load is maintained balanced.
Scheduling	**Static**, automation tasks are scheduled to cores at the application deployment phase. Some data dependency information is not available at compile-time and this may create a conservative parallel scheduling.	**Static/Dynamic**, automation tasks can be scheduled statically or dynamically. In addition to the compile-time information, information from a running program can be used to create more parallelization opportunities.
Configurability	**None**, the program is parallelized according to data dependencies and the user has no control over any performance parameter.	**Flexible**, the user can configure the pipelining for: (1) scan cycle time reduction, (2) increased workload, (3) balance between scan cycle time and increased workload.
Parallelism Granularity	**Application-level**, multiple independent applications (including automation tasks) can be distributed among cores.	**Automation Task-level**, suitable for cyclic control systems handling analog and digital signals.
Engineering Effort	**High**, the user hints the system how to parallelize the program with other system services. Explicit parallel programming primitives are necessary for fine grain parallelism. Compilers, editors, and runtime need to be modified to support parallelism.	**Low/Moderate**, the user simply specifies the functional parameters for the desired scan cycle time and/or additional workload and our pipelining method tunes the program behind the scenes. Identification of critical stateful variables improves the effectiveness of pipelining. Only the runtime requires adjustments to exploit parallelism.

the data dependencies at runtime. Therefore, application-level parallelization techniques perform parallelization only at compile-time and this approach, unfortunately, ignores many other parallelization opportunities hidden in the implicit data dependencies that cause an imbalanced load distribution that inhibit scalability in multi-/many-core cyclic control systems as shown in Figure 2(b). Cyclic control systems call for a light-weight parallelization approach whose run-time performance has a negligible impact in the jitter of the system. Computing explicit data dependencies has a well established practice and therefore we focus on solving the problem of tracking implicit data dependencies because this is ultimately what enables further parallelization opportunities. Paradoxically, rather than tracking the implicit data dependencies in the program to increase the opportunities of parallelization, we rely on the time-based serialization process used by cyclic control systems to identify points in time where other tasks within the same cycle should be preempted. However, rather than using this mechanism for time-sharing multiple tasks in a single CPU, we use it for identifying time-based break-points to create pipeline stages in multiple CPUs. This mechanism guarantees correct execution while honoring both explicit and implicit data dependencies. This light-weight run-time method not only creates valid partitioning of the program into pipeline stages, but also hides the parallelization complexity from the application engineer, the compiler, and the run-time. This is highly beneficial for a cyclic control system because the application engineer does not need to modify existing programs, or to specify parallelization primitives in new programs to take full advantage of the benefits of our pipelining method. Similarly, the compiler remains agnostic from the underlying execution and no significant changes are necessary to generate pipelining-compatible executables. Although the run-time needs to be modified to accommodate pipeline parallelism, the jitter overhead remains the same as in existing systems because the same serialization mechanism is reused to create pipelining opportunities. Figure 6 shows how our algorithm leverages the time-based serialization events generated by the operating system to time-share a single CPU to create pipeline stages and distribute the workload among cores. Notice that while the even numbered iterations (A0, A2, A4, ...) are fired by the original time-based events that dictate the scan cycle time in Core 1, the odd numbered iterations (A1, A3, A5, ...) are fired by an event that occurs exactly a half the scan cycle time in Core 2. Scalability across cores is possible because additional time-based events fire new iterations in different cores. Similarly, scan cycle time reduction and/or workload increase is determined by the points in time when these time-based events occur. Therefore, by leveraging the same serialization events generated by the operating system to exploit concurrency in a single core, we create new pipeline parallelization opportunities without tracking the explicit and implicit data dependencies and maintaining the same execution order as in a uniprocessor implementation. Thus, this existing light-weight mechanism implemented in cyclic control systems can be reused for pipelining applications without incurring in additional and unwanted jitter and performance penalties.

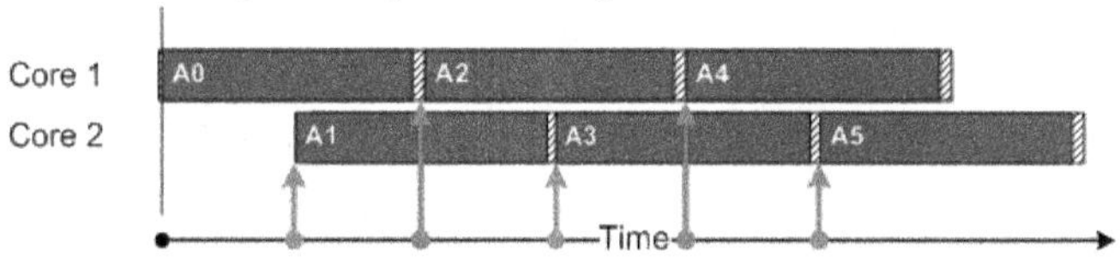

Figure 6: Time-based partitioning of cyclic control applications.

3.2 Process Image Data Consistency

Pipelining a cyclic control application requires special considerations because having multiple iterations being executed simultaneously may cause data inconsistencies in the process image when, for example, data is overwritten by old and invalid data from previous chronological iterations that completes and writes to the process image *after* a future chronological iteration attempts to read that memory location of the process image[6]. Therefore, it is critical that a pipelined cyclic control system relies on the assumption of a *consistent process image during the execution of one cycle* that provides an accurate snapshot of the state of the sensors and actuators in the system. In this Section, we characterize the process image data consistency problem in terms of *where the state of sensors and actuators is maintained* and propose appropriate methods to solve this problem.

3.2.1 Global Process Image

Consider an application whose objective is to detect and count the rising edges of a digital signal. In a given cycle, the state of the input signal (1 or 0) must be compared to the state of the signal on the previous cycle to determine whether there was a rising edge. In other words, on every iteration, the current input's state is stored in memory and used by the next iteration to detect edges. We refer to these type of applications as *stateful* because the state of the external system is maintained within the control application. Similarly, we refer to the variables used to store the state of the system in the application as *stateful variables*. Maintaining the state in memory becomes problematic when the multiple iterations overlap in time because iterations may read old data and cause a miss-detection or a double-detection of a rising edge. Either problem, miss- or double-detection is not acceptable in cyclic control systems and any implementation should guarantee

[6]This situation is a form of a loop carried dependency in dataflow analysis.

that the execution is error-free. Figure 7 shows a pipelined program in 2 cores with 50% overlap where two iterations detect the same rising edge, we refer to this problem as double-detection. Notice that the read of the sensor data occurs at the beginning of the cycle and the write of the state to the global process image occurs at the end of the cycle. In this example, there is a double-detection by A1 and A2 because although A1 identifies a rising edge, it fails to update the global process image before A2 reads the state data. Thus, A2 reads "0" when in reality the state is "1". This problem could be resolved by reducing the gap between the reads/writes of the process image and enforcing that cycles commit data before the following iterations read it. This example shows the importance of maintaining a logically correct process image and also the importance of *the location of process image update within an iteration*.

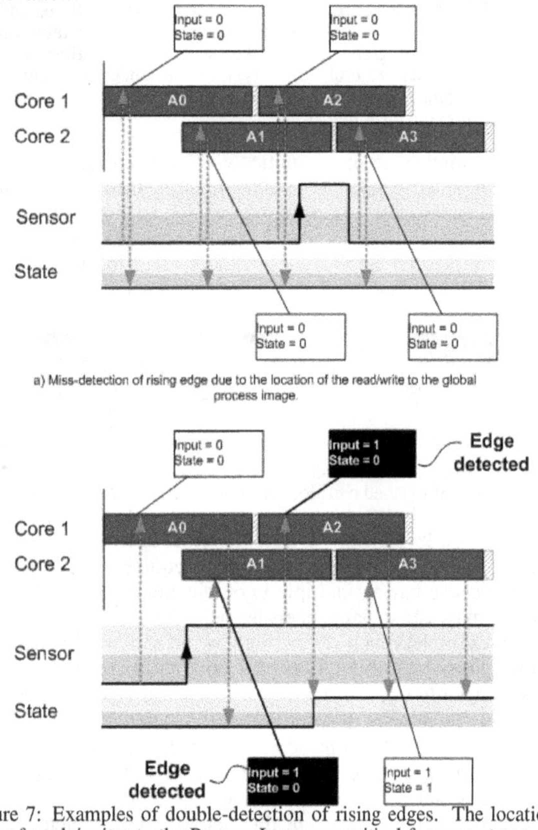
Input = 0
State = 0
Core 1
Core 2
A0
A1
A2
A3
Sensor
State
a) Miss-detection of rising edge due to the location of the read/write to the global process image.
Input = 1
State = 0
Edge detected
Input = 1
State = 1

Figure 7: Examples of double-detection of rising edges. The location in time of reads/writes to the Process Image are critical for correct execution of cyclic control systems.

3.2.2 *Privatized Process Image*

Applications that do not store state information in the program because the state of the physical system is directly read through sensors and modified through actuators are referred to as *stateless* applications. In stateless applications, the state is in the devices (sensors, actuators, and the physical system) and not in the program. One example of a stateless application is an air conditioning control system that periodically reads the temperature of a room (state) to decide whether it is appropriate to turn on or off the cooling system according to a set point. Although the set point is maintained within the application, the state of the system is determined by the sensor data and modified directly through actuator data. Another common example is a motion control application that reads the position of axels (state) from the I/Os to calculate the velocity of the drive for the next cycle. Stateless applications do not present a challenge for pipelining because each iteration is independent from each other and therefore they can be fully parallelized. Although no special consideration is made for pipelining stateless applications, it is important to optimize the process image access of each iteration such that maximum performance in multi-core systems is achieved. The technical challenge for the implementation of pipelining for stateless applications is to feed I/Os as fast as possible to the automation tasks without incurring in unnecessary thread and core contention.

3.3 Time-Based Pipelining for Cyclic Control Systems

Although pipelining is a straightforward concept that allows us to reduce the scan cycle time and/or increase the workload, there are several aspects that need to be considered for an implementation in a cyclic control system. In this Section, we formalize the set of critical parameters related to pipelining in cyclic control systems. These parameters deal with the configuration and correct execution of pipelined cyclic control applications. The `overlap` parameter is used to control the pipelining approach to either minimize the scan cycle time, maximize the workload, or find a compromise between the two. Figure 8 shows how this parameter can accomplish the three pipelining configurations. Figure 8(a) shows that a 0% overlap represents the configuration for maximal workload increase while maintaining the original configured scan cycle time. This is possible because the interleaving of iterations results in idle time that can be used to execute additional workload. The increased workload is proportional to a factor of N×, where N is the number of cores in the system. In this example with 2 cores, the workload increase is 2×. Figure 8(b) shows that (1/N)*100% overlap, where N is the number of cores in the system minimizes the scan cycle time and effectively produces N as many equidistant process image updates. In this example with 2 cores, a 50% overlap produces 2 times more process image updates. To accomplish a compromise between increased workload and cycle time reduction, overlap should be 0 < Overlap < (1/N)*100. This configuration reduces the cycle time and also creates idle periods that can be used to increase the workload. For example, Figure 8(c) shows a 20% overlap in 2 core reduces the scan cycle time and increases the workload that can be executed in the idle times.

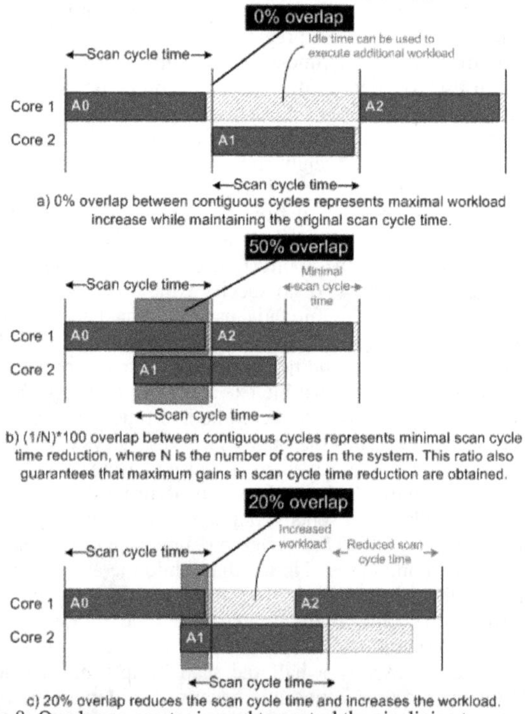
0% overlap
Scan cycle time
Idle time can be used to execute additional workload
Core 1
Core 2
A0
A1
A2
a) 0% overlap between contiguous cycles represents maximal workload increase while maintaining the original scan cycle time.
50% overlap
Minimal scan cycle time
b) (1/N)*100 overlap between contiguous cycles represents minimal scan cycle time reduction, where N is the number of cores in the system. This ratio also guarantees that maximum gains in scan cycle time reduction are obtained.
20% overlap
Increased workload
Reduced scan cycle time
c) 20% overlap reduces the scan cycle time and increases the workload.

Figure 8: Overlap parameter is used to control the pipelining to accomplish (a) maximal workload increase, (b) minimal scan cycle time in N cores while maintaining equidistant updates to the process image, or (c) a combination of both.

As discussed in Section 3.2.1, the update scheme of the process image is very important for stateful applications. Both, the location of the process image update within an iteration, and its relation to the next iteration to avoid breaking the correctness of the process image are critical to avoid double-detections of state changes in the sensor data. As shown in Figure 9, we define two parameters `PI_read` and `PI_update` that control the location within an iteration where sensor data is read from the process image and where the actuator data is written (updated) to the process image, respectively. The location can be anywhere within the cycle and the only rule is to execute the process image reads *before* the process image update. In other words, `PI_read` < `PI_update`.

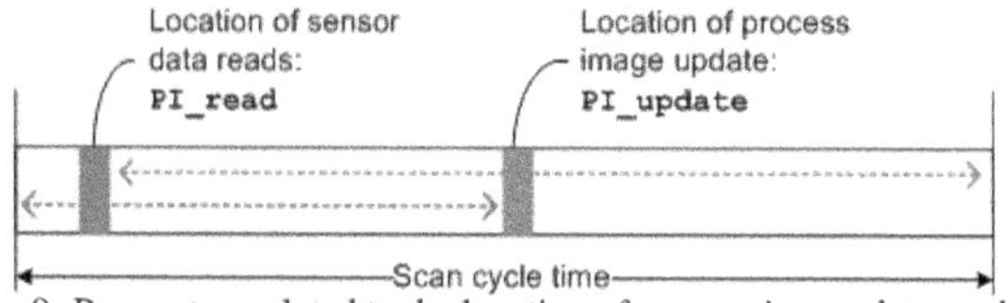

Figure 9: Parameters related to the location of process image data reads and process image update configure the location within a cycle time when these operations occur.

In total, we identify six different pipelined execution schemes that relate to how contiguous iterations interact with each other with respect of the process image access. Figure 10 shows, conceptually, four pipelining schemes (MOST_RECENT_UPDATE, GLOBAL_UPDATE, END_CYCLE_SAME_CORE, and END_CYCLE_DIFFERENT_CORE). These schemes decouple pipeline stages by using a privatized process image for each cycle and forwarding data among cycles' private process images, such that the *next* iteration is correctly *initialized*. What differentiates these pipelining schemes is the location, within a cycle, where the process image data is initialized. The MOST_RECENT_UPDATE scheme in Figure 10(a) defines that a cycle (A3) reads the process image *whenever it is required* (e.g. at the beginning of the cycle) under the assumption that the *producer iteration (e.g. A2) has committed its data*. This scheme performs the process image data reads at the beginning of a cycle and assumes that the previous cycle has committed the most recent process image data. Therefore, this scheme provides limited flexibility to position the process image updates within an iteration because this location should never exceed the overlap limit with respect of the next iteration. The GLOBAL_UPDATE scheme shown in Figure 10(b) is very similar to the MOST_RECENT_UPDATE. The difference is that the GLOBAL_UPDATE uses the shared global memory ("M") instead of the privatized memory copies for each cycle and the read-write accesses to critical variables (i.e. states) are atomic to avoid data dependence violations and incorrect execution. In other words, the overlap is dictated by the read-write chains of critical variables in the global memory where accesses are atomic. This scheme allows for the process image data reads to be anywhere from the beginning of the cycle to the location where the process image is written. The END_CYCLE_SAME_CORE scheme in Figure 10(c) defines that the process image contains the data committed by the *previous iteration allocated to the same core* (e.g. A1→A3). Notice that there is high flexibility to position the process image update anywhere within an iteration because contiguous iterations executed in the same core never overlap and the process image is guaranteed to contain consistent data committed by the previous iteration. The END_CYCLE_DIFFERENT_CORE scheme in Figure 10(d) defines that the process image contains the data committed by the *previous iteration allocated to a different core* (e.g. A2→A3). This provides the strictest scheme because the process image data reads is performed at the beginning of a cycle and the process image update is performed immediately after. However, it still allows for overlapping of contiguous iterations (e.g. A2 and A1).

Two additional pipelining schemes are presented in Figure 11. The MANUAL_UPDATE scheme shown in Figure 11(a) defines that the application programmer determines the overlap between cycles by inserting a marker or annotation in the program that the runtime interprets as a signal to start the next cycle. Thus,

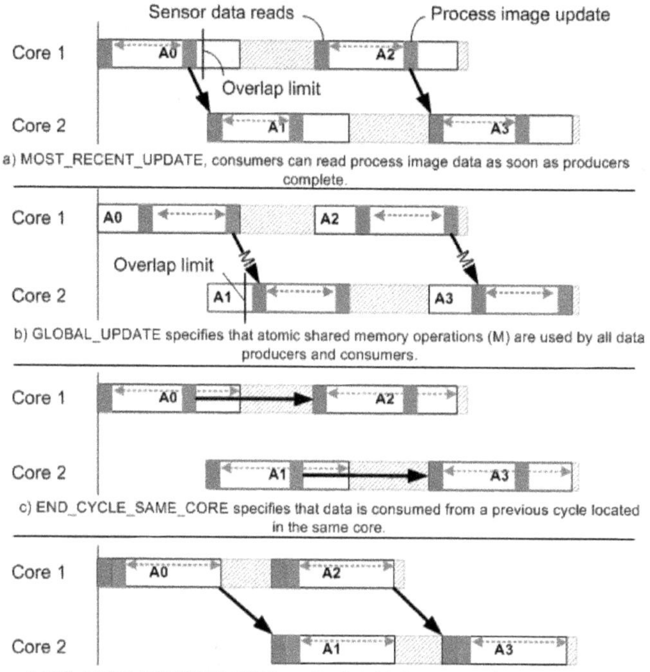

a) MOST_RECENT_UPDATE, consumers can read process image data as soon as producers complete.

b) GLOBAL_UPDATE specifies that atomic shared memory operations (M) are used by all data producers and consumers.

c) END_CYCLE_SAME_CORE specifies that data is consumed from a previous cycle located in the same core.

d) END_CYCLE_DIFFERENT_CORE specifies that data is consumed from a previous cycle located in a different core.

Figure 10: Pipelining execution schemes related to the relationships between contiguous cycles relative to the process image updates.

the MANUAL_UPDATE relies on the manual identification of the stateful variables by the application engineer. We believe it is reasonable to put this burden on the application engineer because no changes in the code are necessary except for the addition of annotations to the typically few and well identified stateful variables. The RESTART_AFTER_VIOLATION scheme defines a speculative execution scheme where cycles are started under the assumption that no data dependence violations exist. However, if a data violation is detected on a critical stateful variable (e.g. A2 in Core1), the incorrect cycles are eliminated from the pipeline (e.g. black A3 in Core 2) and restarted (e.g. A3 restarted in Core 2). Although the eliminated cycle is restarted immediately after the data dependency violation on stateful variables is detected, the overlap between contiguous iterations is reduced and this ultimately affects the cycle time reduction and/or workload increase. In this example, notice that the original overlap is 50% (A0 and A1) and after the violation in A2 occurs, the overlap is reduced to 20% (A2 and A3 (restarted)).

4 Evaluation

This section evaluates our pipelining approach targeting a commercial off-the-shelf quad-core processor. Our run-time implementation is based on a software-PLC[7]. A software-PLC relies on a real-time operating system and real-time multi-threading for the execution of user programs and system services such as communications, scheduler, alarms, and timers. We modified the software-PLC to allow the user program to be executed in multiple threads in such a way that cycles can be assigned to different threads running in different cores. The memory management sub-system of the software-PLC was modified to reflect the read-write policies of the six identified pipelining schemes (See Section 3.3). The time-based pipeline partitioning algorithm was implemented by reutilizing the existing time-sharing mechanism for uniprocessor execution. We tested our implementation in a quad-core x86 running Windows XP with real-time extensions [18], and profiled it using Vtune Amplifier [17]. We use an edge-detection application benchmark

[7]Soft-PLC or software programmable logic controller is based on a personal computer rather than specialized hardware.

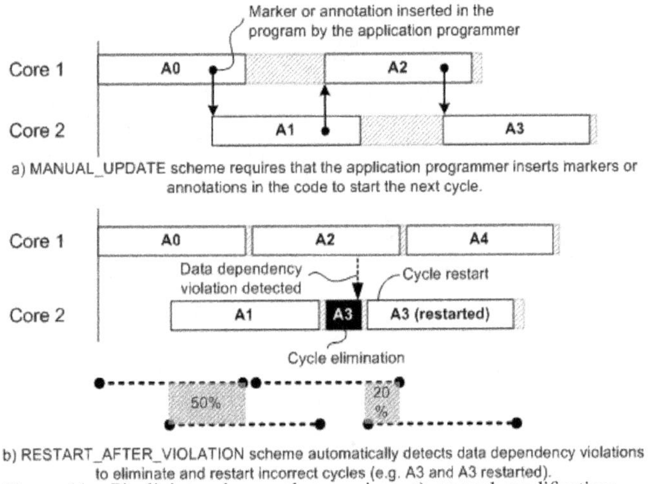
Marker or annotation inserted in the program by the application programmer
Core 1
Core 2
A0
A2
A1
A3
a) MANUAL_UPDATE scheme requires that the application programmer inserts markers or annotations in the code to start the next cycle.
Core 1
Core 2
A0
A2
A4
Data dependency violation detected
Cycle restart
A1
A3
A3 (restarted)
Cycle elimination
50%
20 %
b) RESTART_AFTER_VIOLATION scheme automatically detects data dependency violations to eliminate and restart incorrect cycles (e.g. A3 and A3 restarted).

Figure 11: Pipelining schemes that require: a) manual modifications by the application programmer to signal the start of the next cycle; b) speculative execution where cycles are eliminated and restarted when a data dependence violation is detected at runtime.

very common in industrial applications whose functionality is to count the number of rising edges in a digital signal. One important aspect of this benchmark is that the state of the previous signal is stored in the program and not on the physical device, thus making the cyclic control system a stateful application.

4.1 Performance Envelope of Pipelining Execution Schemes

The amount of overlap, and therefore the potential gains in cycle time reduction and/or workload increase, is determined by the data dependencies in the stateful variables in a program referred to as *stateful dependencies*. Data dependencies can occur at any position within a cycle. For example, a stateful data dependency exactly in the middle of a cycle allows a maximum overlap of 50% within two contiguous iterations. If the dependency is located at the beginning of the cycle then it allows for almost a full overlap, and if located at the end of the cycle then it almost prohibits any overlap. Notice that dependencies are important only when the stateful variables are written and reads can be overlapped without restrictions. It is also important to note that when cycles do not overlap, the gain can be measured in terms of additional workload. In addition to the location of stateful data dependencies, the pipeline execution scheme determines the gains in cycle time and/or workload performance. Consider an application with a single stateful dependency in the middle of the cycle. The MOST_RECENT_UPDATE scheme would allow at most 50% overlap and therefore 2× reduction on cycle time, regardless of the number of available cores. On the other hand, the END_CYCLE_SAME_CORE scheme would allow maximum overlap because the dependency remains within the same core and therefore a N× reduction of cycle time, where N is the number of cores in the system. Figure 12 analyzes the scalability and performance envelope in terms of cycle time reduction of the main four pipelining execution schemes. For every execution scheme, we analyze how the locations of stateful data dependencies affect the cycle time performance gains. For example, Figure 12(a) shows the performance envelope (higher is better) for MOST_RECENT_UPDATE scheme when a stateful dependency occurs in the Middle, Beginning, and End of the cycle. Notice that this location affects the scalability of the scheme. These results show that this scheme scales proportionally to the number of available cores when the stateful dependency occurs at the Beginning but does not improve the cycle time when the dependency occurs at the End of the cycle. Similarly, the maximum cycle time improvement is of about 2× when the dependency occurs in the Middle of the cycle, regardless of the number of available cores. This is due to the fact that a dependency in the middle of a cycle allows for at most 50% overlap. Figure 12(b) shows the results for GLOBAL_UPDATE scheme and are identical to the results of MOST_RECENT_UPDATE scheme because even though the implementation differs, privatized memory and shared global memory, the semantics of execution are the same. Figure 12(c) shows the results for END_CYCLE_SAME_CORE and Figure 12 for END_CYCLE_DIFFERENT_CORE. These results show that these schemes are insensitive to the location of data dependencies and provide the best opportunities for pipeline parallelization and reducing the scan cycle time. This occurs because there will always be a *fully completed cycle* (with the exception of the first iterations that can solve this problem by using the initial values), either in the same or in a different core, and the body of the current cycle can be overlapped completely with its immediate predecessor. These results provide an important insight for the implementation of pipelining algorithms for cyclic control systems because they outline the general strategy to follow in various types of applications where data dependencies occur at different locations within a cycle.

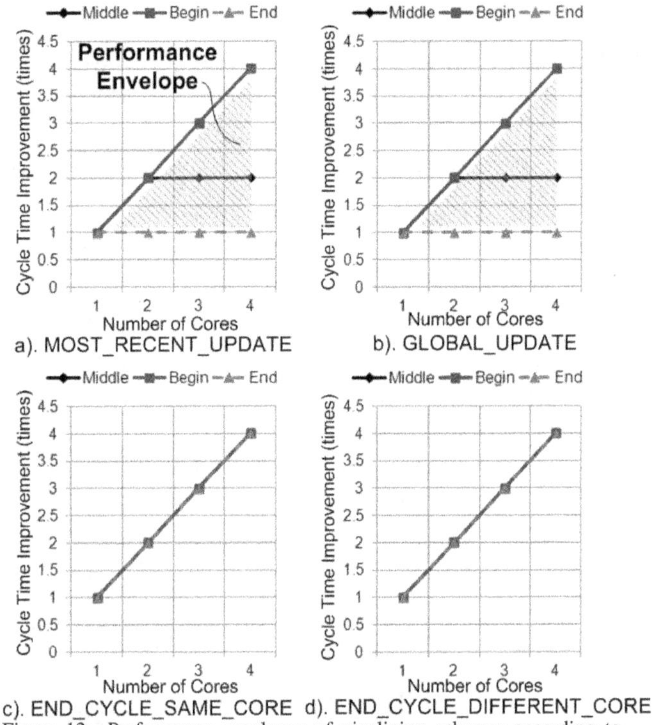
Middle
Begin
End
Performance Envelope
Cycle Time Improvement (times)
Number of Cores
a). MOST_RECENT_UPDATE
b). GLOBAL_UPDATE
c). END_CYCLE_SAME_CORE
d). END_CYCLE_DIFFERENT_CORE

Figure 12: Performance envelopes of pipelining schemes according to different configurations of the process image update (Middle, Beginning, End).

4.2 Resilience of Pipelining Execution Schemes in the Presence of Time Delays

Enforcing correct pipelined execution requires a mechanism to honor the stateful data dependencies in the application. The presence of time delays within a cycle can cause the location of critical stateful dependencies to shift in time and create violations that lead to incorrect execution of the application. It is therefore very important to analyze the resilience of the various pipelining execution schemes when time delays occur and skew these dependencies in time. We analyze the effects of time delays in the pipelining execution schemes, in various configurations of overlapping and location of dependencies, by randomly inserting time delays within the cycles and measuring the number of detected edges in an edge detection benchmark. Figure 13 shows the results of our experiments for GLOBAL_UPDATE, MOST_RECENT_UPDATE, END_CYCLE_DIFFERENT_CORE, and END_CYCLE_SAME_CORE schemes in 2 cores. We use GLOBAL_UPDATE as the baseline for our results because the atomic read-write operations in the global memory guarantee correct executions, as only one

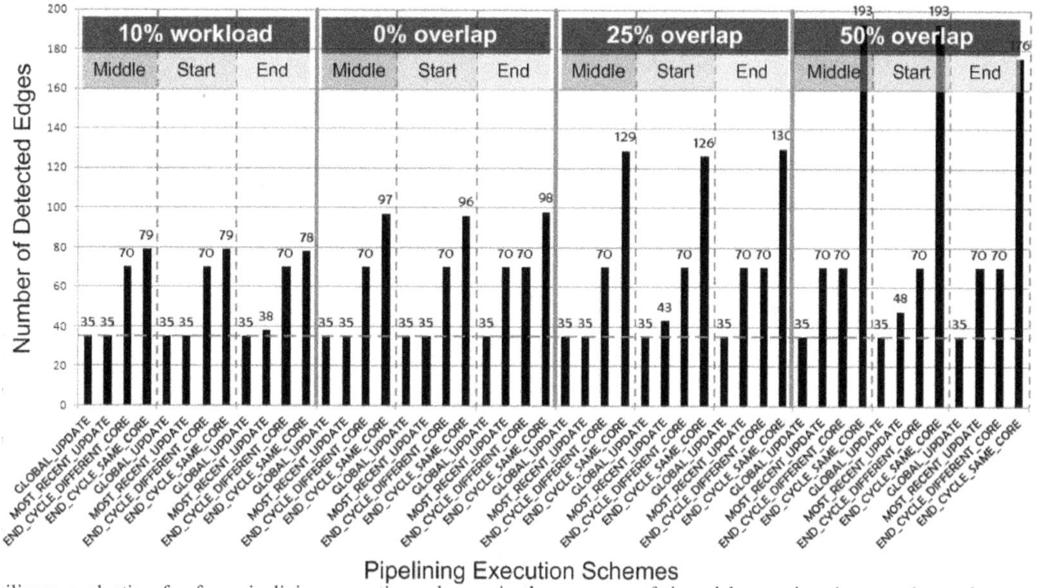

Figure 13: Resilience evaluation for four pipelining execution schemes in the presence of time delays and various overlap and process image update configurations in 2 cores.

core is able to access the critical stateful variables at a time. The graph is divided in four groups representing different pipeline configurations for workload increase (10% workload) and scan cycle time degrees of overlap (0%, 25%, and 50%). Each group contains three sets of results indicating the location of the critical stateful dependency within a cycle (Middle, Start, End). Finally, every set contains the results for the four pipelining execution schemes. The total number of rising edges generated by the simulated physical system is exactly 35 for all experiments. As expected, GLOBAL_UPDATE (our baseline) is the only scheme that produces correct results across all configurations of overlap and location of dependencies. These results show that as the pipelining overlap increases, the execution schemes are more sensitive to time delays that produce double edge detections and therefore incorrect results. In particular, END_CYCLE_DIFFERENT_CORE and END_CYCLE_SAME_CORE are very sensitive to time delays. The most important observation of these results is that the implementation of the pipelining execution schemes calls for a mechanism to prevent critical stateful dependency violations. In the following Section we evaluate two additional execution schemes that prevent stateful data dependence violations in a program.

4.3 Analysis of Manual and Speculative Pipeline Execution Schemes

We identify two schemes that can increase the robustness of our pipelining method against time delays. The MANUAL_UPDATE and RESTART_AFTER_VIOLATION schemes require that the application programmer identifies the stateful variables through annotations in the application code which is a drawback. However, application programmers typically know the application very well and are qualified to take these decisions. The main difference between these two schemes is the way the next cycle is started. MANUAL_UPDATE also requires a marker in the application code inserted by the application programmer that identifies the location within the cycle where the next cycle can be started without causing a stateful data dependency violation. RESTART_AFTER_VIOLATION, on the other hand, speculatively starts the execution of the next cycle according to the configured overlap and assuming that no stateful violations occur. In case that a data dependency violation on a stateful variable is detected, the speculative cycles are eliminated from the pipeline and immediately restarted. It must be noted that several restarts may occur before the stateful variable is free of violations. Data dependency violations in stateful variables are easy to detect by the runtime because the application programmer has annotated these in the source code. In terms of performance, the difference between these two schemes is the CPU utilization. MANUAL_UPDATE, because the location of the start of the next cycle is known a-priori, the idle time can be utilized for executing additional workload (See Figure 11(a)). RESTART_AFTER_VIOLATION, on the other hand, maintains the CPUs utilized at all times with either speculative cycles or restarted cycles (See Figure 11(b)). Figure 14 shows the performance evaluation of MANUAL_UPDATE and RESTART_AFTER_VIOLATION schemes in terms of average edge detection time in our edge detection benchmark. The average edge detection time metric in the edge detection application is equivalent to the scan cycle time reduction. The program is configured with a scan cycle time of 100ms and a single data dependence located exactly at 50ms of every cycle. This location allows for overlaps from 0% to 50% and we measure the performance on 10% increments. As expected, notice that for both schemes, the average detection time decreases as the overlap increases. The RESTART_AFTER_VIOLATION scheme is slightly faster to detect edges when compared to the MANUAL_UPDATE scheme because rather than waiting for a signal to begin the next cycle, it speculatively starts the next cycle and eliminates the need for additional synchronization. In summary, these results show that on 2 cores, these methods can reduce the average edge detection time from 100ms to about 75ms. Notice that in the 50% overlap configuration for RESTART_AFTER_VIOLATION scheme, about 43% of the cycles were eliminated from the pipeline and restarted.

Figure 15 shows the number of restarted cycles in the RESTART_AFTER_VIOLATION scheme when data dependencies occur at the Middle, Beginning, End, and Anywhere (Variable) of the cycle in various overlap configurations from 10% to 90%. The total number of successful cycles is 330 for all experiments. As expected, the restarted cycles correlate to the location of the data dependency. For example, when the data dependency occurs in the Middle, restarts are measured when cycles overlap 50% or more. When the data dependency is located in the Beginning of the cycle, restarts are expected when cycles overlap 90% or more because almost the entire cycle can be overlapped with the next. Similarly, when the data dependency is located at the End of the cycle, restarts are expected when the cycles overlap 20% or more because only a small fragment of a cycle can be overlapped with the next. Additionally, we test the case when the data dependency location is unknown and occurs anywhere in the cycle. These results confirm that when dependencies occur in the Beginning to

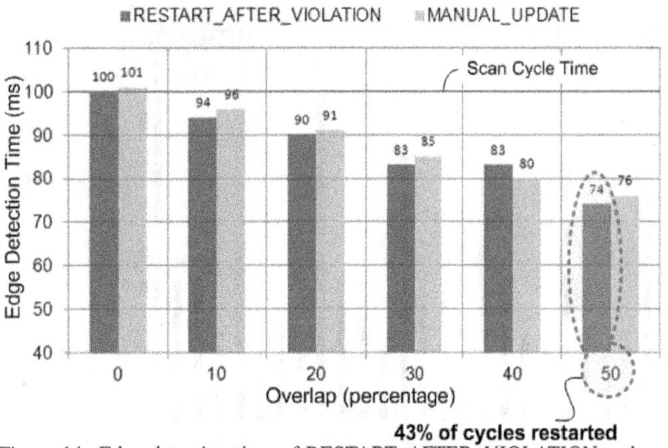

Figure 14: Edge detection time of RESTART_AFTER_VIOLATION and MANUAL_UPDATE schemes.

the Middle of a cycle, not all cycles are restarted and eliminated. In the case where the dependency occurs at the End, on the other hand, every cycle that successfully detects an edge needs to be restarted twice, on average. The variable location of the data dependency presents an interesting case that we will investigate in detail in our future work.

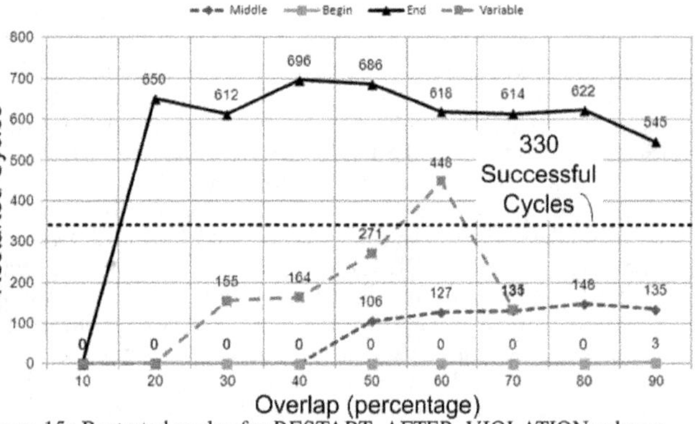

Figure 15: Restarted cycles for RESTART_AFTER_VIOLATION scheme in various configurations of overlap and location of the data dependency.

5 Conclusion

This paper introduced a novel method for enabling parallel processing of cyclic control systems and improve the scan cycle time and/or the execution of additional workload. Our technique uses the pipelining principle to overlap contiguous cycles of cyclic control systems stateful applications over time. To honor the semantics of the program, we presented six pipelining schemes that take into account the dependencies in stateful variables in the application to execute cycles in parallel. Using a realistic edge detection benchmark application and an implementation in a software real-time programmable logic controller on a quad-core processor, we evaluated various aspects of our pipelining methodology and provide the following key insights:

- The scalability of our pipelining approach in terms of improved scan cycle time and/or workload increase is a function of the number of available cores, the execution scheme, and the location of stateful dependencies within the cycle.
- Pipelining in cyclic control systems is very sensitive to delays within the cycle that may cause data dependency violations that, for example, cause double edge detection problems.
- Sensitivity to delays can be mitigated with atomic global memory access but this inhibits parallelization opportunities and therefore scalability.
- Identification of stateful variables in the application code by the application programmer improves pipelining by allowing the programmer to explicitly decide when to start the next cycle, or to speculatively execute the next cycle and eliminate and restart cycles whenever a violation in the stateful variables is detected.
- Speculative pipeline execution is a promising approach because it can be implemented in the runtime system and provides good performance when dependencies occur at the beginning, the middle, or are randomly distributed within a cycle.

6 References

[1] IEC 61499. International Standard for Distributed Systems. http://www.iec61499.de/.

[2] Beckhoff. TwinCAT 3. http://www.beckhoff.com.

[3] M. S. Boggs and A. D. McNutt. Systems, Devices, and/or Methods for Managing Programmable Logic Controller Processing, 03 2009.

[4] A. Bordelon, R. Dye, H. Yi, and M. Flechter. Automatically Creating Parallel Iterative Program Code in a Data Flow Program, 12 2010.

[5] A. Canedo and M. Al-Faruque. Towards Parallel Execution of IEC 61131 Industrial Cyber-Physical Systems Applications. In *Design Automation and Test in Europe (DATE)*, pages 554–558, 2012.

[6] A. Douillet and G. R. Gao. Software-Pipelining on Multi-Core Architectures. In *Proceedings of the 16th International Conference on Parallel Architecture and Compilation Techniques*, PACT '07, pages 39–48, Washington, DC, USA, 2007. IEEE Computer Society.

[7] Freescale. QorIQ Processing Platforms - Industrial. http://www.freescale.com.

[8] T. L. Fulton, L. Trapp, and H. Fuchs. Devices, Systems, and Methods Regarding a PLC System Fault, 07 2010.

[9] B. Gedik, H. Andrade, K.-L. Wu, P. S. Yu, and M. Doo. SPADE: the system s declarative stream processing engine. In *Proceedings of the 2008 ACM SIGMOD international conference on Management of data*, SIGMOD '08, pages 1123–1134, New York, NY, USA, 2008. ACM.

[10] M. I. Gordon, W. Thies, and S. Amarasinghe. Exploiting Coarse-grained Task, Data, and Pipeline Parallelism in Stream Programs. In *Proceedings of the 12th international conference on Architectural support for programming languages and operating systems*, ASPLOS-XII, pages 151–162, New York, NY, USA, 2006. ACM.

[11] J. L. Hennessy and D. A. Patterson. *Computer Architecture, Fourth Edition: A Quantitative Approach*. Morgan Kaufmann Publishers Inc., San Francisco, CA, USA, 2006.

[12] M. Hill and M. Marty. Amdahl's Law in the Multicore Era. *Computer*, 41(7):33 –38, july 2008.

[13] IEC. IEC 61131-3 Edition 2.0, Programmable controllers - Part 3: Programming Languages. http://www.iec.ch, 2011.

[14] Infineon. Infineon automation. www.infineon.com.

[15] T. Instruments. Stellaris MCU for Industrial Automation. www.ti.com.

[16] Intel. Technical Resources for Embedded Designs with Intel Architecture. http://www.intel.com/.

[17] Intel. VTune Amplifier XE 2011. http://software.intel.com/.

[18] IntervalZero. RTX Hard Real-Time Platform. http://www.intervalzero.com.

[19] K.-H. John and M. Tiegelkamp. *IEC 61131-3: Programming Industrial Automation Systems*. Springer, 2010.

[20] E. A. Lee. Disciplined Heterogeneous Modeling. In *MODELS 2010, Keynote Talk*, 2010.

[21] National Instruments. Fundamentals of Motion Control. http://www.ni.com/white-paper/3367/en.

[22] A. Navarro, R. Asenjo, S. Tabik, and C. Cascaval. Analytical Modeling of Pipeline Parallelism. In *Parallel Architectures and Compilation Techniques, 2009. PACT '09. 18th International Conference on*, pages 281 –290, sept. 2009.

[23] I. Pembeci and G. Hager. A Comparative Review of Robot Programming Languages, 2001.

[24] F. M. Proctor and W. P. Shackleford. Real-time Operating System Timing Jitter and its Impact on Motor Control. In *SPIE Sensors and Controls for Intelligent Manufacturing*, pages 10–16, Oct 2001.

[25] Siemens. Isochrone Mode, Function Manual, 2006.

[26] Siemens. Motion Control. http://http://www.automation.siemens.com.

[27] N. Vachharajani, R. Rangan, E. Raman, M. J. Bridges, G. Ottoni, and D. I. August. Speculative Decoupled Software Pipelining. In *Proceedings of the 16th International Conference on Parallel Architecture and Compilation Techniques*, PACT '07, pages 49–59, Washington, DC, USA, 2007. IEEE Computer Society.

[28] H. Wei, J. Yu, H. Yu, M. Qin, and G. R. Gao. Software Pipelining for Stream Programs on Resource Constrained Multi-core Architectures. *IEEE Transactions on Parallel and Distributed Systems*, 99(PrePrints), 2012.

[29] M. Westmijze, M. Bekooij, G. Smit, and M. Schrijver. Evaluation of scheduling heuristics for jitter reduction of real-time streaming applications on multi-core general purpose hardware. In *9th IEEE Symposium on Embedded Systems for Real-Time Multimedia, ESTIMedia 2012*, pages 140–146, 2011.

Optimal CPU Allocation to a Set of Control Tasks with Soft Real–Time Execution Constraints*

Daniele Fontanelli
Dipartimento di Scienza e Ingegneria dell'Informazione,
University of Trento
Via Sommarive, 5
Povo, Trento, Italy
fontanelli@disi.unitn.it

Luigi Palopoli
Dipartimento di Scienza e Ingegneria dell'Informazione,
University of Trento
Via Sommarive, 5
Povo, Trento, Italy
palopoli@disi.unitn.it

Luca Greco
LSS - Supélec
3, rue Joliot-Curie
91192 Gif sur Yvette, France
lgreco@ieee.org

ABSTRACT

We consider a set of control tasks sharing a CPU and having stochastic execution requirements. Each task is associated with a deadline: when this constraint is violated the particular execution is dropped. Different choices of the scheduling parameters correspond to a different probability of deadline violation, which can be translated into a different level for the Quality of Control experienced by the feedback loop. For a particular choice of the metric quantifying the global QoC, we show how to find the optimal choice of the scheduling parameters.

Categories and Subject Descriptors

J.7 [**Computer applications**]: Computers in other systems—*Command and Control, Real–time*

Keywords

Real–time Scheduling, Embedded Control; Stochastic Methods

1. INTRODUCTION

The challenging issues posed by modern embedded control systems are manifold. First, there is a clear trend toward an aggressive sharing of hardware resources, which calls for an efficient allocation of resources. Second, the use of modern sensors such as video cameras and radars, generates a significant computation workload. More importantly, the computation activities required to extract the relevant information from these sensors are heavily dependent on the input data set. Thereby, computation and communication requirements can change very strongly in time. As an example in Figure 1, we show the histograms for the computation time of the different activations of a tracking task used in a visual servoing application. As evident from the picture,

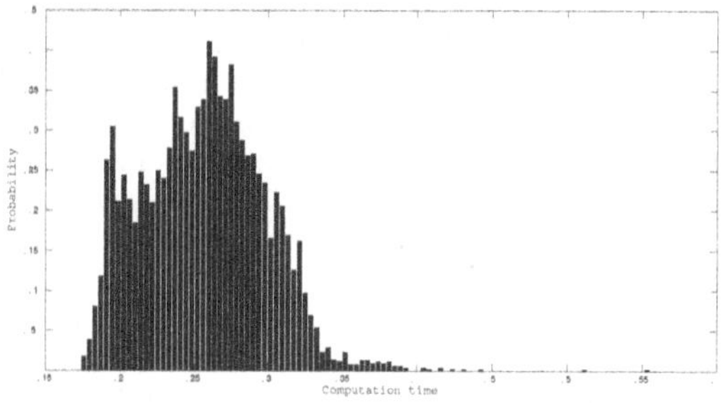

Figure 1: Histrograms for a tracking application

the distribution is centered around 25 ms but has very long tails (some executions required more than 50 ms). Another important element of information is that the level of reliability required to the software implementation of an embedded control system is necessarily very high.

The time varying computation delays and the scheduling interference (i.e., the additional delay that a task suffers for the presence of higher priority tasks) undermine the classical assumptions of digital control: constant sampling time and fixed delay in the feedback loop. The standard solution to this problem relies on the combination of the time–triggered approach [13] and of the real–time scheduling theory. The former nullifies the fluctuations in the loop delay by forcing the communications between plant and controllers to take place at precise points in time. The latter ensures that all activities will *always* be able to deliver the results at the planned instants. This approach is sustainable in terms of resource utilisation if the worst case execution requirements remain close to the average case, which is not a realistic assumption for the new generations of control applications.

Many researchers have challenged the idea that a control system can be correctly designed only if we assume a regular timing behaviour for its implementation. They have investigated on *how to make the design robust* against an irregular timing behaviour of the implementation, focusing on such effects as packet dropout [19, 15], jitter in computation [17, 14] and time varying delays [12]. Other authors have sought suitable ways to modify the scheduling behaviour in overload conditions. In this research line fall the work of Marti [3],

*This work was partially supported by the HYCON2 NoE, under grant agreement FP7-ICT-257462

HSCC'13, April 8–11, 2013, Philadelphia, Pennsylvania, USA.

who proposes to re-modulate the task periods in response to an overload condition. In the same direction goes the work of Lemmon and co-workers [6]. More recently, the onset of a new class of control algorithms, event-triggered [24, 30] or self-triggered [18, 28] dismisses the very idea of periodic sampling and advocates the execution of the control action only when necessary, while innovative models for control applications based on the idea of anytime computing have been explored [11, 2, 23].

In this work, we consider a set of control tasks sharing a CPU and target the problem of finding the scheduling parameters that strikes the best trade–off between the control performance experienced by the different tasks. Our starting point is the consideration that hard real–time scheduling is not necessarily a suitable choice for a large class of control system. This is consistently revealed by both experimental studies [5, 4] and recent theoretical results [10].

Finding the best period assignment that maximises the overall performance of a set of tasks is a well known problem investigated by several authors assuming a fixed priority or an Earliest Deadline First (EDF) scheduler [27, 22, 31]. Other authors have approached the problem of control oriented scheduling of a CPU or of a bus using off-line techniques [26, 20, 29, 16]. In the present work, we assume the presence of a soft real–time scheduling algorithm such as the resource reservations [25, 1]. A key feature of these algorithms is their ability to control the amount of CPU devoted to each task (bandwidth). This allows us to estimate the probability of meeting a deadline for a given bandwidth allocation with fixed periods and, ultimately, to expose the dependence of the Quality of Control (QoC) from the scheduling parameters. We consider as metrics for QoC the asymptotic covariance of the output of the controlled system, assuming that its evolution is affected by process noise. The choice of this metrics enables us to *set up an optimisation problem* where the global QoC is given by the worst QoC achieved by the different control tasks, which is optimised under the constraint that the total allocated bandwidth does not exceed the unity. In a preliminary version of this work [9], we have restricted our analysis to the control of scalar systems with a simplified model of computation. This made for an analytical solution of the problem. In this paper, we extend the reach of our analysis to the case of multi–variable systems controlled with a rather standard model of computation. We characterise the region of feasible choices for the bandwidth of each tasks (choices that guarantee mean square stability). Then we identify conditions under which the problem lends itself to a very efficient numeric search for the optimum. The resulting optimisation algorithm is the key contribution of the paper.

The paper is organised as follows. In Section 2, we offer a description of the problem with a clear statement of assumptions and constraints. In Section 3, we show the computation of the QoC metrics adopted in this paper. In Section 4, we show the closed form expression for the optimal solution, along with the particular form that it takes in some interesting special cases. In Section 5 we collect some numeric examples that could clarify the results of the paper. Finally, in Section 6 we offer our conclusions and announce future work directions.

2. PROBLEM PRESENTATION

We consider a set of real–time tasks sharing a CPU. In our a setting a task $\tau_i, i = 1, \ldots, n$ is a piece of software implementing a controller. Tasks have a cyclic behaviour: when an event occurs at the start time $r_{i,j}$, task τ_i generates a job $J_{i,j}$ (the j-th job). In this work, we will consider periodic activations: $r_{i,j} = jT_i$ with $j \in \mathbb{Z}_{\geq 0}$. Job $J_{i,j}$ finishes at the finishing time $f_{i,j}$, after using the resource for a computation time $c_{i,j}$. If job $J_{i,j}$ is granted an exclusive use of the resource, then $f_{i,j} = r_{i,j} + c_{i,j}$. However, as multiple job requests can be active at the same time, a scheduling mechanism is required to assign, in each moment, the resource to one of them.

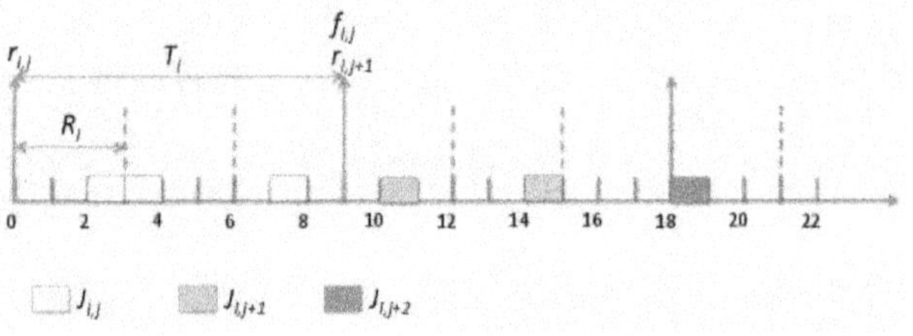

Figure 2: Scheduling mechanism and model of execution.

2.1 Scheduling Mechanism

As a scheduling mechanism, we advocate the use of *resource reservations* [25]. Each task τ_i is associated with a reservation (Q_i, R_i), meaning that τ_i is allowed to execute for Q_i (*budget*) time units in every interval of length R_i (*reservation period*). The budget Q_i is an integer number varying in the interval $[0, \ldots, R_i]$. The bandwidth allocated to the task is $B_i = Q_i/R_i$ and it can be thought of as the fraction of CPU time allocated to the task. In this model, the execution of job $J_{i,j}$ requires at most $\left\lceil \frac{c_{i,j}}{Q_i} \right\rceil R_i$ computation units. For the sake of simplicity, we will assume a small R_i. Therefore, the time required to complete the execution can be approximated as $\frac{c_{i,j}}{B_i}$, which corresponds to a fluid partitioning of the CPU where the task executes as if on a "dedicated" CPU whose speed is a fraction B_i of the speed of the real processor.

For our purposes, it is convenient to choose a reservation period that is an integer sub-multiple of the task period: $T_i = N_i R_i$, $N_i \in \mathbb{N}$. There are several possible implementation of the resource reservations paradigm, one of the most popular being the Constant Bandwidth Server (CBS) [1]. The CBS operates properly as long as the following condition is satisfied.

$$\sum_{i=1}^{n} B_i \leq 1. \tag{1}$$

This constraint captures the simple fact that the cumulative allocation of bandwidth cannot exceed 100% of the computing power. Besides, it also expresses that the CBS is able to utilise the whole availability of computing power. This comes form the fact that the CBS is based on an EDF scheduler [1]. Figure 2 reports a pictorial example of the proposed mechanism for the task τ_i of three consecutive jobs. There are $N_i = 3$ reservation periods. The computation times for the three consecutive jobs are respectively 3, 2 and 1 units.

2.2 Model of Execution

The model of execution we adopt for this paper is time

triggered [13], which is represented in Figure 2 and described as follows. At time $r_{i,j}$ a new sample is collected triggering the activation of a job. The execution of the activity requires the allocation of $c_{i,j}$ time units of the resource time (which needs not be contiguous). When the execution finishes (time $f_{i,j}$), the output data is stored in a temporary variable and is released (i.e., applied to the actuators), only when the deadline $d_{i,j}$ expires. The deadline $d_{i,j}$ is set equal to $r_{i,j} + D_i$ where D_i is the relative deadline that we will set equal to an integer multiple of the reservation period R_i. After being released, the output is held constant until the next update using the customary ZoH approach.

Very important in our scheme is the fact that if the job does not terminate within the deadline (i.e., $f_{i,j} > d_{i,j}$), its execution is cancelled and the output on the actuator remains constant.

2.3 The Control Problem

Each task τ_i is used to control a controllable and observable discrete–time linear system:

$$\begin{aligned} x_i(k+1) &= A_i x_i(k) + F_i u_i(k) + w_i(k) \\ y_i(k) &= C_i x_i(k) \end{aligned} \tag{2}$$

where $x_i(k) \in \mathbb{R}^{n_{x_i}}$ represents the system state, $u_i(k) \in \mathbb{R}^{m_i}$ the control inputs, $w_i(k) \in \mathbb{R}^{n_{x_i}}$ a noise term and $y_i \in \mathbb{R}^{p_i}$ the output functions. One step transition for this discrete time system refers to the evolution of the system step across one temporal unit, which is set equal to the reservation period R_i. The $u_i(k)$ vector is updated at the end of each control task, according to the model of execution described above.

By $P_i(k) = \mathsf{E}\left\{x_i(k)x_i(k)^T\right\}$ we denote the variance of the state resulting from the action of the noise term and of the control action. When $\overline{P}_i = \lim_{k\to\infty} P_i(k) < +\infty$ (i.e., the variance converges) for a given control algorithm, we will say that the closed loop system is *mean square stable* and we will use $\overline{P}_i$ as a QoC metric. Clearly, the smaller is the value of $\overline{P}_i$, the better is the control quality.

2.4 Problem Formulation

The amount of allocated bandwidth (B_i) quantifies the QoS that the task receives and translates into the delays introduced in the feedback loop and in the probability of violating the deadline. Therefore, different values for the bandwidth determine a different value of the QoC given by $\overline{P}_i$. We can identify a critical value $\underline{B}_i$ for the minimum bandwidth that the task has to receive in order for the feedback loop that it implements to be mean square stable.

The objective of this paper is to identify the allocation of resources between the different feedback loops that maximises the system-wide *global Quality of Control.* In mathematical terms, the problem can be set up as follows:

$$\begin{gathered} \min \Phi(\overline{P}_1, \ldots, \overline{P}_n) \\ \text{subject to} \\ \sum_{i=1}^{n} B_i \leq 1 \\ \overline{B}_i \geq B_i \geq \underline{B}_i. \end{gathered} \tag{3}$$

where the upper bound $\overline{B}_i$ is to ensure that the task does not receive more bandwidth than it needs to achieve probability 1 of meeting the deadline. The cost function $\Phi(\cdot)$ gathers the individual QoC performance related to each task into a global performance index. In this paper, we consider as a cost function $\Phi(\overline{P}_1, \ldots, \overline{P}_n) = \max_i \phi_i(\overline{P}_i)$, where $\phi_i(\cdot)$ are suitably defined cost functions used to measure the QoC of each individual task. The constraint $\sum_{i=1}^{n} B_i \leq 1$ comes from the requirement 1 of the CBS scheduling algorithm, whilst the constraint $B_i \geq \underline{B}_i$ is to enforce mean square stability, $\underline{B}_i$ being the critical bandwidth. The analysis presented below could easily be generalised to *weighted* norms. For instance, we could consider $\Phi(\overline{P}_1, \ldots, \overline{P}_n) = \sum_{i=1}^{n} l_i \phi_i(\overline{P}_i)$ where l_i weighs the relevance the i–th feedback loop.

3. COST FUNCTION AND CONSTRAINTS FOR THE OPTIMISATION PROBLEMS

In the time–triggered semantics, the control tasks are activated on a periodic basis: $r_{i,j} = jN_i$, where the $N_i \in \mathbb{N}$ expresses the number of computation units composing a period. This section presents the QoC analysis for a generic task. Since we are referring to a single control task, we will drop the i subscript. For its j-th job, the task receives as an input the state sampled at time jN: $x(jN)$. For notational simplicity, we will henceforth denote this quantity by $\hat{x}(j) \in \mathbb{R}^{n_x}$. The delay introduced by the j-th job will be denoted by $\delta_j = f_j - r_j$. Because we adopt the time–triggered model of computation, the output is released at time $r_j + D$ if $\delta_j \leq D$, and is not released at all if $\delta_j > D$. For the sake of simplicity, in this paper we assume $D = T$. The use of the ZoH model requires holding the data constant until the next output is released, whence the need for an additional state variable $\zeta \in \mathbb{R}^m$ to store the control value. Therefore, the unitary delayed open loop system is given by

$$\begin{aligned} \hat{x}(j+1) &= A^N \hat{x}(j) + \sum_{i=0}^{N-1} A^{N-i-1} F \zeta(j) \\ &\quad + \sum_{i=0}^{N-1} A^{N-i-1} w(Nj+i), \\ \zeta(j+1) &= u(j), \end{aligned}$$

or, in matrix form

$$\begin{bmatrix} \hat{x} \\ \zeta \end{bmatrix}^+ = \begin{bmatrix} A^N & \sum_{i=0}^{N-1} A^{N-i-1} F \\ 0 & 0 \end{bmatrix} \begin{bmatrix} \hat{x} \\ \zeta \end{bmatrix} + \begin{bmatrix} 0 \\ I_m \end{bmatrix} u + \begin{bmatrix} \widetilde{w}(j) \\ 0 \end{bmatrix} \tag{4}$$

with $\widetilde{w}(j) = \sum_{i=0}^{N-1} A^{N-i-1} w(Nj+i)$. Let

$$\begin{aligned} z(j+1) &= A_z z(j) + F_z y(j) \\ u(j) &= C_z z(j) + G_z y(j) \end{aligned}$$

be the stabilizing controller for the system (4), where $z \in \mathbb{R}^{n_z}$ and $y(j) = C\hat{x}(j) \in \mathbb{R}^p$. Define the augmented state $\tilde{x} = [\hat{x}^T, \zeta^T, z^T]^T \in \mathbb{R}^{n_c}$, with $n_c = n_x + n_z + m$. The Schur stable closed loop dynamic matrix of the system can then be written as

$$A_c = \begin{bmatrix} A^N & \sum_{i=0}^{N-1} A^{N-i-1} F & 0 \\ G_z C & 0 & C_z \\ F_z C & 0 & A_z \end{bmatrix} \in \mathbb{R}^{n_c \times n_c},$$

in the nominal condition ($\delta_j \leq T$). On the contrary, if the job is cancelled ($\delta_j > T$), the system evolves with the open loop matrix

$$A_o = \begin{bmatrix} A^N & \sum_{i=0}^{N-1} A^{N-i-1}F & 0 \\ 0 & I_m & 0 \\ 0 & 0 & I_{n_z} \end{bmatrix} \in \mathbb{R}^{n_c \times n_c}.$$

In the following, we assume that the noise $w(\cdot)$ and the computation time $\{c_j\}_{j \in \mathbb{Z}_{\geq 0}}$ are independent identically distributed (i.i.d.) random processes, mutually independent and both independent from the state. The mutual independence is, in our evaluation, a realistic assumption insofar as the tasks are used to implement independent control loops. The i.i.d. nature of $\{c_j\}_{j \in \mathbb{Z}_{\geq 0}}$ is far less obvious. However, in many practical cases of interest (e.g., visual servoing), the analysis of the temporal behaviour of a real–time task that approximates the $\{c_j\}_{j \in \mathbb{Z}_{\geq 0}}$ process as i.i.d. (ignoring its correlation structure) produces a close approximation of the behaviour that is reported in the experiments [21].

The noise $w(\cdot)$ is also assumed to have zero mean and variance $\mathsf{E}\{w(k)w(k)^T\} = W \in \mathbb{R}^{n_x \times n_x}$. Depending on the choice of the bandwidth B, we get different distributions for the i.i.d. process representing the delay. Define $\mu \triangleq \mathsf{P}[\delta_j \leq T]$ the probability of meeting the deadline. The variance $P \in \mathbb{R}^{n_c \times n_c}$ of the state $\tilde{x}$ is given by

$$P(j+1) = \mathsf{E}\left\{\tilde{x}(j+1)\tilde{x}(j+1)^T\right\} =$$
$$\mu \mathsf{E}\left\{(A_c\tilde{x}(j) + v(j))(A_c\tilde{x}(j) + v(j))^T\right\} +$$
$$(1-\mu)\mathsf{E}\left\{(A_o\tilde{x}(j) + v(j))(A_o\tilde{x}(j) + v(j))^T\right\}, \quad (5)$$

where $v(j) \triangleq \left[\tilde{w}(j)^T, 0\right]^T \in \mathbb{R}^{n_c}$. Taking into account the mutual independence of the stochastic processes and the fact that $w(\cdot)$ has null mean and constant variance W, the equation above can be written as

$$P(j+1) = \mu A_c P(j) A_c^T + (1-\mu) A_o P(j) A_o^T + H, \quad (6)$$

where

$$H = \begin{bmatrix} \sum_{i=0}^{N-1} A^{N-i-1}W(A^{N-i-1})^T & 0 \\ 0 & 0 \end{bmatrix}.$$

The relation (6) describes the dynamics of the covariance. Two issues are relevant to this paper: 1) find the values of μ that ensure the convergence of the covariance to a steady state value (we define the infimum of these values as the critical probability), 2) find a measure of the covariance which could effectively be used as a cost function in the optimisation problem.

3.1 Estimating the critical probability

The computation of the critical probability amounts to find the infimum value of μ for which the system is mean square stable. Using Kronecker product properties we can write the dynamics (6) as

$$\text{vec}(P(j+1)) = \left(\mu A_c^{[2]} + (1-\mu)A_o^{[2]}\right)\text{vec}(P(j)) + \text{vec}(H), \quad (7)$$

where $M^{[2]} \triangleq M \otimes M$ and $\text{vec}(\cdot)$ is the linear operator producing a vector by stacking the column of a matrix. This is a discrete–time linear time–invariant system in the state $\text{vec}(P(j)) \in \mathbb{R}^{n_c^2}$. Hence, it admits a steady state solution

---

Algorithm 1 Find Critical Probability
1: **function** FINDMU(A_c,A_o)
2:    $A_C = A_c^{[2]}$
3:    $A_O = A_o^{[2]}$
4:    $\tilde{\mu} = 1$
5:    $\mu_0 = 0$
6:    **while** not$(\tilde{\mu} - \mu_0 < t_\mu \wedge \tilde{\mu} < 1 - t_\mu)$ **do**
7:       $\mu = (\tilde{\mu} - \mu_0)/2 + \mu_0$
8:       $A_2 = (1-\mu) * A_O + \mu * A_C$
9:       **if** SchurStableSegment(A_C, A_2) **then**
10:          $\tilde{\mu} = \mu$
11:       **else**
12:          $\mu_0 = \mu$
13:       **end if**
14:    **end while**
15:    **return** $\tilde{\mu}$
16: **end function**

---

w.r.t. the constant input $\text{vec}(H) \in \mathbb{R}^{n_c^2}$ if

$$\max_i \left|\lambda_i\left(\mu A_c^{[2]} + (1-\mu)A_o^{[2]}\right)\right| < 1, \quad (8)$$

where with $\lambda_i(M)$ we mean the i-th eigenvalue of M. If such a condition is verified, we look for a steady state solution $\bar{P}$ solving the algebraic equation (see (6))

$$\bar{P} = \mu A_c \bar{P} A_c^T + (1-\mu) A_o \bar{P} A_o^T + H. \quad (9)$$

Using again the Kronecker product we can write the unique solution of (9) as

$$\text{vec}(\bar{P}) = \left(I - \mu A_c^{[2]} - (1-\mu)A_o^{[2]}\right)^{-1}\text{vec}(H). \quad (10)$$

It is worth noting that the inverse in (10) exists due to condition (8). Ensuring mean square stability for the system (4) turns to a problem of Schur stability for the matrix pencil $\mu A_c^{[2]} + (1-\mu)A_o^{[2]}$. In other words, we search the critical probability $\tilde{\mu}_c \triangleq \inf_{\bar{\mu} \in [0,1]}\{\bar{\mu} \mid \forall \mu \geq \bar{\mu} \text{ (8) is satisfied}\}$. For our problem, a solution with $\tilde{\mu}_c < 1$ always exists, since A_c is Schur and the eigenvalues are continuous functions of μ. In practice, instead of looking for the infimum in the continuum set $[0, 1]$, we can fix an accuracy level t_μ and look for the minimum in the discrete set $\left\{0, t_\mu, \ldots, t_\mu\left\lfloor\frac{1}{t_\mu}\right\rfloor\right\}$. The Algorithm 1 implements a dichotomic search in the μ space to find $\tilde{\mu} \triangleq \min_{\bar{\mu} \in \left\{0, t_\mu, \ldots, t_\mu\left\lfloor\frac{1}{t_\mu}\right\rfloor\right\}}\{\bar{\mu} \mid \forall \mu \geq \bar{\mu} \text{ (8) is satisfied}\}$. The function SchurStableSegment(A,B) implements the algebraic conditions on a finite number of matrices of [8] and provides a positive answer if all the matrices in the matrix pencil with vertices A and B are Schur.

3.2 Measuring the steady state covariance

As discussed in the next section, expressing the QoC through a measure of the steady state covariance matrix $\bar{P}$ given in the equations (9) which enjoys a monotonicity property w.r.t. the probability μ, significantly simplifies the solution of the optimisation problem.

As a measure of the steady state covariance, we consider the trace of $\bar{P}$ which is a function of the probability μ, i.e., $\phi : \mathbb{R} \to \mathbb{R}$ with

$$\phi(\mu) \triangleq \text{Tr}\{\bar{P}(\mu)\}. \quad (11)$$

The monotonicity of $\phi(\mu)$ can be analysed by studying the sign of $\frac{d\phi(\mu)}{d\mu}$. To this end, we first notice that $\text{Tr}\{AB\} =$

$\text{vec}(A^T)^T\text{vec}(B)$, which leads to $\text{Tr}\{\bar{P}\} = \text{vec}(I)^T\text{vec}(\bar{P})$, from which

$$\frac{d\phi(\mu)}{d\mu} = \text{vec}(I)^T\frac{d\text{vec}(\bar{P}(\mu))}{d\mu},$$

and, plugging (10), yields to

$$\begin{aligned}\frac{d\phi(\mu)}{d\mu} &= \text{vec}(I)^T\frac{d\left(I - \mu A_c^{[2]} - (1-\mu)A_o^{[2]}\right)^{-1}}{d\mu}\text{vec}(H)\\ &= \text{vec}(I)^T\frac{dS(\mu)^{-1}}{d\mu}\text{vec}(H),\end{aligned}$$

with $S(\mu) \triangleq I - \mu A_c^{[2]} - (1-\mu)A_o^{[2]}$. We finally have

$$\begin{aligned}\frac{d\phi(\mu)}{d\mu} &= \text{vec}(I)^T S(\mu)^{-1}\frac{dS(\mu)}{d\mu}S(\mu)^{-1}\text{vec}(H)\\ &= \text{vec}(I)^T S(\mu)^{-1}\left(A_c^{[2]} - A_o^{[2]}\right)S(\mu)^{-1}\text{vec}(H).\end{aligned} \quad (12)$$

Relation (12) leads to express the trace as a polynomial function of μ. The monotonicity of the trace can be assessed by studying the sign of this expression. To this end we can apply Sturms theorem [7] to verify that $\frac{d\phi(\mu)}{d\mu}$ does not have roots in the range $[\widetilde{\mu}, 1]$. The result of this numeric test strongly depends on the specific controller–system pair and analytical results with general validity are difficult to find. However, in the rather broad set of thousands of randomly synthesized open–loop unstable systems, we found out that for more than 95% of them $\text{Tr}\{\bar{P}(\mu)\}$ is a non–increasing function for $\mu \in [\widetilde{\mu}, 1]$ whenever the system is closed in loop with an LQG controller[1]. We remark that a non–increasing function can be made strictly decreasing by simply adding to it a strictly decreasing function of arbitrary small amplitude. We can consider, for instance, $\phi(\mu) \triangleq \text{Tr}\{\bar{P}(\mu)\} + \varepsilon\frac{1-\mu}{1-\widetilde{\mu}}$. for arbitrary small $\varepsilon > 0$. For this reason, henceforth we make the following assumption.

ASSUMPTION 1. *The function $\phi : \mathbb{R} \to \mathbb{R}$ used to measure the QoC of each controlled system is strictly decreasing in the range $[\widetilde{\mu}, 1]$.*

4. OPTIMAL SOLUTION

As assumed in Section 3, the computation time $\{c_{i,j}\}_{j\in\mathbb{Z}_{\geq 0}}$ of the generic task i is a stochastic process, hence its cumulative distribution function $\Gamma_{c_i} : \mathbb{R}_{\geq 0} \to [0,1]$ is non–decreasing. Since $\{c_{i,j}\}_{j\in\mathbb{Z}_{\geq 0}}$ is a computation time, it is perfectly reasonable to assume that its cumulative distribution function is strictly increasing. Let us consider the probability $\mu_{i,j} = \mathsf{P}\{f_{i,j} \geq D_i\}$ of the j-th job of the i-th task of finishing within its deadline. By means of the fluid approximation we can write $\mu_{i,j} = \mathsf{P}\left\{\frac{c_{i,j}}{B_i} \geq D_i\right\} = \mathsf{P}\{c_{i,j} \geq D_iB_i\}$ and using the cumulative distribution $\mu_{i,j} = \Gamma_{c_i}(D_iB_i)$. Due to the fact that $\{c_{i,j}\}_{j\in\mathbb{Z}_{\geq 0}}$ is an i.i.d. process, the probability does not change with the job and we can just drop the subscript j: $\mu_i = \Gamma_{c_i}(D_iB_i)$. Because of the strict monotonicity of the cumulative distribution, its inverse is well defined and we can write

$$B_i = \frac{\Gamma_{c_i}^{-1}(\mu_i)}{D_i},$$

[1]If the test (12) fails for some of the controlled system to be scheduled, they can receive full bandwidth, letting the optimisation work on all the others.

for every $\mu \in [0,1]$. According to such (strictly increasing) relation, we can define the minimum bandwidth required to achieve mean square stability as $\underline{B}_i \triangleq \frac{\Gamma_{c_i}^{-1}(\tilde{\mu}_i)}{D_i}$, where $\tilde{\mu}_i$ is the critical probability computed with the Algorithm 1 in Section 3. Analogously, we can define the maximum bandwidth necessary for the i-th task to finish within its deadline with probability 1 as $\overline{B}_i \triangleq \frac{\Gamma_{c_i}^{-1}(1)}{D_i}$. By means of these definitions, the feasibility set of the optimisation problem (3) can be written as the following polytope

$$\mathcal{B} \triangleq \left\{(B_1,\ldots,B_n) \in \mathbb{R}^n \mid \underline{B}_i \leq B_i \leq \overline{B}_i, \quad \sum_{i=1}^{n} B_i \leq 1\right\}. \quad (13)$$

In our framework the deadlines D_i are fixed, thus the probability μ_i is a strictly increasing function of the bandwidth B_i: $\mu_i = \Gamma_{c_i}(B_i)$. Let us recall that the cost function $\phi_i(\cdot)$ of any task i, introduced in the previous section, is a decreasing function of μ_i. The composition $\phi_i \circ \Gamma_{c_i}(\cdot)$ is then a strictly decreasing function of the bandwidth B_i in the set $[\underline{B}_i, \overline{B}_i]$. With a slight abuse of notation, in what follows we will write $\phi_i(\cdot)$ directly as a function of B_i to refer such a composition.

The optimisation problem (3) can be finally written as

$$\min_{B\in\mathcal{B}} \Phi(B) = \min_{B\in\mathcal{B}} \max_{i\in\{1,\ldots,n\}} \phi_i(B_i) \quad (14)$$

where $B \triangleq (B_1,\ldots,B_n)$ and $\mathcal{B}$ is given in (13).

The optimisation problem (14) can present some special cases that deserve a separate analysis. For example, it is said to be *degenerate* if there exist $i,j \in \{1,\ldots,n\}$ such that $\phi_i(\overline{B}_i) \geq \phi_j(\underline{B}_j)$. In such a case $\phi_i(\cdot)$ is said *to dominate* $\phi_j(\cdot)$. In such a case, due to the strictly decreasing nature of the function $\phi_i(\cdot)$ that attains its minimum for the bandwidth $\overline{B}_i$, $\phi_i(B_i) \geq \phi_j(B_j)$ for any $B_i \in [\underline{B}_i, \overline{B}_i]$ and any $B_j \in [\underline{B}_j, \overline{B}_j]$. For this reason $\Phi(B) = \max_{h\in\{1,\ldots,n\}} \phi_h(B_h) = \max_{h\in\{1,\ldots,n\}\setminus\{j\}} \phi_h(B_h)$. In a degenerate case the bandwidth associated to a dominated function does not influence the cost function, but only the constraints defining the feasibility set. Therefore, it can be fixed to a value ensuring the largest feasibility set. It can be easily verified that, if $\phi_j(\cdot)$ is the dominated function, such a value is $B_j = \underline{B}_j$. It is worth noting that, with this position, the feasibility set is now a $(n-1)$–dimensional facet of the original polytope $\mathcal{B}$. If in the optimisation problem there are more than one dominated function, say $n' < n$ functions with indices in the set $I' \subset \{1,\ldots,n\}$, then the feasibility set is the $(n-n')$–dimensional facet: $\mathcal{B}_{n'} \triangleq \mathcal{B} \setminus \{(B_1,\ldots,B_n) \in \mathbb{R}^n \mid B_h = \underline{B}_h, \forall h \in I'\}$.

Let us now analyse the solution set of the non degenerate optimisation problem (14). First we notice that a function $g : \mathbb{R}^n \to \mathbb{R}$ is said *componentwise strictly decreasing* if for any $x, y \in \mathbb{R}^n$ such that $x <_e y$ (namely $x_i < y_i$ for any $i \in \{1,\ldots,n\}$), then $g(x) > g(y)$. We are now in condition to state the following theorem.

THEOREM 1. *Assume that Assumption (1) holds and that the optimisation problem (14) is not degenerate. Define the optimal solution set as $X^* \triangleq \arg\min_{B\in\mathcal{B}} \Phi(B)$, then we have:*

i) *$X^* \subseteq \partial\mathcal{B}$ where $\partial\mathcal{B}$ is the frontier of the polytope $\mathcal{B}$;*

ii) *there exists $B^* = (B_1^*,\ldots,B_n^*) \in X^*$ such that $\phi_i(B_i^*) = \phi_j(B_j^*)$ for every $i,j \in \{1,\ldots,n\}$;*

iii) *there exists* $\hat{B} = (\hat{B}_1, \dots, \hat{B}_n) \in [\underline{B}_1, \bar{B}_1] \times \cdots \times [\underline{B}_n, \bar{B}_n]$ *unique solution of* $\phi_h(\hat{B}_h) = \min_{i \in 1,\dots,n} \phi_i(\underline{B}_i)$ *for every* $h \in \{1, \dots, n\}$*such that, if* $\sum_{i=1}^n \hat{B}_i \le 1$*, then* $\hat{B} \in X^*$*. Otherwise there exists* B^* *as for point ii) satisfying* $\sum_{i=1}^n B_i^* = 1$*;*

iv) *if* $\left\{(B_1, \dots, B_n) \in \mathbb{R}^n \mid \sum_{i=1}^n B_i = 1\right\} \cap \partial\mathcal{B} \ne \{0\}$ *then* $\left\{(B_1, \dots, B_n) \in \mathbb{R}^n \mid \sum_{i=1}^n B_i = 1\right\} \cap X^* \ne \{0\}$.

PROOF. In order to prove i), let us proceed by contradiction. Assume that there exists $B^* \in X^*$ such that $B^* \in \mathcal{B} \setminus \partial\mathcal{B}$. Being $\mathcal{B} \setminus \partial\mathcal{B}$ an open set, then there exists $\hat{B} \in \mathcal{B}$ such that $B^* <_e \hat{B}$. It is easily to verify that $\Phi(B) = \max_{i \in \{1,\dots,n\}} \phi_i(B_i)$ is a componentwise strictly decreasing function, hence we have $\Phi(\hat{B}) < \Phi(B^*)$, which contradicts the optimality of B^*.

Let us prove the points ii) and iii). Define the real values $\underline{t} \triangleq \max_{i \in 1,\dots,n} \phi_i(\bar{B}_i)$ and $\bar{t} \triangleq \min_{i \in 1,\dots,n} \phi_i(\underline{B}_i)$. In the non degenerate case we have that $\underline{t} \le \bar{t}$. We want to prove that the optimal value $t^* \triangleq \min_{B \in \mathcal{B}} \Phi(B)$ is such that $t^* \in [\underline{t}, \bar{t}]$. By definition of the function $\Phi(\cdot)$, its codomain is the set $[\underline{t}, \max_{i \in 1,\dots,n} \phi_i(\underline{B}_i)]$, hence $t^* \ge \underline{t}$. Moreover, clearly we have $\bar{t} \in [\underline{t}, \max_{i \in 1,\dots,n} \phi_i(\underline{B}_i)]$. Recall that each function $\phi_i : [\underline{B}_i, \bar{B}_i] \to [\phi_i(\bar{B}_i), \phi_i(\underline{B}_i)]$ with $i \in \{1, \dots, n\}$ is strictly monotonic, thus invertible. As a consequence there exists $B \in [\underline{B}_1, \bar{B}_1] \times \cdots \times [\underline{B}_n, \bar{B}_n]$ such that $\Phi(B) = \bar{t} \le \max_{i \in 1,\dots,n} \phi_i(\underline{B}_i)$, hence $t^* \le \bar{t}$ because of its optimality.

Still for the strict monotonicity of any $\phi_i(\cdot)$, we have that for any $t \in [\underline{t}, \bar{t}]$ there exists a unique $B_i \in [\underline{B}_i, \bar{B}_i]$ such that $\phi_i(B_i) = t$. Thus, for any $t \in [\underline{t}, \bar{t}]$ we can find a unique solution $B = (B_1, \dots, B_n) \in [\underline{B}_1, \bar{B}_1] \times \cdots \times [\underline{B}_n, \bar{B}_n]$ such that $\phi_i(B_i) = \phi_j(B_j)$ for each $i, j \in \{1, \dots, n\}$. Let us define in particular $\hat{B} = (\hat{B}_1, \dots, \hat{B}_n)$ as the unique solution of $\phi_i(\hat{B}_i) = \bar{t}$ for each $i \in \{1, \dots, n\}$. If $\sum_{i=1}^n \hat{B}_i \le 1$ then $\hat{B} \in \mathcal{B}$ and it is clearly optimal ($B^* \triangleq \hat{B} \in X^*$). If $\sum_{i=1}^n \hat{B}_i > 1$, recall that $\sum_{i=1}^n \underline{B}_i < 1$ and the continuity of each $\phi_i(\cdot)$ to show that there exists $\tilde{t} \in [\underline{t}, \bar{t}]$ and a unique $\tilde{B} = (\tilde{B}_1, \dots, \tilde{B}_n) \in \mathcal{B}$ such that $\sum_{i=1}^n \tilde{B}_i = 1$ and $\phi_i(\tilde{B}_i) = \phi_j(\tilde{B}_j) = \tilde{t}$ for every $i, j \in \{1, n\}$. It is apparent that $\tilde{B} \in \partial\mathcal{B}$ as $\sum_{i=1}^n \tilde{B}_i = 1$, we must show that $\tilde{B} \in X^*$. In order for any $B = (B_1, \dots, B_n)$ to give $\Phi(B) < \tilde{t}$, it must have at least one component B_i such that $B_i > \tilde{B}_i$ and all the other components such that $B_j \ge \tilde{B}_j$ for every $j \ne i$. But such a B is not a feasible solution as $\sum_{h=1}^n B_h > 1$, hence $B^* \triangleq \tilde{B} \in X^*$.

Concerning point iv), if the solution $\hat{B}$ at point iii) is such that $\sum_{i=1}^n \hat{B}_i > 1$, we have shown how to find a solution $\tilde{B}$ such that $\sum_{i=1}^n \tilde{B}_i = 1$. If instead $\sum_{i=1}^n \hat{B}_i \le 1$, let us define the set $H = \left\{h \in \{1, \dots, n\} \mid (\hat{B}_1, \dots, \hat{B}_h + b, \dots, \hat{B}_n) \notin \mathcal{B} \forall b > 0\right\}$. It is easily to verify that for some $i \in H$ we have $\phi_i(\hat{B}_i) = \Phi(\hat{B})$. Then for any $j \in \{1,.,n\} \setminus H$ we have $\phi_j(\hat{B}_j) \le \Phi(\hat{B})$. If $\left\{(B_1, \dots, B_n) \in \mathbb{R}^n \mid \sum_{i=1}^n B_i = 1\right\} \cap \partial\mathcal{B} \ne \{0\}$ we can easily find a $B = (B_1, \dots, B_n)$ such that $B_i = \hat{B}_i$ for any $i \in H$, $B_j \ge \hat{B}_j$ for any $j \in \{1, \dots, n\} \setminus H$ and $\sum_{i=1}^n B_i = 1$. Due to the strictly decreasing nature of each $\phi_i(\cdot)$, we have $\Phi(B) = \Phi(\hat{B})$. Hence $B \in X^*$.

□

The degenerate case can be addressed by recalling that the facet $\mathcal{B}_{n'}$ is an $(n - n')$–dimensional polytope. We can, thus, apply the previous theorem to such a polytope, with the unique difference that now $\sum_{i \in \{1,\dots,n\} \setminus I'} B_i = 1 - \sum_{i \in I'} \underline{B}_i$.

Algorithm 2 Solve Optimisation

```
function SolveOptimisation(Systems, B̄_i, B_i)
    P^(1) = max_i Tr{P̄_i(B̄_i)}
    B^(1) = [B_1^(1), ..., B_n^(1)] = b(P^(1))
    if Σ B_i^(1) ≤ 1 then
        return B^(1)
    end if
    P^(2) = max_i Tr{P̄_i(B_i)}
    while P^(2) − P^(1) > ε do
        P = (P^(1) + P^(2)) / 2
        B = [B_1, ..., B_n] = b(P)
        if Σ B_i ≤ 1 then P^(2) = P
        else P^(1) = P
        end if
    end while
    return B = [B_1, ..., B_n] = b(P^(2))
end function
function b(P)
    for i = 1, ..., n do
        Bm = B_i; BM = B̄_i
        if Tr{P̄_i(Bm)} ≤ P then B_i = Bm
        else
            while BM − Bm > ε do
                B = (BM+Bm)/2
                if Tr{P̄_i(B)} < P then BM=B
                elseBm=B
                end if
            end while
            B_i = BM
        end if
    end for
    return B = [B_1, ..., B_n]
end function
```

The Algorithm 2 solves Problem (14) in light of Theorem 1. Basically, it applies a binary search until it reaches the conditions expressed by the main theorem.

4.1 A geometric interpretation

A simple geometric interpretation can be useful to understand the meaning of the result above for the more interesting case of non-degenerate problems. Let us focus, for simplicity, on the case of two tasks. Consider the plot in Figure 3.(b). The thick lines represent the constraints (and are dashed when the inequality is strict). The area of the feasible solutions is filled in gray.

The dashed-dotted line represent the level set $\Phi(B_1, B_2) = t$, which is the set of all values for which the infinity norm of the traces of the covariance matrices is fixed to the value t. This set looks like a 90 degree angle, whose vertex is given by the values of the bandwidth B_1 and B_2. The level set $\Phi(B_1, B_2) = \phi_2(B_2)$ (the bottom one in the figure), originates from a feasible point a point in $\mathcal{B}$. The level set $\Phi(B_1, B_2) = \phi_1(B_1)$ is associated with the point $\phi_1(B_1) = \max_i \mathrm{Tr}\left\{\overline{P}_i(\overline{B}_i)\right\}$, which we assumed equal to $\mathrm{Tr}\left\{\overline{P}_1(\overline{B}_1)\right\}$ in the example. Two facts are noteworthy: first, $\phi_1(B_1)$ being a lower bound for $\Phi(B_1, B_2)$, we have $\phi_1(B_1) \le \phi_2(B_2)$; second, the level set $\Phi(B_1, B_2) = \phi_1(B_1)$ is internal to the level set $\Phi(B_1, B_2) = \phi_2(B_2)$ (this is due

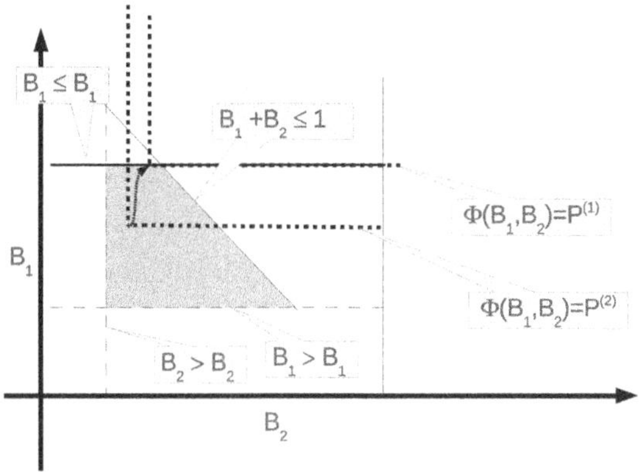

(a)

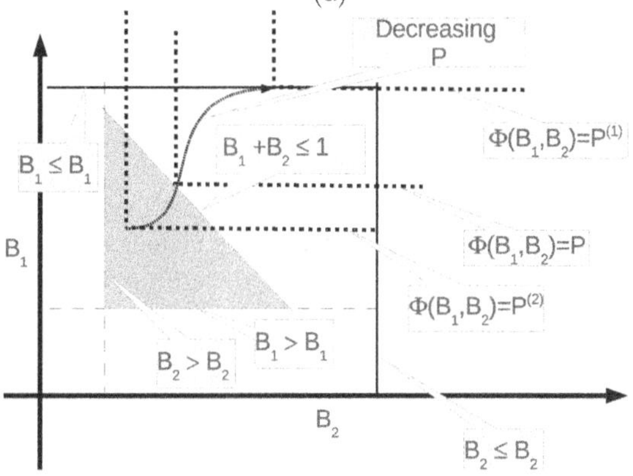

(b)

Figure 3: Graphical representation of the problem addressed by Theorem 1 for the case of two tasks. (a) the first claim of condition iii), (b) the second claim of condition iii).

to the decreasing behaviour of the $\phi_i(B_i)$ functions). Therefore the line joining the vertex of the two level set (represented in dotted notation) remains internal to the level set $\Phi(B_1, B_2) = \phi_2(B_2)$. What is more, it terminates on the vertex of the level $\Phi(B_1, B_2) = \phi_1(B_1)$, since $\phi_1(B_1)$ is a lower bound for $\Phi(B_1, B_2)$. The points on this line remain feasible until it touches the constraint $\sum B_i = 1$, which corresponds to the optimal solution. In this case, the level set becomes tangent to the constraint. The situation just described is the one corresponding to the second claim of part **iii)** of Theorem 1. On the contrary, the case addressed by the first claim of part **iii)** of Theorem 1 is the one depicted in Figure 3.(a). In this case the point $\phi_1(B_1)$ lies inside the feasible area and it is obviously optimal.

5. NUMERIC EVALUATION

In this section, we offer some numeric data that show the results of the optimal allocation of bandwidth algorithm presented in this paper. We consider randomly generated, open-loop unstable, reachable and observable linear continuous time systems subject to a linear combination of continuous time noises. The number of noise sources equals the number of states. The noise processes are normally distributed with zero mean and standard deviation equal to $\sigma = 0.01$.

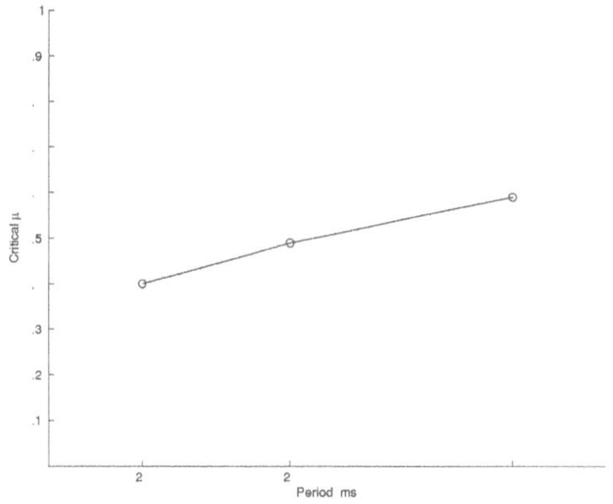

Figure 4: Critical probability versus sampling time for a generic unstable system.

The time varying computation times of the control tasks is described by a probability density function (pdf) f_{C_i}. Two different pdfs are considered: $\mathcal{E}_{\mu_i} = e^{-\frac{c_i}{\mu_i}}$ is exponential with mean value equal to μ_i; $\mathcal{U}_{[a,\,b]}$ is uniform defined in the range $[a, b]$. In accordance with our model, periods are kept constant throughout each simulation run. The relative deadlines were chosen equal to the period, i.e., $D_i = T_i$.

We first show how the sampling period influences the critical probability $\widetilde{\mu}_i$, which increases as the sampling time increases (Fig. 4). This is in accordance with our expectations since the variance of the process noise for the discrete time system increases with the sampling time. This behaviour complicates the analysis for the minimum bandwidth. Indeed, recalling that $\Gamma_{c_i}^{-1}(\cdot)$ is monotonically increasing (since it is the inverse function of the CDF of the computation times) and that $\underline{B_i} = \Gamma_{c_i}^{-1}(\widetilde{\mu}_i)/D_i$, we can see that a larger period increases both the numerator and the denominator. Therefore, the analysis for the minimum bandwidth largely depends on the specific system, while this is not the case for the critical probability $\widetilde{\mu}_i$, which is monotonically increasing for all the tested system.

To expose the relation between the period, the characteristic of the computation time pdfs and the minimum and maximum bandwidth, Fig. 5 depicts the behaviour of the available computation bandwidth for a generic unstable system when the control task computation time has different values of the mean computation times for both uniform and exponential distributions. The bandwidth can be chosen between $\overline{B}_i$ (dashed lines) and $\underline{B}_i$ (solid lines). The value $\underline{B}_i$ is the one associated with the critical probability, while $\overline{B}_i$ guarantees that the loop is always closed at every iteration. This range is shown for different values of the sampling time. Whatever is the computation time distribution, increasing its mean value reduces the available bandwidth. This effect is partially mitigated by the period. For example, from Fig. 5.(a) the maximum and minimum bandwidths are unchanged if the mean value grows from 6 ms to 12 ms and simultaneously the period is doubled from 20 to 40 ms. The benefit is even more evident for the maximum bandwidth. For instance, it is not possible to close the loop at every iteration if the mean value is 14 ms and the period is 20 ms (Fig. 5.(a), however this is possible increasing the period at least up to 28 ms. Similar, discussions can be made for the

T_1	T_2	T_3	f_{C_1}	f_{C_2}	f_{C_3}	$\underline{B}_1$	$\underline{B}_2$	$\underline{B}_3$	B_1^*	B_2^*	B_3^*	B	$\min(\Phi(B))$
20	20	20	$\mathcal{U}_{[4,8]}$	$\mathcal{U}_{[4,12]}$	$\mathcal{U}_{[4,8]}$	0.268	0.296	0.278	0.2912	0.42781	0.28086	1	767.7968
28	28	28	$\mathcal{U}_{[4,8]}$	$\mathcal{U}_{[4,12]}$	$\mathcal{U}_{[4,8]}$	0.2	0.21714	0.20286	0.22879	0.42857	0.20415	0.86152	723.8266
40	40	40	$\mathcal{U}_{[4,8]}$	$\mathcal{U}_{[4,12]}$	$\mathcal{U}_{[4,8]}$	0.149	0.158	0.145	0.18486	0.3	0.14672	0.63158	759.3342
20	20	20	$\mathcal{U}_{[4,8]}$	$\mathcal{U}_{[4,8]}$	$\mathcal{U}_{[4,8]}$	0.268	0.248	0.278	0.29481	0.4	0.28086	0.97567	701.1684
28	28	28	$\mathcal{U}_{[4,8]}$	$\mathcal{U}_{[4,8]}$	$\mathcal{U}_{[4,8]}$	0.2	0.18	0.20286	0.22879	0.28571	0.20415	0.71866	723.8266
40	40	40	$\mathcal{U}_{[4,8]}$	$\mathcal{U}_{[4,8]}$	$\mathcal{U}_{[4,8]}$	0.149	0.129	0.145	0.18486	0.2	0.14672	0.53158	759.3342

Table 1: Table showing numeric examples for the three unstable systems in the case of uniformly distributed computation times. Time values for the distribution and for the period are reported in milliseconds.

T_1	T_2	T_3	f_{C_1}	f_{C_2}	f_{C_3}	$\underline{B}_1$	$\underline{B}_2$	$\underline{B}_3$	B_1^*	B_2^*	B_3^*	B	$\min(\Phi(B))$
20	20	20	$\mathcal{E}_{12}$	$\mathcal{E}_{20}$	$\mathcal{E}_{12}$	0.24931	0.27444	0.29658	0.30429	0.39206	0.30345	1	1240.6107
28	28	28	$\mathcal{E}_{12}$	$\mathcal{E}_{20}$	$\mathcal{E}_{12}$	0.21893	0.21508	0.23345	0.32876	0.4343	0.23645	1	907.4646
40	40	40	$\mathcal{E}_{12}$	$\mathcal{E}_{20}$	$\mathcal{E}_{12}$	0.202	0.17125	0.17935	0.40384	0.41	0.18576	1	887.9284
20	20	20	$\mathcal{E}_{12}$	$\mathcal{E}_{12}$	$\mathcal{E}_{12}$	0.24931	0.16466	0.29658	0.34388	0.34902	0.30688	1	859.701
28	28	28	$\mathcal{E}_{12}$	$\mathcal{E}_{12}$	$\mathcal{E}_{12}$	0.21893	0.12905	0.23345	0.3608	0.40122	0.23795	1	791.3342
40	40	40	$\mathcal{E}_{12}$	$\mathcal{E}_{12}$	$\mathcal{E}_{12}$	0.202	0.10275	0.17935	0.46307	0.34984	0.18656	1	818.1927

Table 2: Table showing numeric examples for the three unstable systems in the case of exponentially distributed computation times. Time values for the distribution and for the period are reported in milliseconds.

exponential distribution (Fig. 5.(b)), although the effect is reduced and no maximum bandwidth is available since the pdf has infinite support.

For an insightful numeric comparison, we report simulation results for three randomly generated open–loop unstable continuous time systems of dimension $n_1 = 2$, $n_2 = 3$ and $n_3 = 4$. The number of inputs is equal to 2, 1 and 2, while the number of outputs is 2, 3 and 4, respectively. The pdf of the computation times is described by f_{C_i}, $i = 1, 2, 3$. The maximum unstable eigenvalues are $e_1 = 3.92$, $e_2 = 0.85$ and $e_3 = 1.81$ for the first, second and third system respectively. In the tables, the first group of three columns reports the task periods T_i for each control task, while the second triplet shows the distribution of the computation times f_{C_i} (expressed in milliseconds). The second and the third groups report the minimum and the optimal bandwidths, $\underline{B}_i$ and B_i^* respectively. The columns annotated with B and $\min(\Phi(B))$ report the overall bandwidth used and the optimal trace. Finally, the last three columns are the periods of the three tasks.

Table 1, summarises the numerical results for uniform distributions for three different periods. In the first three rows the uniform distribution for the second task ranges between one and three reservation periods, while for the second three rows it varies between one and two. The second task is the most critical, since it receives at the optimum the largest amount of bandwidth. Moreover, only for the first row it cannot receive its maximum bandwidth, corresponding to probability one of computing the control action before the deadline. This fact is reflected by the highest maximum trace out of the six cases and by the full utilisation of the computing resources ($B = 1$). The same maximum trace is obtained for the periods $T_i = 28$ ms and $T_i = 40$ ms in both cases, since the optimal bandwidth for the second task always corresponds to its maximum value, while the other two tasks receives exactly the same amount of bandwidth. Finally, for the second set of rows the trace increases with the period, since the noise power also increases. According to our previous discussion on the critical bandwidth behaviour, this is not always true as highlighted by the second set of experiments of Table 2, which shows a similar situation for the exponential pdfs. In this second table, again the second task takes the lion's share, but with a different behaviour since the support of the pdfs is now infinite. This fact is also highlighted by the full utilisation of the computing platform in all cases.

Table 3 reports numeric data in the same situation of Table 1 and Table 2 but it considers the period of each task that minimises the overall bandwidth B or the maximum trace. In the former case, only the uniform distribution represents a valid example. In this case, the period should be increased at most, while the second task should receive its maximum bandwidth. The same picture is obtained if the trace is minimised, but in this second situation the second task has the minimum possible period in order to minimise the noise power. In the case of the exponential distribution, again the task two receives an adequate bandwidth and, in order to leverage the effect of task one (which has the highest eigenvalue hence it is potentially dangerous) the optimal solution leads to minimise its period.

Finally, Table 4 reports a set of experiments in which the optimal bandwidth allocation is computed by increasing the mean value of the exponential distribution of only the first task and fixing the period to 40 ms. The monotonicity of the optimal trace and of the optimal bandwidth associated to the first task is clearly visible. We have found similar results, but we omitted them for the sake of brevity.

6. CONCLUSIONS

In this paper, we have considered an application scenario where multiple tasks are used to implement independent feedback loops. The scheduling decisions determine a different bandwidth reservation for the tasks and have a direct impact on the QoC that they deliver.

The QoC is evaluated using the trace of the steady state covariance as a cost function. We have shown conditions that make the QoC measure a decreasing function of the probability to close the loop, and hence of the bandwidth allocated to the task. We have formulated an optimisation problem where the global cost function is the maximum QoC obtained by the different loops and the constraints are on the minimum bandwidth required to obtain mean square stability and on the sum of the bandwidth allocated to the different tasks. The problems thus obtained lends itself to a

T_1	T_2	T_3	f_{C_1}	f_{C_2}	f_{C_3}	$\underline{B}_1$	$\underline{B}_2$	$\underline{B}_3$	B_1^*	B_2^*	B_3^*	B	$\min(\Phi(B))$
40	40	40	$\mathcal{U}_{[4,8]}$	$\mathcal{U}_{[4,12]}$	$\mathcal{U}_{[4,8]}$	0.149	0.158	0.145	0.18486	0.3	0.14672	0.63158	759.3342
40	20	40	$\mathcal{U}_{[4,8]}$	$\mathcal{U}_{[4,12]}$	$\mathcal{U}_{[4,8]}$	0.149	0.296	0.145	0.19283	0.6	0.14672	0.93955	701.1684
28	40	40	$\mathcal{E}_{12}$	$\mathcal{E}_{20}$	$\mathcal{E}_{12}$	0.21893	0.17125	0.17935	0.34173	0.47151	0.18656	1	852.9288

Table 3: Table showing numeric examples in the same set-up of Table 1 and Table 2 with minimized bandwidth or minimized trace. Time values for the distribution and for the period are reported in milliseconds.

T_1	T_2	T_3	f_{C_1}	f_{C_2}	f_{C_3}	$\underline{B}_1$	$\underline{B}_2$	$\underline{B}_3$	B_1^*	B_2^*	B_3^*	B	$\min(\Phi(B))$
40	40	40	$\mathcal{E}_{12}$	$\mathcal{E}_{20}$	$\mathcal{E}_{12}$	0.202	0.17125	0.17935	0.40384	0.41	0.18576	1	887.9284
40	40	40	$\mathcal{E}_{16}$	$\mathcal{E}_{20}$	$\mathcal{E}_{12}$	0.26934	0.17125	0.17935	0.48126	0.3323	0.18576	1	974.7714
40	40	40	$\mathcal{E}_{20}$	$\mathcal{E}_{20}$	$\mathcal{E}_{12}$	0.33667	0.17125	0.17935	0.53554	0.2797	0.18416	1	1110.2589
40	40	40	$\mathcal{E}_{24}$	$\mathcal{E}_{20}$	$\mathcal{E}_{12}$	0.40401	0.17125	0.17935	0.57396	0.24247	0.18336	1	1335.2052
40	40	40	$\mathcal{E}_{28}$	$\mathcal{E}_{20}$	$\mathcal{E}_{12}$	0.47134	0.17125	0.17935	0.60247	0.21495	0.18256	1	1736.5507

Table 4: Table showing numeric examples for increasing values of the mean of the exponential distribution for the first task. Time values for the distribution and for the period are reported in milliseconds.

very efficient solution algorithm, which is the core contribution of the paper. We have offered numeric examples of the algorithm execution.

We envisage future work directions in the possible use of different cost functions and of different computation models, where a task execution is not dropped as soon as it violates a deadline and an execution delay can be tolerated to some degree.

7. REFERENCES

[1] L. Abeni and G. Buttazzo. Integrating Multimedia Applications in Hard Real-Time Systems. In *Proc. IEEE Real-Time Systems Symposium*, pages 4–13, Dec. 1998.

[2] R. Bhattacharya and G. Balas. Anytime control algorithm: Model reduction approach. *Journal of Guidance Control and Dynamics*, 27:767–776, 2004.

[3] G. Buttazzo, M. Velasco, and P. Marti. Quality-of-Control Management in Overloaded Real-Time Systems. *IEEE Trans. on Computers*, 56(2):253–266, Feb. 2007.

[4] A. Cervin and J. Eker. The control server: A computational model for real-time control tasks. In *ECRTS*, pages 113–120. IEEE Computer Society, 2003.

[5] A. Cervin, D. Henriksson, B. Lincoln, J. Eker, and K.-E. Arzen. How does control timing affect performance? Analysis and simulation of timing using Jitterbug and TrueTime. *IEEE Control Systems Magazine*, 23(3):16–30, June 2003.

[6] T. Chantem, X. S. Hu, and M. Lemmon. Generalized Elastic Scheduling for Real-Time Tasks. *IEEE Trans. on Computers*, 58(4):480–495, April 2009.

[7] H. Dörrie. *100 Great Problems of Elementary Mathematics: Their History and Solutions*, chapter 24, pages 112–116. Dover, 1965.

[8] L. Elsner and T. Szulc. Convex sets of schur stable and stable matrices. *Linear and Multilinear Algebra*, 48(1):1–19, 2000.

[9] D. Fontanelli and L. Palopoli. Quality of service and quality of control in real-time control systems. In *Communications Control and Signal Processing (ISCCSP), 2012 5th International Symposium on*, pages 1 –5, may 2012.

[10] D. Fontanelli, L. Palopoli, and L. Greco. Deterministic and Stochastic QoS Provision for Real-Time Control Systems. In *Proc. IEEE Real-Time and Embedded Technology and Applications Symposium*, pages 103–112, Chicago, IL, USA, April 2011. IEEE.

[11] L. Greco, D. Fontanelli, and A. Bicchi. Design and stability analysis for anytime control via stochastic scheduling. *Automatic Control, IEEE Transactions on*, (99):1–1, 2011.

[12] C.-Y. Kao and A. Rantzer. Stability analysis of systems with uncertain time-varying delays. *Automatica*, 43(6):959–970, June 2007.

[13] H. Kopetz and G. Bauer. The time-triggered architecture. *Proceedings of the IEEE*, 91(1):112–126, 2003.

[14] B. Lincoln and A. Cervin. JITTERBUG: a tool for analysis of real-time control performance. In *Proc. IEEE Conf. on Decision and Control*, pages 1319–1324, Dec. 2002.

[15] Q. Ling and M. Lemmon. Robust performance of soft real-time networked control systems with data dropouts. In *Proc. IEEE Conf. on Decision and Control*, volume 2, pages 1225–1230, Dec. 2002.

[16] R. Majumdar, I. Saha, and M. Zamani. Performance-aware scheduler synthesis for control systems. In *Proceedings of the ninth ACM international conference on Embedded software*, EMSOFT '11, pages 299–308, New York, NY, USA, 2011. ACM.

[17] P. Marti, J. Fuertes, G. Fohler, and K. Ramamritham. Jitter compensation for real-time control systems. In *Proc. IEEE Real-Time Systems Symposium*, pages 39–48, Dec. 2001.

[18] M. Mazo and P. Tabuada. Input-to-state stability of self-triggered control systems. In *Decision and Control, 2009 held jointly with the 2009 28th Chinese Control Conference. CDC/CCC 2009. Proceedings of the 48th IEEE Conference on*, pages 928 –933, dec. 2009.

[19] J. Nilsson and B. Bernhardsson. Analysis of real-time control systems with time delays. In *Proc. IEEE Conf. on Decision and Control*, volume 3, pages 3173–3178, Dec. 1996.

[20] L. Palopoli, A. Bicchi, and A. S. Vincentelli. Numerically efficient control of systems with

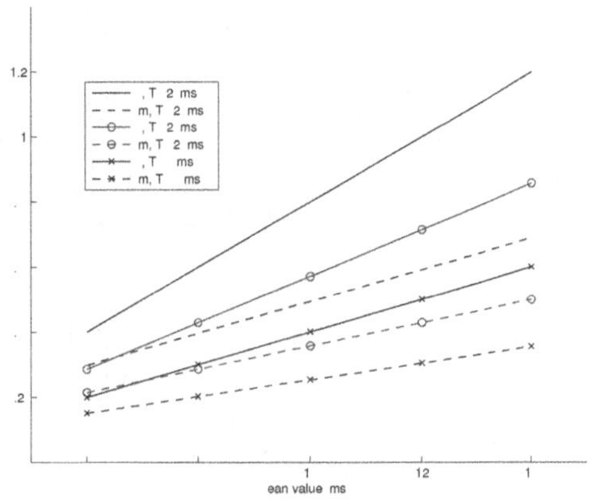

(a)

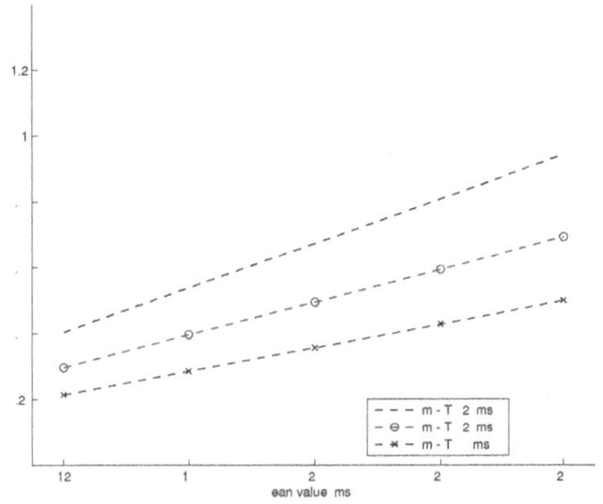

(b)

Figure 5: Available bandwidth for different sampling times as a function of the mean value of the uniform (a) or exponential (b) computation times distributions.

communication constraints. In *Proc. of the IEEE 2002 conference on decision and control (CDC02)*, Las Vegas, Nevada, USA, December 2002.

[21] L. Palopoli, D. Fontanelli, N. Manica, and L. Abeni. An analytical bound for probabilistic deadlines. In *Real-Time Systems (ECRTS), 2012 24th Euromicro Conference on*, pages 179 –188, july 2012.

[22] L. Palopoli, C. Pinello, A. Bicchi, and A. Sangiovanni-Vincentelli. Maximizing the stability radius of a set of systems under real-time scheduling constraints. *Automatic Control, IEEE Transactions on*, 50(11):1790–1795, Nov. 2005.

[23] A. Quagli, D. Fontanelli, L. Greco, L. Palopoli, and A. Bicchi. Designing real-time embedded controllers using the anytime computing paradigm. In *Emerging Technologies & Factory Automation, 2009. ETFA 2009. IEEE Conference on*, pages 1–8. IEEE, 2009.

[24] M. Rabi and K. Johansson. Scheduling packets for event-triggered control. In *Proc. of 10th European Control Conf*, pages 3779–3784, 2009.

[25] R. Rajkumar, K. Juvva, A. Molano, and S. Oikawa. Resource kernels: A resource-centric approach to real-time and multimedia systems. In *Proc. of the SPIE/ACM Conference on Multimedia Computing and Networking*, January 1998.

[26] H. Rehbinder and M. Sanfridson. Integration of off-line scheduling and optimal control. In *Real-Time Systems, 2000. Euromicro RTS 2000. 12th Euromicro Conference on*, pages 137–143. IEEE, 2000.

[27] D. Seto, J. Lehoczky, L. Sha, and K. Shin. On task schedulability in real-time control systems. In *rtss*, page 13. Published by the IEEE Computer Society, 1996.

[28] M. Velasco, P. Martí, and E. Bini. Control-driven tasks: Modeling and analysis. In *IEEE Real-Time Systems Symposium*, pages 280–290. IEEE Computer Society, 2008.

[29] H. Voit, R. Schneider, D. Goswami, A. Annaswamy, and S. Chakraborty. Optimizing hierarchical schedules for improved control performance. In *Industrial Embedded Systems (SIES), 2010 International Symposium on*, pages 9 –17, july 2010.

[30] X. Wang and M. D. Lemmon. Event-triggering in distributed networked control systems. *IEEE Trans. Automat. Contr.*, 56(3):586–601, 2011.

[31] F. Zhang, K. Szwaykowska, W. Wolf, and V. Mooney. Task scheduling for control oriented requirements for cyber-physical systems. In *Real-Time Systems Symposium, 2008*, pages 47 –56, 30 2008-dec. 3 2008.

Safe Schedulability of Bounded-Rate Multi-Mode Systems

Rajeev Alur
University of Pennsylvania,
Philadelphia, USA
alur@seas.upenn.edu

Vojtěch Forejt
Dept. of Computer Science,
University of Oxford, UK
vojfor@cs.ox.ac.uk

Salar Moarref
University of Pennsylvania,
Philadelphia, USA
moarref@seas.upenn.edu

Ashutosh Trivedi
Dept. of Computer Sc. & Eng.,
IIT Bombay, India
trivedi@cse.iitb.ac.in

ABSTRACT

Bounded-rate multi-mode systems (BMS) are hybrid systems that can switch freely among a finite set of modes, and whose dynamics is specified by a finite number of real-valued variables with mode-dependent rates that can vary within given bounded sets. The schedulability problem for BMS is defined as an infinite-round game between two players—the scheduler and the environment—where in each round the scheduler proposes a time and a mode while the environment chooses an allowable rate for that mode, and the state of the system changes linearly in the direction of the rate vector. The goal of the scheduler is to keep the state of the system within a pre-specified safe set using a non-Zeno schedule, while the goal of the environment is the opposite. Green scheduling under uncertainty is a paradigmatic example of BMS where a winning strategy of the scheduler corresponds to a robust energy-optimal policy. We present an algorithm to decide whether the scheduler has a winning strategy from an arbitrary starting state, and give an algorithm to compute such a winning strategy, if it exists. We show that the schedulability problem for BMS is co-NP complete in general, but for two variables it is in PTIME. We also study the discrete schedulability problem where the environment has only finitely many choices of rate vectors in each mode and the scheduler can make decisions only at multiples of a given clock period, and show it to be EXPTIME-complete.

Categories and Subject Descriptors

I.2.8 [**Problem Solving, Control Methods, and Search**]: Control Theory, Scheduling; D.4.7 [**Organization and Design**]: Real-time systems and embedded systems

Keywords

Multi-Mode Systems; Green Scheduling; Controller Synthesis; Invariant Sets; Constrained Control; Stability

HSCC'13, April 8–11, 2013, Philadelphia, Pennsylvania, USA.

1. INTRODUCTION

There is a growing trend towards multi-mode compositional design frameworks [10, 15, 11] for the synthesis of cyber-physical systems where the desired system is built by composing various modes, subsystems, or motion primitives—with well-understood performance characteristics—so as to satisfy certain higher level control objectives. A notable example of such an approach is *green scheduling* proposed by Nghiem et al. [13, 14] where the goal is to compose different modes of heating, ventilation, and air-conditioning (HVAC) installations in a building so as to keep the temperature surrounding each installation in a given comfort zone while keeping the peak energy consumption under a given budget. Under the assumption that the state of the system grows linearly in each mode, Nghiem et al. gave a polynomial algorithm to decide the green schedulability problem. Alur, Trivedi, and Wojtczak [2] studied general constant-rate multi-mode systems and showed, among others, that the result of Nghiem et al. holds for arbitrary multi-mode systems with constant rate dynamics as long as the scheduler can switch freely among the finite set of modes.

In this paper we present *bounded-rate multi-mode systems* that generalize constant-rate multi-mode systems by allowing non-constant mode-dependent rates that are given as bounded polytopes. Our motivations to study bounded-rate multi-mode schedulability are twofold. First, it allows one to model a conservative approximation of green schedulability problem in presence of more complex inter-mode dynamics. Second motivation is theoretical and it stems from the desire to characterize decidable problems in context of design and analysis of cyber-physical systems. In particular, we view a bounded-rate multi-mode system as a two-player extension of constant-rate multi-mode system, and show the decidability of schedulability game for such systems.

Before discussing bounded-rate multi-mode system (BMS) in any further detail, let us review the definition, relevant results, and limitations of constant-rate multi-mode system (CMS). A CMS is specified as a finite set of variables whose dynamics in a finite set of modes is given as mode-dependent constant rate vector. The *schedulability problem* for a CMS and a bounded convex safety set of states is to decide whether there exists an infinite sequence (schedule) of modes and time durations such that choosing modes for corresponding time durations in that sequence keeps the system within the safety set forever. Moreover such schedule is also required to be physically implementable, i.e. the sum of time dura-

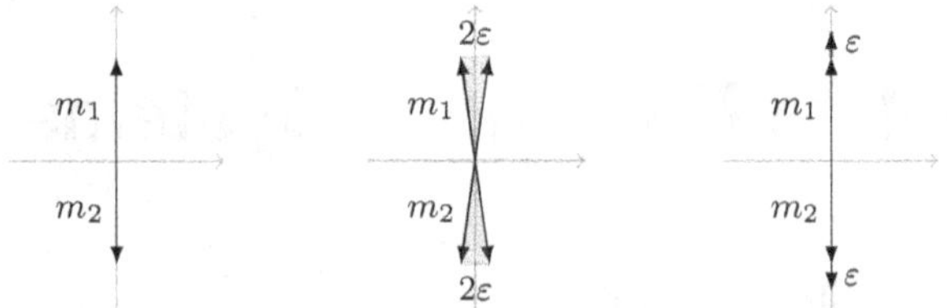

Figure 1: Multi-mode systems with uncertain rates

tions must diverge (the standard non-Zeno requirement [8]). Alur et al. [2] showed that, for the starting states in the interior of the safety set, the necessary and sufficient condition for safe schedulability is the existence of an assignment of dwell times to modes such that the sum of rate vectors of various modes scaled by corresponding dwell time is zero. Intuitively, if it is possible using the modes to loop back to the starting state, i.e. to go to some state other than the starting state and then to return to the starting state, then the same schedule can be scaled appropriately and repeated forever to form a *periodic schedule* that keeps the system inside the interior of any convex safety set while ensuring time divergence. On the other hand, if no such assignment exists then Farkas' lemma implies the existence of a vector such that choosing any mode the system makes a positive progress in the direction of that vector, and hence for any non-Zeno schedule the system will leave any bounded safety set in a finite amount of time. Also, due to constant-rate dynamics such condition can be modeled as a linear program feasibility problem, yielding a polynomial-time algorithm.

EXAMPLE 1. *Consider the 2-dimensional CMS shown in Figure 1 (left) with two modes m_1 and m_2 with rates of the variables as $\vec{r}_1 = (0,1)$ in mode m_1 and $\vec{r}_2 = (0,-1)$ in mode m_2. It is easy to see that the system is schedulable for any starting state (x_0, y_0) in the interior of any bounded convex set S as $\vec{r}_1 + \vec{r}_2 = (0,0)$. The safe schedule consists of the periodic schedule $(m_1, t), (m_2, t)$ for a carefully selected $t \in \mathbb{R}_{>0}$ such that $(x_0, y_0) + \vec{r}_1 t$ stays inside S.*

However, the schedules constructed in this manner are not robust as an arbitrarily small change in the rate can make the schedule unsafe as shown in the following example.

EXAMPLE 2. *Consider a multi-mode system where some environment related fluctuations [8] cause the rate vectors in modes m_1 and m_2 to differ from those in Example 1 by an arbitrarily small $\varepsilon > 0$ as shown in Figure 1 (center). Here, m_1 can have rate-vectors from $\{(0+\delta, 1) : -\varepsilon \le \delta \le \varepsilon\}$, while rate-vectors of m_2 are from $\{(0+\delta, -1) : -\varepsilon \le \delta \le \varepsilon\}$. First we show that the periodic schedule $(m_1, t), (m_2, t)$ proposed in Example 1 is not safe for any t. Consider the case when the rate vector in modes m_1 and m_2 are fixed to $(\varepsilon, 1)$ and $(\varepsilon, -1)$. Starting from the state (x_0, y_0) and following the periodic schedule $(m_1, t), (m_2, t)$ for k steps the state of the system will be $(x_0 + kt\varepsilon, y_0)$ after k steps. Hence it is easy to see that for any bounded safety set the state of the system will leave the safety set after finitely many steps. In fact, for this choice of rate vectors no non-Zeno safe schedule exists at all, since by choosing any mode for a positive time the system makes a positive progress along the X axis.*

We formalize modeling of such multi-mode system under uncertainty as bounded-rate multi-mode systems (BMS). BMSs can also approximate [5] the effect of more complex non-linear, and even time-varying, mode dynamics over a bounded safety set. Formally, a BMS is specified as a finite set of variables whose dynamics in a finite set of modes is given as a mode-dependent bounded convex polytopes of rate vectors. We present the schedulability problem on BMS as an infinite-round zero-sum game between two players, *the scheduler* and *the environment*; at each round scheduler chooses a mode and a time duration, the environment chooses a rate vector from the allowable set of rates for that mode, and the state of the system is evolved accordingly. The recipe for selecting their choices, or *moves*, is formalized in the form of a strategy that is a function of the history of the game so far to a move of the player. A strategy is called *positional* if it is a function of the current state. We say that the scheduler wins the schedulability game, or has a winning strategy, from a given starting state if there is a scheduler strategy such that, irrespective of the strategies of the environment, the state of the system stays within the safety set and time does not converge to any real number. Similarly, we say that the environment has a winning strategy if she has a strategy such that for any strategy of the scheduler the system leaves the safety set in a finite amount of time, or the time converges to some real number. One of the central results of this paper is that the schedulability games on BMS are *determined*, i.e. for each starting state exactly one of the player has a winning strategy. Note that the determinacy of these games could be proved using more general results on determinacy, e.g. [12], however our proof is direct and shows the existence of positional winning strategies.

We distinguish between two kind of strategies of scheduler– the *static strategies*, where scheduler can not observe the decisions of the environment, and the *dynamic strategies*, where scheduler can observe the decisions of the environment so far before choosing a mode and a time. Our use of static vs. dynamic strategies closely corresponds to standard *open-loop* control vs. *closed-loop* control distinction in control theory. Also notice that static strategies correspond precisely to schedules, and we often use these two terms interchangeably. A key challenge in the schedulability analysis of BMS is inadequacy of static strategies as shown below.

EXAMPLE 3. *Consider the BMS of Figure 1 (right) where the rates in mode m_1 and m_2 lie in $\{(0, 1+\delta) : 0 \le \delta \le \varepsilon\}$ and $\{(0, -(1+\delta)) : 0 \le \delta \le \varepsilon\}$, respectively. We hint that there is no static winning strategy of scheduler in this BMS (the formal conditions for the existence of static winning strategies will be analyzed later in the paper). Let us assume, for example, that $\sigma = (m_1, t_1), (m_2, t_2), \ldots$ is a static non-Zeno winning strategy of the scheduler. Moreover consider two strategies π and π' of the environment that differ only in mode m_1 where they propose rates $(1, 0)$ and $(1+\varepsilon, 0)$ respectively. Let ϱ and ϱ' be the sequences of system states and player's choices—what we subsequently refer to as runs—as the game progresses from a starting state (x_0, y_0) where the environment uses strategy π and π', respectively, against scheduler's strategy σ. Let $T_1(i)$ and $T_2(i)$ be the time spent in mode m_1 and m_2, resp., till the i-th round in runs ϱ and ϱ', while T_1 and T_2 be total time spent in mode m_1 and m_2, resp. The state of the system in the runs ϱ and ϱ' after i rounds will be $(x_0, y_0 + T_1(i) - T_2(i))$ and $(x_0, y_0 + T_1(i) - T_2(i) + T_1(i)\varepsilon)$. Hence the distance $T_1(i)\varepsilon$ between states reached after i-rounds in runs ϱ and ϱ' tends to $T_1\varepsilon$ as i tends to ∞. It is easy to see that if σ is a winning strategy then $T_1 = \infty$; since if $T_1 < \infty$ and $T_2 = \infty$ then the system will move in the direction of rates of mode m_2, while if* *both T_1 and T_2 are finite then the strategy is not non-Zeno. Hence system will eventually leave any bounded safety set, contradicting our assumption on σ being a winning strategy.*

The techniques used for schedulability analysis and schedule construction for CMS cannot be generalized to BMS since in a BMS, the scheduler may not have a strategy to loop back to the starting state. In fact, in general scheduler does not have a strategy to revisit any state as is clear from Figure 1 (right)—here the environment can always choose a rate vector in both mode m_1 and m_2 to avoid any previously visited state. However, from our results on BMS it follows that if the scheduler has a winning strategy then he has a strategy to restrict the future states of the system to a ball of arbitrary diameter centered around the starting state.

In order to solve schedulability game for BMS we exploit the following observation[1]: the scheduler has a winning strategy, from all the starting states in the interior of the safety set S, if and only if there is a polytope $P \subseteq S$, such that for every vertex $\overline{v}$ of P there is a mode $m(\overline{v})$ and time $t(\overline{v})$ such that choosing mode $m(\overline{v})$ for time $t(\overline{v})$ from the vertex $\overline{v}$, the line $\overline{v} + \vec{r}t(\overline{v})$ stays within polytope P for all allowable rates $\vec{r}$ of $m(\overline{v})$. In other words, for any vertex of P there is a mode and a time duration such that if the system evolves with any rate vector of that mode for such amount of time, the system stays in P. For a BMS $\mathcal{H}$ we call such a polytope *$\mathcal{H}$-closed.* We show how such a polytope can be constructed for a BMS based on its characteristics. We also analyze the complexity of such a construction. The existence of an $\mathcal{H}$-closed polytope immediately provides a non-Zeno safe dynamic strategy for the scheduler for any starting state in P: find the convex coefficient $(\lambda_1, \lambda_2, \ldots, \lambda_k)$ of the current state $\overline{x}$ with respect to the finite set of vertices $(\overline{x}_1, \overline{x}_2, \ldots, \overline{x}_k)$ of P and choose the mode $m(\overline{x}_i)$ for time $t(\overline{x}_i)\lambda_i$ that maximizes $t(\overline{x}_i)\lambda_i$. Then, for some choice $\vec{r}$ of the environment for $m(\overline{x}_i)$ the system will progress to $\overline{x}' = \overline{x} + t(\overline{x}_i)\lambda_i\vec{r}$. One can repeat this dynamic strategy from the next state $\overline{x}'$ as the current state. We prove that such strategy is both non-Zeno and safe.

An extreme-rate CMS of a BMS $\mathcal{H}$ is obtained by preserving the set of modes, and for each mode assigning a rate which is a vertex of the available rate-set of that mode. The main result of the paper is that an $\mathcal{H}$-closed polytope exists for a BMS $\mathcal{H}$ iff all *extreme-rate* CMSs of $\mathcal{H}$ are schedulable. The "only if" direction of the above characterization is immediate as if some extreme-rate CMS is not schedulable then the environment can fix those rate vectors and win the schedulability game in the BMS. We show the "if" direction by explicitly constructing the $\mathcal{H}$-closed polytope.

EXAMPLE 4. *Consider the BMS $\mathcal{H}$ from Figure 1 (right) with $\varepsilon = 0.5$. The safety set is given as a shaded area in Figure 2 (left) and $\bar{x}_0 = (-1, -0.5)$ is the initial state. Observe that all extreme-rate combinations are schedulable and hence we show a winning strategy. An $\mathcal{H}$-closed polytope for this BMS is the line-segment between the points $(0, 2.5)$ and $(0, -2.5)$ (we explain the construction of such a polytope in Section 3). After translating this line-segment to x_0 and scaling it to fit inside the safety set, we will get the line-segment connecting $\bar{x}_1 = (-1, 1)$ to $\bar{x}_2 = (-1, -2)$, as shown in Figure 2 (left). At vertices $\bar{x}_1$ and $\bar{x}_2$ modes m_2*

[1]Blanchini [3, Theorem 5.2] makes a similar observation to design uniformly stable control for uncertain linear systems.

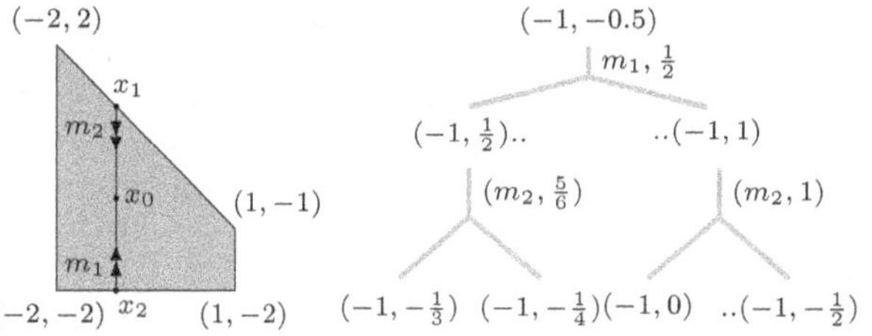

Figure 2: $\mathcal{H}$-closed polytope and dynamic strategy

and m_1, respectively, can be used for 1 time unit. A winning strategy of scheduler is to keep the system's state along the line segment. Our strategy observes the current state $\overline{x}$ and finds the mode to choose by computing convex coefficient $\lambda \in [0, 1]$ s.t. $\overline{x} = \lambda\overline{x}_1 + (1-\lambda)\overline{x}_2$. For instance, at state $\overline{x}_0 = \frac{1}{2}\overline{x}_1 + \frac{1}{2}\overline{x}_2$ the scheduler can choose any of the modes for $\frac{1}{2}$ time units. Assume that it chooses m_1. Based on environment's choice the state of system after $\frac{1}{2}$ time units will be in the set $\{-1, 0.5 + \delta : 0 \leq \delta \leq 0.5\}$. The scheduler observes this new state after $\frac{1}{2}$ time-unit, and chooses mode and time accordingly. For example, if the environment chooses $(0, 1.25)$ and so the next state is $\overline{x} = (-1, 0.75) = \frac{1}{12}\overline{x}_1 + \frac{11}{12}\overline{x}_2$, scheduler can choose mode m_2 for $\frac{11}{12}$ time units. In Figure 2 (right) we show first two rounds of the game. Since, for any point on our line segment scheduler can choose a mode for at least 0.5 time unit and stay on the line segment, such strategy is both safe and non-Zeno.

We also extend the above result to decide the winner starting from arbitrary states, i.e. including those states that lie on the boundary of the safety set. Here we show that the existence of a safe scheduler implies the existence of a safe scheduler which only allows to move from lower-dimensional faces to higher-dimensional ones and not the other way around; this allows us to use an algorithm which traverses the face lattice of the safety set and analyses each face one by one. We also prove co-NP completeness of the schedulability problem, showing the hardness by giving a reduction from 3-SAT to the non-schedulability problem. On a positive note, we show that if the number of variables is two, then the schedulability game can be decided in polynomial time. This is because in such a case we can prove that there is only polynomially many candidates for falsifiers we need to consider, and hence we can check each of them one by one. Finally, we study a discrete version of schedulability games where scheduler can choose time delays only at multiples of a given clock period, while the environment can choose rate vectors from a finite set. We show that discrete schedulability games on BMS are EXPTIME-complete, and that the maximal clock period for which scheduler has a winning strategy can be computed in exponential time. If the system is a CMS, we get a PSPACE algorithm, improving the result of [2] where only an approximation of the maximal clock period for CMS was studied.

For a review of related work on CMS and green scheduling we refer to [14, 13] and [2]. The work closest to this paper is that on *constrained regulator problem* by Blanchini [3]. This work studies the concept of *positively invariant* region that closely corresponds to our definition of $\mathcal{H}$-closed polytope, and present conditions for schedulability of more general linear systems. However, [3] does not settle computational complexity of corresponding problem that form the core of our results. We refer the reader to [4] for an excellent survey on positively invariant sets and their applications in controller synthesis. Another closely related work is by Heymann et al. [8] that considers scheduling problem on BMS where rate-vectors are given as upper and lower rate matrices and the safety set as the entire non-negative orthant. The main result of [8] is that the scheduler wins if he wins in the CMS of the lower rate matrix, and wins only if he wins in the CMS of the upper rate matrix. We study more general BMS and safety sets, and characterize necessary and sufficient condition for schedulability. Finally, to complete the picture, we remark that games on hybrid automata [7, 6], that corresponds to BMS with local invariants and guards, have undecidable schedulability problem.

For the lack of space proofs are either sketched or omitted; full proofs can be found in the technical report [1].

2. PROBLEM DEFINITION

Points and Vectors. Let $\mathbb{R}$ be the set of real numbers. We represent the states in our system as points in $\mathbb{R}^n$ that is equipped with the standard *Euclidean norm* $\|\cdot\|$. We denote points in this state space by $\overline{x}, \overline{y}$, vectors by $\vec{r}, \vec{v}$, and the i-th coordinate of point $\overline{x}$ and vector $\vec{r}$ by $\overline{x}(i)$ and $\vec{r}(i)$, respectively. We write $\vec{0}$ for a vector with all its coordinates equal to 0; its dimension is often clear from the context. The distance $\|\overline{x}, \overline{y}\|$ between points $\overline{x}$ and $\overline{y}$ is defined as $\|\overline{x} - \overline{y}\|$. For two vectors $\vec{v}_1, \vec{v}_2 \in \mathbb{R}^n$, we write $\vec{v}_1 \cdot \vec{v}_2$ to denote their dot product defined as $\sum_{i=1}^{n} \vec{v}_1(i) \cdot \vec{v}_2(i)$.

Boundedness and Interior. We denote a *closed ball* of radius $d \in \mathbb{R}_{\geq 0}$ centered at $\overline{x}$ as $B_d(\overline{x}) = \{\overline{y} \in \mathbb{R}^n : \|\overline{x}, \overline{y}\| \leq d\}$. We say that a set $S \subseteq \mathbb{R}^n$ is *bounded* if there exists $d \in \mathbb{R}_{\geq 0}$ such that for all $\overline{x}, \overline{y} \in S$ we have $\|\overline{x}, \overline{y}\| \leq d$. The *interior* of a set S, $\text{int}(S)$, is the set of all points $\overline{x} \in S$ for which there exists $d > 0$ s.t. $B_d(\overline{x}) \subseteq S$.

Convexity. A point $\overline{x}$ is a *convex combination* of a finite set of points $X = \{\overline{x}_1, \overline{x}_2, \ldots, \overline{x}_k\}$ if there are $\lambda_1, \lambda_2, \ldots, \lambda_k \in [0, 1]$ such that $\sum_{i=1}^{k} \lambda_i = 1$ and $\overline{x} = \sum_{i=1}^{k} \lambda_i \cdot \overline{x}_i$. The *convex hull* of X is then the set of all points that are convex combinations of points in X. We say that $S \subseteq \mathbb{R}^n$ is *convex* iff for all $\overline{x}, \overline{y} \in S$ and all $\lambda \in [0, 1]$ we have $\lambda\overline{x} + (1 - \lambda)\overline{y} \in S$ and moreover, S is a *convex polytope* if it is bounded and there exists $k \in \mathbb{N}$, a matrix A of size $k \times n$ and a vector $\vec{b} \in \mathbb{R}^k$ such that $\overline{x} \in S$ iff $A\overline{x} \leq \vec{b}$. We write *rows*$(M)$ for the number of rows in a matrix M, here *rows*$(A) = k$.

A point $\overline{x}$ is a *vertex* of a convex polytope P if it is not a convex combination of two distinct (other than $\overline{x}$) points in P. For a convex polytope P we write $\mathsf{vert}(P)$ for the finite set of points that correspond to the vertices of P. Each point in P can be written as a convex combination of the points in $\mathsf{vert}(P)$, or in other words, P is the *convex hull* of $\mathsf{vert}(P)$. From standard properties of polytopes, it follows that for every convex polytope P and every vertex $\overline{c}$ of P, there exists a vector $\vec{v}$ such that $\vec{v} \cdot \overline{c} = d$ and $\vec{v} \cdot \overline{x} > d$ for all $\overline{x} \in P \setminus \{\overline{c}\}$ for some d. We call such a vector $\vec{v}$ a *supporting hyperplane* of the polytope P at $\overline{c}$.

2.1 Multi-Mode Systems

A multi-mode system is a hybrid system equipped with finitely many *modes* and finitely many real-valued *variables*. A configuration is described by values of the variables, which change, as the time elapses, at the rates determined by the modes being used. The choice of rates is nondeterministic, which introduces a notion of adversarial behavior. Formally,

DEFINITION 1 (MULTI-MODE SYSTEMS). *A multi-mode system is a tuple $\mathcal{H} = (M, n, \mathcal{R})$ where: M is the finite nonempty set of* modes, *n is the number of continuous variables, and $\mathcal{R} : M \to 2^{\mathbb{R}^n}$ is the rate-set function that, for each mode $m \in M$, gives a set of vectors.*

We often write $\vec{r} \in m$ for $\vec{r} \in \mathcal{R}(m)$ when $\mathcal{R}$ is clear from the context. A finite *run* of a multi-mode system $\mathcal{H}$ is a finite sequence of states, timed moves and rate vector choices $\varrho = \langle \overline{x}_0, (m_1, t_1), \vec{r}_1, \overline{x}_1, \ldots, (m_k, t_k), \vec{r}_k, \overline{x}_k \rangle$ s.t. for all $1 \leq i \leq k$ we have $\vec{r}_i \in \mathcal{R}(m_i)$ and $\overline{x}_i = \overline{x}_{i-1} + t_i \cdot \vec{r}_i$. For such a run ϱ we say that $\overline{x}_0$ is the *starting state*, while $\overline{x}_k$ is its *last state*. An infinite run is defined in a similar manner. We write *Runs* and *FRuns* for the set of infinite and finite runs of $\mathcal{H}$, while *Runs*$(\overline{x})$ and *FRuns*$(\overline{x})$ for the set of infinite and finite runs starting from $\overline{x}$.

An infinite run $\langle \overline{x}_0, (m_1, t_1), \vec{r}_1, \overline{x}_1, (m_2, t_2), \vec{r}_2, \ldots \rangle$ is *Zeno* if $\sum_{i=1}^{\infty} t_i < \infty$. Given a set $S \subseteq \mathbb{R}^n$ of safe states, we say that a run $\langle \overline{x}_0, (m_1, t_1), \vec{r}_1, \overline{x}_1, (m_2, t_2), \vec{r}_2, \ldots \rangle$ is S-safe if for all $i \geq 0$ we have that $\overline{x}_i \in S$ and $\overline{x}_i + t \cdot \vec{r}_{i+1} \in S$ for all $t \in [0, t_{i+1}]$, assuming $t_0 = 0$. Notice that if S is a convex set and $\overline{x}_i \in S$ for all $i \geq 0$, then for all $i \geq 0$ and for all $t \in [0, t_{i+1}]$ we have that $\overline{x}_i + t \cdot \vec{r}_{i+1} \in S$. The concept of S-safety for finite runs is defined in a similar manner. Sometimes we simply call a run safe when the safety set and the starting state is clear from the context.

We formally give the semantics of a multi-mode system $\mathcal{H}$ as a turn-based two-player game between the players, *scheduler* and *environment*, who choose their moves to construct a run of the system. The system starts in a given starting state $\overline{x}_0 \in \mathbb{R}^n$ and at each turn scheduler chooses a timed move, a pair $(m, t) \in M \times \mathbb{R}_{>0}$ consisting of a mode and a time duration, and the environment chooses a rate vector $\vec{r} \in \mathcal{R}(m)$ and as a result the system changes its state from $\overline{x}_0$ to the state $\overline{x}_1 = \overline{x}_0 + t \cdot \vec{r}$ in t time units following the linear trajectory according to the rate vector $\vec{r}$. From the next state $\overline{x}_1$ the scheduler again chooses a timed move and the environment an allowable rate vector, and the game continues forever in this fashion. The focus of this paper is on safe-*schedulability game*, where the goal of the scheduler is to keep the states of the system within a given safety set S, while ensuring that the time diverges (non-Zenoness requirement). The goal of the environment is the opposite, i.e. to visit a state out of the safety set or make the time converge to some finite number.

Given a bounded and convex safety set S, we define *(safe) schedulability objective* $\mathcal{W}^S_{\mathsf{Safe}}$ as the set of S-safe and non-Zeno runs of $\mathcal{H}$. In a schedulability game the winning objective of the scheduler is to make sure that the constructed run of a system belongs to $\mathcal{W}^S_{\mathsf{Safe}}$, while the goal of the environment is the opposite. The choice selection mechanism of the players is typically defined as strategies. A *strategy* σ of scheduler is function $\sigma{:}FRuns{\to}M{\times}\mathbb{R}_{\geq 0}$ that gives a timed move for every history of the game. A strategy π of the environment is a function $\pi : FRuns \times (M \times \mathbb{R}_{\geq 0}) \to \mathbb{R}^n$ that chooses an allowable rate for a given history of the game and choice of the scheduler. We say that a strategy is *positional* if it suggests the same action for all runs with common last state. We write Σ and Π for the set of strategies of the scheduler and the environment, respectively.

Given a starting state $\overline{x}_0$ and a strategy pair $(\sigma, \pi) \in \Sigma \times \Pi$

we define the unique run $Run(\overline{x}_0, \sigma, \pi)$ starting from $\overline{x}_0$ as

$$Run(\overline{x}_0, \sigma, \pi) = \langle \overline{x}_0, (m_1, t_1), \vec{r}_1, \overline{x}_1, (m_2, t_2), \vec{r}_2, \ldots \rangle$$

where for all $i \geq 1$, $(m_i, t_i) = \sigma(\langle \overline{x}_0, (m_1, t_1), \vec{r}_1, \overline{x}_1, \ldots, \overline{x}_{i-1} \rangle)$ and $\vec{r}_i = \pi(\langle \overline{x}_0, (m_1, t_1), \vec{r}_1, \overline{x}_1, \ldots, \overline{x}_{i-1}, m_i, t_i \rangle)$ and $x_i = x_{i-1} + t_i \cdot \vec{r}_i$. The scheduler wins the game if there is $\sigma \in \Sigma$ such that for all $\pi \in \Pi$ we get $Run(\overline{x}_0, \sigma, \pi) \in \mathcal{W}^S_{\mathsf{Safe}}$. Such a strategy σ is *winning*. Similarly, the environment wins the game if there is $\pi \in \Pi$ such that for all $\sigma \in \Sigma$ we have $Run(\overline{x}_0, \sigma, \pi) \notin \mathcal{W}^S_{\mathsf{Safe}}$. Again, π is called *winning* in this case. If a winning strategy for scheduler exists, we say that $\mathcal{H}$ is schedulable for S and $\overline{x}_0$ (or simply *schedulable* if S and $\overline{x}_0$ are clear from the context). The following is the main algorithmic problem studied in this paper.

DEFINITION 2 (SCHEDULABILITY). *Given a multi-mode system $\mathcal{H}$, a safety set S, and a starting state $\overline{x}_0 \in S$, the (safe)* schedulability problem *is to decide whether there exists a winning strategy of the scheduler.*

2.2 Bounded-Rate Multi-Mode Systems

To algorithmically decide schedulability problem, we need to restrict the range of $\mathcal{R}$ and the domain of safety set S in a schedulability game on a multi-mode system. The most general model that we consider is the bounded-rate multi-mode systems (BMS) that are multi-mode systems $(M, n, \mathcal{R})$ such that $\mathcal{R}(m)$ is a convex polytope for every $m \in M$. We also assume that the safety set S is specified as a convex polytope. In our proofs we often refer to another variant of multi-mode systems in which there are only a fixed number of different rates in each mode (i.e. $\mathcal{R}(m)$ is finite for all $m \in M$). We call such a multi-mode system *multi-rate multi-mode systems* (MMS). Finally, a special form of MMS are *constant-rate multi-mode systems* (CMS) [2] in which $\mathcal{R}(m)$ is a singleton for all $m \in M$. We sometimes use $\mathcal{R}(m)$ to refer to the unique element of the set $\mathcal{R}(m)$ in a CMS. The concepts for the schedulability games for BMS and MMS are already defined for multi-mode systems. Similar concepts also hold for CMS but note that the environment has no real choice in this case. For this reason, we can refer to a schedulability game on CMS as a one-player game.

The prime [2] practical motivation for studying CMS was to generalize results on green scheduling problem by Nghiem et al. [14]. We argue that BMS are a suitable abstraction to study green scheduling problem when various rates of temperature change are either uncertain or follow a complex and time-varying dynamics, as shown in the following example.

EXAMPLE 5 (GREEN SCHEDULING). *Consider a building with two rooms A and B. HVAC units in each zone can be in one of the two modes 0 (OFF) and 1 (ON). We write the mode of the combined system as $m_{i,j}$ to represent the fact that rooms A and B are in mode $i \in \{0, 1\}$ and $j \in \{0, 1\}$, respectively. The rate of temperature change and the energy usage for each room is given below.*

Zones	*ON*	*OFF*
A (temp. change rate/ usage)	*-2/2*	*2/1*
B (temp. change/ usage)	*-2/2*	*2/1*

Following [2] we assume that the energy cost is equal to energy usage if peak energy usage at any given point in time is less than or equal to 3 units, otherwise energy cost is 10 times of that standard rate. It follows that to minimize energy cost the peak usage, if possible, must not be higher than 3 units at any given time. We can model the system as a CMS with modes $m_{0,0}$, $m_{0,1}$, and $m_{1,0}$, because these are the only ones that have peak usage at most 3. The variables of the CMS are the temperature of the rooms, while the safety set is the constraint that temperature of both zones should be between 65^oF to 75^oF. The existence of a winning strategy in CMS implies the existence of a switching schedule with energy peak demand less than or equal to 4 units. In Figure 3.(a) we show a graphical representation of such CMS with three modes $m_{0,0}$, $m_{0,1}$ and $m_{1,0}$ and two variables (corresponding to the two axes). The rate of the variables in mode $m_{0,0}$ is $(2, 2)$, in mode $m_{0,1}$ is $(2, -2)$, and in mode $m_{1,0}$ is $(-2, 2)$.

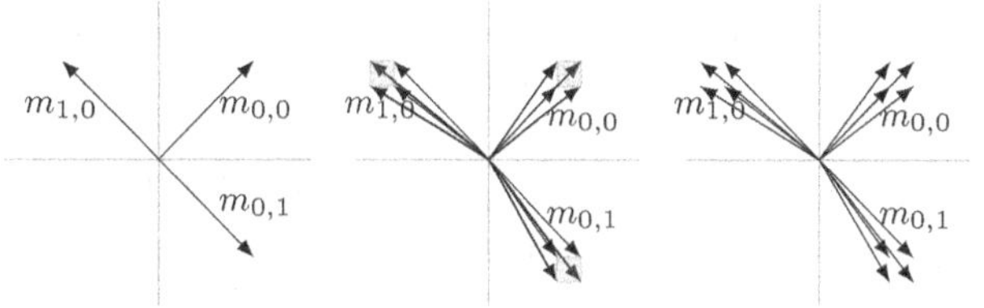

(a) Constant-Rate (b) Bounded-Rate (c) Multi-Rate

Figure 3: Restricted Multi-mode Systems

Now assume that the rate of temperature change in a mode is not constant and can vary within a given margin $\varepsilon > 0$. Schedulability problem for such system can best be modeled as a BMS as shown in Figure 3.(b) where the polytope of possible rate vectors is shown as a shaded region. In Figure 3.(c) we show a MMS where variables can only change with the extreme rates of the BMS in Figure 3.(b).

We say that a CMS $H = (M, n, R)$ is an instance of a multi-mode system $\mathcal{H} = (M, n, \mathcal{R})$ if for every $m \in M$ we have that $R(m) \in \mathcal{R}(m)$. For example, the CMS shown in Figure 3.(a) is an instance of BMS in Figure 3.(b). We denote the set of instances of a multi-mode system $\mathcal{H}$ by $[\![\mathcal{H}]\!]$. Notice that for a BMS $\mathcal{H}$ the set $[\![\mathcal{H}]\!]$ of its instances is uncountably infinite, while for a MMS $\mathcal{H}$ the set $[\![\mathcal{H}]\!]$ is finite whose size is exponential in the size of $\mathcal{H}$. We say that a MMS $(M, n, \mathcal{R}')$ is the *extreme-rate* MMS of a BMS $(M, n, \mathcal{R})$ if $\mathcal{R}'(m) = \mathsf{vert}(\mathcal{R}(m))$. The MMS in Figure 3.(c) is the extreme-rate MMS for the BMS in Figure 3.(b) We write $Ext(\mathcal{H})$ for the extreme-rate MMS of the BMS $\mathcal{H}$.

Notice that for every starting state and winning objective at most one player can have a winning strategy. We say that a game is not *determined* if no player has a winning strategy for some starting state. In the next section we give an algorithm to decide the winner in a schedulability game for an arbitrary starting state. Since for every starting state we can decide the winner, it gives a direct proof of determinacy of schedulability games on BMS. Moreover, it follows from our results that whenever a player has a winning strategy, he has a positional such strategy. These two results together yield the first key results of this paper.

THEOREM 1 (DETERMINACY). *Schedulability games on BMS with convex safety polytopes are positionally determined.*

In Section 4 we analyze the complexity of deciding the winner in a schedulability game. Using a reduction from SAT problem to non-schedulability for a MMS, we prove the following main contribution of the paper.

THEOREM 2. *Schedulability problems for BMS and MMS are co-NP complete.*

On a positive note, we show that schedulability games can be solved in PTIME for BMS and MMS with 2 variables.

3. SOLVING SCHEDULABILITY GAMES

In this section we discuss the decidability of the schedulability problem for BMS. We first present a solution for the case when the starting state is in the interior of a safety set, and generalize it to arbitrary starting states in Section 3.2.

3.1 Starting State in the Interior of Safety Set

Alur et al. [2] presented a polynomial-time algorithm to decide if the scheduler has a winning strategy in a schedulability game on a CMS for an arbitrary starting state. In particular, for starting states in the interior of the safety set, they characterized a necessary and sufficient condition.

THEOREM 3 ([2]). *The scheduler has a winning strategy in a CMS (M, n, R), with convex safety set S and starting state $\overline{x}_0$ in the interior of S, iff there is $\vec{t} \in \mathbb{R}_{\geq 0}^{|M|}$ satisfying:*

$$\sum_{i=1}^{|M|} R(i)(j) \cdot \vec{t}(i) = 0 \text{ for } 1 \leq j \leq n \text{ and } \sum_{i=1}^{|M|} \vec{t}(i) = 1. \quad (1)$$

We call a CMS *safe* if it satisfies (1) and we call H unsafe otherwise. The intuition behind Theorem 3 is that the scheduler has a winning strategy if and only if it is possible to return to the starting state in strictly positive time units. From the results of [2] it also follows that whenever a winning strategy exists, there is a strategy which does not look at a history or even the current state, but only uses a bounded counter of size $\ell \leq |M| - 1$ and after a history of length k makes a decision only based on the number k modulo ℓ. Such strategies are called *periodic*.

It is natural to ask whether the approach of [2] can be generalized to BMS. Unfortunately, Example 3 shows that in a BMS although a winning strategy may exist, it may not be possible to return to the initial state, or indeed visit any state twice. Another natural question to ask is whether a suitable generalization of periodic strategies suffice for BMS. *Static* strategies are BMS analog of periodic strategies that behave in the same manner irrespective of the choices of the environment, i.e. for a static strategy σ we have that $\sigma(\rho) = \sigma(\rho')$ for all runs $\rho = \langle \overline{x}_0, (m_1, t_1), \vec{r}_1, \overline{x}_1, \ldots, (m_k, t_k), \vec{r}_k, \overline{x}_k \rangle$ and $\rho' = \langle \overline{x}_0, (m_1, t_1), \vec{r}'_1, \overline{x}'_1, \ldots, (m_k, t_k), \vec{r}'_k), \overline{x}'_k \rangle$. Static strategies are often desirable in the settings where scheduler can not observe the state of the system. However, we observe [1] that except for the degenerate cases when the BMS contains a subset of modes which induce a safe CMS, scheduler can never win a game on BMS using static strategies. We saw an example of this phenomenon in the Introductory section as Figure 1.(c).

This negative observations imply that to solve the schedulability games for BMS one needs to take a different approach. In the rest of this section, we define the notion of $\mathcal{H}$-closed polytope and show that if such a polytope exists, then for any convex set S we can construct a winning *dynamic* strategy which takes its decisions only based on the last state. We also extend the notion of safety of a CMS to BMS. We say that a BMS $\mathcal{H}$ is *safe* if all instances of its extreme-rate MMS $Ext(\mathcal{H})$ are safe, i.e. all $H \in [\![Ext(\mathcal{H})]\!]$ satisfy (1). Finally, we connect (Lemmas 5 and 6) the existence of $\mathcal{H}$-closed polytope with the safety of the BMS.

Algorithm 1: Dynamic scheduling algorithm

Input: BMMS $\mathcal{H}$, starting state $\overline{x}_0$
Output: non-Terminating Scheduling Algorithm
1 $\gamma :=$ the shortest distance of $\overline{x}_0$ from borders of S;
2 $P := \mathcal{H}$-closed polytope s.t. $P \subseteq B_\gamma(\overline{x}_0)$ and $\overline{x}_0 \in P$;
3 **foreach** $\overline{c} \in \mathsf{vert}(P)$ **do**
4   **foreach** *mode* $m \in M$ **do**
5   **foreach** *extreme rate vector* $\vec{r} \in m$ **do**
6       $t_{\vec{r}} = \max\{t : \overline{c} + \vec{r} \cdot t \in P\}$;
7     $\delta_m = \min_{\vec{r} \in m} t_{\vec{r}}$;
8   $m_* = \arg\max_{m \in M} \delta_m$; $\Delta_{\overline{c}} = \delta_{m_*}$; $m_{\overline{c}} = m_*$;
9 **while** *true* **do**
10   Store current state as $\overline{x}$;
11   Find $(\lambda_{\overline{c}} \geq 0)_{\overline{c} \in \mathsf{vert}(P)}$ where $\overline{x} = \sum_{\overline{c} \in \mathsf{vert}(P)} \lambda_{\overline{c}} \cdot \overline{c}$;
12   Find $\overline{c}_* = \arg\max_{\overline{c} \in \mathsf{vert}(P)} \lambda_{\overline{c}} \cdot \Delta_{\overline{c}}$;
13   Schedule mode $m_{\overline{c}_*}$ for $\lambda_{\overline{c}_*} \cdot \Delta_{\overline{c}_*}$;

Dynamic Scheduling Algorithm. For a BMS $\mathcal{H}$ we call a convex polytope P *$\mathcal{H}$-closed*, if for every vertex of P there exists a mode m such that all the rate vectors of m keep the system in P, i.e. for all $\overline{c} \in \mathsf{vert}(P)$ there exists $m \in M$ and $\tau \in \mathbb{R}_{>0}$ such that for all $\vec{r} \in \mathcal{R}(m)$ we have that $\overline{c} + \vec{r} \cdot t \in P$ for all $t \in [0, \tau]$. An example of a $\mathcal{H}$-closed polytope is given in the Example 4.

Assume that for any $\gamma > 0$ and $\overline{x}_0$ we are able to compute a $\mathcal{H}$-closed polytope which is fully contained in $B_\gamma(\overline{x}_0)$ and contains $\overline{x}_0$. If this is the case, we can use Algorithm 1 to compute a dynamic scheduling strategy. The idea of the algorithm is to build a $\mathcal{H}$-closed polytope which contains the initial state and is fully contained within S, and then construct the strategy based on the modes safe at the vertices of the polytope. The correctness of the algorithm is established by the following proposition.

PROPOSITION 4. *If there exists an $\mathcal{H}$-closed polytope and it can be effectively computed then Algorithm 1 implements a winning dynamic strategy for the scheduler.*

PROOF. Assume that there exists an $\mathcal{H}$-closed polytope and we have an algorithm to compute it. Observe that the strategy is non-Zeno, because $\lambda_{\overline{c}_*} \cdot \Delta_{\overline{c}_*}$ on line 13 is bounded from below by $\frac{1}{|\mathsf{vert}(P)|} \cdot \min_{\overline{c} \in \mathsf{vert}(P)} \Delta_{\overline{c}}$ for any point of P, and $\Delta_{\overline{c}}$ are positive by their construction and the definition of the $\mathcal{H}$-closed polytope. Next, we need to show that under the computed strategy we never leave the convex polytope P. For a state $\overline{x}$ which is of the form $\sum_{\overline{c} \in \mathsf{vert}(P)} \lambda_{\overline{c}} \cdot \overline{c}$, the successor state will be $\overline{x}' = (\sum_{\overline{c} \in \mathsf{vert}(P)} \lambda_{\overline{c}} \cdot \overline{c}) + \lambda_{\overline{c}_*} \cdot \Delta_{\overline{c}_*} \cdot \vec{r}$ where $\vec{r}$ is the rate picked by the environment. We can rewrite $\overline{x}'$ as $(\sum_{\overline{c} \in \mathsf{vert}(P) \setminus \{\overline{c}_*\}} \lambda_{\overline{c}} \cdot \overline{c}) + \lambda_{\overline{c}_*} \cdot (\overline{c}_* + \vec{r} \cdot \Delta_{\overline{c}_*})$. Since $\overline{c}_* + \vec{r} \cdot \Delta_{\overline{c}_*} \in P$, we get that $\overline{x}'$ is a convex combination of points in P and hence lies in P. □

Constructing $\mathcal{H}$-Closed Polytope. We will next show how to implement line 2 of Algorithm 1. We give necessary and sufficient conditions for existence of $\mathcal{H}$-closed polytopes in the following two lemmas. The first lemma shows that an $\mathcal{H}$-closed polytope exists if and only if for any hyperplane (given by its normal vector $\vec{v}$) there exists a mode m such that all its rates stay at one side of the hyperplane.

LEMMA 5. *For a BMS $\mathcal{H}$, a state $\overline{x}_0$ and $\gamma > 0$, there is a $\mathcal{H}$-closed polytope $P \subseteq B_\gamma(\overline{x}_0)$ with $\overline{x}_0 \in P$ if and only if for every $\vec{v}$ there is a mode m such that $\vec{v} \cdot \vec{r} \geq 0$ for all $\vec{r} \in m$.*

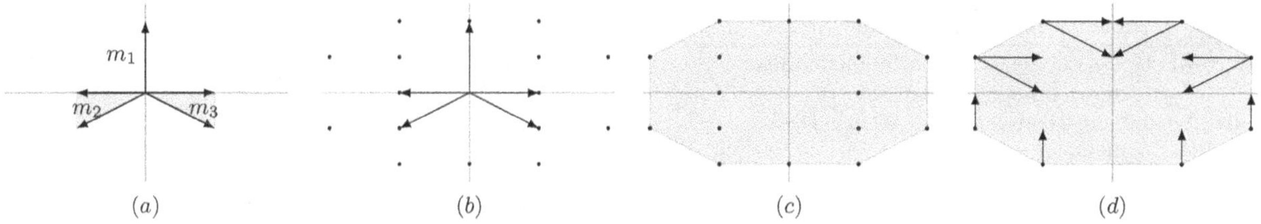

Figure 4: Constructing closed convex polytope

PROOF. Let us fix a BMS $\mathcal{H} = (M, n, \mathcal{R})$. The proof is in two parts. For $\Rightarrow$, assume that the system is schedulable but there exists a vector $\vec{v}$ such that for all modes $m \in M$ there is a rate $\vec{r}_m \in m$ where $\vec{v} \cdot \vec{r}_m < 0$. It implies that if the adversary fixes the rates $\vec{r}_m$ whenever the scheduler chooses m, then the system moves in the direction of vector $-\vec{v}$ (i.e. for all d a state $\overline{x}$ will be reached such that $\vec{v} \cdot \overline{x} < d$), and hence for any bounded safety set and non-Zeno strategy system will leave the safety set. This contradicts with existence of $\mathcal{H}$-closed polytope implying winning scheduler strategy.

To prove the other direction, let $R = \{\vec{r}_1, \ldots, \vec{r}_N\}$ be the set of rates occurring in modes of the extreme-rate MMS of $\mathcal{H}$, i.e. $R = \{\mathcal{R}'(m) : (M, n, \mathcal{R}') \in [\![Ext(\mathcal{H})]\!], m \in M\}$. We claim the following to be the $\mathcal{H}$-closed polytope:

$$P := \{\overline{x}_0 + D \cdot \sum_{i=1}^{N} \vec{r}_i \cdot p_i \mid p_i \in [0,1]\}, \tag{2}$$

where $D = \gamma / \sum_{i=1}^{N} \|\vec{r}_i\|$. Notice that P is a convex polytope since it is a convex hull of points $\overline{x}_0 + D \cdot \sum_{i=1}^{N} \vec{r}_i \cdot p_i$ where $p_i \in \{0,1\}$. Also, due to our choice of D, $P \subseteq B_\gamma(\overline{x}_0)$, and $\overline{x}_0 \in P$. For the sake of contradiction we assume that for every $\vec{v}$ there is a mode m such that all rates $\vec{r}$ of m satisfy $\vec{v} \cdot \vec{r} \geq 0$, but at least one corner $\overline{c}$ of P does not satisfy the defining condition of $\mathcal{H}$-closed polytope, i.e. for all modes i there is a rate vector $\vec{r}_i$ satisfying

$$\overline{c} + t \cdot \vec{r}_i \notin P \text{ for all } t > 0 \tag{3}$$

Let us fix such corner $\overline{c}$. By the supporting hyperplane theorem there is a vector $\vec{v}$ such that, for some d:

$$\vec{v} \cdot \overline{c} \;=\; d \tag{4}$$

$$\vec{v} \cdot \overline{x} \;>\; d, \text{ for all } \overline{x} \in P \setminus \{\overline{c}\} \tag{5}$$

i.e. $\vec{v}$ is supporting P on $\overline{c}$. Let us fix some mode m such that for all rates $\vec{r}$ of m we have $\vec{v} \cdot \vec{r} \geq 0$. Notice that this exists by the assumption. Let $\vec{r}_i$ be a rate of m satisfying (3).

By the definition of P the point $\overline{c}$, a corner of P, is of the form $\overline{x}_0 + D \cdot \sum_{j=1}^{N} \vec{r}_j \cdot p_j$ for some $p_j \in [0,1]$ where $1 \leq j \leq N$ and $\vec{r}_j \in R$. We necessarily have $p_i = 1$, because if $p_i = 1 - \delta$ for some $\delta > 0$, then $\overline{c} + D \cdot \varepsilon \cdot \vec{r}_i \in P$ for any $\varepsilon \leq \delta$ and that will contradict with (3). Notice that for all $k \in [0,1]$ the points $\overline{y}_k = \overline{x}_0 + D \cdot \sum_{j=1}^{N} p_j^k \cdot \vec{r}_j$, where $p_j^k = p_j$ if $j \neq i$ and $p_j^k = k$ otherwise, are all in P. Also notice that point $\overline{y}_1 = \overline{c}$ and for each $k \in [0,1]$ we have that $\overline{y}_k = \overline{y}_0 + D \cdot k \cdot \vec{r}_i$. In particular, $\overline{c} = \overline{y}_1 = \overline{y}_0 + D \cdot \vec{r}_i$. It follows that $\overline{c} - D \cdot \vec{r}_i = \overline{y}_0 \in P$. W.l.o.g. we assume $\vec{r}_i \neq \vec{0}$. Hence, from (5) we get $\vec{v} \cdot (\overline{c} - D \cdot \vec{r}_i) > d$. By rearranging we get $\vec{v} \cdot \overline{c} - D \cdot \vec{v} \cdot \vec{r}_i > d$, and because $\vec{v} \cdot \overline{c} = d$, we get $D \cdot \vec{v} \cdot \vec{r}_i < 0$ which contradicts that $\vec{v} \cdot \vec{r}_i \geq 0$. □

Figures 4.(b)-(c) show how to construct $\mathcal{H}$-closed polytope from (2) for the BMS in Figure 4.(a), while Figure 4.(d) shows that for every corner of the constructed polytope there is a mode that keeps the system inside the polytope.

The following lemma finally gives an algorithmically checkable characterization of existence of $\mathcal{H}$-closed polytope.

LEMMA 6. *Let $\mathcal{H} = (M, n, \mathcal{R})$ be a BMS. We have that for every $\vec{v}$ there is a mode m such that $\vec{v} \cdot \vec{r} \geq 0$ for all $\vec{r} \in m$ if and only if $\mathcal{H}$ is safe.*

PROOF. In one direction, let us assume that $(M, n, R) \in [\![Ext(\mathcal{H})]\!]$ is not safe, and let $Q = \{R(m) \mid m \in M\}$. Then $\vec{0}$ is not a convex combination of points in Q, and so by supporting hyperplane theorem applied to $\vec{0}$ and the convex hull of Q there is $\vec{v}$ and $d > 0$ such that $\vec{v} \cdot R(m) \geq d$ for all $m \in M$. Since $R(m) \in \mathcal{R}(m)$, this direction of the proof is finished. In the other direction, let $\vec{v}$ be such that there is $\vec{r} \in \mathcal{R}(m)$ for all $m \in M$ such that $\vec{v} \cdot \vec{r} < 0$. Then by convexity of $\mathcal{R}(m)$ there is $\vec{r}_m \in \mathsf{vert}(\mathcal{R}(m))$ with the same properties, and we can create a CMS $(M, n, R) \in [\![Ext(\mathcal{H})]\!]$ by putting $R(m) = \vec{r}_m$. This CMS is not safe, because for any strategy, for a sufficiently large time bound a point $\overline{x}$ will be reached such that $(-\vec{v}) \cdot \overline{x}$ is arbitrarily large, and hence any convex polytope will be left eventually. □

Combining Proposition 4 with Lemmas 5 and 6 we get the following main result.

THEOREM 7. *For every BMS $\mathcal{H}$ and the starting state in the interior of a convex and bounded safety set we have that scheduler has a winning strategy if and only if $\mathcal{H}$ is safe.*

Theorem 7 allows us to devise Algorithm 2 and at the same time give its correctness. The reader may have noticed that Theorem 7 bears a striking resemblance to Theorem 3 for

Algorithm 2: Schedulability for Interior States.

Input: BMS $\mathcal{H}$, $\overline{x} \in \mathbb{R}^n$ and $\gamma > 0$

Output: $\mathcal{H}$-closed polytope P contained in $B_\gamma(\overline{x})$ s.t. $\overline{x} \in P$, No if there is no $\mathcal{H}$-closed polytope.

foreach *CMS* $H = (M, n, R)$ *of* $[\![Ext(\mathcal{H})]\!]$ **do**

  Check if there is a satisfying assignment for:

$$\begin{aligned} \textstyle\sum_{m \in M} R(m) \cdot t_m &= \vec{0} \\ \textstyle\sum_{m \in M} t_m &= 1 \\ t_m &\geq 0 \text{ for all } m \in M. \end{aligned} \tag{6}$$

  if *no satisfying assignment exists* **then return** NO

$R := \{\vec{r}_1, \vec{r}_2, \ldots, \vec{r}_N\}$ be the set of rate vectors of $[\![Ext(\mathcal{H})]\!]$;

return the polytope given as convex hull of the points $\overline{x} + \frac{\gamma}{\sum_{i=1}^{N} \|\vec{r}_i\|} \cdot \sum_{i=1}^{N} \cdot p_i \vec{r}_i$ where $p_i \in \{0,1\}$;

Algorithm 3: Schedulability For Arbitrary States

Input: BMS $\mathcal{H}$, a safety set S given by inequalities $A\vec{x} \leq \vec{b}$, and a starting state $\overline{x}_0$.

Output: Yes, if the scheduler wins, No otherwise.

Compute the sequence $\mathcal{I} = \langle I_1, I_2, \ldots, I_\ell \rangle$;
SCHEDULABLE $= \emptyset$, UNSCHEDULABLE $= \emptyset$;
foreach I *in* $\mathcal{I}$ **do**
  if $I' \subseteq I$ *and* $I' \in$ UNSCHEDULABLE **then**
    UNSCHEDULABLE := UNSCHEDULABLE $\cup \{I\}$;
  if $\exists m \in M$ *with only internal rates* **then**
    SCHEDULABLE := SCHEDULABLE $\cup \{(I, \perp)\}$;
  else
    Construct $\mathcal{H}_I$;
    if $\mathcal{H}_I$ *is safe and* P_I *is* $\mathcal{H}_I$*-closed polytope* **then**
      SCHEDULABLE := SCHEDULABLE $\cup \{(I, P_I)\}$;
    else UNSCHEDULABLE:=UNSCHEDULABLE$\cup\{I\}$;
if $\exists I \in$ SCHEDULABLE *and* $\overline{x}_0 \models S|_I$ **then return** Yes;
else return No;

CMS, since the former boils down to checking safety of exponentially many CMS instances. Note, however, that the proof here is much more delicate. While in the case of CMS satisfiability of (1) gives immediately a periodic winning strategy, for BMS this is not the case: even when every instance in $[\![Ext(\mathcal{H})]\!]$ is safe, we cannot immediately see which modes should be used by the winning strategy; this requires the introduction of $\mathcal{H}$-closed polytopes.

3.2 General Case

In this section we present Algorithm 3 that analyses schedulability of arbitrary starting states in S. Notice that a starting state on the boundary of the safety polytope may lie on various faces (planes, edges etc.) of different dimensions. The scheduler may have a winning strategy using modes that let the system stay on some lower dimension face, or there may exists a winning strategy where scheduler first reaches a face of higher dimension where it may have a winning strategy. Before we describe steps of our algorithm, we need to formalize a notion of (open) faces of a convex polytope, a concept critical in Algorithm 3.

Let $Ax \leq b$ be the linear constraints specifying a convex polytope S. We specify a face of S by a set $I \subseteq \{1, \ldots, rows(A)\}$. We write $\overline{x} \models S|_I$, and we say that $\overline{x}$ satisfies $S|_I$, if and only if $A_{1,j}x(1) + \cdots A_{n,j}x(n) = b_j$ for all $j \in I$, and $A_{1,j}x(1) + \cdots A_{n,j}x(n) < b_j$ for all $j \notin I$, i.e. exactly the inequalities indexed by numbers from I are satisfied tightly. Note that every point of S satisfies $S|_I$ for exactly one I. Although there are potentially uncountably many states in every face of S, the following Lemma implies that it is sufficient to analyze only one state in every face.

LEMMA 8. *For a BMS, a convex polytope S, and for all faces I of S, either none or all states satisfying $S|_I$ are schedulable. Moreover, if $I' \subseteq I$ and no point satisfying $S|_{I'}$ is schedulable, then no point satisfying $S|_I$ is schedulable.*

Let $\mathcal{I} = \langle I_1, I_2, \ldots \rangle$ be the sequence of all faces such that $S|_{I_i}$ is satisfied by some state, ordered such that if $I_i \subseteq I_j$, then $i \leq j$. We call a mode m *unusable for* I if there is $\overline{x} \models S|_I$ and $\vec{r} \in \mathcal{R}(m)$ such that $\overline{x} + \vec{r} \cdot \delta \notin S$ for all $\delta > 0$. The rate $\vec{r}$ satisfying this condition is called *external*. A rate $\vec{r}$ is called *internal* if for any $\overline{x}$ such that $\overline{x} \models S|_I$ there is $\delta > 0$ and j such that $I_j \subseteq I$ and $\overline{x} + \vec{r} \cdot \varepsilon \models S|_{I_j}$ for all $0 < \varepsilon \leq \delta$. For a BMS $\mathcal{H}$ and face I we define a BMS $\mathcal{H}_I = (M', n, \mathcal{R}')$ where M' contains all modes of M which are not unusable for I, and $\mathcal{R}'(m)$ is the set of non-internal rates of $\mathcal{R}(m)$.

THEOREM 9. *For every BMS $\mathcal{H}$, a convex polytope safety set S, and a starting state $\overline{x}_0 \in S$, Algorithm 3 decides schedulability problem for $\mathcal{H}$. Moreover, one can construct a dynamic winning strategy using the set* SCHEDULABLE.

PROOF. (Sketch.) Let $\langle I_1, I_2, \ldots \rangle$ be all sets such that $S|_{I_i}$ is satisfied by some state, ordered such that if $I_i \subseteq I_j$, then $i \leq j$. Algorithm 3 analyzes the sets I_i, determining whether the points satisfying $S|_{I_i}$ are schedulable (in which case we call I_i schedulable), or not. Let us assume that I is the first element of the sequence $\langle I_1, I_2 \ldots \rangle$ which has not been analyzed yet. If there is I' such that $I' \subseteq I$ and I' is already marked as not schedulable, then by Lemma 8 one can immediately mark I as non-schedulable. If all modes are unusable, then no point $\overline{x}$ such that $S|_I$ is schedulable. Notice that if there exists an internal rate to face I_j then it must necessarily be the case that I_j is schedulable. If there is a mode m which only has internal rates, there is a winning strategy σ for the scheduler which starts by picking m and a sufficiently small time interval t. This will make sure that after one step a point is reached which is already known to be schedulable and scheduler has a winning strategy.

If none of the previous cases match, the algorithm creates a BMS $\mathcal{H}_I$ and applies Theorem 7 to the system $\mathcal{H}_I$. If there is a $\mathcal{H}_I$-closed polyhedron P, we know that I is schedulable and give a winning scheduler's strategy $\sigma_{\overline{x}}$ for any point $\overline{x} \models S|_I$ as follows. Let $d > 0$ be a number such that for any $\overline{y} \models I_j$ where $j > i$ we have $\|\overline{x}, \overline{y}\| > d$, i.e. d is chosen so that all points of S contained in $B_d(\overline{y})$ satisfy $S|_{I'}$ for $I' \subseteq I$ (this follows from the properties of the sequence $I_1, I_2, \ldots$ and because S is a convex polytope). The strategy $\sigma_{\overline{x}}$ works as follows. If all points in the history satisfy $S|_I$, $\sigma_{\overline{x}}$ mimics $\sigma_{\mathcal{H}_I, \overline{x}, d}$. Otherwise, once a point $\overline{y} \not\models S|_I$ is reached, the strategy $\sigma_{\overline{x}}$ starts mimicking $\sigma_{\overline{y}}$. Note that the strategy $\sigma_{\overline{y}}$ is indeed defined by our choice of d and polytopes stored in SCHEDULABLE set. Although the strategy we obtain in this way may potentially be non-positional, it is a mere technicality to turn it into a positional one.

If $\mathcal{H}_I$ is not schedulable for any set and any point, then it is easy to see that for no point satisfying $S|_I$ there is a schedulable strategy. Indeed, for any strategy σ, as long as σ picks the modes from M', the environment can play a counter-strategy showing that $\mathcal{H}_I$ is not schedulable. When any mode from $m \in M \setminus M'$ is used by σ, we have that m is unusable and so the environment can pick a rate witnessing m's unusability: this will ensure reaching a point outside S. Hence, we can mark I as unschedulable. $\square$

4. COMPLEXITY

In this section we analyze complexity of the schedulability problem for BMS. We begin by showing that in general it is co-NP-complete, however it can be solved in polynomial time if the system has only two variables.

4.1 General Case

PROPOSITION 10. *The schedulability problem for BMS and convex polytope safety sets is in co-NP.*

PROOF (SKETCH). We show that when the answer to the problem of schedulability of a point $\overline{x}$ is No, there is a falsifier that consists of two components:

- a set $I \subseteq \{1, \ldots, rows(A)\}$ s.t. $\overline{x} \models S|_{I'}$ for $I' \supseteq I$, and
- a rate combination $(\vec{r}_m)_{m \in M}$ such that there is a set of modes External $\subseteq M$ where every $\vec{r}_m$ for $m \in$ External is external for I; and the rates $\vec{r}_m$ for $m \notin$ External are neither external, nor internal, and there is a vector $\vec{v}$ such that $\vec{v} \cdot \vec{r}_m > 0$ for all $m \notin$ External.

Let us first show that the existence of this falsifier implies that the answer to the problem is No. Indeed, as long as a strategy of a scheduler keeps using modes $m \notin$ External, the environment can pick the rates $\vec{r}_m$, and a point outside of S will be reached under any non-Zeno strategy, because S is bounded. If the strategy of a scheduler picks any mode $m \in$ External, the environment can win immediately by picking the external rate $\vec{r}_m$ and getting outside of S.

On the other hand, let us suppose that the answer to the problem is No, and let I' be such that $\overline{x} \models S|_{I'}$. Then consider any *minimal non-schedulable* $I \subseteq I'$. We put to External all modes which are unusable, and for every such mode, we pick a rate that witnesses it. Further, there is not any mode with only internal modes and the BMS $\mathcal{H}_I$ must be non-schedulable (otherwise I would be schedulable, or would not be minimal non-schedulable). By Proposition 7 there is an unsafe instance $H = (M', n, R) \in [\![Ext(\mathcal{H}_I)]\!]$. Since M' contains all the modes whose indices are not in External, we can pick the rate from this unsafe instance and we are finished. □

PROPOSITION 11 (CO-NP HARDNESS). *The schedulability problem for MMS is co-NP hard.*

PROOF (SKETCH). The proof for co-NP hardness uses a reduction from the classical NP-complete problem 3-SAT. For a SAT instance ϕ we construct a MMS $\mathcal{H}_\phi$ such that ϕ is satisfiable if and only if $\mathcal{H}_\phi$ is not schedulable for any starting state and bounded convex safety set. . Consider a SAT instance ϕ with k clauses and n variables denoted as $x_1, \ldots, x_n$. The corresponding MMS $\mathcal{H}_\phi = (M, n, \mathcal{R})$ is such that its set of modes $M = \{m_1, \ldots, m_k\}$ corresponds to the clauses in ϕ, and variables are such that variable i corresponds to variable x_i of ϕ. For each variable x_i we define vectors $\vec{p}_i$ and $\vec{n}_i$ such that $\vec{p}_i(i) = 1$, $\vec{n}_i(i) = -1$, and $\vec{p}_i(j) = \vec{n}_i(j) = 0$ if $i \neq j$. The rate-vector function $\mathcal{R}$ is defined such that for each mode m_j and for each SAT variable x_i we have that $\vec{p}_i \in \mathcal{R}(m_j)$ if x_i occurs positively in clause j, and $\vec{n}_i \in \mathcal{R}(m_j)$ if the variable x_i occurs negatively in clause j. The crucial property here is that there is no vector that can have a positive dot product with both $\vec{p}_i$ and $\vec{n}_i$, which allows us to map unsafe rate combinations to satisfying valuations and vice versa. □

The following corollary is immediate.

COROLLARY 12 (CO-NP HARDNESS RESULT FOR BMS). *The schedulability problem for BMS is co-NP hard.*

4.2 BMS with two variables

For the special case of BMS with two variables, we show that the schedulability problem can be solved efficiently.

THEOREM 13. *Schedulability problems for BMS with convex polytope safety sets are in P for systems with 2 variables.*

Algorithm 4: Decide if a two dimension BMS is safe.

Input: BMS $\mathcal{H}$ with two variables.
Output: Return Yes, if $\mathcal{H}$ is safe and No otherwise.

Set R to the set of extreme rate vectors of $\mathcal{H}$;
foreach $\vec{r}_\perp \in R$ **do**
  Set $\vec{u}$ to be a perpendicular vectors to $\vec{r}_\perp$;
  foreach $\vec{v} \in \{\vec{u}, -\vec{u}\}$ **do**
    if *for each* $m \in M$ *there is* $\vec{r} \in m$ *s.t.* $\vec{v} \cdot \vec{r} > 0$ *or there is* $p > 0$ *s.t.* $\vec{r} = p\vec{r}_\perp$ **then return** No;
return Yes

The rest of the section is devoted to the proof of this theorem. The following lemma shows that to check whether a set of rate vectors $R = \{\vec{r}_1, ..., \vec{r}_k\}$ is unsafe it is sufficient to check properties of vectors $\vec{u}$ perpendicular to some vector of R. This observation yields a polynomial time algorithm.

LEMMA 14. *Let R be a set of vectors. There is $\vec{v}$ such that $\vec{v} \cdot \vec{r} > 0$ for all $\vec{r} \in R$ if and only if there are $\vec{u}$ and $\vec{r}_\perp \in R$ satisfying $\vec{u} \cdot \vec{r}_\perp = 0$ and for all $\vec{r} \in R$ either $\vec{u} \cdot \vec{r} > 0$ or $\vec{r} = p \cdot \vec{r}_\perp$ for some $p > 0$.*

PROOF (SKETCH). To obtain $\vec{v}$ we keep changing $\vec{v}$ until it becomes perpendicular to some vector in R. On the other hand, $\vec{v}$ is obtained from $\vec{u}$ by making a sufficiently small change to $\vec{u}$. □

EXAMPLE 6. *Consider an unsafe set of rate vectors $R = \{\vec{r}_1, \vec{r}_2, \vec{r}_3, \vec{r}_4\}$ shown in the following figure in the left side.*

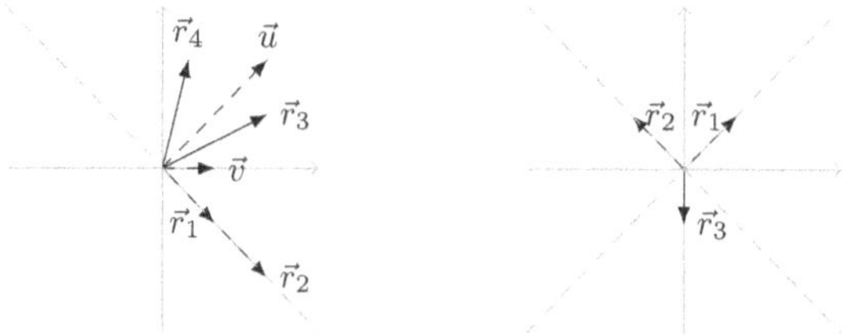

All the rate vectors are on the right side of line $x = 0$ and vector $\vec{v}$ has strictly positive dot product with all of them. As it can be seen in the figure, all the rate vectors are on right-hand side of the line passing through $\vec{r}_1$ and there exists $\vec{u}$ perpendicular to $\vec{r}_1$ such that $\vec{v}' \cdot \vec{r}_i \geq 0$ for all $\vec{r}_i \in R$. Observe that adding a rate vector $\vec{r}_5 = -\vec{r}_1$ to R makes this set of rate vectors safe, and none of rate vectors would satisfy the conditions of Lemma 14.

However, the figure on the right side shows a safe set of rate vectors. As is clear to see that no rate vector has the others on its one side.

The following corollary implies that we can use Lemma 14 to check the safety of a BMS.

COROLLARY 15. *A BMS $\mathcal{H}$ with two variables is not safe if and only if there exists a rate vector $\vec{r}_\perp$ in one of the modes of system and vector $\vec{v}$ perpendicular to it, such that for all modes $m \in \mathcal{H}$: (i) there exists $\vec{r} \in m$ such that $\vec{v} \cdot \vec{r} > 0$; or (ii) $\vec{v} \cdot \vec{r} = 0$ and $\vec{r} = p \cdot \vec{r}_\perp$ for some $p > 0$.*

Algorithm 4 checks whether all the combinations are safe in polynomial time; it chooses a rate vector $\vec{r}_\perp$ at each step and tries to find an unsafe combination using the result of Corollary 15. Note that for any non-zero vector $\vec{r}_\perp$ in two dimensions there are only two vectors which we need to check.

Although there are infinitely many vectors $\vec{v}$ which satisfy conditions of Corollary 15, the conditions we are checking are preserved if we multiply $\vec{v}$ by a positive scalar.

5. DISCRETE SCHEDULABILITY

In this section we discuss the *discrete schedulability problem*, in which a scheduler can only make decisions at integer multiplies of a specified *clock period* Δ and the environment has finitely many choices of rates. Formally, given a MMS $\mathcal{H}$, a closed convex polytope S as safety set, an initial state $x_0 \in S$, the discrete schedulability problem is to decide if there exists a winning strategy of the scheduler where the time delays are multiples of Δ.

THEOREM 16. *Discrete schedulability problem is complete for EXPTIME.*

PROOF. EXPTIME-membership of the problems is shown via discretization of the state space of $\mathcal{H}$. Since the set S is given as a bounded polytope, the size of the discretization can be shown to be at most exponential in the size of $\mathcal{H}$ and Δ, and since the safety games on a finite graph can be solved in P, EXPTIME membership follows. The hardness is shown by a reduction from the countdown games [9]. □

We turn the discrete schedulability problem to an optimization problem, by asking to find supremum of all Δ for which the answer to the discrete schedulability problem is yes. We prove the following, which also improves a result of [2] where only an approximation algorithm was given.

THEOREM 17. *Given a MMS $\mathcal{H}$, a closed convex polytope S and an initial state $\overline{x}_0$, there is an exponential time algorithm which outputs the maximal Δ for which the answer to the discrete schedulability problem is Yes. For a CMS the algorithm can be made to run in polynomial space.*

PROOF (SKETCH). We exploit the fact that as the clock period Δ increases, all the points of the discretization move continuously towards infinity, except for the initial point. This further implies that for Δ to be maximal, there must be a point of the discretization which lies on the boundary of S, since otherwise we could increase Δ by some small number, while preserving the existence of a safe scheduler. By using a lower bound on Δ from Section 3 (obtained as a by-product of the construction of a dynamic strategy), there are only exponentially many candidates for such points, which gives us exponentially many candidates for maximal Δ to consider, and we can check each one by Theorem 16. For the PSPACE bound we don't enumerate the points, but guess them nondeterministically in polynomial space, and utilize [2, Theorem 10] instead of Theorem 16. □

6. CONCLUSION

We investigated systems that comprise finitely many real-valued variables whose values evolve linearly based on a rate vector determined by strategies of the scheduler and the environment. We studied an important schedulability problem for these systems, with application to energy scheduling, that asks whether scheduler can make sure that the values of the variables never leave a given safety set. We showed that when the safety set is a closed convex polytope, existence of non-Zeno winning strategy for scheduler is decidable for any arbitrary starting state. We also showed how to construct such a winning strategy. On complexity side, we showed that the schedulability problem is co-NP complete in general, but for the special case where the system has only two variables, the problem can be decided in polynomial time. Future research includes schedulability problem with respect to more expressive higher-level control objectives including temporal-logic based specification and bounded-rate multi-mode systems with reward functions.

Acknowledgement

This research was partially supported by NSF award CNS 1035715 and NSF Expeditions in Computing award CCF 1138996. Vojtech Forejt was supported by Newton Fellowship of Royal Society.

7. REFERENCES

[1] R. Alur, V. Forejt, S. Moarref, and A. Trivedi. Safe schedulability of bounded-rate multi-mode systems. Technical report. arXiv:1302.0745 [cs.LO].

[2] R. Alur, A. Trivedi, and D. Wojtczak. Optimal scheduling for constant-rate mulit-mode systems. In *HSCC*, pages 75–84. ACM, 2012.

[3] F. Blanchini. Constrained control for uncertain linear systems. *Journal of Optimization Theory and Applications*, 71:465–484, 1991.

[4] F. Blanchini. Set invariance in control. *Automatica*, 35(11):1747–1767, 1999.

[5] T. A. Henzinger. The theory of hybrid automata. In *LICS*, pages 278 –292. IEEE Comp. Soc. Press, 1996.

[6] T. A. Henzinger, B. Horowitz, and R. Majumdar. Rectangular hybrid games. In *CONCUR 99*, pages 320–335. Springer, 1999.

[7] T. A. Henzinger and P. W. Kopke. Discrete-time control for rectangular hybrid automata. *TCS*, 221(1-2):369–392, 1999.

[8] M. Heymann, L. Feng, G. Meyer, and S. Resmerita. Analysis of Zeno behaviors in a class of hybrid systems. *IEEE Tran. on Auto. Ctrl.*, 50:376–383, 2005.

[9] M. Jurdziński, J. Sproston, and F. Laroussinie. Model checking probabilistic timed automata with one or two clocks. *Logical Methods in Comp. Sc.*, 4(3):12, 2008.

[10] J. Le Ny and G.J. Pappas. Sequential composition of robust controller specifications. In *Robotics and Automation*, pages 5190–5195, 2012.

[11] J. Liu, N. Ozay, U. Topcu, and R. Murray. Switching protocol synthesis for temporal logic specifications. In *American Control Conference*, pages 727–734, 2012.

[12] A. Maitra and W. Sudderth. Finitely additive stochastic games with Borel measurable payoffs. *Intl. Journal of Game Theory*, 27(2):257–267, 1998.

[13] T. X. Nghiem, M. Behl, R. Mangharam, and G. J. Pappas. Green scheduling of control systems for peak demand reduction. In *Decision and Control (CDC) and ECC*, pages 5131–5136. IEEE, December 2011.

[14] T. X. Nghiem, M. Behl, G. J. Pappas, and R. Mangharam. Green scheduling: Scheduling of control systems for peak power reduction. *Intl. Green Computing Conf.*, pages 1–8, July 2011.

[15] L. T. X. Phan, I. Lee, and O. Sokolsky. Compositional analysis of multi-mode systems. In *Euromicro Conf. on Real-Time Sys. (ECRTS)*, pages 6–9. IEEE, 2010.

Compositional Heterogeneous Abstraction

Akshay Rajhans

Bruce H. Krogh

Department of Electrical and Computer Engineering
Carnegie Mellon University, Pittsburgh, PA 15213
{ arajhans | krogh }@ece.cmu.edu

ABSTRACT

In model-based development, abstraction provides insight and tractability. Different formalisms are often used at different levels of abstraction to represent the variety of concerns that need to be addressed when designing complex cyber-physical systems. In this paper, we consider the problem of establishing abstraction across heterogeneous formalisms in a compositional manner. We use the framework of behavioral semantics to elucidate the general conditions that must be satisfied to assure that the composition of abstractions for individual components is an abstraction for the composition of the components. The theoretical concepts are illustrated using an example of a cooperative intersection collision avoidance system (CICAS).

Categories and Subject Descriptors

G.4 [**Mathematical Software**]: Verification; I.6.4 [**Simulation and Modeling**]: Model Validation and Analysis

Keywords

Heterogeneous Verification; Compositional Reasoning

1. INTRODUCTION

Model-based development (MBD) refers to the creation of mathematical models of systems under design and checking those models against design specifications using suitable analysis tools. The MBD approach aims to catch errors early in the design process, thereby avoiding costly re-design/re-development cycles. For all but the most trivial cyber-physical systems (CPS), abstraction is essential for making analysis and verification tractable. Different modeling formalisms are often used in various abstractions to facilitate the design of particular aspects of the system. In our previous work, we proposed the use of behavior relations to support abstraction across heterogeneous modeling formalisms [19]. In this paper, we address the problem of establishing heterogeneous abstraction in a compositional manner.

Several tools have been developed to support simulation using heterogeneous models. Ptolemy II, for example, supports hierarchical integration of multiple "models of computation" into a single simulation model based on an actor-oriented formalism [8]. MILAN [17] is an integrated simulation framework that allows different components of a system to be built using different tools. The Metropolis toolchain [5] supports multiple analysis tools for design and simulation. None of these tools deals with abstraction, however, other than in the form of encapsulation for components and subsystems.

Heterogeneous abstraction across particular pairs of modeling formalisms appears often in the literature. Examples include hybrid abstractions of nonlinear systems [13, 10], linear hybrid automata abstractions of hybrid systems with linear continuous dynamics [12], discrete abstractions of hybrid systems [4, 9, 3], and continuous abstractions of hybrid systems [2]. In addition to being specific in the formalisms that are used, many of these methods are not compositional. Compositional methods, such as assume-guarantee reasoning, with abstraction defined by language inclusion [16] and simulation relations [11, 14], are usually defined in the context of a single formalism. Behavior-interaction-priority framework for embedded software uses structured interaction invariants to support compositional analysis but only for transition system models [6]. Our objective is to develop a general framework that elucidates the basic conditions for compositional abstraction between any pair of heterogeneous formalisms.

The notion of tagged signal semantics has been proposed to compare and compose heterogeneous reactive systems [18, 7]. Julius creates a behavioral framework for modeling control as a behavior interconnection problem [15] . These approaches use system trajectories or behaviors as a mathematical framework for creating relations between the semantics of different modeling formalisms. We have used a similar approach to establish abstractions across different formalisms, which can then be used then be used for heterogeneous verification [19]. We use the behavioral framework in this paper to develop compositional heterogeneous abstraction.

The paper is organized as follows. We introduce the notation and the problem description in Sect. 2. Sect. 3 develops a framework for relating the local semantics of a component model with its global semantics for the purposes of composing it with other component models. Sect. 4 develops

HSCC'13, April 8–11, 2013, Philadelphia, Pennsylvania, USA.

compositional heterogeneous abstraction for the case when components at each level share the same local behavioral domain. Sect. 5 extends the development to the more common case where components are developed within their own distinct local behavior domains. Sect. 6 illustrates the theoretical concepts using an example of a cooperative intersection collision avoidance system (CICAS). The concluding section summarizes the results in this paper and discusses directions for future work.

2. MATHEMATICAL PRELIMINARIES

A *model* M is a mathematical description of a system using a *modeling formalism* $\mathcal{M}$, which is a collection of modeling primitives and syntactic rules for building models. Modeling formalisms typically used for CPS include transition systems, hybrid automata, signal-flow models, acausal equation-based models, and queuing networks. The semantics of a model M is defined by a set of *legal behaviors* from a given *behavior domain* B, where the behavior domain is a member of a given *class of behavior domains* $\mathcal{B}$. Behavior classes used to define semantics for CPS models include discrete traces, continuous trajectories and hybrid trajectories. For each behavior class, we assume there exists a syntax, called the *behavior formalism*, which can be used to precisely define behavior domains and individual behaviors. $[\![M]\!]^B$ denotes the set of legal behaviors for a given model M with semantics defined in a given behavior domain B.

Given two models M_0, M_1 with semantic interpretations in the same behavior domain B, model M_1 is called an *abstraction* of model M_0 if $[\![M_0]\!]^B \subseteq [\![M_1]\!]^B$. This is the standard definition of abstraction in the literature, using, for example, language or trace inclusion. We denote abstraction in this case by the notation $M_0 \sqsubseteq^B M_1$. We introduce the following notion of *behavior abstraction functions* as semantic mappings between different behavior domains, perhaps from two different behavior classes.

Definition 1 (Behavior Abstraction Functions) *Given two behavior classes $\mathcal{B}_0$ and $\mathcal{B}_1$ and behavior domains from each behavior class $B_0 \in \mathcal{B}_0$ and $B_1 \in \mathcal{B}_1$, a* behavior abstraction function $\mathcal{A} : B_0 \to B_1$ *maps each behavior in B_0 to a corresponding abstract behavior in B_1.*

Behavior abstraction functions are special cases of behavior relations from [19]. In particular, they are relations that are also functions, i.e., $R \subseteq B_0 \times B_1$ s.t. $(b_0, b_1) \in R$ and $(b_0, b_1') \in R$ only if $b_1 = b_1'$.

Example 1 Consider a behavior domain $B_0 = \mathbb{R}^{\mathbb{R}_+}$ as the set of all 1-d continuous trajectories starting at time 0. Let the variable name for the single dimension be x. Consider another behavior domain $B_1 = \Sigma^* \cup \Sigma^\omega$ defined as the set of all finite or infinite traces with event labels in $\Sigma = \{\alpha, \bar{\alpha}\}$. Consider a usual behavior abstraction technique frequently used in the literature — state-space partitioning, illustrated below. The continuous state-space $\mathbb{R}$ is partitioned in two halves $x \le l_x$ and $x \ge l_x$ at a boundary $x = l_x$ as follows.

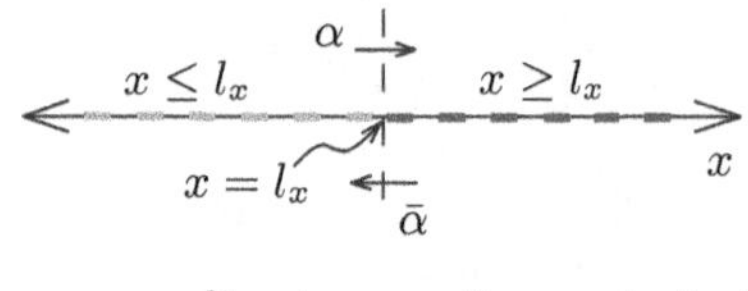

The event corresponding to a continuous trajectory crossing the partition going from $x \le l_x$ to $x \ge l_x$ is associated with the label α and that from $x \ge l_x$ to $x \le l_x$ is associated with the label $\bar{\alpha}$.

Consider $b_0 \in B_0$ and $b_1 \in B_1$ where $b_0 = x(t)$ for $t \in \mathbb{R}_+$ and $b_1 = \sigma_0\sigma_1 \cdots \sigma_N$, for $N \in \mathbb{N} \cup \{\infty\}$. In words, the abstraction function states that $\mathcal{A}(b_0) = b_1$ if (i) $\exists$ event times $t_i \in \mathbb{R}_+$, $i = 0, 1, \ldots, N$ that correspond to the continuous trajectory crossing the boundary in the right direction associated with the label σ_i (i.e., from "$\text{FROM}(\sigma_i)$" to "$\text{TO}(\sigma_i)$" according to the following table) and (ii) there are no crossings between any consecutive event times t_i and t_{i+1}. Mathematically, these conditions can be written as

$$\begin{aligned} &\forall t' \in [0, t_0), && x(t') \in \text{FROM}(\sigma_0), \\ &\forall t' \in [t_{i-1}, t_i), && x(t') \in \text{TO}(\sigma_{i-1}) \cap \text{FROM}(\sigma_i), \\ &\forall t' \ge t_N, && x(t') \in \text{TO}(\sigma_N), \end{aligned}$$

σ	$\text{FROM}(\sigma)$	$\text{TO}(\sigma)$
α	$x \le l_x$	$x \ge l_x$
$\bar{\alpha}$	$x \ge l_x$	$x \le l_x$

Otherwise, $\mathcal{A}(b_0)$ is an empty behavior ε.

With respect to the following picture, $\mathcal{A}(c) = \mathcal{A}(d) = \alpha$ and $\mathcal{A}(f)$ is the infinite string $\alpha\bar{\alpha}\alpha\bar{\alpha}\alpha\bar{\alpha}\cdots$. In contrast, $\mathcal{A}(e) = \varepsilon$ since e never crosses the boundary.

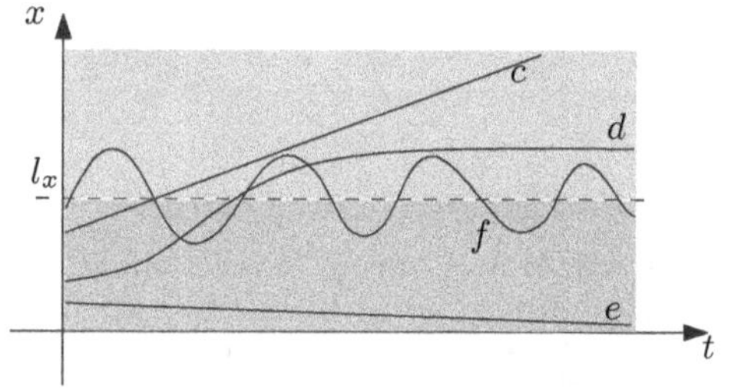

□

Behavior abstraction functions are typically used in the context of models and serve as mappings between the semantics of the two models defined in the two behavior domains under consideration. Therefore, they are usually inferred from relationships between models from given modeling formalisms and associated definitions of the relationships between model primitives and their semantic interpretations.

The set-valued extensions of behavior abstraction functions are defined in the usual way. For a given behavior abstraction function $\mathcal{A} : B_0 \to B_1$ and a set of behaviors $B_1' \subseteq B_1$, the set-valued inverse $\mathcal{A}^{-1}$ is defined as $\mathcal{A}^{-1}(B_1') = \{b_0 \mid \mathcal{A}(b_0) \in B_1'\}$.

Given behavior abstraction functions as semantic mappings, heterogeneous abstraction between two models is defined as follows.

Definition 2 (Heterogeneous Abstraction) *Given heterogeneous behavior classes $\mathcal{B}_0$, $\mathcal{B}_1$, suppose behavior domains $B_0 \in \mathcal{B}_0$ and $B_1 \in \mathcal{B}_1$ are used to define the semantics of models M_0 and M_1, respectively, and that there is an abstraction function $\mathcal{A} : B_0 \to B_1$. Model M_1 is a* heterogeneous abstraction *of M_0 through $\mathcal{A}$, written $M_0 \sqsubseteq^{\mathcal{A}} M_1$, if*

$$[\![M_0]\!]^{B_0} \subseteq \mathcal{A}^{-1}([\![M_1]\!]^{B_1}).$$

This definition asserts that for every behavior of model M_0 in B_0, the abstraction function $\mathcal{A}$ associates a corresponding abstract behavior of model M_1 in B_1.

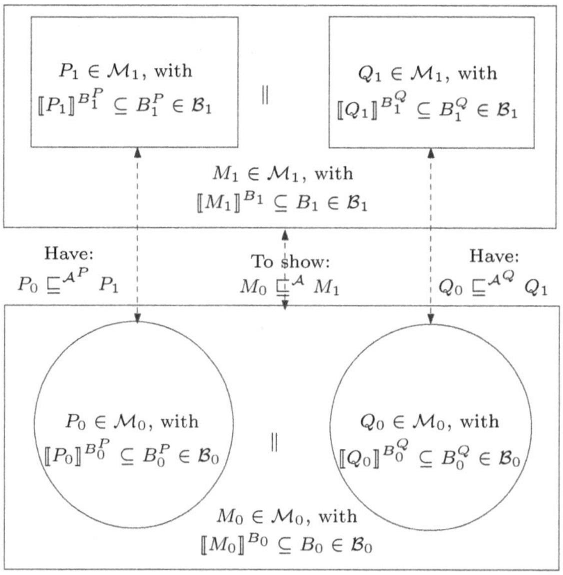

Figure 1: A schematic of compositional heterogeneous abstraction analysis.

Fig. 1 illustrates the compositional heterogeneous abstraction problem considered in this paper. For each of the two levels of abstraction, $i = 0, 1$, we assume there is a modeling formalism $\mathcal{M}_i$ and a behavior class $\mathcal{B}_i$. Component models $P_i, Q_i \in \mathcal{M}_i$ have their semantics defined in terms of *local* behavior domains $B_i^P, B_i^Q \in \mathcal{B}_i$. These local domains include only the variables relevant to the given component. Heterogeneous abstraction between the two models of each component is established via behavior abstraction functions $\mathcal{A}^P$ and $\mathcal{A}^Q$ that are mappings between the respective local behavior domains. To compose the two models to form the system models $M_i \in \mathcal{M}_i$, the local semantics are lifted to global behavior domains $B_i \in \mathcal{B}_i$ to include variables from both components. We seek conditions under which heterogeneous abstraction between component models in their local behavior domains implies heterogeneous abstraction between the composite system models in the global behavior domains.

3. LOCAL VS. GLOBAL SEMANTICS

We begin by defining the relationship between behaviors in local domains for component models and a global domain for the composition.

Definition 3 (Behavior Localization) *Given a behavior class $\mathcal{B}$ and two behavior domains $B, B' \in \mathcal{B}$, an onto function $\downarrow : B \to B'$ (i.e., every element of B' has at least one pre-image in B) is called a* (behavior) localization *of behavior domain B to behavior domain B'.*

Given a localization $\downarrow$ of B to B', for $b \in B$, we will let $b\downarrow$ denote $\downarrow(b)$. The set-valued extension of localization can be defined in the usual way. Next, we consider two common types of variable elimination as examples of behavior localization projections.

Example 2 (Event label removal) Let $L \subseteq \Sigma^*$ be a language over a global alphabet Σ. Let $A \subseteq \Sigma$ be a local alphabet relevant for a component. The localization due to natural projection of L onto the set of strings over A^*, written $L\downarrow_A := \{s\downarrow | s \in L\}$, where $s\downarrow$ is recursively defined as follows.

1. The empty string is projected onto itself, i.e. $\varepsilon\downarrow = \varepsilon$.
2. For any string $s \in \sigma^*$ and $a \in \Sigma$
   - $(s \circ a)\downarrow = (s)\downarrow \circ\ a \ldots$ if $a \in A$
   - $(s \circ a)\downarrow = (s)\downarrow \ldots$ if $a \notin A$ □

Example 3 (Continuous Variable Elimination) Consider a global behavior domain $B = (\mathbb{R}^2)^{\mathbb{R}_+}$ of 2-d continuous trajectories, with the variables along the two dimensions named x and y. Let a local behavior domain be $B' = (\mathbb{R})^{\mathbb{R}_+}$ with the variable name x. Let $\tilde{B} \subseteq B = \{[x(t)\ y(t)]^T | \forall t \in \mathbb{R}_+, x(t) \geq 0, y(t) \in [0, 1]\}$. Then the localization due to elimination of variable y can be written in terms of its existential quantification. $\tilde{B}\downarrow$ can be defined as the set $\{x(t) | \exists y(t) \text{ s.t. } \forall t \in \mathbb{R}_+, x(t) \geq 0, y(t) \in [0, 1]\}$. □

For $b' \in B'$ we will let $b'\uparrow$ denote the set-valued function $\uparrow : B' \to 2^B - \{\emptyset\}$ defined by $\uparrow(b') = \downarrow^{-1}(b') = \{b \in B | b\downarrow = b'\}$. We will call the function $\uparrow$ a *(behavior) globalization* of B' to B. Note that $b'\uparrow$ is always non-empty since the localization function $\downarrow$ is onto.

Behavior localization and globalization are generally inferred from relationships between models from given modeling formalisms and associated definitions of the relationships between model primitives and their semantic interpretations.

Note that in case of compositional heterogeneous analysis as depicted in Fig. 1, there are four different behavior localizations (or globalizations) – one for each component and one at each level of abstraction. We index these with subscripts $i = 0, 1$ for the two levels of abstraction and superscripts $j = 1, 2$ or $j = P, Q$ to distinguish between these wherever necessary.

Given behavior globalizations at the abstract and concrete levels of abstraction, we next define the globalization of a behavior abstraction function between the abstract and concrete local domains.

Definition 4 (Abstraction Globalization) *Given two behavior classes $\mathcal{B}_0$ and $\mathcal{B}_1$, behavior domains from each behavior class: $B_0, B_0' \in \mathcal{B}_0$ and $B_1, B_1' \in \mathcal{B}_1$, localizations $\downarrow_i$ of B_i to B_i' for $i = 1, 2$, and a behavior abstraction function $\mathcal{A}'$ of B_0' to B_1', a behavior abstraction function $\mathcal{A}$ of B_0 to B_1 is said to be a* globalization *of $\mathcal{A}'$ if*

$$\forall b_0 \in B_0 : \mathcal{A}'(b_0\downarrow_0) = \mathcal{A}(b_0)\downarrow_1. \tag{1}$$

In words, the definition of abstraction globalization states that given any global concrete behavior b_0, the abstraction of its localization $b_0\downarrow_0$ at the concrete level 0 through the local abstraction function $\mathcal{A}'$ should be the same as the localization at the abstract level 1 of its corresponding abstract behavior $\mathcal{A}(b_0)$. This concept is illustrated by the following diagram: $\mathcal{A}$ is a globalization of $\mathcal{A}'$ if the diagram commutes.

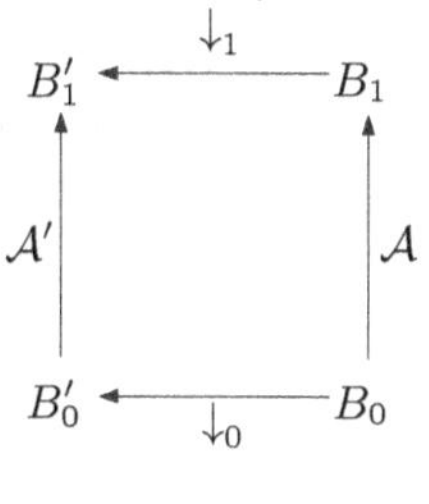

We write $\mathcal{A} = \mathcal{A}'\Uparrow$ if $\mathcal{A}$ is a globalization of $\mathcal{A}'$. We call $\mathcal{A}'$ a *localization* of $\mathcal{A}$, written $\mathcal{A}' = \mathcal{A}\Downarrow$, iff $\mathcal{A} = \mathcal{A}'\Uparrow$.

Note that in case of compositional heterogeneous analysis as depicted in Fig. 1, there are two different abstraction localizations/globalizations – one for each component.

We note the following existence and uniqueness properties of localization/globalization of behavior abstraction functions.

- **Existence of globalization.** For a given local abstraction function $\mathcal{A}'$, it is always possible to construct a globalization $\mathcal{A}'\Uparrow$ s.t. the diagram commutes. This is due to the fact that both localizations $\downarrow_i$, $i = 0, 1$ are onto functions. Therefore, for any local behaviors b'_0 and $b'_1 = \mathcal{A}'(b'_0)$, $b'_0\uparrow_0$ and $b'_1\uparrow_1$ are non-empty. One can then associate every behavior $b_0 \in b'_0\uparrow_0$ with some behavior $b_1 \in b'_1\uparrow_1$, which results in a valid globalization of $\mathcal{A}'$.

- **Non-uniqueness of globalization.** For a given local abstraction function $\mathcal{A}'$, its globalization $\mathcal{A}'\Uparrow$ is not unique. For a b'_0 with $\mathcal{A}'(b'_0) = b'_1$ and $b'_1\uparrow_1 = \{b^0_1, b^1_1\}$, consider a global behavior $b_0 \in b'_0\uparrow_0$. Then $\mathcal{A}^0$ with $\mathcal{A}^0(b_0) = b^0_1$ and $\mathcal{A}^1$ with $\mathcal{A}^1(b_0) = b^1_1$ can both be globalizations of $\mathcal{A}'$. Since localization causes loss of information, its set-valued inverse provides some freedom for creating mappings at the global level; appropriate ones need to be chosen.

- **Non-existence of localization.** For a given global abstraction function $\mathcal{A}$, its localization $\mathcal{A}\Downarrow$ may not exist, i.e., the diagram may not commute for any $\mathcal{A}'$. Consider b^0_0, b^1_0 with $\mathcal{A}(b^0_0) = b^0_1$ and $\mathcal{A}(b^1_0) = b^1_1$, and $b^0_0\downarrow_0 = b^1_0\downarrow_0$, but $b^0_1\downarrow_0 \neq b^1_1\downarrow_0$. For such a case, there can be no $\mathcal{A}'$ s.t. $\mathcal{A}'\Uparrow = \mathcal{A}$.

- **Uniqueness of localization.** For a given global abstraction function $\mathcal{A}$, if $\mathcal{A}\Downarrow$ exists, it is unique. This is simply due to the diagram commuting. $\forall\, b_0$, behaviors $b_0\downarrow_0$, $\mathcal{A}(b_0) =: b_1$, and $b_1\downarrow_1$ are unique. Therefore, for every given mapping $\mathcal{A}(b_0) = b_1$, there is a unique mapping $\mathcal{A}'(b_0\downarrow_0) = b_1\downarrow_1$.

- **Globalization and localization are not necessarily inverse operations.** From the uniqueness of localization and non-uniqueness of globalization, it is straightforward to show that
$$(\mathcal{A}'\Uparrow)\Downarrow = \mathcal{A}'; \tag{2}$$
but $(\mathcal{A}\Downarrow)\Uparrow$ may not be equal to $\mathcal{A}$.

Given the theoretical machinery developed in this section, in the next two sections we find conditions under which compositional heterogeneous abstraction w.r.t. Fig. 1 can be used.

4. HETEROGENEOUS ABSTRACTION IN GLOBAL BEHAVIOR DOMAINS

We start with a simple special-case scenario w.r.t. Fig. 1 in which the semantics of component models P_i and Q_i are defined in the same local behavior domain at each level of abstraction. In this case, the global behavior domains are the same as the local behavior domains , i.e., $B^P_i = B^Q_i = B_i \in \mathcal{B}_i$, $i = 0, 1$. For this special case, only one behavior abstraction function $\mathcal{A}$ is sufficient, as we can set $\mathcal{A}^P = \mathcal{A}^Q = \mathcal{A}$. In this case, we define the *semantic composition* of two component models as follows.

Definition 5 (Semantic Composition) *Given component models P and Q from the same modeling formalism $\mathcal{M}$ with semantics defined in behavior domain B, the composition $P||Q$ is a model in $\mathcal{M}$ s.t.*
$$[\![P||Q]\!]^B = [\![P]\!]^B \cap [\![Q]\!]^B. \tag{3}$$

This definition of composition as the intersection of behavior sets is consistent with the literature for composition using specific behavior domains [15, 18, 7]. For a given modeling formalism $\mathcal{M}$, syntactic techniques may exist for creating a composition, e.g., construction of product automata. We support all such procedures so long as (3) holds.

The following proposition gives conditions for compositional heterogeneous abstraction.

Proposition 1 *For each abstraction level $i = 0, 1$, given component models P_i, Q_i with the semantics of each model interpreted over a behavior domain B_i, and a behavior abstraction function $\mathcal{A} : B_0 \to B_1$, if $P_0 \sqsubseteq^{\mathcal{A}} P_1$ and $Q_0 \sqsubseteq^{\mathcal{A}} Q_1$, then*
$$P_0||Q_0 \sqsubseteq^{\mathcal{A}} P_1||Q_1.$$

PROOF. From $P_0 \sqsubseteq^{\mathcal{A}} P_1$ and $Q_0 \sqsubseteq^{\mathcal{A}} Q_1$, we have $[\![P_0]\!]^{B_0} \subseteq \mathcal{A}^{-1}([\![P_1]\!]^{B_1})$ and $[\![Q_0]\!]^{B_0} \subseteq \mathcal{A}^{-1}([\![Q_1]\!]^{B_1})$. Therefore,
$$\begin{aligned}[\![P_0||Q_0]\!]^{B_0} &= [\![P_0]\!]^{B_0} \cap [\![Q_0]\!]^{B_0}\\ &\subseteq \mathcal{A}^{-1}([\![P_1]\!]^{B_1}) \cap \mathcal{A}^{-1}([\![Q_1]\!]^{B_1})\\ &= \mathcal{A}^{-1}([\![P_1]\!]^{B_1} \cap [\![Q_1]\!]^{B_1})\\ &= \mathcal{A}^{-1}([\![P_1||Q_1]\!]^{B_1}).\end{aligned}$$
■

This proposition states that with global semantics, composition of abstractions is the abstraction of the composition.

Remark 6 (Insufficiency of Behavior Relations) We note that arbitrary behavior relations from [19] that are not functions are not sufficient in even this simple case of compositional heterogeneous abstraction. If a behavior relation $\mathcal{A}$ is not a function, it is possible to have a behavior $b_0 \in [\![P_0]\!]^{B_0} \cap [\![Q_0]\!]^{B_0}$ with $(b_0, p_1) \in \mathcal{A}$, $(b_0, q_1) \in \mathcal{A}$, s.t. $p_1 \in [\![P_1]\!]^{B_1} \backslash [\![Q_1]\!]^{B_1}$ and $q_1 \in [\![Q_1]\!]^{B_1} \backslash [\![P_1]\!]^{B_1}$ but $\nexists\, b_1 \in [\![P_1]\!]^{B_1} \cap [\![Q_1]\!]^{B_1}$ with $(b_0, b_1) \in \mathcal{A}$, as shown in the following Venn diagram.

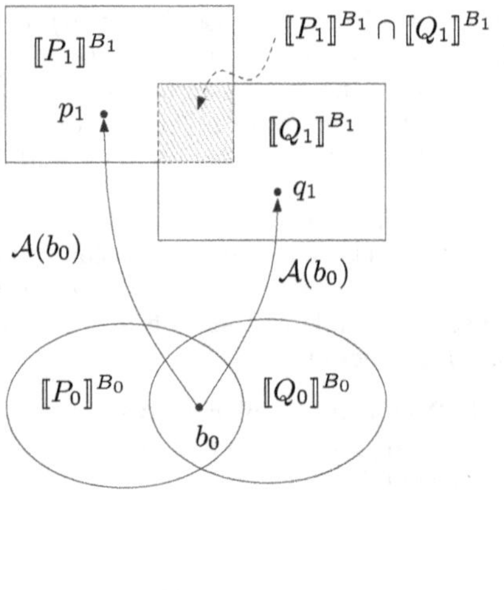

For this b_0, we have $b_0 \in \mathcal{A}^{-1}(\llbracket P_1 \rrbracket^{B_1}) \cap \mathcal{A}^{-1}(\llbracket Q_1 \rrbracket^{B_1})$ but $b_0 \notin \mathcal{A}^{-1}(\llbracket P_1 \rrbracket^{B_1} \cap \llbracket Q_1 \rrbracket^{B_1})$ and therefore the above proof does not hold. The arbitrary mappings that are the source of these counterexamples – those that allow one concrete behavior to be associated with more than one abstract behaviors – are perhaps not necessary in practice. The restriction from behavior relations to functions disallows the possibility of having several abstract behaviors correspond to a single concrete behavior, while still allowing several concrete behaviors to be mapped to a single abstract behavior. □

In the next section, we consider the general case where the local semantics of the two components are defined in terms of distinct behavior domains.

5. HETEROGENEOUS ABSTRACTION IN LOCAL BEHAVIOR DOMAINS

We now consider a more general scenario w.r.t. Fig. 1 in which the component models P_i and Q_i have different local behavior domains B_i^P and B_i^Q in behavior class $\mathcal{B}_i$, for levels of abstraction $i = 0, 1$. In this case, we need to lift the local semantics of the components to common global behavior domains before we can compose them.

Definition 7 (Model Globalization) *Given a global behavior domain B, a model P with its local behavior domain B', and a behavior localization function $\downarrow : B \to B'$, the* (model) globalization *of P is a model P^G s.t. $\llbracket P^G \rrbracket^B = \llbracket P \rrbracket^{B'} \uparrow$.*

For a given modeling formalism $\mathcal{M}$, syntactic approaches for globalization may exist, e.g., addition of self loops for newly added event labels for discrete transition systems, or addition of state variables with unconstrained dynamics for continuous dynamic systems. We support all such procedures that lead to models with the set of behaviors $\llbracket P \rrbracket^{B'} \uparrow$.

Example 4 Consider transition system model P as shown below. The local alphabet is $\Sigma^P = \{\alpha, \beta\}$ and the local behavior domain $B^P = \Sigma^*$. Let the global alphabet be $\Sigma = \{\alpha, \beta, \gamma\}$ and the global behavior domain $B = \Sigma^*$. Let the projection function $\downarrow : B \to B^P$ be defined as per Ex. 2.

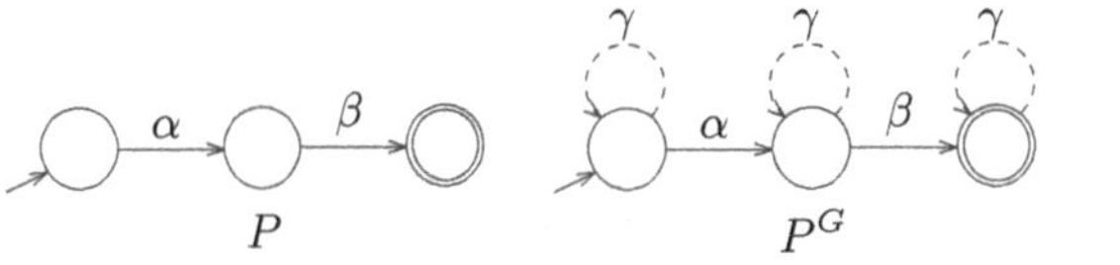

The semantic interpretation of P in the local behavior domain B^P is the set $\{\alpha\beta\}$. The globalization of the set $\{\alpha\beta\}$ in the global behavior domain B is the set $\{\gamma^*\alpha\gamma^*\beta\gamma^*\}$. Note that the syntactic globalization procedure by introducing self loops for the new label γ results in a model P^G shown above, and $\llbracket P^G \rrbracket^B = \llbracket P \rrbracket^{B^P} \uparrow$. □

The following lemma states that heterogeneous abstraction between model globalizations via a global abstraction function is equivalent to heterogeneous abstraction between original models via the localization of the global abstraction function.

Lemma 1 *For abstraction levels $i = 0, 1$, given component models P_i with local behavior domains B_i', behavior localization functions $\downarrow_i : B_i \to B_i'$, let their corresponding globalized models be P_i^G with global behavior domains B_i. If $\mathcal{A} : B_0 \to B_1$ is a global behavior abstraction function and $\mathcal{A}' : B_0' \to B_1'$ is a localization of $\mathcal{A}$, then*

$$P_0^G \sqsubseteq^{\mathcal{A}} P_1^G \Leftrightarrow P_0 \sqsubseteq^{\mathcal{A}'} P_1.$$

PROOF. From the definition of model globalization, we have

$$b_i \in \llbracket P_i^G \rrbracket^{B_i} \Leftrightarrow b_i{\downarrow_i} \in \llbracket P_i \rrbracket^{B_i'} \tag{4}$$

and

$$b_i' \in \llbracket P_i \rrbracket^{B_i'} \Leftrightarrow b_i'{\uparrow_i} \subseteq \llbracket P_i^G \rrbracket^{B_i}. \tag{5}$$

Case I: $P_0^G \sqsubseteq^{\mathcal{A}} P_1^G \Rightarrow P_0 \sqsubseteq^{\mathcal{A}'} P_1$
For any given $b_0 \in \llbracket P_0^G \rrbracket^{B_0}$, let $b_1 := \mathcal{A}(b_0)$. From $P_0^G \sqsubseteq^{\mathcal{A}} P_1^G$, we have $b_1 \in \llbracket P_1^G \rrbracket^{B_1}$. From (1), $\mathcal{A}'(b_0' := b_0{\downarrow_0}) = b_1{\downarrow_1}$. Hence, from (4), we have that $\forall b_0' \in \llbracket P_0 \rrbracket^{B_0'}, \mathcal{A}'(b_0') \in \llbracket P_1 \rrbracket^{B_1'}$, which implies $\llbracket P_0 \rrbracket^{B_0'} \subseteq \mathcal{A}'^{-1}(\llbracket P_1 \rrbracket^{B_1'})$, i.e., $P_0 \sqsubseteq^{\mathcal{A}'} P_1$.

Case II: $P_0^G \sqsubseteq^{\mathcal{A}} P_1^G \Leftarrow P_0 \sqsubseteq^{\mathcal{A}'} P_1$
From $P_0 \sqsubseteq^{\mathcal{A}'} P_1$, we have $b_0' \in \llbracket P_0 \rrbracket^{B_0'} \Rightarrow \mathcal{A}'(b_0') =: b_1' \in \llbracket P_1 \rrbracket^{B_1'}$. From Def. 4 and (5), for any $b_0' \in \llbracket P_0 \rrbracket^{B_0'}$, $b_0 \in b_0'{\uparrow_0} \subseteq \llbracket P_0^G \rrbracket^{B_0} \Rightarrow \mathcal{A}(b_0) =: b_1 \in b_1'{\uparrow_1} \subseteq \llbracket P_1^G \rrbracket^{B_1}$. Therefore, $\llbracket P_0^G \rrbracket^{B_0} \subseteq \mathcal{A}^{-1}(\llbracket P_1^G \rrbracket^{B_1})$, i.e., $P_0^G \sqsubseteq^{\mathcal{A}} P_1^G$. ■

In terms of Fig. 1, the implication of Lemma 1 is the following. When the abstract and concrete models of a component are considered in isolation, it does not matter whether one does the heterogeneous abstraction analysis in the global domains or the local domains.

We now use the result from Lemma 1 in a compositional setting when the component models are composed to form a system model. The following definition generalizes the notion of semantic composition from Def. 5.

Definition 8 (Globalized Semantic Composition) *Given a global behavior domain B, component models P and Q with their corresponding local behavior domains B^P and B^Q, and behavior localizations $\downarrow^P : B \to B^P$ and $\downarrow^Q : B \to B^Q$, the* globalized semantic composition *of P and Q in the global behavior domain B, denoted by $P||^G Q$ is the semantic composition of models P^G and Q^G, which are the globalizations of P and Q respectively, i.e., $P||^G Q = P^G || Q^G$.*

Example 5 Consider two transition system models P and Q as shown below (without the dashed self loops).

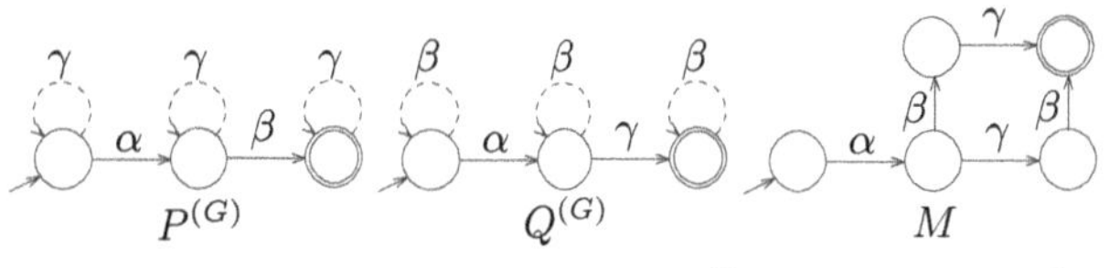

The local alphabets of P and Q are $\Sigma^P = \{\alpha, \beta\}$ and $\Sigma^P = \{\alpha, \gamma\}$, and corresponding behavior domains $B^P = \Sigma^{P^*}$ and $B^Q = \Sigma^{Q^*}$ respectively. Let the global alphabet be $\Sigma = \{\alpha, \beta, \gamma\}$, and the global behavior domain $B = \Sigma^*$. Let $\downarrow^P$ and $\downarrow^Q$ be the natural projections as defined in Ex. 2.

The local sets of behaviors for the two components are $\llbracket P \rrbracket^{B^P} = \{\alpha\beta\}$ and $\llbracket P \rrbracket^{B^P} = \{\alpha\gamma\}$. The semantic globalizations of the two component models yield $\llbracket P \rrbracket^{B^P} \uparrow^P = \{\gamma^*\alpha\gamma^*\beta\gamma^*\}$ and $\llbracket Q \rrbracket^{B^Q} \uparrow^Q = \{\beta^*\alpha\beta^*\gamma\beta^*\}$. The composition $M := P||^G Q$ has corresponding sets of behaviors given by $\llbracket M \rrbracket^B = \llbracket P \rrbracket^{B^P} \uparrow^P \cap \llbracket Q \rrbracket^{B^Q} \uparrow^Q = \{\alpha\beta\gamma, \alpha\gamma\beta\}$.

Note that the syntactic globalization procedure of introducing self loops yields models P^G and Q^G, whose syntactic composition results in a model M as shown above, s.t. $M = P||^G Q$. $\square$

Example 6 Let $B^j := \mathbb{R}^{\mathbb{R}_+}$ be the sets of 1-d continuous trajectories with variable names x_j, $j = p, q$ respectively. Let two components P given by $\dot{x_p} \in [1,2]$ and Q given by $\dot{x_q} \in [3,5]$ respectively have their semantics defined in domains B^j, $j = p, q$. Let $B := (\mathbb{R}^2)^{\mathbb{R}_+}$ be the system behavior domain of 2-d continuous trajectories with variable names along the two dimensions x_p and x_q. The globalizations of P and Q add the missing dimension and leave it unconstrained. Therefore, P^G and Q^G can be obtained as

$$P^G \equiv \begin{bmatrix} \dot{x_p} \\ \dot{x_q} \end{bmatrix} \in \begin{bmatrix} [1,2] \\ (-\infty,\infty) \end{bmatrix}, Q^G \equiv \begin{bmatrix} \dot{x_p} \\ \dot{x_q} \end{bmatrix} \in \begin{bmatrix} (-\infty,\infty) \\ [3,5] \end{bmatrix}.$$

Their composition is $P||^G Q \equiv \begin{bmatrix} \dot{x_p} \\ \dot{x_q} \end{bmatrix} \in \begin{bmatrix} [1,2] \\ [3,5] \end{bmatrix}$. $\square$

For the following discussion, we let models M_i, with the global behavior domains B_i, be the globalized compositions $P_i||^G Q_i$ of component models P_i and Q_i with their local behavior domains B_i^P and B_i^Q, for levels of abstraction $i = 0, 1$ as depicted in Fig. 1. We consider two scenarios in which the source of the abstraction is at the system and component levels respectively.

5.1 Centralized development

First, we consider the case where an abstraction function $\mathcal{A} : B_0 \to B_1$ between the global behavior domains B_0 and B_1 is given. For this case, the following proposition shows that the problem of establishing $M_0 \sqsubseteq^{\mathcal{A}} M_1$ can be reduced to solving two smaller problems $P_0 \sqsubseteq^{\mathcal{A}\Downarrow^P} P_1$ and $Q_0 \sqsubseteq^{\mathcal{A}\Downarrow^Q} Q_1$.

Proposition 2 *For abstraction levels $i = 0, 1$, given component models P_i and Q_i with corresponding local behavior domains B_i^P and B_i^Q, let their globalized semantic compositions be $M_i := P_i||^G Q_i$ in global behavior domains B_i with behavior localizations $\downarrow_i^j : B_i \to B_i^j$, where $j = P, Q$, and a global behavior abstraction function $\mathcal{A} : B_0 \to B_1$. If localizations $\mathcal{A}\Downarrow^P$ and $\mathcal{A}\Downarrow^Q$ of $\mathcal{A}$ exist and $P_0 \sqsubseteq^{\mathcal{A}\Downarrow^P} P_1$ and $Q_0 \sqsubseteq^{\mathcal{A}\Downarrow^Q} Q_1$, then $M_0 \sqsubseteq^{\mathcal{A}} M_1$.*

PROOF. From $P_0 \sqsubseteq^{\mathcal{A}\Downarrow^P} P_1$ and $Q_0 \sqsubseteq^{\mathcal{A}\Downarrow^Q} Q_1$, we know from Lemma 1 that $P_0^G \sqsubseteq^{\mathcal{A}} P_1^G$ and $Q_0^G \sqsubseteq^{\mathcal{A}} Q_1^G$, i.e., that $[\![P_0^G]\!]^{B_0} \subseteq \mathcal{A}^{-1}([\![P_1^G]\!]^{B_1})$ and $[\![Q_0^G]\!]^{B_0} \subseteq \mathcal{A}^{-1}([\![Q_1^G]\!]^{B_1})$. We have,

$$\begin{aligned} [\![P_0||^G Q_0]\!]^{B_0} &= [\![P_0^G]\!]^{B_0} \cap [\![Q_0^G]\!]^{B_0} \\ &\subseteq \mathcal{A}^{-1}([\![P_1^G]\!]^{B_1}) \cap \mathcal{A}^{-1}([\![Q_1^G]\!]^{B_1}) \\ &= \mathcal{A}^{-1}([\![P_1^G]\!]^{B_1} \cap [\![Q_1^G]\!]^{B_1}) \\ &= \mathcal{A}^{-1}([\![P_1||^G Q_1]\!]^{B_1}). \end{aligned}$$

$\blacksquare$

Prop. 2 states that we can establish $M_0 \sqsubseteq^{\mathcal{A}} M_1$ in the global behavior domains by establishing $P_0 \sqsubseteq^{\mathcal{A}\Downarrow^P} P_1$ and $Q_0 \sqsubseteq^{\mathcal{A}\Downarrow^Q} Q_1$ in the local behavior domains of the two components.

Example 7 Consider component models

$$P_0 \equiv \begin{bmatrix} \dot{x} \\ \dot{y} \end{bmatrix} \in \begin{bmatrix} [2,4] \\ [1,2] \end{bmatrix}, \begin{bmatrix} x \\ y \end{bmatrix}(0) \in \begin{bmatrix} [0,l_x) \\ [0,l_y) \end{bmatrix} \text{ and}$$

$$Q_0 \equiv \begin{bmatrix} \dot{x} \\ \dot{z} \end{bmatrix} \in \begin{bmatrix} [3,5] \\ [1,2] \end{bmatrix}, \begin{bmatrix} x \\ z \end{bmatrix}(0) \in \begin{bmatrix} [0,l_x) \\ [0,l_z) \end{bmatrix}.$$

Let P_1 and Q_1 be as follows.

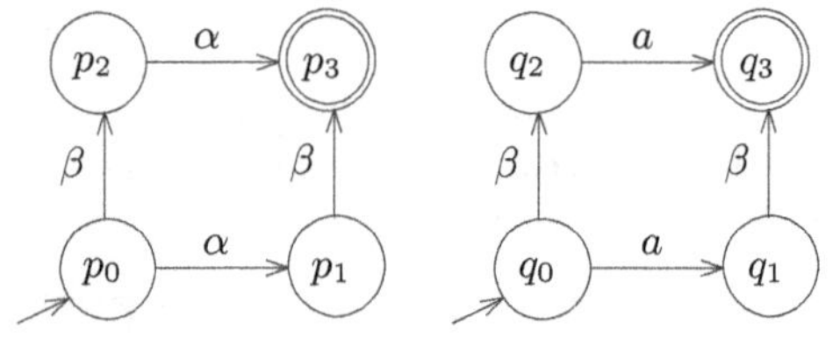

P_1 $\qquad$ Q_1

The compositions are $M_0 := P_0||^G Q_0$ given by

$$\begin{bmatrix} \dot{x} \\ \dot{y} \\ \dot{z} \end{bmatrix} \in \begin{bmatrix} [3,4] \\ [1,2] \\ [1,2] \end{bmatrix}, \begin{bmatrix} x \\ y \\ z \end{bmatrix}(0) \in \begin{bmatrix} [0,l_x) \\ [0,l_y) \\ [0,l_z) \end{bmatrix} \text{ and}$$

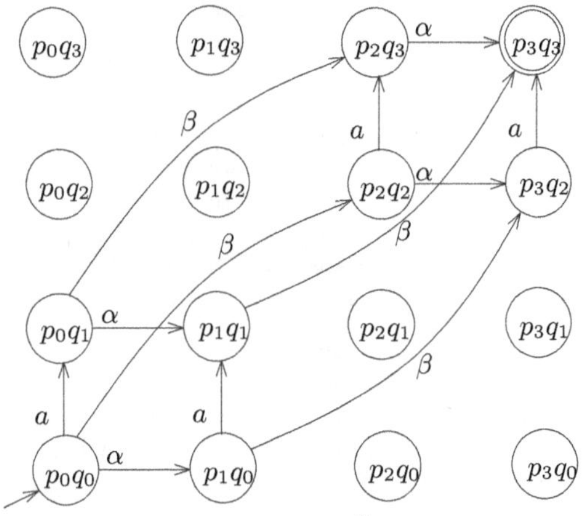

$M_1 := P_1||^G Q_1$.

Global behavior domain $B_0 := (\mathbb{R}^3)^{\mathbb{R}_+}$ for M_0 is the set of 3-d trajectories with variable names x, y, z; while global behavior domain $B_1 := \Sigma^*$ for M_1 is the set of all finite traces over the alphabet $\Sigma = \{\alpha, \bar{\alpha}, \beta, \bar{\beta}, a, \bar{a}\}$. Let the behavior abstraction function $\mathcal{A} : B_0 \to B_1$ be defined by partitioning the continuous state space as follows.

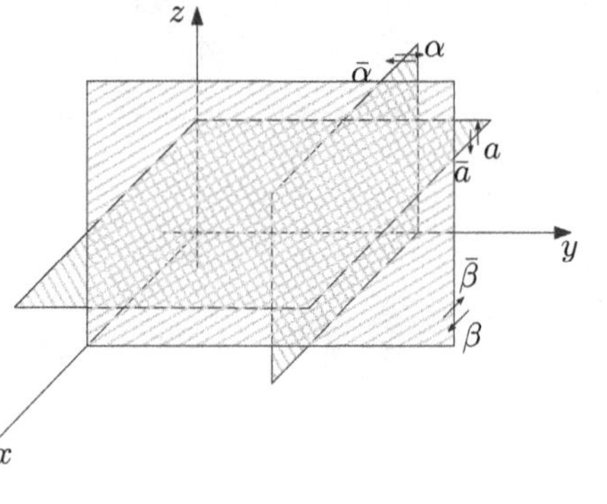

Given $b_0 = [x(t)\ y(t)\ z(t)]^T =: \bar{x}(t)$, $t \in \mathbb{R}_+$ and $b_1 = \sigma_0\sigma_1 \cdots$, $\mathcal{A}^j(b_0) = b_1$ if $\exists$ times $t_i \in \mathbb{R}_+$ s.t.

$$\forall t' \in [0, t_0), \quad \bar{x}(t) \in \text{FROM}(\sigma_0),$$

$$\forall t' \in [t_{i-1}, t_i), \quad \bar{x}(t) \in \text{TO}(\sigma_{i-1}) \cap \text{FROM}(\sigma_i),$$

$$\forall t' \geq t_N \quad \bar{x}(t) \in \text{TO}(\sigma_N)$$

where $i = 1, \ldots, N$ for some $N \in \mathbb{N}$ and FROM($\cdot$) and TO($\cdot$) are given in the following table.

σ	FROM(σ)	TO(σ)
a	$z \leq l_z,\ x, y \in \mathbb{R}$	$z \geq l_z,\ x, y \in \mathbb{R}$
$\bar{a}$	$z \geq l_z,\ x, y \in \mathbb{R}$	$z \leq l_z,\ x, y \in \mathbb{R}$
α	$y \leq l_y,\ x, z \in \mathbb{R}$	$y \geq l_y,\ x, z \in \mathbb{R}$
$\bar{\alpha}$	$y \geq l_y,\ x, z \in \mathbb{R}$	$y \leq l_y,\ x, z \in \mathbb{R}$
β	$x \leq l_x,\ y, z \in \mathbb{R}$	$x \geq l_x,\ y, z \in \mathbb{R}$
$\bar{\beta}$	$x \geq l_x,\ y, z \in \mathbb{R}$	$x \leq l_x,\ y, z \in \mathbb{R}$

Otherwise, $\mathcal{A}(b_0) = \varepsilon$.

The problem of establishing $M_0 \sqsubseteq^{\mathcal{A}} M_1$ for above $\mathcal{A}$ can be reduced to two smaller problems $P_0 \sqsubseteq^{\mathcal{A}\Downarrow^P} P_1$ and $Q_0 \sqsubseteq^{\mathcal{A}\Downarrow^Q} Q_1$ as follows.

Local behavior domains for the two models of component P are $B_0^P = (\mathbb{R}^2)^{\mathbb{R}_+}$ with variable names for the dimensions x and y; and $B_1^P = \Sigma^{P*}$ with $\Sigma^P = \{\alpha, \bar{\alpha}, \beta, \bar{\beta}\}$. Similarly, in case of the two models of component Q, $B_0^Q = (\mathbb{R}^2)^{\mathbb{R}_+}$ with variable names for the dimensions x and z; and $B_1^Q = \Sigma^{Q*}$ with $\Sigma^Q = \{a, \bar{a}, \beta, \bar{\beta}\}$. Let behavior localization functions for the two components at the two levels of abstractions be variable elimination and natural projection as in Ex. 2 and 3. We get the behavior abstraction function localizations $\mathcal{A}^P : B_0^P \to B_1^P$ and $\mathcal{A}^Q : B_0^Q \to B_1^Q$, where $\mathcal{A}^P = \mathcal{A}\Downarrow^P$ and $\mathcal{A}^Q = \mathcal{A}\Downarrow^Q$ are as follows.

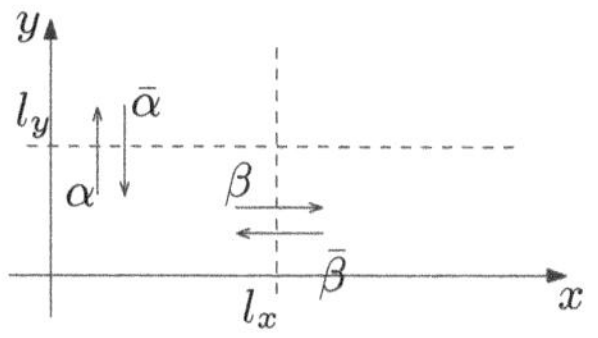

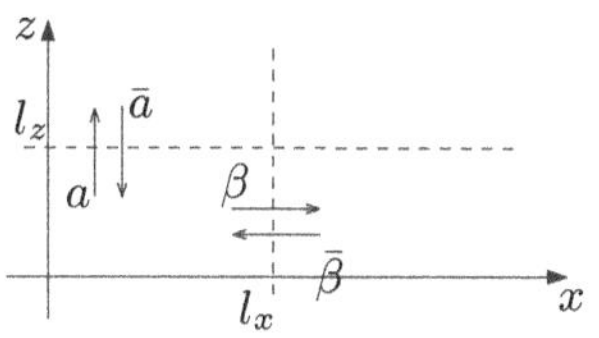

Let $\bar{x}^P = [xy]^T$ and $\bar{x}^Q = [xz]^T$. Given $b_0^P = \bar{x}^P(t)$, $b_0^Q = \bar{x}^Q(t)$ for $t \in \mathbb{R}_+$ and $b_1^j = \sigma_0^j\sigma_1^j \cdots$, $\mathcal{A}^j$, $j = P, Q$, are defined as $\mathcal{A}^j(b_0^j) = b_1^j$ if $\exists$ times $t_i^j \in \mathbb{R}_+$ s.t.

$$\forall t' \in [0, t_0^j), \quad \bar{x}^j(t') \in \text{FROM}^j(\sigma_0^j),$$

$$\forall t' \in [t_{i-1}^j, t_i^j), \quad \bar{x}^j(t') \in \text{TO}^j(\sigma_{i-1}^j) \cap \text{FROM}^j(\sigma_i^j),$$

$$\forall t' \geq t_N^j, \quad \bar{x}^j(t') \in \text{TO}^j(\sigma_N^j),$$

where $i = 1, \ldots, N$ for some $N \in \mathbb{N}$ and FROM$^j(\cdot)$ and TO$^j(\cdot)$ are given in the following tables.

σ	FROM$^P(\sigma)$	TO$^P(\sigma)$
α	$y \leq l_y, x \in \mathbb{R}$	$y \geq l_y, x \in \mathbb{R}$
$\bar{\alpha}$	$y \geq l_y, x \in \mathbb{R}$	$y \leq l_y, x \in \mathbb{R}$
β	$x \leq l_x, y \in \mathbb{R}$	$x \geq l_x, y \in \mathbb{R}$
$\bar{\beta}$	$x \geq l_x, y \in \mathbb{R}$	$x \leq l_x, y \in \mathbb{R}$

σ	FROM$^Q(\sigma)$	TO$^Q(\sigma)$
a	$z \leq l_z, x \in \mathbb{R}$	$z \geq l_z, x \in \mathbb{R}$
$\bar{a}$	$z \geq l_z, x \in \mathbb{R}$	$z \leq l_z, x \in \mathbb{R}$
β	$x \leq l_x, z \in \mathbb{R}$	$x \geq l_x, z \in \mathbb{R}$
$\bar{\beta}$	$x \geq l_x, z \in \mathbb{R}$	$x \leq l_x, z \in \mathbb{R}$

Otherwise, $\mathcal{A}^j(b_0^j) = \varepsilon$.

From the initial conditions and the monotonicity of the dynamics of P_0 (resp. Q_0), we can see that every behavior of the concrete model crosses the $x = l_x$ and $y = l_y$ (resp. $z = l_z$) boundaries in either order and have corresponding behaviors $\alpha\beta$ (resp. $a\beta$) or $\beta\alpha$ (resp. βa) at the discrete level that they map to. Therefore $P_0 \sqsubseteq^{\mathcal{A}^P} P_1$ and $Q_0 \sqsubseteq^{\mathcal{A}^Q} Q_1$. Using Prop. 2, $M_0 \sqsubseteq^{\mathcal{A}} M_1$.

Here, analyzing P_is and Q_is is much easier than analyzing M_is directly. In general, the extent of savings achieved by doing the heterogeneous abstraction analysis compositionally is depends on how much smaller the local behavior domains are compared to the global ones. □

5.2 Decentralized development

Now, we consider the case where the abstraction functions $\mathcal{A}^P : B_0^P \to B_1^P$ and $\mathcal{A}^Q : B_0^Q \to B_0^Q$ between the local behavior domains B_i^P and B_i^Q are given and heterogeneous abstractions of component models $P_0 \sqsubseteq^{\mathcal{A}^P} P_1$ and $Q_0 \sqsubseteq^{\mathcal{A}^Q} Q_1$ are established independently. This is the more common situation in practice, particularly for distributed development. In this case, the following proposition states that if the globalizations of abstraction functions $\mathcal{A}^P\Uparrow^P$ and $\mathcal{A}^Q\Uparrow^Q$ are defined consistently, the heterogeneous abstraction results for the components carry over to their compositions.

Proposition 3 *For abstraction levels $i = 0, 1$, given component models P_i and Q_i with local behavior domains B_i^P and B_i^Q, let their compositions be $P_i||^G Q_i$ in global behavior domains B_i and local behavior abstraction functions be $\mathcal{A}^P : B_0^P \to B_1^P$ and $\mathcal{A}^Q : B_0^Q \to B_1^Q$ s.t. $P_0 \sqsubseteq^{\mathcal{A}^P} P_1$ and $Q_0 \sqsubseteq^{\mathcal{A}^Q} Q_1$. If $\mathcal{A}^P\Uparrow^P = \mathcal{A}^Q\Uparrow^Q =: \mathcal{A}$, then $P_0||^G Q_0 \sqsubseteq^{\mathcal{A}} P_1||^G Q_1$.*

PROOF. The result follows due to $(\mathcal{A}^P\Uparrow^P)\Downarrow^P = \mathcal{A}^P$ and $(\mathcal{A}^Q\Uparrow^Q)\Downarrow^Q = \mathcal{A}^Q$ from (2) and Prop. 2. ■

Prop. 3 states that the heterogeneous abstraction results for component models P_i and Q_i via possibly very different abstraction functions $\mathcal{A}^P$ and $\mathcal{A}^Q$ follow over to the system heterogeneous abstraction so long as $\mathcal{A}^P$ and $\mathcal{A}^Q$ are consistent, i.e., that it is possible to find globalizations$\mathcal{A}^P\Uparrow^P$ and $\mathcal{A}^Q\Uparrow^Q$ that are in agreement with each other. Note from the non-uniqueness of globalization of abstraction functions that there is some design freedom while constructing the semantic mappings at the global behavior domains for the two components such that they agree.

We note the following conditions for agreement of the globalizations of the local abstraction functions from the two components.

- **Disjoint behavior domains.** If the local behavior domains are disjoint (no common variables), the abstraction functions are disjoint. Therefore, when globalized, they are not mutually restrictive and it is always possible to construct globalizations that agree.

- **Agreement in intersection.** For non-disjoint local behavior domains, it is necessary for globalization agreement that the local abstraction functions agree on the "intersection" of the two behavior domains, say $B_i^\cap$, i.e., along the variables common to the two components. If localizations $\mathcal{A}^P\Downarrow^\cap : B_0^\cap \to B_1^\cap$ and $\mathcal{A}^Q\Downarrow^\cap : B_0^\cap \to B_1^\cap$ of $\mathcal{A}^P$ and $\mathcal{A}^Q$ agree, it is always possible to construct globalizations of $\mathcal{A}^P$ and $\mathcal{A}^Q$ that agree due to the fact that variables not common to the two components are not mutually constraining.

We illustrate distributed compositional heterogeneous abstraction analysis in the following section.

6. EXAMPLE

Consider a cooperative intersection collision avoidance system for stop-sign assist (CICAS-SSA) [1] from Fig. 2, which depicts a *subject vehicle* (SV) waiting at a stop-sign-controlled intersection to cross through traffic on a major road. The objective is to augment human judgment of the SV driver about whether a given gap in oncoming traffic is safe by sensing the positions and/or velocities of the oncoming vehicles and doing some computations based on vehicle dynamics, intersection geometry and speed limits. The oncoming vehicle is called the *principal other vehicle* (POV). The SV modeled is allowed to (but doesn't have to) enter the intersection only if the POV is far enough away from the intersection to allow the SV to pass completely through the intersection before the POV has arrived at the intersection. Otherwise the SV has to remain stopped.

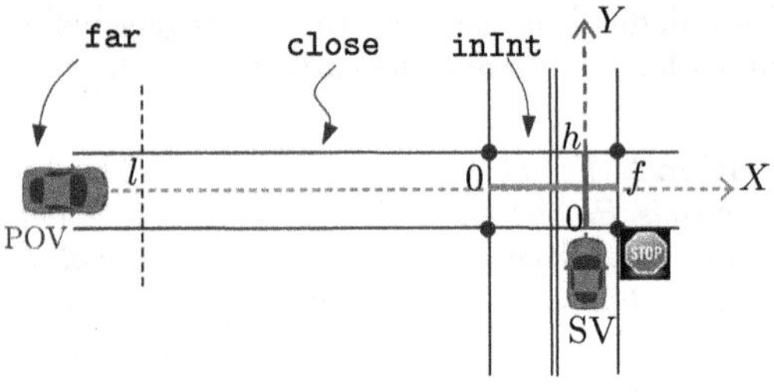

Figure 2: A simple variant of CICAS-SSA.

We model this system at two levels of abstraction. At the detailed level we model the two vehicles using their hybrid dynamics, while at the abstract level, we model simple discrete dynamics. The discrete dynamics can be used to verify protocols such as "if SV hasn't entered in the intersection already, then it doesn't do so once it sees POV get close to the intersection." This protocol verification can then be used in conjunction with other information to construct a hierarchical heterogeneous verification of the system to guarantee safety specifications such as "the two cars are never in the intersection at the same time" as demonstrated in [19]. In this paper, we consider the problem of establishing in a distributed compositional manner that the discrete model of the system used in the protocol verification is a heterogeneous abstraction of the underlying hybrid model of the system. In this distributed compositional heterogeneous abstraction, we use two different kinds of abstraction functions for two components – one using state-space partitioning and another by retaining the discrete transition graph by projecting away all the continuous dynamics.

The POV drives along the major road with its position x increasing over time, at a velocity between a minimum and a maximum, both limits being positive (i.e., it cannot drive in reverse on the highway). We assume that the SV is able to sense the position of the POV to make its decision. Once in the intersection, SV keeps driving with a velocity in the range $[\underline{v}_y, \overline{v}_y]$, both limits assumed positive, eventually clearing the intersection at $y = h$.

6.1 Heterogeneous abstraction for POV

The hybrid model P_0 and the discrete model P_1 for the POV are shown in Fig. 3.

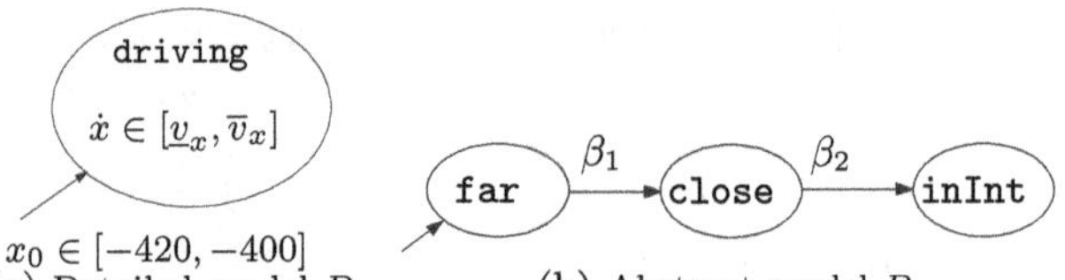

(a) Detailed model P_0 (b) Abstract model P_1

Figure 3: Hybrid and discrete POV models.

The local behavior domains are B_0^{POV}: 1-d hybrid traces, i.e., evolution of the hybrid state $h^{POV} := (l^{POV}, x)$ over time, with $l^{POV} \in \mathcal{L}^{POV} := \{\texttt{driving}\}$ and $x \in \mathbb{R}$; and $B_1^{POV} := \Sigma^{POV*}$ for set of event labels $\Sigma^{POV} = \{\beta_1, \beta_2\}$. The model semantics are $[\![P_0]\!]^{B_0}$: the set of all hybrid traces with the discrete location `driving` and x that starts in the initial condition set $[-420, -400]$ and evolves along any arbitrary derivative in the range $[\underline{v}_x, \overline{v}_x]$, and $[\![P_1]\!]^{B_1}$: the singleton set $\{\beta_1\beta_2\}$.

A behavior abstraction function $\mathcal{A}^{POV} : B_0^{POV} \to B_1^{POV}$ constructed by partitioning the continuous dimension x at boundaries $x = l$ and $x = 0$ is written mathematically as follows. Given $b_0^{POV} = h^{POV}(t) \in B_0^{POV}$ and $b_1^{POV} = \sigma_0\sigma_1 \cdots \in B_1^{POV}$, $\mathcal{A}^{POV}(b_0^{POV}) = b_1^{POV}$ iff $\exists$ times $t_i \in \mathbb{R}_+$ s.t. $\forall t' \in [0, t_0)$, $x(t') \in \text{FROM}(\sigma_0)$; $\forall t' \in [t_{i-1}, t_i)$, $x(t') \in \text{TO}(\sigma_{i-1}) \cap \text{FROM}(\sigma_i)$ for $i = 1, \ldots, N$ for some $N \in \mathbb{N}$; and $\forall t' \geq t_N$, $x(t') \in \text{TO}(\sigma_N)$, where $\text{FROM}(\cdot)$ and $\text{TO}(\cdot)$ are given in the following table.

σ	$\text{FROM}(\sigma)$	$\text{TO}(\sigma)$
β_1	$x \leq l$	$x \in [l, 0]$
β_2	$x \in [l, 0]$	$x \geq 0$

Otherwise, $\mathcal{A}^{POV}(b_0^{POV}) = \varepsilon$.

Suppose the boundary l is at -300. Since the range of velocities is positive, and initial condition is in the range $[-420, -400]$, it is straightforward to show that $\forall b_0^{POV} \in B_0^{POV}$, $\mathcal{A}^{POV}(b^{POV}) = \beta_1\beta_2$. Therefore, $P_0 \sqsubseteq^{\mathcal{A}^{POV}} P_1$. Note that if l is say -410, $\mathcal{A}^{POV}(b^{POV}) = \beta_2$ for some b^{POV} and $P_0 \not\sqsubseteq^{\mathcal{A}^{POV}} P_1$.

6.2 Heterogeneous abstraction for SV

The hybrid model Q_0 and the discrete model Q_1 for the SV are shown in Fig. 4. At the hybrid (respectively, discrete) level, the SV is able to sense the POV position x as a continuous input variable (respectively, the event β_1).

The local behavior domains are B_0^{SV}: the set of 2-d hybrid trajectories $h^{SV}(t)$, where $h^{SV} := (l^{SV}, x, y)$ are the hybrid

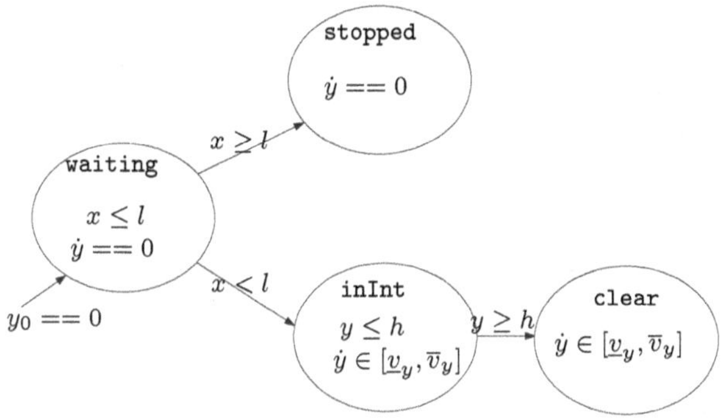

(a) Detailed model Q_0

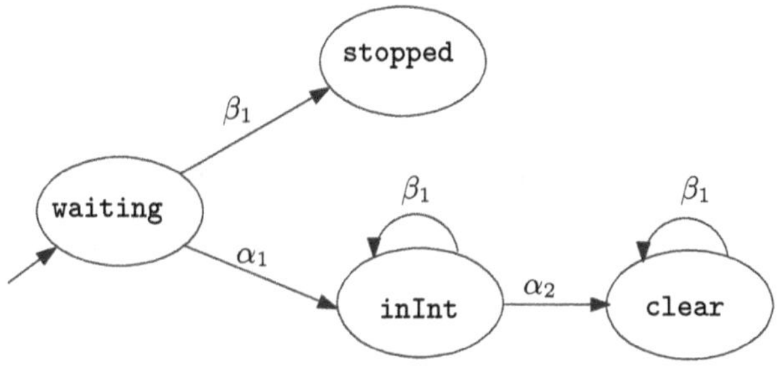

(b) Abstract model Q_1

Figure 4: Hybrid and discrete SV models.

states that take values in $\mathcal{L}^{SV} \times \mathcal{X}^{SV}$, for the discrete set of locations $\mathcal{L}^{SV} := \{\texttt{waiting}, \texttt{stopped}, \texttt{inInt}, \texttt{clear}\}$ and the continuous state space $\mathcal{X}^{SV} := \mathbb{R}^2$; and $B_1^{SV} := \Sigma^{SV*}$ with $\Sigma^{SV} := \{\alpha_1, \alpha_2, \beta_1\}$, where α's signify SV entering and exiting the intersection.

A behavior abstraction function $\mathcal{A}^{SV} : B_0^{SV} \to B_1^{SV}$, constructed by only keeping the discrete part of the hybrid model and adding transition labels, is written formally as follows. Given $b_0^{SV} = h^{SV}(t)$, where $t \in \mathbb{R}_+$ and $h^{SV} = (l^{SV}, x, y)$, and $b_1^{POV} = \sigma_0\sigma_1\cdots$ with states $q_i^{SV} \in \mathcal{L}_i^{SV}$ s.t. $q_i^{SV} \xrightarrow{\sigma_i} q_{i+1}^{SV}$, $\mathcal{A}^{SV}(b_0^{SV}) = b_1^{SV}$ iff $\exists$ times $t_i \in \mathbb{R}_+$ s.t. $\forall\ t' \in [t_i, t_{i+1})$ with $t_0 = 0$, $l^{SV}(t') == q_i^{SV}$. Otherwise, $\mathcal{A}^{POV}(b_0^{POV}) = \varepsilon$.

Because Q_1 has the exact same discrete transition graph as that of Q_0, for every hybrid behavior $b_0^{SV} \in [\![Q_0]\!]^{B_0^{SV}}$, $\mathcal{A}^{SV}(b_0^{SV}) \in [\![Q_1]\!]^{B_1^{SV}}$, i.e., $Q_0 \sqsubseteq^{\mathcal{A}^{SV}} Q_1$.

6.3 Abstraction between compositions

At the discrete level of abstraction, the global unified behavior domain B_1 is Σ^*, where $\Sigma = \Sigma^{POV} \cup \Sigma^{SV} = \{\alpha_1, \alpha_2, \beta_1, \beta_2\}$. Behavior localizations $\downarrow_1^j$, $j = P, Q$ are discrete event projection functions that replace a string not in the local label set by the empty string ε. In this case, the syntactic procedures of adding self loops on the missing labels α_1, α_2 in P_1 and β_2 in Q_1 take care of the globalizations and their composition is simply their product. At the hybrid level, we add an unrestricted continuous variable y in P_0 leaving Q_0 unchanged, and take the parallel composition of the resulting hybrid automata. The resultant system models $M_i := P_i||^G Q_i$, $i = 0, 1$ are as shown in Fig. 5.

The variables common to local behavior domains B_i^{POV} and B_i^{SV} are x and β_1. We have to make sure that the localizations $\mathcal{A}^{POV}\Downarrow^{\cap}$ and $\mathcal{A}^{SV}\Downarrow^{\cap}$ of abstraction functions $\mathcal{A}^{POV}$ and $\mathcal{A}^{SV}$ onto these common variables , i.e., the mappings from behaviors in x to behaviors in $\{\beta_1\}^*$ agree. $\mathcal{A}^{POV}\Downarrow^{\cap}$ is essentially the same as $\mathcal{A}^{POV}$, with the row for β_2 discarded. $\mathcal{A}^{SV}$ puts indirect restrictions on x due to the guard and invariant conditions of the hybrid transitions $(\texttt{waiting}, x) \to (\texttt{stopped}, x)$ that are mapped with the discrete transition $\texttt{waiting} \xrightarrow{\beta_1} \texttt{stopped}$. Such a hybrid transition occurs iff $x \leq l$ and $x \geq l$ hold before and after the transition, i.e., while crossing the boundary $x = l$ in the increasing direction, which agrees with $\mathcal{A}^{POV}\Downarrow^{\cap}$. In the self-loop β_1 transitions, x does not appear and is therefore unrestricted, and in agreement with $\mathcal{A}^{POV}\Downarrow^{\cap}$.

Therefore, using Prop. 3, we can conclude (without having to analyze models M_0 and M_1 directly) that $M_0 \sqsubseteq^{\mathcal{A}} M_1$.

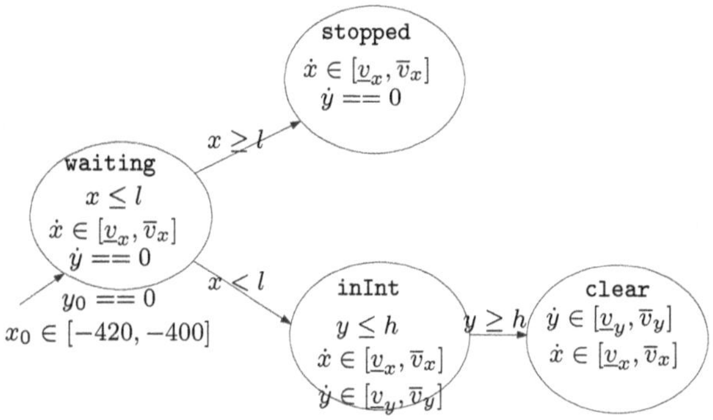

(a) Detailed model $M_0 := P_0||^G Q_0$

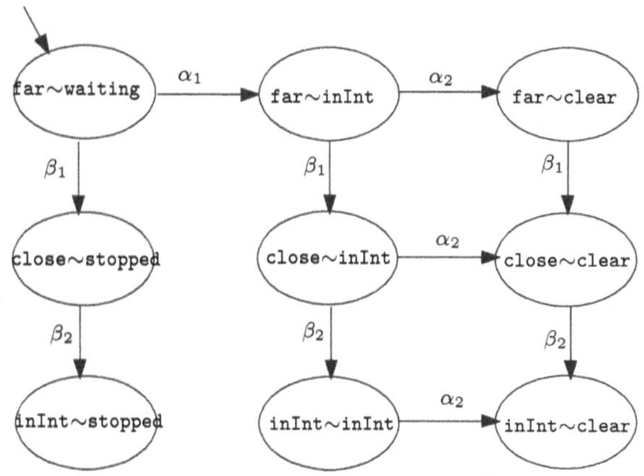

(b) Abstract model $M_1 := P_1||^G Q_1$

Figure 5: Hybrid and discrete system models.

6.4 Need for consistency between abstraction functions

Note that if for some reason, the parameter l is different in models P_0 and Q_0, the consistency condition in Prop. 3 cannot be satisfied and the heterogeneous approach cannot be used. Suppose the reference marker in the POV component is l' rather than l, but the SV thinks it is l. Physically this may correspond to, e.g., a measurement error or parallax for a human SV driver. In this case, there is a disagreement between the two models as to what corresponds to the β_1 event of POV going from **far** to **close**. Since the two abstraction functions disagree on the mapping between behaviors in the variable x and the event β_1 that are common to the local behavior domains of the two components, the design freedom in the non-uniqueness of globalizations while adding the remaining variables and events does not help us resolve this mismatch. Therefore, we cannot find any agreeing globalizations of the two abstraction functions. In such a case, although heterogeneous abstraction still holds for the two components individually, it does not carry over to their composition.

7. CONCLUSION

This paper presents a compositional approach to heterogeneous abstraction. Behavior abstraction functions are proposed to establish semantic associations between heterogeneous formalisms across different levels of abstraction, and localizations/globalizations are used to associate local component behavior domains and global system behavior domains at a given level of abstraction. Sufficient conditions are developed under which heterogeneous abstraction between component models implies heterogeneous abstraction between their compositions. The theoretical concepts are illustrated using the example of a cooperative intersection collision avoidance system (CICAS).

As noted in the paper, abstraction relations as well as globalization and localization typically can be inferred directly from the structure and syntactic rules for constructing models. Future work will address the possibility of heterogeneity between the interacting component models within a given level of abstraction. This requires the development of heterogeneous generalizations of abstraction globalization and globalized semantic composition.

Acknowledgments

The authors gratefully acknowledge support by NSF grants CNS 1035800-NSF and CCF-0926181.

8. REFERENCES

[1] Cooperative intersection collision avoidance systems (CICAS). `http://www.its.dot.gov/cicas/`.

[2] M. Althoff, A. Rajhans, B. H. Krogh, S. Yaldiz, X. Li, and L. Pileggi. Formal Verification of Phase-Locked Loops Using Reachability Analysis and Continuization. In *Proceedings of the IEEE/ACM 2011 International Conference on Computer-Aided Design (ICCAD)*, San Jose, Nov 2011.

[3] R. Alur, T. Dang, and F. Ivancic. Predicate abstraction for reachability analysis of hybrid systems. *ACM Transactions on Embedded Computing Systems*, 5(1):152–199, 2006.

[4] R. Alur, T. A. Henzinger, G. Laffarriere, and G. J. Pappas. Discrete Abstractions of Hybrid Systems. *Proceedings of the IEEE*, 88:971–984, 2000.

[5] F. Balarin, Y. Watanabe, H. Hsieh, L. Lavagno, C. Passerone, and A. Sangiovanni-Vincentelli. Metropolis: an integrated electronic system design environment. *Computer*, 36(4):45–52, april 2003.

[6] S. Bensalem, M. Bozga, T.-H. Nguyen, and J. Sifakis. Compositional verification for component-based systems and application. *IET Software*, 4(3):181–193, 2010.

[7] A. Benveniste, L. P. Carloni, P. Caspi, and A. L. Sangiovanni-Vincentelli. Composing Heterogeneous Reactive Systems. *ACM Transactions on Embedded Computing Systems*, 7(4), July 2008.

[8] S. S. Bhattacharyya, E. Cheong, and I. Davis. PTOLEMY II Heterogeneous Concurrent Modeling and Design in Java. Technical report, University of California, Berkeley, 2003.

[9] A. Chutinan and B. H. Krogh. Verification of Infinite-State Dynamic Systms Using Approximate Quotient Transition Systems. *IEEE Transactions on Automatic Control*, 46:1401–1410, 2001.

[10] T. Dang, O. Maler, and R. Testylier. Accurate Hybridization of Nonlinear Systems. In *Proceedings of the International Conference on Hybrid Systems: Computation and Control (HSCC)*, 2010.

[11] G. Frehse. *Compositional Verification of Hybrid Systems using Simulation Relations*. PhD thesis, Radboud Universiteit Nijmegen, 2005.

[12] G. Frehse. PHAVer: Algorithmic Verification of Hybrid Systems past HyTech. *International Journal on Software Tools for Technology Transfer (STTT)*, 10(3), 2008.

[13] T. A. Henzinger, P.-H. Ho, and H. Wong-Toi. Algorithmic Analysis of Nonlinear Hybrid Systems. *IEEE Transactions on Automatic Control*, 43:225–238, 1998.

[14] T. A. Henzinger, S. Qadeer, S. K. Rajamani, and S. Tasiran. An assume-guarantee rule for checking simulation. *ACM Trans. Program. Lang. Syst.*, 24(1):51–64, Jan. 2002.

[15] A. A. Julius. *On interconnection and equivalence of continuous and discrete systems: a behavioral perspective*. PhD thesis, University of Twente, 2005.

[16] D. Kaynar and N. Lynch. Decomposing Verification of Timed I/O Automata. In *In the Proceedings of the Joint Conference on Formal Modelling and Analysis of Timed Systems (FORMATS) Formal Techniques in Real-Time and Fault Tolerant System (FTRTFT)*, 2004.

[17] A. Ledeczi, J. Davis, S. Neema, and A. Agrawal. Modeling methodology for integrated simulation of embedded systems. *ACM Trans. Model. Comput. Simul.*, 13:82–103, January 2003.

[18] E. A. Lee and A. Sangiovanni-Vincentelli. A Framework for Comparing Models of Computations. *IEEE Transactions on Computer-Aided Design of Integrated Circuits and Systems*, 17(12):1217–1229, Dec 1998.

[19] A. Rajhans and B. H. Krogh. Heterogeneous verification of cyber-physical systems using behavior relations. In *Proceedings of the 15th ACM International Conference on Hybrid Systems: Computation and Control (HSCC)*, 2012.

Certifying the Safe Design of a Virtual Fixture Control Algorithm for a Surgical Robot*

Yanni Kouskoulas
Johns Hopkins University
Applied Physics Lab

David Renshaw
Carnegie Mellon University
Computer Science Dept.

André Platzer
Carnegie Mellon University
Computer Science Dept.

Peter Kazanzides
Johns Hopkins University
Dept. of Computer Science

ABSTRACT

We applied quantified differential-dynamic logic (Qd$\mathcal{L}$) to analyze a control algorithm designed to provide directional force feedback for a surgical robot. We identified problems with the algorithm, proved that it was in general unsafe, and described exactly what could go wrong. We then applied Qd$\mathcal{L}$ to guide the development of a new algorithm that provides safe operation along with directional force feedback. Using KeYmaeraD (a tool that mechanizes Qd$\mathcal{L}$), we created a machine-checked proof that guarantees the new algorithm is safe for all possible inputs.

Categories and Subject Descriptors

F.3.1 [**Logics and the Meaning of Progams**]: Specifying and Verifying and Reasoning about Programs; I.2.9 [**Artificial Intelligence**]: Robotics

Keywords

Quantified differential dynamic logic, medical robotics, formal verification

1. INTRODUCTION

Imagine an operating room in the near future where a surgeon works diligently to excise a tumor at the base of a patient's skull. The physician might employ newly developed robotic technology to more clearly visualize the surgery and more precisely guide the surgical instrument to make the incisions. The ultimate goal: surgery that is safer and more effective than before, providing better patient outcomes.

The robotic machinery to make this future a reality are slowly being developed, but the systems are complex, and can be prone to subtle, unexpected errors. It is easy to see how safety critical such systems are; a bug in the implementation or error in the algorithm that controls the surgical tool might cause it to make the wrong incision, with devastating consequences for the patient.

The usual approach today for ensuring the safety of complex systems is careful design, thoughtful examination of the algorithms, and testing. This approach was applied in [18], where the authors built the system and tested the final product with a surgical procedure on a cadaver. Testing is useful, but only shows the presence of bugs, not their absence.

This paper describes the analysis of one safety property of a skull-base surgery (SBS) robot algorithm, described in [18], to help ensure its safe and predictable operation. Rather than taking a testing approach, we apply formal methods to analyze the control algorithm of interest. This rigorous analysis ensures that the algorithm and the hardware that it controls behave predictably and safely for *all* possible inputs, rather than only for finitely many test cases. The guarantee we seek is much more comprehensive, and can lead to much safer and more predictable systems.

The contribution of this work is that it helps explore how to usefully apply newly developed formal approaches to practical systems. This has two benefits: first, it helps guide the development and refinement of logics and tools, by identifying what is necessary to put these techniques into widespread use; second, it helps the development of practical robotic systems by introducing new formal methods as a powerful and maturing set of design tools.

2. BACKGROUND

The SBS robot in [18] restricts movement of a surgical tool to be within a preoperatively defined surgical site (see Fig. 1), provides force feedback to the surgeon, to indicate when he or she is approaching these boundaries, and aids in fine control of the tool by damping small movements.

To configure this robot, a surgeon describes an operating volume in which to work by a series of planes oriented and positioned in space, called "*virtual fixtures*." Each planar boundary extends infinitely, and the intersection of the volume on the "safe" side of each plane is designed to exclude the areas of anatomy with which the surgeon does not wish to interact. We are considering planar boundaries because that is what was used in the original work.

During surgery, the surgeon holds the tool, (think of it

*This material is based upon work supported by the National Science Foundation under NSF CAREER Award CNS-1054246 and NSF EXPEDITION CNS-0926181.

HSCC'13, April 8–11, 2013, Philadelphia, Pennsylvania, USA.

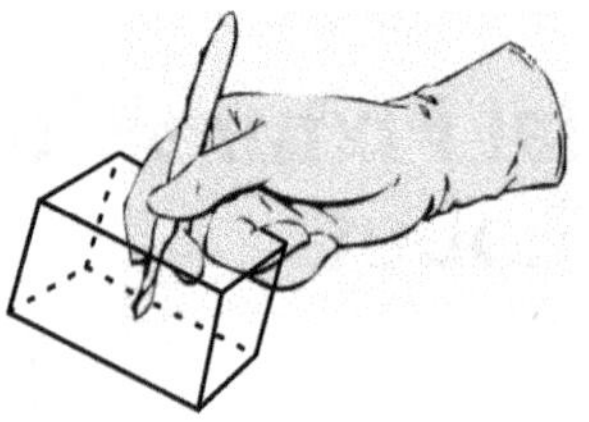

Figure 1: Cooperatively controlled robot enforcing virtual fixtures to restrict a tool-tip to moving within a small volume.

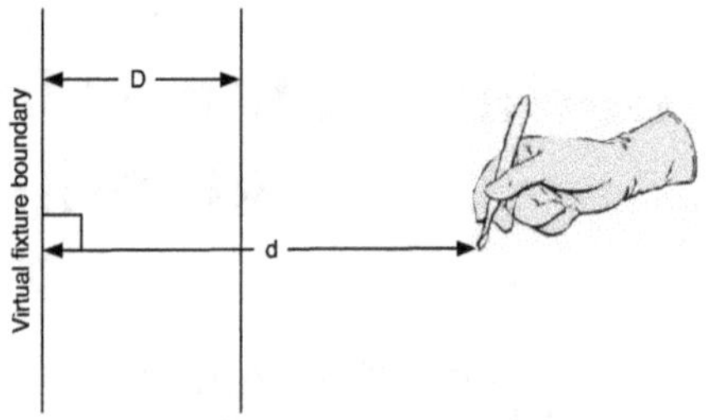

Figure 2: The robot determines which mode it is in depending on the distance from the virtual fixture boundary.

as a scalpel, even though it may be another tool with the same form factor) and applies it to the surgical task at hand, moving it about. The robot is attached to the tool through a rigid mechanical linkage, so as the surgeon exerts forces and torques on the tool, the robot can sense them, and can also exert forces on the tool opposing or aiding the surgeon's movements. This interaction between the SBS robot and surgeon is called cooperative control.

The control algorithm provides three behaviors through different modes of operation: within the center of the safe volume it allows free movement of the tool; close to the surface at the edge of the safe volume it creates an increasing resistance to movement that lets the surgeon know that he or she is close to a virtual boundary; and at the boundary it opposes the movement of the tool, preventing it from crossing the boundary.

Qualitatively, this will produce an invisible boundary that feels soft or mushy. As the tool pushes towards the boundary, the resistance will become progressively firmer; eventually the tool will slow to a stop and "stick" when it reaches the limit of its allowed movement.

The "physics" of the robot are created by an admittance control design, a circuit that converts sensed forces and torques to velocity through a multiplicative factor. If the surgeon exerts force $\bar{f}$ on the tool tip, located at Cartesian position $\bar{p}$, the control circuit translates the input force into a velocity $\bar{p}'$ at the tool tip, given by:

$$\bar{p}' = K(\bar{p})G(\bar{f})\bar{f} \qquad (1)$$

where overbars indicate vectors, prime $'$ indicates a derivative with respect to time, and K and G are 3×3 matrices (when just the position is controlled).

A more detailed version of the admittance control equation can be found in [18], but for our purposes this is an adequate model. G is the scale factor, which in the actual system is a matrix with non-linear (exponential) terms described in [7]. It allows the surgeon to switch between moving rapidly over large areas, and doing precise work in a small area with fine control, without interrupting the workflow. In our study, we simplify G and consider it a constant.

$K(\bar{p})$ is a gain term, used to provide force feedback and enforce the "virtual fixtures" as described above. Its elements change form abruptly depending on the position of the tip of the tool, $\bar{p}$.

Regions of movement are defined in terms of D, a design parameter that indicates a cutoff distance from the boundary in its safe direction, as in Fig. 2. In the free region, i.e., the points whose distance d from the boundary satisfy $d > D$, K is the identity matrix. In the slow region, where $0 \le d \le D$, the system provides force feedback opposing movement in proportion to how close the tool is to the boundary. Past the edges of the boundary in the unsafe zone, K becomes zero; our tool should not reach past the boundary during normal operation, if our algorithm is safe.

During the experiments [18], Eqn. 1 was effective in preventing the tool from penetrating the safety boundary, but had the undesirable property that once the tool was near the boundary, motion in *all* directions was attenuated, including motions away from the boundary. This led to the development of a new control law, where in the slow region ($0 \le d \le D$), the velocity of the tool is attenuated by subtracting a fraction of the component in the direction normal to the boundary. The amount subtracted is proportional to how close to the boundary the tool is at any given point. So if the velocity of the tool is $\bar{p}'$, the distance from the boundary is d, and the unit normal to the boundary is $\hat{n}_1$, the final velocity $\bar{p}_1'$ is given in [5] as:

$$\bar{p}_1' = \bar{p}' - \left(1 - \frac{d}{D}\right)(\bar{p}' \cdot \hat{n}_1)\hat{n}_1 \qquad (2)$$

where $\cdot$ is the dot product of two vectors. The force feedback from each region is applied sequentially, using the attenuated velocity remaining from prior steps. The designers of this algorithm recognize that there is an irregularity in the computation for acute angles, but were not sure what practical effect it had on the system's behavior.

The goal of this research is to ensure that the algorithm safely restricts movement to stay within the virtual fixture boundaries, so that the robot will not adversely impact the safety of the overall system or otherwise unnecessarily endanger the patient.

3. RELATED WORK

There are a number of papers related to the use of virtual fixtures for surgery, such as [10, 1]. The closest to the current work are [4, 9], which take a constrained optimization approach to enforcing more complicated boundaries that are generated by anatomy, and providing an approach that can guide a tool successfully through a specific path. The current work is focused on somewhat orthogonal concerns of improving the fidelity of the model of the interface between discrete program and robotic machinery, (i.e. accurately representing the control mechanism by incorporating realistic delay), and using this improved model to prove safety for all possible inputs, helping to improve the maturity of formal methods. It would be interesting in the future to explore the benefits of applying a formal approach to proving the safety of the more complicated algorithms.

Algorithmic verification approaches such as [2, 3] exist, but they are not appropriate for this work because they could only show safety for a specific number of boundaries, n, i.e. we cannot quantify over elements in the structure in our model. We wish to use one proof to show safety for all possible configurations of any number of boundaries, i.e. $\forall n$, where n is the number of boundaries configured.

At the intersection of formal methods and surgical robotic systems, the authors of this work have also produced related work certifying surgical robot systems in [8, 6], but this prior work is different because it is focused on certifying program implementation for concurrent software, rather than algorithm design for hybrid systems.

4. FORMAL APPROACH

Like all formal methods, the approach we employ has three components: a method for creating a model of the algorithm and the physics of the system, a language for writing a precise specification of its behavior, and a strategy to rigorously prove that the model has exactly the behavior described in the specification. We used two closely related hybrid logics to help us formally verify the safe behavior of the control algorithm used to enforce virtual fixtures: differential-dynamic logic (d$\mathcal{L}$) [11, 13], and quantified differential-dynamic logic [14, 15] (Qd$\mathcal{L}$). Both of these logics use the same approach to modeling the system, specifying behavior, and proving properties.

Modeling a hybrid system in Qd$\mathcal{L}$ and d$\mathcal{L}$ is done via a hybrid program (HP). The hybrid program looks a bit like an imperative programming language, with real numbers and discrete sets comprising its primitive types. The statements available within a hybrid program are: instantaneous, simultaneous assignment of values θ_i to program state variables x_i, e.g. $x_1 := \theta_1, \dots, x_n := \theta_n$; enforcing a logical assertion χ on program state, i.e. $?\chi$; composing program α with β, i.e. $\alpha;\beta$; nondeterministically choosing to execute one of two programs α and β, i.e. $\alpha \cup \beta$; repetition of a program α some nondeterministic number of times, i.e. α^*; and continuous evolution of a set of differential equations, i.e. $x_1' = \theta_1, \dots, x_n' = \theta_n \& \chi$. The last type of statement is the most interesting: it represents the evolution of continuous time in the hybrid system according to the given differential equations, satisfying the first order constraint, χ. The state variables x_i and those contained in terms θ_i, evolve continuously in response to this statement and stop at any time before leaving χ.

Specifying behavior in Qd$\mathcal{L}$ and d$\mathcal{L}$ is done using first-order logic with the usual logical connectives, i.e. $\phi \wedge \psi$, $\phi \vee \psi$, $\neg\phi$, $\phi \rightarrow \psi$, and quantifiers, i.e. $\forall x \phi$, and $\exists x \phi$. The specification language also contains box and diamond modal operators $[\alpha]\phi$ and $\langle\alpha\rangle\phi$; the former is satisfied when ϕ holds after all runs of α, and the latter is satisfied if ϕ holds after at least one run of α.

The proof strategy for these logics depends on applying a set of sound inference rules to transform pieces of the goal into tautologies in real arithmetic.

Qd$\mathcal{L}$ is a generalization of d$\mathcal{L}$, designed to allow quantification of variables used as arguments to other functions, and continue to ensure that the resulting arithmetic is decidable. Decidability is important because when working with these logics the leaves of proof trees are systems of equations devoid of hybrid program constructs; solving them proves or disproves that branch of the proof tree. Quantification over variables used as arguments to other functions becomes useful when we model multiple boundaries.

We are fortunate that there is a mechanization of Qd$\mathcal{L}$, described in [17], that allows us to model and mechanically prove properties of a hybrid system using this new logic.

5. DEVELOPING A SPECIFICATION

We are interested in proving the following critical safety property for our system:

> For any configuration of virtual fixture boundaries, if the surgeon starts the tool at a safe place, for all possible uses of the robot, at each point in time, the tool remains in a safe place.

There are many cases where developing a specification is difficult, because the behavior we want is complicated, or the language we use to describe it is not sufficiently expressive. Neither of these is the case for our particular problem. The property is simple, and our logic is sufficiently expressive. The challenge is how to prove it.

If we call the model of our control algorithm "ctrl," and we have a single boundary i with a unit normal to the boundary $\hat{n}_i$ pointing in the safe direction, a point $\bar{r}_i$ on the boundary, then a tool at position $\bar{p}$ would be safe if $(\bar{p}-\bar{r}_i)\cdot\hat{n}_i \geq 0$. We will call the preceeding expression safe_i.

Thus for the ith boundary, our safety property would be expressed as:

$$\text{safe}_i \rightarrow [\text{ctrl}]\text{safe}_i \tag{3}$$

An implication with a precondition in the implicant, and a modality in the implicand is a common idiom in differential dynamic logic used to represent safety properties. Our safety property says if we start in the safe location for boundary i, and you run the robot using ctrl, then at every point in time, you will also be in a safe location.

If the surgeon specifies 3 virtual fixtures, we could state the safety property as $(\text{safe}_1 \wedge \text{safe}_2 \wedge \text{safe}_3) \rightarrow [\text{ctrl}](\text{safe}_1 \wedge \text{safe}_2 \wedge \text{safe}_3)$. In fact, we considered and verified such properties of the SBS system with a fixed number of virtual fixtures in KeYmaera. But the surgeon can specify any number of virtual fixtures as input, so we would then have to reverify the system if the surgeon ever decided to use 4 or more fixtures. The same problem arises if we consider 10 boundaries or any other fixed number instead.

In Qd$\mathcal{L}$, however, we can develop a single model for an arbitrary number of boundaries represented, e.g., as a linked list. We can traverse boundaries of interest with an uninterpreted function next(), that when applied consecutively to its result, identifies a list of boundaries, and a hybrid program that iterates over the different boundaries in that list: $[i := \text{first}; (i := \text{next}(i))^*]((i \neq \text{end}) \rightarrow \text{safe}_i)$. Unlike most hybrid programs, which represent a model of a hybrid system, this one's sole purpose for existence is to span the list provided to us by the next() function and terminated by a constant symbol end (satisfying next(end) = end).

For a finite, end-terminated list of boundaries defined by next(), we are interested in proving the following property saying that ctrl safely respects all boundaries always:

$$\begin{aligned}&[i := \text{first}; (i := \text{next}(i))^*]((i \neq \text{end}) \rightarrow \text{safe}_i) \rightarrow \\ &\quad [\text{ctrl}][i := \text{first}; (i := \text{next}(i))^*]((i \neq \text{end}) \rightarrow \text{safe}_i)\end{aligned} \tag{4}$$

What remains is developing the model ctrl and proving that it matches this specification.

6. DEVELOPING A MODEL

In this section, we describe the different steps we took to create an accurate model of the system. This means we will start with simplifying assumptions, create a model of our control algorithm, and show how we refine our model to eventually prove the safety of our system.

Why not simply jump straight to the final result? Because the process of developing and refining a model helps identify problems in the design and can lead to changes in the control algorithm we are modeling. Describing the modeling process illustrates the strategy we use to create these models, and provides general lessons that are applicable to many other hybrid system modeling and verification tasks.

Model development is an integral part of formal verification, and how to get to the right model is not always obvious. For this modeling effort we are not sure what the final model will look like, so we will step back and break the problem into two pieces: first, we must create the model of the control algorithm for the safety and force feedback of a single boundary, and second, we must add code to our single-boundary control model that applies that control to many boundaries sequentially, and produces a combined effect that ensures safety for a finite collection of boundaries.

6.1 Event-Driven Continuous Control

The first attempt at modeling our SBS system is a direct translation of the continuous equations in Eqn. 1. This equation implicitly assumes a sort of continuous control, where our system responds infinitely fast and infinitely often; it will lead to what is sometimes called an *event-driven* system, since it responds instantaneously to events.

There are many analog control circuits that do exhibit behavior that is nearly continuous control. In fact, the underlying control circuit that implements admittance control for the SBS robot is one example, assuming its step response is sharp and its settling time is short compared to the other time scales in the system.

Because the admittance control circuit can be modeled by continuous control, it may be tempting to think that a continuously controlled, event-driven model is a good representation for the virtual fixture control algorithm. It is not. The virtual fixture control algorithm cannot be practically implemented as an analog circuit, because K must be set in a manner consistent with both the current geometric state, but also the currently configured set of virtual fixture boundaries. This means the part of our system that computes K will be digital, and can introduce a significant delay in our response.

This is a very general lesson for anyone who is developing complex control algorithms. The lower level control circuits may be analog, and often can be modeled and analyzed very well by continuous control, and the control and circuit theory that we have available in our toolbox. But when building up more complicated control behaviors at higher levels of system abstraction, those behaviors may require more computation, and it may only be practical to implement them using digital components. In this case, modeling the control algorithm using continuous, event-driven control is a poor approximation that may lead to inaccurate conclusions.

Regardless, this sort of approximation is useful because it provides a quick sanity check of our basic concept under ideal conditions; if our control algorithm does not work correctly under continuous control, we cannot expect it to work under a more realistic model that incorporates delay. We will illustrate the utility of this initial step for model development by creating a model of control for a single boundary. Attempting to apply formal methods to this model will force us to refine the model, and eventually identify a problem with our initial control system design.

Our initial modeling attempt begins by writing pseudo code for a hybrid program that represents our control algorithm. There will be a section of code that assigns values to state variables in the system, and we can call this block "input." It is followed by a non-deterministic choice of different statements describing continuous evolution according to a system of differential equations directly taken from the admittance control equation. Each statement corresponds to a mode of operation, describing that mode's continuous behavior, as well as constraints that must hold in that case. Our pseudo-code looks like this:

```
ctrl ≡ (input;
        ((mode 1 diff eqn & mode 1 constraints) ∪
         (mode 2 diff eqn & mode 2 constraints) ∪
         (mode 3 diff eqn & mode 3 constraints) ∪
         ...
         (mode n diff eqn & mode n constraints)))*
```

This defines an event-driven controller, because it will switch from one mode to the other as described by the event of moving from one evolution domain constraint to another. The semantics of Qd$\mathcal{L}$ are nondeterministic, such that the system can switch between overlapping evolution domain constraints at any time. In our system, all evolution domain constraints are disjoint except for their overlapping mode boundaries, so that the system switches over to the other modes exactly at those boundary events.

We can now convert this pseudo-code into an algorithm by adding details. For simplicity, we will first model the controller in two dimensions, and fix a single virtual fixture boundary to the x-axis, with the safe side being quadrants one and two of our cartesian coordinate system, where y is positive. Next, we will fill in the details for "input." From the physician's perspective, the interaction with the system is via simple hand motions, exerting force on the tool to move it around. We model this motion using the nondeterministic assignment to the derivative of the force, creating a piecewise linear model of the force. We use $\bar{f}_p = (f_{xp}, f_{yp})$ to represent the derivative of force with respect to time, and write $f_{xp} := *$ and $f_{yp} := *$ to indicate non-deterministic assignment. (Because our mechanization does not have any way to represent vector quantities, we have to decompose our system into scalar equations.) We relate these quantities to the force by including the appropriate differential equations $f'_x = f_{xp}, f'_y = f_{yp}$ during continuous evolution of our system. Finally, we fill in the differential equations using the basic physics of our admittance control circuit and damping mechanism, given in Eqs. 1 and 2. As in Eqn. 1, $\bar{p}$ represents the position of the tooltip, $\bar{p}'$ is its derivative with respect to time, and $\bar{f}$ represents the force that the surgeon exerts on the system at a given point in time. The logical constraints that are provided during continuous evolution distinguish the different input cases. The refined hybrid program is given in Table 1.

In this simple model, we can see the different cases that our robot controls for, to make a single boundary safe. For

$$
\begin{aligned}
&\text{ctrl}_1 \equiv \big(f_{xp} := *; f_{yp} := *;\\
&/*\ \text{free zone}\ */\\
&(p_x' = Gf_x, p_y' = Gf_y, f_x' = f_{xp}, f_y' = f_{yp} \& (p_y \geq D)) \cup\\
&/*\ \text{slow zone}\ */\\
&(p_x' = Gf_x, p_y' = G\tfrac{p_y}{D} f_y, f_x' = f_{xp}, f_y' = f_{yp} \&\\
&\quad ((0 \leq p_y \leq D) \wedge (f_y \leq 0))) \cup\\
&(p_x' = Gf_x, p_y' = Gf_y, f_x' = f_{xp}, f_y' = f_{yp} \&\\
&\quad ((0 \leq p_y \leq D) \wedge (f_y \geq 0))) \cup\\
&/*\ \text{past boundary}\ */\\
&(p_x' = 0, p_y' = 0, f_x' = f_{xp}, f_y' = f_{yp} \& ((p_y \leq 0) \wedge (f_y \leq 0))) \cup\\
&(p_x' = 0, p_y' = Gf_y, f_x' = f_{xp}, f_y' = f_{yp} \& ((p_y \leq 0) \wedge (f_y \geq 0)))\big)^*
\end{aligned}
$$

Table 1: Simple, event-driven model exhibiting continuous control for a single virtual fixture in two dimensions.

example, there are two cases in the slow zone, one for moving towards the boundary, and one for moving away from it.

For this system, safety means that we do not move past the virtual fixture boundary. We state our safety property as ensuring that the tool tip, located at $\bar{p}$, will never go into the bad region below the x-axis, so $\text{safe}_1 = (p_y \geq 0)$. We also define some sanity requirements for our system, $\text{sane} = (G > 0) \wedge (D > 0)$ to formalize assumptions about the different parameters.

Theorem 1. *The event-driven SBS control system in Table 1 is safe for a single boundary in two dimensions, i.e., the following* d$\mathcal{L}$ *formula is provable:*

$$safe_1 \wedge sane \rightarrow [ctrl_1]safe_1 \tag{5}$$

We were able to mechanize and prove the safety of this system, using KeYmaera, a mechanization of d$\mathcal{L}$.

It is tempting to stop our modeling exercise here, because it seems obvious that this algorithm will be safe for multiple boundaries. Here, we ended up reusing a continuous model appropriate for the design of linear, time-invariant systems, for formal verification of a high level control algorithm that is a hybrid system. For this system, the assumption of continuous control is not appropriate.

Generalizing to Multiple Boundaries.

When we try to extend our formal model to include an arbitrary but finite number of multiple boundaries, we fail; this modeling approach can (with difficulty) be scaled to a higher number of boundaries, but it cannot represent the general case of any number of boundaries in a single model.

To use this modeling approach for an arbitrary number of boundaries, we would have to embed a looping program within different modes describing the continuous dynamics of the system, in order to control for all of the boundaries and their overlapping regions. The semantics of Qd$\mathcal{L}$ and d$\mathcal{L}$ do not allow this, because it would represent the execution of a discrete program instantaneously at every instant of time; this is more than physically impossible to implement.

In the process of formalizing an accurate model of our system, we discover something inaccurate about our model that is dangerous to our conclusions of its safety; our modeling language tells us so by prohibiting what we are trying to do. As discussed above, we have implicitly extended the assumption of continuous control to the implementation of the gain term $K(\bar{p})$, which contains discrete aspects of the system that switch its form, giving the system its hybrid flavor. This system needs a more accurate modeling approach to correctly ensure its safety.

6.2 Improved Delayed-Response Control

To relax the idealized assumptions of the event-driven model, and allow us to represent an arbitrary but unspecified number of boundaries in our system, we will model the control algorithm with delay in the control loop. When we create the software that implements this control algorithm, it will contain a loop that senses its state and the current inputs, makes a decision, and then sends commands to the robotic machinery to respond. This cycle of decision-making introduces some finite (possibly variable) delay, with an upper bound we call ϵ, in the response of the robot. Modeling this delay is appropriate because it more closely matches the system behavior, in that the robot takes some action, and then cannot react again to the external world for some period of time up to ϵ; see Figure 3. For the robot system used in [18], $\epsilon = 18.2$ msec.

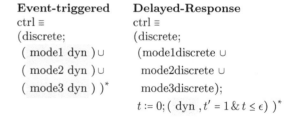

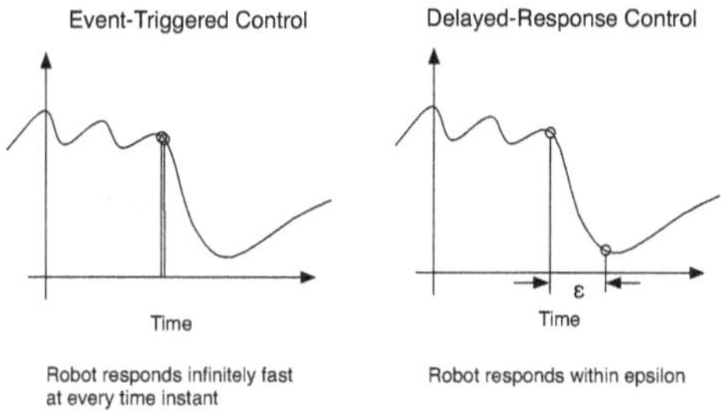

Figure 3: Event-triggered vs. Delayed-Response Control

Converting our event-triggered model to one with a delayed response requires the opposite of what we contemplated doing in the previous section. Rather than embedding a looping program into the continuous dynamics to represent discrete behavior, we must refactor the hybrid program, rewriting it so that the differences in the modes are expressed in the discrete program at the beginning of each continuous evolution. This refactoring is illustrated in Fig. 4.

Once this is done, there will only be one set of differential equations that describes the continuous evolution of our system, and it will simply represent the physics of the system

Event-triggered	**Delayed-Response**
ctrl ≡	ctrl ≡
(discrete;	(discrete;
(mode1 dyn) ∪	(mode1discrete ∪
(mode2 dyn) ∪	mode2discrete ∪
(mode3 dyn))*	mode3discrete);
	$t := 0; (\ \text{dyn}\ , t' = 1 \& t \leq \epsilon)\ )^*$

Figure 4: Converting a model that uses event-triggered control into a model that has a delayed response. Refactoring differences between different modes into a discrete program component, leaving our model with a single term describing the continuous evolution of the system with an extra clock t bounded by ϵ.

and the behavior of our lower-level admittance controller, which should not change regardless of the mode the system is in or the damping decision made by the control algorithm.

6.2.1 First Iteration: Immediate Control

We rewrite Eqn. 1 to match Eqn. 2, the implementation described in [5], for which we wish to prove safety:

$$\bar{p}'(t) = G(\bar{f}(t) - k\hat{n}) \tag{6}$$

In doing so, we define a state variable k that will be used to describe the amount of force-feedback in the direction normal to the boundary. The value of k is computed and set at each step by the discrete program, and used as a constant in the ODE during continuous evolution. Incidentally, this rewriting of Eqn. 1 will become a problem, as it changes the functional form of our equation, and prevents us from generalizing to multiple boundaries.

For our first iteration, we write a discrete program to set the value of k according to the current value of the force and position at the beginning of each step, implement the differential equation above, and use the approach described in Fig. 4 to allow up to ϵ time to elapse before our next control decision. This approach runs into immediate problems, which are evident even without doing a formal proof: Since the system may now be delayed by up to ϵ, we have introduced the possibility that the tool moves so quickly that it crosses the buffer of size D in some time less than ϵ. Whether this occurs in practice will depend on the values of D, ϵ, and the maximum velocity the robot allows. We can prove this if it is not sufficiently obvious.

6.2.2 Second Iteration: Predictive Control

To solve this problem, we must redesign the algorithm so that it anticipates the motion in the time step of duration $\leq\epsilon$, so that it can avoid violating the safety conditions. Rather than reacting only in the fixed forced-feedback zone, we will anticipate, based on the tool's position, its velocity, and its acceleration, whether we need to oppose its motion to prevent it from exceeding the virtual fixture we have defined. Force feedback can be applied if and when the tool is safe, in the slow (force-feedback) zone, and moving slowly enough to provide useful feedback to the user. At this point, our redesign is focused on safety only, since we need to get this right before we add force feedback.

The rest of this section describes the details of this design, which requires: computing the characteristics of the path the tool takes during the time step that represents the robot's delay; enumerating input cases, and computing k so that the robot's behavior remains safe; and proving that the design does in fact ensure the safety we expect.

Calculating the Tool's Path.

In order to create a predictive version of the algorithm, we must calculate the path of the tool with respect to the boundary. The force exerted on the tool at a point in the time step can be written as the initial force at the beginning of a step (subscript 0), plus the derivative of the force times time, and each of these can further be broken into x and y coordinates:

$$\bar{f}(t) = \bar{f}_0 + \bar{f}_p t = (f_{x0} + f_{xp}t)\hat{x} + (f_{y0} + f_{yp}t)\hat{y} \tag{7}$$

where $\hat{x}$ and $\hat{y}$ are unit normals in the x and y directions.

By calculating force components normal to the boundary, we can relate these quantities to the frame of reference of the boundary, and more easily model a boundary at an arbitrary location and an arbitrary orientation. We can write f_n, the force normal to the boundary, as the initial normal force f_{n0} plus the derivative f_{np} of the normal force times time

$$f_n(t) = \bar{f}(t) \cdot \hat{n} = (\bar{f}_0 \cdot \hat{n}) + (\bar{f}_p \cdot \hat{n})t = f_{n0} + f_{np}t \tag{8}$$

By solving the ODE in Eqn. 6, we find the position $\bar{p}$ of the tooltip at any time t during the step, and then we can use this to find the distance $d(t)$ from the boundary:

$$\begin{aligned} \bar{p}(t) &= \bar{p}_0 + G(\bar{f}_0 - k\hat{n})t + \tfrac{1}{2}G\bar{f}_p t^2 \\ d(t) &= (\bar{p}(t) - \bar{r}) \cdot \hat{n} \\ &= d_0 + G(f_{n0} - k)t + \tfrac{1}{2}G f_{np} t^2. \end{aligned} \tag{9}$$

We need to look at the undamped case to know whether we need the robot to apply damping. The discriminant of the quadratic describing the distance will tell us whether our parabolic trajectory without damping intersects the boundary we are interested in; if disc $= (Gf_{n0})^2 - 2Gf_{np}d_0$ is positive, we know that we may intersect the virtual fixture during the next ϵ time step.

Calculating Safe k for Different Input Conditions.

Now we need to use Eqn. 9 to calculate a k that ensures safety throughout the time step of duration $\leq\epsilon$. We want to damp it safely, but not unnecessarily, depending on the specifics of the motion that is being made. For example, if the tool's motion is away from a boundary, we don't need any damping, unless the acceleration eventually reverses its direction and has it intersecting the boundary during this time step, in which case we do. There are various cases for this type of motion, and we have to compute a safe value of k for each case, and then use the results during the execution of our hybrid program.

The cases are shown in Fig. 5, and the calculations for k used in the model for each case follow. How do we know that we have safely addressed all of the cases? This is one of the benefits of a formal approach. We think carefully about the cases, enumerate them, and then calculate a safe k depending on the behavior we seek. Then, we attempt to prove the safety property. If we have missed any cases, or miscalculated k, the proof will fail and lead to a counterexample of whatever cases we have missed. In this way, verification can support the design process for this control algorithm.

There are four cases where the robot should not do anything, because we will not hit the boundary during this time step. These four cases correspond with the subfigures a–d in Fig. 5, and for these cases, the robot will not add any damping; we represent this in our model by setting $k := 0$ for these cases.

(a) We are moving away from the boundary ($f_{n0} \geq 0$), and accelerating away from the boundary ($f_{np} \geq 0$).

(b) We are moving towards the boundary ($f_{n0} \leq 0$), but accelerating away from the boundary ($f_{np} \geq 0$), so that we turn around before we reach it (disc ≤ 0).

(c) We are moving towards the boundary ($f_{n0} \leq 0$), but accelerating away from the boundary ($f_{np} \geq 0$) so that we will intersect it (disc ≥ 0), but not during this time step ($d(\epsilon) \geq 0$ and $f_n(\epsilon) \leq 0$), so no action is necessary yet.

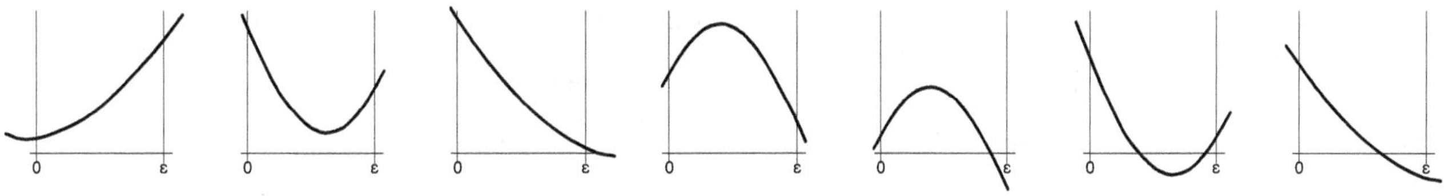

(a) (b) (c) (d) (e) (f) (g)

Figure 5: Each subfigure represents a movement scenario in which the skull-base surgery robot must enforce safety. The y-axis of each subfigure represents the distance of the closest approach of the tool to a single virtual fixture boundary, i.e. $d(t)$. The x-axis of each subfigure represents the progression of time, with a zero-reference being assigned to the beginning of a step, progressing to the maximum delay, ϵ. The final, safe version of the realistic controller contains each of these scenarios as an explicitly controlled case.

(d) We are accelerating towards the boundary ($f_{np} \leq 0$) but we never reach it during our step ($d(\epsilon) \geq 0$).

There are three cases where the robot must interfere, to prevent the tool-tip from crossing the virtual fixture. These remaining three cases correspond with the subfigures e–g in Fig. 5. The logic and calculations for k follow for each case:

(e) We are accelerating towards the boundary ($f_{np} \leq 0$), and we would reach it or exceed it at the end of our time step (i.e. $[k := 0]d(\epsilon) \leq 0$[1]) unless additional damping is applied.

Since we are accelerating towards the boundary, we know the furthest we can reach past the boundary will be at time ϵ. To ensure that we do not cross the boundary, we find a damping value that just slows us enough so that we touch but do not cross the boundary at this time, by solving $d(\epsilon) = 0$ for k. We find $k := f_{n0} + \frac{(d_0 + \frac{1}{2} G f_{np} \epsilon^2)}{(G\epsilon)}$ to ensure safety, so we set that value as k at the beginning of the step for this case.

(f) We are moving towards the bounday ($f_{n0} \leq 0$) but accelerating away from the boundary ($f_{np} \geq 0$), and we will still intersect it (disc ≥ 0) during this time step. We know this because the trajectory has turned around ($f_n(\epsilon) \geq 0$) and reached its minimum.

Here, we need to ensure that the minimum value just touches the boundary, so that no part of the path actually crosses it. We compute the time where our parabolic trajectory reaches its minimum, $t_m = \frac{-f_{n0}}{f_{np}}$; this is the furthest past the boundary that our path reaches. Now we can require that the discriminant of the resulting quadratic equation be zero, and solve for the damping that would be required to create that path. We find $k := f_{n0} - \sqrt{\frac{2 f_{np} d_0}{G}}$ would be the damping required to keep this safe.

(g) We are moving towards the bounday ($f_{n0} \leq 0$) but accelerating away from the boundary ($f_{np} \geq 0$), but we will still intersect it (disc ≥ 0) during this time step. We know this because even though the trajectory is still downward, ($f_n(\epsilon) \leq 0$), we have exceeded the boundary by the end of the step ($[k := 0]d(\epsilon) \leq 0$).

This trajectory is monotonic during our step and moving towards the boundary. The position $[k := 0]d(\epsilon)$ thus represents the furthest possible distance we can go past the boundary during the step, without damping.

By applying damping and opposing the downward trajectory, the minimum will move to an earlier time. Without damping, the time at which our parabolic trajectory reaches a minimum is beyond our time step, $t_m \geq \epsilon$, but by changing the damping k, we may find it has moved so that $t_m < \epsilon$. If the t_m occurs during our ϵ step, we need to produce a force that ensures that the minimum point is raised so that our path does not intersect the boundary. We must solve $d(t_m) = 0$ for k, and in this case, we find that $k := f_{n0} - \sqrt{\frac{2 f_{np} d_0}{G}}$ is a safe damping coefficient, as in case f. If the k computed above still leaves $t_m \geq \epsilon$, then this correction is more than is necessary. We can solve $d(\epsilon) = 0$ for k, and exactly as in case e we find that a damping of $k := f_{n0} + \frac{(d_0 + \frac{1}{2} G f_{np} \epsilon^2)}{(G\epsilon)}$ keeps our trajectory safe.

[1]We use notation from differential dynamic logic to indicate the setting the state variable k to a value in the following calculations. So $[k := 0]d(\epsilon) \leq 0$ is the assertion that with the value of k set to zero, evaluating the equation $d(\epsilon)$ yields a value less than zero.

Proving (Un)Safety.

We have been able to quickly produce a design that accurately represents lag in the controller and the linear damping described in [18, 5], and can be applied to multiple boundaries. The modeling effort and our formal approach has led us to make some modifications to the control algorithm to improve safety, namely we redesigned the algorithm so that it damps its movements predictively, taking into account time lag in the system.

We are able to use KeYmaera to prove the safety of this design for a single boundary, but when we implement multiple boundaries in KeYmaeraD, this algorithm is in general unsafe, and we can prove it with a counterexample. We will present an informal proof by constructing a description of an algorithm that depends upon sequential application of Eqn. 2 for each boundary, and pointing out that any algorithm that makes these assumptions produces a counterexample when one tries to prove its safety.

Consider a set of boundaries, C. For each boundary $i \in C$, there is a a normal $\hat{n}_i$, and a distance of the tool to the closest point on that boundary d_i. We write an assertion,

Q_i, that encodes what it means to be safe for boundary i:

$$Q_i(\bar{w}, \bar{v}) \equiv \left(-\bar{w} \cdot \hat{n}_i \leq -\bar{v} \cdot \hat{n}_i \frac{d_i}{D}\right) \vee (\bar{w} \cdot \hat{n}_i \geq 0) \quad (10)$$

In this assertion, the quantity $\bar{v}$ represents the velocity of the tool tip before we exert any control, and $\bar{w}$ represents the velocity of the tool tip, after we have exerted control. The satisfaction of $Q_i(\bar{w}, \bar{v})$ describes the exerting of safe control for boundary i, when the initial velocity without control would have been $\bar{v}$, by modifying the final velocity to be $\bar{w}$. In other words, $Q_i(\bar{w}, \bar{v})$ is satisfied when both $\bar{w}$ and $\bar{v}$, have components normal to the boundary i that are towards the boundary, and $\bar{w}$'s normal component is attenuated in proportion to its distance from the boundary, *or* when $\bar{w}$ has its component normal to the boundary going away from the boundary.

We assume C_s is a proper subset of C, and $\bar{w}$ is a velocity such that $\forall i \in C_s, Q_i(\bar{w}, \bar{v})$. This means that all forces have been attenuated properly for the boundaries indexed in C_s.

We wish to construct an algorithm which when given $\bar{v}$, and $\bar{w}$, computes a new velocity $\bar{x}$ for each index $j \in C \setminus C_s$ such that,

$$\forall i \in C_s, Q_i(\bar{x}, \bar{v}) \wedge Q_j(\bar{x}, \bar{v}). \quad (11)$$

By induction, such an algorithm could then be applied to each successive boundary to compute the resulting, force-feedback velocity that is safe for all virtual fixture boundaries.

An example of an algorithm that makes these assumptions follows:

$$\bar{x} = \begin{cases} \bar{w} + \left[\left(\frac{d_j}{D}\bar{v} - \bar{w}\right) \cdot \hat{n}_j\right]\hat{n}_j & \text{if } \bar{w} \cdot \hat{n}_j \leq 0 \wedge \bar{v} \cdot \hat{n}_j \leq 0 \wedge \\ & \left(-\bar{w} \cdot \hat{n}_j \geq -\bar{v} \cdot \hat{n}_j \frac{d_j}{D}\right) \\ \bar{w} & \text{if } \bar{w} \cdot \hat{n}_j \leq 0 \wedge \bar{v} \cdot \hat{n}_j \leq 0 \wedge \\ & \left(-\bar{w} \cdot \hat{n}_j \leq -\bar{v} \cdot \hat{n}_j \frac{d_j}{D}\right) \\ \bar{w} & \text{if } \bar{w} \cdot \hat{n}_j \geq 0 \\ \bar{w} - (\bar{w} \cdot \hat{n}_j)\hat{n}_j & \text{if } \bar{w} \cdot \hat{n}_j \leq 0 \wedge \bar{v} \cdot \hat{n}_j \geq 0 \end{cases} \quad (12)$$

The first case simply subtracts out the component of tool velocity normal to the boundary that would be necessary to attenuate it proportional to how close the tool is to the boundary. The second and third cases do nothing, as the normal is already attenuated due to transformations from other boundaries. The fourth case represents conditions where other boundaries are producing attenuating velocity in opposition to what is necessary to attenuate the normal component of velocity with respect to the current boundary.

When attempting a safety proof for this algorithm, we have to prove Eqn. 11 by showing the safety of different cases in Eqn. 12. We ended up disproving one of the cases by finding a counterexample during an attempt to complete the safety proof using KeYmaera. In particular, we found a counterexample in the fourth case where:

$$\begin{aligned} &\{Q_i(\bar{w}, \bar{v}), \bar{w} \cdot \hat{n}_j \leq 0, \bar{v} \cdot \hat{n}_j \geq 0, \hat{n}_i \cdot \hat{n}_j \leq 0\} \nvdash \\ &\quad Q_i(\bar{x}, \bar{v}) \wedge Q_j(\bar{x}, \bar{v}), \end{aligned} \quad (13)$$

violating our specification. The term $\hat{n}_i \cdot \hat{n}_j \leq 0$ describes the subset of the fourth case where the counterexample occurs. The counterexample occurs when boundaries intersect each other at acute angles, and the dot product of their normals is a negative quantity, i.e., they partially face each other.

The existence of this counterexample means that there are geometric configurations of multiple boundaries where the correction introduced by the system for one boundary can push the tool past another virtual fixture boundary into an unsafe location. This means that the tool can "slip" past a virtual fixture boundary, through the edge formed by the intersection of two different virtual fixtures. This also means that for large velocity movements, the process of enforcing one virtual fixture boundary can violate another, sufficiently closely spaced boundary.

For multiple, arbitrarily configured boundaries, an algorithm based on these assumptions is not in general safe.

6.2.3 Third Iteration: Non-linear Damping

We wish to make our system safe, yet produce a directional damping so that motions away from the boundary are not attenuated. Because of the problems described above, we revert back to damping the system with a multiplicative factor as in Eqn. 1. Multiplicative damping ensures that damping from one boundary will not be reversed by further damping from another. Still using the notation from Eqns. 7–8, we will revisit the design using the formal approach we have adopted, which will help us ensure safety as we design the directional damping we seek. As before, we start with

$$\bar{p}' = K(\bar{p})G(\bar{f})\bar{f} \quad (14)$$

but we choose a form for K appropriate to non-linear damping, $K = \left(\begin{smallmatrix} K_x & 0 \\ 0 & K_y \end{smallmatrix}\right)$.

The rest of this section follows the structure of Section 6.2.2, computing the characteristics of the path the tool takes during the time step, enumerating input cases, and computing safe damping coefficient, and finally proving that the design does in fact ensure the safety we expect.

Calculating the Tool's Path.

We can now compute the position of the tool-tip during each step and its distance from a boundary, as before, first solving Eqn. 14, and then computing the distance during the time step as before.

$$\begin{aligned} \bar{p}(t) &= \bar{p}_0 + G(K_x f_{x0}\hat{x} + K_y f_{y0}\hat{y})t + \\ &\quad \tfrac{1}{2}G(K_x f_{xp}\hat{x} + K_y f_{yp}\hat{y})t^2 \\ d(t) &= (\bar{p}(t) - \bar{r}) \cdot \hat{n} \\ &= d_0 + G(K_x f_{x0}n_x + K_y f_{y0}n_y)t + \\ &\quad \tfrac{1}{2}G(K_x f_{xp}n_x + K_y f_{yp}n_y)t^2 \end{aligned} \quad (15)$$

In this case, the discriminant will tell us whether our parabolic trajectory intersects the boundary we are interested in. The discriminant is given by disc $= G^2(K_x f_{x0}n_x + K_y f_{y0}n_y)^2 - 2G(K_x f_{xp}n_x + K_y f_{yp}n_y)d_0$. The turn-around point for this trajectory is at $t_m = \frac{-(K_x f_{x0}n_x + K_y f_{y0}n_y)}{(K_x f_{xp}n_x + K_y f_{yp}n_y)}$ as long as $(K_x f_{xp}n_x + K_y f_{yp}n_y) \neq 0$.

Calculating Safe k for Different Input Conditions.

We go through the exercise of computing the control for each input scenario in Fig. 5, as in Sec. 6.2.2. We assume $k = K_x = K_y$. The only differences are in cases e, f, and g.

(e) Solving for $d(\epsilon) = 0$, we find $k := -\frac{d_0}{G}\left(f_n\epsilon + \frac{1}{2}f_{np}\epsilon^2\right)^{-1}$.

(f) For our system, the minimum point is thus at $t_m = \frac{-(K_x f_{x0}n_x + K_y f_{y0}n_y)}{(K_x f_{xp}n_x + K_y f_{yp}n_y)}$. With $K_x = K_y$, $t_m = \frac{-f_{n0}}{f_{np}}$. We can require that the the discriminant of the resulting quadratic equation be zero, and solve it, finding that $k := \frac{2f_{np}d_0}{Gf_{n0}^2}$ provides appropriate damping to guarantee a safe trajectory.

(g) In general, the trajectory reaches its minimum point at $t_m = \frac{-f_{n0}}{f_{np}}$. It is sufficient to ensure $d(\epsilon) = 0$ exactly as in case e. We find $k := -\frac{d_0}{G}\left(f_n\epsilon + \frac{1}{2}f_{np}\epsilon^2\right)^{-1}$ safely damps the system, as before.

We also add force feedback into this design, as described in [18], defining a slow zone that is between the virtual fixture boundary, and up to a distance D away. We attenuate the tool's movement by a factor proportional to how far into the slow zone the tool has progressed, measured by its distance from the boundary divided by the total thickness of the slow zone, i.e. $\frac{d(t)}{D}$. The tool will slow down more and more as it approaches the boundary. We apply force feedback only when the tool's path primarily moves towards the virtual fixture, and when its starting point is within the slow zone. If after applying attenuation that provides force feedback, we would cross a virtual fixture boundary, we can add further attenuation to those cases, as described above, to ensure that the system remains safe.

Careful modeling of our robot and the application of formal methods have led us to a design of a single-boundary control algorithm that satisfies our original requirements, accurately represents lag, predictively enforces safety, and is *composable* so that it can be applied to multiple boundaries in sequence, and still ensure a safe system.

We can now complete the model of our redesigned control algorithm by taking the single-boundary control program and embedding it in two loops, one nested inside the other. The outer loop models the controller's interaction with the physical world, describing force input from the surgeon, and the evolution of continuous time after the control algorithm has provided force feedback and enforced safety. The nested inner loop models the control algorithm being applied successively to each boundary; each iteration of the inner loop computes and applies force feedback for, and enforces the safety of, a single arbitrarily oriented boundary. The nested inner loop iterates over all of the boundaries. The final model is given in Table 2.

Safety Proof.

After our careful modeling effort, we were able to mechanize the safety proof in KeYmaeraD.[2]

THEOREM 2. *The SBS controller in Table 2 is safe for an arbitrary number of arbitrarily oriented and positioned boundaries in three dimensions, i.e., the following* QdL *formula is valid:*

$$(G > 0) \land (e > 0) \land (D > 0) \land (k \geq 0) \land (\mathit{next}(\mathit{end}) = \mathit{end}) \land (\forall i : B, |\hat{n}(i)|^2 = 1) \land (\forall (h\ g : B), (\mathit{next}(h) \neq \mathit{end}) \rightarrow (h \neq g) \rightarrow (\mathit{next}(h) \neq \mathit{next}(g))) \land [i := \mathit{first}; (i := \mathit{next}(i))^*]((i \neq \mathit{end}) \rightarrow \mathit{safe}_i) \rightarrow [\mathit{ctrl}_2][i := \mathit{first}; (i := \mathit{next}(i))^*]((i \neq \mathit{end}) \rightarrow \mathit{safe}_i)$$

The proof is structured the same way as the model: with two inductive steps, one nested inside the other. Both inductive reasoning steps have the three branches: one representing their base cases, one for the inductive step, and the last for the postcondition. The middle "inductive step" of the first application of induction leads to the second application of loop induction. (This is the location of the nesting.) The second "inductive step" of the second application of inductive reasoning breaks out into one hundred and forty different branches, each representing slightly different conditions of force feedback and safety enforcement.

To see where the branches come from, we can examine the final model in Table 2. For each time step, the system: computes a damping coefficient to provide force feedback (ten cases); independently computes a damping coefficient to ensure safe operation enforcement of virtual fixture boundaries (seven cases discussed in Fig. 5); and then compares the damping coefficients and makes a decision whether to apply safety damping depending on their relative magnitudes. Each of these actions is shown in the model, and the product of these cases, $10 \times 7 \times 2 = 140$, gives us the total number of major subcases in the proof.

The main idea of the quantified proof is to show that for each branch in the inner inductive step, the application of damping for a given boundary, either by leaving the original damping in place, or in applying additional damping to provide force feedback or ensure safety, successfully ensures safety of that boundary. The proof must also show that when additional multiplicative damping factors are added, that additional damping does not impair the safety of the other boundaries that have been controlled-for so far.

The final, completed safety proof has 156,024 proof steps. The proof steps are described by a manually created proof script, that allows the user to describe the structure of the proof. The script is like a program that describes how to conduct the proof, detailing individual steps in parts of the proof, and automating the proving process in other parts, re-using proof strategies as appropriate. On a laptop, the 2D version of the model takes around two minutes to machine-check the proof, confirming its truth, while the full 3D version takes 70 minutes to complete, mostly waiting to automatically solve the arithmetic for certain unoptimized branches of the proof tree.

[2]The final models and proofs are available online at `http://symbolaris.com/pub/medrobot-examples.zip`

7. FUTURE WORK

One avenue of future work might be to explore how to address more complicated boundaries. The current approach could be extended by designing a control algorithm that dynamically adds and removes boundaries depending on where you are in the allowed operating volume. Boundaries can be safely removed at any time, and boundaries can be safely added at the beginning of a time step, providing that the tool tip is in a safe location with respect to the new boundary when it is added. One can conceive of extensions to the current algorithm that would simulate surfaces with convex topologies and cusps, by selectively adding and removing planar boundaries at appropriate times to stay inside one convex connected component at a time. This would be an interesting verification challenge.

8. REFERENCES

[1] J. Abbott, P. Marayong, and A. Okamura. Haptic virtual fixtures for robot-assisted manipulation. In S. Thrun, R. Brooks, and H. Durrant-Whyte, editors, *Robotics Research*, volume 28 of *Springer Tracts in Advanced Robotics*, pages 49–64. Springer, 2007.

[2] M. Fränzle and C. Herde. HySAT: An efficient proof engine for bounded model checking of hybrid systems. *Formal Methods in System Design*, 30:179–198, 2007.

$$
\begin{aligned}
&\text{ctrl}_2 \equiv \Big(f_{xp} := *; f_{yp} := *; f_{zp} := *; k := 1; i := \text{first};\\
&\quad (?i \neq \text{end};\\
&\quad\quad \text{xtr}(i) := *; u(i) := 1;\\
&\quad\quad d_O(i) := (p_x - r_x(i))n_x(i) + (p_y - r_y(i))n_y(i) +\\
&\quad\quad\quad (p_z - r_z(i))n_z(i);\\
&\quad\quad f_{np}(i) := f_{xp}n_x(i) + f_{yp}n_y(i) + f_{zp}n_z(i);\\
&\quad\quad f_n(i) := f_x n_x(i) + f_y n_y(i) + f_z n_z(i);\\
&\quad\quad ((?(d_O(i) \geq d \wedge (3 \geq 3)); // \text{ force feedback section}\\
&\quad\quad\quad u(i) := *; (?u(i)1 = 1)) \cup\\
&\quad\quad (?((0 \leq d_O(i)) \wedge (d_O(i) \leq d));\\
&\quad\quad ((?f_{np}(i) \geq 0 \wedge f_n(i) \geq 0);\\
&\quad\quad\quad u(i) := *; (?u(i)1 = 1)) \cup\\
&\quad\quad ((?f_{np}(i) > 0 \wedge f_n(i) \leq 0 \wedge\\
&\quad\quad (\text{xtr}(i)f_{np}(i) = -1 f_n(i)) \wedge (\text{xtr}(i) \leq (1/2)e));\\
&\quad\quad\quad u(i) := *; (?u(i)1 = 1)) \cup\\
&\quad\quad ((?f_{np}(i) > 0 \wedge f_n(i) \leq 0 \wedge\\
&\quad\quad (\text{xtr}(i)f_{np}(i) = -1 f_n(i)) \wedge (\text{xtr}(i) \geq (1/2)e));\\
&\quad\quad\quad u(i) := *; (?u(i)d = d_O(i))) \cup\\
&\quad\quad ((?f_{np}(i) = 0 \wedge f_n(i) \leq 0);\\
&\quad\quad\quad u(i) := *; (?u(i)d = d_O(i))) \cup\\
&\quad\quad ((?f_{np}(i) < 0 \wedge f_n(i) \geq 0 \wedge\\
&\quad\quad (\text{xtr}(i)f_{np}(i) = -1 f_n(i)) \wedge (\text{xtr}(i) \leq (1/2)e));\\
&\quad\quad\quad u(i) := *; (?u(i)d = d_O(i))) \cup\\
&\quad\quad ((?f_{np}(i) < 0 \wedge f_n(i) \geq 0 \wedge\\
&\quad\quad (\text{xtr}(i)f_{np}(i) = -1 f_n(i)) \wedge (\text{xtr}(i) \geq (1/2)e));\\
&\quad\quad\quad u(i) := *; (?u(i)1 = 1)) \cup\\
&\quad\quad ((?f_{np}(i) = 0 \wedge f_n(i) \geq 0);\\
&\quad\quad\quad u(i) := *; (?u(i)1 = 1)) \cup\\
&\quad\quad ((?f_{np}(i) \leq 0 \wedge f_n(i) \leq 0);\\
&\quad\quad\quad u(i) := *; (?u(i)d = d_O(i)))) \cup\\
&\quad\quad (?((d_O(i) \leq 0) \wedge (3 \geq 3));\\
&\quad\quad\quad u(i) := *; (?u(i)1 = 0)));\\
&\quad\quad \text{dist}(i) := (d_O(i) + Gu(i)(f_n(i)e + f_{np}(i)e^2(1/2)));\\
&\quad\quad \text{disc}(i) := ((Gu(i)f_n(i))^2 - 2Gu(i)f_{np}(i)d_O(i));\\
&\quad\quad ((?(f_{np}(i) \leq 0 \wedge \text{dist}(i) \geq 0); // \text{ safety section}\\
&\quad\quad\quad \text{c} := *; (?\text{c}G = 1); \text{kt} := u(i)) \cup\\
&\quad\quad (?(f_{np}(i) \leq 0 \wedge \text{dist}(i) \leq 0);\\
&\quad\quad\quad \text{c} := *; (?\text{c}G(f_n(i)e + (1/2)f_{np}(i)(e^2)) = 1);\\
&\quad\quad\quad \text{kt} := (-1)d_O(i)\text{c}) \cup\\
&\quad\quad ((?f_{np}(i) \geq 0 \wedge f_n(i) \leq 0 \wedge \text{disc}(i) \leq 0);\\
&\quad\quad\quad \text{c} := *; (?\text{c}G = 1); \text{kt} := u(i)) \cup\\
&\quad\quad ((?f_{np}(i) \geq 0 \wedge f_n(i) \leq 0 \wedge \text{disc}(i) \geq 0\\
&\quad\quad\quad \wedge f_n(i) + f_{np}(i)e \geq 0);\\
&\quad\quad\quad \text{c} := *; (?\text{c}G(f_n(i)^2) = 1);\\
&\quad\quad\quad \text{kt} := 2f_{np}(i)d_O(i)\text{c}) \cup\\
&\quad\quad ((?f_{np}(i) \geq 0 \wedge f_n(i) \leq 0 \wedge \text{disc}(i) \geq 0 \wedge\\
&\quad\quad f_n(i) + f_{np}(i)e \leq 0 \wedge \text{dist}(i) \leq 0);\\
&\quad\quad\quad \text{c} := *; (?\text{c}G(f_n(i)e + (1/2)f_{np}(i)(e^2)) = 1);\\
&\quad\quad\quad \text{kt} := (-1)d_O(i)\text{c}) \cup\\
&\quad\quad ((?f_{np}(i) \geq 0 \wedge f_n(i) \leq 0 \wedge \text{disc}(i) \geq 0 \wedge\\
&\quad\quad f_n(i) + f_{np}(i)e \leq 0 \wedge \text{dist}(i) \geq 0);\\
&\quad\quad\quad \text{c} := *; (?\text{c}G = 1); \text{kt} := u(i)) \cup\\
&\quad\quad ((?f_{np}(i) \geq 0 \wedge f_n(i) \geq 0);\\
&\quad\quad\quad \text{c} := *; (?\text{c}G = 1); \text{kt} := u(i)));\\
&\quad\quad ((?\text{kt} \geq k; k := k) \cup (?\text{kt} < k; k := \text{kt})); i := \text{next}(i)\\
&\quad)^*; (?i = \text{end}); t := 0;\\
&\quad (p_x' = Gf_x k, p_y' = Gf_y k, p_z' = Gf_z k,\\
&\quad f_x' = f_{xp}, f_y' = f_{yp}, f_z' = f_{zp}, t' = 1 \& t \leq e)\Big)^*
\end{aligned}
$$

Table 2: A complete time-triggered model of a redesigned control algorithm that enforces the safety of an arbitrary number of arbitrarily oriented and positioned virtual fixture boundaries, in three dimensions. This model is realistic, provides directional force feedback, and is proven to be safe.

[3] G. Frehse, C. L. Guernic, A. Donzé, S. Cotton, R. Ray, O. Lebeltel, R. Ripado, A. Girard, T. Dang, and O. Maler. SpaceEx: Scalable verification of hybrid systems. In *CAV*, pages 379–395, 2011.

[4] A. Kapoor, M. Li, and R. Taylor. Constrained control for surgical assistant robots. In *ICRA*, volume 1, pages 231–236, 2006.

[5] P. Kazanzides. Virtual fixture computation. Note on combining the effects of multiple virtual fixtures, Dec. 2011.

[6] P. Kazanzides, Y. Kouskoulas, A. Deguet, and Z. Shao. Proving the correctness of concurrent robot software. In *ICRA*, pages 4718–4723. IEEE, 2012.

[7] P. Kazanzides, J. Zuhars, B. Mittelstadt, and R. H. Taylor. Force sensing and control for a surgical robot. Number May, pages 612–617. IEEE, 1992.

[8] Y. Kouskoulas, F. Ming, Z. Shao, and P. Kazanzides. Certifying the concurrent state table implementation in a surgical robotic system (extended version). Technical report, Yale University, 2011.

[9] M. Li, M. Ishii, and R. Taylor. Spatial motion constraints using virtual fixtures generated by anatomy. *Robotics, IEEE Transactions on*, 23(1):4–19, 2007.

[10] S. Park, R. Howe, and D. Torchiana. Virtual fixtures for robotic cardiac surgery. In W. Niessen and M. Viergever, editors, *MICCAI*, volume 2208 of *LNCS*, pages 1419–1420. Springer, 2001.

[11] A. Platzer. Differential dynamic logic for hybrid systems. *J. Autom. Reas.*, 41(2):143–189, 2008.

[12] A. Platzer. Differential-algebraic dynamic logic for differential-algebraic programs. *J. Log. Comput.*, 20(1):309–352, 2010.

[13] A. Platzer. *Logical Analysis of Hybrid Systems: Proving Theorems for Complex Dynamics*. Springer, Heidelberg, 2010.

[14] A. Platzer. Quantified differential dynamic logic for distributed hybrid systems. In A. Dawar and H. Veith, editors, *CSL*, volume 6247 of *LNCS*, pages 469–483. Springer, 2010.

[15] A. Platzer. A complete axiomatization of quantified differential dynamic logic for distributed hybrid systems. *Logical Methods in Computer Science*, 8(4:13):1–44, 2012.

[16] A. Platzer and J.-D. Quesel. KeYmaera: A hybrid theorem prover for hybrid systems. In A. Armando, P. Baumgartner, and G. Dowek, editors, *IJCAR*, volume 5195 of *LNCS*, pages 171–178. Springer, 2008.

[17] D. W. Renshaw, S. M. Loos, and A. Platzer. Distributed theorem proving for distributed hybrid systems. In S. Qin and Z. Qiu, editors, *ICFEM*, volume 6991 of *LNCS*, pages 356–371. Springer, 2011.

[18] T. Xia, C. Baird, G. Jallo, K. Hayes, N. Nakajima, N. Hata, and P. Kazanzides. An integrated system for planning, navigation and robotic assistance for skull base surgery. *International Journal of Medical Robotics*, 4(4):321–330, 2008.

Quantitative Timed Simulation Functions and Refinement Metrics for Real-Time Systems*

Krishnendu Chatterjee
IST Austria (Institute of Science and Technology, Austria)
krishnendu.chatterjee@ist.ac.at

Vinayak S. Prabhu
University of Porto
vinayak@eecs.berkeley.edu

ABSTRACT

We introduce quantitative timed refinement metrics and quantitative timed simulation functions, incorporating zenoness checks, for timed systems. These functions assign positive real numbers between zero and infinity which quantify the *timing mismatches* between two timed systems, amongst non-zeno runs. We quantify timing mismatches in three ways: (1) the maximum timing mismatch that can arise, (2) the "steady-state" maximum timing mismatches, where initial transient timing mismatches are ignored; and (3) the (long-run) average timing mismatches amongst two systems. These three kinds of mismatches constitute three important types of timing differences. Our event times are the *global times*, measured from the start of the system execution, not just the time durations of individual steps. We present algorithms over timed automata for computing the three quantitative simulation functions to within any desired degree of accuracy. In order to compute the values of the quantitative simulation functions, we use a game theoretic formulation. We introduce two new kinds of objectives for two player games on finite state game graphs: (1) eventual debit-sum level objectives, and (2) average debit-sum level objectives. We present algorithms for computing the optimal values for these objectives for player 1, and then use these algorithms to compute the values of the quantitative timed simulation functions.

*This work has been financially supported in part by the European Commission FP7-ICT Cognitive Systems, Interaction, and Robotics under the contract # 270180 (NOPTILUS); by Fundação para Ciência e Tecnologia under project PTDC/EEA-CRO/104901/2008 (Modeling and control of Networked vehicle systems in persistent autonomous operations); by Austrian Science Fund (FWF) Grant No P 23499-N23 on Modern Graph Algorithmic Techniques in Formal Verification; FWF NFN Grant No S11407-N23 (RiSE); ERC Start grant (279307: Graph Games); and the Microsoft faculty fellows award

HSCC'13, April 8–11, 2013, Philadelphia, Pennsylvania, USA.

Categories and Subject Descriptors

F.4 [**Mathematical Logic and Formal Languages**]: Mathematical Logic—*Computability theory*

General Terms

Algorithms, Theory, Verification

Keywords

Timed Automata, Graph Games, Quantitative Simulation Relations, Robustness, Approximation Theories

1. INTRODUCTION

Theories of system approximation for continuous systems are used for analyzing systems that differ to a small extent, as opposed to the traditional boolean yes/no view of system refinement for discrete systems. These theories are necessary as formal models are only approximations of the real world, and are subject to estimation and modelling errors. Approximation theories have been traditionally developed for continuous control systems [2] and more recently for linear and non-linear systems [18, 15, 14], timed systems [16], labeled Markov Processes [12], probabilistic automata [20], quantitative transition systems [11], and software systems [10].

Timed and hybrid systems model the evolution of system outputs as well as the timing aspects related to the system evolution. In this work we develop a theory of system approximation for timed systems by quantifying the *timing differences* between corresponding system events. We first generalize timed refinement relations to metrics on timed systems that quantitatively estimate the closeness of two systems. Given a timed model T_s denoting the abstract specification model, and a model T_r denoting the concrete refined implementation of T_s, we assign a positive real number between zero and infinity to the pair (T_r, T_s) which denotes the quantitative refinement distance between T_r and T_s. Given a trace tr_r of T_r, and a trace tr_s of T_s, we define various distances between the two traces, e.g., the distance being ∞ if the untimed trace sequences differ, and being the supremum of the differences of the matching timepoints for matching events otherwise. Our event times are the *global times*, measured from the start of the system execution, not just the time durations of individual steps. The distance between the systems T_r and T_s is taken to be the supremum of closest matching trace differences from the initial states.

Timed trace inclusion is undecidable on timed automata [1], thus timed refinement is conservatively estimated using *timed simulation relations* [4]. Simulation relations take a branching time view, unlike the linear view

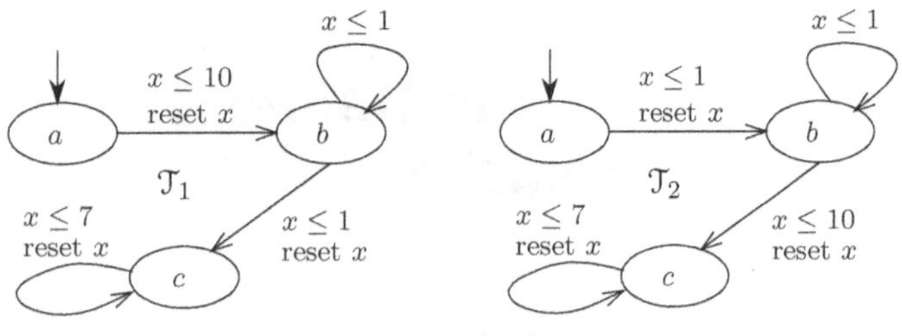

Figure 1: Two timed automata $\mathcal{T}_1, \mathcal{T}_2$.

of refinement relations, and can be defined using two player *games*. We generalize timed simulation relations to quantitative timed simulation functions, and define the values of quantitative timed simulation functions as the real-valued outcome of games played on the corresponding timed graphs.

Zeno runs where time converges is an artifact present in models of timed systems due to model imperfections; such runs are obviously absent in the physical systems which our timed models are meant to represent. We thus exclude Zeno runs in our computation of quantitative timed refinement and quantitative timed simulation relations.

We define three illustrative quantitative timed simulation functions which measure three important system differences. The *maximum time difference* quantitative simulation function denotes the maximum time discrepancy that can arise amongst matching transitions. The *eventual maximum time difference* quantitative simulation function denotes the eventual maximum time discrepancy that arises (ignoring finite time trace prefix discrepancies) amongst matching transitions. This corresponds to the "steady-state" difference between systems, ignoring transient behavior. The *(long-run) average time difference* quantitative simulation function denotes the average time discrepancy amongst matching transitions. This function measures the long-run average time discrepancies, per transition, amongst two timed systems. Ideally, we want all three simulation functions to be as small as possible between the specification and the implementation systems, but minimizing one may lead to increase in values for others. Thus, all three simulation functions give important information about systems. We illustrate the various quantitative timed simulation functions via examples.

Example 1 (Maximum Time Difference)**.** Consider the two timed automata $\mathcal{T}_1$ and $\mathcal{T}_2$ in Figure 1. The locations are labelled with the observations. The starting location of each automaton is the one labelled with the observation a, and the starting value of the clock x is 0. Let us look at the value of the *maximum time difference* quantitative timed simulation function $\mathcal{S}_{\mathsf{MaxDiff}}$ for the state pair $\left(\langle a, x=0\rangle^{\mathcal{T}_1}, \langle a, x=0\rangle^{\mathcal{T}_2}\right)$. The value is (1) infinity if every transition from the state in $\mathcal{T}_1$ cannot be matched by a transition from the matching state in $\mathcal{T}_2$ (and similarly for following steps), that is the state of $\mathcal{T}_1$ *time-abstract simulates* the state of $\mathcal{T}_2$; (2) the maximum time difference between matching transitions of $\mathcal{T}_1$ and $\mathcal{T}_2$ otherwise, amongst *time-divergent* runs. For the two timed automata in Figure 1, it can be checked that $\langle a, x=0\rangle^{\mathcal{T}_1}$ time-abstract simulates $\langle a, x=0\rangle^{\mathcal{T}_2}$, and that the maximum time difference between matching transitions is 9 time units, (*e.g.* between the paths $\langle a, x=0\rangle^{\mathcal{T}_1} \xrightarrow{10} \langle b, x=0\rangle^{\mathcal{T}_1} \xrightarrow{0} \langle c, x=0\rangle^{\mathcal{T}_1} \xrightarrow{5} \langle c, x=0\rangle^{\mathcal{T}_1} \xrightarrow{5} \cdots$ and $\langle a, x=0\rangle^{\mathcal{T}_2} \xrightarrow{1} \langle b, x=0\rangle^{\mathcal{T}_2} \xrightarrow{9} \langle c, x=0\rangle^{\mathcal{T}_2} \xrightarrow{5} \langle c, x=0\rangle^{\mathcal{T}_2} \xrightarrow{5} \cdots$). □

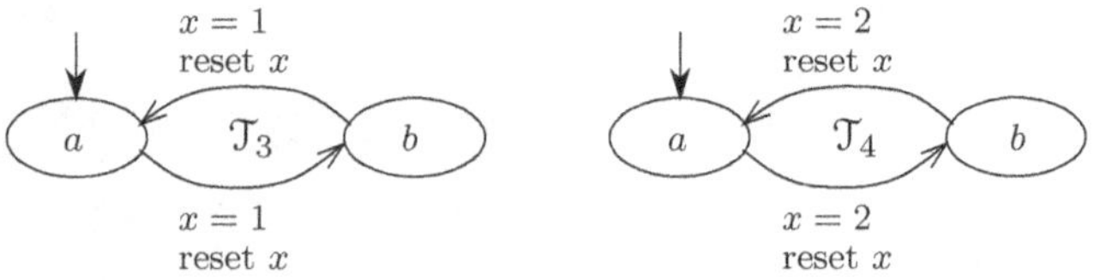

Figure 2: Two timed automata $\mathcal{T}_3, \mathcal{T}_4$.

Example 2 (Global Event Times)**.** Consider the two timed automata in Figure 2. The value of the maximum time difference quantitative timed simulation function $\mathcal{S}_{\mathsf{MaxDiff}}$ for the state pair $\left(\langle a, x=0\rangle^{\mathcal{T}_3}, \langle a, x=0\rangle^{\mathcal{T}_4}\right)$ is ∞, since timing mismatch corresponding to the n-th transition is n (the n-th transition in $\mathcal{T}_3$ occurs at global time n, the n-th transition in $\mathcal{T}_4$ occurs at global time $2 \cdot n$). We depict the timelines in Figure 3. □

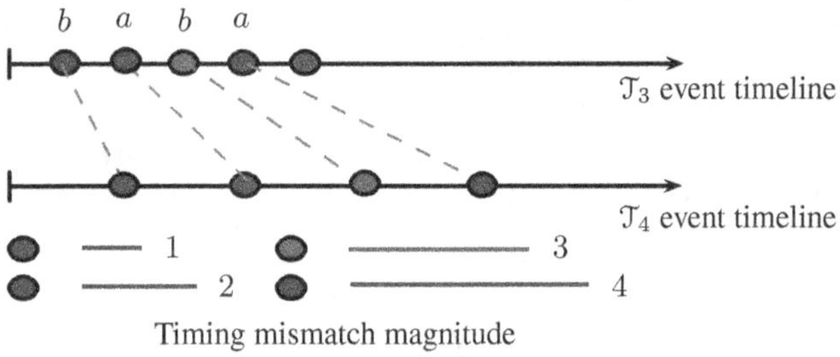

Figure 3: Timeline of $\mathcal{T}_3, \mathcal{T}_4$ events.

Example 3 (Eventual Maximum Time Difference)**.** Consider the two timed automata $\mathcal{T}_1$ and $\mathcal{T}_2$ in Figure 1. Let us look at the value of the *eventual maximum time difference* quantitative timed simulation function $\mathcal{S}_{\mathsf{LimMaxDiff}}$ for the state pair $\left(\langle a, x=0\rangle^{\mathcal{T}_1}, \langle a, x=0\rangle^{\mathcal{T}_2}\right)$. The value is (1) infinity if every transition from the state in $\mathcal{T}_1$ cannot be matched by a transition from the matching state in $\mathcal{T}_2$ (and similarly for following steps), that is the state of $\mathcal{T}_1$ *time-abstract simulates* the state of $\mathcal{T}_2$; (2) the *eventual* maximum time difference between matching transitions of $\mathcal{T}_1$ and $\mathcal{T}_2$ otherwise (ignoring the time differences amongst finite trace prefixes), amongst *time-divergent* runs. In the automata $\mathcal{T}_1, \mathcal{T}_2$, there is a time mismatch only at the transitions from a, and this transition can only occur before time 10. Once the executions reach the location c, the automaton $\mathcal{T}_2$ is able to match the transitions of $\mathcal{T}_1$ at the exact times, with zero time discrepancy. Thus, $\mathcal{S}_{\mathsf{LimMaxDiff}}$ denotes the "steady-state" time discrepancy between $\mathcal{T}_1$, $\mathcal{T}_2$, and this value is zero for the state pair $\left(\langle a, x=0\rangle^{\mathcal{T}_1}, \langle a, x=0\rangle^{\mathcal{T}_2}\right)$, in contrast to the value of 9 for $\mathcal{S}_{\mathsf{MaxDiff}}$ for the state pair. Note that we ignore time-discrepancies for finite *time* (by discarding Zeno runs), not just finite trace prefixes. If we ignore only finite trace prefixes, then we would have obtained a value of 9, as $\mathcal{T}_1$ can loop on the location b by preventing time from progressing (note that the clock x is not reset on the b loop transition). □

Example 4 (Average Time Difference)**.** Consider the two timed automata $\mathcal{T}_5$ and $\mathcal{T}_6$ in Figure 4. Let us look at the value of the (long-run) *average time difference* quantitative timed simulation function $\mathcal{S}_{\mathsf{AvgDiff}}$ for the state pair $\left(\langle a, x=0\rangle^{\mathcal{T}_5}, \langle a, x=0\rangle^{\mathcal{T}_6}\right)$. As usual, for the value to be finite, we require time-abstract simulation. If time-abstract simulation holds, we take the average with respect to the

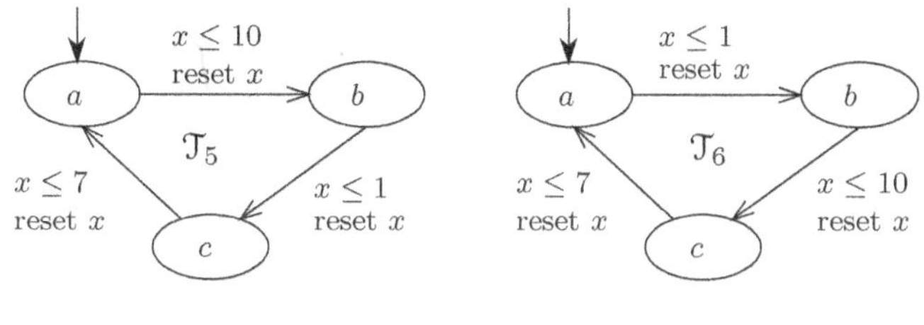

Figure 4: Two timed automata $\mathcal{T}_5, \mathcal{T}_6$.

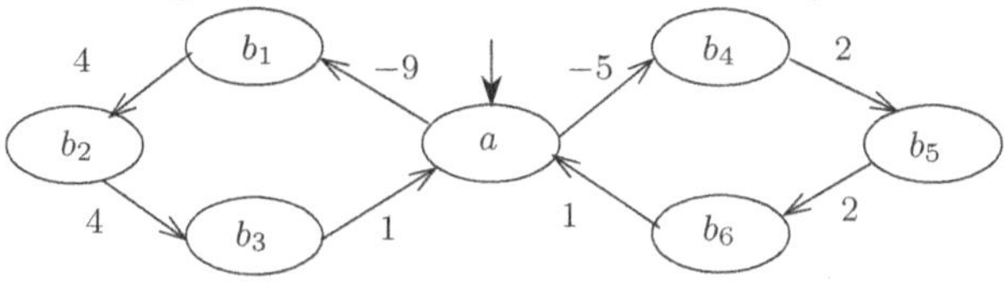

Figure 5: Debit sum-level game

number of transitions (over non-Zeno runs). For the state pair, a time difference of 9 occurs infinitely often, but this difference occurs in only one-third of the transitions (the transitions from a locations). For the transitions from b and c, the time discrepancy is zero. Thus, the value for $\mathcal{S}_{\mathsf{AvgDiff}}$ is $\frac{9+0+0}{3} = 3$. $\square$

To compute the values of the three simulation functions, we use the framework of turn based games on finite-state game graphs. We introduce two new game theoretic objectives (these objectives are required for computing two of the quantitative simulation functions) on these game graphs, namely, *eventual debit-sum level* and *average debit-sum level* objectives, and present novel solutions for both. We need to consider the sums of the weights encountered as in our quantitative simulation functions, the global time is the sum of the time durations of all the preceding transitions.

Eventual debit-sum level and average debit-sum level games are also interesting on their own. We next illustrate average debit-sum level games. These games are played on two-player turn based game graphs. Each transition in the game graph incurs a cost (denoted by a negative weight), or a reward (denoted by a positive weight). These costs can be viewed as monetary losses, or monetary gains. The debit-sum level at a stage in the game denotes the absolute value of the monetary balance, if the balance is negative (the balance is the sum of all the positive and negative costs and rewards). The objective of player 1 is to have the lowest possible average debit-sum level. These games are also applicable, for instance, in financial markets, where banks have to take overnight loans from the Federal Reserve loan windows in case of need (these loans need to be renewed each day the loan is not repaid). It is in the banks interests to minimize the average of the loan amount per day.

Example 5 (Debit Sum-Level Turn Based Games)**.** Consider the turn based game depicted in Figure 5. The only player-1 location is a, the other locations are player-2 locations. The numbers on the edges denote the costs or rewards that player-1 gets when that transition is taken. Positive weights denotes rewards, and negative weights denotes costs. Viewing the weights as monetary transactions, and starting with a monetary balance of zero at a, if player 1 loops around the left loop, then the trace, together with the monetary balances is: $((a, 0)\ (b_1, -9)\ (b_2, -5)\ (b_3, -1))^\omega$, where the numbers denote the accumulated balances during the run of the play. The average negative balance, *i.e*, the average debit-sum level (per unit location visit), is $\frac{0+9+5+1}{4} = \frac{15}{4}$. If player 1 loops around the right loop, then the trace, together with the balances is: $((a, 0)\ (b_4, -5)\ (b_5, -3)\ (b_6, -1)\ )^\omega$. The average negative balance is $\frac{0+5+3+1}{4} = \frac{9}{4}$. Thus the optimum average debit sum-level value for player 1 is $9/4$, and the optimum strategy is to loop around the right-hand side, where it needs to borrow less, on average. $\square$

Our Contributions. Our main contributions in the present work are as follows.

- ⋆ We define three quantitative refinement metrics incorporating Zenoness conditions semantically, that is our refinement metrics ignore artificial Zeno runs present in systems due to modelling artifacts. We also show that these quantitative functions are actually (directed) metrics.
- ⋆ We define quantitative timed simulation functions corresponding to the refinement metrics using a game theoretic formulation. These quantitative simulation functions also incorporate Zenoness conditions for obtaining physically meaningful system differences. As far we know, this is is the first work which handles Zeno runs when computing simulation functions.
- ⋆ We present *decision procedures* for computing all the defined quantitative timed simulation functions to within any desired degree of accuracy for *any* given timed automaton.
- ⋆ We introduce new game theoretic objectives on finite-state game graphs, namely, eventual debit-sum level objectives and average debit-sum level objectives, and present novel solutions for both on finite-state turn based games. These new objectives are required in the computation of the defined quantitative simulation functions.

We have considered the (more challenging) framework of global event times in our quantitative simulation functions. Our solution framework is also applicable where the mismatches are only with respect to transition *durations* (simple algorithms are applicable in this case). Our algorithms can easily be generalized to consider quantitative simulation functions in which an observation σ is allowed to match a different observation σ', but with some matching penalty in case $\sigma \neq \sigma'$ (the penalty being in addition to the timing mismatch of σ, σ'). Thus, our algorithms apply to the computation of quantitative simulation functions which consider the *Skorokhod metric* [17] over mismatches.

Related Work. The most relevant related work is the recent work on the theory of *approximate bisimulation* for continuous and switched systems [15, 14], and [19]. The approximations in [15, 14] are with respect to the real-valued system outputs, and not with respect to the *times* during which the values are output. The simulation relations are constrained to match values at equivalent sample points, thus there is no mechanism to judge the time discrepancies. The work in [19] presents similarity relations where the approximations are with respect to time as well as output values. Computation of similarity relations is reduced to solving a derived game, however, no decidability results are presented for solving these derived games. For timed systems, the work in [16] presented maximum time difference quantitative timed simulation functions, however, Zeno issues were ignored. Our solutions for the new objectives on finite-state game graphs builds on previous work on mean payoff parity games, multi-dimensional mean payoff, and energy games [3, 5, 8, 7, 6]. The new game objectives presented in the present work, that are required for the quantitative timed simulation functions, were previously unstudied, and require new ideas in their solutions. The full details of the paper (omitted for lack of space) can be found in [9].

2. QUANTITATIVE TIMED TRACE DIFFERENCE AND REFINEMENT

We define *quantitative* refinement functions on timed systems. They allow approximate matching of timed traces and generalize timed and untimed refinement relations.

Timed Transition System. A *timed transition system* (TTS) is a tuple $A = \langle S, \Sigma, \rightarrow, \mu, S_0 \rangle$ where

- S is the set of states.
- Σ is a set of atomic propositions (the observations).
- $\rightarrow \subseteq S \times \mathbb{R}^+ \times S$ is the transition relation such that for all $s \in S$ there exists at least one $s' \in S$ such that for some Δ, we have $s \xrightarrow{\Delta} s'$.
- $\mu : S \mapsto 2^{\Sigma}$ is the observation map which assigns a truth value to atomic propositions in each state.
- $S_0 \subseteq S$ is the set of initial states.

We write $s \xrightarrow{t} s'$ if $(s, t, s') \in \rightarrow$. A *state trajectory* is an infinite sequence $s_0 \xrightarrow{t_0} s_1 \xrightarrow{t_1} \ldots$, where for each $j \geq 0$, we have $s_j \xrightarrow{t_j} s_{j+1}$. The state trajectory is *initialized* if $s_0 \in S_0$ is an initial state. A state trajectory $s_0 \xrightarrow{t_0} s_1 \ldots$ induces a *trace* given by the observation sequence $\mu(s_0) \xrightarrow{t_0} \mu(s_1) \xrightarrow{t_1} \ldots$. To emphasize the initial state, we say s_0-trace for a trace induced by a state trajectory starting from s_0. A trace is initialized if it is induced by an initialized state trajectory. Given a trace tr induced by a state trajectory $s_0 \xrightarrow{t_0} s_1 \xrightarrow{t_1} \ldots$, let $\mathsf{time}_{\mathsf{tr}}[i]$ denote $\sum_{j=0}^{i} t_j$, *i.e.* the time of the i-th transition. The trace tr is *time-convergent* or Zeno if $\lim_{i \to \infty} \mathsf{time}_{\mathsf{tr}}[i]$ is finite; otherwise it is time-divergent or non-Zeno. We denote the set of time-divergent initialized traces of a timed transition system A by $\mathsf{Timediv}(A)$, and the set of all initialized traces of A by $\mathsf{Traces}(A)$. A TTS is well formed if from every $s_0 \in S_0$, there exists a s_0-trace in $\mathsf{Timediv}(A)$. We consider only well formed TTS in the sequel. The TTS $A_{\mathfrak{r}}$ *refines* or *implements* the TTS $A_{\mathfrak{s}}$ (the specification) if every initialized trace of $A_{\mathfrak{r}}$ is also an initialized trace of $A_{\mathfrak{s}}$. We first define various quantitative notions of refinement that quantify if the behavior of an implementation TTS is "close enough" to a specification TTS. We begin by defining several metrics on trace differences and refinements.

Maximum Trace Difference Distance. Given two traces $\mathsf{tr} = r_0 \xrightarrow{t_0} r_1 \xrightarrow{t_1} r_2 \ldots$ and $\mathsf{tr}' = s_0 \xrightarrow{t'_0} s_1 \xrightarrow{t'_1} s_2 \ldots$, the maximum trace difference distance $\mathcal{D}_{\mathsf{MaxDiff}}(\mathsf{tr}, \mathsf{tr}')$ is defined by

$$\mathcal{D}_{\mathsf{MaxDiff}}(\mathsf{tr}, \mathsf{tr}') = \begin{cases} \infty & \text{if } r_n \neq s_n \text{ for some } n \\ \sup_n \{ | \mathsf{time}_{\mathsf{tr}}[n] - \mathsf{time}_{\mathsf{tr}'}[n] | \} & \text{otherwise} \end{cases}$$

The distance $\mathcal{D}_{\mathsf{MaxDiff}}(\mathsf{tr}, \mathsf{tr}')$ indicates the maximum time discrepancy between matching observations in the two traces tr and tr'.

Limit-Maximum Trace Difference Distance. Given two traces $\mathsf{tr} = r_0 \xrightarrow{t_0} r_1 \xrightarrow{t_1} r_2 \ldots$ and $\mathsf{tr}' = s_0 \xrightarrow{t'_0} s_1 \xrightarrow{t'_1} s_2 \ldots$, the limit-maximum trace difference distance $\mathcal{D}_{\mathsf{LimMaxDiff}}(\mathsf{tr}, \mathsf{tr}')$ is defined by $\mathcal{D}_{\mathsf{LimMaxDiff}}(\mathsf{tr}, \mathsf{tr}') =$

$$\begin{cases} \infty & \text{if } r_n \neq s_n \text{ for some } n \\ \lim_{M \to \infty} \sup_{n \geq M} \{ | \mathsf{time}_{\mathsf{tr}}[n] - \mathsf{time}_{\mathsf{tr}'}[n] | \} & \text{otherwise} \end{cases}$$

The distance $\mathcal{D}_{\mathsf{LimMaxDiff}}(\mathsf{tr}, \mathsf{tr}')$ indicates the limit-maximum time discrepancy between matching observations in the two traces tr and tr'. That is, it indicates the eventual "steady state" maximum time discrepancy, ignoring any initial spikes in the time discrepancy between the two traces (we still require all observations to be matched).

Limit-Average Trace Difference Distance. Given two traces $\mathsf{tr} = r_0 \xrightarrow{t_0} r_1 \xrightarrow{t_1} r_2 \ldots$ and $\mathsf{tr}' = s_0 \xrightarrow{t'_0} s_1 \xrightarrow{t'_1} s_2 \ldots$, the limit-average trace difference distance $\mathcal{D}_{\mathsf{AvgDiff}}(\mathsf{tr}, \mathsf{tr}')$ is defined by $\mathcal{D}_{\mathsf{AvgDiff}}(\mathsf{tr}, \mathsf{tr}') =$

$$\begin{cases} \infty & \text{if } r_j \neq s_j \text{ for some } j \\ \lim_{M \to \infty} \left(\sup_{n \geq M} \left\{ \frac{\sum_{i=0}^{n} (| \mathsf{time}_{\mathsf{tr}}[i] - \mathsf{time}_{\mathsf{tr}'}[i] |)}{n} \right\} \right) & \text{otherwise} \end{cases}$$

The distance $\mathcal{D}_{\mathsf{AvgDiff}}(\mathsf{tr}, \mathsf{tr}')$ indicates the long run average of the time discrepancies between the two traces.

Proposition 1. *For $\Psi \in \{\mathcal{D}_{\mathsf{MaxDiff}}, \mathcal{D}_{\mathsf{LimMaxDiff}}, \mathcal{D}_{\mathsf{AvgDiff}}\}$, the functions $\Psi()$ are metrics on timed traces.* □

Definition 1 (Refinement Distances). Trace difference metrics $\Psi \in \{\mathcal{D}_{\mathsf{MaxDiff}}, \mathcal{D}_{\mathsf{LimMaxDiff}}, \mathcal{D}_{\mathsf{AvgDiff}}\}$, induce the refinement distances $\mathcal{R}_{\Psi}(A_{\mathfrak{r}}, A_{\mathfrak{s}})$. Formally, given two timed transition systems $A_{\mathfrak{r}}$, $A_{\mathfrak{s}}$, with initial state sets $S_{\mathfrak{r}}, S_{\mathfrak{s}}$ respectively (the "$\mathfrak{r}$" and "$\mathfrak{s}$" subscripts denote "refinement", and "specification" respectively), the refinement distance of $A_{\mathfrak{r}}$ with respect to $A_{\mathfrak{s}}$, induced by Ψ, is given by

$$\mathcal{R}_{\Psi}(A_{\mathfrak{r}}, A_{\mathfrak{s}}) = \inf_{\mathsf{tr}_{q_{\mathfrak{s}}}} \sup_{\mathsf{tr}_{q_{\mathfrak{r}}} \in \mathsf{Timediv}(A_{\mathfrak{r}})} \{ \Psi(\mathsf{tr}_{q_{\mathfrak{r}}}, \mathsf{tr}_{q_{\mathfrak{s}}}) \}$$

where $\mathsf{tr}_{q_{\mathfrak{r}}}$ (respectively, $\mathsf{tr}_{q_{\mathfrak{s}}}$) is a $q_{\mathfrak{r}}$-trace (respectively, $q_{\mathfrak{s}}$-trace) for some $q_{\mathfrak{r}} \in S_{\mathfrak{r}}$ (respectively, $q_{\mathfrak{s}} \in S_{\mathfrak{s}}$). □

Notice that this refinement distance is asymmetric: it is a *directed distance*[11]. The refinement distance $\mathcal{R}_{\Psi}(A_{\mathfrak{r}}, A_{\mathfrak{s}})$ indicates quantitatively how well initialized traces in $A_{\mathfrak{s}}$ match corresponding initialized traces in $A_{\mathfrak{r}}$ with respect to the Ψ trace difference metric.

Proposition 2. *For $\Psi \in \{\mathcal{D}_{\mathsf{MaxDiff}}, \mathcal{D}_{\mathsf{LimMaxDiff}}, \mathcal{D}_{\mathsf{AvgDiff}}\}$, the functions $\mathcal{R}_{\Psi}()$ are directed metrics on* TTS. □

A Note on Zeno-Asymmetry in Refinement Metrics. There appears to be a Zenoness asymmetry in the definitions for refinement metrics, as only Zeno behaviors of $A_{\mathfrak{r}}$ are given special treatment. This is because in case of Zeno behavior by the specification, our definitions automatically give a value of ∞, which is the correct notion. That is, for $\Psi \in \{\mathcal{D}_{\mathsf{MaxDiff}}, \mathcal{D}_{\mathsf{LimMaxDiff}}, \mathcal{D}_{\mathsf{AvgDiff}}\}$, we have $\Psi(\mathsf{tr}_{q_{\mathfrak{r}}}, \mathsf{tr}_{q_{\mathfrak{s}}}) = \infty$ if $\mathsf{tr}_{q_{\mathfrak{r}}}$ is time divergent, and $\mathsf{tr}_{q_{\mathfrak{s}}}$ is time convergent. In the definition of the refinement distances, the "inf" in $\inf_{\mathsf{tr}_{q_{\mathfrak{s}}}} \{ \Psi(\mathsf{tr}_{q_{\mathfrak{r}}}, \mathsf{tr}_{q_{\mathfrak{s}}}) \}$ ensures that time convergent $\mathsf{tr}_{q_{\mathfrak{s}}}$ traces are effectively not considered as they lead to a ∞ value for $\Psi(\mathsf{tr}_{q_{\mathfrak{r}}}, \mathsf{tr}_{q_{\mathfrak{s}}})$. Thus, there is no need to explicitly treat Zeno traces of $A_{\mathfrak{s}}$ differently as a special case.

Timed Simulation Relations. The general trace inclusion problem for timed systems is undecidable [1]; simulation relations allow us to restrict our attention to a computable relation. Let $A_{\mathfrak{r}}$ and $A_{\mathfrak{s}}$be timed transition systems. A binary relation $\preceq \subseteq S_{\mathfrak{r}} \times S_{\mathfrak{s}}$ is a *timed simulation* if $s_{\mathfrak{r}} \preceq s_{\mathfrak{s}}$ implies the following conditions: (i) $\mu(s_{\mathfrak{r}}) = \mu(s_{\mathfrak{s}})$, and (ii) If $s_{\mathfrak{r}} \xrightarrow{t} s'_{\mathfrak{r}}$, then there exists $s'_{\mathfrak{s}}$ such that $s_{\mathfrak{s}} \xrightarrow{t} s'_{\mathfrak{s}}$, and $s'_{\mathfrak{r}} \preceq s'_{\mathfrak{s}}$. The state $s_{\mathfrak{r}}$ is timed simulated by the state $s_{\mathfrak{s}}$ if there exists a timed simulation $\preceq$ such that $s_{\mathfrak{r}} \preceq s_{\mathfrak{s}}$. A binary relation $\equiv$ is a *timed bisimulation* if it is a symmetric timed simulation. Two states $s_{\mathfrak{r}}$ and $s_{\mathfrak{s}}$ are timed bisimilar if there exists a timed bisimulation $\equiv$ with $s_{\mathfrak{r}} \equiv s_{\mathfrak{s}}$. Timed bisimulation is stronger than timed simulation which in turn is stronger than trace inclusion. If state $s_{\mathfrak{r}}$ is timed simulated by state $s_{\mathfrak{s}}$, then every $s_{\mathfrak{r}}$-trace is also a $s_{\mathfrak{s}}$-trace.

Untimed Simulation Relations. *Untimed* simulation and bisimulation relations are defined analogously to timed simulation and bisimulation relations by ignoring the duration of time steps.

Timed simulation and bisimulation require that times be matched exactly. This is often too strict a requirement, especially since timed models are approximations of the real world. On the other hand, untimed simulation and bisimulation relations ignore the times on moves altogether. Analogous to the notions of quantitative refinement presented in Section 2, we will define quantitative notions of simulation functions which lie in between these extremes in Section 4. We will define quantitative simulation functions in a game theoretic framework. The motivation for the game theoretic framework for simulation relations is presented next.

Timed and Untimed Simulation Games. We present an alternative equivalent game theoretic view of timed simulation (a similar view exists for untimed simulation). Given two timed transition systems $A_{\mathfrak{r}}$ and $A_{\mathfrak{s}}$, consider a two player turn-based bipartite timed transition game structure $\mathfrak{G}_t(A_{\mathfrak{r}}, A_{\mathfrak{s}})$ with state space $(S_{\mathfrak{r}} \times S_{\mathfrak{s}} \times \{1\}) \cup (S_{\mathfrak{r}} \times S_{\mathfrak{s}} \times \{2\})$ (the full formal definitions of game structures will be presented in Section 3). The states of player 2 (the antagonist) are $S_{\mathfrak{r}} \times S_{\mathfrak{s}} \times \{2\}$ and the state of player 1 (the protagonist) are $S_{\mathfrak{r}} \times S_{\mathfrak{s}} \times \{1\}$. The transitions are:

Player-2 transitions. $\langle s_{\mathfrak{r}}, s_{\mathfrak{s}}, 2\rangle \xrightarrow{\Delta_{\mathfrak{r}}} \langle s'_{\mathfrak{r}}, s_{\mathfrak{s}}, 1\rangle$ such that $s_{\mathfrak{r}} \xrightarrow{\Delta_{\mathfrak{r}}} s'_{\mathfrak{r}}$ is a valid transition in $A_{\mathfrak{r}}$.

Player-1 transitions. $\langle s_{\mathfrak{r}}, s_{\mathfrak{s}}, 1\rangle \xrightarrow{\Delta_{\mathfrak{s}}} \langle s_{\mathfrak{r}}, s'_{\mathfrak{s}}, 2\rangle$ such that $s_{\mathfrak{s}} \xrightarrow{\Delta_{\mathfrak{s}}} s'_{\mathfrak{s}}$ is a valid transition in $A_{\mathfrak{s}}$.

To decide if $s_{\mathfrak{s}}$ time-simulates $s_{\mathfrak{r}}$, we play the following game. Let $\langle s_{\mathfrak{r}}, s_{\mathfrak{s}}, 2\rangle$ be the initial state such that $\mu(s_{\mathfrak{r}}) = \mu(s_{\mathfrak{s}})$. Player-2 picks a transition of some duration $\Delta_{\mathfrak{r}}$ from this state and moves to some state $\langle s'_{\mathfrak{r}}, s_{\mathfrak{s}}, 1\rangle$. From $\langle s'_{\mathfrak{r}}, s_{\mathfrak{s}}, 1\rangle$, player 1 then picks a transition of duration $\Delta_{\mathfrak{s}}$ such that $\Delta_{\mathfrak{s}} = \Delta_{\mathfrak{r}}$ and moves to $\langle s'_{\mathfrak{r}}, s'_{\mathfrak{s}}, 2\rangle$ such that $\mu(s'_{\mathfrak{s}}) = \mu(s'_{\mathfrak{s}})$. If no such transition exists, then player 1 loses. If the game can proceed forever without player-1 losing, then player 2 loses and player 1 wins. If player 1 has a winning strategy from $\langle s_{\mathfrak{r}}, s_{\mathfrak{s}}, 2\rangle$, then $s_{\mathfrak{s}}$ time-simulates $s_{\mathfrak{r}}$. For untimed simulation, we ignore the time durations of the moves (*i.e.* both players can pick transitions of unequal time duration). We denote the two player turn-based bipartite *untimed* transition game as $\mathfrak{G}_u(A_{\mathfrak{r}}, A_{\mathfrak{s}})$.

3. FINITE-STATE GAME GRAPHS

We will define the values of *quantitative timed simulation functions* in Section 4 through game theoretic formulations of problems for finite-state game graphs. In this section, we first present the basic background on finite-state game graphs, and the relevant known results; then introduce new game theoretic objectives (that were not studied before but are required for quantitative timed simulation functions) and present solutions for the new objectives.

3.1 Basic Definitions and Known Results

In this section we present definitions of finite game graphs, plays, strategies, objectives, notion of winning and the decision problems.

Game graphs. A *game graph* $G = \langle Q, E\rangle$ consists of a finite set Q of states partitioned into player-1 states Q_1 and player-2 states Q_2 (i.e., $Q = Q_1 \cup Q_2$ and $Q_1 \cap Q_2 = \emptyset$), and a set $E \subseteq Q \times Q$ of edges such that for all $q \in Q$, there exists (at least one) $q' \in Q$ such that $(q, q') \in E$. A *player*-1 *game* is a game graph where $Q_1 = Q$ and $Q_2 = \emptyset$. The subgraph of G induced by $S \subseteq Q$ is the graph $\langle S, E \cap (S \times S)\rangle$ (which is not a game graph in general); the subgraph induced by S is a game graph if for all $s \in S$ there exist $s' \in S$ such that $(s, s') \in E$.

Plays and strategies. A game on G starting from a state $q_0 \in Q$ is played in rounds as follows. If the game is in a player-1 state, then player 1 chooses the successor state from the set of outgoing edges; otherwise the game is in a player-2 state, and player 2 chooses the successor state from the set of outgoing edges. The game results in a *play* from q_0, i.e., an infinite path $\rho = q_0 q_1 \ldots$ such that $(q_i, q_{i+1}) \in E$ for all $i \geq 0$. The prefix of length n of ρ is denoted by $\rho(n) = q_0 \ldots q_n$. A *strategy* for player 1 is a function $\pi_1 : Q^* Q_1 \to Q$ such that $(q, \pi_1(\rho \cdot q)) \in E$ for all $\rho \in Q^*$ and $q \in Q_1$. An *outcome* of π_1 from q_0 is a play $q_0 q_1 \ldots$ such that $\pi_1(q_0 \ldots q_i) = q_{i+1}$ for all $i \geq 0$ such that $q_i \in Q_1$. Strategy and outcome for player 2 are defined analogously. A player-1 strategy is *memoryless* if it is independent of the history and depends only on the current state, and hence can be described as a function $\pi_1 : Q_1 \to Q$. Memoryless strategies for player 2 are defined analogously. We denote by Π_1 and Π_2 the set of strategies for player 1 and player 2, respectively. Given a starting state q, a strategy π_1 for player 1 and a strategy π_2 for player 2, we have a unique play $q_0 q_1 q_2 \ldots$, such that $q_0 = q$ and for all $i \geq 0$ (i) if q_i is a player 1 state, then $q_{i+1} = \pi_1(q_0, q_1, \ldots, q_i)$; and (ii) if q_i is a player 2 state, then $q_{i+1} = \pi_2(q_0, q_1, \ldots, q_i)$. We denote the unique play as $\rho(\pi_1, \pi_2, q)$.

Objectives. In this work we will consider both qualitative and quantitative objectives. We first introduce qualitative objectives that we will use in our work. A *qualitative objective* for G is a set $\phi \subseteq Q^\omega$ of winning plays. For a play ρ, we denote by $\mathsf{Inf}(\rho)$ the set of states that occur infinitely often in ρ. We consider Büchi objectives, and its dual coBüchi objectives which are defined as follows. A Büchi objective consists of a set B of Büchi states, and requires that the set B is visited infinitely often. Formally, the Büchi objective defines the following set of winning plays: $\mathsf{Büchi}(B) = \{\rho \mid \mathsf{Inf}(\rho) \cap B \neq \emptyset\}$. Dually the coBüchi objective consists of a set C of coBüchi states and requires that states outside C be visited only finitely often, and defines the set $\mathsf{coBüchi}(C) = \{\rho \mid \mathsf{Inf}(\rho) \subseteq C\}$ of winning plays. When we will consider qualitative objectives, the objective of player 1 will be disjunction of two coBüchi objectives, and the objective of player 2 will be the complement (conjunction of two Büchi objectives). We now introduce several quantitative objectives.

Quantitative objectives. A *quantitative objective* for G is a function $f : Q^\omega \rightarrow \mathbb{R}$ that maps every play to a real-valued number (in contrast a qualitative objective can be interpreted as a function $\phi : Q^\omega \rightarrow \{0, 1\}$ that maps plays to Boolean rewards, with 1 for winning plays). Let $w : E \rightarrow \mathbb{Z}$ be a *weight function* and let us denote by W the largest weight (in absolute value) according to w. For a prefix $\rho(n) = q_0 q_1 \dots q_n$ of a play we denote by $\mathsf{Sum}(w)(\rho(n)) = \sum_{i=0}^{n-1} w(q_i, q_{i+1})$ the sum of the weights of the prefix. The *debit-sum* level at the end of the prefix $\rho(n)$ is defined by: $\mathsf{DebSum}(w)(\rho(n)) = \max(0, -\sum_{i=0}^{n-1} w(q_i, q_{i+1}))$ Note the negative sign in the definition. The debit-sum level denotes the amount by which the accumulated sum of the weights has dipped below 0 at the end of $\rho(n)$ (if the sum of the weights is positive, *i.e.* there is a credit, then the debit-sum level is defined to be 0). We will consider the following objective functions.

Debit-sum level. For a play ρ, the debit-sum level is the maximal debit-sum level that occurs in it. Formally, for a play ρ and the weight function w we have $\mathsf{DebSum}(w)(\rho) = \sup_n \mathsf{DebSum}(w)(\rho(n)) = \inf\{v_0 \mid \forall n \geq 0. v_0 + \mathsf{Sum}(w)(\rho(n)) \geq 0\}$.

Eventual debit-sum level. For a play ρ, the eventual debit-sum level is the maximal debit-sum level that occurs after some point on in the play. Formally, for a play ρ and the weight function w we have $\mathsf{EvDebSum}(w)(\rho) = \limsup_{n \rightarrow \infty} \mathsf{DebSum}(w)(\rho(n)) = \lim_{M \rightarrow \infty} \sup_{n \geq M} \mathsf{DebSum}(w)(\rho(n)) = \inf\{v_0 \mid \exists n_0 \geq 0. \forall n \geq n_0. v_0 + \mathsf{Sum}(w)(\rho(n)) \geq 0\}$.

Average weight. The mean-payoff (or limit-average weight) objective function on a play $\rho = q_0 q_1 \dots$ is the long-run average of the weights of the play, i.e., $\mathsf{Avg}(w)(\rho) = \limsup_{n \rightarrow \infty} \frac{1}{n} \cdot \mathsf{Sum}(w)(\rho(n))$.

Average debit-sum. Along with the previous objective, we introduce a new objective function, which we call the average debit-sum level that assigns to every play the long-run average of the debit-sum levels. Formally, $\mathsf{AvDebSum}(w)(\rho) = \limsup_{n \rightarrow \infty} \frac{\sum_{i=0}^{n} \mathsf{DebSum}(w)(\rho(n))}{n}$. Note that since the debit-sum level is defined to be 0 if the accumulated sum is positive (*i.e.* a positive credit-sum level), a positive credit-sum cannot cancel out a positive debit-sum in the averaging process in $\mathsf{AvDebSum}(w)(\rho)$. Observe that in contrast to mean-payoff objective that is the average of the weights, the average debit-sum has the flavor of the average of the partial sums of the weights.

In the sequel, when the weight function w is clear from context we will omit it and simply write $\mathsf{Sum}(\rho(n))$ and $\mathsf{Avg}(\rho)$, and so on. For each of the above quantitative objectives, we will consider a version of the quantitative objective that is a disjunction with a coBüchi objective. Formally for a quantitative objective f and coBüchi objective $\mathsf{coBüchi}(C)$, the quantitative objective that is the disjunction of the two objectives is defined as follows for a play ρ: if $\rho \in \mathsf{coBüchi}(C)$, then the objective function assigns value 0 to ρ, otherwise it assigns value $f(\rho)$. We will refer to the corresponding version of the quantitative objectives with disjunction with coBüchi objective as $\mathsf{DebSumCB}$, $\mathsf{EvDebSumCB}$, AvgCB, and $\mathsf{AvDebSumCB}$, respectively (and when the weight function w and the coBüchi set C is clear from the context we drop them for simplicity).

Winning strategies, optimal value and optimal strategies. A player-1 strategy π_1 is *winning* (we also say that player 1 is winning, or that q is a winning state) in a state q for a qualitative objective ϕ if $\rho \in \phi$ for all outcomes ρ of π_1 from q. The optimal value for a quantitative objective is the minimal value that player 1 can guarantee against all strategies of player 2. Formally, for a quantitative objective f that maps plays to real-valued rewards, the optimal value $\mathsf{Opt}(f)(q)$ at state q is defined as

$$\mathsf{Opt}(f)(q) = \inf_{\pi_1 \in \Pi_1} \sup_{\pi_2 \in \Pi_2} f(\rho(\pi_1, \pi_2, q)).$$

A strategy for player 1 is an *optimal strategy* if it achieves the optimal value against all strategies of player 2, i.e., a strategy π_1^* is optimal if we have $\mathsf{Opt}(f)(q) = \sup_{\pi_2 \in \Pi_1} f(\rho(\pi_1^*, \pi_2, q))$.

We now present a theorem that summarizes known results about Büchi and coBüchi games, debit sum (minimal initial credit for energy games), and mean-payoff games. The results of Büchi and coBüchi objectives follow from [13], the results for debit sum games credit follows from the results on energy games of [6], and the result for mean-payoff games follows from [8, 3] (also note that in [6, 8, 3] player 1 has a conjunction of energy (or mean-payoff) with parity objectives, whereas in our setting player 1 has the disjunction of energy (or mean-payoff) with parity, and thus the roles of player 1 and player 2 in this work are exchanged as compared to [6, 8, 3]).

Theorem 1. *The following assertions hold for finite-state game graphs.*

1. *The set of winning states in games with disjunction of two coBüchi objectives can be computed in time $O(|Q| \cdot |E|)$, and memoryless winning strategies exist for player 1 and winning strategies of player 2 require one-bit memory (from their respective winning states).*
2. *The optimal values for debit-sum functions with coBüchi disjunctions can be computed in time $O(|Q|^2 \cdot |E| \cdot W)$, and memoryless optimal strategies exist for player 1 and optimal strategies for player 2 require finite memory. If the optimal value is finite, then the optimal value is at most $|Q| \cdot |W|$.*
3. *The optimal values for limit-average functions with coBüchi disjunctions can be computed in time $O(|Q|^2 \cdot |E| \cdot W)$, and memoryless optimal strategies exist for player 1 and the optimal strategies of player 2 may require infinite memory.* □

3.2 New Results and Algorithms

In this section we will present two solutions for problems on finite-state game graphs. The first solution is for games with minimal initial credit for eventual survival, and the second solution for average-sum objectives.

Eventual Debit-Sum Level Objectives. We will solve the problem by a reduction to a coBüchi game. We start with a lemma that is required for the reduction.

Lemma 1. *For all game graphs with a weight function w, the following assertions hold:*

1. *The optimal value of the eventual debit sum level is at most the optimal value of the debit sum level objective* i.e., *for all states q we have*
$$\mathsf{Opt}(\mathsf{EvDebSum})(q) \leq \mathsf{Opt}(\mathsf{DebSum})(q).$$
2. *If the optimal value of the debit sum level objective is infinite, then the optimal value of the eventual debit sum level is also infinite.* □

Theorem 2. *The optimal player-1 strategy and the optimal value* $\mathsf{Opt}(\mathsf{EvDebSumCB})(q)$ *for the eventual debit sum level objective with coBüchi disjunction can be computed in time* $O(|Q|^3 \cdot |E| \cdot W^2 \cdot \log(|Q| \cdot W))$. □

The next example illustrates the difference between debit sum level and eventual debit sum level objectives.

Example 6 (Debit sum vs eventual debit sum level). Consider the game graph G_0 in Figure 6. The game G_0 has

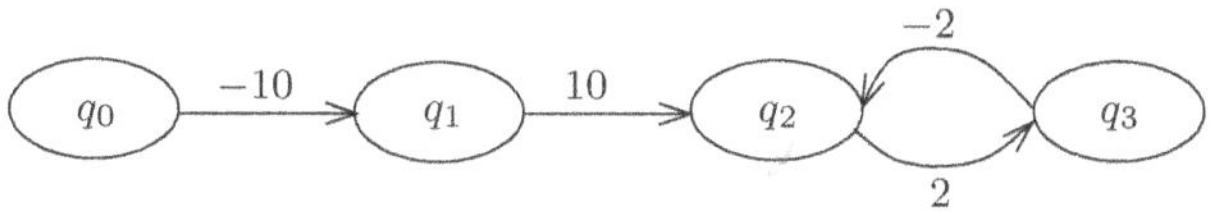

Figure 6: Game Graph G_0

only one play from q_0, namely, $q_0 \to q_1 \to (q_2 \to q_3 \to)^\omega$. It can be seen that $\mathsf{Opt}(\mathsf{DebSum})(q_0)$ is 10 as a debit level of 10 is seen on the transition from q_0 to q_1. However, $\mathsf{Opt}(\mathsf{EvDebSum})(q_0)$ is only 2, as the debit level 10 occurs only once in the play. The debit level 2 however occurs infinitely often in the play. Thus, $\mathsf{Opt}(\mathsf{EvDebSum})(q_0)$ is 2. □

Average Debit-Sum Level Objectives. We start with an example to illustrate average debit-sum level objectives.

Example 7. Consider the game graph G_1 in Figure 7. The game G_1 has only one play from q_0, namely,

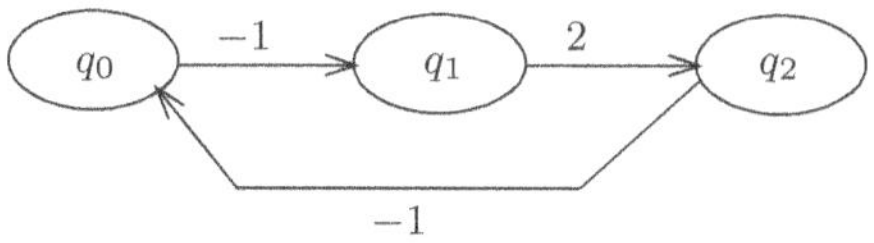

Figure 7: Game Graph G_1

$(q_0 \to q_1 \to q_2 \to)^\omega$ (and similarly only one play from any state). For this play we compute the debit sum and credit sum levels: let $\langle q, d, c\rangle$ denote the state q, and d, c the debit and credit sum levels at that point in the play (note that only either debit sum, or credit sum level can be non-zero, by definition). The play together with debit and credit sum levels is: $\langle q_0, 0, 0\rangle \to (\langle q_1, 1, 0\rangle \to \langle q_2, 0, 1\rangle \to \langle q_0, 0, 0\rangle \to)^\omega$ Thus the average debit sum level $\mathsf{AvDebSum}(w)(q_0) = 1/3$. Now consider the only play from q_2. The play annotated with debit and credit sum levels is: $\langle q_2, 0, 0\rangle \to (\langle q_0, 1, 0\rangle \to \langle q_1, 2, 0\rangle \to \langle q_2, 0, 0\rangle \to)^\omega$ Note that credit levels never rise above 0 in this play. The average debit sum level $\mathsf{AvDebSum}(w)(q_2)$ for this play is 1. Thus, where we "enter" in a cycle affects the value of the average debit sum level. □

Lemma 2. *The following assertions hold given a weight function* w, *and coBüchi objective* $\mathsf{coBüchi}(C)$*:(1) If* $\mathsf{Opt}(\mathsf{DebSumCB})(q) = \infty$, *then* $\mathsf{Opt}(\mathsf{AvDebSumCB})(q) = \infty$. *(2) If* $\mathsf{Opt}(\mathsf{AvgCB})(q) > 0$, *then* $\mathsf{Opt}(\mathsf{AvDebSumCB})(q) = 0$. □

Reduction to mean-payoff coBüchi games. We now use the above lemma to solve the average debit sum problem. Using the above lemma, and solution for $\mathsf{Opt}(\mathsf{DebSumCB})$ and $\mathsf{Opt}(\mathsf{AvgCB})$ we can identify whether $\mathsf{Opt}(\mathsf{AvDebSum})$ is infinite or 0. If $\mathsf{Opt}(\mathsf{DebSumCB})(q)$ is finite, and $\mathsf{Opt}(\mathsf{AvgCB})(q) = 0$, it follows that both players can play strategies to ensure that all cycles formed after their chosen strategies are fixed have sum of weights exactly equal to 0, and has a non coBüchi state. Observe that a positive cycle is only favorable for player 1 for the average debit sum objective, and a negative cycle is favorable for player 2. Hence there exist optimal plays for the average debit sum objective where for all cycles formed along the play the sum of the weights of the cycle will exactly be 0. Thus we reduce the average debit sum problem to solving a larger mean-payoff game as follows: we keep track of the current sum of weights, and since all cycles formed will have exactly 0 sums, we only need to keep track of weights from $-|Q| \cdot W$ to $|Q| \cdot W$. For the limit-average game, we construct a weight function according to the current sum of weights, i.e., if the current sum of weights is ℓ, then the weight function assigns value $\max(-\ell, 0)$. The optimal value of the constructed game with limit-average objective is the optimal value for the average debit sum objective in the original game. The constructed game has $O(|Q|^2 \cdot W)$ states, $O(|E| \cdot |Q| \cdot W)$ edges, and the maximal absolute value of the weight is $O(|Q| \cdot W)$. Thus our reduction and Theorem 1 yield the following result for average debit sum objectives.

Theorem 3. *The optimal player-1 strategy and the optimal value* $\mathsf{Opt}(\mathsf{AvDebSum})(q)$ *for average debit sum objective with coBüchi disjunction can be computed in time* $O(|Q|^6 \cdot |E| \cdot W^4)$. □

From debit-sum to difference-sum. An easy extension of the debit-sum objectives is instead of the sum of the weights, we consider the absolute values of the sum of the weights (*i.e* we consider $|\mathsf{Sum}(\rho(n))|$ values). We call the corresponding version as $\mathsf{DiffSum}$ instead of DebSum. This can be modeled as two weight functions (the original weight function and its negation), and then apply results for two-dimensional energy and mean-payoff games with disjunction with coBüchi objectives. Applying our techniques to solve eventual debit sum, and average debit sum, along with the results of [5, 21, 7] we obtain the following result.

Theorem 4. *The optimal player-1 strategy and the optimal value for difference-sum function with coBüchi disjunction,* $\mathsf{Opt}(\mathsf{DiffSumCB})(q)$, *the optimal value* $\mathsf{Opt}(\mathsf{EvDiffSumCB})(q)$ *for the eventual difference sum level objective with coBüchi disjunction, and the optimal value* $\mathsf{Opt}(\mathsf{AvDiffSumCB})(q)$ *for average difference sum objective with coBüchi disjunction, can all be computed in* $O(poly(Q, E, W))$ *time, where poly is a polynomial function.* □

4. QUANTITATIVE TIMED SIMULATION

We first define quantitative timed simulation functions (QTSFs) for timed transition systems in a game theoretic framework in Subsection 4.1, and then briefly present the model of timed automata. In Subsection 4.2 we present algorithms for computing the quantitative simulation functions for timed automata.

4.1 QTSFs from Timed Games

Timed Transition Game Structures. A *timed transition game structure* is a tuple $\mathfrak{G}_t = \langle S, \to\rangle$ where

- S is the set of states, consisting of player-1 states S_1 and player-2 states S_2 (*i.e.*, $S = S_1 \cup S_2$ and $S_1 \cap S_2 = \emptyset$),
- $\to \subseteq S \times \mathbb{R}^+ \times S$ is the transition relation such that for all $s \in S$ there exists at least one $s' \in S$ such that for some Δ, we have $s \xrightarrow{\Delta} s'$.

Plays, objectives, strategies, outcomes *etc.* are as in finite games (Section 3).

Quantitative Timed Simulation Functions. Analogous to the game theoretic presentation of timed simulation games, we now present a game theoretic definition of quantitative timed simulation functions. Recall the two player turn-based bipartite timed transition game structure $\mathfrak{G}_t(A_\mathfrak{r}, A_\mathfrak{s})$ defined previously in Section 2. Consider a play ρ in $\mathfrak{G}_t(A_\mathfrak{r}, A_\mathfrak{s})$: $\langle s_\mathfrak{r}^0, s_\mathfrak{s}^0, 2\rangle \xrightarrow{\Delta_\mathfrak{r}^0} \langle s_\mathfrak{r}^1, s_\mathfrak{s}^0, 1\rangle \xrightarrow{\Delta_\mathfrak{s}^0} \langle s_\mathfrak{r}^1, s_\mathfrak{s}^1, 2\rangle \xrightarrow{\Delta_\mathfrak{r}^1} \ldots$. Let $\rho(\mathfrak{r})$ be the projection on $A_\mathfrak{r}$, thus $\rho(\mathfrak{r})$ is the $A_\mathfrak{r}$ trajectory $s_\mathfrak{r}^0 \xrightarrow{\Delta_\mathfrak{r}^0} s_\mathfrak{r}^1 \xrightarrow{\Delta_\mathfrak{r}^1} \ldots$. Note that $\rho(\mathfrak{r})$ is a valid trajectory in $A_\mathfrak{r}$. We define $\rho(\mathfrak{s})$ similarly.

Definition 2 (Metric Over Simulation Game Plays). Recall the $\mathcal{D}_{\mathsf{MaxDiff}}, \mathcal{D}_{\mathsf{LimMaxDiff}}, \mathcal{D}_{\mathsf{AvgDiff}}$ trajectory trace difference metrics defined in Section 2. For $\Psi \in \{\mathcal{D}_{\mathsf{MaxDiff}}, \mathcal{D}_{\mathsf{LimMaxDiff}}, \mathcal{D}_{\mathsf{AvgDiff}}\}$, we define $\Psi^{\mathsf{Timediv}}()$ as follows for a play ρ in $\mathfrak{G}_t(A_\mathfrak{r}, A_\mathfrak{s})$:

$$\Psi^{\mathsf{Timediv}}(\rho) = \begin{cases} 0 & \text{if } \rho(\mathfrak{r}) \notin \mathsf{Timediv}(A_\mathfrak{r}) \\ \Psi(\rho(\mathfrak{r}), \rho(\mathfrak{s})) & \text{otherwise} \end{cases} \qquad \square$$

Note that Ψ^{Timediv} can be viewed as a metric over trajectories of the timed transition systems $A_\mathfrak{r}, A_\mathfrak{s}$.

Definition 3 (Quantitative Timed Simulation Functions). Let $A_\mathfrak{r}, A_\mathfrak{s}$ be timed transition systems, and let $\mathfrak{G}_t(A_\mathfrak{r}, A_\mathfrak{s})$ be the two player turn-based bipartite timed transition game structure defined previously. The value of the quantitative simulation function $\mathcal{S}_{\Psi^{\mathsf{Timediv}}}(\langle s_\mathfrak{r}, s_\mathfrak{s}\rangle)$, for $s_\mathfrak{r}$ and $s_\mathfrak{s}$ states of $A_\mathfrak{r}$ and $A_\mathfrak{s}$ respectively, and for $\Psi^{\mathsf{Timediv}} \in \{\mathcal{D}^{\mathsf{Timediv}}_{\mathsf{MaxDiff}}, \mathcal{D}^{\mathsf{Timediv}}_{\mathsf{LimMaxDiff}}, \mathcal{D}^{\mathsf{Timediv}}_{\mathsf{AvgDiff}}\}$, is defined as follows.

$$\mathcal{S}_{\Psi^{\mathsf{Timediv}}}(\langle s_\mathfrak{r}, s_\mathfrak{s}\rangle) = \inf_{\pi_\mathfrak{s} \in \Pi_\mathfrak{s}} \sup_{\pi_\mathfrak{r} \in \Pi_\mathfrak{r}} \Psi^{\mathsf{Timediv}}(\rho(\pi_\mathfrak{r}, \pi_\mathfrak{s}, \langle s_\mathfrak{r}, s_\mathfrak{s}, 2\rangle))$$

where $\rho(\pi_\mathfrak{r}, \pi_\mathfrak{s}, \langle s_\mathfrak{r}, s_\mathfrak{s}, 2\rangle)$ is the trajectory which results given the player-1 strategy $\pi_\mathfrak{s} \in \Pi_\mathfrak{s}$ and the player-2 strategy $\pi_\mathfrak{r} \in \Pi_\mathfrak{r}$. Equivalently, $\mathcal{S}_{\Psi^{\mathsf{Timediv}}}(\langle s_\mathfrak{r}, s_\mathfrak{s}\rangle) = \mathsf{Opt}(\Psi^{\mathsf{Timediv}})(\langle s_\mathfrak{r}, s_\mathfrak{s}, 2\rangle)$. $\square$

Timed Automata. Timed automata [1] suggest a finite syntax for specifying infinite-state timed game structures. A *timed automaton* $\mathcal{T}$ is a tuple $\langle L, \Sigma, C, \mu, \gamma, \rightarrow, S_0\rangle$, where L is the set of locations, Σ is the set of atomic propositions, C is a finite set of clocks, $S_0 \subseteq L \times \mathbb{R}^{+|C|}$ is the set of initial states, $\mu : L \mapsto 2^\Sigma$ is the observation map (it does not depend on clock values), γ is a function which describes the invariants on locations, and $\rightarrow$ is the set of transitions. A *clock valuation* is a function $\kappa : C \mapsto \mathbb{R}_{\geq 0}$ that maps every clock to a nonnegative real. A *state* $s = \langle l, \kappa\rangle$ of the timed automaton $\mathcal{T}$ is a location $l \in L$ together with a clock valuation κ such that the invariant at the location is satisfied, We let S be the set of all states of $\mathcal{T}$. Clock region equivalence, denoted as $\cong$ is an equivalence relation on states of timed automata. The equivalence classes of the relation are called *regions.* Given a state $\langle l, \kappa\rangle$ of $\mathcal{T}$, we denote the *region* containing $\langle l, \kappa\rangle$ as $\mathsf{Reg}(\langle l, \kappa\rangle)$. The region graph $\mathsf{Reg}(\mathcal{T})$ corresponding to $\mathcal{T}$ is the time-abstract bisimulation quotient graph induced by the *region equivalence relation.* We omit the full details for lack of space, these can be found in [9].

4.2 Computing QTSFs on Timed Automata

In this subsection we solve for quantitative simulation functions on timed automata by reducing the problem to games on finite-state graphs. For ease of presentation we assume that all clocks are bounded, *i.e.*, that the invariants of each location can be conjuncted with the clause $\wedge_{x \in C}(x \leq c_{\max})$ for some constant $c_{\max}$. The general case where clocks may be unbounded can be solved using similar algorithms, with some additional bookkeeping.

The solution involves the following steps. We first enlarge the timed game structure corresponding to the timed automaton $\mathcal{T}$ in order to measure elapsed time, and to measure the integer time "ticks" (or integer boundaries) crossed during an execution of $\mathcal{T}$ (a real time Δ corresponds to $\lfloor \Delta \rfloor$ integer boundaries crossed). Then, we define *integer* quantitative simulation functions which depend only on the integer time boundaries crossed, and show that these integer simulation functions are close to the original (real-valued) simulation functions. Next, we show that these integer simulation functions can be computed using finite game graphs, Finally, we present the algorithm which ties all the steps together, and show that we can compute the quantitative simulation functions to within any desired degree of accuracy on timed automata.

Enlarging the Timed Game Structure. Given a timed automata $\mathcal{T}$ where all the clocks are bounded by $c_{\max}$, let $[\![\mathcal{T}]\!]$ denote the timed transition system obtained by adding to $\mathcal{T}$ an extra clock z, which cycles between 0 and 1 for measuring elapsed time, and an integer valued variable *ticks* which takes on values in $\mathbb{N}_{\leq c_{\max}}$, where $\mathbb{N}_{\leq c_{\max}}$ denotes the set $\{0, 1, \ldots, c_{\max}\}$. Formally, the set of states of $[\![\mathcal{T}]\!]$ is $S^{[\![\mathcal{T}]\!]} = S \times \mathbb{R}_{[0,1)} \times \mathbb{N}_{\leq c_{\max}}$, where S is the set of states of $\mathcal{T}$. The state $\langle s, \mathfrak{z}, \mathfrak{d}\rangle$ of $[\![\mathcal{T}]\!]$ has the following components:

- s is the state of the original timed automaton $\mathcal{T}$;
- $\mathfrak{z}$ is the value of the added clock z which gets reset to 0 every time it crosses 1 (i.e., if κ' is the clock valuation resulting from letting time Δ elapse from an initial clock valuation κ, then, $\mathfrak{z} = \kappa'(z) = (\kappa(z) + \Delta) \mod 1$); and
- $\mathfrak{d}$ denotes the value of the integer variable *ticks*, and is equal to the number of integer boundaries crossed by the added clock z since the last transition: if the clock valuation in the previous state was κ, and the transition time duration is Δ, then $\mathfrak{d} = \lfloor \kappa(z) + \Delta \rfloor$ in the current state, where $\lfloor\rfloor$ denotes the integer floor function. Note that since all the clocks in $\mathcal{T}$ are bounded by $c_{\max}$, we have $\mathfrak{d} \leq c_{\max}$, as the maximum duration of a transition is $c_{\max}$, and $\kappa(z) < 1$ in the previous state.

The region equivalence relation can be expanded to $[\![\mathcal{T}]\!]$ states. Two states $\langle\langle l_1, \kappa_1\rangle, \mathfrak{z}_1, \mathfrak{d}_1\rangle$ and $\langle\langle l_2, \kappa_2\rangle \mathfrak{z}_2, \mathfrak{d}_2\rangle$ of $[\![\mathcal{T}]\!]$ are defined to be region equivalent if $\langle l_1, \mathfrak{d}_1\rangle = \langle l_2, \mathfrak{d}_2\rangle$, and $\kappa_1^{z=\mathfrak{z}_1} \cong \kappa_2^{z=\mathfrak{z}_2}$, where $\kappa_i^{z=\mathfrak{z}_i}$ denotes the clock valuation κ_i on C expanded to a clock valuation to $C \cup \{z\}$ by mapping z to $\mathfrak{z}_i$ (we denote the enlarged clock valuation be denoted as $\widehat{\kappa}$). Similar to the region graph $\mathsf{Reg}(\mathcal{T})$, we define an untimed finite state bisimulation quotient graph $\mathsf{Reg}([\![\mathcal{T}]\!])$ for $[\![\mathcal{T}]\!]$.

Given a state s of $\mathcal{T}$, we denote by $[\![s]\!]$ the state $\langle s, 0, 0\rangle$ of $[\![\mathcal{T}]\!]$. For a state trajectory $\mathit{traj} = s_0 \xrightarrow{t_0} s_1 \xrightarrow{t_1} \ldots$, we let $\mathit{traj}[i]$ denote the state s_i. Given a state trajectory traj of the timed automaton $\mathcal{T}$, we denote by $[\![\mathit{traj}]\!]$ the $[\![\mathcal{T}]\!]$ trajectory $[\![\mathit{traj}[0]]\!] \xrightarrow{t_0} \widehat{s}_1 \xrightarrow{t_1} \widehat{s}_2 \ldots$, where $\widehat{s}_i = \langle s_i, \mathfrak{z}_i, \mathfrak{d}_i\rangle$, and $\mathfrak{z}_i, \mathfrak{d}_i$ values are according to the times of the transitions. That is,

$[\![traj]\!]$ denotes the trajectory obtained by adding the clock z, and the integer variable *ticks*, where the values for both the new variables are set to 0 in the starting state $[\![traj[0]]\!]$. The new variables just observe the time, and the integer boundaries crossed for each transition according to the semantics for $[\![\mathcal{T}]\!]$ described previously. The first component of $[\![traj[i]]\!]$ is the same as the state $traj[i]$ for all i. We note that a trajectory is time-divergent iff it satisfies the Büchi constraint $\mathsf{Büchi}\left(\bigvee_{i=1}^{c_{\max}} ticks = i\right)$.

Integer Time. For the trajectory $[\![traj]\!]$, let $\mathsf{time}^{\mathsf{int}}_{[\![traj]\!]}[i]$ denote the number of integer boundaries crossed upto the i-th transition:

$$\mathsf{time}^{\mathsf{int}}_{[\![traj]\!]}[i] = \lfloor \mathsf{time}_{[\![traj]\!]}[i] \rfloor$$

We have the following lemma which expresses $\mathsf{time}^{\mathsf{int}}_{[\![traj]\!]}[i]$ in terms of the of the values of the *ticks* variable in traces. Note that the value of the *ticks* variable is zero in the first state of a valid trajectory $[\![traj]\!]$.

Lemma 3. *Let $traj$ be a trajectory of a timed automaton $\mathcal{T}$ in which all clocks are bounded. We have*

$$\mathsf{time}^{\mathsf{int}}_{[\![traj]\!]}[i] = \sum_{j=0}^{i} \mathfrak{d}_j$$

where $\mathfrak{d}_j$ is the value of the ticks variable in $[\![traj]\!][j]$. □

The Integer Trace Difference Metrics $\mathcal{D}^{\mathsf{int}}_{\mathsf{MaxDiff}}$, $\mathcal{D}^{\mathsf{int}}_{\mathsf{LimMaxDiff}}$, and $\mathcal{D}^{\mathsf{int}}_{\mathsf{AvgDiff}}$. Corresponding to the trace difference metric $\mathcal{D}_{\varphi}()$, for $\varphi = \mathsf{MaxDiff}, \mathsf{LimMaxDiff}, \mathsf{AvgDiff}$, we define the trace difference metric $\mathcal{D}^{\mathsf{int}}_{\varphi}()$, by substituting $\mathsf{time}^{\mathsf{int}}()$ for $\mathsf{time}()$ in the definition of $\mathcal{D}_{\varphi}()$, and using only the location component of $\mathcal{T}$ for matching observations. Lemma 4 shows that $\mathcal{D}^{\mathsf{int}}_{\varphi}()$ closely approximates $\mathcal{D}_{\varphi}()$.

Lemma 4. *Let $traj_1$ and $traj_2$ be two trajectories of a timed automaton $\mathcal{T}$. The following assertions are true for $\varphi \in \{\mathsf{MaxDiff}, \mathsf{LimMaxDiff}, \mathsf{AvgDiff}\}$.*

1. *$\mathcal{D}_{\varphi}([\![traj_1]\!], [\![traj_2]\!]) = \infty$ iff $\mathcal{D}^{\mathsf{int}}_{\varphi}([\![traj_1]\!], [\![traj_2]\!]) = \infty$.*
2. *If both $\mathcal{D}_{\varphi}([\![traj_1]\!], [\![traj_2]\!])$ and $\mathcal{D}^{\mathsf{int}}_{\varphi}([\![traj_1]\!], [\![traj_2]\!])$ are less than ∞, then*

$$|\mathcal{D}_{\varphi}([\![traj_1]\!], [\![traj_2]\!]) - \mathcal{D}^{\mathsf{int}}_{\varphi}([\![traj_1]\!], [\![traj_2]\!])| \leq 1 \quad \square$$

Integer Quantitative Timed Simulation Functions Using $\mathcal{D}^{\mathsf{int}}_{\mathsf{MaxDiff}}$, $\mathcal{D}^{\mathsf{int}}_{\mathsf{LimMaxDiff}}$, and $\mathcal{D}^{\mathsf{int}}_{\mathsf{AvgDiff}}$, we can define integer quantitative simulation functions which approximate $\mathcal{S}_{\Psi^{\mathsf{Timediv}}}$ for $\Psi^{\mathsf{Timediv}} \in \{\mathcal{D}^{\mathsf{Timediv}}_{\mathsf{MaxDiff}}, \mathcal{D}^{\mathsf{Timediv}}_{\mathsf{LimMaxDiff}}, \mathcal{D}^{\mathsf{Timediv}}_{\mathsf{AvgDiff}}\}$.

Definition 4 (Integer Metric over Simulation Game Plays). For $\Lambda \in \{\mathcal{D}^{\mathsf{int}}_{\mathsf{MaxDiff}}, \mathcal{D}^{\mathsf{int}}_{\mathsf{LimMaxDiff}}, \mathcal{D}^{\mathsf{int}}_{\mathsf{AvgDiff}}\}$, we define $\Lambda^{\mathsf{Timediv}}()$ as follows for a play ρ in $\mathfrak{G}_t([\![\mathcal{T}_\mathfrak{r}]\!], [\![\mathcal{T}_\mathfrak{s}]\!])$:

$$\Lambda^{\mathsf{Timediv}}(\rho) = \begin{cases} 0 & \text{if } \rho(\mathfrak{r}) \notin \mathsf{Timediv}([\![\mathcal{T}_\mathfrak{r}]\!]) \\ \Lambda(\rho(\mathfrak{r}), \rho(\mathfrak{s})) & \text{otherwise} \end{cases} \quad \square$$

The *integer quantitative simulation functions* $\mathcal{S}_{\Lambda^{\mathsf{Timediv}}}(\langle s_\mathfrak{r}, s_\mathfrak{s}\rangle)$, can now be defined exactly as in Definition 3, using $\Lambda^{\mathsf{Timediv}}$ instead of Ψ^{Timediv}.

Proposition 3 states that the integer simulation functions closely approximate the original quantitative simulation functions.

Proposition 3 (Integer Simulation Functions Approximate Quantitative Simulation Functions). *Let $\mathcal{T}_\mathfrak{r}, \mathcal{T}_\mathfrak{s}$ be timed automata, with the corresponding enlarged timed transition systems $[\![\mathcal{T}_\mathfrak{r}]\!], [\![\mathcal{T}_\mathfrak{s}]\!]$ respectively, and let $\mathfrak{G}_t([\![A_\mathfrak{r}]\!], [\![A_\mathfrak{s}]\!])$ be the two player turn-based bipartite timed simulation game structure. For $\langle \Lambda, \Psi \rangle$ in $\{\langle \mathcal{D}^{\mathsf{int}}_{\mathsf{MaxDiff}}, \mathcal{D}_{\mathsf{MaxDiff}}\rangle, \langle \mathcal{D}^{\mathsf{int}}_{\mathsf{LimMaxDiff}}, \mathcal{D}_{\mathsf{LimMaxDiff}}\rangle, \langle \mathcal{D}^{\mathsf{int}}_{\mathsf{AvgDiff}}, \mathcal{D}_{\mathsf{AvgDiff}}\rangle\}$, we have the following assertions to be true.*

1. *$\mathcal{S}_{\Lambda^{\mathsf{Timediv}}}(\langle [\![s_\mathfrak{r}]\!], [\![s_\mathfrak{s}]\!]\rangle) = \infty$ iff $\mathcal{S}_{\Psi^{\mathsf{Timediv}}}(\langle [\![s_\mathfrak{r}]\!], [\![s_\mathfrak{s}]\!]\rangle) = \infty$.*
2. *If $\mathcal{S}_{\Lambda^{\mathsf{Timediv}}}(\langle [\![s_\mathfrak{r}]\!], [\![s_\mathfrak{s}]\!]\rangle) < \infty$ and $\mathcal{S}_{\Psi^{\mathsf{Timediv}}}(\langle [\![s_\mathfrak{r}]\!], [\![s_\mathfrak{s}]\!]\rangle) < \infty$, then*

$$|\mathcal{S}_{\Lambda^{\mathsf{Timediv}}}(\langle [\![s_\mathfrak{r}]\!], [\![s_\mathfrak{s}]\!]\rangle) - \mathcal{S}_{\Psi^{\mathsf{Timediv}}}(\langle [\![s_\mathfrak{r}]\!], [\![s_\mathfrak{s}]\!]\rangle)| \leq 1 \quad \square$$

4.2.1 Reduction to Finite Weighted Game Graphs.

We now show how to compute the values of the integer quantitative simulation functions by reductions to finite state weighted games.

The Integer Trace Difference Metrics and Simulation Functions on Untimed Region Graphs. We first lift the integer trace difference metrics $\Lambda^{\mathsf{Timediv}}$ for $\Lambda \in \{\mathcal{D}^{\mathsf{int}}_{\mathsf{MaxDiff}}, \mathcal{D}^{\mathsf{int}}_{\mathsf{LimMaxDiff}}, \mathcal{D}^{\mathsf{int}}_{\mathsf{AvgDiff}}\}$ to (untimed) region graphs. Let $\mathsf{Reg}([\![\mathcal{T}]\!])$ be the region graph corresponding to the enlarged time game structure $[\![\mathcal{T}]\!]$. Let the observation function μ be defined as $\mu(\langle l, \kappa, \mathfrak{z}, \mathfrak{d}\rangle) = \langle \mu(l), \mathfrak{d}\rangle$ (this is done for technical reasons, see [9]).

Given the two timed automata $\mathcal{T}_\mathfrak{r}, \mathcal{T}_\mathfrak{s}$, consider the *untimed* simulation game $\mathfrak{G}_u^{\dagger}(\mathsf{Reg}([\![\mathcal{T}_\mathfrak{r}]\!]), \mathsf{Reg}([\![\mathcal{T}_\mathfrak{s}]\!]))$ defined to be the untimed simulation game $\mathfrak{G}_u(\mathsf{Reg}([\![\mathcal{T}_\mathfrak{r}]\!]), \mathsf{Reg}([\![\mathcal{T}_\mathfrak{s}]\!]))$, but with the observation function $\mu^{\dagger}((\langle l, \kappa, \mathfrak{z}, \mathfrak{d}\rangle) = \mu(l)$. For a play ρ of $\mathfrak{G}_u^{\dagger}$, we define $\rho(\mathfrak{r})$ and $\rho(\mathfrak{s})$ as the projections on $\mathsf{Reg}([\![\mathcal{T}_\mathfrak{r}]\!])$ and $\mathsf{Reg}([\![\mathcal{T}_\mathfrak{s}]\!])$ respectively. For a region graph $\mathsf{Reg}([\![\mathcal{T}]\!])$ we define $\mathsf{Timediv}(\mathsf{Reg}([\![\mathcal{T}]\!])$ as the set of runs satisfying the Büchi condition $\mathsf{Büchi}\left(\bigvee_{i=1}^{c_{\max}} ticks = i\right)$. Note that this has the intended meaning of encoding time divergence. Next we define $\Lambda^{\mathsf{Timediv}}$ for $\Lambda \in \{\mathcal{D}^{\mathsf{int}}_{\mathsf{MaxDiff}}, \mathcal{D}^{\mathsf{int}}_{\mathsf{LimMaxDiff}}, \mathcal{D}^{\mathsf{int}}_{\mathsf{AvgDiff}}\}$ on plays of $\mathfrak{G}_u^{\dagger}(\mathsf{Reg}([\![\mathcal{T}_\mathfrak{r}]\!]), \mathsf{Reg}([\![\mathcal{T}_\mathfrak{s}]\!]))$ using Lemma 3 for defining $\mathsf{time}^{\mathsf{int}}_{\mathsf{Reg}([\![traj]\!])}(i)$ in terms of the sum of the *ticks* variable. Finally, we define the integer simulation functions as for the the timed simulation game $\mathfrak{G}_t$ using the above definitions.

Lemma 5 states that the values of the integer simulation functions of the region graphs are the same as that on timed automata. Note that region graphs are *untimed* structures.

Lemma 5. *Let $\mathcal{T}_\mathfrak{r}, \mathcal{T}_\mathfrak{s}$ be timed automata, and let $\mathsf{Reg}([\![\mathcal{T}_\mathfrak{r}]\!]), \mathsf{Reg}([\![\mathcal{T}_\mathfrak{s}]\!])$ be region graphs of the corresponding enlarged timed game structures $[\![\mathcal{T}_\mathfrak{r}]\!], [\![\mathcal{T}_\mathfrak{s}]\!]$ respectively. For any states $[\![s_\mathfrak{r}]\!]$ of $[\![\mathcal{T}_\mathfrak{r}]\!]$ and $[\![s_\mathfrak{s}]\!]$ of $[\![\mathcal{T}_\mathfrak{s}]\!]$, we have*

$$\mathcal{S}^{\mathfrak{G}_t([\![\mathcal{T}_\mathfrak{r}]\!],[\![\mathcal{T}_\mathfrak{s}]\!])}_{\Lambda^{\mathsf{Timediv}}}\left(\langle [\![s_\mathfrak{r}]\!], [\![s_\mathfrak{s}]\!]\rangle\right) = \mathcal{S}^{\mathfrak{G}_u^{\dagger}(\mathsf{Reg}([\![\mathcal{T}_\mathfrak{r}]\!]),\mathsf{Reg}([\![\mathcal{T}_\mathfrak{s}]\!]))}_{\Lambda^{\mathsf{Timediv}}}\left(\langle \mathsf{Reg}([\![s_\mathfrak{r}]\!]), \mathsf{Reg}([\![s_\mathfrak{s}]\!])\rangle\right)$$

where $\Lambda \in \{\mathcal{D}^{\mathsf{int}}_{\mathsf{MaxDiff}}, \mathcal{D}^{\mathsf{int}}_{\mathsf{LimMaxDiff}}, \mathcal{D}^{\mathsf{int}}_{\mathsf{AvgDiff}}\}$. □

The weighted finite untimed game graph $\mathfrak{F}(\mathsf{Reg}([\![\mathcal{T}_\mathfrak{r}]\!]), \mathsf{Reg}([\![\mathcal{T}_\mathfrak{s}]\!]))$. Now we construct a finite weighted game graph $\mathfrak{F}(\mathsf{Reg}([\![\mathcal{T}_\mathfrak{r}]\!]), ([\![\mathcal{T}_\mathfrak{s}]\!]))$, on which we can use the algorithms of Section 3, to compute the values of the integer quantitative simulation function for $\mathfrak{G}_u^{\dagger}(\mathsf{Reg}([\![\mathcal{T}_\mathfrak{r}]\!]), \mathsf{Reg}([\![\mathcal{T}_\mathfrak{s}]\!]))$. The game structure $\mathfrak{F}$ is essentially the untimed simulation game $\mathfrak{G}_u^{\dagger}$ over the region graphs, where weights are assigned to transitions based on the *tick* values of the region states. The full formal construction is omitted due to lack of space and can be found in [9]. The next lemma states that to compute the values of the integer quantitative simulation functions on the region graphs, we can use the objectives $\mathsf{DiffSumCB}, \mathsf{EvDiffSumCB}, \mathsf{AvDiffSumCB}$ on the weighted finite game $\mathfrak{F}(\mathsf{Reg}(\llbracket \mathcal{T}_{\mathfrak{r}} \rrbracket), \mathsf{Reg}(\llbracket \mathcal{T}_{\mathfrak{s}} \rrbracket))$.

Lemma 6. *Let $\mathcal{T}_{\mathfrak{r}}$ and $\mathcal{T}_{\mathfrak{s}}$ be well-formed timed automata such that all clocks are bounded by $c_{\max}$, and let $\mathfrak{F}(\mathsf{Reg}(\llbracket \mathcal{T}_{\mathfrak{r}} \rrbracket), \mathsf{Reg}(\llbracket \mathcal{T}_{\mathfrak{s}} \rrbracket))$ be the weighted game structure corresponding to $\mathfrak{S}_u^{\dagger}(\mathsf{Reg}(\llbracket \mathcal{T}_{\mathfrak{r}} \rrbracket), \mathsf{Reg}(\llbracket \mathcal{T}_{\mathfrak{s}} \rrbracket))$, as described above. Fix the coBüchi objective $\mathsf{coBüchi}(ticks_{\mathfrak{r}} = 0)$ in the following. For $\langle \Lambda, \Xi \rangle$ equal to $\langle \mathcal{D}^{\mathsf{int}}_{\mathsf{MaxDiff}}, \mathsf{DiffSumCB} \rangle$, or $\langle \mathcal{D}^{\mathsf{int}}_{\mathsf{LimMaxDiff}}, \mathsf{EvDiffSumCB} \rangle$, or $\langle \mathcal{D}^{\mathsf{int}}_{\mathsf{AvgDiff}}, \mathsf{AvDiffSumCB} \rangle$, we have*

$$\mathcal{S}^{\mathfrak{S}_u^{\dagger}(\mathsf{Reg}(\llbracket \mathcal{T}_{\mathfrak{r}} \rrbracket), \mathsf{Reg}(\llbracket \mathcal{T}_{\mathfrak{s}} \rrbracket))}_{\Lambda^{\mathsf{Timediv}}}\Big(\langle \mathsf{Reg}(\llbracket s_{\mathfrak{r}} \rrbracket), \mathsf{Reg}(\llbracket s_{\mathfrak{s}} \rrbracket) \rangle\Big) = \Big(\mathsf{Opt}^{\mathfrak{F}\left(\mathsf{Reg}(\llbracket \mathcal{T}_{\mathfrak{r}} \rrbracket), \mathsf{Reg}(\llbracket \mathcal{T}_{\mathfrak{s}} \rrbracket)\right)}(\Xi)\Big)\Big(\langle \mathsf{Reg}(\llbracket s_{\mathfrak{r}} \rrbracket), \mathsf{Reg}(\llbracket s_{\mathfrak{s}} \rrbracket), 2 \rangle\Big)$$

□

Precision of the Integer Simulation Functions. Given a positive integer $\alpha \geq 1$, and a timed automaton $\mathcal{T}$, let $\alpha \cdot \mathcal{T}$ denote the timed automaton obtained from $\mathcal{T}$ by multiplying every constant by α. Note that if clocks are bounded by $c_{\max}$ in $\mathcal{T}$, then clocks are bounded by $\alpha \cdot c_{\max}$ in $\alpha \cdot \mathcal{T}$. The automaton $\alpha \cdot \mathcal{T}$ is just $\mathcal{T}$ with a blown up timescale. One time unit in $\mathcal{T}$ corresponds to α time units in $\alpha \cdot \mathcal{T}$. We let $\alpha \cdot \llbracket \mathcal{T} \rrbracket = \llbracket \alpha \cdot \mathcal{T} \rrbracket$, and

$$\alpha \cdot \langle l, \kappa, \mathfrak{z}, \mathfrak{d} \rangle = \langle l, \alpha \cdot \kappa, \mathsf{frac}(\alpha \cdot \mathfrak{z}), \lfloor \alpha \cdot \mathfrak{z} \rfloor + \alpha \cdot \mathfrak{d} \rangle$$

where $\mathsf{frac}(\beta)$ denotes the fractional part of β, *i.e.* $\beta - \lfloor \beta \rfloor$ for $\beta \geq 0$. Note that in $\alpha \cdot \llbracket \mathcal{T} \rrbracket$, the clock z *still cycles from 0 to 1.* Thus, we first blow up the time scale of $\mathcal{T}$ to obtain $\alpha \cdot \mathcal{T}$, and *then* take the expanded game structure $\llbracket \alpha \cdot \mathcal{T} \rrbracket$.

Final Algorithms and Results. Applying the previous results, we obtain the following Theorem which states that the values for the (real-valued) quantitative simulation functions $\mathcal{S}_{\Psi^{\mathsf{Timediv}}}$ for $\Psi \in \{\mathcal{D}_{\mathsf{MaxDiff}}, \mathcal{D}_{\mathsf{LimMaxDiff}}, \mathcal{D}_{\mathsf{AvgDiff}}\}$, can be computed to within any desired degree of accuracy using the algorithm in the function $h_{\Psi,\alpha}(s_{\mathfrak{r}}, s_{\mathfrak{s}})$.

Theorem 5. *Let $\mathcal{T}_{\mathfrak{r}}$ and $\mathcal{T}_{\mathfrak{s}}$ be well-formed timed automata such that all clocks are bounded by $c_{\max}$, and let $\alpha \geq 1$ be a positive integer. For $\Psi \in \{\mathcal{D}_{\mathsf{MaxDiff}}, \mathcal{D}_{\mathsf{LimMaxDiff}}, \mathcal{D}_{\mathsf{AvgDiff}}\}$, the function $h_{\Psi,\alpha}()$ is such that for any states $s_{\mathfrak{r}}$ of $\mathcal{T}_{\mathfrak{r}}$ and $s_{\mathfrak{s}}$ of $\mathcal{T}_{\mathfrak{s}}$, we have either (i) $\mathcal{S}_{\Psi^{\mathsf{Timediv}}}(\langle s_{\mathfrak{r}}, s_{\mathfrak{s}} \rangle) = h_{\Psi,\alpha}(s_{\mathfrak{r}}, s_{\mathfrak{s}}) = \infty$; or (ii) both values are finite and $|\mathcal{S}_{\Psi^{\mathsf{Timediv}}}(\langle s_{\mathfrak{r}}, s_{\mathfrak{s}} \rangle) - h_{\Psi,\alpha}(s_{\mathfrak{r}}, s_{\mathfrak{s}})| \leq \frac{1}{\alpha}$.* □

Concluding Remarks. We have presented algorithms for computing the various types of QTSF values (to any desired degree of accuracy) for timed automata. We note that the optimal player-1 strategies in the games required for QTSF values are also computable, and are witnesses to the quantitative simulation function values (dual to simulation relations witnessing the simulation decision problem).

5. REFERENCES

[1] R. Alur and D. L. Dill. A theory of timed automata. *Theor. Comput. Sci.*, 126(2):183–235, 1994.

[2] A. C. Antoulas, D. C. Sorensen, and S. Gugercin. A survey of model reduction methods for large-scale systems. *Contemporary Mathematics*, 280:193–219, 2001.

[3] P. Bouyer, N. Markey, J. Olschewski, and M. Ummels. Measuring permissiveness in parity games: Mean-payoff parity games revisited. In *ATVA*, LNCS 6996, pages 135–149. 2011.

[4] K. Cerans. Decidability of bisimulation equivalences for parallel timer processes. In *CAV*, LNCS 663, pages 302–315. 1992.

Input : States $s_{\mathfrak{r}}, s_{\mathfrak{s}}$ from $\mathcal{T}_{\mathfrak{r}}, \mathcal{T}_{\mathfrak{s}}$ respectively;
$\Psi \in \{\mathcal{D}_{\mathsf{MaxDiff}}, \mathcal{D}_{\mathsf{LimMaxDiff}}, \mathcal{D}_{\mathsf{AvgDiff}}\}$;
α a positive integer

Output: A number approximating $\mathcal{S}_{\Psi^{\mathsf{Timediv}}}(\langle s_{\mathfrak{r}}, s_{\mathfrak{s}} \rangle)$

$\mathsf{Reg}(\llbracket \alpha \cdot \mathcal{T}_{\mathfrak{r}} \rrbracket), \mathsf{Reg}(\llbracket \alpha \cdot \mathcal{T}_{\mathfrak{s}} \rrbracket) :=$ Region graphs of the expanded timed game structures $\llbracket \alpha \cdot \mathcal{T}_{\mathfrak{r}} \rrbracket$ and $\llbracket \alpha \cdot \mathcal{T}_{\mathfrak{s}} \rrbracket$;

$\mathfrak{F} := \mathfrak{F}(\mathsf{Reg}(\llbracket \alpha \cdot \mathcal{T}_{\mathfrak{r}} \rrbracket), \mathsf{Reg}(\llbracket \alpha \cdot \mathcal{T}_{\mathfrak{s}} \rrbracket))$;

`/* Finite weighted turn based game constructed from the region graphs */`

switch Ψ **do**
- **case** $\mathcal{D}_{\mathsf{MaxDiff}}$
  - $\Xi := \mathsf{DiffSumCB}_{\mathsf{coBüchi}(ticks_{\mathfrak{r}}=0)}$;
- **case** $\mathcal{D}_{\mathsf{LimMaxDiff}}$
  - $\Xi := \mathsf{EvDiffSumCB}_{\mathsf{coBüchi}(ticks_{\mathfrak{r}}=0)}$;
- **case** $\mathcal{D}_{\mathsf{AvgDiff}}$
  - $\Xi := \mathsf{AvDiffSumCB}_{\mathsf{coBüchi}(ticks_{\mathfrak{r}}=0)}$;

endsw

return $\alpha^{-1} \cdot \mathsf{Opt}^{\mathfrak{F}}(\Xi)\left(\left\langle \begin{array}{c} \mathsf{Reg}(\llbracket \alpha \cdot s_{\mathfrak{r}} \rrbracket), \\ \mathsf{Reg}(\llbracket \alpha \cdot s_{\mathfrak{s}} \rrbracket), \\ 2 \end{array} \right\rangle\right)$;

Function $h_{\Psi,\alpha}(s_{\mathfrak{r}}, s_{\mathfrak{s}})$

[5] J. Chaloupka. Z-reachability problem for games on 2-dimensional vector addition systems with states is in P. In *RP*, pages 104–119, 2010.

[6] K. Chatterjee and L. Doyen. Energy parity games. In *ICALP*, LNCS 6199, pages 599–610. Springer, 2010.

[7] K. Chatterjee, L. Doyen, T.A. Henzinger, and J.-F. Raskin. Generalized mean-payoff and energy games. In *FSTTCS*, LIPIcs 8, pages 505–516. Schloss Dagstuhl - LZI, 2010.

[8] K. Chatterjee, T.A. Henzinger, and M. Jurdzinski. Mean-payoff parity games. In *LICS*, pages 178–187. IEEE. 2005.

[9] K. Chatterjee and V. S. Prabhu. Quantitative timed simulation functions and refinement metrics for timed systems (full version). *CoRR*, abs/1212.6556, 2012.

[10] S. Chaudhuri, S. Gulwani, and R. Lublinerman. Continuity and robustness of programs. *Commun. ACM*, 55(8):107–115, 2012.

[11] L. d. Alfaro, M. Faella, and M. Stoelinga. Linear and branching system metrics. *IEEE Trans. Soft. Eng.*, 35:258–273, 2009.

[12] J. Desharnais, V. Gupta, R. Jagadeesan, and P. Panangaden. Metrics for labelled markov processes. *TCS*, 318:323–354, 2004.

[13] E.A. Emerson and C. Jutla. Tree automata, mu-calculus and determinacy. In *FOCS'91*, pages 368–377. IEEE, 1991.

[14] A. Girard, A. A. Julius, and G. Pappas. Approximate simulation relations for hybrid systems. *Discrete Event dynamic Systems*, 18(2):163–179, 2008.

[15] A. Girard, G. Pola, and P. Tabuada. Approximately bisimilar symbolic models for incrementally stable switched systems. *Automatic Control, IEEE Transactions*, 55:116 –126, 2010.

[16] T. A. Henzinger, R. Majumdar, and V. S. Prabhu. Quantifying similarities between timed systems. In *FORMATS*, LNCS 3829, pages 226–241. Springer, 2005.

[17] J. Jacod and A. Shiryaev. *Limit Theorems for Stochastic Processes.* Springer, 2003.

[18] G. Pola, P. Pepe, M. D. Di Benedetto, and P. Tabuada. Symbolic models for nonlinear time-delay systems using approximate bisimulations. *Systems & Control Letters*, 59(6):365 – 373, 2010.

[19] J.D. Quesel, M. Fränzle, and W. Damm. Crossing the bridge between similar games. In *FORMATS 2011*, LNCS 6919, pages 160–176, 2011.

[20] F. v. Breugel, M. W. Mislove, J. Ouaknine, and J. Worrell. An intrinsic characterization of approximate probabilistic bisimilarity. In *FoSSaCS*, LNCS 2620, pages 200–215. 2003.

[21] Y. Velner and A. Rabinovich. Church synthesis problem for noisy input. In *FOSSACS*, pages 275–289, 2011.

Formula-free Finite Abstractions for Linear Temporal Verification of Stochastic Hybrid Systems*

Ilya Tkachev
TU Delft - Delft University of Technology
i.tkachev@tudelft.nl

Alessandro Abate
TU Delft - Delft University of Technology
a.abate@tudelft.nl

ABSTRACT

Results on approximate model-checking of Stochastic Hybrid Systems (SHS) against general temporal specifications lead to abstractions that structurally depend on the given specification or with a state cardinality that crucially depends on the size of the specification. In order to cope with the associated issues of generality and scalability, we propose a specification-free abstraction approach that is general, namely it allows constructing a single abstraction to be then used for a whole cohort of problems. It furthermore computationally outperforms specification-dependent abstractions over linear temporal properties, such as bounded LTL (BLTL). The proposed approach unifies techniques for the approximate abstraction of SHS over different classes of properties by explicitly relating the error introduced by the approximation to the distance between transition kernels of abstract and concrete models, and by propagating the error in time over the horizon of the specification. The new technique is compared over a case study to related results in the literature.

Categories and Subject Descriptors

G.3 [**Probability and Statistics**]: Stochastic processes

Keywords

Markov processes, stochastic hybrid systems, formal verification, probabilistic model-checking, linear temporal specifications, finite abstractions, approximate bisimulations.

1. INTRODUCTION

Stochastic Hybrid Systems (SHS) provide a powerful modeling framework for diverse application areas such as systems biology, air traffic control, power networks and telecommunication systems [8, 20]. The reliable employment of SHS models demands solid foundations for their analysis and verification.

One of the most prominent tools for the verification of finite-state systems is model checking. In particular, with focus on the discrete-time case, the verification theory of finite-state probabilistic models known as discrete-time Markov Chains (`dt`-MC) is mature [5]. The formal verification of `dt`-MC is enabled by probabilistic model checking: in this instance verification problems allow for explicit solutions or answers that can be obtained in a numerically efficient way leveraging dedicated probabilistic model checking software [14, 17]. On the other hand, the price to pay for the descriptive generality of SHS, models that are characterized by an uncountable state space, is the lack of explicit solutions and the undecidability for most verification problems [1]. A possible approach to overcome this issue is based on the concept of *abstraction*, namely "a quotient system that preserves some properties of interest, while ignoring details" [19]. Abstractions are ideally *finite*, as this often leads to problem decidability and to explicit solutions. For SHS, finite abstractions are naturally `dt`-MC. Notice however that whenever the original state space is infinite (as for SHS models), it is often only possible to synthesize a finite model that is an *approximate* abstraction of the concrete one [12].

Many properties of interest for SHS can be expressed as PCTL formulae or as linear temporal (LT) specifications [5]. With focus on the former, the work in [20] has formally related the verification of PCTL formulae to the computation of corresponding *value functions* defined over the state space of a SHS. Given an initial state, such a value function represents the probability that the execution of SHS satisfies a given PCTL path formula. Thus one can relate the quality of an approximate abstraction with respect to a given property to the difference between value functions computed respectively over the abstraction and over the concrete model [1].

So far only *formula-dependent* techniques have been developed to find approximate finite abstractions of SHS. The first step of these techniques is to leverage dynamic programming (DP) [6] principles to derive DP-like recursions for the value function related to a given formula. The second step is to build an abstraction in order to numerically compute integrals involved in the DP recursions, with explicit bounds on the approximation error. The work in [1] has developed this approach and applied it to the problem of probabilistic safety (or invariance) within the class of PCTL formulae. Later, [22] has further improved these results by relaxing some of the model assumptions and by finding tighter error bounds, which in turn led to a lower cardinality of the abstraction required to match a given precision. Both works have used DP procedures for bounded-horizon

*This work is supported by the European Commission MoVeS project FP7-ICT-2009-5 257005, by the European Commission Marie Curie grant MANTRAS 249295, by the European Commission NoE HYCON2 FP7-ICT-2009-5 257462, and by the NWO VENI grant 016.103.020.

HSCC'13, April 8–11, 2013, Philadelphia, Pennsylvania, USA.

safety value function developed in [3]. Such recursions have also been developed for the probabilistic reach-avoid problem in [23]. Although PCTL path formulae for safety and reach-avoid are part of a more general class of LT specifications, it has not been clear yet whether DP recursions could be also developed for other LT specifications. Due to this reason, [2] has suggested a new approach for the verification of LT specifications, by reducing the original problem to the safety one defined over a new SHS, the latter being the product between the original SHS and the automaton corresponding to the specification of interest. Let us mention that the approximate abstraction methods discussed above are limited to the verification of bounded-horizon specifications – the work in [25, 26] argued that direct abstraction may not work for infinite-time problems, and developed alternative techniques to tackle them.

Notice that all the described methods require building a brand new approximate abstraction for each given different formula. This contribution is thus challenged to develop *formula-free* finite abstractions over SHS. More precisely: given a SHS $\mathfrak{D}$, a bounded time horizon n and a precision level ε, we provide an explicit way to build a dt-MC $\hat{\mathfrak{D}}$ which allows computing value functions of any n-bounded LT specification with an error that does not exceed ε. This result has several important features. Firstly, no matter how many properties are to be model-checked against $\mathfrak{D}$, one has to construct only a single formula-free abstraction $\hat{\mathfrak{D}}$; one can then use any desired model-checking software to do verification on $\hat{\mathfrak{D}}$ [14, 17]. Secondly, the approach we propose is especially useful when one needs to look into LT specifications that are richer than PCTL path formulae, for example BLTL specifications (their applicative importance was recently emphasized in [16]). For such problems, the only technique available in the literature requires solving the safety problem over the product between a SHS and an automaton expressing the formula [2]. However, the error for the computation of the safety value function depends on the size of the state space, thus the overall error is crucially dependent on the size of the automaton: this is not the case for the proposed new formula-free abstraction method. Lastly, the approach we use to quantify the error of the formula-free abstraction is directly extendable from LT specifications to other verification problems, such as those based on reward properties, and it allows developing a unified technique for the approximate abstraction of SHS over diverse classes of specifications.

For notational convenience, results in this paper are stated for discrete-time Markov processes (dt-MP), a class of models that is more general than discrete-time SHS. The structure of the paper is the following: Section 2 introduces classes of models and specifications of interest, and formalizes the model-checking of LT specifications against dt-MP. Section 3 describes the abstraction technique for BLTL and compares its performance with results from the literature. An extension of the technique from BLTL to other specifications is presented in Section 4. Computational examples are given in Section 5, whereas Section 6 contains the conclusions. Due to space constraints, the proofs of the statements are omitted from this manuscript.

2. MODELS AND SPECIFICATIONS

2.1 Notations

Let us recall some concepts from measure theory – for a detailed exposition the interested reader is referred to the books [7, Chapters 1-3] and [11, Chapters 1-3].

We use $\mathbb{N} = \{1, 2, \dots\}$ to denote the set of natural numbers and write $\mathbb{N}_0 := \mathbb{N} \cup \{0\}$ and $\overline{m,n} = \{m, m+1, \dots, n\}$ whenever $m, n \in \mathbb{N}_0$ and $m < n$. We also use the notation $\mathbb{R}$ for the set of real numbers and $\bar{\mathbb{R}} = \mathbb{R} \cup \{\pm\infty\}$ for the set of extended reals.

For any set X and collection of its subsets $\mathscr{C} \subseteq 2^X$, the σ-algebra generated by $\mathscr{C}$ is denoted by $\sigma(\mathscr{C})$. For example, $\mathscr{B}(\mathbb{R})$ is the *Borel* σ-algebra on $\mathbb{R}$, and is generated by the class of all open subsets of $\mathbb{R}$. We always assume $\mathbb{R}$ to be endowed with its Borel σ-algebra. Given two measurable spaces $(X, \mathscr{X})$ and $(Y, \mathscr{Y})$ the map $f : X \to Y$ is $\mathscr{X}/\mathscr{Y}$*-measurable* if $f^{-1}(A) \in \mathscr{X}$ for any $A \in \mathscr{Y}$. In the case of a $f : X \to \mathbb{R}$ we say that f is $\mathscr{X}/\mathscr{B}(\mathbb{R})$-measurable. For any function $f : X \to \mathbb{R}$ we denote its sup-norm by $\|f\| := \sup_{x \in X} |f(x)|$. We denote by $\mathrm{b}\mathscr{X}$ the space of all bounded $\mathscr{X}$-measurable functions.

If (X, ρ) is a metric space, then $\operatorname{diam}(A) = \sup_{x,y \in A} \rho(x,y)$ denotes the diameter of a set $A \subseteq X$. Let $I \subseteq \mathbb{N}_0$ be some index set and $(X_i, \mathscr{X}_i)$ be a measurable space for any $i \in I$. We denote the corresponding *product measurable space* by $\prod_{i \in I}(X_i, \mathscr{X}_i)$.

We call a function $\mu : \mathscr{X} \to \bar{\mathbb{R}}$ a *measure* on $(X, \mathscr{X})$ if $\mu(\emptyset) = 0$, if μ takes at most one of the values $\pm\infty$, and if $\mu(\bigcup_{n \in \mathbb{N}} A_n) = \sum_{n=1}^{\infty} \mu(A_n)$ for any sequence of disjoint sets $(A_n)_{n \in \mathbb{N}} \subseteq \mathscr{X}$, where the series converges absolutely if $\mu(\bigcup_{n \in \mathbb{N}} A_n)$ is finite. Such measures are also called *signed* measures, in contrast to *positive* measures (namely measures taking values over a subset of $\mathbb{R}_+$). A positive measure $\mu : \mathscr{X} \to \mathbb{R}_+$ is called a *probability* measure (or *distribution*) whenever it holds that $\mu(X) = 1$.

The last notion to be considered is that of a *kernel*: given two measurable spaces $(X, \mathscr{X})$ and $(Y, \mathscr{Y})$, a kernel Q on $(Y, \mathscr{Y})$ given $(X, \mathscr{X})$ is a function $Q : X \times \mathscr{Y} \to \bar{\mathbb{R}}$ such that $Q_x(\cdot)$ is a measure on $(Y, \mathscr{Y})$ for all $x \in X$, and such that the function $x \mapsto Q_x(A)$ is $\mathscr{X}$-measurable for any $A \in \mathscr{Y}$. We say that Q is a stochastic kernel if for any $x \in X$ the measure Q_x is a probability measure. If $(Y, \mathscr{Y}) = (X, \mathscr{X})$ we simply say that Q is a kernel on $(X, \mathscr{X})$, in that case we sometimes write $Q(x, A)$ for $Q_x(A)$.

2.2 Discrete-time Markov processes

This work considers a class of models known as discrete-time Markov processes (dt-MP). Any dt-MP $\mathfrak{D}$ can be uniquely characterized by a triple $(E, \mathscr{E}, P)$, where $(E, \mathscr{E})$ is a measurable space and $P : E \times \mathscr{E} \to [0, 1]$ is a stochastic kernel [21]. The *state space* of $\mathfrak{D}$ is $(E, \mathscr{E})$ and the elements $x \in E$ of the state space are the *states* of $\mathfrak{D}$. P is said to be a *transition kernel* of $\mathfrak{D}$ and the quantity $P(x, A)$ represents the probability of going from the state x to the set $A \in \mathscr{E}$. The work in [3] provides details of the embedding of discrete-time SHS into the dt-MP framework.

The space of trajectories of $\mathfrak{D}$ is given by the product space $(\Omega, \mathscr{F}) := \prod_{k=0}^{\infty}(E, \mathscr{E})$, with a generic trajectory denoted by

$$\omega = (\omega_0, \omega_1, \dots) \in \Omega,$$

where for any $n \in N_0$, $\omega_n \in E$ represents the state of the system modeled by the dt-MP $\mathfrak{D}$ at time epoch n. It further follows from [21, Theorem 2.8] that there exists a unique kernel P defined on $(\Omega, \mathscr{F})$ given $(E, \mathscr{E})$ that satisfies, for any $n \in \mathbb{N}_0$,

$$\mathsf{P}_x\left(\prod_{i=0}^{n} A_i \times \prod_{i=n+1}^{\infty} E\right) = 1_{A_0}(x) \int\limits_{A_1} \dots \int\limits_{A_n} P(x_{n-1}, \mathrm{d}x_n) \dots P(x, \mathrm{d}x_1),$$

where $A_i \in \mathscr{E}$ are arbitrary sets and $i \in \overline{0,n}$. The measure P_x tells which *events* (measurable sets of trajectories) are more probable to happen for $\mathfrak{D}$ than others, given that the initial state is x. By slight abuse of notation, we say that $(\Omega, \mathscr{F}, \mathsf{P})$ as above is a *canonical probability space* for the dt-MP $\mathfrak{D} = (E, \mathscr{E}, P)$.

Any set $F \in \mathscr{F}$ is called an event. We are particularly interested in the following classes of events: given $n \in \mathbb{N}_0$, $F \in \mathscr{F}$ is an *n-horizon* event if $\omega \in F$ together with $\omega_i = \omega'_i$, $i \in \overline{0,n}$ implies $\omega' \in F$. The σ-algebra of n-horizon events is given by

$$\mathscr{F}_n := \sigma\left\{\prod_{i=0}^{n} A_i \times \prod_{i=n+1}^{\infty} E \,\middle|\, A_i \in \mathscr{E},\ i \in \overline{0,n}\right\}, \tag{2.1}$$

and it represents the history of the observations of the dt-MP $\mathfrak{D}$ up to the time epoch n. We say that $(\mathscr{F}_n)_{n\in\mathbb{N}_0}$ is the *natural filtration* of the dt-MP $\mathfrak{D}$. The *Markov property* $\mathsf{P}_x(\omega_{n+1} \in A|\mathscr{F}_n) = P(\omega_n, A)$ suggests that the distribution of the next state of $\mathfrak{D}$ depends on its history only through the current state.

The following concept is important for the abstraction procedure given in Section 3: a dt-MP $\mathfrak{D} = (E, \mathscr{E}, P)$ is called *finitely generated* (f.g.) whenever $\mathscr{E}$ is finite. As an example, we call $\mathfrak{D}$ a discrete-time Markov Chain (dt-MC) if E is finite. Clearly, the finiteness of E implies the finiteness of $\mathscr{E} \subseteq 2^E$, hence we have that a dt-MC is a f.g. dt-MP. However, the inverse statement is not necessarily true: even if $\mathfrak{D}$ is finitely generated, the set E can be uncountable. Such f.g. dt-MP is an artificial object used below as an intermediate step in the abstraction of a general dt-MP into a dt-MC (cf. Figure 1).

2.3 Linear temporal specifications

The class of dt-MP models has been introduced to be model-checked against linear temporal (LT) specifications, which satisfiability can be explicitly decided over any given trajectory $\omega \in \Omega$. Thus, the satisfaction relation is defined over the set $\models \subseteq \Omega \times \mathsf{LT}$ – here the symbol LT designates an abstract class of LT specifications, which will be further detailed below. Each LT specification φ can be characterized by its sat-set as

$$\mathtt{Sat}_\Omega(\varphi) := \{\omega \in \Omega : \omega \models \varphi\},$$

which is the subset of the space Ω containing exactly those trajectories that satisfy the specification φ.

In accordance to [5, Section 10.3], we define the probabilistic model-checking problem over LT specifications as follows. Given a dt-MP $\mathfrak{D} = (E, \mathscr{E}, P)$, an initial state $x \in E$, and an LT specification φ, find the probability that the trajectory of $\mathfrak{D}$ starting from state x satisfies φ. More precisely, one has to evaluate

$$\mathsf{P}_x(\mathtt{Sat}_\Omega(\varphi)). \tag{2.2}$$

Recall that for any initial state $x \in E$ the probability measure P_x is only defined over the σ-algebra $\mathscr{F}$, and not over arbitrary collections of trajectories. Due to this reason, we say that the probabilistic model-checking problem is well posed for $\mathfrak{D}$ if and only if the quantity in (2.2) is defined, that is if $\mathtt{Sat}_\Omega(\varphi) \in \mathscr{F}$.

Whenever the probability in (2.2) is well defined for any φ in a given class of linear temporal specifications LT, one can follow the procedure described in [4, Section 9.1.3] and do equivalently a probabilistic model-checking of *state* specifications, which would yield a **true**/**false** answer instead of an arbitrary number in the interval $[0,1]$, as in the case of (2.2). More precisely, let us define the class of state specifications as $\mathsf{LT}_{\text{state}} = \mathsf{LT} \times 2^{[0,1]}$ with elements (φ, I) where $\varphi \in \mathsf{LT}$ is an LT specification and I is a subset of $[0,1]$. The satisfaction relation can then be defined on the product set $\models \subseteq E \times \mathsf{LT}_{\text{state}}$ by

$$x \models (\varphi, I) \quad \Leftrightarrow \quad \mathsf{P}_x(\mathtt{Sat}_\Omega(\varphi)) \in I.$$

Since the quantity in (2.2) needs to be well defined to be evaluated, we first discuss measurability issues.

Let us focus on a particular class of LT specifications comprising automata [5, Chapter 4] and LTL [5, Chapter 5]. In both cases, specifications are expressed via languages over certain alphabets.[1] Thus, it is sufficient to consider measurability properties of such languages, without focusing on a particular modal logic, thereafter tailoring the developed results to the special cases of LTL or automata, if needed.

We call an *alphabet* some finite set Σ, we call *letters* its elements $\sigma \in \Sigma$ and we call *words* finite or infinite sequences of letters. Let $\mathfrak{S} = \Sigma^{\mathbb{N}_0}$ be the set of all infinite words over the alphabet Σ. The generic element of $\mathfrak{S}$ is denoted by

$$\pi = (\pi_0, \pi_1, \dots) \in \mathfrak{S}, \quad \pi_i \in \Sigma, \quad i \in \mathbb{N}_0.$$

The infinite *language* φ over Σ is an arbitrary collection of infinite words, i.e. $\varphi \subseteq \mathfrak{S}$. We regard words as *traces* of trajectories of a dt-MP, as already done for the case of dt-MC [5, Section 10.3] and non-probabilistic systems [24]. Note that the canonical trajectory space Ω contains only infinite trajectories. It is thus convenient to focus on infinite words and languages, since their finite counterparts can be easily embedded in this framework: to each finite word $\pi' = (\pi'_0, \dots, \pi'_n)$ there corresponds an infinite language $\{\pi'\} \times \prod_{i=n+1}^{\infty} \Sigma$ (we call such a language a *basic* language). The embedding of a finite language into an infinite one can be done in a similar way, word by word. As a result, we shall deal only with infinite words and languages and omit the word "infinite" in both cases.

We regard each language as a specification over a dt-MP as follows. In order to characterize the satisfaction relation $\models$ between trajectories $\omega \in \Omega$ and specifications (or languages) $\varphi \subseteq \mathfrak{S}$, let us introduce the labeling map $\mathsf{L} : E \to \Sigma$. As a result, to each state $x \in E$ of the dt-MP we assign a letter $\mathsf{L}(x) \in \Sigma$. While the system described by a dt-MP evolves in time, it produces a trajectory $\omega_0, \omega_1, \dots$ which in turn produces the word $\mathsf{L}(\omega_0)\mathsf{L}(\omega_1)\dots$ called the *trace* of ω [5, Section 3.2.2]. We say that a trajectory satisfies the specification expressed as an infinite language if its trace belongs to such a language.

More formally, we denote by $\mathsf{L}_* : \Omega \to \mathfrak{S}$ the element-wise extension of L given by $\mathsf{L}_*(\omega_0, \omega_1, \dots) := (\mathsf{L}(\omega_0), \mathsf{L}(\omega_1), \dots)$. We define the satisfaction relation as follows:

$$\omega \models \varphi \quad \Leftrightarrow \quad \mathsf{L}_*(\omega) \in \varphi. \tag{2.3}$$

It follows from (2.3) that $\mathtt{Sat}_\Omega(\varphi) = \mathsf{L}_*^{-1}(\varphi)$ for all $\varphi \in \mathfrak{S}$.

Having characterized sat-sets $\mathtt{Sat}_\Omega$ through the labeling map, we can state the main result about measurability of the sat-sets used in our framework. For this purpose, we introduce the important concept of *measurable language*. Let us endow the alphabet Σ with a discrete σ-algebra 2^Σ, which makes $(\Sigma, 2^\Sigma)$ a measurable space. Hence, $\mathfrak{S}$ can be endowed with its product σ-algebra, which is further denoted by $\mathscr{S}$.

Definition 1. *[18] We say that the language φ over the alphabet Σ (so that $\varphi \subseteq \mathfrak{S}$) is* measurable, *whenever $\varphi \in \mathscr{S}$.*

Obviously, the collection of all measurable languages is just the σ-algebra $\mathscr{S}$, which is closed under intersections and complementations by definition. The following theorem is crucial for our further considerations.

Theorem 1. *If L is a $\mathscr{E}/2^\Sigma$-measurable map, then the sat-set of any measurable language $\varphi \in \mathscr{S}$ is a measurable subset of Ω.*

[1] In our case there is no substantial difference whether to start from a finite set of *atomic propositions* AP and define an alphabet as $\Sigma = 2^{\mathsf{AP}}$, or to start directly from some finite set Σ as an alphabet. For ease of notation we have chosen the latter.

Whenever the map L is $\mathscr{E}/2^{\Sigma}$-measurable, we call a quintuple $(E, \mathscr{E}, P, \Sigma, \mathsf{L})$ a *labeled discrete-time Markov process* and write ldt-MP for short. This notion is different from that of Labeled Markov process (LMP) defined in [9], where the primary goal of using labels is to model the non-determinism in transitions. However, an ldt-MP is similar to a general Labeled Markov Chain [15, Definition 1] with the only difference that in the latter case $\mathsf{L}^{-1} : \Sigma \to \mathscr{E}$ is said to be the labeling map. We say that the ldt-MP is finitely generated (f.g.) if the σ-algebra $\mathscr{E}$ is finite; in particular, if the state space E is finite we use the name *labeled discrete-time Markov Chain* (ldt-MC) in place of ldt-MP.

Theorem 1 states that the model-checking of measurable languages $\varphi \in \mathscr{S}$ over an ldt-MP is a well-posed problem in the sense that (2.2) is well-defined. Although not all infinite languages are measurable [18, Example 8], the important class of ω-regular languages satisfies the measurability property.

Proposition 1. *[18] If $\varphi \subseteq \mathfrak{S}$ is ω-regular, then $\varphi \in \mathscr{S}$.*

2.4 BLTL specifications

Although Proposition 1, together with Theorem 1, implies that the probabilistic model-checking of ldt-MP against ω-regular properties, such as LTL formulae and Büchi automata [5], is a well posed problem, its solution is in general difficult to find: as it was shown in previous work [25, 26], the solution of each particular infinite time-horizon problem depends on structural features of the dt-MP, such as the presence of absorbing sets. Due to this reason, we focus on a general class of bounded time-horizon specifications, which are still very important for applications, for instance in in systems biology [16] and in financial mathematics [27, Part III].

Let us first formalize what the horizon of a specification is. A specification $\varphi \subseteq \mathfrak{S}$ has a horizon equal to $n \in \mathbb{N}_0$ if, for any word $\pi \in \mathfrak{S}$, the value of the letters in π beyond position n does not affect whether $\pi \in \varphi$. More precisely, we call a language $\varphi \subseteq \mathfrak{S}$ *bounded* if there exists $n \in \mathbb{N}_0$ such that

$$(\pi \in \varphi) \wedge \left(\pi_i = \pi'_i,\ i \in \overline{0,n}\right) \quad \Rightarrow \quad (\pi' \in \varphi) \tag{2.4}$$

holds true for all words $\pi, \pi' \in \mathfrak{S}$. Clearly, if φ satisfies (2.4) for some $n \in \mathbb{N}_0$, then it also satisfies it for $n+1$. Thus, it is natural to define the horizon of $\varphi \subseteq \mathfrak{S}$ as follows:

$$\mathsf{H}(\varphi) := \inf\{n \in \mathbb{N}_0 : (2.4) \text{ holds true for } n\}.$$

In other words $\mathsf{H}(\varphi)$ is the smallest $n \in \mathbb{N}_0$ which makes (2.4) hold true for φ, if such n exists, whereas $\mathsf{H}(\varphi) = \infty$ otherwise, where as usual $\inf(\emptyset) := \infty$. As an example, each basic language $\varphi' = \{\pi'\} \times \prod_{i=n+1}^{\infty} \Sigma$, where $\pi' = (\pi'_0, \dots, \pi'_n) \in \Sigma^n$ is bounded, and $\mathsf{H}(\varphi') = n$ whenever Σ has more than one letter. Conversely, it follows from the finite cardinality of the alphabet Σ that each bounded language is a finite union of basic languages. As a result, since any basic language is measurable, so is each bounded language. The equivalent formula for H follows:

$$\mathsf{H}(\varphi) = \inf\left\{n \in \mathbb{N}_0 : \mathsf{L}_*^{-1}(\varphi) \in \mathscr{F}_n\right\}, \tag{2.5}$$

where $\mathscr{F}_n$ is given by (2.1). Thus, $\mathscr{S}_n := \{\varphi \subseteq \mathfrak{S} : \mathsf{H}(\varphi) \le n\}$ – the collection of all languages with an horizon not exceeding n is a sub-σ-algebra of $\mathscr{S}$, and hence it is closed under intersections, unions and complementations.

Clearly, each bounded language can be written via the finite number of basic languages that it contains, which in turn can be written via the corresponding finite words. It is possible to consider some alternative, compact representations of bounded languages. For instance, they appear as accepting languages of Deterministic Finite Automata (DFA) [5] taking only runs that are bounded by some a-priori integer $n \in N_0$ [2]: we give the precise definition later, in Section 3.4.

Another way to compactly encode a bounded language is via BLTL formulae: we now tailor to our study the definition of this logic given for a different class of models in [16]. The syntax of BLTL over alphabet Σ is given by the following grammar:

$$\Phi \quad ::= \quad \sigma \mid \neg\Phi \mid \Phi_1 \wedge \Phi_2 \mid \mathsf{X}\Phi. \tag{2.6}$$

We define the semantics of BLTL by introducing the satisfaction relation between BLTL formulae and infinite words over Σ:

$$\begin{aligned} \pi \models \sigma \quad &\Leftrightarrow \quad \pi_0 = \sigma \\ \pi \models \neg\Phi \quad &\Leftrightarrow \quad \pi \not\models \Phi \\ \pi \models \Phi \wedge \Psi \quad &\Leftrightarrow \quad \pi \models \Phi \text{ and } \pi \models \Psi \\ \pi \models \mathsf{X}\Phi \quad &\Leftrightarrow \quad \theta(\pi) \models \Phi, \end{aligned}$$

where the *shift operator* $\theta : \mathfrak{S} \to \mathfrak{S}$ is given as follows:

$$\theta(\pi_0, \pi_1, \pi_2, \dots) = (\pi_1, \pi_2, \dots).$$

For any BLTL formula Φ, we define its *accepting language* $\mathfrak{L}(\Phi)$ to be the collection of all infinite words that satisfy this formula, namely $\mathfrak{L}(\Phi) := \{\pi \in \mathfrak{S} : \pi \models \Phi\}$.

From the basic BLTL grammar in (2.6) we define the disjunction of two formulae as $\Phi_1 \vee \Phi_2 := \neg(\neg\Phi_1 \wedge \neg\Phi_2)$ and the truth formula as $\texttt{true} := \bigvee_{\Sigma} \sigma$. The "neXt" temporal operator X allows defining the "bounded Until" one. We first introduce powers of X inductively by $\mathsf{X}^0\Phi := \Phi$ and $\mathsf{X}^n\Phi := \mathsf{X}(\mathsf{X}^{n-1}\Phi)$.

We further define $\Phi_1 \mathsf{U}^{\le n} \Phi_2 := \Phi_2 \vee \bigvee_{i=1}^{n} \left(\bigwedge_{j=0}^{i-1} \mathsf{X}^j \Phi_1 \wedge \mathsf{X}^i \Phi_2\right)$ for $n \in N_0$. This formula has the following familiar semantics:

$$\pi \models \Phi_1 \mathsf{U}^{\le n} \Phi_2 \quad \Leftrightarrow \quad \pi \models \Phi_2 \text{ or } \theta^i \pi \models \Phi_2 \text{ for some } i \in \overline{1,n} \text{ and } \theta^j \pi \models \Phi_1 \text{ for all } 0 \le j < i.$$

Other temporal modalities can be defined using the bounded until operator, e.g. "bounded eventually" as $\lozenge^{\le n}\Phi := \texttt{true}\mathsf{U}^{\le n}\Phi$ and "bounded always" as $\square^{\le n}\Phi := \neg(\lozenge^{\le n}\neg\Phi)$. Using BLTL we can then pose well-known verification problems, such as probabilistic reach-avoid (using $\mathsf{U}^{\le n}$), reachability (using $\lozenge^{\le n}$), and safety (using $\square^{\le n}$). As an example, the specification induced by the language $\mathfrak{L}(\square^{\le n}\sigma)$ is equivalent to a finite-horizon safety one [1]. Moreover, BLTL allows to consider more complex properties: let $\Sigma = \{\alpha, \beta, \gamma\}$ and consider the following formula:

$$\Phi = \square^{\le 100}(\alpha \vee \beta) \wedge \lozenge^{\le 50}\square^{\le 50}\alpha.$$

Supposing that $\{\alpha, \beta\}$ corresponds to the safe set and $\{\alpha\}$ to the target set, formula Φ reads as "the system will be safe for at least 100 steps and within the following 50 iterations it will end up spending at least 50 consecutive steps in the target set." For more instances of BLTL formulae see e.g. [16].

Finally, the horizon of accepting languages of BLTL formulae can be found as follows. Clearly, we have that $\mathsf{H}(\mathfrak{L}(\sigma)) = 0$ and the following relations hold

$$\mathsf{H}(\mathfrak{L}(\neg\Phi)) = \mathsf{H}(\mathfrak{L}(\Phi)), \quad \mathsf{H}(\mathfrak{L}(\mathsf{X}\Phi)) = \mathsf{H}(\mathfrak{L}(\Phi)) + 1$$

$$\mathsf{H}(\mathfrak{L}(\Phi_1 \wedge \Phi_2)) = \max(\mathsf{H}(\mathfrak{L}(\Phi_1)), \mathsf{H}(\mathfrak{L}(\Phi_2))).$$

By induction, for any BLTL formula the horizon of its accepting language is finite and hence by (2.5) such a language is measurable, which leads to the well-posedness of probabilistic model checking of BLTL. Note also that for each basic language φ there exists a formula Φ such that $\varphi = \mathfrak{L}(\Phi)$. As a result, BLTL allows describing all possible bounded languages.

3. FINITE ABSTRACTIONS OF LDT-MP

In order to progressively introduce the main results presented in this work, let us first discuss how one can perform verification of BLTL formulae against stochastic models with finite state spaces, specifically `ldt-MC`. Notice that, from (2.6), the grammar of BLTL is a fragment of LTL. Thus, any BLTL formula can be expressed via an automaton, and its verification over an `ldt-MC` is known [5, Chapter 10.3]. On the other hand, any BLTL formula Φ can be directly expressed via the basic components (finite words) of its accepting language $\mathfrak{L}(\Phi)$: one can further compute probabilities of sat-sets for each word and find the sum thereof to obtain the probability of the sat-set for $\mathfrak{L}(\Phi)$.

With focus on the general case of `ldt-MP`, automata model-checking was studied in [2]. However, as it has been mentioned in the introduction, the error for the approximate solution depends on the size of the automaton (cfr. Section 3.4). This is especially important in case of BLTL, which often leads to automata with large state spaces (cfr. Section 5).

To cope with the issues described above, this contribution provides a formula-free abstraction technique made up of two steps. We show that any BLTL model-checking problem over f.g. `ldt-MP` can be explicitly reduced to the same problem over a certain `ldt-MC`. Perhaps not a striking result per se, it motivates looking for finitely generated approximate abstractions of general `ldt-MP`. The overall abstraction scheme is depicted in Figure 1: the general `ldt-MP` is *approximately* abstracted as a f.g. `ldt-MP`, which in turn is *exactly* abstracted as a `ldt-MC`.

3.1 Quotient ldt-MC of a f.g. ldt-MP

A f.g. `ldt-MP` with an infinite state space is an artificial object that is used as an intermediate step between a general `ldt-MP` and a `ldt-MC` in the abstraction procedure. Intuitively, a finitely generated abstraction is useful since it has the same uncountable state space as the original model but only a discrete measurability structure given by its finite σ-algebra. To be more precise, let us first comment on the structure of some arbitrary f.g. `ldt-MP` $\mathfrak{D} = (E, \mathscr{E}, P, \Sigma, \mathsf{L})$. Since the σ-algebra $\mathscr{E}$ is finite, it follows that there exists a finite measurable partition of E which generates $\mathscr{E}$, i.e. there exists a finite collection of disjoint non-empty sets $E_1, \dots, E_N$ satisfying $\bigcup_{i=1}^{N} E_i = E$, and such that $\mathscr{E} = \sigma(E_1, \dots, E_N)$. Although the state space E can still be an uncountable set, the finite structure of $\mathscr{E}$ in particular implies that all measurable maps are constant when restricted to the partition sets. This follows directly from the definition of measurability and the fact that $\mathscr{E}$ is generated by a finite partition. For example, for the stochastic kernel of $\mathfrak{D}$ it holds that

$$P(x', A) = P(x'', A) \quad \forall\, x', x'' \in E_i,\ i \in \overline{1,N},\ A \in \mathscr{E}. \tag{3.1}$$

Moreover, any set $A \in \mathscr{E}$ admits a unique representation of the form $A = \bigcup_{i \in I} E_i$ where $I \subseteq \overline{1,N}$ is some index set, e.g. it is empty for the case $A = \emptyset$. As a result, the stochastic kernel P is uniquely determined by the matrix with entries given by

$$p_{ij} := P(x_i, E_j), \tag{3.2}$$

where x_i can be *any* point in E_i, as it follows from (3.1).

Notice that the construction above means that only the sets E_i, rather than single states $x \in E$ or general subsets of E, are "observable" locations. For example, if $(\Omega, \mathscr{F}, \mathsf{P})$ is a probability space of $\mathfrak{D}$ then the probability that ω_1 belongs to E_i is well-defined and is given by $\mathsf{P}_x(\omega_1 \in E_i) = P(x, E_i)$. However, for any non-empty $E' \subsetneq E_i$ the probability $\mathsf{P}_x(\omega_1 \in E')$ is not defined since $E' \notin \mathscr{E}$. This can be interpreted as follows: we know the one-step transition probability for entering the set E_i, but nothing can be said about the transition probability into a generic subset of E_i.

The above discussion leads to regard the partition sets E_i as equivalence classes of states in E, and to construct a *quotient* `ldt-MC` over the finite state space made up by the collection of such equivalence classes. Such an `ldt-MC` is characterized by transition probabilities derived from the discrete structure of the kernel P given in (3.1). In order to formally present this object, let us introduce the *indexing* map $\mathtt{I} : E \to \overline{1,N}$, defined uniquely by the formula $x \in E_{\mathtt{I}(x)}$, that assigns to each $x \in E$ the index of the partition set that state x belongs to.

Definition 2. *Given a f.g. `ldt-MP` $\mathfrak{D} = (E, \mathscr{E}, P, \Sigma, \mathsf{L})$ we define the quotient `ldt-MC` by $\hat{\mathfrak{D}} = (\hat{E}, \hat{\mathscr{E}}, \hat{P}, \Sigma, \hat{\mathsf{L}})$, where the state space is $\hat{E} = \overline{1,N}$ and $\hat{\mathscr{E}} = 2^{\hat{E}}$; $\hat{P}$ is defined by the stochastic matrix $\hat{P}(i, \{j\}) = p_{ij}$, with p_{ij} given by (3.2); the labeling map is $\hat{\mathsf{L}}(i) = \mathsf{L}(x)$ where x is any element of E_i.*

Note that in Definition 2 the new labeling map $\hat{\mathsf{L}}$ is well-defined since $\mathsf{L} : E \to \Sigma$ is $\mathscr{E}/2^{\Sigma}$-measurable and hence its restriction to any partition set E_i is constant. Let us emphasize that we have used the name *quotient* because $\hat{E}$ can be thought of as a finite collection of equivalence classes of states in the original state space E with an equivalence relation $\sim$ generated by the partition $E_1, \dots, E_N$, i.e. $x' \sim x''$ if and only if $\mathtt{I}(x') = \mathtt{I}(x'')$. Let $(\hat{\Omega}, \hat{\mathscr{F}}, \hat{\mathsf{P}})$ denote the canonical probability space of $\hat{\mathfrak{D}}$. The main result on the quotient `ldt-MC` is stated as follows.

Theorem 2. *For any specification $\varphi \in \mathscr{S}$ it holds that*

$$\mathsf{P}_x\left(\mathtt{Sat}_{\Omega}(\varphi)\right) = \hat{\mathsf{P}}_{\mathtt{I}(x)}\left(\mathtt{Sat}_{\hat{\Omega}}(\varphi)\right).$$

3.2 F.g. abstraction of a general ldt-MP

We have shown that the probabilistic model-checking of a f.g. `ldt-MP` can be reduced to that of its quotient `ldt-MC`. This motivates us to look for finitely generated abstractions of general `ldt-MP`. Obviously, such an abstraction is in general not exact, hence there is no hope for equivalence results as in Theorem 2. The best one can do is constructing an abstract f.g. `ldt-MP` that is designed to approximate the value in (2.2) for the original `ldt-MP`. This leads to the introduction of an appropriate notion of distance between probability measures.

Definition 3. *Let $\mu : \mathscr{X} \to \bar{\mathbb{R}}$ be a signed measure defined on a measurable space $(X, \mathscr{X})$; its* total variation norm *is given by*

$$\|\mu\|_{\mathscr{X}} := \sup_{A \in \mathscr{X}} \left(|\mu(A)| + |\mu(A^c)|\right).$$

If Q is a kernel on $(X, \mathscr{X})$ given $(Y, \mathscr{Y})$, we use the same notation for the induced norm: $\|Q\|_{\mathscr{X}} := \sup_{y \in Y} \|Q_y\|_{\mathscr{X}}$.

Let us now consider an `ldt-MP` $\mathfrak{D} = (E, \mathscr{E}, P, \Sigma, \mathsf{L})$. In order to construct a finitely generated abstraction, we are going to retain its state space E and its logical structure (given by Σ and L): this is done in order to avoid the necessity of abstracting specifications in addition to the model. As a result, the abstraction is obtained modifying $\mathscr{E}$ and P into some finite $\tilde{\mathscr{E}} \subseteq \mathscr{E}$ and $\tilde{P}$, thus resulting in model $\tilde{\mathfrak{D}} = (E, \tilde{\mathscr{E}}, \tilde{P}, \Sigma, \mathsf{L})$. Note that it is always possible to choose some finite $\tilde{\mathscr{E}}$ and $\tilde{P}$, e.g. one can start with $\tilde{\mathscr{E}}$ generated by the labels, then define $\tilde{P}$ such that every label is absorbing. Although this is rarely an optimal choice, it can be further refined as discussed in Section 3.3.

Let $(\Omega, \tilde{\mathscr{F}}, \tilde{\mathsf{P}})$ be the probability space of $\tilde{\mathfrak{D}}$ and let $(\tilde{\mathscr{F}}_n)_{n \in \mathbb{N}_0}$ be its natural filtration. The following result shows that the distance between measures P and $\tilde{\mathsf{P}}$ propagates at most linearly in time via the distance between transition kernels P and $\tilde{P}$.

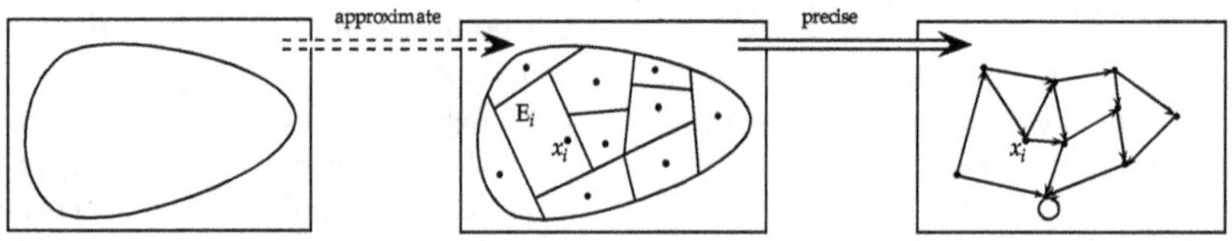

Figure 1: Two-step abstraction procedure: from a general ldt-MP to f.g. ldt-MP, to abstract ldt-MC

LEMMA 1. *For any $n \in \mathbb{N}_0$ the following inequality holds true:*

$$\|\mathsf{P} - \tilde{\mathsf{P}}\|_{\tilde{\mathscr{F}}_n} \le n \cdot \|P - \tilde{P}\|_{\tilde{\mathscr{E}}}.$$

We are now ready to state the main result, which deals with the approximate BLTL model-checking over a general `ldt`-MP using a finite `ldt`-MC abstraction obtained via a f.g. `ldt`-MP.

THEOREM 3. *Let $\mathfrak{D} = (E, \mathscr{E}, P, \Sigma, \mathsf{L})$ be a given `ldt`-MP and let $\tilde{\mathfrak{D}} = (E, \tilde{\mathscr{E}}, \tilde{P}, \Sigma, \mathsf{L})$ be its finitely generated abstraction, and let $\hat{\mathfrak{D}}$ be the quotient `ldt`-MC of $\tilde{\mathfrak{D}}$. Then, for any $x \in E$ and $\varphi \in \mathscr{S}$,*

$$\left|\mathsf{P}_x\left(\mathtt{Sat}_\Omega(\varphi)\right) - \hat{\mathsf{P}}_{\mathtt{I}(x)}\left(\mathtt{Sat}_{\hat{\Omega}}(\varphi)\right)\right| \le \mathsf{H}(\varphi) \cdot \|P - \tilde{P}\|_{\tilde{\mathscr{E}}}.$$

Theorem 3 states that any BLTL probabilistic model-checking problem over an `ldt`-MP can be approximately solved using an appropriate `ldt`-MC abstraction. The derived error bounds are clearly useful only for bounded-horizon specifications φ, whereas for infinite-horizon specifications the distance between kernels P and $\tilde{P}$ in general cannot be employed to control the error. Still, we show in Section 4.1 that for some infinite-horizon specifications bounds on the error can be derived.

The results in Theorem 2 and Theorem 3 can be related to notions of precise and approximate bisimulation, respectively, which have been introduced for `ldt`-MC e.g. in [10].

Let us now focus on the bounded-horizon case $\mathsf{H}(\varphi) < \infty$ and provide an explicit construction of finitely-generated approximations for a given `ldt`-MP $(E, \mathscr{E}, P, \Sigma, \mathsf{L})$. We also show how to upper-bound the distance between kernels. This procedure has two ingredients: the choice of the finite σ-algebra $\tilde{\mathscr{E}}$ and the choice of the corresponding kernel $\tilde{P}$. Consider a finite collection of non-empty $\mathscr{E}$-measurable sets $(E_1, \dots, E_N)$, such that $E_i \cap E_j = \emptyset$ for all $i \neq j$ and $E = \bigcup_{i=1}^N E_i$, and such that for any index $i \in \overline{1,N}$ it holds that $\mathsf{L}|_{E_i} \equiv const$. We define

$$\tilde{\mathscr{E}} = \sigma(E_1, \dots, E_N) \tag{3.3}$$

to be the σ-algebra generated by this partition. To introduce the kernel $\tilde{P}$, we choose representative points $x_i \in E_i$ and define

$$\tilde{P}(x, A) := \sum_{i=1}^{n} 1_{E_i}(x) P(x_i, A) \tag{3.4}$$

for any set $A \in \tilde{\mathscr{E}}$. Note that $\tilde{P}$ given by (3.4) is uniquely determined by the matrix with entries $\tilde{p}_{ij} := P(x_i, E_j)$ (cf. (3.2)). We call a collection $(E_i, x_i)_{i=1}^N$ defined as above a *tagged partition* of the `ldt`-MP $(E, \mathscr{E}, P, \Sigma, \mathsf{L})$. Note that any tagged partition $(E_i, x_i)_{i=1}^N$ generates the pair $(\tilde{\mathscr{E}}, \tilde{P})$ by formulae (3.3), (3.4), hence for a given `ldt`-MP it defines uniquely its finitely generated abstraction $(E, \tilde{\mathscr{E}}, \tilde{P}, \Sigma, \mathsf{L})$.

3.3 Bounds on the distance between kernels

1. General BLTL specifications. Let us now discuss how to find upper bounds on the distance $\|P - \tilde{P}\|_{\tilde{\mathscr{E}}}$, and when is it possible to control it by choice of the tagged partition $(E_i, x_i)_{i=1}^N$. We define $\kappa_i(\tilde{P}) := \sum_{j=1}^N \sup_{x \in E_i} |P(x, E_j) - P(x_i, E_j)|$, $i \in \overline{1,N}$. The next proposition gives bounds on $\|P - \tilde{P}\|_{\tilde{\mathscr{E}}}$ in terms of κ_i.

PROPOSITION 2. *For any tagged partition $(E_i, x_i)_{i=1}^N$:*

$$\|P - \tilde{P}\|_{\tilde{\mathscr{E}}} \le \max_{i \in \overline{1,N}} \kappa_i(\tilde{P}). \tag{3.5}$$

Although bounds in (3.5) are explicit and do not require any assumptions on the model, it may be impractical to find them and to control them by tuning the partition. Due to this reason, let us restrict our attention to the important class of integral kernels.

ASSUMPTION 1. *Let (E, ρ) be a metric space and $\mathscr{E}$ be its Borel σ-algebra. Assume that P is an integral kernel, i.e. that there exists a σ-finite* basis measure *μ on $(E, \mathscr{E})$ and a jointly measurable function $p : E \times E \to \mathbb{R}_+$ such that $P(x, dy) = p(x, y)\mu(dy)$, i.e. $P(x, A) = \int_A p(x, y)\mu(dy)$ for any $x \in E$, $A \in \mathscr{E}$.*

To make our results sharper, we need to assume that the density p satisfies certain Lipschitz-like conditions. The work in [1] raised uniform Lipschitz continuity assumptions in order to control the bound for the computation of value functions characterizing probabilistic safety. This assumption was further relaxed in [22] into local Lipschitz continuity. In this work we further generalize the latter assumption as follows.

ASSUMPTION 2. *Under Assumption 1, for some tagged partition $(E_i, x_i)_{i=1}^N$ there exist measurable functions $\lambda_i : E \to \mathbb{R}_+$ such that $\Lambda_i := \int_E \lambda_i(y)\mu(dy) < \infty$ for all $i \in \overline{1,N}$, and such that*

$$|p(x', y) - p(x'', y)| \le \lambda_i(y)\rho(x', x'') \quad \forall x', x'' \in E_i, \quad \forall y \in E.$$

PROPOSITION 3. *If Assumption 2 is satisfied, then for any index $i \in \overline{1,N}$ it holds that $\kappa_i \le \Lambda_i \delta_i$ where $\delta_i = \mathrm{diam}(E_i)$.*

The latter result together with Theorem 3 leads to:

COROLLARY 1. *If Assumption 2 is satisfied, then for any $x \in E$ and $\varphi \in \mathscr{S}$ the following bound holds true:*

$$\left|\mathsf{P}_x\left(\mathtt{Sat}_\Omega(\varphi)\right) - \hat{\mathsf{P}}_{\mathtt{I}(x)}\left(\mathtt{Sat}_{\hat{\Omega}}(\varphi)\right)\right| \le \mathsf{H}(\varphi) \cdot \max_{i \in \overline{1,N}} \Lambda_i \delta_i.$$

One can notice that the bounds provided in Proposition 3 are similar in shape to those in [22, Theorems 4,6]. This further allows tailoring the sequential and adaptive gridding algorithm in [22, Section V] to our case. Indeed, whenever a precision level ε is fixed, one can start with some partition $(E_i, x_i)_{i=1}^N$ satisfying Assumption 2 and further refine it until the abstraction error becomes smaller than ε. If each refinement reduces the diameter of the partition sets at least by the factor of 2, then clearly the maximum number of refinements necessary to reach the precision level can be upper-bounded (for details see [22]).

It is important to note that results in Proposition 3 hold only if the state space E is bounded with respect to its metric ρ. Indeed, the condition on the finite cardinality of the partition N

that is used for the construction of the *finite* quotient ldt-MC implies that $\mathrm{diam}(E) \le \sum_{i=1}^N \delta_i < \infty$. There are two ways to cope with this restriction: first of all, one can always transform the original metric into an equivalent bounded one, for example $\rho'(x,y) := \frac{\rho(x,y)}{1+\rho(x,y)}$. The transformation of the metric leads to a change in the functions λ_i in Assumption 2: one shall further look for conditions on the original kernel in order to assure that the corresponding integrals Λ_i are bounded. Alternatively, one can introduce an additional assumption on the kernel P in order to deal with unbounded state spaces, as follows. The idea is that the original state space can be approximated with a bounded set, say B_ε, with any precision level $\varepsilon > 0$. More precisely:

ASSUMPTION 3. *Under Assumption 1, assume that there exists $\lambda \in \mathbb{R}$ such that for any points $x', x'', y \in E$ it holds that*

$$|p(x',y) - p(x'',y)| \le \lambda \cdot \rho(x',x''),$$

and for any $\varepsilon > 0$ there exists a bounded set B_ε such that:

$$P(x, B_\varepsilon^c) \le \varepsilon, \quad \forall\, x \in E,$$
$$|p(x',y) - p(x'',y)| \le \varepsilon, \quad \forall\, x', x'' \in B_\varepsilon^c,\ y \in B_\varepsilon.$$

PROPOSITION 4. *If Assumption 3 is satisfied with an $\varepsilon > 0$, let us consider $(E_i, x_i)_{i=1}^N$ to be a tagged partition with $E_N = B_\varepsilon^c$. Denote $\delta = \max_{i \in \overline{1,N-1}} \delta_i$, then*

$$\|P - \tilde{P}\|_{\mathscr{E}} \le \varepsilon + \frac{1}{2} \max\{\varepsilon \cdot \mu(B_\varepsilon), \lambda\delta \cdot \mu(B_\varepsilon)\}. \tag{3.6}$$

Proposition 4 allows reaching any desired precision only if there is a choice of B_ε such that $\lim_{\varepsilon \to 0} \varepsilon \cdot \mu(B_\varepsilon) = 0$. Indeed, in such case it is possible to make ε and $\varepsilon \cdot \mu(B_\varepsilon)$ in (3.6) as small as needed, and then to further construct a partition of B_ε in such a way that $\lambda\delta\mu(B_\varepsilon) < \varepsilon \cdot \mu(B_\varepsilon)$ by tuning δ appropriately.

2. Special case: bounded-horizon probabilistic safety. Let us now tailor the results above for the important case of bounded-horizon safety. Consider an ldt-MP $\mathfrak{D} = (E, \mathscr{E}, P, \Sigma, \mathsf{L})$ with $\Sigma = \{\alpha, \beta\}$. As discussed earlier, we can formulate the probabilistic safety problem via the BLTL formula $\Phi_n = \Box^{\le n}\alpha$, which allows the application of the results above. However, notice that in such a case one has to partition the whole state space, whereas it is known from [1, 22] that it is sufficient to partition only the safe set $A = \mathsf{L}^{-1}(\alpha)$. Let us show how this problem can be studied in the new framework – in other words, below we consider a *formula-dependent* abstraction technique for the safety as a special case of the formula-free one presented above.

Let $\Delta \notin E$ be some auxiliary state introduced to represent set A^c and define a new ldt-MP $\mathfrak{D}' = (E', \mathscr{E}', P', \Sigma, \mathsf{L}')$, where $E' = A \cup \{\Delta\}$ and $\mathscr{E}' = \sigma(\mathscr{E}_A, \{\Delta\})$, where $\mathscr{E}_A = \{B \subseteq A : B \in \mathscr{E}\}$ is the subspace σ-algebra of A. The kernel P' is given by $P'(x,B) = P(x,B)$ for $x \in A$, $B \in \mathscr{E}_A$, and $P'(x,\Delta) = P(x,A^c)$ for $x \in A$, and $P'(\Delta,\{\Delta\}) = 1$. Finally, $\mathsf{L}'(x) = \alpha$ for $x \in A$ and $\mathsf{L}'(\Delta) = \beta$.

Let us denote by $(\Omega', \mathscr{F}', \mathsf{P}')$ the probability space of $\mathfrak{D}'$, then

$$\mathsf{P}_x\left(\mathtt{Sat}_\Omega(\mathfrak{L}(\Phi_n))\right) = \mathsf{P}'_x\left(\mathtt{Sat}_{\Omega'}(\mathfrak{L}(\Phi_n))\right)$$

for all $x \in A$ and $n \in \mathbb{N}_0$, as it follows from the integral representation of the safety probability [1]. As a result, rather than doing verification of safety over $\mathfrak{D}$ we can do this over a simpler ldt-MP $\mathfrak{D}'$, and still obtain the same result. Let us show how Assumption 2 changes in such a case for $\mathfrak{D}$.

First of all, in order to solve the safety problem over $\mathfrak{D}'$ we need to partition E', which reduces to partitioning only A. Suppose now that Assumption 1 holds for $\mathfrak{D}$ and that $\mathrm{diam}(A) < \infty$. We define a metric ρ' over E' by: $\rho'(x,y) = \rho(x,y)$ if $x, y \in A$, and $\rho'(x,\Delta) = \mathrm{diam}(A) + 1$ if $x \in A$. We further derive a new basis measure μ' from μ as $\mu'(B) = \mu(B)$ if $B \in \mathscr{E}_A$ and $\mu'(\{\Delta\}) = 1$. As a result, from $P(x,\mathrm{d}y) = p(x,y)\mu(\mathrm{d}y)$ and $P'(x,\mathrm{d}y) = p'(x,y)\mu'(\mathrm{d}y)$ we obtain the following density p': $p'(x,y) = p(x,y)$ if $x, y \in A$, and $p'(x,\Delta) = P'(x,\{\Delta\})$. Thus, if $\mathfrak{D}$ satisfies Assumption 1 with parameters (ρ, μ, p) then $\mathfrak{D}'$ also satisfies it with parameters (ρ', μ', p') defined above. Thus, for Assumption 2 to hold for $\mathfrak{D}'$, the *original* process $\mathfrak{D}$ only has to satisfy the following relaxed version of this assumption.

ASSUMPTION 4. *Under Assumption 1 for $\mathfrak{D}$: for some partition $(E_i, x_i)_{i=1}^N$ of A there exist measurable functions $\lambda_i' : A \to \mathbb{R}_+$ such that $\Lambda_i' := \int_A \lambda_i'(y)\mu(\mathrm{d}y) < \infty$ for all $i \in \overline{1,N}$, and such that*

$$|p(x',y) - p(x'',y)| \le \lambda_i'(y)\rho(x',x'') \quad \forall x', x'' \in E_i, \quad \forall y \in A.$$

To summarize, the discussion above suggests that in order to solve a probabilistic safety problem over the original ldt-MP $\mathfrak{D}$, one can partition only the safe set A and lump A^c into a single state Δ. Furthermore, Assumption 4 is sufficient to yield bounds as in Corollary 1. Note that although these bounds would hold for any bounded language over $\mathfrak{D}'$, in general only the safety specification over $\mathfrak{D}'$ can be related to that over $\mathfrak{D}$.

3.4 Connections with the literature

1. Discrete-time Stochastic Hybrid Systems. Above we have shown how to bound the quantity $\|P - \tilde{P}\|_{\mathscr{E}}$ needed for the abstraction technique, and which assumptions are sufficient to control this bound. The results have been stated in measure-theoretical terms, for instance dealing with abstract basis measures and densities. In order to further elucidate the meaning of these results over concrete models, as well as to highlight their connection with recent literature, this section focuses on models expressed as discrete-time SHS. We say that an ldt-MP $\mathfrak{D} = (E, \mathscr{E}, P, \Sigma, \mathsf{L})$ is an ldt-SHS if $E = \bigcup_{q \in Q} \{q\} \times D_q$, where D_q are Borel subsets of $\mathbb{R}^{m_q}$ and $m_q \in \mathbb{N}$ for all $q \in Q$. Q is a finite set of *modes* (or locations) and E is a hybrid state space.

We can endow E with a disjoint union topology and choose $\mathscr{E}$ to be its Borel σ-algebra. In other words, any $B \in \mathscr{E}$ can be decomposed uniquely as $B = \bigcup_{q \in Q} \{q\} \times B_q$, where B_q is a Borel subset of D_q. It is thus natural to define a basis measure μ on $(E, \mathscr{E})$ by $\mu(B) = \sum_{q \in Q} \ell^{m_q}(B_q)$, where ℓ^m stands for the Lebesgue measure on $\mathbb{R}^m$. It is common to define the transition kernel of ldt-SHS through its hybrid components [1] as

$$P((q,c), \{q'\} \times \mathrm{d}c') = \begin{cases} T_q(q'|(q,c))T_x(\mathrm{d}c'|(q,c)), & q' = q, \\ T_q(q'|(q,c))T_r(\mathrm{d}c'|(q,c),q'), & q' \ne q, \end{cases}$$

for any $c \in D_q$ and $q \in Q$ and where T_q is a discrete probability law, whereas T_r, T_x are continuous (reset and transition) kernels. The semantical meaning of the conditional distributions T_q, T_r, T_x is given in [1]. Let us now show how the density p can be constructed given the co-product basis measure μ as above, and densities t_x, t_r of T_x, T_r respectively:

$$p((q,c),(q',c')) = \begin{cases} T_q(q'|(q,c))t_x(c'|(q,c)), & q' = q, \\ T_q(q'|(q,c))t_r(c'|(q,c),q'), & q' \ne q. \end{cases}$$

We have just explicitly embedded the densities of a ldt-SHS into the general measure-theoretical framework we use in this contribution. In particular, we obtain that the Lipschitz assumption on T_q, t_x, t_r as per [22, Assumption 2] is indeed a special case of Assumption 4 of this contribution, as it follows from

$$|f(x_1)g(x_1) - f(x_2)g(x_2)| \le \|f\| \cdot |g(x_1) - g(x_2)| + \|g\| \cdot |f(x_1) - f(x_2)|,$$

for any functions $f, g : E \to \mathbb{R}$ and any $x_1, x_2 \in E$. Thus, with focus on the safety problem, Corollary 1 under Assumption 4 in this contribution implies [22, Theorem 6] as a special case, where functions λ'_i have a piecewise-constant shape.

2. Specifications expressed as DFA. Let us discuss the verification of DFA specifications over a general ldt-MP. A DFA is a tuple $\mathscr{A} = (Q, q_0, \Sigma, \mathsf{t}, F)$ where Q is a finite set of states, $q_0 \in Q$ is the initial state, Σ is a finite alphabet, $\mathsf{t} : Q \times \Sigma \to Q$ is a transition function and $F \subseteq Q$ is a set of accepting states. Given an infinite word $\pi \in \mathfrak{S}$, the corresponding trajectory $\eta \in Q^{\mathbb{N}_0}$ is defined by $\eta_0 = q_0$ and $\eta_{i+1} = \mathsf{t}(\eta_i, \pi_i)$ for all $i \in \mathbb{N}_0$. For any $n \in \mathbb{N}_0 \cup \{\infty\}$ we define the accepting language $\mathfrak{L}_n(\mathscr{A}) \subseteq \mathfrak{S}$ as follows: the word $\pi \in \mathfrak{S}$ is n-accepted by $\mathscr{A}$ if its corresponding trajectory $\eta_i \in F$ for some $i \le n$. Clearly, in case $n = \infty$ we obtain the usual accepting condition for DFA, else $\mathfrak{L}_n(\mathscr{A}) \in \mathscr{S}$ and $\mathsf{H}(\mathfrak{L}_n(\mathscr{A})) \le n$. As a result, the model-checking problem is well-posed and we can apply a formula-free abstraction in the case $n < \infty$. Alternatively, we can follow the formula-dependent approach given in [2], which we now recall and compare. Given a ldt-MP $\mathfrak{D} = (E, \mathscr{E}, P, \Sigma, \mathsf{L})$ we define a new ldt-MP $\mathfrak{D}^{\mathscr{A}} = (E^{\mathscr{A}}, \mathscr{E}^{\mathscr{A}}, P^{\mathscr{A}}, \{\alpha, \beta\}, \mathsf{L}^{\mathscr{A}})$ as follows: $E^{\mathscr{A}} = Q \times E$ and $\mathscr{E}^{\mathscr{A}}$ is the corresponding product σ-algebra. Also $\mathsf{L}^{\mathscr{A}}(q, x) = \beta$ if $q \in F$ and $\mathsf{L}^{\mathscr{A}}(q, x) = \alpha$ otherwise. Finally:

$$P^{\mathscr{A}}((q, x), \{q'\} \times \mathrm{d}x') = 1_{\mathsf{t}(q, \mathsf{L}(x))}(q') \cdot P(x, \mathrm{d}x'). \tag{3.7}$$

In this case, for any $n \in \mathbb{N}_0 \cup \{\infty\}$ [2]

$$\mathsf{P}_x(\mathtt{Sat}_\Omega(\mathfrak{L}_n(\mathscr{A}))) = 1 - \mathsf{P}^{\mathscr{A}}_x(\mathtt{Sat}_{\Omega^{\mathscr{A}}}(\mathfrak{L}(\square^{\le n+1}\alpha))), \tag{3.8}$$

where $(\Omega^{\mathscr{A}}, \mathscr{F}^{\mathscr{A}}, \mathsf{P}^{\mathscr{A}})$ denotes the probability space of $\mathfrak{D}^{\mathscr{A}}$. Such an object is called the *product* between the ldt-MP and the DFA, and is alternatively denoted by $\mathfrak{D} \otimes \mathscr{A}$ [2].

To do verification of $\mathscr{A}$ we can either leverage a formula-free abstraction technique to find the left-hand side of (3.8), or a formula-dependent one to evaluate safety in its right-hand side. Note, however, that in the latter case we still have to partition the whole original state space E as α corresponds to $(Q \setminus F) \times E$.

In order to compare the two techniques in more detail, let us suppose that Assumption 2 holds true for $\mathfrak{D}$. By Corollary 1, we have that the error introduced by the formula-free abstraction is equal to $\epsilon_1 = n \cdot \max_i \Lambda_{i,\mathfrak{D}} \delta_i$. Now, with focus on the formula-dependent approach, notice that Assumption 2 for $\mathfrak{D}$ implies Assumption 4 for $\mathfrak{D}^{\mathscr{A}}$. In particular, without having any additional information but Assumption 2 over $\mathfrak{D}$, we can only say that $\lambda'_{i,\mathfrak{D}^{\mathscr{A}}}(q, y) = \lambda_{i,\mathfrak{D}}(y)$ for all pairs (q, y) that belong to the safe set $A = (Q \setminus F) \times E$. As a result, we obtain that

$$\Lambda'_{i,\mathfrak{D}^{\mathscr{A}}} = \int_A \lambda'_{i,\mathfrak{D}^{\mathscr{A}}}(q, y) \mu^{\mathscr{A}}(\{q\} \times \mathrm{d}y) = \sum_{Q\setminus F} \int_E \lambda_i(y) \mu(\mathrm{d}y) = \#(Q \setminus F) \cdot \Lambda_i,$$

where $\#$ denotes the cardinality of a set and $\mu^{\mathscr{A}}$ is a product measure of the measure μ on E and the counting measure on Q. Hence, the error of the formula-dependent approach is

$$\epsilon_2 = (n+1) \cdot \max_i \Lambda'_{i,\mathfrak{D}^{\mathscr{A}}} \delta_i \ge \#(Q \setminus F) \cdot \epsilon_1.$$

Such an error is in most cases larger than the error introduced by the formula-free abstraction – this highlights yet another advantage of the proposed approach. If we fix the precision level and further refine the partition for the formula-dependent abstraction to reach this precision, whenever $\mathfrak{D}$ is an ldt-SHS the cardinality of the corresponding finite abstraction will be $\mathcal{O}(\#(Q \setminus F)^{m+1})$ bigger than that of the formula-free abstraction, where m is the largest dimension of the continuous components of the hybrid state space. To further elucidate this scalability assessment, we provide a concrete example in Section 5.

Clearly, we have supposed that no additional information about the structure of the original system is used. Although such assumption is relevant in many applications where e.g. it is not possible to compute Lipschitz-like functions λ adaptively for any new partition, it motivates exploiting the structure of the DFA $\mathscr{A}$ which possibly may help reducing ϵ_2 – for example one can try using methods from [13]. However, it is by no means clear whether such attempts would in general enable overcoming the factor $\#(Q \setminus F)$, which can be a large integer.

4. FURTHER APPLICATIONS

The previous section has shown how to build a formula-free finite abstraction tailored to the goal of probabilistic BLTL model-checking against ldt-MP models. Let us now discuss how the technique we used to prove the main result in Theorem 3 can lead to other important applications. This technique is given in Lemma 1, which relates the *one-step error* $\|P - \tilde{P}\|_{\tilde{\mathscr{E}}}$ to the *final error* $\|\mathsf{P} - \tilde{\mathsf{P}}\|_{\mathscr{F}_n}$ by showing that the one-step error propagates in time at most linearly. This idea essentially extends a similar method developed for approximate model-checking of particular PCTL specifications, such as safety [1]. One of the possible advantages of the current approach is that the one-step error can be related not only to the final error for safety and BLTL, but also to the final error for other verification problems: such a relation allows working on improvements of the one-step error, rather than on computing the final error for each verification problem. The new results on the one-step error would be then applicable to all verification problems where the relation between the one-step and final errors is known.

To further emphasize the usefulness of the proposed approach, let us recapitulate that the one-step error has been successfully related to the final error in the BLTL model-checking of an ldt-MP. Below we show other examples of verification problems where it is as well worth applying the newly introduced approach based on the one-step error. Note that in each such example this error can be bounded using any of Propositions 2, 3 and 4.

4.1 Infinite-horizon reach-avoid

As already mentioned, the formula-free abstraction technique presented in this work is not directly applicable to general unbounded time instances, since the one-step error cannot be linearly accrued over an infinite horizon. However, the abstraction described above can still be applied over particular instances of infinite-horizon problems.

Consider an ldt-MP $\mathfrak{D} = (E, \mathscr{E}, P, \Sigma, \mathsf{L})$ where $\Sigma = \{\alpha, \beta, \gamma\}$. We are interested in the infinite-horizon reach-avoid problem, which can be stated using the "unbounded Until" operator from LTL or PCTL. Such an operator can be easily expressed using a disjunction over BLTL formulae $\alpha \mathsf{U} \beta := \bigvee_{n=0}^{\infty} \alpha \mathsf{U}^{\le n} \beta$. Note that the infinite disjunction is needed for the definition of U, which is not part of BLTL where only finite disjunctions are allowed. Moreover, the corresponding accepting language $\mathfrak{L}(\alpha \mathsf{U} \beta)$ is not bounded, being an ω-regular language. For this specification we define the following value function $w(x) := \mathsf{P}_x(\mathtt{Sat}_\Omega(\mathfrak{L}(\alpha \mathsf{U} \beta)))$, which is known to be a solution of the Bellman equation [20]

$$w(x) = 1_B(x) + 1_A(x)Pw(x), \qquad (4.1)$$

where $A := \mathsf{L}^{-1}(\alpha)$ and $B := \mathsf{L}^{-1}(\beta)$. Let $r := \sup_{x\in A} P(x,A)$. Whenever $r < 1$, equation (4.1) admits a unique solution [20].

Let now $\tilde{\mathfrak{D}} = (E, \tilde{\mathscr{E}}, \tilde{P}, \Sigma, \mathsf{L})$ be some finitely-generated abstraction of $\mathfrak{D}$ and let us denote the corresponding value function by $\tilde{w}(x) := \tilde{\mathsf{P}}_x(\mathtt{Sat}_\Omega(\mathfrak{L}(\alpha \mathsf{U} \beta)))$.

THEOREM 4. *If $r < 1$ the following bound holds true:*

$$\|w - \tilde{w}\| \leq \frac{\|P - \tilde{P}\|_{\tilde{\mathscr{E}}}}{1 - r}. \qquad (4.2)$$

As in the case of the safety problem in Section 3.3, to solve the reach-avoid problem one can consider a version $\mathfrak{D}'$ of $\mathfrak{D}$ where states corresponding to α and γ are lumped into two single states. Owing to this simpler structure, it is easier for $\mathfrak{D}'$ to satisfy assumptions needed to control the one-step error.

4.2 Rewards

One useful extension of the results in Section 3 stems out of the following observation. Let $\mathfrak{D} = (E, \mathscr{E}, P)$ be some dt-MP and let $\tilde{\mathfrak{D}} = (E, \tilde{\mathscr{E}}, \tilde{P})$ be its finitely-generated abstraction. Let $(\Omega, \mathscr{F}, \mathsf{P})$ and $(\Omega, \tilde{\mathscr{F}}, \tilde{\mathsf{P}})$ represent the canonical probability spaces of $\mathfrak{D}$ and $\tilde{\mathfrak{D}}$ respectively, and let E_x and $\tilde{\mathsf{E}}_x$ denote the corresponding expectations.

THEOREM 5. *If $\xi : \Omega \to \mathbb{R}$ is $\tilde{\mathscr{F}}_n$-measurable then for any $x \in E$*

$$\left|\mathsf{E}_x[\xi] - \tilde{\mathsf{E}}_x[\xi]\right| \leq n \cdot \|\xi\| \cdot \|P - \tilde{P}\|_{\tilde{\mathscr{E}}}$$

This fact can be applied to the approximate computation of reward functionals (cf. [5, Chapter 10.5] for dt-MC), since they are real-valued maps ξ defined over trajectories ω. The $\tilde{\mathscr{F}}_n$-measurability assumption requires the reward ξ to depend only on the first $n+1$ coordinates of ω, i.e. $\omega_0, \dots, \omega_n$. Such an assumption holds for a wide range of problems, for instance in finance [27].

Let us show how to approximately compute the expected value of rewards in the following example. The first hitting time of a set $A \in \mathscr{E}$ is a defined by $\tau_A(\omega) := \inf\{n \geq 0 : \omega_n \in A\}$. For a function $g \in b\mathscr{E}$, set $A \in \mathscr{E}$ and $n \in \mathbb{N}_0$, let us define a reward of the following form: $\xi^n_{g,A}(\omega) := \sum_{k=0}^{\tau_A \wedge n} g(\omega_k)$. For example, if $g = 1_B$ where $B \in \mathscr{E}$ is some set, then $\xi^n_{g,A}$ is the time that a trajectory ω spends in the set B prior to hitting set A and within the time epoch n. Let us show now how to approximately compute $\mathsf{E}_x[\xi^n_{g,A}]$ using finite abstractions. Let $(E_i, x_i)_{i=1}^n$ be a tagged partition that generates $\tilde{\mathfrak{D}}$ – since there is no labeling structure, we only require that $A \in \tilde{\mathscr{E}}$, which is equivalent to $A = \bigcup_{i\in I} E_i$ for some index set i. We cannot apply immediately Theorem 5 to $\xi^n_{g,A}$ as in general it is only $\mathscr{F}_n$-measurable rather than $\tilde{\mathscr{F}}_n$-measurable. Thus, we first approximate the original reward with an $\tilde{\mathscr{F}}_n$-measurable one: we define $\tilde{g}(x) := \sum_{i=1}^N 1_{E_i}(x) g(x_i)$, which is clearly $\tilde{\mathscr{E}}$-measurable, hence $\xi^n_{\tilde{g},A}$ is $\tilde{\mathscr{F}}_n$-measurable.

ASSUMPTION 5. *Let (E, ρ) be a metric space and $\mathscr{E}$ be its Borel σ-algebra. Denote $\delta = \max_{1\leq i\leq N} \operatorname{diam}(E_i)$ and assume that*

$$|g(x') - g(x'')| \leq \eta \cdot \rho(x', x''), \quad \forall x', x'' \in E_i, \quad \forall i \in \overline{1,N},$$

for some constant $\eta > 0$.

THEOREM 6. *If Assumption 5 is satisfied then for any $x \in E$*

$$|\mathsf{E}_x[\xi^n_{g,A}] - \tilde{\mathsf{E}}_x[\xi^n_{\tilde{g},A}]| \leq (n+1)\left(n \cdot \|g\| \cdot \|P - \tilde{P}\|_{\tilde{\mathscr{E}}} + \eta\delta\right).$$

5. CASE STUDY

To elucidate the techniques developed throughout this work, let us consider the following model. Let the state space E be the interval $[0, 10]$, endowed with a Euclidian metric, and let $\mathscr{E}$ be the corresponding Borel σ-algebra. We construct the transition kernel P as an integral kernel with a basis Lebesgue measure $\mu = \ell^1$ and a density p being a weighted sum of two components: $p(x, y) = w(x)p_1(y) + (1 - w(x))p_2(y)$. The weighting function w is chosen to be the relative distance to the center of the interval: $w(x) := \frac{1}{5}|x - 5|$. The function $p_1(y)$ corresponds to a truncated Gaussian distribution given by $p_1(y) := K \cdot e^{-\frac{1}{2}(y-5)^2}$, and $p_2(y) = \frac{1}{10}$ corresponds to the uniform distribution. Here K is a normalization constant defined by $\int_E p_1(y)\mathrm{d}y = 1$, so that $K \approx 0.3989$. The shape of the density p suggests that the closer the current state to the center of the interval, the more impact the truncated Gaussian term has whereas if the current state is far from the center of the interval, the dynamics are affected by the uniform term in p. Note that the dt-MP $(E, \mathscr{E}, P)$ satisfies Assumption 1 by construction.

We introduce the alphabet $\Sigma = \{\alpha, \beta\}$ and the labeling map $\mathsf{L}(x) = \alpha$ if $x \in [4, 6]$ and $\mathsf{L}(x) = \beta$ otherwise. In order to build the formula-free abstraction of the ldt-MP $\mathfrak{D} = (E, \mathscr{E}, P, \Sigma, \mathsf{L})$, we fix a time horizon $T = 100$ and select the precision level to be equal to $\varepsilon = 0.1$. We are going to apply Proposition 3 to find the required size of the partition sets, so we need to check if Assumption 2 holds in this case. It holds that

$$\begin{aligned}|p(x', y) - p(x'', y)| &= |p_2(y) - p_1(y)| \cdot |w(x') - w(x'')| \\ &\leq |p_2(y) - p_1(y)| \cdot |x' - x''|,\end{aligned}$$

so we can select $\lambda(y) = |p_2(y) - p_1(y)|$ as per Assumption 2, regardless of the choice of the partition. In this case it holds that $\Lambda := \int_0^{10} \lambda(y)\mathrm{d}y \approx 1.1422$.

Let us now consider some arbitrary partition $(E_i, x_i)_{i=1}^N$ of $\mathfrak{D}$ and let $\delta = \max_i \operatorname{diam}(E_i)$. It follows from Proposition 3 that the one-step error is given by $\Lambda\delta$ and the final error is equal to $T\Lambda\delta$. As a result, to reach the desired precision level we need to select a partition size $\delta \leq \frac{\varepsilon}{T\Lambda}$ and the cardinality of the partition results in $N_1 = 10/\delta = 11422$. Let us emphasize that this partition leads to a ldt-MC that can be used for the model-checking of any linear temporal property with horizon T.

As a second example, in order to further clarify the statements made for formula-dependent abstraction techniques in Section 3.4, let us consider a particular specification given by a DFA $\mathscr{A} = (Q, q_0, \Sigma, \mathsf{t}, F)$ where $Q = \overline{0, M}$, $q_0 = 0$ and $F = \{M\}$. Further, let the transition function be given by $\mathsf{t}(q, \alpha) = q + 1$ if $q \notin F$, $\mathsf{t}(M, \alpha) = M$ and $\mathsf{t}(q, \beta) = 0$. Such an automaton expresses the BLTL formula $\Phi = \Diamond^{\leq 100-M} \Box^{\leq M} \alpha$ in the sense that $\mathfrak{L}_{M+1}(\mathscr{A}) = \mathfrak{L}(\Phi)$. For simplicity, let us select $M = 50$ from now on. As we have discussed, the formula-free abstraction yields an ldt-MC $\hat{\mathfrak{D}}$ with $N = 11422$ states and introduces the error $\epsilon_1 = 0.1$ over this specification. We can then verify the formula by solving a safety problem on $\hat{\mathfrak{D}}^{\mathscr{A}}$, which requires dealing with $C_1 = \#(Q \setminus F)N + 1 = 50 \cdot 11422 + 1 \approx 6 \cdot 10^5$ states.

As an alternative, we can do formula-dependent abstraction and try to solve the safety problem over $\mathfrak{D}^{\mathscr{A}}$. Based on the discussion in Section 3.4, in order to have the same precision level of 0.1, we need to take partition sets that are of size 50 times smaller, compared to the formula-free abstraction. Since the dimension of the problem is $m = 1$, this results in a cardinality of the obtained Markov Chain equal to $C_2 \approx 50^2 \cdot N_1 = 50 \cdot C_1 \approx 3 \cdot 10^6$ states, which is a number substantially larger than C_1 and as such possibly critical for computations. As a remark, if the dimension of the state space would be bigger, say $m = 2$, then we would have that $C_2 \approx 2500C_1$. If we increased the parameter M (e.g., $M = 70$), as a result we would obtain $C_2 \approx 5 \cdot 10^3 C_1$.

6. CONCLUSIONS

This contribution has presented a formula-free approach for approximate finite abstractions of SHS tailored to BLTL model checking. The work has shown that in a number of problems, for example when dealing with specifications expressed as automata, this approach can mitigate scalability issues related to formula-dependent abstractions – this motivates further need for more tailored and precise formula-dependent abstraction methods. The approach is based on the propagation of the difference in transition kernels introduced by the abstraction (the "one-step error") as a global error over the complete verification problem. Besides the application on formula-free abstractions over BLTL, this technique can also be used over probabilistic safety (PCTL) of SHS, and allows for extensions beyond BLTL.

Results on formula-free abstraction do not hold over general infinite-horizon problems, which calls for specific techniques to tackle these problems [26]. The authors are also looking into extensions of the developed techniques to the controlled case.

7. ACKNOWLEDGMENTS

The authors would like to thank Theo Buehler for the helpful discussions on measure theory, Joost-Pieter Katoen and Manuel Mazo Jr. for comments on model-checking, and Sadegh E.Z. Soudjani for help with the automaton expression for $\Diamond^{\leq n}\Box^{\leq m}$.

8. REFERENCES

[1] A. Abate, J.-P. Katoen, J. Lygeros, and M. Prandini. Approximate model checking of stochastic hybrid systems. *European Journal of Control*, 16:624–641, December 2010.

[2] A. Abate, J.-P. Katoen, and A. Mereacre. Quantitative automata model checking of autonomous stochastic hybrid systems. In *Proceedings of the 14th ACM international conference on Hybrid Systems: Computation and Control*, pages 83–92, 2011.

[3] A. Abate, M. Prandini, J. Lygeros, and S. Sastry. Probabilistic reachability and safety for controlled discrete time stochastic hybrid systems. *Automatica*, 44(11):2724–2734, 2008.

[4] C. Baier. *On algorithmic verification methods for probabilistic systems*. PhD thesis, Universität Mannheim, 1998.

[5] C. Baier and J.-P. Katoen. *Principles of model checking*. The MIT Press, 2008.

[6] D.P. Bertsekas and S.E. Shreve. *Stochastic optimal control: The discrete time case*, volume 139. Academic Press, 1978.

[7] V. I. Bogachev. *Measure theory. Vol. I, II*. Springer-Verlag, Berlin, 2007.

[8] C.G. Cassandras and J. (Eds.) Lygeros. *Stochastic hybrid systems*, volume 24. CRC Press, 2007.

[9] J. Desharnais, V. Gupta, R. Jagadeesan, and P. Panangaden. Metrics for labelled Markov processes. *Theoretical Computer Science*, 318(3):323–354, 2004.

[10] A. D'Innocenzo, A. Abate, and J.P. Katoen. Robust pctl model checking. In *Proceedings of the 15th ACM international conference on Hybrid Systems: Computation and Control*, pages 275 286. ACM, 2012.

[11] G.B. Folland. *Real analysis*. Pure and Applied Mathematics. John Wiley & Sons Inc., New York, second edition, 1999.

[12] A. Girard and G. J. Pappas. Approximation metrics for discrete and continuous systems. *IEEE Transactions on Automatic Control*, 52(5):782 –798, May 2007.

[13] H. Hermanns, B. Wachter, and L. Zhang. Probabilistic CEGAR. In *Computer Aided Verification*, pages 162–175. Springer, 2008.

[14] A. Hinton, M. Kwiatkowska, G. Norman, and D. Parker. PRISM: A tool for automatic verification of probabilistic systems. In H. Hermanns and J. Palsberg, editors, *Tools and Algorithms for the Construction and Analysis of Systems*, volume 3920 of *Lecture Notes in Computer Science*, pages 441–444. Springer Verlag, 2006.

[15] M. Huth. On finite-state approximants for probabilistic computation tree logic. *Theoretical Computer Science*, 346(1):113–134, 2005.

[16] S. Jha, E. Clarke, C. Langmead, A. Legay, A. Platzer, and P. Zuliani. A bayesian approach to model checking biological systems. In *Computational Methods in Systems Biology*, pages 218–234. Springer, 2009.

[17] J.-P. Katoen, M. Khattri, and I. S. Zapreev. A Markov reward model checker. In *Quantitative Evaluation of Systems (QEST)*, pages 243–244, 2005.

[18] N.H. Lâm and L. Vân. Measure of infinitary codes. *Acta Cybernetica*, 11(3):127–137, 1994.

[19] G.J. Pappas. Bisimilar linear systems. *Automatica*, 39(12):2035–2047, 2003.

[20] F. Ramponi, D. Chatterjee, S. Summers, and J. Lygeros. On the connections between PCTL and dynamic programming. In *Proceedings of the 13th ACM international conference on Hybrid Systems: Computation and Control*, pages 253–262, 2010.

[21] D. Revuz. *Markov chains*. North-Holland Publishing, Amsterdam, second edition, 1984.

[22] S.E.Z. Soudjani and A. Abate. Adaptive gridding for abstraction and verification of stochastic hybrid systems. In *Quantitative Evaluation of Systems (QEST)*, pages 59–68, Aachen, DE, 2011.

[23] S. Summers and J. Lygeros. Verification of discrete time stochastic hybrid systems: A stochastic reach-avoid decision problem. *Automatica*, 46(12):1951–1961, 2010.

[24] P. Tabuada. *Verification and control of hybrid systems: A symbolic approach*. Springer Verlag, New York, 2009.

[25] I. Tkachev and A. Abate. On infinite-horizon probabilistic properties and stochastic bisimulation functions. In *Proceedings of the 50th IEEE Conference on Decision and Control and European Control Conference*, pages 526–531, Orlando, FL, December 2011.

[26] I. Tkachev and A. Abate. Regularization of Bellman Equations for Infinite-Horizon Probabilistic Properties. In *Proceedings of the 15th International Conference on Hybrid Systems: Computation and Control*, pages 227–236, Beijing, PRC, April 2012.

[27] P. Wilmott, S. Howison, and J. Dewynne. *The mathematics of financial derivatives: a student introduction*. Cambridge University Press, 1995.

Quantitative Automata-based Controller Synthesis for Non-Autonomous Stochastic Hybrid Systems*

Ilya Tkachev
TU Delft
i.tkachev@tudelft.nl

Alexandru Mereacre
University of Oxford
mereacre@cs.ox.ac.uk

Joost-Pieter Katoen
RWTH Aachen
katoen@cs.rwth-aachen.de

Alessandro Abate
TU Delft
a.abate@tudelft.nl

ABSTRACT

This work deals with Markov processes that are defined over an uncountable state space (possibly hybrid) and embedding non-determinism in the shape of a control structure. The contribution looks at the problem of optimization, over the set of allowed controls, of probabilistic specifications defined by automata – in particular, the focus is on deterministic finite-state automata. This problem can be reformulated as an optimization of a probabilistic reachability property over a product process obtained from the model for the specification and the model of the system. Optimizing over automata-based specifications thus leads to maximal or minimal probabilistic reachability properties. For both setups, the contribution shows that these problems can be sufficiently tackled with history-independent Markov policies. This outcome has relevant computational repercussions: in particular, the work develops a discretization procedure leading into standard optimization problems over Markov decision processes. Such procedure is associated with exact error bounds and is experimentally tested on a case study.

Categories and Subject Descriptors

G.3 [**Mathematics of Computing**]: PROBABILITY AND STATISTICS—*Stochastic processes*

General Terms

Theory, verification

*This work is supported by the European Commission MoVeS project FP7-ICT-2009-5 257005, by the European Commission Marie Curie grant MANTRAS 249295, by the European Commission NoE HYCON2 FP7-ICT-2009-5 257462, and by the NWO VENI grant 016.103.020.

HSCC'13, April 8–11, 2013, Philadelphia, Pennsylvania, USA.

Keywords

stochastic hybrid systems, probabilistic reachability, formal verification, stochastic optimal control, approximate abstractions.

1. INTRODUCTION

Stochastic Hybrid Systems (SHS) are a general and widely applicable mathematical framework involving the interaction of discrete, continuous, and probabilistic dynamics. Because of their generality, SHS have been applied over many areas, including telecommunication networks, manufacturing systems, transportation, and biological systems [8, 9].

SHS can be abstractly regarded as general Markov processes defined over an uncountable (in particular, hybrid) state space and embedded with non-determinism in the sense that they are dependent on a control structure [7]. The allowed control policies are functions of the history: in other words, we allow the control inputs to depend not only on the current state, but also on past trajectory realization and on past choices of control inputs. Markov policies are important special instances of these policies and depend solely on the current state.

SHS models are properly tagged by a labeling function that maps points in the state space to elements of a finite labeling set, which we refer to as an alphabet. SHS are structurally useful in the verification of linear time specifications, for example specifications expressed as deterministic finite-state automata (DFA) with labels. The work in [4] has shown that the verification of linear time specifications (in particular, DFA properties) can be performed by solving a probabilistic reachability problem over a new SHS, which is obtained by taking the cross product between the SHS and the DFA. Probabilistic reachability can then be practically assessed by computing its dual, namely probabilistic invariance [5].

This work generalizes the results in [4] to the case of control-dependent SHS and of DFA specifications. Since the verification of DFA specifications boils down to solving a probabilistic reachability problem, we consider this setup both in its finite- and infinite-horizon formulations. As the models under study are control dependent, we look into the possible maximal and minimal probabilistic reachability formulations: these can be studied as limits of dynamic programming recursions or as solutions of integral equations [7]. Since the cost functions expressing the dynamic programming scheme for probabilistic reachability take a multiplicative form [5], the analysis of their properties is in general difficult. This has lead several works in the literature [18, 10] to try formulating the reach-avoid problem (a generalization of the reachability problem) in terms of additive costs, where the dynamic programming theory is mature [7, 13]. To the best of our knowledge, inspired by the work in [11] the present work provides such a reduction explicitly for the first time. This as a consequence leads to several important results: first of all, we show that Markov policies are sufficient for the optimal probabilistic reachability; additionally, such a technique allows obtaining Bellman recursion and Bellman fixpoint equations for the reachability probability in the most general case, whereas the results in the literature were either focused on Markov policies exclusively [5, 18] or required structural assumptions on the model [10].

Sufficiency of Markov policies has important computational implications, since optimal controls can be synthesized according to the current value of trajectories, rather than based on the entire past history. Further along this computational line, the work introduces a discretization scheme that reduces the original (uncountable) setup to a finite optimization problem over Markov decision processes. Provided some continuity assumptions on the model are valid, the work shows that the discretization scheme can be associated with exact bounds on the introduced error. The obtained error bounds, inspired by [2], are functions of tunable parameters in the model and thus can be made, at the expense of more computations, arbitrarily small.

The article is structured as follows. Section 2 introduces the model syntax and semantics, as well as the class of linear temporal specifications of interest. Section 3 discusses the problem statement (probabilistic reachability) and its alternative formulation via additive cost functions, and derives the main theoretical results in this work: it shows in particular the sufficiency of Markov policies, and elaborates the minimal and maximal optimization problems over finite and infinite horizons. Section 4 puts forward a discretization scheme for the computation of the quantities in Section 3, with an exact quantification of the introduced errors. Finally, Section 5 presents the experimental outcomes over a case study and Section 6 concludes the paper. Due to space constraints, the proofs of the statements are omitted from this manuscript.

2. PRELIMINARIES

2.1 Notations and model definition

In this section we give a brief recap on Borel spaces and related stochastic kernels. It is common in the literature on the theory of controlled discrete-time Markov processes (cdt-MP) to assume that both the state space and the control space are endowed with a certain topological structure [7, 13]. A topological space X is called a Borel space if it is homeomorphic to a Borel subset of a Polish space[1]. Examples of a Borel space are the Euclidean spaces $\mathbb{R}^n$, its Borel subsets endowed with a subspace topology [16], as well as hybrid spaces [5]. Any Borel space X is assumed to be endowed with a Borel σ-algebra, which is denoted by $\mathscr{B}(X)$. We say that a map $f : X \to Y$ is measurable whenever it is Borel measurable.

Given two Borel spaces X, Y the stochastic kernel on X given Y is the map $P : Y \times \mathscr{B}(X) \to [0,1]$ such that $P(\cdot|y)$ is a probability measure on X for any point $y \in Y$, and such that $P(B|\cdot)$ is a measurable function on Y for any set $B \in \mathscr{B}(X)$. Stochastic kernels provide a natural generalization of update laws for deterministic systems, as we further show below.

We adopt the notation from [13] and consider a tuple $\mathfrak{D} = (X, U, \{U(x)\}_{x\in X}, \mathsf{T})$, where X is a Borel space, referred to as the state space of the model, and U is a Borel space to be thought of as the set of controls. Furthermore, $\{U(x)\}_{x\in X}$ is a family of non-empty measurable subsets of U with the property that

$$\mathbb{K} := \{(x,u) : x \in X, u \in U(x)\}$$

is measurable in $X \times U$. Intuitively, $U(x)$ is the set of controls that are feasible at state x. Finally, T is a stochastic kernel on X given $\mathbb{K}$: note that $\mathbb{K}$ is a measurable subset of a Borel space $X \times U$, hence it is itself a Borel space. In order to assure that the set of control policies is not empty we require $\mathbb{K}$ to contain the graph of a measurable function [13, Assumption 2.2.2]. In other words, we assume that there exists a measurable map $\mathsf{k} : X \to U$ such that $\mathsf{k}(x) \in U(x)$ for any $x \in X$.

Definition 1 (cdt-MP). *We call any tuple*

$$\mathfrak{D} = (X, U, \{U(x)\}_{x\in X}, \mathsf{T})$$

that satisfies the assumptions above a cdt-MP.

The following notation is used throughout the paper. We denote the set of positive integers by $\mathbb{N}$ and the set of non-negative integers by $\mathbb{N}_0 := \mathbb{N} \cup \{0\}$. Furthermore, we denote $\overline{m,n} := \{m, m+1, \dots, n-1, n\}$ for $m, n \in \mathbb{N}_0$ such that $m < n$; $\mathbb{R}$ stands for the set of reals.

For any set X we denote by $X^{\mathbb{N}_0}$ the Cartesian product of a countable number of copies of X, i.e. $X^{\mathbb{N}_0} = \prod_{k=0}^{\infty} X$.

2.2 Model semantics

The semantics of a cdt-MP is characterized by its *paths* or executions, which reflect both the history of previous states of the system and of implemented control actions. Paths (often thought as infinite paths) are used to measure the performance of the system, which is done here via model checking methods. Also, a path up to a time epoch n can be used to derive the control action on the next step: these are finite paths that we also call *histories* as in [13, Section 2.2]. Let us finally mention that for technical reasons, when dealing with uncountable state spaces, it is usual to take into consideration also non-admissible paths, i.e. those containing non-feasible controls.

Definition 2 (History). *Given a* cdt-MP $\mathfrak{D}$ *and a number* $n \in \mathbb{N}_0$, *an* n-history *is a finite sequence*

$$h_n = (x_0, u_0, \dots, x_{n-1}, u_{n-1}, x_n), \tag{1}$$

where $x_i \in X$ *are state coordinates and* $u_i \in U$ *are control coordinates of the history. An* n*-history* h_n *is called* admissible *if* $u_i \in U(x_i)$, $i \in \overline{0, n-1}$. *The space of all* n*-histories is denoted by* $\bar{H}_n$, *and its subspace of admissible* n*-histories is denoted by* H_n:

$$\bar{H}_n = (X \times U)^n \times X, \quad H_n = \mathbb{K}^n \times X.$$

Further, we denote projections by $h_n[i] := x_i$ *and* $h_n(i) := u_i$.

Definition 3 (Path). *An infinite* path *of a* cdt-MP $\mathfrak{D}$ *is*

$$\omega = (x_0, u_0, x_1, u_1, \dots), \tag{2}$$

where $x_i \in X$ *and* $u_i \in U$ *for all* $i \in \mathbb{N}_0$. *As above, let us introduce projections* $\omega[i] := x_i$ *and* $\omega(i) := u_i$.

The space of all infinite paths $\Omega = (X \times U)^{\mathbb{N}_0}$ *together with its product* σ*-algebra* $\mathscr{F}$ *is called a* canonical sample space *for*

[1] A Polish space is a topological space which is separable and completely metrizable.

a cdt-MP $\mathfrak{D}$ *[13, Section 2.2]. An infinite path $\omega \in \Omega$ is called* admissible *if $u_n \in U(x_n)$ for any $n \in \mathbb{N}_0$. The space of all infinite admissible paths is denoted by $H_\infty = \mathbb{K}^\infty$ and is a subspace of Ω.*

Given a path ω or a history h_n, we assume below that x_i and u_i are their state and control coordinates respectively, unless otherwise stated. In order to emphasize the control structure, for the path ω as in (2) we also write

$$\omega = x_0 \xrightarrow{u_0} x_1 \xrightarrow{u_1} x_2 \xrightarrow{u_2} \cdots .$$

Clearly, for any infinite path $\omega \in \Omega$ its n-prefix (ending in a state) ω_n is an n-history. However, for notational reasons we aim on denoting finite histories through h_n and paths through ω exclusively, since they usually serve for different purposes. We are now ready to introduce the notion of control policy.

DEFINITION 4 (POLICY). *A* policy *is a sequence $\pi = (\pi_n)_{n\in\mathbb{N}_0}$ of universally measurable stochastic kernels π_n [7, Chapter 8.1], each defined on the control set U given H_n and such that*

$$\pi_n(U(x_n)|h_n) = 1 \tag{3}$$

for all $h_n \in H_n$, $n \in \mathbb{N}_0$. The set of all policies is denoted by Π.

Equation (2) shows that all policies are "admissible" by definition. More precisely, the class of non-admissible policies is of 0 probability: given a policy $\pi \in \Pi$ and an admissible n-history $h_n \in H_n$, the distribution of the next control action u_n given by $\pi(\cdot|h_n)$ is supported on $U(x_n)$. Among the class of all possible policies, we are especially interested in those with a simple structure in that they depend only on the current state, rather than on the whole history.

DEFINITION 5 (MARKOV POLICY). *A policy $\pi \in \Pi$ is called a* Markov *policy if for any $n \in \mathbb{N}_0$ it holds that $\pi_n(\cdot|h_n) = \pi_n(\cdot|x_n)$, i.e. π_n depends on the history h_n only through the current state x_n. The class of all Markov policies is denoted by $\Pi_M \subset \Pi$.*

Although a Markov policy is not history-dependent, it is not necessary *stationary*, i.e. it may depend on the time variable. This fact highlights the difference between Markov policies and memoryless schedulers as defined in [6, Section 10.6]. More precisely, a policy $\pi \in \Pi$ is a memoryless scheduler if and only if it is Markov and stationary, i.e. $\pi_n = \pi_0$ for all $n \in \mathbb{N}_0$.

A central measurability result is that given a cdt-MP $\mathfrak{D}$ and a policy $\pi \in \Pi$, it is always possible to construct a suitable probability measure over the set of paths. Due to technical reasons, such a measure is constructed on the space $(\Omega, \mathscr{F})$ which also contains non-admissible paths, but it is supported on the space H_∞ of admissible paths. More precisely, let a cdt-MP $\mathfrak{D}$, a policy π and a probability measure α on X be given – the latter is referred to be the initial probability distribution of the cdt-MP. By the theorem of Ionescu Tulcea [13], there exists a unique probability measure P^π_α on the canonical sample space $(\Omega, \mathscr{F})$ supported on H_∞, i.e. $\mathsf{P}^\pi_\alpha(H_\infty) = 1$ and such that

$$\mathsf{P}^\pi_\alpha(x_0 \in B) = \alpha(B),$$
$$\mathsf{P}^\pi_\alpha(u_n \in C|h_n) = \pi_n(C|h_n),$$
$$\mathsf{P}^\pi_\alpha(x_{n+1} \in B|h_n, u_n) = \mathsf{T}(B|x_n, u_n)$$

for all $B \in \mathscr{B}(X)$, $C \in \mathscr{B}(U)$ and $h_n \in H_n$, $n \in \mathbb{N}_0$. In the case when the initial distribution is supported on a single point, i.e. $\alpha(\{x\}) = 1$, we write P^π_x in place of P^π_α. Note that the last equation means that T is a transition kernel for the cdt-MP: whenever the current state is known and the control action is chosen, T gives a distribution for the next state.

2.3 Controlled Stochastic Hybrid Systems

Controlled discrete-time Stochastic Hybrid Systems (cdt-SHS) are a particular subclass of cdt-MP with a more explicit structure that distinguishes between continuous and discrete states of the system. The class of cdt-SHS provides rich modeling power and is applicable to various areas [9, 5]. It has been introduced in [2] and further studied in [5, 19]. Here we introduce it directly through the discussed framework: a cdt-SHS is a cdt-MP $\mathfrak{D} = (X, U, \{U(x)\}_{x\in X}, \mathsf{T})$ whose state space is $X = \bigcup_{q\in Q}\{q\} \times D_q$, where Q is a finite set of discrete *modes* and the measurable sets $D_q \in \mathscr{B}(\mathbb{R}^{n(q)})$ are the continuous components of the state space.

The control space U is often taken to be some Borel space. Furthermore, the transition kernel of cdt-SHS is often defined through its hybrid components [5] as:

$$\mathsf{T}(\{q'\}\times dc'|(q,c),u) = \begin{cases} \mathsf{T}_q(q'|(q,c),u)\mathsf{T}_x(dc'|(q,c),u), & q' = q, \\ \mathsf{T}_q(q'|(q,c),u)\mathsf{T}_r(dc'|(q,c,q'),u), & q' \neq q, \end{cases}$$

for any $c \in D_q$ and $q \in Q$ and where T_q is a discrete probability law, whereas T_r, T_x are continuous (reset and transition) kernels. The semantical meaning of the conditional distributions T_q, T_r and T_x is given in [5]. The hybrid structure of the kernel further allows considering the control space U to be a product of the controls affecting the continuous dynamics and of those affecting the discrete dynamics as $U = U_q \times U_c$, where U_q, U_c are some Borel spaces [5].

Markov Decision Processes (MDP) are a subclass of cdt-SHS characterized by finite state and control spaces. Formally, U is finite and each continuous component D_q of a MDP is a singleton, which can be identified with the discrete component q itself. The theory of MDP is mature and allows for explicit solutions of many synthesis problems [6, Section 10.6].

2.4 Deterministic Finite State Automata

We are interested in linear temporal properties of trajectories of a given cdt-MP. For this purpose, in this section we introduce a model known as Deterministic Finite-state Automaton (DFA).

DEFINITION 6 (DFA). *A DFA is a tuple $\mathscr{A} = (Q, q_0, \Sigma, F, \mathsf{t})$, where Q is a finite set of locations, $q_0 \in Q$ is the initial location, Σ is a finite set, $F \subseteq Q$ is a set of accept locations, and $\mathsf{t}: Q \times \Sigma \to Q$ is a transition function.*

We call the set Σ an *alphabet* and its elements $\sigma \in \Sigma$ *letters*. We denote by $\mathfrak{S} = \Sigma^{\mathbb{N}_0}$ (by $\mathfrak{S}_{<\infty}$) the collection of all infinite (finite) words over Σ. A finite word $w = (w[0], \dots, w[n]) \in \mathfrak{S}_{<\infty}$ is accepted by a DFA $\mathscr{A}$ if there exists a finite *run* $z = (z[0], \dots, z[n+1]) \in Q^{n+2}$ such that $z[0] = q_0$, $z[i+1] = \mathsf{t}(z[i], w[i])$ for all $0 \le i \le n$ and $z[n+1] \in F$. Although the verification of DFA is classically based on finite words, it is here more convenient to work with infinite words and define the corresponding accepting languages over infinite words. We say that an infinite word $w \in \mathfrak{S}$ is accepted by a DFA $\mathscr{A}$ if there exists a finite prefix of w accepted by $\mathscr{A}$ as a finite word. This is equivalent to the following statement: an infinite word $w \in \mathfrak{S}$ is accepted by $\mathscr{A}$ if and only if there exists an infinite run $z \in Q^{\mathbb{N}_0}$ such that $z[0] = q_0$, $z[i+1] = \mathsf{t}(z[i], w[i])$ for all $i \in \mathbb{N}_0$ and there exists $j \in \mathbb{N}_0$ such that $z[j] \in F$. Note that $w[i]$ (resp. $z[i]$) denotes the i-th letter (resp. state of the automaton) on w (resp. z). The accepted language of $\mathscr{A}$, denoted $\mathfrak{L}(\mathscr{A})$, is the set of all words accepted by $\mathscr{A}$. We are further interested in other kinds of accepting conditions over the DFA $\mathscr{A}$: we say that the word w is n-accepted by the DFA $\mathscr{A}$ if there exists a *run* $z \in Q^{\mathbb{N}_0}$ such that $z[0] = q_0$, $z[i+1] = \mathsf{t}(z[i], w[i])$ for all $i \in \mathbb{N}_0$ and $z[j] \in F$ for some $j \leq n$. We denote the set of all n-accepted words by $\mathfrak{L}_n(\mathscr{A})$. It is clear that $\mathfrak{L}_\infty(\mathscr{A}) = \mathfrak{L}(\mathscr{A})$ and furthermore that $\mathfrak{L}_n(\mathscr{A}) \subseteq \mathfrak{L}_{n+1}(\mathscr{A})$, where $n \in \mathbb{N}_0$ is arbitrary.

We use the DFA $\mathscr{A}$ to specify properties of the cdt-MP as follows. Let $\mathsf{L} : X \to \Sigma$ be a measurable function which we call the *labeling function* for a cdt-MP $\mathfrak{D}$. To each state $x \in X$ it assigns the letter $\mathsf{L}(x) \in \Sigma$. In the same fashion, each path $\omega = (x_0, u_0, x_1, u_1, x_2, u_2, \dots) \in \Omega$ induces the word $w \in \mathfrak{S}$ given by $w = (\mathsf{L}(x_0), \mathsf{L}(x_1), \mathsf{L}(x_2), \dots)$. We define the function L_Ω which maps paths onto words, i.e., $\mathsf{L}_\Omega(\omega) = w$, as above. Using this function we can introduce the satisfaction relation between paths of $\mathfrak{D}$ and the DFA specification as follows:

$$\omega \models \mathscr{A} \quad \Leftrightarrow \quad \mathsf{L}_\Omega(\omega) \in \mathfrak{L}(\mathscr{A}). \tag{4}$$

As a result, given a policy $\pi \in \Pi$ we can define the probability that a path of $\mathfrak{D}$ satisfies $\mathscr{A}$, i.e. $\mathsf{P}^\pi_\alpha(\omega \models \mathscr{A})$. It can be shown that under the assumption made on the measurability of $\mathsf{L} : X \to \Sigma$ it holds that $\{\omega \in \Omega : \omega \models \mathscr{A}\} \in \mathscr{F}$ and hence such probability is well-defined. For any $n \in N_0$ we also define

$$\omega \models_n \mathscr{A} \quad \Leftrightarrow \quad \mathsf{L}_\Omega(\omega) \in \mathfrak{L}_n(\mathscr{A}). \tag{5}$$

By now it should be clear why we prefer dealing with infinite words: the reason for this is that runs of the cdt-MP are always infinite, i.e. the cdt-MP "never stops". At the same time, the input to the DFA $\mathscr{A}$ is a word $\mathsf{L}_\Omega(\omega)$ which is infinite as well.

The work in [4] studied model-checking of automata specifications against autonomous (i.e. uncontrolled) discrete-time stochastic models over uncountable state spaces. It was shown that the computation of the probability of satisfying a DFA can be restated in terms of a *probabilistic reachability* problem over the product between the original model and the DFA. In this paper we extend this result to the case of cdt-MP. As a result, we need to introduce the probabilistic reachability problem in general terms, and discuss its solution.

3. PROBABILISTIC REACHABILITY

3.1 Problem formulation

In this section we consider a basic and important problem where the probability of reaching a goal set is to be optimized. Let us consider a cdt-MP $\mathfrak{D} = (X, U, \{U(x)\}, \mathsf{T})$ and some goal set $G \in \mathscr{B}(X)$. For any $n \in \mathbb{N}_0$ we define

$$\Diamond^{\leq n} G = \{\omega \in \Omega : x_k \in G \text{ for some } 0 \leq k \leq n\} \tag{6}$$

and further $\Diamond^{\leq \infty} G = \bigcup_{n=0}^\infty \Diamond^{\leq n} G$. We call the event in $\Diamond^{\leq n} G$ an *n-bounded* reachability if $n < \infty$, and and an *unbounded* reachability if $n = \infty$. Since it holds that $\Diamond^{\leq n} G = \bigcup_{k=0}^n \{x_k \in G\}$ for any $n \in \bar{\mathbb{N}}_0 = \mathbb{N}_0 \cup \{\infty\}$, we obtain that $G \in \mathscr{B}(X)$ implies that $\Diamond^{\leq n} G \in \mathscr{F}$ for all $n \in \bar{\mathbb{N}}_0$. Due to this reason, the quantities $\mathsf{P}^\pi_\alpha(\Diamond^{\leq n} G)$ are well-defined for any initial distribution α and any control policy $\pi \in \Pi$. We further refer to these quantities as *reachability probabilities*.

We assume that the target set $G \in \mathscr{B}(X)$ is given and fixed. We are interested in the problem of reachability probability optimization, being either a maximization or a minimization over all possible policies $\pi \in \Pi$. For this purpose we restrict our attention to initial distributions supported at single points and define the following *value functions*: $V^\pi_n(x) := \mathsf{P}^\pi_x(\Diamond^{\leq n} G)$ for all $n \in \bar{\mathbb{N}}_0$. For the problem of maximal reachability, the corresponding value functions are given by

$$V^*_n(x) := \sup_{\pi \in \Pi} V^\pi_n(x) \tag{7}$$

and for the problem of minimal reachability by

$$V_{*,n}(x) := \inf_{\pi \in \Pi} V^\pi_n(x). \tag{8}$$

A policy $\pi^* \in \Pi$ is called optimal for the problem (7) if it satisfies $V^{\pi^*}_n(x) = V^*_n(x)$. Similarly, a policy $\pi_* \in \Pi$ is called optimal for the problem (8) if $V^{\pi^*}_n(x) = V_{*,n}(x)$

The reachability problem introduced above relates to other important problems in the analysis of probabilistic systems. First of all, it is a dual to the probabilistic safety (or invariance) problem, which was studied for cdt-SHS in [5]. We can define such problem as follows: for some $S \in \mathscr{B}(X)$ and any $n \in \mathbb{N}_0$,

$$\Box^{\leq n} S = \{\omega \in \Omega : x_k \in S \text{ for all } 0 \leq k \leq n\} \tag{9}$$

and $\Box^{\leq \infty} S = \bigcap_{n=0}^\infty \Box^{\leq n} S$. The duality between the safety and the reachability is given by the identity $(\Box^{\leq n} S)^c = \Diamond^{\leq n} S^c$, which holds for any $n \in \bar{\mathbb{N}}_0$. As a result, we obtain that the maximal and minimal safety problems can be successfully reformulated in terms of optimal reachability value functions, i.e. if $S = G^c$

$$\begin{aligned} \sup_{\pi \in \Pi} \mathsf{P}^\pi_x(\Box^{\leq n} S) &= 1 - V_{*,n}(x), \\ \inf_{\pi \in \Pi} \mathsf{P}^\pi_x(\Box^{\leq n} S) &= 1 - V^*_n(x), \end{aligned} \tag{10}$$

for any $n \in \bar{\mathbb{N}}_0$. To find the maximal safety one has to look for the minimal reachability over the complement of the goal set.

Another problem related to probabilistic reachability is known as probabilistic reach-avoid [18]. Let us introduce this problem for two given sets $S, G \in \mathscr{B}(X)$ as follows. For $n \in \mathbb{N}_0$ we define

$$S \mathsf{U}^{\leq n} G = \left\{ \omega \in \Omega : \begin{array}{c} x_k \in G \text{ for some } 0 \leq k \leq n \text{ and} \\ x_j \in S \text{ for all } 0 \leq j < k \end{array} \right\}$$

and $S \mathsf{U}^{\leq \infty} G = \bigcup_{n=0}^\infty S \mathsf{U}^{\leq n} G$. The measurability of the defined events is clear. The set S is referred to as the safe set (or the set of legal states) and as above the set G is referred to as the goal set. We further introduce the corresponding reach-avoid value function for $x \in X, n \in \bar{\mathbb{N}}_0$, as

$$W^*_n(x) := \sup_{\pi \in \Pi} \mathsf{P}^\pi_x(S \mathsf{U}^{\leq n} G), \quad W_{*,n}(x) := \inf_{\pi \in \Pi} \mathsf{P}^\pi_x(S \mathsf{U}^{\leq n} G). \tag{11}$$

It is well-known that the reach-avoid problem is more general than the reachability one, since $\Diamond^{\leq n} G = X \mathsf{U}^{\leq n} G$ for any $G \in \mathscr{B}(X)$ and any $n \in \bar{\mathbb{N}}_0$. However, the converse statement also holds true, as we are going to show now: the idea is to state the reach-avoid problem over the cdt-MP $\mathfrak{D}$ as a reachability problem over a new cdt-MP $\check{\mathfrak{D}}$, where the avoid set $A := (S \cup G)^c$ is identified with an auxiliary single state with a loop.

More precisely, given a cdt-MP $\mathfrak{D}$ and two sets $S, G \in \mathscr{B}(X)$ we define $\check{\mathfrak{D}} = (\check{X}, U, \{\check{U}_x\}_{x \in \check{X}}, \check{\mathsf{T}})$ where $\check{X} = S \cup G \cup \{\psi\}$ for some auxiliary state $\psi \notin X$. We further define $\check{U}(x) = U(x)$ for $x \in S \cup G$ and $\check{U}(\psi) = \check{u}$ where $\check{u}$ is an element of U. Finally,

$$\check{\mathsf{T}}(B|x, u) = \begin{cases} \mathsf{T}(B|x, u), & \text{if } B \in \mathscr{B}(S \cup G),\ x \in S \cup G \\ \mathsf{T}(A|x, u), & \text{if } B = \{\psi\},\ x \in S \cup G \\ 1, & \text{if } B = \{\psi\},\ x = \psi. \end{cases}$$

Let us now define the corresponding optimal reachability value functions for the goal set G over the cdt-MP $\check{\mathfrak{D}}$, which we denote by $\check{V}^*_n$ for the maximal reachability and $\check{V}_{*,n}$ for the minimal one. The following result relates the reach-avoid value functions over $\mathfrak{D}$ to the reachability ones over $\check{\mathfrak{D}}$.

PROPOSITION 1. *For any $x \in A$ and any $n \in \bar{\mathbb{N}}_0$ it holds that $W_{*,n}(x) = W_n^*(x) = 0$. Furthermore, for any $x \in S \cup G$ and $n \in \mathbb{N}_0$*

$$W_n^*(x) = \check{V}_n^*(x), \quad W_{*,n}(x) = \check{V}_{*,n}(x).$$

It follows from Proposition 1 that results obtained for the optimization over the reachability probabilities can be directly applied to the case of reach-avoid. Due to this reason, we focus on the former problem and later show explicitly how to extend the obtained results from the reachability to the reach-avoid.

3.2 Formulation with an additive cost

The work in [5] considered the optimization of reachability probabilities over the class of `cdt`-MP where the cost functional took a multiplicative form. Focusing exclusively on Markov policies allowed obtaining DP recursions for value functions. In this work, however, we are interested in a wider class of policies, so the results in [5] are not directly applicable. In particular, an interesting question is the following: is it sufficient to consider only Markov policies in the optimization procedure? In order to answer this question, as well as to derive DP recursions, we are going to reformulate the original optimization problem via an additive cost functional, for which the theory of DP is rather rich [7, 13]. This approach is inspired by the one in [11].

Given a goal set $G \in \mathscr{B}(X)$ we consider a new `cdt`-MP

$$\hat{\mathfrak{D}} := (\hat{X}, U, \{\hat{U}(x,y)\}_{(x,y)\in\hat{X}}, \hat{\mathsf{T}})$$

with an augmented state space $\hat{X} = X \times Y$, where $Y = \{0,1\}$. The states are of the form (x,y) with coordinates being $x \in X$, $y \in Y$. The control space U is the same and we further define $\hat{U}(x,y) := U(x)$. The dynamics of $\hat{\mathfrak{D}}$ are given as follows:

$$\begin{cases} x_{n+1} & \sim \mathsf{T}(\cdot|x_n,u_n) \\ y_{n+1} & = 1_{G^c}(x_n)\cdot y_n, \end{cases}$$

hence the corresponding transition kernel $\hat{\mathsf{T}}$ is given by

$$\hat{\mathsf{T}}\left(B \times \{y'\}|x,y,u\right) := \begin{cases} y \cdot 1_{G^c}(x)\mathsf{T}(B|x,u), & \text{if } y' = 1, \\ \left(1 - y\cdot 1_{G^c}(x)\right)\mathsf{T}(B|x,u), & \text{if } y' = 0. \end{cases}$$

We construct a space of policies $\hat{\Pi}$ and for each $\hat{\pi} \in \hat{\Pi}$ and initial distribution $\hat{\alpha}$ on $\hat{X}$, a probability space $(\hat{\Omega}, \hat{\mathscr{F}}, \hat{\mathsf{P}}_{\hat{\alpha}}^{\hat{\pi}})$ with the expectation $\hat{\mathsf{E}}_{\hat{\alpha}}^{\hat{\pi}}$. We denote by $\hat{\Pi}_M \subset \hat{\Pi}$ the corresponding class of Markov policies for $\hat{\mathfrak{D}}$.

The additive cost structure consists of a cost $c : \hat{X} \to \{0,1\}$ given by $c(x,y) := y \cdot 1_G(x)$ and a functional

$$J_n^{\hat{\pi}}(x,y) := \hat{\mathsf{E}}_{(x,y)}^{\hat{\pi}}\left[\sum_{k=0}^{n} c(x_k,y_k)\right].$$

In order to relate it to the original formulation defined over the `cdt`-MP $\mathfrak{D}$, we first have to establish an explicit relationship between classes of strategies Π and $\hat{\Pi}$. Clearly, we can treat Π as a subset of $\hat{\Pi}$ as any policy $\pi \in \Pi$ for the `cdt`-MP $\mathfrak{D}$ serves also as a policy for the `cdt`-MP $\hat{\mathfrak{D}}$. We let $\iota : \Pi \to \hat{\Pi}$ be the *inclusion* map. On the other hand, we define the *projection* map $\theta : \hat{\Pi} \to \Pi$ by

$$\theta_i(\pi)(du_i|x_0,u_0,\dots,x_i) := \hat{\pi}_i(du_i|x_0,y_0 = 1,u_0,\dots,x_i,y_i = 1)$$

The following result relates the two optimization problems.

THEOREM 1. *For any $n \in \bar{\mathbb{N}}_0$, $\pi \in \Pi$ and $\hat{\pi} \in \hat{\Pi}$, it holds that*

$$J_n^{\hat{\pi}}(x,1) = V_n^{\theta(\hat{\pi})}(x), \quad V_n^{\pi}(x) = J_n^{\iota(\pi)}(x,1). \tag{12}$$

Theorem 1 has several important corollaries. First of all, it can be used to prove that Markov policies are sufficient for the original optimal reachability problem over a bounded time horizon, i.e. in case when $n < \infty$. At the same time, the optimal policy may depend on time and thus is not necessary stationary. Note that for a MDP, a special case of `cdt`-MP, this fact has been already known [6, Section 10.6]. More precisely:

COROLLARY 1. *For any $n \in \mathbb{N}_0$ and $\pi \in \Pi$ there exists $\pi' \in \Pi_M$ such that $V_n^{\pi} = V_n^{\pi'}$, and as a consequence $V_n^*(x) := \sup_{\pi'\in\Pi_M} V_n^{\pi'}(x)$ and $V_{*,n}(x) := \inf_{\pi'\in\Pi_M} V_n^{\pi'}(x)$.*

Moreover, it follows from Theorem 1 that one can do equivalently optimization of the additive cost functionals J to solve the original optimal reachability problem. Let us further define $J_n^*(x,y) := \sup_{\hat{\pi}\in\hat{\Pi}} J_n^{\hat{\pi}}(x,y)$ and $J_{*,n}(x,y) := \inf_{\hat{\pi}\in\hat{\Pi}} J_n^{\hat{\pi}}(x,y)$.

COROLLARY 2. *For any $n \in \bar{\mathbb{N}}_0$, $x \in X$, the following equalities hold true: $J_n^*(x,0) = J_{*,n}(x,0) = 0$ and*

$$V_n^*(x) = J_n^*(x,1), \quad V_{*,n}(x) = J_{*,n}(x,1). \tag{13}$$

Finally, we can exploit DP recursions for the additive cost functionals J_n^* and $J_{*,n}$ to study DP recursion for the optimal reachability value functions. Let us introduce the following operators

$$\mathfrak{I}^* f(x) := 1_G(x) + 1_{G^c}(x) \sup_{u\in U(x)} \int_X f(y)\mathsf{T}(dy|x,u),$$

$$\mathfrak{I}_* f(x) := 1_G(x) + 1_{G^c}(x) \inf_{u\in U(x)} \int_X f(y)\mathsf{T}(dy|x,u),$$

which act on the space of bounded universally measurable functions. These operators can be used to compute optimal value functions recursively, as the following result states.

COROLLARY 3. *For any $n \in \mathbb{N}_0$, the functions V_n^*, $V_{*,n}$ are universally measurable. Moreover, $V_0^* = V_{*,0} = 1_G$ and for any $n \in \mathbb{N}_0$*

$$V_{n+1}^* = \mathfrak{I}^* V_n^*, \quad V_{*,n+1} = \mathfrak{I}_* V_{*,n}. \tag{14}$$

3.3 Infinite time horizon

In the previous section we have successfully restated the original reachability problem as a classical additive-cost problem, which made it possible to derive several important results. In particular, Corollary 3 allows one to compute finite-horizon optimal reachability functions recursively, which can be done using numerical methods based on approximate abstractions: we present them later in Section 4. With focus on the infinite time horizon, it is expected that the solution can be obtained as a limit of solutions of finite-horizon optimization problems, which is a fixpoint of an appropriate operator: either $\mathfrak{I}^*$ for the maximal value function, or $\mathfrak{I}_*$ for the minimal one. However, it is known from the literature that this is not necessarily true [7]. In particular, in our case the fixpoint characterization $V_\infty^* = \mathfrak{I}^* V_\infty^*$ holds in general, whereas additional assumptions are needed to show that $V_{*,\infty} = \mathfrak{I}_* V_{*,\infty}$.

We start with the following result, which describes the properties of reachability probabilities with a fixed policy.

LEMMA 1. *For any $n \in \mathbb{N}_0$, $\pi \in \Pi$ and an initial distribution α,*

$$\mathsf{P}_\alpha^\pi(\lozenge^{\le n} G) \le \mathsf{P}_\alpha^\pi(\lozenge^{\le n+1} G) \le \mathsf{P}_\alpha^\pi(\lozenge^{<\infty} G). \tag{15}$$

Moreover, it holds that

$$\mathsf{P}_\alpha^\pi(\lozenge^{\infty} G) = \lim_n \mathsf{P}_\alpha^\pi(\lozenge^{\le n} G) = \sup_n \mathsf{P}_\alpha^\pi(\lozenge^{\le n} G). \tag{16}$$

The monotonicity result above is crucial for the proof of the fixpoint characterization for the maximal reachability over an infinite time horizon. When considering the limit of functions V_n^* as $n \to \infty$ the key step is to swap the order of $\lim_{n\to\infty}$ and the supremum over the control actions, which comes from $\mathfrak{I}^*$ – this can be done as the limit of an increasing sequence is a supremum itself. This leads us to the following:

LEMMA 2. *For any $x \in X$, $(V_n(x))_{n\in\mathbb{N}_0}$ is a non-decreasing sequence of real numbers and there exists a point-wise limit*

$$V^\star(x) := \lim_n V_n(x) = \sup_n V_n(x), \tag{17}$$

which is the least non-negative fixpoint of $\mathfrak{I}$, i.e. $V^\star = \mathfrak{I}V^\star$ and if there is another fixpoint $f \in \mathscr{B}(X)$ such that $f \geq 0$ then $f \geq V^\star$.

THEOREM 2. *The maximal reachability value function V_∞^* is the least non-negative fixpoint of the operator $\mathfrak{I}^*$.*

REMARK 1. *To our knowledge, this is the first result on the fixpoint characterization of the maximal reachability value function for* cdt-MP. *Although [10, Theorem (2.10)] provides a fixpoint characterization for a (more general) reach-avoid problem, one of the assumptions required there is that under any Markov policy, for any initial condition the set of legal states is left by the path of the* cdt-MP *with a probability 1 in some finite time. This clearly leads to the fact that $V_\infty^* \equiv 1$. Thus, with focus on the reachability problem [10, Theorem (2.10)] can be applied only for the case when the solution is known to be constant.*

Let us now consider the minimal reachability problem: in this case the $\lim_{n\to\infty}$ has to be swapped with the infimum that comes from $\mathfrak{I}_*$, which cannot been done in general. We then tailor a technique in [13, Section 4] to our case. In order to establish the main result we need the following assumption:

ASSUMPTION 1. *The kernel T is strongly continuous: $\mathsf{T}(A|\cdot)$ is a continuous function on $\mathbb{K}$ for any $A \in \mathscr{B}(X)$ [13, Appendix C].*

THEOREM 3. *Under Assumption 1, the minimal reachability function $V_{*,\infty}$ is the least non-negative fixpoint of $\mathfrak{I}_*$.*

Let us discuss some properties of the optimal infinite-horizon reachability value functions and of the fixpoint equations

$$f = \mathfrak{I}^* f, \tag{18}$$

$$f = \mathfrak{I}_* f. \tag{19}$$

First of all, note that $V^*(x) = V_*(x) = 1$ for all $x \in G$, so in case $V^* \equiv 1$ or $V_* \equiv 1$ we say that the corresponding optimization problem has a trivial solution. This case refers to the reachability of the goal state G in some finite time with probability 1 starting from any initial condition. However, by substitution we find that the function $f \equiv 1$ solves both equations (18) and (19), regardless of the shape of the transition kernel T. Due to this reason, we are able to formulate the following result.

PROPOSITION 2. *Equations (18) and (19) have unique solutions if and only if the corresponding optimization problems have trivial solutions.*

Some examples when the solutions of the optimization problems are not trivial can be constructed using appropriate notions of absorbing sets over the cdt-MP.

DEFINITION 7 (STRONGLY AND WEAKLY ABSORBING SETS). *Given the* cdt-MP *$\mathfrak{D} = (X, U, \{U(x)\}_{x\in X}, \mathsf{T})$, the set $A \in \mathscr{B}(X)$ is called* strongly absorbing *if $\mathsf{T}(A|x,u) = 1$ for all $u \in U(x)$ and $x \in A$.*

The set B is called weakly absorbing *if there exists a kernel μ on U given B such that $\mu(U(x)|x) = 1$ for all $x \in B$ and such that*

$$\int_{U(x)} \mathsf{T}(B|x,u)\mu(du|x) = 1, \quad \forall x \in B.$$

Let us briefly comment on the definition above. First of all, every strongly absorbing set A is a weakly absorbing set since the required kernel μ as per Definition 7 in such case can be chosen to be a deterministic one, obtained by the restriction of the map k (defined in Section 2.1) to the set A, i.e.

$$\mu(C|x) = 1_C(\mathsf{k}(x))$$

for any $x \in A$ and $C \in \mathscr{B}(U)$. Furthermore, in the autonomous case when the control set U is identified with a singleton, the notion of weak and strong sets coincide with that of an absorbing set [15]. Intuitively, a strongly absorbing set remains absorbing under any action whereas for a weakly absorbing set there exists a control which makes such set absorbing.

PROPOSITION 3. *If $A \subseteq G^c$ is a strongly (weakly) absorbing set, then $V^*(x) = 0$ ($V_*(x) = 0$) for all $x \in A$.*

The latter result in particular shows that the presence of absorbing subsets on the complement of the goal state violates the uniqueness of the fixpoint equations, and leads to a non-trivial solution of the problem. This fact is already known for the case of autonomous systems [20, 21]. There, it has been shown that under some structural assumptions on the model and on the goal set, the presence of absorbing subsets is not only a sufficient condition for the lack of uniqueness, but is also necessary. In particular, let us further mention that in the second part of [4, Theorem 3], where the infinite-horizon (autonomous) reachability value function is considered, the result holds only under the assumption that the solution of the fixpoint equation is unique – in which case as we have shown the solution is trivial and equal to a constant function 1.

Finally, let us apply the derived results to the probabilistic reach-avoid problem using Proposition 1. We keep G as the goal set, and define $S \in \mathscr{B}(X)$ to be the set of legal states. As above, let W_n^* and $W_{*,n}$ be the optimal reach-avoid value functions as in (11) and define the following operators

$$\mathfrak{R}^* f(x) := 1_G(x) + 1_A(x) \sup_{u\in U(x)} \int_X f(y)\mathsf{T}(dy|x,u),$$

$$\mathfrak{R}_* f(x) := 1_G(x) + 1_A(x) \inf_{u\in U(x)} \int_X f(y)\mathsf{T}(dy|x,u),$$

where as above $A = (S \cup G)^c$ is the set to be avoided. The next result follows immediately from those we have obtained for reachability and from Proposition 1.

THEOREM 4. *It holds that $W_{*,0} = W_0^* = 1_G$ and for any $n \in \mathbb{N}_0$*

$$W_{n+1}^* = \mathfrak{R}^* W_n^*, \quad W_{*,n} = \mathfrak{R}_* W_{*,n}.$$

In particular, Markov policies are sufficient for optimization:

$$W_n^*(x) := \sup_{\pi\in\Pi_M} \mathsf{P}_x^\pi(S\mathsf{U}^{\leq n}G), \quad W_{*,n}(x) := \inf_{\pi\in\Pi_M} \mathsf{P}_x^\pi(S\mathsf{U}^{\leq n}G).$$

Moreover, function W_∞^* is the least non-negative fixpoint of the operator $\mathfrak{R}^*$ and, under Assumption 1, function $W_{*,\infty}$ is the least non-negative fixpoint of the operator $\mathfrak{R}_*$.

Theorem 4 generalizes several results in the literature. First, it extends the work in [18] by considering all possible policies, rather than focusing on Markov policies exclusively. Second, it extends results obtained in [10] over the infinite-time horizon by considering both optimization problems, rather than the maximization one only. Furthermore, we are able to relax the assumptions in [10, Theorem (2.10)] and show that the fixpoint characterization for W^*_∞ holds in a more general case.

3.4 Automata model checking

In section 2.4 we have introduced the following problem: given a cdt-MP $\mathfrak{D}$, a policy $\pi \in \Pi$, an initial distribution α and a DFA $\mathscr{A}$, find the probability $\mathsf{P}^\pi_\alpha(\omega \models \mathscr{A})$. We are further interested in maximizing and minimizing such a probability over all possible policies $\pi \in \Pi$. For this purpose, we are going to reduce this general problem over $\mathfrak{D}$ to the reachability problem studied above over another cdt-MP $\mathfrak{D}\otimes\mathscr{A}$, which we refer to as a product of the cdt-MP $\mathfrak{D}$ and the automaton $\mathscr{A}$. This product is defined as follows:

DEFINITION 8 (PRODUCT BETWEEN cdt-MP AND DFA). *Given a cdt-MP $\mathfrak{D} = (X, U, \{U(x)\}_{x\in X}, \mathsf{T})$, a finite alphabet Σ, a labeling function $\mathsf{L} : X \to \Sigma$, and a DFA $\mathscr{A} = (Q, q_0, \Sigma, F, \mathsf{t})$, we define the product between $\mathfrak{D}$ and $\mathscr{A}$ to be another cdt-MP denoted as $\mathfrak{D} \otimes \mathscr{A} = (\bar{X}, U, \{U(x)\}_{x\in\bar{X}}, \bar{\mathsf{T}})$. Here $\bar{X} = X \times Q$ and*

$$\bar{\mathsf{T}}(A \times \{q'\}|x,q,u) = 1_{\mathsf{t}(q,\mathsf{L}(x))}(q') \cdot \mathsf{T}(A|x,u).$$

We want to show that $\mathsf{P}^\pi_x(\omega \models \mathscr{A})$ can be related to the reachability probability over the cdt-MP $\mathfrak{D} \otimes \mathscr{A}$ with a goal state $G := X \times F$, as it was shown to be the case when the state space is finite [12, Proposition 1]. In order to reformulate the automaton verification as a reachability problem, we are going to follow a procedure similar to the one in Section 3.2, where the reachability problem has been reformulated by an additive cost. We again construct a space of policies $\bar{\Pi}$ for $\mathfrak{D} \otimes \mathscr{A}$, and for each $\bar{\pi} \in \bar{\Pi}$ and initial distribution $\bar{\alpha}$ on $\bar{X}$, a probability space $(\bar{\Omega}, \bar{\mathscr{F}}, \bar{\mathsf{P}}^{\bar{\pi}}_{\bar{\alpha}})$. As in Section 3.2 we need to relate policies in Π to those in $\bar{\Pi}$: again, we can use the fact that $\Pi \subset \bar{\Pi}$ hence any policy $\pi \in \Pi$ over the original cdt-MP $\mathfrak{D}$ is also a policy over $\mathfrak{D} \otimes \mathscr{A}$. By $\iota : \Pi \to \bar{\Pi}$ we can hence denote the corresponding inclusion map. For the other direction, to any policy $\bar{\pi} \in \bar{\Pi}$ we can assign a policy $\theta(\bar{\pi}) \in \Pi$ as follows

$$\theta_i(\pi)(du_i|x_0,u_0,\dots,x_i) = \bar{\pi}_i(du_i|x_0,z_0,u_0,\dots,x_i,z_i),$$

where $z_0 = q_0$ is the initial state of $\mathscr{A}$ and $z_{j+1} = \mathsf{t}(z_j, \mathsf{L}(x_j))$ is defined recursively for $j \in \overline{0, i-1}$. The following technical lemma is necessary to state the main result.

LEMMA 3. *For any $n \in \bar{\mathbb{N}}_0$, $x \in X$ and $\pi \in \Pi$ it holds that*

$$\mathsf{P}^\pi_x(\omega \models_n \mathscr{A}) = \bar{\mathsf{P}}^{\iota(\pi)}_{(x,q_0)}(\lozenge^{\le n+1} G),$$

where $\infty + 1 := \infty$, and for any $\bar{\pi} \in \bar{\Pi}$ it holds that

$$\bar{\mathsf{P}}^{\bar{\pi}}_{(x,q_0)}(\lozenge^{\le n+1} G) = \mathsf{P}^{\theta(\bar{\pi})}_x(\omega \models_n \mathscr{A}).$$

THEOREM 5. *For any $n \in \bar{\mathbb{N}}_0$ and $x \in X$ it holds that*

$$\sup_{\pi\in\Pi} \mathsf{P}^\pi_x(\omega \models_n \mathscr{A}) = \bar{V}^*_{n+1}(x), \qquad \inf_{\pi\in\Pi} \mathsf{P}^\pi_x(\omega \models_n \mathscr{A}) = \bar{V}_{*,n+1}(x), \tag{20}$$

where $\infty + 1 := \infty$ and $\bar{V}^*_n$, $\bar{V}_{*,n}$ are the optimal reachability functions that are defined over the cdt-MP $\mathfrak{D} \otimes \mathscr{A}$.

Notice that in the above theorem we need to consider only Markov policies as was shown in the previous section. However, such policies are Markov with respect to $\mathfrak{D}\otimes\mathscr{A}$ and are not necessarily Markov with respect to $\mathfrak{D}$. In terms of $\mathfrak{D}$ this means that the policy is dependent on the history through the state of the automaton. The actual computation of the optimal reachability value functions can be implemented by discretizing the set of states as well as the set of controls, as elaborated in Section 4.

4. APPROXIMATE ABSTRACTIONS

Since in general there is no hope that the iterations in (14) yield value functions in an explicit form, we introduce an abstraction procedure that leads to numerical methods for the computation of such functions. Moreover, we provide an explicit upper bound on the error caused by the abstraction. We present the results with focus on the maximal reachability problem, with the understanding that a similar procedure applies to the minimal reachability one too.

Let us consider some cdt-MP $\mathfrak{D} = (X, U, \{U(x)\}_{x\in X}, \mathsf{T})$. Since X and U are Borel spaces they are metrizable topological spaces. Let ρ_X and ρ_U be some metrics on X and U respectively, which are consistent with the given topologies of the underlying spaces. We first introduce some technical considerations that are important for the abstraction over the control space. Let $U = \bigcup_{j=1}^M U_j$ be a measurable partition of U, and let $u_j \in U_j$ for $1 \le j \le M$ be arbitrary representative points. Define $\Delta := \max_{1\le j\le M} \mathrm{diam}_U(U_j)$ where the diameter of a subset of U for any $C \subseteq U$ is given by

$$\mathrm{diam}_U(C) = \sup_{u',u''\in C} \rho_U(u',u'').$$

LEMMA 4. *Let $g, \hat{g} : U \to \mathbb{R}$ be two functions. Define two optimization problems: $g^* := \sup_{u\in U} g(u)$ and $\hat{g}^* := \max_{1\le j\le M} \hat{g}(u_j)$. If g is Lipschitz continuous, i.e. if there exists $K > 0$ such that*

$$|g(u') - g(u'')| \le K \cdot \rho_U(u',u'') \tag{21}$$

for all $u', u'' \in U$, then it holds that $|g^* - \hat{g}^*| \le K \cdot \Delta + \|g - \hat{g}\|_U$.

Let us further proceed with the abstraction procedure and choose $G \in \mathscr{B}(X)$ to be a target set. Clearly, the solution of any reachability problem on G is trivial, so we only need to solve these problems on G^c. For this purpose we define $G^c = \bigcup_{i=1}^N G_i$ to be a measurable partition of the set G^c, and we choose representative points $x_i \in G_i$ for $1 \le i \le N$ in an arbitrary way. We further denote $\delta_i := \mathrm{diam}_X(G_i)$.

We abstract the original cdt-MP $\mathfrak{D}$ as an MDP denoted by $\tilde{\mathfrak{D}} = (\tilde{X}, \tilde{U}, \{\tilde{U}(x)\}_{x\in\tilde{X}}, \tilde{\mathsf{T}})$. The finite state and control spaces are $\tilde{X} := \{i\}_{i=1}^N \cup \{\phi\}$ and $\tilde{U} := \{j\}_{j=1}^M$, where $\phi \notin X$ is a "sink" state that corresponds to the target set G of the original cdt-MP. To complete the definition of $\tilde{\mathfrak{D}}$ we have to specify the map $\tilde{U}$ and the kernel $\tilde{\mathsf{T}}$. We choose $\tilde{U}(i) = \tilde{U}$ for any $i \in \tilde{X}$ and

$$\begin{cases} \tilde{\mathsf{T}}(k|i,j) := \mathsf{T}(G_k|x_i,u_j) & \text{for all } 1 \le j \le M,\ 1 \le i,k \le N \\ \tilde{\mathsf{T}}(\phi|i,j) := \mathsf{T}(G|x_i,u_j) & \text{for all } 1 \le j \le M,\ 1 \le i \le N \\ \tilde{\mathsf{T}}(\phi|\phi,j) = 1 & \text{for all } 1 \le j \le M. \end{cases}$$

For $\tilde{\mathfrak{D}}$, let us define $\tilde{V}^*_n$ to be the n-horizon maximal reachability value function of the target state ϕ. Clearly, $\tilde{V}^*_0 = 1_{\{\phi\}}$ and it follows from Corollary 3 that

$$\tilde{V}^*_{n+1}(i) = 1_{\{\phi\}}(i) + 1_{\{\phi\}^c}(i) \max_{1\le j\le M} \sum_{k\in\tilde{X}} \tilde{V}^*_n(k)\,\tilde{\mathsf{T}}(k|i,j).$$

Such value functions can be computed with numerically efficient methods [14]. Clearly, we expect that such functions serve as a good approximation for functions V_n^* defined for the original cdt-MP $\mathfrak{D}$. However, so far they are defined over a different state space, which makes it hard to compare them. Due to this reason, we choose $\hat{V}_n^* : X \to \mathbb{R}$ to be a piece-wise constant interpolation of $\tilde{V}_n^*$, defined in the following manner:

$$\hat{V}_n^*(x) = 1_G(x) + \sum_{i=1}^{N} 1_{G_i}(x)\tilde{V}_n^*(i).$$

Clearly, $\hat{V}_n^*(x) = 1$ for all $x \in G$. Moreover, for all other states x the following result holds true.

LEMMA 5. *For any $1 \le i \le N$ and any $x \in G_i$ it holds that*

$$\hat{V}_{n+1}^*(x) = \max_{1\le j\le M} \int_X \hat{V}_n^*(y)\mathsf{T}(dy|x_i, u_j).$$

To ensure that the difference between V_n^* and $\hat{V}_n^*$ can be made as small as needed by tuning parameters of the partition Δ and δ_i, we need the following assumption.

ASSUMPTION 2. *The action space $U(x)$ does not depend on x, i.e. it holds that $U(x) = U$ for all $x \in X$. In addition, T is an integral kernel, i.e. there exists a σ-finite measure μ on X and a jointly measurable function $t : X \times X \times U \to R$ such that*

$$\mathsf{T}(B|x,u) = \int_B t(y|x,u)\mu(dy), \tag{22}$$

for any $B \in \mathscr{B}(X), x \in X, u \in U$. Moreover, there exist:

1. *measurable functions $\lambda_i : X \to \mathbb{R}$ for $1 \le i \le N$, such that for all $x', x'' \in G_i$ and for all $y \in X, u \in U$:*

$$|t(y|x',u) - t(y|x'',u)| \le \lambda_i(y),$$

where $\Lambda_i := \int_X \lambda_i(y)\mu(dy) < \infty$;

2. *measurable functions $\kappa_i : X \to \mathbb{R}$ for $1 \le i \le N$, such that for all $u', u'' \in U$ and for all $y \in X, x \in G_i$:*

$$|t(y|x,u') - t(y|x,u'')| \le \kappa_i(y),$$

where $K_i := \int_X \kappa_i(y)\mu(dy) < \infty$.

The following result provides upper-bounds on the difference between the value functions of the original model and those computed via the MDP abstraction.

THEOREM 6. *Let V_n^* and $\hat{V}_n^*$ be the functions defined above. Introduce $r := \sup_{x\in G^c, u\in U} \mathsf{T}(G^c|x,u)$ and let Assumption 2 hold true. If $r < 1$, then for all $n \ge 1$ it holds that*

$$\|V_n^* - \hat{V}_n^*\| \le \frac{1-r^n}{1-r} \cdot \max_{1\le i\le N} (\Lambda_i\delta_i + K_i\Delta). \tag{23}$$

If $r = 1$, then for all $n \ge 0$ it holds that

$$\|V_n^* - \hat{V}_n^*\| \le n \cdot \max_{1\le i\le N} (\Lambda_i\delta_i + K_i\Delta). \tag{24}$$

Since the case of autonomous SHS can be regarded as a cdt-SHS with a control space being a singleton, $U = \{u\}$, it is worth commenting on the relation between Theorem 6 and approximation techniques known from the literature on autonomous SHS [3, 17]. First of all, in such a case Assumption 2 is a slight generalization of assumptions on the uniform Lipschitz continuity of kernels in [3] and on the local Lipschitz continuity in [17]. In particular, the application of Theorem 6 under Assumption 2 to the autonomous case implies [17, Theorem 6] as a special case, where the functions λ_i are assumed to be piecewise constant. In turn, [17, Theorem 6] is further known to be a generalization of [3, Theorem 1].

Let us mention that the structure of the bounds allows applying the adaptive gridding procedure developed for the autonomous SHS in [17] over the state space discretization, by computing the local errors Λ_i and choosing the discretization size δ_i accordingly. However, the error introduced by the partition of the control space depends on the global discretization parameter $\max_{1\le i\le N} K_i\Delta$, so the adaptive gridding of the control space may not improve results versus those obtained by the uniform gridding of the control space.

5. CASE STUDY: ENERGY CONTROL

Inspired by [1], we consider a resource allocation problem for an energy network comprised of two subnetworks $i = 1, 2$. The energy provider for each subnetwork is given the choice of generating energy at capacity either by a polluting device (say, a coal plant), or alternatively via local renewables: accordingly, the (normalized) decision variables are $0 \le u^i_{(\cdot)} \le 1, i = 1, 2$, denoting the production of polluting power (u^i_p) and of renewable power (u^i_r) within the ith subnetwork. There is a constraint on the maximal power generation, both globally (over the total generated polluting power) and locally (whatever is not generated via coal can be obtained from the renewables u^i_r), so that

$$u^1_p + u^2_p \le 2T, \qquad u^i_p + u^i_r = T, i = 1, 2. \tag{25}$$

Each of the two subnetworks sets forth a time-varying demand $D_i(k)$, which is not exactly known due to the intrinsic variability of power demand.

The state variables of the model are the time-varying energy levels of each subnetwork, which we denote by E_i for $i = 1, 2$. They are driven by the following dynamics:

$$E_i(k+1) = E_i(k) + P(k)u^i_p(k) + R_i(k)u^i_r(k) - D_i(k), \tag{26}$$

where P is the actual power generated by the coal plant (polluting device) and R_i is the generated power by local renewables. Furthermore, P and R_i are independent random variables, identically distributed in time, and such that $\mathsf{E}[P] \doteq \mu_P > \mathsf{E}[R_i] \doteq \mu_{R_i}$, whereas $\mathsf{Var}[P] \doteq \sigma_P^2 < \mathsf{Var}[R_i] \doteq \sigma_{R_i}^2$. The above relation among parameters is suggested by assuming that the coal plant is a stronger and more reliable (though less desirable) source of power. Variables R_1 and R_2 can be correlated due to the spatial adjacency of the two subnetworks and its effect on the production based on renewables. This, along with the presence of P and of the constraints in (25), couples the two dynamics. We select the demand variables D_i to be independent and identically distributed, and independent from P and R_i, $i = 1, 2$.

We consider two scenarios expressed via specifications, and proceed optimizing over. Let us introduce two additional constraints on the energy levels for the subnetworks, namely:

$$E_1 + E_2 \le S, \tag{27}$$

$$E_i \ge M_i, i = 1, 2, \tag{28}$$

where the threshold S denotes a (constant) limit due to storage, while the second inequality refers to (constant) minimal energy requirements. We are looking at the following problem:

$$\sup_{\pi} \mathsf{P}_x^{\pi} \left\{ \square^{\leq N} (27) \wedge (28) \right\}, \quad (29)$$

which corresponds to the following safety specification: the energy levels have to be above the thresholds and the sum of them must not exceed the storage capacity. Another problem which is interesting to us is whether the energy levels can simultaneously exceed some given value F without ever falling below the zero level beforehand. This problem can be expressed as:

$$E_i \geq F_i, i = 1, 2, \quad (30)$$

$$E_i \geq 0, i = 1, 2, \quad (31)$$

and the related synthesis problem is

$$\sup_{\pi} \mathsf{P}_x^{\pi} \left\{ (31) \mathsf{U}^{\leq N} (30) \right\}, \quad (32)$$

The specification in (29) denotes maximal probabilistic invariance (equivalently, minimal reachability), whereas (32) is a maximal reach-avoid property, which can be reformulated as a maximal probabilistic reachability problem as discussed in Sec. 3.

In the implementation we have used the following model parameters: $T = 1$, $P = 2$ (constant polluting power production), whereas $R_i \sim \mathcal{N}(0.5, 2)$ and $D_i \sim \mathcal{N}(1, 2)$. None of the variables P, R and D depend on the step size k. We have selected the parameters M_i for (29) to be smaller than F_i for (32): for the first synthesis problem we have picked $M_i = 5$ and $S = 30$, whereas for the second one we have picked $F_i = 25$. We have chosen a discretization step $\delta_s = 1$ for the state variables and $\delta_c = 0.05$ for the control variables. All the experiments have been run on a 2.83GHz 4 Core(TM)2 Quad CPU with 3.7Gb of memory. The total running time for the experiments has amounted to 14.37 min.

Fig. 1(a) displays the maximal probabilistic invariance obtained (at the initial time step) for the synthesis problem in (29). It can be seen that the probability decreases close to the boundary of the region defined by the conditions in (30) and (31). This is as expected, since there is a higher probability close to the boundary to falsify the invariance property. The optimal control action at the initial time step for the synthesis problem (29) is plotted in Fig. 1(b). Note that this action suggests using some polluting power generation (rather than relying on the more uncertain renewable power) whenever the energy levels are too low and close to the thresholds M_i. Conversely, whenever the energy levels are high enough, the policy suggests relying on renewables only.

Fig. 2(a) depicts the maximal probabilistic reach-avoid (at the initial time step) for the synthesis problem in (32). As expected, the obtained probability is higher for energy levels close to F_i. The optimal control action for the synthesis problem (32) is given in Fig. 2(b) (here plotted at the final step): since the goal is essentially to maximize the energy levels, it can be seen that the obtained policy selects the reliable coal plant for maximal energy generation (except when already in the goal sets).

The computational results have been obtained by discretizing the state space and computing the discrete value functions, and can be further improved by using an adaptive gridding procedure as in [17]. On the other hand, rather than computing value functions, the obtained MDP abstractions for safety and reach-avoid can be engaged in the verification of the properties of interest using model-checking software [14].

6. CONCLUSIONS AND NEXT STEPS

The contributions of this work are twofold. On the theoretical side, the article has zoomed in on issues related to reachability and invariance optimization over both finite and infinite horizons for non-autonomous stochastic hybrid systems, showing the sufficiency of memoryless optimal policies and tackling the problem of verification of specifications expressed as deterministic, finite-state automata. On the applicative side, a new computational scheme with explicit error quantification has been introduced and applied over a controller synthesis case study from the area of power systems. Both the theoretical and the computational outcomes nicely tailor back to known special models (autonomous, Markovian) from the literature.

The authors are interested in considering models with more complicated control structures (both discrete and continuous), and in looking at verification problems over ω-regular properties expressed as Büchi automata: the latter require non-trivial measure-theoretical results dealing with infinite-horizon problems that go beyond the scope of this work.

7. ACKNOWLEDGMENTS

The authors would like to thank Sadegh E. Z. Soudjani for the discussions on the bounds on the discretization procedure.

8. REFERENCES

[1] MoVeS website. http://www.movesproject.eu.

[2] A. Abate, S. Amin, M. Prandini, J. Lygeros, and S. Sastry. Computational approaches to reachability analysis of stochastic hybrid systems. In *Hybrid Systems: Computation and Control*, pages 4–17. Springer Verlag, 2007.

[3] A. Abate, J.-P. Katoen, J. Lygeros, and M. Prandini. Approximate model checking of stochastic hybrid systems. *European Journal of Control*, 16(6):1–18, 2010.

[4] A. Abate, J.-P. Katoen, and A. Mereacre. Quantitative automata model checking of autonomous stochastic hybrid systems. In *Proceedings of the 14th ACM international conference on Hybrid Systems: Computation and Control*, pages 83–92, Chicago, IL, April 2011.

[5] A. Abate, M. Prandini, J. Lygeros, and S. Sastry. Probabilistic reachability and safety for controlled discrete time stochastic hybrid systems. *Automatica*, 44(11):2724–2734, 2008.

[6] C. Baier and J.-P. Katoen. *Principles of Model Checking*. MIT Press, 2008.

[7] D. Bertsekas and S. Shreve. *Stochastic Optimal Control: The Discrete Time Case*, volume 139. Academic Press, 1978.

[8] H. Blom and J. Lygeros (Eds.). *Stochastic Hybrid Systems: Theory and Safety Critical Applications*. Number 337 in Lecture Notes in Control and Information Sciences. Springer Verlag, Berlin Heidelberg, 2006.

[9] C. Cassandras and J. E. Lygeros. *Stochastic hybrid systems*, volume 24. CRC Press, 2007.

[10] D. Chatterjee, E. Cinquemani, and J. Lygeros. Maximizing the probability of attaining a target prior to extinction. *Nonlinear Analysis: Hybrid Systems*, 5(2):367–381, 2011.

[11] J. Ding, A. Abate, and C. Tomlin. Optimal control of partially observable discrete time stochastic hybrid systems for safety specifications. *Proceedings of the 32nd American Control Conference*, 2013.

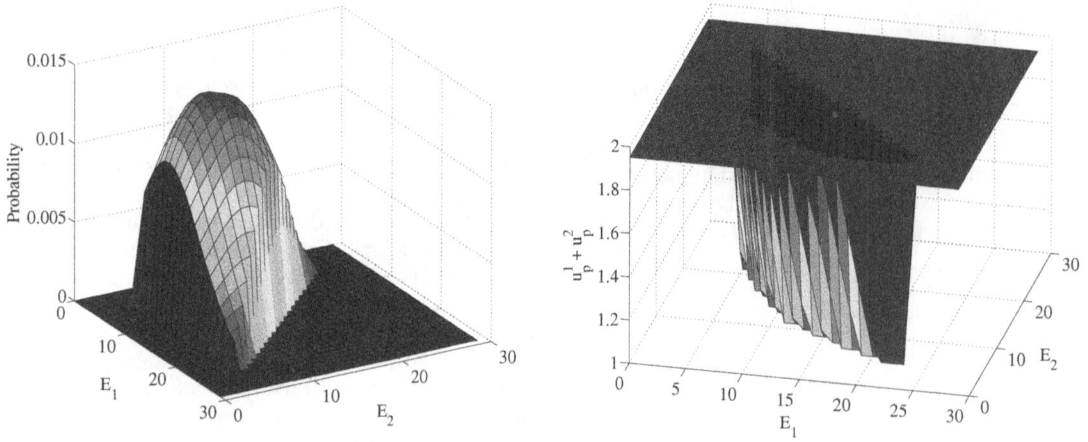

(a) Maximal probabilistic invariance (b) Optimal control action at the initial step

Figure 1: Synthesis problem formulated in (29).

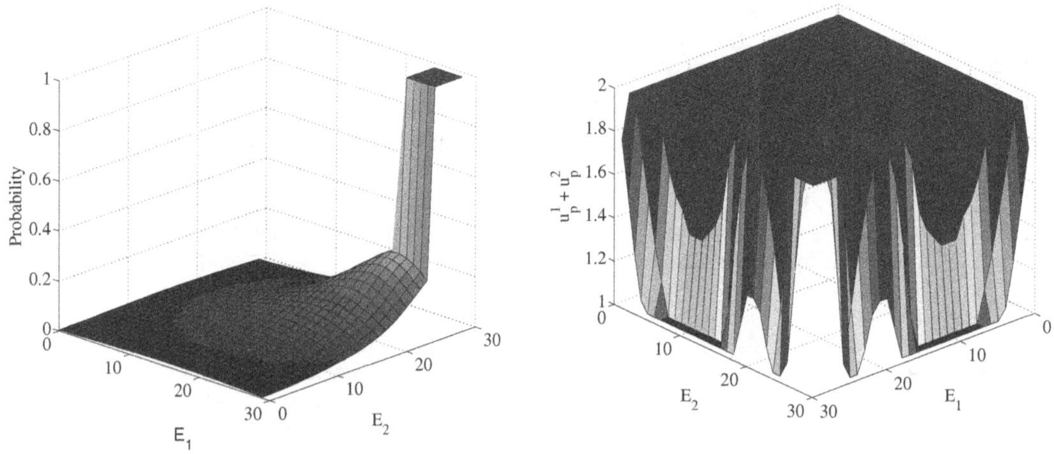

(a) Maximal probabilistic reach-avoid (b) Optimal control action at the final step

Figure 2: Synthesis problem formulated in (32).

[12] V. Forejt, M. Kwiatkowska, G. Norman, and D. Parker. Automated verification techniques for probabilistic systems. *Formal Methods for Eternal Networked Software Systems*, pages 53–113, 2011.

[13] O. Hernández-Lerma and J. B. Lasserre. *Discrete-time Markov control processes*, volume 30 of *Applications of Mathematics (New York)*. Springer Verlag, New York, 1996.

[14] A. Hinton, M. Kwiatkowska, G. Norman, and D. Parker. PRISM: A tool for automatic verification of probabilistic systems. In H. Hermanns and J. Palsberg, editors, *Tools and Algorithms for the Construction and Analysis of Systems*, volume 3920 of *Lecture Notes in Computer Science*, pages 441–444. Springer Verlag, 2006.

[15] S. Meyn and R. Tweedie. *Markov Chains and Stochastic Stability*. Springer Verlag, 1993.

[16] W. Rudin. *Real and complex analysis*. McGraw-Hill Book Co., New York, third edition, 1987.

[17] S. Soudjani and A. Abate. Adaptive gridding for abstraction and verification of stochastic hybrid systems. In *Quantitative Evaluation of Systems (QEST)*, pages 59–68, Aachen, DE, 2011.

[18] S. Summers and J. Lygeros. A Probabilistic Reach-Avoid Problem for Controlled Discrete Time Stochastic Hybrid Systems. In *IFAC Conference on Analysis and Design of Hybrid Systems, ADHS*, Zaragoza, Spain, September 2009.

[19] S. Summers and J. Lygeros. Verification of discrete time stochastic hybrid systems: A stochastic reach-avoid decision problem. *Automatica*, 46(12):1951–1961, 2010.

[20] I. Tkachev and A. Abate. On infinite-horizon probabilistic properties and stochastic bisimulation functions. In *Proceedings of the 50th IEEE Conference on Decision and Control and European Control Conference*, pages 526–531, Orlando, FL, December 2011.

[21] I. Tkachev and A. Abate. Regularization of Bellman Equations for Infinite-Horizon Probabilistic Properties. In *Proceedings of the 15th International Conference on Hybrid Systems: Computation and Control*, pages 227–236, Beijing, PRC, April 2012.

Control Design for Specifications on Stochastic Hybrid Systems

Maryam Kamgarpour, Sean Summers and John Lygeros
{mkamgar,ssummers,lygeros@control.ee.ethz.ch}

Automatic Control Laboratory, ETH Zurich
Physikstrasse 3, 8092, Zurich, Switzerland *

ABSTRACT

We synthesize controllers for discrete-time stochastic hybrid systems such that the probability of satisfying a given specification on the system is maximized. The specifications are defined with finite state automata. It is shown that automata satisfaction is equivalent to a reachability problem in an extended state space consisting of the system and the automaton state spaces. The control policy is defined as a map from this extended state space to the input space. Using existing results on maximizing reachability probability, the control policy is designed to maximize probability of satisfying the specification.

Categories and Subject Descriptors

I.2.8 Problem Solving, Control Methods, and Search

General Terms

Theorey

Keywords

Control Synthesis; Automaton; Stochastic; Hybrid; Reachability; Dynamic Programming

1. INTRODUCTION

In classical control theory specifications on a dynamical system may be formulated as an objective function. The controller would then be designed to minimize or maximize this function. Several desired system properties however may not be readily formulated as an objective function. For example, consider designing an input such that (a) the system trajectory reaches a desired final set after passing through an intermediary set, or (b) the system trajectory visits certain subsets of the state-space in any order. To formulate these and more general specifications, one can use temporal logic.

In Linear Temporal Logic (LTL) operators are defined to quantify events along a single system trajectory. Originally, temporal logic and the tools associated with model checking this logic had been developed for finite transition systems. In systems and control, LTL has also been used for specifications on continuous-state dynamical systems. In this case, one method to verify an LTL specification is through creating a finite state abstraction of the system and then applying existing automata model checking techniques [27, 17]. The class of autonomous systems for which finite state abstractions can be constructed are defined in [18]. Alternatively, an optimization framework for verification is proposed in [3] and timed automata are proposed for specifications on certain classes of autonomous dynamical systems [4].

For stochastic systems, specifications may be defined by requiring that the probability of achieving the desired system property is above a given threshold. For linear stochastic systems, LTL specifications may be verified by creating a finite-state Markov chain abstraction of the system [20]. Probabilistic notions of abstractions [29], randomized approaches [21], and average time to failure [11] have been used for safety verification of autonomous stochastic systems. A probabilistic version of temporal logic, Probabilistic Computation Tree Logic (PCTL), has been developed for property specifications on stochastic systems [13] and more recently extended as PCTL* and applied to finite state Markov chains [5]. Certain specifications such as reaching a goal set while avoiding an unsafe set, can readily be formulated as a PCTL formula or equivalently formulated as a cost function for a general stochastic process and verified via a dynamic programming algorithm [25, 22]. In [19] controller synthesis given PCTL specifications was addressed for finite-state Markov Chains. To the best of our knowledge, previous approaches applied to a continuous state spaces either consider autonomous systems or create a finite abstraction of the controlled system.

We address controller synthesis to satisfy properties specified by automata for discrete-time controlled stochastic hybrid systems (CSHS). Our work is inspired by [1] and is an extension of this work from autonomous to controlled stochastic hybrid systems. In contrast to [1] which associates the specifications with the discrete states of the hybrid stochastic process, we consider a more general framework in which specifications are associated with both the discrete and continuous states of the process and thus the dynamical system is defined independent of a given specification. We

*This work is supported by the European Commission under the MoVeS project FP7-ICT-2009-257005.

HSCC'13, April 8–11, 2013, Philadelphia, Pennsylvania, USA.

create a product of the CSHS with the automaton generating the specification to create an extended CSHS. By formulating the specification as a reachability problem for the extended CSHS, we synthesize a controller that maximizes the probability of satisfying the specifications. In doing so, we also characterize the solution of the infinite horizon reachability for a CSHS. Throughout, we provide several examples to describe how specifications can be formulated with automata and apply the methodology to specifications arising in robotics search and rescue mission.

This paper is organized as follows. In Section 2 we describe the dynamical system under study. In Section 3 we define the formal method for defining specifications and provide several examples. In Section 4 we describe our approach for controller synthesis to maximize the specification probability. In Section 5 we illustrate our approach with a robotic search and rescue mission. In Section 6 we make concluding remarks.

2. SYSTEM MODEL

We start by recalling the definition of a controlled stochastic hybrid system. For any topological space X, let $\mathcal{B}(X)$ denote the Borel $\sigma-$algebra on the space. The dynamical system under consideration is defined as follows [25]:

Definition 1 (CSHS). *A discrete-time controlled stochastic hybrid system (CSHS) is a tuple $\mathcal{H} = (Q^{\mathcal{H}}, X, U, \tau_v, \tau_q, \tau_r)$ with elements defined below:*

- Discrete state-space $Q^{\mathcal{H}} = \{1, 2, \ldots, M\}$;
- Hybrid state-space *defined as* $X = Q^{\mathcal{H}} \times \mathbb{R}^n$;
- Input space U: *a nonempty, compact Borel space;*
- Continuous state transition kernel $\tau_v : \mathcal{B}(\mathbb{R}^n) \times X \times U \to [0, 1]$: *a Borel-measurable stochastic kernel on* $\mathbb{R}^n$ *given* $X \times U$ *which assigns to each* $x = (q, v) \in X$, $u \in U$ *a probability measure* $\tau_v(\cdot|x, u)$ *on the Borel space* $(\mathbb{R}^n, \mathcal{B}(\mathbb{R}^n))$;
- Discrete state transition kernel $\tau_q : Q^{\mathcal{H}} \times X \times U \to [0, 1]$: *a discrete stochastic kernel on* $Q^{\mathcal{H}}$ *given* $X \times U$ *which assigns to each* $x \in X$ *and* $u \in U$ *a probability distribution* $\tau_q(\cdot|x, u)$ *over* $Q^{\mathcal{H}}$;
- Reset transition kernel $\tau_r : \mathcal{B}(\mathbb{R}^n) \times X \times U \times Q^{\mathcal{H}} \to [0, 1]$: *a Borel-measurable stochastic kernel on* $\mathbb{R}^n$ *given* $X \times U \times Q^{\mathcal{H}}$ *which assigns to each* $x \in X$, $u \in U$ *and* $q' \in Q^{\mathcal{H}}$ *a probability measure* $\tau_r(\cdot|x, u, q')$ *on the Borel space* $(\mathbb{R}^n, \mathcal{B}(\mathbb{R}^n))$.

A *Markov policy* in finite time horizon N is defined as a sequence $\mu = (\mu_0, \mu_1, \ldots, \mu_{N-1})$ of Borel measurable maps $\mu_k : X \to U$, $k = 0, 1, \ldots, N-1$. The set of all Markov policies is denoted $\mathcal{M}_N$.

Let $\mathcal{H}$ be a CSHS and $N \in \mathbb{N}$ be a finite horizon. A stochastic process $\mathbf{x} = \{x_k\}_{k=0}^{N}$ with $x_k \in X$ is an *execution* of $\mathcal{H}$ associated with a Markov policy $\mu \in \mathcal{M}_N$ and an initial condition $x_0 \in X$ if at time $k \geq 0$, the state $x_{k+1} = (q_{k+1}, v_{k+1}) \in Q^{\mathcal{H}} \times \mathbb{R}^n$ is obtained as follows: Set the input to $u_k = \mu_k(x_k)$, the discrete state is then updated according to the discrete transition kernel as $q_{k+1} \sim \tau_q(\cdot|x_k, u_k)$. If the discrete state remains the same, namely $q_{k+1} = q_k$, then the continuous state is updated according to the continuous state transition kernel as $v_{k+1} \sim \tau_v(\cdot|x_k, u_k)$. On the other hand, if there is a discrete jump, the continuous state is instead updated according to the reset transition kernel as $v_{k+1} \sim \tau_r(\cdot|x_k, u_k, q_{k+1})$. The execution $\mathbf{x}$ of the CSHS is a time inhomogeneous stochastic process on the sample space $\Omega = X^{N+1}$, endowed with the canonical product topology $\mathcal{B}(\Omega) := \prod_{k=0}^{N} \mathcal{B}(X)$.

Following [2] we compose the transition kernels τ_v, τ_q, τ_r and form a hybrid state transition kernel $\tau : \mathcal{B}(X) \times X \times U \to [0, 1]$ which describes the evolution of the hybrid state $x = (q, v) \in X$:

$$\tau(q', dv'|q, v, u) \qquad (1)$$
$$= \begin{cases} \tau_v(dv'|q, v, u)\tau_q(q|q, v, u) & \text{if } q' = q, \\ \tau_r(dv'|q, v, u, q')\tau_q(q'|q, v, u) & \text{if } q' \neq q. \end{cases}$$

Similarly, we define the infinite horizon execution $\mathbf{x}_\infty = \{x_k\}_{k=0}^{\infty}$, given an infinite horizon Markov policy $\mu = (\mu_0, \mu_1, \ldots)$. Let the set of all infinite horizon Markov policies be denoted by $\mathcal{M}_\infty$.

Given a Markov policy $\mu \in \mathcal{M}_N$ and initial condition $x_0 \in X$, the stochastic kernels $\tau(.|x, \mu_k(x))$, $k = 0, \ldots, N-1$ induce a unique probability measure $P_{x_0}^{\mu}$, on the product space X^{N+1} by Proposition 7.28 of [6]. By the same proposition, the stochastic kernels $\tau(.|x, \mu_k(x))$, for $k = 0, 1, \ldots$ induce a unique probability measure on the sample space $\Omega = \prod_{k=0}^{\infty} X$.

For the results in Section 4, we make the following assumptions.

Assumption 1 (Continuity of stochastic kernels).

- *For each $x = (q, v) \in X$ and $q' \in Q^{\mathcal{H}}$, the function $\tau_q(q'|x, u)$ is continuous in u;*
- *For each $x \in X$ and $E \in \mathcal{B}(\mathbb{R}^n)$, the function $\tau_v(E|x, u)$ is continuous in u;*
- *For each $x \in X$, $q' \in Q^{\mathcal{H}}$, and $E \in \mathcal{B}(\mathbb{R}^n)$, the function $\tau_r(E|x, u, q')$ is continuous in u.*

Note that we only assume continuity of the stochastic kernels in the input space, but not necessarily in the system state. Thus, the model allows for stochastic hybrid systems where transition probabilities change abruptly with changes in the system state. However, at the time of numerical computation, continuity in x may be needed in order to provide bounds on the error arising from the discretization of the continuous state space [2]. If U is finite or countable, then the above assumptions are readily satisfied under the discrete topology on U. Also, if $\tau_v(\cdot|q, v, u)$ (or τ_r) has a density function $f(v'|q, v, u), v' \in \mathbb{R}^n$ for every $q \in Q^{\mathcal{H}}$, and $f(v'|q, v, u)$ is continuous in u, it can be checked that the assumption for τ_v (respectively τ_r) is satisfied.

In the remainder of paper, after defining specifications with automata, we define a stochastic hybrid system whose state space is Cartesian product of the automaton and the dynamical system state spaces. Thus, the stochastic hybrid model serves two purposes. First, it provides a general class of dynamical systems for which specifications may be defined. Second, the product of the automaton with the stochastic hybrid system results in a stochastic hybrid system. Thus the control synthesis tools for this class of systems can be applied to the product system.

3. SPECIFICATIONS

3.1 Automata Specification

Let us first recall the definition of an automaton [8].

Definition 2 (Automaton). *A finite state automaton is a tuple* $\mathcal{A} = (Q^{\mathcal{A}}, Q_0^{\mathcal{A}}, \Sigma, Q_F^{\mathcal{A}}, \Delta)$, *where* $Q^{\mathcal{A}}$ *is a finite set of states;* $Q_0^{\mathcal{A}} \subset Q^{\mathcal{A}}$ *is a set of initial states;* Σ *is a finite alphabet;* $Q_F^{\mathcal{A}} \subset Q^{\mathcal{A}}$ *is a set of final states;* $\Delta \subset Q^{\mathcal{A}} \times \Sigma \times Q^{\mathcal{A}}$ *is a transition relation.*

Let $\mathbf{w} = (w_0, \ldots, w_N)$ with $w_k \in \Sigma$ be a sequence referred to as a *word.* We say that automaton $\mathcal{A}$ *accepts* $\mathbf{w}$ if there exists a sequence of states $(\rho_0, \rho_1, \ldots, \rho_N)$, referred to as a *run*, such that $\rho_0 \in Q_0^{\mathcal{A}}$, $(\rho_k, w_k, \rho_{k+1}) \in \Delta$ for $k = 1, \ldots, N-1$, and $\rho_N \in Q_F^{\mathcal{A}}$. For an infinite word $\mathbf{w}_\infty = (w_0, w_1, \ldots)$, a run denotes a sequence of states ρ, with $\rho_0 \in Q_0^{\mathcal{A}}$, $(\rho_k, w_k, \rho_{k+1}) \in \Delta$ for $k = 1, 2, \ldots$. The simplest accept condition for an infinite word is defined in the framework of nondeterministic Büchi automaton as follows: Let $inf(\rho)$ be the set of states that appear infinitely often in the run ρ when $\mathbf{w}_\infty$ is applied. The automaton is defined to accept $\mathbf{w}_\infty$ if there exists a run ρ such that $Q_F^{\mathcal{A}} \cap inf(\rho) \neq \emptyset$.

In this paper, we consider deterministic automata by requiring that $\forall q \in Q^{\mathcal{A}}$ and $\sigma \in \Sigma$ there exists a unique $q' \in Q^{\mathcal{A}}$ such that $(q, \sigma, q') \in \Delta$. We further assume that for any $q \in Q_F$, $(q, \sigma, q) \in \Delta$ for all $\sigma \in \Sigma$, that is, any finite state is absorbing. This requirement simplifies the accept condition of the automaton for an infinite word.

The link between an automaton and a dynamical system is established by defining the automaton alphabet to be a set of subsets of the state-space of the dynamical system. Recall that X denotes the state space of the stochastic hybrid system. For each automaton state $q \in Q^{\mathcal{A}}$, let A_i^q, $i = 1, \ldots, n_q$ denote Borel measurable subsets of the hybrid state-space X. Define a set of atomic propositions, AP, as union over the automaton states, of all the Borel subsets associated with each automaton state, that is, $AP := \cup_{q \in Q^{\mathcal{A}}} \cup_{i=1}^{n_q} \{A_i^q\}$. To specify system properties, the alphabet of the automaton $\mathcal{A}$ can be defined as sets of subsets of the hybrid space, that is, $\Sigma = 2^{AP}$. In addition, a labeling function $L : X \to 2^{AP}$ assigns to each hybrid state the set of subsets to which it belongs. Thus, given a state $x \in X$ of the dynamical system, a transition from automaton state $q \in Q^{\mathcal{A}}$ to $q' \in Q^{\mathcal{A}}$ occurs if $(q, L(x), q') \in \Delta$. We assume $\cup_{i=1}^{n_q} A_i^q = X$ to ensure that from every automaton state a transition exists.

Definition 3 (Automata specification). *An execution of the CSHS,* $\mathbf{x} = \{x_k\}_{k=0}^N$ *satisfies a specification given by automaton* $\mathcal{A}$ *if there exists a sequence of automaton states* $(\rho_0, \rho_1, \ldots, \rho_N)$ *with* $\rho_0 \in Q_0^{\mathcal{A}}$ *such that:*
(a) $(\rho_k, L(x_k), \rho_{k+1}) \in \Delta$ *for* $k = 1, \ldots, N-1$,
(b) $\rho_N \in Q_F^{\mathcal{A}}$.
Similarly, an infinite execution $\mathbf{x}_\infty = \{x_k\}_{k=0}^\infty$ *satisfies the specification if condition (a) holds and* $inf(\rho) \cap Q_F^{\mathcal{A}} \neq \emptyset$.

Pictorially, an automaton may be represented as a graph in which the nodes correspond to the states and the edges correspond to the transition relation Δ. The initial states are labeled with an incoming arrow and the final states are labeled with a double circle. An example of an automaton is shown in Fig. 1. This automaton, accepts any finite or infinite words of the form $\mathbf{w} = w_0 w_0 \ldots w_0 w_1 \ldots$, with $w_0 = X \setminus G$, $w_1 = G$ and w_0 may repeat a finite number of times until w_1 occurs.

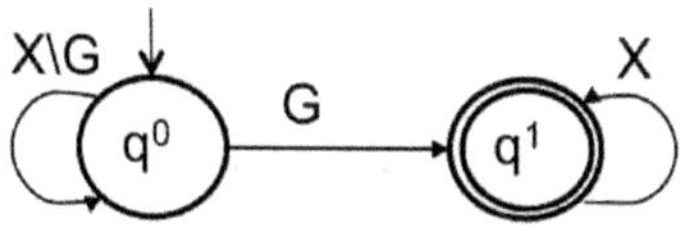

Figure 1: Automaton specifying reaching the set G.

3.2 Connection to Temporal Logic

Linear Temporal Logic (LTL) has been used to define specifications on dynamical systems. Every LTL formula can be converted to an automaton [8]. In general, the resulting automaton is non-deterministic (Büchi automaton) or may have different accept conditions (Rabin automaton) as defined here [5]. Thus, by considering deterministic automata, our framework accommodates specifications defined by a subset of LTL formula.

LTL uses atomic propositions and operators such as conjunction and negation to build specifications as follows. As before, let AP denote a finite set of atomic propositions. The syntax of an LTL formula is defined as follows [8]:

- if $p \in AP$ then p is an LTL formula.
- if ϕ and ψ are LTL formulas then, $\neg\phi$, $\phi \wedge \psi$, $\mathcal{X}\phi$ and $\phi\,\mathcal{U}\,\psi$ are LTL formula.

The semantics of an LTL formula are defined over an execution of the dynamical system under consideration, in this case, a stochastic hybrid system. As before, we assume atomic propositions are described by Borel subsets of the hybrid state-space X. Let $\mathbf{x}_\infty = \{x_k\}_{k=0}^\infty$ denote an infinite execution of the CSHS. The semantics are as follows:

$$
\begin{array}{ll}
\mathbf{x} \models a & a \in L(x_0) \\
\mathbf{x} \models \phi \wedge \psi & \phi \in L(x_0) \wedge \psi \in L(x_0) \\
\mathbf{x} \models \neg\phi & \phi \notin L(x_0) \\
\mathbf{x} \models \mathcal{X}\phi & \phi \in L(x_1) \\
\mathbf{x} \models \phi\,\mathcal{U}\,\psi & \exists N < \infty,\ \psi \in L(x_N) \text{ and} \\
 & \forall j,\ 0 < j < N,\ \phi \in L(x_j)
\end{array}
$$

Expressing complex specifications may be easier with the above syntax and semantics than with creating an equivalent automaton. Meanwhile, since LTL model checking can be done using automata model checking techniques, various algorithms exist that convert an LTL formula to an equivalent automaton.

3.3 Examples

The first problem is a simple reachability specification for which an algorithm is given in Section 4. The other two problems were motivated in the introduction and will be considered in the case studies.

3.3.1 *Reachability*

Assume that we would like to ensure that the hybrid state trajectory reaches a goal set G. We define an automaton with two states $Q^{\mathcal{A}} = \{q^0, q^1\}$, $Q_0^{\mathcal{A}} = q^0$ and $Q_F^{\mathcal{A}} = q^1$. Let $A_1^0 = X \setminus G$ and $A_2^0 = G$. The alphabet is given as $\Sigma = 2^{\{A_1^0, A_2^0\}}$. The transition relation is $\Delta = \{(q^0, A_1^0, q^0), (q^0, A_2^0, q^1), (q^1, X, q^1)\}$. Equivalently, we may use LTL to express this specification as $X\,\mathcal{U}\,G$. The automaton is shown in Fig. 1.

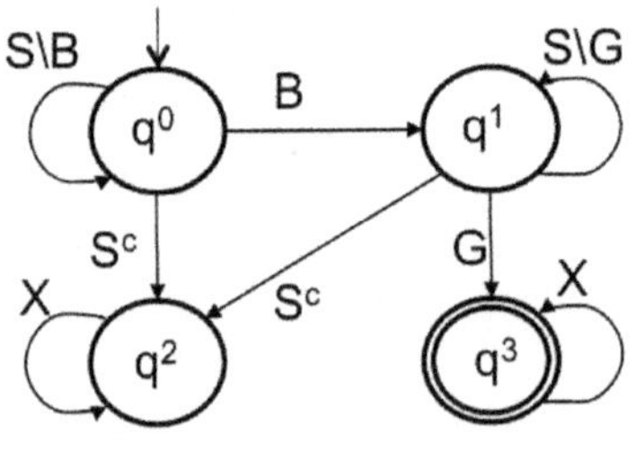

Figure 2: Automaton specifying reaching G, by first passing through B while avoiding S^c at all times.

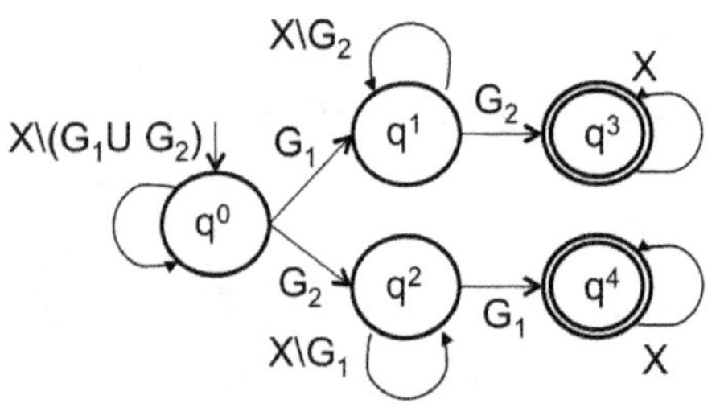

Figure 3: Automaton specifying visiting the sets G_1, G_2 in any order.

3.3.2 Reach-Avoid-Passage

Given three sets $S, B, G \in \mathcal{B}(X)$, consider the specification of reaching the goal set G after passing through B, while staying inside S at all times, as introduced in Problem (a) of Section 1. To capture this, we define $A_1^0 = S \setminus B$, $A_2^0 = B$, $A_3^0 = S^c$, $A_1^1 = S \setminus G$, $A_2^1 = G$, $A_3^1 = S^c$, and $A_1^2 = X$. Equivalently, we can express the specification with the LTL formula $(S \ \mathcal{U}\ B)\ \mathcal{U}\ (S \ \mathcal{U}\ G)$. The associated automaton is shown in Fig. 2.

3.3.3 Traveling Salesman

Consider the specification of reaching two sets, G_1 and G_2, in any order as described in Problem (b) in the introduction. Define $A_1^0 = X \setminus (G_1 \cup G_2)$, $A_2^0 = G_1$, $A_3^0 = G_2$, $A_1^1 = X \setminus G_2$, $A_2^1 = G_2$, $A_1^2 = X \setminus G_1$, $A_2^2 = G_1$. Equivalently, we can express this specification with the LTL formula,
$\neg\big((X\ \mathcal{U}\ G_1)\ \mathcal{U}\ G_2\big) \wedge \neg\big((X\ \mathcal{U}\ G_2)\ \mathcal{U}\ G_1\big)$. The automaton states and the transition relations are shown in Fig. 3.

4. CONTROLLER SYNTHESIS

First, we will discuss how to address the reachability problem for a CSHS.

4.1 Reachability problem

Suppose that the set $G \in \mathcal{B}(X)$ is given as the goal set. The finite-time reachability problem for a CSHS concerns finding a Markov policy which maximizes the probability of the system execution reaching the goal set at some time in the horizon $0, 1, \ldots, N$. This specification was depicted with the automaton of Fig. 1.

The probability that the CSHS trajectory $\mathbf{x} = \{x_k\}_{k=0}^N$ reaches G under a fixed choice of $\mu \in \mathcal{M}_N$ is [25]

$$r_N(x_0, \mu) := P_{x_0}^{\mu}\left(\bigcup_{j=0}^{N} (X \setminus G)^j \times G \times X^{N-j}\right).$$

The control objective is to maximize the reachability probability. Thus, our objective is to compute:

$$r_N^*(x_0) := \sup_{\mu \in \mathcal{M}_N} r_N(x_0, \mu). \tag{2}$$

In addition, we need to find a Markov policy $\mu^* \in \mathcal{M}_N$ which achieves the supremum. The supremum above can be computed via dynamic programming. To state this result, define $\mathcal{F}$ as the set of bounded Borel measurable functions from X to $\mathbb{R}$. For any $f_1, f_2 \in \mathcal{F}$, we denote $f_1 \leq f_2$ if $f_1(x) \leq f_2(x)$ for all $x \in X$.

Define the function $H : X \times U \times \mathcal{F} \to \mathbb{R}$ and the so-called dynamic programming operator $T : \mathcal{F} \to \mathcal{F}$:

$$H(x, u, f) = \int_X f(y)\tau(dy|x, u),$$
$$T[f](x) = \mathbf{1}_G(x) + \mathbf{1}_{X \setminus G}(x) \sup_{u \in U} H(x, u, f). \tag{3}$$

Note that Assumption 1 is sufficient to ensure that operator T preserves Borel measurability [6]. In addition, this assumption and compactness of the set U ensure that for any $J \in \mathcal{F}$, there exists a Borel measurable function such that $f(x) = \arg\sup_{u \in U} H(x, u, J)$ [23]. It is also clear that T preserves boundedness.

Define T^0 as the identity operator on $\mathcal{F}$ and $T^k = T \circ T^{k-1}$ for $k \geq 1$.

Theorem 1 (Finite horizon reachability). *Define the functions $V_k : X \to \mathbb{R}$ by the backward recursion*

$$V_{N-k} = T^k[\mathbf{1}_G].\ k = 0, 1, \ldots, N \tag{4}$$

Then, the following statements hold:

1. *$r_N^*(x_0) = V_0(x_0), \forall x_0 \in X$.*
2. *A policy $\mu^* = (\mu_1^*, \ldots, \mu_{N-1}^*)$ achieving the maximum in (2) exists and is defined $\forall x \in X$ by*

$$\mu_k^*(x) \in \arg\max_{u \in U} \int_X \tau(dy|x, u)V_{k+1}(y). \tag{5}$$

The above theorem is a special case of the result in [15] for a CSHS with two players and thus the proof is omitted. The proof is also provided under slightly different assumption on the stochastic kernels [2, 25].

We also consider infinite horizon reachability. Under a given infinite horizon Markov policy $\mu = (\mu_0, \mu_1, \ldots)$, the infinite horizon reach probability is the probability that $x_N \in G$ for some $0 \leq N < \infty$. This is expressed as

$$r_\infty(x_0, \mu) := P_{x_0}^{\mu}\left(\bigcup_{j=0}^{\infty} (X \setminus G)^j \times G \times X \times \ldots\right).$$

Taking the supremum over all infinite horizon Markov policies, $\mu \in \mathcal{M}_\infty$, of the above probability, we obtain the optimal infinite horizon reachability, denoted by

$$r_\infty^*(x_0) := \sup_{\mu \in \mathcal{M}_\infty} r_\infty(x_0, \mu).$$

Next, we show that the above supremum can be computed as the limit, as $N \to \infty$, of iteration (4) concerning the N-step reachability.

Theorem 2 (Infinite horizon reachability).

1. $r^*_\infty(x_0) = V^*(x_0)$ *where* V^* *is the limit of the finite horizon maximum reachability probability:*
$$V^*(x) = \lim_{N\to\infty} T^N[\mathbf{1}_G](x). \qquad (6)$$
2. V^* *is a fixed point of the dynamic programming operator* T*, that is,* $V^*(x) = T[V^*](x)$.
3. *If* $\mu^* = (\pi, \pi, \dots)$ *is an infinite horizon stationary optimal policy, then* $T_\pi[V^*](x) = T[V^*](x)$*, where the operator* $T_\pi : \mathcal{F} \to \mathcal{F}$ *is defined as*
$$T_\pi[f](x) = \mathbf{1}_G(x) + \mathbf{1}_{X\setminus G}(x)H(x, \pi(x), f).$$

The proof is provided to the Appendix in order to keep the discussion here focused. Note that this theorem provides necessary conditions for optimality of a stationary Markov policy. Sufficient conditions are more difficult to obtain and may be derived under a contraction assumption on the stochastic kernels [6].

Based on the above Theorem, we could derive an algorithm to solve for the fixed point of the operator T. Unfortunately, in general, the dynamic programming operator may not have a unique fixed point. For example, observe that $V^*(x) = 1,\ \forall x \in X$ is always a fixed point. However, if $X \setminus G$ has an absorbing subset, then V^* cannot be equal to one on X. To ensure uniqueness of the fixed point, additional assumptions on the stochastic kernel are needed. Here, we provide one such result.

Let $\bar{\mathcal{F}} \subset \mathcal{F}$ denote the set of non-negative functions such that $f(x) = 1,\ \forall x \in G$, equipped with the supremum norm, denoted here by $\|.\|$. Note that the operator T as defined in Equation (3) maps $\bar{\mathcal{F}}$ to itself.

Proposition 1. *If there exists* γ *such that*
$$\int_{X\setminus G} \tau(dy|x,u) \le \gamma < 1$$
for all $x \in X,\ u \in U$*, then* T *has a unique fixed point in* $\bar{\mathcal{F}}$.

Proof. The proof is an extension of results [22] for the autonomous case. First, for $f \in \bar{\mathcal{F}}$ and $u \in U$ define:
$$T_u[f](x) = \mathbf{1}_G(x) + \mathbf{1}_{X\setminus G}(x)\int_X f(y)\tau(dy|x,u) = \mathbf{1}_G(x)$$
$$+ \mathbf{1}_{X\setminus G}(x)\Big(\int_G \tau(dy|x,u) + \int_{X\setminus G} f(y)\tau(dy|x,u)\Big).$$
Let $f_1, f_2 \in \bar{\mathcal{F}}$. If there exists such γ then for every $u \in U$:
$$\begin{aligned} T_u[f_1](x) - T_u[f_2](x) &= \int_{X\setminus G} \tau(dy|x,u)(f_1(y) - f_2(y)) \\ &\le \int_{X\setminus G} \tau(dy|x,u)\|f_1 - f_2\| \\ &\le \gamma\|f_1 - f_2\|. \end{aligned}$$
So, for every $u \in U$, T_u is a contraction map:
$$\|T_u[f_1] - T_u[f_2]\| \le \gamma\|f_1 - f_2\|$$
Now,
$$\begin{aligned} &\|T_u[f_1] - T_u[f_2]\| \le \gamma\|f_1 - f_2\| \\ \Longrightarrow\ & T_u[f_1](x) \le T_u[f_2](x) + \gamma\|f_1 - f_2\| \\ \Longrightarrow\ & T[f_1](x) \le T[f_2](x) + \gamma\|f_1 - f_2\|. \end{aligned}$$
In the above, the last statement is induced by taking the supremum over $u \in U$. Similarly,
$$\begin{aligned} &\|T_u[f_1] - T_u[f_2]\| \le \gamma\|f_1 - f_2\| \\ \Longrightarrow\ & T_u[f_2](x) \le T_u[f_1](x) + \gamma\|f_1 - f_2\| \\ \Longrightarrow\ & T[f_2](x) \le T[f_1](x) + \gamma\|f_1 - f_2\|. \end{aligned}$$
Thus, for every $x \in X$:
$$|T[f_1](x) - T[f_2](x)| \le \gamma\|f_1 - f_2\|.$$
Taking the supremum over x we get that T is a contraction:
$$\|T[f_1] - T[f_2]\| \le \gamma\|f_1 - f_2\|.$$
Since $T : \bar{\mathcal{F}} \to \bar{\mathcal{F}}$ is a contraction and $\bar{\mathcal{F}}$ is a closed subset of a metric space, by the Contraction Mapping Theorem [23], T has a unique fixed point in $\bar{\mathcal{F}}$. □

Under the condition of Proposition (1), one can solve for the unique fixed point of the operator T to find the optimal infinite horizon reachability probability. However, solving for the fixed point is in general very difficult. A more common method is to obtain V^* using value iteration. By Equation (6) value iteration would converge to V^* even when the fixed point is not unique. In addition, under the assumption of existence of such γ as in Proposition (1) the convergence rate is as follows:
$$\|V^{k+1} - V^*\| = \|T[V^k] - T[V^*]\| \le \gamma\|V^k - V^*\|.$$

In the following section, we show that we can also formulate a linear programming optimization problem to obtain V^*. First, we characterize V^* in the set of fixed points of T, as an extension of the result for autonomous systems [22].

Proposition 2. *Given the set* $\{W_\alpha\}_{\alpha\in I}$ *of all fixed points of the operator* T *for some index set* I*, with* $W_\alpha \in \bar{\mathcal{F}}$*,* $V^*(x)$ *is the point wise minimum among all fixed points, that is,* $V^*(x) = \min_{\alpha\in I} W_\alpha(x)$.

Proof. By induction. Clearly, $\mathbf{1}_G = V_0 \le W_\alpha$ for every α. Suppose for $k \in \mathbb{N}$ we have $V_k \le W_\alpha$ for every α. Then, by monotonicity of the operator T (shown in Lemma 1 of Appendix) we have $T[V_k] \le T[W_\alpha]$. Thus, $V_{k+1} \le W_\alpha$, where the left-hand-side is by definition of V_{k+1} and the right-hand-side holds since W_α is a fixed point of the operator. Since the inequality holds for every k, we have $\lim_{k\to\infty} V_k \le W_\alpha$ for every α. Thus, by Theorem 2, $V^* \le \inf_{\alpha\in I} W_\alpha$. Since V^* satisfies the dynamic programming operator, it follows that the infimum above is achieved and $V^* = \min_{\alpha\in I} W_\alpha$. □

The above characterization enables us to formulate an optimization problem to find V^*.

Proposition 3. *Choose* $c : X \setminus G \to \mathbb{R}_+$ *to be a weight function. Let* V^o *be the solution of the optimization problem:*
$$\min_{V:X\setminus G\to\mathbb{R}_+} \int_{X\setminus G} c(x)V(x)dx$$
$$s.t.\ V(x) \ge \int_G \tau(dy|x,u) +$$
$$\int_{X\setminus G} V(y)\tau(dy|x,u), \quad \forall u \in U, \forall x \in X \setminus G. \qquad (7)$$

Define $V^*(x) = 1$ for $x \in G$ and $V^*(x) = V^o(x)$ for $x \in X \setminus G$. Then, V^* satisfies Equation (6) almost everywhere.

Proof. First, note that the optimization problem has at least one feasible solution, which is V^* defined in (6) restricted to $X \setminus G$. Let V^o be the optimal solution. Since V^o is feasible, we have $T_u[V^*](x) \le V^*(x)$ for every $x \in X$ and thus $T[V^*] \le V^*$. We now show, by contradiction, that the equality must hold almost everywhere.

Suppose $V^*(x) - T[V^*](x) = \epsilon(x) > 0$ for all $x \in A$ where A is a subset of $X \setminus G$ with positive measure. Define $\tilde{V} : \mathcal{F} \to \mathbb{R}$ as follows: $\tilde{V}(x) = V^o(x) - \epsilon(x)$ if $x \in A$ and $\tilde{V}(x) = V^o(x)$ otherwise. One can verify that $\tilde{V}(x)$ also satisfies Constraint (7). In addition, the objective is now reduced since $c(x) > 0$. This contradicts the fact that V^o is the optimal solution. Consequently, V^* is a fixed point of the dynamic programming operator. Now, by the same reasoning, for any fixed point of the dynamic programming operator, W_α, we must have $V^*(x) \le W_\alpha(x)$ almost everywhere, because otherwise the objective function can be decreased while feasibility is maintained. □

The optimization variable of the above problem lives in the linear infinite dimensional space $\mathcal{F}$. It is subject to a linear objective and an infinite number of linear constraints. Thus, it is a linear program in infinite dimensional space. There are several approximation methods for solving this class of problems (see, for example, [14]). Most methods consist of restricting the infinite dimensional decision space with a finite dimensional linear space through approximating the decision variable as a linear combination of a set of basis functions. The infinite constraints may be handled either using sampling approaches [9, 12, 7] or sum-of-squares [28]. Currently we are exploring the application of these methods for infinite horizon dynamic programming [16, 26].

4.2 Automata satisfaction formulated as a reachability problem

In order to maximize the probability of satisfying a specification, we first create an extended hybrid system $\bar{\mathcal{H}}$, whose discrete transitions capture the transitions of the automaton as well as the original CSHS $\mathcal{H}$.

Let $Q = Q^{\mathcal{A}} \times Q^{\mathcal{H}}$. In addition, let $\bar{X} = Q \times \mathbb{R}^n$. The extended CSHS is a tuple $\bar{\mathcal{H}} := (Q, U, \bar{X}, \bar{\tau}_q, \bar{\tau}_r, \bar{\tau}_v)$ with transition kernels defined as follows. Let $q = (q^{\mathcal{A}}, q^{\mathcal{H}}) \in Q, q' = (q'^{\mathcal{A}}, q'^{\mathcal{H}}) \in Q$ and $x \in X$. The discrete-state transition kernel $\bar{\tau}_q : Q \times \bar{X} \times U \to [0,1]$ is:

$$\bar{\tau}_q(q'^{\mathcal{A}}, q'^{\mathcal{H}} | q^{\mathcal{A}}, q^{\mathcal{H}}, v, u)$$
$$= \tau_q(q'^{\mathcal{H}} | q^{\mathcal{H}}, v, u) \mathbf{1}_{\{q'^{\mathcal{A}} \in Q^{\mathcal{A}} | (q^{\mathcal{A}}, L(x), q'^{\mathcal{A}}) \in \Delta\}}.$$

The continuous-state transition kernels are $\bar{\tau}_v : \mathcal{B}(\mathbb{R}^n) \times \bar{X} \times U \to [0,1]$ and $\bar{\tau}_r : \mathcal{B}(\mathbb{R}^n) \times \bar{X} \times U \times Q \to [0,1]$. They are defined for any $q^{\mathcal{A}}, q'^{\mathcal{A}} \in Q^{\mathcal{A}}$ as:

$$\bar{\tau}_v(dv | q^{\mathcal{A}}, q^{\mathcal{H}}, v, u) = \tau_v(dv | q^{\mathcal{H}}, v, u),$$
$$\bar{\tau}_r(dv | q^{\mathcal{A}}, q^{\mathcal{H}}, v, u, q') = \tau_r(dv | q^{\mathcal{H}}, v, u, q'^{\mathcal{H}}).$$

Note that for a non-deterministic automaton the resulting product object $\bar{\mathcal{H}}$, may not be a well-defined hybrid system since the set $\{q'^{\mathcal{A}} \in Q^{\mathcal{A}} | (q^{\mathcal{A}}, L(x), q'^{\mathcal{A}}) \in \Delta\}$ is not singleton. In this case, additional assumptions on the automaton, such as separability [1], may be imposed.

We first focus on finite horizon automata satisfaction according to Definition 3. To address this, we define the policy on the product space $\bar{X}$ as follows:

Definition 4. *An extended Markov policy is a sequence $\bar{\mu} = (\bar{\mu}_0, \bar{\mu}_1, \ldots, \bar{\mu}_{N-1})$ of Borel measurable maps $\bar{\mu}_k : \bar{X} \to U$, $k = 0, 1, \ldots, N-1$. The set of all extended Markov policies is denoted by $\bar{\mathcal{M}}_N$.*

Let $\bar{\mathbf{x}} = \{\bar{x}_k\}_{k=0}^N$ denote a stochastic process with $\bar{x}_k = (q_k^{\mathcal{A}}, x_k) \in Q^{\mathcal{A}} \times X$, $x_k = (q_k^{\mathcal{H}}, v_k)$. We define $\bar{\mathbf{x}}$ as an *execution* of $\bar{\mathcal{H}}$ associated with a Markov policy $\bar{\mu} \in \bar{\mathcal{M}}_N$ and an initial condition $\bar{x}_0 \in Q_0^{\mathcal{A}} \times X$ if for $k \ge 0$, the state $\bar{x}_{k+1} = (q_{k+1}^{\mathcal{A}}, q_{k+1}^{\mathcal{H}}, v_{k+1})$ is obtained as follows: Set the input to $u_k = \bar{\mu}_k(q^{\mathcal{A}}, x_k)$, update the discrete state of the hybrid system according to the discrete transition kernel as $q_{k+1}^{\mathcal{H}} \sim \tau_q(\cdot | x_k, u_k)$. If the discrete state remains the same, namely $q_{k+1}^{\mathcal{H}} = q_k^{\mathcal{H}}$, then the continuous state is updated according to the continuous state transition kernel as $v_{k+1} \sim \tau_v(\cdot | x_k, u_k)$, else, the continuous state is updated according to the reset transition kernel as $v_{k+1} \sim \tau_r(\cdot | x_k, u_k, q_{k+1})$. Finally, the automaton mode is updated by setting it equal to the unique state $q_{k+1}^{\mathcal{A}}$, such that $(q_k^{\mathcal{A}}, L(x_k), q_{k+1}^{\mathcal{A}}) \in \Delta$.

Next, we show that the automata satisfaction problem for $\mathcal{H}$ is equivalently formulated as a reachability problem for $\bar{\mathcal{H}}$. Define $\bar{G} = Q_F^{\mathcal{A}} \times X \subset \bar{X}$ as the goal set. Given $\mathbf{x} = \{x_k\}_{k=0}^N$ as a trajectory of the hybrid dynamical system, we construct $\bar{\mathbf{x}} := \{\bar{x}_k = (q_k^{\mathcal{A}}, x_k)\}_{k=0}^N$ by setting $q_0^{\mathcal{A}} \in Q_0^{\mathcal{A}}$, and for $k \ge 0$, setting $q_{k+1}^{\mathcal{A}}$ as the unique state such that $(q_k^{\mathcal{A}}, L(x_k), q_{k+1}^{\mathcal{A}}) \in \Delta$.

Proposition 4 (Verification in Product Space)**.** *The hybrid system trajectory $\mathbf{x}$, satisfies the specification of the automaton if and only if there exists $k \le N$ such that $\bar{x}_k \in \bar{G}$.*

Proof. $\mathbf{x}$ satisfies the specification of the automaton if and only if the sequence of automaton states $(q_0^{\mathcal{A}}, q_1^{\mathcal{A}}, \ldots, q_k^{\mathcal{A}})$ satisfies $q_0^{\mathcal{A}} \in Q_0^{\mathcal{A}}$, and for $0 \le k \le N-1$, $(q_k^{\mathcal{A}}, L(x_k), q_{k+1}^{\mathcal{A}}) \in \Delta$ and $q_N^{\mathcal{A}} \in Q_F^{\mathcal{A}}$. This run happens if and only if $\bar{\mathbf{x}}$ reaches the goal set $\bar{G}$ (by definition of $\bar{\mathbf{x}}$). □

Similar to the construction in Equation (1) we can define a hybrid transition kernel on the extended state-space $\bar{\tau} : \mathcal{B}(\bar{X}) \times \bar{X} \times U \to [0,1]$. Given a policy $\bar{\mu} \in \bar{\mathcal{M}}_N$ and initial condition $\bar{x}_0 \in \bar{X}$, we have a well-defined probability measure on the product space $\bar{X}^{N+1}$ induced by the stochastic kernels $\bar{\tau}$ as shown in Proposition 7.28 of [6]. Denote this probability measure by $\bar{P}_{\bar{x}_0}^{\bar{\mu}} : \mathcal{B}(\bar{X}^{N+1}) \to [0,1]$. Combining this with the above proposition we conclude that the probability with which the CSHS satisfies an automaton is well-defined and is given by:

$$\bar{r}_N(\bar{x}_0, \bar{\mu}) := \bar{P}_{\bar{x}_0}^{\bar{\mu}} \left(\bigcup_{j=0}^{N} (\bar{X} \setminus \bar{G})^j \times \bar{G} \times \bar{X}^{N-j} \right).$$

Maximizing the probability of automaton satisfaction is thus equivalent to computing $\bar{r}_N^*(\bar{x}_0) := \sup_{\bar{\mu} \in \bar{\mathcal{M}}_N} \bar{r}_N(\bar{x}_0, \bar{\mu})$. It follows from Theorem 1 that $\bar{r}_N^*(\bar{x}_0)$ can be computed with an iterative algorithm, where the iterations are now evaluated in the extended state-space. To apply the policy online, at each time step, one needs to know $(q_k^{\mathcal{A}}, x_k)$, that is, both the state of the automaton and the hybrid system.

Infinite horizon automaton specifications can also be addressed in the product space. In particular, in Proposition 4 the term $k \le N$ would be replaced with $k < \infty$. Reachability computation in the product space would then be addressed based on results of Theorem 2 and Proposition 3.

In general, the iterative algorithm for reachability computation may be done by gridding the continuous state space

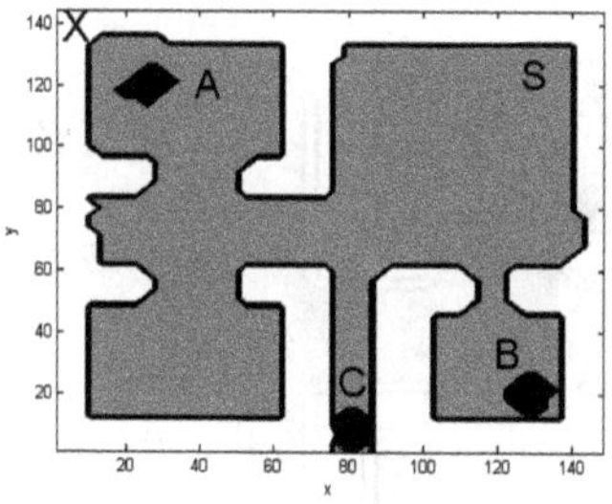

Figure 4: Building map for emergency rescue. The sets labeled A and B represent trapped individuals needing rescue. Set C represents the entrance/exit of the building. The set S represents the safe building terrain.

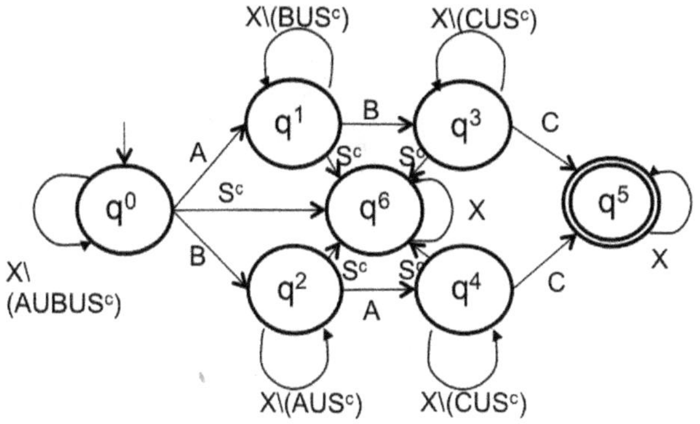

Figure 5: Automaton for the evacuation mission i).

and the input space. The computation is exponential in the dimensions of the continuous state and input spaces and linear in the size of the discrete states of the dynamical system and the automaton. Thus, the limiting factor in computation is the continuous space dimensions. Adaptive gridding [24] as well as approximation methods which do not require gridding [16, 26] have been proposed to alleviate the Curse of Dimensionality.

5. CASE STUDIES

To demonstrate the method introduced in this work, we provide an application in emergency building rescue and evacuation via autonomous rescue robotics. In this scenario, we consider that two individuals are trapped in a building that has suffered structural damage. The damage to the building has made a human rescue mission high risk due to the altered terrain and imminent building collapse expected to happen in exactly 25 minutes. Hence, a robotic solution is preferred.

The potential mission specifications include i) rescue both individuals, ii) rescue only individual 1, and iii) rescue only individual 2. A mission involving the robotic rescue of humans is acceptable if and only if there exists a control strategy such that the specification is satisfied with probability greater than 0.8. In addition, the control strategy for the robot to achieve this probability must be provided.

The building layout is provided in Fig. 4. The labeled sets A and B represent the two individuals that are trapped in the building. The labeled set C represents the building entrance/exit. The three potential mission objectives include the following steps and necessitate completion within 25 minutes and with a success probability greater than 0.8:

1. Mission i) Reach sets A and B in any order. Mission ii) Reach A. Mission iii) Reach B.
2. Return to set C (the exit).
3. Remain throughout in the safe building area S.

More formally, each of the three specifications above can be written as an automaton. The satisfaction of each specification can then be evaluated by solving a stochastic reachability problem on the product space of the corresponding finite state automaton and the stochastic dynamical system representing the robot motion, as described in Section 4.

Mission ii) and iii) can each be represented as the Reach-Avoid-Passage automaton defined in Section 3.3.2 and illustrated in Fig. 2. Mission i), on the other hand, combines the properties of the Reach-Avoid-Passage automaton and the Traveling Salesmen automaton defined in Section 3.3.2 and Section 3.3.3. The automaton for Mission i) is illustrated in Figure 5.

The rescue robot dynamics are given by stochastic difference equations representing the evolution of the robot position and the robot pose:

$$x_{k+1} = x_k + \delta(v\cos(\theta_k)) + \omega_x$$
$$y_{k+1} = y_k + \delta(v\sin(\theta_k)) + \omega_y$$
$$\theta_{k+1} = \theta_k + \delta u + \omega_\theta.$$

The states x and y represent the spatial location of the robot (meters) and the state θ represents the heading angle (radians). The input variables v and u represent the linear velocity of the robot and the angular velocity of the robot, respectively. In the case study below, the angular velocity takes values $u \in [-6\ 6]$ radians per minute and the linear velocity v is assumed constant at $v = 24$ meters per minute. The time step is $\delta = 0.0625$ minutes, hence 25 minutes necessitates a time horizon of $N = 400$. The terms $\omega_x \sim \mathcal{N}(0, 0.4^2)$, $\omega_y \sim \mathcal{N}(0, 0.4^2)$, and $\omega_\theta \sim \mathcal{N}(0, 0.125^2)$ are disturbance perturbations due to the uncertain terrain of the damaged building.

To evaluate the feasibility of each mission objective, and consequently choose a single mission for the robotic rescue vehicle, we solved the dynamic program of Section 4 for the resulting product stochastic hybrid systems, where the automaton associated with each mission is combined with the rescue robot dynamics. The rescue robot itself is not hybrid but the product system is a stochastic hybrid system. The result was an optimal decision strategy for each mission. An example strategy, which maps the initial state of the robot to an angular velocity, is illustrated in Fig. 6. In addition, the set of all initial robot states that allowed for the logic specification satisfaction with probability greater than 0.8 is computed. The resulting sets are illustrated in Figures 7 and 8. The solution was obtained by gridding the continuous state on a $50 \times 50 \times 20$ grid. The computation time was 8000 seconds, thus 20 seconds per dynamic programming iteration, on a dual-core processor of 2.66 GHz.

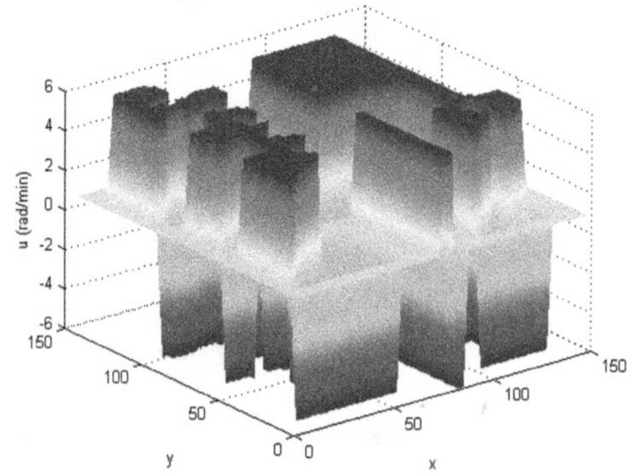

Figure 6: Example optimal strategy synthesized for the robot rescue mission using the dynamic programming approach introduced for mission i), for a fixed automata state and robot pose θ.

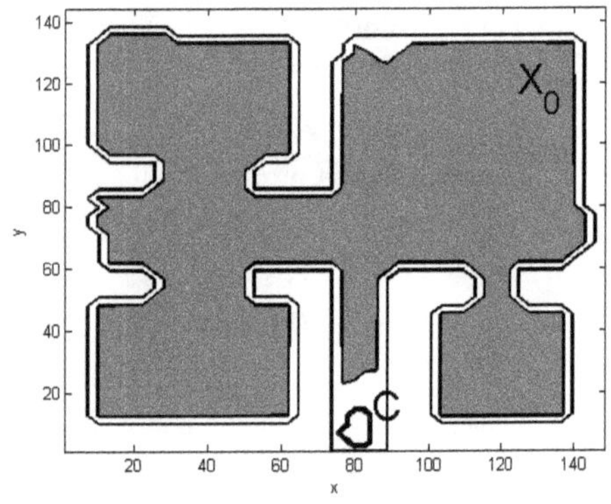

Figure 7: The colored set X_0 represents the set of all initial robot positions for which there exists a θ_0 and a controller such that the rescue of both individuals can be accomplished within 25 minutes with probability greater than 0.8. Note that the set of robot entrance points does not intersect X_0. Hence, a mission involving the rescue of both individuals will not succeed with sufficiently high probability.

6. CONCLUSIONS

We showed that the problem of controller synthesis to satisfy specifications defined by finite state deterministic automaton on stochastic hybrid systems can be achieved by solving a reachability problem. The reachability problem was solved in the product space of the automaton and the hybrid system using a dynamic programming algorithm. The synthesized controller achieves the upper bound on the probability of satisfying the specification. Future work includes extensions to specifications defined by more general non-deterministic automata, approximation methods for solving the infinite dimensional linear program arising due to infinite horizon reachability and numerical implementations.

7. ACKNOWLEDGEMENTS

We would like to thank A. Abate, M. Prandini, F. Ramponi for discussions.

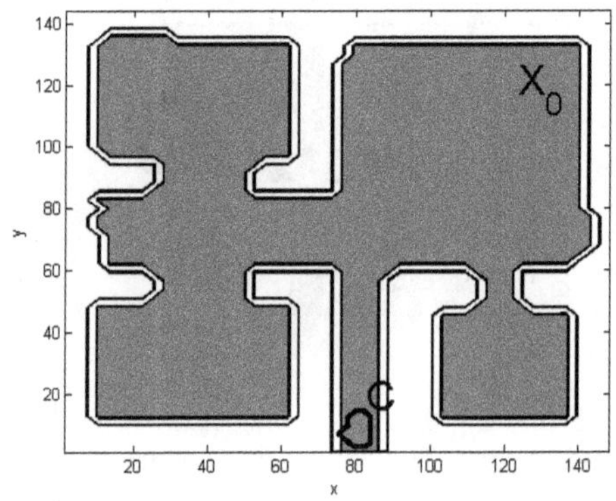

Figure 8: The colored set X_0 represents the set of all initial robot positions for which there exists a θ_0 and a controller such that the rescue of one individual can be completed with probability greater than 0.8 and within 25 minutes. The set of robot entrance points is within X_0 for both individual rescue scenarios. Hence, a rescue mission for a single individual (either 1 or 2) is feasible given the optimal control strategy is implemented.

8. REFERENCES

[1] A. Abate, J.-P. Katoen, and A. Mereacre. Quantitative automata model checking of autonomous stochastic hybrid systems. In E. Frazzoli and R. Grosu, editors, *Hybrid systems: computation and control*, Lecture Notes in Computer Sciences, pages 83–92. ACM, 2011.

[2] A. Abate, M. Prandini, J. Lygeros, and S. Sastry. Probabilistic reachability and safety for controlled discrete time stochastic hybrid systems. *Automatica*, 44(11):2724–2734, November 2008.

[3] H. Abbas and G. Fainekos. Linear hybrid system falsification through local search. In T. Bultan and P.-A. Hsiung, editors, *Automated Technology for Verification and Analysis*, volume 6996 of *Lecture Notes in Computer Science*, pages 503–510. Springer Berlin / Heidelberg, 2011.

[4] R. Alur and D. L. Dill. A theory of timed automata. *Theoretical Computer Science*, 126(2):183 – 235, 1994.

[5] C. Baier, J. Katoen, et al. *Principles of model checking*. MIT press, 2008.

[6] D. P. Bertsekas and S. E. Shreve. *Stochastic Optimal Control: The Discrete-Time Case*. Athena Scientific, 2007.

[7] G. Calafiore and M. Campi. Uncertain convex programs: randomized solutions and confidence levels. *Mathematical Programming*, 102(1):25–46, 2005.

[8] E. M. Clarke, O. Grumberg, and D. A. Peled. *Model Checking*. The MIT Press, 2001.

[9] D. de Farias and B. Van Roy. The linear programming approach to approximate dynamic programming. *Operations Research*, 51(6):850–865, 2003.

[10] G. B. Folland. *Real Analysis: Modern Techniques and Their Applications*. Wiley, second edition, 1999.

[11] M. Fränzle, T. Teige, and A. Eggers. Satisfaction meets expectations. In *Integrated Formal Methods*, pages 168–182. Springer, 2010.

[12] C. Guestrin, D. Koller, R. Parr, and S. Venkataraman. Efficient solution algorithms for factored MDPs. *J. Artif. Intell. Res. (JAIR)*, 19:399–468, 2003.
[13] H. Hansson and B. Jonsson. A logic for reasoning about time and reliability. *Formal Aspects of Computing*, 6:512–535, 1994.
[14] O. Hernández-Lerma and J. Lasserre. Approximation schemes for infinite linear programs. *SIAM Journal on Optimization*, 8(4):973–988, 1998.
[15] M. Kamgarpour, J. Ding, S. Summers, , A. Abate, J. Lygeros, and C. Tomlin. AStochastic Games Framework for Verification and Control of Discrete Time Stochastic Hybrid Systems. *Automatica*, 2012. to appear.
[16] N. Kariotoglou, S. Summers, T. H. Summers, M. Kamgarpour, and J. Lygeros. Approximate dynamic programming for stochastic reachability. In *European Control Conference*, 2012. submitted.
[17] M. Kloetzer and C. Belta. A fully automated framework for control of linear systems from temporal logic specifications. *Automatic Control, IEEE Transactions on*, 53(1):287–297, feb 2008.
[18] G. Lafferriere, G. Pappas, and S. Sastry. O-minimal hybrid systems. *Mathematics of Control, Signals, and Systems (MCSS)*, 13(1):1–21, 2000.
[19] M. Lahijanian, S. Andersson, and C. Belta. Control of Markov decision processes from PCTL specifications. In *American Control Conference (ACC), 2011*, pages 311–316. IEEE, 2011.
[20] M. Lahijanian, S. B. Andersson, and C. Belta. A Probabilistic Approach for Control of a Stochastic System from LTL Specifications. In *IEEE Conference on Decision and Control*, 2009.
[21] M. Prandini and A. Abate. Approximate abstractions of stochastic systems: a randomized method. In *Proceedings of the IEEE Conference on Decision and Control and European Control Conference*, 2011.
[22] F. Ramponi, D. Chatterjee, S. Summers, and J. Lygeros. On the connections between PCTL and dynamic programming. In K. H. Johansson and W. Yi, editors, *Hybrid Systems: Computation and Control*, volume 3927 of *Lecture Notes in Computer Science*, pages 253–262. ACM, 2010.
[23] W. Rudin. *Principles of Mathematical Analysis*. McGraw-Hill New York, 1964.
[24] S. E. Z. Soudjani and A. Abate. Adaptive gridding for abstraction and verification of stochastic hybrid systems. In *Proceedings of the Eighth International Conference on Quantitative Evaluation of SysTems*, pages 59–68, 2011.
[25] S. Summers and J. Lygeros. Verification of discrete time stochastic hybrid systems: A stochastic reach-avoid decision problem. *Automatica*, 46(12):1951 – 1961, 2010.
[26] T. H. Summers, N. Kariotoglou, M. Kamgarpour, S. Summers, and J. Lygeros. Approximate Dynamic Programming via Sum of Squares Programming. In *European Control Conference*, 2012. submitted.
[27] P. Tabuada and G. J. Pappas. Model Checking LTL over Controllable Linear Systems Is Decidable. In O. Maler and A. Pnueli, editors, *Hybrid Systems: Computation and Control*, volume 2623 of *Lecture Notes in Computer Science*, pages 498–513. Springer, 2003.
[28] Y. Wang and S. Boyd. Approximate dynamic programming via iterated Bellman inequalities. Manuscript preprint, 2010.
[29] L. Zhang, Z. She, S. Ratschan, H. Hermanns, and E. Hahn. Safety verification for probabilistic hybrid systems. In *Computer Aided Verification*, pages 196–211. Springer, 2010.

9. APPENDIX

Here, we prove Theorem 2 on infinite horizon reachability by referring to established results in [6]. First, we have the following Lemma:

Lemma 1. *The operator T satisfies a monotonicity property: If $f, f' : X \to \mathbb{R}$ are Borel-measurable functions such that $f \leq f'$, then $T[f] \leq T[f']$.*

Proof. From non-negativity of the stochastic kernel τ it follows that for every $u \in U$, $x \in X$, $\int_X f(y)\tau(dy|x,u) \leq \int_X f'(y)\tau(dy|x,u)$. Thus

$$\sup_{u \in U} \int_X f(y)\tau(dy|x,u) \leq \sup_{u \in U} \int_X f'(y)\tau(dy|x,u)$$

and the result follows. □

Under a given policy $\mu = (\mu_0, \mu_1, \dots)$ the probability of reaching the goal set G in infinite time horizon is

$$\begin{aligned} r_\infty(x_0, \mu) &:= P^{\mu}_{x_0}\left(\bigcup_{j=0}^{\infty}(X \setminus G)^j \times G \times X \times \cdots \times X\right) \\ &= \sum_{j=0}^{\infty} P^{\mu}_{x_0}\left((X \setminus G)^j \times G \times X \times \cdots \times X\right) \\ &= \lim_{N\to\infty} \sum_{j=0}^{N} P^{\mu_{0\to N}}_{x_0}\left((X \setminus G)^j \times G \times X^{N-j}\right) \\ &= \lim_{N\to\infty} r_N(x_0, \mu_{0\to N}). \end{aligned}$$

where $\mu_{0\to N} = (\mu_0, \dots, \mu_{N-1})$ and the first equality holds due to mutual exclusiveness of events in the union. Denote $V^* : X \to \mathbb{R}$ to be the optimal infinite horizon reachability, that is: $V^*(x_0) = \sup_{\mu \in \mathcal{M}_\infty} r_\infty(x_0, \mu)$. To facilitate application of results in [6], we transform the maximization into a minimization by defining the function $J^* : X \to \mathbb{R}$:

$$J^*(x_0) = \inf_{\mu \in \mathcal{M}_\infty} -r_\infty(x_0, \mu).$$

In addition, $J_N : X \to \mathbb{R}$ is recursively defined, with $J_0 = -\mathbf{1}_G$ and

$$J_N(x_0) = \inf_{\mu_{0\to N} \in \mathcal{M}_N} -r_N(x_0, \mu_{0\to N}),\ N \in \mathbb{N}$$

Note that $J_N = -V_N$, where V_N was the optimal value function defined in Equation (4) of Theorem 1.

Lemma 2. *For each $x_0 \in X$, the sequence $\{J_N(x_0)\}_{N=0}^{\infty}$ converges.*

Proof. We equivalently show that $\forall x_0 \in X$, $V_N(x_0)$ defined in Equation (4) converges. For each $N \geq 0$, $V_N(x_0)$ is the optimal finite horizon reach probability. Thus, for every $x_0 \in X$ and $N \geq 0$, $V_N(x_0) \in [0,1]$. In addition, by

Theorem 1, $V_N(x_0) = T^N[\mathbf{1}_G](x_0)$, $\forall x_0 \in X$. From the definition of T it is clear that $\mathbf{1}_G \leq T[\mathbf{1}_G]$. Given this and monotonicity of T as shown in Lemma 1, we conclude $T^N[\mathbf{1}_G] \leq T^{N+1}[\mathbf{1}_G]$ for every $N \geq 0$. Thus, $\forall x_0 \in X$, the sequence $\{V_N(x_0)\}_{N=1}^{\infty}$ is bounded and monotonically increasing, and hence converges (see for example Theorem 3.14 of [23]). □

Since T preserves Borel measurability [15] it follows that V_N and thus J_N are also Borel measurable for each N. Thus, $\lim_{N\to\infty} J_N$ is the limit of a sequence of Borel-measurable functions, and is also Borel-measurable (Proposition 2.7 of [10]).

Next, we define the function $\tilde{H} : X \times U \times \mathcal{F} \to \mathbb{R}$ and the operator $\tilde{T} : \mathcal{F} \to \mathcal{F}$ as:

$$\tilde{H}(x,u,J) = -\mathbf{1}_G(x) + \mathbf{1}_{X\setminus G}(x)\int_X J(y)\tau(dy|x,u),$$
$$\tilde{T}[J](x) = -\mathbf{1}_G(x) + \mathbf{1}_{X\setminus G}(x)\inf_{u\in U}\tilde{H}(x,u,J).$$

It follows from Theorem 1 that $J_N = \tilde{T}^N[-\mathbf{1}_G]$.

Proposition 5. *Define $J_0(x) = -\mathbf{1}_G(x)$. The map $\tilde{H}$ satisfies the following properties:*
(a) $\tilde{H}(x,u,J_0) \leq J_0$, $\forall u \in U$
(b) if $\{J_k\}_{k=0}^{\infty} \in \mathcal{F}$ is a sequence satisfying $J_{k+1} \leq J_k \leq J_0$ then $\forall u \in U$

$$\lim_{k\to\infty} \tilde{H}(x,u,J_k) = \tilde{H}(x,u,\lim_{k\to\infty} J_k)$$

Proof. (a) Since τ is a stochastic Kernel and $-1 \leq J_0(x) \leq 0$ for all x, we have $-1 \leq \int_X J_0(y)\tau(dy|x,u) \leq 0$ and by multiplying the inequality by $\mathbf{1}_G$ we get $-\mathbf{1}_{X\setminus G}(x) \leq \mathbf{1}_{X\setminus G}(x)\int_X J_0(y)\tau(dy|x,u) \leq 0$. Then, by adding $-\mathbf{1}_G(x)$ to both sides of the right-hand-side inequality, we get $-1 \leq -\mathbf{1}_G(x) + \mathbf{1}_{X\setminus G}(x)\int_X J_0(y)\tau(dy|x,u) \leq -\mathbf{1}_G(x)$ and the result on $\tilde{H}$ follows.
(b) Since $J_{k+1} \leq J_k \leq 0$ and for each k, J_k is Borel measurable, by the Monotone Convergence Theorem, we have that $\lim_{k\to\infty} J_k(x)$ exists and that

$$\int_X \lim_{k\to\infty} J_k(y)\tau(dy|x,u) = \lim_{k\to\infty}\int_X J_k(y)\tau(dy|x,u)$$

The result on $\tilde{H}$ follows by its definition. □

It follows from part (a) and (b) of the above proposition that assumptions D, $D.1$ of Proposition 5.3 [6] respectively hold. We conclude (by Proposition 5.3, [6]) that J^* is a fixed point of the operator $\tilde{T}$, that is, $J^* = \tilde{T}[J^*]$, and can be found from the dynamic programming recursion, that is $J^* = \lim_{N\to\infty} J_N$. Given that $V^* = -J^*$ and $V_N = -J_N = T^N[\mathbf{1}_G]$ we conclude parts (1) and (2) of Theorem 2:

$$V^* = \lim_{N\to\infty} T^N[\mathbf{1}_G],$$
$$V^* = T[V^*].$$

Part (3) follows from Proposition 5.5 in [6].

Rewarding Probabilistic Hybrid Automata

Ernst Moritz Hahn
University of Oxford, United Kingdom*

Holger Hermanns
Saarland University, Germany

ABSTRACT

The joint consideration of randomness and continuous time is important for the formal verification of many real systems. Considering both facets is especially important for wireless sensor networks, distributed control applications, and many other systems of growing importance. Apart from proving the quantitative safety of such systems, it is important to analyse properties related to resource consumption (energy, memory, bandwidth, etc.) and properties that lie more on the economical side (monetary gain, the expected time or cost until termination, etc.). This paper provides a framework to decide such *reward* properties effectively for a generic class of models which have a discrete-continuous behaviour and involve both probabilistic as well as nondeterministic decisions. Experimental evidence is provided demonstrating the applicability of our approach.

Categories and Subject Descriptors: I.6.4 [Computing Methodologies]: Simulation and Modelling - Model Validation and Analysis; G.3 [Mathematics of Computing]: Probability and Statistics.

General Terms: Performance; Reliability; Verification.

Keywords: probabilistic hybrid automaton; abstraction; model checking; expected rewards; probabilistic automaton; performance evaluation; performability; continuous time; nondeterminism; simulation relation.

1. INTRODUCTION

The inclusion of stochastic phenomena in the hybrid systems framework is crucial for a spectrum of application domains, ranging from wireless communication and control to air traffic management and to electric power grid operation [13, 23]. As a consequence, many different stochastic hybrid system models have been proposed [2, 35, 11, 12, 1, 28], together with a vast body of mathematical tools and techniques.

Recently, model checkers for stochastic hybrid systems have emerged [36, 40, 17]. In this context, the model of probabilistic hybrid automata [35] is of particular interest, since it pairs expressiveness and modelling convenience in a way that model checking is indeed possible. In particular, it enables to piggyback [40, 17] the solution of quantitative probabilistic model checking problems on qualitative model checking approaches for hybrid systems. Solvers such as HSOLVER [31], PHAVER [19], or SPACEEX [20] can be employed for the latter. Thus far, the prime focus in this context has been put on approximating or bounding reach probabilities in probabilistic hybrid automata. This is appropriate for quantifying system safety and reliability, but not for availability, survivability, throughput and resource consumption questions.

In this paper, we aim to overcome this restriction in a framework that is as general as possible, while retaining the idea of piggybacking on existing hybrid system solvers. Taking up initial ideas [22], we decorate probabilistic hybrid automata with *rewards*, which can be considered as costs or bonuses. We discuss a method for handling properties that quantify expected rewards. The properties we consider are either the *minimal or maximal expected total accumulated reward over all executions of a model* or the *minimal or maximal expected time-average reward over all executions of the model.* We will need to postpone the precise formalisation of these notions (cf. Definition 15), until we have defined the semantics of our models. Using appropriate reward structures and property types, this approach allows us to reason about the cost accumulated until termination, the long-run cost of system operation, system availability [16] and survivability [14], time or cost until stabilisation, and many other properties of interest. Proofs backing up the results presented in this paper can be found in [21].

Related Work. Reward properties for classical (i.e. nonstochastic) timed automata have been considered by Bouyer et al. [10, 9]. Rutkowski et al. [32] considered a controller synthesis problem for average-reward properties in classical hybrid automata. Discrete-time stochastic hybrid automata have been considered for the analysis of reward properties [37, 36, 18], and have been studied with importance sampling techniques recently [41]. Methods which approximate continuous-time stochastic hybrid automata by Markov chains [29, 24] also allow for an extension to reward-based properties. To the best of our knowledge, the present paper is the first to address reward-based properties of probabilistic hybrid automata involving nondeterminism, stochastic behaviour as well as continuous time in full generality, harvesting well-understood and effective methods originally developed for the verification of classical hybrid automata.

*Part of this work was done while Ernst Moritz Hahn was with Saarland University.

HSCC'13, April 8–11, 2013, Philadelphia, Pennsylvania, USA.

2. PROBABILISTIC HYBRID AUTOMATA

This section introduces the notion of probabilistic hybrid automata we are going to use, and describes how rewards are integrated into the model. To get started, we first define a generic multi-dimensional *post operator*, which will be used to describe the continuous behaviour of our model. In this operator, we reserve the first two dimensions for the accumulation of reward, respectively the advance of time. In the context of hybrid systems [3, 4], post operators are often described by differential (in)equations. However, our notion is independent of the formalism used.

DEFINITION 1. *A k-dimensional* post operator *with $k \in \mathbb{N}$ and $k \geq 2$ is a function*

$$Post: \mathbb{R}^k \to 2^{\mathbb{R}^k}.$$

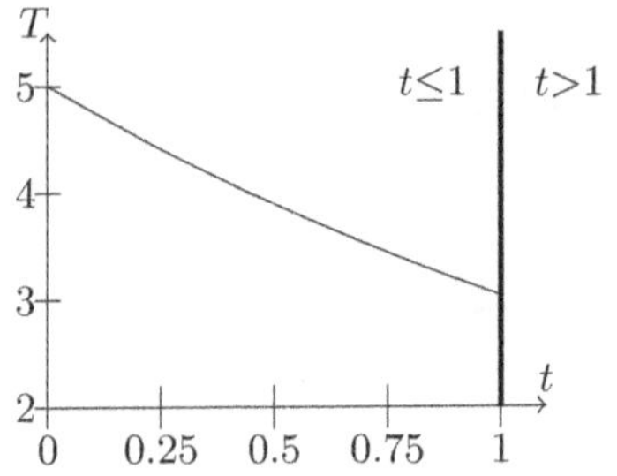

Figure 1: Post operator.

$Post(r, t, v)$ will be used to describe the possible values of the continuous variables after a timed transition. This implies an update of the reward and the time dimension. If there is a constant $c \in \mathbb{R}_{\geq 0}$ satisfying that for any $r, t \in \mathbb{R}$, $v \in \mathbb{R}^{k-2}$,

$$Post(r, t, v) \subseteq \{(r + c\mathrm{t}, t + \mathrm{t}) \mid \mathrm{t} \in \mathbb{R}_{\geq 0}\} \times \mathbb{R}^{k-2},$$

we call the post operator *reward affine*. Models which use only reward affine post operators will turn out to allow for abstractions which are particularly precise.

EXAMPLE 1. *Consider* $Post_{\mathrm{Check}} \colon \mathbb{R}^3 \to 2^{\mathbb{R}^3}$ *with*

$$\begin{aligned} &Post_{\mathrm{Check}}(r, t, T) \\ &\stackrel{\mathrm{def}}{=} \{(r, t + \mathrm{t}, T \exp(-0.5\mathrm{t})) \mid \mathrm{t} \in \mathbb{R}_{\geq 0} \wedge \mathrm{t} + t \leq 1\}. \end{aligned}$$

For $(r, t, T) = (0, 0, 5)$, *the behaviour is depicted in Figure 1. The graph denotes the set of points which can be reached by a timed transition. The axis labelled with* t *denotes both the values of the time passed as well as the continuous variable* t *(and here also the value of variable* r*). The axis* T *displays the third dimension. After time* 0.25, *it has a value of* ≈ 4.41.

Post operators will appear in the definition of the probabilistic hybrid automaton model we consider. As a preparation, we first define classical hybrid automata.

DEFINITION 2. *A* classical hybrid automaton (HA) *is a tuple*

$$\mathcal{H} = (M, k, \overline{m}, \langle Post_m \rangle_{m \in M}, Cmds, Rew), \text{ where}$$

- M *is a finite set of* modes,
- $k \in \mathbb{N}$ *with* $k \geq 2$ *is the* dimension,
- $\overline{m} \in M$ *is the* initial mode,
- $Post_m$ *is a* k*-dimensional post operator for each* m,
- *Cmds is a finite set of* guarded commands *of the form*

$$g \to u, \text{ where}$$

  - $g \subseteq M \times \mathbb{R}^k$ *is a* guard,
  - $u \colon (M \times \mathbb{R}^k) \to 2^{M \times \mathbb{R}^k}$ *is an* update function *with*
  - $u(s) \subseteq M \times \{(0, 0)\} \times \mathbb{R}^{k-2}$,
  - *if* $s \in g$ *then* $u(s) \neq \emptyset$,
- *for each* $s = (m, v)$ *with* $Post_m(v) = \emptyset$, *there is a command with guard* g *with* $s \in g$, *and*
- $Rew \colon ((M \times \mathbb{R}^k) \times Cmds) \to \mathbb{R}_{\geq 0}$ *is a* reward structure.

The continuous-time behaviour of an HA in a given mode m is determined by the corresponding post operator. Whenever the guard of a guarded command is satisfied, the command can be executed in zero time. If executed, a nondeterministic choice over successor updates and modes results. Multiple guards of commands may be satisfied at the same time, implying a nondeterministic selection over these commands.

Another obvious concept needed for the setting considered is that of a probability distribution.

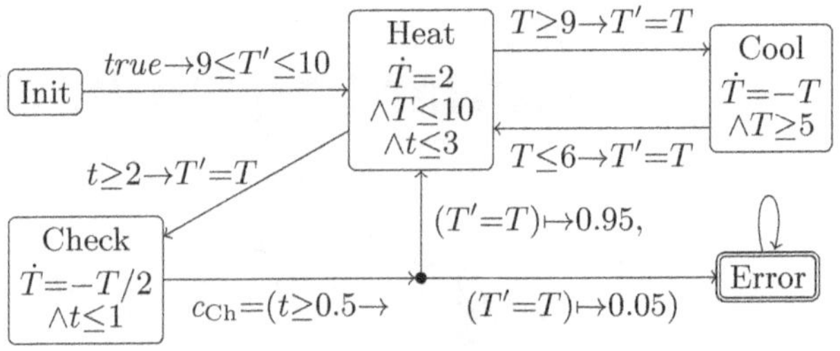

Figure 2: PHA modelling a thermostat.

DEFINITION 3. *A* finite probability distribution *over a set* Ω *is a function* $\mu \colon \Omega \to [0, 1]$, *where there are only finitely many* $a \in \Omega$ *with* $\mu(a) > 0$, *and it is* $\sum_{a \in \Omega} \mu(a) = 1$. *In a* Dirac *probability distribution* μ, *there is only a single* $a \in \Omega$ *with* $\mu(a) = 1$. *With* $Distr(\Omega)$, *we denote the set of all finite probability distributions over* Ω. *Given* n *pairwise different elements* $a_i \in \Omega$ *and probabilities* $p_i \geq 0$, $1 \leq i \leq n$ *with* $\sum_{i=1}^{n} p_i = 1$, *we use* $[a_1 {\mapsto} p_1, \ldots, a_n {\mapsto} p_n]$ *to denote the probability distribution* μ *with* $\mu(a_i) = p_i$.

With this extension, we can now specify probabilistic hybrid automata (similar to Sproston [35, Section 2]).

DEFINITION 4. *A* probabilistic hybrid automaton (PHA) *is a tuple*

$$\mathcal{H} = (M, k, \overline{m}, \langle Post_m \rangle_{m \in M}, Cmds, Rew)$$

where all components of this tuple are as in Definition 2, and satisfy the same constraints, except for

- *Cmds, which is a finite set of* probabilistic guarded commands *of the form*

$$g \to [u_1 {\mapsto} p_1, \ldots, u_n {\mapsto} p_n], \text{ where}$$

  - $g \subseteq M \times \mathbb{R}^k$ *is a* guard,
  - $u_i \colon (M \times \mathbb{R}^k) \to 2^{M \times \mathbb{R}^k}$ *is an* update function,
  - $u_i(s) \subseteq M \times \{(0, 0)\} \times \mathbb{R}^{k-2}$,
  - *if* $s \in g$ *then* $u_i(s) \neq \emptyset$ *for* $1 \leq i \leq n$.

$\mathcal{H}$ *is* reward affine *if all its post operators are reward affine and if* $Rew(\cdot, c)$ *is constant for all* $c \in Cmds$. *Under these assumptions we can consider Rew as a function*

$$Rew \colon Cmds \to \mathbb{R}_{\geq 0}.$$

Probabilities are incorporated in this definition as part of the commands that update mode and variables according to the probabilities p_i associated with the ith update option u_i.

A HA can now be viewed as a PHA where each guarded command has only a single update option, to be chosen by a Dirac distribution, or a (possibly uncountable) nondeterministic choice over Dirac distributions.

EXAMPLE 2. *Figure 2 depicts a PHA model of a simple unreliable thermostat, where* $\dot{r} = 0$ *and* $\dot{t} = 1$ *in each mode. The model can switch between modes* Heat *and* Cool *to adjust the temperature of its environment. At certain occasions the system may enter a* Check *mode. The post operator of* Check *has been described in Example 1, the other ones are similar. On execution of the command* c_{Ch}, *the system moves to mode* Heat *with probability* 0.95, *and to mode* Error *with probability* 0.05. *We thus have*

$$c_{\mathrm{Ch}} = (g \to [u_{\mathrm{ChH}} {\mapsto} 0.95, u_{\mathrm{ChE}} {\mapsto} 0.05]), \text{ where}$$

- $g = \{\mathrm{Check}\} \times \mathbb{R} \times [0.5, \infty) \times \mathbb{R}$,
- $u_{\mathrm{ChH}}(m, r, t, T) = \{(\mathrm{Heat}, 0, 0, T)\}$, *and*
- $u_{\mathrm{ChE}}(m, r, t, T) = \{(\mathrm{Error}, 0, 0, T)\}$.

The formalisation of the other commands is similar, but does not include nontrivial probabilities.

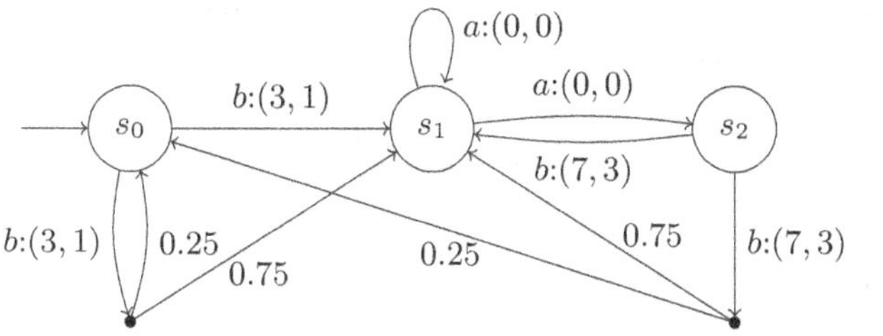

Figure 3: Probabilistic automaton.

In the post operators, we have already integrated means to refer to reward and time. Indeed, the first two dimensions of a PHA are used to record reward accumulation, and time advances as the system lifetime time progresses. We will "collect" the reward and times whenever a command is executed, and therefore reset these dimensions whenever executing commands. The component *Rew* associates rewards to discrete transitions of a PHA.

As time and reward are present as explicit dimensions in our construction, guards and invariants can relate to them. This is in contrast to e.g. priced timed automata [10], where this is forbidden, so as to avoid crossing the undecidability boundary. In the setting considered here, this makes no difference because we build on machinery (for HAs) that is developed for undecidable theories, in the form of heuristics.

EXAMPLE 3. *Consider the thermostat of Figure 2. Here it is not possible to leave the* Error *mode once entered. We define a reward structure* Rew_{acc} *with* $Rew_{\mathrm{acc}}(c) \stackrel{\mathrm{def}}{=} 0$ *if* $c \in \{c_{\mathrm{E}}, c_{\mathrm{IH}}\}$ *and* $Rew_{\mathrm{acc}}(c) \stackrel{\mathrm{def}}{=} 1$ *else. The system thus earns a reward for executing any command, except for the one of* Error, *and the one to initialise the system. With this reward structure, the minimal sum of reward values accumulated, expresses the minimal expected number of commands executed until an error happens.*

Now, assume we have extended mode Error *so that the system can recover after a certain time (e.g. by adding a reset transition back to the initial state). In such a system, it makes sense to consider the long-run behaviours. We can for instance look at a reward structure* Rew_{lra} *assigning constant 0 to each command. If in addition we modify the post operator in such a way that r is increased by* 1 *per time unit in mode* Error, *we can use this to reason about the percentage of time the system is not operational on the long run.*

3. PHA SEMANTICS

In this section, we describe the semantics of PHAs. The semantics maps on variations of infinite-state Markov decision processes [30], known as probabilistic automata [33].

DEFINITION 5. *A* probabilistic automaton (PA) *is a tuple*

$$\mathcal{M} = (S, \overline{s}, Act, \mathcal{T}), \text{ where}$$

- S *is a set of* states,
- $\overline{s} \in S$ *is the* initial state,
- *Act is a set of actions, and the*
- transition matrix $\mathcal{T}\colon (S \times Act) \to 2^{Distr(S)}$ *assigns sets of probability distributions to state-action pairs.*

For each $s \in S$, *we require* $\{a \in Act \mid \mathcal{T}(s,a) \neq \emptyset\} \neq \emptyset$.
PAs contain a (possibly uncountable) set of states, whereof one is initial. In each $s \in S$, there is a nondeterministic choice of actions $a \in Act$ and distributions $\mu \in \mathcal{T}(s,a)$ over successor states.

EXAMPLE 4. *In Figure 3, we depict a finite example PA* $\mathcal{M} \stackrel{\mathrm{def}}{=} (S, \overline{s}, Act, \mathcal{T})$. *Here, we have* $S \stackrel{\mathrm{def}}{=} \{s_0, s_1, s_2\}$, $\overline{s} \stackrel{\mathrm{def}}{=} s_0$, $Act = \{a, b\}$, *and*

$$\mathcal{T}(s_0,a) \stackrel{\mathrm{def}}{=} \emptyset, \mathcal{T}(s_0,b) \stackrel{\mathrm{def}}{=} \{[s_0 \mapsto 0.25, s_1 \mapsto 0.75], [s_1 \mapsto 1]\},$$
$$\mathcal{T}(s_1,a) \stackrel{\mathrm{def}}{=} \{[s_1 \mapsto 1], [s_2 \mapsto 1]\}, \mathcal{T}(s_1,b) \stackrel{\mathrm{def}}{=} \emptyset,$$
$$\mathcal{T}(s_2,a) \stackrel{\mathrm{def}}{=} \emptyset, \mathcal{T}(s_2,b) \stackrel{\mathrm{def}}{=} \{[s_1 \mapsto 1], [s_0 \mapsto 0.25, s_1 \mapsto 0.75]\}.$$

DEFINITION 6. *A* finite path *of a PA* $\mathcal{M} = (S, \overline{s}, Act, \mathcal{T})$ *is a tuple*

$$\beta_{\mathrm{fin}} = s_0 a_0 \mu_0 \ldots s_{n-1} a_{n-1} \mu_{n-1} s_n \in (S \times Act \times Distr(S))^* \times S,$$

where $s_0 = \overline{s}$ *and for all* i *with* $0 \leq i < n$ *it is* $\mu_i \in \mathcal{T}(s_i, a_i)$. *An* infinite path *is a tuple*

$$\beta_{\mathrm{inf}} = s_0 a_0 \mu_0 \ldots \in (S \times Act \times Distr(S))^\omega,$$

where $s_0 = \overline{s}$ *and* $\mu_i \in \mathcal{T}(s_i, a_i)$ *holds for all* $i \geq 0$. *By* $Path^{\mathrm{fin}}_{\mathcal{M}}$, *we denote the set of all finite paths and by* $Path^{\mathrm{inf}}_{\mathcal{M}}$ *we denote the set of all infinite paths of* $\mathcal{M}$.

We let $\beta_{\mathrm{fin}}[i] \stackrel{\mathrm{def}}{=} \beta_{\mathrm{inf}}[i] \stackrel{\mathrm{def}}{=} s_i$ *denote the* $(i+1)$*th state of a finite or infinite path (for the i-s defined). By* $\mathrm{last}(\beta_{\mathrm{fin}}) \stackrel{\mathrm{def}}{=} s_n$ *we denote the last state of a finite path. For* $\beta, \beta' \in Path^{\mathrm{fin}}_{\mathcal{M}} \uplus Path^{\mathrm{inf}}_{\mathcal{M}}$ *we write* $\beta \leq \beta'$ *in case either* $\beta = \beta'$ *or if* β *is a finite prefix of* β'.

By $\mathrm{trace}(\beta_{\mathrm{fin}}) = a_0 a_1 \ldots a_{n-1}$ *we denote the* trace *of a finite path, and accordingly for infinite paths. The sets of all finite and infinite traces are defined as* $Trace^*_{\mathcal{M}} \stackrel{\mathrm{def}}{=} Act^*$ *and* $Trace^\omega_{\mathcal{M}} \stackrel{\mathrm{def}}{=} Act^\omega$. *Given* $\gamma = a_0 a_1 \ldots \in Trace^*_{\mathcal{M}} \uplus Trace^\omega_{\mathcal{M}}$, *we define* $\gamma[i] \stackrel{\mathrm{def}}{=} a_i$ *as the* $(i+1)$*th action on the trace.*

Consider a subset $Act_{\mathrm{fair}} \subseteq Act$ *of the actions of* $\mathcal{M}$. *We consider a path* $\beta \in Path^{\mathrm{inf}}_{\mathcal{M}}$ *as* Act_{fair}-fair *if there are infinitely many* $i \geq 0$ *with* $\mathrm{trace}(\beta)[i] \in Act_{\mathrm{fair}}$. *By* $Path^{Act_{\mathrm{fair}}}_{\mathcal{M}}$ *we denote the set of all* Act_{fair}*-fair paths of* $\mathcal{M}$.

EXAMPLE 5. *A finite path in the PA of Figure 3 is*

$$\beta_{\mathrm{fin}} \stackrel{\mathrm{def}}{=} s_0 b [s_0 \mapsto 0.25, s_1 \mapsto 0.75] s_0,$$

and with $Act_{\mathrm{fair}} \stackrel{\mathrm{def}}{=} \{b\}$, *an* Act_{fair}*-fair infinite path is*

$$\beta_{\mathrm{inf}} \stackrel{\mathrm{def}}{=} s_0 (b [s_0 \mapsto 0.25, s_1 \mapsto 0.75] s_0)^\omega.$$

We have

$$\mathrm{trace}(\beta_{\mathrm{fin}}) = b, \mathrm{trace}(\beta_{\mathrm{inf}}) = b^\omega,$$
$$\mathrm{last}(\beta_{\mathrm{fin}}) = s_0, \beta_{\mathrm{fin}}[0] = s_0, \beta_{\mathrm{inf}}[15] = s_0.$$

Path sets are by themselves not sufficient to describe the properties of PAs. This is because nondeterministic behaviour is intertwined with probabilistic behaviour. This asks for instances to resolve the nondeterminism. These instances, called *schedulers*, induce a purely probabilistic behaviour, which can be subjected to stochastic analyses.

DEFINITION 7. *A* scheduler *for PA* $\mathcal{M} = (S, \overline{s}, Act, \mathcal{T})$ *is a function* $\rho\colon Path^{\mathrm{fin}}_{\mathcal{M}} \to Distr(Act \times Distr(S))$. *For* $\beta \in Path^{\mathrm{fin}}_{\mathcal{M}}$, *we require* $\rho(\beta)(a,\mu) > 0$ *implies* $\mu \in \mathcal{T}(\mathrm{last}(\beta), a)$. *With* $Sched_{\mathcal{M}}$ *we denote the set of schedulers of* $\mathcal{M}$.

A scheduler ρ *is called* simple *if it only maps to Dirac distributions and if for all* $\beta, \beta' \in Path^{\mathrm{fin}}_{\mathcal{M}}$ *with* $\mathrm{last}(\beta) = \mathrm{last}(\beta')$ *we have* $\rho(\beta) = \rho(\beta')$. *We can interpret it as being of the form* $\rho\colon S \to (Act \times Distr(S))$.
In this paper, we develop results valid for general schedulers, thus not restricted to simple schedulers. But since simple schedulers are simpler to describe and understand, they appear in illustrating examples. We are now in the position to define probability measures on paths.

DEFINITION 8. *We define* $Pr_{\mathcal{M},\rho}\colon Path^{\mathrm{fin}}_{\mathcal{M}} \to [0,1]$ *for PA* $\mathcal{M} = (S, \overline{s}, Act, \mathcal{T})$ *and scheduler* $\rho\colon Path^{\mathrm{fin}}_{\mathcal{M}} \to Distr(Act \times$

$Distr(S))$: Given $\beta = s_0 a_0 \mu_0 s_1 a_1 \mu_1 \ldots s_n \in Path_{\mathcal{M}}^{\text{fin}}$, let

$$Pr_{\mathcal{M},\rho}(\beta) \stackrel{\text{def}}{=}$$

$$\rho(s_0)(a_0,\mu_0)\mu_0(s_1)\rho(s_0 a_0 \mu_0 s_1)(a_1,\mu_1)\mu_1(s_2)\cdots\mu_{n-1}(s_n).$$

We define the cylinder of $\beta \in Path_{\mathcal{M}}^{\text{fin}}$

$$Cyl(\beta) \stackrel{\text{def}}{=} \{\beta' \in Path_{\mathcal{M}}^{\text{inf}} \mid \beta \leq \beta'\}$$

as the set of infinite paths which start with β. Then, we use the generated σ-algebra

$$\Sigma_{\mathcal{M}} \stackrel{\text{def}}{=} \sigma(\{Cyl(\beta) \mid \beta \in Path_{\mathcal{M}}^{\text{fin}}\}),$$

to obtain the measurable space $(Path_{\mathcal{M}}^{\text{inf}}, \Sigma_{\mathcal{M}})$.

There is a unique extension [27] of $Pr_{\mathcal{M},\rho}\colon Path_{\mathcal{M}}^{\text{fin}} \to [0,1]$ to $Pr_{\mathcal{M},\rho}\colon \Sigma_{\mathcal{M}} \to [0,1]$ where for all $\beta \in Path_{\mathcal{M}}^{\text{fin}}$ it is

$$Pr_{\mathcal{M},\rho}(Cyl(\beta)) \stackrel{\text{def}}{=} Pr_{\mathcal{M},\rho}(\beta).$$

Using the definition of a fair path and the probability of paths, we can define fair schedulers.

DEFINITION 9. *A scheduler $\rho \in Sched_{\mathcal{M}}$ of a PA $\mathcal{M} = (S, \bar{s}, Act, \mathcal{T})$ is called Act_{fair}-fair for $Act_{\text{fair}} \subseteq Act$ if*

$$Pr_{\mathcal{M},\rho}(Path_{\mathcal{M}}^{Act_{\text{fair}}}) = 1.$$

$Sched_{\mathcal{M}}^{Act_{\text{fair}}}$ denotes the set of Act_{fair}-fair schedulers of $\mathcal{M}$.

We define two stochastic processes associated to a PA.

DEFINITION 10. *Let $\mathcal{M} = (S, \bar{s}, Act, \mathcal{T})$ be a PA. We define the state process and the action process of $\mathcal{M}$ as*

$$X^{\mathcal{M}}\colon (Path_{\mathcal{M}}^{\text{inf}} \times \mathbb{N}) \to S \text{ with } X^{\mathcal{M}}(\beta,n) \stackrel{\text{def}}{=} \beta[n],$$

$$Y^{\mathcal{M}}\colon (Path_{\mathcal{M}}^{\text{inf}} \times \mathbb{N}) \to Act \text{ with } Y^{\mathcal{M}}(\beta,n) \stackrel{\text{def}}{=} \text{trace}(\beta)[n]$$

for $\beta \in Path_{\mathcal{M}}^{\text{inf}}$ and $n \in \mathbb{N}$.

We equip our PAs with reward structures. We remark that rew_{num} and rew_{den} of the same reward structure rew in the following definition are *not* meant to denote lower or upper bounds on the specification of rewards.

DEFINITION 11. *Given PA $\mathcal{M} = (S, \bar{s}, Act, \mathcal{T})$, a reward structure is a pair $(rew_{\text{num}}, rew_{\text{den}})$ of two functions*

$$rew_{\text{num}}, rew_{\text{den}}\colon (S \times Act) \to \mathbb{R}_{\geq 0}.$$

Given a reward structure $(rew_{\text{num}}, rew_{\text{den}})$, we say that it is affine, if for all $a \in Act$ there are $mul_a \in \mathbb{R}_{\geq 0}$ and $add_a \in \mathbb{R}_{\geq 0}$, where for all $s \in S$ with $\mathcal{T}(s,a) \neq \emptyset$ it is

$$rew_{\text{num}}(s,a) = mul_a rew_{\text{den}}(s,a) + add_a.$$

We will use reward structures $rew = (rew_{\text{num}}, rew_{\text{den}})$ to specify two different reward-based properties of PAs. For the definition of one of them, we will use both the functions rew_{num} and rew_{den}, for the other one we only need rew_{num}.

EXAMPLE 6. *In Figure 3 we depict a PA along with a reward structure $rew \stackrel{\text{def}}{=} (rew_{\text{num}}, rew_{\text{den}})$. We have*

$$rew_{\text{num}}(s_0,b) \stackrel{\text{def}}{=} 3, rew_{\text{den}}(s_0,b) \stackrel{\text{def}}{=} 1,$$

$$rew_{\text{num}}(s_1,a) \stackrel{\text{def}}{=} 0, rew_{\text{den}}(s_1,a) \stackrel{\text{def}}{=} 0,$$

$$rew_{\text{num}}(s_2,b) \stackrel{\text{def}}{=} 7, rew_{\text{num}}(s_2,b) \stackrel{\text{def}}{=} 3.$$

The reward structure is affine, and for action b we have the factors $mul_b = 2$ and $add_b = 1$.

We define properties based on these reward structures.

DEFINITION 12. *Given a PA $\mathcal{M} = (S, \bar{s}, Act, \mathcal{T})$ together with a reward structure $rew = (rew_{\text{num}}, rew_{\text{den}})$ and $\rho \in Sched_{\mathcal{M}}$, the accumulated reward is the expectation*

$$\text{val}_{\mathcal{M},rew,\text{acc}}^{\rho} \stackrel{\text{def}}{=} E_{\mathcal{M},\rho}\left[\lim_{n \to \infty} \sum_{i=0}^{n} rew_{\text{num}}(X_i^{\mathcal{M}}, Y_i^{\mathcal{M}})\right]$$

under the probability measure $Pr_{\mathcal{M},\rho}$. The fractional long-run average reward is defined as

$$\text{val}_{\mathcal{M},rew,\text{lra}}^{\rho} \stackrel{\text{def}}{=} E_{\mathcal{M},\rho}\left[\lim_{n \to \infty} \frac{\sum_{i=0}^{n} rew_{\text{num}}(X_i^{\mathcal{M}}, Y_i^{\mathcal{M}})}{\sum_{i=0}^{n} rew_{\text{den}}(X_i^{\mathcal{M}}, Y_i^{\mathcal{M}})}\right],$$

in case we have $Pr_{\mathcal{M},\rho}(A) = 1$, where A is the set of paths on which the property is well-defined. In the above, we let $\frac{0}{0} \stackrel{\text{def}}{=} 0$ and $\frac{x}{0} \stackrel{\text{def}}{=} \infty$ for $x \neq 0$. For $Act_{\text{fair}} \subseteq Act$, we define the Act_{fair}-fair maximal value

$$\text{val}_{\mathcal{M},rew,\text{lra}}^{+,Act_{\text{fair}}} = \sup_{\rho \in Sched_{\mathcal{M}}^{Act_{\text{fair}}}} \text{val}_{\mathcal{M},rew,\text{lra}}^{\rho},$$

and accordingly for accumulated rewards and minimal values. For $\text{val}_{\mathcal{M},rew,\text{lra}}^{+,Act_{\text{fair}}}$ ($\text{val}_{\mathcal{M},rew,\text{lra}}^{-,Act_{\text{fair}}}$) we only take the supremum (infimum) over the schedulers ρ for which $Pr_{\mathcal{M},\rho}(A) = 1$.

There are more complicated notions of fractional long-run averages which are well-defined on all paths [38]. They agree with the definition above if it exists, which we use for clarity. We will later on use reward-extended PAs as the semantics of PHAs. When considering accumulated reward properties, we add up all rewards we come across along a certain path. The value we consider then is the expected value over all paths. Properties of this kind can for instance be used to reason about the expected time until system termination or the number of steps until an error is reached. Fractional long-run average values specify a value that is reached in the long-run operation of a system. The numerator will later on describe the value of which we want to obtain the average. The denominator will describe the time which has passed in a PHA. It is necessary to use a variable denominator here rather than to assume that each step takes one unit of time, because in the semantics of such models, not all steps take the same duration of time to be performed.

Later on, the timed transitions will correspond to the non-fair actions. If a scheduler would be allowed to choose an infinite number of timed actions, it could let the time flow stop at a moment in which a very high or very low value has been reached. In this case, this value would then form the long-run average value, which we consider as being unrealistic. Fairness can prevent this problem.

EXAMPLE 7. *Reconsider the PA of Figure 3. In this model, the accumulated reward would be infinite, so that we only consider the fractional long-run average reward. We derive for $Act_{\text{fair}} \stackrel{\text{def}}{=} \{b\}$ that*

$$\text{val}_{\mathcal{M},rew,\text{lra}}^{+,Act_{\text{fair}}} = \frac{12}{5} = 2.4, \text{val}_{\mathcal{M},rew,\text{lra}}^{-,Act_{\text{fair}}} = \frac{7}{3} \approx 2.333,$$

which is for instance obtained by the simple schedulers ρ^+ and ρ^- with

$$\rho^+(s_0) \stackrel{\text{def}}{=} (b, [s_0 \mapsto 0.25, s_1 \mapsto 0.75]), \rho^+(s_1) \stackrel{\text{def}}{=} (a, [s_2 \mapsto 1]),$$

$$\rho^+(s_2) \stackrel{\text{def}}{=} (b, [s_0 \mapsto 0.25, s_1 \mapsto 0.75]),$$

$$\rho^-(s_0) \stackrel{\text{def}}{=} (b, [s_1 \mapsto 1]), \rho^-(s_1) \stackrel{\text{def}}{=} (a, [s_2 \mapsto 1]),$$

$$\rho^-(s_2) \stackrel{\text{def}}{=} (a, [s_1 \mapsto 1]).$$

We now define the semantics of PHAs as PAs formally.

DEFINITION 13. *For PHA $\mathcal{H} = (M, k, \overline{m}, \langle Post_m \rangle_{m \in M}, Cmds, Rew)$, the semantics is a PA*

$$[\![\mathcal{H}]\!] \stackrel{\text{def}}{=} (S, \bar{s}, Act, \mathcal{T}), \text{ where}$$

- $S \stackrel{\text{def}}{=} M \times \mathbb{R}^k$,
- $\bar{s} \stackrel{\text{def}}{=} (\overline{m}, 0, \ldots, 0)$,
- $Act \stackrel{\text{def}}{=} Cmds \uplus \{\tau\}$,
- *for $s = (m,v) \in S$ we have that*

– for $c = (g \to [u_1 \mapsto p_1, \dots, u_n \mapsto p_n]) \in Cmds$ it is $\mathcal{T}(s, c) \stackrel{\text{def}}{=} \emptyset$ if $s \notin g$ and else:

$$\mathcal{T}(s, c) \stackrel{\text{def}}{=} \{\mu \in Distr(S) \mid \exists s'_1 \in u_1(s), \dots, s'_n \in u_n(s). \\ \forall s' \in S.\ \mu(s') = \sum_{s'_i = s'} p_i\},$$

– it is $\mathcal{T}(s, \tau) \stackrel{\text{def}}{=} \{[(m, v') \mapsto 1] \mid v' \in Post_m(v)\}$.

The semantics is similar to usual notions of HAs, which are usually given in terms of *labelled transition systems* [26]. The difference is in the probabilistic guarded commands, where we can have a probabilistic choice over successor states in addition to nondeterminism.

EXAMPLE 8. *Consider the PHA $\mathcal{H}$ of Figure 2. Then*

$$[\![\mathcal{H}]\!] = (S, \overline{s}, Act, \mathcal{T}),\ \textit{with}$$

- $S = M \times \mathbb{R}^3$,
- $\overline{s} = (\text{Init}, 0, 0, 0)$,
- $Act = \{c_{\text{IH}}, c_{\text{HCo}}, c_{\text{HCh}}, c_{\text{CoH}}, c_{\text{E}}, c_{\text{Ch}}, \tau\}$,
- $\mathcal{T} \colon (S \times Act) \to 2^{Distr(S)}$.

For $s = (\text{Check}, r, t, T) \in \{\text{Check}\} \times \mathbb{R}^3$, *we have*

$$\mathcal{T}(s, c_{\text{IH}}) = \mathcal{T}(s, c_{\text{HCo}}) = \mathcal{T}(s, c_{\text{HCh}}) \\ = \mathcal{T}(s, c_{\text{CoH}}) = \mathcal{T}(s, c_{\text{E}}) = \emptyset,$$

it is $\mathcal{T}(s, c_{\text{Ch}}) = \emptyset$ *if* $t < 0.5$ *and we have* $\mathcal{T}(s, c_{\text{Ch}}) = \{[(\text{Heat}, 0, 0, T) \mapsto 0.95, (\text{Error}, 0, 0, T) \mapsto 0.05]\}$ *else. Further,*

$$\mathcal{T}(s, \tau) = \{[(\text{Check}, r, t + \mathsf{t}, T\exp(-0.5\mathsf{t})) \mapsto 1] \\ \mid \mathsf{t} \in \mathbb{R}_{\geq 0} \wedge \mathsf{t} + t \leq 1\}.$$

With these preparations, we can define the semantics of reward structures of PHAs.

DEFINITION 14. *Given PHA* $\mathcal{H} = (M, k, \overline{m}, \langle Post_m \rangle_{m \in M}, Cmds, Rew)$ *with semantics* $[\![\mathcal{H}]\!] = (S, \overline{s}, Act, \mathcal{T})$, *the* reward semantics *is the reward structure* $\text{rew}(\mathcal{H}) \stackrel{\text{def}}{=} (rew_{\text{num}}, rew_{\text{den}})$ *associated to* $[\![\mathcal{H}]\!]$. *For* $s = (m, r, t, v) \in S$ *and* $c \in Cmds$, *let*

$$rew_{\text{num}}(s, c) \stackrel{\text{def}}{=} Rew(s, c) + r, rew_{\text{den}}(s, c) \stackrel{\text{def}}{=} t,$$

and $rew_{\text{num}}(s, \tau) \stackrel{\text{def}}{=} rew_{\text{den}}(s, \tau) \stackrel{\text{def}}{=} 0$.

In this definition, whenever a command is executed, the reward of this command is obtained. Additionally, the timed rewards and the time accumulated until the execution of this command become effective here. As in our model only paths of infinitely many commands are relevant, this is equivalent to a reward semantics in which rewards and passage of time are attached directly to timed transitions. By postponing the collection of timed rewards to the execution of subsequent commands, we will be able to simplify the computation of abstractions of PHAs.

We are now in the position to define the values of reward properties, the technical core of our approach, using the semantics of PHAs and their reward structures.

DEFINITION 15. *Given PHA* $\mathcal{H} = (M, k, \overline{m}, \langle Post_m \rangle_{m \in M}, Cmds, Rew)$, *we define the* maximal and minimal time-average reward *as*

$$\text{val}^{+}_{\mathcal{H},\text{lra}} \stackrel{\text{def}}{=} \text{val}^{+,Cmds}_{[\![\mathcal{H}]\!],\text{rew}(\mathcal{H}),\text{lra}}\ \textit{and}\ \text{val}^{-}_{\mathcal{H},\text{lra}} \stackrel{\text{def}}{=} \text{val}^{-,Cmds}_{[\![\mathcal{H}]\!],\text{rew}(\mathcal{H}),\text{lra}}$$

and define the accumulated rewards *accordingly as*

$$\text{val}^{+}_{\mathcal{H},\text{acc}} \stackrel{\text{def}}{=} \text{val}^{+,Cmds}_{[\![\mathcal{H}]\!],\text{rew}(\mathcal{H}),\text{acc}}\ \textit{and}\ \text{val}^{-}_{\mathcal{H},\text{acc}} \stackrel{\text{def}}{=} \text{val}^{-,Cmds}_{[\![\mathcal{H}]\!],\text{rew}(\mathcal{H}),\text{acc}}.$$

We only optimise over fair schedulers, because otherwise, we could assign a relevant probability mass to paths which are *time convergent* [8, Chapter 9], that is their trace will end in a sequence $\tau\tau\tau\dots$ corresponding to time durations $\mathsf{t}_0\mathsf{t}_1\mathsf{t}_2, \dots$ with $\sum_{i=0}^{\infty} \mathsf{t}_i < \infty$. This way, time effectively stops, which means that only the reward up to this point of time will be taken into account, which is unrealistic. Now assume that infinitely many commands are executed, and that the automaton is *structurally (strongly) nonzeno* [6][5, Definition 6], that is the guards are defined so that they cannot be executed without a minimal fixed delay. Then this ensures the time divergence of the path.

One point is worth noting. One might be interested in models in which it is a legal behaviour to eventually reside in a mode m of the model without further executing any commands. However, the definition above requires that infinitely many commands are executed on each legal path. Because of this, such models have to be adapted accordingly. This can be done, e.g. by adding a new auxiliary command c_m which can be executed infinitely often after a given delay whenever residing in m.

3.1 Expressing Properties

Table 1 provides an overview of common properties that are expressible using the mechanisms described. Here, **F** denotes the set of failed states of the PHA, and **T** describes states in which operation has terminated. The *availability* [16] of a system can then be expressed as a time-average reward value, by specifying the reward of the PHA under consideration so that in each mode in which the system is available the reward increases with rate 1 per time unit and zero else. *Survivability* [14] is the ability of a system to recover a certain quality of service level in a timely manner after a disaster. Here, we consider the maximal expected time needed to recover from an error condition. This can be expressed using expected total rewards, by maximising over all states of the system (or, an abstraction of the system) in which it is not available.

4. ABSTRACTION

In this section, we develop the necessary tools for the abstraction for PHAs and their reward structures. For the theoretical justification of our abstraction method in its entirety we use simulation relations. The relations are actually never constructed during the verification process, just like the full semantics of PHAs, which is our reference semantics, but its construction is prohibitive.

To prove the validity of abstractions of PHAs for reward-based properties, we extend the definition of simulation relations [34, 33] to take into account reward structures. A simulation relation requires that every successor distribution of a state of a simulated PA $\mathcal{M}$ is related to a successor distribution of its corresponding state of a simulating PA $\mathcal{M}_{\text{sim}}$ using a *weight function* [25, Definition 4.3].

DEFINITION 16. *Let* $\mu \in Distr(S)$ *and* $\mu_{\text{sim}} \in Distr(S_{\text{sim}})$ *be two distributions. For a relation* $R \subseteq S \times S_{\text{sim}}$, *a* weight function *for* (μ, μ_{sim}) *with respect to* R *is a function* $w \colon (S \times S_{\text{sim}}) \to [0, 1]$ *with*

1. $w(s, s_{\text{sim}}) > 0$ *implies* $(s, s_{\text{sim}}) \in R$,
2. $\mu(s) = \sum_{s_{\text{sim}} \in S_{\text{sim}}} w(s, s_{\text{sim}})$ *for* $s \in S$, *and*
3. $\mu_{\text{sim}}(s_{\text{sim}}) = \sum_{s \in S} w(s, s_{\text{sim}})$ *for* $s_{\text{sim}} \in S_{\text{sim}}$.

We write $\mu \sqsubseteq_R \mu_{\text{sim}}$ *if and only if there exists a weight function for* (μ, μ_{sim}) *with respect to* R.

Using weight functions, we define simulations.

DEFINITION 17. *Given the two PAs* $\mathcal{M} = (S, \overline{s}, Act, \mathcal{T})$ *and* $\mathcal{M}_{\text{sim}} = (S_{\text{sim}}, \overline{s}_{\text{sim}}, Act, \mathcal{T}_{\text{sim}})$, *we say that* $\mathcal{M}_{\text{sim}}$ simulates $\mathcal{M}$, *denoted by* $\mathcal{M} \preceq \mathcal{M}_{\text{sim}}$, *if and only if there exists a relation* $R \subseteq S \times S_{\text{sim}}$, *which we will call* simulation relation *from now on, where*

	property	Rew	$\dot{r}$	type	remark	
(a)	time until failure	0	$\begin{cases} 0 & s \in \mathbf{F} \\ 1 & \text{else} \end{cases}$	minimum accumulated reward	$\mathbf{F}$ absorbing	[7]
(b)	cost until termination	$\begin{cases} 0 & s \in \mathbf{F} \\ \text{any} & \text{else} \end{cases}$	$\begin{cases} 0 & s \in \mathbf{T} \\ \text{any} & \text{else} \end{cases}$	maximum accumulated reward	$\mathbf{T}$ absorbing	[7]
(c)	long-run cost of operation	any	any	maximum long-run average reward	-	[7]
(d)	system availability	0	$\begin{cases} 0 & s \in \mathbf{F} \\ 1 & \text{else} \end{cases}$	minimum long-run average reward	-	[16]
(e)	survivability	0	$\begin{cases} 1 & s \in \mathbf{F} \\ 0 & \text{else} \end{cases}$	maximum accumulated reward	maximum over $s \in \mathbf{F}$	[14]

Table 1: Overview of expressible reward properties.

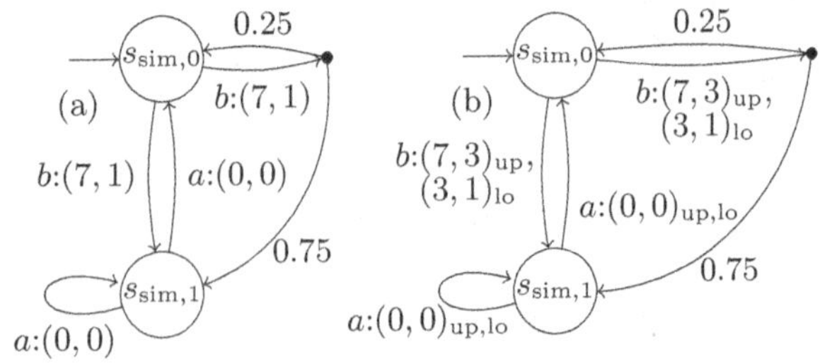

Figure 4: (a) PA simulating the one of Figure 3. (b) PA affinely simulating the one of Figure 3.

1. *it is* $(\overline{s}, \overline{s}_{\text{sim}}) \in R$,
2. *for each* $(s, s_{\text{sim}}) \in R$, $a \in Act$, *and* $\mu \in \mathcal{T}(s, a)$, *there is a distribution* $\mu_{\text{sim}} \in Distr(S_{\text{sim}})$ *with* $\mu_{\text{sim}} \in \mathcal{T}_{\text{sim}}(s_{\text{sim}}, a)$ *and* $\mu \sqsubseteq_R \mu_{\text{sim}}$.

For our purposes, we must consider rewards of PAs.

DEFINITION 18. *Consider two PAs* $\mathcal{M} = (S, \overline{s}, Act, \mathcal{T})$ *and* $\mathcal{M}_{\text{sim}} = (S_{\text{sim}}, \overline{s}_{\text{sim}}, Act, \mathcal{T}_{\text{sim}})$ *with reward structures* $rew = (rew_{\text{num}}, rew_{\text{den}})$, $rew_{\text{sim}} = (rew_{\text{sim,num}}, rew_{\text{sim,den}})$, *and a simulation relation R between the PAs. We say that R is* upper-bound compatible, *if in case we have* $(s, s_{\text{sim}}) \in R$ *then for all* $a \in Act$ *it is*

$$rew_{\text{num}}(s, a) \leq rew_{\text{sim,num}}(s_{\text{sim}}, a),$$
$$rew_{\text{den}}(s, a) \geq rew_{\text{sim,den}}(s_{\text{sim}}, a).$$

If there exists such a relation R, we write

$$(\mathcal{M}, rew) \overset{\text{up}}{\preceq} (\mathcal{M}_{\text{sim}}, rew_{\text{sim}}).$$

We define lower-bound compatible *simulations R accordingly by swapping* $\leq$ *and* $\geq$ *above and write*

$$(\mathcal{M}, rew) \overset{\text{lo}}{\preceq} (\mathcal{M}_{\text{sim}}, rew_{\text{sim}}).$$

With simulations, we can establish upper and lower bounds on the reward properties of simulated models by considering the corresponding property in the simulating model.

LEMMA 1. *For PAs* $\mathcal{M}$ *and* $\mathcal{M}_{\text{sim}}$ *with reward structures* rew *and* rew_{sim}, *if* $(\mathcal{M}, rew) \overset{\text{up}}{\preceq} (\mathcal{M}_{\text{sim}}, rew_{\text{sim}})$ *then*

$$\text{val}^{+,Act_{\text{fair}}}_{\mathcal{M},rew,\text{lra}} \leq \text{val}^{+,Act_{\text{fair}}}_{\mathcal{M}_{\text{sim}},rew_{\text{sim}},\text{lra}},$$

accordingly for the accumulated rewards. In case $(\mathcal{M}, rew) \overset{\text{lo}}{\preceq} (\mathcal{M}_{\text{sim}}, rew_{\text{sim}})$, *we obtain lower bounds for minimal values.*

We can thus bound the maximal (minimal) reward values from above (below). The principle idea of this simulation is, that the simulating automaton can mimic the behaviour of the simulated one, while overapproximating or underapproximating respectively rew_{num} and rew_{den}.

EXAMPLE 9. *In Figure 4 (a) we give a PA with corresponding reward structures which simulates the one of Figure 3 by an upper-bound compatible simulation relation. The maximal fractional long-run average reward of the latter is indeed much higher than the former, namely, 7 rather than* $\frac{12}{5} = 2.4$. *To obtain a lower-bound compatible simulation relation, we would replace the rewards* (7, 1) *by* (3, 3), *thus to obtain a reward of* 1 *which is considerably lower than the minimal reward* $\frac{7}{3} \approx 2.333$ *of the original model.*

In the case of affine reward structures, we can define a different simulation relation to obtain more precise results.

DEFINITION 19. *Consider two PAs* $\mathcal{M} = (S, \overline{s}, Act, \mathcal{T})$ *and* $\mathcal{M}_{\text{sim}} = (S_{\text{sim}}, \overline{s}_{\text{sim}}, Act, \mathcal{T}_{\text{sim}})$ *between which there is a simulation relation R. Consider affine reward structures*

$$rew = (rew_{\text{num}}, rew_{\text{den}}),$$
$$rew_{\text{sim,up}} = (rew_{\text{sim,up,num}}, rew_{\text{sim,up,den}}),$$
$$rew_{\text{sim,lo}} = (rew_{\text{sim,lo,num}}, rew_{\text{sim,lo,den}}).$$

We require that rew, $rew_{\text{sim,up}}$ *and* $rew_{\text{sim,lo}}$ *are affine with the same factors* mul_a, add_a *(cf. Definition 11) for each action a, that is for* $s \in S$ *and* $\mathbf{z} \in \mathbf{A}$ *it is*

$$rew_{\text{num}}(s, a) = mul_a rew_{\text{den}}(s, a) + add_a,$$
$$rew_{\text{sim,up,num}}(\mathbf{z}, a) = mul_a rew_{\text{sim,up,den}}(\mathbf{z}, a) + add_a,$$
$$rew_{\text{sim,lo,num}}(\mathbf{z}, a) = mul_a rew_{\text{sim,lo,den}}(\mathbf{z}, a) + add_a.$$

Then, we define R as affine compatible *if for all* $(s, s_{\text{sim}}) \in R$ *and* $a \in Act$ *it is*

$$rew_{\text{sim,lo,num}}(s_{\text{sim}}, a) \leq rew_{\text{num}}(s, a) \leq rew_{\text{sim,up,num}}(s_{\text{sim}}, a),$$
$$rew_{\text{sim,lo,den}}(s_{\text{sim}}, a) \leq rew_{\text{den}}(s, a) \leq rew_{\text{sim,up,den}}(s_{\text{sim}}, a).$$

If there exists such a relation, we write

$$(\mathcal{M}, rew) \overset{\text{aff}}{\preceq} (\mathcal{M}_{\text{sim}}, rew_{\text{sim,up}}, rew_{\text{sim,lo}}).$$

As before, affine simulations maintain reward properties.

LEMMA 2. *Consider the PAs* $\mathcal{M} = (S, \overline{s}, Act, \mathcal{T})$ *with the reward structure* rew *and* $\mathcal{M}_{\text{sim}} = (S_{\text{sim}}, \overline{s}_{\text{sim}}, Act, \mathcal{T}_{\text{sim}})$ *with reward structures* $rew_{\text{sim,up}}$ *and* $rew_{\text{sim,lo}}$ *with* $(\mathcal{M}, rew) \overset{\text{aff}}{\preceq} (\mathcal{M}_{\text{sim}}, rew_{\text{sim,up}}, rew_{\text{sim,lo}})$. *We define*

$$\mathcal{M}_{\text{aff}} \overset{\text{def}}{=} (S_{\text{sim}}, \overline{s}_{\text{sim}}, Act \times \{\text{up}, \text{lo}\}, \mathcal{T}_{\text{aff}}),$$
$$rew_{\text{aff}} \overset{\text{def}}{=} (rew_{\text{aff,num}}, rew_{\text{aff,den}}), \text{ where}$$

- *for* $s \in S_{\text{sim}}$ *and* $a \in Act$ *it is* $\mathcal{T}_{\text{aff}}(s, (a, \text{up})) \overset{\text{def}}{=} \mathcal{T}_{\text{aff}}(s, (a, \text{lo})) \overset{\text{def}}{=} \mathcal{T}_{\text{sim}}(s, a)$,
- *for* $s \in S_{\text{sim}}$ *and* $a \in Act$ *it is* $rew_{\text{aff,num}}(s, (a, \text{up})) \overset{\text{def}}{=} rew_{\text{sim,up,num}}(s, a)$, *accordingly for* $rew_{\text{aff,num}}(s, (a, \text{lo}))$, $rew_{\text{aff,den}}(s, (a, \text{lo}))$ *and* $rew_{\text{aff,den}}(s, (a, \text{lo}))$.

Then we have

$$\text{val}^{+,Act_{\text{fair}}}_{\mathcal{M},rew,\text{lra}} \leq \text{val}^{+,Act_{\text{fair}}}_{\mathcal{M}_{\text{aff}},rew_{\text{aff}},\text{lra}},$$

and accordingly for the accumulated rewards and the minimising cases.

Similar to upper-bound and lower-bound compatible simulations, the affinely simulating automaton $\mathcal{M}_{\text{sim}}$ can mimic the behaviours of the simulated one. In $\mathcal{M}_{\text{aff}}$ then, it also mimics the behaviours of the original model, but can use randomised choices over (a, up) and (a, lo) to obtain exactly the same reward as when choosing a in the original model. The reason that we will obtain results which are more precise is, intuitively, that for nonaffine reward structures we had to bound rew_{num} and rew_{den} from opposite directions.

EXAMPLE 10. *In Figure 4 (b) we give a PA which affinely simulates the one of Figure 3. Maximal and minimal long-run averages are* 3 *and* $\frac{7}{3} \approx 2.333$*, which is more precise than the values obtained from Figure 4 (a) in Example 9.*

We describe abstract state spaces to subsume uncountably many states of the infinite semantics of PHAs.

DEFINITION 20. *An* abstract state space *of dimension k for a set of modes M is a finite set* $\mathbf{A} = \{\mathbf{z}_1, \dots, \mathbf{z}_n\}$ *where* $\mathbf{z}_i = (m_i, \zeta_i) \in M \times 2^{\mathbb{R}^k}$ *and it is* $\bigcup_{(m,\zeta)\in\mathbf{A}} \zeta = \mathbb{R}^k$ *for all* $m \in M$*. We identify* (m, ζ) *with the set* $\{m\} \times \zeta$ *which allows us to apply the usual set operations on abstract states, and we will for instance write* $s \in (m, \zeta)$.

We do not require $\mathbf{A}$ to be a *partitioning* of $M \times \mathbb{R}^k$, that is we do allow overlapping states. This way, one concrete state may be contained in several abstract states. We need to allow this, because in several hybrid system solvers from which we obtain these abstractions, these cases indeed happen. For instance, in the tool HSOLVER [31] we may have overlapping borders, whereas for PHAVER [19] we may also have common interiors of abstract states.

An abstraction of a PHA is defined as follows. There, we will need to transfer probability distributions over the states of the PHA semantics to the states of abstractions.

DEFINITION 21. *Consider an arbitrary PHA* $\mathcal{H} = (M, k, \overline{m}, \langle Post_m \rangle_{m\in M}, Cmds, Rew)$*, and abstract state space* $\mathbf{A} = \{\mathbf{z}_1, \dots, \mathbf{z}_n\}$ *of corresponding dimension and modes. We say that*

$$\mathcal{M} = (\mathbf{A}, \overline{\mathbf{z}}, Cmds \uplus \{\tau\}, \mathcal{T})$$

is an abstraction *of $\mathcal{H}$ using $\mathbf{A}$ if*

- $(\overline{m}, 0, \dots, 0) \in \overline{\mathbf{z}}$,
- *for all* $\mathbf{z} \in \mathbf{A}$, $s \in \mathbf{z}$, $c = (g \to [u_1 \mapsto p_1, \dots, u_n \mapsto p_n]) \in Cmds$*, if* $s \in \mathbf{z} \cap g$*, then for all*

  $$(s_1', \dots, s_n') \in u_1(s) \times \dots \times u_n(s)$$

  there are

  $$(\mathbf{z}_1', \dots, \mathbf{z}_n') \in \mathbf{A}^n \text{ with } s_i \in \mathbf{z}_i, 1 \le i \le n$$

  so that there is $\mu \in Distr(\mathbf{A})$ *with* $\mu(\mathbf{z}') = \sum_{\mathbf{z}'=\mathbf{z}_i'} p_i$ *and* $\mu \in \mathcal{T}(\mathbf{z}, c)$,
- *for all* $\mathbf{z} \in \mathbf{A}$, $s = (m, v) \in \mathbf{z}$ *and all* $s' = (m, v') \in \{m\} \times Post_m(v)$*, we require that there is* $\mathbf{z}' \in \mathbf{A}$ *with* $s' \in \mathbf{z}'$ *and* $[\mathbf{z}' \mapsto 1] \in \mathcal{T}(\mathbf{z}, \tau)$.

By $Abs(\mathcal{H}, \mathbf{A})$ *we denote the set of all such abstractions.*

Next, we equip PHA abstractions with rewards.

DEFINITION 22. *Let $\mathcal{H}$ be a PHA with rewards Rew and consider* $\mathcal{M} = (\mathbf{A}, \overline{\mathbf{z}}, Cmds \uplus \{\tau\}, \mathcal{T}) \in Abs(\mathcal{H}, \mathbf{A})$*. The* abstract upper-bound reward structure *is defined as*

$$\text{absup}(\mathcal{H}, \mathcal{M}) \stackrel{\text{def}}{=} (rew_{\text{num}}, rew_{\text{den}}), \text{ where}$$

- *for all* $\mathbf{z} \in \mathbf{A}$ *it is* $rew_{\text{num}}(\mathbf{z}, \tau) \stackrel{\text{def}}{=} rew_{\text{den}}(\mathbf{z}, \tau) \stackrel{\text{def}}{=} 0$,
- *for all* $\mathbf{z} \in \mathbf{A}$ *and* $c = (g \to [u_1 \mapsto p_1, \dots, u_n \mapsto p_n]) \in Cmds$ *it is*

$$rew_{\text{num}}(\mathbf{z}, c) \stackrel{\text{def}}{=} \sup_{s=(m,r,t,v)\in\mathbf{z}} r + Rew(s, c),$$
$$rew_{\text{den}}(\mathbf{z}, c) \stackrel{\text{def}}{=} \inf_{(m,r,t,v)\in\mathbf{z}} t.$$

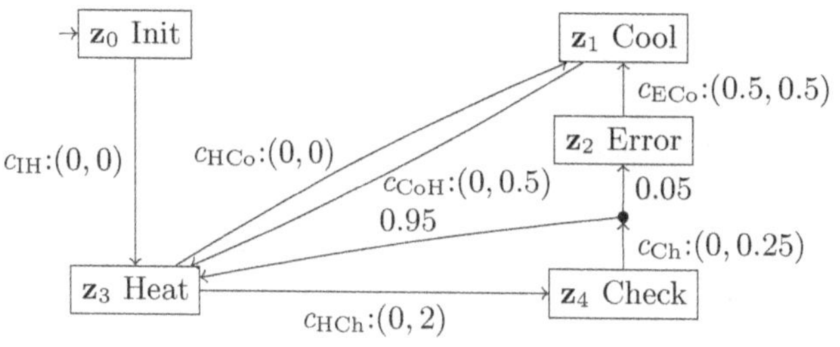

Figure 5: PHA abstraction for rewards.

The abstract lower-bound reward structure absl0 *is defined accordingly by swapping* sup *and* inf.

We can use these reward structures to safely bound the reward values of PHAs semantics.

THEOREM 1. *Consider a PHA $\mathcal{H}$ with reward structure Rew and* $\mathcal{M} = (\mathbf{A}, \overline{\mathbf{z}}, Cmds \uplus \{\tau\}, \mathcal{T}) \in Abs(\mathcal{H}, \mathbf{A})$ *with* $rew_{\text{up}} \stackrel{\text{def}}{=} \text{absup}(\mathcal{H}, \mathcal{M})$*. Then*

$$\text{val}^{+}_{\mathcal{H},\text{lra}} \le \text{val}^{+,Cmds}_{\mathcal{M},rew_{\text{up}},\text{lra}},$$

and accordingly for accumulated rewards and minimal values.

The theorem follows by Lemma 1, because abstractions simulate the semantics of PHAs.

EXAMPLE 11. *In Figure 5 we sketch an abstraction of Figure 2 with an abstract upper-bound reward structure for the PHA reward structure* Rew_{lra} *of Example 3. Thus, we have a reward of* 1 *per time in* Error *and* 0 *else. In the abstract states, we left out constraints to restrict to states which are actually reachable. Consider the abstract state* $\mathbf{z}_4$*. As the mode of this state is* Check*, we obtain a reward of* 0 *when executing* c_{Ch}*. According to the guard of this command (which we slightly modify when using this reward structure), we have to wait at least until* $t \ge 0.25$ *to execute it. Now consider* $\mathbf{z}_2$*. We can leave this state at point* $t = 0.25$*, and we thus obtain a reward and time of* 0.25.

In case we are given reward affine PHAs, we can use more precise reward structures in the abstraction.

DEFINITION 23. *Consider a reward affine PHA $\mathcal{H}$ with commands $Cmds$ and the reward structure Rew and* $\mathcal{M} = (\mathbf{A}, \overline{\mathbf{z}}, Cmds \uplus \{\tau\}, \mathcal{T}) \in Abs(\mathcal{H}, \mathbf{A})$*. We define the* affine abstraction

$$\mathcal{M}_{\text{aff}} \stackrel{\text{def}}{=} (\mathbf{A}, \overline{\mathbf{z}}, Act_{\text{aff}}, \mathcal{T}_{\text{aff}}),$$

where $Act_{\text{aff}} \stackrel{\text{def}}{=} \{(\tau, \text{up}), (\tau, \text{lo})\} \uplus \{(c, \text{up}) \mid c \in Cmds\} \uplus \{(c, \text{lo}) \mid c \in Cmds\}$*, and define* $\mathcal{T}_{\text{aff}}$ *as* $\mathcal{T}_{\text{aff}}(\mathbf{z}, (a, \text{up})) \stackrel{\text{def}}{=} \mathcal{T}_{\text{aff}}(\mathbf{z}, (a, \text{lo})) \stackrel{\text{def}}{=} \mathcal{T}(\mathbf{z}, a)$*. Then, the* abstract affine reward structure *is defined as* $rew_{\text{aff}} = (rew_{\text{num}}, rew_{\text{den}})$ *where*

- *for all* $\mathbf{z} \in \mathbf{A}$*, let*

  $$rew_{\text{num}}(\mathbf{z}, (\tau, \text{up})) \stackrel{\text{def}}{=} rew_{\text{num}}(\mathbf{z}, (\tau, \text{lo}))$$
  $$\stackrel{\text{def}}{=} rew_{\text{den}}(\mathbf{z}, (\tau, \text{up})) \stackrel{\text{def}}{=} rew_{\text{den}}(\mathbf{z}, (\tau, \text{lo})) \stackrel{\text{def}}{=} 0,$$

- *for all* $\mathbf{z} \in \mathbf{A}$ *with mode m and* $c = (g \to [u_1 \mapsto p_1, \dots, u_n \mapsto p_n]) \in Cmds$*, let*

  $$rew_{\text{num}}(\mathbf{z}, (c, \text{up})) \stackrel{\text{def}}{=} cv_{\text{sup}} + Rew(c),$$
  $$rew_{\text{den}}(\mathbf{z}, (c, \text{up})) \stackrel{\text{def}}{=} v_{\text{sup}},$$

$$rew_{\mathrm{num}}(\mathbf{z},(c,\mathrm{lo})) \stackrel{\mathrm{def}}{=} cv_{\mathrm{inf}} + Rew(c),$$
$$rew_{\mathrm{den}}(\mathbf{z},(c,\mathrm{lo})) \stackrel{\mathrm{def}}{=} v_{\mathrm{inf}},$$

with the factor c of Definition 1 and

$$v_{\mathrm{sup}} \stackrel{\mathrm{def}}{=} \sup\{t \mid (m,r,t,v) \in \mathbf{z}\},$$
$$v_{\mathrm{inf}} \stackrel{\mathrm{def}}{=} \inf\{t \mid (m,r,t,v) \in \mathbf{z}\}.$$

We then define $\mathrm{absaff}(\mathcal{H},\mathcal{M}) \stackrel{\mathrm{def}}{=} (\mathcal{M}_{\mathrm{aff}}, rew_{\mathrm{aff}})$.

THEOREM 2. *Consider a reward affine PHA* $\mathcal{H}$ *with reward structure Rew and abstraction* $\mathcal{M} = (\mathbf{A}, \overline{\mathbf{z}}, Cmds \uplus \{\tau\}, \mathcal{T}) \in Abs(\mathcal{H}, \mathbf{A})$, *with* $\mathrm{absaff}(\mathcal{H},\mathcal{M}) = (\mathcal{M}_{\mathrm{aff}}, rew_{\mathrm{aff}})$. *Then*

$$\mathrm{val}^{+}_{\mathcal{H},\mathrm{lra}} \leq \mathrm{val}^{+,Cmds}_{\mathcal{M}_{\mathrm{aff}},rew_{\mathrm{aff}},\mathrm{lra}},$$

accordingly for accumulated rewards and minimal values.

The theorem follows by Lemma 2.

EXAMPLE 12. *If using an affine abstraction, we replace the actions of Figure 5 by*

$$(c_{\mathrm{IH}},\mathrm{up}){:}(0,0), (c_{\mathrm{IH}},\mathrm{lo}){:}(0,0),$$
$$(c_{\mathrm{CoH}},\mathrm{up}){:}(0,2), (c_{\mathrm{CoH}},\mathrm{lo}){:}(0,0.5),$$
$$(c_{\mathrm{ECo}},\mathrm{up}){:}(0.5,0.5), (c_{\mathrm{ECo}},\mathrm{lo}){:}(0.5,0.5),$$
$$(c_{\mathrm{HCo}},\mathrm{up}){:}(0,3), (c_{\mathrm{HCo}},\mathrm{lo}){:}(0,0),$$
$$(c_{\mathrm{Ch}},\mathrm{up}){:}(0,0.5), (c_{\mathrm{Ch}},\mathrm{lo}){:}(0,0.25).$$

4.1 Computing Abstractions

A practical recipe to compute abstractions of PHAs has been developed in earlier work [39], and is implemented in the tool PROHVER. The abstraction process is piggybacked on solvers for HAs (as in Definition 2). Concretely, the tool applies PHAVER for this purpose. Thus far however there was no need and no means to compute reward structures for the abstraction at hand.

According to Definition 22 and Definition 23, we have to find suprema and infima of the variables for rewards and time. How this can be done depends on the hybrid systems solver used. PHAVER uses polyhedra to represent abstract states, which can be represented as sets of linear inequations. Because of this, we can use a linear programming tool to find the minimal and maximal values of reward and time variables.

In some cases, this construction can be simplified. If we do not have time-dependent rewards and only a constant reward value for each command, and only want to consider the expected accumulated reward, we do not need to compute infima or suprema at all. For affine abstractions, we only need a variable to remember the time since a mode change, because we can compute the rewards from these values; in Definition 23 the values r are not used.

In usual abstractions of HAs, the information about the time which a continuous transition takes is lost. This is the main reason why we encode reward and time into the first two dimension of a PHA, rather than assigning them directly to the timed transitions.

4.2 Algorithmic Considerations

After we have obtained a finite abstraction and have computed the according reward structure using linear programming, it remains to compute expected accumulated or long-run average rewards in the abstraction. There are several algorithms which we can apply for this purpose. For accumulated rewards we can use algorithms based on policy iteration or linear programming [30]. In case we want to compute time-average average values, there are also algorithms using linear programming [15] or policy iteration [38].

len.	PHAVER	S	uptime constr.	uptime ana.	uptime res.	commands constr.	commands ana.	commands res.
–	0	7	0	0	0.00	0	0	42.00
1	1	102	0	0	45.33	0	0	42.00
0.05	151	24593	6	65	62.02	1	103	48.84
0.03	2566	93979	25	584	62.72	2	722	50.78

Table 2: Accumulated rewards in thermostat.

len.	PHAVER	S	time in error constr.	time in error ana.	time in error result	time in error (lin) constr.	time in error (lin) ana.	time in error (lin) result
1	0	126	0	0	0.013	0	0	0.013
0.5	1	382	0	0	0.011	1	0	0.011
0.1	28	7072	4	9	0.009	6	14	0.009
0.05	185	29367	8	124	0.009	25	313	0.009

Table 3: Long-run average rewards in thermostat.

5. EXPERIMENTS

We implemented the analysis methods to compute expected accumulated and time-average rewards in our tool PROHVER, and have applied them on two case studies. Experiments were run on an Intel(R) Core(TM)2 Duo CPU with 2.67 GHz and 4 GB RAM.

5.1 Thermostat

The first properties we consider for our running thermostat example are the minimal expected time and the expected number of commands (except from Init to Heat) it takes until the error mode is reached, as in Example 3.

Results are given in Table 2. The *constraint length* (len.) is a parameter which influences the precision with which PHAVER builds the abstract state space. It splits the abstract states along a given variable, in our case T. We also provide the times in seconds which PHAVER needed to build the abstraction (PHAVER), the number of states (S), the time we needed to compute the reward structures (constr.) and the computed value bound (result).

Initially, we performed these experiments splitting on t. Doing so, we obtained much worse reward bounds, indeed always 42 for the expected number of commands until error. Manual analysis showed that this refinement only excluded the first direct transition from Heat to Check without a previous visit to Cool. The resulting lower bound on the expected number of commands therefore is

$$Rew(c_{\mathrm{HCo}}) + Rew(c_{\mathrm{CoH}}) + (Rew(c_{\mathrm{HCh}}) + Rew(c_{\mathrm{ChH}}))\frac{1}{0.05} = 1 + 1 + (1+1)\frac{1}{0.05} = 42.$$

Table 2 shows that this is not the ultimate answer to the minimal expected number of executed commands. This implies that at later points of time a direct transition from Heat to Check without visiting Cool in between is not always possible.

The time needed to construct the reward structures for the expected number of commands is much lower than the one for the expected time, because in the first case the reward is constant, and we thus can avoid solving a large number of linear programming problems. The time for the analysis is higher, though.

Next, we consider the maximal expected time fraction f spent in the Error mode of the modified thermostat variant in which this mode can be left. This allows us to obtain a lower bound for the system long-run availability $1 - f$. We also provide results when using the more precise method taking advantage of the fact that the reward structure is affine, by using a reward structure as in Example 12. Results are given in Table 3. As seen, using affine reward structures

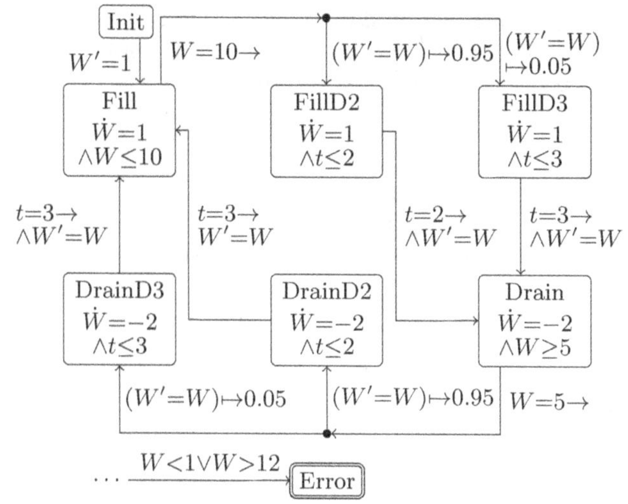

Figure 6: Water level control automaton.

len.	PHAVER	S	uptime constr.	uptime ana.	uptime result	commands constr.	commands ana.	commands result
–	1	10	0	0	0.0	0	0	40
1	0	53	0	0	137.0	0	0	40
0.05	5	814	1	1	164.8	0	2	40
0.01	31	4294	1	42	166.1	0	46	40

Table 4: Accumulated rewards in water control.

does not improve the result: This was expected by comparing Example 11 and Example 12. Some of the actions of Example 12 have the same values as actions in Example 11 (e.g. $(c_{\mathrm{CoH}}, \mathrm{lo}) : (0, 0.5)$ corresponds to $c_{\mathrm{CoH}} : (0, 0.5)$), while the remaining ones are suboptimal choices for a maximising scheduler (e.g. $(c_{\mathrm{CoH}}, , \mathrm{up}) : (0, 2)$). Thus, a scheduler in the affine abstraction basically takes the same choices as in the previous abstraction.

By the definition of affine simulation, it is necessary to solve twice as many linear optimisation problems to obtain an abstraction as when using an abstract upper-bound reward structure, so the time needed by ProHVer is larger.

5.2 Water Level Control

We consider a model of a water level control system (extended from the one of Alur et al. [3]) which uses wireless sensors. Values submitted are thus subject to probabilistic delays, due to the unreliable transport medium. A sketch of the model is given in Figure 6. The water level W of a tank is controlled by a monitor. Its change is specified by an affine function. Initially, the water level is $W = 1$. When no pump is turned on (Fill), the tank is filled by a constant stream of water ($\dot{W}$). When a water level of $W = 10$ is seen by a sensor of the tank, the pump should be turned on. However, the pump features a certain delay, which results from submitting control data via a wireless network. With a probability of 0.95 this delay takes 2 time units (FillD2), but with a probability of 0.05 it takes 3 time units (FillD3). The delay is modelled by the timer t. After the delay has passed, the water is pumped out with a higher speed than it is filled into the tank ($\dot{W} = -2$ in Drain). There is another sensor to check whether the water level is below 5. If this is the case, the pump must turn off again. Again, we have a distribution over delays here (DrainD2 and DrainD3). Similar to the thermostat case, we considered the minimal expected time and number of command executions until the Error mode is reached. As this model features only affine continuous dynamics, we successfully obtained results using constraints on t. Results are given in Table 4.

len.	PHAVER	S	avg. energy constr.	avg. energy ana.	avg. energy result	avg. energy (lin.) constr.	avg. energy (lin.) ana.	avg. energy (lin.) result
1	0	51	0	0	1.985	0	0	1.814
0.1	2	409	0	0	1.656	0	0	1.640
0.02	11	2149	0	0	1.630	1	0	1.627
0.01	24	4292	0	1	1.627	2	1	1.625

Table 5: Long-run average rewards in water control.

For the second property, we remove Error and assume that the operational bounds of the system are safe. We are interested in the average energy consumption of the system. We assume that no energy is spent in the modes where the pump is switched off. While the pump is running, 2 units of energy are consumed per time unit, starting the pump takes 10 units of energy and switching it off again takes 6 units. The reward structure thus features both rewards obtained from timed transitions as well as those obtained from command executions. Results are given in Table 5. As seen, using affine reward structures improves the result more than in the thermostat case. This happens because in a larger percentage of the time a nonzero reward in the numerator is obtained. In the thermostat case study, this was only possible in mode Error. As the time spent in that mode in the thermostat setting is small, the induced difference between the lower and upper values with the two kinds of reward abstractions is also rather small.

6. CONCLUSION

We have presented a framework to handle probabilistic hybrid automata decorated with costs or bonuses. The resulting abstraction and verification approach considerably enriches the spectrum of properties that are amenable to model checking. It is now possible to check properties including cost accumulated until termination, the long-run cost of system operation, and other popular quantities. This includes long-run availability, as mentioned in Section 5, and many other quantities of interest. For instance, we are now in the position to compute the long-run survivability [14] of a probabilistic hybrid system. This quantity asks for a time bounded reach analysis [40] nested inside a long-run cost analysis.

To arrive at this powerful framework, we have decorated PHAs with reward structures associated to taking a transition, have discussed how these structures carry over to the semantical model (PAs) and how properties of PHAs are defined. These properties can reason about the expected total or long-run average reward.

On the level of the semantical model, we have extended our abstraction framework to work well with reward structures, and this required some nontrivial considerations. Probabilistic simulation relations serve as means to guarantee the correctness of the abstractions we build, and they needed to be augmented to properly cover reward-based properties.

To allow for the automatic analysis of reward-based properties, we extended our framework accordingly. In case we have rewards depending on the time, as for instance the average time the system is operational, it has turned out necessary to take a closer look at the form of the abstract states, to find out about minimal or maximal reward values. The effectivity of our approach has been demonstrated by using it to compute reward-based values on two case studies.

Acknowledgements. This work was supported by the Transregional Collaborative Research Centre SFB/TR 14 AVACS, the NWO-DFG bilateral project ROCKS, the European Union Seventh Framework Programme under grant agreement numbers 295261 (MEALS), and 318490 (SENSATION), and the ERC Advanced Grant VERIWARE.

7. REFERENCES

[1] A. Abate, M. Prandini, J. Lygeros, and S. Sastry. Probabilistic reachability and safety for controlled discrete time stochastic hybrid systems. *Automatica*, 44(11):2724–2734, 2008.

[2] E. Altman and V. Gaitsgory. Asymptotic optimization of a nonlinear hybrid system governed by a Markov decision process. *SICON*, 35(6):2070–2085, 1997.

[3] R. Alur, C. Courcoubetis, N. Halbwachs, T. A. Henzinger, P.-H. Ho, X. Nicollin, A. Olivero, J. Sifakis, and S. Yovine. The algorithmic analysis of hybrid systems. *TCS*, 138(1):3–34, 1995.

[4] R. Alur, T. Dang, and F. Ivančić. Predicate abstraction for reachability analysis of hybrid systems. *ACM TECS*, 5(1):152–199, 2006.

[5] E. Asarin, O. Bournez, T. Dang, O. Maler, and A. Pnueli. Effective synthesis of switching controllers for linear systems. *Proc. IEEE*, 88(7):1011–1025, 2000.

[6] E. Asarin, O. Maler, and A. Pnueli. Symbolic controller synthesis for discrete and timed systems. In *Hybrid Systems*, pages 1–20, 1994.

[7] C. Baier, L. Cloth, B. R. Haverkort, H. Hermanns, and J.-P. Katoen. Performability assessment by model checking of Markov reward models. *Formal Methods in System Design*, 36:1–36, 2010.

[8] C. Baier and J.-P. Katoen. *Principles of Model Checking*. MIT Press, 2008.

[9] P. Bouyer, U. Fahrenberg, K. G. Larsen, and N. Markey. Quantitative analysis of real-time systems using priced timed automata. *Commun. ACM*, 54(9):78–87, 2011.

[10] P. Bouyer, N. Markey, J. Ouaknine, and J. Worrell. The cost of punctuality. In *LICS*, pages 109–120, 2007.

[11] M. L. Bujorianu. Extended stochastic hybrid systems and their reachability problem. In *HSCC*, pages 234–249, 2004.

[12] M. L. Bujorianu, J. Lygeros, and M. C. Bujorianu. Bisimulation for general stochastic hybrid systems. In *HSCC*, pages 198–214, 2005.

[13] C. Cassandras and J. Lygeros. *Stochastic Hybrid Systems*, volume 24. CRC Press and IEEE Press, 2006.

[14] L. Cloth and B. R. Haverkort. Model checking for survivability. In *QEST*, pages 145–154, 2005.

[15] L. de Alfaro. *Formal Verification of Probabilistic Systems*. PhD thesis, Stanford University, 1997.

[16] E. de Souza e Silva and H. R. Gail. Calculating availability and performability measures of repairable computer systems using randomization. *J. ACM*, 36(1):171–193, 1989.

[17] M. Fränzle, E. M. Hahn, H. Hermanns, N. Wolovick, and L. Zhang. Measurability and safety verification for stochastic hybrid systems. In *HSCC*, pages 43–52, 2011.

[18] M. Fränzle, T. Teige, and A. Eggers. Satisfaction meets expectations - computing expected values of probabilistic hybrid systems with SMT. In *IFM*, pages 168–182, 2010.

[19] G. Frehse. PHAVer: Algorithmic verification of hybrid systems past HyTech. *STTT*, 10(3):263–279, 2008.

[20] G. Frehse, C. L. Guernic, A. Donzé, S. Cotton, R. Ray, O. Lebeltel, R. Ripado, A. Girard, T. Dang, and O. Maler. SpaceEx: Scalable verification of hybrid systems. In *CAV*, pages 379–395, 2011.

[21] E. M. Hahn. *Model Checking Stochastic Hybrid Systems*. PhD thesis, Saarland University, 2013. To appear.

[22] E. M. Hahn, G. Norman, D. Parker, B. Wachter, and L. Zhang. Game-based abstraction and controller synthesis for probabilistic hybrid systems. In *QEST*, pages 69–78, 2011.

[23] A. Hartmanns, H. Hermanns, and P. Berrang. A comparative analysis of decentralized power grid stabilization strategies. In *WSC*, 2012.

[24] J. Hu, J. Lygeros, and S. Sastry. Towards a theory of stochastic hybrid systems. In *HSCC*, pages 160–173, 2000.

[25] B. Jonsson and K. G. Larsen. Specification and refinement of probabilistic processes. In *LICS*, pages 266–277, 1991.

[26] R. M. Keller. Formal verification of parallel programs. *Commun. ACM*, 19(7):371–384, 1976.

[27] J. Kemeny, J. Snell, and A. Knapp. *Denumerable Markov Chains*. D. Van Nostrand Company, 1966.

[28] A. Platzer. Stochastic differential dynamic logic for stochastic hybrid programs. In *CADE*, pages 446–460, 2011.

[29] M. Prandini and J. Hu. A stochastic approximation method for reachability computations. In H. Blom and J. Lygeros, editors, *Stochastic Hybrid Systems*, volume 337 of *LNCIS*, pages 107–139. Springer, 2006.

[30] M. L. Puterman. *Markov Decision Processes: Discrete Stochastic Dynamic Programming*. Wiley, 1994.

[31] S. Ratschan and Z. She. Safety verification of hybrid systems by constraint propagation-based abstraction refinement. *ACM TECS*, 6(1), 2007.

[32] M. Rutkowski, R. Lazić, and M. Jurdziński. Average-price-per-reward games on hybrid automata with strong resets. *STTT*, 13(6):553–569, 2011.

[33] R. Segala. *Modeling and Verification of Randomized Distributed Real-time Systems*. PhD thesis, MIT, 1995.

[34] R. Segala and N. A. Lynch. Probabilistic simulations for probabilistic processes. *NJC*, 2(2):250–273, 1995.

[35] J. Sproston. Decidable model checking of probabilistic hybrid automata. In *FTRTFT*, pages 31–45, 2000.

[36] T. Teige. *Stochastic Satisfiability Modulo Theories: A Symbolic Technique for the Analysis of Probabilistic Hybrid Systems*. PhD thesis, Carl von Ossietzky Universität Oldenburg, 2012.

[37] I. Tkachev and A. Abate. Regularization of Bellman equations for infinite-horizon probabilistic properties. In *HSCC*, pages 227–236, 2012.

[38] C. von Essen and B. Jobstmann. Synthesizing efficient controllers. In *VMCAI*, pages 428–444, 2012.

[39] L. Zhang, Z. She, S. Ratschan, H. Hermanns, and E. M. Hahn. Safety verification for probabilistic hybrid systems. In *CAV*, pages 196–211, 2010.

[40] L. Zhang, Z. She, S. Ratschan, H. Hermanns, and E. M. Hahn. Safety verification for probabilistic hybrid systems. *European Journal of Control*, 18:572–587, 2012.

[41] P. Zuliani, C. Baier, and E. M. Clarke. Rare-event verification for stochastic hybrid systems. In *HSCC*, pages 217–226, 2012.

Approximating Acceptance Probabilities of CTMC-Paths on Multi-Clock Deterministic Timed Automata *

Hongfei Fu †
Lehrstuhl für Informatik 2, RWTH Aachen
Ahornstraße 55, 52074 Aachen, Germany
hongfeifu@cs.rwth-aachen.de

ABSTRACT

We consider the problem of approximating the probability mass of the set of timed paths under a continuous-time Markov chain (CTMC) that are accepted by a deterministic timed automaton (DTA). As opposed to several existing works on this topic, we consider DTA with multiple clocks. Our key contribution is an algorithm to approximate these probabilities using finite difference methods. An error bound is provided which indicates the approximation error. The stepping stones towards this result include rigorous proofs for the measurability of the set of accepted paths and the integral-equation system characterizing the acceptance probability, and a differential characterization for the acceptance probability.

Categories and Subject Descriptors

D.2.4 [**Software Engineering**]: Software/Program Verification—*Model Checking*

General Terms

Algorithms, Verification

Keywords

continuous-time Markov chains; deterministic timed automata; model checking

1. INTRODUCTION

Continuous-time Markov chains (CTMCs) [16] are one of the most prominent models for performance and dependability analysis of real-time stochastic systems. They are the semantical backbones of Markovian queueing networks,

*A full version of this paper is available at `http://arxiv.org/abs/1210.4787`

†Supported by a CSC (China Scholarship Council) Scholarship

HSCC'13, April 8–11, 2013, Philadelphia, Pennsylvania, USA.

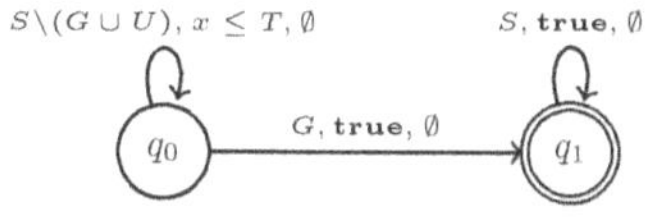

Figure 1: DTA $\mathcal{A}_1$

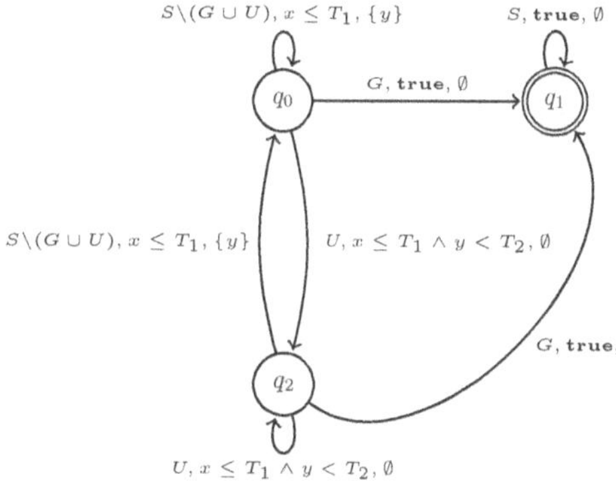

Figure 2: DTA $\mathcal{A}_2$

stochastic Petri nets and calculi for system biology and so forth. The desired behaviour of these systems is specified by various measures such as reachability with time information, timed logics such as CSL[3, 19], mean response time, throughput, expected frequency of errors, etc.

Verification of continuous-time Markov chains has received much attention in recent years [4]. Many applicable results have been obtained on time-bounded reachability [3, 15], CSL model checking [3, 19], and so on. In this paper, we focus on verifying CTMC against timed automata specification. In particular we consider approximating the probabilities of sets of CTMC-paths accepted by a deterministic timed automata (DTA) [1, 10]. In general, DTA represents a wide class of linear real-time specifications. For example, we can describe time-bounded reachability probability "to reach target set $G \subseteq S$ within time bound T while avoiding unsafe states $U \subseteq S$" ($G \cap U = \emptyset$) by the single-clock DTA $\mathcal{A}_1$ (Fig. 1), and the property "to reach target set $G \subseteq S$ within time bound T_1 while successively remaining in unsafe states $U \subseteq S$ for at most T_2 time" ($G \cap U = \emptyset$) by the two-clock DTA $\mathcal{A}_2$ (Fig. 2), both with initial configuration $(q_0, \vec{0})$. (We omit redundant locations that cannot reach the accepting state.)

The problem to verify CTMC against DTA specifications is first considered by Donatelli *et al.* [14] where they enriched CSL with an acceptance condition of one-clock DTA to obtain the logic CSL^{TA}. In their paper, they proved that CSL^{TA} is at least as expressive as CSL and asCSL [3, 2], and is strictly more expressive than CSL. Moreover, they presented a model-checking algorithm for CSL^{TA} using Markov regenerative processes. Chen *et al.* [12] systematically studied the DTA acceptance condition on CTMC-paths. More specifically, they proved that the set of CTMC-path accepted by a DTA is measurable and proposed a system of integral equations which characterizes the acceptance probabilities. Moreover, they demonstrated that the product of CTMC and DTA is a piecewise deterministic Markov process [13], a dynamic system which integrates both discrete control and continuous evolution. Afterwards, Barbot *et al.* [5] put the approximation of DTA acceptance probabilities of CTMC-paths into practice, especially the algorithm on one-clock DTA which is first devised by Donatelli *et al.* [14] and then rearranged by Chen *et al.* [12]. Later on, Chen *et al.* [11] proposed approximation algorithms for time-bounded verification of several linear real-time specifications, where the restricted time-bounded case, in which the time guard $x < T$ with a fresh clock x and a time bound T is enforced on each edge that leads to some final state of the DTA, is covered. Very recently, Mikeev *et al.* [17] applies the notion of DTA acceptance condition on large structured CTMCs. It is worth noting that Brázdil *et al.* [10] also studied DTA specifications. However they focused on semi-Markov processes as the underlying continuous-time stochastic model and limit frequencies of locations (in the DTA) as the performance measures, rather than path-acceptance.

Our contributions are as follows. We start by providing a rigorous proof for the measurability of CTMC paths accepted by a DTA, correcting the proof provided by Chen *et al.* [12]. We confirm the correctness of the integral equation system characterizing acceptance probabilities provided by Chen *et al.* [12] by providing a formal proof, and derive a differential characterization. This provides the main basis for our algorithm that approximates acceptance probabilities. We provide tight error bounds for our approximation algorithm. Whereas other works [5, 17, 14] focus on single-clock DTA, our approximation scheme is applicable to any multi-clock DTA. To our knowledge, this is the first such approximation algorithm with error bounds. Barbot *et al.* [5] suggested an approximation scheme, but did not provide any error bounds.

The paper is organized as follows. In Section 2 we introduce some preliminaries. In Section 3 we prove the measurability of accepted paths, and prove the integral equations [12] that characterize the acceptance probability. In Section 4 we develop several tools useful to our main result. In Section 5 we propose a differential characterization for the family of acceptance probability functions. Based on these results, we establish and solve our approximation scheme in Section 6 by using finite difference methods [18], which is the main result of the paper. Section 7 concludes the paper and discusses some possible future works.

All integrals in this paper should be basically understood as Lebesgue Integral. A full version of this paper can be found at `http://arxiv.org/abs/1210.4787`.

2. PRELIMINARIES

In this section we introduce continuous-time Markov chains [16] and deterministic timed automata [1, 10, 12].

2.1 Continuous-Time Markov Chains

Definition 1. A *continuous-time Markov chain* (CTMC) is a tuple $(S, L, \mathbf{P}, \lambda, \mathcal{L})$ where

- S is a finite set of *states*, and L is a finite set of *labels*;
- $\mathbf{P} : S \times S \mapsto [0, 1]$ is a *transition matrix* such that $\sum_{u \in S} \mathbf{P}(s, u) = 1$ for all $s \in S$;
- $\lambda : S \mapsto \mathbb{R}_{>0}$ is an *exit-rate function*, and $\mathcal{L} : S \mapsto L$ is a *labelling function*.

Intuitively, the running behaviour of a CTMC is as follows. Suppose s is the current state of a CTMC. Firstly, the CTMC stays at s for t time units where the dwell-time t observes the negative exponential distribution with rate $\lambda(s)$. Then the CTMC changes its current state to some state u with probability $\mathbf{P}(s, u)$ and continues running from u, and so forth. The one-step probability of the transition from s to u whose dwell time lies in the interval I equals $\mathbf{P}(s, u) \cdot \int_{t \in I} \lambda(s) \cdot e^{-\lambda(s)t}\, dt$. Besides, the labelling function $\mathcal{L}$ assigns each state s a label which indicates the set of atomic properties that hold at s.

To ease the notation, we denote the probability density function of the negative exponential distribution with rate $\lambda(s)$ by Λ_s, i.e., $\Lambda_s(t) = \lambda(s) \cdot e^{-\lambda(s) \cdot t}$ when $t \geq 0$ and $\Lambda_s(t) = 0$ when $t < 0$. It is worth noting that under our definition, we restrict ourselves such that the rates of all states are positive. CTMCs which contain states with rate 0 (i.e. *deadlock* states without outgoing transitions) can be adjusted to our case by (i) changing the rate of a deadlock state s to any positive value and (ii) setting $P(s, s) = 1$ and $P(s, u) = 0$ for all $u \neq s$, i.e., by making a self-loop on s.

Below we formally define a probability measure on sets of CTMC-paths, following the definitions from [3]. Suppose $\mathcal{M} = (S, L, \mathbf{P}, \lambda, \mathcal{L})$ be a CTMC. An $\mathcal{M}$-*path* π is an infinite sequence $s_0 t_0 s_1 t_1 \dots$ such that $s_n \in S$ and $t_n \in \mathbb{R}_{\geq 0}$ for all $n \in \mathbb{N}_0$. In other words, the set of $\mathcal{M}$-paths, denoted by $\text{Path}(\mathcal{M})$, is essentially $(S \times \mathbb{R}_{\geq 0})^\omega$. Given an $\mathcal{M}$-path $\pi = s_0 t_0 s_1 t_1 \dots$, we denote s_n and t_n by $\pi[n]$ and $\pi\langle n \rangle$, respectively.

A *template* θ is a finite sequence $s_0 I_0 \dots s_{l-1} I_{l-1} s_l$ such that $l \geq 1$, $s_n \in S$ for all $0 \leq n \leq l$ and I_n is an interval in $\mathbb{R}_{\geq 0}$ for all $0 \leq n \leq l-1$. Given a template $\theta = s_0 I_0 \dots s_{l-1} I_{l-1} s_l$, we define the *cylinder set* $R_\theta \subseteq \text{Path}(\mathcal{M})$ by: $\pi \in R_\theta$ iff $\pi[n] = s_n$ for all $0 \leq n \leq l$, and $\pi\langle n \rangle \in I_n$ for all $0 \leq n \leq l-1$.

An *initial distribution* Θ is a function $S \mapsto [0, 1]$ such that $\sum_{s \in S} \Theta(s) = 1$. The *probability space* $(\Omega, \mathcal{F}, \mathcal{B}_\Theta)$ over $\mathcal{M}$-paths with initial distribution Θ is defined as follows:

- $\Omega = \text{Path}(\mathcal{M})$;
- $\mathcal{F} \subseteq 2^\Omega$ is the smallest σ-algebra generated by the *cylindrical family* $\{R_\theta \mid \theta \text{ is a template}\}$ of subsets of Ω.
- $\mathcal{B}_\Theta : \mathcal{F} \mapsto [0, 1]$ is the unique probability measure such that $\mathcal{B}_\Theta(R_\theta) = \Theta(s_0) \cdot \prod_{n=0}^{l-1} \left\{ \mathbf{P}(s_n, s_{n+1}) \cdot \int_{I_n} \Lambda_{s_n}(t)\, dt \right\}$ for every template $\theta = s_0 I_0 \dots s_{l-1} I_{l-1} s_l$.

Intuitively, the probability space $(\Omega, \mathcal{F}, \mathcal{B}_\Theta)$ is generated by all cylinder sets R_θ, where $\mathcal{B}_\Theta(R_\theta)$ is the product of the initial probability and those one-step probabilities specified in Θ and θ. The uniqueness of $\mathcal{B}_\Theta$ is guaranteed by Carathéodory's Extension Theorem [8].

When Θ is a Dirac distribution on $s \in S$ (i.e., $\Theta(s) = 1$), we simply denote $(\Omega, \mathcal{F}, \mathcal{B}_\Theta)$ by $(\Omega, \mathcal{F}, \mathcal{B}_s)$. In this paper, we focus on the computation of $\mathcal{B}_s$, since any $\mathcal{B}_\Theta$ can be expressed as a linear combination of $\{\mathcal{B}_s\}_{s \in S}$.

2.2 Deterministic Timed Automata

Suppose $\mathcal{X}$ be a finite set of *clocks*. A (*clock*) *valuation* over $\mathcal{X}$ is a function $\eta : \mathcal{X} \mapsto \mathbb{R}_{\geq 0}$. We denote by $\mathrm{Val}(\mathcal{X})$ the set of valuations over $\mathcal{X}$. Sometimes we will view a clock valuation as a vector with an implicit order on $\mathcal{X}$.

A *guard* (or *clock constraints*) over a finite set of clocks $\mathcal{X}$ is a finite conjunction of basic constraints of the form $x \bowtie c$, where $x \in \mathcal{X}$, $\bowtie \in \{<, \leq, >, \geq\}$ and $c \in \mathbb{N}_0$. We denote the set of guards over $\mathcal{X}$ by $\Phi(\mathcal{X})$. For each $\eta \in \mathrm{Val}(\mathcal{X})$ and $g \in \Phi(\mathcal{X})$, the satisfaction relation $\eta \models g$ is defined by: $\eta \models x \bowtie c$ iff $\eta(x) \bowtie c$, and $\eta \models g_1 \wedge g_2$ iff $\eta \models g_1$ and $\eta \models g_2$. Given $g \in \Phi(\mathcal{X})$, we may also refer g to the set of valuations that satisfy g: this may happen in the context such as $g_1 \cap g_2$, etc. Given $X \subseteq \mathcal{X}$, $\eta \in \mathrm{Val}(\mathcal{X})$ and $t \in \mathbb{R}_{\geq 0}$, the valuations $\eta[X := 0]$, $\eta + t$, and $\eta - t$ are defined as follows:

1. if $x \in X$ then $\eta[X := 0](x) := 0$, otherwise $\eta[X := 0](x) := \eta(x)$;
2. $(\eta + t)(x) := \eta(x) + t$ for all $x \in \mathcal{X}$;
3. $(\eta - t)(x) := \eta(x) - t$ for all $x \in \mathcal{X}$, provided that $\eta(x) \geq t$ for all $x \in \mathcal{X}$.

Intuitively, $\eta[X := 0]$ is obtained by resetting all clocks of X to zero on η, and $\eta + t$ resp. $\eta - t$ is obtained by delaying resp. backtracking t time units from η.

Definition 2. [1, 12, 10] A *deterministic timed automaton* (DTA) is a tuple $(Q, \Sigma, \mathcal{X}, \Delta, F)$ where

- Q is a finite set of *locations*;
- $F \subseteq Q$ is a set of *final locations*;
- Σ is a finite *alphabet* of *signatures*;
- $\mathcal{X}$ is a finite set of *clocks*;
- $\Delta \subseteq Q \times \Sigma \times \Phi(\mathcal{X}) \times 2^{\mathcal{X}} \times Q$ is a finite set of *rules* such that
  1. Δ is *deterministic*: whenever $(q_1, a_1, g_1, X_1, q_1')$, $(q_2, a_2, g_2, X_2, q_2') \in \Delta$, if $(q_1, a_1) = (q_2, a_2)$ and $g_1 \cap g_2 \neq \emptyset$ then $(g_1, X_1, q_1') = (g_2, X_2, q_2')$.
  2. Δ is *total*: for all $(q, a) \in Q \times \Sigma$ and $\eta \in \mathrm{Val}(\mathcal{X})$, there exists $(q, a, g, X, q') \in \Delta$ such that $\eta \models g$.

Given $q \in Q$, $\eta \in \mathrm{Val}(\mathcal{X})$ and $a \in \Sigma$, we define the triple $(\mathbf{g}^\eta_{q,a}, \mathbf{X}^\eta_{q,a}, \mathbf{q}^\eta_{q,a})$ such that $(q, a, \mathbf{g}^\eta_{q,a}, \mathbf{X}^\eta_{q,a}, \mathbf{q}^\eta_{q,a})$ is the unique rule satisfying $\eta \models \mathbf{g}^\eta_{q,a}$.

Let $\mathcal{A} = (Q, \Sigma, \mathcal{X}, \Delta, F)$ be a DTA. A *configuration* of $\mathcal{A}$ is a pair (q, η), where $q \in Q$ and $\eta \in \mathrm{Val}(\mathcal{X})$. A *timed signature* is a pair (a, t) where $a \in \Sigma$ and $t \in \mathbb{R}_{\geq 0}$. We define the *one-step transition function*[1, 12]

$$\kappa^{\mathcal{A}} : (Q \times \mathrm{Val}(\mathcal{X})) \times (\Sigma \times \mathbb{R}_{\geq 0}) \mapsto Q \times \mathrm{Val}(\mathcal{X})$$

by: $\kappa^{\mathcal{A}}((q, \eta), (a, t)) = (\mathbf{q}^{\eta+t}_{q,a}, (\eta + t)[\mathbf{X}^{\eta+t}_{q,a} := 0])$. We may represent $\kappa^{\mathcal{A}}((q, \eta), (a, t)) = (q', \eta')$ by the more intuitive notation "$(q, \eta) \xrightarrow{(a,t)}_{\mathcal{A}} (q', \eta')$". We omit "$\mathcal{A}$" if the context is clear.

Intuitively, the configuration $\kappa((q, \eta), (a, t))$ is obtained as follows: firstly we delay t time-units at (q, η) to obtain $(q, \eta + t)$; then we find the unique rule $(q, a, g, X, q') \in \Delta$ such that $\eta + t \models g$; finally, we obtain $\kappa((q, \eta), (a, t))$ by changing the location to q' and resetting $\eta + t$ with X. The determinism and the totality of Δ together ensures that κ is a function.

Definition 3. [12] Let $\mathcal{A} = (Q, \Sigma, \mathcal{X}, \Delta, F)$ be a DTA. A *timed word* is an infinite sequence of timed signatures. The *run* of $\mathcal{A}$ on a timed word $w = \{(a_n, t_n)\}_{n \in \mathbb{N}_0}$ with *initial configuration* (q, η), denoted by $\mathcal{A}_{q,\eta}(w)$, is the unique infinite sequence $\{(q_n, \eta_n)(a_n, t_n)\}_{n \in \mathbb{N}_0}$ which satisfies that $(q_0, \eta_0) = (q, \eta)$ and $(q_{n+1}, \eta_{n+1}) = \kappa^{\mathcal{A}}((q_n, \eta_n), (a_n, t_n))$ for $n \geq 0$. A timed word w is *accepted by* $\mathcal{A}$ with initial configuration (q, η) (*abbr.* "w accepted by $\mathcal{A}_{q,\eta}$") iff $\mathcal{A}_{q,\eta}(w) = \{(q_n, \eta_n)(a_n, t_n)\}_{n \in \mathbb{N}_0}$ satisfies that $q_n \in F$ for some $n \geq 0$. Moreover, w is *accepted* by $\mathcal{A}_{q,\eta}$ *within* k *steps* ($k \geq 0$) iff $\mathcal{A}_{q,\eta}(w) = \{(q_n, \eta_n)(a_n, t_n)\}_{n \in \mathbb{N}_0}$ satisfies that $q_n \in F$ for some $0 \leq n \leq k$.

3. MEASURABILITY AND THE INTEGRAL EQUATIONS

In this section, we sketch our proofs for the measurability of the set of CTMC-paths accepted by a DTA and the system of integral equations that characterizes the acceptance probability. The notion of acceptance follows the previous ones in [12].

We fix a CTMC $\mathcal{M} = (S, L, \mathbf{P}, \lambda, \mathcal{L})$ and a DTA $\mathcal{A} = (Q, \Sigma, \mathcal{X}, \Delta, F)$ such that $\Sigma = L$. The notion of acceptance is defined as follows.

Definition 4. [12] The set of $\mathcal{M}$-paths *accepted* by $\mathcal{A}$ w.r.t $s \in S$, $q \in Q$ and $\eta \in \mathrm{Val}(\mathcal{X})$, denoted by $\mathsf{Path}^{\mathcal{M}\otimes\mathcal{A}}(s, q, \eta)$, is defined as the set of $\mathcal{M}$- paths π such that $\pi[0] = s$ and $\mathcal{L}_\pi$ is accepted by $\mathcal{A}_{q,\eta}$, where $\mathcal{L}_\pi$ is the timed word whose n-th signature is $(\mathcal{L}(\pi[n]), \pi\langle n\rangle)$ for all $n \geq 0$. Moreover, the set of $\mathcal{M}$-paths accepted by $\mathcal{A}$ w.r.t s, q and η *within* k*-steps* ($k \geq 0$), denoted by $\mathsf{Path}_k^{\mathcal{M}\otimes\mathcal{A}}(s, q, \eta)$, is defined as the set of $\mathcal{M}$-paths π such that $\pi[0] = s$ and $\mathcal{L}_\pi$ is accepted by $\mathcal{A}_{q,\eta}$ within k steps.

We omit "$\mathcal{M} \otimes \mathcal{A}$" in "$\mathsf{Path}^{\mathcal{M}\otimes\mathcal{A}}$" and "$\mathsf{Path}_k^{\mathcal{M}\otimes\mathcal{A}}$" if the underlying context is clear. Note that $\mathcal{L}_\pi$ specifies the behaviour of $\mathcal{M}$ observable by an outside observer. By definition, we have $\bigcup_{k \geq 0} \mathsf{Path}_k(s, q, \eta) = \mathsf{Path}(s, q, \eta)$.

Remark 1. We point out the main error in the measurability proof by Chen *et al.* [12]. The error appears on Page 11 under the label "(1b)" which handles the equality guards in timed transitions. In (1b), for an timed transition e emitted from q with guard $x = K$, four DTA $\mathcal{A}_e, \overline{\mathcal{A}}_e, \mathcal{A}_e^>, \mathcal{A}_e^<$ are defined w.r.t the original DTA $\mathcal{A}$. Then it is argued that

$$Paths^C(\mathcal{A}_e) = Paths^C(\overline{\mathcal{A}}_e) \backslash (Paths^C(\mathcal{A}_e^>) \cup Paths^C(\mathcal{A}_e^<))$$

This is incorrect. The left part $Paths^C(\mathcal{A}_e)$ excludes all timed paths which involve both the guard $x > K$ and the guard $x < K$ (from q). However the right part does not. So the left and right part are not equal.

We describe our main results in this section as follows. Let $\mathcal{V} := S \times Q \times \mathrm{Val}(\mathcal{X})$. We abbreviate

$$\left(\mathbf{g}^\eta_{q,\mathcal{L}(s)}, \mathbf{X}^\eta_{q,\mathcal{L}(s)}, \mathbf{q}^\eta_{q,\mathcal{L}(s)}\right)$$

as $(\mathbf{g}^{\eta}_{q,s}, \mathbf{X}^{\eta}_{q,s}, \mathbf{q}^{\eta}_{q,s})$. Given $(s,q,\eta) \in \mathcal{V}$ and $u \in S$, we define $(s,q,\eta)_u := (u, \mathbf{q}^{\eta}_{q,s}, \eta[\mathbf{X}^{\eta}_{q,s} := 0]) \ (\in \mathcal{V})$. We denote by $\langle J \rangle$ the characteristic function of a set J.

THEOREM 1. *For all $(s,q,\eta) \in \mathcal{V}$ and $k \geq 0$, $\mathsf{Path}_k(s,q,\eta)$ is measurable under Ω. Furthermore, the family*

$$\{\mathsf{prob}_k : \mathcal{V} \mapsto [0,1]\}_{k \geq 0} ,$$

where $\mathsf{prob}_k(s,q,\eta)$ is the probability mass of $\mathsf{Path}_k(s,q,\eta)$ under $\mathcal{B}_s$, satisfies the following properties: $\mathsf{prob}_0(s,q,\eta) = \langle F \rangle(q)$; If $q \in F$ then $\mathsf{prob}_{k+1}(s,q,\eta) = 1$, otherwise

$$\mathsf{prob}_{k+1}(s,q,\eta) =$$

$$\int_0^{+\infty} \left\{ \Lambda_s(t) \cdot \left[\sum_{u \in S} \mathbf{P}(s,u) \cdot \mathsf{prob}_k\left((s,q,\eta+t)_u\right) \right] \right\} \mathrm{d}t$$

We prove this result by decomposing $\mathsf{Path}_k(s,q,\eta)$ into disjoint path sets according to the first $k+1$ states of π and the first k rules from $\mathcal{A}_{q,\eta}(\mathcal{L}(\pi))$, for some $\pi \in \mathsf{Path}_k(s,q,\eta)$. From Theorem 1, one can prove the following corollary:

COROLLARY 1. *For all $(s,q,\eta) \in \mathcal{V}$, $\mathsf{Path}(s,q,\eta)$ is measurable under Ω. Moreover, the function $\mathsf{prob} : \mathcal{V} \mapsto [0,1]$, where $\mathsf{prob}(s,q,\eta)$ is the probability mass of $\mathsf{Path}(s,q,\eta)$ under $\mathcal{B}_s$, satisfies the following system of integral equations: If $q \in F$ then $\mathsf{prob}(s,q,\eta) = 1$, otherwise*

$$\mathsf{prob}(s,q,\eta) =$$

$$\int_0^{+\infty} \left\{ \Lambda_s(t) \cdot \left[\sum_{u \in S} \mathbf{P}(s,u) \cdot \mathsf{prob}\left((s,q,\eta+t)_u\right) \right] \right\} dt$$

4. EQUIVALENCE RELATIONS, PRODUCT REGION GRAPH AND LIPSCHITZ CONTINUITY

In this section we prepare several tools to derive the differential characterization for the function prob. In detail, we review several equivalence relations on clock valuations [1] and the product region graph between CTMC and DTA [12], and derive a Lipschitz Continuity of the function prob.

Below we fix a CTMC $\mathcal{M} = (S, L, \mathbf{P}, \lambda, \mathcal{L})$ and a DTA $\mathcal{A} = (Q, L, \mathcal{X}, \Delta, F)$. We denote by $T^{\mathcal{A}}_x$ the largest number c that appears in some guard $x \bowtie c$ of $\mathcal{A}$ on clock x, by $T^{\mathcal{A}}_{\max}$ the number $\max_{x \in \mathcal{X}} T^{\mathcal{A}}_x$, and define $\lambda^{\mathcal{M}}_{\max} := \max\{\lambda(s) \mid s \in S\}$. We omit $\mathcal{M}$ or $\mathcal{A}$ if the context is clear.

4.1 Equivalence Relations

Definition 5. [1] Two valuations $\eta, \eta' \in \mathrm{Val}(\mathcal{X})$ are *guard-equivalent*, denoted by $\eta \equiv_{\mathrm{g}} \eta'$, if they satisfy the following conditions:

1. for all $x \in \mathcal{X}$, $\eta(x) > T_x$ iff $\eta'(x) > T_x$;
2. for all $x \in \mathcal{X}$, if $\eta(x) \leq T_x$ and $\eta'(x) \leq T_x$, then (i) $\mathrm{int}(\eta(x)) = \mathrm{int}(\eta'(x))$ and (ii) $\mathrm{frac}(\eta(x)) > 0$ iff $\mathrm{frac}(\eta'(x)) > 0$.

where int(), frac() are the integral and fractional part of a real number, respectively. Moreover, η and η' are *equivalent*, denoted by $\eta \sim \eta'$, if (i) $\eta \equiv_{\mathrm{g}} \eta'$ and (ii) for all $x, y \in \mathcal{X}$, if $\eta(x), \eta'(x) \leq T_x$ and $\eta(y), \eta'(y) \leq T_y$, then $\mathrm{frac}(\eta(x)) \bowtie \mathrm{frac}(\eta(y))$ iff $\mathrm{frac}(\eta'(x)) \bowtie \mathrm{frac}(\eta'(y))$ for all $\bowtie \in \{<, =, >\}$. We will call equivalence classes of $\sim$ *regions*. Given a region $[\eta]_\sim$, we say that $[\eta]_\sim$ is *marginal* if $\eta(x) \leq T_x$ and $\mathrm{frac}(\eta(x)) = 0$ for some clock x.

In other words, equivalence classes of $\equiv_{\mathrm{g}}$ are captured by a boolean vector over $\mathcal{X}$ which indicates whether $\eta(x) > T_x$, an integer vector which indicates the integral parts on $\eta(x) \leq T_x$ and a boolean vector which indicates whether $\eta(x)$ is an integer when $\eta(x) \leq T_x$; equivalence classes of $\sim$ is further captured by a linear order on the set $\{x \in \mathcal{X} \mid \eta(x) \leq T_x\}$ w.r.t $\mathrm{frac}(\eta(x))$. Below we state some basic properties of $\equiv_{\mathrm{g}}$ and $\sim$.

PROPOSITION 1. *[1] The following properties on $\equiv_{\mathrm{g}}$ and $\sim$ hold:*

1. *Both $\equiv_{\mathrm{g}}$ and $\sim$ is an equivalence relation over clock valuations, and has finite index;*
2. *if $\eta \equiv_{\mathrm{g}} \eta'$ then they satisfy the same set of guards that appear in $\mathcal{A}$;*
3. *If $\eta \sim \eta'$ then*
   - *for all $t > 0$, there exists $t' > 0$ such that $\eta + t \sim \eta' + t'$, and*
   - *for all $t' > 0$, there exists $t > 0$ such that $\eta + t \sim \eta' + t'$.*
4. *If $\eta \sim \eta'$, then $\eta[X := 0] \sim \eta'[X := 0]$ for all $X \subseteq \mathcal{X}$. Moreover, for all $\eta \in \mathrm{Val}(\mathcal{X})$ and $X \subseteq \mathcal{X}$, $\{\eta'[X := 0] \mid \eta' \in [\eta]_\sim\}$ is a region.*

Besides these two equivalence notions, we define another finer equivalence notion as follows.

Definition 6. Two valuations $\eta, \eta' \in \mathrm{Val}(\mathcal{X})$ are *bound-equivalent*, denoted by $\eta \equiv_{\mathrm{b}} \eta'$, if for all $x \in \mathcal{X}$, either $\eta(x) > T_x$ and $\eta'(x) > T_x$, or $\eta(x) = \eta'(x)$.

It is straightforward to verify that $\equiv_{\mathrm{b}}$ is an equivalence relation. The following proposition specifies the relation between $\equiv_{\mathrm{b}}$ and prob, see Barbot *et al.* [5].

PROPOSITION 2. *Let $s \in S$, $q \in Q$ and $\eta, \eta' \in \mathrm{Val}(\mathcal{X})$. If $\eta \equiv_{\mathrm{b}} \eta'$, then $\mathsf{prob}(s,q,\eta) = \mathsf{prob}(s,q,\eta')$.*

In the following, we further introduce a useful proposition.

PROPOSITION 3. *For each $\eta \in \mathrm{Val}(\mathcal{X})$, there exists $t_1 > 0$ such that $\eta + t \sim \eta + t'$ for all $t, t' \in (0, t_1)$. For each $\eta \in \mathrm{Val}(\mathcal{X})$ such that $\eta(x) > 0$ for all $x \in \mathcal{X}$, there exists $t_2 > 0$ such that $\eta - t \sim \eta - t'$ for all $t, t' \in (0, t_2)$. Moreover, if $[\eta]_\sim$ is non-marginal, then for all $t \in (0, t_1)$ and $t' \in (0, t_2)$, $\eta \sim \eta + t$ and $\eta \sim \eta - t'$.*

We denote η^+ to be a representative in $\{\eta + t \mid t \in (0, t_1)\}$, and η^- to be a representative in $\{\eta - t \mid t \in (0, t_2)\}$, where t_1, t_2 are specified in Proposition 3. The choice among the representatives will be irrelevant because they are equivalent under $\sim$. Note that if $[\eta]_\sim$ is not marginal, then $[\eta]_\sim = [\eta^+]_\sim = [\eta^-]_\sim$.

4.2 The Product Region Graph

We define a qualitative variation of the product region graph proposed by Chen *et al.* [12], mainly to derive a qualitative property of the function prob.

Definition 7. The *product region graph*

$$G^{\mathcal{M} \otimes \mathcal{A}} = (V^{\mathcal{M} \otimes \mathcal{A}}, E^{\mathcal{M} \otimes \mathcal{A}})$$

of $\mathcal{M}$ and $\mathcal{A}$ is a directed graph defined as follows: $V = S \times Q \times (\mathrm{Val}(\mathcal{X})/\sim)$, $((s,q,r),(s',q',r')) \in E$ iff (i) $\mathbf{P}(s,s') > 0$ and (ii) there exists $\eta \in r, \eta' \in r'$ and $t > 0$ such that $[\eta + t]_\sim$ is not a marginal region and $(q', \eta') = \kappa((q,\eta), (\mathcal{L}(s), t))$. We say that a vertex $(s,q,r) \in V$ is *final* if $q \in F$.

We will omit $\mathcal{M} \otimes \mathcal{A}$ in $G^{\mathcal{M}\otimes\mathcal{A}} = (V^{\mathcal{M}\otimes\mathcal{A}}, E^{\mathcal{M}\otimes\mathcal{A}})$ if the context is clear. The following proposition states the relationship between prob and the product region graph.

PROPOSITION 4. *For all $(s,q,\eta) \in \mathcal{V}$, $\mathsf{prob}(s,q,\eta) > 0$ iff the vertex $(s,q,[\eta]_\sim)$ can reach some final vertex in G.*

4.3 Lipschitz Continuity

Below we prove a Lipschitz Continuity property of the function prob. More specifically, we prove that all functions that satisfies a boundness condition related to $\equiv_\mathrm{b}$ and the system of integral equations specified in Corollary 1 are Lipschitz continuous. The Lipschitz continuity will be fundamental to our differential characterization and the error bound of our approximation result.

THEOREM 2. *Let $h : \mathcal{V} \mapsto [0,1]$ be a function which satisfies the following conditions for all $s \in S$, $q \in Q$ and $\eta, \eta' \in \mathrm{Val}(\mathcal{X})$:*

- *if $\eta \equiv_\mathrm{b} \eta'$ then $h(s,q,\eta) = h(s,q,\eta')$;*
- *if $q \in F$ then $h(s,q,\eta) = 1$, otherwise $h(s,q,\eta)$ is equal to the integral*

$$\int_0^{+\infty} \left\{ \Lambda_s(t) \cdot \left[\sum_{u \in S} \mathbf{P}(s,u) \cdot h\left((s,q,\eta+t)_u\right) \right] \right\} \mathrm{d}t$$

Then for all $s \in S$, $q \in Q$ and $\eta, \eta' \in \mathrm{Val}(\mathcal{X})$, if $\|\eta - \eta'\|_\infty < 1$ then

$$|h(s,q,\eta) - h(s,q,\eta')| \le M_1 \cdot \|\eta - \eta'\|_\infty ,$$

where $M_1 := |\mathcal{X}| \cdot \lambda_{\max} T_{\max} \cdot e^{\lambda_{\max} T_{\max}}$.

PROOF. (Sketch) If $q \in F$, then the result follows from $h(s,q,\eta) = h(s,q,\eta') = 1$. From now on we suppose that $q \notin F$. To prove the theorem, it suffices to prove that

$$|h(s,q,\eta) - h(s,q,\eta')| \le \lambda_{\max} T_{\max} \cdot e^{\lambda_{\max} T_{\max}} \cdot \|\eta - \eta'\|_\infty$$

if $\|\eta - \eta'\|_\infty < 1$ and η, η' differ only on one clock, i.e.

$$|\{x \in \mathcal{X} \mid \eta(x) \neq \eta'(x)\}| = 1 .$$

To this end we define $\delta(\epsilon)$ for each $\epsilon \in (0,1)$ as follows:

$$\begin{aligned}\delta(\epsilon) := \sup \{ & |h(s,q,\eta) - h(s,q,\eta')| \mid \\ & (s,q) \in S \times Q, \eta, \eta' \in \mathrm{Val}(\mathcal{X}), \|\eta - \eta'\|_\infty \le \epsilon \\ & \text{and } \eta, \eta' \text{ differ only on one clock}\}\end{aligned}$$

Note that for all $\eta, \eta' \in \mathrm{Val}(\mathcal{X})$ and $X \subseteq \mathcal{X}$:

- if η and η' differ at most on one clock, then so are $\eta[X := 0]$ and $\eta'[X := 0]$;
- $\|\eta[X := 0] - \eta'[X := 0]\|_\infty \le \|\eta - \eta'\|_\infty$.

Suppose $(s,q) \in S \times Q$ and $\eta, \eta' \in \mathrm{Val}(\mathcal{X})$ which satisfies $\|\eta - \eta'\|_\infty \le \epsilon < 1$ and differ only on the clock x, i.e., $\eta(x) \neq \eta'(x)$ and $\eta(y) = \eta'(y)$ for all $y \neq x$. W.l.o.g we can assume that $\eta(x) < \eta'(x)$. We clarify two cases below.

Case 1: $\mathrm{int}(\eta(x)) = \mathrm{int}(\eta'(x))$. Then by $\eta(x) < \eta'(x)$, we have $\mathrm{frac}(\eta(x)) < \mathrm{frac}(\eta'(x))$. Consider the "behaviours" of $\eta + t$ and $\eta' + t$ when t goes from 0 to ∞. We divide $[0, \infty)$ into open integer intervals $(0,1), (1,2), \ldots, (T_{\max} - 1, T_{\max})$ and $(T_{\max}, \infty)$. For each $n < T_{\max}$, we further divide the interval $(n, n+1)$ into the following open sub-intervals:

$$\begin{aligned}&(n, n+1-\mathrm{frac}(\eta'(x))), (n+1-\mathrm{frac}(\eta(x)), n+1)\\ &\qquad (n+1-\mathrm{frac}(\eta'(x)), n+1-\mathrm{frac}(\eta(x)))\end{aligned}$$

One can observe that for $t \in (n, n+1-\mathrm{frac}(\eta'(x))) \cup (n+1-\mathrm{frac}(\eta(x)), n+1)$, we have $\eta + t \equiv_\mathrm{g} \eta' + t$, which implies that $\eta + t$ and $\eta' + t$ satisfies the same set of guards in $\mathcal{A}$. However for $t \in (n+1-\mathrm{frac}(\eta'(x)), n+1-\mathrm{frac}(\eta(x)))$, it may be the case that $\eta + t \not\equiv_\mathrm{g} \eta' + t$ due to their difference on clock x. Thus the total length for t within $(n, n+1)$ such that $\eta + t \not\equiv_\mathrm{g} \eta' + t$ is smaller than $|\eta(x) - \eta'(x)|$. Thus we have (†):

$$\begin{aligned}\delta_n &:= \left| \int_n^{n+1} \Lambda_s(t) \cdot \left\{ \sum_{u \in S} \left[\mathbf{P}(s,u) \cdot h\Big((s,q,\eta+t)_u\Big) \right.\right.\right.\\ &\qquad \left.\left.\left. - \mathbf{P}(s,u) \cdot h\Big((s,q,\eta'+t)_u\Big) \right] \right\} \mathrm{d}t \right|\\ &\le \int_n^{n+1} \left\{ \lambda(s) e^{-\lambda(s)t} \cdot \delta(\epsilon) \right\} \mathrm{d}t\\ &\qquad + \lambda(s) e^{-\lambda(s)n} \cdot |\eta(x) - \eta'(x)|\\ &\le \delta(\epsilon) \cdot \int_n^{n+1} \left\{ \lambda(s) e^{-\lambda(s)t} \right\} \mathrm{d}t + \lambda(s) e^{-\lambda(s)n} \cdot \epsilon\end{aligned}$$

Note that for all $t \in (T_{\max}, \infty)$ and $X \subseteq \mathcal{X}$,

$$(\eta + t)[X := 0] \equiv_\mathrm{b} (\eta' + t)[X := 0] .$$

This implies $h((s,q,\eta+t)_u) = h((s,q,\eta'+t)_u)$. Therefore we have (‡):

$$\begin{aligned}& |h(s,q,\eta) - h(s,q,\eta')|\\ \le\ & \sum_{n=0}^{T_{\max}-1} \delta_n\\ \le\ & \delta(\epsilon) \cdot \int_0^{T_{\max}} \left\{ \lambda(s) e^{-\lambda(s)t} \right\} \mathrm{d}t + \lambda(s)\epsilon \cdot \sum_{n=0}^{T_{\max}-1} e^{-\lambda(s)n}\\ \le\ & \delta(\epsilon) \cdot (1 - e^{-\lambda(s) T_{\max}}) + \epsilon \cdot \lambda(s) \cdot T_{\max}\\ \le\ & \delta(\epsilon) \cdot (1 - e^{-\lambda_{\max} T_{\max}}) + \epsilon \cdot \lambda_{\max} \cdot T_{\max}\end{aligned}$$

Case 2: $\mathrm{int}(\eta(x)) < \mathrm{int}(\eta'(x))$. By $|\eta(x) - \eta'(x)| < 1$, we have $\mathrm{int}(\eta(x)) + 1 = \mathrm{int}(\eta'(x))$ and $\mathrm{frac}(\eta'(x)) < \mathrm{frac}(\eta(x))$. By similar arguments we can also obtain that

$$|h(s,q,\eta) - h(s,q,\eta')| \le \delta(\epsilon) \cdot (1 - e^{-\lambda_{\max} T_{\max}}) + \epsilon \cdot \lambda_{\max} \cdot T_{\max}$$

Thus by the definition of $\delta(\epsilon)$, we obtain

$$\delta(\epsilon) \le \delta(\epsilon) \cdot (1 - e^{-\lambda_{\max} T_{\max}}) + \epsilon \cdot \lambda_{\max} \cdot T_{\max}$$

which implies $\delta(\epsilon) \le \epsilon \cdot e^{\lambda_{\max} T_{\max}} \cdot \lambda_{\max} \cdot T_{\max}$. By letting $\epsilon = \|\eta - \eta'\|_\infty$, we obtain the desired result. □

Then by Corollary 1, Proposition 2 and Theorem 2, we obtain:

COROLLARY 2. *For all $(s,q) \in S \times Q$ and $\eta, \eta' \in \mathrm{Val}(\mathcal{X})$, if $\|\eta - \eta'\|_\infty < 1$ then*

$$|\mathsf{prob}(s,q,\eta) - \mathsf{prob}(s,q,\eta')| \le M_1 \cdot \|\eta - \eta'\|_\infty$$

where M_1 is defined as in Theorem 2.

5. DIFFERENTIAL CHARACTERIZATION

In this section we present a differential characterization for the function prob. We fix a CTMC $\mathcal{M} = (S, L, \mathbf{P}, \lambda, \mathcal{L})$ and a DTA $\mathcal{A} = (Q, L, \mathcal{X}, \Delta, F)$. We extend $\lambda(.)$ and $\mathbf{P}(.,.)$ to a triple $(s,q,\eta) \in \mathcal{V}$ by: $\lambda(s,q,\eta) = \lambda(s)$ and $\mathbf{P}((s,q,\eta),u) =$

$\mathbf{P}(s,u)$ for $u \in S$. We also extend $+$ to a triple $(s,q,\eta) \in \mathcal{V}$ by: $(s,q,\eta)+t=(s,q,\eta+t)$.

Below we introduce our notion of derivative, which is a directional derivative as follows.

Definition 8. Given a function $h:\mathcal{V}\mapsto[0,1]$, we denote by $\nabla_{\mathbf{1}}^{+}h$ and resp. $\nabla_{\mathbf{1}}^{-}h$ the right directional derivative and resp. the left directional derivative of h along the direction $\mathbf{1}$ if the derivative exists. Formally, we define

- $\nabla_{\mathbf{1}}^{+}h(s,q,\eta) := \lim_{t\to 0^+}\frac{1}{t}\cdot(h(s,q,\eta+t)-h(s,q,\eta))$, if the limit exists.
- $\nabla_{\mathbf{1}}^{-}h(s,q,\eta) := \lim_{t\to 0^+}\frac{1}{t}\cdot(h(s,q,\eta)-h(s,q,\eta-t))$, if $\eta(x)>0$ for all $x\in\mathcal{X}$ and the limit exists.

for each $(s,q,\eta)\in\mathcal{V}$.

Below we calculate these directional derivatives. Given a triple $(s,q,\eta)\in\mathcal{V}$ and $u\in S$, we denote by $(s,q,\eta)_u^{+}$ the triple $(u,\mathbf{q}_{q,s}^{\eta^+},\eta[\mathbf{X}_{q,s}^{\eta^+}:=0])$ $(\in\mathcal{V})$, and by $(s,q,\eta)_u^{-}$ the triple $(u,\mathbf{q}_{q,s}^{\eta^-},\eta[\mathbf{X}_{q,s}^{\eta^-}:=0])$ when $\eta(x)>0$ for all clocks x.

THEOREM 3. *For all $(s,q,\eta)\in\mathcal{V}$ with $q\notin F$, the directional derivative $\nabla_{\mathbf{1}}^{+}\mathsf{prob}(s,q,\eta)$ exists, and $\nabla_{\mathbf{1}}^{-}\mathsf{prob}(s,q,\eta)$ exists given that $\eta(x)>0$ for all $x\in\mathcal{X}$. Furthermore, for all $v\in\mathcal{V}$,*

$$\nabla_{\mathbf{1}}^{+}\mathsf{prob}(v)=\lambda(v)\cdot\mathsf{prob}(v)-\lambda(v)\cdot\sum_{u\in S}\mathbf{P}(v,u)\cdot\mathsf{prob}(v_u^{+})$$

and

$$\nabla_{\mathbf{1}}^{-}\mathsf{prob}(v)=\lambda(v)\cdot\mathsf{prob}(v)-\lambda(v)\cdot\sum_{u\in S}\mathbf{P}(v,u)\cdot\mathsf{prob}(v_u^{-})$$

whenever $\nabla_{\mathbf{1}}^{-}\mathsf{prob}(v)$ exists.

PROOF. We first prove the case for $\nabla_{\mathbf{1}}^{+}\mathsf{prob}(v)$. Denote $v=(s,q,\eta)$. By Corollary 1,

$$\mathsf{prob}(v+t)=$$
$$\int_0^{+\infty}\left\{\Lambda_s(\tau)\cdot\left[\sum_{u\in S}\mathbf{P}(s,u)\cdot\mathsf{prob}((v+(t+\tau))_u)\right]\right\}\mathrm{d}\tau$$

for $t\ge 0$. Note that the integrand function is Riemann integratable since it is piecewise continuous on τ. By the variable substitution $\tau'=t+\tau$, we have for $t\ge 0$,

$$\mathsf{prob}(v+t)=e^{\lambda(s)\cdot t}\cdot$$
$$\int_t^{+\infty}\left\{\Lambda_s(\tau)\cdot\left[\sum_{u\in S}\mathbf{P}(s,u)\cdot\mathsf{prob}((v+\tau)_u)\right]\right\}\mathrm{d}\tau$$

Then we have

$$\begin{aligned}
&\mathsf{prob}(v+t)-\mathsf{prob}(v)\\
=\;& e^{\lambda(s)\cdot t}\cdot\int_t^{+\infty}\left\{\Lambda_s(\tau)\cdot\left[\sum_{u\in S}\mathbf{P}(s,u)\cdot\mathsf{prob}((v+\tau)_u)\right]\right\}\mathrm{d}\tau\\
&-\int_0^{+\infty}\left\{\Lambda_s(\tau)\cdot\left[\sum_{u\in S}\mathbf{P}(s,u)\cdot\mathsf{prob}((v+\tau)_u)\right]\right\}\mathrm{d}\tau\\
=\;&(e^{\lambda(s)\cdot t}-1)\cdot\int_t^{+\infty}\left\{\Lambda_s(\tau)\cdot\left[\sum_{u\in S}\mathbf{P}(s,u)\cdot\mathsf{prob}((v+\tau)_u)\right]\right\}\mathrm{d}\tau\\
&-\int_0^{t}\left\{\Lambda_s(\tau)\cdot\left[\sum_{u\in S}\mathbf{P}(s,u)\cdot\mathsf{prob}((v+\tau)_u)\right]\right\}\mathrm{d}\tau
\end{aligned}$$

By Proposition 3, there exists $t_1>0$ such that $\mathbf{q}_{q,s}^{\eta+\tau}$ and $\mathbf{X}_{q,s}^{\eta+\tau}$ does not change for $\tau\in(0,t_1)$. Thus the integrand function in the integral

$$\int_0^t\left\{\Lambda_s(\tau)\cdot\left[\sum_{u\in S}\mathbf{P}(s,u)\cdot\mathsf{prob}((v+\tau)_u)\right]\right\}\mathrm{d}\tau$$

is continuous on τ when $t\in(0,t_1)$. Thus by L'Hôspital's Rule and the fact that $\lim_{t\to 0^+}(v+t)_u=v_u^{+}$, we obtain

$$\nabla_{\mathbf{1}}^{+}\mathsf{prob}(v)=\lambda(v)\cdot\mathsf{prob}(v)-\lambda(v)\cdot\sum_{u\in S}\mathbf{P}(v,u)\cdot\mathsf{prob}(v_u^{+})$$

The case for $\nabla_{\mathbf{1}}^{-}\mathsf{prob}(v)$ can be handled analogously. □

By some further analysis, we can obtain the following theorem.

THEOREM 4. *The function* prob *is the unique solution of the following system of differential equations on $h:\mathcal{V}\mapsto[0,1]$: given any $s\in S, q\in Q$ and $\eta,\eta'\in\mathrm{Val}(\mathcal{X})$,*

1. *if $\eta\equiv_{\mathrm{b}}\eta'$ then $h(s,q,\eta)=h(s,q,\eta')$;*
2. *$h(s,q,\eta)=0$ if $(s,q,[\eta]_\sim)$ cannot reach a final vertex in G, and $h(s,q,\eta)=1$ if $q\in F$;*
3. *if $(s,q,[\eta]_\sim)$ can reach a final vertex in G and $q\notin F$, then*
$$\nabla_{\mathbf{1}}^{+}h(v)=\lambda(v)\cdot h(v)-\lambda(v)\cdot\sum_{u\in S}\mathbf{P}(v,u)\cdot h(v_u^{+})$$
where $v:=(s,q,\eta)$, and
$$\nabla_{\mathbf{1}}^{-}h(v)=\lambda(v)\cdot h(v)-\lambda(v)\cdot\sum_{u\in S}\mathbf{P}(v,u)\cdot h(v_u^{-})$$
when $v:=(s,q,\eta)$ and $\eta(x)>0$ for all $x\in\mathcal{X}$.

6. FINITE DIFFERENCE METHODS

In this section, we deal with the approximation of the function prob through finite approximation schemes. We establish our approximation scheme based on Theorem 4 and by finite difference methods [18]. Then we prove that our approximation scheme converges to prob with a derived error bound.

We fix a CTMC $\mathcal{M}=(S,L,\mathbf{P},\lambda,\mathcal{L})$ and a DTA $\mathcal{A}=(Q,L,\mathcal{X},\Delta,F)$. For computational purpose we assume that all numerical values in $\mathcal{M}$ are rational.

Given valuation η and $t\ge 0$, we define $\eta\oplus t\in\mathrm{Val}(\mathcal{X})$ by: $(\eta\oplus t)(x)=\min\{T_x,\eta(x)+t\}$ for all $x\in\mathcal{X}$. Note that $\eta\oplus 0=\eta$ iff $\eta(x)\le T_x$ for all clocks x.

We extend $\oplus$ and $[\centerdot]_\sim$ to a triple $(s,q,\eta)\in\mathcal{V}$ as follows: $(s,q,\eta)\oplus t=(s,q,\eta\oplus t)$ and $[(s,q,\eta)]_\sim=(s,q,[\eta]_\sim)$. Moreover, we say that $[(s,q,\eta)]_\sim$ is *marginal* if $[\eta]_\sim$ is marginal. Note that by Lipschitz Continuity and Proposition 2, we have $\mathsf{prob}(v+t)=\mathsf{prob}(v\oplus t)$ for all $v\in\mathcal{V}$ and $t\ge 0$.

6.1 Approximation Schemes

We establish our approximation scheme in two steps: firstly, we discretize the hypercube $\prod_{x\in\mathcal{X}}[0,T_x]\subseteq\mathrm{Val}(\mathcal{X})$ into small grids; secondly, we establish our approximation scheme by building constraints between these discrete values through finite difference methods. From Lipschitz Continuity and Proposition 2, we do not need to consider valuations outside the hypercube $\prod_{x\in\mathcal{X}}[0,T_x]$. The discretization is as follows.

Definition 9. Let $m \in \mathbb{N}$. A clock valuation η is *on m-grid* if $\eta(x) \in [0, T_x]$ and $\eta(x) \cdot m$ is an integer for all clocks x. The set of discrete values D_m of concern is defined as follows:

$$\mathsf{D}_m = \{\mathsf{h}[(s,q,\eta)] \mid (s,q,\eta) \in \mathcal{V} \text{ and } \eta \text{ is on } m\text{-grid}\} .$$

Below we fix a $m \in \mathbb{N}$ and define $\rho := m^{-1}$. Based on Theorem 4, we establish our basic approximation scheme Γ_m, as follows.

Definition 10. The approximation scheme Γ_m consists of the discrete values D_m and a system of linear equations on D_m. The system of linear equations contains one of the following equations for each $\mathsf{h}[v] \in \mathsf{D}_m$:

- $\mathsf{h}[v] = 0$ if $[v]_\sim$ (as a vertex of G) cannot reach a final vertex in G;
- $\mathsf{h}[v] = 1$ if $[v]_\sim$ is a final vertex in G;
- If $[v]_\sim$ can reach a final vertex in G however itself is not a final vertex, then

$$\frac{\mathsf{h}[v \oplus \rho] - \mathsf{h}[v]}{\rho} = \lambda(v) \cdot \mathsf{h}[v] - \lambda(v) \cdot \sum_{u \in S} \mathbf{P}(v,u) \cdot \mathsf{h}[v_u^+] .$$

In other words, we relate elements of D_m by using $\nabla_1^+\mathsf{h}$ in Theorem 4. Note that $\mathsf{h}[v]$ is in essence v. Sometimes we will not distinguish between $\mathsf{h}[v]$ and v.

We note that Γ_m does not have initial values from which we can approximate prob incrementally. One fundamental problem is whether Γ_m has a solution, or even a unique solution. Another fundamental problem is the error bound $\max\{|h^*[v] - \mathsf{prob}(v)| \mid \mathsf{h}[v] \in \mathsf{D}_m\}$ provided that h^* is the unique solution of Γ_m.

Below we first derive the *error bound* of Γ_m which is the error bound of each linear equality when we substitute all $\mathsf{h}[v]$ by $\mathsf{prob}(v)$. Note that generally the error bound of an approximation scheme does not imply any information of the error bound of the solution to the approximation scheme.

Proposition 5. *For all $\mathsf{h}[v] \in \mathsf{D}_m$, if $[v]_\sim$ is not a final vertex and can reach some final vertex in G then*

$$\left|\frac{1}{\rho} \cdot (\mathsf{prob}(v \oplus \rho) - \mathsf{prob}(v)) - \nabla_1^+\mathsf{prob}(v)\right| < M_2 \cdot \rho ,$$

where $M_2 := 2\lambda_{\max} \cdot M_1$.

Proof. Suppose $\mathsf{h}[v] \in \mathsf{D}_m$ such that $[v]_\sim$ is not a final vertex and can reach some final vertex in G. Since $\mathsf{h}[v] \in \mathsf{D}_m$, the function $f[v](t) : t \mapsto \mathsf{prob}(v \oplus t)$ is continuous on $[0, \rho]$ and is differentiable on $(0, \rho)$. By Lagrange's Mean Value Theorem, there exists $\rho' \in (0, \rho)$ such that

$$\frac{1}{\rho} \cdot (\mathsf{prob}(v \oplus \rho) - \mathsf{prob}(v)) = \frac{\mathrm{d}}{\mathrm{d}t} f[v](\rho') .$$

By Theorem 3, we have

$$\frac{\mathrm{d}}{\mathrm{d}t} f[v](\rho') = \lambda(v) \cdot \mathsf{prob}(v + \rho') - \lambda(v) \cdot \sum_{u \in S} \mathbf{P}(v,u) \cdot \mathsf{prob}((v + \rho')_u^+)$$

and

$$\nabla_1^+\mathsf{prob}(v) = \lambda(v) \cdot \mathsf{prob}(v) - \lambda(v) \cdot \sum_{u \in S} \mathbf{P}(v,u) \cdot \mathsf{prob}(v_u^+)$$

Then by Corollary 2, we obtain the desired result. □

To analyze Γ_m, we further define several auxiliary approximation schemes. Below we define $\mathsf{B}_m, \mathsf{B}_m^{\max}$ as follows:

$$\mathsf{B}_m = \{\mathsf{h}[v] \in \mathsf{D}_m \mid [v]_\sim \text{ is not final and can reach some final vertex in } G\}$$
$$\mathsf{B}_m^{\max} = \{\mathsf{h}[v] \in \mathsf{B}_m \mid v = (s,q,\eta) \text{ and } \eta(x) = T_x \text{ for all } x \in \mathcal{X}\}$$

For each $\mathsf{h}[v] \in \mathsf{B}_m$, we denote by $N_v \in \mathbb{N}_0$ the minimum number such that either $\mathsf{h}\,[v \oplus (N_v \cdot \rho)] \in \mathsf{B}_m^{\max}$ or $[v \oplus (N_v \cdot \rho)]_\sim$ cannot reach some final vertex in G. We first transform Γ_m into an equivalent form.

Definition 11. The approximation scheme Γ'_m consists of the discrete values D_m, and the system of linear equations which contains one of the following linear equalities for each $\mathsf{h}[v] \in \mathsf{D}_m$:

- $\mathsf{h}[v] = 0$ if $[v]_\sim$ cannot reach a final vertex in G;
- $\mathsf{h}[v] = 1$ if $[v]_\sim$ is a final vertex of G.
- If $\mathsf{h}[v] \in \mathsf{B}_m \backslash \mathsf{B}_m^{\max}$, then

$$\mathsf{h}[v] = \frac{1}{1 + \rho \cdot \lambda(v)} \cdot \mathsf{h}[v \oplus \rho] + \frac{\rho \cdot \lambda(v)}{1 + \rho \cdot \lambda(v)} \cdot \sum_{u \in S} \mathbf{P}(v,u) \cdot \mathsf{h}[v_u^+] \quad (1)$$

- if $\mathsf{h}[v] \in \mathsf{B}_m^{\max}$ then $\mathsf{h}[v] = \sum_{u \in S} \mathbf{P}(v,u) \cdot \mathsf{h}[v_u^+]$.

It is clear that Γ'_m is just a re-formulation of Γ_m. Note that the case $\mathsf{h}[v] \in \mathsf{B}_m^{\max}$ is derived from $v \oplus \rho = v$. The error bound of Γ'_m is as follows.

Proposition 6. *For all $\mathsf{h}[v] \in \mathsf{B}_m^{\max}$,*

$$\mathsf{prob}(v) = \sum_{u \in S} \mathbf{P}(v,u) \cdot \mathsf{prob}(v_u^+) .$$

For all $\mathsf{h}[v] \in \mathsf{B}_m \backslash \mathsf{B}_m^{\max}$, $|\mathsf{prob}(v) - z| < M_2\rho^2$, where z equals

$$\frac{1}{1 + \rho \cdot \lambda(v)} \cdot \mathsf{prob}(v \oplus \rho) + \frac{\rho \cdot \lambda(v)}{1 + \rho \cdot \lambda(v)} \cdot \sum_{u \in S} \mathbf{P}(v,u) \cdot \mathsf{prob}\,(v_u^+)$$

Proof. The case $\mathsf{h}[v] \in \mathsf{B}_m^{\max}$ is due to the fact that $\nabla_1^+\mathsf{prob}(v) = 0$. The case $\mathsf{h}[v] \in \mathsf{B}_m \backslash \mathsf{B}_m^{\max}$ can be directly derived from the statement of Proposition 5. □

Below we unfold Γ'_m into another equivalent form Γ''_m.

Definition 12. The approximation scheme Γ''_m consists of the discrete values D_m, and one of the following linear equality for each $\mathsf{h}[v] \in \mathsf{D}_m$:

- $\mathsf{h}[v] = 0$ if $[v]_\sim$ cannot reach some final vertex in G, and $\mathsf{h}[v] = 1$ if $[v]_\sim$ is a final vertex;
- if $\mathsf{h}[v] \in \mathsf{B}_m \backslash \mathsf{B}_m^{\max}$, then

$$\mathsf{h}[v] = \sum_{l=0}^{N_v - 1} \left\{ \left(\frac{1}{1 + \rho \cdot \lambda(v)}\right)^l \cdot \frac{\rho \cdot \lambda(v)}{1 + \rho \cdot \lambda(v)} \cdot \sum_{u \in S} \mathbf{P}(v,u) \cdot \mathsf{h}\left[(v \oplus (l \cdot \rho))_u^+\right] \right\} + \left(\frac{1}{1 + \rho \cdot \lambda(v)}\right)^{N_v} \cdot f(v) \quad (2)$$

where $f(v) := 0$ if $[v \oplus (N_v \cdot \rho)]_\sim$ cannot reach some final vertex in G and $f(v) := \sum_{u \in S} \mathbf{P}(v,u) \cdot \mathsf{h}[(v \oplus (N_v \cdot \rho))_u^+]$ if $\mathsf{h}\,[v \oplus (N_v \cdot \rho)] \in \mathsf{B}_m^{\max}$;

- if $\mathsf{h}[v] \in \mathsf{B}_m^{\max}$ then $\mathsf{h}[v] = \sum_{u \in S} \mathbf{P}(v,u) \cdot \mathsf{h}[v_u^+]$.

Intuitively, Γ''_m is obtained by unfolding $\mathsf{h}[v \oplus \rho]$ further in Equation (1) whenever $v \oplus \rho \in \mathsf{B}_m \backslash \mathsf{B}_m^{\max}$. It is not hard to prove that Γ'_m and Γ''_m are equivalent.

PROPOSITION 7. *Γ'_m and Γ''_m are equivalent, i.e., they have the same set of solutions.*

The following proposition can be derived from Proposition 6.

PROPOSITION 8. *The error bound of Γ''_m is $M_3 \cdot \rho$, where $M_3 = T_{\max} \cdot M_2$.*

6.2 Analysis of the Approximation Schemes

Below we analyse the approximation schemes proposed in the previous subsection. We fix some $m \in \mathbb{N}$ and $\rho := m^{-1}$. We define $\lambda_{\min} := \min\{\lambda(s) \mid s \in S\}$ and $p_{\min} := \min\{\mathbf{P}(s,u) \mid s, u \in S, \mathbf{P}(s,u) > 0\}$. Note that $\lambda_{\min} > 0$.

We describe Γ''_m by a matrix equation $\mu = \mathbf{A}\mu + \mathbf{b}$ where μ is the vector over B_m to be solved, $\mathbf{b} : \mathsf{B}_m \mapsto \mathbb{R}$ is a vector and $\mathbf{A} : \mathsf{B}_m \times \mathsf{B}_m \mapsto \mathbb{R}$ is a matrix. More specifically, the row $\mathbf{A}(\mathsf{h}[v], -)$ is specified by the coefficents on $\mathsf{h}[v'] \in \mathsf{B}_m$ in Equation (2); the value $\mathbf{b}(\mathsf{h}[v])$ is the sum of the values over $\mathsf{D}_m \backslash \mathsf{B}_m$ in Equation (2). The exact permutation among B_m is irrelevant. Analogously, we describe Γ'_m by a matrix equation $\mu = \mathbf{C}\mu + \mathbf{d}$.

Below we analyse the equation $\mu = \mathbf{A}\mu + \mathbf{b}$. To this end, we first reproduce (on CTMC and DTA) the notions of δ-*seperateness* and δ-*wideness*, which is originally discovered by Brazdil *et al.* [10] on semi-Markov processes and DTA.

Below we define the transition relation $\xrightarrow{t}$ over $\mathcal{V}$ by: $(s,q,\eta) \xrightarrow{t} (u,q',\eta')$ iff $\mathbf{P}(s,u) > 0$ and $(q,\eta) \xrightarrow{\mathcal{L}(s),t} (q',\eta')$. Let $\mathcal{R}_\eta := \{0,1\} \cup \{\mathrm{frac}(\eta(x)) \mid x \in \mathcal{X} \text{ and } \eta(x) \le T_x\}$ for each $\eta \in \mathrm{Val}(\mathcal{X})$. Intuitively, $\mathcal{R}_\eta$ captures the fractional values on η.

Definition 13. A clock valuation η is δ-*separated* if for all $d_1, d_2 \in \mathcal{R}_\eta$, either $d_1 = d_2$ or $|d_1 - d_2| \ge \delta$. A transition $(s,q,\eta) \xrightarrow{t} (u,q',\eta')$ is δ-*wide* if $t \ge \delta$ and for all $\tau \in (t-\delta, t+\delta)$, $\eta + \tau \sim \eta + t$. Furthermore, a transition path

$$(s_0,q_0,\eta_0) \xrightarrow{t_1} (s_1,q_1,\eta_1) \dots \xrightarrow{t_n} (s_n,q_n,\eta_n) \quad (n \ge 1)$$

is δ-*wide* if all its transitions are δ-wide.

Intuitively, A transition is δ-wide if one can adjust the transition by up to δ time units, while keeping the DTA rule used on this transition. The following result is the counterpart of the one on semi Markov processes and DTA [10].

PROPOSITION 9. *For all $(s,q,\eta) \in \mathcal{V}$, if η is δ-separated and $(s,q,[\eta]_\sim)$ can reach some final vertex in G however $q \notin F$, then there exists an at most $|V|$-long, $\delta/|V|$-wide transition path from (s,q,η) to some (s',q',η') with $q' \in F$.*

Then we study the linear equation system $\mu = \mathbf{A}\mu + \zeta$ where $\zeta : \mathsf{B}_m \mapsto \mathbb{R}$ is an arbitrary real vector. Below we define $\zeta_{\max} : \mathsf{B}_m \mapsto \mathbb{R}$ such that $\zeta_{\max}(\mathsf{h}[v]) = M_3 \cdot \rho$ for all $\mathsf{h}[v] \in \mathsf{B}_m$. We denote by $|\zeta|$ the vector such that $|\zeta|(\mathsf{h}[v]) = |\zeta(\mathsf{h}[v])|$ for all $\mathsf{h}[v] \in \mathsf{B}_m$. We extend $\le$ to vectors over B_m in a pointwise fashion.

PROPOSITION 10. *Suppose $m > 2|V|^2$. Let ζ be a vector over B_m such that $|\zeta| \le \zeta_{\max}$. Then the matrix series $\sum_{n=0}^{\infty} \mathbf{A}^n\zeta$ converges. Moreover, we have*

$$\| \sum_{n=0}^{\infty} \mathbf{A}^n \zeta \|_\infty \le |V| \cdot \mathfrak{c}^{-|V|} \cdot (M_3 \cdot \rho) \ ,$$

where

$$\mathfrak{c} := e^{-\lambda_{\max} T_{\max}} \cdot p_{\min} \cdot \frac{\lambda_{\min}}{2|V|^2 + \lambda_{\min}} \ .$$

PROOF. (Sketch) Let $\delta := |V|^{-2}$ and $k := \lfloor m/|V|^2 \rfloor$. We analyse $(\sum_{n=0}^{\infty} \mathbf{A}^n\zeta)(\mathsf{h}[v^*])$ for each $\mathsf{h}[v^*] \in \mathsf{B}_m$. Firstly, we consider the case when $\zeta = \zeta_{\max}$ and $\mathsf{h}[v^*] \in \mathsf{B}_m^{\max}$. Denote $v^* = (s^*, q^*, \eta^*)$. By definition, η^* is 1-separated. Then by Proposition 9, there exists a shortest $|V|^{-1}$-wide path

$$(s^*, q^*, \eta^*) = (s_1, q_1, \eta_1) \xrightarrow{t_1} \dots \xrightarrow{t_{n-1}} (s_n, q_n, \eta_n)$$

with $1 < n \le |V|$, $q_i \notin F$ for $1 \le i \le n-1$ and $q_n \in F$.

Then we define $\{\mathsf{B}'_i\}_{1 \le i \le n}$ with each $\mathsf{B}'_i \subseteq \mathsf{D}_m$ by:

$$\mathsf{B}'_1 = \{\mathsf{h}[v^*]\} \text{ and } \mathsf{B}'_{i+1} = \bigcup \{\mathrm{Post}_i(v) \mid \mathsf{h}[v] \in \mathsf{B}'_i\},$$

where for each $v \in \mathcal{V}$:

- $\mathrm{Post}_i(v) := \{\mathsf{h}[(v')^+_{s_{i+1}}] \mid v' \in \mathrm{Delay}_i(v)\}$;
- $\mathrm{Delay}_i(v) := \{v \oplus \tau \mid \tau \in [t_i, t_i + \delta), \mathsf{h}[v \oplus \tau] \in \mathsf{D}_m\}$.

We have $\bigcup_{i=1}^{n-1} \mathsf{B}'_i \subseteq \mathsf{B}_m$. We prove by induction on $i \ge 1$ that for all $v \in \mathsf{B}'_{n-i}$, $|(\mathbf{A}^i\zeta_{\max})(v)| \le (1 - \mathfrak{c}^i) \cdot M_3\rho$. Note that $\mathbf{A}\zeta_{\max} \le \zeta_{\max}$ and $\mathbf{A}\zeta_1 \le \mathbf{A}\zeta_2$ for all $\vec{0} \le \zeta_1 \le \zeta_2$. It follows that $\mathbf{A}^j\zeta_{\max} \le \mathbf{A}^i\zeta_{\max}$ for all $0 \le i \le j$.

Base Step: $i = 1$. Consider any $v \in \mathsf{B}'_{n-1}$. By $\mathsf{B}'_n \subseteq \mathsf{V}'_n$, we know that $[v']_\sim$ is final in G for all $v' \in \mathrm{Post}_{n-1}(v)$. If $v \oplus (N_v \cdot \rho) \in \mathrm{Delay}_{n-1}(v)$, then from Γ''_m we have

$$\begin{aligned} 1 - \sum_{\mathsf{h}[v'] \in \mathsf{B}_m} \mathbf{A}(\mathsf{h}[v], \mathsf{h}[v']) &\ge p_{\min} \cdot \left(\frac{1}{1 + \rho \cdot \lambda(v)}\right)^{N_v} \\ &\ge p_{\min} \cdot \left(\frac{1}{1 + \rho \cdot \lambda(v)}\right)^{T_{\max}/\rho} \\ &\ge p_{\min} \cdot e^{-\lambda_{\max} T_{\max}} \\ &\ge \mathfrak{c} \end{aligned}$$

Otherwise, there are at least $\lfloor \delta/\rho \rfloor = k$ distinct elements in $\mathrm{Delay}_{n-1}(v)$. Note that $k\rho \ge \frac{1}{2}|V|^{-2}$. We have

$$\begin{aligned} & 1 - \sum_{\mathsf{h}[v'] \in \mathsf{B}_m} \mathbf{A}(\mathsf{h}[v], \mathsf{h}[v']) \\ \ge\ & \left(\frac{1}{1 + \rho \cdot \lambda(v)}\right)^{T_{\max}/\rho} \cdot \frac{k\rho \cdot \lambda(v)}{1 + \rho \cdot \lambda(v)} \cdot p_{\min} \\ \ge\ & e^{-\lambda_{\max} T_{\max}} \cdot \frac{\frac{1}{2}|V|^{-2} \cdot \lambda(v)}{1 + \frac{1}{2}|V|^{-2} \cdot \lambda(v)} \cdot p_{\min} \\ \ge\ & e^{-\lambda_{\max} T_{\max}} \cdot p_{\min} \cdot \frac{\lambda_{\min}}{2|V|^2 + \lambda_{\min}} \end{aligned}$$

Thus we have $|(\mathbf{A}\zeta_{\max})(v)| \le (1 - \mathfrak{c}) \cdot M_3\rho$.

Inductive Step: Suppose $(\mathbf{A}^i\zeta_{\max})(v) \le (1 - \mathfrak{c}^i) \cdot M_3\rho$ for all $v \in \mathsf{B}'_{n-i}$. We prove the case for $i+1$. Fix some $v \in \mathsf{B}'_{n-(i+1)}$. If $v \oplus (N_v \cdot \rho) \in \mathrm{Delay}_{n-(i+1)}(v)$, then

$$\begin{aligned} (\mathbf{A}^{i+1}\zeta_{\max})(v) &\le \left(1 - p_{\min} \cdot \left(\frac{1}{1 + \rho \cdot \lambda(v)}\right)^{N_v} \cdot \mathfrak{c}^i\right) \cdot M_3\rho \\ &\le (1 - \mathfrak{c}^{i+1}) \cdot M_3\rho \end{aligned}$$

Otherwise, there are at least $\lfloor \delta/\rho \rfloor = k$ distinct elements in $\mathrm{Delay}_{n-(i+1)}(v)$. By $\mathrm{Post}_{n-(i+1)}(v) \subseteq \mathsf{B}'_{n-i}$ and $\mathbf{A}^i \zeta_{\max} \leq \zeta_{\max}$, we obtain that

$$\begin{aligned} & (\mathbf{A}^{i+1}\zeta_{\max})(v) \\ \leq \; & M_3\rho \cdot \left\{1 - \left(\frac{1}{1+\rho\cdot\lambda(v)}\right)^{N_v} \cdot \frac{k\rho\cdot\lambda(v)}{1+\rho\cdot\lambda(v)} \cdot p_{\min} \cdot \mathfrak{c}^i\right\} \\ \leq \; & M_3\rho \cdot \left\{1 - e^{-\lambda_{\max}T_{\max}} \cdot p_{\min} \cdot \frac{\lambda_{\min}}{2|V|^2+\lambda_{\min}} \cdot \mathfrak{c}^i\right\} \\ = \; & M_3\rho \cdot (1-\mathfrak{c}^{i+1}) \end{aligned}$$

Then the result follows. Then we obtain that

$$\begin{aligned} (\mathbf{A}^{|V|-1}\zeta_{\max})(v) & \leq (\mathbf{A}^i\zeta_{\max})(v) \\ & \leq (1-\mathfrak{c}^i)\cdot M_3\rho \\ & \leq (1-\mathfrak{c}^{|V|-1})\cdot M_3\rho \end{aligned}$$

for all $1 \leq i \leq n-1$ and $v \in \mathsf{B}'_{n-i}$. Thus $\mathbf{A}^{|V|-1}\zeta_{\max}(v^*) \leq (1-\mathfrak{c}^{|V|-1}) \cdot M_3\rho$ for all $v^* \in \mathsf{B}_m^{\max}$.

Now consider an arbitrary $v \in \mathsf{B}_m$ while $\zeta = \zeta_{\max}$. If either $v \oplus (N_v\rho) \notin \mathsf{B}_m^{\max}$ or $\mathsf{h}[(v \oplus (N_v\rho))_u^+] \notin \mathsf{B}_m$ for some $u \in S$ such that $\mathbf{P}(v,u) > 0$, then we have

$$\begin{aligned} (\mathbf{A}^{|V|}\zeta_{\max})(v) & \leq (\mathbf{A}\zeta_{\max})(v) \\ & \leq \left(1 - \left(\frac{1}{1+\rho\lambda(v)}\right)^{T_{\max}/\rho} \cdot p_{\min}\right) \cdot M_3\rho \\ & \leq (1 - e^{-\lambda_{\max}T_{\max}} \cdot p_{\min}) \cdot M_3\rho \\ & \leq (1-\mathfrak{c}^{|V|}) \cdot M_3\rho \end{aligned}$$

Otherwise we have

$$\begin{aligned} & (\mathbf{A}\cdot\mathbf{A}^{|V|-1}\zeta_{\max})(v) \\ \leq \; & \left(1 - \left(\frac{1}{1+\rho\lambda(v)}\right)^{T_{\max}/\rho} \cdot \mathfrak{c}^{|V|-1}\right) \cdot M_3\rho \\ \leq \; & (1-\mathfrak{c}^{|V|}) \cdot M_3\rho \end{aligned}$$

Then we have $\mathbf{A}^{|V|}\zeta_{\max} \leq (1-\mathfrak{c}^{|V|}) \cdot \zeta_{\max}$. It follows that $\mathbf{A}^{i|V|}\zeta_{\max} \leq (1-\mathfrak{c}^{|V|})^i \zeta_{\max}$ for all $i \in \mathbb{N}$. Thus by the monotonicity of $\mathbf{A}$, we have $\sum_{i=0}^{\infty} \mathbf{A}^i\zeta_{\max}$ converges since $\sum_{i=0}^{\infty} \mathbf{A}^i\zeta_{\max}$ is bounded by $|V| \cdot \mathfrak{c}^{-|V|} \cdot \zeta_{\max}$.

Finally we consider any ζ such that $|\zeta| \leq \zeta_{\max}$. Note that $|\mathbf{A}^i\zeta| \leq \mathbf{A}^i\zeta_{\max}$ since all entries of $\mathbf{A}^i$ are non-negative. Thus by Cauchy Criterion, it follows that $\sum_{i=0}^{\infty}\mathbf{A}^i\zeta$ converges and $\| \sum_{i=0}^{\infty}\mathbf{A}^i\zeta \|_\infty \leq |V| \cdot \mathfrak{c}^{-|V|} \cdot M_3\rho$. □

By Proposition 10, the system of linear equation $\mu = \mathbf{A}\mu + \zeta$ has a solution for all $|\zeta| \leq \zeta_{\max}$ when $m > 2|V|^2$. Below we assume that $m > 2|V|^2$. The following propositions show that the linear equation has a unique solution.

Proposition 11. *For all solutions μ of $\mu = \mathbf{A}\mu + \zeta$ with $\| \zeta \|_\infty < M_3\rho$, we have $|\mu| \leq \mu^*$, where $\mu^* := \sum_{i=0}^{\infty}\mathbf{A}^i\zeta_{\max}$.*

Proof. Let μ be an arbitrary solution of $\mu = \mathbf{A}\mu + \zeta$. Define $\mu' = \mu^* - \mu$. By the fact that $\mu^* = \mathbf{A}\mu^* + \zeta_{\max}$, we have $\mu'(\mathsf{h}[v]) > (\mathbf{A}\mu')(\mathsf{h}[v])$ for all $\mathsf{h}[v] \in \mathsf{B}_m$. Suppose there is some $\mathsf{h}[v] \in \mathsf{B}_m$ such that $\mu'(\mathsf{h}[v]) < 0$. W.l.o.g we assume that $\mu'(\mathsf{h}[v])$ is the least element of $\{\mu'(\mathsf{h}[v']) \mid \mathsf{h}[v'] \in \mathsf{B}_m\}$. Denote $c := \sum_{\mathsf{h}[v'] \in \mathsf{B}_m} \mathbf{A}(\mathsf{h}[v], \mathsf{h}[v']) \in [0,1]$. We have $\mu'(\mathsf{h}[v]) > c \cdot \mu'(\mathsf{h}[v])$, which implies $c > 1$. Contradiction. Thus $\mu' \geq \vec{0}$. Similar arguments holds if we define $\mu' = \mu^* + \mu$. Thus we have $|\mu| \leq \mu^*$. □

Proposition 12. *The system of linear equations*

$$\mu = \mathbf{A}\mu + \zeta$$

has a unique solution for all ζ such that $\|\zeta\|_\infty < M_3\rho$. It follows that $\mathbf{I} - \mathbf{A}$ is invertible where $\mathbf{I}$ is the identity matrix.

Proof. By Proposition 10, the system $\mu = \mathbf{A}\mu + \zeta$ has a solution. And by Proposition 11, all solutions of $\mu = \mathbf{A}\mu + \zeta$ are bounded by μ^*. Suppose it has two distinct solutions. Then the homogeneous system of linear equations $\mu = \mathbf{A}\mu$ has a non-trivial solution, which implies that the solutions of $\mu = \mathbf{A}\mu + \zeta$ cannot be bounded. Contradiction. Thus $\mu = \mathbf{A}\mu + \zeta$ has a unique solution and $\mathbf{I} - \mathbf{A}$ is invertible. □

Now we analyse Γ'_m (Γ_m). In the following theorem which is the main result of the paper, we prove that the equation $\mu = \mathbf{C}\mu + \mathbf{d}$ has a unique solution (i.e. $\mathbf{I} - \mathbf{C}$ is invertible), and give the error bound between the unique solution and the function prob.

Theorem 5. *The matrix equation $\mu = \mathbf{C}\mu + \mathbf{d}$ (for Γ'_m) has a unique solution $\overline{\mu}$. Moreover,*

$$\max_{\mathsf{h}[v] \in \mathsf{B}_m} |\overline{\mu}(\mathsf{h}[v]) - \mathsf{prob}(v)| \leq |V| \cdot \mathfrak{c}^{-|V|} \cdot M_3\rho \; .$$

Proof. (Sketch) We first prove that $\mu = \mathbf{C}\mu + \mathbf{d}$ has a unique solution. Let $\mu = \mathbf{C}\mu + \zeta$ be a matrix equation such that $\| \zeta \|_\infty < M_2\rho^2$. We can equivalently expand this equation into some equation $\mu = \mathbf{A}\mu + \zeta'$ with $\| \zeta' \|_\infty < M_3 \cdot \rho$. Since $\mu = \mathbf{A}\mu + \zeta'$ has a unique solution, we have $\mu = \mathbf{C}\mu + \zeta$ also has a unique solution. Thus $\mathbf{I} - \mathbf{C}$ is invertible and $\mu = \mathbf{C}\mu + \mathbf{d}$ has a unique solution.

Now we prove the error bound between $\overline{\mu}$ and prob. Define the vector μ' such that $\mu'(\mathsf{h}[v]) = \overline{\mu}(\mathsf{h}[v]) - \mathsf{prob}(v)$ for all $\mathsf{h}[v] \in \mathsf{B}_m$. By Proposition 6, μ' is the unique solution of $\mu = \mathbf{C}\mu + \zeta$ for some $\| \zeta \|_\infty < M_2\rho^2$. Then μ' is also the unique solution of the equation $\mu = \mathbf{A}\mu + \zeta'$ for some $\| \zeta' \|_\infty < M_3\rho$. By Proposition 10, $\| \mu' \|_\infty \leq |V| \cdot \mathfrak{c}^{-|V|} \cdot M_3\rho$. □

By Theorem 5 and the Lipschitz Continuity (Corollary 2), we can approximate $\mathsf{prob}(s,q,\eta)$ as follows: given $\epsilon \in (0,1)$, we choose m sufficiently large and some $\mathsf{h}[v] \in \mathsf{D}_m$ such that $|\mathsf{prob}(v) - \mathsf{prob}(s,q,\eta)| < \frac{1}{2}\epsilon$ and $M_3|V|\mathfrak{c}^{-|V|} \cdot \rho < \frac{1}{2}\epsilon$. Then we solve the system Γ'_m to obtain $\overline{\mu}(\mathsf{h}[v])$.

7. CONCLUSION AND FUTURE WORK

We have shown an algorithm to approximate the acceptance probabilities of CTMC-paths by a multi-clock DTA under finite acceptance condition. Unlike the result by Barbot *et al.* [5], we are able to derive an approximation error. Chen *et al.* [12] demonstrated that computing the acceptance probability of CTMC-paths by a multi-clock DTA under Muller acceptance condition can be reduced to the one under finite acceptance condition. Thus our result can also be applied to Muller acceptance conditions. One future direction is to refine our approximation algorithm by importing zone-based techniques [6]. Another future direction is to extend this result to continuous-time Markov decision processes (CTMDP) [7] or continuous-time Markov games (CTMG) [9, 15]. A more challenging task would be to consider the acceptance probabilities of CTMC-paths by a non-deterministic timed automaton.

8. ACKNOWLEDGEMENTS

I thank Prof. Joost-Pieter Katoen for suggestions on the writing of this paper, and anonymous referees for helpful comments. The author is supported by a CSC (China Scholarship Council) scholarship.

9. REFERENCES

[1] R. Alur and D. L. Dill. A theory of timed automata. *Theor. Comput. Sci.*, 126(2):183–235, 1994.

[2] C. Baier, L. Cloth, B. R. Haverkort, M. Kuntz, and M. Siegle. Model checking Markov chains with actions and state labels. *IEEE Trans. Software Eng.*, 33(4):209–224, 2007.

[3] C. Baier, B. R. Haverkort, H. Hermanns, and J.-P. Katoen. Model-checking algorithms for continuous-time Markov chains. *IEEE Trans. Software Eng.*, 29(6):524–541, 2003.

[4] C. Baier, B. R. Haverkort, H. Hermanns, and J.-P. Katoen. Performance evaluation and model checking join forces. *Commun. ACM*, 53(9):76–85, 2010.

[5] B. Barbot, T. Chen, T. Han, J.-P. Katoen, and A. Mereacre. Efficient CTMC model checking of linear real-time objectives. In P. A. Abdulla and K. R. M. Leino, editors, *TACAS*, volume 6605 of *Lecture Notes in Computer Science*, pages 128–142. Springer, 2011.

[6] J. Bengtsson and W. Yi. Timed automata: Semantics, algorithms and tools. In J. Desel, W. Reisig, and G. Rozenberg, editors, *Lectures on Concurrency and Petri Nets*, volume 3098 of *Lecture Notes in Computer Science*, pages 87–124. Springer, 2003.

[7] N. Bertrand and S. Schewe. Playing optimally on timed automata with random delays. In M. Jurdzinski and D. Nickovic, editors, *FORMATS*, volume 7595 of *Lecture Notes in Computer Science*, pages 43–58. Springer, 2012.

[8] P. Billingsley. *Probability and Measure.* John Wiley & Sons, New York, NY, USA, 1995.

[9] T. Brázdil, J. Krcál, J. Kretínský, A. Kučera, and V. Rehák. Stochastic real-time games with qualitative timed automata objectives. In P. Gastin and F. Laroussinie, editors, *CONCUR*, volume 6269 of *Lecture Notes in Computer Science*, pages 207–221. Springer, 2010.

[10] T. Brázdil, J. Krcál, J. Kretínský, A. Kučera, and V. Rehák. Measuring performance of continuous-time stochastic processes using timed automata. In M. Caccamo, E. Frazzoli, and R. Grosu, editors, *HSCC*, pages 33–42. ACM, 2011.

[11] T. Chen, M. Diciolla, M. Z. Kwiatkowska, and A. Mereacre. Time-bounded verification of CTMCs against real-time specifications. In U. Fahrenberg and S. Tripakis, editors, *FORMATS*, volume 6919 of *Lecture Notes in Computer Science*, pages 26–42. Springer, 2011.

[12] T. Chen, T. Han, J.-P. Katoen, and A. Mereacre. Model Checking of Continuous-Time Markov Chains Against Timed Automata Specifications. *Logical Methods in Computer Science*, 7(1), 2011.

[13] M. Davis. *Markov Models and Optimizations.* Chapman & Hall, New York, NY, USA, 1993.

[14] S. Donatelli, S. Haddad, and J. Sproston. Model checking timed and stochastic properties with CSL^{TA}. *IEEE Trans. Software Eng.*, 35(2):224–240, 2009.

[15] J. Fearnley, M. Rabe, S. Schewe, and L. Zhang. Efficient approximation of optimal control for continuous-time Markov games. In S. Chakraborty and A. Kumar, editors, *FSTTCS*, volume 13 of *LIPIcs*, pages 399–410. Schloss Dagstuhl - Leibniz-Zentrum fuer Informatik, 2011.

[16] W. Feller. *An Introduction to Probability Theory and Its Applications.* John Wiley & Sons, New York, NY, USA, 1966.

[17] L. Mikeev, M. R. Neuhäußer, D. Spieler, and V. Wolf. On-the-fly verification and optimization of DTA-properties for large Markov chains. *Form. Methods Syst. Des.*, 2012.

[18] J. Thomas. *Numerical Partial Differential Equations: Finite Difference Methods.* Springer-Verlag, New York, NY, USA, 1995.

[19] L. Zhang, D. N. Jansen, F. Nielson, and H. Hermanns. Automata-based CSL model checking. In L. Aceto, M. Henzinger, and J. Sgall, editors, *ICALP (2)*, volume 6756 of *Lecture Notes in Computer Science*, pages 271–282. Springer, 2011.

Specification-Guided Controller Synthesis for Linear Systems and Safe Linear-Time Temporal Logic*

Matthias Rungger
Department of Electrical Engineering
UCLA
rungger@ee.ucla.edu

Manuel Mazo Jr.
Delft Center for Systems and Control
Delft University of Technology
m.mazo@tudelft.nl

Paulo Tabuada
Department of Electrical Engineering
UCLA
tabuada@ee.ucla.edu

ABSTRACT

In this paper we present and analyze a novel algorithm to synthesize controllers enforcing linear temporal logic specifications on discrete-time linear systems. The central step within this approach is the computation of the maximal controlled invariant set contained in a possibly non-convex safe set. Although it is known how to compute approximations of maximal controlled invariant sets, its exact computation remains an open problem. We provide an algorithm which computes a controlled invariant set that is guaranteed to be an under-approximation of the maximal controlled invariant set. Moreover, we guarantee that our approximation is at least as good as any invariant set whose distance to the boundary of the safe set is lower bounded. The proposed algorithm is founded on the notion of sets adapted to the dynamics and binary decision diagrams. Contrary to most controller synthesis schemes enforcing temporal logic specifications, we do not compute a discrete abstraction of the continuous dynamics. Instead, we abstract only the part of the continuous dynamics that is relevant for the computation of the maximal controlled invariant set. For this reason we call our approach specification guided. We describe the theoretical foundations and technical underpinnings of a preliminary implementation and report on several experiments including the synthesis of an automatic cruise controller. Our preliminary implementation handles up to five continuous dimensions and specifications containing up to 160 predicates defined as polytopes in about 30 minutes with less than 1GB memory.

Categories and Subject Descriptors

I.2.8 [**Problem Solving, Control Methods and Search**]: Control Theory; I.2.2 [**Automatic Programming**]: Program Synthesis

*The work of the first and last authors was partially supported by the NSF award 1035916 and by the NSF Expeditions in Computing project ExCAPE: Expeditions in Computer Augmented Program Engineering.

Keywords

Controller Synthesis, LTL Specification, Set Invariance

1. INTRODUCTION

Traditional control systems design is concerned with rather simple specifications such as asymptotic stability, *i.e.*, reaching some set as time tends to infinity. Although simple, these specifications are often hard to enforce due to the complexity of the models employed in control theory and, in particular, due to the infinite nature of the state space over which those models are defined. In sharp contrast is the work on formal verification of systems or in discrete-event controller synthesis. In those two realms, the problem specification is typically much more complex, involving not only specifications regarding the limiting behavior of the system at hand, but also sequencing of actions, choices between alternative actions, etc. These specifications, are typically formalized in a temporal logic [25] such as linear temporal logic (LTL), which lets the user specify both logical and temporal properties. In those areas of research more complex specifications can be tackled because of the finite state nature of the system models. The use of systems evolving on finite state spaces allows for the algorithmic treatment of problems and therefore the use of computers for the computation of its solution or be it the verification of a given property, or the design of a control strategy to enforce a property.

With the advent of ubiquitous computing platforms to sense and control physical systems, there has been an increasing level of concern regarding the safety of such systems. This has, in turn, sparked the interest of computer scientists to verify the correctness of software interacting with the physical world, as well as from control scientists aiming at synthesizing software that guarantees the safe control of a given physical plant. Until recently, most of the approaches to these problems relied on extensive testing to prove correctness of the designs at hand. In the past few years, however, there has been a surge of research aimed at the synthesis of correct-by-design control software. Under this new paradigm, testing or verification is no longer needed as the software is designed taking into account all possible contingencies described by the models of the system to be controlled.

As already mentioned, when the models employed to describe the system at hand have finitely many states, automated techniques are already available to synthesize controllers. Thus, the most common approach to provide correct-by-design synthesis techniques is to convert the infinite state models, usually employed in the modeling of physical processes, into finite state models. This translation, referred to as abstraction, needs to be carefully done so that properties guaranteed in the simpler finite state model carry through to the original infinite state system. While numerous theoretical results are available supporting such abstraction procedures [31], direct application of them, imposing uniform grids on the state space, usually results in computationally intractable models. This is especially the case when dealing with higher dimensional systems due to the *curse of dimensionality* resulting in an exponential growth of the finite models with the dimension of the system. There exists a great variety of different approaches [23, 33, 16, 12, 10, 22, 1, 27, 36, 20, 40, 35] and tools, *e.g.* LTL-Con [15], LTLMoP [11], TuLiP [41], Pessoa [21]. All of them with their different virtues. The systems considered range from simple double integrator dynamics [10], over linear dynamics [33, 16, 40] or nonlinear dynamics [22, 27, 20, 35] to hybrid dynamics [12] and stochastic systems [1, 36]. Different notion of abstractions like behavioral containment [23, 27], exact (bi)similar relations [33] or approximate notions [12, 22, 20] have been analyzed. Also the regarded specification language ranges from full LTL [33, 10], restricted versions of LTL [22, 40], to special cases of reach-avoid problems [27, 35]. However, most of them suffer from the curse of dimensionality, as they follow the same two step approach by first computing a finite abstraction and performing controller synthesis afterwards. Alternative approaches are thus needed in order to make the synthesis of correct-by-design controllers practical.

In the present work, we focus on an algorithm that avoids the computation of a finite state abstraction of the continuous system to be controlled. A similar approach, which also connects the computation of the abstraction to the specification is outlined in [13], where the authors focus on co-safe LTL specifications. We consider discrete-time linear time-invariant systems and a subset of the full linear temporal logic (LTL) called safe-LTL. Safe-LTL formulas always specify a safety property. These are properties whose violation can be checked by looking at a finite prefix of the violating run. The well-known automata theoretic approach to synthesize controllers that enforce LTL specifications, proceeds as follows, see [17, 19]. First, the negated specification is translated into a finite state automaton, which is then synchronized with the system. If the synchronized system has an accepting run, the property given by the specification does not hold. Hence, the objective of a controller is to prevent runs from being accepted which can be formalized as safety game or as the computation of the maximal controlled invariant set.

Therefore, we revisit the well-known fixed point iteration to compute the maximal controlled invariant subset (MCIS) of a given *safe* set K, see *e.g.* [3] or for a more detailed exposition the monographs [2, 5]. When the system is linear and the set K is polyhedral, all the intermediate computations can be performed exactly by computing polyhedral projections and intersections, see *e.g.* [37]. Thus, provided that the iterations terminate, we can solve the synthesis problem. Unfortunately, termination is not guaranteed in general. There exists some results which address this problem for a convex safe set K. The authors in [4, 5] exploit the contraction property of an asymptotically stabilizable system, and show that the MCIS can be under-approximated with arbitrary accuracy in finite time. The authors in [9], initialize the iteration with a controlled invariant set, and are able to provide an invariant under-approximation of the MCIS in every step of the iteration. Moreover, they show that their approximation approaches the MCIS in the limit. Unfortunately, in our case, we cannot assume the safe set K to be convex as it is given by unions and intersections of the polyhedral sets that correspond to the atomic propositions in a safe-LTL formula.

The fixed point iteration for a non-convex safe set is often carried out for piecewise linear systems [14, 26, 24, 42]. However, none of these approaches ensure finite termination. Furthermore, it is not straightforward to apply the solutions for convex safe sets to non-convex safe sets. Also, from a practical point of view, the exact computation of the MCIS with non-convex safe sets might not be a wise choice since the number of polytopes might grow combinatorially over the iterations. This constitutes a serious problem as illustrated in Section 6.2.

We propose a practical implementation of the MCIS computation for which we can guarantee termination. This is attained at the price of providing an under-approximation of the actual MCIS which is, nevertheless, a controlled invariant subset. The proposed technique is based on the notion of sets adapted to the dynamics of a linear system, see [31]. These are sets which in controllable normal form coordinates, also known as Brunvosky normal form, are Cartesian products of intervals. For (unions of) such sets, the computation of the MICS is relatively simple, and can be computed in a finite number of iterations. This was already observed in [37] and [38], and later generalized in [32, 33] and [28]. However, one rarely encounters problems in which the relevant sets are already adapted to the dynamics as assumed in [32, 33] and [28]. The idea we explore in this paper is to under-approximate the relevant sets by adapted sets and then perform the fixed point computation for the approximating sets. The approximations of the sets at hand are represented much in the same flavor as proposed in [6] and are stored in the form of binary decision diagrams [39, 30] (BDD). BDDs are selected as a data structure because of their efficiency in representing and manipulating binary functions or equivalently discrete sets.

One aspect of our work is that we do not require the computation of the complete discrete abstraction of the continuous dynamics. Instead, we abstract only the part of the continuous dynamics that plays a role in the computation on the MCIS. The computational implications of this are reflected on the experimental results, where we can solve problems up to five continuous variables, which was previously not possible, see Section 6.2. A second aspect of the present work is a certain completeness result for our approach: we show that our approximation is at least as good as any invariant set whose distance to the boundary of the safe set is lower bounded, see Section 4.2. A statement that is rarely found in the abstraction based synthesis community.

2. PRELIMINARIES

In what follows, we mostly operate in a space $X = Q \times \mathbb{R}^n$, given as the product of a discrete space Q and the Euclidean space $\mathbb{R}^n$. The projection of X onto Q and $\mathbb{R}^n$ is denoted by the mappings $\pi_Q : X \to Q$ and $\pi_{\mathbb{R}^n} : X \to \mathbb{R}^n$, respectively. Given a subset $K \subseteq X$ we use $K(q)$ to denote the set $K(q) = \{x \in \mathbb{R}^n \mid (q, x) \in K\}$. We define the Hausdorff distance d_H on X with respect to the product metric given by the discrete metric on Q and the Euclidean distance d in $\mathbb{R}^n$. A cube in $\mathbb{R}^n$ with center $c \in \mathbb{R}^n$ and radius $r \in \mathbb{R}^n$, is denote defined by $\mathsf{B}(c,r) := \{x \in \mathbb{R}^n \mid \forall_{i\in\{1,\dots,n\}} : -r_i \leq x_i - c_i \leq r_i\}$.

3. THE SYNTHESIS PROBLEM

In this section, we introduce the synthesis problem, which is composed of three entities: a linear control *system* that describes the physical process we want to control; a set of *atomic propositions* that represent the sets of states that are of interest for the control task; and the safe-LTL *formula* that describes the desired closed-loop behavior of the system in terms of the atomic propositions.

3.1 The System

We consider discrete-time linear control systems:

$$x^+ = Ax + Bu, \tag{1}$$

given by the matrices $A \in \mathbb{R}^{n\times n}$ and $B \in \mathbb{R}^{n\times m}$. Throughout the paper, we assume the system (1) to be controllable.

3.2 The Specification

Let $\mathcal{P} \subseteq 2^{\mathbb{R}^n}$ denote the set of *atomic propositions*, with each element $P_i \in \mathcal{P}$ given by a convex polytope:

$$P_i = \mathcal{H}(C_i, c_i). \tag{2}$$

Here $\mathcal{H}(C_i, c_i) = \{x \in \mathbb{R}^n : C_i x \leq c_i\}$ denotes the set of solutions of the system of inequalities $C_i x \leq c_i$ with $C_i \in \mathbb{R}^{r_i \times n}$ and $c_i \in \mathbb{R}^{r_i}$. We assume that the sets $P_i \subseteq \mathbb{R}^n$ are bounded. The atomic propositions are linked to the system (1) through the map $h : \mathbb{R}^n \to 2^{\mathcal{P}}$ associating to each state $x \in \mathbb{R}^n$ the set of atomic propositions that are satisfied at x, *i.e.*, $h(x) = \{P_i \in \mathcal{P} : x \in P_i\}$. These atomic propositions are used to construct safe-LTL formulas formalizing the desired specification to be enforced by the controller to be synthesized. For a definition of the syntax and semantics of LTL and safe-LTL we refer the readers to [17, 19]. Here, we simply mention that safe-LTL formulas always define safety specifications and we provide a concrete example illustrating its usefulness as a specification formalism.

Cruise control example.

Consider a truck with a trailer as depicted in Figure 1. We want to design a controller ensuring that the highway speed limits are always satisfied. The longitudinal dynamics of the truck and trailer are given by the continuous-time linear control system

$$\dot{x} = \begin{bmatrix} 0 & -1 & 1 \\ \frac{k_s}{m} & -\frac{k_d}{m} & \frac{k_d}{m} \\ 0 & 0 & 0 \end{bmatrix} x + \begin{bmatrix} 0 \\ 0 \\ 1 \end{bmatrix} u,$$

where the entries of the state vector $x = (d, v_1, v_2) \in \mathbb{R}^3$ correspond to the distance d between the truck and the trailer, the velocity of the trailer v_2 and the velocity of the truck v_1, respectively. A more detailed explanation of this model is given in Section 6.1.

Let the truck be driving on a highway on which three speed limits $v_a = 15.6\,\mathrm{m/s}$ (ca. 35 mph), $v_b = 24.5\,\mathrm{m/s}$ (ca. 55 mph) and $v_c = 29.5\,\mathrm{m/s}$ (ca. 66 mph) are imposed. We would like to design a controller guaranteeing that the truck obeys the velocity limits. We assume that each velocity limit sign on the highway is equipped with a radio transmitter

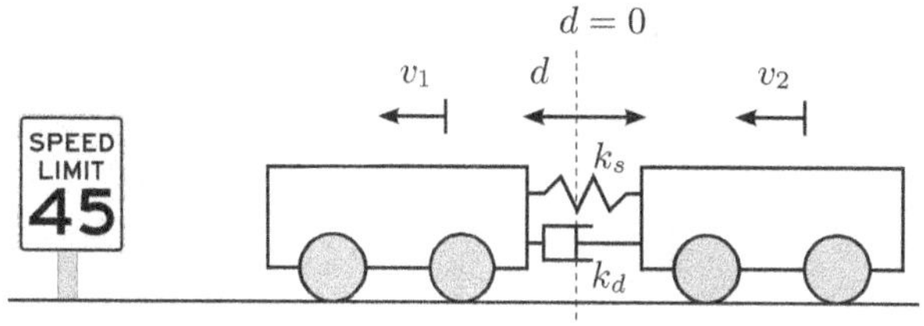

Figure 1: The truck with trailer.

that automatically transmits the limit to the truck. Then we would like to ensure that the truck obeys the current speed limit at most after $T \in \mathbb{N}$ time steps after the truck receives the information that the speed limit has changed.

For simplicity we consider only the limits v_a and v_b. A safe-LTL formula encoding the specification is given by:

$$\Box(\varphi_a \wedge \varphi_b) \tag{3}$$

where φ_a and φ_b are defined as:

$$m_a \implies \Diamond_{\leq T}(t_a W m_b) \quad \text{and} \quad m_b \implies \Diamond_{\leq T}(t_b W m_a)$$

respectively. The atomic proposition m_i, $i \in \{a, b\}$ encodes the fact that limit v_i is in place while the atomic proposition t_i encodes the satisfaction of the speed limit, that is, $v_1 \leq v_i$. The operator $\Diamond_{\leq T}$ requires $t_a W m_b$ to be satisfied in T or less steps while the formula $t_a W m_b$ requires t_a to be satisfied unless m_b becomes true, *i.e.*, the speed limit changes from m_a to m_b. We return to this example in Section 6.1 where we compute the maximal controlled invariant set.

3.3 The synthesis problem

There exists a well known automata theoretic approach to synthesize controllers with respect to LTL formulas and specifically, with respect to safe-LTL formulas. Safe-LTL formulas represent a fragment of full LTL formulas that specifies safety properties, while full LTL formulas can also specify liveness properties. Loosely speaking, a formula φ defines a safety property, if its violation can be checked by looking at a finite prefix of a trajectory. This property allows the use of finite automata in the synthesis of controllers enforcing safe-LTL instead of Büchi automata that require more complex algorithms, see *e.g.* [19]. Kupferman and Vardi [17] show that a *bad-prefix finite state automaton* can be constructed from a safe-LTL formula φ. This automaton accepts at least one bad prefix for every trajectory that violates the specification φ.

Let $A_{\neg\varphi} = (Q, Q_0, F, \delta, g, 2^{\mathcal{P}})$ denote the bad-prefix finite-state automaton with respect to the formula φ where Q is the set of states, $F \subseteq Q$ is the set of accepting states, $\delta \subset Q \times Q$ is the transition relation, $g : Q \to 2^{\mathcal{P}}$ is the output function and $2^{\mathcal{P}}$ is the output set. We define the transition system:

$$S_\varphi = (X, X_0, U, \delta_\varphi, Y, H) \tag{4}$$

as the synchronous product of system (1) and the bad-prefix automaton $A_{\neg\varphi}$ by:

- the set of states $X = \{(q, x) \in Q \times \mathbb{R}^n \mid g(q) = h(x)\}$;
- the initial states $X_0 = \{(q, x) \in Q_0 \times \mathbb{R}^n \mid g(q) = h(x)\}$;
- the set of inputs $U = \mathbb{R}^m$;

- the transition relation $\delta_\varphi \subseteq X \times X$;
- the set of outputs $Y = 2^{\mathcal{P}}$;
- the output map $H(q,x) = g(q)$ for each $(q,x) \in X$.

The transition relation δ_φ of the synchronous product is implicitly defined by $((q,x),(q',x')) \in \delta_\varphi$ iff there exists $u \in U$ such that

$$x' = Ax + Bu \wedge (q,q') \in \delta.$$

A *run* of the transition system S_φ associated to the initial state $(q,x) \in X_0$ is an infinite sequence $\zeta : \mathbb{N}_0 \to X$ with $\zeta(0) = (q,x)$ that satisfies $(\zeta(t), \zeta(t+1)) \in \delta_\varphi$ for all times $t \in \mathbb{N}_0$. Sometimes we will denote $\zeta(t)$ simply by ζ_t.

We refer the reader to [17, 19] for a detail description of the bad-prefix automaton and in particular to [13] where the very same bad-prefix automaton is used.

It is well-known that the controller enforcing φ on (1) corresponds to the controller forcing the system S_φ to stay within the *safe set*

$$K = \bigcup_{q \in Q \setminus F} \{q\} \times h^{-1} \circ g(q) \tag{5}$$

for all times. As long as trajectories remain in K, no finite-prefix of a trajectory violating the specification is reached and thus no violation of the specification ever occurs. Moreover, it is known that this controller is memoryless in the sense that it is given by a map $\mu : X \to 2^U$ and it can be chosen to be maximal, *i.e.*, any other controller $\mu' : X \to 2^U$ that enforces φ satisfies $\mu'(q,x) \subseteq \mu(q,x)$, see *e.g.* [31].

In summary, the synthesis of a controller enforcing a safe-LTL specification is reduced to the computation of a controlled invariant set for the system S_φ. Therefore, in the remaining parts of the paper, we solely focus on the computation of the maximal controlled invariant subset of K.

4. COMPUTATION OF INVARIANT SETS

Let us introduce the algorithm to compute the maximal controlled invariant subset (MCIS) of K defined by:

$$\begin{aligned}\mathcal{K}(K) = \big\{(q,x) \in X \,\big|\, &\text{there exists a run } \zeta \text{ of } S_\varphi \text{ with} \\ &\zeta_0 = (q,x) \text{ and } \zeta_t \in K \text{ for all } t \in \mathbb{N}_0\big\}.\end{aligned}$$

When no confusion arises we will denote the MCIS simply by $\mathcal{K}$. This set is computed by the well-known Algorithm 1, see *e.g.* [3]. The algorithm is initialized with the set K and proceeds iteratively by computing the set K_{i+1} as the intersection of K_i with the set of states that reach K_i in one step. Assuming that the algorithm terminates after $i \in \mathbb{N}_0$ steps, the output satisfies $K_i = \mathcal{K}(K)$, see *e.g.* [14, Lemma 2.1].

Algorithm 1 Computation of $\mathcal{K}(K)$

Input: K
Init: $K_0 := K$
while $K_i \neq K_{i+1}$ **do**

$$K_{i+1} := K_i \cap \{z \in X \mid \exists z' \in K_i : (z,z') \in \delta_\varphi\} \tag{6}$$

end while
Output: K_i

4.1 Approximation of the Maximal Controlled Invariant Set

For the following analysis, we assume the system (1) to be given in *special Brunovsky normal form*, see [7], *i.e.*, the system matrices are of form

$$A = \begin{bmatrix} A_{\mu_1} & 0 & \dots & 0 \\ 0 & A_{\mu_2} & \dots & 0 \\ \vdots & \vdots & \ddots & \vdots \\ 0 & 0 & 0 & A_{\mu_m} \end{bmatrix}, B = \begin{bmatrix} b_{\mu_1} & 0 & \dots & 0 \\ 0 & b_{\mu_2} & \dots & 0 \\ \vdots & \vdots & \ddots & \vdots \\ 0 & 0 & 0 & b_{\mu_m} \end{bmatrix}$$

for some $\mu_1, \dots, \mu_m \in \mathbb{N}$ with $\sum_{i=1}^m \mu_i = n$, and $A_{\mu_i} \in \mathbb{R}^{\mu_i \times \mu_i}$, $b_{\mu_i} \in \mathbb{R}^{\mu_i \times 1}$ are of the form

$$A_{\mu_i} = \begin{bmatrix} 0 & 1 & 0 & \dots & 0 \\ 0 & 0 & 1 & \dots & 0 \\ \vdots & \vdots & \vdots & \ddots & \vdots \\ 0 & 0 & 0 & \dots & 1 \\ 0 & 0 & 0 & \dots & 0 \end{bmatrix}, \quad b_{\mu_i} = \begin{bmatrix} 0 \\ 0 \\ \vdots \\ 0 \\ 1 \end{bmatrix}.$$

The original Brunovsky normal form is similarly defined, only the last rows of the matrices A_{μ_i} contain possible non-zero entries. However, such rows can be cancelled out by a simple input transformation. Given that the system is controllable, all these transformations are carried out without loss of generality and automatically computed by our implementation. See [28] for a more detailed treatment.

In our approximation of the safe set K, we use the fact that for all $q \in Q$, the set $K(q)$ results from a finite union and intersection of atomic propositions $P_i \in \mathcal{P}$. Therefore, $K(q)$ is given as a finite union of polytopes

$$K(q) = \bigcup_{i=1}^{p_q} \mathcal{H}(C_i, c_i)$$

for some $C_i \in \mathbb{R}^{r_i \times n}$ and $c_i \in \mathbb{R}^{r_i}$. We now proceed by approximating these sets based on a uniform grid in $\mathbb{R}^n$ denoted by $[\mathbb{R}^n]_\rho$ and parameterized by $\rho \in \mathbb{R}_+$:

$$[\mathbb{R}^n]_\rho = \{x \in \mathbb{R}^n : x = \rho k, k \in \mathbb{Z}^n\}.$$

Let $\mathsf{C}_\rho(\xi)$ denote the *cell* associated to the grid point $\xi \in [\mathbb{R}^n]_\rho$:

$$\mathsf{C}_\rho(\xi) = \{x \in \mathbb{R}^n \mid \forall_{i \in \{1,\dots,n\}} : \xi_i \leq x_i \leq \xi_i + \rho\}.$$

and note that a cell is a Cartesian product of intervals and is thus adapted to the dynamics [31]. We define the over-approximation $\hat{K}$ and the under-approximation $\check{K}$ of the safe set K by

$$\begin{aligned}\hat{K}(q) &= \{x \in \mathbb{R}^n \mid \exists \xi \in [\mathbb{R}^n]_\rho : x \in \mathsf{C}_\rho(\xi) \cap K(q) \neq \varnothing\}, \\ \check{K}(q) &= \{x \in \mathbb{R}^n \mid \exists \xi \in [\mathbb{R}^n]_\rho : x \in \mathsf{C}_\rho(\xi) \subseteq K(q)\}.\end{aligned} \tag{7}$$

These approximations, being a union of adapted sets, are also adapted sets. It is straightforward to verify that $\check{K} \subseteq K \subseteq \hat{K}$ holds. Moreover, it follows immediately from the definition of MCIS that

$$\mathcal{K}(\check{K}) \subseteq \mathcal{K}(K) \subseteq \mathcal{K}(\hat{K}).$$

THEOREM 1. *Suppose system* (1) *is in special Brunovsky normal form and let $\check{K}$ and $\hat{K}$ be the sets as defined in* (7). *Then Algorithm 1, with input $\check{K}$ and $\hat{K}$ terminates in a finite number of iterations and the output K_i is the maximal controlled invariant subset of $\check{K}$ and $\hat{K}$, respectively.*

Although the proof of this theorem follows from the results in [33] we include it here for the sake of completeness. Before we prove finite termination of the algorithm, we need a technical result for the set

$$\mathsf{Pre}(D) = \{x \in \mathbb{R}^n \mid \exists u \in \mathbb{R}^m : Ax + Bu \in D\} \quad (8)$$

and the following definition.

DEFINITION 1. *We call a set $D \subseteq \mathbb{R}^n$ finitely representable on $[\mathbb{R}^n]_\rho$ if there exist $\xi^1, \ldots, \xi^p \in [\mathbb{R}^n]_\rho$ such that $D = \cup_i \mathsf{C}_\rho(\xi^i)$.*

LEMMA 1. *Suppose (1) is in special Brunovsky normal form. If D and D' are finitely representable sets then $D' \cap \mathsf{Pre}(D)$ is finitely representable.*

PROOF OF THEOREM 1. We show the assertion for the under-approximation $\mathcal{K}(\check{K})$, *i.e.*, Algorithm 1 is invoked with $\check{K}$. The statement for $\hat{K}$ follows analogously. We can write equation (6) as

$$K_{i+1} = K_i \cap \bigcup_{(q,q')\in\delta, q'\in\pi_Q(K_i)} \{q\} \times \mathsf{Pre}(K_i(q')).$$

Note that $K_0(q)$ is finitely representable for all $q \in Q$. Thus, we can use induction and invoke Lemma 1 to conclude that that the sets $K_i(q)$ are finitely representable for every $i \in \mathbb{N}_0$ and $q \in Q$.

Now we identify every $K_i(q)$ with its representing grid points $K_i(q) \cap [\mathbb{R}^n]_\rho$ and define the operator

$$G : 2^{\check{K}_\rho} \to 2^{\check{K}_\rho}, K_i \cap Q \times [\mathbb{R}^n]_\rho \mapsto K_{i+1} \cap Q \times [\mathbb{R}^n]_\rho$$

with $\check{K}_\rho = K \cap Q \times [\mathbb{R}^n]_\rho$ and K_i and K_{i+1} given in (6). Note that $(2^{\check{K}_\rho}, \subseteq)$ is a complete lattice and $2^{\check{K}_\rho}$ is finite. In addition, it is straightforward to see that G is monotone. Then, we apply Tarski's fixed point theorem, see *e.g.* [34] or [31, Corollary A.6], and conclude that there exists $i \in \mathbb{N}_0$ such that the maximal fixpoint $\mathcal{K} = G(\mathcal{K})$ of G is given by $\mathcal{K} = G^i(\check{K}_\rho)$, where G^i denotes the ith-fold composition of G with itself. □

4.2 Approximation Analysis

In this subsection we discuss how well we can approximate $\mathcal{K}(K)$ with $\mathcal{K}(\check{K})$. It is well-known that the MCIS can be arbitrarily well over-approximated [29]. However, this is in strong contrast to under-approximations. It is shown in [8] that replacing K with an arbitrarily small under-approximation may lead to an empty MCIS. Nevertheless, if K contains a controlled invariant set I "strictly inside", then by taking $\check{K}$ to be a sufficiently close approximation of K we can capture I in the sense that $I \subseteq \mathcal{K}(\check{K})$. In other words, if the synthesis problem has a solution for a safe set that is "strictly inside" K, we are guaranteed to find that solution. We now make these ideas precise.

DEFINITION 2. *A set $I \subseteq X$ is called controlled invariant with respect to S_φ if for every $z \in I$ there exists a run ζ of S_φ such that $\zeta_0 = z$ and $\zeta_t \in I$ for all $t \in \mathbb{N}_0$.*

We now formalize the meaning of "strictly inside" by requiring $\mathsf{d_H}(I, \partial K) \geq \gamma$ for some $\gamma > 0$ where $\mathsf{d_H}$ is the Hausdorff distance between the set I and the boundary of K denoted by ∂K. Whenever $\mathsf{d_H}(I, \partial K) \geq \gamma$ holds, we can construct an under-approximation $\check{K}$ of K containing I and thus we have the guarantee $I \subseteq \mathcal{K}(\check{K})$ by maximality of $\mathcal{K}(\check{K})$.

In the following, we use $\check{K}_\rho$ to indicate that the under-approximation $\check{K}$ is obtained with respect to the grid $[\mathbb{R}^n]_\rho$.

THEOREM 2. *Let $I \subseteq X$ be a controlled invariant set with respect to S_φ contained in $I \subseteq K \subseteq X$ satisfying $\mathsf{d_H}(I, \partial K) \geq \gamma$ for some $\gamma > 0$. Then, there exists $\rho \in \mathbb{R}_+$ such that $I \subseteq \mathcal{K}(\check{K}_\rho)$.*

PROOF. Suppose the assertion is not true. Then there exists a run ζ of S_φ with initial state $\zeta_0 \in I \backslash (\cup_{\rho\in\mathbb{R}_+} \mathcal{K}(\check{K}_\rho))$ with $\zeta_t \in I$ for all $t \in N_0$. Therefore, there exists $\gamma \in \mathbb{R}_+$ with $\mathsf{d_H}(\pi_{\mathbb{R}^n}(\zeta_t), \partial K(\pi_Q(\zeta_t))) \geq \gamma$. Moreover, there exists $\rho_0 \in \mathbb{R}_+$ such that for all $\rho \in \mathbb{R}_+$, $\rho \leq \rho_0$ we have $\mathsf{d_H}(\check{K}_\rho, K) \leq \gamma/2$, which implies $\zeta_t \in \check{K}_\rho$ for all $t \in \mathbb{N}_0$ and therefore $\zeta_0 \in \mathcal{K}(\check{K}_\rho)$. A contradiction. □

4.3 Input Constraints

So far we have assumed that there are no constraints on the inputs of the linear system. However, it is often the case that the control inputs are restricted to a subset $V \subseteq U$ of the input space $U = \mathbb{R}^m$. If that is the case we can aim at computing an input constrained version of the MCIS. That is, in the definition of the transition relation δ_φ, see (4), we enforce the inputs to be elements in $U = V$ instead of $U = \mathbb{R}^m$. Let $\mathcal{K}_V(K)$ denote the MCIS with respect to the constrained input space V. We incorporate such input constraints by extending the state space of the system so as to incorporate the inputs. The system matrices of the extended system are given by

$$A' = \begin{bmatrix} A & B \\ 0 & 0 \end{bmatrix}, \quad B' = \begin{bmatrix} 0 \\ I \end{bmatrix} \quad (9)$$

where I is the identity matrix in $\mathbb{R}^m$. Each element P'_i of the set of atomic propositions $\mathcal{P}'$ is correspondingly modified to

$$P'_i = P_i \times V$$

which in turn results in a modified safe set K' according to (5). The following lemma ensures the completeness and soundness of this scheme.

LEMMA 2. *Let S_φ and S'_φ be the transition systems obtained as the synchronous product of (1) and (9) with the bad-prefix automaton, respectively, and let K and K' denote the corresponding safe sets. Then, the input constrained MCIS of S_φ, denoted by $\mathcal{K}_V(K)$, coincides with the MCIS of S'_φ, denoted by $\mathcal{K}'(K')$, on X:*

$$\mathcal{K}_V(K) = \pi_X(\mathcal{K}'(K')),$$

where π_X is the projection from $X \times V$ to X.

PROOF. Let $(q, x) \in \mathcal{K}_V(K)$. This implies, that there exists a run ζ of S_φ with initial state (q, x) such that $\zeta_t \in K$ for all $t \geq 0$. Let $\upsilon : \mathbb{N}_0 \to V$ denote the associated input sequence. We define the sequence $\upsilon' : \mathbb{N}_0 \to V$ by a shift operation $\upsilon'_t := \upsilon_{t+1}$. By the definition of the extended system (9) we are able to chose a run ζ' of S'_φ with initial state $\zeta'_0 = (q, x, \upsilon_0)$ that satisfies

$$\zeta'_t = (\zeta_t, \upsilon_{t-1}).$$

As S_φ obeys the constraints $\zeta_t \in X$ and $\upsilon_t \in V$ it follows that $\zeta'_t \in X \times V$ for all $t \geq 0$ which implies that $(q, x, \upsilon_0) \in \tilde{\mathcal{K}}'$ and thus $\mathcal{K}_V \subset \pi_X(\mathcal{K}')$.

For the reverse inclusion, let $(q', x', u') \in \mathcal{K}'$. This implies that there exists a run ζ' of S'_φ with initial sate (q', x', u') that satisfies $\zeta'_t \in X \times V$ for all $t \geq 0$. Let π_V be the projection from $X \times V$ to V and define the sequence $\upsilon_t := \pi_V(\zeta'_t)$. We see, again by the definition of the extended system (9), that we can pick a run ζ of S_φ with initial state (q', x') that satisfies $(\zeta_t, \upsilon_t) = \zeta'_t$, which implies that $(q, x) \in \mathcal{K}_V$. $\square$

5. SYMBOLIC IMPLEMENTATION

In this section, we describe the implementation of our approach using binary decision diagrams (BDDs) and algebraic decision diagrams (ADDs). Decision diagrams are data structures that efficiently represent binary functions $f_b : \mathbb{B}^n \to \mathbb{B}$ with binary codomain ($\mathbb{B} = \{0, 1\}$) and binary functions $f_a : \mathbb{B}^n \to \mathbb{R}$ with real codomain. In particular, we focus on the two operations that are unique to our approach: the under-approximation of the atomic propositions and the computation of the set iterates (6). In what follows, we show how these two operations can be performed in a symbolic manner, *i.e.*, without iterating over the grid points. Moreover, all the operations are implemented using ordinary BDD manipulations like conjunction, disjunction, variable reordering and quantifier elimination as well as ADD manipulations like thresholding, addition and subtraction. Standard BDD packages as CUDD [30] provide efficient implementations for all of those operations.

5.1 Set Encoding

Our implementation of Algorithm 1 starts with the encoding of the safe set K, see (5), in terms of BDDs on a binary domain $\mathbb{B}^{Nn}$. In doing so, we assume that the sets $K(q) \subseteq \mathbb{R}^n$ are contained in the unit-cube $\mathsf{B}(0, 1)$. If that is not the case, we scale the sets and the continuous dynamics accordingly.

Once the grid $[\mathbb{R}^n]_\rho$ is fixed, the dimension Nn of the binary domain $\mathbb{B}^{Nn}$ corresponds to the product of the state space dimension $n \in \mathbb{N}$ and the number of bits $N \in \mathbb{N}$ that are required to binary encode the finite number of grid points of $[\mathsf{B}(0, 1)]_\rho = \mathsf{B}(0, 1) \cap [\mathbb{R}^n]_\rho$.

Given a grid point $\xi = (\xi_1, \ldots, \xi_n) \in [\mathsf{B}(0, 1)]_\rho$ we denote its binary encoding by $\xi_\mathbb{B} = (\xi_{1_\mathbb{B}}, \ldots, \xi_{n_\mathbb{B}}) \in \mathbb{B}^{Nn}$ with each $\xi_{i_\mathbb{B}} \in \mathbb{B}^N$.

5.2 Symbolic Approximation of Polytopes

Every polytope is obtained as a finite intersection of half-spaces. Henceforth, we focus on the approximation of half-spaces of the form

$$R = \{x \in \mathbb{R}^n \mid Cx \leq c\}$$

with $C \in \mathbb{R}^{1 \times n}$ and $c \in \mathbb{R}$. Once we have the BDD representation of the half-spaces, computing the intersection can be realized by computing the conjunction of the two BDDs representing these sets.

The under-approximation and the over-approximation of the set R is represented by the BDDs $\check{r} : \mathbb{B}^{Nn} \to \mathbb{B}$ and $\hat{r} : \mathbb{B}^{Nn} \to \mathbb{B}$, respectively, defined as:

$$\check{r}(\xi_\mathbb{B}) = 1 \quad \Leftrightarrow \quad \mathsf{C}_\rho(\xi) \subseteq R, \tag{10}$$

$$\hat{r}(\xi_\mathbb{B}) = 1 \quad \Leftrightarrow \quad \mathsf{C}_\rho(\xi) \cap R \neq \varnothing. \tag{11}$$

Rather than using the definition of $\check{r}$ and $\hat{r}$ we compute these BDDs by making use of the ADDs $\check{f} : \mathbb{B}^{Nn} \to \mathbb{R}$ and $\hat{f} : \mathbb{B}^{Nn} \to \mathbb{R}$ given by

$$\check{f}(\xi_\mathbb{B}) = \max_{x \in \mathsf{C}_\rho(\xi)} Cx \quad \text{and} \quad \hat{f}(\xi_\mathbb{B}) = \min_{x \in \mathsf{C}_\rho(\xi)} Cx.$$

The functions $\check{f}$ and $\hat{f}$ associate to each grid point ξ the maximum and minimum value of Cx with x restricted to $x \in \mathsf{C}_\rho(\xi)$. Notice that we have the following implications

$$\check{f}(\xi_\mathbb{B}) \leq c \Leftrightarrow \mathsf{C}_\rho(\xi) \subseteq R \quad \text{and} \quad \hat{f}(\xi_\mathbb{B}) \leq c \Leftrightarrow \mathsf{C}_\rho(\xi) \cap R \neq \varnothing.$$

Therefore, we are able to define the approximating BDDs $\check{r}$ and $\hat{r}$ simply by truncating the ADDs $\check{f}$ and $\hat{f}$. In particular, we obtain all grid points $\xi_\mathbb{B} \in \mathbb{B}^{Nn}$ that satisfy $\check{r}(\xi_\mathbb{B}) = 1$ respectively $\hat{r}(\xi_\mathbb{B}) = 1$ by

$$\check{r}^{-1}(1) = \{\xi_\mathbb{B} \in \mathbb{B}^{Nn} \mid \check{f}(\xi_\mathbb{B}) \leq c\}$$
$$\hat{r}^{-1}(1) = \{\xi_\mathbb{B} \in \mathbb{B}^{Nn} \mid \hat{f}(\xi_\mathbb{B}) \leq c\}.$$

The ADDs $\check{f}$ and $\hat{f}$ can in turn be efficiently computed as we outline in the reminder of this subsection.

To provide a clearer presentation, we consider the one-dimensional case $n = 1$ and assume that $C \geq 0$. We fix the particular encoding scheme as follows: a grid point $\xi \in [B(0, 1)]_\rho$ is obtained by a binary element $\xi_\mathbb{B} = b_0 \ldots b_{N-1}$ of $\mathbb{B}^N$ by

$$\xi = \sum_{i=0}^{N-1} 2^{-i} b_i - 1.$$

With this encoding scheme in mind, we compute $\check{f}$ iteratively over the number of bits by the functions $\check{f}^i : \mathbb{B}^i \to \mathbb{R}$ obtained from

$$\check{f}^0(b_0) = Cb_0$$
$$\check{f}^{i+1}(b_0 \ldots b_{i+1}) = \check{f}^i(b_0 \ldots b_i) - 2^{-i}C(1 - b_{i+1}).$$

It is easy to verify that $\check{f} = \check{f}^{N-1}$. We proceed in an analogous way for the general case, as well as to compute $\hat{f}$.

5.3 Implementation of the Set Iterates

In this subsection, we describe the implementation of the main computation in Algorithm 1, *i.e.*, the iteration (6) given by

$$K \cap K'$$

with $K' = \{z \in X \mid \exists z' \in K : (z, z') \in \delta_\varphi\}$. In particular we focus on how to obtain a BDD representation of K' given a BDD representation of K. The set K' can be written in terms of

$$K' = \{(q, x) \in X \mid \exists q' \in \pi_Q(K) \wedge (q, q') \in \delta \wedge x \in \mathsf{Pre}(K(q))\}.$$

with the Pre operator as defined in (8). Notice that the elements in $(q, x) \in K'$ do not need to be output synchronized, *i.e.*, we do not require $g(q) = h(x)$. The synchronization of the elements in K' is achieved by the intersection of K' with the synchronized elements K in (6).

In the following we describe, how we obtain K' in two steps. First, we compute the set $\tilde{K}$ from K by

$$\tilde{K} = \{(q, x) \in X \mid q \in \pi_Q(K) \wedge x \in \mathsf{Pre}(K(q))\}$$

and subsequently K' from $\tilde{K}$ by

$$K' = \{(q, x) \in X \mid \exists q' \in \pi_Q(\tilde{K}) : (q, q') \in \delta \wedge x \in \tilde{K}(q')\}.$$

It is not difficult to see that the computation of K' via $\tilde{K}$ can be obtained through standard BDD operations like existential abstraction and conjunction.

In the remainder of this subsection, we focus on the computation of $\tilde{K}$ from K, which basically corresponds to the implementation of the Pre operator. For the sake of presentation, let us focus on the single input case.

Let $k : \mathbb{B}^{|Q|Nn} \to \mathbb{B}$ denote the BDD representation of the set K given by

$$k(q_{\mathbb{B}}, \xi_{\mathbb{B}}) = 1 \Leftrightarrow \{q\} \times \mathsf{C}_\rho(\xi) \subseteq K.$$

Notice that we assume that all sets $K(q)$ are obtained from the above approximation procedure. Hence, the sets $K(q)$ are finitely representable and therefore, k is an exact representation of K.

As detailed in [28], whenever the linear system is given in Brunovsky normal form, the Pre computation is implemented by the following simple existential abstraction and variable shift:

$$\tilde{k}(q_{\mathbb{B}}, \xi_{\mathbb{B}_1}, \ldots, \xi_{\mathbb{B}_n}) = 1 \Leftrightarrow$$
$$\exists u_{\mathbb{B}} : k(q_{\mathbb{B}}, \xi_{\mathbb{B}_2}, \ldots, \xi_{\mathbb{B}_{n-1}}, u_{\mathbb{B}}) = 1.$$

We refer the reader to [28] for a more detailed explanation and the multi-dimensional input case.

Notice that all our computations, including the approximation of polytopes, are performed symbolically, *i.e.*, we avoid any iteration over the grid points $\xi_{\mathbb{B}}$. Moreover, none of the operations in our implementation includes the costly-to-obtain abstraction, *i.e.*, a BDD representation of the transition relation $\delta_\varphi \subseteq X \times X$, which is the case for several known approaches, *e.g.* [21, 10, 27, 35, 40] just to mention a few. In the proposed scheme, the computation of the abstraction is restricted to the safe set K. As a result, we are able to save memory as well as computation time.

6. EXPERIMENTAL RESULTS

In this section, we demonstrate the performance of the developed algorithm for some examples. In the first subsection, we synthesize a cruise controller for a truck on a highway. In the second part, we compare the performance of Algorithm 1 with the polyhedral approach, described in [14] and [26] in terms of an example from [24].

The accuracy of the computed solutions will be estimated by the upper bound $\hat{e}$ of the relative error between the volume of the MCIS and the volume of the computed under-approximation:

$$\hat{e} = \frac{\mathsf{vol}\,\mathcal{K}(\hat{K}) - \mathsf{vol}\,\mathcal{K}(\check{K})}{\mathsf{vol}\,\mathcal{K}(\check{K})} \geq \frac{\mathsf{vol}\,\mathcal{K}(K) - \mathsf{vol}\,\mathcal{K}(\check{K})}{\mathsf{vol}\,\mathcal{K}(K)}.$$

We implemented all the algorithms in C using the decision diagram library from the University of Colorado (CUDD) [30]. All computations are performed on an Intel Core i7 1.8GHz processor using 4GBytes of memory.

6.1 Cruise Controller for a Truck

We return to the example that was briefly described in Section 3.

Recall that we seek to design a controller for the truck with a trailer depicted in Figure 1. The objective of the controller is to enforce the highway speed limits. We start with the continuous-time linear model:

$$\dot{x} = \begin{bmatrix} 0 & -1 & 1 \\ \frac{k_s}{m} & -\frac{k_d}{m} & \frac{k_d}{m} \\ 0 & 0 & 0 \end{bmatrix} x + \begin{bmatrix} 0 \\ 0 \\ 1 \end{bmatrix} u,$$

where the entries of the state vector $x = (d, v_2, v_1) \in \mathbb{R}^3$ correspond to the distance d between the truck and the trailer, the velocity of the trailer v_2 and the velocity of the truck v_1, respectively. Here, we model the connection between the truck and the trailer by a spring-damper system with constants $k_s = 4500\,\mathrm{N/kg}$ and $k_d = 4600\,\mathrm{Ns/m}$. The mass of the trailer is fixed to $m = 1000\,\mathrm{kg}$. The input of our model is the constrained acceleration $u \in [-4, 4]\,\mathrm{m/s^2}$ of the truck.

We consider three speed limits $v_a = 15.6\,\mathrm{m/s}$ (ca. 35 mph), $v_b = 24.5\,\mathrm{m/s}$ (ca. 55 mph) and $v_c = 29.5\,\mathrm{m/s}$ (ca. 66 mph) and would like to design a controller, that guarantees that the truck obeys the effective velocity constraint. In particular, we would like to ensure that the truck obeys the current speed limit at most $T \in \mathbb{N}$ time steps after the controller receives the information that the speed limit has changed. The number of steps T, after which we enforce the speed limit is a parameter that we vary throughout the experiments.

Truck without trailer. We provide a complete description of a simplified version of this problem. However, we present the experimental results for the original example. The simplified version consists of the truck without a trailer subject only to two speed limits v_a and v_b. As a first step, we discretize the continuous-time system by sampling the solution of the system under piecewise constant inputs, with sampling rate $h = 0.4\,\mathrm{s}$. In the second step, we extend the state space with the input space so that we are able to account for the input constraints $u \in [-4, 4]$. The resulting system is given by

$$z^+ = \begin{bmatrix} 1 & h \\ 0 & 0 \end{bmatrix} z + \begin{bmatrix} 0 \\ h \end{bmatrix} v \qquad (12)$$

with $z = (v_1, u)$ and unconstrained input v. In addition to the model of the truck we introduce an automaton as a model for the environment, *i.e.*, a model for the change of speed limits over time. The automaton is sketched in Figure 2. It has two states, m_a and m_b, one for each speed limit and it can arbitrarily change between these states. The full plant model is given by the composition of the linear system (12) and the automaton in Figure 2.

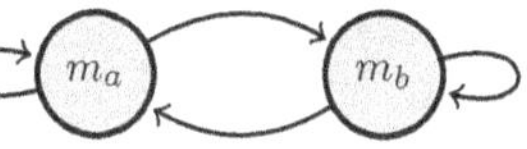

Figure 2: Model of the highway.

Notice that so far we assumed that our plant is given by a linear system and not as a switched system with linear dynamics. However, as we will describe below, for this particular example, it is an easy task to modify Algorithm 1 accordingly.

We define the atomic propositions by $P_0 = \mathcal{H}(C_0, c_0)$, $P_1 = \mathcal{H}(C_1, c_1)$ and $P_2 = \mathcal{H}(C_1, c_2)$ with

$$C_0 = \begin{bmatrix} -1 & 0 \\ 1 & 0 \\ 0 & -1 \\ 0 & 1 \end{bmatrix}, \quad c_0 = \begin{bmatrix} 0 \\ 35 \\ 4 \\ 4 \end{bmatrix}$$

and $C_1 = \begin{bmatrix} 1 & 0 \end{bmatrix}$, $c_1 = 15.6\,\mathrm{m/s}$, $c_2 = 24.5\,\mathrm{m/s}$. We use P_0 to specify our operating speed range $5 \leq v_1 \leq 35\,\mathrm{m/s}$ as well as to account for the input constraints $-4 \leq u \leq 4\,\mathrm{m/s^2}$. The propositions P_1 and P_2 are used to enforce the speed limits v_a and v_b, respectively.

The specification is given by the safe-LTL formula (3) and the corresponding bad-prefix finite state automaton for $T = 2$ is depicted in Figure 3.

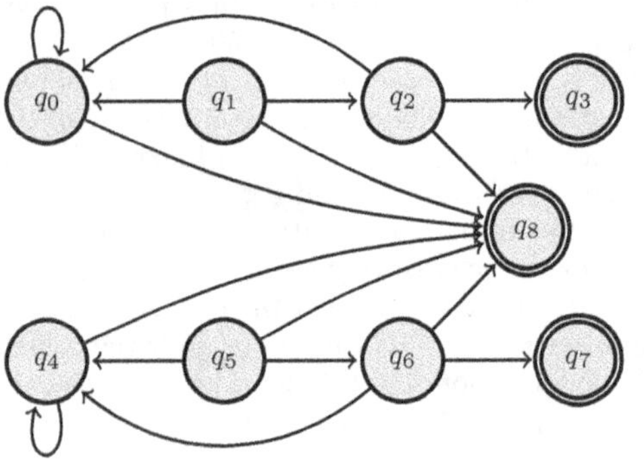

Figure 3: The bad-prefix finite state automaton accepting the prefixes of the undesired behavior.

The output function $g : Q \to Y$ is given by

$$\begin{aligned} g(q_0) &= \{m_a, P_0, P_1\}, & g(q_1) &= \{m_a, P_0\}, \\ g(q_2) &= \{m_a, P_0\}, & g(q_3) &= \{m_a, P_0\}, \\ g(q_4) &= \{m_b, P_0, P_2\}, & g(q_5) &= \{m_b, P_0\}, \\ g(q_6) &= \{m_b, P_0\}, & g(q_7) &= \{m_b, P_0\}, \\ g(q_8) &= \varnothing. \end{aligned}$$

The states q_0 and q_4 of the bad-prefix automaton represent the states where the speed limits v_a and v_b, respectively are met by the truck. For the states q_1 and q_5 the truck is allowed to violate the speed limit, but has to reach q_0 or q_4 in two steps. The same holds for q_2 and q_6 only that the truck needs to meet the speed limits in one step. Otherwise, the system ends up in one of the marked states, which represent the undesired behavior.

We now form the synchronous composition of the bad prefix finite state automaton with the plant. Recall that the plant is in fact the synchronous composition of the continuous dynamics in (12) with the automaton in Figure 2 that has $M = \{m_a, m_b\}$ as set of states. The set that we want to render invariant is given by $K_0 = Q_0 \times M \times \mathbb{R}^2$ with $Q_0 = \{q_0, q_1, q_2, q_4, q_5, q_6\}$. Note that we do not have control over the change of the speed limit. Hence, we need to compute a MCIS that is robust with respect to uncontrollable changes in speed limits. This is achieved by replacing the sets K_i in each iteration just before the computation of (6) in Algorithm 1 with

$$K_i' = \{(q, m, x) \in K_i \mid (q, m_a, x) \in K_i \wedge (q, m_b, x) \in K_i\}.$$

This replacement ensures that the elements in $\mathcal{K}$ are independent of the mode $m \in M$. A more general approach to handle the action of the environment is to consider a game whose solution would require replacing Algorithm 1 with a version that is robust with respect to environment actions.

We approximately computed the MCIS $\mathcal{K}$ where we used 10 bits in each dimension to approximate the atomic propositions. The continuous part of $\mathcal{K}$, associated to the states q_0 and q_4 is illustrated in the left subplots of Figure 4.

The cutoff corners of the sets clearly indicate the integrator dynamics of the truck. Notice that, even though for q_4 the truck is allowed to drive up to $v_b = 24.5\,\mathrm{m/s}$, it stays close to $v_a = 15.6\,\mathrm{m/s}$. That results from the fact, that the truck needs to be able to reduce its velocity to v_b within $Th = 0.8\,\mathrm{s}$ at all times. We show the approximation of the MCIS $\check{\mathcal{K}}$ for the case $T = 10$ in the right subplots of Figure 4. As expected, the truck is now allowed to drive at a much higher speed in mode m_b compared to the case $T = 2$. The run-time t_r of the algorithm to compute $\check{\mathcal{K}}$ for both cases is below a second and the error is bounded by $\hat{e} \leq 0.2$.

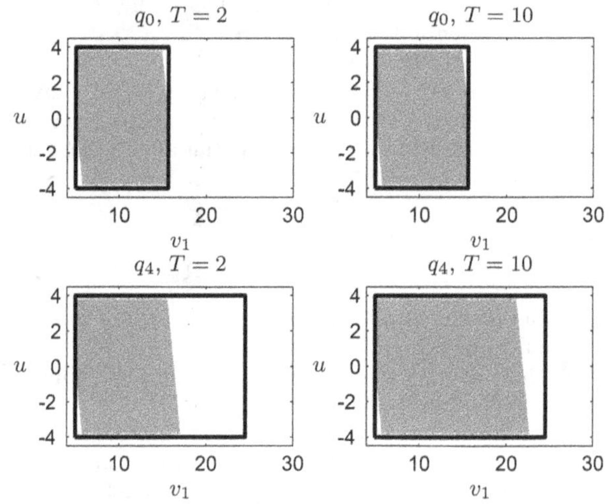

Figure 4: The set $\check{\mathcal{K}}$ associated to q_0 and q_4 using $N = 10$ bits per dimension for $T = 2$ and $T = 10$.

The run-times of Algorithm 1 and error estimates for various parameters $T \in \{2, 10\}$ and $N \in \{10, 11, 12\}$ are listed in Table 1. We conducted the computations with all three speed limits, which results in a maximum number of 33 discrete states on the bad-prefix automaton.

N\T	2		10	
	t_r	$\hat{e}$	t_r	$\hat{e}$
10	0.2s	0.24	0.2s	0.27
11	0.3s	0.12	0.3s	0.13
12	0.3s	0.06	0.3s	0.06

Table 1: Run-times of Alg. 1 and error bounds $\hat{e}$.

Truck with trailer. Now we consider the truck with trailer. Following the same steps as above, we sample the solution of the continuous-time system under piecewise constant inputs, in order to obtain a discrete-time system. Additionally, we augment the state space with the input space which results in the four-dimensional state vector

$$z = [d, v_2, v_1, u]^\top.$$

We use the same bad-prefix automaton to specify the desired system behavior. For this case, however, in addition to the speed limits we constrain the distance between the two trucks to the interval

$$-1/2\,\mathrm{m} \leq d \leq 1/2\,\mathrm{m}.$$

Thus, the polytopes associated to the atomic propositions $P_0 = \mathcal{H}(C_0, c_0)$, $P_1 = \mathcal{H}(C_1, c_1)$ and $P_2 = \mathcal{H}(C_1, c_2)$ are given by

$$C_0 = \begin{bmatrix} -1 & 0 & 0 & 0 \\ 1 & 0 & 0 & 0 \\ 0 & -1 & 0 & 0 \\ 0 & 1 & 0 & 0 \\ 0 & 0 & -1 & 0 \\ 0 & 0 & 1 & 0 \\ 0 & 0 & 0 & -1 \\ 0 & 0 & 0 & 1 \end{bmatrix}, \quad c_0 = \begin{bmatrix} 1/2 \\ 1/2 \\ 5 \\ 35 \\ 5 \\ 35 \\ 4 \\ 4 \end{bmatrix}$$

and $C_1 = \begin{bmatrix} 0 & 1 & 0 & 0 \end{bmatrix}$, $c_1 = 15.6\,\mathrm{m/s}$, $c_2 = 24.5\,\mathrm{m/s}$.

The times required to compute the MCISs are listed in Table 2. Compared to the previous case, we can clearly observe the burden of the higher dimensional state space in terms of the run-times and tightness of the approximation. Also, for this example the number of discrete states is observable in the computation times, *i.e.*, as we increase $T = 2$ to $T = 10$ the bad prefix automaton becomes larger and so do the computation times.

N\T	2		10	
	t_r	$\hat{e}$	t_r	$\hat{e}$
10	1m39s	2.31	2m40s	2.38
11	4m09s	1.01	4m31s	1.04
12	6m48s	0.58	7m52s	0.62
13	10m38s	0.43	16m01s	0.46

Table 2: Run-times of Alg. 1 and error bounds $\hat{e}$.

6.2 Comparison with an Example from the Literature

In this subsection, we show some evidence of the previously mentioned claim, that the combinatorial complexity within the known polyhedral approach [14, 26] constitutes a serious problem when applied to controller synthesis for safe-LTL specifications.

For this purpose, we consider Example 5.1 introduced in [24]. This example describes a linear system with a constrained three dimensional state space and a constrained two dimensional input space. The set of atomic propositions is formulated in terms of subsets of the state space $Y = \mathsf{B}(\mathbf{0}, \mathbf{30})$, $O_1 = \mathsf{B}(\mathbf{10}, \mathbf{5})$, $O_2 = \mathsf{B}(\mathbf{-5}, \mathbf{5})$ and $O_3 = \mathsf{B}(\mathbf{-15}, \mathbf{10})$ together with subsets of the input space $V = \mathsf{B}(\mathbf{0}, \mathbf{2})$, $W_1 = \mathsf{B}(\mathbf{-3/2}, \mathbf{1/2})$, $W_2 = \mathsf{B}(\mathbf{-1/4}, \mathbf{1/4})$ and $W_3 = \mathsf{B}(\mathbf{2/5}, \mathbf{1/5})$. Here we use the bold notation, to indicate that the number corresponds to a vector in the appropriate dimension, e.g., for $Y = \mathsf{B}(\mathbf{0}, \mathbf{30})$ we have $\mathbf{0} = [0,0,0]^\mathsf{T}$ and $\mathbf{30} = [30, 30, 30]^\mathsf{T}$.

We interpret the sets Y and V as the domain of the problem, while the sets O_i and W_i represent obstacles, in the state space and input space, respectively. We computed the MCIS $\mathcal{K}$ for the following specifications of increasing complexity

$$\begin{aligned}
\varphi_0 &= \square(Y \times V) \\
\varphi_1 &= \square((Y \wedge \neg O_1) \times V) \\
\varphi_2 &= \square(Y \times (V \wedge \neg W_1)) \\
\varphi_3 &= \square((Y \wedge \neg O_1) \times (V \wedge \neg W_1)) \\
\varphi_4 &= \square((Y \wedge_{i=1}^{2} \neg O_i) \times (V \wedge_{i=1}^{2} \neg W_i)) \\
\varphi_5 &= \square((Y \wedge_{i=1}^{3} \neg O_i) \times (V \wedge_{i=1}^{2} \neg W_i)) \\
\varphi_6 &= \square((Y \wedge_{i=1}^{3} \neg O_i) \times (V \wedge_{i=1}^{3} \neg W_i))
\end{aligned}$$

In all of the conducted computations the polyhedral algorithm terminated. We show the run-times of the computations in comparison with our symbolic implementation of Algorithm 1 in Figure 5. The subplot inside the main plot shows the outcome for $\varphi_0, \ldots, \varphi_4$. Clearly, for simple specifications the polyhedral approach outperforms the symbolic scheme. This simply results from the fact that the safe set first has to be approximated, which takes more than ninety percent of the time. However, as the problem becomes more complex the approximation effort lessens in comparison with the effort to compute the set iterates. For example, for the problem with two obstacles in the state space and input space, *i.e.*, φ_4, we can observe that the proposed symbolic approach starts to outperform the polyhedral approach.

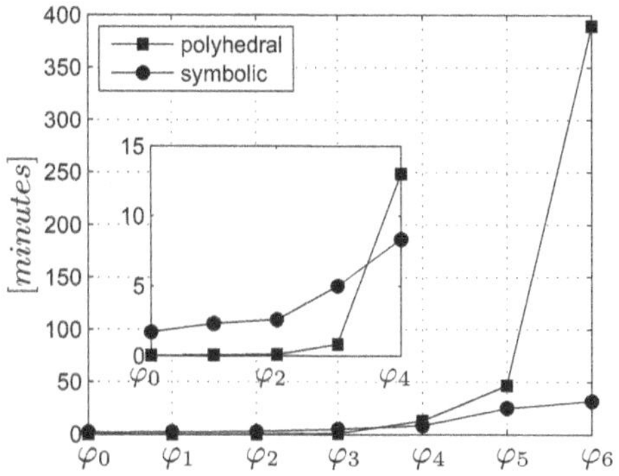

Figure 5: Comparison of the polyhedral approach with the symbolic scheme in terms of run-times.

We implemented the polyhedral approach using MATLAB and the multi-parametric toolbox [18]. The toolbox provides several methods to compute intersections and projections of polytopes. Specifically, we used the mex interface to the "fast" C implementation of the Fourier-Motzkin quantifier elimination to compute polyhedral projections, see `help polytope/projection`.

7. DISCUSSION

In this paper we presented an approach to the synthesis of controllers enforcing safe-LTL specifications on discrete-time linear systems. It was shown through examples that it scales up to 5 continuous variables thus placing it at the forefront of the existing controller synthesis techniques enforcing temporal logic specifications on continuous systems. We are currently working to increase the size of the systems that can be handled by improving the approximation of sets which constitutes the bottleneck of the current implementation.

8. REFERENCES

[1] A. Abate, J. P. Katoen, J. Lygeros, and M. Prandini. Approximate model checking of stochastic hybrid systems. *European Journal of Control*, 16:624, 2010.

[2] J. P. Aubin. *Viability Theory.* Systems & Control: Foundations & Applications. Birkhäuser, 1991.

[3] D. Bertsekas and I. B. Rhodes. On the minimax reachability of target sets and target tubes. *Automatica*, 7:233–247, 1971.

[4] F. Blanchini. Ultimate boundedness control for uncertain discrete-time systems via set-induced lyapunov functions. In *Proc. of the 30th IEEE CDC*, pages 1755–1760, 1991.

[5] F. Blanchini and S. Miani. *Set-Theoretic Methods in Control.* Systems & Control: Foundations & Applications. Birkhäuser, 2008.

[6] O. Bournez, O. Maler, and A. Pnueli. Orthogonal polyhedra: Representation and computation. In *HSCC*, LNCS, pages 46–60. Springer, 1999.

[7] P. Brunovský. A classification of linear controllable systems. *Kybernetika*, 6:173–188, 1970.

[8] P. Collins. Optimal semicomputable approximations to reachable and invariant sets. *Theory of Computing Systems*, 41:33–48, 2007.

[9] E. De Santis, M. D. Di Benedetto, and L. Berardi. Computation of maximal safe sets for switching systems. *IEEE TAC*, 49:184–195, 2004.

[10] G. E. Fainekos, A. Girard, H. Kress-Gazit, and G. J. Pappas. Temporal logic motion planning for dynamic robots. *Automatica*, 45:343–352, 2009.

[11] C. Finucane, G. Jing, and H. Kress-Gazit. LTLMoP website. `http://ltlmop.github.com/`, 2010.

[12] A. Girard and G. J. Pappas. Hierarchical control system design using approximate simulation. *Automatica*, 45:566–571, 2009.

[13] E. A. Gol, M. Lazar, and C. Belta. Language-guided controller synthesis for discrete-time linear systems. In *HSCC*, pages 95–104. ACM, 2012.

[14] E. C. Kerrigan. *Robust Constraint Satisfaction: Invariant Sets and Predictive Control.* PhD thesis, Dep. of Eng., University of Cambridge, 2000.

[15] M. Kloetzer and C. Belta. LTL-Con website. `http://iasi.bu.edu/Software.html`, 2006.

[16] M. Kloetzer and C. Belta. A fully automated framework for control of linear systems from temporal logic specifications. *IEEE TAC*, 53:287–297, 2008.

[17] O. Kupferman and M. Y. Vardi. Model checking of safety properties. *Formal Methods in System Design*, 19:291–314, 2001.

[18] M. Kvasnica, P. Grieder, M. Baotić, and M. Morari. Multi-parametric toolbox (mpt). In *HSCC*, volume 2993, pages 121–124. Springer, 2004.

[19] T. Latvala. Efficient model checking of safety properties. In *In Model Checking Software. 10th International SPIN Workshop*, pages 74–88, 2003.

[20] R. Majumdar and M. Zamani. Approximately bisimilar symbolic models for digital control systems. In *CAV*, volume 7358, pages 362–377. Springer, 2012.

[21] M. Mazo Jr., A. Davitian, and P. Tabuada. Pessoa website.. `http://www.cyphylab.ee.ucla.edu/pessoa`, 2009.

[22] M. Mazo Jr., A. Davitian, and P. Tabuada. Pessoa: A tool for embedded controller synthesis. In *CAV*, volume 6174 of *LNCS*, pages 566–569. Springer, 2010.

[23] T. Moor, J. Raisch, and S. O'Young. Discrete supervisory control of hybrid systems based on l-complete approximations. *Discrete Event Dynamic Systems*, 12:83–107, 2002.

[24] E. Pérez, C. Ariño, F. X. Blasco, and M. A. Martínez. Maximal closed loop admissible set for linear systems with non-convex polyhedral constraints. *Journal of Process Control*, pages 529 – 537, 2011.

[25] A. Pnueli. The temporal logic of programs. In *Proc. of 18th Annual Symp. on Foundations of Computer Science*, pages 46–57, 1977.

[26] S. V. Rakovic, P. Grieder, M. Kvasnica, D. Q. Mayne, and M. Morari. Computation of invariant sets for piecewise affine discrete time systems subject to bounded disturbances. In *Proc. of the 43rd IEEE CDC*, pages 1418–1423, 2004.

[27] G. Reißig. Computing abstractions of nonlinear systems. *IEEE TAC*, 56:2583–2598, 2011.

[28] M. Rungger, M. Mazo Jr., and P. Tabuada. Scaling up controller synthesis for linear systems and safety specifications. In *Proc. of the 51th IEEE CDC*, 2012.

[29] P. Saint-Pierre. Approximation of the viability kernel. *Applied Math & Optimization*, 29:187–209, 1994.

[30] F. Somenzi. *CUDD: CU Decision Diagram Package. Release 2.5.0.* University of Colorado at Boulder, 2012. `http://vlsi.colorado.edu/~fabio/CUDD/`.

[31] P. Tabuada. *Verification and Control of Hybrid Systems - A Symbolic Approach.* Springer, 2009.

[32] P. Tabuada and G. J. Pappas. Model checking LTL over controllable linear systems is decidable. In *HSCC*, pages 498–513. Springer, 2003.

[33] P. Tabuada and G. J. Pappas. Linear time logic control of discrete-time linear systems. *IEEE TAC*, 51:1862–1877, 2006.

[34] A. Tarski. A lattice-theoretical fixpoint theorem and its applications. *Pacific Journal of Mathematics*, 5:285–309, 1955.

[35] Y. Tazaki and J. Imura. Discrete abstractions of nonlinear systems based on error propagation analysis. *IEEE TAC*, 57:550–564, 2012.

[36] I. Tkachev and A. Abate. On infinite-horizon probabilistic properties and stochastic bisimulation functions. In *Proc. of the 50th IEEE CDC and ECC*, pages 526–531, 2011.

[37] R. Vidal, S. Schaffert, J. Lygeros, and S. Sastry. Controlled invariance of discrete time systems. In *HSCC*, pages 437–451. Springer, 2000.

[38] R. Vidal, S. Schaffert, O. Shakernia, J. Lygeros, and S. Sastry. Decidable and semi-decidable controller synthesis for classes of discrete time hybrid systems. In *Proc. of the 40th IEEE CDC*, pages 1243–1248, 2001.

[39] I. Wegener. *Branching Programs and Binary Decision Diagrams - Theory and Applications.* SIAM Monographs on Discrete Mathematics and Applications, 2000.

[40] T. Wongpiromsarn, U. Topcu, and R. M. Murray. Receding horizon temporal logic planning. *IEEE TAC*, 57:2817–2830, 2012.

[41] T. Wongpiromsarn, U. Topcu, N. Ozay, H. Xu, and R.M. Murray. TuLiP website. `http://sourceforge.net/apps/mediawiki/tulip-control/`, 2010.

[42] B. Yordanov, J. Tůmová, I. Černá, J. Barnat, and C. Belta. Formal analysis of piecewise affine systems through formula-guided refinement. *Automatica*, 2012.

Temporal Logic Model Predictive Control for Discrete-Time Systems

Ebru Aydin Gol
Boston University
Boston, MA 02215, USA
ebru@bu.edu

Mircea Lazar
Eindhoven University of Technology
Den Dolech 2, 5600MB,
Eindhoven, The Netherlands.
m.lazar@tue.nl

ABSTRACT

This paper proposes an optimal control strategy for a discrete-time linear system constrained to satisfy a temporal logic specification over a set of linear predicates in its state variables. The cost is a quadratic function that penalizes the distance from desired state and control trajectories. The specification is a formula of syntactically co-safe Linear Temporal Logic (scLTL), which can be satisfied in finite time. It is assumed that the reference trajectories are only available over a finite horizon and a model predictive control (MPC) approach is employed. The MPC controller solves a set of convex optimization problems guided by the specification and subject to progress constraints. The constraints ensure that progress is made towards the satisfaction of the formula with guaranteed satisfaction by the closed-loop trajectory. The algorithms proposed in this paper were implemented as a software package that is available for download. Illustrative case studies are included.

Categories and Subject Descriptors

I.2.8 [**Artificial Intelligence**]: Problem Solving, Control Methods, and Search—*Control theory*; D.2.4 [**Software Engineering**]: Software/Program Verification—*Formal methods*

Keywords

Linear temporal logic, Formal synthesis, Model predictive control

1. INTRODUCTION

In recent years, there has been increasing interest in formal synthesis of control strategies for dynamical systems [1–10]. Unlike "classical" control problems, in which the specifications are stability or closeness to a reference point or trajectory possibly coupled with safety, the above works allow for richer specifications that translate to formulas of temporal logics such as Linear Temporal Logic (LTL) [11] and fragments of LTL such as GR(1) [12] and syntactically co-safe LTL (scLTL) [13]. The main challenge in the above works is to abstract the dynamics of the system to a finite-state transition graph, for which automata game techniques can be used for control synthesis [11].

In this work, we focus on synthesis of optimal control strategies from temporal logic specifications. While some results exist for finite systems such as deterministic transition systems and Markov Decision Processes [14, 15], this problem is largely open for systems with infinitely many states. We treat the temporal logic specifications as constraints in the optimal control problem and propose a model predictive control (MPC) solution, as the natural approach for such constrained problems. The main challenge in this work is the integration of the temporal logic specifications in the MPC problem.

Within the field of hybrid systems, MPC has been studied quite extensively. MPC control schemes were developed for rich subclasses of hybrid systems including piece-wise affine systems [16], hybrid dynamical systems with continuous and discrete states [17], and mixed logical dynamical systems, which include hybrid automata [18]. However, in these works, only classical control specifications were considered, such as asymptotic stability [16, 17], attractivity [18] and safety [17, 18].

MPC under temporal logic specifications was first studied in [5], where discrete-time linear systems and specifications in the form of a formula of GR(1) were considered. A receding horizon controller was derived for a finite abstraction of the system and then refined to the original system. The satisfaction of the specification was guaranteed by assuming that the specification automaton had a known partial order structure. In our recent work on MPC of finite-state transition systems [14], the satisfaction of the specification of an arbitrary LTL formula was guaranteed by a Lyapunov-type function rather than assuming a particular structure of the automaton. In this work, we extend this technique to dynamical systems with infinitely many states.

In this paper we consider the following set-up, which was not considered by the above-mentioned works: MPC of discrete-time linear systems subject to scLTL formulas over linear predicates in the state variables. The cost is a quadratic function that penalizes the distance between the actual and desired state and control trajectories over a finite time horizon. The goal is to find a control strategy such that the trajectory of the closed-loop system originating from a given initial state satisfies the formula and minimizes the cost. Note that the syntactically co-safe fragment of LTL is rich enough to express a wide spectrum of finite-time properties of dynamical systems, *e.g.*, "Go to A or B and avoid C for all times before reaching T. Do not go to D unless E was visited before." The results from this paper can be extended to discrete-time piecewise affine systems as briefly outlined in Section 6 - the focus on linear systems is for simplicity of presentation.

Our approach to the above problem consists of two main steps.

HSCC'13, April 8–11, 2013, Philadelphia, Pennsylvania, USA.

First, by using the framework developed in [7], we perform an iterative partitioning of the state space guided by an automaton enforcing the satisfaction of the scLTL formula. Second, we design an MPC controller over the automaton and the state space of the system. Essentially, we use the automaton to translate the formula into a type of constraint that can be embedded into the MPC problem. In contrast to [5], this allows us to combine the automaton path generation and the control synthesis processes to produce a more efficient and less conservative exploration of the state space. The proposed MPC controller produces an optimal control sequence with respect to the available reference trajectory by solving a set of quadratic programs (QPs). The first control is applied and the process is repeated until a final state of the automaton is reached. The constraints of the optimization problem guarantee that the produced trajectory follows an automaton path while making progress towards a final state. As a result, the closed-loop trajectory satisfies the specification while the cost over the available finite-horizon reference trajectories is minimized at each time step.

The remainder of the paper is organized as follows. In Section 2, we introduce some notation and preliminaries necessary throughout the paper. We formulate the problem and outline our approach in Section 3. The specification-guided partition of the state space and the definition of the function that will be used to enforce the satisfaction are presented in Section 4. The main result of the paper is included in Section 5, where we show that the MPC control strategy is guaranteed to satisfy the specification. The extension to discrete-time PWA systems is discussed in Section 6. We present two case studies in Section 7 before we conclude with final remarks in Section 8.

2. NOTATION AND PRELIMINARIES

For a set $\mathcal{S}$, $\mathrm{Co}(\mathcal{S})$, $\#\mathcal{S}$, and $2^{\mathcal{S}}$ stand for its convex hull, cardinality, and power set, respectively. We use $\mathbb{R}$, $\mathbb{R}_+$, $\mathbb{Z}$, and $\mathbb{Z}_+$ to denote the sets of real numbers, non-negative reals, integer numbers, and non-negative integers. For $m,n \in \mathbb{Z}_+$, we use $\mathbb{R}^n$ and $\mathbb{R}^{m\times n}$ to denote the set of column vectors and matrices with n and $m \times n$ real entries, respectively.

A polyhedron (polyhedral set) in $\mathbb{R}^n$ is the intersection of a finite number of open and/or closed half-spaces. A polytope is a compact polyhedron. We use $\mathcal{V}(\mathcal{P})$ to denote the set of vertices of a polytope $\mathcal{P}$.

In this work, the control specifications are given as formulas of syntactically co-safe linear temporal logic (scLTL):

Definition 2.1 [19] An scLTL formula over a set of atomic propositions P is inductively defined as follows:

$$\Phi := p|\neg p|\Phi\vee\Phi|\Phi\wedge\Phi|\Phi\ \mathcal{U}\ \Phi|\bigcirc\Phi|\diamondsuit\Phi, \tag{1}$$

where p is an atomic proposition, $\neg$ (negation), $\vee$ (disjunction), $\wedge$ (conjunction) are Boolean operators, and $\bigcirc$ ("next"), $\mathcal{U}$ ("until"), and $\diamondsuit$ ("eventually") are temporal operators.

An infinite word w over a finite set Σ is an infinite sequence $w = w_0w_1\ldots$, where $w_i \in \Sigma$ for all $i \in \mathbb{Z}_+$. Similarly, a finite word w over a finite set Σ is a finite sequence $w = w_0\ldots w_d$, where $d \in \mathbb{Z}_+$ and $w_i \in \Sigma$ for all $i = 0,\ldots,d$.

The semantics of scLTL formulas is defined over infinite words over 2^P as follows:

Definition 2.2 The satisfaction of an scLTL formula Φ at position $i \in \mathbb{Z}_+$ of a word w over 2^P, denoted by $w_i \models \Phi$, is recursively defined as follows:

- $w_i \models p$ if $p \in w_i$,
- $w_i \models \neg p$ if $p \notin w_i$,
- $w_i \models \Phi_1 \vee \Phi_2$ if $w_i \models \Phi_1$ or $w_i \models \Phi_2$,
- $w_i \models \bigcirc\Phi$ if $w_{i+1} \models \Phi$,
- $w_i \models \Phi_1\ \mathcal{U}\ \Phi_2$ if there exists $j \geq i$ such that $w_j \models \Phi_2$ and for all $i \leq k < j$ $w_k \models \Phi_1$,
- $w_i \models \diamondsuit\Phi$ if there exists $j \geq i$ such that $w_j \models \Phi$.

A word w satisfies an scLTL formula Φ, written as $w \models \Phi$, if $w_0 \models \Phi$.

An important property of scLTL formulas is that, even though they have infinite-time semantics, their satisfaction is guaranteed in finite time. Explicitly, for any scLTL formula Φ over P, any satisfying infinite word over 2^P contains a satisfying finite prefix[1]. We use $\mathcal{L}_\Phi$ to denote the set of all (finite) prefixes of all satisfying infinite words.

Definition 2.3 A deterministic finite state automaton (FSA) is a tuple $\mathcal{A} = (Q,\Sigma,\rightarrow_{\mathcal{A}},Q_0,F)$, where Q is a finite set of states, Σ is a set of symbols, $\rightarrow_{\mathcal{A}}\subseteq Q\times\Sigma\times Q$ is a deterministic transition relation, $Q_0 \subseteq Q$ is a set of initial states, and $F \subseteq Q$ is a set of final states.

An accepting run $r_{\mathcal{A}}$ of an automaton $\mathcal{A}$ on a finite word $w = w_0\ldots w_d$ over Σ is a sequence of states $r_{\mathcal{A}} = q_0\ldots q_{d+1}$ such that $q_0 \in Q_0$, $q_{d+1} \in F$ and $(q_i,w_i,q_{i+1}) \in\rightarrow_{\mathcal{A}}$ for all $i = 0,\ldots,d$. The set of all words corresponding to all of the accepting runs of $\mathcal{A}$ is called the language accepted by $\mathcal{A}$ and is denoted as $\mathcal{L}_{\mathcal{A}}$.

For any scLTL Φ formula over P, there exists an FSA $\mathcal{A}$ with input alphabet 2^P that accepts the prefixes of all the satisfying words, *i.e.*, $\mathcal{L}_\Phi$ [19]. There are algorithmic procedures and off-the-shelf tools, such as *scheck2* [20], for the construction of such an automaton.

Definition 2.4 Given an FSA $\mathcal{A} = (Q,\Sigma,\rightarrow_{\mathcal{A}},Q_0,F)$, its dual automaton is a tuple $\mathcal{A}^D = (Q^D,\rightarrow^D,\Gamma^D,\tau^D,Q_0^D,F^D)$, where

$$\begin{aligned}
&Q^D = \{(q,\sigma,q') \mid (q,\sigma,q') \in\rightarrow_{\mathcal{A}}\},\\
&\rightarrow^D = \{((q,\sigma,q'),(q',\sigma',\bar{q})) \mid (q,\sigma,q'),(q',\sigma',\bar{q}) \in\rightarrow_{\mathcal{A}}\},\\
&\Gamma^D = \Sigma,\\
&\tau^D : Q^D \rightarrow \Gamma^D, \quad \tau^D((q,\sigma,q')) = \sigma,\\
&Q_0^D = \{(q_0,\sigma,q) \mid q_0 \in Q_0\},\\
&F^D = \{(q,\sigma,q') \mid q' \in F\}.
\end{aligned}$$

Informally, the states of the dual automaton $\mathcal{A}^D$ are the transitions of $\mathcal{A}$. There is a transition between two states of $\mathcal{A}^D$ if the corresponding transitions are connected by a state in $\mathcal{A}$. Γ^D is a set of output symbols and it is the same as the set of symbols of $\mathcal{A}$. τ^D is an output function. For a state of $\mathcal{A}^D$, τ^D produces the symbol that enables the transition in $\mathcal{A}$. The set of initial states Q_0^D of $\mathcal{A}^D$ is the set of all transitions that leave an initial state in $\mathcal{A}$. Similarly, the set of final states F^D of $\mathcal{A}^D$ is the set of transitions that end in a final state of $\mathcal{A}$.

An accepting run $r_{\mathcal{A}^D}$ of a dual automaton is a sequence of states $r_{\mathcal{A}^D} = q_0\ldots q_d$ such that $q_0 \in Q_0^D$, $q_d \in F^D$ and $(q_i,q_{i+1}) \in\rightarrow_{\mathcal{A}^D}$ for all $i = 0,\ldots,d-1$. An accepting run $r_{\mathcal{A}^D}$ produces a word $w = w_0\ldots w_d$ over Γ^D such that $\tau(q_i) = w_i$, for all $i = 0,\ldots,d$. The output language $\mathcal{L}_{\mathcal{A}^D}$ of a dual automaton $\mathcal{A}^D$ is the set of all words that are generated by accepting runs of $\mathcal{A}^D$. The construction of a dual automaton $\mathcal{A}^D$ from an FSA $\mathcal{A}$ guarantees that any word produced by $\mathcal{A}^D$ is accepted by $\mathcal{A}$:

[1] We abuse the terminology and say that a finite word satisfies a formula if it contains a satisfying finite prefix.

Proposition 2.5 *The output language of the dual automaton $\mathscr{A}^D$ coincides with the language accepted by the automaton $\mathscr{A}$, i.e., $\mathscr{L}_{\mathscr{A}} = \mathscr{L}_{\mathscr{A}^D}$.*

3. PROBLEM FORMULATION

Consider a discrete-time linear control system of the form

$$x_{k+1} = Ax_k + Bu_k, \quad x_k \in \mathbb{X},\ u_k \in \mathbb{U}, \tag{2}$$

where $A \in \mathbb{R}^{n\times n}$ and $B \in \mathbb{R}^{n\times m}$ describe the system dynamics, $\mathbb{X} \subset \mathbb{R}^n$ and $\mathbb{U} \subset \mathbb{R}^m$ are polyhedral sets, and $x_k \in \mathbb{X}$ and $u_k \in \mathbb{U}$ are the state and the applied control at time $k \in \mathbb{Z}_+$, respectively.

Let $x_0^r, x_1^r \ldots$ and $u_0^r, u_1^r, \ldots$ denote a reference trajectory and a reference control sequence, respectively. We assume that, for some N, at time k the reference trajectory of length $N+1$, $x_k^r, \ldots x_{k+N}^r$, and the reference control sequence of length N, $u_k^r, \ldots, u_{k+N-1}^r$, are known. At time $k \in \mathbb{Z}_+$, the cost of a finite trajectory $x_k, \ldots, x_{k+N}$ originating at x_k and generated by the control sequence $\mathbf{u}_k = u_k, \ldots, u_{k+N-1}$ is defined with respect to the available reference trajectory and control sequence as follows:

$$\begin{aligned} C(x_k, \mathbf{u}_k) &= (x_{k+N} - x_{k+N}^r)^\top L_N (x_{k+N} - x_{k+N}^r) \\ &+ \sum_{i=0}^{N-1} \Big\{ (x_{k+i} - x_{k+i}^r)^\top L (x_{k+i} - x_{k+i}^r) \\ &+ (u_{k+i} - u_{k+i}^r)^\top R (u_{k+i} - u_{k+i}^r) \Big\}, \end{aligned} \tag{3}$$

where $L, L_N \in \mathbb{R}^{n\times n}$ and $R \in \mathbb{R}^{m\times m}$ are positive definite matrices.

Remark 3.1 As it will become clear later in the paper, unlike classical MPC, the terminal cost weight matrix L_N no longer plays a role in guaranteeing recursive feasibility, and, as such, it can be chosen freely based on performance or other reasons.

Let $P = \{p_i\}_{i=0,\ldots,l}$ for some $l \geq 1$ be a set of atomic propositions given as linear inequalities in $\mathbb{R}^n$. Each atomic proposition p_i induces a half-space

$$[p_i] := \{x \in \mathbb{R}^n \mid c_i^\top x + d_i \leq 0\},\ c_i \in \mathbb{R}^n, d_i \in \mathbb{R}. \tag{4}$$

A trajectory $x_0, x_1, \ldots$ of system (2) produces a word $P_0P_1\ldots$ where $P_i \subseteq P$ is the set of atomic propositions satisfied by x_i, *i.e.*, $P_i = \{p_j \mid x_i \in [p_j]\}$. scLTL formulas over the set of predicates P can therefore be interpreted over such words (see Section 2). A system trajectory satisfies an scLTL formula over P if the word produced by the trajectory satisfies the corresponding formula. The main problem considered in this paper can be formulated as follows:

Problem 3.2 Given an scLTL formula Φ over a set of linear predicates P, a dynamical system as defined in (2), and an initial state $x_0 \in \mathbb{X}$, find a feedback control strategy such that the closed-loop trajectory originating at x_0 satisfies Φ while minimizing the cost (3).

We propose a two-step solution to Problem 3.2. In the first step, by using existing tools [7], we construct an automaton from the specification formula. The states of the automaton correspond to polyhedral subsets of the state space of system (2), and any satisfying trajectory of system (2) follows a sequence of polyhedral sets defined by an accepting run of the automaton. In the second step, we design an MPC controller that minimizes the cost over the available reference trajectory, while ensuring that the resulting trajectory satisfies the specification. While the automaton is constructed "offline", at each stage, the MPC controller solves an optimization problem "online" and produces the control action. The constraints of the optimization problem ensure that the produced trajectory lies within an automaton path and makes progress towards a final state, and hence, the produced trajectory satisfies the specification while the cost over the available reference trajectory is minimized.

As it will be established, the designed controller is recursively feasible, meaning that if the MPC optimization problem is feasible for the initial state at the initial time instant, then it remains feasible until the specification is satisfied. The correctness of the solution will therefore be guaranteed.

There are two main motivations for the proposed MPC scheme. First, we assume that the reference trajectories are available over a finite horizon. This allows us to incorporate a dynamic environment, in which case, the dynamic nature of the environment can be captured in the cost function. Second, even though scLTL specifications regard only finite executions, the length, $\overline{N}$, of the optimal satisfying trajectory is unknown. As it will become clear later in the paper, $\overline{N}$ can be found by solving optimization problems for all satisfying automaton runs. However, this computation is very expensive for a large $\overline{N}$, therefore, it also motivates the proposed MPC approach.

4. AUTOMATON GENERATION

In Section 4.1, we summarize a result from [7], which is the first step of our solution to Problem 3.2. In Section 4.2, we define a Lyapunov-type function that is used to enforce the satisfaction of the specification in the MPC controller.

4.1 Language-Guided Control

All words that satisfy the specification formula Φ over the set of linear predicates P are accepted by an FSA $\mathscr{A} = (Q, 2^P, \rightarrow_{\mathscr{A}}, Q_0, F)$. The dual automaton $\mathscr{A}^D = (Q^D, \rightarrow^D, \Gamma^D, \tau^D, Q_0^D, F^D)$ is constructed by interchanging the states and the transitions of $\mathscr{A}$ (Definition 2.4). As the transitions of $\mathscr{A}$ become states of $\mathscr{A}^D$, elements from 2^P label the states and define polyhedral sets within the state-space of system (2). For a dual automaton state $q \in Q^D$, $\mathscr{P}_q \subset \mathbb{X}$ is used to denote the corresponding polyhedral set.

In [7], we developed a procedure for iterative refinement of the dual automaton and the corresponding polyhedral partition of the state space of system (2) with the goal of finding initial states and corresponding feedback control strategies producing satisfying trajectories. Starting with the initial dual automaton, at each iteration, we checked whether feedback controllers could be designed for the original system to "match" the transitions of the dual automaton. We presented two methods, one based on vertex interpolation and the other one on polyhedral Lyapunov functions, to solve so-called $\mathscr{P}_q - to - \mathscr{P}_{q'}$ control problems, in which the goal was to design a state-feedback controller driving all the states from $\mathscr{P}_q$ to $\mathscr{P}_{q'}$ in finite time while remaining in $\mathscr{P}_q$ until they reached $\mathscr{P}_{q'}$. At each iteration, each transition (q, q') was labeled with a cost $J((q, q'))$ that equaled the minimum number of discrete time steps necessary for all states in $\mathscr{P}_q$ to reach $\mathscr{P}'_q$ under the determined state feedback law. If no controller could be found, then the cost was set to infinity. The cost of a state q was defined as the shortest path cost from q to a final state on the graph of the automaton weighted with transition costs.

The refinement algorithm proposed in [7] iteratively partitions the regions with infinite cost, *i.e.*, the regions for which there do not exist sequences of feedback controllers driving all the corresponding states to a region corresponding to a final state in the automaton. This procedure results in a monotonically increasing, with respect to set inclusion, set of initial states of system (2) for which an admissible control strategy can be found. As we showed in [7] (Theorem 6.1), if the refinement algorithm terminates, then all the satisfying trajectories of system (2) originate in the resulting set of initial states, denoted by $\mathbb{X}_0^\Phi$. In this paper, as our goal is to find a control strategy for a given initial state x_0, we terminate the algorithm at the ith iteration if $x_0 \in \mathbb{X}_{0,i}^\Phi$, where $\mathbb{X}_{0,i}^\Phi \subseteq \mathbb{X}$ is the union of the regions corresponding to start states of the automaton with finite path costs obtained at the ith iteration.

Remark 4.1 If we do not stop the refinement algorithm described above at the ith iteration when $x_0 \in \mathbb{X}_{0,i}^\Phi$, and it terminates in finite time, then we obtain $\mathbb{X}_0^\Phi$, and the method developed in this paper provides a solution to Problem 3.2 for each initial state $x_0 \in \mathbb{X}_0^\Phi$.

Example 4.2 To illustrate the construction of the refined dual automaton, we consider the double integrator dynamics with sampling time of 1 sec., which are described by (2) with

$$A = \begin{bmatrix} 1 & 1 \\ 0 & 1 \end{bmatrix}, \quad B = \begin{bmatrix} 0.5 \\ 1 \end{bmatrix}. \tag{5}$$

We assume that the control constraint set is $\mathbb{U} = \{u \mid -2 \leq u \leq 2\}$ and the initial state is $x_0 = \left[\begin{smallmatrix} 1.4 \\ -2.8 \end{smallmatrix}\right]$. The specification is to visit region $\mathbb{A}$ or region $\mathbb{B}$, and then the target region $\mathbb{T}$, while always avoiding obstacles $\mathbb{O}_1$ and $\mathbb{O}_2$, and staying inside a safe region $\mathbb{X}$, where

$$\begin{aligned}
\mathbb{A} &= \{x \mid -6 \leq x_1 \leq -5, 1 \leq x_2 \leq 2\},\\
\mathbb{B} &= \{x \mid -5 < x_1 \leq -4, -3 \leq x_2 \leq -2\},\\
\mathbb{T} &= \{x \mid -0.5 \leq x_i \leq 0.5, i = 1,2\},\\
\mathbb{O}_1 &= \{x \mid -10 \leq x_i \leq -5, i = 1,2\},\\
\mathbb{O}_2 &= \{x \mid -5 < x_1 \leq 1.85, -4 < x_2 < -3\},\\
\mathbb{X} &= \{x \mid -10 \leq x_1 \leq 1.85, -10 \leq x_2 \leq 2\}.
\end{aligned}$$

These polyhedral regions, together with the linear predicates used in their definitions, are shown in Figure 1. By using these predicates, the specification can be written as the following scLTL formula:

$\Phi_1 = ((p_0 \wedge p_1 \wedge p_2 \wedge p_3 \wedge \neg(p_4 \wedge p_5) \wedge \neg(\neg p_5 \wedge \neg p_6 \wedge \neg p_7))\ \mathcal{U}\ (p_8 \wedge p_9 \wedge p_{10} \wedge p_{11})) \wedge (\neg(p_8 \wedge p_9 \wedge p_{10} \wedge p_{11})\ \mathcal{U}\ ((p_5 \wedge p_{12} \wedge p_{13}) \vee (\neg p_5 \wedge p_7 \wedge p_{14} \wedge p_{15})))$.

A vertex interpolation method is used to synthesize the transition controllers. The initial state x_0 is covered by $\mathbb{X}_{0,81}^{\Phi_1}$ at iteration 81. If we continue the refinement process, it stops at iteration 183, and produces the maximal set of satisfying initial states $\mathbb{X}_0^{\Phi_1}$. The refined dual automaton obtained at iteration 183 has 228 finite cost states and 879 finite cost transitions. The sets $\mathbb{X}_{0,81}^{\Phi_1}$ and $\mathbb{X}_0^{\Phi_1}$ are shown in Figure 1 (b) and (c), respectively. The computation of $\mathbb{X}_{0,81}^{\Phi_1}$ and $\mathbb{X}_0^{\Phi_1}$ took 201 and 298 *seconds*, respectively, on an iMac with a Intel Core i5 processor at 2.8GHz with 8GB of memory. □

In the remainder of the paper, for simplicity of notation, we use

$$\mathscr{A}^D = (Q^D, \rightarrow^D, \Gamma^D, \tau^D, Q_0^D, F^D) \tag{6}$$

to denote the (refined) dual automaton obtained at the last iteration of the algorithm presented above. The corresponding transition cost function will be denoted by $J : \rightarrow^D \longrightarrow \mathbb{Z}_+$. We use $\mathscr{P}_q \subset \mathbb{X}$ to denote the polyhedral region of state $q \in Q^D$.

4.2 Distance Function

In control theory, control Lyapunov functions are used to enforce closed-loop stability of an equilibrium point. In this paper, we define a Lyapunov-type function to enforce the satisfaction of the accepting condition of an automaton. The idea of using such a real positive function to enforce a Büchi acceptance condition on the trajectories of a finite deterministic transition system was introduced in [14]. Here, we extend this concept to discrete-time linear systems and focus on acceptance conditions of finite state automata.

Definition 4.3 A function $V : \bigcup_{q \in Q^D} \{\{q\} \times \mathscr{P}_q\} \rightarrow \mathbb{Z}_+$ is called a distance function for a system (2) and a dual automaton (6) if it satisfies:

(i) $V(q,x) = 0$ for all $q \in F^D$.

(ii) For each $(q,x) \in \bigcup_{q \in Q^D} \{\{q\} \times \mathscr{P}_q\}$, it holds that if $V(q,x) \neq 0$ and $V(q,x) \neq \infty$, then there exists a control $u \in \mathbb{U}$ such that $x' = Ax + Bu$, $x' \in \mathscr{P}_{q'}$, $(q,q') \in \rightarrow^D$, and $V(q',x') < V(q,x)$.

Informally, the distance function at (q,x), $q \in Q^D, x \in \mathscr{P}_q$ is defined as an upper bound for the time required to reach $\mathscr{P}_{q_f}$ from x by applying the corresponding polytope-to-polytope feedback controllers along a shortest path $qq_1 \ldots q_f$ on the graph of the automaton, where q_f is a final state of the automaton. In the rest of this section, we formalize this description and then show that this function satisfies the properties of Definition 4.3.

Let $\mathbf{P}_q$ denote the set of all finite paths (not necessarily acyclic) from q to F^D, *i.e.*,

$$\mathbf{P}_q = \{\mathbf{q} = q_0 q_1 \ldots q_d \mid d \in \mathbb{Z}_+, (q_i, q_{i+1}) \in \rightarrow^D, i = 0, \ldots, d-1, \; q_d \in F^D, q_0 = q\}. \tag{7}$$

The cost $J^p(\mathbf{q})$ of an automaton path $\mathbf{q} = q_0 \ldots q_d$ is defined as the sum of the corresponding transition costs:

$$J^p(\mathbf{q}) = \sum_{i=0}^{d-1} J((q_i, q_{i-1})).$$

The cost $J^s(q)$ of a state $q \in Q^D$ is the cost of the shortest path from q to a final state, *i.e.*,

$$J^s(q) = \min_{\mathbf{q} \in \mathbf{P}_q} J^p(\mathbf{q}).$$

The successor $S(q)$ of a state $q \in Q^D$ is the state that succeeds q in the shortest path from q to F^D, *i.e.*,

$$qS(q) \ldots = \arg\min_{\mathbf{q} \in \mathbf{P}_q} J^p(\mathbf{q}).$$

We define the function $J^T : \bigcup_{(q,q') \in \rightarrow^D} \{\{(q,q')\} \times \mathscr{P}_q\} \longrightarrow \mathbb{Z}_+$ such that $J^T((q,q'),x)$ is an upper bound on the number of steps required to reach $\mathscr{P}_{q'}$ from the state $x \in \mathscr{P}_q$ by system (2). Later in this section we provide a method to determine $J^T((q,q'),x)$ given the feedback controllers used to solve the corresponding $\mathscr{P}_q - to - \mathscr{P}_{q'}$ control problem.

We define the distance function at (q,x) as

$$V(q,x) = \begin{cases} 0 & \text{if } q \in F^D, \\ J^T((q,S(q)),x) + J^s(S(q)) & \text{otherwise.} \end{cases} \tag{8}$$

The following properties are used to define $J^T((q,q'),x)$ and to prove that the function defined above is indeed a distance function according to Definition 4.3. *It is important to note that both $\mathscr{P}_q - to - \mathscr{P}_{q'}$ control methods from [7] satisfy these properties.* The detailed description of the controllers for solving $\mathscr{P}_q - to - \mathscr{P}_{q'}$ control problems can be found in Section 5 of [7].

Property 4.4 Let $g : \mathscr{P}_q \rightarrow \mathbb{U}$ be a feedback control law that solves the $\mathscr{P}_q - to - \mathscr{P}_{q'}$ control problem. Let $k^* = J((q,q'))$ and assume

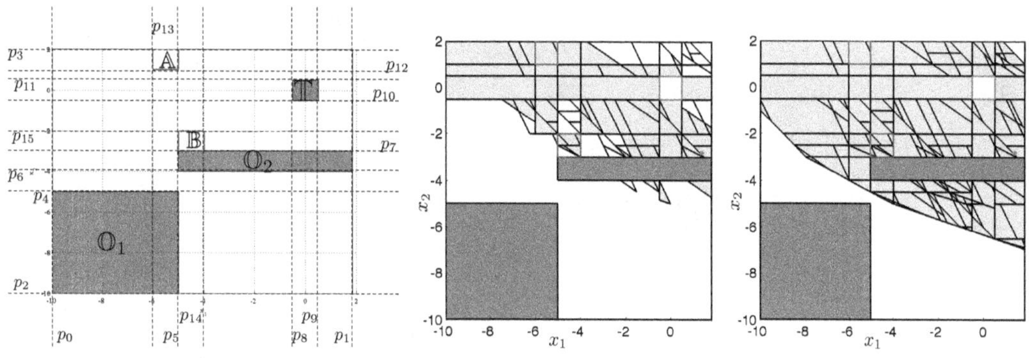

(a) (b) (c)

Figure 1: (a) The regions and the corresponding linear predicates for the specification from Example 4.2. The predicates are shown in the half planes where they are satisfied; (b) and (c) The set of satisfying initial states (shown in yellow) at iterations 81 and 183. The initial state is marked with a red point.

that $k^* \geq 1$. Let $\{x^i_j\}_{j=0,\ldots,k^*}$ be the trajectory generated by the feedback control $g(\cdot)$ from the vertex $v^i \in \mathscr{V}(\mathscr{P}_q)$, *i.e.*, $x^i_0 := v^i$. The feedback control $g(\cdot)$ satisfies the following properties:

(i) For each vertex $v^i \in \mathscr{V}(\mathscr{P}_q)$:

$$x^i_j \in \mathscr{P}_q, j = 0,\ldots,k^*-1, \text{ and } x^i_{k^*} \in \mathscr{P}_{q'}.$$

(ii) If $x \in Co(\{x^i_k\}_{i=1,\ldots,\#\mathscr{V}(\mathscr{P}_q)})$, $k < k^*$, it holds that

$$Ax + Bg(x) \in Co(\{x^i_{k+j}\}_{i=1,\ldots,\#\mathscr{V}(\mathscr{P}_q)}),$$

for some $1 \leq j \leq k^* - k$.

(iii) If $x \in \{x' \in \mathscr{P}_q \mid \exists u \in \mathbb{U} : Ax' + Bu \in \mathscr{P}_{q'}\}$, then

$$x \in Co(\{x^i_{k^*-1}\}_{i=1,\ldots,\#\mathscr{V}(\mathscr{P}_q)}).$$

If Property 4.4 holds, we define $J^T(\cdot,\cdot)$ as

$$J^T((q,q'),x) = k^* - \arg\max\{j \in \{0,\ldots,k^*\} \mid x \in Co(\{x^i_j\}_{i=1,\ldots,\#\mathscr{V}(\mathscr{P}_q)})\}. \quad (9)$$

Proposition 4.5 *The function defined in* (8) *is a distance function according to Definition 4.3.*

PROOF. The property *(i)* of Definition 4.3 is satisfied trivially by the function $V(\cdot,\cdot)$ (8). To prove that the function satisfies the property *(ii)*, we consider two cases: $J^T((q,S(q)),x) = 1$ and $J^T((q,S(q)),x) > 1$. Note that $J^T((q,S(q)),x) = 1$ and (9) imply that $x \in Co(\{x^i_{k^*-1}\}_{i=1,\ldots,\#\mathscr{V}(\mathscr{P}_q)})$. From Property 4.4-*(i)* and *(ii)*, it follows that

$$x' = Ax + Bg(x) \in Co(\{x^i_{k^*}\}_{i=1,\ldots,\#\mathscr{V}(\mathscr{P}_q)}) \subseteq \mathscr{P}_{S(q)}.$$

As such, in the case $J^T((q,S(q)),x) = 1$ the claim holds as

$$V(x',S(q)) \leq J^s(S(q)) \text{ for all } x' \in \mathscr{P}_{S(q)}.$$

In the case when $J^T((q,S(q)),x) > 1$, by (9),

$$x \in Co(\{x^i_k\}_{i=1,\ldots,\#\mathscr{V}(\mathscr{P}_q)})$$

for some $k < k^* - 1$. By Property 4.4-*(ii)*, one of the following holds:
(a) $Ax + Bg(x) \in \mathscr{P}_{q'}$,
(b) $Ax + Bg(x) \in Co(\{x^i_{k+j}\}_{i=1,\ldots,\#\mathscr{V}(\mathscr{P}_q)}), 1 \leq j < k^* - k$.

The claim is true for *(a)* by the same argument given for the $J^T((q,S(q)),x) = 1$ case. For *(b)*, as $j < k^* - k$, $Ax + Bg(x) \in \mathscr{P}_q$. From (9), it follows that

$$J^T((q,S(q)),Ax + Bg(x)) < J^T((q,S(q)),x).$$

Hence, $V(q,Ax + Bg(x)) < V(q,x)$. □

5. MPC STRATEGY

In this section, we present an MPC scheme to solve Problem 3.2 for a given dual automaton $\mathscr{A}^D = (Q^D, \rightarrow^D, \Gamma^D, \tau^D, Q^D_0, F^D)$ and a transition cost function $J : \rightarrow^D \longrightarrow \mathbb{Z}_+$. We formulate an MPC optimization problem over $\bigcup_{q \in Q^D} \{\{q\} \times \mathscr{P}_q\}$ to be solved "online" at each time step. The constraints of the optimization problem guarantee that the resulting closed-loop trajectory of system (2) follows an automaton path and makes progress towards the final states of the automaton.

Definition 5.1 An automaton-enabled finite trajectory

$$\mathbf{T} = (q_0,x_0),\ldots,(q_N,x_N)$$

is a sequence of automaton (6) and system (2) state pairs such that

(i) for each $k = 0,\ldots,N-1$ there exists $u_k \in \mathbb{U}$ such that $x_{k+1} = Ax_k + Bu_k$,
(ii) $x_k \in \mathscr{P}_{q_k}$, for all $k = 0,\ldots,N$,
(iii) $(q_k, q_{k+1}) \in \rightarrow^D$, for all $k = 0,\ldots,N-1$.

The definition of an automaton-enabled trajectory implies that the projection $\gamma_{\mathscr{A}}(\mathbf{T}) = q_0 \ldots q_N$ of the trajectory onto the automaton states is an automaton path and the projection $\gamma_X(\mathbf{T}) = x_0 \ldots x_N$ onto the state space of system (2) is a trajectory of system (2) that follows the sequence of polyhedra defined by the automaton path. The automaton-enabled trajectory is related to the hybrid trajectory of a hybrid system [21], in the sense that the trajectory is defined over a hybrid state space, *i.e.*, $\bigcup_{q \in Q^D} \{\{q\} \times \mathscr{P}_q\}$.

The construction of the dual automaton $\mathscr{A}^D$ from Section 4.1 and Definition 5.1 guarantee that for any satisfying trajectory $\mathbf{x} = x_0,\ldots,x_d, d \in \mathbb{Z}_+$ of system (2) originating at x_0, there exists an automaton-enabled trajectory $\mathbf{T}$ such that $\gamma_X(\mathbf{T}) = \mathbf{x}$ and $\gamma_{\mathscr{A}}(\mathbf{T})$ is an accepting run of $\mathscr{A}^D$. Therefore, in the MPC controller design, we restrict our attention to the control sequences that generate automaton-enabled trajectories. We use $\mathbf{U}_N(q,x)$ to denote the set of all control sequences of length N that produce automaton-enabled trajectories starting from (q,x) as characterized in Definition 5.1. By following the standard MPC notation, we use

$$\mathbf{T}_k = (q_{0|k}, x_{0|k}) \dots (q_{N|k}, x_{N|k}),$$

to denote a *predicted* automaton-enabled trajectory originating at (q_k, x_k), *i.e.*, $q_{0|k} = q_k$, $x_{0|k} = x_k$, at time $k \in \mathbb{Z}_+$.

At each time-step $k \in \mathbb{Z}_+$, we solve an optimization problem over $\mathbf{U}_N(q_k, x_k)$ and find the optimal automaton-enabled trajectory

$$\mathbf{T}^*_k = (q^*_{0|k}, x^*_{0|k}) \dots (q^*_{N|k}, x^*_{N|k})$$

generated by the optimal control sequence $\mathbf{u}^*_k \in \mathbf{U}_N(q_k, x_k)$, until a final automaton state is reached, *i.e.*, $q_k \in F^D$. As the first control of the optimal control sequence is applied, we have the following relation:

$$x_{k+1} = x^*_{1|k}, \qquad q_{k+1} = q^*_{1|k}, \qquad k = 0, 1, 2, \dots. \tag{10}$$

The MPC optimization problem is formulated as follows:

$$\min_{\mathbf{u}_k \in \mathbf{U}_N(q_k, x_k)} C(x_k, \mathbf{u}_k),$$

$$s.\,t. \begin{cases} V(q_{N|k}, x_{N|k}) < V(q_0, x_0), & \text{if } k = 0. \\ V(q_{N|k}, x_{N|k}) < V(q^*_{N|k-1}, x^*_{N|k-1}), & \text{if } k \geq 1. \end{cases} \tag{11a}$$

For $k \geq 1$, the optimal predicted trajectory obtained at the previous time step is used to enforce the satisfaction of the specification. As the predicted trajectory at time k must end in a state that is closer to a final state than the last state of $\mathbf{T}^*_{k-1}$, the resulting trajectory eventually reaches a final automaton state.

The optimization problem formulation is analogous to the classical MPC formulation, where $C(\cdot,\cdot)$, N, and the constraint given in (11a) are called the objective function, the prediction horizon, and the terminal state constraint, respectively [22]. There are, however, significant differences due to the definition of an automaton-enabled trajectory. Specifically, the terminal set defined by the constraint given in (11a) is a subset of $\bigcup_{q \in Q^D} \{\{q\} \times \mathscr{P}_q\}$ and the set of admissible control sequences $\mathbf{U}_N(q_k, x_k)$ is not necessarily convex.

Next, we show that the optimal solution of the MPC problem given in (11) can be found by solving a set of convex optimization problems. Specifically, we propose to solve an optimization problem for each automaton path from the set

$$\mathbf{P}^N_{q_k} = \{\mathbf{q}' = q_{0|k} q_{1|k} \dots q_{N|k} \mid q_{0|k} := q_k, \exists d \in \mathbb{Z}_+ s.t.\, N \leq d,$$
$$\mathbf{q} = q_{0|k} \dots q_{d|k}, \mathbf{q} \in \mathbf{P}_{q_k}\}, \tag{12}$$

where $\mathbf{P}_{q_k}$ is defined as in (7). Since Q^D and $\rightarrow^D$ are finite sets, $\mathbf{P}^N_{q_k}$ is a finite set. The definition of an automaton-enabled trajectory $\mathbf{T}_k$ of horizon N (Definition 5.1) implies that $\gamma_{\mathscr{A}}(\mathbf{T}_k) \in \mathbf{P}^N_{q_k}$ for any trajectory that can be produced by a control sequence from the set $\mathbf{U}_N(q_k, x_k)$.

Given a finite automaton path $\mathbf{q}_k \in \mathbf{P}^N_{q_k}$, let $\mathbf{U}^{\mathbf{q}_k}_N(q_k, x_k)$ denote the set of all control sequences that produce an automaton-enabled trajectory $\mathbf{T}_k$ with $\gamma_{\mathscr{A}}(\mathbf{T}_k) = \mathbf{q}_k$. Essentially, $\mathbf{U}^{\mathbf{q}_k}_N(q_k, x_k)$ is the set of all control sequences that produce trajectories of system (2) that originate at x_k and follow the sequence of polyhedra defined by $\mathbf{q}_k$. Then, it is straightforward to see that

$$\mathbf{U}_N(q_k, x_k) = \bigcup_{\mathbf{q}_k \in \mathbf{P}^N_{q_k}} \mathbf{U}^{\mathbf{q}_k}_N(q_k, x_k). \tag{13}$$

Consider a path $\mathbf{q}_k = q_{0|k} \dots q_{N|k} \in \mathbf{P}^N_{q_k}$ and the following optimization problem in the variables $\mathbf{u}_k = u_{0|k}, \dots, u_{N-1|k}$:

$$\min C(x_k, \mathbf{u}_k),$$

subject to

$$x_{i|k} \in \mathscr{P}_{q_{i|k}}, \qquad i = 1, \dots, N, \tag{14a}$$

$$u_{i|k} \in \mathbb{U}, \qquad i = 0, \dots, N-1, \tag{14b}$$

$$\begin{cases} V(q_{N|k}, x_{N|k}) < V(q_0, x_0), & \text{if } k = 0. \\ V(q_{N|k}, x_{N|k}) < V(q^*_{N|k-1}, x^*_{N|k-1}), & \text{if } k \geq 1. \end{cases} \tag{14c}$$

The set of control sequences that satisfy constraints (14a) and (14b) is $\mathbf{U}^{\mathbf{q}_k}_N(q_k, x_k)$. Therefore, the optimal solution of the MPC problem given in (11) can be found by solving an optimization problem as given in (14) for each $\mathbf{q}_k \in \mathbf{P}^N_{q_k}$.

Next, we show that the set of states of system (2) that satisfies the progress constraint given in (14c) is a convex subset of $\mathscr{P}_{q_{N|k}}$. By the definition of the distance function given in (8), the progress constraint given in (14c) at time $k \geq 1$ takes the following from:

$$J^T((q_{N|k}, S(q_{N|k})), x_{N|k}) < V(q^*_{N|k-1}, x^*_{N|k-1}) - J^s(S(q_{N|k})). \tag{15}$$

Equation (15) implies that the trajectory originating at $x_{N|k}$ must reach $\mathscr{P}_{S(q_{N|k})}$ within

$$\bar{k} := V(q^*_{N|k-1}, x^*_{N|k-1}) - 1 - J^s(S(q_{N|k})) \tag{16}$$

discrete time instants with respect to the feedback controller $g(\cdot)$ that is used to compute $J((q_{N|k}, S(q_{N|k})))$. The recursive application of Property 4.4-*(ii)* implies that the inequality given in (15) is satisfied if $\bar{k} \geq 1$ and

$$x_{N|k} \in Co(\{x^i_{k^*-\bar{k}}\}_{i=1,\dots,\#\mathscr{V}(\mathscr{P}_q)}). \tag{17}$$

As shown in (17), the set of states that satisfy the inequality given in (15) is a polyhedral subset of $\mathscr{P}_{q_{N|k}}$. By observing that at time $k = 0$, a subset of $\mathscr{P}_{q_{N|k}}$ can be defined with respect to $V(q_0, x_0)$ in the same way, we conclude that the progress constraint can be implemented by linear inequalities.

Remark 5.2 The costs of the transitions are not considered while constructing the set $\mathbf{P}^N_{q_k}$. A transition (q, q') with infinite cost means that the synthesis method failed to find a feedback control law that solves the corresponding $\mathscr{P}_q - to - \mathscr{P}_{q'}$ control problem. However, the problem might have a solution for subsets of $\mathscr{P}_q$ and $\mathscr{P}_{q'}$.

As shown above, the solution of the MPC optimization problem given in (11) can be found by solving a set of convex optimization problems for a given prediction horizon. To guarantee that the resulting closed-loop trajectory of system (2) reaches a region $\mathscr{P}_{q_f}$, where $q_f \in F^D$; at each time-step the prediction horizon is determined with respect to the predicted trajectory obtained at the previous step. Specifically, if the predicted trajectory obtained at the previous step does not include a final automaton state, then N, the length of the observed reference trajectory, is used as the prediction horizon. However, if the predicted trajectory obtained at the previous step visits a final state at position j for the first time, then $j-1$ is used as the prediction horizon. The following function is used to determine the prediction horizon for a given trajectory $\mathbf{T}_k = (q_{0|k}, x_{0|k}) \dots (q_{N|k}, x_{N|k})$:

$$I(\mathbf{T}_k) = \begin{cases} N & \text{if } 0 < V(q_{i|k}, x_{i|k}), \forall i = 0, \dots, N \\ j-1 & \text{if } 0 < V(q_{i|k}, x_{i|k}), \forall i = 0, \dots, j-1, \\ & \quad V(q_{j|k}, x_{j|k}) = 0. \end{cases} \tag{18}$$

If $V(q_{j|k}, x_{j|k}) = 0$ for some $j \le N$, then $V(q_{j-1|k}, x_{j-1|k}) = 1$ and $q_{j|k} \in F^D$ by Definition 4.3 and Property 4.4. Consequently, when $V(q_{j|k}, x_{j|k}) = 0$ for some $j \le N$ and $I(\mathbf{T}_k)$ is used as the prediction horizon, the progress constraint enforces the predicted trajectory at time $k+1$ to visit an accepting state at position $j-1$. Moreover, function $I(\cdot)$ allows us to optimize the cost until the specification is satisfied, *i.e.*, until a final automaton state is reached.

Algorithm 1 Automaton-guided MPC

Input: $\mathscr{A}^D = (Q^D, \to^D, \Gamma^D, \tau^D, Q_0^D, F^D)$, transition cost function $J : \to^D \to \mathbb{Z}_+$, an initial condition $x_0 \in \mathscr{P}_{q_0}$ for some $q_0 \in Q_0^D$ with finite cost, MPC horizon N.

1: Set $k := 0$, $I_k := N$. (Initialization)
2: $V(q^*_{I_k|k-1}, x^*_{I_k|k-1}) := V(q_0, x_0)$. (For the first iteration)
3: **while** $q_k \notin F^D$ **do**
4: $\quad OptCost = \infty$, $\mathbf{u}_k^* = \emptyset$, $\mathbf{T}_k^* = \emptyset$.
5: $\quad$ Compute $\mathbf{P}^{I_k}_{q_k}$.
6: $\quad$ **for all** $\mathbf{q} = q_{0|k} \dots q_{I_k|k} \in \mathbf{P}^{I_k}_{q_k}$ **do**
7: $\quad\quad$ **if** $J^s(S(q_{I_k|k})) < V((q^*_{I_k|k-1}, x^*_{I_k|k-1})) - 1$ **then**
8: $\quad\quad\quad c = \min_{\mathbf{u}_k = u_{0|k} \dots u_{I_k-1|k}} C(x_k, \mathbf{u}_k)$ subject to (14a), (14b) and (17)
9: $\quad\quad\quad$ **if** $c < OptCost$ **then**
10: $\quad\quad\quad\quad OptCost := c$, set $\mathbf{u}_k^*$ and $\mathbf{T}_k^*$ with respect to the solution of the optimization problem (line 8).
11: $\quad\quad\quad$ **end if**
12: $\quad\quad$ **end if**
13: $\quad$ **end for**
14: $\quad$ Apply $\mathbf{u}^*_{0|k}$, $q_{k+1} := q^*_{1|k}$, $x_{k+1} := x^*_{1|k}$.
15: $\quad I_{k+1} := I(\mathbf{T}_k^*)$.
16: $\quad k := k+1$.
17: **end while**

The proposed MPC controller is summarized in Algorithm 1, where a set of optimization problems is solved at each time step until a final automaton state is reached (line 3). At each time step, the linear quadratic optimization problem given in line 8 is solved for each automaton path $\mathbf{q} \in \mathbf{P}^{I_k}_{q_k}$ (12) which satisfies the condition given in line 7. Note that if an automaton path does not satisfy the condition given in line 7, the problem given in (14) becomes infeasible. When the loop over $\mathbf{P}^{I_k}_{q_k}$ (line 6) is terminated, the first element of the optimal control sequence $\mathbf{u}_k^*$ is applied and the state (q_{k+1}, x_{k+1}) is computed.

Assumption 5.3 The length of any satisfying trajectory of system (2) originating at x_0 is lower bounded by N.

Assumption 5.3 is made to simplify the presentation of the following results. The length, $\overline{N}$, of the shortest satisfying trajectory of system (2) originating at x_0 can be found by solving a set of optimization problems of the form (14). Then, the assumption is not necessary if $\overline{N}$ is used as the initial prediction horizon in the case when $\overline{N} < N$.

Lemma 5.4 *Suppose that Assumption 5.3 holds, $V(\cdot,\cdot)$ and $J^T(\cdot,\cdot)$ are defined as in (8) and (9), respectively, and there exists $q_0 \in Q^D$ such that $x_0 \in \mathscr{P}_{q_0}$ and $J^s(q_0) < \infty$. Then, the optimization problem given in line 8 of Algorithm 1 is feasible for some $\mathbf{q}_0 \in \mathbf{P}^N_{q_0}$ at the initial condition (q_0, x_0).*

PROOF. *$V(q_0, x_0) < \infty$, since $V(q_0, x) \le J^s(q_0)$ for all $x \in \mathscr{P}_{q_0}$. By Proposition 4.5 and Assumption 5.3, there exists a control sequence $\mathbf{u} = u_0, \dots, u_{N-1}$ and an automaton path $\mathbf{q} = q_0, \dots, q_N$ such that the value of the distance function strictly decreases along the trajectory $\mathbf{T} = (q_0, x_0) \dots (q_N, x_N)$ generated by $\mathbf{u}$ from the initial condition (q_0, x_0), and hence, $V(q_N, x_N) < V(q_0, x_0)$. As such, in the case $k = 0$, the optimization problem is feasible for $\mathbf{q}$.* □

Theorem 5.5 *Suppose that Assumption 5.3 holds, $V(\cdot,\cdot)$ and $J^T(\cdot,\cdot)$ are defined as in (8) and (9), respectively, and there exists $q_0 \in Q^D$ such that $x_0 \in \mathscr{P}_{q_0}$ and $J^s(q_0) < \infty$. Then:*

(i) If the optimization problem given in (14) is feasible for some $\mathbf{q}_k \in \mathbf{P}^{I_k}_{q_k}$ at time k for state (q_k, x_k), then there exists $\mathbf{q}_{k+1} \in \mathbf{P}^{I_{k+1}}_{q_{k+1}}$ such that the problem is feasible for $\mathbf{q}_{k+1}$ and state (q_{k+1}, x_{k+1}).

(ii) The trajectory of system (2) produced by the closed-loop system satisfies the specification.

PROOF. *(i)* Let $\mathbf{T}_k^* = (q^*_{0|k}, x^*_{0|k}) \dots (q^*_{I_k|k}, x^*_{I_k|k})$ be the trajectory generated by the optimal control sequence $\mathbf{u}_k^* = u^*_{0|k} \dots u^*_{I_k|k}$ at step k. From (18) and the progress constraint (15), we have that $I_{k+1} \le I_k$. By Proposition 4.5, there exists a control $u \in \mathbb{U}$ and a state $q' \in Q^D$ such that $x' = Ax_{I_{k+1}|k} + Bu \in \mathscr{P}_{q'}$, $(q_{I_{k+1}|k}, q') \in \to^D$ and

$$V(q', x') < V(q_{I_{k+1}|k}, x_{I_{k+1}|k}).$$

As such, the control sequence $u^*_{1|k} \dots u^*_{I_k|k} u'$ is a feasible solution of the optimization problem at step $k+1$ for $q^*_{1|k} \dots q^*_{I_k|k} q'$.

(ii) We first show that the produced trajectory reaches a final automaton state in finite time. Let V^* be the value of the distance function at the initial condition, $V^* = V(q_0, x_0)$. By Lemma 5.4, the optimization problem is feasible at time $k = 0$ for some $\mathbf{q}_0 \in \mathbf{P}_{q_0}$. From Theorem 5.5-*(i)* it follows that an optimal predicted trajectory $\mathbf{T}_k^*$ exists at each time step until a final state is reached. The sequence of optimal trajectories satisfies the following inequality until a final automaton state appears in a trajectory:

$$V^* = V(q_0, x_0) > V(q^*_{N|0}, x^*_{N|0}) > V(q^*_{N|1}, x^*_{N|1}) \dots \quad (19)$$

The strict decrease implies that there exists $v \le V^*$ such that the trajectory $\mathbf{T}_v^*$ visits a final automaton state. The optimization horizon at step $v+1$ satisfies $I_{v+1} < N$ and the predicted trajectory $\mathbf{T}^*_{v+1}$ must end in a final state. Therefore, the sequence of optimization horizons satisfies the following set of inequalities for some $w \le N$:

$$N = I_v > I_{v+1} > I_{v+2} > \dots > I_{v+w} = 0.$$

$I_{v+w} = 0$ implies that $q_{v+w} \in F^D$, and hence the trajectory produced by the closed-loop system reaches a final automaton state within $v + w \le N + V^*$ steps.

The proposed MPC controller produces a finite trajectory (q_0, x_0) $, \dots,$ $(q_l, x_l), l \le N + V^*$. Next, we show that the projected system trajectory $x_0, \dots, x_l$ satisfies Φ. It is assumed that $x_0 \in \mathscr{P}_{q_0}$ and $q_0 \in Q_0^D$. By the definition of $\mathbf{P}^{I_k}_q$, $(q_i, q_{i+1}) \in \to^D$ for all $i = 0, \dots, l-1$ and by the termination condition $q_l \in F^D$. Consequently, $q_0 \dots q_l$ is an accepting automaton run. The constraints of the optimization problem given in line 8 ensure that $x_k \in \mathscr{P}_{q_k}$ for all $i = 0, \dots, l$. Hence, the system trajectory $x_0 \dots x_l$ satisfies the specification. □

Remark 5.6 Unlike the classical terminal constraint set and cost function method for guaranteeing stability of MPC, in the proposed approach the formula satisfaction and recursive feasibility are decoupled from the optimization of the cost function (3). Nevertheless, the progress constraint (14c), which guarantees recursive feasibility, and thus, formula satisfaction by the properties of the constructed automaton path, is a constraint on the terminal state. As such, it can be regarded as an analogy of the terminal constraint method of standard, stabilizing MPC. Previously, such Lyapunov-type constraints have been proposed in [17] for stabilizing MPC of hybrid systems with general discrete dynamics and in [14] for MPC of finite state deterministic transition systems.

Remark 5.7 (Complexity) The number of paths of length d originating at a node of a graph is upper bounded by b^d, where b is the branching factor of the graph, *i.e.,* the maximum number of outgoing edges from a node. Therefore, the size of $\mathbf{P}^N_{q_k}$ is upper bounded by b^N. Consequently, to find the optimal solution of the MPC problem (11), b^N optimization problems should be solved in the worst case. However, the experiments (see Table 1) show that this exponential bound is rarely reached as the number of outgoing transitions is significantly less then b for most of the dual automaton states. Moreover, as the algorithm iterates, the number of paths that fail the feasibility check in line 7 increases, since the distance to a final automaton state decreases.

6. EXTENSION TO PWA SYSTEMS

In this section, we discuss the extension of the presented temporal logic model predictive control approach to piece-wise affine systems. A discrete-time PWA control system is described by

$$x_{k+1} = A_i x_k + B_i u_k + c_i, \quad x_k \in \mathbb{X}_i,\ u_k \in \mathbb{U},\ i = 1,\dots,l, \tag{20}$$

where l is the number of modes (different dynamics), $\mathbb{X}_i, i = 1,\dots,l$ provide a polyhedral partition of $\mathbb{X} = \bigcup_{i=1,\dots,l} \mathbb{X}_i \subset \mathbb{R}^n$, such that $\mathbb{X}_i \cap \mathbb{X}_j = \emptyset$ for all $i \neq j$, $\mathbb{U}$ is a polyhedral set in $\mathbb{R}^m$, and A_i, B_i, and c_i are matrices of appropriate sizes.

The extension of the method from the linear dynamics (2) to the PWA dynamics (20) requires an additional refinement procedure that takes into account the particular structure of the state space $\mathbb{X} = \bigcup_{i=1,\dots,l} \mathbb{X}_i$. This procedure is summarized in Algorithm 2.

Algorithm 2 Refinement of $\mathscr{A}^D$ with respect to $\mathbb{X} = \bigcup_{i=1,\dots,l} \mathbb{X}_i$

$Q_{in} := Q^D$
for all $q \in Q_{in}$ **do**
  $q_Q := \{q_j \mid \mathscr{P}_{q_j} = \mathscr{P}_q \cap \mathbb{X}_j, \mathscr{P}_{q_j} \neq \emptyset\}$
  $\rightarrow^D := \rightarrow^D \cup \left\{\bigcup_{(q,q') \in \rightarrow^D} \{(q_j, q') \mid q_j \in q_Q\}\right\} \setminus \{(q,q') \mid q' \in Q^D\}$
  $\rightarrow^D := \rightarrow^D \cup \left\{\bigcup_{(q',q) \in \rightarrow^D} \{(q', q_j) \mid q_j \in q_Q\}\right\} \setminus \{(q',q) \mid q' \in Q^D\}$
  $Q_0^D := \{Q_0^D \setminus \{q\}\} \cup q_Q$ if $q \in Q_0^D$
  $F^D := \{F^D \setminus \{q\}\} \cup q_Q$ if $q \in F^D$
end for

Algorithm 2 preserves the dual automaton language while ensuring that for each dual automaton state q there exists $\alpha(q) \in \{1,\dots,l\}$ such that $\mathscr{P}_q \subseteq \mathbb{X}_{\alpha(q)}$. Consequently, for each dual automaton state q there exist an associated system dynamics ($A_{\alpha(q)}$, $B_{\alpha(q)}$, $c_{\alpha(q)}$) that can be used to solve $\mathscr{P}_q - to - \mathscr{P}_{q'}$ control synthesis problems for transitions leaving the state q. Therefore, to obtain the refined dual automaton, the algorithms proposed in [7] can directly be used for PWA systems after the refinement step.

The MPC problem given in (11) still reduces to solving a finite number of QPs, that only depends on the size of the set $\mathbf{P}^N_{q_k}$ due to the fact that only one mode is active for any $\mathscr{P}_q$ in a path. Therefore, once a refined dual automaton is obtained as explained above, Algorithm 1 can be used to control a PWA system by adapting the dynamics in the optimization problem given in (14) as follows:

$$x_{j+1|k} = A_{\alpha(q_{j|k})} x_{j|k} + B_{\alpha(q_{j|k})} u_{j|k} + c_{\alpha(q_{j|k})}, \text{for all } j = 1,\dots,N,$$

The use of this extension to PWA systems is illustrated in Case Study 2 in Section 7.

Remark 6.1 The active dynamics within a polyhedral set $\mathscr{P}_q$ is uniquely defined, because each $\mathscr{P}_q$ corresponds to a unique automaton state. In turn, each automaton state is associated with a unique mode of the PWA system (20).

Remark 6.2 It is of further interest to investigate how the tools developed in this paper can be employed to reduce the complexity associated with solving mixed integer convex optimization problems (MIQP or MILP) that arise in hybrid MPC, *i.e.,* MPC of hybrid systems such as PWA systems or more general, MLD systems.

7. IMPLEMENTATION AND CASE STUDIES

The computational framework developed in this paper was implemented as a Matlab software package LanGuiMPC (Language-Guided Model Predictive Control), which is freely downloadable from `hyness.bu.edu/software`. The toolbox takes as input an scLTL formula over a set of linear predicates, the matrices of a PWA system, the control constraint set and state regions corresponding to the modes of the system, an initial state, a reference trajectory, and a horizon length, and produces a solution to Problem 3.2. The tool, which uses *scheck2* [20] for the construction of the FSA and MPT [23] for polyhedral operations, also allows for simulating the trajectories of the closed-loop system for 2D or 3D state-spaces.

The case studies described below were generated using LanGuiMPC running on the same computer as in Example 4.2.

7.1 Case Study 1 : Double Integrator

Consider the double integrator dynamics and the specification given in Example 4.2. The cost function is defined as in (3) with

$$L_N = L = \begin{bmatrix} 0.5 & 0 \\ 0 & 0.5 \end{bmatrix}, \quad R = 0.8. \tag{21}$$

To illustrate the main results of the paper, we use both satisfying and violating reference trajectories. The satisfying reference trajectory $\mathbf{x}^{r_1}$ is generated by the reference control sequence $\mathbf{u}^{r_1} = 1.8, 1.4, 1.8, 1.8, 1, 0, -0.7, 0, 0, 0, 0.1, -1.2, 0$ from the initial condition $\left[\begin{smallmatrix} 1 \\ -6 \end{smallmatrix}\right]$. The reference trajectory and two simulated trajectories of the closed-loop system originating at $\left[\begin{smallmatrix} 1 \\ -6 \end{smallmatrix}\right]$ for optimization horizons $N = 2$ and $N = 4$ are shown in Figure 2 (a-b).

Second, we use a violating reference trajectory generated by the reference control sequence $\mathbf{u}^{r_2} = 0.6, 0.6, 0.8, 0.8, 1, 0.4, 0.4, -0.6, -0.6, -0.6, -0.4, -0.6, 1.0, 0$ from the initial condition $\left[\begin{smallmatrix} 1.4 \\ -2.8 \end{smallmatrix}\right]$. The reference trajectory and the trajectory generated by Algorithm 1 for optimization horizon $N = 6$ are shown in Figure 2(c). The controlled trajectory closely follows the reference trajectory for the first 2 steps, then the distance between them increases as the controlled trajectory visits region B to satisfy the specification.

The average number of QPs solved to generate the closed-loop trajectories are given in Table 1. The data is consistent with Remark 5.7 since the number of QPs solved is significantly less then the theoretical bound.

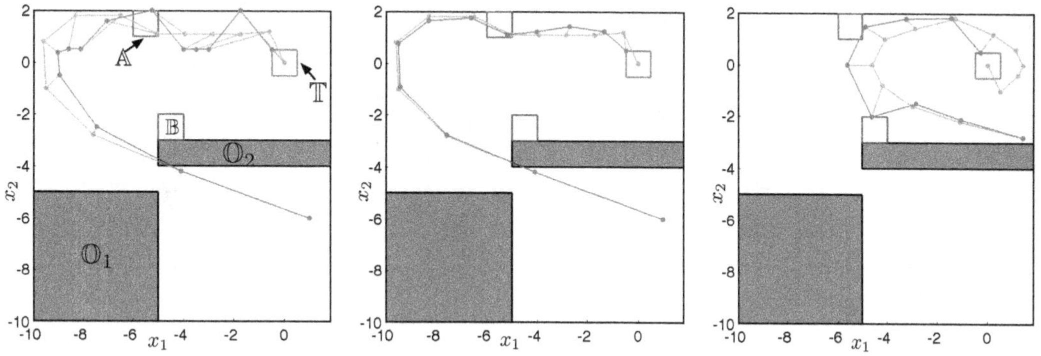

(a) (b) (c)

Figure 2: Case Study 1: The reference trajectory (green, satisfying in (a) and (b), and violating in (c)) and the trajectory of the controlled system (red). Pairs of points on the reference and controlled trajectory corresponding to the same time are connected by a shaded red line. (a) $N = 2$, total cost $= 29.688$. (b) $N = 4$, total cost $= 0.886$. (c) $N = 6$, total cost $= 5.12$.

Case study #	1	1	1	2	2
Horizon length N	2	4	6	6	7
Worst case (b^N)	24^2	24^4	24^6	7^6	7^7
# QP	3	20	2795	171	609

Table 1: The number of QPs required to be solved in the worst case, and the average number of QPs solved at each iteration for the trajectories shown in Figure 2 and Figure 3.

7.2 Case Study 2 : Toggle Switch

In this case study, we apply the proposed method to a PWA system, which models a synthetic biological repressor network known as the toggle switch [24]. The system and the specification are adapted from [10], where the goal was to find a constant control for each operating region from a specification given as an LTL formula over the regions. The state space of the PWA system is partitioned into 36 polyhedral operating regions that are shown in Figure 3. We assume that the controls are constrained to set $\mathbb{U} = \{u \in \mathbb{R}^2 \mid -2.5 \leq u_i \leq 2.5, i = 1,2\}$ and we omit the dynamics due to the space constraints. Some trajectories of the uncontrolled PWA system, *i.e.*, $\mathbf{u}_k = \left[\begin{smallmatrix}0\\0\end{smallmatrix}\right]$ for all $k \in \mathbb{Z}_+$, are shown in Figure 3(a), where trajectories go towards one of two possible stable equilibria.

We consider the following specification for the PWA system:

"A system trajectory never enters the interior of region $\mathbb{O}$, eventually visits a target region $\mathbb{T}$, and stays inside a safe set $\mathbb{X}$ before it reaches $\mathbb{T}$", where

$$\begin{aligned}\mathbb{T} &= \{x \mid 0 \leq x_1 \leq 20, 75 \leq x_2 \leq 100\},\\ \mathbb{X} &= \{x \mid 0 \leq x_i \leq 100, i = 1,2\},\\ \mathbb{O} &= \{x \mid 40 < x_1 < 80, 20 < x_2 < 50\}.\end{aligned}$$

These polyhedral regions and linear predicates used to define these regions are shown in Figure 3. The specification is formally stated as the following scLTL formula:

$\Phi_2 = (p_0 \wedge p_1 \wedge p_2 \wedge p_3 \wedge \neg(\neg p_4 \wedge \neg p_5 \wedge \neg p_6 \wedge \neg p_7)) \;\mathcal{U}\; (p_0 \wedge p_8 \wedge p_3 \wedge p_9)$.

Algorithm 2 is used to refine the initial dual automaton with respect to the operating regions of the PWA system. The dual automaton is not further refined since all the states have finite costs. The transition controllers and the corresponding costs are computed with polyhedral Lyapunov functions method.

The matrices of the cost function (3) are

$$L_N = L = \begin{bmatrix} 0.5 & 0 \\ 0 & 0.5 \end{bmatrix} \text{ and } R = \begin{bmatrix} 0.8 & 0 \\ 0 & 0.8 \end{bmatrix}. \tag{22}$$

The reference control sequence $\mathbf{u}^{r_3}$ and the reference trajectory $\mathbf{x}^{r_3}$ are defined as

$$\mathbf{u}_i^{r_3} = \begin{bmatrix} 0 \\ 0 \end{bmatrix}, \mathbf{x}_i^{r_3} = \begin{bmatrix} 10 \\ 90 \end{bmatrix}, \quad \text{for all } i = 0, \ldots, 99. \tag{23}$$

Note that the reference trajectory corresponds to a point in the interior of the target region. The reference control sequence and the reference trajectory allow us to minimize the magnitude of the control applied and the distance to the target region. As shown in Figure 3, the simulated trajectories advance towards the target region diagonally for the first steps, and then they follow the edge of the region $\mathbb{O}$, and finally reach the target region. To prevent trajectories from reaching the boundaries of the region $\mathbb{O}$, standard techniques of motion planning such as *obstacle inflation* can be used.

8. CONCLUSIONS

We showed that optimal control of discrete-time linear systems can be implemented in a receding-horizon fashion with guaranteed satisfaction of temporal logic specifications. We focused on quadratic costs penalizing the distance from reference state and control trajectories over finite horizons and formulas over linear state predicates in a fragment of LTL called syntactically co-safe LTL. The main contribution of this work is to show that the satisfaction of the formula can be guaranteed eventually based on constraints added to the optimization problem that is solved at each time step. This can be seen as an extension to classical MPC approaches, in which such constraints, called terminal constraints, are used to enforce stability. The proposed algorithms were implemented as a user-friendly software tool that is available for download.

9. ACKNOWLEDGEMENTS

This work was partially supported at Boston University by the ONR under grants MURI 014-001-0303-5 and MURI N00014-10-10952 and by the NSF under grant CNS-0834260.

The authors would like to acknowledge Saša V. Raković from University of Oxford for his suggestions on preliminary versions of this manuscript.

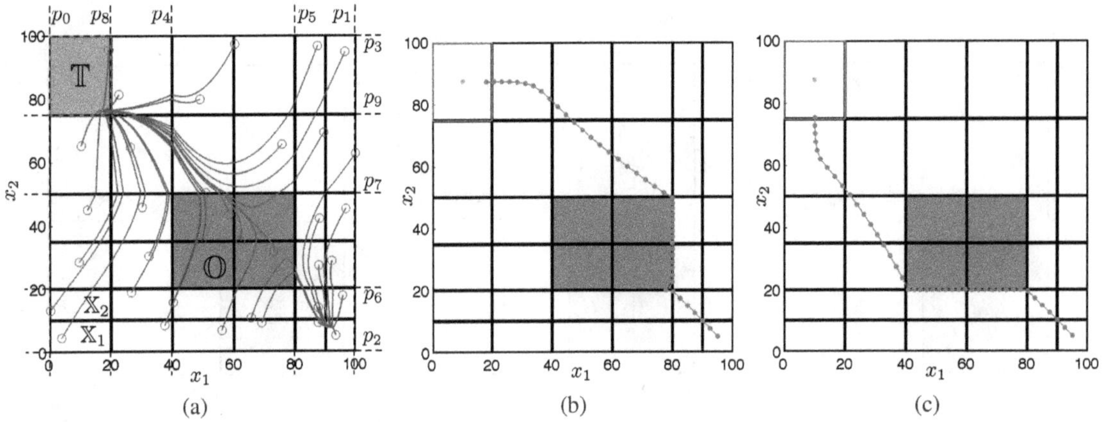

Figure 3: Case study 2: (a) Some trajectories of the uncontrolled PWA system, the regions and the corresponding linear predicates. The predicates are shown in the half planes where they are satisfied. The operating regions of the PWA system are shown with thick black borders and two of them are labeled. The trajectories go towards one of two possible stable equilibria (initial states are marked by circles). (b-c) The point $[10, 90]^\top$ is marked with a green dot. The trajectory of the controlled system originating at $[95, 5]^\top$. (b) $N = 6$, total cost $= 124479$. (c) $N = 7$, total cost$= 121096$.

10. REFERENCES

[1] M. Kloetzer and C. Belta, "A fully automated framework for control of linear systems from temporal logic specifications," *IEEE Transactions on Automatic Control*, vol. 53, no. 1, pp. 287 –297, 2008.

[2] P. Tabuada and G. Pappas, "Model checking LTL over controllable linear systems is decidable," ser. Lecture Notes in Computer Science, O. Maler and A. Pnueli, Eds. Springer-Verlag, 2003, vol. 2623.

[3] A. Girard, "Synthesis using approximately bisimilar abstractions: state-feedback controllers for safety specifications," in *Hybrid Systems: Computation and Control*. ACM, 2010, pp. 111–120.

[4] H. K. Gazit, G. Fainekos, and G. J. Pappas, "Where's Waldo? Sensor-based temporal logic motion planning," in *IEEE Conference on Robotics and Automation*, Rome, Italy, 2007, pp. 3116–3121.

[5] T. Wongpiromsarn, U. Topcu, and R. M. Murray, "Receding horizon temporal logic planning for dynamical systems," in *IEEE Conf. on Decision and Control*, Shanghai, China, 2009, pp. 5997–6004.

[6] A. Bhatia, L. E. Kavraki, and M. Y. Vardi, "Motion planning with hybrid dynamics and temporal goals," in *IEEE Conf. on Decision and Control*, Atlanta, GA, 2010, pp. 1108–1115.

[7] E. A. Gol, M. Lazar, and C. Belta, "Language-guided controller synthesis for discrete-time linear systems," in *Hybrid Systems: Computation and Control*. ACM, 2012, pp. 95–104.

[8] C. Sloth and R. Wisniewski, "Complete abstractions of dynamical systems by timed automata," *Nonlinear Analysis: Hybrid Systems*, vol. 7, no. 1, pp. 80 – 100, 2013.

[9] S. Karaman and E. Frazzoli, "Sampling-based motion planning with deterministic μ-calculus specifications," in *IEEE Conf. on Decision and Control*, Shanghai, China, 2009, pp. 2222 –2229.

[10] B. Yordanov, J. Tumova, C. Belta, I. Cerna, and J. Barnat, "Temporal logic control of discrete-time piecewise affine systems," *IEEE Transactions on Automatic Control*, vol. 57, no. 6, pp. 1491 –1504, 2012.

[11] C. Baier and J. Katoen, *Principles of model checking*. The MIT Press, 2008.

[12] N. Piterman and A. Pnueli, "Synthesis of reactive(1) designs," in *Verification, Model Checking, and Abstract Interpretation*. Springer, 2006, pp. 364–380.

[13] A. P. Sistla, "Safety, liveness and fairness in temporal logic," *Formal Aspects of Computing*, vol. 6, pp. 495–511, 1994.

[14] X. C. Ding, M. Lazar, and C. Belta, "Receding horizon temporal logic control for finite deterministic systems," in *American Control Conference*, Montreal, Canada, 2012, pp. 715 –720.

[15] X. C. Ding, S. L. Smith, C. Belta, and D. Rus, "Mdp optimal control under temporal logic constraints," in *IEEE Conf. on Decision and Control*, Orlando, FL, 2011, pp. 532 –538.

[16] M. Lazar, W. P. M. H. Heemels, S. Weiland, and A. Bemporad, "Stabilizing model predictive control of hybrid systems," *IEEE Transactions on Automatic Control*, vol. 51, no. 11, pp. 1813–1818, 2006.

[17] S. Di Cairano, M. Lazar, A. Bemporad, and W. P. M. H. Heemels, "A control Lyapunov approach to predictive control of hybrid systems," in *Hybrid Systems: Computation and Control*, ser. Lecture Notes in Computer Science, vol. 4981. St. Louis, MO: Springer Verlag, 2008, pp. 130–143.

[18] A. Bemporad and M. Morari, "Control of systems integrating logic, dynamics, and constraints," *Automatica*, vol. 35, pp. 407–427, 1999.

[19] O. Kupferman and M. Y. Vardi, "Model checking of safety properties," *Formal Methods in System Design*, vol. 19, pp. 291–314, 2001.

[20] T. Latvala, "Efficient model checking of safety properties," in *In Model Checking Software. 10th International SPIN Workshop*. Springer, 2003, pp. 74–88.

[21] J. Hespanha, "Modelling and analysis of stochastic hybrid systems," *Control Theory and Applications, IEEE Proceedings*, vol. 153, no. 5, pp. 520 – 535, 2006.

[22] J. B. Rawlings and D. Q. Mayne, *Model predictive control theory and desing*. Nob Hill Pub, 2009.

[23] M. Kvasnica, P. Grieder, and M. Baotić, "Multi-Parametric Toolbox (MPT)," 2004. [Online]. Available: http://control.ee.ethz.ch/~mpt/

[24] T. Gardner, C. Cantor, and J. Collins, "Construction of a genetic toggle switch in escherichia coli," *Nature*, vol. 403, no. 6767, pp. 339–342, 2000.

Iterative Temporal Motion Planning for Hybrid Systems in Partially Unknown Environments

Matthew R. Maly
mmaly@rice.edu

Morteza Lahijanian
morteza@rice.edu

Lydia E. Kavraki
kavraki@rice.edu

Hadas Kress-Gazit
hadaskg@cornell.edu

Moshe Y. Vardi
vardi@cs.rice.edu

ABSTRACT

This paper considers the problem of motion planning for a hybrid robotic system with complex and nonlinear dynamics in a partially unknown environment given a temporal logic specification. We employ a multi-layered synergistic framework that can deal with general robot dynamics and combine it with an iterative planning strategy. Our work allows us to deal with the unknown environmental restrictions only when they are discovered and without the need to repeat the computation that is related to the temporal logic specification. In addition, we define a metric for satisfaction of a specification. We use this metric to plan a trajectory that satisfies the specification as closely as possible in cases in which the discovered constraint in the environment renders the specification unsatisfiable. We demonstrate the efficacy of our framework on a simulation of a hybrid second-order car-like robot moving in an office environment with unknown obstacles. The results show that our framework is successful in generating a trajectory whose satisfaction measure of the specification is optimal. They also show that, when new obstacles are discovered, the reinitialization of our framework is computationally inexpensive.

Categories and Subject Descriptors

I.2.9 [**Artificial Intelligence**]: Robotics

General Terms

Algorithms, Verification

Keywords

Motion Planning, Temporal Logic, Formal Synthesis

1. INTRODUCTION

In "classical" motion planning, robots with dynamics along with basic point-to-point robotic tasks are considered. These tasks are specified as, "go from A to B and avoid obstacles."

To allow planning for more complex robots with complicated missions, various computational frameworks for planning with temporal logic specifications have been developed in the recent years (e.g., [10, 16, 17, 27, 32]). The increased expressivity as the result of the employment of temporal logics accommodates complex tasks that, for instance, require reaching one of a set of goals ("go to A or B"), visiting targets sequentially ("go to A, B, and C in this order"), and temporal conditions in reachability of targets ("first go to A or B and then eventually to C. If D is ever visited, then avoid B"). In this paper, we focus on planning with temporal goals for systems with general dynamics that can be realistically modeled as hybrid systems.

Most of the existing works in motion planning with temporal logic specifications consider static workspaces with full knowledge of their maps. Such assumptions, however, do not usually hold in real-world scenarios. For instance, a mobile robot in a warehouse setting may not be aware of a fallen box from a shelf that has blocked an aisle, or a mobile robot in an office environment may not know about the states of the office doors before its deployment. In such scenarios, it is reasonable to assume some information about the environment (e.g., the floor plan of the warehouse or the office building), but the motion-planning framework needs to have the capability of dealing with unforeseen obstacles in the environment. In fact, with complex specifications, it is imperative to consider the cases where the environment changes.

Recent works in synthesis-based approaches have begun to consider such cases (e.g., [25, 29]). In these works, synthesis involves the creation of a control strategy that can account for every possible environmental uncertainty. However, for the cases in which the number of environmental uncertainties is large, the problem becomes too complex. In these cases, it is advantageous to adopt an iterative temporal planning approach, in which online replanning is performed when unexpected environmental features are discovered. The difference between the above two approaches is discussed further in Section 1.1. The iterative planning approach poses the extra question of what to do in case the specification cannot be met due to newly discovered environmental restrictions. This gives rise to the need for a formal definition for a measure of satisfaction of a specification, a topic which is partially addressed in this paper.

Consider, for instance, a janitor robot in the office building whose schematic representation is shown in Figure 1. The office environment consists of a lobby, which includes an obstacle shown as a black rectangle, and five rooms each

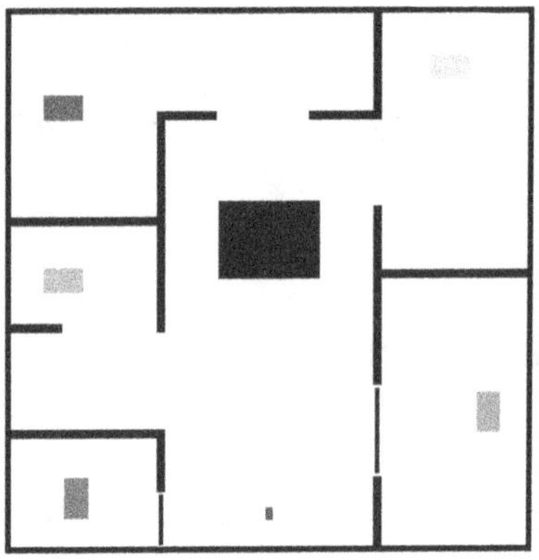

Figure 1: A schematic representation of an office building consisting of a lobby and five rooms. The lobby includes a black rectangular obstacle (center of figure), and each room has a door. The doors of three rooms are open and two are closed. The robot is shown as a blue rectangle in the lobby. The properties of interest in this environment are represented by the red, orange, purple, yellow, and green rectangles.

with a door. As the robot moves in the environment, its dynamics are subject to a set of restrictions associated with the features in the rooms (e.g., floor material and existence of sensitive objects). The properties of interest in this office environment are shown as orange, purple, red, green, and yellow rectangles. They represent office objects such as plants, a desk, a coffee maker, a blackboard, and a supply cabinet in the rooms. An example of a motion specification is as follows.

> "Visit the red region (to water the plants), go to the green region (to turn off the coffee maker), go to the yellow region (to clean the blackboard), and go to the purple region (to pick up the duster from supply cabinet) before visiting the orange region (to dust the desk). Do these tasks in any order and always avoid obstacles."

In this example, the robot initially has no knowledge of the obstacle in the lobby and the state of the office doors. It discovers them as it moves in the environment. Note that (for the instance captured in this figure) the robot will not be able to visit the red and green regions since the doors of their rooms are closed. Thus, there are parts of the above specification that cannot be satisfied. Nevertheless, for such tasks, we should allow the robot to continue with the mission even if it fails to satisfy parts of the specification due to unknown environmental constraints.

In this paper, we consider such realistic scenarios of planning in a partially unknown workspace for robots with complex and nonlinear dynamics. We focus on hybrid systems as they reveal the generality of our approach and lead to interesting application scenarios. Furthermore, we study the meaning of partial satisfaction of a temporal logic specification and how to measure it. Thus, the following problem emerges as a key challenge:

> "Given a mission expressed as a temporal logic specification for a mobile robot in a partially unknown workspace, find a plan for the robot to satisfy the mission's goals as closely as possible."

The main contribution of this work is a framework for iterative temporal motion planning. This framework allows for motion planning to be performed in a partially unknown environment. It enables a robot with nonlinear hybrid dynamics to modify its plan online, when it discovers obstacles that were not initially known, to satisfy a temporal logic specification without having to return to its original point and replan from scratch. The method is designed to avoid the recomputation of the automaton that is associated with the satisfaction of the temporal specification.

Another contribution of this paper is a scheme to measure closeness of satisfaction of a temporal logic specification and a method to maximize that measure. Thus, the framework can handle and provide guarantees of maximum satisfaction of the specification for the cases in which the discovered obstacles block regions of interest and render the temporal specification unsatisfiable. Then, instead of aborting the mission, the robot modifies its plan such that the measure of closeness of satisfaction of the specification is maximized. For a typical robotic task specification, optimizing this measure corresponds to satisfying as many requirements in the specification as the environment allows.

We tested this framework on a second-order car-like robot with hybrid dynamics in an office environment where regions of interest are known *a priori*. The robot was given a temporal specification requiring it to visit the regions of interest, without initial knowledge of obstacles that make some of the regions unreachable. In our simulation experiments, we varied the number of regions the robot should visit as well as the accuracy of the robot's initial map of the workspace. In all cases, our framework successfully found a trajectory for the robot to satisfy the specification as closely as possible.

The remainder of the paper is organized as follows. Section 1.1 contains related work. Section 2 describes our hybrid robot model and the type of temporal logic that we use. Section 3 details the problem we consider and gives a high-level overview of our approach to solving it. We present our iterative temporal motion-planning framework in Section 4. In Section 5, we introduce our simulation experiment and demonstrate results. The paper concludes with final remarks and a discussion of future work in Section 6.

1.1 Related Work

Much work has been done toward the problem of planning for robotic systems to satisfy high-level temporal logic specifications. For general robot models with nonlinear dynamics, static workspaces, and temporal goals, motion planning approaches have been proposed to solve the problem using deterministic μ-calculus specifications [14] and co-safe LTL specifications [3–5, 27].

Synthesis-based approaches to such problems require strong assumptions on the robot's dynamics and a construction of a finite discrete model for the motion of the robot in its workspace with a simulation relation to the continuous model. A provably correct hybrid controller can then be generated as a state machine that encodes the robot actions necessary to satisfy the task [18]. The synthesis of this hybrid controller requires time and space polynomial in the size of the reachable state space of the system; this is often called the state explosion problem [19]. One way to address this problem is to use a coarser abstraction, which requires a stronger assumption on controllers. Other suggestions include receding horizon techniques [32]. Work has also been done to address the issue of controller uncertainty, modeling the robot as a Markov decision process [9,22,23].

The issue of synthesis from high-level specifications in an unknown or dynamic environment has been studied both for abstract systems (e.g., [6]) and specifically for robotics (e.g., [7, 15, 25, 29]). If the geometry of the environment changes, whether due to an unknown region becoming reachable [29] or a known region becoming unreachable [25], then the hybrid controller must be updated to incorporate the change. As global resynthesis of the hybrid controller is expensive, there exist approaches to locally patch the controller to incorporate the changes in less time; still, initial work in this area has shown that patching the hybrid controller can still require significant time to complete [25]. In addition, the work in [12] addresses the issue of motion planning for a mobile robot to move from a start region to a goal region, where secondary regions of interest can potentially be discovered along the way. When a secondary target is discovered, the robot replans a new trajectory to visit the target along its way to the goal region.

Our work is most closely related to [3–5, 27] in that we are taking a motion-planning approach instead of using synthesis. Synthesis is a process that generates a strategy to ensure system correctness in all possible scenarios. Such approaches require knowledge of all possible environmental uncertainties. However, in real-world applications, it may not be feasible to obtain enough information of the environment for synthesis. Also, when there are too many unknowns, synthesizing a plan can be unsuccessful. In such cases, iterative planning methods are more natural and viable to employ since they do not require a full knowledge of the environmental restrictions. Therefore, in this paper, we propose an online iterative planning approach to deal with partially unknown environments. Rather than accounting for everything that can go wrong, we plan based on what we know and deal with new restrictions only when they are discovered. In other words, we plan a trajectory given the currently known state of the environment. During execution of the trajectory, if an unforeseen problem is encountered, we replan a new trajectory from the current state on-the-fly. This framework is inspired by replanning scenarios in robotics [1,2].

A key advantage of our approach lies in the high-level structure through which we guide a low-level continuous motion planner. This high-level structure is a product of the abstraction of the hybrid system and an automaton that derives from the temporal logic specification. The abstraction is achieved quickly by a geometric partition of the robot's workspace. However, the automaton from the specification can be very expensive to compute. By keeping the automaton and the abstraction separate, we prevent changes to the environment from requiring us to recompute the automaton. Changes to the environment simply require modifications to the decomposition of the workspace and, hence, to the abstraction, which is an inexpensive operation to perform. This is in contrast to other works that use synthesis approaches to plan for robots to satisfy temporal logic specifications. In these works, typically the task specification and assumptions on the environment must be encoded into a single hybrid controller that can be expensive to change [18,25]. Another advantage of our framework is the use of motion planning, which supports systems with any type of high-dimensional hybrid (possibly nonlinear) dynamics. Synthesis-based approaches deal with a restricted class of robot systems that typically involve linear dynamics. When the dynamics of the system are sufficiently complex, it is difficult (if not impossible) to synthesize provably correct controllers [5].

The issue of what to do when a specification is determined to be unsatisfiable has been explored before. In [28], an algorithm was defined to report a reason as to why a GR(1) LTL specification is unrealizable. The work in [15] presents a method of changing an unsatisfiable nondeterministic Büchi automaton into the "closest" satisfiable one, where all actions of the robot are represented using a finite state machine. Our approach to unsatisfiable specifications differ in that we do not change the automaton; instead, we provide a simple metric to define partial satisfaction that is meaningful for an interesting set of scenarios.

2. PRELIMINARIES

2.1 Robot Hybrid Model

In this paper, we consider a general mobile robot whose dynamics are subject to restrictions in the regions of a partially unknown environment. We describe its motion in such an environment by the hybrid system $H = (S, s_0, \text{INV}, \text{SENSE}, E, \text{GUARD}, \text{JUMP}, U, \text{FLOW}, \Pi, L)$, where

- $S = Q \times X$ is the hybrid state space that is a product of a set of discrete modes, $Q = \{q_1, q_2, \ldots, q_m\}$ for some finite $m \in \mathbb{N}$, by a set of continuous state spaces $X = \{X_q \subseteq \mathbb{R}^{n_q} : q \in Q\}$;
- $s_0 \in S$ is the initial state;
- $\text{INV} = \{\text{INV}_q : q \in Q\}$, is the set of invariants, where $\text{INV}_q : X_q \to \{\top, \bot\}$;
- $\text{SENSE} : X_q \to \{\top, \bot\}$, is the sensing function that returns *true* if an unknown obstacle is detected;
- $E \subseteq Q \times Q$ describes discrete transitions between modes in Q;
- $\text{GUARD} = \{\text{GUARD}_{q_i,q_j} : (q_i, q_j) \in E\}$, where $\text{GUARD}_{q_i,q_j} : X_{q_i} \times \{\top, \bot\} \to \{\top, \bot\}$ is a guard function that enables transitions between different modes given the continuous state of the robot and the unknown-obstacle detector readings (i.e., output of SENSE);
- $\text{JUMP} = \{\text{JUMP}_{q_i,q_j} : (q_i, q_j) \in E\}$, where $\text{JUMP}_{q_i,q_j} : X_{q_i} \to X_{q_j}$ is the jump function. In this paper, we assume that each JUMP_{q_i,q_j} is the identity function;
- $U = \{U_q \subset \mathbb{R}^{m_q} : q \in Q\}$ is the set of input spaces;
- $\text{FLOW} = \{\text{FLOW}_q : q \in Q\}$, where $\text{FLOW}_q : X_q \times U_q \times \mathbb{R}^{\geq 0} \to X_q$ is the flow function that describes the continuous dynamics of the system through a set of differential equations;
- Π is a set of atomic propositions;
- $L : S \to 2^{\Pi}$ is a labeling function assigning to each hybrid state possibly several elements of Π.

A pair $s = (q, x) \in S$ denotes a hybrid state of the system. $\text{FLOW}_q(x, u, t)$ gives the continuous state of the system when the input u is applied for t time units starting from state x.

2.2 Syntactically Co-safe LTL

We use syntactically co-safe LTL to write the specifications of robotic tasks. Its syntax and semantics are defined below.

DEFINITION 1 (SYNTAX). *Let $\Pi = \{\pi_1, \pi_2, \ldots, \pi_N\}$ be a set of Boolean atomic propositions. A syntactically co-safe LTL formula over Π is inductively defined as following:*

$$\phi := \pi \,|\, \neg\pi \,|\, \phi \vee \phi \,|\, \phi \wedge \phi \,|\, \mathcal{X}\phi \,|\, \phi\mathcal{U}\phi \,|\, \mathcal{F}\phi$$

where $\pi \in \Pi$, $\neg$ (negation), $\vee$ (disjunction), and $\wedge$ (conjunction) are Boolean operators, and $\mathcal{X}$ ("next"), $\mathcal{U}$ ("until"), and $\mathcal{F}$ ("eventually") are temporal operators.

DEFINITION 2 (SEMANTICS). *The semantics of syntactically co-safe LTL formulas are defined over infinite traces over 2^Π. Let $\sigma = \{\tau_i\}_{i=0}^{\infty}$ with $\tau_i \in 2^\Pi$ be an infinite trace and $\sigma^i = \tau_i, \tau_{i+1}, \ldots$ and $\sigma_i = \tau_0, \tau_1, \ldots, \tau_{i-1}$. σ_i is a prefix of the trace σ. $\sigma \models \phi$ indicates that σ satisfies formula ϕ and is recursively defined as following:*

- $\sigma \models \pi$ *if* $\pi \in \tau_0$;
- $\sigma \models \neg\pi$ *if* $\pi \notin \tau_0$;
- $\sigma \models \phi_1 \vee \phi_2$ *if* $\sigma \models \phi_1$ *or* $\sigma \models \phi_2$;
- $\sigma \models \phi_1 \wedge \phi_2$ *if* $\sigma \models \phi_1$ *and* $\sigma \models \phi_2$;
- $\sigma \models \mathcal{X}\phi$ *if* $\sigma^1 \models \phi$;
- $\sigma \models \phi_1\mathcal{U}\phi_2$ *if* $\exists k \geq 0$, *s.t.* $\sigma^k \models \phi_2$, *and* $\forall i \in [0, k)$, $\sigma^i \models \phi_1$;
- $\sigma \models \mathcal{F}\phi$ *if* $\exists k \geq 0$, *s.t.* $\sigma^k \models \phi$.

An important property of syntactically co-safe LTL formulas is that, even though they have infinite-time semantics, finite traces are sufficient to satisfy them. Hence, syntactically co-safe LTL is an appropriate specification language to describe robotic tasks which are required to be realized in finite horizon.

From a syntactically co-safe LTL formula ϕ, a deterministic finite automaton (DFA) can be constructed that accepts precisely all of the formula's satisfying finite traces [21]. Such a DFA is given by a tuple $\mathcal{A}_\phi = (Z, \Sigma, \delta, z_0, F)$, where

- Z is a finite set of states;
- $\Sigma = 2^\Pi$ is the input alphabet, where each input symbol is a truth assignment to the propositions in Π;
- $\delta : Z \times \Sigma \to Z$ is the transition function;
- $z_0 \in Z$ is the initial state;
- $F \subseteq Z$ is the set of accepting states.

A finite run of $\mathcal{A}_\phi$ is a sequence of states $\omega = \omega_0\omega_1 \ldots \omega_n$, where $\omega_0 = z_0$ and $\omega_i \in Z$ for $i = 1, \ldots, n$. ω is called an accepting run if $\omega_n \in F$. An input trace that realizes an accepting run is a ϕ-satisfying trace.

3. PROBLEM DESCRIPTION AND OVERALL APPROACH

In this paper, we consider a mobile robot with complex and possibly nonlinear hybrid dynamics moving in an environment. Some of the features and obstacles in the environment may not be known before the deployment of the robot. We make the natural assumption that the robot can detect an unknown obstacle when it is within the robot's obstacle-detector range. This is an assumption commonly made in related work [2]. The regions in the environment impose different sets of restrictions on the dynamics of the robot. Each point in the environment holds a set of properties (propositions). Let Π denote the set of all environmental propositions. We assume that while the robot has full information of the propositions and their locations in the environment, it has only partial *a priori* knowledge of the regions (obstacles) of the environment. We are interested in deploying this robot in such a partially unknown environment with temporal logic specifications.

Due to possible unknown obstacles in the environment, the satisfaction of the specification cannot be guaranteed. Nevertheless, we do not want the robot to abort the mission if it realizes that fragments of the specification cannot be met. Instead, we require the robot to satisfy the specification as closely as possible. We envision many scenarios where this can be an advantageous approach (e.g., the janitor robot example in Section 1). We formally define and discuss the definition of satisfying a specification as closely as possible below and in Section 4.3. We now focus on the following problem.

PROBLEM: *Given a partially unknown environment and a task specification expressed as a syntactically co-safe LTL formula ϕ over Π, find a robot motion plan that satisfies ϕ as closely as possible.*

We model the motion of the robot in the environment as a hybrid system (Section 2.1). This allows us to capture the changes in the robot dynamics by the transitions between the modes of the hybrid system. We assume that the robot moves in a two-dimensional environment. We choose the continuous states of the hybrid system in mode q, $x \in X_q$, such that its first two components refer to the position of the robot. Let $Pr(A)$ denote projection of a Euclidean set A onto $\mathbb{R}^2$. We define the workspace of the robot as $W = \{(q, W_q) : W_q = Pr(X_q), q \in Q\}$ and denote the set of (polygonal) obstacles in mode q by $W_{q,obs}$. The robot has partial *a priori* knowledge of the obstacles in its workspace. Thus, $W_{q,obs} = W^k_{q,obs} \cup W^u_{q,obs}$ where $W^k_{q,obs}$ and $W^u_{q,obs}$ refer to the sets of known and unknown obstacles in mode q, respectively. We also assume that the robot can detect an unknown obstacle when it comes within some proximity of it. This is represented by SENSE in the hybrid model. We establish the relationship between the hybrid state of the robot $s = (q, x)$ and its workspace by function $h_H : S \to W$. Similarly, given $\mathrm{w} \in W$, we define $h_H^{-1}(\mathrm{w}) = \{s \in S : h_H(s) = \mathrm{w}\}$. We choose Π as the set of atomic propositions for the hybrid system and assign them to the hybrid states according to their positions in the workspace. Hence, the problem is now reduced to designing a motion plan for H that satisfies ϕ.

We employ a multi-layered synergistic framework [4, 27] to solve the motion planning problem by using the initial knowledge of the workspace. The framework consists of three main layers: a high-level search layer, a low-level search layer, and a synergy layer that facilitates the interaction between the high-level and the low-level search layers (see Figure 2). The high-level planner uses an abstraction of the hybrid system and the specification formula ϕ to suggest high level plans. The low-level planner uses the dynamics of the hybrid system and the suggested high-level plans to explore the state-space for feasible solutions. In our work, the low-level layer is a sampling-based planner and does not assume the existence of a controller [26].

To satisfy a specification in an undiscovered environment, an iterative high-level planner is employed. That is, every time an unknown obstacle is encountered, the high-level planner modifies the coarse high-level plan online by accounting for the geometry of the discovered obstacle, the path traveled to that point, and the remaining segment of the specification that is yet to be satisfied. This replanning is achieved in four steps. First, a "braking" operation is applied to prevent the robot from colliding with the newly discovered obstacle. Simultaneously, a new abstraction of the hybrid system is computed through a new decomposition of the modified environment map (workspace) [3]. Next, the traveled path is mapped on the new abstraction model. Finally, a new satisfying plan is generated as a continuation of the explored portion of the old plan. Thus, the robot does not need to reinitialize (return to its starting point) every time it encounters an unknown environmental feature. Moreover, the robot's progress in satisfying the specification is preserved. This iterative motion-planning framework is discussed in detail in Section 4.

Recall that from ϕ, a DFA can be constructed that accepts all of the formula's satisfying finite traces [21, 27]. We use this DFA to design a satisfying high-level plan. We also utilize the DFA to define a metric to measure the "distance-to-satisfaction" of a specification in cases in which the specification is unsatisfiable. This measure is used to produce a high-level plan that satisfies the specification as closely as possible. We formally define this metric in Section 4.3.

In general, a contingency maneuver can be used instead of a "braking" operation as the first step of the approach. Our framework is by no means limited to a stopping maneuver, and the exploration for the "best" contingency plan is left for future work. Moreover, it is important to note that our method of generating a new high-level plan is fast. This is due to the following two reasons: (1) we are not recomputing the DFA, which does not need to change since the specification formula does not change following discovery of an obstacle, and (2) we generate the abstraction of the hybrid system by decomposing the workspace through triangulation, which has been shown to be computationally inexpensive [5]. For instance, the computation time for high-level replanning for the janitor robot example moving in the office environment shown in Figure 1 is in the order of a fraction of a second.

4. PLANNING FRAMEWORK

In this section, we describe our iterative planning framework, which consists of three main layers: a high-level planner, a low-level search layer, and a synergy layer as shown in Figure 2. The high-level planner generates a set of coarse satisfying plans by searching over a structure called a product automaton (Section 4.2). This structure is the product of discrete abstraction $\mathcal{M}$ of the hybrid system (Section 4.1) and automaton $\mathcal{A}_\phi$ corresponding to the formula ϕ. Each of these plans is a sequence of the states of the product automaton which can be mapped back to the states of $\mathcal{M}$. The low-level search layer produces continuous trajectories that follow a satisfying high-level plan. This is achieved by expanding a sampling-based motion tree in the direction of a suggested high-level plan in the hybrid state-space. The synergy layer facilitates the two-way interaction between the high-level and the low-level search layers (Sections 4.2 and 4.3). Algorithm 1 contains the framework pseudocode; it relies on subroutines detailed in Algorithms 2, 3, and 4.

4.1 Abstraction

To produce a high-level plan, we first abstract the hybrid system H to a discrete model $\mathcal{M} = (D, d_0, \rightarrow_D, \Pi, L_D)$, where D is a set of discrete states, $d_0 \in D$ is the initial state, $\rightarrow_D \subseteq D \times D$ is the transition relation, and $L_D : D \rightarrow 2^\Pi$ is a labeling function. We refer to the model $\mathcal{M}$ as the *abstraction* of the hybrid system. To construct $\mathcal{M}$, we first partition each workspace W_q (for each discrete mode q) into a set of *regions* (i.e. $W_q = \bigcup_{i=1}^{N_r^q} r_i^q$). Specifically, we use a geometry-based conforming Delaunay triangulation of W_q that respects the propositional regions and the boundaries of the obstacles.

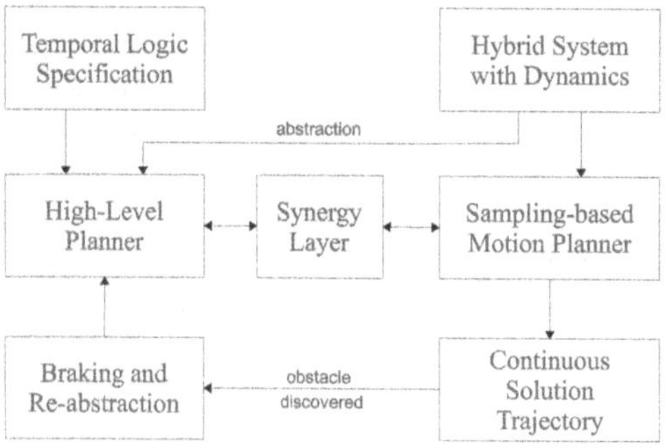

Figure 2: Multi-layered synergistic motion planning framework.

Recall from Section 3 that $W_q = Pr(X_q)$, where X_q is the domain of the continuous states of the hybrid system in mode q. Thus, the decomposition of W_q induces a partition in the hybrid state space. Let S_i^q denote the set of all hybrid states that correspond to the region r_i^q in W_q (i.e. $S_i^q = \{(q, x) \in S \mid x \in Pr^{-1}(r_i^q),\, r_i^q \subseteq W_q,\, q \in Q\}$). Then,

$$S = \bigcup_{q \in Q} \bigcup_{i=1}^{N_r^q} S_i^q.$$

We associate a unique discrete state $d \in D$ to each S_i^q. We model this correspondence with a family of maps $\{\Upsilon_q : S^q \rightarrow D \mid S^q = q \times X_q,\, q \in Q\}$; then the previous sentence can be written as $\Upsilon_q(S_i^q) = d$. Moreover, D can be written as $D = \cup_{q \in Q}\{\Upsilon_q(S_i^q) \mid 1 \leq i \leq N_r^q\}$.

We construct the transition relation $\rightarrow_D$ to include geometric adjacencies between regions of a given workspace as well as adjacencies between discrete modes enabled by GUARD functions of the hybrid system. Specifically, for each pair of geometrically adjacent regions r_j^q and r_k^q in W_q, we add the transition $(\Upsilon_q(S_j^q), \Upsilon_q(S_k^q))$ to $\rightarrow_D$; furthermore for each pair of sets S_l^q and $S_m^{q'}$ between which a discrete jump is possible, we add the transition $(\Upsilon_q(S_l^q), \Upsilon_q(S_m^{q'}))$ to $\rightarrow_D$. All the hybrid states in S_i^q have the same labels since the triangulation of each workspace W_q respects the propositional regions. Hence, the labeling function L_D corresponds to the labeling function L from the definition of H; that is, $L_D(\Upsilon_q(S_i^q)) = L(s)$ for every $s \in S_i^q$. For further details, we refer the reader to our previous work [3].

It should be noted that the initial construction of $\mathcal{M}$ is based on the initial knowledge of the environment map. As the robot discovers unknown obstacles, the map is updated and a new abstraction is generated. Given that this method is based on a triangulation of a two-dimensional space, obtaining a new abstraction is fast. Furthermore, we initially assume transitions between all adjacent partitions of the workspace are realizable even though the dynamics of the robot may prevent some transitions. This does not create a problem in our planning framework because the synergistic framework will bias its discrete search against unrealizable transitions. In fact, one of the advantages of our planning framework is that it does not require a bisimilar abstraction and, hence, allows for inexpensive and fast construction of an approximate abstraction model.

4.2 Initializing the Product Automaton

The structure we use to guide the tree of system trajectories is a product automaton, which is computed as $\mathcal{P} = \mathcal{M} \times \mathcal{A}_\phi$. In line 2 of Algorithm 1, we compute the minimal DFA $\mathcal{A}_\phi$ corresponding to the formula ϕ [21, 24]. Though this translation can require time doubly exponential with respect to the number of propositions in ϕ, we only compute $\mathcal{A}_\phi$ once, so the translation can be seen as an offline step. We refer to elements of $\mathcal{P}$ as *high-level states*. $\mathcal{P}$ is a directed graph in which there exists an edge from high-level state (d_1, z_1) to (d_2, z_2) iff d_1 and d_2 are adjacent in $\mathcal{M}$ and $\mathcal{A}_\phi.\delta(z_1, \mathcal{M}.L_D(d_2)) = z_2$, where $\mathcal{A}_\phi.\delta$ is the deterministic transition function for $\mathcal{A}_\phi$. The latter condition means that there exists a transition in $\mathcal{A}_\phi$ from z_1 to z_2 whose label is satisfied by the set of propositions $\mathcal{M}.L_D(d_2)$ that hold true at d_2. We call a high-level state $(d, z) \in \mathcal{P}$ an *accepting* (or *goal*) state iff z is an accepting state in $\mathcal{A}_\phi$.

For each high-level state $(d, z) \in \mathcal{P}$, we assign a weight defined by

$$w(d, z) = \frac{(\text{COV}(d, z) + 1) \cdot \text{VOL}(d)}{\text{DISTFROMACC}(z) \cdot (\text{NUMSEL}(d, z) + 1)^2} \quad (1)$$

where $\text{COV}(d, z)$ is the number of tree vertices (generated by the low-level planner) associated with (d, z) (an estimate of coverage), $\text{VOL}(d)$ is the area of the workspace corresponding to the abstraction state d, $\text{DISTFROMACC}(z)$ is the minimum distance from automaton state z to an accepting state in the automaton, and $\text{NUMSEL}(d, z)$ is the number of times (d, z) has been selected for tree expansion in line 2 of Algorithm 4. Then, to each directed edge $e = ((d_1, z_1), (d_2, z_2))$ in $\mathcal{P}$, we assign the weight

$$w(e) = \frac{1}{w(d_1, z_1) \cdot w(d_2, z_2)}. \quad (2)$$

The values in (1) and (2) are continually updated as the planning framework progresses. For example, whenever the low-level planner creates a tree vertex associated with a high-level state (d, z), the value of $\text{COV}(d, z)$ is incremented by one. The estimates in (1) and (2) have been shown to work well in previous work [4]. In general, a weighing scheme that incorporates more than just number-of-edge distance is useful to promote expansion in unexplored areas (i.e., where COV and NUMSEL are both small) and to discourage expansion in areas where attempts at exploration have repeatedly failed (i.e., where NUMSEL $\gg$ COV).

4.3 Planning

Once the product automaton has been computed, line 4 of Algorithm 1 computes a trajectory for the system that satisfies the formula ϕ as closely as possible. The details of this approach are given in Algorithm 2. Many details are similar to the framework discussed in past works [3–5]. We differ from them by allowing for online replanning in light of newly discovered obstacles in Algorithm 1, and by partially satisfying an unsatisfiable specification when computing a lead in Algorithm 3.

The core loop of our planning algorithm is shown in lines 4, 5, and 6 of Algorithm 2. The subroutine COMPUTELEAD in Algorithm 3 creates leads that reach as close as possible to an accepting state. Each lead computed in line 4 is a suggested sequence of contiguous high-level states through which EXPLORE attempts to guide the tree of motions.

Measure of Satisfiability.

We present a measure of satisfiability that uses the graph-based distance to an accepting state in the DFA. Each high-level state (d, z) is annotated with the graph-based distance value $\text{DISTFROMACC}(z)$ corresponding to the automaton state z. Our framework computes trajectories that end in a high-level state (d_g, z_g) such that $\text{DISTFROMACC}(z_g)$ is minimized. The function DISTFROMACC is an intuitive measure on the automaton that translates to a reasonable high-level plan for many formulas that we have encountered, such as the example specification in Section 1. For such a specification, a trajectory that minimizes DISTFROMACC takes the robot to all reachable regions of interest, while a non-optimal trajectory with respect to DISTFROMACC would miss some reachable regions. The topic of "approximating" temporal properties is a subject of ongoing research. Generally, it requires making the satisfaction relation quantitative rather than qualitative. For example, the satisfaction value can be an arbitrary lattice element rather than a Boolean value; cf. [20]. In addition, the authors in [31] describe a synthesis algorithm to minimize quantitative satisfaction error given a set of contradictory specifications.

Algorithm 1 Framework for planning for hybrid systems with LTL specifications in a partially unknown environment

Input: A robot model described by a hybrid system $H = (S, s_0, \text{INV}, \text{SENSE}, E, \text{GUARD}, \text{JUMP}, U, \text{FLOW}, \Pi, L)$,
a bounded workspace $W \subset \mathbb{R}^2$,
a set of initially known obstacles $O \subset W$,
a co-safe LTL formula ϕ defined over $H.\Pi$,
and a time bound $t_{\max}$.

Output: Returns `true` if successful in moving the robot through the workspace to satisfy ϕ; returns `false` otherwise.

```
1: M ← COMPUTEABSTRACTION(W, O, H.Π, H.L)
2: A_φ ← COMPUTEMINDFA(φ, W, H.L)
3: P ← COMPUTEPRODUCT(M, A_φ, H.Π, H.L)
4: {x_i}_{i≥0} ← PLAN(H, W, O, H.Π, H.L, P, t_max)
5: j ← 1
6: while j < |{x_i}| do
7:    Move system from state x_{j−1}.s to state x_j.s
8:    if H.SENSE(x_j.s) = ⊤ then
9:       Apply braking operation to reach stopped robot state s'
10:      H.s_0 ← s'
11:      Add discovered obstacle to O
12:      M ← COMPUTEABSTRACTION(W, O, H.Π, H.L)
13:      P ← COMPUTEPRODUCT(M, A_φ, H.Π, H.L)
14:      {x_i}_{i≥0} ← PLAN(H, W, O, H.Π, H.L, P, t_max)
15:      if PLAN was unsuccessful then
16:         return false
17:      j ← 1
18:   j ← j + 1
19: return true
```

COMPUTELEAD computes a lead that ends in a high-level state (d, z) such that the number of transitions from z to an accepting state in the automaton, given by $\text{DISTFROMACC}(z)$, is minimized. If the specification ϕ is satisfiable in the current environment, then $\text{DISTFROMACC}(z) = 0$, i.e., (d, z) is an accepting state. On the other hand, if the specification ϕ is unsatisfiable, then z is as close as possible to an accepting state in the automaton. In many cases, there are multiple candidate high-level states that tie under the DISTFROMACC metric. To break ties, we choose the high-level state with minimal edge-weight distance from the starting high-level state, using the edge-weight function defined in (2).

The subroutine ComputeAvailableCells in line 5 of Algorithm 2 computes the set of high-level states that exist in the current lead and are nonempty. A high-level state (d, z) is *nonempty* if there exists at least one tree vertex associated with it, i.e., if $\text{COV}(d, z) > 0$. In the first few iterations of Plan, the only nonempty high-level state will be (d_0, z_0), where d_0 is the abstraction region containing the initial system state $H.s_0$, and z_0 is the initial state of the DFA. As the algorithm progresses, the tree planner reaches more high-level states, and the set C of available cells grows larger. To promote progress, we favor high-level states that are closest to the end of the lead. Specifically, moving backwards along the lead, for each nonempty high-level state (d, z) we encounter, we add (d, z) to the set C of available cells and then quit early with probability 0.5.

The subroutine Explore, given in Algorithm 4, corresponds to the low-level search layer of our framework. This function promotes tree expansion in high-level states from the set C. In line 2 of Explore, a high-level state (d, z) is sampled from C with probability $w(d, z)/\sum_{(d',z')\in C} w(d', z')$. Then, in line 3, we perform one iteration of the low-level tree planner to promote expansion from the set of tree vertices associated with the high-level state (d, z) and obtain a new tree vertex v. Any tree-based motion planner can be used in this step; in our approach, we are using an EST-like approach [13]. If z is an accepting state of the automaton, then v is returned as the endpoint of a solution trajectory. This trajectory is reconstructed by Plan in line 8 by following parent vertices back to the root of the tree. Otherwise, if the new vertex v corresponds to a newly reached high-level state that is in the current lead, then the high-level state is added to the set of available cells in line 8 of Explore to be considered in the future.

4.4 Discovering an Obstacle and Replanning

Once a system trajectory that satisfies ϕ is computed, we begin moving the robot along the trajectory. At each state in the trajectory, we query the robot's range sensor in line 8 of Algorithm 1. We assume that the robot's range sensor checks for obstacles within radius ρ of the center of the robot and reports a polygonal model of any previously unknown obstacle that it finds. If no new obstacles are discovered along the trajectory, then the robot reaches the final state of the planned trajectory and stops, having completed its mission. If an obstacle is discovered by the range sensor from some state s along the trajectory, then we apply a braking operation to the robot to reach some stopped state s'. The braking operation should respect the dynamics of the system. In the general case, the robot should perform a contingency maneuver to avoid the newly discovered obstacle [1, 11]. The radius ρ of the range sensor is assumed to be large enough for the braking or contingency maneauver to safely be performed. Once the braking maneuver is complete, we recompute the discrete abstraction $\mathcal{M}$ to ignore the new obstacle, recompute the product automaton $\mathcal{P}$, and replan a trajectory from s', following the same planning approach described in Section 4.3. Once a new trajectory is found by the planner, we resume moving the robot from s' along the new trajectory. It is important to note that because we recompute the entire discrete abstraction from scratch, all of the previous edge weights in $\mathcal{P}$ are lost and recomputed in the next planning iteration.

Algorithm 2 Plan: Temporal planning algorithm for hybrid systems

Input: A robot model described by a hybrid system $H = (S, s_0, \text{INV}, \text{SENSE}, E, \text{GUARD}, \text{JUMP}, U, \text{FLOW}, \Pi, L)$,
a bounded workspace $W \subset \mathbb{R}^2$,
a set of known obstacles $O \subset W$,
a product automaton $\mathcal{P}$,
and a time bound $t_{\max}$.
Output: Returns a sequence of triplets, each containing hybrid system state, control, and corresponding high-level state, representing a system trajectory that satisfies the specification. Reports an error and aborts if no such trajectory could be found within time $t_{\max}$.

1: $\mathcal{T} \leftarrow$ InitializeTree(s_0)
2: `solved` $\leftarrow$ `false`
3: **while** Time Elapsed $< t_{\max}$ **do**
4: $K = ((d_1, z_1), \ldots, (d_k, z_k)) \leftarrow$ ComputeLead($\mathcal{P}, H.s_0$)
5: $C \leftarrow$ ComputeAvailableCells(K)
6: $v \leftarrow$ Explore($H, W, O, \mathcal{T}, C, K, \mathcal{P}, \Delta t$)
7: **if** $v \neq$ `NULL` **then**
8: Follow `v.parent` to construct trajectory $\{x_i\}_i$
9: **return** $\{x_i\}_i$
10: Report unsuccessful and exit

Algorithm 3 ComputeLead: Subroutine to compute high-level guides

Input: A product automaton $\mathcal{P}$ (product of DFA and workspace decomposition) and a starting high-level state (d_0, z_0).
Output: Returns a lead, which is a sequence of high-level states beginning with the given start (d_0, z_0) and ending as close as possible to an accepting state.

1: $F \leftarrow \arg\min_{(d,z)\in\mathcal{P}}\{\text{DistFromAcc}(z)\}$
2: Run Dijkstra's all-pairs shortest-path algorithm on $\mathcal{P}$ with source (d_0, z_0); store parent map `parent` and weight map `weight`
3: $(d_g, z_g) \leftarrow \arg\min_{(d,z)\in F}\{\texttt{weight}[(d, z)]\}$
4: Construct lead $K = ((d_0, z_0), \ldots, (d_g, z_g))$ using `parent` map
5: **return** K

5. EXPERIMENTS

To test our approach, we have created an experiment for a second-order car with hybrid dynamics to explore an office-like environment. The full map of the office is shown in Figure 3(a). Geometrically, the robot is modeled as a rectangle with length $l = 0.2$ and width $w = 0.1$. The robot state has a continuous component (x, y, θ, v, ψ), which includes the planar position $(x, y) \in [0, 10]^2$, heading $\theta \in [-\pi, \pi]$, forward velocity $v \in [-1/6, 1]$, and steering angle $\psi \in [-\pi/6, \pi/6]$. The robot state also includes a discrete component $g \in \{1, 2, 3\}$, corresponding to the gear of the car. The car is controlled with the input pair $u = (u_0, u_1)$, where u_0 is the forward acceleration and $u_1 \in [-\pi/18, \pi/18]$ is the steering angle velocity. Given the current gear g, we bound the acceleration input so that $u_0 \in [-1/6, g/6]$. The dynamics of the car are given by $\dot{x} = v\cos(\theta)$, $\dot{y} = v\sin(\theta)$, $\dot{\theta} = v\tan(\psi)/l$, $\dot{v} = u_0$, and $\dot{\psi} = u_1$. We model each gear as a separate discrete mode of the hybrid system. We define guards and jumps on the dynamics of the robot so that the robot switches gears as follows. If, when in gear $g < 3$, the car achieves velocity $v > g/6$, then the car switches to gear $g + 1$. If, when in gear $g > 1$, the car achieves velocity

Algorithm 4 EXPLORE: Tree-exploration subroutine

Input: A robot model described by a hybrid system $H = (S, s_0, \text{INV}, \text{SENSE}, E, \text{GUARD}, \text{JUMP}, U, \text{FLOW}, \Pi, L)$,
a bounded workspace $W \subset \mathbb{R}^2$,
a set of known obstacles $O \subset W$,
a tree of motions $\mathcal{T}$,
a set of available high-level states C,
a lead K,
a product automaton $\mathcal{P}$,
and an exploration time Δt.

Output: Returns a tree vertex that reaches the goal high-level state if one was found; returns NULL otherwise.

```
while TIME ELAPSED < Δt do
    (d, z) ← C.sample()
    v ← SELECTANDEXTEND(T, H, (d, z), W, O, P, PROP)
    if v.z ≠ ∅ then
        if v.z.isAccepting() then
            return v
        if (v.d, v.z) ∉ C ∧ (v.d, v.z) ∈ L then
            C ← C ∪ {(v.d, v.z)}
return NULL
```

$v < (g-1)/6$, then the car switches to gear $g-1$. Immediately following a gear switch, the acceleration input bounds are updated accordingly. Furthermore, we define guards in the office so that in small rooms (rooms containing the red and orange propositions), the car is restricted to first gear. In larger rooms (rooms containing the yellow, purple, and green propositions), the car is restricted to first and second gears. Specifically, the guards in the rooms prevent the robot from switching outside of the allowable gears by restricting the robot's velocity, so that the guard conditions to switch gears are never satisfied. The car is given a sensing radius of 1. If, when executing a solution trajectory, it discovers a new obstacle within its sensing radius, the car switches to an "emergency" mode in which it applies an emergency deceleration sufficient to reduce its velocity to $\epsilon > 0$ before colliding with the obstacle. It then rebuilds the abstraction, recomputes the product automaton, switches out of the emergency mode into the mode corresponding to gear $g = 1$, and computes a new trajectory to follow.

In this experiment, the robot is asked to visit N randomly chosen regions of interest in any order, where $N \in \{1, \ldots, 5\}$. Formally, the robot is given the co-safe LTL specification

$$\phi_N = \bigwedge_{i=0}^{N-1} \mathcal{F}p_i, \tag{3}$$

where each p_i corresponds to a propositional region in the office environment.

The robot's initial map includes the walls of the office. However, the robot does not know the current status of the doors into each room, nor does it know whether there are any obstacles in the central lobby. Specifically, the robot is unaware that the doors to two of the rooms are closed (we model this as rectangular obstacles filling the doorways), and there is a large rectangular obstacle in the center of the lobby. Figure 3(a) contains the actual map of the office, and Figure 3(b) contains the robot's initial map. We include the triangulations in the maps in Figure 3 to demonstrate granularity. A triangulation always respects the currently known obstacles and the geometry of the propositional regions.

We have implemented our framework and experiments in C++ using the Open Motion Planning Library (OMPL) [8]. For the co-safe LTL formulas considered in our experiments,

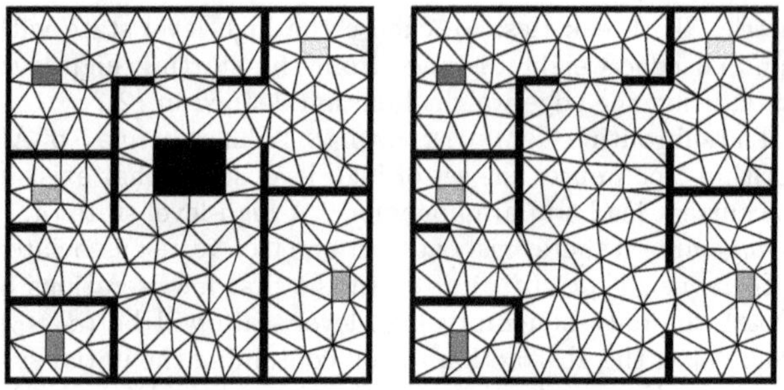

(a) (b)

Figure 3: (a) an office-like environment with propositional regions of interest; (b) the robot's initial map, in which 3 obstacles are unknown.

Table 1: Experimental data for office experiment with a full initial map and a partial initial map

Initial Map	N	Solution Time	Time Computing Product $\mathcal{P} = \mathcal{M} \times \mathcal{A}_\phi$
Full	1	1.36	0.003
	2	4.05	0.003
	3	11.19	0.004
	4	21.49	0.006
	5	34.85	0.008
Partial	1	3.78	0.009
	2	19.32	0.037
	3	54.83	0.088
	4	257.82	0.206
	5	549.86	0.415

we have converted them to minimal DFA's by using scheck [24]. To triangulate environments, we use Triangle [30]. All experiments were run on the Shared University Computing Grid at Rice. Each experiment used a 2.83 GHz Intel Xeon processor with 16 GB RAM. For each set of input parameters, we average our timing measurements over 50 independent runs.

Table 1 contains experimental data for satisfying the coverage formula ϕ_N in the office environment, comparing the full initial map (Figure 3(a)) to a partial initial map (Figure 3(b)). With a fully accurate initial map, the robot does not encounter any unanticipated obstacles, and so our method behaves equivalently to the past method presented in [3–5]. We are including data for the full initial map for comparison. For the partial map, planning times increase significantly with the number of regions of interest in the coverage formula. Visiting more regions causes the robot

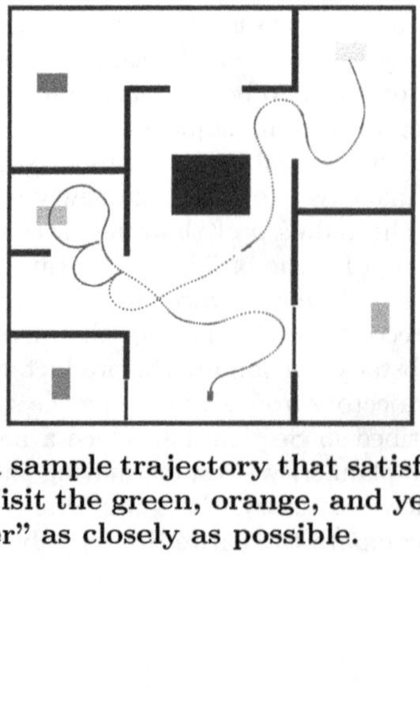

Figure 4: A sample trajectory that satisfies the specification "Visit the green, orange, and yellow regions in any order" as closely as possible.

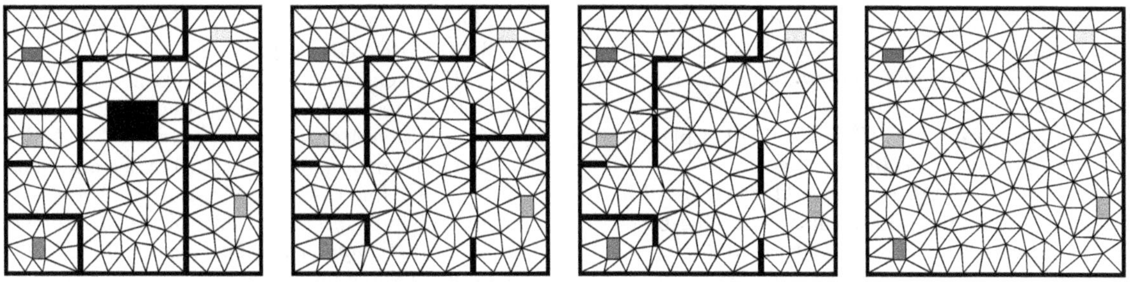

Figure 5: Office-like environments in which (a) all obstacles are known; (b) 3 obstacles are unknown; (c) 6 obstacles are unknown; (d) all 16 obstacles are unknown.

Table 2: Experimental data for office experiment with formula ϕ_5 when varying the initial map

Number of Initially Unknown Obstacles	Solution Time	Time Computing Product $\mathcal{P} = \mathcal{M} \times \mathcal{A}_\phi$
0	34.85	0.008
3	594.86	0.415
6	1430.71	0.952
16	3001.96	1.966

to discover more unknown obstacles, each of which requires the robot to brake. Every time the robot comes to a stop near a newly discovered obstacle, planning a solution trajectory from that stopped point is often time-consuming for the low-level motion-planning layer. This is due to the close proximity of the robot and the obstacle. With longer-range sensors, this problem can be alleviated. For all experiments, times spent recomputing the product automaton $\mathcal{P}$ remain very small (see Tables 1 and 2). It should be noted that the times spent recomputing $\mathcal{P}$ in Tables 1 and 2 also include the time to regenerate the abstraction $\mathcal{M}$. Figure 4 contains an example trajectory for the robot, given a specification to visit three regions of interest (green, orange, and yellow), one of which is unreachable (green) due to a closed door. First, the robot drives toward the room containing the green region. When it encounters the door, it brakes and recomputes the abstraction and the product automaton. The planning framework uses our measure of satisfiability to generate another trajectory that satisfies the specification as closely as possible, which is to visit the two remaining regions. A similar replanning step occurs when the robot encounters the large obstacle in the lobby. When the robot is in the room with the orange region, it stays only in first gear as required.

To test the importance of the initial map, we have also run experiments with initial maps of varying accuracy, ranging from a completely known environment to a completely unknown environment in which no obstacles or walls are initially known by the robot (except for the bounding box of the environment). The four types of initial maps are shown in Figure 5. As before, all propositional regions are initially known. Table 2 contains data for this set of experiments. We focus on solving ϕ_5, the most difficult of the formulas to consider. As seen in Table 2, the time spent building and recomputing $\mathcal{P}$ and $\mathcal{M}$ is negligible compared to the time spent planning solution trajectories.

6. CONCLUSION

In this paper, we have presented an iterative motion planning framework for a hybrid system with complex and possibly nonlinear dynamics given a temporal logic specification and a partially unknown environment. We have also presented a metric of satisfiability which we can optimize in cases where obstacles in the environment prevent full satisfaction of the given temporal logic specification; in such cases, the robotic system satisfies the specification as closely as possible.

For future work, we plan to change the replanning step to retriangulate only the part of the workspace that has changed upon discovery of a new obstacle, instead of retriangulating the entire workspace. This would allow the framework to keep many of the edge weights in the product automaton, so that not all information from the synergy layer is lost. In addition, we plan to add support for obstacles to disappear from the robot's initial map (the current framework only supports obstacles appearing). We could also assume a probabilistic distribution on where and when obstacles will appear, and then generate trajectories that maximize probability of successful satisfaction of the specification. Finally, we would like to consider a "greedy" temporal motion planning approach that begins executing a partial trajectory along a lead in the product automaton. This is to prevent the framework from wasting time generating an entire solution trajectory for a large specification, only to discover an obstacle early in that trajectory, stop, and recompute another solution trajectory. As seen in our experiments, when the robot's initial map is sufficiently inaccurate, this wasted planning time can add up.

7. ACKNOWLEDGMENTS

The authors would like to acknowledge Devin Grady, Ryan Luna, and Mark Moll of Rice University for their helpful feedback and suggestions. Work on this project by Maly, Kavraki, Kress-Gazit, and Vardi has been supported in part by NSF Expeditions 1139011. Maly, Lahijanian, Kavraki, and Vardi have also been supported in part by NSF CCF 1018798. Kavraki has been supported in part by the U.S. Army Research Laboratory and the U.S. Army Research Office under grant number W911NF-09-1-0383. This work was also supported in part by the Shared University Grid at Rice funded by NSF under Grant EIA-0216467 and a partnership between Rice University, Sun Microsystems, and Sigma Solutions, Inc.

8. REFERENCES

[1] K. E. Bekris, D. K. Grady, M. Moll, and L. E. Kavraki. Safe distributed motion coordination for second-order systems with different planning cycles. *Intl. J. of Robotics Research*, 31(2):129–149, Feb. 2012.

[2] K. E. Bekris and L. E. Kavraki. Greedy but safe replanning under kinodynamic constraints. In *IEEE Intl. Conf. on Robotics and Automation*, pages 704–710, 2007.

[3] A. Bhatia, L. Kavraki, and M. Vardi. Motion planning with hybrid dynamics and temporal goals. In *Decision and Control, IEEE Conf. on*, pages 1108–1115, 2010.

[4] A. Bhatia, L. Kavraki, and M. Vardi. Sampling-based motion planning with temporal goals. In *Robotics and Automation, IEEE Int. Conf. on*, pages 2689–2696, May 2010.

[5] A. Bhatia, M. Maly, L. Kavraki, and M. Vardi. Motion planning with complex goals. *Robotics Automation Magazine, IEEE*, 18(3):55 –64, Sep. 2011.

[6] R. Bloem, K. Greimel, T. Henzinger, and B. Jobstmann. Synthesizing robust systems. In *Formal Methods in Computer-Aided Design*, pages 85–92, 2009.

[7] Y. Chen, J. Tumova, and C. Belta. LTL robot motion control based on automata learning of environmental dynamics. In *Robotics and Automation, IEEE Int. Conf. on*, pages 5177 –5182, May 2012.

[8] I. A. Şucan, M. Moll, and L. E. Kavraki. The open motion planning library. *IEEE Robotics & Automation Magazine*, 19:72–82, December 2012.

[9] X. Ding, S. Smith, C. Belta, and D. Rus. MDP optimal control under temporal logic constraints. In *Decision and Control and European Control Conf. (CDC-ECC), IEEE Conf. on*, pages 532–538, 2011.

[10] G. Fainekos, A. Girard, H. Kress-Gazit, and G. J. Pappas. Temporal logic motion planning for dynamic robots. *Automatica*, 45:343–352, 2009.

[11] T. Fraichard. A short paper about motion safety. In *Proc. 2007 IEEE Intl. Conf. on Robotics and Automation*, pages 1140–1145, Apr. 2007.

[12] D. K. Grady, M. Moll, C. Hegde, A. C. Sankaranarayanan, R. G. Baraniuk, and L. E. Kavraki. Multi-objective sensor-based replanning for a car-like robot. In *IEEE Intl. Symp. on Safety, Security, and Rescue Robotics*, 2012.

[13] D. Hsu, J. Latombe, and R. Motwani. Path planning in expansive configuration spaces. *Intl. J. of Computational Geometry and Applications*, 9(4-5):495–512, 1999.

[14] Karaman and Frazzoli. Sampling-based motion planning with deterministic μ-calculus specifications. In *IEEE Conf. on Decision and Control*, 2009.

[15] K. Kim and G. Fainekos. Approximate solutions for the minimal revision problem of specification automata. In *Proc. of the IEEE Int. Conf. on Intelligent Robots and Systems*, pages 265–271, 2012.

[16] M. Kloetzer and C. Belta. A fully automated framework for control of linear systems from temporal logic specifications. *Automatic Control, IEEE Transactions on*, 53(1):287–297, 2008.

[17] H. Kress-Gazit, G. Fainekos, and G. Pappas. Where's waldo? Sensor-based temporal logic motion planning. In *Robotics and Automation, 2007 IEEE Int. Conf. on*, pages 3116 –3121, Apr. 2007.

[18] H. Kress-Gazit, G. Fainekos, and G. Pappas. Temporal-logic-based reactive mission and motion planning. *Robotics, IEEE Transactions on*, 25(6):1370–1381, Dec. 2009.

[19] H. Kress-Gazit, T. Wongpiromsarn, and U. Topcu. Correct, reactive, high-level robot control. *Robotics Automation Magazine, IEEE*, 18(3):65 –74, Sep. 2011.

[20] O. Kupferman and Y. Lustig. Lattice automata. In *Proc. 8th Int. Conf. on Verification, Model Checking, and Abstract Interpretation*, volume 4349 of *Lecture Notes in Computer Science*, pages 199–213, 2007.

[21] O. Kupferman and M. Y. Vardi. Model checking of safety properties. *Formal Methods in System Design*, 19:291 – 314, 2001.

[22] M. Lahijanian, S. B. Andersson, and C. Belta. Temporal logic motion planning and control with probabilistic satisfaction guarantees. *IEEE Transactions on Robotics*, 28(2):396–409, Apr. 2012.

[23] M. Lahijanian, J. Wasniewski, S. Andersson, and C. Belta. Motion planning and control from temporal logic specifications with probabilistic satisfaction guarantees,. In *IEEE Int. Conf. on Robotics and Automation*, pages 3227–3232, Alaska, 2010.

[24] T. Latvala. Efficient model checking of safety properties. In *Model Checking Software*, pages 74–88. Springer, 2003.

[25] S. C. Livingston, R. M. Murray, and J. W. Burdick. Backtracking temporal logic synthesis for uncertain environments. In *IEEE Intl. Conf. on Robotics and Automation*, pages 5163–5170, 2012.

[26] E. Plaku, L. Kavraki, and M. Vardi. Motion planning with dynamics by a synergistic combination of layers of planning. *IEEE Trans. on Robotics*, 26(3):469–482, Jun. 2010.

[27] E. Plaku, L. E. Kavraki, and M. Y. Vardi. Falsification of LTL safety properties in hybrid systems. In *Proc. of the Conf. on Tools and Algorithms for the Construction and Analysis of Systems (TACAS 2009)*, York, UK, 2009.

[28] V. Raman and H. Kress-Gazit. Analyzing unsynthesizable specifications for high-level robot behavior using LTLMoP. In *Proc. of the 23rd Int. Conf. on Computer Cided Verification*, CAV'11, pages 663–668, Berlin, Heidelberg, 2011. Springer-Verlag.

[29] S. Sarid, B. Xu, and H. Kress-Gazit. Guaranteeing high-level behaviors while exploring partially known maps. In *Proc. of Robotics: Science and Systems*, Sydney, Australia, July 2012.

[30] J. Shewchuk. Triangle: Engineering a 2D quality mesh generator and Delaunay triangulator. In *Applied Computational Geometry Towards Geometric Engineering*, volume 1148 of *Lecture Notes in Computer Science*, chapter 23, pages 203 – 222. Springer-Verlag, Berlin/Heidelberg, 1996.

[31] P. Černý, S. Gopi, T. A. Henzinger, A. Radhakrishna, and N. Totla. Synthesis from incompatible specifications. In *Proc. of the tenth ACM Int. Conf. on Embedded software*, EMSOFT '12, pages 53–62, New York, NY, USA, 2012. ACM.

[32] T. Wongpiromsarn, U. Topcu, and R. Murray. Receding horizon control for temporal logic specifications. In *Proc. of the 13th ACM Int. Conf. on Hybrid Systems: Computation and Control*, pages 101–110. ACM, 2010.

HSCC'13 Posters

Nikolaos Athanasopoulos *(Eindhoven University of Technology)* and Mircea Lazar *(Eindhoven University of Technology)*: **On stability analysis of switched linear systems via stability theory in the space of compact sets.**

Martin Clochard *(ENS Pairs)*, Swarat Chaudhuri *(Rice University)*, and Armando Solar-Lezama *(Massachusetts Institute of Technology)*: **Stochastic Program Synthesis with Smoothed Numerical Search.**

Yi Deng *(Rensselaer Polytechnic Institute)*, Akshay Rajhans *(Carnegie Mellon University)*, and A. Agung Julius *(Rensselaer Polytechnic Institute)*: **STRONG: A Trajectory-based Verification Toolbox for Hybrid Systems.**

Charles Freundlich *(Steven Institute of Technology)*, Philippos Mordohai *(Stevens Institute of Technology)*, and Michael M. Zavlanos *(Duke University)*: **A Hybrid Control Approach to the Next-Best-View Problem using Stereo Vision.**

Jie Fu *(University of Delaware)*, Herbert G. Tanner *(University of Delaware)*, and Jeffrey Heinz *(University of Delaware)*: **Adaptive Symbolic Design with Grammatical Inference.**

Shahab Kaynama *(University of California at Berkeley)*, Ian M. Mitchell *(University of British Columbia)*, Meeko Oishi *(University of New Mexico)*, and Guy A. Dumont *(University of British Columbia)*: **Safety-Preserving Control of High-Dimensional Continuous-Time Uncertain Linear Systems.**

Fei Miao *(University of Pennsylvania)*, Miroslav Pajic *(University of Pennsylvania)*, and George J. Pappas *(University of Pennsylvania)*: **A stochastic game formulation for system against replay attack.**

Andrew K. Winn *(Rensselaer Polytechnic Institute)* and A. Agung Julius *(Rensselaer Polytechnic Institute)*: **Optimization-based Improvements for Safety Controller Synthesis via Human Generated Trajectories.**

HSCC'13 Tool Demonstrations

Stanley Bak *(University of Illinois at Urbana-Champaign)* and Marco Caccamo *(University of Illinois at Urbana-Champaign)*: **Computing Reachability for Nonlinear Systems with HyCreate.**

Goran Frehse *(VERIMAG)*: **A Comparison of the STC and LGG Reachability Algorithms in SpaceEx.**

Sicun Gao *(Carnegie Mellon University)*, Soonho Kong *(Carnegie Mellon University)*, and Edmund M. Clarke *(Carnegie Mellon University)*: **dReach: A Reachability Analysis Tool for Nonlinear Hybrid Systems.**

Arnd Hartmanns *(Saarland University)* and Holger Hermanns *(Saarland University)*: **Modeling and Analysis of Stochastic, Hybrid and Real-Time Systems with the Modest Toolset.**

Taylor T. Johnson *(University of Illinois at Urbana-Champaign)* and Sayan *Mitra (University of Illinois at Urbana-Champaign)*: **The Passel Verification Tool for Hybrid Automata Networks.**

Stefan Mitsch *(Carnegie Mellon University)* and André Platzer *(Carnegie Mellon University)*: **Modeling and Verification of Hybrid Systems with Sphinx and KeYmaera.**

Yash Vardhan Pant *(University of Pennsylvania)*, Harsh Jain *(University of Pennsylvania)*, Abhijeet Mulay *(University of Pennsylvania)*, and Rahul Mangharam *(University of Pennsylvania)*: **ProtoDrive: An Experimental Platform for Electric Vehicle Energy Scheduling and Control.**

Author Index

www.ingramcontent.com/pod-product-compliance
Lightning Source LLC
Chambersburg PA
CBHW080150060326
40689CB00018B/3924